HANDBOOK OF

HUMAN FACTORS and ERGONOMICS in HEALTH CARE and PATIENT SAFETY

SECOND EDITION

Human Factors and Ergonomics

Series Editor

Gavriel Salvendy

Professor Emeritus
School of Industrial Engineering
Purdue University

Chair Professor & Head
Dept. of Industrial Engineering
Tsinghua Univ., P.R. China

Published Titles

Conceptual Foundations of Human Factors Measurement, *D. Meister*

Content Preparation Guidelines for the Web and Information Appliances:
Cross-Cultural Comparisons, *H. Liao, Y. Guo, A. Savoy, and G. Salvendy*

Designing for Accessibility: A Business Guide to Countering Design Exclusion, *S. Keates*

Handbook of Cognitive Task Design, *E. Hollnagel*

The Handbook of Data Mining, *N. Ye*

Handbook of Digital Human Modeling: Research for Applied Ergonomics and Human Factors Engineering
V. G. Duffy

Handbook of Human Factors and Ergonomics in Health Care and Patient Safety, Second Edition
P. Carayon

Handbook of Human Factors in Web Design, Second Edition, *R. Proctor and K. Vu*

Handbook of Occupational Safety and Health, *D. Koradecka*

Handbook of Standards and Guidelines in Ergonomics and Human Factors,
W. Karwowski

Handbook of Virtual Environments: Design, Implementation, and Applications,
K. Stanney

Handbook of Warnings, *M. Wogalter*

Human-Computer Interaction: Designing for Diverse Users and Domains, *A. Sears
and J. A. Jacko*

Human-Computer Interaction: Design Issues, Solutions, and Applications, *A. Sears
and J. A. Jacko*

Human-Computer Interaction: Development Process, *A. Sears and J. A. Jacko*

The Human-Computer Interaction Handbook: Fundamentals, Evolving Technologies, and Emerging
Applications, Second Edition, *A. Sears and J. A. Jacko*

Human Factors in System Design, Development, and Testing, *D. Meister
and T. Enderwick*

Introduction to Human Factors and Ergonomics for Engineers, *M. R. Lehto and J. R. Buck*

Macroergonomics: Theory, Methods and Applications, *H. Hendrick and B. Kleiner*

Practical Speech User Interface Design, *James R. Lewis*

Smart Clothing: Technology and Applications, *Gilsoo Cho*

Theories and Practice in Interaction Design, *S. Bagnara and G. Crampton-Smith*

The Universal Access Handbook, *C. Stephanidis*

Usability and Internationalization of Information Technology, *N. Aykin*

User Interfaces for All: Concepts, Methods, and Tools, *C. Stephanidis*

Forthcoming Titles

Computer-Aided Anthropometry for Research and Design, *K. M. Robinette*

Cross-Cultural Design for IT Products and Services, *P. Rau*

Foundations of Human–Computer and Human–Machine Systems, *G. Johannsen*

Introduction to Human Factors and Ergonomics for Engineers, Second Edition,
  M. R. Lehto

The Human–Computer Interaction Handbook: Fundamentals, Evolving Technologies, and Emerging
  Applications, Third Edition, *J. A. Jacko*

The Science of Footwear, *R. S. Goonetilleke*

Human Performance Modeling: Design for Applications in Human Factors
  and Ergonomics, *D. L. Fisher, R. Schweickert, and C. G. Drury*

HANDBOOK OF
HUMAN FACTORS and ERGONOMICS in HEALTH CARE and PATIENT SAFETY

SECOND EDITION

Edited By
Pascale Carayon

CRC Press
Taylor & Francis Group
Boca Raton London New York

CRC Press is an imprint of the
Taylor & Francis Group, an **informa** business

CRC Press
Taylor & Francis Group
6000 Broken Sound Parkway NW, Suite 300
Boca Raton, FL 33487-2742

© 2012 by Taylor & Francis Group, LLC
CRC Press is an imprint of Taylor & Francis Group, an Informa business

No claim to original U.S. Government works

Printed in the United States of America on acid-free paper
Version Date: 20110803

International Standard Book Number: 978-1-4398-3033-8 (Hardback)

Library of Congress Cataloging-in-Publication Data

Handbook of human factors and ergonomics in health care and patient safety / [edited by] Pascale Carayon. -- 2nd ed.
    p. cm. -- (Human factors and ergonomics)
  Includes bibliographical references and index.
  ISBN 978-1-4398-3033-8 (hardback)
  1. Medical errors--Prevention--Handbooks, manuals, etc. 2. Human engineering--Handbooks, manuals, etc. 3. Patients--Safety measures--Handbooks, manuals, etc. 4. Health facilities--Design and construction--Handbooks, manuals, etc. I. Carayon, Pascale.

R729.8.H34 2011
610.28'9--dc23   2011028092

Visit the Taylor & Francis Web site at
http://www.taylorandfrancis.com

and the CRC Press Web site at
http://www.crcpress.com

This handbook is dedicated to Hal Hendrick, the leader in human factors

and ergonomics and the father of macroergonomics.

Contents

Series Preface xiii
Foreword xv
Preface... xvii
Acknowledgments xix
Editor... xxi
Contributors xxiii

Section I   Introduction

1. **Human Factors and Ergonomics in Health Care and Patient Safety** 3
   Pascale Carayon

2. **Opportunities and Challenges in the Pursuit of Patient Safety**.................................... 17
   Kerm Henriksen

3. **Embedded Risks and Challenges of Modern Health Care and the Urgent Call for Proactive Human Factors** 27
   Daniel Gopher and Yoel Donchin

Section II   Macroergonomics and Systems

4. **Historical Perspective and Overview of Macroergonomics** 45
   Hal W. Hendrick

5. **Work System Design in Health Care**... 65
   Pascale Carayon, Carla J. Alvarado, and Ann Schoofs Hundt

6. **Error as Behavior and Murphy's Law: Implications for Human Factors and Ergonomics**............................. 81
   Marilyn Sue Bogner

7. **Organizational Learning in Health Care** 97
   Ann Schoofs Hundt

8. **The Relationship between Physician Professionalism and Health Care Systems Change**........................... 109
   Maureen A. Smith and Jessica M. Bartell

Section III   Job and Organizational Design

9. **The Effect of Workplace Health Care Worker Stress and Burnout on Patient Outcomes** 119
   Eric S. Williams, Jonathon R.B. Halbesleben, Linda Baier Manwell, Julia E. McMurray, Joseph Rabatin, Ayesha Rashid, and Mark Linzer

10. **Safety Culture in Health Care** 133
    Kenji Itoh, Henning Boje Andersen, and Marlene Dyrløv Madsen

11. Handoffs and Transitions of Care..163
Robert L. Wears, Shawna J. Perry, and Emily S. Patterson

12. High-Reliability Organizations in Health Care...173 ✓
Vinit Desai, Peter M. Madsen, and Karlene H. Roberts

13. The Relation between Teamwork and Patient Safety..185 ✓
David P. Baker, Eduardo Salas, James B. Battles, and Heidi B. King

14. Health-Care Work Schedules...199
Ann E. Rogers

Section IV Physical Ergonomics

15. The Physical Environment in Health Care..215
Carla J. Alvarado

16. Assessment and Evaluation Tools for Health-Care Ergonomics: Musculoskeletal Disorders and Patient Handling..235
Sue M. Hignett, Michael J. Fray, and Mary Matz

Section V Technology

17. Human Factors and Ergonomics of Health Information Technology Implementation.................249
Ben-Tzion Karsh, Richard J. Holden, and Calvin K. L. Or

18. Human–Computer Interaction Design in Health Care..265
Frank A. Drews and Heidi S. Kramer

19. Trust in Health Technologies..281
Enid Montague and John D. Lee

20. Human Factors in Telemedicine..293
Peter Hoonakker

21. Using Health IT to Improve Health Care and Patient Safety.............................305
James M. Walker

Section VI Human Error

22. Human Error in Health Care..323 ✓
Frank A. Drews

23. Medical Failure Taxonomies..341
Bruce R. Thomadsen

24. Human Error Reduction Strategies in Health Care...385 ✓
René Amalberti and Sylvain Hourlier

25. Communicating about Unexpected Outcomes, Adverse Events, and Errors.........401 ✓
James W. Pichert, Gerald B. Hickson, Anna Pinto, and Charles Vincent

26. **Human Factors Engineering of Health Care Reporting Systems**...423 ✓
    Christopher W. Johnson

27. **Error Recovery in Health Care**..449 ✓
    Tosha B. Wetterneck

Section VII Human Factors and Ergonomics Methodologies

28. **Cognitive Work Analysis in Health Care**..465
    Michelle L. Rogers, Emily S. Patterson, and Marta L. Render

29. **Human Factors Risk Management for Medical Products**...475
    Edmond W. Israelski and William H. Muto

30. **Human Factors Analysis of Workflow in Health Information Technology Implementation**......................507
    *Pascale Carayon, Randi Cartmill, Peter Hoonakker, Ann Schoofs Hundt, Ben-Tzion Karsh,
    Daniel Krueger, Molly L. Snellman, Teresa N. Thuemling, and Tosha B. Wetterneck*

31. **Video Analysis: An Approach for Use in Health Care**...523
    Colin F. Mackenzie and Yan Xiao

32. **Usability Evaluation in Health Care**...543
    John Gosbee and Laura Lin Gosbee

33. **Medical Simulation**..557
    Mark W. Scerbo and Brittany L. Anderson

34. **Simulation-Based Training for Teams in Health Care: Designing Scenarios,
    Measuring Performance, and Providing Feedback**...573
    Michael A. Rosen, Eduardo Salas, Scott I. Tannenbaum, Peter J. Pronovost, and Heidi B. King

Section VIII Human Factors and Ergonomics Interventions

35. **Ergonomics Programs and Effective Implementation**..597
    Michael J. Smith

36. **Work Organization Interventions in Health Care**..611
    Kari Lindström, Gustaf Molander, and Jürgen Glaser

37. **Team Training for Patient Safety**..627 ✓
    Eduardo Salas, Sallie J. Weaver, Michael A. Rosen, and Megan E. Gregory

38. **Human Factors Considerations in Health IT Design and Development**...649
    Marie-Catherine Beuscart-Zéphir, Sylvia Pelayo, Elizabeth Borycki, and Andre Kushniruk

39. **Human Factors and Ergonomics in Patient Safety Management**...671
    Tommaso Bellandi, Sara Albolino, Riccardo Tartaglia, and Sebastiano Bagnara

Section IX Specific Applications

40. **Human Factors and Ergonomics in Intensive Care Units**...693
    Ayse P. Gurses, Bradford D. Winters, Priyadarshini R. Pennathur, Pascale Carayon, and Peter J. Pronovost

41. **Human Factors and Ergonomics in the Emergency Department** 709
Shawna J. Perry, Robert L. Wears, and Rollin J. Fairbanks

42. **Human Factors and Ergonomics in Pediatrics** 723
Matthew C. Scanlon and Paul Bauer

43. **Human Factors and Ergonomics in Home Care** 743
Teresa Zayas-Cabán and Rupa S. Valdez

44. **Human Factors and Ergonomics in Primary Care** 763
Tosha B. Wetterneck, Jamie A. Lapin, Ben-Tzion Karsh, and John W. Beasley

45. **In Search of Surgical Excellence: A Work Systems Approach** 775
Douglas A. Wiegmann, Sacha Duff, and Renaldo Blocker

46. **Human Factors and Ergonomics in Medication Safety** 785
Elizabeth Allan Flynn

47. **Human Factors and Ergonomics in Infection Prevention** 793
Carla J. Alvarado

48. **Human Factors in Anesthesiology** 803
Matthew B. Weinger

Index 825

Series Preface

In the past five years, since the publication of this handbook, great debates and increased emphasis have been placed on providing high-quality, affordable health care to all segments of society around the world. While this handbook predominantly deals with the human and organizational issues of health care and patient safety, these approaches can lead to significant reductions in health-care costs related to improvements in patient safety, which can reduce or eliminate the need to perform additional procedures to correct the mishaps caused by poorly designed and implemented health-care systems and processes. Application of human factors to health care also increases patient satisfaction and well-being, which can accelerate patient recovery and improve preventive care, therefore reducing health-care costs further. Effective human factors and ergonomic interventions also result in significant cost savings.

A series of reports by the Institute of Medicine (IOM) has described the multiple quality of care and patient safety problems and the need for human factors participation in analyzing and redesigning health-care systems and processes. For instance, in the year 2000, the IOM published a report (*To Err Is Human: Building a Safer Health System*) that attracted much attention from health-care institutes from the United States and all over the world. According to data reported by IOM, between 44,000 and 98,000 Americans die each year as the result of medical errors. A 2005 report by the IOM and the National Academy of Engineering (*Building a Better Delivery System: A New Engineering/Health Care Partnership*) emphasizes the need for industrial and systems engineering, including human factors, to partner with health-care experts and organizations. This handbook presents state-of-the-art knowledge on the application of human factors and ergonomics to medical errors and quality and safety of care.

With the rapid introduction of highly sophisticated computers, telecommunications, new ways of delivering patient care services, and innovative organizational approaches in the health-care industry, a major shift has occurred in the way patients receive care and in the way health-care staff use technology to deliver care. The objective of this book series on human factors and ergonomics is to provide researchers and practitioners a platform where important issues related to these changes can be discussed. Methods and recommendations are presented that will ensure that emerging technologies and work organizations will provide increased productivity quality, satisfaction, safety, and health in a new workplace and in the information society.

This handbook provides a comprehensive account on how to manage and improve health-care delivery in order to minimize medical errors and maximize patient satisfaction and well-being in an economically responsive way. Hence, the effective use of this handbook may result in fewer medical errors, which could reduce malpractice suits and thus reduce the cost of health-care delivery. The handbook also provides invaluable information for reducing stress and workload and increasing the well-being of health-care personnel.

The 98 authors of this handbook include 72 from academia, 20 from the health-care industry, and 6 from government agencies. These individuals are some of the very best and most respected in their field. The 121 tables, 98 figures, and 4000 references provide an insightful presentation of the subject with pointers for further research and development. The handbook is especially invaluable to health-care employees and managers, human factors specialists, hospital administrators and leaders, and health-care policy makers.

Gavriel Salvendy
Series Editor
Purdue University/Tsinghua University, China

Foreword

Over the last decade patient safety has been a focal point of health care. Hospitals are identifying hazards, investigating mistakes, implementing interventions to reduce hazards, and training staff on patient safety. Despite these efforts, there is limited empiric evidence of decreases in patient harm. Although estimates for the number of patients harmed from medical errors varies between 2% and 33%, far too many patients suffer preventable harm.

Researchers and patient safety experts advocate the use of human factors and systems engineering principles, theories, tools, and methods to improve safety. In 2005, the Institute of Medicine and the National Academy of Engineering identified human factors engineering as an important tool for designing better health-care systems. Nonetheless, human factors and systems engineering is incorporated in few areas of health care. Health care is the only high-risk industry where the operators (e.g., physicians, nurses), and sometimes lawyers and administrators, investigate adverse events and develop solutions without the experts who understand people, systems, and how they interact to impact safety. Not surprisingly, "re-educating staff" or stressing "vigilance" are the most common recommendations emerging from these error investigations. While mostly ineffective, collectively across the globe, these types of interventions consume significant resources.

Although errors are more visible now, our patients continue to suffer preventable harm, making patients, regulators, accreditators, and caregivers increasingly frustrated. Newspaper articles and congressional hearings routinely highlight examples of patients who were harmed from their interaction with our health-care system. While there is broad consensus that faulty systems rather than faulty people are the cause of most errors, the health-care community struggles to find practical and scientifically sound ways to address, surface, and mitigate hazards.

Over health care's decade-long efforts to improve patient safety, we have seen work that is more independent than interdependent, more competitive than cooperative, more focused on activities than on results. We have learned much, despite little evidence that patient harm has decreased. Perhaps most importantly, we learned that health care must embrace the science of patient safety. There are no quick fixes. Efforts to improve patient safety must be guided by theory, rooted in evidence, and robustly evaluated. This science must draw from diverse disciplines, linking clinicians and administrators with human factors engineers, systems engineers, psychologists, and sociologists. Such interdisciplinary research groups are still rare, but slowly growing.

We are increasingly aware that health-care information technology (HIT) and devices, thought by many to be a panacea for patient safety, are a double-edged sword. These technologies, while defending against some mistakes, invariably introduce new mistakes and new harms. This reality does not impugn technology. Rather, it impugns the methods by which we develop and introduce technology or other system changes into health care. Any system change may defend against some mistakes but invariably introduce new ones. The underlying problem with technology may lie not in the operation but in the design that is often invisible to the operator of the equipment.

The second edition of *Handbook of Human Factors and Ergonomics in Health Care and Patient Safety*, edited by Dr. Pascale Carayon, provides a conceptual framework and practical strategies to more fully apply human factors principles in health care to improve patient safety. This book gives us the details about human factors research, theories, and tools and demonstrates through examples how the theories and tools have been applied. The authors are experienced researchers and practitioners of human factors engineering in health care, and their collective wisdom is an excellent "field guide" on human factors engineering. The book provides important information for health-care safety and quality leaders, researchers, and practitioners. Perhaps, as the principles and tools discussed in this book become widely applied in health care, we will begin to realize significant improvements in patient safety. Indeed, we may proactively design technologies and a health-care system that prevents most errors and defends or minimizes the harmful events that continue to occur.

The work of improving patient safety has proven difficult. As we understand the sciences underlying patient safety and begin to unveil the systems upon which health care is based, we recognize how vulnerable our systems are. So much safety is built upon the backs of experts and dedicated clinicians who perform heroic acts every day. If not for these heroes, far more patients would suffer preventable harm.

Patients and these clinicians are counting on you. They need you to read this book, apply these principles, and make patients safer. In our work to reduce bloodstream infections, we have seen what is possible when patient safety efforts are guided by science yet made ruthlessly practical. I hope you put the knowledge from this book to use.

Peter J. Pronovost

Preface

In 2007, when the first edition of the handbook was published, there was burgeoning interest in human factors and ergonomics in health care. The interest of healthcare practitioners, decision makers, and policy makers for human factors and ergonomics has continued to grow. The human factors and ergonomics discipline has responded to the challenge with enthusiasm and dedication. The body of human factors and ergonomics knowledge specifically relevant to health care and patient safety has grown steadily. Therefore, the second edition of the *Handbook of Human Factors and Ergonomics in Health Care and Patient Safety* adds significant contributions to the first edition of the handbook. Several new topics are addressed in the second edition, including work schedules, error recovery, telemedicine, workflow analysis, simulation, health information technology development, and design, patient safety management, and a greater range of application domains such as medication safety, surgery, anesthesia, and infection prevention. Readers of the first edition of the handbook will want to read updated versions of chapters published in the first edition and many other new chapters on highly important topics for health-care quality and patient safety.

The handbook represents a complete reference book on human factors and ergonomics research, concepts, theories, models, methods, and interventions that have been (or can be) applied in health care. Special emphasis is put on the contributions of human factors and ergonomics to the improvement of patient safety and quality of care. The handbook also covers human factors and ergonomics issues specific to a range of health-care domains or applications; this section of the handbook was significantly enhanced in the second edition.

Human factors and ergonomics professionals and researchers can help healthcare organizations, professionals, and institutions to improve the design of healthcare systems and processes and, therefore, enhance quality of care and reduce patient safety problems. These efforts require active collaboration between human factors professionals, on the one hand, and health-care providers, professionals, and leaders, on the other hand. Both groups can benefit from reading this handbook by learning about human factors and ergonomics in health care and patient safety, human factors and ergonomics tools and intervention strategies, as well as applications of human factors and ergonomics in specific healthcare settings and contexts.

The book is a multiauthored handbook. It contains 48 chapters organized in 9 sections: (I) Introduction, (II) Macroergonomics and Systems, (III) Job and Organizational Design, (IV) Physical Ergonomics, (V) Technology, (VI) Human Error, (VII) Human Factors and Ergonomics Methodologies, (VIII) Human Factors and Ergonomics Interventions, and (IX) Specific Applications. The sections cover theory, research, tools and methods, applications, and various perspectives on human factors and ergonomics in health care and patient safety. The chapter authors are highly respected individuals from academia, health care, industry, and government from Canada, China, Denmark, Finland, France, Germany, Israel, Italy, Japan, the United Kingdom, and the United States. I would like to thank the authors for their contribution to this handbook and their commitment to improving health-care quality and patient safety.

Acknowledgments

Eight years ago, Mike Smith, my mentor, collaborator, and friend for more than 25 years, suggested the idea of this handbook. After the success of the first edition and the completion of the second edition, I want to thank Mike for his suggestion as well as his contribution to the first and second editions of the handbook. Thank you, Mike, for your continued support; you have been a great role model.

I would like to thank Gavriel Salvendy of Purdue University and Tsinghua University for his encouragement and support during the initial development of the handbook as well as with the second edition of this handbook. As the Human Factors and Ergonomics series editor for Taylor & Francis, he was instrumental in facilitating the publication of this handbook. During my entire career, Gavriel has challenged me to excel in my research and has provided tremendous help.

I would also like to thank my colleagues and students at the Center for Quality and Productivity Improvement at the University of Wisconsin–Madison. A special thank you goes to Teresa Thuemling, who was very helpful in the final stage of putting together this handbook. I also want to thank Carla Alvarado and Ann Schoofs Hundt for their support and friendship; I have learned a lot (and am still learning) from them regarding the specificity of health care and patient safety. Thank you to both of them for interesting conversations about the current and future challenges experienced by health care. Many of my colleagues in the Department of Industrial and Systems Engineering and in the School of Medicine and Public Health at the University of Wisconsin–Madison are research collaborators and have contributed to this second edition; thank you to all!

This handbook has also benefited from the many ideas, input, and feedback from my health-care collaborators at various hospitals, health-care systems, and schools of medicine, pharmacy, and nursing at universities in the United States and across the world.

I would also like to recognize the support of funding agencies, foundations, and organizations that has allowed me and my colleagues to do the research reported in this handbook.

I could not have completed the major task of putting together this handbook without the support and patience of my husband, Peter. You are there for me—thanks, schatje! Last but not least, I want to thank my parents, Michèle and Robert, Laurence, Christian, Sophie, Camille, Clémence, Marie, Clara, and Pierre-Dominique. *Merci de tout coeur. Dank u wel.*

Pascale Carayon
March 2011; Madison, Wisconsin

Editor

Pascale Carayon is Procter & Gamble Bascom Professor in Total Quality in the Department of Industrial and Systems Engineering and the director of the Center for Quality and Productivity Improvement (CQPI) at the University of Wisconsin–Madison. She leads the *Systems Engineering Initiative for Patient Safety (SEIPS)* at the University of Wisconsin–Madison (http://cqpi. engr.wisc.edu/seips_home). She received her diploma in engineering from the Ecole Centrale de Paris, France, in 1984, and her PhD in industrial engineering from the University of Wisconsin–Madison in 1988. Her research examines systems engineering, human factors and ergonomics, sociotechnical engineering, and occupational health and safety and has been funded by the Agency for Healthcare Research and Quality, the National Science Foundation, the National Institutes for Health (NIH), the National Institute for Occupational Safety and Health, the Department of Defense, various foundations, and private industry. She is the North American editor for *Applied Ergonomics*, the editor of the SocioTechnical System Analysis Department of *IIE Transactions on Healthcare Systems Engineering*, and a member of the editorial boards of the *Journal of Patient Safety*, *Behaviour and Information Technology*, *Work and Stress*, and *AHRQ WebM&M*. She is a fellow of the Human Factors and Ergonomics Society and a fellow of the International Ergonomics Association. Dr. Carayon is also a member of the National Research Council Board on Human-Systems Integration.

Contributors

Sara Albolino
Center for Clinical Risk
    Management and Patient Safety
Regione Toscana
Florence, Italy

Carla J. Alvarado
Center for Quality and Productivity
    Improvement
University of Wisconsin–Madison
Madison, Wisconsin

René Amalberti
Haute Autorité de Santé
Saint-Denis, France

Henning Boje Andersen
Management Engineering
    Department
Technical University of Denmark
Kongens Lyngby, Denmark

Brittany L. Anderson
Old Dominion University
Norfolk, Virginia

Sebastiano Bagnara
Faculty of Architecture
University of Sassari
Alghero, Italy

David P. Baker
IMPAQ International
Columbia, Maryland

Jessica M. Bartell
Group Health Cooperative of South
    Central Wisconsin
Madison, Wisconsin

James B. Battles
Center for Quality Improvement
    and Patient Safety
Agency for Healthcare Research
    and Quality
Rockville, Maryland

Paul Bauer
Children's Hospital of Wisconsin
Medical College of Wisconsin
Milwaukee, Wisconsin

John W. Beasley
Department of Family Medicine
University of Wisconsin School of
    Medicine and Public Health
Madison, Wisconsin

Tommaso Bellandi
Center for Clinical Risk
    Management and Patient Safety
Regione Toscana
Florence, Italy

Marie-Catherine Beuscart-Zéphir
Evalab—Clinical Investigation
    Centre for Innovative
    Technologies (CIC-IT),
Centre Hospitalier Universitaire
    (CHU)
and
Université Lille–Nord de France
Lille, France

Renaldo Blocker
Department of Industrial and
    Systems Engineering
University of Wisconsin–Madison
Madison, Wisconsin

Marilyn Sue Bogner
Institute for the Study of Human
    Error LCC
Bethesda, Maryland

Elizabeth Borycki
School of Health Information
    Science
University of Victoria
Victoria, British Columbia, Canada

Pascale Carayon
Department of Industrial and
    Systems Engineering
and
Center for Quality and Productivity
    Improvement
University of Wisconsin–Madison
Madison, Wisconsin

Randi Cartmill
Center for Quality and Productivity
    Improvement
University of Wisconsin–Madison
Madison, Wisconsin

Vinit Desai
University of Colorado–Denver
Denver, Colorado

Yoel Donchin
Head, Patient Safety Unit
Hadassah Hebrew University
    Medical Center
Jerusalem, Israel

Frank A. Drews
University of Utah
and
VA Center for Human Factors in
    Patient Safety
and
Informatics, Decision Enhancement
    and Surveillance Center
Salt Lake City, Utah

Sacha Duff
Department of Industrial and
    Systems Engineering
University of Wisconsin–Madison
Madison, Wisconsin

Rollin J. Fairbanks
National Center for Human Factors
  Engineering in Healthcare
MedStar Health

and

Georgetown University School of
  Medicine
Washington, DC

Elizabeth Allan Flynn
Department of Pharmacy Care
  Systems
Auburn University
Auburn, Alabama

and

School of Pharmacy
University of Florida
Gainesville, Florida

Michael J. Fray
Healthcare Ergonomics and Patient
  Safety Unit
Loughborough University
Leicestershire, United Kingdom

Jürgen Glaser
Department of Medicine
Institute for Occupational Medicine
University of Munich
Munich, Germany

and

Department of Psychology
University of Konstanz
Konstanz, Germany

Daniel Gopher
Director, Research Center for Work
  Safety and Human Engineering
Faculty of Industrial Engineering
  and Management Technion
Haifa, Israel

John Gosbee
Red Forest Consulting, LLC

and

University of Michigan Health
  System
Ann Arbor, Michigan

Laura Lin Gosbee
Red Forest Consulting, LLC
Ann Arbor, Michigan

Megan E. Gregory
University of Central Florida
Orlando, Florida

Ayse P. Gurses
Quality and Safety Research Group
Department of Anesthesiology and
  Critical Care Medicine
Johns Hopkins University School of
  Medicine

and

Johns Hopkins University
  Bloomberg School of Public
  Health
Baltimore, Maryland

Jonathon R.B. Halbesleben
Department of Management and
  Marketing
Culverhouse College of Commerce
  and Business Administration
University of Alabama
Tuscaloosa, Alabama

Hal W. Hendrick
University of Southern California
Los Angeles, California

Kerm Henriksen
Center for Quality Improvement
  and Patient Safety
Agency for Healthcare Research
  and Quality
Rockville, Maryland

Gerald B. Hickson
Center for Patient and Professional
  Advocacy
Vanderbilt University School of
  Medicine
Nashville, Tennessee

Sue M. Hignett
Healthcare Ergonomics and Patient
  Safety Unit
Loughborough University
Leicestershire, United Kingdom

Richard J. Holden
Department of Medicine and
  Biomedical Informatics

and

Center for Research and Innovation
  in Systems Safety
Vanderbilt University
Nashville, Tennessee

Peter Hoonakker
Center for Quality and Productivity
  Improvement
University of Wisconsin–Madison
Madison, Wisconsin

Sylvain Hourlier
Thales Avionics
Bordeaux, France

Ann Schoofs Hundt
Center for Quality and Productivity
  Improvement
University of Wisconsin–Madison
Madison, Wisconsin

Edmond W. Israelski
Abbott Laboratories
Abbott Park, Illinois

Kenji Itoh
Department of Industrial
  Engineering and Management
Tokyo Institute of Technology
Tokyo, Japan

Christopher W. Johnson
School of Computing Science
University of Glasgow
Glasgow, Scotland, United
  Kingdom

Ben-Tzion Karsh
Department of Industrial and
  Systems Engineering

and

Department of Family Medicine

and

Center for Quality and Productivity
  Improvement
University of Wisconsin–Madison
Madison, Wisconsin

Heidi B. King
Department of Defense Patient
   Safety Program
Office of the Assistant Secretary of
   Defense (Health Affairs)
TRICARE Management Activity
Falls Church, Virginia

Heidi S. Kramer
University of Utah
Salt Lake City, Utah

Daniel Krueger
Bellin Health Systems
Green Bay, Wisconsin

Andre Kushniruk
School of Health Information
   Science
University of Victoria
Victoria, British Columbia, Canada

Jamie A. Lapin
Department of Industrial and
   Systems Engineering
and
Center for Quality and Productivity
   Improvement
University of Wisconsin–Madison
Madison, Wisconsin

John D. Lee
Department of Industrial and
   Systems Engineering
University of Wisconsin–Madison
Madison, Wisconsin

Kari Lindström
Finnish Institute of Occupational
   Health
The Centre of Expertise for Work
   Organizations
Helsinki, Finland

Mark Linzer
Hennepin County Medical Center
Minneapolis, Minnesota

Colin F. Mackenzie
Shock Trauma Anesthesiology
   Research Organized Research
   Center
University of Maryland School of
   Medicine
Baltimore, Maryland

Marlene Dyrløv Madsen
Danish Institute of Medical
   Simulation
Herlev University Hospital
Herlev, Denmark

Peter M. Madsen
Brigham Young University
Provo, Utah

Linda Baier Manwell
Center for Women's Health and
   Research
University of Wisconsin–Madison
Madison, Wisconsin

Mary Matz
Veterans' Health Administration
Office of Public Health
Occupational Health Strategic
   Healthcare Group
Tampa, Florida

Julia E. McMurray
Meriter Medical Group
Madison, Wisconsin

Gustaf Molander
Department of Social Policy
University of Helsinki
Helsinki, Finland

Enid Montague
Department of Industrial and
   Systems Engineering
University of Wisconsin–Madison
Madison, Wisconsin

William H. Muto
Abbott Laboratories
Irving, Texas

Calvin K. L. Or
Department of Industrial and
   Manufacturing Systems
   Engineering
University of Hong Kong
Pokfulam, Hong Kong

Emily S. Patterson
Division of Health Information
   Management Systems
School of Allied Medical
   Professions
College of Medicine
Ohio State University
Columbus, Ohio

Sylvia Pelayo
Evalab—Clinical Investigation
   Centre for Innovative
   Technologies (CIC-IT)
Centre Hospitalier Universitaire
   (CHU)
and
Université Lille–Nord de France
Lille, France

Priyadarshini R. Pennathur
School of Medicine
Johns Hopkins University
Baltimore, Maryland

Shawna J. Perry
Department of Emergency
   Medicine
School of Medicine
Virginia Commonwealth University
Richmond, Virginia

James W. Pichert
Center for Patient and Professional
   Advocacy
Vanderbilt University School of
   Medicine
Nashville, Tennessee

Anna Pinto
Clinical Safety Research Unit
Department of Surgery and Cancer
Imperial College London
London, United Kingdom

Peter J. Pronovost
Department of Anesthesiology and
   Critical Care Medicine
Quality and Safety Research Group
Johns Hopkins University School of
   Medicine
and
Johns Hopkins University
   Bloomberg School of Public
   Health
and
Johns Hopkins University School of
   Nursing
Baltimore, Maryland

Joseph Rabatin
Alpert Medical School
Bellvue Hospital
Brown University
New York, New York

Ayesha Rashid
Hennepin County Medical Center
University of Minnesota
Minneapolis, Minnesota

Marta L. Render
Division of Pulmonary and Critical
    Care
Cincinnati VAMC Getting at Patient
    Safety Center
University of Cincinnati
Cincinnati, Ohio

Karlene H. Roberts
Haas School of Business
University of California–Berkeley
Berkeley, California

Ann E. Rogers
Nell Hodgson Woodruff School of
    Nursing
Emory University
Atlanta, Georgia

Michelle L. Rogers
School of Information Science and
    Technology
Drexel University
Philadelphia, Pennsylvania

Michael A. Rosen
Armstrong Institute for Patient
    Safety and Quality
and
Department of Anesthesiology and
    Critical Care Medicine
Johns Hopkins University School of
    Medicine
Baltimore, Maryland

Eduardo Salas
Department of Psychology
and
Institute for Simulation and
    Training
University of Central Florida
Orlando, Florida

Matthew C. Scanlon
Children's Hospital of Wisconsin
Medical College of Wisconsin
Milwaukee, Wisconsin

Mark W. Scerbo
Old Dominion University
Norfolk, Virginia

Maureen A. Smith
Department of Population Health
    Sciences, Family Medicine, and
    Surgery
and
Health Innovation Program
School of Medicine and Public
    Health
University of Wisconsin–Madison
Madison, Wisconsin

Michael J. Smith
Department of Industrial and
    Systems Engineering
and
Center for Quality and Productivity
    Improvement
University of Wisconsin–Madison
Madison, Wisconsin

Molly L. Snellman
Department of Industrial and
    Systems Engineering
University of Wisconsin–Madison
Madison, Wisconsin

Scott I. Tannenbaum
The Group for Organizational
    Effectiveness, Inc.
Albany, New York

Riccardo Tartaglia
Center for Clinical Risk
    Management and Patient Safety
Regione Toscana
Florence, Italy

Bruce R. Thomadsen
Department of Medical Physics,
    Human Oncology, Engineering
    Physics, Biomedical Engineering
    and Industrial and Systems
    Engineering
School of Medicine and Public
    Health
University of Wisconsin–Madison
Madison, Wisconsin

Teresa N. Thuemling
Center for Quality and Productivity
    Improvement
University of Wisconsin–Madison
Madison, Wisconsin

Rupa S. Valdez
Department of Industrial and
    Systems Engineering
University of Wisconsin–Madison
Madison, Wisconsin

Charles Vincent
Clinical Safety Research Unit
Department of Surgery and Cancer
Imperial College London
London, United Kingdom

James M. Walker
Geisinger Health System
Danville, Pennsylvania

Robert L. Wears
University of Florida Health Science
    Center
Jacksonville, Florida
and
Clinical Safety Research Unit
Imperial College
London, United Kingdom

Sallie J. Weaver
Department of Psychology and
    Institute for Simulation and
    Training
University of Central Florida
Orlando, Florida

Matthew B. Weinger
Geriatric Research Education and
    Clinical Center
Tennessee Valley VA Healthcare
    System
and
Center for Research and Innovation
    in Systems Safety
Vanderbilt University Medical
    Center
and
Department of Anesthesiology and
    Biomedical Informatics
Vanderbilt University School of
    Medicine
Nashville, Tennessee

Tosha B. Wetterneck
Department of Medicine
School of Medicine and Public
    Health

and

Center for Quality and Productivity
    Improvement
Systems Engineering Initiative for
    Patient Safety
University of Wisconsin–Madison
Madison, Wisconsin

Douglas A. Wiegmann
Department of Industrial and
    Systems Engineering
University of Wisconsin–Madison
Madison, Wisconsin

Eric S. Williams
Department of Management and
    Marketing
Culverhouse College of Commerce
    and Business Administration
University of Alabama
Tuscaloosa, Alabama

Bradford D. Winters
School of Medicine
Johns Hopkins University
Baltimore, Maryland

Yan Xiao
Baylor Health Care System
Dallas, Texas

Teresa Zayas-Cabán
Center for Primary Care,
    Prevention, and Clinical
    Partnerships
Agency for Healthcare Research
    and Quality
Rockville, Maryland

Section I

Introduction

1

Human Factors and Ergonomics in Health Care and Patient Safety

Pascale Carayon

CONTENTS

1.1 Definition of HFE..4
    1.1.1 Domains of HFE...4
    1.1.2 From Micro- to Macroergonomics...5
    1.1.3 Complexity in Health Care and HFE..6
1.2 HFE: Design for and by People..7
    1.2.1 Physical, Cognitive, and Psychosocial Characteristics of People............7
    1.2.2 Diversity of People in Health Care...8
    1.2.3 Design by People in Health Care..9
1.3 HFE Systems Approaches...9
    1.3.1 System Boundaries, Interactions, and Levels...10
    1.3.2 Macroergonomic Model of Patient Safety..10
    1.3.3 System Design and Life Cycle...11
1.4 HFE Contribution to Healthcare Transformation..11
1.5 Conclusion..12
Appendix A...13
References..13

In the United States, the recent healthcare reform has generated discussion and debate about how to provide access to high-quality health care to the majority of the population at a reasonable cost. The question has thus arisen about how to design healthcare systems and processes to meet this ambitious objective. Major efforts are underway and are being proposed to answer this important question. For instance, The Centers for Medicare & Medicaid Services has formally established the new Center for Medicare and Medicaid Innovation (CMMI) (Feder 2010). As stated on its Web site (http://www.innovations.cms.gov/), the goal of CMMI is to develop, test, evaluate, and deploy new care and payment models. An example of a new care model suggested by the U.S. healthcare reform and the CMMI is the patient-centered medical home. Different approaches and definitions of the medical home have been proposed by, for instance, the NCQA (National Committee for Quality Assurance) standards (www.ncqa.org) and AHRQ or the Agency for Healthcare Research and Quality (www.ahrq.gov). In a review paper, Stange et al. (2010) define the patient-centered medical home as "a team of people embedded in the community who seek to improve the health and healing of the people in that community" and who "work to optimize the fundamental attributes of primary care

combined with evolving new ideas about organizing and developing practice and changing the larger health care and reimbursement systems" (p. 602). The human factors and ergonomics (HFE) discipline can contribute to the design and implementation of the medical home by, for instance, providing principles for effective teamwork or contributing to the design of usable and useful health information technologies used by healthcare professionals and patients involved in the medical home. Therefore, HFE can help to design and implement new care models in the context of the U.S. healthcare reform.

As suggested by Paget (2004), clinical work is an "error-ridden activity." A number of reports and studies in the United States and other countries confirm the importance of patient safety and medical errors (Kohn et al. 1999; Baker et al. 2004). Across the world, the issue of patient safety is attracting increasing attention from governments, regulators, and other governmental agencies. For instance, the World Health Organization (WHO) launched the World Alliance for Patient Safety in 2004 (http://www.who.int/patientsafety). The World Alliance has focused its campaigns on hand hygiene and safety in surgery. It has also invested in educational programs, research, and other resources aimed at improving the implementation of patient safety programs and activities across the world. For instance, the WHO has invested in the development of an international classification for patient safety (Runciman et al. 2009). The classification includes many HFE concepts, such as types of human error and violations, system failures, resilience, error detection, and recovery, which are discussed in this book. HFE can contribute to patient safety improvements across the world in both developed and developing countries.

The need to enhance patient safety training in medical schools and other health sciences educational and training programs has also been recognized. Recently, the WHO's World Alliance for Patient Safety issued a set of recommendations and guidelines for patient safety curriculum in medical schools (Walton et al. 2010). The 11 topics of the WHO curriculum are as follows: (1) patient safety, (2) human factors engineering and its importance to patient safety, (3) systems and the impact of complexity on patient care, (4) effective teamwork, (5) understanding and learning from errors, (6) understanding and managing clinical risk, (7) introduction to quality improvement methods, (8) engaging with patients and carers, (9) infection control, (10) patient safety and invasive procedures, and (11) improving medication safety. The second topic directly addresses human factors engineering; other topics of the WHO curriculum are also addressed in this book, including systems and complexity, teamwork, organizational learning, human error, and the human factors aspects of infection control and medication safety.

The second edition of the *Handbook of Human Factors and Ergonomics in Health Care and Patient Safety* addresses the challenges described in this introduction as well as the challenge offered by Lucian Leape (2004) for human factors professionals and researchers to tackle the difficult patient safety and healthcare quality problems. This second edition presents and discusses a variety of HFE issues, concepts, and methods that can help understand, identify, mitigate, and remove the obstacles to safe health care and ultimately improve the quality of health care. It includes 9 sections with 48 chapters: (1) Introduction, (2) Macroergonomics and systems, (3) Job and organizational design, (4) Physical ergonomics, (5) Technology, (6) Human error, (7) Human factors and ergonomics methodologies, (8) Human factors and ergonomics interventions, and (9) Specific applications.

1.1 Definition of HFE

According to the International Ergonomics Association (IEA), ergonomics or human factors can be defined as follows:

> Ergonomics (or human factors) is the scientific discipline concerned with the understanding of interactions among humans and other elements of a system, and the profession that applies theory, principles, data and methods to design in order to optimize human well-being and overall system performance. (www.iea.cc)

In this definition, the "system" represents the physical, cognitive, and organizational artifacts that people interact with. The system can be a technology, a software, or a medical device; a person, a team, or an organization; a procedure, policy, or guideline; or a physical environment. Interactions between people and the systems are tasks. Ergonomists focus on the *design* of systems so that they fit the needs, abilities, and limitations of people (International Ergonomics Association [IEA] 2004; Meister and Enderwick 2001). As suggested by Leape et al. (1995) and many other experts and practitioners in health care and patient safety (Bogner 1994; Cook et al. 1998; Carayon 2007; Vincent 2010), the discipline of human factors can contribute to the safe *design* of healthcare systems by considering the various needs, abilities, and limitations of people involved in those systems. The quality and safety of care provided by healthcare systems are, of course, dependent on the patients' risk factors and the technical skills and knowledge of the healthcare staff, but they are also influenced by various characteristics of the system (Vincent et al. 2004); these system characteristics can be redesigned, changed, and improved by applying HFE principles and methods (Carayon et al. 2006; Carayon 2007). In health care, the "people involved" are diverse and include healthcare providers and workers, patients, and their families. People have varied needs, abilities, and limitations that change over time. The diverse theories, models, concepts, and methods of the HFE discipline can be used to address those individual needs, abilities, and limitations in system design.

1.1.1 Domains of HFE

According to the International Ergonomics Association (2000), the discipline of HFE covers three major domains: (1) physical ergonomics concerned with physical activity, (2) cognitive ergonomics concerned with mental processes, and (3) organizational ergonomics (also called macroergonomics) concerned with sociotechnical system design. Physical ergonomics focuses primarily on the physical characteristics of the person, cognitive ergonomics on the cognitive characteristics of the person, and organizational ergonomics on the psychosocial characteristics of the person. Important physical HFE issues in health care and patient safety, such as the design of hospital facilities, the design of physical environment, patient handling, patient room design, noise, and alarms, are addressed in this book. Physical ergonomic issues are also relevant for specific applications in health care, such as intensive care environments and emergency rooms. Methods for assessing physical ergonomic factors in health care and for designing and implementing interventions aimed at reducing physical stress are addressed in Sections IV and VII.

Human error is a cognitive ergonomic issue that has received much attention in health care and patient safety. Section VI deals with the complexity of patient care, taxonomies of medical failures, human factors in the design of reporting systems, and communication about unexpected errors and outcomes. Various approaches for human error reduction strategies are described in Section VIII. There is increasing understanding among healthcare leaders and managers of the contribution of latent failures (i.e., those failures that lay dormant for a long time and that are removed in space and time from active failures [Reason 1990]) and organizational factors to medical errors and patient safety. The HFE discipline has had a major impact in health care; healthcare leaders and managers have a better understanding of the human mechanisms involved in medical errors (e.g., levels and types of human performance and human error—skill-, rule-, and knowledge-based performance), and the influence of system characteristics on human behavior, performance, and error (Bogner 1994; Vincent et al. 1998; Carayon et al. 2007a).

The design and implementation of technologies in health care have raised various cognitive ergonomic issues, such as usability of medical devices, as well as macroergonomic issues related to the implementation and use of those technologies (e.g., training, participation in the change process). These issues are described in Section V.

Organizational ergonomic issues, such as job stress and burnout of healthcare workers, organizational culture and learning, teamwork, work schedules, and organizational design (e.g., transitions of care), are addressed in Section III. Numerous organizational ergonomic issues relevant to specific healthcare settings are addressed in Section IX. For instance, the challenges related to the family as a complex organization, such as roles and responsibilities for health information management, are highlighted in Chapter 43. The diversity of clinical needs experienced by patients and the organizational design consequences, such as difficulty in standardizing processes and the diversity of people, locations, and systems involved in the care of patients, are described in Chapter 44. Organizational ergonomic issues, such as human factors risk management in medical products, assessment of safety culture and climate, and incident analysis, are addressed in Section VII.

1.1.2 From Micro- to Macroergonomics

The diversity of topics addressed by HFE researchers and practitioners is clearly demonstrated in this book. Historically, the HFE discipline has developed to encompass an increasing number and type of interactions between people and systems (Wilson 2000). Hendrick (Hendrick 1991, 1997, 2008; Hendrick and Kleiner 2001) has described five "human-system interface technologies" of the HFE discipline:

- Human–machine interface technology, or hardware ergonomics
- Human–environment interface technology, or environmental ergonomics
- Human–software interface technology, or cognitive ergonomics
- Human–job interface technology, or work design ergonomics
- Human–organization interface technology, or macroergonomics

Table 1.1 includes health care and patient safety examples for each interface level. In health care, hardware ergonomics is concerned with the design of technologies, work layouts, and facilities. Hardware ergonomics draws on knowledge, concepts, and methods of physical and cognitive ergonomics primarily. Environmental ergonomic issues in health care include noise, temperature, humidity, airflow, and

TABLE 1.1

Human–System Interfaces in Health Care and Patient Safety

Human–System Interfaces	Examples in Health Care and Patient Safety
Human–machine interface, or hardware ergonomics	Design of controls (e.g., controls in a telemetry monitoring unit), displays (e.g., display of an anesthesia machine), workspaces (e.g., nursing station, patient room), and facilities (e.g., hospital, nursing home)
Human–environment interface, or environmental ergonomics	Noise, temperature, humidity, airflow, and vibration
Human–software interface, or cognitive ergonomics	Usability of medical devices and information technologies Information overload
Human–job interface, or work design ergonomics	Work schedule and work content
Human–organization interface, or macroergonomics	Teamwork, organizational culture and learning, work system, sociotechnical system, and HROs

vibration, and draw mainly from the physical ergonomics domain. Human–software interface issues are varied given the increasing number and diversity of devices, equipment, and technologies designed, implemented, and used in health care. They include usability, usefulness, workload, and error recovery. Work design ergonomics in health care is concerned with the way work is designed and organized, such as the content of work and task allocation. Macroergonomic issues in health care and patient safety are varied, as demonstrated, in particular, in Sections II and III. They include teamwork, organizational culture and learning, work system design, and high-reliability organizations (HROs).

1.1.3 Complexity in Health Care and HFE

There are many aspects of healthcare that are predictable and routine, for example, much of the care for chronic conditions; however, in other areas of health care, there are complex aspects that challenge the designers of healthcare work systems and processes

(Vincent 2010). Table 1.2 describes and compares different dimensions and characteristics of the complexity of healthcare systems as proposed by various authors (Plsek and Greenhalgh 2001; Effken 2002; Carayon 2006). Four major themes emerge from this analysis of complexity in health care. First, healthcare systems involve a range of people with varied backgrounds and perspectives who interact, communicate, collaborate, and make decisions. The healthcare systems are "fluid" as people move in and out of the systems (or sub-systems) and change over time. Therefore, healthcare systems should be designed to facilitate and support self-organization, adaptation, and resilience. Second, healthcare systems are dynamic; they evolve over time, respond to disturbances and contingencies, and adapt over time. Behaviors of individuals in those dynamic healthcare systems are often based on internal mental models that also evolve over time. Therefore, the design of healthcare systems should foster individual and organizational abilities to change, learn, and adapt. Third, health care involves uncertainty related to patient care and trajectory; this requires knowledge

TABLE 1.2

Characteristics and Dimensions of Complexity in Health Care

Health Care as a Complex Sociotechnical System (Carayon 2006)	Health Care as a Complex Dynamic Sociotechnical System (Effken 2002)	Health Care as a Complex Adaptive System (Plsek and Greenhalgh 2001)
Social system with heterogeneous perspectives in a distributed system	Patient care involves groups of people that cooperate and need to communicate on a continuous basis. Decision making is distributed and performed by people with different perspectives	Boundaries are blurred (fuzzy), rather than fixed, and well defined. Members of the system change, and people may belong to multiple systems Interaction leads to continually emerging, new behaviors, which can produce new creative solutions Self-organization is inherent through simple locally applied rules
Dynamic system and disturbances with both tight and loose coupling	Dynamic system	Agents' actions are based on internalized rules such as mental models, which evolve over time Agents and system(s) can adapt to local contingencies over time Systems are embedded within other systems and evolve over time Systems have patterns of behavior that can be identified, such as attractor patterns
Uncertainty	Patient care involves "emerging contingencies that require ad hoc, pragmatic responses" Value conflicts between not-for-profit and volunteer organizations, and financial objectives, and between local patient care and organizational financial goals	Unpredictability is inherent in the system Nonlinearity is inherent in the system Tension and paradox are inherent, and do not necessarily need to be resolved Competition and cooperation are often synergistic
Large problem spaces (e.g., large number of illnesses and possible diagnoses)		
Hazardous system such as medical errors		
Automation and interaction mediated through technologies		

and flexibility among healthcare professionals. Fourth, healthcare systems tackle a large number of high-risk problems (e.g., illnesses, diagnoses) in which people (e.g., physician, patient) use diverse medical, informational, and communication technologies and devices. Hazards are multiple in healthcare systems and cannot be easily predicted, anticipated, or eliminated. Therefore, healthcare systems need to be continuously evaluated and redesigned (Carayon 2006).

The many facets of complexity in health care also emphasize the need to consider the local context. Similar to the principle of deference to expertise advocated by HRO experts (Weick and Sutcliffe 2001), healthcare systems and processes should be designed to provide autonomy and flexibility at the lowest level possible; healthcare workers at the front line should have the tools and resources to deal with disturbances, contingencies, and uncertainties. Their work systems should be designed to facilitate their performance. Chapter 5 addresses this issue in more detail.

Given the complexity of health care, HFE interventions that do not consider the organizational context are unlikely to have a significant, sustainable impact on patient safety and quality of care. For instance, improving the physical design of a medical device or the cognitive interface of health information technology is important; but without understanding the organizational context in which these technologies are used, the tools may not be used safely. For example, the research on work-around and patient safety risks related to the use of bar coding medication administration (BCMA) technology has identified a range of system and organizational issues that may defeat the safety features of the technology (Patterson et al. 2002; Carayon et al. 2007b; Koppel et al. 2008). Therefore, macroergonomics (or organizational ergonomics) should play a major role in health care and patient safety.

1.2 HFE: Design for and by People

HFE professionals aim at understanding the interactions among humans and other system elements; they use their knowledge, models, and methods to improve the design, implementation, and use of various systems (International Ergonomics Association, IEA 2000). "Humans" in healthcare systems include the providers and workers that are either directly or indirectly involved in patient care as well as patients and their families and friends. Three categories of human characteristics need to be considered in any HFE analysis: (1) physical, (2) cognitive, and (3) psychosocial individual characteristics.

1.2.1 Physical, Cognitive, and Psychosocial Characteristics of People

Working environments in health care can impose many physical, cognitive, or psychosocial stressors on the health care providers and workers as well as on patients and families. According to the Bureau of Labor Statistics (BLS), nursing is an occupation at high risk for work-related musculoskeletal injuries (Bureau of Labor Statistics (BLS) 2000). Designing the healthcare workplace for optimal human performance and using technologies such as lifting devices to minimize the need for human strength can help reduce the physical stressors experienced by nurses. The design of healthcare systems and processes also needs to consider the cognitive characteristics of people. Often people in health care are overwhelmed by the number of tasks to perform (e.g., high patient flow in an emergency department, limited time allocated for a primary care physician to visit with a patient), need to work under high time pressure (e.g., quick decision making during a code), or have a hard time coping with information load (e.g., need to keep up with medical knowledge, information overload related to poor design of electronic health record technology). These various aspects of workload can be addressed by understanding the cognitive limitations and abilities of individuals, as well as the larger organizational system that may either add to workload (e.g., poor supply chain management that requires nurses to search for supplies) or may be able to reduce the workload (e.g., teamwork to facilitate task distribution and allocation and communication).

Psychosocial work stress and burnout are becoming increasingly important in healthcare organizations. The 2004 report by the Institute of Medicine identifies a range of working conditions that contribute to job stress and burnout of nurses, which can also be related to patient safety (Institute of Medicine Committee on the Work Environment for Nurses and Patient Safety 2004). Berland et al. (2008) interviewed critical care nurses who described how high job demands, lack of job control, and low social support can produce hazards and affect patient safety. Many of the psychosocial work issues of importance to providers in stressful working environments, such as work schedules and time pressure, are described in Section III.

To better design healthcare systems from a physical viewpoint, several physical human characteristics need to be understood: height, weight, reach envelop, physical strength, physical movement, and sensory characteristics (e.g., vision, hearing). For instance, a technology such as BCMA is often implemented on a Personal Digital Assistant (PDA). Several physical characteristics of the BCMA technology users should be considered, such as the size of their hands for holding the PDA and their visual capacity in reading the small font on the PDA screen. Such technologies should be designed for the full range of physical human characteristics. Physical human characteristics are also important for designing the physical layout of hospital units, patient rooms, exam rooms, or nursing stations. The following ergonomic principles should be considered in healthcare workspace layout design: to minimize detection and perception time, to minimize decision time, to minimize manipulation time, and to optimize opportunity for movement (Carayon et al. 2003). These principles consider physical characteristics of people, such as reach envelop in the design of a workstation to allow opportunity for movement and reduce static physical loading. These principles also rely on knowledge of the cognitive characteristics of humans, such as information processing and reaction time to minimize perception time and decision time.

Cognitive human characteristics of importance in health care and patient safety include information processing and decision making, knowledge and expertise, and human error. Information processing and decision making models (Klein et al. 1993; Rasmussen et al. 1994; Wickens et al. 2004) provide useful information on cognitive human characteristics, for example, limited attention capacity, decision making heuristics, and characteristics of short- and long-term memory. The usability of medical devices and associated instructions and training materials relies very much on knowledge regarding the cognitive characteristics of end users. Various usability methods that can be used to evaluate medical devices and other healthcare technologies are described in Chapter 32. In addition to considering information processing and other cognitive characteristics of users, usability methods can also identify problems related to physical interactions. For instance, a screen that is too small may not be readable by people with visual problems.

Cognitive and psychosocial human characteristics are also important to consider when designing training programs and materials. For instance, mental models of end users can provide valuable information on how end users perceive specific tasks. The introduction of a technology changes the way tasks are performed, therefore requiring end users to develop a new mental model for their work. This need for the development of a new mental model should be addressed in the training offered to end users regarding the new technology (Salas and Cannon-Bowers 1997). For more information, see Chapter 37.

In the previous section, we discussed how the HFE discipline evolved over time from micro- to macroergonomics and has integrated organizational issues in system design. Motivation and satisfaction are important to consider when implementing any organizational

and technological change in healthcare institutions. For instance, a key HFE principle is user involvement. This principle has been clearly highlighted in the area of information technology design and implementation (Eason 1988, 2001; Carayon and Karsh 2000; Karsh 2004). Understanding how end users can be involved in an organizational or technological change to foster and support their motivation and satisfaction is critical for the success and sustainability of health care and patient safety improvement activities.

Understanding why humans make errors has benefited from theories and models based on cognitive and organizational ergonomics (Reason 1990; Wickens et al. 2004). The well-known taxonomy of slips-lapses and mistakes reflects two different cognitive mechanisms, one more routine, sometimes automatic mechanism that is susceptible to distractions and breaks in the routine, and another mechanism that involves higher-cognitive processing and decision making processes (Reason 1990; Rasmussen et al. 1994; Wickens et al. 2004). More recently, experts have also highlighted the role of organizational factors in creating the conditions (i.e., latent failures) for human errors and accidents (Reason 1997). This book proposes system models of human error and safety, for instance the Artichoke Systems Approach for Error Prevention.

Important knowledge, concepts, and methods that consider social needs and characteristics in, for instance, the design of teamwork have been produced by HFE. A range of best practices and guidelines for implementing teamwork in health care is described in Chapter 37. Given the collaborative nature of most health care and patient safety activity, it is important to understand the social fabric of healthcare work systems. For instance, the physical design of a hospital needs to consider not only the physical needs and requirements of the end users (e.g., patients, nurses, physicians), but also their social needs, such as need for communication, teamwork, and social interaction. HFE can provide the basic principles and tools for designing healthcare systems that support and encourage the social needs of healthcare providers, patients, and their families.

1.2.2 Diversity of People in Health Care

The definition of human factors (or ergonomics) by the IEA does not make any specific assumption about the "humans." A lot of HFE research and practice target the workers, for example, nurses, physicians, pharmacists, and other healthcare providers and staff (e.g., technicians, orderlies, maintenance personnel, biomedical engineers). This book describes numerous HFE issues for the workers in healthcare systems. On the other hand, HFE knowledge, concepts, and methods can equally apply to the design of systems with which patients and their families interact. For instance, the viewpoint of both the providers and the patients and their families is addressed in Chapter 25. Several HFE issues of relevance to children as patients are listed in Chapter 42.

Healthcare workers, patients, and their families all have physical, cognitive, and psychosocial characteristics that need to be considered when studying and designing the systems with which they interact. Sometimes different groups with different characteristics and different objectives exist in the same environment or interact with the same technologies. For instance, an infusion pump is used by a nurse who sets the medicine and the IV. The same IV pump is also part of the system in which the patient's medical condition evolves. The IV pump may be attached to a pole that limits the mobility of the patient (e.g., going to the bathroom). Healthcare system design is therefore challenging because of the variety of end user groups and their different system interactions in order to achieve various objectives (i.e., various tasks).

1.2.3 Design by People in Health Care

Healthcare workers and professionals, patients, and their families are not passive members of the healthcare system; they are active participants who process, communicate, and share information, make decisions, and adapt to contingencies. They also participate in detecting and analyzing errors and other failures in the systems, and they often help in designing and improving systems and processes.

The HFE discipline has increasingly paid attention to the way HFE knowledge, concepts, and methods are being adopted, implemented, and used. In conceptualizing the introduction of HFE in health care as an innovation, various barriers can be identified, and recommendations for enhancing the adoption and use of HFE have been proposed (Carayon 2010). It is not sufficient to understand the physical and cognitive interactions between people and systems. Although such knowledge is important for understanding the positive and negative system characteristics, from a human factors viewpoint, to successfully change systems and achieve better performance, enhanced safety, and improved quality of working life, knowledge about psychosocial human characteristics and methods such as participatory ergonomics are necessary. Participatory ergonomics was originally developed to facilitate the design and implementation of HFE programs and activities (Noro and Imada 1991). Participatory ergonomics includes various methods in which end users are involved in the identification and analysis of risk factors in their work system as well as the design and implementation of ergonomic solutions (Noro and Imada 1991; Wilson and Haines 1997).

Participatory ergonomics programs vary on the following dimensions: permanency (is the participatory ergonomics effort a one-shot effort or part of an on-going organizational effort?), involvement (whether people participate directly or indirectly), level of influence or the organizational level that is targeted by the participatory ergonomics program (e.g., department, entire organization), decision making power (who has the power to make decisions? e.g., individual consultation versus group delegation), composition of the group involved in the participatory process, requirement for participation (i.e., voluntary or compulsory), focus (e.g., analysis of a task, workstation, or job), remit (i.e., involvement in various phases of the participatory process), and role of the ergonomist (Haines et al. 2002). Participatory ergonomics programs are varied and use a range of methods, such as design decision groups (Wilson 1991), quality circles and other quality improvement methods (Noro and Imada 1991; Zink 1996), and representative ergonomic facilitators (Haims and Carayon 1998).

Evanoff and colleagues (Bohr et al. 1997; Evanoff et al. 1999) have conducted several studies of participatory ergonomics in health care to address physical stressors that contribute to work-related musculoskeletal disorders. A key challenge identified in these studies was the high workload and time pressure related to patient care. These conditions make the involvement of workers, in particular ICU nurses, more difficult and challenging. Participatory ergonomics approaches need to be developed and tested to effectively and efficiently involve healthcare workers as well as patients. Those innovative methods need to consider the psychosocial benefits of participation (e.g., motivation and satisfaction of healthcare workers and patients), while at the same time not overburdening healthcare workers and patients who are already experiencing high workload and time pressure. Healthcare workers have a primary commitment to patient care, and they may not have the time and other resources to be active participants in system analysis and redesign. The involvement of patients and their families in healthcare system design is in its infancy and needs to be encouraged and further developed.

1.3 HFE Systems Approaches

An important element of the IEA definition of HFE relates to system design. This book has an entire section dedicated to HFE systems approaches in health care and patient safety. This section describes various systems approaches and frameworks: work system design, macroergonomics and sociotechnical system design, system model of the factors contributing to errors (the "Artichoke" model), and organizational learning as a system characteristic. It also addresses the conflicts and possible compromises between the professional medical model and systems approaches.

1.3.1 System Boundaries, Interactions, and Levels

A key issue in system design is the definition of the system, that is, the objectives of the system and the boundaries of the system. The boundaries of the system can be physical (e.g., a patient room as the system), organizational (e.g., a hospital or a department as the system), or temporal (e.g., beginning and end of a patient visit). Guidance for understanding the various elements and characteristics of systems is provided by many of the systems approaches and frameworks included in this book. For instance, the work system model developed by Smith and Carayon (Smith and Carayon-Sainfort 1989; Carayon and Smith 2000; Smith and Carayon 2000) assumes that an individual (e.g., a healthcare worker, a patient) performs a variety of tasks using tools and technologies in a particular physical environment and location under organizational conditions. Specific facets of the five work system elements are described in this book (Chapter 5). The work system model is used to describe HFE issues related to specific applications (e.g., home care). Intervention and redesign approaches based on the work system model are proposed in Chapter 35.

Many HFE principles exist that can help design the systems, the system elements, and the interactions between system elements to produce high-quality safe care and to ensure the well-being, safety, and performance of end users (Carayon et al. 2007a). Increasingly, researchers and healthcare practitioners have begun to understand that interactions and interfaces between connected, dependent systems for patient care produce challenges for quality and safety of care. For example, medication reconciliation for patients being admitted to a hospital involves the admitting team of the hospital but may require communication with the primary care physician and the pharmacy where the patient gets his/her medications, both being located elsewhere. This example shows that the care provided to patients depends on several systems (e.g., admitting team at hospital, primary care physician, pharmacy) that are connected to each other and depend on each other to produce high-quality safe care. Issues related to the interactions and interfaces between systems and organizations are addressed in this book. Chapter 11 highlights HFE issues of transitions of care, such as communication and sense making between providers

during a shift change. Patient safety problems arise or are detected and solved at the interface between systems. How effectively interactions between systems can contribute to the discovery of errors therefore preventing patient harm is addressed in chapters of Section IX such as Chapter 41 as well as in Chapter 11.

According to Plsek and Greenhalgh (2001), the boundaries of healthcare systems can be blurry and fuzzy (see Table 1.2). Therefore, defining the specific system to analyze and redesign may be challenging. For instance, physicians may be working in a hospital but not be directly employed by the hospital, although they work alongside hospital colleagues who are employees of the hospital. This may pose unique challenges when trying to introduce changes throughout the hospital organization.

Another challenging aspect of system design are the various system levels and their interactions (Rasmussen 2000; Vicente 2003). The 2001 IOM (Institute of Medicine) report on *Crossing the Quality Chasm* defines four levels at which interventions are needed to improve the quality and safety of care (Institute of Medicine Committee on Quality of Health Care in America 2001; Berwick 2002): Level A—experience of patients and communities, Level B—microsystems of care (that is, the small units of work that actually give the care that the patient experiences), Level C—healthcare organizations, and Level D—healthcare environment. Designing a safe system at levels A or B may be limited because of constraints at levels C or D. Much discussion about how to redesign systems and processes for improving patient safety and quality of care recognizes the hierarchy of system levels. The Agency for Healthcare Research and Quality is providing funding to research projects targeted at redesign and change at multiple levels; see, for example, the recent AHRQ program announcement on patient safety and medical liability reform. Rasmussen (1997) and Moray (1994) have proposed models for describing the hierarchy of systems and their interactions. A key issue for safety, effectiveness, and efficiency is the alignment of systems (Vicente 2003); incentives at one level may negate safety and other improvement efforts at a lower level. Therefore, objectives and incentives of different inter-connected system levels need to be congruent.

1.3.2 Macroergonomic Model of Patient Safety

The Systems Engineering Initiative for Patient Safety (SEIPS) model (Carayon et al. 2006) is based on the macroergonomic work system model developed by Smith and Carayon (Smith and Carayon-Sainfort 1989; Carayon 2009; Carayon and Smith 2000; Smith and Carayon 2000), and incorporates the Structure-Process-Outcome model of healthcare quality (Donabedian 1978). The key characteristics of the SEIPS model are as follows: (1) description of the work system elements (this generic system model can be easily used to describe healthcare systems), (2) incorporation of the well-known quality of care model developed by Donabedian (this increases the acceptance of the SEIPS model and HFE concepts by the healthcare community), (3) care process (care process, care pathway, patient journey, and workflow are terms often used in health care; our model explains how the work system can affect care processes), and (4) integration of patient outcomes and organizational/employee outcomes (this element of the model emphasizes the importance of quality of working life and occupational safety and health, in addition to patient safety; we also describe links between worker safety and patient safety).

The SEIPS model has been used by numerous healthcare researchers, professionals, and educators. We have used the SEIPS model in our own research to study the implementation of various forms of health information technology (e.g., EHR, CPOE, BCMA, smart infusion pump, tele-ICU) and to examine patient safety in multiple care settings (e.g., ICU, pediatrics, primary care, outpatient surgery, cardiac surgery, transitions of care). Other researchers have used the SEIPS model to study the timeliness of follow-up of abnormal test results in outpatient settings (Singh et al. 2009), to examine the safety of EHR technology (Sittig and Singh 2009), and to evaluate ways of improving electronic communication and alerts (Hysong et al. 2009). The SEIPS model has been adopted by patient safety leaders such as Peter Pronovost from Johns Hopkins University (Pronovost et al. 2009). For more information on the SEIPS model, see Chapter 5.

1.3.3 System Design and Life Cycle

In pursuit of quality, healthcare systems and processes are often analyzed and redesigned; for instance, when renovating a hospital unit or designing a new healthcare facility, when deploying a new healthcare information technology or purchasing a new medical device, when implementing a new care process, practice, or guideline, or when redesigning daily rounds and improving teamwork. HFE should be considered in all of these projects as they all deal with system redesign and behavior change. For instance, Weinert and Mann (2008) recommend the use of HFE to facilitate the adoption of new processes in the intensive care unit. HFE should be a major contributor to implementation science, dissemination research, or type 2 translational research in health care.

There are three types of system design project (Carayon et al. 2003; Meister and Enderwick 2001): (1) design of a new system (e.g., construction of a new hospital), (2) updating a system (e.g., technological change in medication administration such as BCMA), and (3) system redesign (e.g., effort aimed at redefining the objectives of a clinical microsystem, its systems, and its processes). The design of a new system versus the updating or redesign of an existing system poses different kinds of challenges. In the design of a new system, there may not be any historical data or previous experience to build on and to use to determine the system design characteristics. On the other hand, this may facilitate the creation and implementation of new structures and processes because of lack of resistance to change. Updating or redesigning a system should involve significant planning work to address not only the system design itself (i.e., content of the change), but also the implementation organization (i.e., process of the change). Many factors need to be considered at the stages of system implementation, such as involvement and participation of end users, information about the change communicated to end users and other stakeholders, training and learning, feedback, and project management (Korunka et al. 1993; Korunka and Carayon 1999). The transition period during which the new system is being implemented can be a source of uncertainty and stress; therefore providing adequate support (e.g., "super users" or other experts available for consultation) and sufficient time for learning and adaptation are important.

Designing a new healthcare system or process and updating or redesigning a system or process may be considered as a "project," that is, it has a beginning and an end; but the effects and consequences of the change may develop over time, even long after the change has been implemented. Therefore, it is important to consider issues related to episodic change as well as continuous change (Weick and Quinn 1999). In a study of the implementation of smart IV pump technology in an academic hospital, we described the longitudinal changes that occurred after the introduction of the new technology (Carayon et al. 2008). This case highlights the need to understand technology implementation in a longitudinal manner. Once a technology is implemented, several changes can subsequently occur in the technology itself and/or in the tasks and processes associated with the technology; users of the technology adapt to it and adapt the technology (Tyre and Orlikowski 1994). Therefore, the implementation of a technology in health care can have characteristics of both an episodic change and a continuous change (Weick and Quinn 1999).

1.4 HFE Contribution to Healthcare Transformation

This chapter began with a description of the U.S. healthcare reform and the systemic changes engendered by the reform. HFE can (and should) contribute to healthcare reform and healthcare transformation. HFE knowledge and skills are needed to improve patient safety and solve other quality of care problems; this need has been recognized and articulated in particular by the healthcare community (Leape et al. 2002). However, sometimes HFE is misused and misunderstood. For instance, a "human factors" curriculum for surgical clerk students has recently been proposed (Cahan et al. 2010); however, the curriculum was actually about psychosocial issues and communication, not about HFE (Hu and Greenberg 2010). We need to take every opportunity to explain and "sell" HFE.

Collaborating with healthcare professionals and organizations is critical to ensure that HFE has relevant impact. HFE professionals and researchers may not have the knowledge and skills to identify the issues that are important and relevant to patient safety and quality of care (Carayon 2010). HFE leaders continue to remind HFE professionals and researchers of the need to collaborate with domain experts (Moray 2000; Rasmussen 2000), and health care is a prime example of this need for HFE to engage in close collaboration with domain experts (e.g., healthcare professionals, health services researchers and other healthcare researchers, and healthcare organizations). HFE leaders in health care and patient safety research have also emphasized the need for meaningful long-term collaborations between HFE and health care (Gopher 2004).

In order to address the challenge of implementing HFE in health care and patient safety, it is useful to consider HFE as an innovation and to identify the issues related to adoption, dissemination, implementation, and sustainability of HFE innovations (Carayon 2010). Examples of HFE innovations in health care include the following: (1) use of HFE tools and methods, (2) education and training in HFE knowledge, and (3) HFE professionals employed by healthcare organizations (Carayon 2010). All three approaches are necessary and should be encouraged, supported, and deployed. First, HFE researchers need to develop reliable and valid tools and methods, ensure their usability in the healthcare context, and make them available to healthcare organizations (as an example of the successful translation of HFE to health care, see the AHRQ TeamSTEPSS program; http://teamstepps.ahrq.gov/). Second, increasing HFE knowledge in healthcare organizations requires well-designed training and educational programs (e.g., the SEIPS short course on human factors engineering and patient safety; http://cqpi.engr.wisc.edu/short-course_home). However, these two approaches may not be sufficient, particularly for complex patient safety and quality of care problems that require in-depth HFE knowledge; these problems are better addressed through in-house HFE expertise. This requires that healthcare organizations hire HFE professionals or "biculturals," that is, healthcare professionals who receive in-depth HFE training (Carayon and Xie 2011). The in-house HFE experts will know the healthcare organization, its operations, and culture; the combination of organizational knowledge and HFE expertise can address complex patient safety and quality of care problems. To my knowledge, very few healthcare organizations have actually hired HFE professionals or have encouraged healthcare professionals to become "biculturals." Some healthcare organizations have relied on research collaboration and consulting partnerships with HFE professionals in order to compensate for the lack of in-house HFE expertise; in this partnership scenario, the sustainability of HFE knowledge within the healthcare organization is probably tenuous and depends on the internal organizational capacity for adopting HFE, particularly as part of standard operating procedures (e.g., purchasing of equipment, IT projects).

1.5 Conclusion

This book represents the diversity of the HFE discipline and its various applications to health care and patient safety. Lucian Leape (2004) has called for HFE researchers and practitioners to take on the challenges of designing high-quality, safe healthcare systems and processes. This book provides important information to HFE practitioners engaged in efforts to improve systems and processes in health care. It also clearly highlights the health care and patient safety gaps in HFE knowledge that HFE researchers need to consider.

Vincent (2010) contrasts two approaches to patient safety improvement: (1) removing sources of error and (2) focusing on resilience and organizational culture. He then suggests that we need to combine both methods. This book addresses both approaches; some chapters focus on one or the other approach, while other chapters propose ways of combining the two methods. The HFE discipline aims at improving the interactions between humans and various system elements (Wilson 2000). Whether health care is a system has been debated; however, systems approaches are necessary to understand the multiple, complex interactions between humans and systems as well as between systems. This book can help practitioners in healthcare organizations become more familiar with HFE systems approaches, concepts, and methods. It provides many methods and tools as well as the underlying frameworks necessary to adequately apply HFE knowledge.

Appendix A

Bibliography on Human Factors and Ergonomics

Salvendy, G. (ed.). 2006. *Handbook of Human Factors and Ergonomics*, 3rd edn. New York: John Wiley & Sons.

Stanton, N., A. Hedge, K. Brookhuis, E. Salas, and H.W. Hendrick (eds.). 2004. *Handbook of Human Factors and Ergonomics Methods*. Boca Raton, FL: CRC Press.

Wilson, J.R. and Corlett, N. (eds.). 2005. *Evaluation of Human Work*, 3rd edn. Boca Raton, FL: CRC Press.

Key Journals in the Area of Human Factors and Ergonomics

- *Applied Ergonomics*
- *Ergonomics*
- *Human Factors*

Selected Journals in the Area of Health Care and Patient Safety

- *Journal of the American Medical Association (JAMA)*
- *Quality and Safety in Health Care*
- *Journal of Patient Safety*
- *Joint Commission Journal of Quality & Safety*

Key National and International Organizations in the Area of Health Care and Patient Safety

- Agency for Healthcare and Research Quality (www.ahrq.gov)
- Institute for Healthcare Improvement (www.ihi.org)
- Institute of Medicine (www.iom.edu/)
- Institute for Safe Medication Practices (www.ismp.org)
- National Patient Safety Foundation (www.npsf.org)
- The Joint Commission (www.jointcommission.org/)
- U.S. Food and Drug Administration (www.fda.gov)
- VA National Center for Patient Safety (http://www.patientsafety.gov)
- WHO World Alliance for Patient Safety (www.who.int/patientsafety)

References

Baker, G.R., P.G. Norton, V. Flintoft, R. Blais, A. Brown, J. Cox, E. Etchells, W.A. Ghali, P. Hebert, S.R. Majumdar, M.O'Beirne, L. Palacios-Derflingher, R.J. Reid, S. Sheps, and R. Tamblyn. 2004. The Canadian adverse events study: The incidence of adverse events among hospital patients in Canada. *Journal of the Canadian Medical Association* 170(11):1678–1686.

Berland, A., G.K. Natvig, and D. Gundersen. 2008. Patient safety and job-related stress: A focus group study. *Intensive and Critical Care Nursing* 24(2):90–97.

Berwick, D.M. 2002. A user's manual for the IOM's 'Quality Chasm' report. *Health Affairs* 21(3):80–90.

Bogner, M.S. (ed.). 1994. *Human Error in Medicine*. Hillsdale, NJ: Lawrence Erlbaum Associates.

Bohr, P.C., B.A. Evanoff, and L. Wolf. 1997. Implementing participatory ergonomics teams among health care workers. *American Journal of Industrial Medicine* 32(3):190–196.

Bureau of Labor Statistics (BLS). 2000. *Lost-Worktime Injuries and Illnesses: Characteristics and Resulting Time Away From Work, 1998*. Washington, DC: Bureau of Labor Statistics, U.S. Department of Labor.

Cahan, M.A., A.C. Larkin, S. Starr, S. Wellman, H.L. Haley, K. Sullivan, S. Shah, M. Hirsh, D. Litwin, and M. Quirk. 2010. A human factors curriculum for surgical clerkship students. *Archives of Surgery* 145(12):1151–1157.

Carayon, P. 2006. Human factors of complex sociotechnical systems. *Applied Ergonomics* 37:525–535.

Carayon, P. 2007. *Handbook of Human Factors in Health Care and Patient Safety*. Mahwah, NJ: Lawrence Erlbaum Associates.

Carayon, P., C. Alvarado and A.S. Hundt. 2007a. Work design and patient safety. *TIES—Theoretical Issues in Ergonomics Science* 8(5):395–428.

Carayon, P. 2009. The balance theory and the work system model. Twenty years later. *International Journal of Human-Computer Interaction* 25(5):313–327.

Carayon, P. 2010. Human factors in patient safety as an innovation. *Applied Ergonomics* 41(5):657–665.

Carayon, P., C. Alvarado, and A.S. Hundt. 2003. *Reducing Workload and Increasing Patient Safety Through Work and Workspace Design*. Washington, DC: Institute of Medicine.

Carayon, P., C.J. Alvarado, et al. 2003. Work system and patient safety. *Human Factors in Organizational Design and Management-VII*. H. Luczak and K.J. Zink. Santa monica, CA, IEA Press: 583–589.

Carayon, P., A.S. Hundt, B.T. Karsh, A.P. Gurses, C.J. Alvarado, M. Smith, and P.F. Brennan. 2006. Work system design for patient safety: The SEIPS model. *Quality and Safety in Health Care* 15(Suppl. I):i50–i58.

Carayon, P. and B. Karsh. 2000. Sociotechnical issues in the implementation of imaging technology. *Behaviour and Information Technology* 19(4):247–262.

Carayon, P. and M.J. Smith. 2000. Work organization and ergonomics. *Applied Ergonomics* 31:649–662.

Carayon, P., T.B. Wetterneck, A.S. Hundt, M. Ozkaynak, J. DeSilvey, B. Ludwig, P. Ram, and S.S. Rough. 2007b. Evaluation of nurse interaction with bar code medication administration technology in the work environment. *Journal of Patient Safety* 3(1):34–42.

Carayon, P., T.B. Wetterneck, A.S. Hundt, S. Rough, and M. Schroeder. 2008. Continuous technology implementation in health care: The case of advanced IV infusion pump technology. In *Corporate Sustainability as a Challenge for Comprehensive Management*, K. Zink (ed.). New York: Springer.

Carayon, P. and A. Xie. 2011. Decision making in healthcare system design: When human factors engineering meets health care. In *Cultural Factors in Decision Making and Action*, R. W. Proctor, S. Y. Nof, and Y. Yih (eds.). London, U.K.: Taylor & Francis.

Cook, R.I., D.D. Woods, and C. Miller. 1998. *A Tale of Two Stories: Contrasting Views of Patient Safety*. Chicago, IL: National Patient Safety Foundation.

Donabedian, A. 1978. The quality of medical care. *Science* 200:856–864.

Eason, K. 1988. *Information Technology and Organizational Change*. London, U.K.: Taylor & Francis.

Eason, K. 2001. Changing perspectives on the organizational consequences of information technology. *Behaviour and Information Technology* 20(5):323–328.

Effken, J.A. 2002. Different lenses, improved outcomes: A new approach to the analysis and design of healthcare information systems. *International Journal of Medical Informatics* 65(1):59–74.

Evanoff, V.A., P.C. Bohr, and L. Wolf. 1999. Effects of a participatory ergonomics team among hospital orderlies. *American Journal of Industrial Medicine* 35:358–365.

Feder, J. L. 2010. Your mission, should you choose to accept it. *Health Affairs* 29(11):1985.

Gopher, D. 2004. Why is it not sufficient to study errors and incidents: Human factors and safety in medical systems. *Biomedical Instrumentation and Technology* 38(5):387–409.

Haims, M.C. and P. Carayon. 1998. Theory and practice for the implementation of 'in-house', continuous improvement participatory ergonomic programs. *Applied Ergonomics* 29(6):461–472.

Haines, H., J.R. Wilson, P. Vink, and E. Koningsveld. 2002. Validating a framework for participatory ergonomics (the PEF). *Ergonomics* 45(4):309–327.

Hendrick, H.W. 1991. Human factors in organizational design and management. *Ergonomics* 34:743–756.

Hendrick, H.W. 1997. Organizational design and macroergonomics. In *Handbook of Human Factors and Ergonomics*, G. Salvendy (ed.). New York: John Wiley & Sons.

Hendrick, H.W. 2008. Applying ergonomics to systems: Some documented "lessons learned". *Applied Ergonomics* 39(4):418–426.

Hendrick, H.W. and B.M. Kleiner. 2001. *Macroergonomics—An Introduction to Work System Design*. Santa Monica, CA: The Human Factors and Ergonomics Society.

Hu, Y.Y. and C.C. Greenberg. 2010. What are human factors? *Archives of Surgery* 145(12):1157.

Hysong, S., M. Sawhney, L. Wilson, D. Sittig, A. Esquivel, M. Watford, T. Davis, D. Espadas, and H. Singh. 2009. Improving outpatient safety through effective electronic communication: A study protocol. *Implementation Science* 4(1):62.

Institute of Medicine Committee on Quality of Health Care in America. 2001. *Crossing the Quality Chasm: A New Health System for the Twenty-First Century*. Washington, DC: National Academy Press.

Institute of Medicine Committee on the Work Environment for Nurses and Patient Safety. 2004. *Keeping Patients Safe: Transforming the Work Environment of Nurses*. Washington, DC: The National Academies Press.

International Ergonomics Association (IEA). 2004. *The Discipline of Ergonomics 2000* [cited August 22, 2004].

Karsh, B.T. 2004. Beyond usability: Designing effective technology implementation systems to promote patient safety. *Quality and Safety in Health Care* 13:388–394.

Klein, G.A., J. Orasanu, and R. Calderwood (eds.). 1993. *Decision Making in Action: Models and Methods*. Norwood, NJ: Ablex Publishing.

Kohn, L.T., J.M. Corrigan, and M.S. Donaldson (eds.). 1999. *To Err Is Human: Building a Safer Health System*. Washington, DC: National Academy Press.

Koppel R., T. Wetterneck, J.L. Telles, and B.T. Karsh. 2008. Workarounds to barcode medication administration systems: Their occurrences, causes, and threats to patient safety. *Journal of the American Medical Informatics Association* 15:408–423.

Korunka, C. and P. Carayon. 1999. Continuous implementations of information technology: The development of an interview guide and a cross-national comparison of Austrian and American organizations. *The International Journal of Human Factors in Manufacturing* 9(2):165–183.

Korunka, C., A. Weiss, and B. Karetta. 1993. Effects of new technologies with special regard for the implementation process per se. *Journal of Organizational Behavior* 14(4):331–348.

Leape, L. 2004. Human factors meets health care: The ultimate challenge. *Ergonomics in Design* 12(3):6–12.

Leape, L.L., D.W. Bates, D.J. Cullen, J. Cooper, H.J. Demonaco, T. Gallivan, R. Hallisey, J. Ives, N. Laird, G. Laffel, R. Nemeskal, L.A. Petersen, K. Porter, D. Servi, B.F. Shea, S.D. Small, B.J. Sweitzer, B.T. Thompson, M. Vander Vliet, and ADE Prevention Study Group. 1995. Systems analysis of adverse drug events. *Journal of the American Medical Association* 274(1):35–43.

Leape, L.L., D.M. Berwick, and D.W. Bates. 2002. What practices will most improve safety? Evidence-based medicine meets patient safety. *Journal of the American Medical Association* 288(4):501–507.

Meister, D. and T.P. Enderwick. 2001. *Human Factors in System Design, Development, and Testing*. Mahwah, NJ: Lawrence Erlbaum Associates.

Moray, N. 1994. Error reduction as a systems problem. In *Human Error in Medicine*, M.S. Bogner (ed.). Hillsdale, NJ: Lawrence Erlbaum Associates.

Moray, N. 2000. Culture, politics and ergonomics. *Ergonomics* 43(7):858–868.

Noro, K. and A. Imada. 1991. *Participatory Ergonomics*. London, U.K.: Taylor & Francis.

Paget, M.A. 2004. *The Unity of Mistakes*. Philadelphia, PA: Temple University Press.

Patterson, E.S., R.I. Cook, and M.L. Render. 2002. Improving patient safety by identifying side effects from introducing bar coding in medication administration. *Journal of the American Medical Informatics Association* 9:540–553.

Plsek, P. and T. Greenhalgh. 2001. Complexity science: The challenge of complexity in health care. *British Medical Journal* 323:625–628.

Pronovost, P.J., C.A. Goeschel, J.A. Marsteller, J.B. Sexton, J.C. Pham, and S.M. Berenholtz. 2009. Framework for patient safety research and improvement. *Circulation* 119(2):330–337.

Rasmussen, J. 1997. Risk management in a dynamic society: A modelling problem. *Safety Science* 27(2/3):183–213.

Rasmussen, J. 2000. Human factors in a dynamic information society: Where are we heading? *Ergonomics* 43(7):869–879.

Rasmussen, J., A.M. Pejtersen, and L.P. Goodstein. 1994. *Cognitive Systems Engineering, Wiley Series in Systems Engineering*. New York: Wiley.

Reason, J. 1990. *Human Error*. Cambridge U.K.: Cambridge University Press.

Reason, J. 1997. *Managing the Risks of Organizational Accidents*. Burlington, VT: Ashgate.

Runciman, W., P. Hibbert, R. Thomson, T.V.D. Schaaf, H. Sherman, and P. Lewalle. 2009. Towards an international classification for patient safety: Key concepts and terms. *International Journal for Quality in Health Care* 21(1):18–26.

Salas, E. and J.A. Cannon-Bowers. 1997. Methods, tools, and strategies for team training. In *Training for a Rapidly Changing Workforce: Applications of Psychological Research*, M.A. Quinones and A. Ehrenstein (eds.). Washington, DC: American Psychological Association.

Singh, H., E.J. Thomas, S. Mani, D. Sittig, H. Arora, D. Espadas, M.M. Khan, and L.A. Petersen. 2009. Timely follow-up of abnormal diagnostic imaging test results in an outpatient setting: Are electronic medical records achieving their potential? *Archives of Internal Medicine* 169(17):1578–1586.

Sittig, D.F. and H. Singh. 2009. Eight rights of safe electronic health record use. *Journal of the American Medical Association* 302(10):1111–1113.

Smith, M.J. and P. Carayon. 2000. Balance theory of job design. In *International Encyclopedia of Ergonomics and Human Factors*, W. Karwowski (ed.). London, U.K.: Taylor & Francis.

Smith, M.J. and P. Carayon-Sainfort. 1989. A balance theory of job design for stress reduction. *International Journal of Industrial Ergonomics* 4(1):67–79.

Stange, K.C., P.A. Nutting, W.L. Miller, C.R. Jaen, B.F. Crabtree, S.A. Flocke, and J.M. Gill. 2010. Defining and measuring the patient-centered medical home. *Journal of General Internal Medicine* 25(6):601–12.

Tyre, M.J. and W.J. Orlikowski. 1994. Windows of opportunity: Temporal patterns of technological adaptation in organizations. *Organization Science* 5(1):98–188.

Vincent, C. 2010. *Patient Safety*. Chichester, U.K.: Wiley-Blackwell.

Vincent, C., K. Moorthy, S.K. Sarker, A. Chang, and A.W. Darzi. 2004. Systems approaches to surgical quality and safety—From concept to measurement. *Annals of Surgery* 239(4):475–482.

Vincent, C., S. Taylor-Adams, and N. Stanhope. 1998. Framework for analysing risk and safety in clinical medicine. *BMJ* 316(7138):1154–1157.

Vicente, K.J. 2003. What does it take? A case study of radical change toward patient safety. *Joint Commission Journal on Quality and Safety* 29(11):598–609.

Walton, M., H. Woodward, S.V. Staalduinen, C. Lemer, F. Greaves, D. Noble, B. Ellis, L. Donaldson, B. Barraclough, for, and as Expert Lead for the Sub-Programme on behalf of the Expert Group convened by the World Alliance of Patient Safety. 2010. The WHO patient safety curriculum guide for medical schools. *Quality and Safety in Health Care* 19(6):542–546.

Weick, K.E. and R.E. Quinn. 1999. Organizational change and development. *Annual Review of Psychology* 50:361–386.

Weick, K.E. and K.M. Sutcliffe. 2001. *Managing the Unexpected: Assuring High Performance in an Age of Complexity*. San Francisco, CA: Jossey-Bass.

Weinert, C.R. and H.J. Mann. 2008. The science of implementation: Changing the practice of critical care. *Current Opinion in Critical Care* 14(4):460–465.

Wickens, C.D., J.D. Lee, Y. Liu, and S.E.G. Becker. 2004. *An Introduction to Human Factors Engineering*. 2nd ed. Upper Saddle River, NJ: Prentice Hall.

Wilson, J.R. 1991. Participation—A framework and a foundation for ergonomics? *Journal of Occupational Psychology* 64:67–80.

Wilson, J.R. 2000. Fundamentals of ergonomics in theory and practice. *Applied Ergonomics* 31(6):557–567.

Wilson, J.R. and H.M. Haines. 1997. Participatory ergonomics. In *Handbook of Human Factors and Ergonomics*, G. Salvendy (ed.). New York: John Wiley & Sons.

Zink, K.J. 1996. Continuous improvement through employee participation: Some experiences from a long-term study. In *Human Factors in Organizational Design and Management-V*, O. Brown Jr. and H.W. Hendrick (eds.). Amsterdam, the Netherlands: Elsevier.

2

Opportunities and Challenges in the Pursuit of Patient Safety

Kerm Henriksen

CONTENTS

2.1 Opportunities 17
    2.1.1 Simulation 17
    2.1.2 Evidence-Based Design 18
    2.1.3 Diagnostic Error 19
    2.1.4 Adaptive Environments 20
    2.1.5 Expanding the Human–System Interface 21
2.2 Challenges 22
    2.2.1 Understanding the Problem 22
    2.2.2 Interventions and the Traditional Approach 23
    2.2.3 Defining Our Methods 23
    2.2.4 Neglecting the Impact of Contextual Factors 23
    2.2.5 Defining Our Metrics 24
    2.2.6 Final Note 24
References 25

Healthcare reform is upon us. The United States has adopted an incremental approach to change rather than vastly overhauling the system, and, hence, the reforms are likely to be occurring for some time into the future. Periods of change typically offer new opportunities, and given the investment that the human factors and ergonomics (HF&E) community already has made in health care, the future looks bright for human factors researchers and practitioners with an interest in health care, especially those willing to stretch their comfort zones a bit.

2.1 Opportunities

Several research domains are discussed that have been receiving increasing attention because of their potential to enhance patient safety and improve the quality of care. HF&E professionals, in conjunction with their clinical counterparts, have an opportunity of playing an important role in these areas; many already have. The selection of domains is based simply on the author's own interests and observations at the time of writing rather than a more robust and systematic overview of the opportunities available. For a quick capture of the latter, the reader need do no more than peruse the titles of the chapters of this book.

2.1.1 Simulation

One of medicine's moral dilemmas has been that of putting today's patients at risk for the training of tomorrow's practitioners (Ziv et al. 2003). However, recent developments in healthcare simulation have started to reduce the need for practicing on patients as a way for clinicians to become experienced and skilled. A national focus on patient safety, advances in medical simulation technology, a progressive lowering of associated costs, shortening of hours for residents (thereby reducing exposure to infrequent and difficult cases), incorporating lessons learned from other high-risk industries, and greater recognition of the performance-based component of clinical competency are some of the converging trends that have started to move the medical training pendulum from a predominant focus on knowledge acquisition to performance accountability (Henriksen and Patterson 2007). The advantages of simulation for health care have received wide comment—safety of patients during a period when providers are inexperienced and learning new procedures, optimal control of the training conditions to promote new skill acquisition, the integration of multiple components and skills for fluid individual and team performance, and return on investment in the form of costs avoided from complications, injury, and death. Less frequently mentioned, but just as relevant, is the role that simulation and modeling can play in serving as a platform or test bed for new technologies and their impact on work processes, for usability testing and iterative design improvement, for assessment of clinical performance in critical care domains, and for prediction of factors that can threaten patient safety and impede the quality of care.

Potential advantages, of course, are not the same thing as actual advantages. Having access to a tool does not mean one knows how to use it effectively. Aviation's impressive use of simulation goes back more than 80 years when Edwin Link, in 1929, built a device that replicated an aircraft cockpit with maneuverable controls. Link's "pilot-maker" or "blue-box" subsequently evolved into the high-volume, generic simulators to train military pilots during World War II and then the sophisticated wide-field visual systems and six degree-of-freedom motion platforms of today (Flexman and Stark 1987; L3 Communications). A similar long-term research and development effort is needed in health care for simulation's actual benefits to be fully realized and sustained. New initiatives frequently begin with great enthusiasm and then disillusionment sets in once the hyperbole dies down and serious work needs to get underway. The real stalwarts recognize that basic questions need to be asked if a better understanding of the optimal learning conditions that underlie skilled clinical performance is to be gained. What provider tasks, patient conditions, or system anomalies need to be simulated? How are acceptable levels of performance established and measured? Are there agreed-upon nomenclatures, taxonomies, and metrics (Satava et al. 2003)? How are the visual, auditory, and tactile requirements of the various user-simulation interfaces determined and represented?

Simulations that strive for high physical realism are touted as "high fidelity"—a term that imparts a misguided message. A common mistake is to equate high fidelity with high training effectiveness. To be sure, many clinical and surgical tasks require highly faithful representations (e.g., rendering visual images faithfully for the detection of skin melanoma or receiving accurate haptic feedback such as feeling tissue resistance when using a cutting tool). However, for the learning of many complex new skills such as acquiring the necessary eye–hand coordination that underlies skillful execution of many minimally invasive surgical procedures, a departure from high fidelity is indeed required. Before the aspiring laparoscopic surgeon can

make a minimal and precise incision, he or she must first learn how to navigate in a 3D operational space by looking at a 2D image with diminished depth cues and manipulate geometric objects with grasping tools, none of which resembles actual anatomy, organs, or tissue. It is not until trainees have progressed to more advanced levels that the simulated representation becomes more realistic and they are expected to perform more intricate tasks that actually warrant a more faithful simulation.

Many slow-onset disease processes and conditions that are difficult to diagnose because their subtle cues and manifestations unfold gradually over time can be compressed by the simulation to unfold faster than in real time. This allows exposure to a wider range of patient conditions than what a resident would encounter during a typical rotation. It is precisely this property of departing from "high fidelity" and selectively controlling the conditions for practice that is not possible in the clinical setting that gives simulation such great learning potential. The challenge before us is a more mindful exploration of appropriate fidelity for new skill promotion and test bed research, not a mindless preoccupation with "high fidelity" (Beaubien and Baker 2004; Henriksen and Patterson 2007; Scerbo 2007).

2.1.2 Evidence-Based Design

In parallel fashion to evidence-based medicine and the rapidly expanding patient safety and quality literature, there is an emerging evidence base from healthcare architecture, health design, environmental psychology, medical and nursing specialty areas, human factors, and industrial engineering—collectively known as evidence-based design (EBD)—that has focused on the physical environment and how its design can serve to facilitate or impede the quality of care that patients receive as well as the work-life quality of their providers. Evidence-based design strives to use the best available information from credible research to construct patient rooms, improve air quality, improve lighting, reduce noise, encourage hand hygiene, improve patient vigilance, reduce falls, reduce walking distances, incorporate nature, and accommodate needs of families. (Hamilton 2004; Henriksen et al. 2007; Joseph and Rashid 2007; Ulrich et al. 2008).

As an example, the emerging evidence suggests that single-bed rooms have several advantages over double rooms and open bays, and are a significant step in achieving the Institute of Medicine's (2001) quality aims of patient-centeredness, safety, effectiveness, and efficiency. Single-bed rooms leave patients less exposed to both airborne and contact transmission routes of pathogens. Reducing the occurrence of preventable hospital-acquired infections is a very worthy aim. Other advantages include easier and quicker access to equipment and supplies from standard room layouts, fewer patient transfers and associated disruptions of care, shorter lengths of stay, reduced noise levels, better patient–staff communication, natural light from wider windows, and better privacy and satisfaction of care by patients and families. The design, layout, physical structures, and equipment in patient rooms also have a direct bearing on patient falls, the majority of which occur as patients make their way from the bed to the bathroom. In some new patient room designs, the bathroom is placed behind the headwall to reduce distance to the bathroom, maximum use is made of handrails and grab bars, and unobstructed lines of sight are important design features, enabling nurses to have greater visibility of patients without disturbing them yet assisting them as needed (Reiling 2007).

Quality and safety of care are not the only concerns for those involved in the design of new healthcare facilities. Somehow these needs have to be balanced and integrated with the needs for environmental safety (i.e., use of toxic-free materials and products) and sustainability that enables the meeting of today's resource requirements without compromising the needs of future generations (Guenther and Vittori 2008). In addition to healthcare designs that transition us toward carbon neutral and zero waste facilities, designs that provide for the capture and reuse of water, another diminishing resource, will be needed. As an industry, health care is a large-scale user of potable water, and as with fossil fuel, the days of insatiable consumption are behind us. Assuming that the current hospital building boom in the United States and other countries will continue for another 10–15 years, there is an opportunity for HF&E professionals to be more involved in EBD research, to apply the tools and methods of the trade as well as develop new ones, and to be more closely aligned with the planning and design phases of healthcare facilities building projects.

2.1.3 Diagnostic Error

Although the Institute of Medicine's *To Err is Human* (2000) report identified diagnosis as a problematic area in its discussion of the most common types of preventable error, the seminal patient safety report actually had very little to say about diagnostic error. In putting a strong spotlight on some of the more readily identifiable system factors, diagnostic error seemed to be left in the shadows (Wachter 2010). Just as physicians traditionally have viewed system-based failures as an institutional problem, hospital executives have viewed diagnostic error as an individual physician matter. Both

views are shortsighted. It did not take too long, however, for the neglect of diagnostic error to be noticed as investigators started to issue alarms to clinicians and their healthcare organizations of the need for a better understanding of diagnostic performance (Graber et al. 2002; Graber 2005). Among medical specialties serving as fertile ground for deeper inquiry are emergency medicine, radiology, pediatrics, surgery, internal and family medicine, laboratory medicine, and pathology. Given its chaotic mix of system-based, cognitive, affective, and variable patient factors, emergency medicine is a specialty where instances of "flesh and blood" diagnostic error come to mind in their most compelling form. Premature closure (failure to consider alternative diagnoses) can occur quickly in the emergency department (ED), and once the patient with the "stomach ache" gets triaged to the waiting room and labeled a "routine case of gastroenteritis," the likelihood of discovering his rupturing aortic aneurysm becomes less and less. The gatekeeping functions of internal and family medicine with their high-volume stream of patients likewise can facilitate similar cognitive biases.

Research on biases in decision making under conditions of uncertainty through the uncritical use of shorthand rules of thumb (heuristics) was well underway by cognitive psychologists in the mid-1970s (Tversky and Kahneman 1974; Fischhoff 1975). By the 1990s, the research was having a profound influence on economic theory, giving rise to the new discipline of behavioral economics and a subsequent Nobel Prize in 2002 for one of its leading researchers. The significance of the research extends well beyond economics, however. It extends to any sphere of human activity where decisions have to be made under conditions of uncertainty, where the problem-solving domain is complex, and where individuals tacitly and intuitively rely on underlying heuristics that serve them well much of the time, but not all of the time. Making decisions under conditions of uncertainty is an apt description of much that occurs in medicine, and, thus, it is not surprising that investigators of diagnostic error have an interest in the cognitive biases that underlie diagnostic performance (Croskerry 2003; Redelmeier 2005).

In addition to the uncritical use of heuristics or cognitive dispositions that lead to systematic and predictable errors, affective dispositions to respond (i.e., our affective state and the influence of aversions and attractions on decision making) also are likely to be operating in clinical settings (Croskerry 2005, 2009). Patients who are viewed as just not taking care of themselves—the obese, the alcoholic, the substance abuser, the frequent complainer—might receive less time and concentrated attention, especially when the waiting room is full. Groopman (2007) provides cogent examples of how a physician's aversion or liking for a patient can influence the diagnostic work up. Yet, affective dispositions to respond have received less attention than the cognitive dispositions. Both are in need of further empirical study.

Likewise, the disciplines of radiology, pathology, and dermatology are vulnerable to yet another response tendency—what might be called perceptual dispositions to respond—given their dependence on orienting and perceptual processes and the ways these processes can lead us down erroneous paths. Perceptual dispositions are intricately interwoven with the cognitive and affective dispositions. This bundle of inherent response dispositions is influenced, in turn, by a host of contextual and system-related factors that can further influence what is looked for and what is found. How many computed tomography (CT) images can a radiologist look at over the course of a busy weekend emergency room (ER) shift before he or she becomes "blind" to what should have been detected? How many images get skipped? How much specific patient history should be given to the radiologist by the referring clinician? Are there dual dangers of giving too little information and leaving the radiologist without sufficient context or providing too much which may bias and limit the examination too severely, allowing something else to go unnoticed? Under what conditions, can specificity's gain be sensitivity's loss?

For human factors researchers with keen interests in cognition, perception, emotion, and system-based factors and who appreciate the intricate interactions that can occur among these components, the study of diagnostic processes and accuracy can be quite engaging. Not all investigators have chosen to focus on inherent response dispositions. Some have made a case for concentrating on system-based factors (Newman-Toker and Pronovost 2009). Some investigators have focused on system glitches that occur during different stages of the diagnostic workup (Hickner et al. 2005), on how the use of information technology and computerized aids can improve communication and reduce omission errors (Singh et al. 2008), and on how follow-up and feedback mechanisms can be devised so that physicians can track patient outcomes better and recalibrate their diagnostic processes where needed (Schiff 2008). Still other investigators have examined the use of providers to enhance resilience and serve as ameliorators to interrupt errors that would otherwise cascade through the system and harm patients (Parnes et al. 2007). In brief, the diagnostic error literature is rapidly expanding, revealing a richness and complexity characteristic of the patient safety literature writ large.

2.1.4 Adaptive Environments

Increases in longevity, a growing elderly population, and a steady migration of medical devices and technologies into care facilities (i.e., homes, assisted living quarters, and nursing homes) are placing new challenges on those that design and care for the aging population. Homes, in particular, vary considerably and have not been designed as a provider workplace or with prolonged healthcare services in mind (Henriksen et al. 2009; National Research Council 2010). As the elderly population strives to age in place with dignity, the home typically is where they intend to remain for as long as possible. In addition to incompatible space layouts and structural obstacles such as narrow door frames and hallways, there are many home environmental features that can serve as hazards to elderly patients (Pinto et al. 2000; McCullagh 2006). These include but are not limited to poor lighting, stairs, unstable furniture, loose rugs, dust, and debris, high-risk areas such as bathroom, and bacteria-ridden surfaces (Gershon et al. 2008). A number of socio-technical factors appear to be converging: the growing elderly population with its multiple chronic and acute healthcare needs, efforts to reduce healthcare costs with shorter hospital stays, the continued migration of medical devices, technology, and therapies into the home, the variable competency of lay and professional home healthcare providers, and the questionable supportiveness of the home environment for the delivery of healthcare services. As these factors converge in their collective interactivity and complexity, there is concern that new threats to patient safety and less than adequate quality of care may be the result.

A host of challenges exist for providers and caregivers in the home setting. How do providers protect themselves from musculoskeletal injury when lifting patients or transporting medical equipment in restricted spaces? To whom can they turn when an extra pair of hands is needed? How prepared are they to dispose of sharps, dressings, and other medical waste products? How knowledgeable and skillful are they with a wide range of care activities: coaching in self-help skills; lifting and assisting with ambulation; helping with medication management; monitoring of vital signs; ensuring the operation and maintenance of an assortment of devices and technologies that may or may not have been designed for home use; and assisting with dialysis, chemotherapy, respiratory, and infusion procedures, formerly limited to the inpatient setting. Skin lesions, open wounds, and pressure ulcers may need to be attended to, problems with elimination checked, and changes in affect/ behavioral status detected. In many instances, elderly lay caregivers with their own chronic care limitations are taking care of elderly spouses (Donelan et al. 2002). Under these circumstances, it is important to understand the sensory, physical, cognitive, and mobility capabilities of caregivers as well as patients. With a sound understanding of all the major home healthcare factors in play, potential threats to safety and quality can be identified by examining the mismatches among the cognitive and physical capabilities of patients and their providers and the medical therapies and care processes that need to be performed, the medical devices and technologies in use, and the physical environment in which care processes take place.

While much of the EBD research discussed earlier pertained to the hospital setting, there is a need to apply EBD findings, universal design principles (Lidwell et al. 2003), and usability practices to long-term care facilities and the home as well. Some of these practices entail intelligent use of space for improved mobility and reduced steps, ground floor adjacencies (e.g., bedroom/bathroom), affordances for ease of use (e.g., step-in showers), elevator/chair lift systems, information display and telemonitoring systems, automated shelving, variable height countertops, wide door widths, passageways, and ramps, ergonomic considerations given to portability, placement, and storage of equipment, adjustable and recessed lighting, adjustable height toilet seats, no step entrances, and level access to green areas (e.g., gardens, patio). Planners and builders of new and remodeled construction envision an expanding market for adaptable housing as the percentage of the population that is 65 years and older continues to increase in the decades ahead. The benefits of adaptable and universal design considerations accrue not just to the elderly but also to sick children, disabled adults, and veterans undergoing rehabilitation from injuries. Actually, the benefits accrue to everyone.

2.1.5 Expanding the Human–System Interface

At a policy level, health information technology (health IT) has been embraced as part of the nation's solution to address rising healthcare costs while attempting simultaneously to improve safety and quality of care. Within the past decade, the Agency for Healthcare Research and Quality (AHRQ) has made a major investment in the issuance of multiple grants and contracts to better understand how assorted technologies—electronic health records, computerized provider order entry (CPOE), bar-coded medication administration, and telehealth, among others—can be cost-effectively implemented. While there have been successful demonstrations of these technologies, they come with many challenges and lessons learned that probably were not fully appreciated by the manufacturers, early advocates, and initial implementers.

Given a desired output or goal, the *human–system interface* traditionally is the convenient label that is applied to any of the interactions that occur between humans and the equipment, and technology and work environments to which they are exposed. Healthcare providers

that use medical devices and equipment extensively have considerable first-hand experience with the poor fit that frequently exists between the design of the devices' controls and displays and the capabilities and knowledge of users. Receiving less recognition is the realization that devices do not exist by themselves. They are coupled and connected to other system components, serve multiple users, and need to be integrated compatibly into an existing unit of care (micro-system) as well as a broader organizational or socio-technical system. Four levels to the *human–system interface* are identified and examined here—patient-device, provider-device, microsystem-device, and socio-technical-device—to give us a fuller understanding of the context of use. Mismatches between devices and people not only exist horizontally within these levels, each with their own sources of confusion, but also as a result of vertical misalignments between levels (Rasmussen 1997; Vicente 2002; Henriksen, 2007).

In adopting a patient-centric model of care, it is best to start with the patient and see what questions might be raised by recognizing a *patient–device interface*. Does the device or accessory attachment need to be fitted or adapted to the patient? What physical, cognitive, and affective characteristics of patients need to be taken into account in the design and use of devices and accessory items? Given economic pressures to move patients out of hospitals as soon as possible and the potential need to continue some device-based medical therapy in the home, what sort of understanding does the patient need to have of device monitoring, operation, and maintenance? In what form should this understanding be provided? In brief, the role of the patient in relation to the device and its immediate environment deserves careful consideration.

Providers, of course, are subject to a similar set of device and technology issues. Technological innovations place new demands on providers as sophisticated monitoring systems, smart infusion pumps, electronic medical records, computerized physician order entry, and bar coding become more common. With respect to the *provider–device interface*, how do we facilitate the provider's ability to operate, maintain, and understand the overall functionality of the device as well as its connections and functioning in relation to other system components? In addition to displays and controls that need to be designed with perceptual and motor capabilities in mind, the device needs to be designed in a way that enables the nurse or physician to determine quickly the underlying state of the device. Increasing miniaturization of computer-controlled devices has increased their opaque quality, leaving many providers at a loss in terms of determining the full functionality of the device or even what state it is in. With a poor understanding of device functionality, providers are at a further loss when the device malfunctions and when swift decisive action may be critical for patient care. The design challenge is in creating provider–device interfaces that facilitate the formation of appropriate mental models of device functioning and that encourage meaningful dialogue and sharing of tasks between user and device.

The next interface level in our progression of interfaces is the *microsystem–device interface*. At the microsystem level (i.e., contained organizational units such as operating rooms [ORs] or intensive care units [ICUs], it is recognized that medical equipment and devices frequently do not exist in stand-alone form but are tied into and coupled with other components and accessories that collectively are intended to function as a seamless, integrated system. Providers, on the other hand, are quick to remind us that this is frequently not the case given the amount of time they spend looking for appropriate cables, lines, connectors, and other accessories. In many ORs and ICUs, there continue to be an eclectic mix of monitoring units from different vendors that interface with other devices in different ways that only furthers the cognitive workload placed on provider personnel. Another microsystem interface problem, as evidenced by several alerts from healthcare safety organizations, are medical gas mix-ups where nitrogen and carbon dioxide have been mistakenly connected to the oxygen supply system. Gas system safeguards using incompatible connectors have been overridden on occasion with adapters and other retrofitted connections. The lesson for those concerned about safety is to be mindful that the very need for a retrofitted adaptor is a warning signal that a connection is being sought that may not be intended by the device manufacturer and that it may be incorrect and harmful (ECRI 2006).

Yet other device-related concerns are sociotechnical in nature and hence we refer to a *sociotechnical–device interface*. How well are the technical requirements for operating and maintaining the device supported by the physical and socio-organizational environment of the user? Are the facilities and workspaces where the device is used adequate? Are quality assurance procedures in place to ensure proper operation and maintenance of the device? What sort of training do providers receive in device operation before using the device with patients? Are chief operating officers and nurse managers committed to safe device use as an integral component of patient safety? As devices and various forms of health IT play an increasing role in efforts to improve patient safety and quality of care, greater scrutiny needs to be directed at discerning the optimal and less-than-optimal conditions in the sociotechnical environment for proper support, maintenance, and safe use.

2.2 Challenges

With opportunities, of course, come challenges and responsibilities. It is not unusual for scientific disciplines, especially applied disciplines, to have a body of basic principles that they espouse publicly, yet violate in practice. Given repeated and frequent departures, the core principles lose their relevance and meaning and the discipline's practitioners are left with empty clichés for guidance. Everyone agrees, for example, that representative samples are needed if we wish to generalize to the population from which the samples are taken (and taking a simple random sample is one way of achieving this), yet the further the investigator moves from a laboratory to a practice setting, this fundamental principle is violated frequently—often out of necessity—and a convenience sample is used. If the results derived from the sample cannot be regarded as unbiased with confidence, one cannot be comfortable generalizing them to the population. Five other challenges of applied human factors research deserve mention. While they are cast within the context of patient safety, they are germane to a wide swath of human factors research domains.

2.2.1 Understanding the Problem

The author once worked for a vice president who had earned the moniker of "No Toes Tulley" (out of sense of respect for anonymity, he is fictitiously called Tom Tulley here). While his name is changed, his problem-solving mode as described is true. Tom would no sooner be presented with a potential problem; and before the sentence could be completed, he would miraculously have a solution; that is, he would be so eager to solve the problem that he would shoot off his toes in the process of withdrawing his figurative six-shooter from his holster to plug away at the problem. When solutions were so readily available to Tom, it was as though he did not need to gain a deeper and fuller understanding of the problem.

Is it possible that human factors investigators, like vice presidents, could fall victim to accepting a problem as stated or read somewhere too readily, and then proceed down a path of misdirected research? Do we sometimes err by defining the problem in terms of a too readily available solution with which we are too comfortable? By focusing on only a limited aspect of what is already known, there may be a danger of missing the larger, multi-factorial problem. It may be tempting to focus on simple fixes—the low hanging fruit—rather than address fundamental underlying issues that take a more prolonged period of study. To be sure, after close to two decades of only sporadic success in trying to address the needless injury and deaths that occur in the treatment and care of patients, it is understandable that many patient safety advocates are frustrated with the limited progress to date and are looking for solutions today. In fact, there are many threats to patient safety that have knowable solutions, but because of the pull of the status quo, organizational–cultural barriers, incompatible infrastructures, or the operation of perverse incentives, the actionable implementation of these solutions remains an uphill struggle. There are other knowable patient safety problems that have only partial or yet untested solutions. Adding to the complexity further are the unknown patient safety problems beyond our realm of awareness for which we currently know nothing about. Rather than shooting from the hip and losing our toes, we may find it less risky to spend more time in expanding our understanding of the problem space that impacts our daily activities. There is no shortage of human factors methodologies and front-end analytic techniques for assisting us in this pursuit.

2.2.2 Interventions and the Traditional Approach

Considerable effort seems to be expended in healthcare and other applied disciplines on research that compares the latest technology, tool, or best practice with the traditional approach. Frequently, not much that is new is learned from this sophomoric approach to research other than the researchers were probably premature in wanting to rush into an intervention-control group comparison in the first place. Suppose that a patient safety researcher with an interest in simulation shows that residents exposed to simulation-based training perform better on certain measures than those residents exposed to the traditional approach (which is typically a "hit or miss" apprenticeship experience, derisively referred to as "see one, do one, teach one"). Should we conclude from the study that simulation-based training is more effective than the apprenticeship model? The most that such a study shows is that a particular exemplar (i.e., set of simulation exposures) of a treatment class called simulation is more effective than another exemplar (i.e., set of apprenticeship experiences) of the treatment class referred to as apprenticeship or the traditional approach. Are the exemplars comparable or are we comparing an excellent set of simulation exposures with a run-of-the-mill set of apprenticeship experiences? We could just as easily compare a very rich set of apprenticeship experiences with a set of poorly executed and tedious simulation encounters. The lesson here is straightforward. When developing and perfecting a new intervention, we'd be wise to resist the temptation of seeking immediate scientific gratification by comparing our premature intervention with some hapless control group. Instead, we can use our time more productively by first engaging in an extensive period of

development, iterative testing, and refinement with representative users.

2.2.3 Defining Our Methods

A significant contributor to the conflicting state of affairs, in fair measure, is inadequate definition of our methods. The typical practice is to label one group that continues doing what they have done all along as the traditional method or control group, while the other group that receives most of the attention and effort is the fortunate intervention group. Since the traditional group serves as little more than a convenient straw man that is given short shrift, we receive very little information about it in the journal that publishes the non-reproducible results. With little or no operational clarification as to what the traditional group encounters, and with only slightly better clarification about the intervention group, prospects for replicating or building upon the reported research are very slim indeed.

Neglecting to adequately define and be fully attuned to our interventional and control methods is a recipe for unrecognized variation and non-replicable results. Geis (1976) cogently put it this way: the problem of definition and subsequent execution becomes transparent when one arm of the comparison, such as the traditional method, is so poorly defined that two exemplars of the traditional method that range from the inept to excellent could be used to restate the experimental question. The same holds true also for the intervention arm of the comparison when variation from the inept to the excellent exists. Wouldn't it be more useful to simply test and tease apart the critical differences that underlie two levels of proficiency in implementing the intervention? Hence, the question whether our favorite intervention is more effective than the traditional approach becomes almost meaningless when we know that one investigator is more or less skillful at implementing the intervention and its associated protocols than another investigator, and such is usually the case. Given such sloppiness in defining and executing our methods, should we expect anything other than the conflicting results that permeate the literature?

2.2.4 Neglecting the Impact of Contextual Factors

Related to the challenge of rendering clear definitions of methods is the challenge of understanding the importance of contextual factors in implementing patient safety interventions. In administering a therapeutic agent such as an antibiotic, it does not make much difference whether it is administered in a well-resourced teaching hospital in the Northeast or a small community hospital in the Southwest. In either setting, it is likely to have its intended effect. Such is not the case with many patient safety interventions (Shekelle et al. 2011). Patient safety interventions are often multi-faceted and very sensitive to the organizational context or setting in which they are carried out (Henriksen 2010). They involve not only the core intervention itself, such as a checklist to reduce central line infections or a decision support tool to improve diagnostic performance, but also the purposeful control and alignment of several factors—leadership buy-in, fully supplied work area, trained personnel, willing patient safety climate—for the intervention to be implemented in a uniform and consistent fashion for the duration of the study and, if successful, to be sustained thereafter. What gets implemented is a "bundle" of work-system considerations along with the intervention, making it sometimes difficult to separate the core intervention from the enabling contextual factors that are purposefully controlled and bundled. It can be argued that the core intervention plus the controlled contextual factors is, in fact, the intervention (Pronovost et al. 2006, 2010). Inability to replicate bundles across disparate sites is what makes what is possible in the Northeast sometimes not possible in the Southwest and vice versa. It is not unusual for novice researchers to focus mainly on the core intervention and to neglect the contextual factors while veteran researchers have learned the hard way to be especially vigilant to contextual factors. Hospitals are dynamic and fluid places and no one expects the researcher to be able to control all the factors that can disrupt the intervention. However, to keep surprises to a reasonable minimum, the importance of taking organizational and work-system factors into account and considering them as an integral part of the intervention cannot be overstated. By working together, HF&E researchers and clinicians can identify and learn to optimize those work-system factors that play a very significant role in ensuring successful patient safety interventions.

2.2.5 Defining Our Metrics

Just as insufficient detail is given frequently to defining our methods, there also are lapses in defining dependent measures. What does better mean when someone asks the following question: "Is treatment X better than treatment Y?" How is better determined? There can be just as many dependent measures as there are investigators. Are we referring to some metric of safety, effectiveness, or timeliness? Or does better mean fewer resources, less costs, or greater user acceptance? It is not unusual for a given metric to be more faithfully aligned, derived, or to make more sense for a particular treatment compared to another treatment since different treatments are initially developed for specific purposes. Metric y may be ideally suited for evaluating treatment Y, but not treatment X. This

gives rise to the question as to what metrics should be used for a fair comparison if that is the intent of the research. Given that investigators make personal investments in developing treatment approaches and as designers of the research select and develop the metrics as well, care has to be taken to avoid the risk of wittingly or unwittingly creating the conditions for self-fulfilling prophesies when resorting to hypotheses testing. Heedful attention to the metric development process and the use of multiple dependent measures help to mitigate potential selective metric bias; otherwise, the results of comparisons between treatments indeed may be spurious.

2.2.6 Final Note

The opportunities described earlier and those the reader finds in other chapters of the book are the cause for celebration. On many healthcare fronts, the human factors and ergonomics community is making valuable contributions in conjunction with its clinical partners. At the same time, there is a need to sharpen our self-examination capabilities if we are to continue to have even greater impact on improving patient safety and quality of care. A few areas for critical self-examination have been discussed: (a) a need exists to more fully understand the nature of multi-faceted problems before we choose to draw our weapons and dispense of them; (b) when so much variation exists in ability to implement our favorite intervention and the traditional approach, why choose the path of folly and compare the two; (c) authors, reviewers, and editors of journal articles need to do a better job of ensuring clear definition and operational clarification of the intervention and control methods deployed; (d) the importance of understanding the role of contextual factors as integral components of the intervention is a lesson many have learned the hard way; and (e) care needs to be taken with the selection and development of metrics so as not to favorably bias a particular treatment approach. The discussed pitfalls and areas in need of improvement are by no means exhaustive; others can surely add to the list. Improvement, in many instances, is a matter of attending to sound research basics. In other instances, investigators may experience pressure by their institutions or funding agencies to demonstrate the effectiveness of their interventions long before they have sufficient time, experience, or practice in implementing them effectively. Rather than rush in and prematurely compare the latest intervention to the traditional approach, more substantive progress is likely to be made by first identifying and learning how to control the critical variables so that their combined effectiveness can be optimized in relation to other work system factors.

References

Beaubien, J.M. and D.P. Baker. 2004. The use of simulation for training teamwork skill in healthcare: How low can you go? *Qual Saf Health Care* 13(Suppl 1): i51–i56.

Croskerry, P. 2003. The importance of cognitive errors in diagnosis and strategies to minimize them. *Acad Med* 78(8): 775–780.

Croskerry, P. 2005. Diagnostic failure: A cognitive and affective approach. In *Advances in Patient Safety: From Research to Implementation*. Vol. 2. Concepts and Methodology, eds. K. Henriksen, J.B. Battles, E. Marks et al., pp. 241–254. AHRQ Publication No. 05-0021-2. Rockville, MD: Agency for Healthcare Research and Quality.

Croskerry, P. 2009. Cognitive and affective dispositions to respond. In *Patient Safety in Emergency Medicine*, eds. P. Croskerry, K.S. Cosby, S.M. Schenkel et al., pp. 219–227. Philadelphia, PA: Wolters Kluwer-Lippincott Williams & Wilkins.

Donelan, K., C.A. Hill, C. Hoffman et al. 2002. Challenged to care: Informal caregivers in a changing health system. *Health Aff* 21: 222–231.

ECRI. 2006. Preventing misconnections of lines and cables. *Health Devices* 35(3): 81–95.

Fischhoff, B. 1975. Hindsight ≠ foresight: The effect of outcome knowledge on judgment under uncertainty. *J Exp Psychol Human* 104: 288–299.

Flexman, R.E. and E.A. Stark. 1987. Training simulators. In *Handbook of Human Factors*, ed. G. Salvendy, pp. 1012–1038. New York: John Wiley & Sons.

Geis, G. 1976. Trick or treatment. *Learn Dev* 7(3): 1–6.

Gershon, R.R.M., A.N. Canton, V.H. Raveis et al. 2008. Household related hazardous conditions with implications for patient safety in the home health care sector. *J Patient Saf* 4: 277–234.

Graber, M. 2005. Diagnostic errors in medicine: A case of neglect. *Jt Comm J Qual Patient Saf* 31(2): 106–113.

Graber, M., R. Gordon, and N. Franklin. 2002. Reducing diagnostic errors in medicine: What's the goal? *Acad Med* 77(10): 981–992.

Groopman, J. 2007. *How Doctors Think*. Boston, MA: Houghton Mifflin Company.

Guenther, R. and G. Vittori. 2008. *Sustainable Healthcare Architecture*. Hoboken, NJ: John Wiley & Sons.

Hamilton, D.K. 2004. Four levels of evidence-based practice. The American Institute of Architects. Retrieved June 29, 2011 from http//: www.sereneview.com/pdf/4levels_ebd.pdf

Henriksen, K. 2007. Human factors and patient safety: Continuing challenges. In *Handbook of Human Factors and Ergonomics in Health Care and Patient Safety*, ed. P. Carayon, pp. 21–37. Mahwah, NJ: Lawrence Erlbaum Associates.

Henriksen, K. 2010. Partial truths in the pursuit of patient safety. *Qual Saf Health Care* 19(Suppl 3): i3–i7.

Henriksen, K., S. Issacson C. Zimring et al. 2007. The role of the physical environment in crossing the quality chasm. *Jt Comm J Qual Patient Saf* 33(11): 68–80.

Henriksen, K., A. Joseph, and T. Zayas-Caban. 2009. The human factors of home health care: A conceptual model for examining safety and quality concerns. *J Patient Saf* 5: 229–236.

Henriksen, K. and M.D. Patterson. 2007. Simulation in health care: Setting realistic expectations. *J Patient Saf* 3(3): 127–134.

Hickner, J.M., D.H. Fernald D.M. Harris et al. 2005. Communicating critical test results: Issues and initiatives in the testing process in primary care physician offices. *Jt Comm J Qual Patient Saf* 31(2): 81–89.

Institute of Medicine. 2000. *To Err Is Human: Building a Safer Health System*. Washington, DC: National Academy Press.

Institute of Medicine. 2001. *Crossing the Quality Chasm: A New Health System for the 21st Century*. Washington, DC: National Academy Press.

Joseph, A. and M. Rashid. 2007. The architecture of safety: Hospital design. *Curr Opin Crit Care* 13: 714–719.

Lidwell, W., K. Holden, and J. Butler. 2003. *Universal Principles of Design*. Gloucester, MA: Rockport Publishers, Inc.

L3 Communications. Link Simulation & Training. Setting the standard for over 75 years. Available at http://www.link.com/history.html (accessed February 15, 2011).

McCullagh, M.C. 2006. Home modifications: How to help patients make their homes safer and more accessible as their abilities change. *Am J Nurs* 106(10): 54–63.

National Research Council. 2010. *The Role of Human Factors in Home Health Care: Workshop Summary*. Washington, DC: The National Academies Press.

Newman-Toker, D.E. and P.J. Pronovost. 2009. Diagnostic errors—The next frontier for patient safety. *JAMA* 301(10): 1060–1062.

Parnes, B., D. Fernald J. Quintela et al. 2007. Stopping the error cascade: A report on ameliorators from the ASIPS collaborative. *Qual Saf Health Care* 16: 12–16.

Pinto, M.R., S. DeMedici C. Van Sant et al. 2000. Ergonomics, gerontechnology, and design of the home-environment. *Appl Ergon* 106(10): 54–63.

Pronovost, P.J., C.A. Gpeschel E. Colantuoni et al. 2010. Sustaining reductions in catheter related bloodstream infections in Michigan intensive care units. *BMJ* 340: c309.

Pronovost, P.J., D. Needham, and S. Berenholtz et al. 2006. An intervention to decrease catheter-related bloodstream infections in the ICU. *N Engl J Med* 355(26): 2725–2732.

Rasmussen, J. 1997. Risk management in a dynamic society: A modeling problem. *Saf Sci* 27: 183–213.

Redelmeier, D.A. 2005. The cognitive psychology of missed diagnoses. *Ann Intern Med* 142: 115–120.

Reiling, J. 2007. *Safe by Design: Designing Safety in Health Care Facilities, Processes, and Culture*. Oakbrook Terrace, IL: Joint Commission Resources.

Satava, R.M., A.G. Gallagher, and C.A. Pellegrini. 2003. Surgical competence and surgical proficiency: Definitions, taxonomy, and metrics. *J Am Coll Surg* 196: 933–937.

Scerbo, M.W. and S. Dawson. 2007. High fidelity, high performance? *Simul Healthcare* 2(4): 224–230.

Schiff, G.D. 2008. Minimizing diagnostic error: The importance of follow-up and feedback. *Am J Med* 121(5A): S38–S42.

Shekelle, P.G., P.J. Pronovost, R.M. Wachter et al. 2011. Advancing the science of patient safety. *Ann Intern Med*, 154: 699–701.

Singh, H., A.D. Naik R. Rao et al. 2008. Reducing diagnostic errors through effective communication: Harnessing the power of information technology. *J Gen Intern Med* 23(4): 489–494.

Tversky, A. and D. Kahneman. 1974. Judgment under uncertainty: Heuristics and biases. *Science* 185: 1124–1131.

Ulrich, R.S., C. Zimring X. Zhu et al. 2008. A review of the research literature on evidence-based healthcare design. *HERD J* 1: 61–125.

Vicente, K.J. 2002. From patients to politicians: A cognitive engineering view of patient safety. *Qual Saf Health Care* 11: 302–304.

Wachter, R.M. 2010. Why diagnostic errors don't get any respect-and what can be done about them. *Health Aff* 29(9): 1605–1610.

Ziv, A., P.R. Wolpe S.D. Small et al. 2003. Simulation-based medical education: An ethical imperative. *Acad Med* 78: 783–788.

3

Embedded Risks and Challenges of Modern Health Care and the Urgent Call for Proactive Human Factors

Daniel Gopher and Yoel Donchin

CONTENTS

3.1 Nature and Embedded Risks of Modern Health Care 28
    3.1.1 Focus on Hospitals and Secondary Care Environments 29
3.2 Aspects and Dimensions of Human Factors in Hospital and Secondary Care Environments 30
    3.2.1 Inclusive Task Analysis and System Evaluation 30
        3.2.1.1 Administration of Magnesium Sulfate 30
        3.2.1.2 Performance of a Surgical Procedure 31
    3.2.2 Technology and Care Systems Are Not Only Robots, Computer Displays, Imaging, and
        Sophisticated Devices 32
    3.2.3 Medical Tasks Are Multiprocess and Multistage 32
    3.2.4 High Workload Is the Product of Multiple Determinants and Work Characteristics 33
3.3 Team Work: Its Cognitive and Human Performance Constituents 35
3.4 Urgent Need for Developing Appropriate Reference Databases 36
3.5 Call for Human Factors Engineering in Modern Health Care 39
References 40

The focus of this chapter is on the urgent need to develop and adopt a proactive, comprehensive, and integrative human factors design approach to modern health care, with a special emphasis on hospitals and secondary (professional) healthcare environments. The main claim is that significant improvements in the quality of care and enhancement of patient safety require and are contingent on a major shift of the present medical error–driven efforts and investments to an emphasis on proactive human factors involvement and on the development of corresponding conceptual approaches and methods for its systematic implementation. There are four main contributing factors to this argument:

1. Modern healthcare systems are highly susceptible to user errors.
2. Major investments have been made in the development of direct patient care capabilities, but not in the development of human factors and user-friendly work environment for care providers.
3. The public and professional focus on understanding, studying, and collecting data on errors and the emphasis on determining the accountability of medical team members have led to the emergence of blame culture, while diverting attention and impairing the ability of professionals to obtain and develop the much-needed knowledge on human performance and human factors.
4. The emerging recognition in the potential for and importance of human factors, the need for a shift in quality of care and patient safety efforts, and a move away from an emphasis on personal accountability aspects have still to be conceptualized and defended. Not enough has been done to explain and describe the global dimensional structure of human factors in medical care, as well as the prospective requirements for systematic undertakings of proactive human factors.

The chapter examines and discusses each of these four points and elaborates on the common general dimensions and critical aspects of the required human factors involvement.

3.1 Nature and Embedded Risks of Modern Health Care

Modern health care is powerful and continues to progress exponentially. The healthcare industry has been revolutionized by novel diagnostic, monitoring, and intervention techniques. Contemporary medicine is able to diagnose, provide pharmacological solutions, and carry out surgical interventions and therapeutic procedures of the type and in cases that were deemed impossible only a few years ago. Notwithstanding, it should be recognized that these improved capabilities made health care much more complex and expensive. All modern developed nations invest and spend an increasingly high percentage of their yearly Global Domestic Product (GDP) on healthcare costs. The June 2010 statistical report of the international Organization for Economic Co-operation and Development (OECD) shows the comparative percentages of the yearly GDP invested on health care in the more developed countries. For 2008, the United States leads with 16.0%, followed by France with 11.2%, Switzerland with 10.7%, Germany, Canada Austria, and the Netherlands, all spending yearly more than 10% of their GDP on health care. Furthermore, the relative costs per person have increased considerably over the years. In all of the aforementioned countries these costs more than doubled and even tripled in 1990–2006. Across the 33 OECD-surveyed countries, the average percent of GDP invested in health care increased from 7.8% to 9% in 2000–2009. In parallel, there has been an increase in the rate and severity of medical errors and deterioration in patient safety. Early 1990s studies proposed an estimate of 98,000–120,000 deaths per year attributed to medical errors only in the United States. (Leape et al., 1994). More recent studies reaffirmed and speculated a raise in this estimate (Thomas et al., 1999). In terms of economic impact, the 2010 report of the U.S. Society of Actuaries' Health Section provides a staggering number of $19 billion for the extra costs of medical errors (Shreve et al., 2010). The increasing frequency and costs of medical error have been frequently referred to as the "healthcare medical paradox": highly trained practitioners, widespread state of the art technology, health expenditures greater than 15% of the GDP, and unparalleled biomedical research. Yet, medical errors, overuse, underuse, and misuse are common, serious, and systemic in nature, but largely preventable. What might be the determinants of this paradox and how can it be resolved?

One unfortunate concomitant of the high and increasing cost of improved health care is that it cannot be made available to all. In modern democratic societies where equal opportunities and social welfare are important values, this consequence creates a major economic and value conflict and concern. The struggle to preserve these values on one hand and to keep costs under control on the other hand often forces policy makers and healthcare management to increase systems' intake load and cut corners to save costs. High workload, exceeded capacity, overwork, and shortage of resources have been major problems of modern health care in many developed countries. These problems, in turn, increase the probability of accidents and adverse events.

Other major functional consequences of the high power of health care have been a dramatic increase in the technological complexity and richness of its working environment, rapid accumulation of massive volumes of information for individual patient cases, and the involvement of heterogeneous teams of professionals to carry out most care procedures (Carayon, 2007; Donchin and Gopher, 2011). These developments have increased by default the susceptibility of the care process to human error. It is not only that each workstation has become loaded with technology and operational procedures, but medical care procedures have developed into a multistep, multistage relay competition in complex environments and require team members to transfer responsibility and information for each relay (Cook et al., 2000).

Considering the aforementioned high investment in medical research and care processes, it should also be recognized that investments have been made and are still being made primarily for the study and development of direct medical interventions: drug administration, surgery (e.g., robotic surgery), and diagnostic procedures (e.g., development of technologies and diagnostic tools that enable the physician to go into a blood vessel or to the eye). In contrast, and despite the increased awareness of the growing frequency and severity of medical adverse events and errors, little attention has been devoted to the need of developing an inclusive, integrative, user-oriented approach to the design of medical systems and working environments, which will match and cope with the task requirements of the new emerging systems and healthcare procedures. Consequently, from the perspective of human factors engineering, care givers (physicians, nurses, and associated personnel) are in most cases forced to work in a hostile and poor human factors designed environment (Donchin et al., 1995; Donchin and Gopher, 2011; Scanlon and Karsh, 2010). Such environments are bound to increase load, reduce efficiency, and are highly conductive to errors. This is, in our view, the prime cause and contributor to human errors in medical systems, rather than low motivation or negligence of care givers. Only recently there has been a growing recognition of the severity and potential consequences of this problem, leading to a call for increased focus on aspects of human factors and user-friendly design of medical systems (Leape and

Berwick, 2005; Carayon, 2007; Moreno et al., 2009). The focus of this chapter is on an attempt to explain and characterize the general dimensions and contributors of these demands as well as the challenges for future human factors work in medical systems.

The need for a comprehensive human factors and user-friendly design to enable efficient and safe use of the new healthcare systems was late to be recognized and considered because error reduction and enhanced patient safety were dominated by a competing perspective. The growing evidence for human (performer) related medical errors, accidents, and adverse events have led to increasing public criticism, the condemnation of medical teams and care providers, and the formation of what has been commonly termed the "blame culture." Individual care givers, physicians, and nurses have been accused, held personally responsible, and deemed accountable for the occurrence of errors. This position has led the professional community to focus on the study of medical errors and human error in general and advocate the collection of data on adverse events, incident reports, and errors as its primary source of information (Bogner, 1994; Kohn et al., 1999; Reason, 2000). This emphasis has been accompanied and fortified by thriving malpractice law suits and insurance business. It has shifted attention and directed efforts away from the human factors viewpoint. What has been a long time starting point and design pillar in aviation, space, industry, and power plant operations has been a late arrival to modern health care and is still in the stage of forming up and fighting its justification and position. A major emphasis for better system design can be considered as a shift of present focus and scope for the major efforts in the improvement of patient safety and quality of health care, as they assign the main blame to the contribution of bad designs rather than incompatible or negligent performers. Such a shift calls for taking major steps to improve the overall design of systems, care procedures, and work environments before evaluating the responsibility and accountability of individual performers. The dimensions and richness of the human factors requirements in modern health care are still to be fully recognized and conceptualized.

3.1.1 Focus on Hospitals and Secondary Care Environments

When considering human factors work in medical systems it is important to distinguish between two categories or two types of context. One is concerned with medical work in hospitals and secondary care environments, in which multiple tasks and complex procedures are carried out by professionals and medical teams. The second includes the use of medical devices developed for personal use of individuals to enable self

measurement of essential parameters (e.g., blood pressure, glucose level), or layperson use in public emergency situations (e.g., defibrillators, manual artificial ventilation devices). The development of medical health technology and its increased availability, as well as the remote capability for information transfer and supervision, have greatly increased the feasibility, frequency, and proliferation of individual self use devices. The decision to focus discussion in the chapter on medical health care in hospitals, in which care is the most intensive and provided by medical teams and extensive technological environment, stems from the observation that this context has been the major contributor to the number and adverse consequences of incidents, accidents, and errors in health care. In addition, because most tasks combine multiple technologies and coordinated team work, they call for an inclusive and integrative human factors approach that extends much beyond the human factors and user centered single system, single user design. We feel that this approach and its needs have been less recognized and conceptually developed. By comparison, human factors of single systems as well as the general category of medical devices for personal and emergency applications have been almost immediately recognized and associated with the more general category of consumer products. This domain has already identified human factors, user-centered design, and usability testing orientation as a prime design and marketing requirement. Corresponding with this approach are several of the recently developed human factors design standards for medical devices and the steps made by the Association for the Advancement of Medical Instrumentation (AAMI) and the U.S. Federal Drug Administration (FDA) to include them in the evaluation of new products. In April 2010 AAMI published a new human factors standard for the medical device industry entitled AAMI/HE75: 2009—Human Factors in Medical Device Design. While industrial standards for the design of medical devices are important and have their own value and contribution, we believe that a main concern is the absence of sufficient conceptualization and complementary standards, which consider the global human factors requirements, accompanied by a systematic, institutionalized investment in meeting these requirements at the hospital and professional care levels. These are the levels at which technologies, care procedures, and professional teams combine to serve a highly variable patient population.

In the following sections we attempt to describe dominant human factors and performance dimensions of healthcare systems that can characterize and guide the development of a general proactive human factors approach. These dimensions emerge from the review of contemporary literature and from our leadership in over three dozens of projects conducted by the Technion team in different hospital wards in Israel (Donchin and Gopher, 2011). The review and evaluation focus exclusively on human factors and cognitive psychology variables and considerations. While, as clearly stated and demonstrated by several authors, an inclusive model of healthcare systems should also include social, organizational, management, and policy perspectives (Carayon et al., 2007), this chapter does not discuss these factors (see other chapters in this handbook for a discussion of these topics).

3.2 Aspects and Dimensions of Human Factors in Hospital and Secondary Care Environments

3.2.1 Inclusive Task Analysis and System Evaluation

A central requirement for an integral and inclusive human factors approach to the design of tasks, systems, and work procedures is a need to obtain a comprehensive and integrative perspective on the performed tasks. We need to develop a framework and approach for such analysis and its implications for the design of systems and work procedures. Surgeons, anesthesiologists, and nurses in operating rooms, medical teams in different intensive care units or on the hospital wards, teams at imaging facilities, all perform a wide variety of routine and emergency tasks within multielement complex environments. Each task may comprise many procedural segments and stages and involve multiple technologies and engineering systems. Some examples include preparation and administration of medications, calculating and delivering radiation therapy, and information retrieval and recording. Of key importance to proper task design are a systematic description of all task components, demands, and involved activities; conceptualization of each performer role(s); and the execution of a task in its environment from start to end.

To illustrate the concept and scope of a comprehensive task analysis let us examine a simple task of medication administration in a pediatric ward and a complex task of conducting surgery. Both examples are taken from projects conducted at the Hadassah Hebrew University medical school in Jerusalem, Israel. They are summarized in Donchin and Gopher, Chapters 4 and 6 (2011).

3.2.1.1 Administration of Magnesium Sulfate

Magnesium is an intracellular electrolyte needed for the activation of enzymes and many metabolic reactions as well in the transmission of nerve impulse. Low levels (hypomagnesaemia) are the result of deficient intake or the effect of drug administration or the influence of a disease. The way to supply it is by infusion of special commercially available solutions. Nurses are commonly responsible for medication administration to the patients based on physician orders. An overdose has deleterious effect and may lead to death.

Let us review what is performed by a nurse who is following a physician written order, to administer 625 mg magnesium sulfate in IV, over a period of 2 h to a child in a pediatric ward; the main steps to be performed are as follows:

1. Detect and read the order (order form and doctor handwriting).
2. Locate the magnesium solution bottle in the medication cabinet packed with bottles and reading the concentration of magnesium in the bottle (medication cabinet, solution bottles and content of information labels).
3. Compute the amount to be extracted to contain 625 mg (mental arithmetic).
4. Use the proper syringe to withdraw the accurate volume from the bottle (injector, pump, volume measures).
5. Dissolve the medication in a 500 or 1000 mL of Hartman's solution (solution containers and information on labels).
6. Add the solution to a "voluset" and set up a special automatic syringe pump.
7. Prepare an Intravenous Automatic Controller or special automatic pump.
8. Compute the rate of drops per minute for the required administration period (IV rates indicators and controllers).
9. Connect or insert the Intravenous (IV) tubing to patient port and set the IVAC regulator (pipes and skin needles, connectors).
10. Update physician order form upon execution (order form and update format).

Note that although following up and administering a doctor single medication order may appear at the outset to be a basic simple task, the actual task includes 10 steps and requires information extraction from congested, poorly designed labels, mental arithmetic, general knowledge, procedural knowledge, control, operation, and adaptation of several instruments and devices. The technology components include the doctor's order form, the magnesium bottle, medication information labels, solution bag, syringe, IV control machine, IV lines, and connectors. All of these performance and technology components are important constituents of successful performance. Errors or failure of the whole process can be attributed to the influence of one or several of the components (Thornburg et al., 2006). Hence, inclusive task design should consider all elements and their interrelations.

3.2.1.2 Performance of a Surgical Procedure

The operating room (OR) is one of the more complex healthcare environments and the performance of a surgery is in most cases a complex, intensive, long duration, and coordinated effort of a heterogeneous professional team. The basic surgery team includes a senior surgeon, an assisting surgeon, an anesthesiologist (sometimes two), a scrub nurse, and a circulating nurse. In many cases there may be a technician, additional physicians and nurses, and imaging and special systems experts. Each team member has a role and specific responsibilities and performs multiple tasks (e.g. Weinger and Englund, 1990). However, performance should be supported by and coordinated with other team members for the successful joint effort, that is, the surgical procedure. As a work environment, the OR is filled with high level technologies: surgical instruments and devices, anesthesia machines, monitors, imaging systems, as well as multiple low-level technologies and manual devices: hand and mechanical tools, syringes, medications, pads, forms, etc. Given the technology and human components of the task, how can a comprehensive and integrative human approach to OR and surgery design be applied? From this brief analysis it is clear that a human factors analysis of individual technology and system is important to assure that its design complies with human factors principles and standards. However, this is not enough. There are multiple, different, sophisticated and simple engineering systems, a relatively large and diversified team of professionals performing a wide variety of tasks in a dynamically changing situation, where the cost of incidents and error is high. These are all interrelated and interdependent components that need to be considered in a global integrative design approach. To enable such an approach the human factors expert is required to acquire a deep and detailed understanding of the objectives, goals, and uses of each of the engineering components, as well as the roles and responsibilities of and division of labor between the different surgery team members. Also required is a detailed task analysis of the different subtasks performed by each member of the team. Examples for such tasks include moving the patient from the gurney to the OR bed, carrying out the induction of anesthesia, monitoring vital signs during surgery, delivering the proper instruments and auxiliary materials by the circulating nurse upon demands while updating inventory records, counting inserted and returned pads and instruments, etc. In a manner similar to the example of the preparation and administration of magnesium, each of these tasks comprises multiple steps and activities that have

to be completed in a certain order and that need to be monitored. It is important to note that in both the medication administration and the surgery examples, the direct medical operations (administration of magnesium, or surgery and anesthesia) are surrounded and complemented by multiple indirect activities, such as preparation, monitoring, maintenance, recording, and report (we further discuss this point in the next section). It is crucial to asses how all these tasks are implemented in the general physical environment, coordinated, and synchronized when there is a single performer or when other team members collaborate in the care procedure.

The two examples clarify why a formulated model for assessing the whole task and the interrelations between its elements takes the human factors involvement and responsibilities much over and beyond a limited user-centered examination of each single procedure or technology component. While each of the separate elements at a patient bedside in intensive care may be well conceived and properly human factors justified, the global collection of elements, layout, and arrangement around the bed should be seemingly well coordinated and consistent. If not conceived and designed to support performance of the integral care task and its dominant variations, even if single composites may be optimal, their combination in a single work environment may create many difficulties and inconsistencies and lead to increased workload for the attending nurse or a physician who attempts to obtain a quick update on the patient's medical status. By analogy, think of the design of an airplane flight cockpit, driver instrument and control panels, air traffic control work stations, or supervisory monitoring panels of industrial processes. These work stations have all been constructed and configured with an integral and comprehensive unified perspective of the work environment and the multiplicity of tasks and cross tasks to be performed; this applies to many environments such as flying, air traffic control, or process supervision and monitoring. Hence, the composition, relationship, and global arrangement of all elements are planned together to afford the best overall interaction and global performance of the task. The challenges for coherent design already exist in environments with a single performer. They are considerably magnified when there is a team of performers who need to coordinate effort under time pressure and high costs of error.

The importance of integral comprehensive design, which is well understood in the domains mentioned earlier (e.g., air traffic control), has not been formally agreed upon, conceptualized, and adopted by healthcare providers and the healthcare community. Very few if any intensive care units, operating theaters, hospitalization institutes, and chronic care units have been carefully human factors designed as integrative unified functional entities. Very frequently healthcare workstations and performance procedures gradually emerge and develop like a quilt blanket, depending on space, budget constraints, donations, and management policy. This situation creates large functional variability in the performance of the same task even within the same unit; it further increases task difficulty, workload, and probability of errors.

A major drawback in the ability to develop and propose a comprehensive human factors design approach to healthcare work is that we do not have yet good enough descriptions and functional task models of the major tasks performed in the medical environment using the language, terminology, and functional descriptors of human factors, human performance, and cognitive psychology. Using the terminologies and models of these domains, what are the tasks and the environment of a surgeon, circulating nurse, or a pediatric nurse preparing a medication for IV administration? Such an analysis is mandatory in order to apply an integrative human factors design approach and propose evaluation recommendations. It is important to describe and explain the various procedures, subtasks, and subsystems that have to be attended to, coordinated, organized, and synchronized for carrying out each task. Recognition of the importance of these aspects and the steps to uncover and describe them may be a key first call for major progress of human factors contribution to health care.

3.2.2 Technology and Care Systems Are Not Only Robots, Computer Displays, Imaging, and Sophisticated Devices

The progress of modern health care is clearly linked with the development of sophisticated technology. Surgery innovations, computing, imaging, measurement and diagnostic, information display and transfer, robotic interventions, and other are all instantiated in innovative healthcare interfaces. Their development and implementation call for a wide span of human factors involvement in their design and application (Alvarado, 2007; Cao and Rogers, 2007; Clarkson, 2007; Karsh et al., 2006; Nemeth et al., 2007). Nonetheless, the incorporation and use of such systems in the conduct of the vast majority of medical procedures and care tasks is surrounded by and interweaved with numerous low-technology devices and manual task components. The latter are integral components of performing the task and may be equal contributors to its efficient and safe conduct. Human factors of this category and type of components should also be examined and properly designed. Surgery involves multiple hand tools and mechanical devices, pads, forms, inventory lists, medical records, and reporting systems; reading labels, counting used tools and pad, and use of storage and disposal devices are all integral parts of surgery. The preparation and

administration of medications involve very little high technology but a large number of low-technology elements and a number of procedural steps, control skills, and computation requirements. This type of low technology and assisting activities are integral parts of the majority of medical care processes, even when the most advanced technologies are available and incorporated in patient care. Assessment and proper design of low-technology elements such as forms and medication labels is important. Inadequate content or poor design can be a source of performance difficulty and a cause of errors. Forms and patient records still appear to have an important role in modern healthcare processes as they were in the past with traditional and less developed care systems. It is hence functionally important that they be properly designed, whether they are manually filled paper forms or converted to computerized versions. Similarly, the wide assembly of information labels should be identified and made distinguishable, informative, and easy to read. An inclusive, comprehensive, and integrative human factors examination of a care procedure should therefore include high- and low-level technology elements, manual and computerized tools, and paper and paperless environments. Their combined requirements and demands characterize the task to be performed and determine its overall difficulty and susceptibility to errors.

3.2.3 Medical Tasks Are Multiprocess and Multistage

This section underlines an important common denominator of the cognitive structure and cognitive demands of medical care tasks with implications for training, skill acquisition, and accurate performance. It should be recognized that the vast majority of tasks performed by physicians and nurses are multiprocess and multistage (Cook et al., 1998). In most tasks the sequence order of activities and the completeness of performing the process and stage are important; a surgeon in surgery, a physician examining a patient in ward, a scrub nurse in the OR, a nurse preparing a medication, an x-ray or ultrasound technician, and a radiotherapy operator are some illustrations of multiprocess and multistage tasks. The significance of this observation from a cognitive psychology and human performance point of view is that the training, acquisition, and execution of procedural skills is a distinguished topic in training and memory research that specifies modes of introduction, organization, training, and maintenance (Anderson, 1982; Meyer and Kieras, 1999; Gopher, 2006, 2007). Analysis of medical tasks with this perspective may lead to alternative design of tasks, development of better work procedures, instruction, training, and introduction of alerts and cuing performers. Three general guiding principles are clear explication of task steps and stages, their transitions, and order; development of assisting mnemonic and memorization aids; inserting compulsory review and sequence validation steps. A cognitive human performance examination of the sequential structure of medical care procedures is mostly absent and much needed to assist performance and reduce errors in modern health care.

3.2.4 High Workload Is the Product of Multiple Determinants and Work Characteristics

Overload and high workload are typical and dominant complaints raised by medical managers and personnel when describing their work. Human factors experts are many times called upon to propose solutions. Complaints of high workload, overloaded systems, and job and task demands that are hard to cope with are frequent and recurrent across the range of hospital and secondary care environments. A prevalent accompanying solution is to increase the number of staff members and/or decrease the number of patients to reduce load. The construct of workload has been widely discussed, theorized, and studied in the domains of cognitive processes and human performance (Gopher, 1994; Wickens and Hollands, 2000). Within this literature, there is no doubt that the number of performers (members of the medical team) relative to the volume and pace of incoming patients is an important determinant of overall workload. Nonetheless, it is also well established that for a given care task there may be additional factors that increase load, which can be equally or even stronger contributors to load. For example, a questionnaire administered by Gurses and Carayon to 265 nurses in 17 different intensive care units, and structured interviews conducted in another group of nurses, in which they were asked to identify workload elevating aspects in their daily work. Both revealed four main clusters of performance obstacles associated with increased workload: Equipment, information transfer and communication, help from others, and intra-hospital transport (Carayon and Gürses, 2005; Gurses and Carayon, 2007, 2009; Carayon, 2008; Gurses et al., 2009). Listed and briefly discussed in the following are work aspects leading to higher workload, which are different from and added to the effect of care team to patient number ratio:

1. *Lack of standardization and consistency of equipment*: Hospital wards normally include multiple patient care stations (e.g., incubators in a neonatal units, special intensive care beds and surroundings in ICU, number of beds in cardiac internal ward). When an attending staff member is responsible for several patients, and when responsibilities change and alternate, overall workload is considerably increased if there is lack of standardization and consistency of types, models, and functional control of systems and care devices, either within or between care stations. When moving from one patient or one bed to another, the difficulty of adjustment, reorientation, and shift further increase task demands when the equipment changes or the same equipment with a different model is used. The requirement to shift, change, or adopt new orientations results in higher load and increased probability of errors. Unfortunately such changes between patient rooms and bedsides are very common in hospitals and secondary care environments. Equally prevalent are control differences and information display mismatches within the same care station (Donchin et al., 1995; Carayon and Curses, 2008).

2. *Lack of standardization and consistency of the physical layout*: Increased workload is also caused by lack of consistency and standardization of the location, order, and space arrangement of systems, equipment, and storage facilities in functionally similar or even identical work sites. If the location and configuration of monitors, controllers, and storage cabinets change from one operating theater to another, from one incubator or intensive care bed to another, the need to readjust and adapt to the changed setting slows down performance, and increases workload and the probability of performance errors, even if those tasks are done daily and repeatedly.

3. *Very spacey or too small and congested work spaces*: The overall physical size of the unit and its design may also be a contributor to performers' increased workload. Large and spacey, multiroom care units may require the attending medical staff, physicians, nurses, and attending personnel to walk on a daily basis for large distances and continually move between spaces and rooms. Also, when a frequently used common facility or storage cabinet (e.g., medication cabinet) is located in an outside room or spatially remote from the patient bedside, the recurrent requirement to use it for different patients at different times increases work hustles and workload. In two different projects in which we were involved, emergency wards were moved to newly designed spaces that were twice to three times larger than the old ward. The medical staff had many difficulties in coping with the mobilization and attention requirements of the large distances and spread of facilities and

patients. They reported increased time load, attention load, and physical fatigue. Both units had to be readjusted and reduced to a physical size that the medical team felt comfortable with. Another example comes from a students' observational study conducted at a spacey and well equipped intensive care unit, in which a computerized central medication cabinet was placed in a free space at the end corner of the unit. Nurses who attended patients in beds located at the other end of the unit tried to save time and effort by checking out medications for several patients at the same time in anticipation of several administration passes rather than for a single patient and a single medication pass. This shortcut was developed to reduce load, but it increased the difficulty of maintaining the separation between patients and administration orders.

On the other hand, if a multi-patient care unit is too small, crowded, and congested, workload is also significantly increased for the attending staff needed to safely negotiate their path between beds and perform efficiently at each bedside. In a study conducted in a neonatal ward of a large hospital, we observed that due to space limitations incubators were only 90–100 cm apart from each other, rather than the recommended 150–200 cm separation (Standards Subcommittee Intensive Care Society, 1983). Given the large overall number of monitor devices, cables, and pipes for each incubator and their different layout around incubators due to physical constraints, attending staff had hard time to perform their tasks efficiently and accurately and to move comfortably among units.

4. *Difficulties in accessing and recording information*: Retrieval and integration of information, as well as keeping records of decisions and orders are significant activities and time segments of contemporary care procedures. Medical information files of patients are rich and accumulate rapidly due to the ability to measure many parameters, the availability of blood constituents, imaging results, and others. Similarly the degrees of freedom for interventions and care recommendations have much increased. Consequently, attending physicians and nurses during patient rounds, consultations, and regular care are required to invest considerable time and effort in retrieval and review of information records and summaries, as well as filling reports and records (Koppel et al., 2005).

The ease and smoothness with which these important activities can be carried out is a crucial contributor of elevating or reducing task workload in medical care. Analysis and evaluation of this aspect is an important challenge for the human factors expert. Well-designed, user-friendly, information technology systems can have an important role in the reduction of this type of task workload in modern health care. Even when no computerized paper forms and medical folders are being used, workload can considerably be eased with good human factors design (Sellen and Harper, 2003; Sela et al., 2005). When information retrieval, review, and recording capabilities are poor, workload and error rates are substantially elevated (Reed, 1996; Brown, 1997; Bogenstätter et al., 2009).

5. *Insufficient work structure—clarity of work procedure, work structure format, division of labor, and responsibilities*: Because most medical procedures are carried out by several performers and are mostly based on coordinated team work, confusion, misunderstanding, and higher workload for each team member can result if the global task and the involved stages and procedures are not well structured. Healthcare work in surgery, trauma, emergency units, regular wards, and intensive care follow this work structure.

There are several cognitive and human performance aspects of a well planned work structure: task objectives should be described and known to all performers, all work procedures should be developed and clarified, and there should be a clear and justified division of labor and responsibilities between team members of the same and different profession (e.g., among nurses; between nurses and physicians). Similar clarification and formalization is required for task segments performed by external units such as laboratories and pharmacies. Insufficient clarity or ambiguity in any of the task segments or involved procedures elevates immediately workload, and increases confusion and the probability of error.

6. *Communication and information transfer—none formulated and insufficient information exchange and communication routines*: Another important cognitive aspect of working in teams associated with workload is the existence of formatted and sufficient oral communication, and formal and informal exchange of information between team members. If team members are not well informed and do not

properly communicate with other team members, the task workload on each team member is considerably increased. This is because each member's performance may be carried out with only partial or incomplete information on all aspects relevant to his own performed segments or on concurrent, supplementary, or complementary accomplishments of other members in the team. Incomplete and ambiguous task knowledge may lead to delayed response, reduced confidence, and elevation of task load. A variety of formal communication and information exchange instances have been incorporated in healthcare procedures. Examples are briefings before surgery, doctor rounds, and shift changes. However, many of these already incorporated formal routines are not sufficiently planned and well attended to. Moreover, there are many other less formal operations and transitions in the care procedure that should be communicated and informed among team members. The quality of this type of information exchange is an important determinant of the quality of care and the workload on each team member. A good example for recognizing the importance of communication is the effort by World Health Organization (WHO) and the Joint Commission to introduce pre-surgery briefing and check list routines into the OR (Joint Commission on Accreditation of Healthcare Organizations, JCAHO, 2007). Many hospital wards have developed local formats and patterns for doctor round and the conduct of shift changes. This is a good beginning, but it is far from being enough or complete. A systematic assessment and development of solutions is another call for human factors contribution to improve the quality of care and reduce load.

To summarize this discussion on possible determinants of and contributors to workload, it is clear that the workload of medical staff members is determined by much more than just the size of the team and the number and flow rate of patients. Workload can be the joined product of several cognitive and human performance factors associated with the engineering, physical and procedural formulation, and cognitive structure of the care task. The review of possible contributing factors underlines the importance of a careful and detailed assessment of workload in any given work environment along the dimensions and aspects discussed in this section. There are multiple ways by which a simple and small investment can ease load, reduce errors, and improve the quality of care.

3.3 Team Work: Its Cognitive and Human Performance Constituents

As already indicated in several of the previous sections, a notable feature of modern health care is the collaborative involvement of teams in the vast majority of care procedures. Team members may have similar or very different educational, professional, and specialty backgrounds. It is hence important that the cognitive and human factors design implications and matching requirements of team work be examined and attended to as part of human factors contribution to the improvement of care quality and patient safety. Team work efficiency and ways to improve it is a central topic in contemporary behavioral and management science both in research and application. Nonetheless, there is a clear distinction between social, organizational, and management perspectives that focus on motivation, leadership, creativity, decision behavior, management, and organizational climate topics (Cook and Wood, 1994; Cooke et al., 2004; Baker et al., 2007). The human factors viewpoint examines the cognitive, information sharing, transfer, and coordination modes required to enhance a single member's task performance and the overall team functionality. The present discussion is devoted exclusively to the latter category of factors—cognitive and human factors aspects of team work in medical care. Teams associated with the performance of a task or a procedure can be generally categorized as synchronous or asynchronous. Synchronous teams are those in which members collaborate concurrently within a single-focused and dynamic care event. Examples are the conduct of a surgery in which anesthesiologists, surgeons, nurses, and other professionals collaborate in an operation on a patient; or a trauma care case in which a combined team of physicians and nurses work together to give primary help, stabilize, and diagnose a rushed-in accident survivor. Asynchronous teams are those in which the care process requires the collaborative involvement of several members, but each of them interacts individually and intermittently with the patient, prior to or following other members of the team. In health care there are many representatives for this type of teams: regular care in a hospital ward by the medical team, nursing shift change, handover of a ward from the day shift to the night shift, or the on-call physician. In elective surgery all the personnel involved in the different steps, stages, and transition of patient, starting from the hospital ward, to reception and waiting rooms, surgery theater, recovery room, and back to the hospital ward.

The importance of distinguishing between these two categories of teams is that the key cognitive and human factors requirements in each type are somewhat different. In the case of synchronous teams, the key component is the existence of a sufficiently detailed and updated cognitive map of patient medical status and the flow of the ongoing process that should be presented and shared by all. The situation is quite analogous to the general battle map display in military operation rooms or the air traffic display monitor for controllers in the flight control center. Quick response and efficient performance of each member in a surgery and overall team coordination and efficiency benefit and improve if members have in common an advanced knowledge of patient data, surgery plans, and known routine requirements (Artman, 2000). Such common knowledge and sharing is indeed the heart and justification of the contemporary demand for a routine conduct of pre-surgery briefings (World Joint Organization's Patient safety checklist for surgery, 2008) and its more developed forms (Einav et al., 2010). Equally important for efficient team work is a good real time and well integrated situation map of the patient that can be observed by all members, along with a clear statement of the present stage of process and the move from one stage to another when it takes place (Mohammed and Dumville, 2001). Segments of these requirements have already been incorporated in medical settings, but these are only preliminary steps and much more needs to be extracted and modeled (Cannon-Bowers et al., 1993; Lingard et al., 2005).

With regard to the work of asynchronous teams, from a cognitive and human performance viewpoint the most significant factor to focus upon is the existence of multiple transitions and transfer points (Cook et al., 2000), some of which are routine, timed, and formal; others are less planned and scheduled. Examples of routine and scheduled transitions are shift change of the nursing staff, doctor rounds, and doctors' handover of wards from the day to the night attendants and on-call doctor. Examples of more variable transitions and transfers are the moves of a patient during a surgical procedure from the hospital ward to the OR and back, the intermittent care activities of nurses, doctors and laboratory tests of a patient in a hospital ward, and patient treatment in emergency wards (Artman and Wearn, 1999). What is common to all transitions and transfers is that in each of them there should be transfer of information and responsibility from the current to next or anticipated team member. A transition is successful only if both information and responsibility transfer are carried out successfully. An important challenge for human factors is to develop the format and content of planned formal and variable transitions. Human factors principles and cognitive psychology knowledge should be drawn upon to develop the format, content, and mode of shift change, doctors' rounds, and handover activities as well as to guide the less formal transition. Such a design and human factors involvement is an important contributor to efficient team work and the quality of care. An

interesting variant of a formal transition and change is a case in which the medical record of the patient is used by multiple users to guide treatment plans and monitor patient progress along the procedure. A good example is the patient record file in radiotherapy (Sela et al., 2005). This file may be the main record and situation map that accompany a patient during several months of radiation treatment. It is being accessed and used by doctors, nurses, physicists, dosimetries, social workers, and others. This is an asynchronous team that accompanies a patient throughout his treatment. It is necessary that the record include the information required by all specialists, in a format that can be easily accessed and updated. The design of process records for a team of healthcare professionals with multiple backgrounds and roles is another unique call for team work related aspect of human factors work in health care.

3.4 Urgent Need for Developing Appropriate Reference Databases

It is important to recognize that an appropriate and trustful evaluation of the human factors needs in each of the topics discussed in this chapter, and prioritizing the investment in corrective efforts, depend on the availability of credible databases to refer to and the ability to collect valid reference information. To our knowledge, at this point of time such databases do not exist and they are not developed as a routine part of projects. Present efforts and data collection are mainly directed to the collection and study of medical errors, adverse events, and incidents reports (e.g., American Hospital Association, 1999; Leape, 2002; Gawande et al., 2003; Weinger et al., 2003; Nemeth et al., 2006). Hence, this is the main existing collective effort and global information source on patient safety, healthcare quality, and their economic costs. While being a good instigator for public and political awareness, errors and incidents studies cannot serve the required role of data sources and guide for a proactive and anticipatory Human Factors involvement and prevention efforts. The major drawbacks of error-based studies as systematic and scientific source of information have been recently described and discussed by Gopher (2004). According to this paper, data collections based on error and incidents reports suffer from four main problems: a nonrepresentative sample of problem coverage, quality and completeness of reported information, absence of a proper comparative reference bases, and the wisdom of hindsight.

The present general estimate is that at the best only about 5%–7% of the total actual number of errors and adverse events are documented and reported (Bogner,

1994; Kohn et al., 1999). A recent example for this state of affairs is a study conducted in 2007 by the state of Indiana in the United States to evaluate the usefulness of a compulsory reporting system of incidents and adverse events. Data were collected for only a total of 105 reporting of mishaps over 1 year period in 291 medical care facilities (1). Avoidance of filling reports is not difficult to understand given concerns about possible consequences to the reporter. Nonetheless, the low percentage of events actually reported and the selective reporting are likely to make this whole collection a small and non representative sample of the actual body of existing errors. A related problem is the quality and completeness of information in the reports. This problem arises because reports are based on subjective and biased information (similar to eye witness reports problems), and mostly based on retrospective memory recounts, which are bound to be partial and incomplete.

A third type of problem with incident and adverse events reports is the absence, in most cases, of a proper comparative reference base for their evaluation. What is missing is the ability to compare or compute the relative frequency of errors of a certain type (e.g., medications administration), as a percentage of the total number (population) of activities of this type in the evaluated unit(s) work. What is usually compared is the relative frequency of errors of a certain type from the overall number of errors collected in the unit(s). This is an incorrect and misleading comparison. The correct comparison should be done relative to the total number of activities of the same type (e.g., total number of medication administrations) in the studied units during the relevant time period. It is quite possible that a certain type of error will be 10% of all collected errors in the unit, but the activity involved may represent 25% of all activities in the evaluated unit, or vice versa. A good example of such a possible computation problem and fallacy is reported by Donchin et al. (1995) who conducted an observational study at an intensive care unit. Physicians were found to contribute only 45% of errors while nurses accounted for 55% of a total number of 348 errors observed and recorded during a 3 month observation period. However, during this period nurses were recorded to perform more than 87% of the total number of activities, while doctors were involved in less than 5%. Which professional population is then more susceptible to and is involved in more errors? The answer is quite clear and underlines the impossibility of judging the seriousness and relative frequency of a type of error without having the frequency of activities (correct and erroneous) as a reference base.

Last but not least is the problem that conclusions drawn from error investigations or incident reports are always the "wisdom of hindsight." Namely, they all use an adverse event as a starting point and trace back the sequence of events to reveal its possible sources and

determinants. The problem with this type of hindsight, for example, in root cause analyses, is that while the analysis may be logical and explanatory, it lacks predictive capability and generalization power. Retrospective analysis of events and decision points of a multiple choice dynamic process does not enable to generate valid predictive schemes. To illustrate, think for example that a person (including the present authors) who progressed in his professional career can probably give a good and well justified account of the trajectory that he followed, which located him in his present job. However, questioning each of the decision and selection points from the start of his career, including education, professional training and work experience, what is the likelihood that the present position could have been predicted at each point? This is the exact difference between hindsight and foresight explanations. It is the latter which is the target of scientific work. Error studies and incident reporting systems are all hindsight wisdom and are not able to provide insight for future errors. There is therefore an urgent call for the search and development of alternative information sources and collection formats that may fill the gap and provide adequate pointers, support, and feedback to guide human factors work efforts. This type of data collection should evaluate and collect human factors performance relevant information on difficulties and user-oriented design flaws along the categories, dimensions, and topics discussed in this chapter. Information and analysis should cover all tasks and modes of system/unit operation, independent of and prior to the occurrence of specific errors or adverse events. The traditional tools and techniques for task and system analysis that were developed in the Human Factors and Ergonomic domain (Salvendy, 2006) are applicable for the present purpose, although they should be adjusted and customized to cover the additional categories and specific topics signifying healthcare tasks and working environments.

One main difficulty in adopting such an approach is that it is hard to apply widely and robustly across many units and work environments. This is because the traditional data collection and task analysis are commonly performed by human factors professionals, and typically do not involve more than one unit at a time. By comparison, incident and adverse event reports can be collectively filed and updated concurrently by all staff members across the whole span of tasks and segments of one or more units. There is a need to try and develop similar formats and methods of data collection in health care for the type of information and data relevant to our human factors objectives.

An example of a possible candidate for an alternative data collection approach has been recently investigated in hospitals in Israel by our team (Donchin and Gopher, 2011; Morag and Gopher, 2006). It is based on short problem reports filled by hospital wards medical and administrative staff members. Members are asked to report recurrent problems, difficulties, and hazards in the conduct of their daily work, associated with human factors and human performance work aspects. It is argued that this type of problems and difficulties are good indicators and contributing factors that can be directly linked to reduced care quality and increased probability of errors. Furthermore, such reports are advanced accounts of daily difficulties, hassles, and hazards. They are completed independently and prior to the occurrence of errors or adverse events and hence do not carry concerns of accountability, negligence, blame, or guilt. Five types of one-page report forms have been developed, each covering a different work aspect. The five are instruments and equipment, physical space and environmental conditions, work structure and division of labor, medication administration, and forms and recording files. Table 3.1 presents a condensed summary of the five report forms.

Reports were distributed and collected over a period of three months in four wards of two tertiary care hospitals in Israel: Rambam Haifa and Hadassah Jerusalem. A total of 359 reports were collected during this period, twice the number collected with the existing compulsory incidents reporting system over a five year period. Moreover, reports were contributed by both doctors and

TABLE 3.1

Five Potential Problem Areas Associated with Human Factors Aspects with Their Contributing Factors

Work Patterns and Procedures	
☐ Carrying out a procedure (knowledge, understanding, ability to follow)	☐ Difficulty transferring information among staff members of the same group
☐ Absence of procedures	☐ Work structure with relation to support units
☐ Division of work between different profession staff members	☐ Information exchange with support units (consultants, pharmacy, etc.)
☐ Division of work between members of the same profession (e.g., nurses)	☐ Difficulty transferring information between members of different groups
☐ Task load	☐ Other _____

Physical space and layout	
☐ Department design and space division	☐ Work and preparation areas
☐ Lack of function specific rooms	☐ Environmental factors causing danger
☐ Incompatibility of designated rooms	☐ Crowding
☐ Storage space	☐ Other _____

Medication administration	
☐ Drug information (dosage, effect, sensitivity, drug family)	☐ Positive or negative reactions with other drugs
☐ Drug administration method (including preparation, giving rate)	☐ Drug administration time and schedule
☐ Drug recording (place, convenience, copying)	☐ Calculation and computation problems
☐ Drug prescriber's instructions	☐ Drug administration procedure
☐ Drug nomenclature or commentary	☐ Problems with auxiliary equipment
☐ Similarity in name/packing with other medication	☐ Work surfaces and drug preparation
☐ Label misleading information (concentration, units)	☐ Drugs storage
☐ Lack of information	☐ Other _____

Instruments and equipment	
☐ Knowledge on equipment operation	☐ Equipment reliability or calibration failure
☐ Reading displays and indications	☐ Absence of equipment
☐ Ergonomic problems (awkward posture, force, weight)	☐ Equipment storage area
☐ User-unfriendly operation and use	☐ Equipment organization and arrangement
☐ Operation mistakes	☐ Equipment variability (lack of standardization in performing similar operation)
☐ Maintenance	☐ Other _____

Data management	
☐ Need to recopy (record the same data several times on the same form)	☐ No uniformity of field names (size, abbreviations, etc.)
☐ Lacking space to record the necessary data	☐ Problems in reading the information on the form
☐ No place on the form for patient identity information	☐ Problems of extracting information
☐ No identification of time and source of recorded data	☐ Form storage space
☐ Hard to locate the required information in the whole form	☐ Other _____

nurses, while incident reports with the existing compulsory system were exclusively filled by the nursing staff. The problem categories and signifying aspects as presented in Table 3.1 correspond well with the dimensions and topics discussed in the previous sections of the chapter. They are markedly different from the topics and issues contained in the incident reports of the existing system. It is not difficult to see how this new type of information basis can serve and guide human factors engineering design and work restructuring efforts. It can also be used for before and after comparative evaluations. Whether this specific approach or an alternative format will be adopted, the difference in the nature and potential value of this type of information as compared to the back tracking accounts of incident reports clarifies the need and urgency for developing new reference bases (see also Weigann and Dunn, 2010).

3.5 Call for Human Factors Engineering in Modern Health Care

The common denominator of all topics and care aspects presented in this chapter is the claim that the format and embedded risks of modern health care require the inclusion of human factors considerations from the earliest stages of design and development. Taken together, technology, physical layout, organizational structure, work procedures, and overall load on care providers combine to instigate a strong justification for continuous and systematic involvement, supervision, and inspiration of human factors domain specialists who have broad background, knowledge, and understanding of the contemporary healthcare work environment and procedures. The driving goals of their involvement are enhancement of work efficiency, reduction of workload, improvement of care quality, increased safety, and reduction of errors.

We consider the dimensions and topics discussed in this chapter to be representative and general enough to delineate the nature and scope of required efforts. They also provide a preliminary formula for the development of a general framework and focus on a proactive and comprehensive human factors approach to health care in hospitals and secondary care environments. To recapitulate, the topics discussed are as follows:

1. Inclusive task analysis and system evaluation

2. Technology and care systems are not only computer displays, imaging, and sophisticated devices

3. Medical tasks are multiprocess and multistage

4. High workload is the product of multiple determinants and work characteristics

5. Team work—its cognitive and human performance constituents

6. The critical need of developing valid and credible reference databases

Attending to each of these topics by itself has important implications for improved care quality and safety enhancement. The combined influence of these topics is much more powerful and substantial. A discussion of difficulties and challenges associated with each care aspect brings forward the need to incorporate human factors–based considerations and decisions as early as possible, maintain monitoring and supervision throughout system employment life cycle, and add focal efforts when problems are observed, or changes are considered and introduced. Note that while errors and incidents may also be recorded and provide supplementary information they are not the focus, the trigger, or the drivers of this work. Data collection, design, and performance evaluation should start much earlier, be more inclusive, examine a wider scope of implications, and include stronger anticipatory predictions. This is why we believe that the existing emphasis on the study of errors and the continuing efforts to increase reporting compliance and collect and analyze incidents are diverting attention and shifting investments away from the required and potentially effective human factors efforts (see also Leape, 2004).

In the effort to overcome difficulties and improve health care, aviation has been one task domain whose evaluation approach and work protocols have been strongly argued for, copied, and drawn upon, in particular, its safety protocols incident investigation and error reporting culture (Bogner, 1994; Helmreich, 2000; Vashdi et al., 2007). Both civil and military aviation traditions have been considered role models to be adopted and copied. We do not want at this point to elaborate on the difficulties experienced and reported by the aviation safety authorities in achieving only low reporting percentage and report biases of incidents and near miss events, which are similar to the healthcare community. It is also beyond the scope of this chapter to describe the clear distinctions between the much simpler and constrained tasks and system environments of pilots and air crews as compared with those of health care and healthcare providers. We do want, however, to bring forward and emphasize the fact that aviation flight safety efforts and incident report traditions are applied as a starting point on aviation systems, cockpits, airplanes, and flight procedures that have been the subject of intensive human factors involvement from their early design stages onward (Wiener and Nagel, 1988; Tsang and Vidulich, 2003; Weigman and Shappell, 2003). Aviation systems have been a major topic of human factors work from its early days. Human factors has been introduced to the aviation industry and community following the Second World War and its involvement and contributions have intensified and proliferated since then. Airplane cockpits, cabin crew work stations, air traffic control stations, and control towers undergo proactive, comprehensive, and integrative human factors design and evaluation from their earliest design stages and are continually monitored. Only then at systems daily operational phases, data on safety procedures and protocols, near miss, incident, and error reports are collected and studied. These analyses are done only in a second phase and are not the primary effort to develop user-centered efficient and safe interfaces and task performance. Let us not reverse the order in medical systems and change the scope of human factors work in health care. Aviation safety begins with human factors efforts; aviation specialists conduct the required analysis and collect the relevant data. Incidents and accidents reports and investigations come later as a supportive and not as the formative source of information. This is a logical order and an approach that should be equally preached and maintained in the field of health care. At the present state of affairs, we have no doubt that the most cost effective efforts to improve healthcare quality and reduce errors should focus attention and efforts first on human factors work, while error studies and improvements in the collection of incident reports should come second.

Another important observation appears to emerge from the review of the human factors challenges in health care and the format in which care environments are integrated and developed. The existing large differences and diversities among units within the same hospital and between different hospitals and care environments require that a human factors team be associated or reside within a hospital or care unit. This team will be responsible for the human factors related unit design configurations, consistency of devices and work stations, and the formulation of work procedures. The chapter shows very clearly that the design optimization of the human factors and usability characteristics of individual systems or work units are not sufficient. Many and different work stations are included in each work environment. It is important to understand their interrelations and combine such an assembly in a way that will match with unit task goals and work structure. This work can only be properly done from the inside, that is, from within the hospital or in close association with the hospital. The call is to have a human factors team working in or closely associated with large hospitals or care units, in much the same way that specially trained architects plan, guide, and supervise hospital construction plans and space design (American Institute of Architects, 2001). To our knowledge, at present such teams are rare or none existent. It is common to establish in-house hospital risk management and legal units,

which sometimes employ large teams (lawyers and analysts), while not having even a single human factors specialist. Our view is that healthcare improvement and safety, as well as a reduction in the number of malpractice suits, can greatly benefit from including a human factors team within the care unit organizational framework. This is an additional important call for improving health care through the involvement of human factors.

References

Alvarado, C. 2007 The physical environment in health care. In *Handbook of Human Factors and Ergonomics in Health Care and Patient Safety*, Ed. P. Carayon, pp. 287–307. Mahwah, NJ: Lawrence Erlbaum Associates, Inc.

American Hospital Association. 1999. *Hospital Statistics*. Chicago, IL.

American Institute of Architects AIA. 2001. *Guidelines for Design and Construction of Hospital and Health Care Facilities*. Washington, DC: AIA Press.

Anderson, J. 1982. Acquisition of cognitive skill. *Psychological Review*, 89, 369–406.

Artman, H. 2000. Team situation assessment and information distribution. *Ergonomics*, 43(8), 1111–1128.

Artman, H. and Wearn, Y. 1999. Distributed cognition in an emergency coordination center. *Cognition, Technology and Work*, 1, 237–246.

Baker, P. D., Salas, E., Barach, P., Battles, J., and King, H. 2007. The relation between teamwork and patient safety. In *Handbook of Human Factors and Ergonomics in Health Care and Patient Safety*, Ed. P. Carayon, pp. 259–271. Mahwah, NJ: Lawrence Erlbaum Associates, Inc.

Barach, P. and Small, S. D. 2000. Reporting and preventing medical mishaps: Lessons from non-medical near miss reporting system. *British Medical Journal*, 320, 759–763.

Bogner, M. S. (Ed.). 1994. *Human Error in Medicine*. Hillsdale, NJ: Lawrence Erlbaum Associates, Inc.

Bogenstätter, Y., Tschan, F., Semmer, N. K., Spychiger, M., Breuer, M., and Marsch, S. 2009. How accurate is information transmitted to medical professionals joining medical emergency? A simulator study. *Human Factors*, 51, 115–125.

Brown, N. R. 1997. Context memory and the selection of frequency estimation strategies. *Journal of Experimental Psychology: Journal of Experimental Psychology: Learning, Memory, and Cognition*, 23, 898–914.

Cannon-Bowers, J. A., Salas, E., and Converse, S. A. 1993 Shared mental models in expert team decision making. In *Current Issues in Individual and Group Decision Making*, Ed. N. J. Jr. Castellan, pp. 221–246. Hillsdale, NJ: Erlbaum.

Cao, G. L. and Rogers, G. 2007. Robotics in health care: HF issues in surgery. In *Handbook of Human Factors and Ergonomics in Health Care and Patient Safety*, Ed. C. Carayon, pp. 411–421. Hillsdale, NJ: Lawrence Erlbaum.

Carayon, P. (Ed.). 2007. *Handbook of Human Factors and Ergonomics in Health Care and Patient Safety*. Hillsdale, NJ: Lawrence Erlbaum Associates, Inc.

Carayon, P. 2008. Nursing workload and patient safety—A human factors engineering perspective. In *Safety and Quality: An Evidence-Based Handbook for Nurses*, Ed. R. Hughes, pp. 1–14. Rockville, MD: Agency for Healthcare Research and Quality. Electronically available at http://www.ahrq.gov/qual/nurseshdbk

Carayon, P. and Gürses, A. P. 2005. A human factors engineering conceptual framework of nursing workload and patient safety in intensive care units. *Intensive and Critical Care Nursing*, 21, 284–301.

Carayon, P. and Gurses, A. P. 2008. Nursing workload and patient safety—A human factors engineering perspective. In *Patient Safety and Quality: An Evidence-Based Handbook for Nurses*, Ed. R. G. Hughes, pp. 203–213. Rockville, MD: Agency for Healthcare Research and Quality.

Carayon, P., Kosseff, A., Borgsdorf, A., and Jacobsen, K. 2007. Collaborative initiatives for patient safety. In *Handbook of Human Factors and Ergonomics in Healthcare and Patient Safety*, Ed. P. Carayon, pp. 147–158. Mahwah, NJ: Lawrence Erlbaum Associates.

Clarkson, W. J. 2007. Human factors engineering and the design of medical devices. In *Handbook of Human Factors and Ergonomics in Health Care and Patient Safety*, Ed. P. Carayon, pp. 367–382. Hillsdale, NJ: Lawrence Erlbaum Associates, Inc..

Cook, R., Render, M., and Woods, D. D. 2000. Gaps in the continuity of care and progress on patient safety. *British Medical Journal*, 320, 791–794.

Cook, R. I. and Woods, D. D. 1994. Operating at the sharp end: The complexity of human error. In *Human Error in Medicine*, Ed. M. S. Bogner, pp. 255–310. Hillsdale, NJ: Lawrence Erlbaum Associates, Inc.

Cook, R. I., Woods, D. D., and Miller, C. (Eds.). 1998. *A Tale of Two Stories: Contrasting Views on Patient Safety*. Chicago, IL: National Health Care Safety Council of the National Patient Safety Foundation at the AMA.

Cooke, N. J., Salas, E., Kiekel, P. A., and Bell, B. 2004. Advances in measuring team cognition. In *Team Cognition: Process and Performance at the Inter- and Intra-Individual Level*, Eds. E. Salas, S. M. Fiore, and J. A. Cannon-Bowers, pp. 83–106. Washington, DC: American Psychological Association.

Donchin, Y. 2010. The role of ergonomics in modern medicine. In *ESCIM Year Book*, Eds. H. Flatten and R. P. Moreno, pp. 251–257. Berlin, Germany: MWV Publication.

Donchin, Y. and Gopher, D. (2011). *Around the Patient Bed: human Factors and Safety in Health care*. Carta, Jerusalem, Israel.

Donchin, Y., Gopher, D., Olin, M., Badihi, Y., Beisky, M., Sprung, C. L. et al. 1995. A look into the nature and causes of human error in the intensive care unit. *Critical Care Medicine*, 23, 294–300.

Einav, Y., Gopher, D., Kara, I., Ben-Yosef, O., Lawn, M., Laufer, N. et al. 2010. Preoperative briefing in the operating room: Shared cognition, team work, and patient safety. *Chest*, 137, 443–449.

Frey, B., Buettiker, V., Hug, M. I., Katharina, W., Gessler, P., Ghelfi, D. et al. 2002. Does critical incident reporting contribute to medication error prevention? *European Journal of Pediatrics*, 161, 594–599.

Gawande, A. A., Zinner, M. J., Studdert, D. M., and Brennan, T. A. 2003. Analysis of errors reported by surgeons at three teaching hospitals. *Surgery*, 133, 614–621.

Gopher, D. 1994. Analysis and measurement of mental load. In *International Perspectives on Cognitive Psychology*, Vol. II, Eds. G. d'Yde Walle, P. Eelen, and P. Bertelson, Chap. 13, pp. 265–291. London, U.K.: Lawrence Erlbaum.

Gopher, D. 2004. Why it is not sufficient to study errors and incidents—Human factors and safety in medical systems. *Biomedical Instrumentation and Technology*, 45, 387–391.

Gopher, D. 2006. Control processes in the formation of task units. In *Psychological Science around the World, Volume 2, Social and Applied Issues*, Ed. Q. Jing, pp. 385–404. London, U.K.: Oxford Psychology Press. A chapter based on a keynote address given at the 28th International Congress of Psychology.

Gopher, D. 2007. Emphasis change as a training protocol for high demands tasks. In *Attention: From Theory to Practice*, Eds. A. Kramer, D. Wiegman, and A. Kirlik. Oxford: Psychology Press.

Gurses, A. P. and Carayon, P. 2007. Performance obstacles of intensive care nurses. *Nursing Research*, 56, 185–194.

Gurses, A. P., Carayon, P., and Wall, M. 2009. Impact of performance obstacles on intensive care nurses' workload, perceived quality and safety of care, and quality of working life. *Health Services Research*, 44, 422–443.

Gurses, A. P. 2009. Exploring performance obstacles of intensive care nurses. *Applied Ergonomics*, 40, 509–518.

Gurses, A. P. and Carayon, P. 2009. Exploring performance obstacles of intensive care nurses. *Applied Ergonomics*, 40, 509–518.

Helmreich, R. L. 2000. On error management. Lessons from aviation. *British Medical Journal*, 320, 781–785.

Henriksen, K. 2007. Human factors and patient safety: Continuing challenges. In *Handbook of Human Factors and Ergonomics in Health Care and Patient Safety*, Ed. P. Carayon, pp. 21–41. Mahwah, NJ: Lawrence Erlbaum Associates, Inc.

Joint Commission on Accreditation of Healthcare Organizations (JCAHO). 2007. *Universal Protocol for Preventing Wrong Site, Wrong Procedure, Wrong Person Surgery*. Oakbrook Terrace, IL: JCAHO. Available at: www.jointcommission.org/PatientSafety/UniversalProtocol

Karsh, B. T., Holden, R. J., Alper, S. J., and Or, C.K.L. 2006. Safety by design a human factors engineering paradigm for patient safety: Designing to support the performance of the healthcare professional. *Quality & Safety in Health Care*, 15, i59–i65.

Kohn, L. T., Corrigan, J. M., and Donaldson, J. C. (Eds.). 1999. *To Err is Human: Building a Safer Health System. Institute of Medicine*. Washington, DC: National Academy Press.

Koppel, R. K., Metlay, J. P., Cohen, A., Abaluck, B., Localio, A. R., Kimmel, S. E. et al. 2005. Role of computerized physician order entry systems in facilitating medication errors. *Journal of the American Medical Association*, 293, 1197–1203.

Leape, L. L. 1994. Error in medicine. *The Journal of the American Medical Association*, 272, 851–857.

Leape, L. L. 2002. Reporting of adverse events. *New England Journal of Medicine*, 347, 1633–1638.

Leape, L. L. 2004. Human factors meets health care: The ultimate challenge. *Ergonomics in Design*, 12, 6–13.

Leape, L. L., Brennan, T. A., Laird, N., Lawthers, A. G., Localio, A. R., Barnes, B. A., Hebert, L., Newhouse, J. P., Weiler, P. C., and Hiatt, H. 1991. The nature of adverse events in hospitalized patients, results of the Harvard Medical Practice Study II. *The New England Journal of Medicine*, 7, 324(6), 377–384.

Leape, L. L. and Berwick, D. M. 2005. Five years after to err is human: What have we learned? *Journal of the American Medical Association*, 293, 2384–2390.

Lingard, L., Espin, S., Rubin, B., Whyte, S., Colmenares, M., Baker, G. R. et al. 2005. Getting teams to talk: Development and pilot implementation of a checklist to promote safer operating room communication. *Quality and Safety in Health Care*, 14, 340–346.

Meyer, D. E. and Kieras, D. 1999. Précis to a practical unified theory of cognition and action. In *Attention and Performance XVII*, Eds. D. Gopher and A. Koriat, pp. 15–88. Cambridge, MA: MIT Press.

Mohammed, S. and Dumville, B. 2001. Team mental models in a team knowledge framework: Expanding theory and measurement across disciplinary boundaries. *Journal of Organizational Behavior*, 22, 89–106.

Morag, I. and Gopher, D. 2006. A reporting system of difficulties and hazards in hospital wards as a guide for improving human factors and safety. *Proceeding of the 50th Annual Meeting of the Human Factors and Ergonomic Society*, pp. 1014–1018. San Francisco, CA: HFES.

Moreno, R. P., Rhodes, A., Donchin, Y. 2009, Patient safety in intensive care medicine: The declaration of Vienna. *Journal of Intensive Care Medicine*, 35(10), 1667–1672.

Nemeth, C., Cook, R. I., Dierks, M., Donchin, Y., Patterson, Emily., Bitan, Y., Crowley, J., McNee, S., and Powell, T., (2006). Learning from investigation: Experience with understanding health care adverse events. In *Proceedings of the Human Factors and Ergonomics Society 50th Annual Meeting*, Santa Monica, CA, pp. 914–917.

Nemeth, C., Cook, R. I., Dierks, M., Donchin, Y., Patterson, E., Bitan, Y. et al. 2006. Learning from investigation: Experience with understanding health care adverse events. *Proceedings of the Human Factors and Ergonomics Society 50th Annual Meeting*, pp. 914–917. San Francisco, CA: HFES.

Organization for Economic Cooperation and Development (OECD).2010.*HealthDataReport*.http://www.oecd.org/document/30/0,3343,en_2649_34631_12968734_1_1_1_1,00.html

Reason, J. 2000. Human error: Models and management. *British Medical Journal*, 72, 768–770.

Reed, S. K. 1996. *Cognition: Theory and Application*. Pacific Grove, CA: Thomson Publishing Inc.

Salvendy, G. 2006. *Handbook of Human Factors and Ergonomics*. Hoboken, NJ: John Wiley.

Scanlon, M. C. and Karsh, B. T. 2010. Value of human factors to medication and patient safety in the intensive care unit. *Critical Care Medicine*, 38, 590–596.

Sela, Y., Auerbach, Y., Straucher, Z., Rogachov, M., Klimer, O., Gopher, D. et al. 2005. A cognitive user centered redesign of a radiotherapy chart. *Proceedings of the 49th Annual Meeting of the Human Factors and Ergonomics Society*, pp. 989–993. Orlando, FL: HFES.

Sellen, A. J. and Harper, R. E. R. 2003. *The Myth of the Paperless Office*. Cambridge, MA: MIT Press.

Shreve, J., Van Den Bos, J., Gray, T., Halford, M., Rustagi, K., and Ziemkiewicz, E. 2010. *The Economic Measurement of Medical Errors*. Schaumburg, IL: The Society of Actuaries, Milliman.

Sinreich, D., Gopher, D., Ben-Barak, S., Marmur, Y., and Menchel, R. 2005. Mental models as a practical tool in the engineer's toolbox. *International Journal of Production Research*, 43, 2977–2996.

Standards Subcommittee Intensive Care Society. 1983. *Standards for Intensive Care Units*. London, U.K.: Biomedica Ltd.

Thomas, E. J, Studdert, D. M, Newhouse, J. P, Zbar, B. I., Howard, K. M., Williams, E. J. et al. 1999. Costs of medical injuries in Utah and Colorado. *Inquiry*, 36, 255–264.

Thornburg, K., Geb, T., and Daraper, S. 2006. Categorizing adverse medical device and medication event frequency. *Proceedings of the Human Factors and Ergonomics Society 50th Annual Meeting*, pp. 1009–1013. San Francisco, CA: HFES.

Tsang, P. and Vidulich, M. 2003. *Principles and Practice of Aviation Psychology*. Mahwah, NJ: Lawrence Erlbaum Associate Publishers.

Vashdi, D. R., Bamberger, P. A., Erez, M., and Meilik, A. 2007. Briefing—Debriefing: Using a reflexive organizational learning model from the military to enhance the performance of surgical teams. *Human Resource Management*, 64, 115–142.

Weinger, M. B. and Englund, C. E. 1990. Ergonomics and human factors affecting anesthetic vigilance and monitoring performance in the operating room environment. *Anesthesiology*, 73, 995–1021.

Weinger, M. B., Slagle, J., Jain, S., and Ordonez, N. 2003 Retrospective data collection and analytical techniques for patient safety studies. *Journal of Biomedical Informatics*, 36, 106–119.

Wickens, D. C. and Hollands, J. G. 2000. *Engineering Psychology and Human Performance* 3rd edn. Upper Saddle River, NJ: Prentice-Hall Publishers.

Wiegmann, D. A. and Dunn, W. F. 2010. Changing culture: A new view of human error and patient safety. *Chest*, 137, 250–252.

Wiegmann, D. A. and Shappell, S. A. 2003. *Human Error Approach to Aviation Accident Analysis: The Human Factors Analysis and Classification System*. Burlington, VT: Ashgate Press.

Wiener, E. L. and Nagel, D. C. 1988. *Human Factors in Aviation*. New York: Elsevier.

World Health Organization's patient-safety checklist for surgery. 2008. *Lancet*. July 5; 372(9632): PubMed Index: 18603137.

Section II

Macroergonomics and Systems

4

Historical Perspective and Overview of Macroergonomics*

Hal W. Hendrick

CONTENTS

4.1 Introduction 46
4.2 Beginning: Select Committee on Human Factors Futures, 1980–2000 46
4.3 Integrating ODAM with Ergonomics 48
    4.3.1 Definition of Macroergonomics 49
    4.3.2 Purpose of Macroergonomics 49
4.4 Theoretical Basis of Macroergonomics 49
    4.4.1 Sociotechnical Systems Theory and the Tavistock Studies 49
    4.4.2 Tavistock Studies 49
    4.4.3 Joint Causation and Subsystem Optimization 50
    4.4.4 Joint Optimization vs. Human-Centered Design 50
    4.4.5 General Systems Theory 51
4.5 Implementing Macroergonomics 51
    4.5.1 Macroergonomics vs. Industrial and Organizational Psychology 51
4.6 Pitfalls of Traditional Approaches to Work System Design 51
    4.6.1 Technology-Centered Design 51
    4.6.2 "Left-Over" Approach to Function and Task Allocation 52
    4.6.3 Failure to Consider the System's Sociotechnical Characteristics 52
    4.6.4 Criteria for an Effective Work System Design Approach 53
4.7 Macroergonomic Analysis of Structure 53
    4.7.1 Complexity: Differentiation 53
    4.7.2 Complexity: Integration 54
    4.7.3 Formalization 54
    4.7.4 Centralization 55
4.8 Sociotechnical System Considerations in Work System Design 55
    4.8.1 Technology: Perrow's Knowledge-Based Model 55
    4.8.2 Personnel Subsystem 56
    4.8.3 Degree of Professionalism 56
    4.8.4 Psychosocial Characteristics 56
    4.8.5 Environment 57
    4.8.6 Integrating the Results of the Separate Assessments 58
4.9 Relation of Macro- to Microergonomic Design 58
4.10 Synergism and Work System Performance 59
    4.10.1 When Systems Have Incompatible Organizational Designs 59
    4.10.2 When Systems Have Effective Macroergonomic Designs 59
4.11 Macroergonomics Results 59
4.12 Macroergonomics Methodology 60
    4.12.1 Methods Developed for Macroergonomics Application 60
        4.12.1.1 Macroergonomic Analysis and Design 60
        4.12.1.2 HITOP 60
        4.12.1.3 Top MODELER 60

---

* Adapted, updated, and modified from Hendrick, H. W. and Kleiner, B. M. (2001). *Macroergonomics: An Introduction to Work System Design.* Santa Monica, CA: Human Factors and Ergonomics Society. Minor revisions were made by Pascale Carayon.

      4.12.1.4 CIMOP 60
      4.12.1.5 Systems Analysis Tool 60
      4.12.1.6 Anthropotechnology 60
    4.12.2 Methods Adapted for Macroergonomics Application 60
      4.12.2.1 Kansai Ergonomics 61
      4.12.2.2 Cognitive Walkthrough Method 61
      4.12.2.3 Participatory Ergonomics 61
4.13 Macroergonomics as a Key to Improving Health Care and Patient Safety 61
References 61

The subdiscipline of macroergonomics is traced from the inception of ergonomics in the 1940s to formal consideration of organizational design and management (ODAM) factors in work system design over the past three decades. The concept of macroergonomics, its theoretical basis, and its methodology are described. Results of two typical macroergonomic interventions are summarized. The author concludes by speculating on the potential of macroergonomics for improving health care and patient safety.

4.1 Introduction

Ergonomics, or human factors as the discipline also is known, is considered to have had its formal inception in the late 1940s. Ergonomics initially focused on the design of human–machine interfaces and physical environment factors as they affect human performance. This emphasis on optimizing the interfaces between individual operators and their immediate work environment, or what herein will be referred to as *microergonomics*, characterized the ergonomics/human factors discipline for the first three decades of its formal existence.

In North America, this initial microergonomic focus was on the human factors (e.g., perception, response time, transfer of training, etc.) associated with the design of such sociotechnical systems as aircraft cockpits, gun sites, and other military systems, and the field accordingly was labeled as *human factors*. In Europe, following the World War II, the primary focus was on rebuilding war-torn industries and the primary emphasis was on applying the science of work, or *ergonomics* (ergo = work; nomics = study of), to the design of factory workstations. While the primary emphasis was somewhat different, the objective of both human factors and ergonomics was to apply scientific knowledge about human capabilities, limitations, and other characteristics to the design of operator controls, displays, tools, seating, workspace arrangements, and physical environments to enhance health, safety, comfort, and productivity, and

to minimize design-induced human error (Chapanis, 1988; Hendrick, 2000). In more recent decades, not only has human–machine interface design been applied to a broader number of systems, but also to a progressively increasing number of consumer products to make them safer and more usable, including medical devices.

Beginning with the development of the silicon chip, the rapid development of computers and automation occurred, and along with this development came an entirely new set of human–system interface problems. As a result, a new subdiscipline of ergonomics emerged which centered on software design and became known as *cognitive ergonomics*. The emphasis of this new subdiscipline was on how humans think and process information and how to design software to have dialogue with humans in the same manner. As a direct result of the development of the cognitive ergonomics subdiscipline, the human factors/ergonomics discipline grew about 25% during the 1980s. However, the emphasis remained on individual human–system interfaces.

In the late 1970s, ergonomists began to realize that our traditional predominate focus on individual human–system interfaces, or *microergonomics*, was not achieving the kinds of improvements in overall total system functioning that both managers and ergonomists felt should be possible. Ergonomists began to realize that they could effectively design human–machine, human–environment, and human–software interfaces, and still have a poorly designed work *system*. Out of this realization developed the subdiscipline of *macroergonomics*.

4.2 Beginning: Select Committee on Human Factors Futures, 1980–2000

As with most new subdisciplines, one can identify a number of precursors within the larger discipline, or related disciplines, and this is the case with macroergonomics. However, the most direct link leading to the formal development of macroergonomics as a distinct subdiscipline can be traced back to the U.S. Human

Factors Society's Select Committee on Human Factors Futures, 1980–2000. In the late 1970s, many dramatic changes were occurring in all aspects of industrialized societies and their built environments. Arnold Small, Professor Emeritus of the University of Southern California (USC) and former President of The Human Factors Society, noted these changes and believed that traditional human factors/ergonomics would not be adequate to effectively respond to these trends. At Professor Small's urging, in 1978 the Human Factors Society (now, the Human Factors and Ergonomics Society) formed a "Select Committee on Human Factors Futures, 1980–2000" to study these trends and determine their implications for the human factors discipline. Arnold was appointed chair of this Select Committee. He, in turn, appointed me to that committee and specifically charged me to research trends related to the management and organization of work systems.

After several years of intensive study and analysis, the committee members reported on their findings in October 1980 at the HFS Annual Meeting in Los Angeles, California. Among other things, I noted the following six major trends as part of this report.

- *Technology.* Recent breakthroughs in the development of new materials, micro-miniaturization of components, and the rapid development of new technology in the computer and telecommunications industries would fundamentally alter the nature of work in offices and factories during the 1980–2000 time frame. In general, we were entering a true information age of automation that would profoundly affect work organization and related human–machine interfaces.

- *Demographic shifts.* The average age of the work populations in the industrialized countries of the world will increase by approximately 6 months for each passing year during the 1980s and most of the 1990s. Two major factors account for this "graying" of the workforce. First, the aging of the post-World War II "baby boom" demographic bulge that now has entered the workforce. The second factor was the lengthening of the average productive life span of workers because of better nutrition and health care. In short, during the next two decades, the workforce will become progressively more mature, experienced, and professionalized. As the organizational literature has shown (e.g., see Robbins, 1983), as the level of professionalism (i.e., education, training, and experience) increases, it becomes important for work systems to become less formalized (i.e., less controlled by standardized procedures, rules, and detailed job descriptions), tactical

decision-making to become decentralized (i.e., delegated to the lower-level supervisors and workers), and management systems to similarly accommodate. These requirements represent profound changes to traditional bureaucratic work systems and related human–system interfaces.

- *Value changes.* Beginning in the mid-1960s and progressing into the 1970s, a fundamental shift occurred in the value systems of workforces in the United States and Western Europe. These value system changes and their implications for work systems design were noted by a number of prominent organizational behavior researchers, and were summarized by Argyris (1971). In particular, Argyris noted that workers now both valued and expected to have greater control over the planning and pacing of their work, greater decision-making responsibility, and more broadly defined jobs that enable a greater sense of both responsibility and accomplishment. Argyris further noted that, to the extent organizations and work system designs do not accommodate these values, organizational efficiency and quality of performance will deteriorate. These value changes were further validated in the 1970s by Yankelovich (1979), based on extensive longitudinal studies of workforce attitudes and values in the United States. Yankelovich found these changes to be particularly dramatic and strong among those workers born after the World War II. Additionally, Yankelovich particularly noted that these baby boom workers expected to have meaningful tasks and the opportunity for quality social relationships on the job.

- *Ergonomics-based litigation.* In the United States, litigation based on the lack of ergonomic safety design of both consumer products and the workplace is increasing, and awards of juries often have been high. The message from this litigation is clear: managers *are* responsible for ensuring that adequate attention is given to the ergonomic design of both their products and their employees' work environments to ensure safety. [Note: I also could have included responsibility for customer health and safety while on the employer's premises.]

One impact of this message, as well as from the competition issue, noted earlier, is that ergonomists are likely to find themselves functioning as true management consultants. A related implication of equal importance is that ergonomics education programs will need to provide academic courses in organizational theory, behavior, and management to prepare their students for this consultant role.

- *World competition.* Progressively, U.S. industry is being forced to compete with high quality products from Europe and Japan; and other countries, such as Taiwan and Korea, soon will follow. Put simply, the post-World War II dominance by U.S. industry is gone. In light of this increasingly competitive world market, the future survival of most companies will depend on their efficiency of operation and production of state-of-the-art products of high quality. In the final analysis, the primary difference between successful and unsuccessful competitors will be the quality of the *ergonomic* design of their products and of their total work organization, and the two are likely to be interrelated.

- *Failure of traditional (micro) ergonomics.* Early attempts to incorporate ergonomics into the design of computer work stations and software have resulted in improvement, but have been disappointing in terms of (a) reducing the work system productivity costs of white collar jobs (b) improving intrinsic job satisfaction, and (c) reducing symptoms of high job stress. [Note: I also could have added reducing the level of work-related musculoskeletal disorders to a greater degree than that achieved through work station redesign alone.]

As I noted several years later, we had begun to realize that it was entirely possible to do an outstanding job of ergonomically designing a system's components, modules, and subsystems, yet fail to reach relevant systems effectiveness goals because of inattention to the *macro-ergonomic* design of the overall work *system* (Hendrick, 1984). Investigations by Meshkati (1986) and Meshkati and Robertson (1986) of failed technology transfer projects, and by Meshkati (1990) of major system disasters (e.g., Three Mile Island and Chernobyl nuclear power plants and the Bhopal chemical plant) have all led to similar conclusions.

I concluded in my 1980 report that, "for the human factors/ergonomics profession to truly be effective, and responsive to the foreseeable requirements of the next two decades and beyond, there is a strong need to integrate organizational design and management (ODAM) factors into our research and practice" (p. 5). It is interesting to note that all of these predictions from 1980 have come to pass, and are continuing. These needs, and the outstanding success of a number of macroergonomic interventions, would appear to account for the

rapid growth and development of macroergonomics that since has occurred.

4.3 Integrating ODAM with Ergonomics

As a direct response to my report, in 1984 an ODAM technical group was formed within the Human Factors Society. Similar groups were formed that year in both the Japan Ergonomics Research Society and the Hungarian society, and less formal interest groups were formed in other ergonomics societies internationally. In 1985, the International Ergonomics Association (IEA) formed a Science and Technology Committee comprising eight technical committees. The first eight committees formed were based on the input from the various Federated Societies as to what areas could most benefit from an international level technical committee. One of those first eight was an ODAM Technical Committee (TC). This TC consistently has been one of the IEA's most active. For example, the IEA ODAM TC has helped organize the highly successful biennial IEA International Symposia on Human Factors in ODAM, with the proceedings of each being commercially published by either North-Holland or the IEA Press.

In 1988, in recognition of its importance to the ergonomics discipline, ODAM was made one of the five major themes of the 10th IEA Triennial Congress in Sidney, Australia. In recognition of both its importance and rapid growth, it was 1 of 12 themes for the 11th Triennial Congress in Paris, France in 1990. At the 12th Triennial Congress in Toronto, Canada in 1994, and again at the 13th Congress in Tempare, Finland in 1997, a major multi-session symposium on Human Factors in ODAM was organized. For both of these Congresses, more papers were received on macroergonomics and ODAM than on any other topic. A similar symposium was held during the 14th Triennial Congress in the year 2000 where ODAM/macroergonomics was one of the three areas with the most presentations. A similar trend was evident at the 15th Triennial Congress in Seoul, Korea, and subsequent IEA triennial Congresses in The Netherlands and China.

By 1986, sufficient conceptualization of the ergonomics of work systems had been developed to identify it as a separate subdiscipline, which became formally identified as *macroergonomics* (Hendrick, 1986). In 1998, in response to the considerable methodology, research findings, and practice experience that had developed during the 1980s and 1990s, the Human Factors and Ergonomics Society's ODAM TG changed its name to the "Macroergonomics Technical Group" (METG).

4.3.1 Definition of Macroergonomics

Conceptually, macroergonomics can be defined as a top-down, sociotechnical systems approach to work system design, and the carry-through of that design to the design of jobs and related human–machine and human–software interfaces (Hendrick, 1997; Hendrick and Kleiner, 2001).

Although top-down conceptually, in practice, macroergonomics is top-down, middle-out, and bottom-up. *Top-down,* an overall work system structure may be prescribed to match an organization's sociotechnical characteristics. *Middle out,* an analysis of subsystems and work processes can be assessed both up and down the organizational hierarchy from intermediate levels, and changes made to ensure a harmonized work system design. *Bottom-up,* most often involves identification of work system problems by employees and lower-level supervisors that result from higher-level work system structural or work processes. Most often, a true macroergonomic intervention involves all three strategies. The macroergonomics design process tends to be nonlinear and iterative, and usually involves extensive employee participation at all organizational levels.

4.3.2 Purpose of Macroergonomics

The ultimate purpose of macroergonomics is to ensure that work systems are fully harmonized and compatible with their sociotechnical characteristics. In terms of systems theory, such a fully harmonized and compatible system can result in synergistic improvements in various organizational effectiveness criteria, including health, safety, comfort, and productivity (Zint, 2002).

4.4 Theoretical Basis of Macroergonomics

Macroergonomics is soundly grounded in both empirically developed and validated sociotechnical systems theory and general systems theory.

4.4.1 Sociotechnical Systems Theory and the Tavistock Studies

The sociotechnical model of work systems was empirically developed in the late 1940s and 1950s by Trist and Bamforth (1951) and their colleagues at the Tavistock Institute of Human Relations in the United Kingdom. Follow-on research was carried out by Katz and Kahn at the Survey Research Center of the University of Michigan and many others. This follow-on research served to further confirm and refine the sociotechnical systems model.

The sociotechnical systems model views organizations as transformation agencies, transforming inputs into outputs. Sociotechnical systems bring three major elements to bear on this transformation process: a *technological subsystem, personnel subsystem, and work system design* comprising the organization's structure and processes. These three elements interact with each other and a fourth element, the relevant aspects of the ex*ternal environment* upon which the work system is dependent for its survival and success. Insight into sociotechnical systems theory is provided by the classic studies of Welsh deep seam coal mining in the UK by the Tavistock Institute.

4.4.2 Tavistock Studies

The origin of sociotechnical systems theory, and the coining of the term, *sociotechnical systems,* can be traced to the studies of Trist and Bamforth relative to the effects of technological change in a deep seem Welsh coal mine (DeGreene, 1973). Prior to the introduction of new machine technology, the traditional system of mining coal in the Welsh mines was largely manual in nature. Small teams of miners who functioned relatively autonomously carried out the work. To a large extent, the group itself exercised control over work. Each miner could perform a variety of tasks, thus enabling most jobs to be interchangeable among the workers. The miners derived considerable satisfaction from being able to complete the entire "task." Further, through their close group interaction, workers could readily satisfy social needs on the job. As a result of these work system characteristics, the psychosocial and cultural characteristics of the workforce, the task requirements, and the work system's design were *congruent.*

The technological change consisted of replacing this more costly manual, or *shortwall* method of mining with mechanical coal cutters. Miners no longer were restricted to working a short face of coal. Instead, they now could extract coal from a long wall. Unfortunately, this new and more technologically efficient *longwall* system resulted in a work system design that was not congruent with the psychosocial and cultural characteristics of the workforce. Instead of being able to work in small, close-knit groups, shifts of 10–20 workers were required. The new jobs were designed to include only a narrowly defined set of tasks. The new work system structure severely limited opportunities for social interaction, and job rotation no longer was possible. The revised work system required a high degree of interdependence among the tasks of the three shifts. Thus, problems from one shift carried over to the next, thereby holding up labor stages in the extraction process. This complex and rigid work system was highly sensitive to both productivity and social disruptions. Instead of achieving the expected improved productivity, low production, absenteeism, and inter-group rivalry became common (DeGreene, 1973).

As part of follow-on studies of other coal mines by the Tavistock Institute (Trist et al., 1963), this conventional longwall method was compared with a new, *composite* longwall method. The composite work system design utilized a combination of the new technology and features of the old psychosocial work structure of the manual system. As compared with the conventional longwall system, the composite work system's design reduced the interdependence of the shifts, increased the variety of skills utilized by each worker, permitted self-selection by workers of their team members, and created opportunities for satisfying social needs on the job. As a result of this work system redesign, production became significantly higher than for either the conventional longwall or the old manual system; and absenteeism and other measures of poor morale and dissatisfaction dropped dramatically (DeGreene, 1973).

Prior to the Tavistock studies, there was a widely held belief in *technological determinism*. It incorporated the basic concept of Taylorism that there is one best way to organize work. It was a belief that the "one best way" for designing the work system is determined by the nature of the technology employed. Based on the Tavistock Institute studies, Emery and Trist (1960) concluded that *different organizational designs can utilize the same technology*. As demonstrated by the Tavistock studies, including the mining study, the key is to select a work system design that is compatible with the characteristics of (a) the people who will constitute the personnel portion of the system and (b) the relevant external environment, and then to (c) employ the available technology in a manner that achieves congruence.

4.4.3 Joint Causation and Subsystem Optimization

As noted earlier, sociotechnical systems theory views organizations as open systems engaged in transforming inputs into desired outcomes (DeGreene, 1973). *Open* means that work systems have permeable boundaries exposed to the environments in which they exist. These environments thus permeate the organization along with the inputs to be transformed. There are several primary ways in which environmental changes enter the organization: Through its marketing or sales function, through the people who work in it, and through its materials or other input functions (Davis, 1982).

As transformation agencies, organizations continually interact with their external environment. They receive inputs from their environment, transform these into desired outputs, and export these outputs back to their environment. In performing this transformation process, work systems bring two critical factors to bear

on the transformation process: Technology in the form of a *technological subsystem*, and people in the form of a *personnel subsystem*. The design of the technological subsystem tends to define the *tasks* to be performed, including necessary tools and equipment, whereas the design of the personnel subsystem primarily prescribes the *ways* in which tasks are performed. The two subsystems interact with each other at every human–machine and human–software interface. Thus, the technological and personnel subsystems are *mutually interdependent*. They both operate under *joint causation*, meaning that both subsystems are affected by causal events in the external environment.

Joint causation underlies the related sociotechnical system concept of *joint optimization*. The technological subsystem, once designed, is relatively stable and fixed. Accordingly, it falls to the personnel subsystem to adapt further to environmental change. Because the two subsystems respond jointly to causal events, optimizing one subsystem and fitting the second to it results in suboptimization of the joint work *system*. Consequently, maximizing overall work system effectiveness requires jointly optimizing *both* subsystems. Thus, in order to develop the best possible fit between the two, joint optimization requires the *joint design* of the technical and personnel subsystems, given the objectives and requirements of each, and of the overall work system (Davis, 1982). Inherent in this joint design is developing an optimal structure for the overall work system.

4.4.4 Joint Optimization vs. Human-Centered Design

At first glance, the concept of joint optimization may appear to be at odds with human-centered interface design. It might seem that human-centered design would lead to maximizing the personnel subsystem at the expense of the technological subsystem and, thus, to suboptimization of the work system. In fact, this has proven not to be the case. In human centered design, the goal is to make optimal use of humans by the *appropriate* employment of technology so as to optimize jointly the capabilities of each, and their interfaces. When a human-centered design approach is not taken, invariably the capabilities of the technological subsystem are maximized at the expense of the personnel subsystem. In recent decades, this often has been exemplified by automating and giving the human the leftover functions to perform, which suboptimizes the personnel subsystem and, thus, the total work system. To achieve the appropriate balance then, joint optimization is operationalized through (a) joint design, (b) a human-centered approach to function and task allocation and design, and (c) attending to the organization's sociotechnical characteristics.

4.4.5 General Systems Theory

General systems theory holds that all systems are more than the simple sum of their individual parts, that the parts are interrelated, and that changing one part will affect other parts. Further, when the individual parts are harmonized, the performance effects will be more than the simple sum of the parts would indicate. Thus, like biological and other complex systems, the whole of sociotechnical systems is more than the simple sum of its parts. When the parts are fully harmonized with each other and with the work system's sociotechnical characteristics, improvements in work system effectiveness can be much greater than the simple sum of the parts would indicate, such as will be illustrated in the case examples described later.

4.5 Implementing Macroergonomics

A true macroergonomic effort is most feasible when a major work system change already is to take place—for example, when changing over to a new technology, replacing equipment, or moving to a new facility. Another opportunity is when there is a major change in the goals, scope, or direction of the organization. A third situation in which management is likely to be receptive is when the organization has a costly chronic problem that has *not* proven correctable with a purely microergonomic effort, or via other intervention strategies. Recently, the desire to reduce lost time accidents and injuries, and related costs, has led senior managers in some organizations to support a true macroergonomics intervention. As will be illustrated with an actual case later, many of these efforts have achieved dramatic results.

Frequently, a true macroergonomic change to the work system is not possible initially. Rather, the ergonomist begins by making microergonomic improvements, which yield positive results within a relatively short period of time. This sometimes is referred to as "picking the low hanging fruit." When managers see these positive results, they become interested in, and willing to support further ergonomic interventions. In this process, if the ergonomist has established a good working rapport with the key decision-makers, the ergonomist serves to raise the consciousness level of the decision-makers about the full scope of ergonomics and its potential value to the organization. Thus, over time, senior management comes to support progressively larger ergonomics projects—ones that actually change the nature of the work system as a whole. Based on my personal experience and observations, this process typically takes about 2 years from the time one has established the necessary working rapport and gained the initial confidence of the key decision-maker(s).

4.5.1 Macroergonomics vs. Industrial and Organizational Psychology

Industrial and organizational (I/O) psychology can be viewed conceptually as the flip side of the coin from ergonomics. Whereas ergonomics focuses on designing work systems to fit people, I/O psychology primarily is concerned with selecting people to fit work systems. This is especially true of classical industrial psychology as compared with microergonomics.

Organizational psychology can be viewed as the opposite side of the coin from macroergonomics, although there is greater overlap than in the case of industrial psychology and microergonomics. Both organizational psychology and macroergonomics are concerned with the design of organizational structures and processes, but the focus is somewhat different. In the case of organizational psychology, improving employee motivation and job satisfaction, developing effective incentive systems, enhancing leadership and organizational climate, and fostering teamwork are common objectives. While these objectives are important and are considered in macroergonomics interventions, the primary focus of macroergonomics is to design work systems that are compatible with the organization's sociotechnical characteristics, and then ensure that the microergonomic elements are designed to harmonize with the overall work system structure and processes.

4.6 Pitfalls of Traditional Approaches to Work System Design

Over a 20 year period, I was involved either by myself or with my graduate students in assessing more than 200 organizational units. Based on these assessments, I identified three highly interrelated work system design practices that frequently underlie dysfunctional work system development and modification efforts. These are *technology-centered design, a "left-over" approach to function and task allocation,* and *a failure to consider an organization's sociotechnical characteristics* and integrate them into its work system design (Hendrick, 1995).

4.6.1 Technology-Centered Design

In exploiting technology, designers incorporate it into some form of hardware or software to achieve some desired purpose. To the extent that those who must operate or maintain the hardware or software are considered, it usually is in terms of what skills, knowledge, and training will be required. Often, even this kind of consideration is not thought through ergonomically. As a result, the intrinsic motivational aspects of jobs, psychosocial characteristics of the workforce, and other related work system design factors rarely are considered. Yet, these are the very factors that can significantly improve work system effectiveness.

Typically, if ergonomic aspects of design are considered, it usually is after the equipment or software *already is designed*. Then, the ergonomist may be called in to modify some of the human–system interfaces to reduce the likelihood of human error, eliminate awkward postures, or improve comfort. Often, even this level of involvement does not occur until testing of the newly designed system reveals serious interface design problems. At this point in the design process, because of cost and schedule considerations, the ergonomist is severely limited in terms of making fundamental changes to improve the work system. Instead, the ergonomist is restricted to making a few "band aid" fixes of specific human–machine, human–environment, or human–software interfaces. Thus, the ultimate outcome is a suboptimal work system.

There is a well-known relationship between when professional ergonomics input occurs in the design process and the value of that input in terms of system performance: the earlier the input occurs in the design process, the greater and more economical is the impact on system effectiveness.

A related impact of a technology-centered approach to redesigning existing work systems is that employees are not actively involved throughout the planning and implementation process. As the organizational change literature frequently has shown, the result often is not only a poorly designed work system, but also a lack of commitment and, not infrequently, either overt or passive-aggressive resistance to the changes. From the author's personal observations, when a technology centered approach is taken, if employees are brought into the process at all, it is only after the work system changes have been designed—the employee's role being to do informal usability testing of the system. As a result, when employees find serious problems with the changes (as often happens), cost and schedule considerations prevent any major redesign to eliminate or minimize the identified deficiencies.

Given that most of the so-called *reengineering* efforts of the early 1990s used a technology-centered approach, it is not surprising that a large majority of them have been unsuccessful (Keidel, 1994). As Keidel has noted, these efforts failed to address the "soft" (i.e., human) side of engineering and often ignored organizational effects.

4.6.2 "Left-Over" Approach to Function and Task Allocation

A technology-centered approach often leads to treating the persons who will operate and maintain the system as impersonal components. The focus is on assigning to the "machine" any functions or tasks which its technology enables it to perform. Then, what is left over is assigned to the persons who must operate or maintain it, or to be serviced by it. As a result, the function and task allocation process fails to consider the characteristics of the workforce and related external environmental factors. Most often, the consequence is a poorly designed work system that fails to make effective use of its human resources.

As noted earlier from the sociotechnical systems literature, effective work system design requires *joint design* of the technical and personnel subsystems (DeGreene, 1973). Put in ergonomic terms, joint optimization requires a *human-centered* approach. In terms of function and task allocation, Bailey (1989) refers to it as a *humanized task* approach. He notes, "This concept essentially means that the ultimate concern is to design a job that *justifies* using a person, rather than a job that merely can be done by a human. With this approach, functions are allocated and the resulting tasks are designed to make full use of human skills and to compensate for human limitations. The nature of the work itself should lend itself to internal motivational influences. The left over functions are allocated to computers" (p. 190).

4.6.3 Failure to Consider the System's Sociotechnical Characteristics

As previously noted, from the sociotechnical systems literature, four major characteristics or elements of sociotechnical systems can be identified. These are the (1) *technological subsystem*, (2) *personnel subsystem*, (3) *external environment*, and (4) *organizational design* of the work system. These four elements interact with one another, so a change in any one affects the other three (and, if not planned for, often in dysfunctional or unanticipated ways). Because of these interrelationships, characteristics of each of the first three elements affect the fourth: the organizational design of the work system. Empirical models have been developed of these relationships, which can be used to determine the optimal work system structure (i.e., the optimal degrees of vertical and horizontal differentiation, integration, formalization, and centralization to design into the work system) and have been integrated into the Macroergonomic Analysis of Structure (MAS) method.

Unfortunately, as first was documented by the Tavistock studies of coal mining more than four decades ago a technology centered approach to the organizational design of work systems does *not* adequately consider the key characteristics of the other three sociotechnical system elements. Consequently, the resulting work system design most often is *suboptimal*. (Trist and Bamforth, 1951; Emery and Trist, 1960).

4.6.4 Criteria for an Effective Work System Design Approach

Based on the pitfalls cited previously, several criteria can be gleaned for selecting an effective work system design approach.

- *Joint design*. The approach should be *human-centered*. Rather than designing the technological subsystem and requiring the personnel subsystem to conform to it, the approach should require design of the personnel subsystem jointly with the technological subsystem. Further, it should allow for extensive employee participation throughout the design process.

- *Humanized task approach*. The function and task allocation process should first consider whether there is a *need* for a human to perform a given function or task before making the allocation to humans or machines. Implicit in this criterion is a systematic consideration of the professionalism requirements (i.e., education and training), and of the cultural and psychosocial characteristics of the personnel subsystem.

- *Consider the organization's sociotechnical characteristics*. The approach should systematically evaluate the organization's sociotechnical system characteristics, and then integrate them into the work system's design.

One approach that meets all three of these criteria is macroergonomics. As was noted earlier, conceptually, macroergonomics is a top-down sociotechnical systems approach to work system design, and the carry-through of the overall work system design to the design of human–job, human–machine, human–environment, and human–software interfaces. It is a top-down sociotechnical systems approach in that it begins with an analysis of the relevant sociotechnical system variables, and then systematically utilizes these data in designing the work system's structure and related processes. Macroergonomics is human-centered in that it systematically considers the worker's professional and psychosocial characteristics in designing the work system; and then carries the work system design through to the ergonomic design of specific jobs and related hardware and software interfaces. Integral to this human-centered design process is the joint design of the technical and personnel subsystems, using a humanized task approach in allocating functions and tasks. A primary methodology of macroergonomics is *participatory ergonomics* (Noro and Imada, 1991).

4.7 Macroergonomic Analysis of Structure

Macroergonomics involves the development and application of human–organization interface technology; and this technology is concerned with optimizing the organizational structure and related processes of work systems. Accordingly, an understanding of macroergonomics requires an understanding of the key dimensions of organizational structure.

The organizational structure of a work system can be conceptualized as having three core dimensions. These are *complexity, formalization,* and *centralization* (Robbins, 1983; Bedeian and Zammuto, 1991; Stevenson, 1993).

Complexity refers to the degree of *differentiation* and *integration* that exist within a work system. Differentiation refers to the extent to which the work system is segmented into parts; integration refers to the number of mechanisms that exist to integrate the segmented parts for the purposes of communication, coordination, and control.

4.7.1 Complexity: Differentiation

Work system structures employ three common types of differentiation: *Vertical, horizontal,* and *spatial*. Increasing any one of these three increases a work system's complexity.

Vertical differentiation. Vertical differentiation is measured in terms of the number of hierarchical levels separating the chief executive position from the jobs directly involved with the system's output. In general, as the size of an organization increases, the need for greater vertical differentiation also increases (Mileti et al., 1977). For example, in one study, size alone was found to account for 50%–59% of the variance (Montanari, 1975). A major reason for this strong relationship is the practical limitations of *span of control*. Any one manager is limited in the number of subordinates that he or she can direct effectively (Robbins, 1983). Thus, as the number of first-level employees increases, the number of first-line supervisors also must increase. This, in turn, requires more supervisors at each successively higher level, and ultimately results in the creation of more hierarchical levels in the work system's structure.

Although span of control limitations underlie the size-vertical differentiation relationship, it is important to note that these limitations can vary considerably, depending on a number of factors. Thus, for an organization of a given size, if large spans of control are appropriate, the number of hierarchical levels will be fewer than if small spans of control are required. A major factor affecting span of control is the *degree of professionalism* (education and skill requirements) designed into employee jobs. Generally, as the level of professionalism increases, employees are able to function more autonomously, and thus need less supervision. Consequently, the manager can supervise effectively a larger number of employees. Other factors which affect span of control are the degree of formalization, type of technology, psychosocial variables, and environmental characteristics. (See Hendrick, 1997 or Hendrick and Kleiner, 2001, 2002 for a more detailed discussion.)

Horizontal differentiation. Horizontal differentiation refers to the degree of departmentalization and specialization within a work system. Horizontal differentiation increases complexity because it requires more sophisticated and expensive methods of control. In spite of this drawback, specialization is common to most work systems because of the inherent efficiencies in the division of labor. Adam Smith (1970/1876) demonstrated this point over 200 years ago. Smith noted that 10 workers, each doing particular tasks (job specialization), could produce about 48,000 pins per day. On the other hand, if the 10 workers each worked separately and independently, performing all of the production tasks, they would be lucky to make 200.

Division of labor creates groups of specialists, or *departmentalization*. The most common ways of designing departments into work systems are on the basis of (1) function, (2) simple numbers, (3) product or services, (4) client or client class served, (5) geography, and (6) process. Most large corporations will use all six (Robbins, 1983).

Two of the most common ways to determine whether or not a work group should be divided into one or more departments are the degree of commonality of *goals*, and of *time orientation*. To the extent that subgroups either differ in goals, or have differing time orientations, they should be structured as separate departments. For example, sales persons differ from R&D employees on both of these dimensions. Not only do they have very different goals, but also the time orientation of sales personnel usually is *short* (1 year or less), whereas it usually is *long* (3 or more years) for R&D personnel. Thus, they clearly should be departmentalized separately, and usually are (Robbins, 1983).

Spatial dispersion. Spatial dispersion refers to the degree an organization's activities are performed in multiple locations. There are three common measures of spatial dispersion. These are (1) the number of geographic locations comprising the total work system, (2) the average distance of the separated locations from the organization's headquarters, and (3) the proportion of employees

in these separated units in relation to the number in the headquarters (Hall et al., 1967). In general, complexity increases as any of these three measures increases.

4.7.2 Complexity: Integration

As noted earlier, *integration* refers to the number of mechanisms designed into a work system for ensuring communication, coordination, and control among the differentiated elements. As the differentiation of a work system increases, the need for integrating mechanisms also increases. This occurs because greater differentiation increases the number of units, levels, etc. that must communicate with one another, coordinate their separate activities, and be controlled for efficient operation. Some of the more common integrating mechanisms that can be designed into a work system are formal rules and procedures, committees, task teams, liaison positions, and system integration offices. Computerized information and decision support systems also can be designed to serve as integrating mechanisms. Vertical differentiation, in itself, is a primary form of integrating mechanism (i.e., a manager at one level serves to coordinate and control the activities of several lower-level groups).

Once the differentiation aspects of a work system's structure have been determined, a major task for the macroergonomics professional is to then determine the kinds and number of integrating mechanisms to design into the work system. Too few integrating mechanisms will result in inadequate coordination and control among the differentiated elements; too many integrating mechanisms stifle efficient and effective work system functioning, and usually increase costs. A systematic analysis of the type of technology, personnel subsystem factors, and characteristics of the external environment can all be used to help determine the optimal number and types of integrating mechanism (see the MAS method, described later).

4.7.3 Formalization

From an ergonomics design perspective, formalization can be defined as the degree to which jobs within the work system are standardized. Highly formalized designs allow for little employee discretion over what is to be done, when it is to be done, or how it is to be accomplished (Robbins, 1983). In highly formalized designs, there are explicit job descriptions, extensive rules, and clearly defined procedures covering work processes. Ergonomists can contribute to formalization by designing jobs, machines, and software so as to standardize procedures and allow little opportunity for operator decision discretion. By the same token, human–job, human–machine, and human–software interfaces can be ergonomically designed to permit greater flexibility and scope to employee decision-making (i.e., low formalization). When there is low formalization, employee behavior is relatively unprogrammed and the work system allows for considerably greater use of one's mental abilities. Thus, greater reliance is placed on the employee's professionalism and jobs tend to be more intrinsically motivating.

In general, the simpler and more repetitive the jobs to be designed into the work system, the higher should be the level of formalization. However, caution must be taken not to make the work system so highly formalized that jobs lack any intrinsic motivation, fail to effectively utilize employee skills, or degrade human dignity. Invariably, good macroergonomic design can avoid this extreme. The more nonroutine or unpredictable the work tasks and related decision-making, the less amenable is the work system to high formalization. Accordingly, reliance has to be placed on designing into jobs a relatively high level of professionalism.

4.7.4 Centralization

Centralization refers to the degree to which formal decision-making is concentrated in a relatively few individuals, group, or level, usually high in the organization. When the work system structure is highly centralized, lower-level supervisors and employees have only minimal input into the decisions affecting their jobs (Robbins, 1983). In highly decentralized work systems, decisions are delegated downward to the lowest level having the necessary expertise.

It is important to note that work systems carry out two basic forms of decision-making, *strategic* and *tactical*, and that the degree of centralization often is quite different for each. *Tactical* decision-making has to do with the day-to-day operation of the organization's business; *strategic* decision-making concerns the long range planning for the organization. Under conditions of low formalization and high professionalism, tactical decision-making may be highly decentralized, whereas strategic decision-making may remain highly centralized. Under these conditions, it also is important to note that the information required for strategic decision-making often is controlled and filtered by middle management or even lower level personnel. Thus, to the extent that these persons can reduce, summarize, selectively omit, or embellish the information that gets fed to top management, the less is the actual degree of centralization of strategic decision-making.

In general, centralization is desirable (1) when a comprehensive perspective is needed, (2) when it provides significant economies, (3) for financial, legal, or other decisions which clearly can be done more efficiently when centralized, (4) when operating in a highly stable and predictable external environment, and (5) where the

decisions have little effect on employees' jobs or are of little employee interest. Decentralized decision-making is desirable (1) when an organization needs to respond rapidly to changing or unpredictable conditions at the point where change is occurring, (2) when "grass roots" input to decisions is desirable, (3) to provide employees with greater intrinsic motivation, job satisfaction, and sense of self worth, (4) when it can reduce stress and related health problems by giving employees greater control over their work, (5) to more fully utilize the mental capabilities and job knowledge of employees, (6) to gain greater employee commitment to, and support for, decisions by involving them in the process, (7) when it can avoid overtaxing a given manager's capacity for human information processing and decision-making, and (8) to provide greater training opportunity for lower-level managers.

4.8 Sociotechnical System Considerations in Work System Design

As noted in our coverage of sociotechnical systems theory, the design of a work system's structure (which includes how it is to be managed) involves consideration of the key elements of three major sociotechnical system components: (a) the technological subsystem, (b) the personnel subsystem, and (c) the relevant external environment. Each of these three major sociotechnical system components has been studied in relation to its effect on the fourth component, organizational structure, and empirical models have emerged that can be used to optimize a work system's organizational design. The models of each of these components that I have found most useful have been integrated into a macroergonomics method that I have labeled Macroergonomics Analysis of Structure (MAS).

4.8.1 Technology: Perrow's Knowledge-Based Model

Perhaps the most thoroughly validated and generalizable model of the technology-organization design relationship is that of Perrow (1967) which utilizes a *knowledge-based* definition of technology. In his classification scheme, Perrow begins by defining technology by the action a person performs upon an object in order to change that object. Perrow notes that this action always requires some form of technological knowledge. Accordingly, technology can be categorized by the required knowledge base. Using this approach, he has identified two underlying dimensions of knowledge-based technology. The first of these is *task variability* or the number of exceptions encountered in one's work.

TABLE 4.1

Perrow's Knowledge-Based Technology Classes

		Task Variability	
		Routine with Few Exceptions	High Variety with Many Exceptions
Problem analyzability	Well defined and analyzable	Routine	Engineering
	Ill defined and unanalyzable	Craft	Nonroutine

For a given technology, these can range from routine tasks with few exceptions to highly variable tasks with many exceptions.

The second dimension has to do with the type of search procedures one has available for responding to task exceptions, or *task analyzability*. For a given technology, the search procedures can range from tasks being well defined and solvable by using logical and analytical reasoning to being ill-defined with no readily available formal search procedures for dealing with task exceptions. In this latter case, problem-solving must rely on experience, judgment, and intuition. The combination of these two dimensions, when dichotomized, yields a 2 × 2 matrix as shown in Table 4.1. Each of the four cells represents a different knowledge-based technology.

1. Routine technologies have few exceptions and well-defined problems. Mass production units most frequently fall into this category. Routine technologies are best accomplished through standardized procedures, and are associated with high formalization and centralization.

2. Nonroutine technologies have many exceptions and difficult to analyze problems. Aerospace operations often fall into this category. Most critical to these technologies is flexibility. They thus lend themselves to decentralization and low formalization.

3. Engineering technologies have many exceptions, but they can be handled using well-defined rational-logical processes. They therefore lend themselves to centralization, but require the flexibility that is achievable through low formalization.

4. Craft technologies typically involve relatively routine tasks, but problems rely heavily on experience, judgment, and intuition for decision. Problem-solving thus needs to be done by those with the particular expertise. Consequently, decentralization and low formalization are required for effective functioning.

4.8.2 Personnel Subsystem

At least two major aspects of the personnel subsystem are important to organizational design. These are the degree of professionalism and the psychosocial characteristics of the workforce.

4.8.3 Degree of Professionalism

Degree of professionalism refers to the education and training requirements of a given job and, presumably, possessed by the incumbent. Robbins (1983) notes that formalization can take place either on the job or off. When done on the job, formalization is *external* to the employee; rules, procedures, and the human–machine and human–software interfaces are designed to limit employee discretion. Accordingly, this tends to characterize unskilled and semiskilled positions. When done off the job, it is done through the professionalization of the employee. Professionalism creates formalization that is *internal* to the worker through a socialization process that is an integral part of formal professional education and training. Thus, values, norms, and expected behavior patterns are learned *before* the employee enters the organization.

From a macroergonomics design perspective, there is a trade-off between formalizing the organizational structure and professionalizing the jobs and related human–machine and human–software interfaces. As positions in the organization are designed to require persons with considerable education and training, they also should be designed to allow for considerable employee discretion. If low education and training requirements characterize the design of the positions, then the work system should be more highly formalized.

4.8.4 Psychosocial Characteristics

In addition to careful consideration of cultural differences (e.g., norms, values, mores, role, and expectations), I have found the most useful integrating model of psychosocial influences on organizational design to be that of *cognitive complexity*. Harvey, Hunt, and Schroder (1961) have identified the higher-order structural personality dimension of concreteness-abstractness of thinking, or cognitive complexity, as underlying different conceptual systems for perceiving reality. We all start out in life relatively concrete in our conceptual functioning. As we gain experience we become more abstract or complex in our conceptualizing, and this changes our perceptions and interpretations of our world. In general, the degree to which a given culture or subculture (a) provides through education, communications, and transportation systems an opportunity for *exposure* to diversity, and (b) encourages through its child-rearing and educational practices and *active* exposure to this diversity (i.e., an active openness to learning from exposure to new experiences), the more cognitively complex the persons of that particular group will become. An active exposure to diversity increases the number of conceptual categories that one develops for storing experiential information, and number of "shades of gray" or partitions within conceptual categories. In short, one develops greater *differentiation* in one's conceptualizing. With an active exposure to diversity one also develops new rules and combinations of rules for *integrating* conceptual data and deriving more insightful conceptions of complex problems and solutions. Note from our earlier review, that these same two dimensions of "differentiation" and "integration" also characterize *organizational* complexity. Relatively concrete adult functioning consistently has been found to be characterized by a relatively high need for structure and order and for stability and consistency in one's environment, closedness of beliefs, absolutism, paternalism, and ethnocentrism. Concrete functioning persons tend to see their views, values, norms, and institutional structures as relatively unambiguous, static, and unchanging. In contrast, cognitively complex persons tend to have a relatively low need for structure and order or stability and consistency, and are open in their beliefs, relativistic in their thinking, and have a high capacity for empathy. They tend to be more people-oriented, flexible, and less authoritarian than their more concrete colleagues, and to have a dynamic conception of their world: they *expect* their views, values, norms, and institutional structures to change (Harvey et al., 1961; Harvey, 1963).

In light of the preceding text, it is not surprising that I have found evidence to suggest that relatively concrete managers and workers function best under moderately high centralization, vertical differentiation, and formalization. In contrast, cognitively complex workgroups and managers seem to function best under relatively low centralization, vertical differentiation, and formalization.

Although only weakly related to general intelligence, cognitive complexity is related to education. Thus, within a given culture, educational level can sometimes serve as a relative estimate of cognitive complexity.

Of particular importance is the fact that since the post-World War II "baby boomers" have entered the workforce in industrially developed societies, the general complexity level of educated or highly trained employees has become moderately *high*. This can be traced to the greater exposure to diversity, and less authoritarian and absolutist child rearing practices these adults experienced while growing up, as compared with those who experienced childhood prior to World War II (Harvey et al., 1961; Harvey, 1963). As a result, successful firms are likely to be those having work system designs that respond to the guidelines given herein for more

cognitively complex workforces—particularly if their workforces are highly professionalized.

4.8.5 Environment

Critical to the success, and indeed, the very survival of an organization, is its ability to adapt to its external environment. In open systems terms, organizations require monitoring and feedback mechanisms to follow and sense changes in their specific task environment, and the capacity to make responsive adjustments. "Specific task environment" refers to that part of the organization's external environment that is made up of the firm's critical constituencies (i.e., those that can positively or negatively influence the organization's effectiveness). Negandhi (1977), based on field studies of 92 industrial organizations in five different countries, has identified five external environments that significantly impact on organizational functioning. These are *socioeconomic* including the nature of competition and the availability of raw materials; *educational* including both the availability of educational programs and facilities and the aspirations of workers; *political* including governmental attitudes toward business, labor, and control over prices; *legal*; and *cultural* including the social class or caste system, values, and attitudes.

Of particular importance to us is the fact that specific task environments vary along two dimensions that strongly influence the effectiveness of a work system's design: These are the degrees of *environmental change* and *complexity*. The degree of change refers to the extent to which a specific task environment is dynamic or remains stable over time; the degree of complexity refers to whether the number of relevant specific task environments is few or many in number. As illustrated in Table 4.2, these two environmental dimensions, in combination, determine the *environmental uncertainty* of an organization. Of all the sociotechnical system factors that impact the effectiveness of a work system's design, environmental uncertainty repeatedly has been shown to be the most important (Burns and Stalker, 1961; Emery and Trist, 1965; Lawrence and Lorsch, 1969; Duncan, 1972; Negandhi, 1977). With a high degree of uncertainty, a premium is placed on an organization's ability to be flexible and rapidly responsive to change. Thus, the greater the environmental uncertainty, the

TABLE 4.2

Environmental Uncertainty of Organizations

		Degree of Change	
		Stable	**Dynamic**
Degree of complexity	Simple	Low uncertainty	Mod. high uncert.
	Complex	Mod. low uncert.	High uncertainty

more important it is for the work system's structure to have relatively low vertical differentiation, decentralized tactical decision-making, low formalization, and a high level of professionalism among its work groups. By contrast, highly certain environments are ideal for high vertical differentiation, formalization, and centralized decision-making, such as found in classical bureaucratic structures. Of particular note is the fact that, today, most high technology corporations are operating in highly dynamic and complex environments. From my observations, although many of these corporations have increased the level of professionalism of their employees, they have not yet fully adapted their work system's design to their environments. I believe this is an issue to be thoroughly assessed in health systems.

4.8.6 Integrating the Results of the Separate Assessments

The separate analyses of the key characteristics of a given organization's technological subsystem, personnel subsystem, and specific task environment each should have provided guidance about the structural design for the work system. Frequently, these results will show a natural convergence. At times, however, the outcome of the analysis of one sociotechnical system element may conflict with the outcomes of the other two. When this occurs, the macroergonomics specialist is faced with the issue of how to reconcile the differences. Based both on the suggestions from the literature and my personal experience in evaluating over 200 organizational units, the outcomes from the analyses can be integrated by weighting them approximately as follows: If the technological subsystem analysis is assigned a weight of "1," give the personnel subsystem analysis a weight of "2," and the specific task environment analysis a weight of "3." For example, let's assume that the technological subsystem falls into Perrow's "routine" category, the personnel subsystem's jobs call for a high level of professionalism and external environment has moderately high complexity. Weighting these three as suggested earlier would indicate that the work system should have moderate formalization and centralization. Accordingly, the results would indicate that most jobs should be redesigned to require a somewhat lower level of professionalism, and attendant hardware and software interfaces should be designed/redesigned to be compatible.

It is important to note that the specific functional units of an organization may differ in the characteristics of their technology, personnel, and specific task environments—particularly within larger organizations. Therefore, the separate functional units may, themselves, need to be analyzed as though they were separate organizations, and the resultant work systems designed accordingly.

4.9 Relation of Macro- to Microergonomic Design

Through a macroergonomic approach to determining the optimal design of a work system's structure and related processes, many of the characteristics of the jobs to be designed into the system, and of the related human–machine and human–software interfaces, have already been prescribed. Some examples are as follows (Hendrick, 1991).

1. Horizontal differentiation decisions prescribe how narrowly or broadly jobs must be designed and, often, how they should be departmentalized.

2. Decisions concerning the level of formalization and centralization will dictate (1) the degree of routinization and employee discretion to be ergonomically designed into the jobs and attendant human–machine and human–software interfaces, (2) the level of professionalism to be designed into each job, and (3) many of the design requirements for the information, communications, and decision support systems, including what kinds of information are required by whom, and networking requirements.

3. Vertical differentiation decisions, coupled with those concerning horizontal differentiation, spatial dispersion, centralization, and formalization will prescribe many of the design characteristics of the managerial positions, including span of control, decision authority and nature of decisions to be made, information and decision support requirements, and qualitative and quantitative educational and experience requirements.

In summary, effective macroergonomic design drives much of the microergonomic design of the system, and thus insures *optimal ergonomic compatibility* of the components with the work system's overall structure. In sociotechnical system terms, this approach enables joint optimization of the technical and personnel subsystems from top to bottom throughout the organization. The result of this is greater assurance of optimal *system* functioning and effectiveness, including productivity, safety and health, comfort, intrinsic employee motivation, and quality of work life.

4.10 Synergism and Work System Performance

From the preceding text, it should be apparent that macroergonomics has the potential to improve the ergonomic design of organizations by ensuring that their respective work system's designs harmonize with the organizations' critical sociotechnical characteristics. Equally important, macroergonomics offers the means to ensure that the design of the entire work system, down to each individual job and workstation, harmonizes with the design of the overall work system. As we noted earlier, a widely accepted view among general system theorists and researchers is that organizations are synergistic—that the whole is more than the simple sum of its parts. Because of this synergism, it is my experience that the following tend to occur in our complex organizations.

4.10.1 When Systems Have Incompatible Organizational Designs

When a work system's structures and related processes are grossly incompatible with their sociotechnical system characteristics, and/or jobs and human–system interfaces are incompatible with the work system's structure, the whole is *less than* the sum of its parts. Under these conditions, we can expect the following to be poor: (a) productivity, especially *quality* of production, (b) lost time accidents and injuries, and (c) motivation and related aspects of quality of work life (e.g., stress), and for human error incidents to be relatively high. Needless to say, these are literally life or death concerns for health systems.

4.10.2 When Systems Have Effective Macroergonomic Designs

When a work system has been effectively designed from a macroergonomics perspective, and that effort is carried through to the microergonomic design of jobs and human–machine, human–environment, and human–software interfaces, then production, safety and health, and quality of work life will be much *greater* than the simple sum of the parts would indicate.

4.11 Macroergonomics Results

A major reason for the rapid spread and development of macroergonomics has been the dramatic results that have been achieved when it was applied to improving an organization's work system. Based

TABLE 4.3

Results of a Macroergonomic Intervention in a Petroleum Company

	Reduction after	
	2 Years (%)	9 Years (%)
Motor vehicle accidents	51	63
Industrial accidents	54	70
Lost workdays	94	97
Off the job injuries	84	69

$60,000 savings in petroleum delivery costs per year.

TABLE 4.4

Designing a New University College

Results (as Compared with Old USC Academic Unit)
27% savings in operating expenses
23% reduction in campus staffing requirements
20% reduction in off campus study center administrative time (30 centers)
67% reduction in average processing time for student registrations, grades, etc., from off campus locations

on both sociotechnical and general systems theory, at ODAM-III in 1990, I predicted that 50%–90% or greater improvements would be typical of macroergonomics interventions having strong management and labor support in comparison to the 10%–20% gains expected from microergonomics. These results subsequently have been achieved around the world in a number of interventions. A good example is the work of my former USC colleague, Andy Imada. Working with a large petroleum distribution company in the southwestern United States, Andy achieved the following results (Table 4.3).

Imada is convinced that taking a macroergonomic approach and using participatory ergonomics at all levels of the organization enabled not only the initial improvements, but the integration of safety and ergonomics into the organizational culture, which accounts for the sustained results over time. (See Imada, 2002, for a full description of this intervention.)

In a macroergonomics project, which involved designing a new university college at the University of Denver, my colleagues and I achieved the results shown in Table 4.4, as compared with our prior similar organization at the University of Southern California. These results are particularly impressive in that the USC academic unit already was an efficient, well functioning organization, refined by over 20 years of experience. Of particular note, a true humanized task approach was used in which we first determined what we needed humans to do, designed the jobs accordingly, and only then selected the equipment and designed or purchased the software to support the humans. (See Hendrick and Kleiner, 2002, for a full description.)

4.12 Macroergonomics Methodology

4.12.1 Methods Developed for Macroergonomics Application

Since the first Human Factors in ODAM symposium in 1984, in addition to MAS described earlier, a number of important new macroergonomic methodologies have been developed and successfully applied, including the following [in addition to the references cited hereafter for some of these methods, see Part 5 of Stanton et al. (2004) for more detailed descriptions.]

4.12.1.1 Macroergonomic Analysis and Design

Macroergonomic analysis and design (MEAD) is a 10-step process for systematically analyzing a work system's processes that was developed by Brian Kleiner of Virginia Tech. Through MEAD's application, deficiencies in work system processes can be identified and corrected (see Hendrick and Kleiner, 2001, 2002 for a detailed description).

4.12.1.2 HITOP

Developed by Ann Majchrzak of the University of Southern California and her colleagues, HITOP is a step-by-step manual procedure for implementing technological change. The procedure is designed to enable managers to be more aware of the organizational and human implications of their technology plans, and thus better able to integrate technology within its organizational and human context. It has been successfully applied in a number of manufacturing organizations.

4.12.1.3 Top MODELER

Top MODELER is a decision support system that also was developed by Ann Majchrzak of the University of Southern California and her colleagues. Its purpose is to help manufacturing organizations identify the organizational changes required when new process technologies are being considered.

4.12.1.4 CIMOP

Computer-Integrated Manufacturing, Organization and People (CIMOP) is a knowledge-based system for evaluating computer integrated manufacturing, organization, and people system design. It was developed by J. Kantola and Waldemar Karwowski of the University of Louisville. The intended users of CIMOP are companies designing, redesigning, or implementing a CIM system. (See Hendrick and Kleiner, 2002, for a more detailed description.)

4.12.1.5 Systems Analysis Tool

Systems analysis tool (SAT) is a method developed by Michelle Robertson of the Liberty Mutual Research Center for conducting systematic trade-off evaluations of work system intervention alternatives. It is an adaptation, elaboration, and extension of the basic steps of the scientific method. SAT has proven useful in enabling both ergonomists and managerial decision makers to determine the most appropriate strategy for making work system changes (see Hendrick and Kleiner, 2002 for a detailed description).

4.12.1.6 Anthropotechnology

Although not intentionally developed as a macroergonomics tool, anthropotechnology is, in effect, a macroergonomics methodology. It deals specifically with analysis and design modification of systems for effective technology transfer from one culture to another. It was developed by the distinguished French ergonomist, Alain Wisner. Wisner and others, such as Philippe Geslin of the Institut National de la Recherche Agronomique (INRA), have been highly successful in applying anthropotechnology to systems transferred from industrially developed to industrially developing countries. (See Wisner, 1995, for a detailed description.)

4.12.2 Methods Adapted for Macroergonomics Application

In addition, most of the traditional research methods have been modified and adopted for macroergonomic application. These include the following: Macroergonomic Organizational Questionnaire Surveys (MOQS)—an adaptation of the organizational questionnaire survey method by Pascale Carayon and Peter Hoonakker (2004) of the University of Wisconsin, laboratory experiment, field study, field experiment (perhaps the most widely used conventional method in macroergonomic studies—make work system changes to one part of the organization and if it improves organizational effectiveness, then implement throughout the work system), interview survey, and focus groups. Brian Kleiner has done a number of interesting laboratory studies in his macroergonomics laboratory at Virginia Tech to investigate sociotechnical variables as they relate to macroergonomics. (See Hendrick and Kleiner, 2001, 2002 for some examples.) Carayon and Hoonakker have successfully applied MOQS in a variety of interventions.

Several methods initially developed for microergonomics also have been adapted for macroergonomic application. Most notable are the following.

4.12.2.1 Kansai Ergonomics

This method initially was developed for consumer product design by Mitsuo Nagamachi and his colleagues at Hiroshima University, but can be applied to evaluating worker affective responses to work system design changes.

4.12.2.2 Cognitive Walkthrough Method

The cognitive walkthrough method (CWM) is a usability inspection method that rests on the assumptions that evaluators are capable of taking the perspective of the user and can apply this user perspective to a task scenario to identify design problems. As applied to macroergonomics, evaluators can assess the usability of conceptual designs of work systems to identify the degree to which a new work system is harmonized or the extent to which workflow is integrated.

4.12.2.3 Participatory Ergonomics

Participatory ergonomics (PE) is an adaptation of participatory management that was developed for both micro- and macroergonomic interventions. When applied to evaluating the overall work system, the employees work with a professional ergonomist who serves as both the facilitator and resource person for the group. A major advantage of this approach is that the employees are in the best position to both know the problem symptoms and to identify the macroergonomic intervention approach that will be most acceptable to them. Equally important, having participated in the process, the employees are more likely to support the work system changes—even if their own preferred approach is not adopted. Finally, the participatory approach has proven to be particularly effective in establishing an ergonomics and safety culture, which sustains performance and safety improvements that initially result from the macroergonomic intervention. Participatory ergonomics is the most widely used method in successful macroergonomic interventions. It often is used in combination with other methods, such as those described previously.

4.13 Macroergonomics as a Key to Improving Health Care and Patient Safety

All sociotechnical systems are organizations that exist to accomplish some purpose, be it manufacturing, refining, product distribution, sales, or to provide some kind of service. Organizations accomplish their respective purposes by transforming some kind of input into output, that is, a product or service. As noted earlier, all

sociotechnical systems comprise a personnel subsystem, technological subsystem, and a work system design (structure and processes), and interact with relevant aspects of their external environment which permeate the organization, and upon which the system must be responsive to survive and be successful. It thus is not surprising that macroergonomics has successfully been applied to a wide spectrum of organizations. For example, during the first five IEA International Symposia on Human Factors in Organizational Design and Management (1984–1996), successful macroergonomic interventions were reported for 23 manufacturing, 8 construction, 2 community planning and development, 9 interorganizational networks, and 38 service organizations. The service organizations included various governmental organizations, architectural design firms, telecommunications companies, banks, credit card companies, and an advertising agency, printing business, insurance company, and industrial product supplier. They also included 4 hospitals.

As with the cases mentioned earlier, all types of healthcare organizations are complex sociotechnical systems. They share with these other organizations the same issues of employee health, safety, comfort, productivity, and the need to eliminate system-induced human error. Unlike most of the organizational types, mentioned previously, the services healthcare systems provide to their customers may be truly life and death matters, as well as matters of enhancing customer quality of life. An employee medical error potentially can result in a patient's death. Poor or inefficient patient care can drastically degrade that customer's quality of life. In addition, inefficient healthcare systems increase healthcare costs. Accordingly, the potential of macroergonomics for enabling healthcare work systems to better meet their goals is of greater importance than for most other types of organizations. With real top management, and employee commitment at all organizational levels, macroergonomic interventions—including systematic carry-through to the microergonomic design/modification of jobs and related human–machine, human–environment, and human–software interfaces—can result in the same dramatic improvements in organizational effectiveness that have been experienced in many other types of sociotechnical systems (Hendrick, 2008).

References

Argyris, C. (1971). *Management and Organizational Development.* New York: McGraw-Hill.
Bailey, R. W. (1989). *Human Performance Engineering* (2nd edn.). Englewood Cliffs, NJ: Prentice-Hall.

Bedeian, A. G. and Zammuto, R. F. (1991). *Organizations: Theory and Design*. Chicago, IL: Dryden.

Burns, T. and Stalker, G. M. (1961). *The Management of Innovation*. London, U.K.: Tavistock.

Carayon, P. and Hoonakker, P. L. T. (2004). Macroergonomics organizational questionnaire survey (MOQS). In: N. Stanton, A. Hedge, K. Brookhuis, E. Salas, and H. W. Hendrick (Eds.), *Handbook of Human Factors and Ergonomics Methods* (pp. 76–1/76–10). Boca Raton, FL: CRC Press.

Chapanis, A. (1988). To communicate the human factors message you have to know what the message is and how to communicate it. Keynote address to the HFAC/ACE Conference, Edmonton, Alberta, Canada [reprinted in the *Human Factors Society Bulletin*, 1991, 34(11), 1–4, part 1; 1992, 35(1), 3–6, part 2].

Davis, L. E. (1982). Organizational design. In: G. Salvendy (Ed.), *Handbook of Industrial Engineering* (pp. 2.1.1–2.1.29). New York: John Wiley & Sons.

Degreene, K. (1973). *Sociotechnical Systems*. Englewood Cliffs, NJ: Prentice-Hall.

Duncan, R. B. (1972). Characteristics of organizational environments and perceived environmental uncertainty. *Administrative Science Quarterly*, 17, 313–327.

Emery, F. E. and Trist, E. L. (1960). Sociotechnical systems. In: C. W. Churchman and M. Verhulst (Eds.), *Management Sciences: Models and Techniques* (pp. 83–97). Oxford: Pergamon.

Emery, F. E. and Trist, E. L. (1965). The causal texture of organizational environments. *Human Relations*, 18, 21–32.

Hall, R. H., Haas, J. E., and Johnson, N. J. (1967). Organizational size, complexity, and formalization. *Administrative Science Quarterly*, June, 303.

Harvey, O. J. (1963). System structure, flexibility and creativity. In: O. J. Harvey (Ed.), *Experience, Structure and Adaptability*. New York: Springer.

Harvey, O. J., Hunt, D. E., and Schroder, H. M. (1961). *Conceptual Systems and Personality Organization*. New York: Wiley.

Hendrick, H. W. (1984). Wagging the tail with the dog: Organizational design considerations in ergonomics. In: *Proceedings of the Human Factors and Ergonomics Society 28th Annual Meeting* (pp. 899–903). Santa Monica, CA: Human Factors and Ergonomics Society.

Hendrick, H. W. (1986). Macroergonomics: A concept whose time has come. *Human Factors Society Bulletin*, 30, 1–3.

Hendrick, H. W. (1991). Human factors in organizational design and management. *Ergonomics*, 34, 743–756.

Hendrick, H. W. (1995). Future directions in macroergonomics. *Ergonomics*, 38, 1617–1624.

Hendrick, H. W. (1997). Organizational design and macroergonomics. In: G. Salvendy (Ed.), *Handbook of Human Factors and Ergonomics* (pp. 594–636). New York: Wiley.

Hendrick, H. W. (2000). Human factors and ergonomics converge—A long march. In: I. Kuorinka (Ed.), *History of the International Ergonomics Association: The First Quarter of a Century* (pp. 125–142). Geneva, Switzerland: IEA Press.

Hendrick, H. W. (2008). Applying ergonomics to systems: Some documented "lessons learned." *Applied Ergonomics*, 39, 418–426.

Hendrick, H. W. and Kleiner, B. M. (2001). *Macroergonomics: An Introduction to Work System Design*. Santa Monica, CA: Human Factors and Ergonomics Society.

Hendrick, H. W. and Kleiner, B. M. (Eds.) (2002). *Macroergonomics: Theory, Methods and Applications*. Mahwah, NJ: Lawrence Erlbaum.

Imada, A. S. (2002). A macroergonomic approach to reducing work-related injuries. In: H. W. Hendrick and B. M. Kleiner (Eds.), *Macroergonomics: Theory, Methods and Applications* (pp. 347–358). Mahwah, NJ: Lawrence Erlbaum.

Keidel, R. W. (1994). Rethinking organizational design. *Academy of Management Executive*, 8(4), 12–30.

Lawrence, P. R. and Lorsch, J. W. (1969). *Organization and Environment*. Homewood, IL: Irwin.

Meshkati, N. (1986). Human factors considerations in technology transfer to industrially developing countries: An analysis and proposed model. In: O. Brown, Jr. and H. W. Hendrick (Eds.), *Human Factors in Organizational Design and Management II* (pp. 351–368). Amsterdam, the Netherlands: North-Holland.

Meshkati, N. (1990). Human factors in large-scale technological system's accidents: Three Mile Island, Bhopal, and Chernobyl. *Industrial Crisis Quarterly*, 5, 133–154.

Meshkati, N. and Robertson, M. M. (1986). The effects of human factors on the success of technology transfer projects to industrially developing countries: A review of representative case studies. In: O. Brown, Jr. and H. W. Hendrick (Eds.), *Human Factors in Organizational Design and Management II* (pp. 343–350). Amsterdam, the Netherlands: North-Holland.

Mileti, D. S., Gillespie, D. S., and Haas, J. E. (1977). Size and structure in complex organizations. *Social Forces*, 56, 208–217.

Montanari, J. R. (1975). An expanded theory of structural determinism: An empirical investigation of the impact of managerial discression on organizational structure. Unpublished doctorial dissertation, University of Colorado, Boulder, CO.

Negandhi, A. R. (1977). A model for analyzing organization in cross-cultural settings: A conceptual scheme and some research findings. In: A. R. Negandhi, G. W. England, and B. Wilpert (Eds.), *Modern Organizational Theory* (285–312). Kent State, OH: University Press.

Noro, K. and Imada, A. S. (1991). *Participatory Ergonomics*. London, U.K.: Taylor & Francis.

Perrow, C. (1967). A framework for the comparative analysis of organizations. *American Sociological Review*, 32, 194–208.

Robbins, S. R. (1983). *Organizational Theory: The Structure and Design of Organizations*. Englewood Cliffs, NJ: Prentice-Hall.

Smith, A. (1970). *The Wealth of Nations*. London, U.K.: Penguin. (Originally published in 1876.)

Stanton, N., Hedge, A., Brookhuis, K., Salas, E., and Hendrick, H. (Eds.) (2001). *Handbook of Human Factors and Ergonomics Methods*. London, U.K.: Taylor & Francis.

Stevenson, W. B. (1993). Organizational design. In: R. T. Golembiewski (Ed.), *Handbook of Organizational Behavior* (pp. 141–168). New York: Marcel Dekker.

Trist, E. L. and Bamforth, K. W. (1951). Some social and psychological consequences of the longwall method of coal-getting. *Human Relations*, 4, 3–38.

Trist, E. L., Higgin, G. W., Murray, H., and Pollock, A. B. (1963). *Organizational Choice*. London: Tavistock.

Wisner, A. (1995). Situated cognition and action: Implications for ergonomic work analysis and anthoroptechnology. *Ergonomics*, 38(8), 1542–1557.

Yankelovich, D. (1979). *Work, Values and the New Breed*. New York: Van Norstrand Reinhold.

Zink, K. J. (2002). A vision of the future of macroergonomics. In: H. W. Hendrick and B. M. Kleiner (Eds.), *Macroergonomics: Theory, Methods and Applications* (pp. 347–358). Mahwah, NJ: Lawrence Erlbaum.

5

Work System Design in Health Care

Pascale Carayon, Carla J. Alvarado, and Ann Schoofs Hundt

CONTENTS

5.1 Model of Work System 66
    5.1.1 Elements of the Work System Model 66
    5.1.2 Examples of Work Systems 68
    5.1.3 Work System Interactions 69
5.2 Work System and Patient Safety 70
    5.2.1 Pathways between Work System and Patient Safety 70
    5.2.2 SEIPS Model of Work System and Patient Safety 71
    5.2.3 Applications of the SEIPS Model 72
5.3 Human Factors and Ergonomics in Work System Design 73
    5.3.1 Work System Design Process 73
    5.3.2 Implementation of Work System Redesign 74
    5.3.3 Continuous Work System Design 75
    5.3.4 Work System Analysis 75
5.4 Conclusion 75
References 76

The system of healthcare delivery across the world is in great need of redesign. The 1999 report by the Institute of Medicine raised the issue of patient safety to the forefront of the public, policy makers, and healthcare leaders (Kohn et al. 1999). Citing data from studies conducted in the United States, the report indicated that between 44,000 and 98,000 American hospital patients were dying from preventable medical errors every year. Other studies, for instance in the United States (Rothschild et al. 2005), Canada (Baker et al. 2004), the United Kingdom (Vincent et al. 2008), and the Netherlands (van den Bemt et al. 2002), have continued to demonstrate the significance and extent of patient safety problems. The World Health Organization has also taken on the issue of patient safety by creating the World Alliance for Patient Safety (http://www.who.int/patientsafety/en/). The World Alliance for Patient Safety has focused its initiatives on surgical safety (see, e.g., the results of an international study looking at the effectiveness of surgical checklists by Haynes et al. [2009]) and hand hygiene. The system of healthcare delivery experiences many other problems, in addition to patient safety. Working conditions for healthcare professionals and workers are often sources of stress, burnout, and dissatisfaction. For instance, the following problematic working conditions of nurses have been highlighted by the IOM (Institute of Medicine Committee on the Work Environment for Nurses and Patient Safety 2004): high demands (e.g., demands for documentation), poorly designed physical space, stressful work schedules, and interruptions and distractions. Improvements in working conditions are also necessary for many other healthcare job categories, as demonstrated, for example, by job dissatisfaction, stress, and burnout experienced by physicians (McMurray et al. 1999; Wetterneck et al. 2002). Poor work conditions can contribute to negative outcomes for healthcare organizations, such as high turnover and injuries (Cohen-Mansfield 1997). Healthcare providers may experience a range of negative feelings and emotions, and suffer from health problems and injuries when their work is not ergonomically designed (Stubbs et al. 1983; Linzer et al. 2000). Patients and their families may be indirectly affected by poor design of the work environment in which care is provided to them (Agency for Healthcare Research and Quality 2002; Institute of Medicine Committee on the Work Environment for Nurses and Patient Safety 2004). A recent study shows a clear relationship between work schedules of nurses and patient outcomes, such as pneumonia deaths (Trinkoff et al. 2011). Concerns about the work schedules for medical residents have also been expressed for both patient safety and the safety and well-being of residents (Ulmer et al. 2008). Finally, society at large may be affected by poor design of healthcare working environments. For instance, nursing shortage has been linked to difficulty in attracting and keeping nurses because of poor working conditions (Wunderlich and Kohler 2001; Institute of Medicine Committee on the Work Environment for Nurses and Patient Safety 2004). Therefore, designing the work systems of healthcare providers, professionals, and workers is an important issue for healthcare organizations, the providers themselves, patients and their families, and society at large.

Understanding the characteristics and principles of work system design provides the foundation for healthcare organizations to engage in work improvements that can ultimately lead to a range of positive outcomes for the organizations themselves (e.g., reduced turnover), the workers (e.g., increased job satisfaction), patients and their families (e.g., improved quality and safety of care), and society (e.g., decreased nursing shortage) (Kovner 2001; Sainfort et al. 2001; Carayon et al. 2003a,b). However, although the need for system redesign has been clearly recognized (Institute of Medicine Committee on Quality of Health Care in America 2001), it is unclear whether it has been taken seriously as suggested by Shortell and Singer (2008). This chapter will contribute to the systems-based approach advocated by many to redesign healthcare systems. We first describe the work system model proposed by Smith and Carayon (Smith and Carayon-Sainfort 1989; Carayon and Smith 2000; Smith and Carayon 2000; Carayon 2009), and then explain how the design of work systems can be related to patient safety. The SEIPS (Systems Engineering Initiative for Patient Safety) model of work system and patient safety (Carayon et al. 2006) is then described, and various applications of the work system model and the SEIPS model are presented. Finally, principles for using the work system approach are presented.

5.1 Model of Work System

In the healthcare literature, the concept of "system" has been extensively discussed as a key element for improving patient safety and quality of care (Donabedian 1980; Leape 1997; Institute of Medicine Committee on Quality of Health Care in America 2001; Berwick 2002; Buckle et al. 2003; Vincent et al. 2004; Shortell and Singer 2008). In this chapter, the concept of "system" is used in a very specific sense: a system represents the various elements of work that a healthcare person uses, encounters, and experiences in order to perform his/her job. The concept of "work" system is extended to include systems experienced by patients and their families and friends.

Designing or redesigning work systems requires a systematic approach that takes into account the various elements of work. The work system model developed by Smith and Carayon is a very useful model to understand the many different elements of work (Smith and Carayon-Sainfort 1989; Carayon and Smith 2000; Smith and Carayon 2000; Carayon 2009). The work system is comprised of five elements: the *person* performing different *tasks* with various *tools and technologies* in a *physical environment* under certain *organizational conditions* (see Figure 5.1).

5.1.1 Elements of the Work System Model (see Table 5.1)

The *person* at the center of the work system can be a physician, a nurse, a pharmacist, a respiratory therapist, a unit clerk, a receptionist, or any other healthcare professional or worker. The person can also be a patient and his/her families and friends. The person has physical, cognitive, and psychosocial characteristics. Physical characteristics include, for instance, height, weight, physical strength, and reach envelop. Cognitive characteristics include, for instance, information processing capacity, knowledge, and expertise. Psychosocial characteristics include, for instance, motivation, needs, and tolerance for ambiguity. The work system model applies to all of the people involved in health care, including the patient (Carayon et al. 2003).

The *tasks* performed by the individual represent the specific interactions between the person and his/her work in order to accomplish particular objectives. Task is a key concept in human factors and ergonomics (HFE) (Kirwan and Ainsworth 1992). HFE makes the distinction between the "prescribed task" and the "activity" or the task actually carried out (Hackman 1969; Leplat 1989). The prescribed task is defined by the organization: it is what is expected of the worker, and what she/he is supposed to do. What the worker actually does may be different from the prescribed task. The activity or the task being carried out is what the worker actually does and can be characterized by work system analysis (e.g., direct observation, interview with the worker) (Leplat 1989). The task actually carried out is influenced by a redefinition process that considers the context and circumstances in which the work is being performed (Hackman 1969). Various dimensions can be used to characterize the task: difficulty or challenge, content, variety, repetitiveness, skill utilization, autonomy and job control, clarity, uncertainty, demands, contact with others, and feedback (Carayon et al. 2001). Several chapters in this book describe various aspects of the tasks of healthcare workers (see, e.g., Chapter 9).

The person performing a task uses various *tools and technologies*, which can be simple (e.g., paper and pencil),

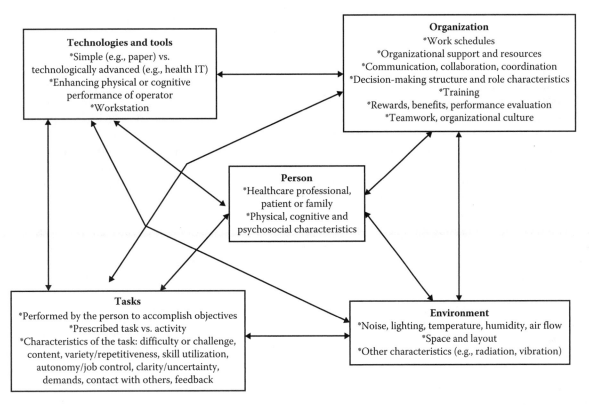

FIGURE 5.1
Model of the work system. (From Smith, M.J. and Carayon-Sainfort, P., *Int. J. Ind. Ergon.*, 4(1), 67, 1989; Carayon, P. and Smith, M.J. *Appl. Ergon.*, 31, 649, 2000.)

TABLE 5.1

Elements of the Work System Model

Work System Elements	Examples
Person	Physician, nurse, pharmacist, respiratory therapist, technician, unit clerk, receptionist, etc…
	Physical, cognitive, and psychosocial characteristics
Task	Distinction between prescribed task and activity or task actually carried out
	Difficulty or challenge, content, variety, repetitiveness, skill utilization, autonomy and job control, uncertainty, demands, contact with others, and feedback
Tools and technologies	Physical, cognitive, and psychosocial dimensions Usability and support to healthcare work
Physical environment	Noise, lighting, temperature-humidity-air flow, space, layout, vibration, and radiation
Organizational conditions	Work schedules, organizational support (e.g., social support from supervisors and managers, resources provided during a technological or organizational change), communication, collaboration, coordination, decision-making structure, role characteristics such as ambiguity and conflict, training, rewards, benefits, performance evaluation, teamwork, and organizational culture

can be technologically advanced (e.g., bar coding medication administration technology, MRI), require the use of hands (e.g., hand tools such as surgical endoscopic tools), may involve remote sensing (e.g., robotic surgery, telemedicine), enhance the capacity of the operator (e.g., lifting aids, remote monitoring), and support other tools and technologies (e.g., nursing workstation where other equipment is located and used). The tools and technologies interact with the physical, cognitive, and psychosocial characteristics of the person, and other elements of the work system. For instance, studies on health information technologies have identified various human factors problems: workflow issues in the design and implementation of electronic health record (EHR) technology in ambulatory care (Carayon et al. 2010), poor coordination between nurses and physicians when a bar coding medication administration (BCMA) system was implemented at VA hospitals (Patterson et al. 2002), and poor usability of a computerized provider order entry system (Koppel et al. 2005). Tools and technologies should be designed to support the work of healthcare individuals and teams in performing the tasks necessary to accomplish quality and safety of care objectives.

The *physical environment* can be characterized on the following dimensions: noise, lighting, temperature-humidity-air flow, space and layout, and other characteristics (e.g., vibration, radiation). Much is known about how to design the various dimensions of the physical environment (Parsons 2000) (see, e.g., Chapter 15). It is important to recognize that the physical environment interacts with all of the characteristics of the person. Physical characteristics of the person need to be considered when designing, for instance, the height of a workstation and the placement of various pieces of equipment. The design of the physical environment can also affect the cognitive characteristics of the person, such as providing cues for attracting the attention of healthcare providers. Psychosocial characteristics of the person are also important to consider when designing a physical environment. For instance, the physical layout of a unit can facilitate communication between workers by creating a space where various providers can meet and discuss.

A range of *organizational conditions* govern and influence the way a person performs tasks using tools and technologies in a specific physical environment: work schedules, organizational support (e.g., social support from supervisors and managers, resources provided during a technological or organizational change), communication, collaboration, coordination, decision-making structure, role characteristics such as ambiguity and conflict, training, rewards, benefits, performance evaluation, teamwork, and organizational culture. A number of chapters in this book describe these organizational conditions, for instance, Chapters 7 and 10.

5.1.2 Examples of Work Systems

The work system model is a generic model that can be used to describe elements of work for any type of healthcare work performed either by an individual or by a team. For instance, the work system of intensive care unit (ICU) nurses has been characterized as follows. ICU nurses often have critical care credentials, such as AACN (American Association of Critical-Care Nurses) certification. An ICU nurse performs a range of activities, including direct nursing care (e.g., taking vital signs, patient assessment, patient treatment), indirect nursing care (e.g., conversation with patient, calibration of monitors), documentation (e.g., nursing notes), administrative activities (e.g., staff meetings), housekeeping (e.g., cleaning bed), and other miscellaneous activities (Wong et al. 2003). ICU nurses use a range of tools and technologies to perform these activities such as monitors, medical devices (e.g., infusion pump), other equipment (e.g., BCMA technology), and health information technologies (e.g., electronic medication administration technology, EHR). Most tasks performed by ICU nurses are performed in the ICU, a physical environment characterized by high levels of noise, varying levels of lighting, and other environmental stressors. When an ICU nurse "takes a patient on the road", that is, takes the patient outside of the ICU for an x-ray, then the physical environment changes: some tools and technologies may no longer be easily accessible (e.g., ventilator) and support staff is limited. A range of organizational conditions influence the work system of ICU nurses, such as work schedules (e.g., 12 h shift, rules for overtime), communication and collaboration with ICU physicians, and ICU nursing management style.

The work system model can be used to describe the elements of the ICU nursing work system; it can also be used to identify the work system elements that either hinder (i.e., barriers) or facilitate (i.e., facilitators) the ability of ICU nurses to do their job. Gurses and Carayon (Gurses 2005; Gurses and Carayon 2007, 2009) have identified the following key work system barriers as experienced by ICU nurses:

- Task: Spending time to address family needs and to teach family

- Tools and technologies: Lack of equipment availability, patient rooms not well-stocked, time searching for supplies and equipment

- Environment: Distractions and numerous phone calls from family members, insufficient space

- Organizational conditions: Delay in getting medications from pharmacy and in seeing new medical orders, time searching for patient chart

As indicated earlier, the work system model can also be used to describe the "work" of patients in their various contacts with the healthcare system. Teresa Zayas-Caban and Rupa Valdez describe the work system as applied to home care in Chapter 43. Henriksen et al. (2009) provide information about various work system elements, such as home healthcare tasks (i.e., assisting with ambulation, medication management, monitoring vital signs, administering respiratory/infusion therapies, handling lesions/wounds/pressure ulcers, and assisting with medical devices), tools and technologies (i.e., understanding functionality, ease of use, adjustability, maintenance and troubleshooting, training and cognitive aids, and dealing with power outages) and physical environment (i.e., space layout, room adjacencies, interior features, ramps and door widths, lighting and HVAC, indoor/outdoor transitions, and gardens). Or et al. (2009) also focus on the home care work system, but from the viewpoint of health information technology. They identified a range of work system issues in the design and implementation of health information technology into the practice of home care nursing.

Another example of the work system can be found in Figure 5.2, which describes the work-educational system of residents, in particular in light of duty hours.

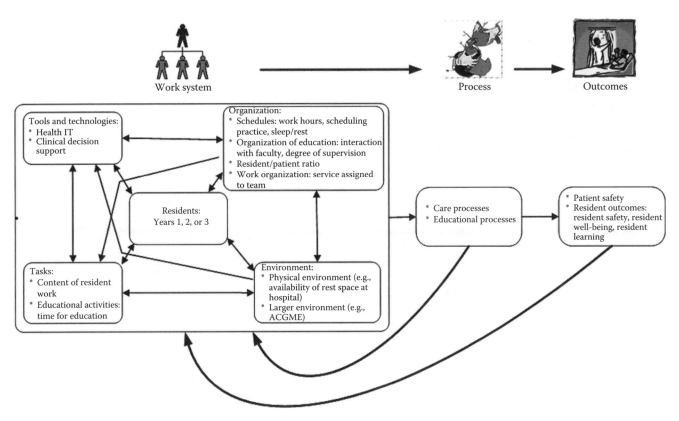

FIGURE 5.2
Work-educational system of medical residents.

For instance, the organizational element describes organizational issues related to schedules (i.e., work hours, scheduling practice, sleep/rest) and organizational issues related to education (i.e., interaction with faculty, degree of supervision, resident/patient ratio, work organization).

5.1.3 Work System Interactions

It is important to underline the "system" nature of work, that is, the work system comprises elements that interact with each other. Consider the example of the task of central venous cannulation in the ICU. It is an essential part of patient management in a variety of clinical settings, for example, for critically ill patients requiring long-term venous access and patients with cancer requiring chemotherapy. Unfortunately, patient complications of CVC cannulation are very common. Noise and other sensory disruptions abound in the modern ICU, creating stress, masking communication, patient responses and alarms (characteristics of the environment and the task). The environment is crowded and messy and physically difficult to work in (characteristics of the physical environment that interact with task performance). Crowds of people are waiting to get a moment of the physician's and/or nurse's time, all making it more difficult for them to carry out their task without

interruptions (organizational conditions that affect task performance). Hospital supply stocking schedules and storage often make supplies difficult to locate or not available when needed (organizational condition and characteristic of the physical environment). The hospital administration may not budget for or approve new devices or technology for the tasks (organizational conditions and technology). To correct a perceived problem with task performance, the administration may introduce additional new tasks for a group of employees. With the introduction of nurses performing peripherally inserted central venous catheters (CVC-PICC), there is a possibility of physician workload reduction. Though an effective solution to balance the physician workload, it is important not to create negative aspects for the nursing workload. This example demonstrates how various elements of the work system interact with each other, and how elements of the work system of one worker (e.g., the physician) can affect elements of the work system of another worker (e.g., the nurse). Too often because of a lack of understanding of the entire structure and design of the overall work system and of how various components affect healthcare worker performance, a solution may be proposed that does not consider the potential for creating "imbalance" within the system work factors and, therefore, lead to increases in workload on the person.

In the context of increasing technological change in health care, we need to understand the interactions between tools and technologies and other elements of the work system. The following example describes the implementation of smart intravenous pumps in a hospital setting and their impact on the work system of different healthcare providers. Given the range of intravenous pump users and the variety of patient care environments and clinical circumstances surrounding their use, one might reasonably expect that organizations would consider this variation in "smart" IV pump use and in turn how it could influence implementation of smart IV pumps organization-wide. For example, features associated with smart pumps including alarm thresholds and drug libraries require careful consideration of user needs and expectations. Because more potent drugs are included in drug libraries, consideration must be given to the type and extent of training an ICU versus general medical/surgical versus pediatric nurse would receive. Use in the operating room by certified registered nurse anesthetists and anesthesiologists requires careful assessment of clinical practices in order to arrive at a consensus when determining upper and lower dose limits associated with unsafe levels of IV drug administration. Consensus in this case may be difficult to achieve.

According to Shortell and Singer (2008), "To take systems seriously requires the implementation of systems thinking throughout the organization and across organizations" (p. 444). Systems thinking emphasizes the need to not only describe the system elements, but to also understand interactions between system elements, and between various systems and sub-systems. The system interactions are critical to address performance and safety issues (Wilson 2000). Understanding system interactions is particularly important when some type of change is implemented, such as a new technology. This change will impact other work system elements in either positive or negative ways. So-called unanticipated consequences are often due to the lack of understanding of systems thinking and system interactions.

5.2 Work System and Patient Safety

A range of safety models have been proposed that describe the contribution of work-related factors (e.g., performance shaping factors) to accidents (Rasmussen 1982; Reason 1990; Vincent et al. 1998). Whereas human error and safety models describing the "chain of events" leading to accidents are interesting and useful for analyzing and evaluating specific accidents, other models and approaches need to be considered when redesigning

work (Vincent 2010). These provide information on *how* to change work and on *what* needs be changed in the work system (Parker and Wall 1998). Before one can redesign work systems to improve patient safety, it is important to understand the relationship between the work system and patient safety.

5.2.1 Pathways between Work System and Patient Safety

We have identified four different pathways or mechanisms between work (re)design and patient safety (see Table 5.2).

First, redesigning work may directly target the causes or sources of patient safety problems. For instance, a number of studies have shown that excessive working hours and poorly designed work schedules of nurses (Smith et al. 1979; Rogers et al. 2004) and physicians (Landrigan et al. 2004; Lockley et al. 2004) can lead to various patient safety problems. Several proactive risk assessment methods, such as FMEA and RCA, allow the identification of errors and specific work system factors that contribute to patient safety problems (Carayon et al. 2003, 2009). These methods can benefit from the use of the work system model in order to systematically address all elements of work that can contribute to an actual incident or event (e.g., RCA) or to anticipate the work factors that may contribute to vulnerabilities in a process (e.g., FMEA). For instance, Faye et al. (2010) identified the work system elements that contributed to failures in the nursing medication administration process. Studies performed in manufacturing have shown relationships between ergonomic deficiencies (e.g., poor working postures) and poor quality of assembly (Eklund 1995; Axelsson 2000). In a similar manner, redesigning work in health care may remove the sources of possible error and improve quality of work, that is, patient safety. These sources of error can be contained in any of the

TABLE 5.2

Pathways between Work System Redesign and Patient Safety

1. Work redesign may directly target the causes or sources of patient safety problems

2. Work redesign may lead to improved efficiencies by removing performance obstacles, therefore freeing up time and reducing workload for nurses to provide better, safer patient care

   Work redesign may lead to the reexamination of who does what, that is, the objectives of work, and indirectly improve quality and safety of care

3. Work design can be considered as part of the 'Structure' element of Donabedian's model of quality of care. Therefore, improving work can improve care processes, and therefore patient outcomes, including patient safety. [SEIPS model of work system and patient safety [Carayon et al. 2006]]

work system elements or can be the product of interactions between elements of the work system.

Second, work redesign may lead to improved efficiencies. Inefficiencies can be considered as "performance obstacles" (Peters et al. 1982; Brown and Mitchell 1991; Park and Han 2002), that is, aspects of the work system that hinder healthcare workers' ability to provide good safe patient care (Carayon et al. 2005; Gurses and Carayon 2009). Removing inefficiencies may directly or indirectly improve patient safety. Inefficiencies may represent sources of error or conditions that enhance the likelihood of error. For example, reliance on paper copies of patients' medical records can result in delays and inefficiencies when there is a problem locating them. As a result unwise (and subsequently unsafe) clinical decisions may be made in the interest of averting delays when a provider determines sufficient (although not necessarily ideal) information is available. Sources of performance obstacles or inefficiencies can be categorized according to the work system model: task-related inefficiencies (e.g., lack of job-related information, interruptions), tools and technologies (e.g., poor functioning or inaccurate equipment, inadequate materials and supplies), organizational inefficiencies (e.g., inadequate staffing, lack of support, lack of time), and physical environment (e.g., noisy, crowded environment) (Peters and O'Connor 1980; Peters et al. 1982; Carayon et al. 2005). Removing performance obstacles and inefficiencies may free up time and reduce workload (e.g., more efficient use of currently employed nurses), therefore reducing the likelihood of error.

Research has been conducted to identify performance obstacles in health care, specifically among staff of outpatient surgery centers and intensive care nurses (Carayon et al. 2005; Gurses and Carayon 2009). This research highlights the variety of work factors that can hinder performance of healthcare providers, that is, their capacity to provide high-quality safe care. For instance, among staff in several outpatient surgery centers, conflict between nurses and physicians, lack of staffing, lack of staff training, and poor work schedules were identified as major performance obstacles. Among ICU nurses, we found that the major performance obstacles were related to patient requirements and inadequate staffing (e.g., poor performance, inadequate skills, insufficient number of people) (Carayon et al. 2005; Gurses and Carayon 2009).

Third, work redesign may lead to improved effectiveness. Whereas efficiency is the degree to which work is performed well, effectiveness relates to the content of work itself: are we doing the right things? Work redesign may lead an organization to ask questions regarding task allocation (who is doing what?) and job content (what tasks should be performed that

are not performed and what tasks are performed that should not be performed?). For instance, Hendrickson et al. (1990) conducted a study on how hospital nurses spend their time. Work sampling techniques were used to evaluate time spent in various activity categories (i.e., with patient, patient chart, preparation of therapies, shift change activities, professional interaction, miscellaneous activities, checking physician's orders, unit-oriented in-service, paperwork, phone communications, supplies, miscellaneous nonclinical). Results show that within a typical 8h shift nurses spend an average of 31% of their time with patients. Such results may lead to the reexamination of the tasks and content of nursing jobs. Work redesign can contribute to increased efficiency, that is, the elimination of unnecessary activities, may free up time for healthcare workers to provide better, safer patient care, and may lead to a more efficient effective distribution of tasks.

Fourth, Donabedian's framework for assessing quality of care can fit the work system model and provide justification for another pathway between work design and patient safety (Donabedian 1978, 1980, 1988). In Donabedian's model, the *structure* includes the following: *the organizational structure* (work system model element = organization); *the material resources* (work system model elements = environment, tools/technology); and *the human resources* (work system model elements = worker, tasks). Donabedian's two other means of assessing quality include evaluating the *process(es)* of care—how worker tasks and clinical processes are both organized and performed (e.g., Was the care provided in compliance with the clinical pathway? Was medically indicated care provided?), and evaluating the *outcome(s)* of care—assessing the clinical results and impacts of and patient satisfaction with the care provided. Donabedian (1980) proposes that a direct relationship exists between structure, process, and outcome. The structure of an organization affects how safely care is provided (the process); and the means of caring for and managing the patient (the process) affects how safe the patient is upon discharge (outcome). We have developed a patient safety model (SEIPS model) that incorporates the work system model in Donabedian's framework (Carayon et al. 2003a). See the following for further discussion of the SEIPS model of work system and patient safety.

5.2.2 SEIPS Model of Work System and Patient Safety

According to the SEIPS model of work system and patient safety (Carayon et al. 2006), the *work system* in which care is provided affects both the work and clinical *processes*, which in turn influence the patient and organizational *outcomes* of care. Changes to any aspect of the work system will, depending on how the change or improvement is implemented, either negatively or positively affect the work and clinical processes and the consequent patient and organizational outcomes. Redesigning a work system requires careful planning and consideration to ensure that neither the quality of care nor patient or worker safety is compromised because of a lack of consideration to all of the elements of the work system. Likewise, subsequent review of the success of change in an organization cannot be fully accomplished without assessing the entire work system, the clinical and work processes, and both the clinical and organizational outcomes of patient care.

Figure 5.2 describes an example of the SEIPS model applied to the work and educational system of residents. Two processes are highlighted: care process and educational process. The outcomes of importance are (1) patient safety and (2) resident outcomes, that is, resident safety and well-being.

According to Donabedian (1978), the care process is what is done to or for the patient. According to the SEIPS model, the care process includes a range of tasks done by different individuals using various tools and technologies in a physical and organizational environment. Because the Donabedian's model of healthcare quality has been primarily used for measuring and evaluating quality of care, the concept of care process is often described as a series of process measures, such as time from diagnosis to treatment, or percentage of patients who have been immunized or who are receiving other recommended health services. These process measures rarely provide information about "what is done to or for the patient"; they are summary measures that do not indicate exactly what was involved in the process. The work system model and its extension, the SEIPS model of work system and patient safety can help to understand what exactly is involved in the care process, therefore providing for richer and deeper description of care processes. The SEIPS model can therefore help healthcare leaders and policy makers to begin to specify and clarify the relationship between structure (or work system) and processes and patient outcomes. Describing the care process in light of the work system also provides opportunities to identify the barriers and facilitators to patient care. Focusing the work system analysis on specific high-risk care processes (e.g., central line insertion, medication ordering or administration, bedside rounds, patient discharge from hospital) is critical (Walker and Carayon 2009).

5.2.3 Applications of the SEIPS Model

The SEIPS model has been extensively used by researchers at the University of Wisconsin-Madison. For instance, the work system model has been used to identify factors that impact the IV medication administration process

(Wetterneck et al. 2006) and the outpatient surgery process (Carayon et al. 2004), to describe performance obstacles and facilitators experienced by healthcare providers in outpatient surgery centers (Carayon et al. 2006) and in ICUs (Carayon et al. 2005; Gurses and Carayon 2007), and to examine the work systems of cardiac surgical care (Wiegmann et al. 2009).

The SEIPS model has also been used to analyze nursing processes (Boston-Fleischhauer 2008a,b), such as the nursing clinical documentation process (Boston-Fleischhauer 2008). Using the SEIPS model, Catchpole and McCulloch (2010) describe issues in critical care work systems that can affect patient safety. They also use the SEIPS model to propose a range of solutions for improving patient safety and quality.

The SEIPS model has helped to identify the system factors that can influence nurses' use of BCMA technology. In a study of inpatient nurses' interaction with BCMA technology, Carayon et al. (2007) described the following work system factors as influencing nursing performance and potentially medication safety: potentially unsafe task sequences including the typical work-around of documentation occurring before administration of the medication (Task element of the work system model), automation surprises and alarms (Technology element), interruptions (Organization element), various environmental factors such as low lighting and noise (Environment element), and patient in contact isolation (Person element). Koppel et al. (2008) specifically examined work-arounds to the use of BCMA technology, using various sources of data including our own research (Carayon et al. 2007). They categorized the probable causes to BCMA work-arounds using the work system model. *Technology*-related causes included need for multiple scans to read barcode, size of computer cart or scanner (also related to the size of the patient room, i.e., part of the *environment*), and scanning difficulties. The standard BCMA scanning *task* may be slower or more difficult than other methods. A large range of *organizational* factors that contributed to BCMA work-arounds were identified, such as use of non-formulary medication, medication order not in system, medications without bar codes, patients without wristband barcode, inadequate staffing, and lack of training. *Patient* barcodes may not be usable or accessible, and therefore, the nurse may not be able to use the BCMA technology.

Hysong et al. (2009) proposed to use the SEIPS model to examine three specific processes: electronic alerting for abnormal test results, electronic referral requests, and provider-pharmacy communication via CPOE. The main work system elements for these processes include: (1) person: providers, nurses and clerks; (2) task: alert processing; (3) technology: view alert system; (4) environment: ambulatory clinic; and (5) organization: VA medical center. They plan to use task analysis and focus groups in order to analyze communication activities involved in the three processes, the work system barriers and facilitators to communication, as well as strategies for improvement in electronic communication through an EMR.

Sittig and Singh (2010) have proposed a sociotechnical model specifically designed for studying health information technology. Their model is based on the SEIPS model, and extends the technology element of the work system model to include hardware and software computing infrastructure, clinical content, and human–computer interface. Their concept of "task" is focused on workflow and communication in order to emphasize collaboration in health care.

5.3 Human Factors and Ergonomics in Work System Design

The strength of the work system model is its comprehensiveness; it is generic and can be applied to analyze all types of healthcare work systems and processes. However, because the work system model is broad and comprehensive, a major challenge in designing or redesigning work systems is the breadth of knowledge required by the analysts and the decision makers. For instance, the Handbook of Human Factors and Ergonomics edited by Gavriel Salvendy (2006) has about 1700 pages. Chapter 5 of the first edition of this book provides a series of tables describing where to find information specific to each work system element (Carayon et al. 2007). The breadth and depth of HFE knowledge necessary to analyze and redesign healthcare work systems represents a major challenge for both the users of the knowledge (healthcare professionals and managers) and the developers of the knowledge (i.e., researchers).

A question raised by ergonomists is how to ensure that HFE knowledge and criteria are considered in the early stage of work system design (Clegg 1988; Slappendel 1994; Luczak 1995; Dekker and Nyce 2004). Johnson and Wilson (1988) discuss two approaches for taking into account ergonomics in work system development: (1) provision of guidelines, and (2) ergonomics input within collaborative design. Because of the voluminous body of HFE knowledge and the varied contexts in which healthcare work systems can be found, it is unlikely that specific guidelines could be developed. Therefore, the second strategy proposed by Johnson and Wilson (1988) may be more easily implemented. This necessitates the close collaboration of ergonomists and healthcare professionals at various stages of work system design.

5.3.1 Work System Design Process

Designing or redesigning work systems can represent a major organizational investment requiring the involvement of numerous people, substantial time to conduct evaluations, analyze data and design and implement solutions, adequate resources, and sufficient expertise and knowledge. Like any major organizational or technological change, work (re)design needs to be "managed." A process needs to be implemented for coordinating all the personnel, activities, and resources involved in the work (re)design project. The collaborative work (re)design process involves the following:

- A series of steps and activities logically and chronologically organized, such as the Plan-Do-Check-Act (PDCA) cycle.
- A "toolbox" of tools and methods that one can use to evaluate the work system and design and implement solutions (Carayon et al. 2003). A number of chapters in this book describe various HFE tools and methods, such as cognitive work analysis (Chapter 28).
- A set of overarching principles that can guide an organization when embarking into a work (re)design project (e.g., participation and involvement of workers and various stakeholders, feedback).

Work system redesign may be a specific objective for a healthcare organization; it is also possible that work systems may need to be redesigned in the implementation of other types of change. For instance, when designing a new healthcare facility, decisions are made that affect not only the physical environment, but also the rest of the work system such as support to communication and collaboration, and access to supplies and equipment. A large variety of projects can benefit from HFE input and expertise. This raises issues regarding the availability of HFE expertise within healthcare institutions, as well as the need to allocate adequate resources so that HFE criteria are considered as early as possible in the project timeline (Carayon 2005, 2010). A variety of jobs and functions within healthcare organizations can benefit from HFE knowledge, principles, and tools (Carayon 2005). This HFE body of knowledge and expertise needs to be translated and transferred to the people involved in those jobs and functions.

One healthcare area in which work system design issues are receiving increasing attention is related to the recent investment in health information technology in the U.S. healthcare system; this has prompted numerous questions about the sociotechnical impact of these technologies. When designing and implementing this type of technology, Clegg (1988) suggests that decisions need to be made regarding the following factors:

- *The type and level of technology.* Different elements can impact decisions regarding the type and level of technology: resources available, expected return on investment, and technology "push."
- *The allocation of functions between humans and machines.* In general, the human aspect is considered late in the design of technology, therefore leaving the "leftover" tasks to people.
- *The roles of humans in the system.* Once allocation of function decisions have been made, the various tasks need to be organized into job designs for the future users of the technology.
- *The organizational structures to support workers.* Healthcare organizations that are introducing new health information technology should ensure that adequate organizational structures are available to support workers through the transition phase as well as afterward (e.g., refresher or continuous training).
- *The way in which people participate in their design.* The type, extent, and timing of worker participation in the design of work systems are all important aspects to consider (Smith and Carayon 1995; Haines et al. 2002).

Usability has been identified as a major issue in the implementation of health information technology. Usability evaluation, aside from evaluating aspects of learnability and memorability (Nielsen 1993), can likewise be used when assessing specific health information technology applications for selection and implementation by including worker input in the selection process. In turn this user feedback can be used to aid in the design of new user training modules. Once the technology is implemented, the healthcare organization must provide adequate support to monitor and sustain a successful implementation by being responsive to user needs, questions, and difficulties. Likewise it is important to monitor for (un)anticipated problems and/or patient safety concerns arising from the technology or provider interaction with it and, as appropriate, share findings with users.

5.3.2 Implementation of Work System Redesign

In the phase of work system implementation, the question arises as to the methods and processes to use in order to facilitate the change process, and rapidly achieve the expected outcomes (i.e., improved quality and safety of care). The way change is implemented

(i.e., process implementation) is central to the successful adaptation of organizations to changes. A "successful" work system implementation from the human factors viewpoint is defined by its "human" and organizational characteristics: reduced/limited negative impact on people (e.g., stress, dissatisfaction) and on the organization (e.g., delays, costs, medication errors), and increased positive impact on people (e.g., acceptance of change, job control, enhanced individual performance) and on the organization (e.g., efficient implementation process, safe patient care). Success also includes decreasing medical errors and improving quality of care. Several authors have recognized the importance of the process of implementation in achieving a successful organizational change (Korunka et al. 1993; Tannenbaum et al. 1996).

Participatory ergonomics is a powerful method for implementing work system changes (Wilson 1995). Participation has been used as a key method for implementing various types of organizational changes, such as physical ergonomic programs (Haims and Carayon 1996; Wilson and Haines 1997), continuous improvement programs (Zink 1996), and technological change (Eason 1988; Carayon and Karsh 2000). Noro and Imada (1991) define participatory ergonomics as a method in which end-users of ergonomics (e.g., workers, nurses) take an active role in the identification and analysis of ergonomic risk factors as well as the design and implementation of ergonomic solutions. Evanoff and his colleagues have conducted studies on participatory ergonomics in health care (Bohr et al. 1997; Evanoff et al. 1999). One study examined the implementation of participatory ergonomics teams in a medical center. Three groups participated in the study: a group of orderlies from the dispatch department, a group of ICU nurses, and a group of laboratory workers. Overall, the team members for the dispatch and the laboratory groups were satisfied with the participatory ergonomics process, and these perceptions seem to improve over time. However, the ICU team members expressed more negative perceptions. The problems encountered by the ICU team seem to be related to the lack of time and the time pressures due to the clinical demands. A more in-depth evaluation of the participatory ergonomics program on orderlies showed substantial improvements in health and safety following the implementation of the participatory ergonomics program (Evanoff et al. 1999). The studies by Evanoff and colleagues demonstrate the feasibility of implementing participatory ergonomics in health care, but highlight the difficulty of the approach in a high-stress high-pressure environment, such as an ICU, where patient needs are critical and patients need immediate or continuous attention. More research is needed in order to develop ergonomic methods for implementing work system changes that lead to improvements in human and organizational outcomes, as well as improved quality and safety of care. This research should consider the high-pace, high-pressure work environment in healthcare organizations.

5.3.3 Continuous Work System Design

The implementation of any new work system always engenders problems and concerns. The process by which these problems and concerns are resolved is important from a human factors point of view, but also from a quality of care and patient safety point of view. It is necessary to have the capability and tools to identify potential HFE problems, and quality and safety of care problems in a timely manner.

How does one ensure that continuous improvements in work system design and important outcomes (quality of care and patient safety, as well as human and organizational outcomes) are achieved? Various models and approaches to quality improvement and management have been proposed and implemented in health care (e.g., Shortell et al. 1995). This research would benefit from an HFE point of view in order to simultaneously optimize work system design and improve quality and safety of care. Ergonomic approaches to quality management and improvement have emphasized the importance of job and organizational design and quality of working life (Carayon et al. 1999), the link between ergonomic deficiencies and quality deficiencies (Eklund 1995; Axelsson 2000), and the importance of management approaches for improved safety and health (Zink 2000). All of these ergonomic approaches have much to offer in order to design continuous improvement systems and processes in health care. The goal of the improvement systems and processes would be to improve human and organizational outcomes, and as well as quality of care and patient safety; this is a key element of the SEIPS model of work system and patient safety (Carayon et al. 2006).

5.3.4 Work System Analysis

The analysis of work systems relies on a range of data collection and analysis methods that are typically used in HFE (Wilson and Corlett 2005). Observations can provide detailed information about tasks as they are performed in the real healthcare environment. Interviews or focus groups can complement observations by providing additional information that may not be readily available to the observers. A combination of observation and interviews was used in a study of nurses' interaction with BCMA technology (Carayon et al. 2007). As described earlier, this analysis identified a range of work system factors that affected nurses' ability to administer medications using the BCMA technology.

Data collected on the work system using observations, interviews, focus groups, or other methods (e.g.,

questionnaire, review of documentation) can be analyzed to identify problems, performance obstacles, or vulnerabilities. For instance, proactive risk assessment methods use work system data as input to identify vulnerabilities in care processes, for example, the nursing medication management process (Faye et al. 2010).

The combination of quantitative and qualitative data is also important for a work system analysis. For instance, interviews were used to gather in-depth information about performance obstacles perceived by 15 ICU nurses (Gurses and Carayon 2009); these qualitative data were complemented by quantitative data collected via a survey of 272 ICU nurses (Gurses and Carayon 2007). Some of the work system barriers identified by ICU nurses during the interviews (e.g., problem related to patient transport) were actually experienced by very few nurses. A quantitative modeling of the impact of work system obstacles allowed the identification of a small number of work system factors as key predictors of perceived workload and perceived safety and quality of care (Gurses et al. 2009). The combination of quantitative and qualitative data can therefore enhance the quality of the work system analysis.

5.4 Conclusion

Describing the work system, its elements, and their interactions is only the first step in work design (see Table 5.3 for a set of questions to ask in order to describe the work system). We also need to evaluate the work system and its elements, that is, to define the positive and negative aspects of the work system elements. The evaluation of the positive and negative aspects relies on knowledge

TABLE 5.3

Questions to Ask for Describing the Work System

1. What are the characteristics of the *individual* performing the work? Does the individual have the musculoskeletal, sensory, and cognitive abilities to do the required task? If not, can any of these be accommodated for the task?
2. What *tasks* are being performed and what are characteristics of the tasks that may contribute to unsafe patient care? What in the nature of the tasks allows the individual to perform them safely or assume risks in the process?
3. What in the *physical environment* can be sources of error or promotes safety? What in the physical environment insures safe behavior or leaves room for unsafe behavior?
4. What *tools and technologies* are being used to perform the tasks and do they increase or decrease the likelihood of untoward events?
5. What in the *organization* prevents or allows exposure to hazard? What in the organization promotes or hinders patient safety? What allows for assuming safe or unsafe behavior by the individual?

in the fields of ergonomics, job stress, and job/organizational design (Carayon et al. 2007). These three fields of research have developed a number of theories, models, and principles that characterize "good" work (Carayon et al. 2007). This information is important for not only evaluating the work system, but also designing solutions for improving the work system and subsequent outcomes, such as performance and patient safety. Once the positive and negative aspects of the work system and its elements have been defined, we need to come up with solutions that will improve the work design and increase patient safety.

Solutions for improving work design and patient safety may involve the elimination of negative aspects of the work system. As an example, proper healthcare provider hand hygiene is the corner stone of any infection control effort. However, hand hygiene is performed less than half of the time it is required for patient care. Healthcare providers state negative aspects for this lack of hand hygiene, such as handwashing agents causing irritation and dryness, sinks inconveniently located or lack of sinks, lack of soap and paper towels, and too busy and insufficient time. Since eliminating these negative aspects may not be feasible, solutions for improving work design may involve other elements of the work system that compensate for or balance out the negative aspects. Placement of waterless alcohol-based hand hygiene agents in patient care areas or personal bottles of the agent for each healthcare provider eliminates the need for sinks, soap, and paper towels. The waterless alcohol hand rub is fast acting and can be used while performing other tasks such as communicating with patients, family members, and professional colleagues. This compensatory effect is the essence of the Balance Theory proposed by Carayon and Smith (Smith and Carayon-Sainfort 1989; Carayon and Smith 2000). In many situations, it may not be possible to eliminate negative aspects of the work system because of constraints due to the physical design, due to the infrastructure, due to cost, due to recruitment constraints, etc... In those situations, it is important to come up with work design solutions that take into account those negative aspects and try to minimize their impact on workers and patient safety.

The five elements of work (i.e., the individual, tasks, tools and technologies, physical environment, and organizational conditions) represent a system: they influence each other, and interact with each other. A change in any one element of the work system can have effects on other elements. This system concept has a number of consequences for work (re)design:

- We cannot look at one work element in isolation.
- Whenever there is a change in work, we need to consider the effects on the entire work system.

- The activity of work (re)design necessitates knowledge and expertise in a variety of domains, for example, environmental design such as lighting and noise, job design such as autonomy and work demands, and organizational design such as safety culture, teamwork, and communication.
- Work redesign aims at removing the "negative" aspects of work, i.e., the aspects contributing to poor performance and unsafe patient care. When this is not feasible, work redesign involves building or relying on other elements of work to compensate for or balance out the negative aspects of work.

References

Agency for Healthcare Research and Quality. 2002. *Impact of Working Conditions on Patient Safety*. Rockville, MD: Agency for Healthcare Research and Quality.

Axelsson, J.R.C. 2000. Quality and ergonomics—Towards successful integration. PhD dissertation, Department of Mechanical Engineering, Linkoping University, Linkoping, Sweden.

Baker, G.R., P.G. Norton, V. Flintoft, R. Blais, A. Brown, J. Cox, E. Etchells, W.A. Ghali, P. Hebert, S.R. Majumdar, M. O'Beirne, L. Palacios-Derflingher, R.J. Reid, S. Sheps, and R. Tamblyn. 2004. The Canadian adverse events study: The incidence of adverse events among hospital patients in Canada. *Journal of the Canadian Medical Association* 170 (11):1678–1686.

Berwick, D.M. 2002. A user's manual for the IOM's 'Quality Chasm' report. *Health Affairs* 21 (3):80–90.

Bohr, P.C., B.A. Evanoff, and L. Wolf. 1997. Implementing participatory ergonomics teams among health care workers. *American Journal of Industrial Medicine* 32 (3):190–196.

Boston-Fleischhauer, C. 2008a. Enhancing healthcare process design with human factors engineering and reliability science, Part 1: Setting the context. *Journal of Nursing Administration* 38 (1):27–32.

Boston-Fleischhauer, C. 2008b. Enhancing healthcare process design with human factors engineering and reliability science, Part 2: Applying the knowledge to clinical documentation systems. *Journal of Nursing Administration* 38 (2):84–89.

Brown, K.Q. and T.R. Mitchell. 1991. A comparison of just-in-time and batch manufacturing: The role of performance obstacles. *Academy of Management Journal* 34 (4):906–917.

Buckle, P., P.J. Clarkson, R. Coleman, R. Lane, D. Stubbs, J. Ward, J. Jarrett, and J. Bound. 2003. *Design for Patient Safety*. London, U.K.: Department of Health Publications.

Carayon, P. 2005. Top management's view on human factors and patient safety: Do they see it? In *Healthcare Systems Ergonomics and Patient Safety*, eds. R. Tartaglia, S. Bagnara, T. Bellandi, and S. Albolino. Florence, Italy: Taylor & Francis.

Carayon, P. 2009. The balance theory and the work system model… Twenty years later. *International Journal of Human-Computer Interaction* 25 (5):313–327.

Carayon, P. 2010. Human factors in patient safety as an innovation. *Applied Ergonomics* 41 (5):657–665.

Carayon, P., C.J. Alvarado, P. Brennan, A. Gurses, A. Hundt, B.T. Karsh, and M. Smith. 2003a. Work system and patient safety. In *Human Factors in Organizational Design and Management-VII*, eds. H. Luczak and K.J. Zink. Santa Monica, CA: IEA Press.

Carayon, P., C. Alvarado, Y. Hsieh, and A.S. Hundt. 2003. A macroergonomic approach to patient process analysis: Application in outpatient surgery. Paper read at *XVth Triennial Congress of the International Ergonomics Association*, August 24–29, 2003, Seoul, Korea.

Carayon, P., C. Alvarado, and A.S. Hundt. 2003b. *Reducing Workload and Increasing Patient Safety through Work and Workspace Design*. Washington, DC: Institute of Medicine.

Carayon, P., C.J. Alvarado, and A.S. Hundt. 2007. Work system design in healthcare. In *Handbook of Human Factors and Ergonomics in Healthcare and Patient Safety*, ed. P. Carayon. Mahwah, NJ: Lawrence Erlbaum Associates.

Carayon, P., C.J. Alvarado, A.S. Hundt, S. Springman, and P. Ayoub. 2006. Patient safety in outpatient surgery: The viewpoint of the healthcare providers. *Ergonomics* 49 (5–6):470–485.

Carayon, P., H. Faye, A.S. Hundt, B.T. Karsh, and T. Wetterneck. 2009. Patient safety and proactive risk assessment. In *Handbook of Healthcare Delivery Systems*, ed. Y. Yuehwern. London, U.K.: Taylor & Francis.

Carayon, P., A.P. Gurses, A.S. Hundt, P. Ayoub, and C.J. Alvarado. 2005. Performance obstacles and facilitators of healthcare providers. In *Change and Quality in Human Service Work*, eds. C. Korunka and P. Hoffmann. Munchen, Germany: Hampp Publishers.

Carayon, P., M.C. Haims, and C.L. Yang. 2001. Psychosocial work factors and work organization. In *The International Encyclopedia of Ergonomics and Human Factors*, eds. W. Karwowski. London, U.K.: Taylor & Francis.

Carayon, P., A.S. Hundt, B.-T. Karsh, A.P. Gurses, C.J. Alvarado, M. Smith, and P.F. Brennan. 2006. Work system design for patient safety: The SEIPS model. *Quality & Safety in Health Care* 15 (Supplement I):i50–i58.

Carayon, P. and B. Karsh. 2000. Sociotechnical issues in the implementation of imaging technology. *Behaviour and Information Technology* 19 (4):247–262.

Carayon, P., B.T. Karsh, R. Cartmill, P. Hoonakker, A.S. Hundt, D. Krueger, T. Thuemling, and T. Wetterneck. 2010. *Incorporating Health Information Technology into Workflow Redesign-Summary Report*. Rockville, MD: Agency for Healthcare Research and Quality.

Carayon, P., F. Sainfort, and M.J. Smith. 1999. Macroergonomics and total quality management: How to improve quality of working life? *International Journal of Occupational Safety and Ergonomics* 5 (2):303–334.

Carayon, P., K. Schultz, and A.S. Hundt. 2004. Righting wrong site surgery. *Joint Commission Journal on Quality and Safety* 30 (7):405–410.

Carayon, P. and M.J. Smith. 2000. Work organization and ergonomics. *Applied Ergonomics* 31:649–662.

Carayon, P., T.B. Wetterneck, A.S. Hundt, M. Ozkaynak, J. DeSilvey, B. Ludwig, P. Ram, and S.S. Rough. 2007. Evaluation of nurse interaction with bar code medication administration technology in the work environment. *Journal of Patient Safety* 3 (1):34–42.

Catchpole, K. and P. McCulloch. 2010. Human factors in critical care: Towards standardized integrated human-centred systems of work. *Current Opinion in Critical Care* 16 (6):618–622.

Clegg, C. 1988. Appropriate technology for manufacturing: Some management issues. *Applied Ergonomics* 19 (1):25–34.

Cohen-Mansfield, J. 1997. Turnover among nursing home staff. *Nursing Management* 28 (5):59–64.

Dekker, S.W.A. and J.M. Nyce. 2004. How can ergonomics influence design? Moving from research findings to future systems. *Ergonomics* 47 (15):1624–1639.

Donabedian, A. 1978. The quality of medical care. *Science* 200:856–864.

Donabedian, A. 1980. *The Definition of Quality and Approaches to its Assessment*. Ann Arbor, MI: Health Administration Press.

Donabedian, A. 1988. The quality of care. How can it be assessed? *Journal of the American Medical Association* 260 (12):1743–1748.

Eason, K. 1988. *Information Technology and Organizational Change*. London, U.K.: Taylor & Francis.

Eklund, J.A.E. 1995. Relationships between ergonomics and quality in assembly work. *Applied Ergonomics* 26 (1):15–20.

Evanoff, V.A., P.C. Bohr, and L. Wolf. 1999. Effects of a participatory ergonomics team among hospital orderlies. *American Journal of Industrial Medicine* 35:358–365.

Faye, H., A.J. Rivera-Rodriguez, B.T. Karsh, A.S. Hundt, C. Baker, and P. Carayon. 2010. Involving intensive care unit nurses in a proactive risk assessment of the medication management process. *The Joint Commission Journal on Quality and Patient Safety* 36 (8):376–384.

Gurses, A.P. 2005. *ICU Nursing Workload: Causes and Consequences—Final Report*. Rockville, MD: Agency for Healthcare Research and Quality.

Gurses, A.P. and P. Carayon. 2007. Performance obstacles of intensive care nurses. *Nursing Research* 56 (3):185–194.

Gurses, A. and P. Carayon. 2009. A qualitative study of performance obstacles and facilitators among ICU nurses. *Applied Ergonomics* 40 (3):509–518.

Gurses, A., P. Carayon, and M. Wall. 2009. Impact of performance obstacles on intensive care nurses workload, perceived quality and safety of care, and quality of working life. *Health Services Research* 44 (2 Pt 1):422–443.

Hackman, J.R. 1969. Toward understanding the role of tasks in behavioral research. *Acta Psychologica* 31:97–128.

Haims, M.C. and P. Carayon. 1996. Implementation of an 'in-house' participatory ergonomics program: A case study in a public service organization. In *Human Factors in Organizational Design and Management*, eds. O.J. Brown and H.W. Hendrick. Amsterdam, the Netherlands: Elsevier.

Haines, H., J.R. Wilson, P. Vink, and E. Koningsveld. 2002. Validating a framework for participatory ergonomics (the PEF). *Ergonomics* 45 (4):309–327.

Haynes, A.B., T.G. Weiser, W.R. Berry, S.R. Lipsitz, A.H. Breizat, E.P. Dellinger, T. Herbosa, S. Joseph, P.L. Kibatala, M.C. Lapitan, A.F. Merry, K. Moorthy, R.K. Reznick, B. Taylor, and A.A. Gawande. 2009. A surgical safety checklist to reduce morbidity and mortality in a global population. *New England Journal of Medicine* 360 (5):491–499.

Hendrickson, G., T.M. Doddato, and C.T. Kovner. 1990. How do nurses use their time? *Journal of Nursing Administration* 20 (3):31–37.

Henriksen, K., A. Joseph, and T. Zayas-Caban. 2009. The human factors of home health care: A conceptual model for examining safety and quality concerns. *Journal of Patient Safety* 5 (4):229–236.

Hysong, S., M. Sawhney, L. Wilson, D. Sittig, A. Esquivel, M. Watford, T. Davis, D. Espadas, and H. Singh. 2009. Improving outpatient safety through effective electronic communication: A study protocol. *Implementation Science* 4 (1):62.

Institute of Medicine Committee on the Work Environment for Nurses and Patient Safety. 2004. *Keeping Patients Safe: Transforming the Work Environment of Nurses*. Washington, DC: The National Academies Press.

Institute of Medicine Committee on Quality of Health Care in America. 2001. *Crossing the Quality Chasm: A New Health System for the 21st Century*. Washington, DC: National Academy Press.

Johnson, G.I. and J.K. Wilson. 1988. Future directions and research issues for ergonomics and advanced manufacturing technology (AMT). *Applied Ergonomics* 19 (1):3–8.

Kirwan, B. and L.K. Ainsworth. 1992. *A Guide to Task Analysis*. London, U.K.: Taylor & Francis.

Kohn, L.T., J.M. Corrigan, and M.S. Donaldson (eds.). 1999. *To Err is Human: Building a Safer Health System*. Washington, DC: National Academy Press.

Koppel, R., J.P. Metlay, A. Cohen, B. Abaluck, A.R. Localio, S.E. Kimmel, and B.L. Strom. 2005. Role of computerized physician order entry systems in facilitating medications errors. *Journal of the American Medical Association* 293 (10):1197–1203.

Koppel, R., T. Wetterneck, J.L. Telles, and B.T. Karsh. 2008. Workarounds to barcode medication administration systems: Their occurrences, causes, and threats to patient safety. *Journal of the American Medical Informatics Association* 15:408–423.

Korunka, C., A. Weiss, et al. (1993). Effects of new technologies with special regard for the implementation process per se. *Journal of Organizational Behavior* 14(4): 331–348.

Kovner, C. 2001. The impact of staffing and the organization of work on patient outcomes and health care workers in health care organizations. *The Joint Commission Journal on Quality Improvement* 27 (9):458–468.

Landrigan, C.P., J.M. Rothschild, J.W. Cronin, R. Kaushal, E. Burdick, J.T. Katz, C.M. Lilly, P.H. Stone, S.W. Lockley, D.W. Bates, and C.A. Czeisler. 2004. Effect of reducing interns' work hours on serious medical errors in intensive care units. *New England Journal of Medicine* 351 (18):1838–1848.

Leape, L.L. 1997. A systems analysis approach to medical error. *Journal of Evaluation in Clinical Practice* 3 (3):213–222.

Leplat, J. 1989. Error analysis, instrument and object of task analysis. *Ergonomics* 32 (7):813–822.

Linzer, M., T.R. Konrad, J. Douglas, J.E. McMurray, D.E. Pathman, E.S. Williams, M.D. Schwartz, M. Gerrity, W. Scheckler, J.A. Bigby, and E. Rhodes. 2000. Managed care, time pressure, and physician job satisfaction: Results from the physician worklife study. *Journal of General Internal Medicine* 15 (7):441–450.

Lockley, S.W., J.W. Cronin, E.E. Evans, B.E. Cade, C.J. Lee, C.P. Landrigan, J.M. Rothschild, J.T. Katz, C.M. Lilly, P.H. Stone, D. Aeschbach, and C.A. Czeisler. 2004. Effect of reducing interns' weekly work hours on sleep and attentional failures. *New England Journal of Medicine* 351 (18):1829–1837.

Luczak, H. 1995. Macroergonomic anticipatory evaluation of work organization in production systems. *Ergonomics* 38 (8):1571–1599.

McMurray, J.E., M. Linzer, J. Douglas, and T.R. Konrad. 1999. Burnout in US women physicians: Assessing remediable factors in worklife. Paper read at *22nd Annual Meeting of the Society of General Internal Medicine*, April 29–May 1, 1999, San Francisco, CA.

Nielsen, J. 1993. *Usability Engineering*. Amsterdam, the Netherlands: Morgan Kaufmann.

Noro, K. and A. Imada. 1991. *Participatory Ergonomics*. London, U.K.: Taylor & Francis.

Or, C.K.L., R.S. Valdez, G.R. Casper, P. Carayon, L.J. Burke, P.F. Brennan, and B.T. Karsh. 2009. Human factors and ergonomics in home care: Current concerns and future considerations for health information technology. *Work—A Journal of Prevention Assessment & Rehabilitation* 33 (2):201–209.

Park, K.S. and S.W. Han. 2002. Performance obstacles in cellular manufacturing implementation—Empirical investigation. *International Journal of Human Factors and Ergonomics in Manufacturing* 12 (1):17–29.

Parker, S. and T. Wall. 1998. *Job and Work Design*. Thousand Oaks, CA: Sage Publications.

Parsons, K.C. 2000. Environmental ergonomics: A review of principles, methods and models. *Applied Ergonomics* 31:581–594.

Patterson, E.S., R.I. Cook, and M.L. Render. 2002. Improving patient safety by identifying side effects from introducing bar coding in medical administration. *Journal of the American Medical Informatics Association* 9 (5):540–533.

Peters, L.H., M.B. Chassie, H.R. Lindholm, E.J. O'Connor, and C.R. Kline. 1982. The joint influence of situational constraints and goal setting on performance and affective outcomes. *Journal of Management* 8:7–20.

Peters, L.H. and E.J. O'Connor. 1980. Situational constraints and work outcomes: The influences of a frequently overlooked construct. *Academy of Management Review* 5 (3):391–397.

Rasmussen, J. 1982. Human errors. A taxonomy for describing human malfunction in industrial installations. *Journal of Occupational Accidents* 4:311–333.

Reason, J. 1990. *Human Error*. Cambridge U.K.: Cambridge University Press.

Rogers, A.E., W.T. Hwang, L.D. Scott, L.H. Aiken, and D.F. Dinges. 2004. The working hours of hospital staff nurses and patient safety. *Health Affairs* 23 (4):202–212.

Rothschild, J.M., C.P. Landrigan, J.W. Cronin, R. Kaushal, S.W. Lockley, E. Burdick, P.H. Stone, C.M. Lilly, J.T. Katz, C.A. Czeisler, and D. W. Bates. 2005. The Critical Care Safety Study: The incidence and nature of adverse events and serious medical errors in intensive care. *Critical Care Medicine* 33:1694–1700.

Sainfort, F., B. Karsh, B.C. Booske, and M.J. Smith. 2001. Applying quality improvement principles to achieve healthy work organizations. *Journal on Quality Improvement* 27 (9):469–483.

Salvendy, G. (ed.). 2006. *Handbook of Human Factors and Ergonomics*, 3rd edn. New York: John Wiley & Sons.

Shortell, S.M., J.L. O'Brien, J.M. Carman, R.W. Foster, E.F.X. Hughes, H. Boerstler, and E.J. O'Connor. 1995. Assessing the impact of continuous quality improvement/total quality management: Concept versus implementation. *Health Services Research* 30 (2):377–401.

Shortell, S.M. and S.J. Singer. 2008. Improving patient safety by taking systems seriously. *Journal of the American Medical Association* 299 (4):445–447.

Sittig, D. F. and H. Singh. 2010. A new sociotechnical model for studying health information technology in complex adaptive healthcare systems. *Quality & Safety in Health Care* 19 (3):i68–i74.

Slappendel, C. 1994. Ergonomics capability in product design and development: An organizational analysis. *Applied Ergonomics* 25 (5):266–274.

Smith, M.J. and P. Carayon. 1995. New technology, automation, and work organization: Stress problems and improved technology implementation strategies. *The International Journal of Human Factors in Manufacturing* 5 (1):99–116.

Smith, M.J. and P. Carayon. 2000. Balance theory of job design. In *International Encyclopedia of Ergonomics and Human Factors*, ed. W. Karwowski. London, U.K.: Taylor & Francis.

Smith, M.J. and P. Carayon-Sainfort. 1989. A balance theory of job design for stress reduction. *International Journal of Industrial Ergonomics* 4 (1):67–79.

Smith, M.J., M.J. Colligan, I.J. Frockt, and D.L. Tasto. 1979. Occupational injury rates among nurses as a function of shift schedule. *Journal of Safety Research* 11 (4):181–187.

Stubbs, D.A., P. Buckle, M.P. Hudson, P.M. Rivers, and C.J. Worringham. 1983. Back pain in the nursing profession. I. Epidemiology and pilot methodology. *Ergonomics* 26:755–765.

Tannenbaum, S.I., E. Salas, et al. (1996). Promoting team effectiveness. *Handbook of Work Group Psychology*. M.A. West. New York, John Wiley: 503–530.

Trinkoff, A.M., M. Johantgen, C.L. Storr, A.P. Gurses, Y. Liang, and H. Kihye. 2011. Nurses' work schedule characteristics, nurse staffing, and patient mortality. *Nursing Research* 60 (1):1–8.

Ulmer, C., D.W. Wolman, and M.E. Johns (eds.). 2008. *Resident Duty Hours: Enhancing Sleep, Supervision, and Safety*. Committee on Optimizing Graduate Medical Trainee (Resident) Hours and Work Schedule to Improve Patient Safety. Washington, DC: The National Academies Press.

van den Bemt, P.M., R. Fijn, P.H. van der Voort, A.A. Gossen, T.C. Egberts, and J.R. Brouwers. 2002. Frequency and determinants of drug administration errors in the intensive care unit. *Critical Care Medicine* 30 (4):846–850.

Vincent, C. 2010. *Patient Safety*. Chichester, U.K.: Wiley-Blackwell.

Vincent, C., P. Aylin, B.D. Franklin, A. Holmes, S. Iskander, A. Jacklin, and K. Moorthy. 2008. Is health care getting safer? *British Medical Journal* 337 (7680):1205–1207.

Vincent, C., K. Moorthy, S.K. Sarker, A. Chang, and A.W. Darzi. 2004. Systems approaches to surgical quality and safety—From concept to measurement. *Annals of Surgery* 239 (4):475–482.

Vincent, C., S. Taylor-Adams, and N. Stanhope. 1998. Framework for analysing risk and safety in clinical medicine. *BMJ* 316 (7138):1154–1157.

Walker, J. and P. Carayon. 2009. From tasks to processes: The case for changing health information technology to improve health care. *Health Affairs* 28 (2):467.

Wetterneck, T.B., M. Linzer, J.E. McMurray, J. Douglas, M.D. Schwartz, J.A. Bigby et al. 2002. Worklife and satisfaction of general internists. *Archives of Internal Medicine* 169:649–656.

Wetterneck, T.B., K.A. Skibinski, T.L. Roberts, S.M. Kleppin, M. Schroeder, M. Enloe, S.S. Rough, A.S. Hundt, and P. Carayon. 2006. Using failure mode and effects analysis to plan implementation of Smart intravenous pump technology. *American Journal of Health-System Pharmacy* 63:1528–1538.

Wiegmann, D.A., A.A. Eggman, A.W. ElBardissi, S.E. Henrickson, and T.M. III Sundt. 2009. Improving cardiac surgical care: A work systems approach. *Applied Ergonomics* to be published.

Wiegmann, D.A., A.A. Eggman, et al. (2010). Improving cardiac surgical care: A work system approach. *Applied Ergonomics* 41(5): 701–712.

Wilson, J.R. 1995. Ergonomics and participation. In *Evaluation of Human Work*, eds. J.R. Wilson and E.N. Corlett. London, U.K.: Taylor & Francis.

Wilson, J.R. 2000. Fundamentals of ergonomics in theory and practice. *Applied Ergonomics* 31 (6):557–567.

Wilson, J.R. and N. Corlett (eds.). 2005. *Evaluation of Human Work*, 3rd edn. Boca Raton, FL: CRC Press.

Wilson, J.R. and H.M. Haines. 1997. Participatory ergonomics. In *Handbook of Human Factors and Ergonomics*, ed. G. Salvendy. New York: John Wiley & Sons.

Wong, D.H., Y. Gallegos, M.B. Weinger, S. Clack, J. Slagle, and C.T. Anderson. 2003. Changes in intensive care unit nurse task activity after installation of a third-generation intensive care unit information system. *Critical Care Medicine* 31 (10):2488–2494.

Wunderlich, G.S. and P.O. Kohler (eds.). 2001. *Improving the Quality of Long-Term Care*. Washington, DC: National Academy Press.

Zink, K.J. 1996. Continuous improvement through employee participation: Some experiences from a long-term study. In *Human Factors in Organizational Design and Management*, eds. O. Brown, Jr. and H.W. Hendrick. Amsterdam, the Netherlands: Elsevier.

Zink, K. 2000. Ergonomics in the past and the future: From a German perspective to an international one. *Ergonomics* 43 (7):920–930.

6

Error as Behavior and Murphy's Law: Implications for Human Factors and Ergonomics

Marilyn Sue Bogner

CONTENTS

6.1 Introduction 82
    6.1.1 Error Definitions 82
    6.1.2 Information from Health Care Error Definitions 82
6.2 Problem 83
    6.2.1 Stop Rule 83
    6.2.2 Where Does the Cause of Error Belong? 83
6.3 Nature of Error 84
    6.3.1 James Bond Syndrome 84
6.4 Artichoke Systems Approach 85
6.5 Error, the Artichoke, and Murphy's Law 87
6.6 Expansion of Focus in Product Design 87
    6.6.1 Medical Devices Are Consumer Products 87
    6.6.2 Proactive Error Prevention 88
    6.6.3 Hard-Wired Human Characteristics 88
    6.6.4 Range of Anticipated Contexts of Use 88
6.7 Care Provider, Product, and Conditions of Use Misfits 88
    6.7.1 Monitor Alarms 89
    6.7.2 Snooze Alarm 89
    6.7.3 Infusion Pumps 89
    6.7.4 Artichoke Approach to Error Prevention 90
    6.7.5 Artichoke Orientation of Observation 90
    6.7.6 Error Prevention through Product Design 90
6.8 Artichoke as a Lens 90
    6.8.1 Reactive Approach to Design to Prevent Errors 91
6.9 Product Design 91
    6.9.1 Excessive Therapeutic Radiation Incidents through the Decades 91
    6.9.2 Device Testing 93
    6.9.3 Simulation of Real-World Health Care 93
6.10 Designing for the Real-World through the Lens of the Artichoke 93
References 94

Reducing the incidence of error in health care is illusive. Despite over a decade of concerted efforts focused on the health care provider as the source of error, there has been no discernable reduction in errors except for those situations in which the incidence of error in a given facility or a given unit of a facility is reduced as the result of Herculean efforts by one individual. When that individual leaves the reduction rarely if ever persists. Because targeting the care provider as the sole source of error has not been effective, it is time to try another approach. Human error is a real-world issue; so reducing the incidence of error requires a practical real-world approach—an approach that considers the seemingly unfailing operation of Murphy's law that whatever can go wrong will go wrong. Such a practical approach however cannot be launched without understanding the nature of the problem—the nature of human error.

The term "human error" strikes fear in the hearts of people as well as being a ready explanation for an action with adverse consequences. The phrase "to err is human" has become an accepted although unsubstantiated human characteristic; that phrase is the first part of a literary quote "To err is human; to forgive, divine" from Alexander Pope's 1711 *Essay on Criticism* (1711/2004). Examples abound that committing errors is not a uniquely human trait; herding dogs, most particularly Australian sheepdogs, herd people and cats as well as sheep, thus committing an error. More poignantly many varieties of flies lay eggs on carrion so the larvae have food; certain plants emit carrion odor which attracts the flies. The flies lay their eggs on the odor-emitting plant, which does not provide their larvae with the food that carrion would; so as a result of the fly's error, the larvae die (Mach, 1905).

The omnipresent presumption of a human trait to commit errors not only connotes blame, which jeopardizes the respect of a person associated with an error in the professional community as well as in their own self-concept; but it also restricts the scope of consideration of possible causes of error, which impacts reducing the likelihood of their occurrence. This discussion provides an understanding of what human error is and how that understanding can be incorporated into product design to prevent errors and decrease the likelihood of the operation of Murphy's law and the often resultant worse case scenarios.

This is accomplished by presenting typical definitions of error that actually are descriptions of errors which are not useful in preventing error. A discussion follows that defies the unsupported presumption that to err is human by describing error as what it actually is—behavior—which provides insights into the nature of error. These insights are incorporated into an evidence-based approach that addresses the role of contextual factors in error—an approach that removes the person as sole focus as cause of error hence the target for blame. This enables care providers to apply their unique knowledge of error inducing factors to enhance safety with noteworthy implications for the practice of human factors and ergonomics.

6.1 Introduction

Human error is a universal issue that occurs throughout all aspects of life and across industries. Health care error arguably is the most onerous because everyone experiences health care, hence the possibility of dire, often irreversible consequences. The latter point was brought home by the Institute of Medicine (IOM) of the U.S. National Academy of Sciences report *To Err Is Human* (Kohn et al. 1999). That report stated 44,000–98,000 hospitalized people die annually due to medical error. This

is more than all the annual deaths from highway accidents suggesting that receiving health care is a high-risk activity. Those data caused such a hue and cry from constituencies that Congress appropriated money to fund research to support the report's recommendation that programs be developed to hold health care providers accountable for their errors. The goal of that funding was to reduce the incidence of errors by 50% in 5 years. The vast majority of that research resulted in computerized programs for health care providers to report their errors.

At the end of the 5 years, in 2004, it was found that the products of the research not only did not meet the goal of a 50% reduction in errors, but also the impact of those efforts on error was negligible (Commonwealth Fund, 2004). More cogently expressed "... little data exist showing progress and researchers are still debating not how to save lives, but what to measure" (Zwillich, 2004). Despite over a decade of ongoing efforts to reduce the incidence of error targeted at the care provider, there has been no appreciable progress. The lack of progress in addressing this serious social issue suggests that there is merit in considering what is being addressed as human error.

6.1.1 Error Definitions

The majority of health care error definitions relate to specific aspects of health care: technical errors reflecting skill failures, or judgmental errors that involve the selection of an incorrect strategy of treatment. Normative errors occur when the social values embedded within medicine as a profession are violated. Errors are defined in terms of the stage in the process of health care in which they occur: errors of missed diagnosis, mistakes during treatment, medication mistakes, inadequate postoperative care, and mistaken identity. Also errors are defined in terms of a process per se: failure of a planned action to be completed as intended (i.e., error of execution) or the use of a wrong plan to achieve an aim (i.e., error of planning).

In addition health care errors are defined in terms used across industries such as being skill-based, rule-based, and knowledge-based. Errors also are defined according to the stage of cognitive activity at which they occur: mistakes, lapses, and slips happen at the planning, storage, and execution stages respectively. In addition errors are defined in terms of the degree of activity involved as in latent errors and active errors.

6.1.2 Information from Health Care Error Definitions

Reports of health care errors based on the aforementioned definitions provide data for trending—a stated desire in the 1999 IOM report—and for the colorful bar graphs and pie charts displayed on computer screens as indicating the usefulness of software presented by vendors at professional meetings. What of the usefulness the data from those definitions have in addressing the ultimate goal of preventing errors in the real world? Deming (1986), an American pioneer in quality who had considerable influence in industry, pointed out that figures (statistics) on accidents do nothing to reduce the frequency of accidents.

6.2 Problem

A commonality across the definitions of error as well as the consideration of error in the IOM report is that the source of the error is the person, the health care provider. It is unclear how those definitions are useful in preventing errors. Identifying an error as reflecting skill failures points to a problem with the person associated with the error. From this the only apparent approach to correct the problem is remedial skill training, but there is no indication of what skill needs remediation; so the only avenue to pursue would be retraining all skills involved in the task—hardly a cost-effective approach. Similarly identifying an error as a slip provides no information about how to address that error, no target for actions except to hold the health care provider accountable—and what does that actually mean with respect to a slip? Further analysis finds that these definitions do not explicate what occurs in an error so that actions might be taken to prevent it; rather they are descriptions of what occurred and provide no indication of a target for change—change that must occur to prevent recurrence of the error and possible adverse health care outcomes.

6.2.1 Stop Rule

The insistent attribution of the cause of error to the person associated with it can be explained in terms of a seemingly human need to find a reason why something happened, in this discussion, an error. People engage in a subjectively pragmatic cognitive search to identify an acceptable cause for an occurrence and conjecture a reasonable explanation—an explanation that can be addressed. This is known as the Stop Rule because people employing that cognitive process in causal explanations stop when a conjectured cause is identified and terminate the search without considering other possible causes (Rasmussen, 1990). The care provider associated with an error is a ready explanation for the error and illustrates the Stop Rule in action.

The potency of this Rule is apparent in the persistence of the interpretation that when a care provider is associated with an error it is caused by that person often attributing carelessness or direct defiance of rules. Following that, stern responses to error by health care providers have been suggested. One such approach is that physicians involved in an error who do not respond to initial warnings and counseling, receive punishment of short-term denial of hospital privileges with possible "... stronger auditing methods (for compliance with patient safety practices) such as video surveillance, computerized triggers, and unannounced, secret monitoring of compliance by hospital personnel" (Wachter and Pronovost, 2009). These proposed draconian activities are yet another attempt to hold health care providers accountable for their errors—attempts that have not met with success in reducing errors.

Not only have approaches directed toward the care provider as the sole cause of error not been successful, there is a lack of any promise of their success. Continuing an approach that has not worked is analogous to repeating an action on a computer that does not do what one intends. If it does not work, repeating it will not make it do so. Similarly if an approach is not productive, more applications of it, even aggressive applications as in strongly striking a computer key, will not make it work. Persisting with an approach that has not worked not only is not productive but in health care lives are also affected. It is past time to stop doing what does not work and change the approach.

6.2.2 Where Does the Cause of Error Belong?

Deming states that there are two kinds of mistakes: one in which the cause is assigned to a person and one of a special common cause in which the cause belongs to the system—"The split is possibly 99% system, 1% from carelessness" (1986, p. 479). He continues that the first step in reducing accidents is to determine if the cause is a person or a set of conditions of the system. He observes that the cause usually is attributed to the person and this leads to an ineffective solution; accidents from a common cause will continue until the accidents are understood and the set of conditions in the context of the accident that induce error are corrected. Applied to health care this provides a critical message for addressing errors—that accidents/errors must be understood and factors in the environmental context of care corrected to reduce the incidence of and prevent errors.

The definitions of health care errors presented earlier in this discussion all consider the care provider as the source of the error, which Deming notes leads to an ineffective solution. In addition, as noted previously, those definitions are descriptions rather than definitions. The actual definition of error explicates the nature of error—what occurs in error regardless of where it occurs.

6.3 Nature of Error

For there to be an error, an act must occur; for there to be a change from one state to another, a behavior must occur. This may seem to be stating the obvious because the term behavior often is used in discussing error; however behavior in the actual meaning of the word typically is not addressed in either error analysis or product design to avoid errors. In those activities the emphasis is on the person and the manifestation of the error rather than what actually constitutes an error—the nature of error.

Insight into the nature of error is obtained from the dictionary definition of behavior as "One's reactions under specified circumstances" (Webster, 1994, p. 163). The insight is that behavior, indeed the act that is considered an error, is not caused solely by the person involved; rather it reflects the reaction of the person with respect to stimuli. Centuries of empirical and theoretical literature of the discipline that studies how people act, psychology, and the millennia of writing in that discipline's parent, philosophy, as well as the literature of all the physical and social sciences attest to that definition.

To study and understand the behavior of the entity under consideration regardless of the discipline, the context in which that entity exists must be considered. Thus, the behavior of performing a task indeed of committing an error (B) is a function (f) of a person (P) interacting (×) with factors in the environment (E) expressed as $B = f \{P \times E\}$ (Lewin, 1943/1964). Considering only the care provider performing a task as the sole source of error is incomplete and misleading.

As much as people may want to believe that they are in command of their activity and that they alone initiate a behavior, they do not. This is a subtle but critical point in considering the nature of error (Bogner, 2000). Behavior is activated by cues that evoke an action—cues from the physical environment or cognitive stimuli triggered by an aspect of the environment such as a comment. This has been manifest for centuries in the theater. The script for each character sets out the actor's reaction to other actors as well as to props and aspects of the sets. So it is in health care.

6.3.1 James Bond Syndrome

Typically when a situation occurs such as misadministrations of the drugs Capastat and Cepastat, the nurse involved is chastised and removed from caring for the patient requiring the drug and replaced by another nurse. Another misadministration occurs involving the replacement nurse. This is an example of the James Bond syndrome of health care error. James Bond always gets the girl no matter which actor is James Bond. This happens because the actors have the James Bond script. For James Bond not to get the girl, the script must be changed—the script of factors in the context of the performance. To reduce the incidence of error such as the drug administration example, the script for that activity must be changed; however before any change occurs—no matter how reasonable the change may seem to a person other than the involved care provider—it is necessary to identify the aspect(s) of the script that induce the error behavior of the care provider so the change will be effective.

To reduce the incidence of misadministration of Cepastat and Capastat, the problem must be identified and the script changed. One aspect of the problem in this example is obvious—the similarity of the names. Further exploration which will be described in the next section of this discussion finds that the vials containing the drugs are of the same shape with similar labels. It is well documented in the psychological literature that similar items are easily perceived as the same. Given that the vials of the drug are not easily differentiated, the appearance of one of the vials must be changed sufficiently to be easily differentiated from the other such as by marking with colored tape. This changes the script affecting the actions of the person administering the drug so the appropriate drug is given—James Bond does not get the girl.

Identifying environmental context factors that can affect the care provider may seem an insurmountable task given the complexity and variety of health care environments; however, that task can be accomplished thanks to studies on error conducted in industries such as nuclear power and manufacturing (Rasmussen, 1982; Senders and Moray, 1991; Moray, 1994). The findings of those studies which are analogous across industries identify eight evidence-based hierarchical categories of interacting factors—systems of factors that affect task performance.

The hierarchical categories of factors in health care terms are: provider characteristics—psychological as well as physical, patient characteristics, means of providing care—in this discussion medical devices and health care products, ambient conditions, physical environment, social environment, organizational factors, and the overarching legal-regulatory-reimbursement-national culture factors (Bogner, 2002). These interacting categories of characteristics operationally define systems of factors envisioned as concentric circles which comprise the supra-ordinate system of the script of the environmental context in which care is provided.

To aid in conceptualizing the care provider in the context of factors, the supra-ordinate system of the environment is represented as an Artichoke with the concentric circles of the hierarchical systems of factors in

FIGURE 6.1
The Artichoke model of the environmental context of task performance. (Reproduced from Bogner, M.S., The Artichoke Systems Approach for identifying the why of error, in P. Carayon (Ed.), *Handbook of Human Factors and Ergonomics in Health Care and Patient Safety*, Lawrence Erlbaum Associates, Inc., Mahwah, NJ, pp. 109–126. Copyright 2007 by Lawrence Erlbaum Associates, Inc. With permission of Taylor & Francis via Copyright Clearance Center.)

the environmental context represented as the layers of leaves (Figure 6.1). The care provider in the center of the Artichoke encased in the leaves illustrates the influence of the environmental context script. This concept is the basis for the Artichoke Systems Approach to addressing human error in health care. (For a discussion of the evidence-based systems of factors that affect the provider and the Artichoke model, see Bogner, 2007.)

6.4 Artichoke Systems Approach

The Artichoke Systems Approach addresses error in five steps.

Step 1 determines the context in which the behavior considered as an error occurred. Typically the focus of attention is on the outcome of the error and the person associated with that outcome; however, those entities cannot be addressed without an environmental context because of the nature of error being behavior. The lack of consideration of the environmental context changes $B = f\{P \times E\}$ to $B = f\{P \times \_\_\}$ which does not define a behavior.

The Artichoke model represents the care provider's perspective of the environmental factors that contribute to error. The concept is analogous to the actor's script of a character. As a script for an actor's character is the perspective of only that character, the Artichoke model of a care provider represents the perspective of that one care provider. If it is desired to address the perspective of another person, the Artichoke of that other person

must be considered just as in a play each character is represented by a unique script that determines the perspective and behavior of the actor. If there were not this differentiation of roles in a play, the behavior of the characters would become muddled and the story compromised.

It is the same in considering the behavior of care providers; if the perspective of each provider, that is, the Artichoke of factors of the context in which each person functions were not considered individually the perspectives become intertwined and the analysis of error is compromised if not unable to be performed. It is emphasized that this approach is not concerned with what the care provider did *per se*—the act associated with an error because that does not indicate what caused the behavior. Rather the purpose of the approach is to identify the factors that contributed to if not induced the act—factors in the script. Focusing on the act alone provides no means of identifying contextual factors that contribute to error so they might be addressed.

Step 2 involves identifying the factors associated with the error including precursor events—events which occur prior to the error act that establish conditions favorable to the error occurring, that is, events that set the stage for an error. It is to be noted that by definition precursor events happen prior to the error and as such cannot directly affect the care provider; however factors from the outcome of those events are present in the Artichoke and can contribute to error. Factors contributing to an error can be identified by considering the error guided by the Artichoke Systems Check List. That check list provides lines for each of the Artichoke systems of factors on which the care provider can note factors he or she perceives influenced his or her error behavior that led to an error (Table 6.1).

It is best if the care provider records the factors as soon as possible after the incident. If that is not feasible, the incident should be described. It is emphasized that the focus is the incident; the care provider can be

TABLE 6.1

Artichoke Systems Check List

Artichoke Systems
Check List
Error-Inducing Factors/Characteristics
Patient _____
Means of providing care _____
Care provider _____
Ambient factors _____
Physical factors _____
Social factors _____
Organization factors _____
Legal-regulatory-reimbursement-cultural factors _____

anonymous on the check list so he or she cannot be blamed for the incident. The important information is the factors the provider experienced that contributed to an error. The check list also is to be used to record factors that lead to a near miss or an error that almost happened, an error that occurred without an adverse outcome, and hazards or errors waiting to happen. Because those situations are much more numerous than errors with adverse outcomes, they can provide a wealth of information.

Step 3 determines the targets for remediation by applying the 5 Why technique to each of the identified factors and precursor events with the technique being continued until an actionable item is identified for each of the factors. An actionable item is a factor upon which actions can be taken to change it and reduce the likelihood of that factor contributing to the recurrence of an error. The actions may be physical as in modifying a cabinet or procedural as in altering work schedules. In general the 5 Why technique is to ask why something occurred, record the response then ask why that occurred. That response is recorded and the process continues drilling down by asking why of each response until an actionable item is identified.

The systems of factors on the check list can be used as prompts. The number of Whys asked may be more or fewer than 5—however many are necessary to identify an actionable item. It is not the number of Whys that is important; it is the product of that repeated questioning. It is emphasized that the product of that technique must be actionable or there can be no impact on reducing error.

If multiple factors are identified from the initial Why, each factor should be addressed individually asking why it occurred using the check list as appropriate, recording the responses and drilling down to an actionable item for that factor. The provider then returns to each of the multiple factors and drills through the responses and questions until an actionable item is identified for *each* factor.

It is again noted that in this approach the person's name need not appear on the check list—the purpose of the analysis is to identify those factors that contributed to the error so they can be changed not to hold the person involved accountable. The use of the prompts of characteristics on the check list stimulates consideration of the multiple categories of factors, hence serves to prevent the Stop Rule from focusing on the first seemingly reasonable factor as the explanation. An example of a surgical incident demonstrates the value of this simple technique. The incident is a patient's liver was lacerated during laparoscopic surgery.

The surgeon conducted a 5 Why analysis of the incident: Why did I lacerate Mrs. Patient's liver? *Because of the fulcrum effect of the long handles of the instruments for laparoscopic surgery.* Why did the fulcrum effect occur? *My body lurched from being off-balance.* Why was I off balance? *I inadvertently kicked the foot pedal off the step stool.* Why was the pedal on a step stool? *I needed additional height to perform the procedure.* Why did I need the additional height? *I am somewhat short; the patient weighed 400 lb so had considerable body mass and the instruments have long shafts.* Why didn't I lower the operating table? *It was as low as it could be which was not low enough.* This technique identifies the actionable items of the step stool platform, the foot pedal, and the vertical mobility of the operating table—items to be addressed in Step 4.

Step 4 implements the remediation of those identified aspects of factors and events that contributed to error. Management should designate a specific person to receive and act on the identified error inducing factors. This should not be a person with responsibilities that could compromise changing the identified factors such as might occur in the office of risk management because that orientation might not be amenable to changing the identified items. A reasonable focal point would be the office of the safety engineer or the department of biomedical engineering.

The actionable items from Step 3 are addressed. The problem of the step stool in the example can be reduced by purchasing stools with larger platforms and securing the foot pedal to the step stool platform by a C clamp. The other factor contributing to the incident cannot be so readily remediated. Operating room table manufacturers should be notified of the need for increased vertical mobility; however the realization of that change may take considerable time and when the desired table is available the cost may prohibit implementing that change.

Although not a factor identified by the application of the 5 Why technique in the example, patient characteristics are a system of factors in the Artichoke and merit consideration. Certainly morbid obesity contributed to the adverse incident. Such obesity cannot be alleviated as can an extremely anxious patient by administrating a calming drug; however, the height from the body mass could be accommodated if a tall surgeon were available to perform the procedure. Such sensitivity to potential error inducing conditions can be developed by care providers integrating the Artichoke System Approach into their consideration of the context in which they perform tasks.

Step 5 is the evaluation of the effectiveness of the remediation activities. It is important that the care providers do this because they experience the conditions of task performance and can identify what affects them. The success of this approach can be measured by the extent of effective remediation of factors identified by the care providers as contributing to an error—the measure being in terms of reducing or preventing error.

When such effective remediation occurs, the value of the approach is multiplied by implementing the changes in comparable conditions throughout the facility such as the larger platform with secured foot pedals in all facility operating rooms (OR) to reduce the likelihood of recurrence of the adverse incident.

The discussion of error when adverse outcomes are considered addresses problems after there has been harm or a near miss. It is a reactive approach. A goal of human factors/ergonomics is to design products to prevent errors and adverse outcomes, and this approach involves more than identifying factors that contributed to an error. Such factors pertain to the error behavior that was addressed. The relevance of those factors could be applied to preventing the occurrence of that error in the future use of the product and to reducing the likelihood of error when using products affected by the identified factors.

Because the reactive approach is concerned with what has already occurred, considering only those factors to prevent error by design narrows the range of factors to be considered and dampens creativity in preventing error through product design. This tie to what has passed presents a particular challenge when designing new products or modifications of existing devices with the omnipresent threat of Murphy's law haunted by the specter of a worse case scenario. A proactive approach to preventing error is needed; the Artichoke is the basis for that approach.

6.5 Error, the Artichoke, and Murphy's Law

Error is after the fact. Recurrence of an error can be prevented by addressing factors identified as contributing to error in the product design. This also may be applicable for similar products to a limited degree. We like to believe that error would not happen if the product were designed appropriately, that is, factors contributing to error are addressed in the design of the product to proactively prevent errors. The all important and challenging task is determining the factors that are likely to contribute to error when a product is used so those factors can be addressed in product design. These are real-world factors as contrasted to those in hypothetical research situations with questionable real-world validity.

Real-world validity is necessary because the real world is the context in which health care products are used. Thus the task in proactively preventing error through product design is determining real-world factors that are likely to contribute to error in the use of a yet-to-be-developed medical device. In other words the

task is to determine how a product might be designed to prevent what has not happened.

Emphasizing designing for the real-world in a discussion of human factors/ergonomics might be considered redundant because human factors and ergonomics are applied disciplines. Being applied, however, does not ensure that human factors/ergonomic considerations address real-world factors optimally or even appropriately. As with so much, the devil is in the details—details of the context, the Artichoke of factors, in which care is provided. Those details become all the more important when designing to address the adage that products should function appropriately in a worse case scenario—a scenario which reflects the operation of Murphy's law.

6.6 Expansion of Focus in Product Design

Rather than focus nearly exclusively on the person-product interface, efforts to prevent error by product design must consider the real-world contexts in which the product will be used. This can be accomplished by considering the systems of factors on the Artichoke Check List and identifying probable error inducing factors in all projected contexts of use. Once factors that are likely to contribute to error have been determined, proactive efforts can be directed to addressing them in the product design to prevent those errors from happening.

6.6.1 Medical Devices Are Consumer Products

Device use is behavior which by definition is influenced by factors in the context of the activity. Because environmental factors affecting device use are not obviously considered in design or included in the adverse outcome reports, the term behavior typically is accepted as referring to the health care provider and the device void of context. As the meaning of a sentence out of context can be unclear, the behavior of using a medical device designed without considering the environmental factors becomes a function of the person, alone, $B = f \{P \times \_\}$, which as discussed previously provides no viable information for preventing error.

It is understandable that those who design medical devices and modifications of such devices design from their own perspective; that is, designing for use by someone who is similar to them in physical anthropomorphic characteristics and senses as well as drawing on their own experiences using a product similar to that under consideration or learning about it in any number of ways. This works well for the development of consumer products that fit with the designers' experiences which may be for all products except those used in health care. Unless the product designer currently delivers health care in the real world, he or she likely is unaware of nuances in behavior affected by design when a medical device is used in such situations.

6.6.2 Proactive Error Prevention

To effectively reduce errors proactively health care products must be designed to accommodate characteristics of the two interacting elements that constitute error behavior: projected user populations and the full range of anticipated contexts of use. Without concerted consideration of those factors the device design can elicit behavior that forges links in a chain of events that ultimately activates Murphy's law to create a worse case scenario. Both of the elements of error behavior however are relative strangers to device design. One of the elements, characteristics of the projected user populations, can be found by consulting the psychological literature.

6.6.3 Hard-Wired Human Characteristics

Although anthropomorphic characteristics of the projected user populations are addressed in product design sometimes more apparently than others, there are a number of other less obvious hard-wired human characteristics that are critical in the design of safe and effective medical devices. Among those characteristics are: aversion to noise, cognitive processes compromised by stress and fatigue, similar words confused/perceived as the same, similar objects confused/perceived as the same, and thinking interrupted by distraction. These characteristics are inherent to the person and not amenable to change, hence referred to as hard-wired.

Because error is a function of the person interacting with factors in the environment, designing a medical product to accommodate user characteristics addresses but one aspect, albeit a critical one, of what must be considered in designing a product that is safe and effective. Factors from the range of anticipated contexts of use also must be addressed in proactive product design.

6.6.4 Range of Anticipated Contexts of Use

Consideration of the contexts of use acknowledges the importance of designing a product to address likely real-world conditions in which it will be is used—conditions that can affect the use of the product. Examples of such conditions are the presence of small children with the possibility of their playing with the product, the presence of pet hair, cramped space for the device, and a dearth of electric outlets. A product should be designed to provide safe and effective care not only in clinical settings but also in disaster and emergency situations. Situations without electricity and in whatever weather and altitude comprise contexts for providing care as well as for self-care of chronic conditions in environments such as at sports fields, campsites, and fashion shows.

The first decade of the twenty-first century finds an ever increasing number of lay persons providing health care often doing so with sophisticated medical products. Those lay persons include patients from children to frail elderly who engage in self-care and those caring for others in home care. The medical devices typically given to lay individuals to provide care are not different from those used by health care professionals in hospitals. That is beginning to change but not without monetary cost. In the best of worlds, medical products would be designed to be used by both lay care providers and health care professionals; however, this is not the best of worlds and cost is a major factor, often a prohibiting factor in health care.

Consideration of the projected population of users for proactive product design brings with it not only an expansion of user capabilities to be addressed in designing products but also an expansion of contexts in which the products will be used. The Artichoke Check List provides guidance in addressing both aspects of that expansion. Concern for preventing errors can be addressed in the design of those devices that increasingly are used to provide health care to others as well as oneself. It is important, not only for ease of use by lay providers, that products to provide such care are designed to be used by care providers of varying knowledge and abilities from health care professionals to those lay care provides. This necessitates designing simple easy to use products that can be used in providing sophisticated treatments.

It is well documented that stress and fatigue compromise a person's cognitive processes. Knowing that, to reduce the likelihood of error devices to be used in conditions that are stressful and fatigue inducing, such as emergency rooms (ER), OR, and intensive care units (ICU), should be designed so their use is intuitive and simple. It is not so—the most technologically sophisticated devices are those to be used in the ER, OR, and ICU, thus establishing conditions for error and a worse case scenario.

6.7 Care Provider, Product, and Conditions of Use Misfits

The importance of design that addresses the hard-wired characteristics of the user is illustrated by examples of medical device designs that do not consider those characteristics or the effect of contextual factors in product design. For example, a rudimentary consideration of factors in the context of use in the design of displays in medical devices kept at patients' bedsides such as the dim light in patients' room during the night could have prevented errors from interpreting the number 1 as 7 or the converse. The lack of consideration of user hard-wired characteristics in the design of monitor alarms and infusion pumps profoundly impact the viability of the design of those devices although such misfits between design and conditions of use *per se* are not mentioned in the literature

6.7.1 Monitor Alarms

The consequence of the design of a medical device that did not address hard-wired user characteristics and real-world device use factors was reported in the *Boston Globe*, February 21, 2010: "MGH (Massachusetts General Hospital) death spurs review of patient monitors. Heart alarm was off; device issues spotlight a growing national problem." The very serious health care issue expressed in those few words "death spurs review of patient monitors. Heart alarm was off" points to the critical need to consider in device design the hard-wired human characteristics of care providers as well as the systems of factors affecting them.

It is well documented that people find intense noise offensive and people remove what is offensive. Given that the cacophony created by multiple monitors in hospital units such as the OR, ER, and ICUs can be offensive to care providers, it is predictable that an alarm will be disabled for a respite from noise. It is also known that people can be distracted from a task so it is predictable that the intention to re-enable the alarm might be diverted and the alarm remains silent. This information could have foretold the likelihood of the death at MGH and had it been heeded it would have prevented the contribution to what the article states is "a growing national problem." That this increasingly happens when an analogous situation has been successfully addressed by a non-health care product—a snooze alarm—is particularly distressing.

6.7.2 Snooze Alarm

A consumer product designed to address the human proclivity of turning off an alarm and not rearming it, the snooze alarm, provides a lesson to be learned for the design of alarms used in health care. Prior to snooze alarms people accommodated their tendency to turn off an alarm and not do the intended task whether the task was getting out of bed or turning off an oven by setting second or even a third alarm to sound at varying time intervals. With the advent of the snooze alarm, the clock rearms the alarm to sound at a later time and repeatedly does so until it actually is turned off. Some hotels acknowledge this proclivity even with snooze alarms in the guest rooms by asking when a wake-up call is requested if a second wake-up call is desired.

The snooze alarm is a concept that can save lives if it were modified and incorporated in alarming medical devices. The device would be modified so it would not be possible to disable the alarm for more than a specified period of time at which point the alarm automatically would rearm. Had the heart monitor been designed to address the well known human proclivity of being distracted and not completing an intended task, the MGH death as well as deaths from similar incidents over the years, indeed over decades might have been prevented.

It is incumbent on medical device manufacturers and their consultants in the human factors/ergonomics community to demonstrate in their products the sensitivity to inherent user characteristics of those who produce snooze alarms. Another example of the design of a medical device that does not address a hard-wired human characteristic of the user is the infusion pump.

6.7.3 Infusion Pumps

Knowing that stress and fatigue compromise a person's cognitive processes, to reduce the likelihood of error devices to be used in stressful and fatigue inducing contexts of care, devices for use in the ER, OR, and ICUs should be designed so that their use is intuitive and simple. This is not the case for a device that is used in all those settings—the infusion pump.

An infusion pump is a programmable device that regulates the flow of an intravenous (IV) fluid into a patient. The device is attached to a pole with a bag of fluid hanging near the top. The fluid is delivered to the patient from the bag via plastic tubing threaded through the pump which controls the rate of flow. The programming of the pump is complex; at times the concentration of the drug to be delivered has to be calculated—a process that increases the complexity in using the device.

The programming of such pumps is performed by a nurse in a hospital context replete with noise and multiple distractions not unlike a ringing bell. When the nurse is interrupted or otherwise distracted during programming, he or she may not remember at what point the distraction occurred and could resume the programming at an inappropriate place resulting in an unintentional rate of flow. This can be avoided by identifying factors in the environmental context that are likely to adversely affect the use of the device and designing the device to be immune to those factors.

That a person's attention and ability to concentrate are compromised by conditions such as those in the context of a hospital room is well documented in the psychological literature and might be avoided or at least reduced by addressing those conditions in the product design.

Programming errors are a concern and have been addressed by studying the process of programming infusion pumps in usability laboratories. That laboratory setting consists of controlled conditions, in which the programming activity is observed in isolation so that only the programming is observed, typically videotaped, and design recommendations made based on those data. The design of the pump might be modified according to those findings and marketed for use in hospitals.

Having been studied in isolation, the design of the pump likely is ill prepared for the cacophony of factors in the environmental conditions of use and prone to error because an inherent defining factor in behavior has not been addressed in the design of that product. Although when a programming error occurs in the hospital, the source of the error typically is attributed to the nurse; actually the source is a device that is not designed to withstand the vicissitudes of the real world. This can exacerbate the impact of interruptions and distractions and hinder recovery to the appropriate programming. Thus because the pump is used in a context alien to its design a nurse using it can open the door for the invasion of Murphy's law and errors ensue.

6.7.4 Artichoke Approach to Error Prevention

The Artichoke Systems Approach to error prevention was previously described with respect to an error that has occurred providing guidance for the identification and remediation of factors in the context of medical device use that contributed to the error. Addressing such factors only in considering errors is one aspect of error prevention. Error remediation might absolve some factors from their error inducing qualities; however, there is no shortage of factors in the many and varied care providing contexts of the use of health care products—factors that typically are not considered in designing such products with the possible exception of the anthropomorphic characteristics of the intended users. Factors to be considered can be identified by honing the Artichoke orientation of observation.

Not considering Artichoke factors during product development to determine factors that are likely to induce error in the anticipated contexts of care and addressing them in the product design does not negate their impact on the care provider and his or her behavior. Rather not considering such factors results in a product design that is vulnerable to error inducing factors in the context in which it is used. This perpetuates the cycle of error by opening the door to error and extending an invitation to a worse case scenario courtesy of Murphy's law.

6.7.5 Artichoke Orientation of Observation

The Artichoke orientation of observation is not to the technical aspects of using the product such as the construction of a keyboard. Rather, the orientation is to systematically observe and record factors in the context of use of the product or a similar product that affect the care provider to contribute to errors using the technical as well as other aspects of a product. The observation should be guided by the Artichoke Systems Check List.

Designing a product without addressing factors in the context of use denies the consideration of the nature of error in the design and by so doing does not protect the user from error inducing factors nor the patient from adverse outcomes should those factors come into play.

6.7.6 Error Prevention through Product Design

The Artichoke proactive approach to error prevention is to guide product design to be in harmony with the hard-wired as well as other general characteristics of the person using the product, the P of the previously stated expression of behavior, and factors that affect the person in the projected real-world contexts of use, the E of that equation.

6.8 Artichoke as a Lens

Given that use of a medical product is behavior and behavior involves the person interacting with factors in the context of care, then not only do such factors contribute to if not induce error, but such factors known to affect performance when addressed in designing a product can prevent error in the use of the device and by so doing thwart Murphy's law and avoid worse case scenarios.

The Artichoke serves as a lens, one side of which guides the reactive identification of factors that contributed to an error in the supra-ordinate system of the context. The other side of the lens is the proactive guidance of the Artichoke systems of factors in determining real-world factors in the projected supra-system contexts of use that are likely to contribute to error. Addressing those factors in the design of a medical device or health care product proactively prevents error.

To ensure that a device can be safely and effectively used across contexts, various contexts should be envisioned and the systems of factors on the Artichoke Check List adjusted for each context should be reviewed for factors that would likely contribute to error and those factors noted to be addressed in the product design. This Artichoke proactive approach can result in health care products that can be used safely and effectively in all reasonably possible real-world contexts of use.

6.8.1 Reactive Approach to Design to Prevent Errors

A critical aspect of the Artichoke Systems Approach is that errors must be understood to effectively be addressed. This necessitates identification and analysis of the act that is an error. This is in contrast to acting to correct an error based on an assumption, an opinion of what occurred or the product of an analysis conducted days if not weeks after the error. This analysis typically is conducted by someone(s) who knows about the error from a report which often is only a statement of what occurred that is deemed an error, typically presuming the care provider is the sole source of the error. This can and does result in unsatisfactory resolution of the problem and the likelihood of that error continues to occur.

Implicit in the designation of the care provider as the focus of this discussion is an expanded target for design. The design target, rather than being the care provider *per se* or the static existence of a product such as a medical device, is the behavior of the care provider using a product in the real world to enhance the health of another person or one's self. This concept is critical to the analysis of health care error as well as in the design of medical devices to prevent error. Without it, there will be misfits between the product design and the conditions of use.

6.9 Product Design

When considering design, a health care need should be identified, a concept for addressing that need developed, and factors necessary for the real-world use of the product from that concept determined and addressed during the development of the product. This is of critical importance because without addressing real-world factors that affect the use of the product—factors that can contribute to error using a product—medical devices can and in many instances are developed and marketed that induce error when used.

Following the definition of error as behavior, the typical focus of developing a product in isolation from factors impacting its use must be expanded and the product developed to proactively address such factors in the contexts of use. That can be accomplished by using the characteristics of the Artichoke Check List as prompts to determine factors that influence the behavior of using

the device by members of the target population and addressing those factors in the design of the product.

Real-world conditions typically are not considered when adverse outcomes involving medical devices are addressed. The usual industry response to an adverse event focuses only on the user, which ignores the interaction of the user and factors in the context of care including characteristics of the target device. The ongoing issue of therapeutic radiation incidents involving linear accelerators and Computerized Tomography (CT) brain profusion scans supports the point that medical devices are consumer products and must be treated as such by involving actual users in their design. It is critically important to consider real-world error inducing factors in the development and modifications of such products through the Artichoke Systems Approach.

When considering design, a health care need should be identified, a concept for addressing that need developed, and factors necessary for the real-world use of the product from that concept determined and addressed during the development of the product. This is of utmost importance because without addressing real-world factors that affect their use, medical devices that induce error when used can and in many instances are developed and marketed sometimes with catastrophic consequences. This is evident in linear accelerators.

6.9.1 Excessive Therapeutic Radiation Incidents through the Decades

Linear accelerators generate beams of radiation that pass through normal tissue to radiate deeper lying cancerous areas. Until the technological revolution the conditions of the operation of therapeutic radiation devices such as the linear accelerator were known and directly controlled by the operator. This changed rapidly when the ubiquitous computer chip and other technological innovations invaded health care; the radiation beam and its focus then came under the exclusive control of software. The software has the capability of increased accuracy in targeting tumors however its complexity creates hidden avenues for errors. It is to be noted that software although internal to the device is an aspect of the design.

Software issues in the delivery of therapeutic radiation were documented in 1985 with the Therac-25 linear accelerator fatal radiation burn incident. That was not the first to occur; six similar incidents had happened previously. This incident is notable because the apparent cause was identified. Despite being fired from her position after the incident, the operator involved in the fatal Therac-25 overdose together with the hospital medical physicist persevered in analyzing the conditions that lead to the overdose until they discovered that the apparent cause was the operator's skill in keying the necessary dose information into the machine.

The operator entered the dose in less than 8 s. Because it took 8 s for the software to ready the beam, the intensity of the beam was not at zero and the conditions were ripe for Murphy's law. With no indication to the operator, subsequent entered doses were added by the software to the programmed intensity. Those precursor events established the conditions for the worse case scenario of a massive radiation overdose. Although the manufacturer later acknowledged that data entry in less than 8 s was known to cause a problem, it was assumed no one could enter data that quickly and that information was not provided to the user (Leveson and Turner, 1993).

It is mind boggling that despite the potential for a deadly overdose of radiation however it might occur, the device had no fail-safe protection or even an alarm or other type of warning to indicate to the operator that an aberration of the radiation dose was occurring. Indeed, in this incident the message displayed was one that had appeared in other treatment situations so there was no hint of a problem. The worse case scenario was realized when the massive overdose treatment was delivered and the patient received a severe radiation burn. He died 4 months later from complications of the overdose having suffered inordinately in the interim.

The software issues of the Therac-25 and earlier accelerators are not unique. In 2010 ongoing problems with radiation therapy were brought to the attention of the public by a series of articles in the *New York Times*. Those articles report details of linear accelerator problems and the deaths of individuals caused by the radiation that was to cure their health issues. This is not a problem that has affected a small number of people. More than 1000 reports of errors in radiation therapy have been filed with the Food and Drug Administration (FDA) from 2000–2010. Seventy-four percent of those errors were caused by linear accelerators. The most frequently cited cause of error was computer software (Bogdanich, 2010b).

Radiation overdoses also have occurred with CT brain profusion scans. Those scans are given to apparent stroke victims to identify blood clots in the brain that can be dissolved to prevent further damage. Evidence of radiation overdoses began to emerge in the summer of 2009 occurring in hospitals throughout the country—hospitals with excellent reputations such as Cedars Sinai in Los Angeles (269 incidents) as well as in less well known facilities (Bogdanich, 2010a). Because there is no process for reporting incidents, except to the FDA, and symptoms of excessive radiation may not appear for months, the actual number of deaths and serious injuries from excessive radiation is far more than 1000. Given the rule of thumb that 1 in 10 incidents are reported, the number of adverse outcomes could be as many as 10,000. This is a far from trivial number of radiation overdoses.

In 2009 a large wound care company treated 3000 radiation injuries, most of which were sufficiently severe to require treatment in a hyperbaric chamber (Bogdanich, 2010d). Linear accelerators continue to deliver massive overdoses of radiation seemingly randomly—the apparent result of the dose entered into the computer being compromised by software programming inherent in device as well as issues in design of user interface and contextual factors. Among those issues is that the operator is not notified that the dose entered has become an overdose or is ineffectively notified by a message on the display located where it is likely not to be seen by the operator. Not being warned of a problem, the operator does his or her job and delivers the radiation often in multiple short doses.

The patient receives the treatment in a room containing the accelerator; the door is shut so the operator has no visual contact with the patient hence cannot observe patient distress from receiving an overdose of radiation. Despite the software changing the radiation dose unbeknownst to the operator and the operator not receiving notification of the impending catastrophe or seeing the patient's reaction to an overdose so he or she might stop, the cause of the incident typically is attributed to the operator who more often than not is fired. Such is the faith in technology and the power of the Stop Rule.

The problems with the internal software programming are exacerbated when an accelerator is modified to enhance the capability for focusing the beam for stereotactic radiosurgery (SRS). This modification consists of retrofitting the accelerator with devices made by different companies. One such focusing device comprises heavy metal jaws that can be moved to narrow or enlarge the opening emitting the beam to pinpoint focus the beam of high-intensity radiation to the target site through a cone attached to the jaws. When no message about the closeness of fit of the cone within the square holding jaws appears on the display and a visual check is not possible due to the design of the device, the operator assumes the entered dose is what will be delivered.

A patient in Evanston, Illinois received treatment from a SRS device that focused the beam on a tiny nerve near the brain stem which was causing facial pain. The intense beam leaked through the space between the cone and the corners of the jaws and radiated an area of the brain four times larger than intended. As a result of that radiation the patient is in a nursing home nearly comatose not able to speak, eat, or walk; her young children are cared for by their father (Bogdanich and Rebelo, 2010).

Radiation overdoses also result from errors when installing the SRS device. In 2004 and 2005 an installation error in a Florida hospital resulted in 77 brain cancer patients receiving 50% more radiation than prescribed.

Another cause of SRS radiation overdoses is the calibration of the device. Over half the patients undergoing SRS in a Missouri hospital were overdosed about 50% because of a calibration error that eluded routine checks for 5 years (Bogdanich and Ruiz, 2010).

These instances of radiation overdose and their all too often catastrophic consequences reflect, as noted by Dr. Howard Amol chief of clinical physics at Memorial Sloan-Kettering Cancer Center, that hospitals often are too trusting of new radiation related computer systems and software relying on them as if they had been tested when they have not (Bogdanich, 2010c). Testing is an important aspect of the development of consumer products. Medical devices are consumer products, yet devices such as the linear accelerator that can do such catastrophic harm are not tested as rigorously as coffee makers.

6.9.2 Device Testing

Testing of medical devices occurs in clinical trials typically conducted by research groups under contract to the device manufacturer. Despite that many adverse outcomes result from the use of medical devices. Adverse events such as those in the use of therapeutic radiation might have been prevented had the devices been tested by experienced device operators with real-world user characteristics such as speed of dose entry and other real-world factors that contribute to error been identified and addressed. The Therac-25 incident likely would have been prevented. Similarly the lack of fit of the cone within the jaw of the SRS device probably would have been identified and the tragic outcomes for radiation overdoses avoided.

It is acknowledged that real-world testing of linear accelerators and other medical devices would put subjects acting as patients at risk so those products are marketed without appropriate testing in the real-world contexts of use. Risk should not be an acceptable reason for not testing a device. Lessons can be learned for safe and effective design by testing in a simulation facility.

6.9.3 Simulation of Real-World Health Care

Often the term simulation refers to computer based modeling or other part task exercises; however to actually test the operation of a device, particularly a device that has the potential for profound adverse outcomes, the simulation must involve an actual care provider—not a retired care provider consultant—in a real-world context of intended use not just the physical context but with typical distractions such as a physician changing the order at the last minute and the lack of telephone contact with the patient in the room where radiation is received—a condition in the Therac-25 incident—and

a mannequin that can record the outcome of the treatment such as radiation dosage.

To avoid possible unintended bias, the testing facility should be an independent entity with industry paying to have devices tested. Hospitals and professional organizations as well as medical device firms can build files of worse case scenarios from de-identified reports of incidents using a specific device and the conditions of use to identify factors to be addressed in testing. Scenarios also can be developed that are counter to human behavior proclivities in the psychological literature—as discussed previously.

The device prototype should be tested in a neutral simulation testing facility rather than under the auspices of industry as in clinical trials or funded by industry. The testing facility provides factors from actual real-world contexts—factors guided by the Artichoke—including noise and distractions in the facility's real-world settings of use of the device. Testing can involve as many and as varied incident scenarios as are available. This affords a rigorous testing of the use of the device by executing in a few hours a myriad of worse case scenarios that may not occur in years of use. Thus the viability of the device to perform as required by the user is assessed and safety issues are identified that are provided to the manufacturer to be addressed by modifying the design.

The modified prototype design is tested in the simulated real-world contexts. Feedback again is provided to industry. Testing continues until it is demonstrated that the identified issues have been successfully addressed and the device is safe and effective. Details and pictures of such a testing facility that has the capability to record what transpires during testing appear in Matern (2010).

The importance of testing medical devices by actual care providers in real-world simulated settings cannot be overemphasized. It is imperative that the design of medical products be demonstrated to be safe and effective in the real world prior to marketing. The consequences of inappropriate design are too profound not to do so. This is indicated in the previous discussions of infusion pumps, alarms, and particularly radiation therapy devices.

6.10 Designing for the Real World through the Lens of the Artichoke

Given a lack of abundance of crystal balls, anticipating what might go wrong has all the earmarks of an impossible task. This is far from actuality, however. The Artichoke guides the determination of factors in real-world context of use that together with consideration of the providers' hard-wired human characteristics can anticipate Murphy's law and thwart its occurrence both proactively as well as reactively. Prevention is more expeditious than reaction in terms of potential for harm as well the cost to industry and society; however, incidents occur that provide opportunities for lessons to be learned—opportunities that must be seized if safe and effective medical devices are to be developed.

The Artichoke can serve as lens through which insights can be obtained for designing health care products including medical devices. Being considered a lens essentially melds the reactive and proactive approaches to factors that contribute to error in the use of health care products so they can be addressed as appropriate in product design. The melding affords insights into product design issues.

The time has come to proactively design medical devices for use in providing care in the real world. This involves addressing not only the physical characteristics of the target user population including human characteristics such as reaction to stress and lack of tolerance of noise as well as characteristics of patients such as obesity that affect the care provider in performing a task. Lessons in designing for the real world also can be learned from applying the Artichoke approach to consumer products because medical devices are consumer products, perhaps the ultimate consumer products due to their life sustaining use and as such should be designed with the care and testing as other consumer products. The possibility of Murphy's law creating a worse case scenario should be a driving force in all device design.

That there are not more adverse outcomes in using medical devices is testimony to the skill and ingenuity of the users. Such skill and ingenuity, however, may be stretched beyond limits in meeting the formidable challenges of devices of increasing technological sophistication as well as unprecedented changes in the characteristics of the patient population that can confound device use such as morbid obesity. Murphy's law will not go away, but its operation can be reduced if not avoided by employing the Artichoke to assist in determining factors in the projected contexts of care including the hard-wired human qualities that affect behavior to be addressed in the proactive design of health care products including medical devices.

Factors in the context of care that are relevant to providing health care are operative whether they are addressed in the device design or not. If they are, then the device might be designed to be used safely and effectively. If they are not, the stage is set for Murphy's law to come into play and the provider is Artichoked with the conditions of a worse case scenario. It is incumbent upon the human factors and ergonomics communities to expand the focus from the user-system interface void of overt consideration of the full range of contextual factors that impact the task performance and

work diligently to determine error inducing factors in the real world context of care and proactively address them when designing the most critical of all consumer products—those used in providing health care.

These efforts can bridge and eventually close the chasm between the needs of the full range of care providers from the most highly trained and skilled professionals to those with abilities limited by disease or medication who provide care in the real-world contexts of their everyday lives and the products developed and marketed to meet those needs.

References

Bogdanich, W. (2010a). After stroke scans, patients face serious health risks. *New York Times*, July 31. www.nytimes.com/2010/08/01/health/01radiation.html. Accessed February 16, 2011.

Bogdanich, W. (2010b). FDA toughens process for radiation equipment. *New York Times*, April 8. www.nytimes.com/2010/04/09/health/policy/09radiation.html. Accessed February 16, 2011.

Bogdanich, W. (2010c). As technology surges, radiation safeguards lag. *New York Times*, January 27, 2010. www.nytimes.com/2010/01/27/us/27radiation.html. Accessed February 16, 2011.

Bogdanich, W. (2010d). Radiation offers new cures and ways to do harm. *New York Times*, January 24. www.nytimes.com/2010/01/24/health/policy/09radiation.html. Accessed February 16, 2011.

Bogdanich, W. and Rebelo, K. (2010). A pinpoint beam strays invisibly, harming instead of helping. *New York Times*, December 28. www.nytimes.com/2010/12/29/health/29radiation.html. Accessed February 16, 2011.

Bogdanich, W. and Ruiz, R. (2010). Radiation errors reported in Missouri. *New York Times*, February 24. www.nytimes.com/2010/02/25/us/25radiation.html. Accessed February 16, 2011.

Bogner, M. S. (2000). A systems approach to medical error. In C. Vincent and B. DeMol (Eds.), *Safety in Medicine* (pp. 83–100). Amsterdam, the Netherlands: Pergamon.

Bogner, M. S. (2002). Stretching the search for the "why" of error: The systems approach. *Journal of Clinical Engineering*; 27: 110–115.

Bogner, M. S. (2007). The Artichoke systems approach for identifying the why of error. In P. Carayon (Ed.), *Handbook of Human Factors and Ergonomics in Health Care and Patient Safety* (pp. 109–126). Mahwah, NJ: Lawrence Erlbaum Associates, Inc.

Commonwealth Fund (2004). The end of the beginning: Patient safety five years after *to err is human* http://www.cmwf.org/publications/publications_show.htm?doc_id=250749. Accessed February 16, 2011.

Deming, W. E. (1986). *Out of the Crisis*. Cambridge, MA: Massachusetts Institute of Technology Center for Advanced Engineering Study.

Kohn, L. T., Corrigan, J. M., and Donaldson, M. S. (Eds.). (1999). *To Err Is Human: Building a Safer Health System.* Washington, DC: National Academy Press.

Levenson, N. and Turner, C. (1993). An investigation of the Therac-25 accidents. *IEEE Computer*; 26 (7): 18–41. http://ei.cs.vt.edu/~cs3604/lib/Therac_25/Therac_1.html. Accessed February 16, 2011.

Lewin, K. (1964). Defining the "field at a given time." In D. Cartwright (Ed.), *Field Theory in Social Science* (pp. 43–59). New York: Harper & Row. (Original work published 1943.)

Mach, E. (1905). *Knowledge and Error.* Dordrecht, the Netherlands: D. Reidel Publishing Co.

Matern, U. (2010). Experimental OR & ergonomic facility. www.wwH-c.com. Accessed February 16, 2011.

Moray, N. (1994). Error reduction as a systems problem. In M. S. Bogner (Ed.), *Human Error in Medicine* (pp. 67–92). Hillsdale, NJ: Lawrence Erlbaum Associates, Inc.

Pope, A. (2004). *An Essay on Criticism.* Whitefish, MT: Kessinger. (Original work published 1711.)

Rasmussen, J. (1982). Human errors: A taxonomy for describing human malfunction in industrial installations. *Journal of Occupational Accidents*; 4: 311–333.

Rasmussen, J. (1990). Human error and the problem of causality in analysis of accidents. *Philosophical Transactions of the Royal Society of London*; 337: 449–462.

Senders, J. W. and Moray, N. P. (1991). *Human Error: Cause, Prediction, and Reduction.* Mahwah, NJ: Lawrence Erlbaum Associates, Inc.

Wachter, R. and Pronovost, P. (2009). Balancing "no blame" with accountability in patient safety. *The New England Journal of Medicine*; 361: 1401–1406.

Webster (1994). *New Riverside University Dictionary.* Boston, MA: Houghton Mifflin.

Zwillich, T. (2004). *Little Progress Seen in Patient Safety Measures.* Washington, DC: Reuters Health Information. November 2004.

7

Organizational Learning in Health Care

Ann Schoofs Hundt

CONTENTS

7.1 Organizational Learning Research 98
    7.1.1 Some Early Research 98
    7.1.2 Synthesis and Expansion of Organizational Learning Concepts 99
    7.1.3 Learning Organizations 100
    7.1.4 Individual and Organizational Learning 101
7.2 Organizational Learning Research Applied to Health Care 102
    7.2.1 Organizational Structure Considerations and Levels of Learning 102
    7.2.2 Learning (or Not) from Failure 103
    7.2.3 Leadership 104
7.3 Risk Assessments as a Means for Promoting Organizational Learning 104
    7.3.1 Risk Assessment Requirements in Health Care 105
    7.3.2 Risk Assessment in Other Industries 105
7.4 Conclusion 106
References 106

Learning is not at all foreign to health care. Individual practitioners, and practitioners collectively through their respective professional associations, continually attempt to improve the means of providing care to patients albeit technically, technologically, and/or interpersonally. *Organizational* learning, however, poses challenges that require a willingness for individuals, groups or teams, and organizations to change when and where change is needed. The "organizational" aspect requires development of shared mental models, a well-articulated mission, and a willingness to surrender personal and group goals in the interest of the organization. This must continuously occur between all providers, managers, and others working within an organization, despite the fact that these individuals may not be accustomed to working together, nor do they perceive any professional relationship or interdependence. These groups may, in fact, represent various cultures within the organization (Carroll and Quijada 2007) and may also have conflicting goals or perspectives.

Organizational learning requires that everyone working in a health care organization understands that they must function as a system comprising numerous, oftentimes-complicated, interdependent parts. And to the extent possible, they must understand the system in which they work. Those providing care, because of the type and extent of training they receive, as well as their status in society, frequently practice in a fashion that makes communication, teamwork, and other aspects of effectively functioning systems a challenge. The varying philosophical underpinnings that promote individual accountability, as suggested by Smith and Bartell in this volume, can also present barriers to understanding organizations as systems and ultimately achieving organizational learning.

Tremendous progress has been made over the years in elevating the standard of health care and offering more options for care. As a result, providers have generally been receptive to adopting proven clinical alternatives that result in better patient outcomes. This is the one common indisputable goal of all working in health care, regardless of their role, be they housekeepers, technicians, surgeons, administrators, or nurses (Bohmer and Edmondson 2001): every patient should be provided the best and safest care possible. From this, we suggest that the commitment to individual learning, and in some instances group learning, that already exists in health care can provide the motivation for management to espouse and promote organizational learning.

This is especially important today as health care undergoes significant scrutiny by individuals receiving care, those paying for care, agencies accrediting health care organizations, and government legislating health care. *Clinical microsystems* in health care organizations that espouse learning have been shown to (a) produce better outcomes through ongoing measurement and process improvement, (b) exhibit greater efficiency (with less waste), (c) utilize varying levels of information technology to promote ongoing communication, (d) function as a team to manage their patients, (e) maintain a focus on both staff and the patient, (f) have a healthy balance between top-management support and individual autonomy, and (g) display a strong and cohesive culture (Nelson et al. 2002). A valid question to raise here is as follows: if learning can accomplish such significant results at a lower (i.e., microsystem) level, what can be done to accomplish equally—or possibly more—significant results at the organization level? What must be done to achieve learning at a higher level that then permeates the organization? In other words, how do we achieve organizational learning in health care?

In this chapter, we review the concept of organizational learning and aspects of it that have been studied in various areas of research. We review a selection of papers that are considered seminal works, frequently cited, or especially pertinent to the topics we later develop. A summary of the discussion is found in Table 7.1. We also present an overview of organizational learning in health care as well as a how risk assessments can complement and facilitate organizational learning. More thorough reviews of the full body of research on organization learning exist elsewhere (Crossan et al. 1995; Crossan and Guatto 1996; Huber 1991; Leavitt and March 1988; Shrivastava 1983; Popper and Lipshitz 1998; Wang and Ahmed 2003).

7.1 Organizational Learning Research

One challenge to organizational learning, be it in health care or any other industry, is determining how to define and characterize it. Cyert and March (1963) are generally credited as being among the first to introduce the notion of organizational learning through their presentation of the *behavioral theory of the firm*. Since then, others have offered varying definitions and conceptual frameworks that describe their perspective on organizational learning. Some of these follow.

7.1.1 Some Early Research

In their early work, Cyert and March (1963) presented organizational learning as one of the four major underpinnings of business decision making (p. 116). They characterized organizations as being adaptive (much like individuals) that respond to changes that are both internal and external to their environment. Cyert and

TABLE 7.1

Summary of Organizational Learning Research

Researchers	Tenets/Findings
Cyert and March (1963)	• "Behavioral theory of a firm" • Organizations are adaptive • Learning occurs from experience • Organizations have short-term focus
Cangelosi and Dill (1965)	• "Adaptation" (from Cyert and March) results from stress • Learning is sporadic and stepwise • Failure leads to change • Need to consider individual learning
Argyris and Schon (1978)	• "Theory of action" • Learning must be communicated • Introduce "single- and double-loop learning"
Duncan and Weiss (1979)	• Focus on knowledge • Knowledge must be communicated
Shrivastava (1983)	• Pointed out fragmented and multidisciplinary approach to organizational learning to date • Proposed two-dimensional typology based on origin and the "how" of learning
Fiol and Lyles (1985)	• Change does not imply learning • Learning results in improved performance • Need balance between stability and change • Four contextual factors to organizational learning: culture, strategy, structure, environment
Leavitt and March (1988)	• Learning is based on experience, history-dependent, attempts to meet targets
Senge (1990)	• Learning organization • Leadership is key
Simon (1991)	• Individual learning in organizations is a "social not solitary phenomenon"
Crossan et al. (1999)	• "4I"s—intuiting, interpreting, integrating and institutionalizing—progression of learning in organizations
Tucker et al. (2000s)	• First- and second-order problem solving: learning occurs with second-order problem solving but is considerably less common • Research focus: nurse problem-solving behavior
Chuang et al. (2000s)	• Organizational learning depends on structure and norms of organization • Organizational learning between organizations depends on extent of interdependence • Research focus: health care
Tamuz et al. (2011)	• More likely to learn from errors with grave consequence at one's own as well as those at another organization

March also posited that learning results from an organization's experience and, given the short-term focus in many organizations, that the current and pressing issues are most readily and frequently addressed. When change is being considered, preference for attention in many organizations is given to routine actions, standard operating procedures, and similar existing structures within the organization. This perspective can hamper creativity and negatively influence recognition of what would otherwise pose significant learning opportunities.

Shortly afterward, Cangelosi and Dill (1965) presented an early experimental case study based on premises of Cyert and March. In the experiment, graduate students were given a management simulation project that required them to develop an organizational structure along hierarchical functionally specialized lines for a fictitious company. Findings of the experiment suggested that (a) adaptation results from stress (frequently due to an unanticipated negative outcome), (b) learning is "sporadic and stepwise rather than continuous and gradual" (p. 203), and (c) determining what to learn requires an understanding of how to learn. They also determined that failure to meet goals led to change. Using a system as the basis, Cangelosi and Dill suggested that there are three types of stress: those stimulating subsystem (e.g., individual and/or group) learning, those stimulating total-system learning, and those stimulating both types of learning. As a result of their work they proposed the need to (a) evaluate organizational learning in light of individual learning (or an interaction of the two), (b) identify facets of an organization—primarily its environment—that define the necessary learning tasks and the ability required to achieve them, and (c) determine when learning—and what is learned—actually occurs.

Argyris and Schon's seminal work (1978) presented organizational learning through a *theory of action* framework. Explaining their theory in light of systems terminology, individual workers or groups of workers receive *inputs* that may or may not coincide with their understanding of the system's premises, structure, or strategic plans. These inputs are then *processed* by the workers who then respond to the input by either changing or maintaining the way they do things to conform to the system's requirements. As a result, any response or *output* (either affirming or negative) must be communicated and then stored in the organization's memory to ensure that the organization learned from the experience. If this does not occur, Argyris and Schon pointed out that "the individual will have learned but the organization will not have done so" (p. 19). This notion is later described by Shrivastava (1983) as *assumption sharing*. Like many following them, Argyris and Schon (1978) placed significant emphasis on the relationship between individual and organizational learning: "Just as individuals are the agents of organizational action, so they are the agents for organizational learning" (p. 19). We will continue to discuss the relationship between individual and organizational learning, as well as group learning, throughout this chapter.

Another concept Argyris and Schon (1978) presented in their work is that of single-loop and double-loop learning. *Single-loop learning* occurs within the constraints of an existing organizational structure, policies, and procedures. The issue (in this case the learning opportunity) in question undergoes minimal in-depth scrutiny despite the fact that the best solution may require a significant change in the way the organization or unit operates. This type of learning is generally less dramatic and frequently does not result in organizational learning. It coincides with what March (1991) later termed *exploitation*.

Conversely the inherent nature of *double-loop learning*, which requires examination of organizational norms, operations, and strategies prior to making a decision, generally results in organizational learning. Conflict between individuals or groups frequently occurs in double-loop learning because of the level of scrutiny that occurs, yet this conflict is not considered to be negative. Thus the goal (of organizational learning) is achieved. March refers to double-loop learning as *exploration* (1991) due to the creativity and risk-taking involved in solving problems that ultimately lead to organizational learning.

Other early work conducted by Duncan and Weiss (1979) recognized the value that organizational learning offered by helping researchers understand how organizations and the people in them function over time. They defined organizational learning as "the process within the organization by which *knowledge* about action–outcome relationships and the effect of the environment on these relationships is developed" (p. 84). They pointed out, however, that this knowledge should produce a change in behavior. Access to and use of the knowledge is central to organizational learning. Like Argyris and Schon (1978), Duncan and Weiss (1979) stated that regardless of its acceptance and change (or not), any knowledge acquired must be clearly, regularly, and widely communicated to achieve a *system of learning* (p. 89).

7.1.2 Synthesis and Expansion of Organizational Learning Concepts

The fact that the research was "fragmented and multidisciplinary" was noted by Shrivastava (1983) in his review of past organizational learning research. After his review of the work already summarized here, he found the concepts to be complementary but in want of empirical research. This led him to propose a two-dimensional typology of organizational learning. One dimension spanned the *origin* of learning: individual—organizational (resulting in three components, accounted for by a "middle" element representing groups or teams); the other spanned the *how* of learning: evolutionary—design (based on how spontaneous versus intentional the learning is). As a result, his model offered six categories of organizational learning. Interestingly the origin concept continues to undergo considerable discussion whereas the facets related to how learning is achieved have received little attention. Later we will discuss that lack of attention to the "how" of organizational learning may, at least in part, be responsible for the fact that organizations have been reluctant or slow to fully and successfully adopt organizational learning.

When distinguishing between learning and adaptation, Fiol and Lyles (1985) suggested that change alone does not imply learning. A different outcome may occur by chance or be temporarily achieved if a necessary change in processes does not become institutionalized. Fiol and Lyles stated that no consensus existed that related to the definition of organizational learning, at least in part because various divergent bodies of research assessed different aspects of the concept. They pointed out that general agreement existed around the notion that learning resulted in improved performance. In turn they proposed a definition for organizational learning that incorporated their belief: "the process of improving actions through better knowledge and understanding" (p. 803). They (like others such as Argyris and Schon, 1978; Duncan and Weiss, 1979) emphasized that knowledge, when communicated properly, resulted in organizational learning. This type of learning is more than the sum of individuals' (within the organization) learning. They also asserted that organizations require a balance between stability versus the changes associated with learning.

Fiol and Lyles's synthesis of past work resulted in proposing that four contextual factors affect the likelihood that (organizational) learning occurs. This included a *culture* that represented shared beliefs, ideologies, and norms that affect an organization's actions; deliberate *strategy* that drives organizational goals and objectives by providing "bounds and context" for decision making as well as "momentum" that promotes learning (p. 805); a *structure* within which decisions are made that may vary by or within the unit (i.e., limited vs. extensive) depending on the context of the decision; and *environments*, both internal and external to the organization, that have varying degrees of complexity and require varying degrees of change. This coincides with Senge (1990a, b) and other learning organization researchers whom we will discuss later in this chapter.

Leavitt and March (1988) reviewed the organizational learning literature and summarized it by building on behavioral studies of organizations. Using this framework, they stated that behavior (i.e., learning) in organizations is based on *experience* that (a) is either direct or interpretive and sometimes results in drawing a conclusion from a limited number of experiences (referred to as *superstitious learning*); (b) is *history-dependent*, drawing from memory that is both internal and external to the organization; and (c) is geared

to meeting *targets* that require a form of intelligence, implying that organizations can learn from their own limitations and/or mistakes. Leavitt and March also built on the diffusion of innovation literature (Rogers 2003) to suggest how the transfer of knowledge from one organization to the other occurs (Leavitt and March 1988).

Organization Science devoted its first issue of 1991 to the topic of organizational learning. In this issue, March, Sproull, and Tamuz presented their work on how organizations can learn from samples of one or fewer (1991). Both error and high reliability literature also build on this notion where catastrophic incidents in high-risk work environments (e.g., aviation, nuclear power, military, etc.) are relatively rare. Yet, the entire industry responds when an event of grave consequence occurs. In addition, they point out the value of and need to review and understand historical events to promote organizational learning. Events posing similar risk or having comparable outcomes must be recognized. This work supports the careful review of near misses and hypothetical consequences of actions as identified through risk assessment (Krouwer 2004; Solomon and Petosa 2001) that has been introduced to health care in recent years. (Interestingly, the Tamuz et al. paper was reprinted 12 years later, in the 2003, volume 12, *Quality and Safety in Health Care*, pp. 465–472.) Another perspective included in the 1991 issue of *Organization Science* includes Weick's (1991) description of strategies for conducting organizational learning research. One strategy suggested retaining the traditional definition of organizational learning that is based primarily on the concept of *stimulus–response* and thus would focus on behavior change. The other strategy suggested greater emphasis on knowledge acquisition and thus would not require a change in behavior.

7.1.3 Learning Organizations

The concept of learning organizations as presented by Peter Senge (1990a, b) brought the notion of organizational learning to a more functional level for managers. Learning organizations offer a way of managing an organization by espousing and implementing rather clear, although not simple, tenets. The distinction between learning organizations and organizational learning stems from the cognitive basis for adaptation and improvement that exists in learning organization research. Learning organizations require that shared mental models are developed. This is much like the "stated and known shared paradigm" proposed by Duncan and Weiss (1979) and also the fact that organizations are systems (Edmondson and Moingeon 1998). Learning organizations offer yet another perspective on organizational learning.

Senge proposed that *leadership* is the distinguishing factor that influences whether or not a learning organization can be achieved. The leader is a *designer* who clearly and consistently states the organization's purpose and priorities, a *teacher* to help all in the organization (leader included) develop clear views of "current reality" that result in shared mental models (Senge 1990b), and a *steward* of those they lead and those they serve as defined in the organization's mission. When serving these roles, the leader helps build shared vision; identify, challenge, and test mental models (much like double-loop learning as defined by Argyris and Schon, 1978); and promote systems thinking.

Subsequently, leaders motivate everyone in the organization to recognize and understand the big picture and how workers contribute to it. The leaders are responsible for developing an environment that promotes and facilitates learning, and then ensures that idea (information) exchange occurs within and between all levels of the organization (Garvin 1993).

Given these premises we suggest that unit and group leaders serve as the conduits for organizational learning between top-level managers and others working within the organization. Considerable time and effort is required to achieve effective communication between individuals who function at varying levels within the organization, yet multilevel, multidirectional communication is required for learning to occur. Leaders must recognize that "whichever direction ideas flow through the organization, it is clear that nothing will happen unless they do flow" (Simon 1991).

One must be careful to not oversimplify organizational learning. Explanations for why organizational learning has not been achieved point to the strong focus on top management to successfully promote and oversee the implementation of organizational learning processes. Yet leaders alone cannot ensure that organizational learning occurs. Likewise, the relatively scant research and lack of tools and guidelines on how to successfully achieve organizational learning all pose challenges for leaders. Finally, those leaders of organizations that aspire to achieve organizational learning discover that there are no objective means of measuring the success of it (Garvin, Edmondson, and Gino 2008). What top management can do is ensure that the "building blocks" (p. 110) to support organizational learning are solid and secure. This requires that top management create and maintain a supportive learning environment, ensure that specific knowledge-sharing processes and practices are developed and maintained, and reinforce learning through their own words and actions. Workers should be free of fear if they challenge or disagree with top management, while leaders should recognize and promote differences and new ideas. Keep in mind though that organizational learning

cannot be achieved and perpetuated without cooperation and a willingness by everyone in the organization, regardless of their role, to espouse such learning.

7.1.4 Individual and Organizational Learning

Building on the tenets of learning organizations, Stata (1989) provided a useful distinction between individual and organizational learning. He suggested that unlike individual learning, organizational learning requires shared insights, knowledge, and mental models among all major decision makers. He cautioned, however, that learning occurs at the rate at which the "slowest link learns" (p. 64). He also emphasized that learning builds on previous knowledge and experience that together constitute organizational memory. He saw systems thinking as the mechanism by which both individual and organizational learning occur.

This distinction is further developed by Kim (1993) who offered two aspects to his definition of learning: an operational aspect and a conceptual aspect—the "know-how" and "know-why" of learning, respectively (p. 38). He stated that a balance between the two must exist for organizational learning to occur and that the *transfer* of learning "is at the heart of organizational learning" (p. 37).

Simon (1991) went so far as to suggest that *individual* learning in organizations does not occur at the one-person level. He sees individual learning in organizations as a "social, not a solitary, phenomenon" (p. 125). The transfer of information from one individual (or group) to another individual (or group) is critical to learning. Once this learning is part of the organization's memory, it becomes organizational learning. These ideas coincide with his view of organizations as being systems comprising interrelated roles.

More recent work by Crossan and her group (Crossan, Lane, and White 1999; Vera and Crossan 2004) offered a means of learning in organizations that progresses through a "4I" cycle comprising intuiting, interpreting, integrating, and institutionalizing, similar to the framework proposed by Huber (1991). This research offers a link between the individual, the group or team, and the organization levels and explains the manner in which learning occurs, either (a) through a feedback mechanism (much like March's, 1991 exploitation), that promotes learning based on current organizational procedures, structures, and norms, or (b) a feed forward mechanism (exploration) where individual and/or group learning affects the organizational level in a manner that may require examination and subsequent changes within the organization, sometimes at the highest levels. Others build on this work by introducing the impact of politics in organizations (Lawrence et al. 2005).

7.2 Organizational Learning Research Applied to Health Care

Interestingly, chapter 1, paragraph 1 of Argyris and Schon's seminal work (1978) reads:

> There has probably never been a time in our history when members, managers, and students of organizations were so united on the importance of organizational learning. Costs of *health care …* have risen precipitously and we urge agencies concerned with these services to learn to increase their productivity and efficiency (p. 8).

Little did they know that costs were to be one of many issues (that now include quality and safety) that draws significant public attention and more scrutiny to health care than existed in the 1970s (Kohn et al. 2000). Unfortunately, relatively little has been accomplished in the health care industry to promote or achieve organizational learning. Leaders, known to be key to the change process (Senge 1990b), have not expressed sufficient or significant willingness to change and lead the change; this must be more than lip service. One way to accomplish this, which builds on values all in health care espouse, is to promote organizational learning as a way of both improving the system in which providers function (in Donabedian's (1978) terms: both the structure and processes of care) and then achieving the outcomes they so intently hope to achieve. In other words, we need to mine the "know what" and "know how" that results from individual and group learning efforts and then communicate this learning throughout the organization. Simon's (1991) "flow of ideas" must constantly occur.

7.2.1 Organizational Structure Considerations and Levels of Learning

"Human beings are naturally programmed to learn, organizations are not" (Carroll and Edmondson 2002). As we know health care organizations are collections of individuals who sometimes work alone and other times work in groups or teams that are generally aligned by their department or the service provided. There is a fluid nature to the work of those in health care that requires constant adaptation. In addition, some of those practicing in health care organizations (i.e., the physicians) are generally not employees of it. This poses challenges to building administrative structures and strategic management that promotes open dialog between independent, while also dependent, groups. Given these realities, how can a health care organization be managed to promote change and

learning and achieve better outcomes for patients and staff alike?

As stated previously, a characteristic of everyone who works in health care is an unselfish commitment to the customer: the patient. Our current scientific environment affords patients and providers alike continually changing and continuously improving health services. When improvement is clearly demonstrated professional societies, peers, and manufacturers offer common avenues for communicating the new information (i.e., learning) to their audiences. It is then the responsibility of the individual or groups to introduce the new practices and learning to their peers and/or the organization.

An example of this is discussed in various papers by Edmondson and others (Edmondson 2003; Edmondson et al. 2001; Pisano et al. 2001) who studied the introduction of minimally invasive cardiac surgery in 16 hospitals. This technology required considerable changes to the routines of those participating on the cardiac surgery team in the operating room. The hospitals studied were similar (all were among top-ranked cardiac surgery programs) yet they demonstrated widely varying levels of success when implementing the new technology. From their study, the researchers found that the extent of success depended heavily on factors related to individual, team, and organizational learning, including: (a) the manner in which team members were selected, their roles defined, and learning planned; (b) the team's training and the extent of practice and the willingness of its members to incorporate the changes required into their respective practices; (c) the openness of the team leader to request and accept input, and then make necessary changes to standard operating procedures; and (d) a willingness of the team to review data related to the team's perceived performance and make improvements when it was deemed appropriate.

In the successful implementations, the research team observed double-loop learning demonstrated by the individual team members and the group as a whole. Both were willing to forego previous standardized work procedures, develop new work processes, and promote learning. There was ongoing multidirectional *communication* within the teams and an expressed desire to *improve* and gain more *knowledge*. This example demonstrates more than local learning (Carroll and Edmondson 2002). It goes beyond March's (1991) exploitation to exploration and is similar to the performance improvements demonstrated by clinical microsystems (Mohr and Batalden 2002; Nelson et al. 2002).

A similar discussion on health care reengineering, in this case with the implementation of a medical home, is presented by Steele and his colleagues (2010). The success of their medical home model is credited, at least in part, to the commitment to continuous learning that occurs at all levels in the organization. Thus, changes to the model regularly occur as the organization learns from its experience. Patient outcomes have also improved as a result of the model.

At a different level of analysis, Tucker reported on cases of individual nurses solving problems—opportunities for learning—based on what they incurred in their daily practice (Tucker et al. 2002; Tucker and Edmondson 2002; Tucker 2004). In nearly every case observed, the respective nurse dealt with his job obstacle as a unique occurrence, never saw it in light of the system in which he was practicing, and rarely informed his manager of the specific event, or any pattern of similar events he experienced or observed. An opportunity for organizational learning was lost in nearly every one of these cases. Why? One reason may be due to the fact that these individuals did not appear to understand that they worked within a system of highly interdependent functions and work processes. Nor did they appear to work in a culture that promoted individual input as a means of identifying opportunities for system improvements. Finally, the nurses' potentially high workload may have influenced their ability to influence—or even consider—any sort of (organizational) learning.

7.2.2 Learning (or Not) from Failure

We now introduce the concept of learning from failure. Although Cangelosi and Dill (1965) found failure to consistently result in learning in their case study, this was not supported in early research conducted in health care (Edmondson 1996; Tucker and Edmondson 2003). In one study (Edmondson 1996), the detection of drug errors, subsequent reporting of them, and the likelihood that learning would occur from the errors, varied both between and within two hospitals. The factor that most affected error identification was the respective unit's willingness and ability to openly discuss mistakes. Although the researchers could only speculate that more open units with accessible nurse managers were likely to perform better (i.e., experience less medication errors), the researchers were able to prove that error detection and reporting was a function of the unit's leadership.

In a study of 26 nurses at nine hospitals of varying sizes and patient population, Tucker and Edmondson (2003) observed 196 "failures." Of these, the vast majority (86%) of the failures were, upon examination, determined to be problems rather than errors. The research team investigated the type of problem solving that occurred for each problem. Coined *first-order problem solving*, the team found that most frequently (93% of the time) the nurses independently dealt with the problems by simply obtaining the missing information or

supply and then proceeding with their work. In light of a system, this form of problem solving is considered to be counterproductive due to the fact that little if any communication (or learning) occurs. The worker fails to report an event that has the potential to result in a system improvement or an opportunity for learning. In isolated instances (the remaining 7% of the time) *second-order problem solving* occurred. After the worker addressed the respective problem, she then communicated it to the responsible individual or department and also conveyed the incident to her manager. Once the manager was involved, the likelihood for greater analysis and possible change increased.

The more common instances of first-order problem solving did not encourage organizational learning to occur. Kim (1993) referred to these as instances of situational learning, considered by many to not even constitute learning. Tucker suggested three reasons why first-order problem solving occurred: (a) there is an emphasis on individual vigilance in health care—each provider is personally responsible for solving their own problems; (b) concerns of unit efficiency conflict with the idea that workers perform outside the realm of direct patient care; and, finally (c) more often today, health care workers are empowered to solve problems because frontline managers are frequently distanced from the local aspect of their job and are inaccessible to their staff to address issues related to daily operations. Instead, many managers deal with issues at a different level and then lose touch with the issues experienced by individuals who report to them. As demonstrated in the instances presented, most of the opportunities for organizational learning that arose from the various problems workers incurred were lost.

Research in health care and other domains is now beginning to support the premise that organizations learn from each other. Studies have thus far demonstrated the following: (a) The extent of organizational learning depends on the level of dependency between organizations (Chuang et al. 2007). (b) Organizations are more likely to learn from others' big failures than they are to learn from successes or failures having little or no consequence (Madsen 2010), near misses, or threats (Tamuz et al. 2011). (c) Organizations are also learning from their own as well as outside (e.g., regional, national) reporting systems (Tamuz et al. 2011). (d) Benchmarking between work groups and/or organizations has influenced organizational learning.

So, how do organizations learn from their own or others' failure? How can opportunities for organizational learning be recognized and fostered? Much of the current literature supports the overwhelmingly influential role of leadership. We briefly discuss this in the next section.

7.2.3 Leadership

"Leadership must be distributed broadly if organizations are to increase their capacity for learning and change" (Carroll and Edmondson 2002). Much of the focus of patient safety and organizational learning in health care literature points to the need for leadership to promote and instill learning within the organization at all levels and to then communicate the learning that occurred throughout the organization (Mohr et al. 2002; Batalden et al. 2002; Batalden and Splaine 2002). Therefore learning must be managed (Bohmer and Edmondson 2001).

Chuang and others (Chuang et al. 2004; Berta and Baker 2004) were among the first to assess patient safety in light of organizational learning. In their research, they posited that varying factors affect the extent of responsiveness to patient safety issues within an organization that in turn affects the extent of organizational learning that occurs. They identified event-specific, group, organizational, and inter-organizational factors that influence an organization's willingness and ability to respond to adverse events and then possibly learn from them.

A more recent work by this research team (Chuang et al. Berta 2007) suggested that the extent of organizational learning depends on the structure and norms of health care organizations. Organizational learning is a "multi-level" (p. 332) interdependent process whereby ongoing feedback and reinforced learning between individuals, groups, and management must occur for learning to become institutionalized. Carroll and Fahlbruch (2011) also found that the relationship between teams and managers is crucial for team and organizational learning to be achieved. Especially pertinent to health care is the recognition that tight coupling of groups and units has also been shown to influence the extent to which group learning extends to organizational learning (Argote 1999).

Learning must be part of an organization's mission. Change, resulting from organizational learning, must be allowed to occur and result in improved work processes that then have a positive impact on the patient, worker, and organizational outcomes. One aspect of the continuous improvement movement that was also emphasized by the Institute of Medicine's *To Err is Human* (Kohn et al. 2000) is the fact that health care must function better as a system. In many instances, errors occur due to the nature of the system and must be addressed, resolved, learned from, and then communicated system-wide. Top-down (feedback) and bottom-up (feed forward) learning must occur to achieve organizational learning (Crossan et al. 1999; Vera and Crossan 2004). The politics of organizations must be recognized but not be allowed to interfere with the organization's ability to learn (Lawrence et al. 2005). There must be a shared vision and shared mental models.

Also, we must learn from the science of training (Salas et al. in this volume; Baker et al. in this volume) to recognize that it provides opportunities for learning by using proven training methods. All too often, learning occurs but the communication and ongoing monitoring of what changed as a result of learning does not. In these instances, the knowledge gained never becomes a part of the organization's memory; it is not institutionalized.

The aforementioned characteristics are a combination of the structure, values, and skills that Carroll and Edmondson (2002) suggest enhance organizational learning. Amid the countless priorities in health care, one cannot overlook the influence, value, and power organizational learning offers health care organizations, those providing care and services, and, most important, those receiving the services.

7.3 Risk Assessments as a Means for Promoting Organizational Learning

To this point we have discussed theories of organizational learning, findings from limited research that explain when and why organizational learning is sometimes achieved in health care and elsewhere, and the unique challenges health care poses for organizational learning. More recently, we observe that learning from failure has been successful in those instances where the consequences of error have been most negative. In this section, we discuss what has been learned by using risk assessment methods (both prospectively and retrospectively) to promote organizational learning. Discussion of specific methods and philosophies is found elsewhere (Carayon et al. 2010) and in this handbook (Israelski and Muto in this volume).

Can health care, where external accrediting bodies and regulating agencies require regular risk assessments, utilize these methods to achieve more than compliance with external requirements? Can organizations successfully conduct risk assessments, implement necessary changes, thoroughly communicate the outcomes, and achieve the same level of learning other industries that use these methods have achieved? Carroll and Fahlbruch (2011) suggest that risk assessments "... offer an opportunity to learn about safe and unsafe operations, generate productive conversations across engaged stakeholders, and bring about beneficial changes to technology, organization, and mental models" (p. 2).

7.3.1 Risk Assessment Requirements in Health Care

The risk assessment methods most commonly used in health care include failure mode and effects analysis (FMEA) and its associated health care failure mode and effects analysis (HFMEA), root cause analysis (RCA), fault tree analysis (FTA), and probabilistic risk assessment (PRA). Cause and effect diagrams (also know as Ishikawa and fishbone diagrams) are also used to identify and evaluate known or potential causes of error or problems. Many health care organizations conduct risk assessments to comply with external accrediting and licensing requirements. We question, however, how frequently these tools are recognized and used as a means of educating individuals, groups, and ultimately the organization. When outcomes of risk assessments result in changes to work, standardization, or more clearly written procedures, how often and how well is the motivation for the change communicated? Do health care providers and others affected by change(s) that results from a risk assessment fully understand the reason for the change? If they do not, and therefore see the change as unwarranted, do workers (including clinicians and other providers) create potentially unsafe workarounds to minimize the impact of the change on their work? All too often workers return to past practices and the learning opportunity that results from a risk assessment is lost.

One of the greatest challenges health care poses for successful use of risk assessment methods stems from the complexity of health care (Leveson 2011) at both the organizational and process level. Many of the risk assessment methods were developed for use in industries that incur little change or change slowly over an extended period of time. The numerous interactions and interdependencies of the processes and systems in health care make it difficult for participants in health care risk assessment efforts to identify all of the possible interactions and outcomes. Yet identifying all such interactions and outcomes is key to the success of risk assessment. Work by Wetterneck and her colleagues (2009) proposes guidelines for planning and conducting risk assessments based on interviews of individuals who served on FMEA teams. Clearly defining the team's objective and limiting the scope of the FMEA were two key guidelines identified.

Likewise, Leveson (2011) stresses the need to question the way risk analyses have traditionally occurred and ensure that teams think in terms of systems. This is especially relevant to health care. Ramanujam and Goodman (2011) also suggest that the risk assessment method a team utilizes is not as important as the quantity and quality of open communication and sharing that occurs between team members. Wetterneck and colleagues (2009) also identified the importance of team composition. Participants and leaders of risk

assessments must be careful, however, to not oversimplify the process being assessed or recommendations made as a result of the risk assessment.

Not surprisingly, research findings confirm (Ramanujam and Goodman 2011; Carroll and Fahlbruch 2011) that simply conducting a risk assessment does not ensure that a group learns. For learning to occur, any knowledge gained and changes made to processes must be clearly linked to the group's effort. All learning must be shared to ensure that the individuals and the group learn. What was learned must also be stored and be readily accessible over the long term. One means of achieving this is by regularly monitoring for subsequent events, similar to or related to the process the risk assessment analyzed. Monitoring can help ensure that known or suspected problems are immediately identified or do not reoccur, and can also provide a feedback loop. Simultaneously the learning is reinforced by relating change(s) made to what was learned. This is something health care has not been successful in developing. Feedback systems are not inherent to daily operations in many health care organizations. Also, communication and reinforcement of what is learned at multiple levels—individual, group, and organization—is not commonplace in health care.

7.3.2 Risk Assessment in Other Industries

Ramanujam and Goodman (2011) suggests that learning must be an explicit goal when conducting risk assessments. Group members must go beyond "just" analyzing risks and ensure that learning has also occurred. Group members must be encouraged to reflect on what the team has learned, what changes occurred as a result of the team's effort (linking the solution(s) with the identified cause(s)), and determine how the group will "remember" what was learned. Feedback mechanisms can and should be developed to reinforce organizational memory. The role of managers as both team members and representatives of top management demonstrates how crucial their role in organizational learning is. They must simultaneously learn and consistently communicate what was learned to all directions in the organization.

Using the nuclear industry as an example, Wahlstrom (2011) describes what has been done to promote organizational learning. He points out the long tradition within the industry to share knowledge. Although health care is a larger industry in the United States, knowledge sharing has been and continues to be key to improving health and health care systems and processes. The nuclear industry also functions under various oversight industries that support peer review. Likewise, health care organizations and providers function under regular oversight by licensing, regulating, and payment systems as well as internal peer review mechanisms. Finally, Wahlstrom points out that the nuclear industry has generally developed and implemented effective mechanisms that facilitate organizational learning. Is this where health care differs? What has health care accomplished as an industry to develop effective feedback loops at all levels that communicate learning and ensure that learning is not lost? Espousing organizational learning requires organizational change. Is organizational learning "lost" because health care organizations constantly tend to more frequently change in response to external mandates?

Wahlstrom (2011) also reminds us that although risk assessments can and should result in organizational learning, and that organizational learning implies something better, changes implemented as a result of a risk assessment can produce problems. For example, changes not carefully planned and/or communicated can produce unintended consequences (Ash et al. 2004). By designing effective monitoring systems the frequency and impact of these problems can be lessened. Another consideration for risk assessments is to recognize that solutions that are appropriate in one organization may not translate well to another. Adopting processes from one organization may require adaptation to successfully address issues identified from a risk assessment. Finally, organizations must recognize and ensure that the initial costs associated with the change should have long-term benefits. If this is not the case, the risk assessment did not achieve one of the fundamental goals: to lessen or eliminate risks associated with current system and process design.

7.4 Conclusion

Briefly we have reviewed seminal organizational learning research and presented it in light of health care. Given the extensive scrutiny health care organizations receive, upper and middle managers and others working in health care organizations must recognize the need to espouse and practice organizational learning. Organizational learning offers a powerful means of improving both the quality and safety of care as well as the skills and work environment for those working in health care. We are beginning to observe more instances of learning through failure, both within organizations that experience failure and when an organization's failure is observed by other health care organizations. Risk assessments, now required by many accrediting and regulating bodies, offer a powerful means of responding to and anticipating known or potential failure and risk. We must emphasize that

learning at all levels should be one of the primary objectives of risk assessments. Communicating and monitoring the outcomes, lessons learned, and changes made as a result of risk assessments will provide one significant means of promoting and demonstrating organizational learning.

References

Argote, L. 1999. *Organizational Learning: Creating, Retaining and Transferring Knowledge*. Boston: Kluwer Academic Publishers.

Argyris, C. and D.A. Schon. 1978. *Organizational Learning: A Theory of Action Perspective*. Reading: MA: Addison-Wesley Publishing Company.

Ash, J.S., M. Berg, and E. Coiera. 2004. Some unintended consequences of information technology in health care: The nature of patient care information system-related errors. *Journal of the American Informatics Association* 11 (2):104–112.

Baker, D., Salas. E., P. Barach, J. Battles, and H. King. 2011. The relation between teamwork and patient safety. In *Handbook of Human Factors and Ergonomics in Health Care and Patient Safety*, ed. P. Carayon. Mahwah, NJ: Lawrence Erlbaum Associates.

Batalden, P.B. and M. Splaine. 2002. What will it take to lead the continual improvement and innovation of health care in the twenty-first century?. *Quality Management in Health Care* 11 (1):45–54.

Batalden, P.B., D.P. Stevens, and K.W. Kizer. 2002. Knowledge for improvement: Who will lead the learning?. *Quality Management in Health Care* 10 (3):3–9.

Berta, W.B. and R. Baker. 2004. Factors that impact the transfer and retention of best practices for reducing error in hospitals. *Health Care Management Review* 29 (2):90–97.

Bohmer, R.M.J. and A.C. Edmondson. 2001. Organizational learning in health care. *Health Forum Journal* 44 (2):32–35.

Cangelosi, V.E. and W.R. Dill. 1965. Organizational learning: Observations toward a theory. *Administrative Science Quarterly* 10 (2):175–203.

Carayon, P., H. Faye, A.S. Hundt, B.-T. Karsh, and T.B. Wetterneck. 2010. Patient safety and proactive risk assessment. In *Handbook of Healthcare Delivery Systems*, ed. Y. Yih. Boca Raton, FL: Taylor & Francis.

Carroll, J. S. and B. Fahlbruch. 2011. "The gift of failure: New approaches to analyzing and learning from events and near-misses." Honoring the contributions of Bernhard Wilpert. *Safety Science* 49 (1):1–4.

Carroll, J.S. and A. Edmondson. 2002. Leading organizational learning in health care. *Quality and Safety in Health Care* 11 (1):51–56.

Carroll, J.S. and M.A. Quijada. 2007. Tilting the culture in health care: Using cultural strengths to transform organizations. In *Handbook of Human Factors and Ergonomics in Health Care and Patient Safety*, ed. P. Carayon. Mahwah, NJ: Lawrence Erlbaum Associates.

Chuang, Y.-T., L. Soberman-Ginsburg, and W.B. Berta. 2004. Why others do, but you don't? A multi-level model of responsiveness to adverse events. Paper read at *Academy of Management Annual Meeting*, 2004, New Orleans, LA.

Chuang, Y., L. Ginsburg, and W. B. Berta. 2007. Learning from preventable adverse events in health care organizations: Development of a multilevel model of learning and propositions. *Health Care Management Review* 32:330–340.

Crossan, M. and T. Guatto. 1996. Organizational learning research profile. *Journal of Organizational Change Management* 9 (1):107.

Crossan, M.M., H.W. Lane, and R.E. White. 1999. An organizational learning framework: From intuition to institution. *Academy of Management Review* 24 (3):522–537.

Crossan, M.M., H.W. Lane, R.E. White, and L. Djurfeldt. 1995. Organizational learning: Dimensions for a theory. *The International Journal of Organizational Analysis* 3 (4):337–360.

Cyert, R.M. and J.G. March. 1963. *A Behavioral Theory of the Firm*. Englewood Cliffs: NJ: Prentice-Hall, Inc.

Donabedian, A. 1978. The quality of medical care: Methods for assessing and monitoring the quality of care for research and for quality assurance programs. *Science* 200:856–864.

Duncan, R. and A. Weiss. 1979. Organizational learning: Implications for organizational learning. *Research in Organizational Behavior* 1:75–123.

Edmondson, A. and B. Moingeon. 1998. From organizational learning to the learning organization. *Management Learning* 29 (1):5–20.

Edmondson, A.C. 1996. Learning from mistakes is easier said than done: Group and organizational influences on the detection and correction of human error. *The Journal of Applied Behavioral Science* 32 (1):5–28.

Edmondson, A.C. 2003. Speaking up in the operating room: How team leaders promote learning in interdisciplinary action teams. *Journal of Management Studies* 40 (6):1419–1452.

Edmondson, A.C., R.M. Bohmer, and G.P. Pisano. 2001. Disrupted routines: Team learning and new technology implementation in hospitals. *Administrative Science Quarterly* 46:685–716.

Fiol, C.M. and M.A. Lyles. 1985. Organizational learning. *Academy of Management Review* 10 (4):803–813.

Garvin, D.A., A.C. Edmondson, and F. Gino. 2008. Is yours a learning organization? *Harvard Business Review*:109–116.

Garvin, D.A. 1993. Building a learning organization. *Harvard Business Review*:78–91.

Huber, G. 1991. Organizational learning: The contributing processes and the literatures. *Organization Science* 2 (1):88–115.

Israelski, E.W. and W.H. Muto. 2011. Human factors risk management for medical products. In *Handbook of Human Factors and Ergonomics in Health Care and Patient Safety*, edited by P. Carayon. Mahwah, NJ: Lawrence Erlbaum Associates.

Kim, D.H. 1993. The link between individual and organizational learning. *Sloan Management Review* 35 (1):37–50.

Kohn, L.T., J.M. Corrigan, and M.S. Donaldson. 2000. *To Err is Human: Building a Safer Health System*. Edited by. Washington, D.C.: National Academy Press.

Krouwer, J.S. 2004. An improved failure mode effects analysis for hospitals. *Archives of pathology and laboratory medicine* 128 (6):663–667.

Lawrence, T.B., M.K. Mauws, B. Dyck, and R.F. Kleysen. 2005. The politics of organizational learning: Integrating power into the 4I framework. *Academy of Management Review* 30 (1):180–191.

Leavitt, B. and J.G. March. 1988. Organizational learning. *Annual Review of Sociology* 14:319–340.

Leveson, N. 2011. Applying systems thinking to analyze and learn from events. *Safety Science* 49 (1):55–64.

Madsen, P. and Desai, V. 2010. Failing to learn? The effects of failure and success on organizational learning in the global orbital launch vehicle industry. *Academy of Management Journal* 53 (3):451–476.

March, J.G. 1991. Exploration and exploitation in organizational learning. *Organization Science* 2 (1):71–87.

March, J.G., L.S. Sproull, and M. Tamuz. 1991. Learning from samples of one or fewer. *Organization Science* 2 (1):1–13.

Mohr, J.J., H.T. Abelson, and P. Barach. 2002. Creating effective leadership for improving patient safety. *Quality Management in Health Care* 11 (1):69–78.

Mohr, J.J. and P.B. Batalden. 2002. Improving safety on the front lines: The role of clinical microsystems. *Quality and Safety in Health Care* 11 (1):45–50.

Nelson, E.C., P.B. Batalden, T.P. Huber, J.J. Mohr, M.M. Godfrey, L.A. Headrick, and J.H. Wasson. 2002. Microsystems in health care: Part 1. Learning from high-performing frontline clinical units. *The Joint Commission Journal on Quality Improvement* 28 (9):472–493.

Pisano, G.P., R.M.J. Bohmer, and A.C. Edmondson. 2001. Organizational differences in rates of learning: Evidence from the adoption of minimally invasive cardiac surgery. *Management Science* 47 (6):752–768.

Popper, M. and R. Lipshitz. 1998. Organizational learning mechanisms: A structural and cultural approach to organizational learning. *Journal of Applied Behavioral Science* 34 (2):161–179.

Ramanujam, R. and P.S. Goodman. 2011. The challenge of collective learning from event analysis. *Safety Science* 49 (1):83–89.

Rogers, Everett M. 2003. *Diffusion of innovations*. 5th ed. New York: Free Press.

Salas, E., K.A. Wilson-Donnelly, D.E. Sims, C.S. Burke, and H.A. Priest. 2011. Teamwork training for patient safety: Best practices and guiding principles. In *Handbook of Human Factors and Ergonomics in Health Care and Patient Safety*, edited by P. Carayon. Mahwah, NJ: Lawrence Erlbaum Associates.

Senge, P.M. 1990a. *The Fifth Discipline: The Art and Practice of the Learning Organization*. New York: Currency Doubleday.

Senge, P.M. 1990b. The leader's new work: Building learning organizations. *Sloan Management Review*:7–23.

Shrivastava, P. 1983. A typology of organizational learning systems. *Journal of Management Studies* 20 (1):7–28.

Simon, H.A. 1991. Bounded rationality and organizational learning. *Organization Science* 2 (1):125–134.

Smith, M. and J. Bartell. 2011. The relationship between physician professionalism and health care systems change. In *Handbook of Human Factors and Ergonomics in Health Care and Patient Safety*, edited by P. Carayon. Mahwah, NJ: Lawrence Erlbaum Associates.

Solomon, R. and L. Petosa. 2001. Appendix F: Quality improvement and proactive hazard analysis models: Deciphering a new Tower of Babel. In *Patient Safety: Achieving a New Standard of Care*: Institute of Medicine.

Stata, R. 1989. Organizational learning: The key to management innovation. *Sloan Management Review* 30 (3):63–74.

Steele, G.D., J.A. Haynes, D.E. Davis, J. Tomcavage, W.F. Stewart, T.R. Graf, R.A. Paulus, K. Weikel, and J. Shikles. 2010. How Geisinger's advanced medical home model argues the case for rapid-cycle innovation. *Health Affairs* 29 (11):2047–2053.

Tamuz, M., K.E. Franchois, and E.J. Thomas. 2011. What's past is prologue: Organizational learning from a serious patient injury. *Safety Science* 49 (1):75–82.

Tucker, A.L. 2004. The impact of operational failures on hospital nurses and their patients. *Journal of Operations Management* 22:151–169.

Tucker, A.L. and A.C. Edmondson. 2002. Managing routine exceptions: A model of nurse problem solving behavior. *Advances in Health Care Management* 3:87–113.

Tucker, A.L. and A.C. Edmondson. 2003. Why hospitals don't learn from failures: Organizational and psychological dynamics that inhibit system change. *California Management Review* 45 (2):55–72.

Tucker, A.L., A.C. Edmondson, and S. Spear. 2002. When problem solving prevents organizational learning. *Journal of Organizational Change Management* 15 (2):122–137.

Vera, D. and M. Crossan. 2004. Strategic leadership and organizational learning. *Academy of Management Review* 29 (2):222–240.

Wahlstrom, B. 2011. Organisational learning – Reflections from the nuclear industry. *Safety Science* 49:65–74.

Wang, C.L. and P.K. Ahmed. 2003. Organizational learning: A critical review. *The Learning Organization* 10 (1):8–17.

Weick, K.E. 1991. The nontraditional quality of organizational learning. *Organization Science* 2 (1):116–124.

Wetterneck, T.B., A.S. Hundt, and P. Carayon. 2009. FMEA team performance in health care: A qualitative analysis of team member perceptions. *Journal of Patient Safety* 5 (2):102–108.

8

The Relationship between Physician Professionalism and Health Care Systems Change

Maureen A. Smith and Jessica M. Bartell

CONTENTS

8.1 Physician Professionalism 110
8.2 Organizational Incentives and Agency Theory 110
8.3 Systems Perspective: An Engineering Approach...111
8.4 Medical Approach to Systems Change..111
8.5 New Approach 112
    8.5.1 Disease Management Programs..113
    8.5.2 Multidisciplinary Patient Care Teams 113
8.6 Accountability 113
8.7 Conclusions..114
References..114

The idea of physician professionalism dates back to the Hippocratic oath, which instructed physicians to practice medicine for "the benefit of patients and abstain from whatever is deleterious and mischievous" (Shortell et al., 1998). Patient trust in the physician–patient relationship is based on the idea that physicians have responsibility and control over medical decision making and that physicians prioritize the needs of patients over all other considerations (Mechanic and Schlesinger, 1996). In recent years, the issue of professionalism among physicians has been the central focus of numerous national and international medical conferences, medical journal publications, and public initiatives by professional organizations (American Board of Internal Medicine Foundation, 2003). Concern by these groups over the effects of a changing U.S. health care delivery system on physicians' primary dedication to patients has spawned a call for renewed commitment to the principles of the primacy of patient welfare, patient autonomy, and social welfare (ABIM Foundation, ACP-ASIM Foundation, and European Federation of Internal Medicine, 2002). In part, this professionalism "movement" has been motivated by perceived negative effects of the efforts by health care organizations to improve quality and decrease costs on the ability of physicians to serve as advocates for their patients (Shortell et al., 1998). In the current era in which competition between health care organizations is increasingly based on reducing inappropriate variations in care (Wennberg, 1998) and improving quality

of care and patient satisfaction (Enthoven and Vorhaus, 1997), the perceived conflict between the goals of physicians as professionals and organizational management efforts may help to explain why so many organizational attempts to change physician behavior meet with failure (Grimshaw et al., 2001).

There has been increasing recognition that quality medical care is a property of systems not just of individuals. In 2001, the Institute of Medicine issued its report on health care quality, *Crossing the Quality Chasm*, that stressed the importance of system design in creating health care environments that are both safe and that produce quality health-related outcomes (Institute of Medicine, 2001). Human factors engineering, a concept adapted from the field of industrial engineering that promotes system design as a method to improve the interactions between the worker and the work environment, is increasingly used to address issues of work efficiency and safety (Helander, 1997). Systems design has also been identified as an important component of quality improvement and patient safety. Research has shown system improvements that decrease reliance on individual memory and attention, decrease error rates, and improve the quality of care (Institute of Medicine, 2001). This systems approach to changing physician behavior may not only be a more effective approach, but may be better accepted by physicians than traditional organizational incentives to decrease costs and improve quality. Furthermore, this approach may be more reconcilable with the concept of physician professionalism because of its more indirect effects on physician behavior.

In this chapter, we examine the tension between physician professionalism and the financial and nonfinancial incentives that currently dominate the relationships between physicians and organizations. We examine the roles of system redesign in health care quality improvement and discuss the implications of a systems approach to health care organizational management for physician job satisfaction, professionalism, accountability, and the quality of patient care.

8.1 Physician Professionalism

The concept of professionalism implies that professionals are bound by the ethics of their professions to serve in their clients' best interests. Professionals are drawn to their professions because of an altruistic desire to serve, and this altruism restrains them from responding primarily to their own economic self-interest. They are also embedded in a community of peers with similar altruistic intentions and values that serve to further restrain their self-interest (Sharma, 1997). Historically, the physician was seen as the central force from which healing power originated and was controlled. With this conceptualization of physician professionalism came great responsibility and great respect. However, it has been noted that physician professionalism also fosters a "culture of blame" when things go wrong because, if physicians are responsible for the entirety of the medical process, then they are also exclusively to blame for poor quality care (Leape, 1994). It is this desire to maintain autonomy and responsibility related to professional status that may, in part, explain physician resistance to incentives connected with health care organizational management.

8.2 Organizational Incentives and Agency Theory

Multiple and sometimes conflicting organizational incentives have created a sense among physicians that their role as professionals may be threatened (Edwards et al., 2002). These organizational incentives include financial as well as nonfinancial incentives to manage utilization, decrease inappropriate variations in care, encourage evidence-based practice of medicine, and increase productivity. Examples of these incentives include salary withholds and bonuses, utilization review, practice guidelines, treatment reminders and prompts, and peer comparisons (Flynn et al., 2002). These incentives are increasing in prevalence and in influence: a study of the arrangement that physicians make with managed care organizations demonstrated that extensive utilization review occurred in 62% of physician practices, 63% of physicians used practice guidelines, and 68% received personalized profiles of their practice patterns (Gold et al., 1995). These incentives, which are often viewed as a form of bureaucratic control, may be resented by practicing physicians who see control and authority over their practice decisions as their professional responsibility. As a result, these incentives have often been less than fully successful in motivating actual changes in physician behavior. Even with additional training and incentives, changing physician behavior is recognized as a very difficult process (Grimshaw et al., 2001).

The role and power of these incentives—as well as physicians' resistance to them—can be explained by agency theory. Agency theory was developed in the context of corporate interactions to describe the relationship between those who contract for services (principals) and those who perform such services (agents). This theory has been applied in the health care setting to describe the interactions between health care organizations (i.e., principals) and physicians (i.e., agents; Flood and Fennell, 1995). Agency theory says that financial and nonfinancial incentives are offered by principals in order to induce behaviors in agents that are consistent with the goals of the principal (Sappington, 1991). These incentives cause physicians to balance the interests of their patients (physicians' primary concern as professionals) with another set of interests, namely the interests of the organization with which the physicians are associated. This concept has been coined "double agency" and refers to the tension between these different influences and loyalties (Shortell et al., 1998). Because physician professionalism suggests that physicians should act as "perfect agents" for their patients by holding patient goals paramount (Shortell et al., 1998), the uncomfortable balance of having to act as agent for both patients and organizations may explain some of the physician resistance to organizational incentives.

There is growing distress among physicians, which appears largely related to the incentives present in medical practice associated with managed care. This distress may be contributing to the growing crisis of confidence in physicians among patients (Mechanic and Schlesinger, 1996) as well as to increased career dissatisfaction among physicians involved in managed care (Linzer et al., 2000). Furthermore, the inherent conflicts between organizational incentives and physician professionalism may undermine effectiveness of health care organizational quality improvement initiatives. For these reasons, a new approach to managing medical practice and physician behavior is needed. A systems perspective, which focuses on the structure of health care, health care processes, and the interface between workers and the work environment, may be more compatible with physician professionalism than are direct incentives. It may also be a more effective method of quality improvement and may help to improve physician acceptance of organizational change efforts.

8.3 Systems Perspective: An Engineering Approach

The Institute of Medicine (2001), in its report *Crossing the Quality Chasm*, called on health care organizations to take a systematic approach to quality improvement. The report pointed to evidence that systems changes can effectively change practice and improve health care quality by reducing medical errors and decreasing inappropriate variations in care. Although the safety and efficiency arguments are convincing, it is becoming increasingly apparent that system redesign may also help physicians to improve their worklives and to better meet their patients' needs while functioning within the required constraints of health care delivery systems (Morrison and Smith, 2000).

The balance theory of human factors engineering, which focuses on system redesign to optimize the interactions between workers and the system they work in, stresses the interactions between the individual, the task, technology, environment, and organizational factors (Smith and Sainfort, 1989; Carayon and Smith, 2000). The balance model places the individual in the center of the system with his or her work affected by the other aspects of the work system (i.e., the task, technology, the work environment, and organizational factors). The individual then makes decisions that are also affected by stress and other modulating factors such as individual competence, age, motivation, and gender (Helander, 1997). Finally, the individual manifests a response output. The interaction between this action and the work system produces an outcome. In designing systems for better efficiency and safety, human factors engineers concentrate on redesigning environmental and workplace factors rather than selection and training of individuals. Training is utilized to ease organizational changes, especially with the implementation of new technology.

Human factors engineering has been increasingly employed in order to change the way that health care is delivered (Gosbee, 1999). For example, human factors engineering has been employed in anesthesiology in the development of engineered safety devices such as a system of gas connectors that does not allow a gas hose or cylinder to fit the wrong site. This type of advance, combined with new technologies, standards, and guidelines, and an emphasis on a "culture of safety," has helped to dramatically decrease the rate of death due to anesthesia (Gaba, 2000). Another example of system change to improve safety is the institution of computerized physician order entry, which has been shown to reduce physician and nursing transcription errors (Mekhjian et al., 2002). Both of these examples illustrate a new approach to health care: the redesign of the environment and important tools in order to optimize the performance of the operator (i.e., the physician).

Efficiency in health care has similarly been helped by systems improvements. The enormous expansion of knowledge and technology in medicine in recent years has made it impossible for physicians to apply all available preventive screening, convey all known information about relevant medical conditions, and provide all available treatments for each patient in the increasingly limited time available for an individual office appointment. This situation, in which physicians find themselves on a hypothetical treadmill and "running faster just to stand still," has been referred to as "hamster health care" (Morrison and Smith, 2000, p. 1541). Systems redesign, such as computerized screening protocols and electronic communications, may relieve the physician of some of the less demanding functions of health care and therefore save time for more meaningful patient care.

Quality improvement efforts that incorporate system redesign may also be more successful in changing physician behavior, maintaining physician job satisfaction, and preserving physician professionalism than traditional incentive approaches. System redesign, which is based on the idea of making it "easy to do things right and hard to do things wrong," may be a more indirect, easier approach to changing physician behavior (Leape, 1994). Well-designed systems may be less resented than direct financial and nonfinancial incentives set forth by health care organizations. Often, systems changes are quite transparent, so they are not in conflict with physician autonomy and control. However, involvement of physicians in the process of system design is important to both success of the system and to physician acceptance (Leape, 1994). Maintenance of a sense of job control and participation in the process of organizational change may improve physician acceptance of and adherence to organizational quality improvement programs (Edwards et al., 2002).

8.4 Medical Approach to Systems Change

Avedis Donabedian (1988) was instrumental in developing a theory and a language for talking about quality improvement in health care. His model of quality assessment—the structure-process-outcomes (SPO) model—foreshadowed many of the ideas of human factors engineering. First, the SPO model acknowledges that the structural factors such as material resources, human resources, and organizational structure likely influence the quality of care. Second, amenities of care, which include convenience, comfort, quiet, and privacy (all characteristics of the environment surrounding the health care interaction), are emphasized as important components of quality. Third, the interactions between the provider and patient are prioritized.

These components of the SPO model parallel the components of human factors engineering (HFE) models of quality assessment (see Table 8.1). Where the structure/environment is represented by material, organizational, and human resources in the SPO model, it is represented by organizational factors, work environment, individual characteristics, and technology features in HFE. In the SPO model, process or tasks are represented by physician technical quality and interpersonal relationship; in the HFE model, it is represented by aspects of the task and the process surrounding the individual's decision making. Finally, outcomes are defined as effects of health care on health status of individuals and populations according to the SPO model and as both negative and positive effects such as errors, accidents, injuries, stress, productivity, time, and quality in the HFE model.

There are, however, a number of differences between the SPO model and a HFE approach. These differences stem largely from the fact that physician professionalism,

which influences the medical approach to quality assessment, is highly reflected in the SPO model. First, the primary emphasis in the SPO model is on the performance of practitioners. Donabedian (1988) wrote "Our power, our responsibility and our vulnerability all flow from the fact that we are the foundation for that ladder, the focal point for that family of concentric circles" (p. 1743) that define quality health care. Because of the centrality of the physician role, the SPO model places the responsibility for other elements of quality within the health care system (the physician–patient interaction, the "amenities of care," and the structural elements) mostly on physicians and minimally on patients or health care systems. In Donabedian's view, structural elements should not carry much weight in quality assessment because prior evidence has demonstrated a weak link between the structure and processes of health care (Palmer and Reilly, 1979) and because structural characteristics are a "rather blunt instrument" in quality assessment (Donabedian, 1988, pp. 1745–1746). Because of this lack of emphasis on structure as a determinant of quality health care, the physician remains responsible for quality of care.

Another difference between the SPO model and the HFE model is that researchers who use the SPO model have become quite sophisticated in understanding the relationships of process to patients' health-related outcomes and in measuring these outcomes (Brook et al., 1996; Clancy and Eisenberg, 1998). Human factors engineering, on the other hand, focuses primarily on the interaction of the individual and his or her work surroundings, including structure and process, and individual and system outcomes; but HFE has only recently started to focus on patient outcomes.

8.5 New Approach

Any attempt to apply a systems perspective to health care must take into account the current medical environment, which has been shaped by physician professionalism and the tension between physician professionalism and the current organizational incentives structure. Because of the historical centrality of the physician in the design and conceptualization of health care systems, in the past, there has been a lack of focus within quality improvement efforts on the structural aspects of health care. There has instead been a focus on clinical processes of care carried out by the physician and other practitioners. If health care organizations are going to both improve the quality of health care and acknowledge the difficulties that physicians have with the change process due to their belief in physician

Table 8.1

Structure-Process-Outcome Model versus Human Factor Engineering Perspective

Domain	Structure-Process-Outcome	Human Factors Engineering
Structure/ environment	1. Material resources responsibilities, training, coworkers 2. Human resources 3. Organizational structure motivation, gender, stress 4. Technology features	1. Organization communication 2. Ambient noise, climate, illumination 3. Operator competence, age
Process/tasks	1. Physician technical quality and feedback 2. Physician interpersonal relationships	1. Task composition, allocation 2. Operator perception, decision making, and response
Outcomes	1. Effects of health care on health status of individuals and populations	1. Negative outcomes: errors, accidents, injuries, physiological stress 2. Positive outcomes: productivity, time quality

professionalism, a new approach to quality improvement is needed, one that incorporates more of a systems perspective and emphasizes the structure of health care delivery and reengineers the processes of care to relieve physicians of exclusive responsibility for the quality of health care provided.

There have been a number of developments in the way that health care is provided that focus on incorporating a systems approach to quality improvement. These approaches to health care have largely been developed by organizations looking for ways to provide quality health care while balancing the needs of physicians, patients, and health care organizations (Berwick, 1996). These developments include disease management programs and collaborative multidisciplinary patient care teams. These approaches, although still in the early stages of development and evaluation, show great potential for incorporating a systems approach to health care while preserving physician professionalism.

8.5.1 Disease Management Programs

Disease management is a concept that first appeared in the U.S. health care field approximately 15 years ago. It is a system for managing chronic conditions at a health care system level in order to improve the effectiveness and efficiency of care across the continuum of care. High prevalence, high-cost conditions are selected. Disease maps and care maps are established to determine how the disease is managed within the health care organization and what drives the costs of care. The most expensive and highest risk patients are identified and interventions to improve care for these patients are evaluated. The programs are monitored and adjusted as necessary using a continuous quality improvement framework. Some examples of these successful disease management programs include focused discharge planning and postdischarge care for patients with congestive heart failure (Phillips et al., 2004), intensive monitoring, education, and follow-up for patients with diabetes mellitus by nurses and nurse case managers (Sidorov et al., 2002), and tertiary prevention of coronary heart disease risk factors (DeBusk et al., 1994).

Disease management programs take advantage of system-level analyses to identify the need for interventions and implement changes in the structure and processes of care around certain high-prevalence, high-cost diseases. Interventions may include physician incentives such as education, bonuses, and profiles, but these incentives are established as only a part of multiple interventions contributing to system redesign (Weingarten et al., 2002). The responsibility for management of the condition is therefore spread more equally across the many participants in the delivery of health care—physicians, other practitioners, employees more indirectly involved

in patient care (e.g., receptionists, administrators, other clinic personnel), and patients (Bodenheimer, 1999). If implemented well, disease management programs should improve physician worklife through better coordination and support of care (Bodenheimer et al., 2002).

8.5.2 Multidisciplinary Patient Care Teams

The development of multidisciplinary patient care teams within health care organizations takes advantage of system-level work redesign to provide effective health care. These teams have been especially effective for managing the health care of patients with chronic diseases (Wagner, 2000). By taking advantage of team members' individual skill sets, patient care teams can create a system-level health care delivery system in which individuals' roles and functions are optimized and effectiveness of care is maximized. This type of quality improvement intervention has the potential to improve care through population management, protocol-based regulation of medication, self-management support, and intensive follow-up (Wagner, 2000). Similar to other system-level approaches, this type of quality improvement effort does not take away from the professional role of physicians or create pressure on physicians to change the way they practice medicine with financial incentives or negative consequences for noncompliance. Instead, these teams create a collaborative, positive environment to supplement physician expertise in the day-to-day management of chronic illness.

8.6 Accountability

One of the major issues left to be resolved in this new approach to health care systems change is that of accountability. Accountability is a crucial component of any health care system because, ultimately, someone or something must be responsible for the activities, actions, and outcomes of health care. Under the professional model of health care, accountability is firmly established: physicians are responsible to their professional colleagues and organizations and to their individual patients. In addition, with the current structure of U.S. health care in which health care organizations operate as economic entities within a larger marketplace, accountability occurs when patient-consumers show disapproval of a physician or health plan through "exit" (i.e., switching providers or plans; Emanuel and Emanuel, 1996). These different accountabilities are not entirely compatible, but at least they are well defined. Under the systems model, however, accountability is much less clear. The systems model emphasizes the multifactorial nature of error and a non-punitive approach (Wears et al., 2000). Accountability under this model occurs through a method of root cause analysis, which evaluates the many contributors to error. Although this more diffuse structure of accountability has many advantages, health system change that employs a systems approach needs to carefully define which entities are responsible for which sets of activities and who is responsible for explaining or answering for which actions (Emanuel and Emanuel, 1996).

8.7 Conclusions

Incorporation of a systems perspective into quality improvement efforts may help to improve physician acceptance of quality management activities and to decrease the tension between these activities and physician professionalism. Formation of disease management programs and multidisciplinary teams are examples of system-level interventions that may be effective in improving quality while preserving physician professionalism. More research is needed to determine which system-level interventions are most effective and to what extent these interventions truly have positive effects on physician job satisfaction, physician attitudes toward quality improvement efforts, and the ability of physicians to maintain a sense of professionalism. HFE has much to add to the medical approach to health care quality improvement, but the optimal approach and methodology that combines a systems approach, takes into account physician professionalism, and clearly establishes accountability has yet to be defined.

References

ABIM Foundation, ACP-ASIM Foundation, and European Federation of Internal Medicine. (2002). Medical professionalism in the new millennium: A physician charter. *Annals of Internal Medicine, 136*(3), 243–246.

American Board of Internal Medicine Foundation. (2003). *The Medical Professionalism Project and the Physician Charter*, from http://www.abimfoundation.org/professional.html

Berwick, D. M. (1996). A primer on leading the improvement of systems. *British Medical Journal, 312*(7031), 619–622.

Bodenheimer, T. (1999). Disease management—Promises and pitfalls. *New England Journal of Medicine, 340*(15), 1202–1205.

Bodenheimer, T., Wagner, E. H., and Grumbach, K. (2002). Improving primary care for patients with chronic illness: The chronic care model, Part 2. *Journal of the American Medical Association, 288*(15), 1909–1914.

Brook, R. H., McGlynn, E. A., and Cleary, P. D. (1996). Quality of health care. Part 2: Measuring quality of care. *New England Journal of Medicine, 335*(13), 966–970.

Carayon, P. and Smith, M. J. (2000). Work organization and ergonomics. *Applied Ergonomics, 31*(6), 649–662.

Clancy, C. M. and Eisenberg, J. M. (1998). Outcomes research: Measuring the end results of health care. *Science, 282*(5387), 245–246.

DeBusk, R. F., Miller, N. H., Superko, H. R., Dennis, C. A., Thomas, R. J., Lew, H. T. et al. (1994). A case-management system for coronary risk factor modification after acute myocardial infarction. *Annals of Internal Medicine, 120*(9), 721–729.

Donabedian, A. (1988). The quality of care. How can it be assessed? *Journal of the American Medical Association, 260*(12), 1743–1748.

Edwards, N., Kornacki, M. J., and Silversin, J. (2002). Unhappy doctors: What are the causes and what can be done? *British Medical Journal, 324*(7341), 835–838.

Emanuel, E. J. and Emanuel, L. L. (1996). What is accountability in health care? *Annals of Internal Medicine, 124*(2), 229–239.

Enthoven, A. C. and Vorhaus, C. B. (1997). A vision of quality in health care delivery. *Health Affairs, 16*(3), 44–57.

Flood, A. B. and Fennell, M. L. (1995). Through the lenses of organizational sociology: The role of organizational theory and research in conceptualizing and examining our health care system. *Journal of Health and Social Behavior, 35*(Special Issue), 154–169.

Flynn, K. E., Smith, M. A., and Davis, M. K. (2002). From physician to consumer: The effectiveness of strategies to manage health care utilization. *Medical Care Research and Review, 59*(4), 455–481.

Gaba, D. M. (2000). Anaesthesiology as a model for patient safety in health care. *British Medical Journal, 320*(7237), 785–788.

Gold, M. R., Hurley, R., Lake, T., Ensor, T., and Berenson, R. (1995). A national survey of the arrangements managed-care plans make with physicians. *New England Journal of Medicine, 333*(25), 1678–1683.

Gosbee, J. and Lin, L. (1999). The role of human factors engineering in medical device and medical system errors. In: C. Vincent (Ed.), *Clinical Risk Management: Enhancing Patient Safety* (pp. 309–330). London: BMJ Publishing.

Grimshaw, J. M., Shirran, L., Thomas, R., Mowatt, G., Fraser, C., Bero, L. et al. (2001). Changing provider behavior: An overview of systematic reviews of interventions. *Medical Care, 39*(8 Suppl 2), II2–45.

Helander, M. G. (1997). The human factors profession. In: G. Salvendy (Ed.), *Handbook of Human Factors and Ergonomics* (pp. 3–16). New York: Wiley.

Institute of Medicine. (2001). *Crossing the Quality Chasm: A New Health System for the 21st Century*. Washington, DC: National Academy Press.

Leape, L. L. (1994). Error in medicine. *Journal of the American Medical Association, 272*(23), 1851–1857.

Linzer, M., Konrad, T. R., Douglas, J., McMurray, J. E., Pathman, D. E., Williams, E. S. et al. (2000). Managed care, time pressure, and physician job satisfaction: Results from the physician worklife study. *Journal of General Internal Medicine, 15*(7), 441–450.

Mechanic, D. and Schlesinger, M. (1996). The impact of managed care on patients' trust in medical care and their physicians. *Journal of the American Medical Association, 275*(21), 1693–1697.

Mekhjian, H. S., Kumar, R. R., Kuehn, L., Bentley, T. D., Teater, P., Thomas, A. et al. (2002). Immediate benefits realized following implementation of physician order entry at an academic medical center. *Journal of the American Medical Informatics Association, 9*(5), 529–539.

Morrison, I. and Smith, R. (2000). Hamster health care. *British Medical Journal, 321*(7276), 1541–1542.

Palmer, R. H. and Reilly, M. C. (1979). Individual and institutional variables which may serve as indicators of quality of medical care. *Medical Care, 17*(7), 693–717.

Phillips, C. O., Wright, S. M., Kern, D. E., Singa, R. M., Shepperd, S., and Rubin, H. R. (2004). Comprehensive discharge planning with postdischarge support for older patients with congestive heart failure: A meta-analysis. *Journal of the American Medical Association, 291*(11), 1358–1367.

Sappington, D. E. M. (1991). Incentives in principal-agent relationships. *The Journal of Economic Perspectives, 5*(2), 45–66.

Sharma, A. (1997). Professional as agent: Knowledge asymmetry in agency exchange. *Academy of Management Review, 22*(3), 758–798.

Shortell, S. M., Waters, T. M., Clarke, K. W., and Budetti, P. P. (1998). Physicians as double agents: Maintaining trust in an era of multiple accountabilities. *Journal of the American Medical Association, 280*(12), 1102–1108.

Sidorov, J., Shull, R., Tomcavage, J., Girolami, S., Lawton, N., and Harris, R. (2002). Does diabetes disease management save money and improve outcomes? A report of simultaneous short-term savings and quality improvement associated with a health maintenance organization-sponsored disease management program among patients fulfilling health employer data and information set criteria. *Diabetes Care, 25*(4), 684–689.

Smith, M. J. and Sainfort, P. C. (1989). A balance theory of job design for stress reduction. *International Journal of Industrial Ergonomics, 4*(1), 67–79.

Wagner, E. H. (2000). The role of patient care teams in chronic disease management. *British Medical Journal, 320*(7234), 569–572.

Wears, R. L., Janiak, B., Moorhead, J. C., Kellermann, A. L., Yeh, C. S., Rice, M. M. et al. (2000). Human error in medicine: Promise and pitfalls, part 1. *Annals of Emergency Medicine, 36*(1), 58–60.

Weingarten, S. R., Henning, J. M., Badamgarav, E., Knight, K., Hasselblad, V., Gano, A., Jr. et al. (2002). Interventions used in disease management programmes for patients with chronic illness-which ones work? Meta-analysis of published reports. *British Medical Journal, 325*(7370), 925.

Wennberg, D. E. (1998). Variation in the delivery of health care: the stakes are high. *Annals of Internal Medicine, 128*(10), 866–868.

Section III

Job and Organizational Design

9

The Effect of Workplace Health Care Worker Stress and Burnout on Patient Outcomes

Eric S. Williams, Jonathon R.B. Halbesleben, Linda Baier Manwell, Julia E. McMurray,
Joseph Rabatin, Ayesha Rashid, and Mark Linzer

CONTENTS

9.1 Prevalence of Stress and Burnout in Health Care Providers...119
9.2 Conservation of Resources Theory 120
9.3 Relationship of Stress and Burnout with Patient Outcomes 121
    9.3.1 Relationship of Stress and Burnout with Patient Satisfaction...121
    9.3.2 Relationship of Stress and Burnout with Quality of Care..122
9.4 Implications for Future Research 123
9.5 Implications for Stress and Burnout Management...125
9.6 Conclusion 128
References...128

Huge changes have occurred in the health care milieu over the last decade, including an exponential growth in information technology, major redesigns in care delivery systems and organizations, shortages among key health care personnel, changing patient demographics, and the rise of consumerism. The increased workload and demands necessitated by these changes have exacerbated job stress and burnout for health care professionals, and impacted the populations they serve.

This chapter focuses on the relationship between stress and burnout among health care workers and their impact on quality of patient care. Early literature was limited to reports on the prevalence and severity of stress and burnout with limited linkage to patient outcomes. More recent literature has progressed beyond descriptive studies to explore multiple outcomes, notably patient satisfaction and various aspects of quality of care. This newer body of literature not only confirms the previously noted high levels of burnout for health care professionals, but begins to assess the consequences of stress and burnout not only for patients, but also for organizations and health care delivery systems.

With this in mind, we begin by summarizing the prevalence of stress and burnout among health care workers. We continue by presenting an emerging theory that provides a framework for understanding how health care worker stress and burnout may affect patients, and use it to examine the literature on how stress and burnout

among health care workers affects patient outcomes. We conclude by recommending directions for future research and discussing interventions for reducing stress and burnout.

9.1 Prevalence of Stress and Burnout in Health Care Providers

Stress and burnout have long been considered occupational risks for nurses, with approximately 40% reporting higher levels of burnout than other health care professionals (Aiken et al. 2001). Burnout is more prevalent among nurses younger than 30 years of age (Erickson and Grove 2010) and in nursing specialties that have less work control (Browning et al. 2007). Research indicates that sources of stress have shifted over the last several decades. Stressors from individual-related issues of patient care such as increased decision-making and responsibility (Menzies 1960) have given way to stressors from organizational restructuring, new technologies, increased management responsibilities, and heavier patient loads (Aiken et al. 2001). Additionally, younger nurses' expectations of less bureaucracy and more autonomy have resulted in intergenerational friction among nursing personnel (Green 2010).

Burnout rates of 25%–60% have been found in varying groups of physicians (Embriaco et al. 2007), with the highest levels reported for subspecialists in emergency departments (Goldberg et al. 1996), critical care (Embriaco et al. 2007), surgery (Shanafelt Balch et al. 2009), and oncology (Trufelli et al. 2008). Rates among primary care clinicians are somewhat lower, ranging from 18% to 27% (Shugerman et al. 2001; Linzer et al. 2009). Burnout tends to be higher for physicians in academia; a recent study found that 34% of faculty members in a department of internal medicine met criteria for burnout (Shanafelt West et al. 2009). Women physicians seem particularly susceptible to burnout. McMurray et al. (2000) reported that female physicians are 60% more likely than their male counterparts to report signs or symptoms of burnout.

Rates of burnout among medical residents vary widely from 18% to 82%, with some data indicating higher rates in surgical residents (Shanafelt et al. 2002; Garza et al. 2004; Thomas 2004). Rates of burnout for medical students hover around 45% (Dyrbye et al. 2005, 2006). In a study assessing the association between burnout and dropping out of medical school, the investigators found that "each one-point increase in emotional exhaustion and depersonalization score, and each one-point decrease in personal accomplishment score was associated with a 7% increase in the odds of serious thoughts of dropping out during the following year" (Dyrbye et al. 2010).

The growth of intensive care in hospitals has increased work stressors for intensive care unit (ICU) personnel who now report some of the highest rates of psychological distress. In a recent survey of 115 Australian physician intensivists, 80% reported signs of psychological stress and discomfort (Shehabi et al. 2008). A study of 978 French intensivists reported that 47% endorsed a high level of burnout (Embriaco et al. 2007). In a multi-hospital study comparing ICU nurses with general nurses, 24% of the ICU nurses reported symptoms of post-traumatic stress disorder (PTSD) related to their work environment compared to 14% of the general nurses (Mealer et al. 2007).

While one study of pharmacists noted a reasonably high level of job satisfaction (67.2%), pharmacists reported significant levels of job stress and work overload, especially in chain and mass pharmacies and hospital settings (Mott et al. 2004). With regard to pharmacy students, Henning et al. (1998) surveyed 477 medical, dental, nursing, and pharmacy students for psychological distress, perfectionism, and imposter feelings. Nearly 28% of all students were experiencing psychiatric levels of distress; the pharmacy students, however, were at greatest risk with half reporting distress levels similar to those reported by psychiatric populations.

For dentists, burnout prevalence rates range from 10% to 47% (Ahola and Hakanen 2007; Alemany Martinez et al. 2008). A longitudinal study of dentists in Finland noted that 23% of respondents developed depression over a 3-year period, and 63% of those who were depressed at baseline reported significant burnout. Thus, the effects between burnout and depression seemed to be reciprocal: "occupational burnout predicted new cases of depressive symptoms and depression predicted new cases of burnout" (Ahola and Hakanen 2007).

Other health professionals, such as occupational therapists have also been shown to have significant levels of burnout (40%), especially for those who work in chronic care facilities (Painter et al. 2003). One study of physical and occupational therapists reported that 94% had negative attitudes about their work and clients and 97% reported low feelings of personal accomplishments (Balogun et al. 2002). One study of Canadian nursing home personnel found that health care aides were significantly more exhausted than registered nurses and registered practical nurses (Ross et al. 2002). Work-related stressors for these caregivers include lack of autonomy, peer identification, challenges for new skills and recognition, and positive feedback about performance (Carr and Kazanowski 1994).

9.2 Conservation of Resources Theory

Before examining the effect of stress and burnout on patient outcomes, we turn to an examination of the theory that underpins this work. Conservation of Resources (COR) (Hobfoll 1988, 1989, 1998) suggests that "people strive to retain, protect, and build resources and that what is threatening to them is the potential or actual loss of those valued resources" (Hobfoll 1988). Thus, stress occurs when (1) resources are threatened, (2) resources are lost, or (3) resources are not adequately gained following a significant investment. Resources can be put into four categories: (1) objective, (2) personal, (3) condition, or (4) energy. Objective resources are actual assets, valued because of their physical nature in providing for basic human needs (housing, safety, nutrition, etc). Personal resources are skills (occupational, life) and characteristics (hardiness, self-efficacy) that are valued to the extent that they enable a person to manage life's challenges. Condition resources lay a foundation for the acquisition of other resources. These structures or states facilitate access to pools of other resources such as health, employment, seniority, and marriage. For example, employment provides self-esteem and the ability to purchase insurance or other resource protection strategies. Energy resources include time, money, and knowledge. Their value rests on their ability to aid in the acquisition of other resources.

Beyond resources, there are key two principles in understanding COR. The first is that resource loss is more salient than resource gain. This argument draws from research showing that people place much greater emphasis on loss in their decision-making than on gains (Tversky and Kahneman 1974). The second is that people must invest resources in order to protect against resource loss, recover from losses, and gain resources. COR also presents two important corollaries. The first is that those with greater resources are less vulnerable to resource loss and more capable of orchestrating resource gain. Conversely, those with fewer resources are more vulnerable to resource loss and less capable of achieving resource gains. The second corollary is that those who lack resources are not only more vulnerable to resource loss, but that initial loss begets future loss. Both of these speak to loss and gain spirals or, put another way, the rich get richer and the poor get poorer.

Hobfoll and Freedy (1993) extended the COR model to burnout through the idea that burnout results from inadequate return for work resources invested across time. This is based on Shirom's (1989) observation that resource depletion is a central facet of job burnout. That is, work resources are continually invested, but are depleted (or, at least, placed at risk) as they do not generate sufficient return. Resource depletion is particularly insidious due to COR's first principle that resource loss has greater psychological import than resource gain (Tversky and Kahneman 1974). As people burn out, they experience emotional exhaustion due to dwindling resources. Hobfoll (2001) argued that as resources are depleted, individuals take a defensive position over investing remaining resources. For example, resource management strategies such as withdrawal or depersonalization may be employed. This defensive positioning to protect remaining resources is the key to linking provider stress and burnout with patient outcomes. When health care professionals experience high levels of stress and burnout, they are more strategic in how they invest their remaining resources so as to minimize any further resource loss. In particular, providers may be reluctant to invest resources in areas of their career that are less likely to provide resources in return. For example, provider–patient relationships are a significant source of burnout due to perceived inequities in resource investment (e.g., the provider perceives that he/she invests more in the relationship than the patient; (Van Dierendonck et al. 1994; Bakker et al. 2000; Halbesleben 2006). Thus, we expect that providers would likely avoid a significant resource investment in patient relationships.

Shanafelt et al. (2002) provide an illustration of this concept in their report that burned out health care providers are more likely to provide suboptimal care (e.g., withholding information, not considering patients' desires in treatment). They found that the depersonalization component of burnout is associated with provision of suboptimal care, supporting the COR-based prediction that providers shy away from relationships that cause resource loss without resultant reciprocity. It is perhaps no surprise that burnout is associated with multiple negative patient outcomes such as lower patient satisfaction, reduced patient safety, and inferior quality of care (Halbesleben and Rathert 2008; Williams et al. 2009). From a COR theory perspective, as resources are lost, health care professionals are less likely to invest resources in areas that led to those losses (patient relationships), leading to worse outcomes for patients. In the following section, we review the literature in this area, using COR theory to explain the mechanism linking stress and burnout to patient outcomes.

9.3 Relationship of Stress and Burnout with Patient Outcomes

9.3.1 Relationship of Stress and Burnout with Patient Satisfaction

The literature relating stress and burnout to patient outcomes falls into two major categories: patient satisfaction and quality of care. Turning first to patient satisfaction, Leiter et al. (1998) examined the relationship between unit-level nurse burnout and patient satisfaction in 16 hospital units. They found emotional exhaustion and depersonalization to be negatively associated with patient satisfaction. In a cross-sectional study of over 800 nurses and 600 patients, Vahey et al. (2004) found that emotional exhaustion and personal accomplishment were associated with lower patient satisfaction, but depersonalization was not. Interestingly, Garman et al. (2002) found much stronger relationships between emotional exhaustion and patient satisfaction than depersonalization or personal accomplishment.

Exploring anesthesiologists and anesthesiology, nurses working in surgical suites at six different facilities, Capuzzo et al. (2007) reported that staff burnout was not associated with patient satisfaction. However, their study brings up an important point that dovetails with Halbesleben and Rathert's study. Capuzzo et al. (2007) note that the time spent with conscious patients is minimal among anesthesiology staff; this, in combination with the context of the interaction (i.e., just before surgery) suggests that patients are not in the best position to observe burnout among this group of professionals. A similar study investigated staff burnout and patient satisfaction with the quality of dialysis care among 68 nephrologists, 334 nurses, and 695 of their patients (Argentero et al. 2008). They found relationships between two of the three burnout dimensions (emotional exhaustion and lack of personal accomplishment) and patient satisfaction.

Studying physicians, Halbesleben and Rathert (2008) found only the depersonalization subscale to be significantly related to patient satisfaction. They also examined patients' perceptions of the physician's depersonalization and found a very high correlation between patient perceived depersonalization and patient satisfaction ($r = -58$). A very unusual study examined stress among 26 English general practitioners related to nights "on call" and the subsequent impact on patient satisfaction (French et al. 2001). They found that GPs were stressed right before, during, and after night call and those patients who were seen during the period immediately before and after were less satisfied.

9.3.2 Relationship of Stress and Burnout with Quality of Care

The literature in this area can be divided into two groups of studies; those that depend on health care worker self-reported quality of care and other, more objective measures of care quality. Beginning with worker self-reports, Schaufeli et al. (1995) linked ICU unit-level nurse burnout with ICU performance, finding that burnout was associated with nurse perceptions of lower effectiveness in the unit. However, when examining a more objective measure of effectiveness (standard mortality ratio), they found no significant relationships with burnout. A study by Van Bogaert et al. (2009) reported that burnout accounted for 46% of the variance in nurses' perceptions of care quality across 31 units in two hospitals. Using data from focus groups of nurses working with critically ill patients, Berland et al. (2008) found that nurses believed that job-related stress could be associated with a higher risk for patients.

Laschinger and Leiter (2006, 2009) developed the Nursing Worklife Model which predicts an association between burnout and the increased likelihood of adverse events. They found support for this prediction in a sample of over 8500 Canadian nurses. In a study of nurses that integrated the Nursing Worklife Model and the Conservation of Resources theory, Halbesleben et al. (2008) found burnout to be associated with lower perceptions of patient safety. However, burnout was not associated with error reports by the nurses and it was negatively associated with reports of near misses, thereby indicating that burned out nurses were less

likely to report near misses. Although this study focused on nurses' responses and did not include patient data, it still has implications for patient safety because near miss reports are an important driver for organizational learning around safety (Roberto et al. 2006).

Finally, Redfern et al. (2002) reported that staff stress was associated with lower perceptions of quality of care as rated by nursing home residents. Although stress was measured directly from the staff, this study extends the notion that patients may be sensitive to staff stress and perceive it as lower quality of care. Particularly noteworthy was that they found this statistically significant finding with a very small sample (one home with 44 staff member surveys and 22 resident surveys).

Turning to physicians, Shanafelt et al. (2002) surveyed 115 medical residents and found that (1) three-quarters exhibited symptoms of burnout, (2) burned out physicians reported more suboptimal patient care practices, and (3) only the depersonalization scale was associated with suboptimal patient care. Using the same measure of self-reported suboptimal care practices, Williams et al. (2007) found that both burnout and stress were related to perceived self-reported suboptimal patient care. In sample of Canadian physicians, Williams et al. (2007) found that stress was negatively related to perceptions of quality patient care.

Shanafelt, et al. (2009) examined error reporting among 7905 surgeons and found that nearly 9% reported a major error in the last 3 months. All three components of burnout were associated with a greater likelihood of reporting an error, even after controlling for personal and professional factors. In an earlier examination of physician stress, Firth-Cozens and Greenhalgh (1997) surveyed 225 English GPs and hospital physicians and found that 82 reported incidents where symptoms of stress had impacted patient care. Stress attributions included fatigue (57%), overwork (28%), depression or anxiety (8%), or the effects of alcohol (5%). Shirom et al. (2006) examined the impact of burnout on self-reported quality of care in 890 Israeli physicians. They found, using a different formulation of burnout, that physical fatigue, cognitive weariness, and emotional exhaustion were all related to quality of care. In another international study, Prins, et al. (2009) examined burnout in 2115 physicians in the Netherlands and found that burnout was associated with more error reporting. In a second international study, Kazmi et al. (2008) examined stress and job performance among 55 Pakistani physician house officers and found that job stress was negative related to all four measures of perceived job performance: knowledge, skills, attitude, and effectiveness/job quality. Finally, Wallace and Lemaire (2009) report a fascinating qualitative study exploring physicians' perceptions of the link between physician wellness and quality of care. Data from 42 physician interviews indicate that the link is not at the forefront of physicians' awareness, likely due to both a medical culture that sees physicians as unbreakable workhorses and that overwhelming workloads prevent physicians from thinking about their own well-being.

Now turning to studies that include objective quality of care measures, Dugan et al. (1996) assessed the effect of job stress on turnover, absenteeism, injuries, and patient incidents in a sample of hospital nurses. Zero order correlations were found between a perceived stress index and patient incidents overall ($r = 0.43$), medication errors ($r = 0.40$), patient falls ($r = 0.33$), but not IV errors ($r = 0.15$). However, a second stress measure composed of reported stress symptoms found more modest, nonsignificant correlations. Units reporting higher levels of nurse stress were consistently more likely to exhibit increased medication errors and patient falls. In their investigation of nurses in Dutch ICU units, Keijsers et al. (1995) found linkages between all three burnout measures and standard mortality ratio, an objective measure of unit performance. Somewhat counter to these findings, Ida, et al. (2009) found no relationship between stress and a medical errors-based measure of job performance among their sample of 502 Japanese nurses.

In their study of physicians, Halbesleben and Rathert (2008) found that patient-reported recovery time after hospitalization was positively associated with the depersonalization component of burnout ($r = 0.23$), but not with either of the other two burnout components. Again they found a stronger relationship when examining patient perceptions of depersonalization ($r = 0.32$), though the pattern was not as stark as when examining satisfaction. Linzer et al. (2009) found that working conditions affected physician stress and burnout as well as some objective measures of quality of care. However, neither stress nor burnout was found to be predictive of quality of care.

In a somewhat different vein, Van den Hombergh et al. (2009) investigated the performance of Dutch general medical practices and reported relationships between job stress and physician availability. While they did not directly link this to patient outcomes, less physician accessibility can result in difficulty obtaining adequate care. In a multipart study, Jones, et al. (1988) examined the critical relationship between stress and medical malpractice. First, they compared 91 hospital departments for malpractice risk. Those at highest risk (e.g., sued within the last year) were comprised of employees who reported significantly higher levels of job stress across three of four subscales: job stress, organizational stress, job dissatisfaction, and personal stress. Then, using a different set of 61 hospitals, the investigators assessed malpractice

frequency and the same four stress subscales. They found strong correlations ($r = 0.39-0.56$) with three of the four measures. After controlling for the number of hospital beds, organizational and job stress remained significant. The average level of personal stress in the organization was not related to malpractice risk or frequency.

9.4 Implications for Future Research

The COR model and literature discussed in this chapter contain a number of potential directions for future research as well as one area where additional research efforts may yield diminishing returns. This latter arena involves reports of the prevalence and predictors of job stress and burnout for particular groups. While we recognize that documenting stress and burnout levels across time is useful for tracking purposes, current databases yield thousands of articles on these topics and continued focus on this area is becoming redundant. More fertile ground involves looking at job stress and burnout as independent variables. As this paper examines the link between stress and burnout with patient outcomes, we will confine our remaining discussion to four promising directions for future research.

The first arena involves identifying intervening mechanisms to explain how health care worker stress affects patient outcomes. Williams et al. (2009) suggest a linkage between stress and burnout with patient satisfaction and trust. They draw on COR theory (Hobfoll 1989) and process models of burnout (Cherniss 1980; Golembiewski and Munzenrider 1988; Leiter and Maslach 1988) which suggest that as health care workers burn out, they tend to cope by being increasingly careful about how they invest their resources. This can result in depersonalization in interactions with patients as well as cynicism and withdrawal. For physicians, withdrawal and depersonalization are manifested by the communication style that is most familiar and comfortable, that is, the biomedical style in which the physician controls the conversation, delivering information and advice to a passive patient (Roter et al. 1997). Of course, such a style violates the norm of reciprocity central to the medical encounter (Blau 1964; Bakker et al. 2000; Halbesleben 2006). Not surprisingly, patients see this violation of the norm of reciprocity and react with less satisfaction and trust (Hall and Dornan 1988; Bertakis et al. 1991; Stewart 1995; Roter et al. 1997).

Future research should examine *specific* behaviors exhibited by health care workers when they are burned out and engaging in depersonalization; following the methodology developed by Roter and colleagues would be useful. Such a design should include several elements: (1) measurement of emotional exhaustion and depersonalization; (2) use of an analysis technique specifically designed for this purpose (e.g., Roter Interactive Analysis System for physician–patient interactions); and (3) audio and videotaping of the encounters, so that raters can analyze both verbal and nonverbal aspects of the medical encounters. Future research may also want to link such a study to patient outcomes such as satisfaction and trust, or to clinical outcomes similar to those used by Halbesleben and Rathert (2008).

Ratanawongsa et al. (2008) used several of these elements, but found that physician burnout was not significantly related to physician or patient affect, patient-centeredness, verbal dominance, length of the encounter, or patient ratings of satisfaction, confidence, or trust. While valuable, this effort was limited by a focus on burnout in general, a long latency between burnout measurement and patient encounters (15 months on average) and small sample sizes (40 physicians with 235 patients). Future research could build on this work, employing larger samples, less time between burnout measurement and patient encounters, measures of emotional exhaustion and depersonalization, and videotapes and audiotapes of the encounters.

A second, related, arena for future research involves the linkage between health care worker stress and burnout with patient safety, quality of care, and medical errors. We know that the relationship between stress and performance follows the Yerkes–Dodson inverted U where the highest level of performance is at moderate levels of stress and lowest levels of performance are associated with low and high levels of stress (Jex 1998). Related research has also shown, for example, that role overload is negatively correlated with nurse supervisory ratings of nurse performance, job motivation, and quality of patient care (Jamal 1984). Research on physicians has shown that excessive workload is negatively related to performance on tasks requiring monitoring or constant attention (Spurgeon and Harrison 1989). Further, sleep deprivation leads to poor performance because it impairs the ability to sustain attention for long periods of time (Kjellberg 1977). What has not been thoroughly investigated is the question of the role that stress and burnout play in medical errors. That is, do stressed workers actually commit more medical errors than nonstressed workers? The general literature concerning stress and mistakes at work suggests a potential link (Jex et al. 1991, Jex 1998), however much of the literature is based on self-reported perceptions of error and that literature using more objective measures seems mixed. An AHRQ report notes that existing work is suggestive that job stress may play some role in medical errors, but the evidence is not definitive (Hickam et al. 2003).

As with much of the previous discussion, it is unlikely that stress and burnout have a direct impact on errors, but that some intervening process can explain how stress leads to errors. Halbesleben and colleagues's (Halbesleben et al. 2008, 2010a,b) work on the notion of workarounds leading to safety concerns (both patient safety and worker safety) may be a starting point. They propose, again based on COR theory, that when faced with stress, health care professionals may not be able to fully carry out intended safety procedures which may expose patients to unsafe conditions. For example, in their study of four ICUs, Halbesleben et al. (2010b) documented examples where nurses bypassed safety procedures associated with medication administration because of blocks they encountered in the process (see also Koppel et al. 2008). While research is needed to empirically establish the relationship between workarounds and medical errors, this appears to be a natural extension of the stress-patient outcomes literature.

A third arena for future research lies at the organizational level. Most of the research into stress (and stress interventions as discussed below) has occurred at the individual level, leaving an opportunity to investigate job stress and related variables at the unit or organizational levels. Keijsers and colleagues (1995) explored the linkage between burnout and unit performance in 20 Dutch ICUs. Performance was measured subjectively through the perception of unit nurses and objectively using the APACHE III Standard Mortality Ratio. Interestingly, both emotional exhaustion and depersonalization were moderately negatively correlated with perceived unit performance and perceived personal performance, but slightly positively correlated with objective unit performance. A structural equation analysis reversed the causal chain suggested in the manuscript. Among four competing models, a model suggesting that perceived unit performance was indirectly related to emotional exhaustion and depersonalization through its effect on perceived personal performance had the best fit. An additional exemplar of this work is that of Jones et al. (1988) which is discussed in detail elsewhere in this chapter.

Another stream of scholarship at the organizational level worthy of extension comes from the "healthy organizations" movement. Murphy (1995) offered four directions for future research on occupational stress management, three of which merit comment here. The first suggests that stress interventions must target not only the individual, but also the job and organizational characteristics. Such an approach requires substantial knowledge of job redesign and organizational change. A second direction involves improvement in measuring stressors. As all firms have accounting systems to measure financial health, likewise there should be a system to assess "organizational health." A third direction involves ensuring that workers are key participants in organizational change efforts (see also Halbesleben 2009). That is, organizational-level stress management must proceed from a grassroots base rather than being dictated from the top. Karasek's and Theorell's (1990) book, *Healthy Work: Stress, Productivity, and the Reconstruction of Work Life*, is a great resource on redesigning work and organizations to be healthier.

9.5 Implications for Stress and Burnout Management

The implications for management of stress and burnout mainly lie in the various interventions that can be used to manage stress. From the perspective of COR theory, interventions should be based on (1) enhancing resources, (2) eliminating vulnerability to resource loss, and (3) activating gain spirals. Successful interventions need to provide health care professionals with resources that allow them to meet the demands of their job. Not surprisingly, the literature on individual stress management is much more developed than the literature focusing on organizational interventions (Murphy 1995). Individual interventions are easier to implement and do not require substantial investments of organizational resources. While they can be effective, their effects seem to be more short term when compared to interventions that effect some sort of systemic change within the organization (Awa et al. 2010), perhaps because organizational interventions are more effective at providing longer-term and more universal resources to employees. Figure 9.1 illustrates our conceptual model and the areas where both individual and organizational interventions can have an impact. The remainder of this section will discuss some individual level stress management techniques and then turn to a discussion of the importance and implications of further research into organizational level interventions.

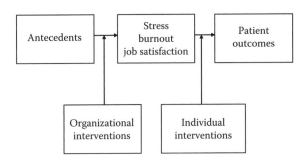

FIGURE 9.1
Intervention model.

Ways for managing stress are as varied as the stressors themselves. Some stress management techniques focus on the physical body incorporating exercise and physical relaxation techniques such as deep breathing exercises, yoga, massage, and progressive muscular relaxation. Other stress management techniques focus on the mind and involve meditation, self-hypnosis, and embracing solitude. A third set of stress management techniques involve time management (organizing, prioritizing), developing better communication skills, talking with others, finding a hobby, minimizing competition, and maintaining a journal about stressors. Despite this variability, there is a common theme of providing personal resources for those who participate in the intervention.

Most stress management techniques for physicians target the individual. McCue and Sachs (1991) tested the efficacy of a stress management workshop with 43 medicine and pediatric residents, and compared them to a control group of 21 residents. The workshop taught interpersonal skills, prioritization, and positive outlook skills as well as techniques to increase stamina and meet self-care needs. A stress measure and the Maslach Burnout Inventory were administered to all subjects 2 weeks before, and 6 weeks after the workshop. Results showed that the overall level of stress declined for the intervention group while the control group scores increased. The intervention group also experienced decreased emotional exhaustion and depersonalization compared to the control group.

Some interventions target specific populations of physicians. For example, female physicians face unique stressors. One study reports that they are 60% more likely than male physicians to report signs or symptoms of burnout (McMurray et al. 2000). While this is due in part to the difficulty of balancing domestic responsibilities with career demands (Spickard et al. 2002), there are also stressors related to patient expectations. Female physicians are expected to spend more time listening which leads to increased time pressure during office visits (McMurray et al. 2000). Adjusting panel sizes (patient lists) by patient gender is one step that can be taken to mitigate work stress for female physicians.

Garside (1993) organized a mutual aid group as a stress management tool for physicians who provide primary health care to people with AIDS. The themes identified and discussed by group members included anger, helplessness, lack of job satisfaction, the pain of seeing young patients die, and the pain of repeatedly delivering crushing messages to patients. Garside concluded that, "the commitment of the physicians to the group over a prolonged period and their own testimony indicated that with appropriate facilitation, physicians under stress from providing care to people with AIDS can benefit from the experience of mutual aid that has been so beneficial to other caregivers."

Similarly, much of the nursing literature is dedicated to common stress management techniques such as exercise, development of coping skills, and relaxation techniques. Nursing, however, is also moving toward alternative therapies such as philophonetics, reflexivity, and personality hardiness. For example, Sherwood and Tager (2000), applied philophonetics counseling to nurses who scored high on a self-reported burnout scale. Philophonetics involves the development of individually tailored, self-controlling, emotional, and cognitive strategies that extend conversational counseling and psychotherapy to include body awareness, movement/gesture, visualization, and sound. These strategies address feelings of victimization, disorientation, loss of decision-making power, lack of interpersonal boundaries, and disconnection from one's inner being and internal resources. Participating nurses reported significant reductions on all of the burnout scale items.

Encouraging reflexivity in professionals has long been recognized as a positive educational exercise. Reflecting on actions after an event allows the practitioner to review the situation and to gain insights for future practice (Hewitt 2003). Reflexivity can also be used as a stress management technique to explore personal values and beliefs around a particular issue. For example, Peerson and Yong (2003) focused on four issues relative to nursing clinical practices: seeking technological solutions to health and ill-health, moving from the nurse–patient relationship to the patient–healer relationship, utilizing critical pathways, and supporting evidence-based nursing. Based on their experience, Peerson and Yong (2003) urged nurses to "engage in reflexivity and not to lose sight of their selves (knowledge, expertise and skills)… in their contribution to health care."

Enhancing personality hardiness is another technique used to manage stress in the nursing milieu. Hardiness is defined as the interrelatedness of three factors controlled by the individual through lifestyle: control of the environment, commitment to self-fulfilling goals, and reasonable levels of challenge in daily life. Thomsens et al. (1999) found that these traits serve as buffers and seem to protect individuals from the psychological repercussions of stress. Nikou (2003), designed a study to investigate the relationships among hardiness, stress, and health-promoting behaviors in students attending a nursing student conference. The results indicated that hardiness was inversely related to stress and positively related to health-promoting behaviors.

Given that interpersonal interaction tends to be a common source of stress for health care professionals, a recent intervention was developed to target that specific problem. Scanera et al. (2009) developed a training program for mental health professions that targeted improvement in interpersonal relationship management as an intervention to reduce burnout and found

that it was effective in reducing the depersonalization component, both immediately following the training and after a longer-term follow up. While their findings are a bit tentative due to design concerns (e.g., no control group), it provides initial evidence that targeting interpersonal processes may be a fruitful avenue for intervention programs.

Brake et al. (2001) studied a group of Dutch dentists to determine the long-term effects of a burnout intervention program. Using the Maslach Burnout Inventory, they identified a group of high-risk subjects who received feedback on their scores and were invited to participate in an intervention program. Two surveys were administered post-intervention; one several weeks afterward and one a year later. In the first survey, the 92 respondents indicated an improvement in all elements of the MBI. The second survey indicated substantial relapse. Interestingly, a group of controls showed no changes at the first survey, but a subgroup that had specifically pursued some strategy to reduce burnout demonstrated a reduction in burnout symptoms.

Organizational interventions to reduce stress have taken on a new importance given the emphasis on "systems thinking" in the IOM report "Crossing the Quality Chasm" (Institute of Medicine 2001). However, few organizational interventions are reported in the health care literature (van der Klink et al. 2001) even though many good recommendations have been made (Ivancevich and Matteson 1987; Murphy 1995). While this literature is still emerging, it has been suggested that it may be critical for addressing the issues of stress and burnout in health care organizations (Halbesleben 2009). Therefore, we review it here with the goal of sparking new research and thinking about organizational interventions.

Karasek's model (1981) notes that control over the workplace is the major moderator between extreme job demands and stress. Participatory decision-making is a primary way to enhance perceptions of employee control. Jackson (1983) proposed a conceptual model in which participation in decision-making decreases role ambiguity and conflict. Decreases in role stressors in turn were hypothesized to increase job satisfaction and decrease emotional distress, absenteeism, and turnover intentions. Using an experimental design (Solomon four group design) with a sample of 126 nurses, nursing assistants, and unit clerks, Jackson found strong support for the effectiveness of increasing participation in decision-making. Manipulations of participation decreased role conflict and ambiguity and increased perceived influence, which in turn resulted in greater job satisfaction and lower levels of stress, absence frequency, and turnover intentions.

Landsbergis and Vivona-Vaughan (1995) drew on work by both Karasek (1981) and Jackson (1983) in their development of a conceptual model and employee participation intervention. Specifically, they proposed the creation of employee problem-solving committees to reduce stressful working conditions and enhance department functioning. The committees identified significant (but different) organizational stressors (e.g., communication, workloads) and a number of specific proposals were implemented. Unfortunately, one department showed little change while another showed mixed results. The author attributed the lack of results to substantial problems in the implementation of the interventions.

Dunn et al. (2007) reported on a natural intervention resulting from the dissolution of a multispecialty medical group. The new organization developed a program which aimed to improve physician control, promote orderly practice, and create meaning in clinical work life. Data were collected yearly from 2000 to 2005 (except for 2004). Results were generally, but universally supportive. Emotional exhaustion and several components of physician satisfaction decreased significantly while depersonalization, lack of personal accomplishment, and turnover remained stable.

Similarly, Heaney et al. (1995) evaluated the effectiveness of a multi-faceted program called the Caregiver Support Program. The two goals of the program were to train workers to (1) develop and tap into workplace support systems, and to (2) develop and use participatory problem-solving techniques in work team meetings. They used a simple experimental design with a sample of 1375 direct care staff from group homes for mentally ill adults. Results showed that those who had participated in the program reported better team climate, more supervisory support, better coping abilities, and reduced depressive symptoms.

Reid et al. (2009) presented the results of the Medical Home demonstration project. The medical home model was implemented in one group practice within the larger Group Health system and was compared against two comparable control practices. The intervention reduced patient panels, expanded visit length to 30 min, allocated additional staff time for coordination and outreach activities, and added additional staff for clinicians. Additionally, there were a variety of structural, team, point-of-care, patient outreach, and management changes. Adjusting for baseline, the demonstration clinic reported high quality patient experience (6 or 7 ratings), lower staff burnout, marginally higher quality, but no differences in overall cost. Patients of the demonstration clinic used more e-mail and phone services that the clinic offered, had more specialty care, but fewer specialty care visits. The authors concluded that the medical home is a viable concept for practice design as it can increase clinician well-being and enhance the patient experience and quality without increasing overall cost.

As part of their impressive four-study manuscript, Jones et al. (1988) presented the results of two organization-wide stress management programs. The first used a sample of 700 hospital employees and looked at the impact of a stress management program on the reduction of medical errors. Responses to an employee survey and a series of discussions about how to manage stress resulted in a series of policy changes and a stress management training program. Results showed a significant reduction in medication errors. The second study examined the impact of an organization-wide stress management program on malpractice frequency. A set of 22 hospitals participated in the intervention and was matched with a control group of 22 similar hospitals. The results of this quasi-experiment showed that the experimental group experienced a statistically significant decline in malpractice frequency.

Recently, Taner (2009) applied the Six Sigma approach to reduce ER physician turnover by reducing burnout and stress. They developed initiatives aimed at improving doctor's working conditions and found them to be effective in reducing stress and burnout. Their approach may serve as a solid model for future burnout interventions, not only because it worked, but because it is based on a popular movement within the health care industry.

Organizational interventions to reduce stress can have an effect, but are costly. More importantly, the management of stress in organizations seems to be an activity that occurs on an infrequent ad-hoc basis. This may be explained, in part, by the finding that physicians who experience high levels of stress appear less likely to engage in the quality improvement activities that might actually help address the stress (Quinn et al. 2009). For organizational stress management to have a real impact, it must be incorporated into the everyday activities of the organization. Fortunately, the concept of strategic human resource management (SHRM) has gained currency in recent years (Ulrich 1997). Quite simply, it argues that an organization's human resources practices and policies must be aligned with the company's strategy for the company to receive maximum benefit from its costliest resource—its people. Managing employee attitudes, including stress, is one of the more important aspects of human resource management. Both SHRM and the "system thinking" espoused in the IOM reports are based on the idea that only by changing organizational systems and strategies will a safer and less stressful health care workplace be achieved. As Halbeslebeben (2009) noted, if people were good at managing their own stress, stress would not be such a big problem for health care organizations. Instead, we must address the environment within which the stressed-out professionals work if we are to see real progress in reducing stress.

9.6 Conclusion

Given the massive changes in our health care system, it is imperative that we understand the impact of these changes on the health care worker and on their most important product—patient care. Continuing to ignore the impact of changes in the health care system on providers will, inevitably, exacerbate the shortages felt in many health care professions as qualified professionals leave for less stressful environs. The result will be a continuation of the current shortage of workers and a more subtle and pernicious decline in the quality of care despite the energy spent on engineering "system" solutions. Until policy makers realize that the health care worker is the center of the health care "system," any efforts toward long-term improvements in patient safety are doomed to failure.

References

Ahola, K. and J. Hakanen. 2007. Job strain, burnout, and depressive symptoms: A prospective study among dentists. *Journal of Affective Disorders* 104 (1–3):103–110.

Aiken, L. H., S. P. Clarke, D. M. Sloane, J. A. Sochalski, R. Busse, H. Clarke, P. Giovannetti, J. Hunt, A. M. Rafferty, and J. Shamian. 2001. Nurses' reports on hospital care in five countries. *Health Affairs (Millwood)* 20 (3):43–53.

Alemany Martinez, A., L. Berini Aytes, and C. Gay Escoda. 2008. The burnout syndrome and associated personality disturbances. The study in three graduate programs in Dentistry at the University of Barcelona. *Medicina Oral Patologia Oraly Cirugia Bucal* 13 (7):E444–450.

Argentero, P., B. Dell'Olivo, and M. S. Ferretti. 2008. Staff burnout and patient satisfaction with the quality of dialysis care. *American Journal of Kidney Diseases* 51 (1):80–92.

Awa, W. L., M. Plaumann, and U. Walter. 2010. Burnout prevention: A review of intervention programs. *Patient Education and Counseling* 78 (2):184–190.

Bakker, A., W. Schaufeli, H. Sixma, W. Bosveld, and D. Dierendonck. 2000. Patient demands, lack of reciprocity, and burnout: A five-year longitudinal study among general practitioners. *Journal of Organizational Behavior* 21 (4):425–441.

Balogun, J. A., V. Titiloye, A. Balogun, A. Oyeyemi, and J. Katz. 2002. Prevalence and determinants of burnout among physical and occupational therapists. *Journal of Allied Health* 31 (3):131–139.

Berland, A., G. K. Natvig, and D. Gundersen. 2008. Patient safety and job-related stress: A focus group study. *Intensive and Critical Care Nursing* 24 (2):90–97.

Bertakis, K., D. Roter, and S. Putnam. 1991. The relationship of physician medical interview style to patient satisfaction. *The Journal of Family Practice* 32 (2):175–181.

Blau, P.M. 1964. *Exchange and Power in Social Life*. New York: Wiley.

Brake, H. T., H. Gorter, J. Hoogstraten, and M. Eijkman. 2001. Burnout intervention among Dutch dentists: Long-term effects. *European Journal of Oral Science* 109 (6):380–387.

Browning, L., C. S. Ryan, S. Thomas, M. Greenberg, and S. Rolniak. 2007. Nursing specialty and burnout. *Psychology Health and Medicine* 12 (2):248–254.

Capuzzo, M., G. Gilli, L. Paparella, G. Gritti, D. Gambi, M. Bianconi, F. Giunta, C. Buccoliero, and R. Alvisi. 2007. Factors predictive of patient satisfaction with anesthesia. *Anesthesia and Analgesia* 105 (2):435–442.

Carr, K. K. and M. K. Kazanowski. 1994. Factors affecting job satisfaction of nurses who work in long-term care. *Journal of Advanced Nursing* 19 (5):878–883.

Cherniss, C. 1980. *Professional Burnout in Human Services Organizations*. New York: Praeger.

Dugan, J., E. Lauer, Z. Bouquot, B. K. Dutro, M. Smith, and G. Widmeyer. 1996. Stressful nurses: The effect on patient outcomes. *Journal of Nursing Care Quality* 10 (3):46–58.

Dunn, P. M., B. B. Arnetz, J. F. Christensen, and L. Homer. 2007. Meeting the imperative to improve physician well-being: Assessment of an innovative program. *Journal of General Internal Medicine* 22 (11):1544–1552.

Dyrbye, L. N., M. R. Thomas, D. V. Power, S. Durning, C. Moutier, F. S. Massie, Jr., W. Harper, A. Eacker, D. W. Szydlo, J. A. Sloan, and T. D. Shanafelt. 2010. Burnout and serious thoughts of dropping out of medical school: A multi-institutional study. *Academic Medicine* 85 (1):94–102.

Dyrbye, L. N., M. R. Thomas, and T. D. Shanafelt. 2005. Medical student distress: Causes, consequences, and proposed solutions. *Mayo Clinic Proceedings* 80 (12):1613–1622.

Dyrbye, L. N., M. R. Thomas, and T. D. Shanafelt. 2006. Systematic review of depression, anxiety, and other indicators of psychological distress among U.S. and Canadian medical students. *Academic Medicine* 81 (4):354–73.

Embriaco, N., E. Azoulay, K. Barrau, N. Kentish, F. Pochard, A. Loundou, and L. Papazian. 2007. High level of burnout in intensivists: Prevalence and associated factors. *American Journal of Respiratory and Critical Care Medicine* 175 (7):686–692.

Erickson, R. J. and W. J. C. Grove. 2010. Why emotional matter: Age, agitation, and burnout among registered nurses. *Online Journal of Issues in Nursing* (1), http://www.nursingworld.org/MainMenuCategories/ANAMarketplace/ANAPeriodicals/OJIN/FunctionalMenu/AboutOJIN.aspx.

Firth-Cozens, J. and J. Greenhalgh. 1997. Doctors' perceptions of the links between stress and lowered clinical care. *Social Science and Medicine* 44 (7):1017–1022.

French, D.P., R. K. McKinley, and A. Hastings. 2001. GP stress and patient dissatisfaction with nights on call: An exploratory study. *Scandinavian Journal of Primary Health Care* 19 (3):170–173.

Garman, A. N., J. M. Corrigan, and S. Morris. 2002. Staff burnout and patient satisfaction: Evidence of relationships at the care unit level. *Journal of Occupational Health Psychology* 7 (3):235–241.

Garside, B. 1993. Mutual aid group: A response to AIDS-related burnout. *Health and Social Work* 18 (4):259–267.

Garza, J. A., K. M. Schneider, P. Promecene, and M. Monga. 2004. Burnout in residency: A statewide study. *Southern Medical Journal* 97 (12):1171–1173.

Goldberg, R., R. W. Boss, and L. Chan. 1996. Burnout audit's correlates in emergency room physicians: Four years' experience with a wellness booth. *Academy of Emergency Medicine* 3:1156–1164.

Golembiewski, R. and R. Munzenrider. 1988. *Phases of Burnout: Developments in Concepts and Applications.* New York: Praeger Publishers.

Green, J. 2010. What nurses want: Different generations, different expectations. *Hospitals and Health Networks*, http://www.hhnmag.com/hhnmag_app/hospitalconnect/search/article.jsp?dcrpath=HHNMAG/PubsNewsArticle/data/0503HHN_FEA_CoverStory&domain=HHNMAG.

Halbesleben, J. R. 2006. Patient reciprocity and physician burnout: What do patients bring to the patient-physician relationship? *Health Services Management Research* 19 (4):215–222.

Halbesleben, J. R. B. 2009. *Managing Stress and Preventing Burnout in the Healthcare Workplace.* Chicago: Health Administration Press.

Halbesleben, J. R. B. 2010a. The role of exhaustion and work-arounds in predicting occupational injuries: A cross-lagged panel study of health care professionals. *Journal of Occupational Health Psychology* 15 (1):1–16.

Halbesleben, J. R. B., G. T. Savage, D. S. Wakefield, and B. J. Wakefield. 2010b. Rework and workarounds in nurse medication administration processes. *Health Care Management Review* 35 (2):124–133.

Halbesleben, J. R. and C. Rathert. 2008. Linking physician burnout and patient outcomes: Exploring the dyadic relationship between physicians and patients. *Health Care Management Review* 33 (1):29–39.

Halbesleben, J. R., B. J. Wakefield, D. S. Wakefield, and L. B. Cooper. 2008. Nurse burnout and patient safety outcomes: Nurse safety perception versus reporting behavior. *Western Journal of Nursing Research* 30 (5):560–577.

Hall, J. and M. Dornan. 1988. Meta-analysis of satisfaction with medical care: Description of research domain and analysis of overall satisfaction levels. *Social Science and Medicine* 27 (6):637–644.

Heaney, C. A., R. H. Price, and J. Rafferty. 1995. Increasing coping resources at work: A field experiment to increase social support, improve work team functioning, and enhance employee mental health. *Journal of Organizational Behavior* 16 (4):335–352.

Henning, K., S. Ey, and D. Shaw. 1998. Perfectionism, the imposter phenomenon and psychological adjustment in medical, dental, nursing and pharmacy students. *Medical Education* 32 (5):456–464.

Hewitt, B. E. 2003. The challenge of providing family-centered care during air transport: An example of reflection on action in nursing practice. *Contemporary Nurse* 15 (1–2):118–124.

Hickam, D., S. Severance, A. Feldstein, L. Ray, P. Gorman, S. Schuldheis, W. R. Hersh, K. P. Krages, and M. Helfand. 2003. *The Effect of Health Care Working Conditions on Patient Safety.* Washington, DC.: Agency for Healthcare Quality and Research.

Hobfoll, S.E. 1988. *The Ecology of Stress.* New York: Hemisphere.

Hobfoll, S. E. 1989. Conservation of resources. A new attempt at conceptualizing stress. *American Psychologist* 44 (3):513–524.

Hobfoll, S.E. 1998. *Stress, Culture and Community: The Psychology and Philosophy of Stress.* New York: Plenum Press.

Hobfoll, S. E. 2001. The influence of culture, community, and the self in the stress process: Advancing conservation of resources theory. *Applied Psychology an International Review* 50 (3):337–421.

Hobfoll, S. E. and J. Freedy. 1993. Conservation of resources: A general stress theory applied to burnout. In: *Professional Burnout: Recent Development in Theory and Research,* edited by W. Schaufeli, C. Maslach and T. Marek. Washington, DC: Taylor and Francis.

Ida, H., M. Miura, M. Komoda, N. Yakura, T. Mano, T. Hamaguchi, Y. Yamazaki, K. Kato, and K. Yamauchi. 2009. Relationship between stress and performance in a Japanese nursing organization. *International Journal of Health Care Quality Assurance* 22 (6):642–656.

Institute of Medicine. 2001. *Crossing the Quality Chasm: A New Health System for the 21st Century.* Washington, DC: National Academy Press.

Ivancevich, J.M. and M. T. Matteson. 1987. Organizational level stress management interventions: A review and recommendations. *Journal of Organizational Behavior Management* 8 (2):229–248.

Jackson, S. E. 1983. Participation in decision making as a strategy for reducing job-related strain. *Journal of Applied Psychology* 68 (1):3–19.

Jamal, M. 1984. Job stress and job performance controversy: An empirical assessment. *Organizational Behavior and Human Decision Processes* 33 (1):1–21.

Jex, S. M. 1998. *Stress and Job Performance: Theory, Research, and Implications.* Thousand Oaks, CA: Sage.

Jex, S., D. Baldwin, P. Hughes, C. Storr, S. Conard, and D. Sheehan, 1991. Behavioral consequences of job-related stress among resident physicians: The mediating role of psychological strain. *Psychological Reports* 69:339–349.

Jones, J., B. Barge, B. Steffy, L. Fay, L. Kunz, and L. Wuebker. 1988. Stress and medical malpractice: Organizational risk assessment and intervention. *Journal of Applied Psychology* 73 (4):727–735.

Karasek, R, D. Baker, F. Marxer, A. Ahlbom, and T. Theorell. 1981. Job decision latitude, job demands and cardiovascular disease: A prospective study of Swedish men. *American Journal of Public Health* 71 (1):694–705.

Karasek, R. and T. Tores. 1990. *Healthy Work: Stress, Productivity, and the Reconstruction of Working Life.* Chicago, IL: Basic Books.

Kazmi, R., S. Amjad, and D. Khan. 2008. Occupational stress and its effect on job performance. A case study of medical house officers of district Abbottabad. *Journal of Ayub Medical College Abbottabad* 20 (3):135–139.

Keijsers, G. J., W. B. Schaufeli, P. M. LeBlanc, C. Zwerts, and D. R. Miranda. 1995. Performance and burnout in intensive care units. *Work and Stress* 9 (4):513–527.

Kjellberg, A. 1977. Sleep deprivation and some aspects of performance: II. Lapses and other attentional effects. *Waking and Sleeping* 1:145–148.

Koppel, R., T. Wetterneck, J. L. Telles, and B. Karsh. 2008. Workarounds to barcode medication administration systems: Their occurrences, causes, and threats to patient safety. *Journal of the American Medical Informatics Association* 15 (4):409–423.

Landsbergis, P. A. and E. Vivona-Vaughan. 1995. Evaluation of an occupational stress intervention in a public agency. *Journal of Organizational Behavior* 16 (1):29–48.

Laschinger, H. K. S. and M. P. Leiter. 2006. The impact of nursing work environments on patient safety outcomes: The mediating role of burnout. *Journal of Nursing Administration* 36 (5):259–267.

Laschinger, H. K. S. and M. P. Leiter. 2009. The nursing worklife model: The role of burnout in mediating work environment's relationship with job satisfaction. In: *Handbook of Stress and Burnout in Health Care*, edited by J. R. B. Halbesleben. New York: Nova Science.

Leiter, M., P. Harvie, and C. Frizzell. 1998. The correspondence of patient satisfaction and nurse burnout. *Social Science and Medicine* 47 (10):1611–1617.

Leiter, M. and C. Maslach. 1988. The impact of interpersonal environment on burnout and organizational commitment. *Journal of Organizational Behavior* 9 (4):297–308.

Linzer, M., L. B. Manwell, E. S. Williams, J. A. Bobula, R. L. Brown, A. B. Varkey, B. Man, J. E. McMurray, A. Maguire, B. Horner-Ibler, and M. D. Schwartz. 2009. Working conditions in primary care: Physician reactions and care quality. *Annals of Internal Medicine* 151 (1):28–36, W6–9.

McCue, J. D. and C. L Sachs. 1991. A stress management workshop improves residents' coping skills. *Archives of Internal Medicine* 151 (11):2273–2277.

McMurray, J., M. Linzer, T. Konrad, J. Dougas, R. Shugerman, and K. Nelson. 2000. The work lives of women physicians results from the physician work life study. *Journal of General Internal Medicine* 15 (6):372–380.

Mealer, M. L., A. Shelton, B. Berg, B. Rothbaum, and M. Moss. 2007. Increased prevalence of post-traumatic stress disorder symptoms in critical care nurses. *American Journal of Respiratory and Critical Care Medicine* 175 (7):693–697.

Menzies, I. E. P. 1960. Nurses under stress. *International Nursing Review* 7:9–16.

Mott, D. A., W. R. Doucette, C. A. Gaither, C. A. Pedersen, and J. C. Schommer. 2004. Pharmacists' attitudes toward worklife: Results from a national survey of pharmacists. *Journal of American Pharmaceutical Association* 44 (3):326–336.

Murphy, L. R. 1995. Occupational stress management: Current status and future directions. In: *Trends in Organizational Behavior, 2*, edited by C. L. Cooper and D. M. Rousseau. New York: John Wiley & Son.

Nikou, V. R. 2003. The relationships of hardiness, stress, and health-promoting behaviors in undergraduate female nursing students. Paper read at Promoting Students'

Success, 14th International Nursing Research Congress, Sigma Theta Tau International, at St. Thomas, U.S. Virgin Islands.

Painter, J., D. Akroyd, S. Eliott, and R. D. Adams. 2003. Burnout among occupational therapists. *Occupational Therapy and Health Care* 17 (1):63–77.

Peerson, A. and V. Yong. 2003. Reflexivity in nursing: Where is the patient? Where is the nurse? *Australian Journal of Holistic Nursing* 10 (1):30–45.

Prins, J. T., F. M. van der Heijden, J. E. Hoekstra-Weebers, A. B. Bakker, H. B. van de Wiel, B. Jacobs, and S. M. Gazendam-Donofrio. 2009. Burnout, engagement and resident physicians' self-reported errors. *Psychology and Health Medicine* 14 (6):654–666.

Quinn, M. A., A. Wilcox, E. J. Orav, D. W. Bates, and S. R. Simon. 2009. The relationship between perceived practice quality and quality improvement activities and physician practice dissatisfaction, professional isolation, and work-life stress. *Medical Care* 47 (8):924–928.

Ratanawongsa, N., D. Roter, M. C. Beach, S. L. Laird, S. M. Larson, K. A. Carson, and L. A. Cooper. 2008. Physician burnout and patient-physician communication during primary care encounters. *Journal of General Internal Medicine* 23 (10):1581–1588.

Redfern, S., S. Hannan, I. Norman, and F. Martin. 2002. Work satisfaction, stress, quality of care and morale of older people in a nursing home. *Health and Social Care in the Community* 10 (6):512–517.

Reid, R. J., P. A. Fishman, O. Yu, T. R. Ross, J. T. Tufano, M. P. Soman, and E. B. Larson. 2009. Patient-centered medical home demonstration: A prospective, quasi-experimental, before and after evaluation. *American Journal of Management Care* 15 (9):e71–87.

Roberto, M. A., R. M. Bohmer, and A. C. Edmondson. 2006. Facing ambiguous threats. *Harvard Business Review* 84 (11):106–113, 157.

Ross, M., A. Carswell, and W. B. Danziel. 2002. Staff burnout in long-term care facilities. *Geriatrics Today* 5 (3):132–135.

Roter, D., M. Stewart, S. Putnam, M. Lipkin, W. Stiles, and T. Inui. 1997. Communication patterns of primary care physicians. *Journal of the American Medical Association* 277 (4):350–356.

Scarnera, P., A. Bosco, E. Soleti, and G. E. Lancioni. 2009. Preventing burnout in mental health workers at interpersonal level: An Italian pilot study. *Community Mental Health Journal* 45 (3):222–227.

Schaufeli, W. B., G. J. Keijsters, and D. R. Miranda. 1995. Burnout, technology use, and ICU performance. In: *Organizational Risk Factors for Job Stress*, edited by S. L. Sauter and L. R. Muraphy. Washington, DC: American Psychological Association.

Shanafelt, T., K. Bradley, J. Wipf, and A. Back. 2002. Burnout and self-reported patient care in an internal medicine residency program. *Annals of Internal Medicine* 136 (5):358–367.

Shanafelt, T. D., C. M. Balch, G. J. Bechamps, T. Russell, L. Dyrbye, D. Satele, P. Collicott, P. J. Novotny, J. Sloan, and J. A. Freischlag. 2009. Burnout and career satisfaction among American surgeons. *Annals of Surgery* 250 (3):463–471.

Shanafelt, T. D., C. M. Balch, G. Bechamps, T. Russell, L. Dyrbye, D. Satele, P. Collicott, P. J. Novotny, J. Sloan, and J. Freischlag. 2009. Burnout and medical errors among American Surgeons. *Annals of Surgery* 251(6):995–1000.

Shanafelt, T. D., C. P. West, J. A. Sloan, P. J. Novotny, G. A. Poland, R. Menaker, T. A. Rummans, and L. N. Dyrbye. 2009. Career fit and burnout among academic faculty. *Achieves of Internal Medicine* 169 (10):990–995.

Shehabi, Y., G. Dobb, I. Jenkins, R. Pascoe, N. Edwards, and W. Butt. 2008. Burnout syndrome among Australian intensivists: A survey. *Critical Care and Resuscitation* 10 (4):312–315.

Sherwood, P. and Y. Tagar. 2000. Experience awareness tools for preventing burnout in nurses. *Australian Journal of Holistic Nursing* 7 (1):15–20.

Shirom, A. 1989. Burnout in work organizations. In: *International Review of Industrial and Organizational Psychology*, edited by C. L. Cooper and I. T. Robertson. New York: Wiley.

Shirom, A., N. Nirel, and A. D. Vinokur. 2006. Overload, autonomy, and burnout as predictors of physicians' quality of care. *Journal of Occupational Health Psychology* 11 (4):328–342.

Shugerman, R., M. Linzer, K. Nelson, J. Douglas, R. Williams, and R. Konrad. 2001. Pediatric generalists and subspecialists: Determinants of career satisfaction. *Pediatrics* 108 (3):E40.

Spickard, A., Jr., S. G. Gabbe, and J. F. Christensen. 2002. Mid-career burnout in generalist and specialist physicians. *Journal of the American Medical Association* 288 (12):1447–1450.

Spurgeon, P. and J. M. Harrison. 1989. Work performance and the health of junior hospital doctors: A review of the literature. *Work and Stress* 3 (2):117–128.

Stewart, M. 1995. Effective physician-patient communication and health outcomes: A review. *Canadian Medical Association Journal* 152 (9):1423–1433.

Taner, M. 2009. An application of Six Sigma methodology to turnover intentions in health care. *International Journal of Health Care Quality Assurance* 22 (3):252–265.

Thomas, N. K. 2004. Resident burnout. *JAMA* 292 (23):2880–2889.

Thomsens, S., B. Arnetz, P. Nolan, J. Soares, and J. Dallander. 1999. Individual and organizational well-being in psychiatric nursing. *Journal of Advanced Nursing* 30 (3):749–757.

Trufelli, D. C., C. G. Bensi, J. B. Garcia, J. L. Narahara, M. N. Abrao, R. W. Diniz, C. Miranda Vda, H. P. Soares, and A. Del Giglio. 2008. Burnout in cancer professionals: A systematic review and meta-analysis. *European Journal of Cancer Care (Engl)* 17 (6):524–531.

Tversky, A. and D. Kahneman. 1974. Judgment under uncertainty: Heuristics and biases. *Science* 185 (4157):1124–1131.

Ulrich, D. 1997. *Human Resource Champions: The Next Agenda for Adding Value and Delivering Results*. Boston, MA: Harvard Business School Press.

Vahey, D., L. Atken, D. Sloane, S. Clarke, and D. Vargas. 2004. Nurse burnout and patient satisfaction. *Medical Care* 42 (2 Suppl):II-57-II-66.

Van Bogaert, P., H. Meulemans, S. Clarke, K. Vermeyen, and P. Van de Heyning. 2009. Hospital nurse practice environment, burnout, job outcomes and quality of care: Test of a structural equation model. *Journal of Advanced Nursing* 65 (10):2175–2185.

van der Homberg, P., B. Kunzi, G. Elwyn, J. van Doremalen, R. Akkermans, R. Grol, M. Wensing. 2009. High workload and job stress are associated with lower practice performance in general practice: An observation study in 239 general practices in the Netherlands. *BMC Health Services Research* 9:118–125.

van der Klink, J. J. L., R. W. B. Blonk, A. H. Schene, and F. J. H. van Dijk. 2001. The benefits of interventions for work-related stress. *American Journal of Public Health* 91 (2):270–276.

Van Dierendonck, D., W. B. Schaufeli, and H. J. Sixma. 1994. Burnout among general practitioners: A perspective from equity theory. *Journal of Social and Clinical Psychology* 13 (1):86–100.

Wallace, J. E. and J. Lemaire. 2009. Physician well being and quality of patient care: An exploratory study of the missing link. *Psychology Health and Medicine* 14 (5):545–52.

Williams, E. S., E. R. Lawrence, K. S. Campbell, and S. Spiehler. 2009. The effect of emotional exhaustion and depersonalization on physician-patient communication: A theoretical model, implications, and directions for future research. *Advances in Health Care Management* 8:3–20.

Williams, E. S., L. Manwell, T. R. Konrad, and M. Linzer. 2007. The relationship of organizational culture, stress, satisfaction, and burnout with physician-reported error and suboptimal patient care: Results from the MEMO study. *Health Care Management Review* 32 (3):203–212.

Williams, E. S., K. V. Rondeau, Q. Xiao, and L. H. Francescutti. 2007. Heavy physician workloads: Impact on physician attitudes and outcomes. *Health Service Management Research* 20 (4):261–269.

10

Safety Culture in Health Care

Kenji Itoh, Henning Boje Andersen, and Marlene Dyrløv Madsen

CONTENTS

10.1 Introduction: Safety Culture and Its Roles to Patient Safety 134
   10.1.1 Safety Culture versus Climate 134
   10.1.2 Definition of Safety Culture 135
   10.1.3 Overlapping Types of Safety Culture 136
   10.1.4 Links to Safety Management 137
10.2 Approaches to Safety Culture: What, Where, and How Should We Measure? 138
   10.2.1 Dimensions of Safety Culture 138
   10.2.2 Methods of Study and Assessment 139
      10.2.2.1 Questionnaire Survey 139
      10.2.2.2 Focus Group Interviews 140
      10.2.2.3 Critical Incident Technique 140
      10.2.2.4 Safety Audit 140
10.3 Assessment of Safety Culture: Requirements, Methods, and Tools 141
   10.3.1 Requirements for Safety Culture Tools 141
   10.3.2 Measurements of Safety Outcomes 142
   10.3.3 Association with Safety-Related Outcomes 143
   10.3.4 Assessment Tools in Health Care 143
      10.3.4.1 ORMAQ 146
      10.3.4.2 SAQ 146
      10.3.4.3 SCS 146
      10.3.4.4 HSOPS 146
      10.3.4.5 Stanford/PSCI 146
   10.3.5 Considerations When Conducting Surveys 147
10.4 Staff Perceptions of Safety Culture: How Do Employees Perceive Safety-Related Issues? 147
   10.4.1 Safety Culture Perceived by Health-Care Professionals 147
   10.4.2 Differences between Clinical Specialties/Wards 149
   10.4.3 Differences in Organizational Safety Culture 151
   10.4.4 Multinational Comparisons in Safety Culture 151
   10.4.5 Climate Changes in Health Care 152
10.5 Correlations with Safety Outcome: How Does Safety Culture Contribute to Patient Safety? 154
   10.5.1 Meaning of Incident Reporting Rates 154
   10.5.2 Comparing Safety Culture Indices with Reporting Rates 155
   10.5.3 Continuous Tracking of Incident Rates and Safety Performance Level 157
10.6 Conclusion: Perspectives for Improving Patient Safety 158
Acknowledgments 158
References 159

10.1 Introduction: Safety Culture and Its Roles to Patient Safety

It is widely recognized that human factors play a crucial role for safety in modern workplaces, not only in high-tech, human–machine operation domains (e.g., Bryant, 1991; Amalberti, 1998; Hollnagel, 1998) such as aviation, maritime, and nuclear power plants, but also in health care (Kohn et al., 1999). The recognition that operational safety (and patient safety in health care) depends on our abilities to control human error does not mean that efforts should be directed exclusively to the psychological mechanisms underlying human error. Rather, effective safety management should be directed at factors that are conducive to human failure (Rasmussen, 1986)—in particular factors that are within the direct control of the organization. Thus, *organizational factors* have long been acknowledged to be of critical importance for safety in human–machine system operations (Griffiths, 1985; Reason, 1993), and in Reason's thesis it has been pointed out that organizational problems are frequently *latent causal factors* that contribute or even lead to the occurrence of human error made by frontline personnel and has become part of the industry standard in this field (Reason, 1997). Indeed, the majority of contributing causes to major accidents may be attributed to the organizations themselves (ibid.). For example, it has been reported that 40% of incidents in the Dutch steel industry were caused by organizational failures (van Vuuren, 2000). Similarly, based on studies in aviation and maritime operations, it has been suggested that the quality and safety, by which the operators accomplish their tasks, are affected not only by their professional and technical competence and skills, but also by their attitudes to and perceptions of their job roles, their organization, and management (Helmreich and Merritt, 1998).

The term *safety culture* was introduced in the late 1980s when, in the aftermath of the Chernobyl nuclear power accident in 1986, the International Atomic Energy Agency (IAEA) issued several reports analyzing and describing the causes of the disaster, arguing forcefully that the root causes of the accident were to be found in the safety culture that existed in the organization running the power plants (IAEA, 1991). Thus, the concept of safety culture was invoked to explain an organizational mindset that tolerated gross violations and individual risk taking behaviors (INSAG, 1991). The INSAG (1991) defined the notion as follows: "Safety culture is that assembly of characteristics and attitudes in organizations and individuals which establishes that, as an overriding priority, nuclear plant safety issues receive the attention warranted by their significance."

Since the publication of the INSAG reports, a large number of studies have developed models, measures and tools, and instruments for safety culture. While we return to the definitions of safety culture in the subsequent section, it should be noted that there have also appeared a great number of studies of safety culture in health care, building methods, and techniques which can be well adapted to this safety critical domain, and carrying practical projects for measuring and diagnosing safety culture in specific organizations and work units since early 1990s (e.g., Helmreich et al., 1998; Gershon et al., 2000; Singer et al., 2003).

After publishing the first edition of this handbook in 2006, an increasing number of projects have been conducting to measure and diagnose safety culture or climate in health care. For example, according to a survey made by Mannion et al. (2009) in the United Kingdom, 97% of acute and primary care health-care organizations viewed safety culture as a central task for clinical management, and one third of them have been using safety culture instruments as a part of clinical management activities. There have also been large-scale surveys conducted for measuring current states in health-care safety culture in many countries, e.g., United States (Singer et al., 2009; Sorra et al., 2009) and Japan (Itoh and Andersen, 2008). For example, the Agency for Healthcare Research and Quality (AHRQ) continuously collected safety culture data, and the 2009 report (Sorra et al., 2009) stated the analysis results from a total of about 200,000 staff responses from 622 U.S. hospitals, including trending results over time by comparing with the 2007 and 2008 data. Thus, nowadays, safety culture assessment has been widespread in health care not only for scientific work, but also as a management tool in individual organizations for measurement and diagnosis of their safety status.

This chapter describes, first, how the notions of safety culture/climate are defined and used in the human factors literature about nonmedical domains as well as in the patient safety literature. Then, based on our experiences of safety culture studies not only in health care, but also in maritime, railway, process industry, and other industrial domains, we illustrate some main themes of safety culture by reviewing results of recent questionnaire surveys. Finally, we review the effects of safety culture on patient safety, and discuss some issues in developing and maintaining a positive safety culture.

10.1.1 Safety Culture versus Climate

Since the 1930s (Krause et al., 1999)—and long before the term "safety culture" was introduced—it has been acknowledged that differences in organizational factors have an impact on the rate of occupational injuries. Thus, organizational factors—such as psychosocial stressors, work team atmosphere, perceptions of leadership—have been studied for decades in relation to occupational accidents, usually under the heading of "organizational climate" or "safety climate."

The distinction between safety culture and safety climate has been discussed by a number of authors. Safety culture is most commonly conceptualized as a three-layer structure, following Schein's (1992) work on organizational culture, as summarized in Table 10.1: (1) an inner layer of "basic assumptions," i.e., a core of largely tacit underlying assumptions taken for granted by the entire organization; (2) a middle layer of "espoused values," i.e., values and norms that are embraced and adopted; and (3) an outer layer of "artifacts" that include tangible and overt items and acts such as procedures, inspections, and checklists. Safety culture is seen as part of the overall culture of the organization that affects members' attitudes and perceptions related to hazards and risk control (Cooper, 2000). Thus, culture concerns shared symbolic and normative structures that (a) are largely tacit (implicit, unconscious), (b) are largely stable over time, and (c) can be assigned a meaning only by reference to surrounding symbolic practices of the cultural community. For these properties, it is difficult to address culture by quantitative methods such as questionnaire surveys since they cannot fully represent the underlying elements of safety culture, and therefore it must be measured by qualitative methods (Flin, 2007).

In contrast, the distinct but closely related notion of *safety climate* is viewed as governed by safety culture and contextual and possibly local issues; so climate refers to employees' context-dependent attitudes and perceptions about safety-related issues (Flin et al., 2000; Glendon and Stanton, 2000). Safety climate has been characterized as reflecting the surface manifestation of culture: "the workforce's attitudes and perceptions at a given place and time. It is a snapshot of the state of safety providing an indicator of the underlying safety culture of an organization," and it may be appropriate to use the term,

safety climate for dimensions elicited by a questionnaire-based survey, which could obtain transient, surface features at a given point in time (Mearns et al., 1998).

As such, culture and climate can theoretically be distinguished in terms of how stable, tacit, and interpretable these shared values, attitudes, etc. are. Still, we will sometimes use the term climate to emphasize the need for referring to local, changeable and explicit attitudes, and perceptions. However, since the term safety culture is much more commonly used than safety climate, we shall, having no need for distinguishing between "underlying" and "overt" attitudes and perceptions, mostly use the term safety culture in its inclusive sense. In the rest of this chapter, following this convention, we shall, unless precision is required, refer to "safety culture and climate."

10.1.2 Definition of Safety Culture

Numerous definitions have been proposed of the concepts of safety culture and climate (e.g., Zohar, 1980; Pidgeon and O'Learry, 1994; Flin et al., 2000). The most widely accepted and most often quoted safety culture definition is the one put forward by the Advisory Committee on the Safety of Nuclear Installations (ACSNI, 1993): "The safety culture of an organization is the product of individual and group values, attitudes, perceptions, competencies and patterns of behavior that determine the commitment to, and the style and proficiency of, an organization's health and safety management. Organizations with a positive safety culture are characterized by communications founded on mutual trust, by shared perceptions of the importance of safety and by confidence in the efficacy of preventive measures."

According to this well-known definition, an organization's safety culture involves thus the shared *values, attitudes, perceptions, competencies,* and *patterns of behavior* of its members; when an organization has a positive safety culture, there is a mutual and high level of trust

TABLE 10.1

Schein's Three Levels of "Culture" Adapted to Medicine

Levels of Culture	Characteristics	Examples in Medicine	Levels of Interpretation and Methods
Artifacts (cultural symptoms)	Visible artifacts, objects, and behavior; changeable, context dependent, but often difficult to decipher	Equipment, procedures, communication routines, standard services, alarms, dress code, and hierarchical structures	Climate/quantitative methods
Espoused values (and attitudes)	Official and unofficial policies, norms and values (in/not in accordance with underlying assumptions)	Mission statement, team norms, "learn from mistakes"	Culture/qualitative methods
Basic assumptions	Unconscious beliefs, values, and expectations shared by individuals (taken as given), implicit (tacit), relatively persistent, cognitive and normative structures	The "primum non nocere" creed, the natural science paradigm	

Source: Schein, E.H., *Organizational Culture and Leadership*, 2nd edn., Jossey-Bass, San Francisco, CA, 1992.

among employees, and they share the belief that safety is important and can be controlled.

But a number of other, though not very dissimilar, definitions have been offered. For example, Hale defines safety culture as "the attitudes, beliefs and perceptions shared by natural groups as defining norms and values, which determine how they act and react to risks and risk control systems" (Hale, 2000). The Confederation of British Industry (CBI, 1991) defined safety culture as, "the ideas and beliefs that all members of the organization share about risk, accidents and ill health."

In a comprehensive review, Guldenmund (2000) cites and discusses 18 definitions of safety culture and climate appearing in the literature from 1980 to 1997. This proliferation of definitions has led to some difficulty in interpretations. It is not surprising therefore that Reason has noted—in his discussion of safety culture and organizational culture—that the latter notion "has as much definitional precision as a cloud" (Reason, 1997).

Still, the most often cited definitions are similar in that they refer to normative beliefs, and *shared* values and attitudes about safety-related issues by members of an organization. Nearly all analysts agreed that safety culture is a relatively stable, multidimensional, holistic construct that is shared by organizational members (e.g., Guldenmund, 2000).

At the same time, safety culture is regarded as being relatively *stable* over time—for example, De Cock et al. (1986; cited in Guldenmund, 2000) found no significant change of *organizational* culture over a 5-year interval. The content of safety culture—the norms and assumptions it is directed at—consists of underlying factors or dimensions—such as perceptions of commitment, leadership involvement, and willingness to learn from incidents. Moreover, safety culture has a *holistic* nature, and is something shared by people, members, or groups of members within an organization. Indeed, the holistic or shared aspect is stressed in most definitions, involving terms such as "molar" (Zohar, 1980), "shared" (Cox and Cox, 1991), "group" (Brown and Holms, 1986), "set" (Pidgeon, 1991), and "assembly" (IAEA, 1991). When members successfully *share* their attitudes or perceptions, "the whole is more than the sum of the parts" (ACSNI, 1993). Finally, safety culture is *functional* in the sense that it guides members to take adaptive actions for their tasks, as Schein (1992) regards organizational culture as "the way we do things around here."

10.1.3 Overlapping Types of Safety Culture

Viewing safety culture from a different perspective, we may distinguish between four overlapping spheres of culture, each of which may distinguish the perceptions and attitudes of different organizational units: national,

domain-specific, professional, and organizational culture. Thus, the safety culture shared by a given team of doctors and nurses will be shaped not only by their organization (e.g., hospital), but also by their different national cultures (see Helmreich et al., 1998, for a review of a number of studies of especially aviation personnel). Again, when comparing different safety critical domains—health care, aviation, process industry, nuclear power, etc.—we sometimes find considerable differences in attitudes and perceptions to safety relevant aspects (Gaba et al., 2003). Finally, within the same domain, we should expect that there might well be—possibly task related—differences between professions, such as between doctors and nurses or pilots and cabin staff (Helmreich et al., 1998). These different and sometimes overlapping layers of cultures are intertwined to shape the safety culture of any particular organizational unit, and each of them can have both positive and negative impact on patient safety (Helmreich et al., 2001).

Regarding the national culture, Hofstede (1991, 2001) built a famous model based on responses about values and norms across different countries involving four dimensions: power distance, individualism–collectivism, uncertainty avoidance, and masculinity–femininity. Hofstede carried out a number of questionnaire surveys with 20,000 IBM employees working in more than 50 countries, mapping the different national cultures in terms of these four dimensions. Adapting parts of this paradigm, Helmreich and his collaborators (e.g., Helmreich et al., 1998 2001) have been conducting comparative studies on national cultures in aviation, finding national differences similar to Hofstede's results.

Shifting our attention to cultures within an organization, we may sometimes observe *local* variations of culture (we may even suppose, for simplicity, that the organization is characterized by the same national, domain-level, and professional culture). Recently, Zohar and colleagues (Zohar, 2003, 2008; Zohar et al., 2007) argued that safety climate can be constructed at multiple hierarchical levels, i.e., hospital climate and unit climate, which have been associated with managerial commitment of both senior manager and supervisor as its core elements. Similarly, Singer et al. (2007) proposed a multiple-level framework of safety climate, which included three factors at the organizational level, i.e., senior managers' engagement, organizational resources, and overall emphasis on safety; two at the unit level, i.e., unit safety norms, and unit recognition and support for safety; and three at the individual level, i.e., fear of shame, fear of blame, and learning.

Connecting to the aforementioned issue, Itoh et al. (2001) identified local differences in safety culture characterizing each of the different contractors of a railway track maintenance company. The study also found a relationship between the cultural types and accident/incident rates. Such variations of culture within a single organization are often called local cultures or subcultures. The variations in subcultures may possibly depend on both task and domain related characteristics (e.g., differences in workload, department and specialties, exposure to risks, and work shift patterns) and on demographic factors (e.g., age, experience, gender, seniority, and positions in the organization). They may also, it is natural to speculate, be influenced by individual factors related to local leadership and the team atmosphere defined by the most dominating and charismatic team members.

10.1.4 Links to Safety Management

The rationale behind studying the safety culture of a given organization and groups of members is, briefly put, that by measuring and assessing safety culture we may be able to identify "weak points" in the attitudes, norms, and practices of the target groups and organizations; in turn, knowledge of "weak points" may be used to guide the planning and implementation of intervention programs directed at enabling the target groups and organizations to develop improved patient safety practices and safety management mechanisms. In addition, one of the greatest advantages of safety culture is its applicability to diagnosing safety status *proactively*, that is, safety culture indices can be measured *before* an actual accident takes place. Moreover, it is well known that only a subset of all incidents is reported in health care (Antonow et al., 2000). Therefore, safety culture assessment may serve as an independent means of estimating the risk level, and identifying and ultimately controlling so-called "latent conditions" in specific organizations and units (Reason, 1997).

Adapting the approach of Nieva and Sorra (2003), we can summarize the objectives of safety culture assessment for a given organization, ward, or department as follows:

1. Profiling (diagnosis): An assessment may aid in determining the specific safety culture or climate profile of the work unit; including the identification of "strong" and "weak" points.

2. Awareness enhancement: It may serve to raise staff awareness, typically when conducted in parallel with other staff oriented patient safety initiatives.

3. Measuring change: Assessment may be applied and repeated over time to detect changes in perceptions and attitudes, possibly as part of a "before-and-after-intervention" design.

4. Benchmarking: It may be used to evaluate the standing of the unit in relation to a reference sample (comparable organizations and groups).

5. Accreditation: It may be part of a, possibly mandated, safety management review or accreditation program.

As can be seen from the definitions mentioned previously, safety culture is coupled not only to employees' beliefs and attitudes about safety-related issues—for instance, motivation, morale, risk perceptions, and attitudes toward management and factors that impact on safety, e.g., fatigue, risk taking, and violations of procedures, all of which can be partially controlled or are influenced by efforts of safety management—but also to safety management issues such as management's commitment to safety, its communication style and the overt rules for reporting errors, etc. (Andersen, 2002). In addition, as will be further discussed in Section 10.2.1 (dimensions of safety culture), an organization's safety culture reflects its policies about error management and sanctions against employees who commit errors or violations, the openness of communications between management and operators, and the level of trust between individuals and senior management (Helmreich, 2000a).

It is well known that the probability of human error is influenced by various work environment factors such as training, task frequency, human–machine interfaces, quality of procedures, supervision/management quality, work load, production and time pressure, fatigue, etc. Such error inducing factors are called "performance shaping factors" (PSFs)—a term name, now widely used, that was introduced in the early 1980s by human reliability assessment (HRA) analysts who seek to structure and, ultimately, quantify the probability of human failure in safety critical domains and in particular nuclear power production (Swain and Gottman, 1983; Hollnagel, 1998). On the HRA approach, different types of relevant PSFs have been identified and analyzed for a given set of tasks. In fact, the traditionally cited PSFs (confer preceding text) are tightly related to safety culture in so far that a "good" safety culture will reflect a shared understanding of the importance and the means to control the factors that have an impact on human reliability.

To put it briefly, while effective safety management consists in identifying the factors that impact on the safety of operations and having the means available to implement control mechanisms, safety culture comprises the mutual awareness of and supportive attitudes directed at controlling risks (Duijm et al., 2004). There is, therefore, a growing awareness that applications of human reliability analysis may contribute to a focus for the strengthening of safety culture when developing practices and strategies that help health-care staff to control the potentially detrimental effects of PSFs (Kohn et al., 1999; Department of Health, 2000).

10.2 Approaches to Safety Culture: What, Where, and How Should We Measure?

10.2.1 Dimensions of Safety Culture

A number of dimensions (scales, components, or aspects) of safety culture have been proposed ranging from psychosocial aspects (motivation, morale, team atmosphere, etc.) to behavioral and attitudinal factors regarding management, job, incident reporting, and others. Many of these dimensions have been elicited by applying multivariate analysis techniques such as factor analysis and principal component analysis to questionnaire data. Some dimensions proposed by different researchers are quite similar—though the terms may differ—while others differ very much. The number of dimensions also differs among studies, ranging from 2 to 16 according to Guldenmund (2000) that conducted a literature survey up to 1997. Additional dimensions of safety culture have been proposed since this review, but the diversity of dimensions still remains (Guldenmund, 2007).

The variation in safety culture dimensions has several sources: first, researchers have employed different questionnaires (so, they have different questions to respondents); second, surveys have been made of quite different fields or domains with quite different hazard levels, recruitment criteria, training requirements, regulation regimes (e.g., nuclear power production, aircraft carriers, airline piloting, construction and building industry, offshore oil production platforms, shipping, railways, health-care organizations, and units); and third, when aggregating a group of question items, the choice of a label is essentially a subjective interpretation. However, although the labels of dimensions will often vary, it is clear that similar dimensions (according to the meaning of the labels) can be found across the different sets of dimensions proposed by different research groups.

In his seminal study of Israeli manufacturing workers Zohar (1980) proposed an eight-dimensional model that comprised (1) the importance of safety training, (2) management attitudes toward safety, (3) effects of safe conduct on promotion, (4) level of risk at the work place, (5) effects of required work pace on safety, (6) status of safety officer, (7) effects of safe conduct on social status, and (8) status of the safety committee. The eight-dimensional model was aggregated by Zohar into a two-dimensional model: (1) perceived relevance of safety to job behavior; and (2) perceived management attitude toward safety. Cox and Flin (1998) identified the following five dimensions, which they called "emergent factors": (1) management commitment to safety, (2) personal responsibility, (3) attitudes to hazards, (4) compliance with rules, and (5) workplace conditions. Pidgeon

et al. (1994) proposed four dimensions to capture "good" safety culture: (1) senior management commitment to safety; (2) realistic and flexible customs and practices for handling both well-defined and ill-defined hazards; (3) continuous organizational learning through practices such as feedback system, monitoring, and analyzing; and (4) care and concern for hazards which is shared across the workforce. These four dimensions seem to relate to the dimensions proposed both by Zohar (1980) and by Cox et al. (1998), and they have broader aspects of safety-related organizational issues such as a concept of organizational learning, which was not overtly included in Zohar's (1980) original factors.

In health care, a number of studies have also been seeking to develop dimensions of safety culture. For example, Gershon et al. (2000) extracted six dimensions from questionnaire responses with a specific focus on blood-borne pathogen risk management: (1) senior management support for safety programs, (2) absence of workplace barriers to safe work practices, (3) cleanliness and orderliness of the work site, (4) minimal conflict and good communication among staff members, (5) frequent safety-related feedback/training by supervisors, and (6) availability of personal protective equipment and engineering controls. Sorra and Nieva (2004) elicited 12 dimensions from the U.S. hospital settings—a similar set of dimensions was also extracted by applying the same instrument to Dutch hospital respondents (Smits et al., 2008): (1) teamwork across hospital units, (2) teamwork within units, (3) hospital handoffs and transitions, (4) frequency of event reporting, (5) nonpunitive response to error, (6) communication openness, (7) feedback and communication about error, (8) organizational learning—continuous improvement, (9) supervisor/manager expectations and action promoting patient safety, (10) hospital management support for patient safety, (11) staffing, and (12) overall perceptions of safety.

The examples mentioned earlier, on the one hand, illustrate the variation in safety culture factors we have referred to. In general, the criterion for including a putative dimension as a safety culture factor is that there is evidence (or reason to believe) that it correlates with safety performance. A given presumed safety culture dimension should be included only if a more "positive" score on this dimension is correlated with a higher safety record. If two comparable organizations or organizational units have different safety outcome rates (e.g., accident and incident rates), we should expect that a high-risk unit has a "lower" score on the dimension in question. However, using this criterion also means that safety culture dimensions may not directly relate to safety (e.g., Andersen et al., 2004a). In fact, it would be wrong (and naive) to expect that the dimensions of safety culture that have the strongest link with safety outcome are ones that most directly are "about" safety.

On the contrary, there is evidence that the strongest indicators of safety performance are ones that relate to dimensions that have an indirect link to safety, such as motivation, morale, and work team atmosphere (ibid.).

Among the aforementioned dimensions, on the other hand, several elements overlap or are similar, and therefore there have been attempts to explore "core" dimensions of safety culture by generalizing from diverse dimensions that have appeared in the literature. Flin et al. (2000) reviewed studies on safety climate instruments used in industrial sectors up to the late 1990s, and identified 6 "common" themes in 18 papers: (1) management/supervision, (2) safety systems, (3) risk, (4) work pressure, (5) competence, and (6) procedures/rules. Flin and her research group also conducted a literature review in health care and extracted 10 features of safety culture—some of them were overlapped with or similar to those in the industry mentioned earlier while others are not—from results of 12 instruments (Flin et al., 2006): (1) management/supervision, (2) safety systems, (3) risk perception, (4) job demands, (5) reporting/speaking up, (6) safety attitudes/behaviors, (7) communication and feedback, (8) teamwork, (9) personal resources, and (10) organizational factors. In particular, they identified *management commitment to safety* as the most frequently measured safety culture dimension in health care, and suggested the following three "core" themes of health-care safety culture: management/supervisor commitment to safety, safety systems, and work pressure (ibid.). Similarly, Singla et al. (2006) elicited six broad categories of safety culture factors—which are overlapped or quite similar to the already mentioned "common" themes of industrial culture elicited by Flin et al. (2000)—from 23 individual dimensions by survey results of 13 instruments: management/supervision, risk, work pressure, competence, rules, and miscellaneous.

10.2.2 Methods of Study and Assessment

When we investigate safety culture of a specific organization, a couple of metrics should be measured: *level* and *strength* (Zohar et al., 2007), although most safety climate/culture studies so far have focused on the former metric. The level of climate/culture refers to the degrees of positive or negative value for a specific dimension in an organization or work unit as a whole, i.e., low to high, or percentage of positive or negative responses. The strength indicates coherence, homogeneity, congruence of a culture and the extent to which employees share the same values, perceptions, and attitudes to a specific dimension of culture/climate, and is often measured by variation or standard deviation of data related to the dimension when using a quantitative method.

Several approaches have been applied to investigating the safety culture of particular organizations: safety audits, peer reviews, performance indicator measures, some interview techniques, including structured interviews with management and employees, focus group, critical incident interviews, behavioral observations and field studies, and questionnaire surveys. Some methods that have been applied relatively frequently are described briefly.

10.2.2.1 Questionnaire Survey

A safety culture/climate questionnaire is a tool or instrument for measuring safety culture or climate of organizations in safety critical domains such as health care. Developers and users of a questionnaire will seek to identify the level and profile of safety culture in a target organization and possibly in its groups through elicitation of employee views and attitudes about safety-related issues. In particular, emphasis is assigned to the perception of employee groups regarding their organization's safety systems, their attitudes to and perceptions of management and, more generally, factors that are believed to impact on safety. Respondents are provided with a set of fixed (closed ended) response options, often in terms of rank-based responses on a Likert-type scale. It is not uncommon to include in the questionnaire open-ended questions that prompt respondents to provide responses in their own words, and thus, qualitative data of this type must be interpreted and possibly categorized by the researchers.

Just as there are variations in the themes and dimensions included by different theorists under "culture" and "climate," there is, in a similar vein, only moderate consistency—though some overlap is typically found—across different safety climate/culture survey tools in the dimensions they cover (Collins and Gadd, 2002). Several attempts and suggestions have been made to identify emerging themes (Flin et al., 2000; Wiegmann et al., 2004) for a common classification system to reduce the general number of dimensions (Guldenmund, 2000). Dimensions are either determined a priori (e.g., reproduced from previous questionnaires) or through survey iteration and refinement of items and dimensions. Dimensions are validated statistically using a measure of reliability (e.g., Cronbach's alpha)—this is discussed in more details in Section 10.3.1—and some forms of multivariate statistical method (e.g., factor analysis or principal component analysis). Still, not all questionnaires are validated statistically. It is important, however, to note that even though it is possible to perform a statistical evaluation of the identification of dimensions, the labeling of the dimensions will always remain subjective. We review some specific instruments applying questionnaires that have been frequently used for measuring organizational culture in health care in Section 10.3.

10.2.2.2 *Focus Group Interviews*

This type of interviews normally involves five to eight interviewees and a pair of interviewers. The interviewers ask the interviewees to react to a few open-ended, related issues. The aim of focus group interviews is to establish an informal forum where the interviewees are invited to articulate their own attitudes, perceptions, feelings, and ideas about the issues under scrutiny (Kitzinger, 1995; Marshall and Rossman, 1999). It is one of the biggest advantages of the focus group interview that themes, viewpoints, and perspectives are brought up, which might not otherwise have been thought of. It is important to promote a free exchange of viewpoints among the interviewees, and this is most often done using a prompting technique that resembles semi-structured interviews.

A semi-structured interview is characterized by the use of an interview guide structured by clusters of themes that may be injected with probes. The role of the interviewer is to usher the interviewees through the selected themes and, importantly, to ensure that all participants get a chance to voice their opinions and that no single person gets to dominate the others. Focus group interviews can be especially helpful when developing a questionnaire because they provide researchers with perspectives and views—and even terminology and phrases—related to themes that might otherwise be missed. In addition, these interviews also work very well as follow up on results from a survey, providing the researchers with "reasons" and background for the data collected.

The focus group interview with the aforementioned characteristics is ideal for employee groups, whereas the management group might be too small or too hierarchical to lend itself to this type of interview. Therefore, upper management representatives will, therefore, often be asked to participate in semi-structured interviews (one or two interviewees).

10.2.2.3 *Critical Incident Technique*

Another interview technique often used when studying human factors issues in safety critical domains is critical incident technique (CIT) (Flanagan, 1954; Carlisle, 1986). While the semi-structured interview is intended to draw a wide and comprehensive picture of the operator attitudes and values, the CIT focuses on the operators' narratives about specific incidents or accidents in which they themselves have been involved. They are asked to recall a critical incident and to talk about what happened; how they reacted; what the consequences were to themselves, to others, or to their work environment; what went well and not so well; and what they or others might have learned from the incident. In particular,

interviewees are asked to recall and recount the precursors—the contributing and possibly exacerbating factors behind the event—and factors that, if in place, could have contributed to resolving the incident.

This technique is very useful to identify human factors issues, and to understand and provide a basis for possibly planning a change of the states in "performance shaping factors," e.g., procedures, training, team interaction guidelines, human–machine interfaces, and workplace redesign. Thus, when a number of interviewees have offered their recalls of specific incidents, the data may reflect strengths and weaknesses in the current safety culture and climate.

10.2.2.4 *Safety Audit*

Safety audits employ methods that combine qualitative interviews and factual observations of (missing) documentation, (faulty) protections and barriers, actual behaviors, etc. in order to identify hazards existing in a workplace. Several safety audit tools have been developed comprising safety performance indicators to determine whether a given safety management delivery system (say, the mechanism to ensure that staff is trained to operate their equipment safely or the mechanism to ensure that learning lessons are derived from incidents) is fully implemented, or only partially so, or possibly only at a rudimentary level (e.g., Duijm et al., 2004). Audits are particularly useful for estimating the extent to which the organization's policies and procedures have been defined and are being followed and, to some extent—how they might be improved. On the behavioral observation (or field study) approach, employees' behaviors and activities are observed during their task performance in order to identify potentially risky behaviors.

Among these approaches the most frequent and widespread method for safety culture assessment is the questionnaire-based survey. This is a useful and, depending on the quality of the questionnaire tool, a reliable method of collecting response data from larger groups—and it is especially time efficient for large groups of respondents. However, some researchers believe that questionnaire surveys are unsuitable to fully uncover culture—refer to the distinction between "culture" and "climate" in Section 10.1.1. For example, Schein argues that it is only through iterative in-depth interviews that values and assumptions of organizational members may be revealed, and that it is doubtful whether questionnaires may be capable of exposing values and hidden cultural assumptions. According to Schein (2000), "culture questionnaire scores do correlate with various indices of organizational performance, but these measures are more appropriately measures of climate than of culture." While most will agree that questionnaires are best suited to elicit (explicit) attitudes and perceptions—and therefore "climate"—not everyone will agree that climate measures cannot illuminate culture. Indeed, to the extent that a successful factor analysis may identify underlying factors behind overt responses to items that might not, on their surface, appear similar, this type of approach can be said to uncover normative or attitudinal structures.

When one has to choose among methods of assessment, one may wish to consider that culture can have several levels of expression within an organization, and each method has its strengths and weaknesses. If the aim is to obtain a comprehensive picture of the organizational culture, one should preferably apply both quantitative and qualitative methods. If resources are scarce and the aim may be solely to provide an empirical basis for planning and selecting a limited set of interventions or revisions of safety management mechanisms, it may not be practical to aim for a comprehensive picture of the safety culture. Considering the resources and possibly the time available, the user should try to choose the method according to the overall safety aims behind the study under consideration as well as any available prior knowledge. For example, qualitative methods are not as easily adopted without prior knowledge or experience, as is the case with quantitative methods, where guidelines for use often follow the tools available. Additionally quantitative results can be benchmarked with other work units and hospitals, and be repeated to detect the effects of intervention programs.

In summary, there are several methods available for investigating safety culture in an individual organization or work unit: safety audits, peer reviews, performance indicator measures, interview techniques such as focus group and semi-structured interviews, behavioral observations, questionnaire surveys, and so forth. Each method has its strengths and weaknesses, its highly useful and less useful applications, etc. Among these methods, some tools involving questionnaire-based survey—which will be mentioned in the next section—can produce safety culture index values that have been shown to correlate with safety performance. It is important to select appropriate methods and tools that match the purposes of safety culture assessment as well as the characteristics of an organization under study.

10.3 Assessment of Safety Culture: Requirements, Methods, and Tools

10.3.1 Requirements for Safety Culture Tools

There are a number of quality requirements for safety culture tools that should be considered when selecting

for a given application. In the following we refer to these requirements as "selection criteria." Generally speaking, the requirements contained in these criteria raise issues about whether a given tool can be used to measure what it intends to measure (content validity), correlates with safety performance (external validity or criterion validity), yields consistent results (reliability), covers the safety culture dimensions that the user wants to have covered (relevance and comprehensiveness), is practical to administer (usability), is culturally referenced and tested in environments demographically and culturally much different from the user's (universality), and is targeted at the user's respondent groups (group targeting).

It is worthwhile pointing out some of these criteria have to be balanced against each other. For example, no tool can be rated highly on both comprehensiveness and usability, since the more comprehensive a questionnaire is with respect to themes it covers, the greater number of items it must include and, therefore, the lower it will score on usability. These selection criteria are described more in detail as follows.

1. *Validity*: The issue of validity concerns whether the tool may successfully be used to measure what it is supposed to measure. There are four components of validity that are relevant:

   a. *Pilot testing*: A pilot test will serve to identify items that are ambiguous, hard to understand, or are understood in ways that differ from what the developers had in mind. No questionnaire should be used as a survey instrument unless it has been thoroughly pilot tested.

   b. *Content validity*: This validity is related to how relevant items of a tool or questionnaire are to the targeted construct for the assessment purpose. This can be judged based on various sources such as relevant theory, empirical literature, and expert judgment (Flin et al., 2006).

   c. *Consistency* (internal validity): A questionnaire containing items purporting to probe different underlying factors or dimensions should be tested as inter-item reliability. Internal validity means that the items that address the same underlying factor correlate with each other. A widely used measure of this criterion is Cronbach's alpha (Cronbach, 1951; Pett et al., 2003). A high value of inter-item reliability does not ensure that the purported factor may not consist of subfactors (to be determined by various factor analyses or, in general, multivariate methods).

A commonly-accepted level, 0.7 or higher, for the alpha coefficient has been used (Spiliotopoulou, 2009).

   d. *Criterion validity* (external validity): Criterion validity of safety culture scales should be assessed by correlations of their scores with safety outcome data, which are preferably collected by some *other* methods from the safety culture questionnaire (Flin et al., 2006). Possible safety outcome data in health care could be worker behaviors, worker injuries, patient injuries, or other organizational outcomes (e.g., litigation costs) (Flin, 2007).

2. *Reliability* (test-retest reliability): This means that the survey tool will yield the same result if the same population is surveyed repeatedly (with the same techniques and in the same circumstances). In practice, this requirement is often examined as "test-retest" reliability by use of the data collected twice using the same tool in a short-term interval, taking advantage of characteristics of safety culture that is stable over time, as mentioned in Section 10.1. However, this criterion is sometimes difficult or even impossible to establish, since attitudes and perceptions are liable to change over time.

3. *Relevance and comprehensiveness*: We deal with these, in principle, distinct features at the same time. Relevance refers to whether a tool seeks to measure important—or relevant—dimensions of safety culture. Comprehensiveness refers to the extent to which a tool covers all the (relevant) dimensions.

4. *Practicality* (usability): This criterion refers to ease of use with which a given tool may be administered and it includes considerations of length and the time required for respondents to complete it. Also relevant in this respect are considerations about statistical analysis of results.

5. *Nonlocality*: This criterion is related to universality or possible cultural bias of a tool, e.g., is it tied to a specific regional or national culture? Users who consider using a questionnaire developed and tested within their own national or ethnic culture need not worry about this requirement.

6. *Job orientation and setting*: This refers to the types of staff for which a tool has been designed and tested (nurses, hospital physicians, pharmacists, etc.). It is also connected to the setting for which a tool is developed, e.g., is it for hospital staff or is it developed for another setting such as nursing homes? Questionnaires may often require considerable adaptation when they are transferred across work domains.

7. *Documentation*: This refers to the documentation available about a given tool: Whether there are sources availability that describe a tool in terms of the aforementioned selection criteria as well as history of its development and use.

10.3.2 Measurements of Safety Outcomes

To examine criterion validity of safety culture tools, attempts have been made to compare "scores" of safety culture dimensions against safety-related outcome data. Several methods of measuring safety performance have been suggested. The most intuitive and strong measure of safety outcome involves the *accident* or *incident rate* of an organization. This type of data can be evaluated repeatedly and regularly for an entire organization and its work units such as clinical departments and wards, but it may also allow a comparison between work units or between organizations in the same domain, e.g., hospital wards. At the same time, there are many reasons why we should be wary of using accident data—even within one and the same domain and the same type of tasks. First, such data may be essentially dependent on external factors and may not reflect internal processes—for example, university hospital clinics may be more likely to admit patients who are more ill and may therefore experience a greater rate of adverse events (Baker et al., 2004). Second, accident data may be of dubious accuracy due to underreporting by some organizations and overreporting by others (Glendon and McKenna, 1995).

Reporting of near misses and incidents may be a useful measure of safety performance—though its overriding goal is not to derive reliable statistics about rates of different types of incidents, but to enable organizations to learn from such experiences (Barach and Small, 2000). Thus, Helmreich (2000b) stresses that one should not seek to derive rates from incident reporting but rather focus on the valuable lessons they contain. But the wide variation observed across, for example, hospitals and individual departments may have much more to do with local incentives and local "reporting culture" than with actual patient risks (Cullen et al., 1995). Therefore, when making comparative studies between organizations—and even between work units within a single organization—we have serious difficulties in interpreting the incident rate. As will be discussed later with illustration of a case study in Section 10.5, this may either be interpreted as a measure of risk—the greater the rate of reported incidents of a given type, the greater is the likelihood that a patient injury may take place; *or* it may be taken as an index of the inverse of risk, that is, as an index of safety—that is, the more that staff are demonstrating willingness to report, the greater is their sensitivity to errors and learning potential, and so, the greater is the safety in their department (Edmondson, 1996, 2004).

There are two types of reporting rates about incidents. One is an actual rate which is usually calculated from the number of incident reports submitted by staff under study. The other is a self-reported rate of incidents, which are elicited as staff responses to (a subjective measure of) safety performance. When testing criterion validity of safety culture scales, as mentioned in the last section, it is recommended to obtain safety outcome data independently by using different methods from the one used for measuring a safety culture. In this sense, the actual incident data are more preferable to use when examining the criterion validity of the scales. However, it seems very few studies (so far) have used actual safety performance data in this sense of validation. Instead, self-reported data are more common sources for validating safety culture tools (Cooper, 2000; Gershon et al., 2004).

10.3.3 Association with Safety-Related Outcomes

Among the requirements for safety culture instruments mentioned in Section 10.3.1, only a few studies yet have been successfully addressed to the criterion validity, i.e., association with safety outcome (Cooper and Phillips, 2004). In health care, for example, Clarke (2006) identified only a weak association of safety culture with safety performance using a meta-analysis of related studies. Most of health-care studies which tried to identify correlations with safety outcomes have used self-reported data pertaining to incidents (Sorra and Nieva, 2003) and to safety behaviors (Gershon et al., 2000; Neal et al., 2000). Only a few projects used actual safety outcome data rather than self-report, and most of these studies employed occupational injuries (Felknor et al., 2000; Vredenburgh, 2002). We have also tried to identify relationships of safety culture factors with safety performance by use of actual incident reports submitted by health-care staff, and one of such case studies (Itoh, 2007) will be introduced in Section 10.5.

Outside health care, however, there are occasionally opportunities for acquiring safety performance data that may be less susceptible to local and random vagaries. Still, most studies that have compared safety culture measures with safety performance have referred to self-reported accidents and incidents (Cooper, 2000; Gershon et al., 2004). For example, Diaz and Cabrera (1997), using questionnaire responses of three comparable companies, argued that rank orders of employees' attitudes toward safety coincided with those of employees' perceived (self-reported) safety level. A limited number of studies have examined the relations between safety culture measures and accident risk or ratio based on *actual* accident-related data. Among such a small number of studies addressing actual accident and incident rates, Sheehy and Chapman (1987), however, could not find an evidence of a correlation between employee attitudes and accident risk based on objective data on accidents. In contrast, Johnson (2007) identified a correlation of safety culture with lost work days per employees in a heavy manufacturing company.

Itoh et al. (2004), focusing on operators of track maintenance trains, identified a correlation of the train operators' morale and motivation with actual incident rates for each of the five branches belonging to a single contractor of a Japanese high-speed railway. A branch that employed train operators having higher morale and motivation exhibited a lower incident rate. They also found the very same correlation for company-based responses collected from all track maintenance companies working for the high-speed railway. Andersen et al. (2004b) found that two daughter companies—involved in the same line of production, operating under the same procedures and regulations—had different lost time incident rates as well as self-reported incident rates and that the "high-rate" company showed more "negative" attitudes and perceptions on 56 out of 57 items on which they showed significantly different responses.

10.3.4 Assessment Tools in Health Care

We introduce some safety culture assessment tools—which have been frequently used in health care—and briefly describe them here in order to illustrate the variety and scope of these to measure safety culture and climate, and more extensive overviews of assessment tools and recommendations for their use can be referred to review articles (e.g., Davies et al., 1999; Guldenmund, 2000, 2007; Scott et al., 2003; Gershon et al., 2004; Colla et al., 2005; Flin et al., 2006; Singla et al., 2006; Flin, 2007; Flin et al., 2009).

As most frequently used or well-known assessment tools for health-care safety culture we selected the following five tools: Operating Room Management Attitudes Questionnaire (ORMAQ), Safety Attitudes Questionnaire (SAQ), Safety Climate Survey (SCS), Hospital Survey on Patient Safety (HSOPS), and Stanford/Patient Safety Center of Inquiry (PSCI). We outline these tools in Table 10.2 in terms of the general objectives, content and construct of the tools,

TABLE 10.2

Assessment Tables of Safety Culture Frequently Used in Health Care

	ORMAQ (Helmreich et al., 1998)	SAQ (Sexton et al., 2006)	SCS (Pronovost et al., 2003)	HSOPS (Sorra et al., 2004)	Stanford/PSCI Culture Survey (Singer et al., 2003)
Objectives and description of tool	To survey health care (operating room; OR) staff attitudes to and perceptions of safety-related issues, e.g., teamwork, communication, stress, error, and error management	To survey health-care staff attitudes concerning error, stress, and teamwork. This tool has been also frequently used to compare cross-sectional staff attitudes in health care and with those of airline cockpit crews (by use of the data from the FMAQ)	"To gain information about the perceptions of safety of front line clinical staff and management commitment to safety." Can show variations across different departments and disciplines; can be repeated to assess impact on interventions. A possible measure to develop, improve, or monitor changes in the culture of safety	"To help hospitals assess the extent to which its cultural components emphasize the importance of patient safety, facilitate openness in discussing error, and create an atmosphere of continuous learning and improvement rather than a culture of covering up mistakes and punishment"	"To measure and understand fundamental attitudes towards patient safety culture and organizational culture and ways in which attitudes vary by hospital and between different types of healthcare personnel." A general tool for assessing safety culture/climate across different hospitals settings and personnel
General/specific setting	Originally for operating room and ICU staff Also for variants for general (e.g., Itoh et al., 2008), and for other specialties, e.g., for anesthesia (Flin et al., 2003)	Several versions for specific settings, e.g., ICU (Thomas et al., 2003) and OR (Sexton et al., 2006)	Originally for ICU	General	General
Construct and development	ORMAQ is an FMAQ type questionnaire for human factors attitudes of OR personnel. The initial "attitudes" questionnaire was expanded to include items about safety practices, interaction with other employee groups, perceptions of management communications and feedback, and some open-ended questions. Some modified versions of ORMAQ were developed	Refinement of the Intensive Care Unit Management Attitudes Questionnaire, which was devised from a questionnaire widely used in aviation, the FMAQ (Flight Management Attitudes Questionnaire), by discussion with health care providers and subject matter experts. From the original pool of over 100 items, 60 items were selected with reference to relevant theories and models	Adapted from the FMAQ, and shortened version of the SAQ	Developed on the basis of literature pertaining to safety, error, and accidents, and error reporting; review of existing safety culture surveys; interviews with employees and managers in different hospitals; a number of transfusion service survey items and scales with demonstrated internal consistency reliability have been modified and included in the hospital-wide survey (informed by earlier work by Nieva and Sorra, 2003)	Developed from five different survey tools, including ORMAQ. Also, adapted from five former US questionnaires (aviation and health care). The preliminary instrument was tested in pilot studies in VA (Veterans Administration) facilities. The questionnaire was revised as a 45-item version called Patient Safety Climate in Health care Organizations (PSCHO) (Gaba et al., 2003)

	ORMAQ	(60 items)	(10 items)	(42 items)	(82 items)
No. of items and dimensions and Safety culture dimensions (excluding demographics)	The original ORMAQ comprises four parts: (1) Operating room management attitudes (57 items), (2) Leadership style (2 items), (3) Work goal (15 items), and (4) Teamwork (8 items). Clear description of dimensions was not given in the original article. Flin et al. (2003) proposed eight categories from Part 1 items: (1) Leadership-structure (2) Confidence-Assertion (3) Information sharing (4) Stress and fatigue (5) Team work (6) Work values (7) Error (8) Organizational climate	60 items and 6 dimensions: (1) Teamwork climate (2) Job satisfaction (3) Perception of management (4) Safety climate (5) Working conditions (6) Stress recognition	10 items, in particular concerning perceptions of strong and proactive organizational commitment to patient safety. Dimensions are not well identified in the original article (Pronovost et al., 2003); 3 dimensions according to review article (Flin et al., 2006): (1) Supervisor and management commitment to safety (2) Knowledge of how to report adverse event (3) Understanding of systems as the cause of adverse events	42 items and 12 dimensions, grouped into four levels: unit level, hospital level, outcome variables, and other measures. The dimensions are (1) communication openness; (2) feedback and communication about error; (3) frequency of events reported; (4) handoffs and transitions; (5) management support for patient safety; (6) nonpunitive response to error; (7) organizational learning—continuous improvement; (8) overall perceptions of patient safety; (9) staffing; (10) supervisor/manager expectations and actions promoting safety; (11) teamwork across units; and (12) teamwork within units	82 items and 5 dimensions (elicited from 30 items): (1) Organization (2) Department (3) Production (4) Reporting/seeking help (5) Shame/self-awareness
Tested in original usage and applied	In one of earlier work on ORMAQ (Helmreich and Schaefer, 1994), 156 staff responses were collected from 53 surgeons, 45 anesthesiologists, 32 surgical nurses, and 22 anesthesia nurses in Switzerland	10,843 responses in 203 clinical areas, including OR, ICU, inpatient settings and ambulatory clinics in three countries: US, UK, and New Zealand	Many hospitals in both US and Europe	Piloted in 12 hospitals in 2003 in the US	15 hospitals in California, sample of 6312, response rate 47% overall (62% excluding physicians)
Test history and results	Interpersonal and communication issues give impacts to inefficiencies, errors, and frustrations in OR. Organizational interventions and formal training in human factors aspects of team performance contribute to tangible improvements in performance and substantial reduction in human error	Applying the confirmatory factor analysis to responses to 30 out of 60 items, the aforementioned 6 factor model was confirmed in a general satisfactory level. Results of cross-organizational and cross-sectional comparisons show great variations in provider attitudes within and among organizations. The developers concluded possible use of the tool in benchmarking of safety attitudes among organizations	Staff perceived supervisors having a greater commitment to safety than senior leaders. Nurses had more positive perceptions of safety than physicians Results of resurveying show that improvement in staff perceptions of the safety climate is linked to decreases in actual errors, patient length of stay, and employment turnover	Psychometric analysis provides solid evidence supporting 12 dimensions and 42 items, plus additional background questions (originally 20 dimensions) Psychometric analysis consisting of: item analysis, content analysis, exploratory and confirmatory factor analysis, reliability analysis, composite score construction, correlation analysis, and analysis of variance	After first revision 82 items, final revision 30 items plus demographics, leaving at least 1–2 items per dimension. Analyzing data in "problematic responses": Clinicians gave more problematic responses than nonclinicians. Nurses were most pessimistic among clinicians Senior managers gave fewer problematic responses than frontline employees

types (professional groups) of health-care personnel assessed, dimensions covered, and brief description of survey results.

10.3.4.1 ORMAQ

The ORMAQ was developed by Robert Helmreich and his research team as the operating room (OR) version of the famous, widely used safety culture tool in aviation, named Flight Management Attitudes Questionnaire (FMAQ; Helmreich et al., 1998). The original version of ORMAQ had four sections as well as two open-ended questions about teamwork and job satisfaction: (1) operating room management attitudes—which is a main part related to safety culture—(57 close-ended items responded on a Likert-type scale), (2) leadership style (style preferred and style most frequently seen by respondents among four styles, from autocratic to democratic type), (3) work goals (15 questions rated the degrees of importance), and (4) teamwork (perceptions of quality of teamwork and cooperation with different professional members). There have been several variants adapted from the ORMAQ to specific areas and to specific countries (language versions), e.g., hospital staff in general (Itoh et al., 2002, 2008)—which will be mentioned in Section 10.4—and anesthesiologists (Flin et al., 2003). The original ORMAQ questionnaire can be seen in Appendix of Helmreich et al. (1998).

10.3.4.2 SAQ

The SAQ (Sexton et al., 2006) is one of the most widely used assessment tools of health-care safety culture/climate. It is a refinement of an ICU (intensive care unit) version of the FMAQ (Sexton et al., 2000). There are several variants corresponding to specific work units or clinical settings such as ICU, OR, and inpatient setting as well as general focus survey tools. The ICU version of SAQ comprises 60 items and covers 19 dimensions, which can be grouped into six safety culture factors: teamwork climate, job satisfaction, perceptions of management, safety climate, working conditions, and stress recognition. The questionnaire can be downloaded with registration of the user from the developer's web page (access May 22, 2010): http://www.uth.tmc.edu/schools/med/imed/patient_safety/questionnaires/registration.html.

10.3.4.3 SCS

The SCS (Pronovost et al., 2003) is based on the SAQ, and moreover is like its shortened version. Therefore, the SCS consists of 10 items—no description was given about the total number of items and of dimensions in Pronovost et al. (2003). The instrument has been tested

in many countries, the authors note. It is not comprehensive, however, leaving out a number of potentially revealing dimensions, and primarily focusing on management commitment to safety. It would be a sensible choice if the user wants a tool that imposes few demands on staff's time and to track changes over time. The questionnaire could be downloaded, but recently has been removed from the web site of the Institute for Healthcare Improvement, and therefore it is not available at the moment.

10.3.4.4 HSOPS

The AHRQ developed a questionnaire for the survey named HSOPS to assess hospital staff opinions about patient safety issues in 2003 (Sorra et al., 2004). Since then, the AHQR has been continuously collecting data from a great number of U.S. hospitals applying this questionnaire, and published an annual report about the current states in safety culture, analyzing, and compiling the data from their huge database, the HSOPS Culture Comparative Database (Sorra et al., 2009). This questionnaire has been used in many studies not only in the United States but in other countries, e.g., the Netherlands (Smits et al., 2008) and Turkey (Bodur and Filtz, 2010). This tool may be a good choice for a comprehensive measure of climate and culture, and to know the level of safety culture in a specific organization or country by comparing with other studies. It has strong content validity, is well structured, and since it also considers outcome measures it facilitates an external (criterion) validation. One of the objectives of this tool is to provide feedback to staff to strengthen awareness of patient safety and the importance of reinforcing a positive culture. We find this to be the tool of choice if the user wants to establish a basis for planning an intervention program.

10.3.4.5 Stanford/PSCI

The questionnaire used for Stanford/PSCI culture survey (Singer et al., 2003) is a general tool to assess safety culture across different hospital settings and personnel. This questionnaire was directed at a broad range of specialties and work settings, and it includes 82 close-ended items covering 16 topics related to safety culture. Among these items, 30 items comprise a core part of the questionnaire from which five dimensions were elicited by the factor analysis. Collected data were analyzed as "problematic" responses to each question item. The first survey was conducted to measure and compare safety culture between 15 hospitals in California. The results show that problematic responses varied widely between the hospitals and clinicians—i.e., physicians, nurses, pharmacists, and technicians—especially nurses were likely to give more problematic responses than nonclinicians.

10.3.5 Considerations When Conducting Surveys

There are several important things that we should consider when conducting a safety culture survey. The user should first make clear about the focus and objectives of the safety culture assessment and the resources required. This includes the time individual respondents need to fill out in the questionnaire, resources for data collection, entry, analysis, interpretation, reporting and—importantly—feedback of results to management, safety managers, and staff. The work needed to make a useful and successful survey should not be underestimated, but such work experience quickly accumulates and a subsequent survey will be much quicker to run.

It will probably be the most relevant, in particular, for health-care staff to run a self-administered survey (and not a phone interview or personal interview survey). Nowadays, web-based tools may be a possibility, though availability of web-linked PCs and staff familiarity with IT should be considered. Carayon and Hoonakker (2004) reviewed studies that used mail surveys and different forms of electronic surveys, summarizing advantages and disadvantages and what to consider when choosing a survey method. Their key findings show that mail surveys tend to lead to a higher response rate than electronic survey (but web-based ones are getting better results), whereas the latter tend to yield more completed questionnaires, greater likelihood of answers to open-ended questions and in general a higher response quality. In addition, electronic surveys are of course much easier and cheaper to administer.

Involvement of key stakeholders is of critical importance to obtaining a high response rate from all relevant groups, which, in turn, is necessary to reduce the risk of sampling bias. A low response rate (50% or lower) will necessarily bring speculation that respondents may not be representative of the target group. In our experience, most people want to know what the survey will be used for and they want to receive firm assurances about anonymity. This also means that respondents may be reluctant to supply potentially revealing demographic information (e.g., age, position, department, and length of employment in current department). It might, though, be helpful if the survey is administered by an independent, reputable research or survey organization that issues guarantee that data will only be reported to the host hospital and the departments at an aggregate level. Even when staff is encouraged to fill out a questionnaire during working time, they may often not feel they can take the time to complete the survey. Finally, management and department leaders may feel that some items get "too close" and that the survey invites respondents to criticize their superiors. All of these considerations make it necessary for a local survey leader to obtain explicit support from management, local leaders, and employee representatives. Low response rates are not uncommon. For example, Singer et al. (2003) report a response rate of just 47%. Andersen et al. (2002) similarly obtained a response rate of just 46% for doctors and 53% for nurses (total of 51%). In contrast, in Itoh et al.'s (2008) survey, for which the organizational agreements were obtained in advance, the mean response rate of nurses over 82 hospitals was 88% although that of doctors was almost as low as the aforementioned surveys, i.e., 51%.

10.4 Staff Perceptions of Safety Culture: How Do Employees Perceive Safety-Related Issues?

10.4.1 Safety Culture Perceived by Health-Care Professionals

In this section, we illustrate characteristics of safety culture in health care, based on the results of our questionnaire survey of Japanese hospitals (Itoh et al., 2008). We describe health-care staff responses and summarize results at the level of organizations (hospitals), professional groups, and specialties and wards. Using multinational data from New Zealand, Iceland, Nigeria, and Japan (Itoh et al., 2005), we also briefly mention cross-national differences in health-care safety culture.

The section of safety culture related issues in our questionnaire was adapted from Helmreich's "ORMAQ" (Helmreich at al., 1998; see also in Section 10.3.4), which contains 57 five-point Likert-type items about perceptions of and attitudes to job, teamwork, communication, hospital management, as well as other safety-related issues. From responses to all items in this section, we elicited twelve dimensions by the principal component analysis with 44% of cumulative variance accounted for. The safety culture dimensions elicited were (1) recognition of communication, (2) morale and motivation, (3) power distance, (4) recognition of stress effects on own performance, (5) trust in management, (6) safety awareness, (7) awareness of own competence, (8) collectivism–individualism, (9) cooperativeness, (10) recognition of stress management for team members, (11) seniority dependency, and (12) recognition of human error.

A summary of approximately 22,000 health-care staff responses (including about 1000 doctors, 18,000 nurses, 540 pharmacists, and 1900 technicians) collected from

84 Japanese hospitals in 2006 is shown in Table 10.3 (the mean response rate of this sample was 84%) based on four professional groups. In the table, the percentage [dis]agreement is referred to as the percentage of respondents who had positive [negative] responses to a meaning of particular dimension title, excluding neutral responses. As can be seen in this table, there were significant differences between doctors, nurses, pharmacists, and technicians for all the safety culture dimensions. Yet, some common trends of safety culture were observed across these professional groups. Most Japanese health-care employees indicate positive perceptions of communication within their organizations as well as of stress management for team members. For example, about 90% of respondents agreed that team members should be monitored for signs of stress and fatigue during task.

All respondents in any professional group still showed modest recognition of stress effects on their own performance in contrast with the aforementioned strong awareness of stress management for team members.

In particular, less than 20% of doctors and technicians exhibited realistic recognition of stress effects during work. In contrast, the sample results also indicate that a large part of health-care staff has realistic recognition of human error (high agreement with this indicates realistic recognition). That is, all four professional groups well recognize that "human error is inevitable," and they do not agree with the item "errors are a sign of incompetence."

All four groups perceive rather small power distance (higher agreement with this index indicates a large power distance)—the term derived from Hofstede (1991) refers to the psychological distance between superiors and subordinate members and a small distance means that leaders and their subordinates have open communication initiated not only by leaders but also by juniors. Previous studies (e.g., Spector et al., 2001) have shown that Japanese are around the "upper middle" when compared with other nations in terms of power distance—so, while not at the extreme high end (e.g., Arab countries and Malaysia) the Japanese are not at

TABLE 10.3

Percentage Agreements and Disagreements for Safety Culture Dimensions

SC Dimensions	Dr. (%)	Ns. (%)	Phar. (%)	Tech. (%)	p
I. Recognition of communication	95	95	96	98	***
	0	0	0	0	
II. Morale and motivation	65	47	54	60	***
	9	19	15	11	
III. Power distance	1	1	1	2	***
	84	82	87	78	
IV. Recognition of stress effects	19	24	28	17	***
	19	10	9	14	
V. Trust in management	56	58	44	43	***
	14	11	23	21	
VI. Safety awareness	57	73	59	58	***
	3	1	3	3	
VII. Awareness of own competence	50	32	35	34	***
	6	12	12	10	
VIII. Collectivism–individualism	52	58	53	53	***
	4	2	3	3	
IX. Cooperativeness	15	23	18	25	***
	34	19	25	20	
X. Recognition of stress management	84	81	76	80	***
	1	1	2	2	
XI. Seniority dependency	85	55	69	69	***
	1	7	2	4	
XII. Recognition of human error	90	81	88	83	***
	1	4	2	3	

Source: Itoh, K. and Andersen, H.B., A national survey on health-care safety culture in Japan: Analysis of 20,000 staff responses from 84 hospitals, *Proceedings of the International Conference on Healthcare Systems Ergonomics and Patient Safety, HEPS 2008*, Strasbourg, France, June 2008 (CD-ROM).
Upper row: % agreement, Lower row: % disagreement.
p between four professional groups ***: $p < 0.001$.

the extreme low end either (e.g., Denmark and Ireland). We will discuss about a relative level of safety culture in Japanese health care later based on results of cross-national comparisons.

Among the differences between the professional groups, we would notice that doctors indicated a higher level of morale and motivation, awareness of own competence, and attitudes of seniority dependency than the other groups. Also, nurses' safety awareness was far stronger than that of the other groups, and they were more liable to express collective or team-oriented attitudes.

10.4.2 Differences between Clinical Specialties/Wards

Percentage agreements and disagreements are shown based both on specialties of the doctors—i.e., internal medicine, surgery, and anesthesia—in Table 10.4 and on the types of wards/workplaces of nurses in Table 10.5. Nurses are classified into eight work unit groups: internal medicine, surgery, ICU, outpatient,

pediatrics, psychiatrics, mixed ward, and OR. The statistical analysis involved rank-based tests (Kruskal–Wallis) to each question item and to each dimension. There are nonsignificant differences between doctors' specialties for most safety culture indices except for recognition of stress effects, trust in management, and collectivism–individualism. There is an overall trend of differences across doctors' specialties in that anesthesiologists have slightly more "negative" attitudes and perceptions with regard to safety culture; that is, their recognition of stress effects on their own performance is less realistic; they were less liable to take team-oriented attitudes; and their trusts in management were weaker than physicians and surgeons. The results seem to indicate that doctors belonging to different specialties may develop different types of safety culture and even local subcultures.

In contrast to the results from doctors, there are significant differences for all twelve safety culture dimensions across the work unit groups of nurses (this is to be expected when the sample is so large). Among the

TABLE 10.4

Doctor's Specialty-Based Comparisons for Safety Culture Dimensions

SC Dimensions	Physician (%)	Surgeon (%)	Anesthesiologist (%)	Total (%)	p
I. Recognition of communication	94	95	91	95	
	0	0	0	0	
II. Morale and motivation	64	66	57	65	
	9	8	16	9	
III. Power distance	1	1	2	1	
	86	84	82	84	
IV. Recognition of stress effects	22	16	7	19	*
	17	24	22	19	
V. Trust in management	55	61	40	56	*
	12	14	23	14	
VI. Safety awareness	53	61	60	57	
	3	2	4	3	
VII. Awareness of own competence	47	51	60	50	
	7	5	2	6	
VIII. Collectivism–individualism	54	51	36	52	*
	5	4	2	4	
IX. Cooperativeness	14	15	16	15	
	34	33	32	34	
X. Recognition of stress management	84	84	91	84	
	1	1	0	1	
XI. Seniority dependency	84	86	82	85	
	1	0	4	1	
XII. Recognition of human error	90	91	88	90	
	1	2	0	1	

Source: Itoh, K. and Andersen, H.B., A national survey on health-care safety culture in Japan: Analysis of 20,000 staff responses from 84 hospitals, *Proceedings of the International Conference on Healthcare Systems Ergonomics and Patient Safety, HEPS 2008,* Strasbourg, France, June 2008 (CD-ROM).

Upper row: % agreement, Lower row: % disagreement.

p between three specialty groups *: *p* < 0.05.

TABLE 10.5

Nurse's Work Unit–Based Comparisons for Safety Culture Dimensions

SC Dimensions	Internal Medicine (%)	Surgery (%)	ICU (%)	OR (%)	Out-patient (%)	Psychiatrics (%)	Pediatrics (%)	Mixed Ward (%)	Total (%)	p
I. Recognition of communication	95	95	96	96	97	95	95	95	95	***
	0	0	0	1	0	0	0	0	0	
II. Morale and motivation	45	44	41	44	57	52	45	45	47	***
	21	19	24	22	13	16	17	21	19	
III. Power distance	1	1	1	1	1	1	0	1	1	***
	83	82	83	82	79	81	81	86	82	
IV. Recognition of stress effects on own performance	26	27	27	25	14	20	25	25	24	***
	9	8	10	11	15	13	11	10	10	
V. Trust in management	61	60	54	51	52	58	66	59	58	***
	10	9	13	17	14	13	7	11	11	
VI. Safety awareness	72	73	71	70	75	73	77	75	73	***
	1	1	1	2	1	1	1	1	1	
VII. Awareness of own competence	29	31	32	30	35	28	29	31	32	***
	13	12	10	11	10	17	11	11	12	
VIII. Collectivism–individualism	58	60	59	54	59	51	60	58	58	***
	2	2	2	4	2	4	2	2	2	
IX. Cooperativeness	23	23	22	24	22	27	23	23	23	***
	20	19	20	21	21	13	16	21	19	
X. Recognition of stress management for team members	80	80	82	83	84	77	79	82	81	***
	1	2	1	1	1	2	2	1	1	
XI. Seniority dependency	53	56	59	57	55	44	58	57	55	***
	7	6	6	6	7	11	6	5	7	
XII. Recognition of human error	81	82	86	83	75	76	80	83	81	***
	4	3	3	4	6	7	3	3	4	

Source: Itoh, K. and Andersen, H.B., A national survey on health-care safety culture in Japan: Analysis of 20,000 staff responses from 84 hospitals, *Proceedings of the International Conference on Healthcare Systems Ergonomics and Patient Safety, HEPS 2008*, Strasbourg, France, June 2008 (CD-ROM).

Upper row: % agreement, Lower row: % disagreement.

p between eight ward groups ***: $p < 0.001$.

eight groups, two extremes stood out as remarkable in terms of responses to these dimensions, as can be seen in Table 10.5, mapping of all eight work unit groups to a coordinate plain having two combined dimensions— (1) power distance, and recognition of stress effects and human error, (2) morale and motivation, and safety awareness—in Figure 10.1. One type is nurses working for outpatients. Compared with most of the other groups, they indicated a higher level of morale and motivation, and stronger safety awareness, and they expressed stronger awareness of communication, stress management for team members, and their own competence; but in contrast, they perceived larger power distance and exhibited less realistic acknowledgement of human error, and, finally, less realistic recognition of their own performance limitations under stress and a relatively lower level of trust in management. Nurses working in the psychiatric ward showed themselves similar to the outpatient group: they exhibited higher

level of morale and motivation, stronger awareness of safety and communication, relatively larger power distance perceived, and less realistic recognition of human error and stress effects. However, their awareness of own competence and stress management for team members was not strong unlike the outpatient nurses.

The other extreme type is ICU nurses. In contrast to the outpatient nurses, their morale and motivation, safety awareness, and recognition of importance of communication were lower than most of the other ward groups, but they perceived smaller power distance, and expressed more realistic recognition of human error and stress effects on own performance. The OR nurse group shared largely the same attitudes and perceptions. However, their awareness of stress management for team member was relatively high unlike the ICU group. In addition, among the eight work unit groups, the OR nurses gave the lowest scores for trust in management as well as safety awareness.

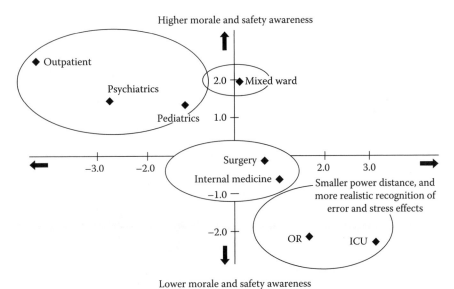

FIGURE 10.1
Mapping of each work unit group of nurses in perceived safety culture. (From Itoh, K. and Andersen, H.B., A national survey on health-care safety culture in Japan: Analysis of 20,000 staff responses from 84 hospitals, *Proceedings of the International Conference on Healthcare Systems Ergonomics and Patient Safety, HEPS 2008*, Strasbourg, France, June 2008 (CD-ROM).)

Besides these two extreme groups, nurses working in internal medicine and surgery wards are quite similar to one another in their perceptions of safety culture, and they are characterized in the center of each dimension. These results might in part reflect differences in tasks and work conditions, e.g., more technical work (operating room) versus more clerical work (outpatients) and work with babies and infants (pediatrics) versus work with many elderly patients (internal medicine, etc.). Such variations in task characteristics may in turn give rise to work unit–based subcultures in a hospital.

10.4.3 Differences in Organizational Safety Culture

In Table 10.6, we summarized hospital-based responses of nurses—a trend of organizational differences is shared with doctors though we do not mention here—in terms of mean, maximum, minimum, and its range of percentage agreement and disagreement for each dimension over 82 hospitals surveyed. Differences across the 82 hospitals are much larger than across work units—the latter reviewed in the last sections—for most indices. Such great organizational variation in safety culture has also been observed in US hospitals (Singer et al., 2009). In our Japanese sample, for most of the safety culture indices, the ranges of percentage agreement and disagreement across the hospitals are more than three times greater than those across nurses' work unit groups. Such variations among health-care organizations are specifically large for dimensions

related to respondent "attitudes", such as morale and motivation, and trust in management. Organizational differences are relatively small for dimensions related to "practices" and "procedures" such as recognition of communication and of stress management for team members as well as power distance.

From these survey results, we would conclude that there are large variations in safety culture across Japanese hospitals—probably any country in the world—and somewhat smaller variations across specialties and work units. One may speculate that similar findings might be found in other settings. In addition, one may speculate about the reasons why hospital differences are greater than work unit and specialty differences. At the very least, the latter result would seem to indicate that the organizational culture of the individual hospitals plays a greater role than the variation in work and task conditions across specialties and work units.

10.4.4 Multinational Comparisons in Safety Culture

In this section, we review some cross-national differences and similarities in safety culture, comparing the Japanese data with three country samples collected in 2002–2003 using a similar questionnaire which included the same ORMAQ type "safety culture" section (Itoh et al., 2005). The three country samples we compared are Iceland (83 responses; including 29 doctors and 54 nurses), New Zealand (197; 57, 142), and Nigeria (223; 164, 59). Table 10.7 shows percentage agreements and disagreements of the doctor and the

TABLE 10.6

Variations of Safety Culture Dimensions across 82 Hospitals (from Nurse Sample)

SC Dimensions		Mean (%)	Max. (%)	Min. (%)	Range (%)
I. Recognition of communication	% Agree	96	100	89	11
	% Disagree	0	2	0	2
II. Morale and motivation		48	71	25	46
		18	32	4	28
III. Power distance		1	4	0	4
		83	95	68	28
IV. Recognition of stress effects on own performance		23	35	8	27
		11	31	4	27
V. Trust in management		57	86	30	56
		12	42	2	40
VI. Safety awareness		74	92	58	34
		1	3	0	3
VII. Awareness of own competence		33	60	17	43
		12	23	3	20
VIII. Collectivism–individualism		58	69	43	26
		2	10	0	10
IX. Cooperativeness		23	36	11	25
		19	30	11	19
X. Recognition of stress management for team members		81	89	69	20
		2	8	0	8
XI. Seniority dependency		54	74	34	40
		7	18	1	17
XII. Recognition of human error		81	92	63	30
		4	9	0	9

Source: Itoh, K. and Andersen, H.B., A national survey on health-care safety culture in Japan: Analysis of 20,000 staff responses from 84 hospitals, *Proceedings of the International Conference on Healthcare Systems Ergonomics and Patient Safety, HEPS 2008*, Strasbourg, France, June 2008 (CD-ROM).
Upper row: % agreement, lower row: % disagreement.

nurse group on the safety culture dimensions for each of the four countries.

In Table 10.7, we can see significant cultural differences between the countries for most safety culture dimensions, and some of them are rather big—these differences may include not only "actual" difference but also some "psychological" response bias, for example, Japanese people are often stated more likely to avoid giving extreme responses, e.g., 1 or 5 on a five-point scale. Comparative results appearing in the table may imply remarkable differences in some dimensions: Japanese staff's morale and motivation, safety awareness and competence awareness, both doctors and nurses, are very low, compared with the other countries' professionals. We mentioned in Section 10.4.1 that a power distance in Japanese health care was rather small. But compared with the three countries, the power distance in Japan is not relatively small, in particular, much larger than Iceland and New Zealand. This may exhibit an actual state in national culture (Hofstede, 1991, 2001; Spector et al., 2001)

We can see a common trend in some dimensions among the four countries although there are significant differences because of the big Japanese sample. On the one hand, the importance of communication is well recognized in health care, and human error is realistically recognized except in Nigeria. But on the other hand, recognition of stress effects on one's own performance have not been realistic enough.

10.4.5 Climate Changes in Health Care

In this section, we use the term "climate" instead of "culture" since we now focus on changes over a short term, i.e., 5 years. To ascertain possible changes in the 5-year interval we performed comparative analyses, using a dataset for which the nurse responses were selected from six hospitals that participated in two surveys conducted in 2002 and 2006. The sample included 1625 nurse responses (94% response rate) in 2006 and 1703 (88%) in the 2002 survey. Comparative results are summarized in Table 10.8 in terms of mean scores of

TABLE 10.7

Cross-National Comparisons of Safety Culture in Health Care

SC Dimensions	Doctors					Nurses				
	JP (%)	IS (%)	NZ (%)	NI (%)	*p*	JP (%)	IS (%)	NZ (%)	NI (%)	*p*
I. Recognition of communication	95	96	94	94	**	95	100	99	93	***
	0	0	0	2		0	0	0	4	
II. Morale and motivation	65	86	93	83	***	47	87	84	93	***
	9	0	2	3		19	2	2	2	
III. Power distance	1	0	0	15	**	1	0	0	0	***
	84	93	93	76		82	92	97	93	
IV. Recognition of stress effects	19	3	29	33	***	24	8	15	22	***
	19	17	11	7		10	32	26	19	
V. Trust in management	56	72	57	49		58	65	67	42	***
	14	10	14	23		11	12	11	25	
VI. Safety awareness	57	72	71	87	***	73	94	90	93	***
	3	3	0	1		1	0	0	2	
VII. Awareness of own competence	50	100	95	96	***	32	98	93	91	***
	6	0	0	0		12	0	1	2	
VIII. Collectivism–individualism	52	59	84	61	***	58	71	80	50	***
	4	0	0	1		2	2	1	4	
IX. Cooperativeness	15	21	29	27	*	23	24	45	41	***
	34	31	25	28		19	24	14	19	
X. Recognition of stress management	84	71	77	94	***	81	71	73	95	***
	1	4	4	2		1	6	4	2	
XI. Seniority dependency	85	90	89	91	***	55	60	67	74	***
	1	0	0	0		7	2	4	12	
XII. Recognition of human error	90	90	93	77	**	81	77	81	62	*
	1	7	4	8		4	9	6	22	

JP:Japan, IS: Iceland, NZ: New Zealand, NI: Nigeria.
Upper row: % agreement; Lower row: % disagreement.
*: *p* < 0.05, **: *p* << 0.01, ***: *p* < 0.001 between the four countries.

each safety "climate" dimension in 2006 and 2002 when using aggregated data from the entire dataset as well as each of the six hospitals.

The results from the aggregated data of the hospitals that were surveyed twice indicate significant differences in most safety climate dimensions in the 5-year interval. It must be noticed that each hospital exhibited the same trend of climate change for each dimension—becoming more positive or more negative—as that of the aggregated data of the twice-surveyed hospital sample. Therefore, we believe that the safety climate change in the recent 5-year interval elicited here is a general trend happening in Japanese health care.

Overall changes between 2002 and 2006 in Japanese nurses' safety climate have taken place as follows: Staff motivation and morale have decreased, and recognition of the importance of communication and safety awareness has become weaker. Moreover, nurses are more likely to behave individualistically—though many of them are still teamwork oriented; and the perceived

power distance has become larger. However, all dimensions of safety climate have not gone negative for the last 5 years. Nurses' recognition of human error and the effects of stress on their own performance have become more realistic.

These changes may have largely been due to intensive organization-wide activities pertaining to patient safety, particularly safety training and establishment of safety-related rules and procedures including error reporting. We hypothesized as reasons for these changes in health-care safety climate that safety training and rules and procedures, in a way which most Japanese health-care organizations have introduced or reinforced, on the one hand, may make their employees recognize more realistically human factors' aspects such as errors, stress, and workload. On the other hand, these organizational changes may lead to a blame culture that in turn contributes to lower motivation and morale and larger power distance within a hospital or a department.

TABLE 10.8

Changes in Japanese Nurses' Safety Climate in 5-Year Interval (2002/2006)

SC Dimensions	Hospitals						All Six Hospitals
	A	**B**	**C**	**D**	**E**	**F**	
I. Recognition of communication	4.41*	4.34**	4.17***	4.13***	4.17***	4.23*	4.23***
	4.51	4.47	4.42	4.35	4.36	4.30	4.39
II. Morale and motivation	3.40***	3.08	3.15***	3.08**	3.29**	3.26**	3.19***
	3.85	3.21	3.47	3.30	3.45	3.44	3.42
III. Power distance	2.02	2.08	2.05*	2.16***	2.03	2.23	2.22***
	2.05	2.04	1.94	1.95	2.01	2.22	2.13
IV. Recognition of stress effects on own performance	3.08*	3.20**	3.05**	3.30	3.14	3.25*	3.18**
	2.95	3.07	3.15	3.25	3.08	3.15	3.12
V. Trust in management	3.62	3.32	3.39***	3.55*	3.60	3.45	3.47
	3.62	3.30	3.70	3.36	3.58	3.34	3.48
VI. Safety awareness	3.90	3.80	3.79*	3.68	3.81	3.73	3.78***
	3.90	3.89	3.89	3.72	3.86	3.75	3.84
VII. Awareness of own competence	3.28	3.17	3.22*	3.13	3.19	3.18	3.19***
	3.43	3.24	3.31	3.15	3.24	3.21	3.25
VIII. Collectivism–individualism	3.64	3.55*	3.38***	3.40	3.59	3.56	3.55***
	3.73	3.64	3.55	3.67	3.61	3.62	3.62
IX. Cooperativeness	3.11*	2.90	2.99**	3.03*	3.10	2.99	3.00**
	2.96	2.93	2.87	2.88	3.00	2.99	2.94
X. Recognition of stress management	3.86	3.93	3.71**	3.78	3.79	3.81	3.81
	3.91	3.86	3.84	3.75	3.80	3.81	3.82
XI. Seniority dependency	3.59	3.54	3.73	3.74	3.57	3.54	3.62**
	3.50	3.43	3.72	3.62	3.56	3.46	3.55
XII. Recognition of human error	3.97	3.95*	3.95**	4.00	4.04**	3.91	3.96***
	3.85	3.82	3.79	3.98	3.86	3.84	3.85

Source: Itoh, K. and Andersen, H.B., A national survey on health-care safety culture in Japan: Analysis of 20,000 staff responses from 84 hospitals, *Proceedings of the International Conference on Healthcare Systems Ergonomics and Patient Safety, HEPS 2008*, Strasbourg, France, June 2008 (CD-ROM).

Upper row: Mean score of safety climate dimension in 2006; Lower row: in 2002.

*: $p < 0.05$, **: $p << 0.01$, ***: $p < 0.001$ between 2002 and 2006.

10.5 Correlations with Safety Outcome: How Does Safety Culture Contribute to Patient Safety?

10.5.1 Meaning of Incident Reporting Rates

As mentioned earlier, there are so far scant "hard" data that support the presumption of a causal link from safety culture to safety outcome. The lack of solid empirical evidence for the crucial link is not specific to health care, and we reviewed briefly the status in terms of evidence in other domains in Section 10.3.3. In this section, we mention an empirical link of safety culture to safety outcome based on results of a Japanese study (Itoh, 2007). Before we describe the results of correlation analysis between safety outcome and safety culture indices, we briefly discuss interpretation of the rate of incident reports submitted to the hospital's reporting system.

For some time, we have raised the question of whether one may interpret the rate of incident reporting of a given unit as an indicator of *safety* or *risk*. This question is particularly hard to answer when comparing rates obtained from different organizations or different work units that may possibly have different procedures, criteria, and practices for error reporting as well as different safety cultures. Making an error is likely to be different from documenting an error, which is based on capturing the error, discussing it, etc., which, in turn, leads to learning from such an experience (Edmondson, 1996, 2004); therefore, it may not be possible to identify this difference only by the data of incident reporting. We believe the interpretation of the reporting rate becomes possible when we discuss the comparative results of incident reporting, for instance, between work units in a specific organization, with consideration of the units' safety cultures. In principle, there are two opposite possibilities. On the one hand, if we assume that all employees apply identical criteria for submitting a report to their hospital's reporting system (and assume that the tasks and patient distribution are comparable) the reporting rate will of course predict the likelihood of errors or incidents. Thus, it is reasonable to consider that more accidents must take place when errors are made more frequently (if tasks and patient profiles are comparable). Therefore, on these hypothetical assumptions, the incident reporting rate would be an index of "risk."

On the other hand, if we have no reason to uphold these assumptions about similar reporting criteria, the rate of incident reporting may reflect the staff's safety awareness or sensitivity to errors or incidents (again, assuming that we may keep the impact of possible variations in tasks and patient profiles constant). Thus, it is natural to hold that staff who has a higher sensitivity to errors—and the importance of learning from errors—will be more likely to submit a report of an incident that might be undetected or unreported by staff who has a lower sensitivity or perhaps a lower appreciation of the importance of learning from experience. In this case, the rate of incident reporting can be interpreted as a measure of "safety." That is, units that have relatively high rates of reporting will be indicative of greater attention to errors and learning—so, we should expect these units to actually be safer (Edmondson, 1996).

For the case study illustrated here (Itoh, 2007), quantitative indices of actual incident reporting were produced using statistical summary of 550 incident reports submitted by nurses during a whole year in a hospital; the nurses also participated in the safety culture survey mentioned in the last section. The hospital is public, belonging to a local municipality about 100 km far from

Tokyo. It covers almost all clinic areas and employs 57 full-time doctors and 259 nurses with 355 beds at the time of the survey (in 2002)—the hospital employs more doctors and nurses now.

The indices of incident reporting are (1) reporting rate of all incidents to the system, i.e., the rate of incidents, including all level cases, submitted to the hospital's reporting system per nurse in a year, and (2) the rate of Level 3 or higher (Level 3+) cases, i.e., the rate of reported cases whose severity levels are higher than or equal to 3 per nurse in a given year. Regarding the latter index, many Japanese hospitals classify accidents and incidents into 6 levels of severity, ranging from 0 (near miss) to 5 (death). Usually, cases at Levels 0 and 1 are regarded as no-effect incidents. A majority of reports submitted by nurses in this hospital (95%)—also any hospital in Japan—were at Levels 0 and 1. Those resulting in a temporary, small effect such as slight fever, headache, or bad mood are at Level 2. Cases are assigned to Level 3 (or higher) only if an additional treatment is required during a prolonged period of hospitalization due to causes not related to the underlying disease of the patient. Therefore, events classified at Level 3 or higher may be considered as indicating risk of accident as compared to lower levels.

10.5.2 Comparing Safety Culture Indices with Reporting Rates

We shall here discuss the effects of safety culture on reporting behavior and, ultimately, on patient safety. In Figure 10.2 are depicted results of correlating each of the two reporting rates with one of the safety culture indices, *power distance*, where all five work units are plotted in the geometric plane. As can be seen in the figure (a),

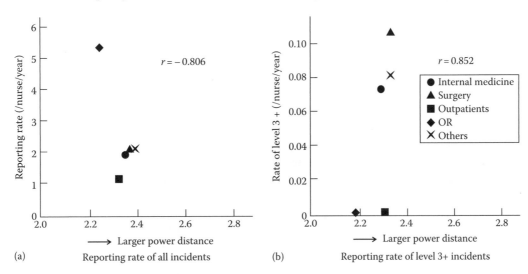

FIGURE 10.2
Correlations of power distance with incident reporting rates. (a) Reporting rate of all incidents; (b) Reporting rate of Level 3+ incidents. (From Itoh, K., *JMS*, 3(1), 3, 2007 (in Japanese). With permission.)

the actual reporting rate of all incidents is negatively correlated with power distance ($r = -0.81$; though a coefficient is not meaningful to present due to the small number of work units).

Thus, it may be suggested that a small power distance contributes to a higher rate of incident reporting in this hospital. As mentioned previously, this safety culture dimension means a psychological distance between leaders and subordinates, and small power distance leads to open communication between team members and superiors. Therefore, it may be natural to interpret that a smaller power distance leads to a better organizational culture having potentially smaller accident risk and, in turn, to greater patient safety. If we could assume this relationship, the actual reporting rate of all incidents could work as a *safety* measure in *this* hospital; that is, the greater number of incident cases nurses are likely to report, the safer that particular ward or work unit becomes. This assumption will be further checked based on a correlation with another dimension.

As can be seen in Figure 10.2b, our sample showed a positive correlation between power distance and the reporting rate of Level 3+ cases ($r = 0.85$). This implies that the larger power distance exists in a work unit, the more frequently incidents at higher levels are made. As suggested in the last section, the reporting rate of higher severity cases is closely related to the occurrence of accidents, and therefore, this index seems to measure *risk* rather than safety. This conjecture is also supported by the hospital's continuous activities to increase patient safety. A risk manager of the hospital has, during an interview, stressed that hospital-wide initiatives for patient safety were initiated 5 years before our survey and that special attention has been devoted to designing the incident reporting system with a view to supporting efficient report submission and follow-up in terms of counter measures. The risk manager also stated that there is no case that *nurses*—though she had no idea about doctors—hold back submitting incident reports for events at Level 2 or higher—our index is more conservative, i.e., Level 3 or higher. Therefore, there is a reason to believe that the actual rate of Level 3+ incidents serves to measure accident *risk* in *this* hospital (again, when we consider comparable departments in terms of tasks and patient profiles).

Another safety culture dimension, *recognition of human error*, is shown in its relation to the actual rates of incident reporting in Figure 10.3. Looking at the figure (a) first, we can see that this safety culture index seems to have a positive correlation ($r = 0.83$) with the reporting rate of all incidents—which was assumed to serve as a safety measure previously. From this positive correlation, we may say that the more realistic recognition toward human errors becomes, the more frequently an incident report is brought up. In addition, applying the same assumption of this reporting index, it makes sense because its interpretation also suggests positive effects of realistic recognition of human error on safety outcome as power distance. Therefore, it seems that the reporting rate of all incidents worked as a *safety* measure in *this* hospital at the time of the survey although whether this reporting index serves as a safety or risk measure may depend on the hospital and on the time—moreover on the level of maturity of the hospital's reporting culture at a given time—according to our experiences of safety culture projects.

As shown in Figure 10.3b, realistic recognition of human error seems to be negatively correlated—but a weak correlation—with the reporting rate of Level 3+ incidents ($r = -0.42$). In this figure, the outpatient nurses

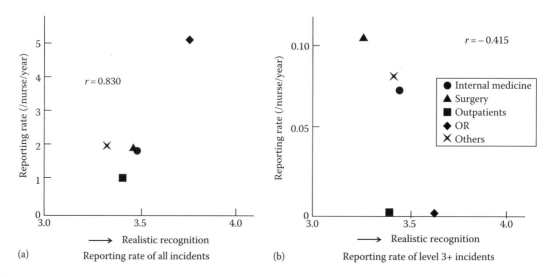

FIGURE 10.3

Correlations of recognition of human error with incident reporting rates. (a) Reporting rate of all incidents; (b) Reporting rate of Level 3+ incidents. (From Itoh, K., *JMS*, 3(1), 3, 2007 (in Japanese). With permission.)

seem to be an outlier from the other work unit groups, and excluding this group the correlation becomes much stronger. It might be argued that the outlier should in fact be removed on the ground that outpatient nurses (the outlier group) may be expected to have a fewer likelihood of involving themselves in adverse events because of more clerical tasks and less interactions with patients than ward nurses. As we noted in the last section, outpatient nurses' subculture was characterized as one extreme among the other groups, e.g., higher morale and motivation, and safety awareness. These characteristics of safety culture may also impact safety performance measures.

Applying the interpretation about the reporting rate of Level 3+ incidents to the aforementioned results, health-care staff's realistic recognition of human error may contribute to a *lower* risk of adverse events. At the same time, the results may suggest that the actual reporting rate is affected not only by the safety culture index in the workplace but also by task characteristics and related procedures, e.g., outpatients versus others, although we need more cases and results to derive a sound conclusion.

10.5.3 Continuous Tracking of Incident Rates and Safety Performance Level

From the same hospital in which correlation analysis between safety culture and safety outcome was applied (see the previous section), we obtained more data of incident reports between 2000 and 2004. In Figure 10.4, transitions of the two reporting indices, i.e., reporting rate of all incidents (a) and that of Level 3+ incidents (b), are depicted for each of the five work unit groups. First, looking at the figure presenting the Level 3+ incidents (b), we can see that the rate has been continuously

decreasing in all work units. According to the interviews with the risk manager, as mentioned before, she was sure that every nurse has submitted any event at Level 2 or higher to the hospital's reporting system even before the period of our survey (at latest in 2000). Therefore, it may be possible to assume that the reporting rate of Level 3+ incidents measures accident risk throughout the period in the figure, i.e., between 2000 and 2004, in this hospital. She also stated that the hospital implemented patient safety initiatives and its staff intensively tackled patient safety activities during that period. Accepting these statements, it is suggested that the hospital's safety level had been increasing during this period. In addition, more importantly, the improvement of safety level may appear as a visible outcome of continuous safety activities in this hospital.

Regarding transitions of the reporting rate of all incidents, which were shown in Figure 10.4a, the index increased in all the work units except in the OR up to 2002. Subsequently, the reporting rate went down with no exceptional unit group. According to the discussion about the transition of Level 3+ incidents, again it seems that safety level in every work unit has improved year by year. Therefore, it may be natural to infer that something about incident reporting was changed (probably a reporting standard was fixed) in or near that year—it may also be interpreted that the OR unit came to this point a year earlier than the other units. We can guess that the reporting system had been stable and efficient for some time, in particular for nurses; therefore we may say that nursing staff has almost an identical submission criterion—which events they should report or should not—and they no longer held back reporting of such events. If this "guess" can be true, more reporting means more errors made in this hospital, and therefore the reporting rate of all incidents was changed to serve

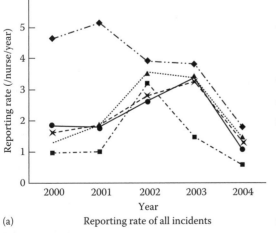

(a) Reporting rate of all incidents

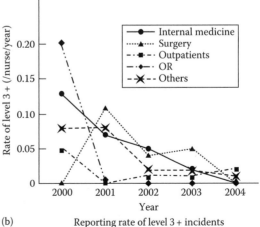

(b) Reporting rate of level 3 + incidents

FIGURE 10.4
Safety performance estimation from transitions of incident reporting rates. (a) Reporting rate of all incidents; (b) Reporting rate of Level 3+ incidents. (From Itoh, K., *JMS*, 3(1), 3, 2007 (in Japanese). With permission.)

as a *risk* measure after 2002. Combined with the interpretation from the transitions of Level 3+ incidents, we could see that the safety level went up in every workplace of nurses in this hospital even after 2000.

Honestly speaking, we are not 100% sure of the aforementioned "guess" about this hospital. However, we believe that in-depth analyses of results of safety culture survey like those mentioned in the last section allows hospital leaders and risk managers to interpret the meaning of incident reporting, and moreover to diagnose more certainly and more reliably a safety level in a specific organization and its work units.

10.6 Conclusion: Perspectives for Improving Patient Safety

In this chapter, we have reviewed the concepts of safety culture and safety climate, some widely quoted definitions and dimensions of safety culture, and methods and tools of safety culture assessment. We have also illustrated how some recent survey results of safety culture dimensions may be correlated to other indicators of safety outcomes. We have shown that much larger safety culture differences exist across different organizations or hospitals than across different departments, wards, or specialties within any single hospital. Comparing safety culture related attitudes between Iceland, New Zealand, Nigeria, and Japan, we have observed large differences in some dimensions and common trends in other between these four countries. A determinant of these differences is no doubt tied to institutional, legal, and regulatory aspects—which we have merely alluded to.

We have demonstrated a link of safety culture to actual safety performance based on the correlation between some of the reviewed safety culture indices and the rate of error reporting (cf. Section 10.5.2 and 10.5.3), although the survey results reviewed in this chapter do not allow us to conclude that there *is* sound evidence for such a causal link. We therefore suggest that further studies should be made to seek to ascertain the possible strength—and modifiers—of such a link. Also, as described in Section 10.5.1, there are considerable difficulties in interpreting the rate of incident reporting. In our survey (e.g., Itoh, 2007), based on several relating facts and the interview with the risk manager, we concluded that the rate of incident reporting (at all severity levels) *may* be regarded as a "safety" measure at the time of the survey but be changed to a "risk" measure in *that* hospital just a few years later. Thus, we do not wish to argue that this putative link between the rate of incident reporting

and risk measure is universal for any health-care organization or even within the same hospital over time. We have been conducting a similar investigation aiming at examining relationships of the incident reporting rate with safety measures for a larger number of hospitals (e.g., Itoh, 2011). Indeed, we have identified some hospitals where the rate of incident reporting may function as a "safety" measure, similar to results of Edmondson's (1996) study conducted in the United States in the mid-1990s. At the same time, there were also other hospitals where this index may be a "risk" measure like the one we illustrated in this chapter. In addition, there were also some hospitals for which we could not determine with any confidence at all whether the index works as a measure of safety or a measure of risk.

One of the greatest advantages of safety culture assessment is its potential for supporting *proactive* patient safety activities. Results of survey should be used prospectively and, in combination with a proactive regime, support the identification of points at which a specific local safety culture may need to be strengthened; equally, survey tools can be used to measure changes over time, including effects of intervention programs within departments or hospitals.

Finally, we would like to stress the importance of participation and involvement of all members of an organization in safety activities, including front line staff, technicians, orderlies, leaders in departments, and senior managers. Commitment to safety is required not only by management but by all the members within an organization—as illustrated in the stress on "shared" norms described in Section 10.1. Through such hospital-wide participation, it is possible to tackle safety activities continuously; to build upon the experience of such initiatives, their outcomes, the knowledge and techniques involved, and the repeated applications to new cases; and, finally, to help in reinforcing safety awareness of health-care practices with frontline staff and management.

Acknowledgments

Some studies mentioned in this chapter were in part supported by Grant-in-Aid for Scientific Research A(2) (No. 18201029), Japan Society for the Promotion of Science, funding provided to the first author. We are grateful to members of the Danish patient safety survey project group, particularly Niels Hermann, Doris Østergaard, and Thomas Schiøler for their contributions to the elaboration of survey instruments and the carrying out of questionnaire surveys.

References

ACSNI (1993). Advisory committee on the safety of nuclear installations: Human Factors Study Group Third Report: Organising for safety. HSE Books, Sheffield, U.K.

Amalberti, R. (1998). Automation in aviation: A human factors perspective. In: Garland, D. Wise, J. and Hopkin, D. (Eds.), *Aviation Human Factors* (pp. 173–192). Hillsdale, NJ: Lawrence Erlbaum Associates.

Andersen, H.B. (2002). Assessing safety culture. Technical Report R-1459, Risø National Laboratory, Roskilde, Denmark.

Andersen, H.B., Hermann, N., Madsen, M.D., Østergaard, D., and Schiøler, T. (2004a). Hospital staff attitudes to models of reporting adverse events: Implications for legislation. *Proceedings of the 7th International Conference on Probabilistic Safety Assessment and Management* (In: Spitzer, C., Schmocker, U., and Dang, V.N. (Eds.). London: Springer Verlag), pp. 2720–2725, Berlin, Germany, June.

Andersen, H.B., Madsen, M.D., Hermann, N., Schiøler, T., and Østergaard, D. (2002). Reporting adverse events in hospitals: A survey of the views of doctors and nurses on reporting practices and models of reporting. *Proceedings of the Workshop on the Investigation and Reporting of Incidents and Accidents*, pp. 127–136, Glasgow, U.K., July.

Andersen, H.B., Nielsen, K.J., Carstensen, O., Dyreborg, J., Guldenmund, F., Hansen, O.N., Madsen, M.D., Mikkelsen, K.L., and Rasmussen, K. (2004b). Identifying safety culture factors in process industry. *Proceedings of Loss Prevention 2004 11th International Symposium*. Loss Prevention and Safety Promotion in the Process Industries, Prague, Czech Republic, May.

Antonow, J.A., Smith, A.B., and Silver, M.P. (2000). Medication error reporting: A survey of nursing staff. *Journal of Nursing Care Quality*, 15(1): 42–48.

Baker, G.R., Norton, P.G., Flintoft, V., Blais, R., Brown, A., Cox, J., Etchells, E., Ghali, W.A., Herbert, P., Majumdar, S.R., O'Beirne, M., Palacious-Derflingher, L., Reid, R.J., and Tamblyn, R. (2004). The Canadian adverse events study: The incidence of adverse events among hospital patients in Canada. *Canadian Medical Association Journal*, 170(11): 1678–1686.

Barach, P. and Small, S. (2000). Reporting and preventing medical mishaps: Lessons from non-medical near miss reporting systems. *British Medical Journal*, 320: 759–763.

Bodur, S. and Filtz, E. (2010). Validity and reliability of Turkish version of "Hospital Survey on Patient Safety Culture" and perception of patient safety in public hospitals in Turkey. *BMC Health Services Research*, 10: 28.

Brown, R.L. and Holmes, H. (1986). The use of a factor-analytic procedure for assessing the validity of an employee safety climate model. *Accident Analysis and Prevention*, 18(6): 455–470.

Bryant, D.T. (1991). *The Human Element in Shipping Causalities*. London, U.K.: HMSO.

Carayon, P. and Hoonakker, P.L.T. (2004). Macroergonomics organizational questionnaire survey (MOQS). In: Stanton, N., Hedge, A., Brookhuis, K., Salas, E., and Hendrick, H. (Eds.), *Handbook of Human Factors and Ergonomics Methods* (pp. 76–1/76–10). Boca Raton, FL: CRC Press.

Carlisle, K.E. (1986). *Analysing Jobs and Tasks*. Englewood Cliffs, NJ: Educational Technology Publications.

CBI (1991). *Developing a Safety Culture*. London, U.K.: Confederation of British Industry.

Clarke, S. (2006). The relationship between safety climate and safety performance: A meta-analytic review. *Journal of Occupational Health Psychology*, 11(4): 315–327.

Colla, J.B., Bracken, A.C., Kinney, L.M., and Weeks, W.B. (2005). Measuring patient safety climate: A review of surveys. *Quality and Safety in Health Care*, 14: 364–366.

Collins, A.M. and Gadd, S. (2002). *Safety Culture: A Review of the Literature*. Sheffield, U.K.: Health and Safety Laboratory, An Agency of the Health and Safety Executive.

Cooper, M.D. (2000). Towards a model of safety culture. *Safety Science*, 36: 111–136.

Cooper, M.D. and Phillips, R.A. (2004). Exploratory analysis of the safety climate and safety behavior relationship. *Journal of Safety Research*, 35(5): 497–512.

Cox, S.J. and Cox, T. (1991). The structure of employee attitudes to safety: An European example. *Work and Stress*, 5(2): 93–106.

Cox, S.J. and Flin, R. (1998). Safety culture: Philosopher's stone or man of straw? *Work and Stress*, 12(3): 189–201.

Cronbach, L.J. (1951). Coefficient alpha and the internal structure of tests. *Psychometrika*, 16(3): 297–334.

Cullen, D., Bates, D., Small, S., Cooper, J., Nemeskal, A., and Leape, L. (1995). The incident reporting system does not detect adverse events: A problem for quality improvement. *Journal of Quality Improvement*, 21(10): 541–548.

Davies, F., Spencer, R., and Dooley, K. (1999). Summary guide to safety climate tools. Offshore Technology Report 1999/063, London, U.K.: HSE Books.

Department of Health (2000). An organisation with a memory. Report of an expert group on learning from adverse events in the NHS, The Stationery Office, London, U.K.

Diaz, R.I. and Cabrera, D.D. (1997). Safety climate and attitude as evaluation measures of organizational safety. *Accident Analysis and Prevention*, 29(5): 643–650.

Duijm, N.J., Andersen, H.B., Hale, A., Goossens, L., and Hourtolou, D. (2004). Evaluating and managing safety barriers in major hazard plants. *Proceedings of the 7th International Conference on Probabilistic Safety Assessment and Management* (In: Spitzer, C. Schmocker, U., and Dang, V.N. (Eds.). London, U.K.: Springer Verlag), pp. 110–115. Berlin, Germany, June.

Edmondson, A.C. (1996). Learning from mistakes is easier said than done: Group and organizational influences on the detection and correction of human error. *Journal of Applied Behavioral Science*, 32(1): 5–28.

Edmondson, A.C. (2004). Learning from failure in health care: Frequent opportunities, pervasive barriers. *Quality and Safety in Health Care*, 13(Suppl. 2): ii3–ii9.

Felknor, S.A., Aday, L.A., Burau, K.D., Delclos, G.L., and Kapadia, A.S. (2000). Safety climate and its association with injuries and safety practices in public hospitals in Costa Rica. *International Journal of Occupational and Environmental Health*, 6(1): 18–25.

Flanagan, J.C. (1954). The critical incident technique. *Psychological Bulletin*, 51(4): 327–359.

Flin, R. (2007). Measuring safety culture in healthcare: A case for accurate diagnosis. *Safety Science*, 45: 653–667.

Flin, R., Burns, C., Mearns, K., Yule, S., and Robertson, E.M. (2006). Measuring safety climate in health care. *Quality and Safety in Health Care*, 15: 109–115.

Flin, R., Fletcher, G., McGeorge, P., Sutherland, A., and Patey, R. (2003). Anaesthesiologists' attitudes to teamwork and safety. *Anaesthesia*, 58: 233–242.

Flin, R., Mearns, K., O'Connor, P., and Bryden, R. (2000). Measuring safety climate: Identifying the common features. *Safety Science*, 34(1–3): 177–192.

Flin, R., Winter, J., Sarac, C., and Raduma, M. (2009). Human factors in patient safety: Review of topics and tools. Report for Methods and Measures Working Group of WHO Patient Safety, World Health Organization, Geneva, Switzerland.

Gaba, D.M., Singer, S.J., Sinaiko, A.D., Bowen, J.D., and Ciavarelli, A.P. (2003). Differences in safety climate between hospital personnel and naval aviators. *Human Factors*, 45(2): 173–185.

Gershon, R.R.M., Karkashian, C.D., Grosch, J.W., Murphy, L.R., Escamilla-Cejudo, A., Flanagan, P.A., Bernacki, E., Kasting, C., and Martin, L. (2000). Hospital safety climate and its relationship with safe work practices and workplace exposure incidents. *American Journal of Infection Control*, 28(3): 211–221.

Gershon, R.R.M., Stone, P.W., Bakken, S., and Larson, E. (2004). Measurement of organizational culture and climate in healthcare. *Journal of Nursing Administration*, 34(1), 33–40.

Glendon, A.I. and McKenna, E.F. (1995). *Human Safety and Risk Management*. London, U.K.: Chapman & Hall.

Glendon, A.I. and Stanton, N.A. (2000). Perspectives on safety culture. *Safety science*, 34: 193–214.

Griffiths, D.K. (1985). Safety attitudes of management. *Ergonomics*, 28: 61–67.

Guldenmund, F.W. (2000). The nature of safety culture: A review of theory and research. *Safety Science*, 34: 215–257.

Guldenmund, F.W. (2007). The use of questionnaires in safety culture research: An evaluation. *Safety Science*, 45: 723–743.

Hale, A.R. (2000). Culture's confusions. Editorial for the special issue on safety culture and safety climate. *Safety Science*, 34: 1–14.

Helmreich, R.L. (2000a). Culture and error in space: Implications from analog environments. *Aviation, Space, and Environmental Medicine*, 71(9): A133–A139.

Helmreich, R.L. (2000b). On error management: Lessons from aviation. *British Medical Journal*, 320: 781–785.

Helmreich, R.L. and Merritt, A.C. (1998). *Culture at Work in Aviation and Medicine: National, Organizational and Professional Influences*. Aldershot, U.K.: Ashgate.

Helmreich, R.L. and Schaefer, H. (1994). Team performance in operating room. In: Bogner, M.S. (Ed.), *Human Error in Medicine* (pp. 225–253). Hillsdale, NJ: Lawrence Erlbaum Associates.

Helmreich, R.L., Wilhelm, J.A., Klinect, J.R., and Merritt, A.C. (2001). Culture, error, and crew resource management. In: Salas, E., Bowers, C.A., and Edens, E. (Eds.), *Improving Teamwork in Organizations* (pp. 305–331). Hillsdale, NJ: Lawrence Erlbaum Associates.

Hofstede, G. (1991). *Cultures and Organizations: Software of the Mind*. London, U.K.: McGraw-Hill.

Hofstede, G. (2001). *Culture's Consequences: Comparing Values, Behaviours, Institutions and Organisations across Nations (2nd ed.)*. Thousand Oaks, CA: Sage Publications.

Hollnagel, E. (1998). *Cognitive Reliability and Error Analysis Method (CREAM)*. London, U.K.: Elsevier Science.

IAEA (1986). *Summary Report on the Post-Accident Review Meeting on the Chernobyl Accident*. Vienna, Austria: International Safety Advisory Group.

IAEA (1991). *Safety Culture, Safety Series No. 75-INSAG-4*. Vienna, Austria: International Atomic Energy Agency.

INSAG (1991). *Safety Culture. Safety Series No. 75-INSAG-4*. Vienna, Austria: IAEA, International Atomic Energy Agency.

Itoh, K. (2007). Human factors approach to healthcare risk management: Towards patient-centered management and well-established safety culture (Part I). *Journal of Medical Safety*, 3(1): 3–11 (in Japanese).

Itoh, K. (2011). Does incident reporting rate indicate risk or safety in healthcare? Implications from correlations with safety climate scores. *Proceedings of the International Conference on Healthcare Systems Ergonomics and Patient Safety*, HEPS 2011, pp. 201–204, Oviedo, Spain, June.

Itoh, K., Abe, T., and Andersen, H.B. (2002). A survey of safety culture in hospitals including staff attitudes about incident reporting. *Proceedings of the Workshop on the Investigation and Reporting of Incidents and Accidents*, pp. 144–153, Glasgow, U.K., July.

Itoh, K. and Andersen, H.B. (2008). A national survey on healthcare safety culture in Japan: Analysis of 20,000 staff responses from 84 hospitals. *Proceedings of the International Conference on Healthcare Systems Ergonomics and Patient Safety*, HEPS 2008. Strasbourg, France, June (CD-ROM).

Itoh, K. and Andersen, H.B. (2010). Dimensions of healthcare safety climate and their correlation with safety outcomes in Japanese hospitals. *Proceedings of the European Safety and Reliability Conference 2010, ESREL 2010*, pp. 1655–1663, Rhodes, Greece, September.

Itoh, K., Andersen, H.B., Madsen, M.D., and Abe, T. (2005). Comparative results of cross-national surveys on hospital safety culture: The views and attitudes of healthcare staff towards safety-related issues and reporting of adverse events and errors. Technical Report, Department of Industrial Engineering and Management, Tokyo Institute of Technology, No. 2005-1, Tokyo Institute of Technology, Tokyo, Japan.

Itoh, K., Andersen, H.B., and Seki, M. (2004). Track maintenance train operators' attitudes to job, organisation and management and their correlation with accident/incident rate. *Cognition, Technology, and Work*, 6(2): 63–78.

Itoh, K., Andersen, H.B, Seki, M., and Hoshino, H. (2001). Safety culture of track maintenance organisations and its correlation with accident/incident statistics. *Proceedings of the 20th European Annual Conference on Human Decision Making and Manual Control*, pp. 139–148, Copenhagen, Denmark, June.

Johnson, S.E. (2007). The predictive validity of safety climate. *Journal of Safety Research*, 38: 511–521.

Kitzinger, J. (1995). Introducing focus groups. *British Medical Journal*, 311: 299–302.

Kohn, L.T., Corrigan, J.M., and Donaldson, M.S. (Eds.) (1999). *To Err is Human: Building a Safer Health System*. Washington, DC: National Academy Press.

Krause, T.R., Seymour, K.J., and Sloat, K.C.M. (1999). Long-term evaluation of a behavior-based method for improving safety performance: A meta-analysis of 73 interrupted time-series replications. *Safety Science*, 32: 1–18.

Mannion, R., Konteh, F.H., and Davies, H.T.O. (2009). Assessing organisational culture for quality and safety improvement: A national survey of tools and tool use. *Quality and Safety in Health Care*, 18: 153–156.

Marshall, C. and Rossman, G.B. (Eds.) (1999). *Designing Qualitative Research*. Thousand Oaks, CA: Sage Publications.

Mearns, K., Flin, R., Gordon, R., and Fleming, M. (1998). Measuring safety climate on offshore installations. *Work and Stress*, 12: 238–254.

Neal, A., Griffin, M.A, and Hart, P.M. (2000). The impact of organizational climate on safety climate and individual behaviour. *Safety Science*, 34: 99–109.

Nieva, V.F. and Sorra, J. (2003). Safety culture assessment: A tool for improving patient safety in healthcare organizations. *Quality and Safety in Health Care*, 12(Suppl. 2): ii17–ii23.

Pett, M.A., Lackey, N.R., and Sullivan, J.J. (2003). *Making Sense of Factor Analysis: The Use of Factor Analysis for Instrument Development in Health Care Research*. Thousand Oaks, CA: Sage Publications.

Pidgeon, N.F. (1991). Safety culture and risk management in organizations. *Journal of Cross-Cultural Psychology*, 22(1): 129–140.

Pidgeon, N.F. and O'Learry, M. (1994). Organizational safety culture: Implications for aviation practice. In: N.A. Johnston, N. McDonald, and R. Fuller (Eds.), *Aviation Psychology in Practice*, pp. 21–43. Aldershot, U.K.: Avebury Technical Press.

Pronovost, P.J., Weast, B., Holzmueller, C.G., Rosenstein, B.J., Kidwell, R.P., Haller, K.B., Feroli, E.R., Sexton, J.B., and Rubin, H.R. (2003). Evaluation of the culture of safety: Survey of clinicians and managers in an academic medical center. *Quality and Safety in Health Care*, 12: 405–410.

Rasmussen, J. (1986). *Information Processing and Human-Machine Interaction: An Approach to Cognitive Engineering*. New York: Elsevier/North Holland.

Reason, J. (1993). Managing the management risk: New approaches to organizational safety. In: Wilpert, B. and Qvale, T. (Eds.), *Reliability and Safety in Hazardous Work Systems* (pp. 7–22). Hove, U.K.: Lawrence Erlbaum Associates.

Reason, J. (1997). *Managing the Risk of Organizational Accidents*. Aldershot, U.K.: Ashgate.

Schein, E.H. (1992). *Organizational Culture and Leadership*, 2nd edn. San Francisco, CA: Jossey-Bass.

Schein, E.H. (2000). Sense and nonsense about culture and climate. In: Ashkanasy, N.M., Wilderom, C.P.M., and Peterson, M.F. (Eds.), *Handbook of Organizational Culture and Climate* (pp. xxiii–xxx). Thousand Oaks, CA: Sage Publications.

Scott, T., Mannion, R., Davies, H., and Marshall, M. (2003). The quantitative measurement of organizational culture in health care: A review of the available instruments. *Health Services Research*, 38(3): 923–945.

Sexton, J.B., Helmreich, R.L., Nielands, T.B., Rowan, K., Vella, K., Boyden, J., Roberts, P.R., and Thomas, E.J. (2006). The safety attitudes questionnaire: Psychometric properties, benchmarking data, and emerging research. *BMC Health Services Research*, 6: 44.

Sexton, J.B., Thomas, E.J., and Helmreich, R.L. (2000). Error, stress, and teamwork in medicine and aviation: Cross sectional surveys. *British Medical Journal*, 320: 745–749.

Sheehy, P.N. and Chapman, A.J. (1987). Industrial accidents. In: C.L., Cooper and Robertson, I.T. (Eds.), *International Review of Industrial and Organizational Psychology* (pp. 201–227). New York: John Wiley & Sons.

Singer, S.J., Gaba, D.M., Falwell, A., Lin, S., Hayes, J., and Baker, L. (2009). Patient safety climate in 92 US hospitals: Differences by work area and discipline. *Medical Care*, 47(1): 23–31.

Singer, S.J., Gaba, D.M., Geppert, J.J., Sinaiko, A.D., Howard, S.K., and Park, K.C. (2003). The culture of safety in California hospitals. *Quality and Safety in Health Care*, 12(2): 112–118.

Singer, S., Meterko, M., Baker, L., Gaba, D., Falwell, A., and Rosen, A. (2007). Workplace perceptions of hospital safety culture: Development and validation of the patient safety climate in healthcare organizations survey. *Health Service Research*, 42(5): 1999–2021.

Singla, A.K., Kitch, B.T., Weissman, J.S., and Campbell, E.G. (2006). Assessing patient safety culture: A review and synthesis of the measurement tools. *Journal of Patient Safety*, 2(3): 112–118.

Smits, M., Christiaans-Dingelhoff, I., Wagner, C., van der Wal, G., and Groenewegen, P.P. (2008). The psychometric properties of 'Hospital Survey on Patient Safety Culture' in Dutch hospitals. *BMC Health Services Research*, 8: 230.

Sorra, J., Famolaro, T., Dyer, N., Nelson, D., and Khanna, K. (2009). Hospital survey on patient safety culture: 2009 comparative database report. Agency for Healthcare Research and Quality, AHRQ Publication No. 09-0030, Rockville, MD.

Sorra, J.S. and Nieva, V.F. (2003). Psychometric analysis of the hospital survey on patient safety. Final Report to Agency for Healthcare Research and Quality (AHRQ), Washington, DC.

Sorra, J.S. and Nieva, V.F. (2004). *Hospital Survey on Patient Safety Culture*. Rockville, MD: Agency for Healthcare Research and Quality.

Spector, P.E., Cooper, C.L., and Sparks, K. (2001). An international study of the psychometric properties of the Hofstede Values Survey Module 1994: A comparison of individual and country/province level results. *Applied Psychology—An International Review*, 50(2): 269–281.

Spiliotopoulou, G. (2009). Reliability reconsidered: Cronbach's alpha and paediatric assessment in occupational therapy. *Australian Occupational Therapy Journal*, 56: 150–155.

Swain, A.D. and Guttmann, H.E. (1983). *Handbook of Human Reliability Analysis with Emphasis on Nuclear Power Plant Applications*. Washington, DC: NUREG-1278, U.S. Nuclear Regulatory Commission.

Thomas, E.J., Sexton, J.B., Bryan, J., and Helmreich, R.L. (2003). Discrepant attitudes about teamwork among critical care nurses and physicians. *Critical Care Medicine*, 31(3): 956–959.

van Vuuren, W. (2000). Cultural influences on risks and risk management: Six case studies. *Safety Science*, 34 (1–3): 31–45.

Vrendenburgh, A.G. (2002). Organizational safety: Which management practices are most effective in reducing employee injury rates? *Journal of Safety Research*, 33: 259–276.

Wiegmann, D.A., Zhang, H., von Thaden, T.L., Sharma, G., and Gibbons, A.M. (2004). Safety culture: An integrative review. *International Journal of Aviation Psychology*, 14(2): 117–134.

Zohar, D. (1980). Safety climate in industrial organizations: Theoretical and applied implications. *Journal of Applied Psychology*, 65: 96–101.

Zohar, D. (2003). Safety climate: Conceptual and measurement issues. In: Quick, J. and Tetrick, L. (Eds.), *Handbook of Occupational Health Psychology* (pp. 123–142). Washington, DC: American Psychological Association.

Zohar, D. (2008). Safety climate and beyond: A multi-level multi-climate framework. *Safety Science*, 46: 376–387.

Zohar, D., Livne, Y., Tenne-Gazit, O., Admi, A., and Donchin, Y. (2007). Healthcare climate: A framework for measuring and improving patient safety. *Critical Care Medicine*, 35(5): 1312–1317.

11

Handoffs and Transitions of Care

Robert L. Wears, Shawna J. Perry, and Emily S. Patterson

CONTENTS

11.1 Overview...163
11.2 Conventional Wisdom and Unspoken Assumptions..164
    11.2.1 Handoffs Are Handoffs Are Handoffs..164
    11.2.2 Handoffs Are Hazardous..164
    11.2.3 Handoffs Are Information Transfers...164
    11.2.4 Handoffs Are "Low Hanging Fruit"..164
    11.2.5 Standardization as an Independent Good..164
    11.2.6 Training Settings Are Good Models for Practice Settings.................................164
11.3 Challenges to the Conventional Wisdom...164
    11.3.1 Handoffs Are Heterogeneous...165
        11.3.1.1 Common Ground..165
        11.3.1.2 Probability of Interaction...165
        11.3.1.3 Point in Trajectory of Care...165
        11.3.1.4 Number of Patients..165
    11.3.2 Handoffs Are Sources of Rescue..165
    11.3.3 Handoffs Are Not (Perhaps Not Even Mostly) about Information..................166
        11.3.3.1 Responsibility and Authority..166
        11.3.3.2 Understanding..166
        11.3.3.3 Handoffs Are Not Isolated...166
    11.3.4 All the Fruit Is High...166
    11.3.5 Requisite Variety versus Mindless Standardization...167
    11.3.6 Residency Is Not Practice...167
11.4 Recent Insights into Handoffs...167
11.5 Possible Framings for Handoffs and Their Implications...168
11.6 Summary..169
References..170

11.1 Overview

Hospitals operate as a continuous system—24 h a day, 7 days a week, 365 days a year—because patient care cannot be temporarily suspended (Randell et al., 2010). In addition, modern healthcare requires a diverse, highly specialized variety of skills and knowledge that cannot be found in a single individual. Each of these two characteristics separately makes handoffs of care necessary, to allow for both continuity of care across time and across professional specialties or disciplines. In addition, the interaction of these two requirements affects handoffs in different ways. Handoffs have commonly been viewed as risky by health professionals, but in general, the degree of insight and understanding of handoffs among health professionals has been superficial and less than satisfactory (Cohen and Hilligoss, 2010; Philibert and Leach, 2005).

In this chapter, we will review disparate views of handoffs in healthcare, discuss some recent studies shedding new light on handoffs, and conclude with a discussion of possible framings under which handoffs might be usefully viewed. These framings might be considered the beginnings of a theoretical model to enable a better understanding of handoffs, and better support for them.

11.2 Conventional Wisdom and Unspoken Assumptions

11.2.1 Handoffs Are Handoffs Are Handoffs

There has been a general presumption in healthcare that handoffs are basically all the same thing, and this has led to a search for single, "silver bullet" solutions that can be applied across the board to fix all handoffs in all circumstances. This approach is exemplified in the Joint Commission's 2006 National Patient Safety Goals, which required delivery organizations to develop a "standardized approach to handoff communications" in care delivery organizations (Joint Commission on Accreditation of Healthcare Organizations, 2006).

11.2.2 Handoffs Are Hazardous

The common view among health professionals is that handoffs are unmitigated hazards, necessary evils that should be avoided if at all possible. For example, a commonly used text on the organization and management of the emergency department (ED) notes that "shift change is well known as a high risk period" and (wisely) advises physicians to be careful (Salluzo et al., 1997). (It is also interesting to note that of almost 900 pages, a total of one half page is devoted to the discussion of the shift change handoff.) Although there are many reports of patient injuries involving handoffs (Gandhi, 2005; Horwitz et al., 2008; JCAHO, 2002; Kitch et al., 2008; Vidyarthi, 2004), it is rare to see in a discussion of those reports any mention of the problems of hindsight bias, outcome bias, or length-biased sampling problems that bedevil meaningful studies of the handoff. (Briefly, the hindsight and outcome biases concern the well-established tendency of observers to rate the quality of a process more favorably when they know the outcome was favorable, and vice versa; and a similar tendency to assume that information known in retrospect to have been important would have been easily recognized as salient in prospect [Caplan et al., 1991; Fischhoff, 2003; Henriksen and Kaplan, 2003]. Length-biased sampling is the problem that more complex patients, who are more likely to experience adverse events, are also more likely to undergo handoffs because of longer length of stay or a greater number of specialties involved in their care [Akman et al., 2007; Westbrook et al., 2010].)

11.2.3 Handoffs Are Information Transfers

It is commonly assumed that handoffs are straightforward transfers of information (typically envisioned as discrete data elements). The problem here is not that handoffs do not involve transferring information, but rather that information transfer is viewed as a sufficient model for handoffs, that it encompasses the totality of all that is important about them. One sees this assumption clearly in the Joint Commission's statement that the primary objective of the handoff is "to provide accurate information" (The Joint Commission, 2008).

11.2.4 Handoffs Are "Low Hanging Fruit"

The rhetoric about handoffs often paints them as poorly structured haphazard almost random events, and implies that they could be quite easily fixed by various forms of standardization, such as standardized data sets, checklists, procedural tools such as Situation, Background, Assessment, Recommendation (SBAR) (Haig et al., 2006), technological interventions such as voicemail (Horwitz et al., 2009c), or computerized handoff support based on an electronic medical record (Petersen et al., 1998).

11.2.5 Standardization as an Independent Good

An unspoken assumption, implicit in many discussions, about handoffs is that variation is bad and that standardization is good in virtually any circumstance. Most initial attempts at improving handoffs began with some form of standardization of the data content (Snow et al., 2009). The SBAR protocol has been commonly promoted as a means of standardizing the handoff (Haig et al., 2006), (although it seems designed for situation communication up an authority gradient [Leonard et al., 2004]).

11.2.6 Training Settings Are Good Models for Practice Settings

This assumption may well be recognized by many of the participants in discussions about handoffs, but it is seldom acknowledged as a limitation, and there has been even less discussion that the needs and capabilities of trainees may not reasonably reflect the needs and capabilities of experienced practitioners.

11.3 Challenges to the Conventional Wisdom

One thing that seems quite remarkable about the conventional wisdom regarding handoffs is that it contains virtually no discussion of what the purpose of the handoff is (Hilligoss and Cohen, 2010). We will take the view here that the purpose of the handoff is to assist in concisely preparing the next practitioner(s) to act safely and effectively in their care of this patient (or these patients) (Cohen and Hilligoss, 2010).

11.3.1 Handoffs Are Heterogeneous

If one looks across the handoffs literature from setting to setting, or observes a variety of handoffs in a single setting, it seems readily apparent that there is a great deal of variety in the setting, the participants, and the constraints that affect the handoff. In observational studies of handoffs in the ED, we noted four factors that seemed to characterize different types of handoffs (Behara et al., 2005; Cheung et al., 2010).

11.3.1.1 Common Ground

The degree of common ground held by the participants in the handoff is an important characteristic explaining some of the observed variability in handoff processes. Shift change handoffs are almost always "like to like" (e.g., emergency physician to emergency physician, internist to internist, ICU nurse to ICU nurse, etc.). The large area of context, information, and understanding held in common in these turnovers allows them to be extremely concise, using densely encoded language, symbols, and gestures to efficiently build shared understanding. In addition, all parties in these handoffs commonly have experience on both the oncoming and off going (or sending and receiving) sides of the interaction.

In contrast, in handovers that across organizational boundaries (e.g., ED to ward, OR to PACU, floor to ICU, etc.), participants share only a moderate amount of common ground based on their all being health professionals, but typically from different specialties. In these handoffs, each participant typically has experience on only one side of the exchange. A subelement of this dimension of handoffs is whether or not the receiving party is obligated to accept the patient (for example, the handoff from an ambulance crew to the ED staff) or whether they must be persuaded to accept (i.e., they have an option to refuse, for example, in most handoffs involving hospital admission, and in many handoffs involving transfers to another service).

Finally, in the handoff that occurs when patients are discharged from the hospital there is very little common ground at all in that the exchange is typically between a health professional and a layperson.

11.3.1.2 Probability of Interaction

Handoffs differ in the probability that the receiving party will actually have to use the information and understanding gained in the handoff in their care of the patient. In many shift change handoffs, the probability is close to 100%, while in the handoff for nighttime coverage, the probability is typically much lower; a physician may take a handoff for overnight coverage of a large number of patients, but is only likely to be called to engage with a handful of them.

11.3.1.3 Point in Trajectory of Care

Many handoffs occur at a logical point in the patient's trajectory of care; for example, the handoff from the OR to the PACU, or from the ED to the ICU. Handoffs at some logical transition point lend themselves well to structured sets of data since there will be many commonalities among different patients about which elements of recent care are relevant, and which variables are important to attend to in the future. In contrast, handoffs that occur at an arbitrary point in time (e.g., shift change or night-time coverage handoffs) have no clear relationship to what is going on with the patient at a time. Therefore, the relevance of recent events, and the salience of certain variables for the future cannot be prespecified, but can only be determined in context.

11.3.1.4 Number of Patients

Finally, handoffs differ dramatically in the number of patients involved. Most handoffs across organizational boundaries involve small numbers of patients, typically only one. In contrast, shift change handoffs involve large numbers of patients, typically in the 10s to 20s or higher. This increases the constraint on the time that can be invested in the handoff and consequently increases the importance of both conciseness and prioritization. In particular, prioritization among patients is seldom an issue in handoffs across organizational boundaries, but almost always an issue in shift change handoffs.

11.3.2 Handoffs Are Sources of Rescue

From a control engineering point of view it seems reasonable that handoff should pose hazards (because having two controllers in a process always raises the risk of conflict, poor coordination, or miscommunication). However, prospective observational studies of handoffs (which do not carry the risks of the hindsight and outcome biases noted earlier) have found that handoffs sometimes serve to correct a course of events that was misdirected and would have resulted in tragedy had it continued (Feldman, 2003; Wears et al., 2003). This property of handoffs has been noted in the literature since at least the 1980s (Cooper, 1989; Cooper et al., 1982). In addition, it is also well known that bringing a "fresh set of eyes" to a problem is virtually the only way to overcome fixation errors (Woods and Cook, 1999). And finally, the principle of "requisite variety" would further support the idea that handoffs can improve care by bringing different points of view to the same problem (Ashby, 1957). Given these findings, it seems curious that handoff should have developed such a bad reputation in healthcare. While even prospective observations of handoffs have shown they pose risks as well as benefits to patients, one wonders whether the emphasis on the ill effects of handoffs stems as much from the hindsight and outcome biases as it does from their actual hazards. Understanding handoffs as sources of rescue has important implications for intervention, since one common approach to "the handoff problem" is to reduce the number of patients handed off, for example, by overlapping shifts. Such a strategy could reduce hazards related to miscommunication, but may increase hazards related to premature closure and would forgo the opportunity for recovery in such cases. In addition, an upcoming handoff acts as a trigger to develop a synthesis of events, data, and information that have accumulated as isolated fragments during previous work, in a sense forcing a period of reflection to develop understanding that may not have occurred previously.

11.3.3 Handoffs Are Not (Perhaps Not Even Mostly) about Information

Perhaps the most pernicious notion in the conventional wisdom is the idea that the information transfer paradigm captures all that is important about handoffs. There are three important drawbacks in limiting thinking about handoffs to the communication or information transfer framework.

11.3.3.1 Responsibility and Authority

Characterizing handoffs solely as communication exercises aimed at transferring information leads to a great deal of ambiguity about what exactly constitutes a handoff. For example, many activities in clinical work (for example, radiology reports from imaging studies, or grand rounds, or morning report) are clearly communications transferring information, but not something we would consider handoffs. Handoffs are unique because in addition to information, they also transfer both responsibility and authority (Lardner, 1996). Limiting the thinking about handoffs to information transfer misses two important dimensions (i.e., responsibility and authority) that differentiate handoffs from other types of clinical communications.

11.3.3.2 Understanding

The second problem with using information transfer as the defining characteristic of a handoff is that the transfer of information, even completely and accurately, does not guarantee the transfer of understanding, and in fact, in some ways understanding can be transferred despite inaccuracies or omissions in the discrete data points. Because it is understanding that is important in guiding future care, a narrow focus on data—ensuring that every pebble put into one end of the pipe comes out the other—is at best insufficient to ensure understanding, and at worst, may drown out meaning in a litany of marginally relevant facts. "A wealth of information creates a poverty of attention" (Simon, 1971)—an emphasis on data tends to value comprehensiveness over salience. In addition, the handoff is (or should be) only one modality in preparing the receiving party; the handoff is particularly well suited to answer the question "To what should I attend during my period of responsibility?" while other sorts of informational resources (the chart, ward logbooks, the status board, etc.) would seem better suited to support the retrieval of an arbitrarily specific datum.

11.3.3.3 Handoffs Are Not Isolated

The handoff episode does not stand alone as the sole means for transferring information or assuring continuity, in two important senses. First (as discussed further later), handoffs are typically preceded and followed by specific activities. Preparatory activity begins prior to the handoff—information is assembled and synthesized, lists are converted to stories. After the handoff, there is often a period of post-handoff consolidation, where "loose ends" are identified and resolved. Thus to focus on the handoff conversations in isolation can miss important segments of the work of assuring continuity of care.

Second, handoffs are not always the best vehicle for some forms of information transfer. Many of the adverse events glibly labeled as handoff or communication failures might better be reconstrued as failures to provide adequate informational support for clinical work. In these cases, the handoff is being used as a crutch to compensate for deficiencies deeper in the organization; modifying the handoff so that it better compensates for gaps better remedied in other ways risks damaging those aspects of handoffs that are central to its effectiveness and do not lend themselves to other forms of informational support.

11.3.4 All the Fruit Is High

The reader should already be convinced by the foregoing that the complexity of handoffs precludes any sort of easy, "one-size-fits-all," silver bullet interventions. Richard Cook has famously pointed out that all the easily solvable problems have been solved already. Handoffs are exquisitely situated in a complex context, and although there may be general principles that might make them better, the key to success is developing a detailed understanding of how those principles will play out in each specific context, to use a military analogy, a strategy of "defeat in detail" rather than "defeat in the main." In addition, it is at least possible if not likely that focusing on improving the handoff as an isolated act could be misdirected. If we adopt the position that the goal of the handoff is to *assist* in preparing the receiving party to act safely and effectively, rather than to be the sole means by which that preparation occurs, then we would begin to focus on other forms of informational support in addition to the handoff exchange itself, and perhaps reach a greater level of success.

11.3.5 Requisite Variety versus Mindless Standardization

Standardization brings many benefits. It supports efficiency in communication and coordination by allowing parties to assume that what is omitted is known to be unimportant; it aids memory in ensuring that important items are not omitted; it decreases mental workload by providing a well understood and agreed framework that does not have to be constructed anew on every occasion. However, variety is also important; every controller of a process must possess at least as much variety as that process (Ashby, 1958). In addition, variety is needed for adaptation to change and for organizational learning (March, 1991). The question is not so much standardization versus variety as it is to identify where standardization might be beneficial and where it might be harmful.

When standardization is mentioned in discussions of handoffs in healthcare it almost always occurs in two contexts: a complaint that handoffs are highly unstandardized, and the notion that standardization of data would improve them. Both ideas are limited.

First, observational studies of handoffs have shown that, despite their surface variety, they do possess an underlying structure, i.e., an underlying standardization. For example, handoffs tend to share a common four-phase structure (pre-handoff preparation, general discussion, specific handover, post-handoff cleanup) (Behara et al., 2005; Durso et al., 2007; Grusenmeyer, 1995). They generally use a constant order among patients, typically a geographical (e.g., by room number) which is often conflated with acuity (generally, sickest first). Within patients, they use a standardized order of presentation (typically the traditional case presentation format). They frequently make use of informal artifacts (whiteboards, logbooks, "cheat sheets"), often to the exclusion of formal ones (e.g., the chart). They are most commonly performed by single profession groups (doctors to doctors, nurses to nurses). And finally, they have a general abstract quality of translating stories about patients to lists of things to be done; said lists to be translated back to stories at the next transition. Thus despite their superficial variety and admitted shortcomings, there is a fair degree of standardization to handoffs already—they are not just haphazard, random events.

The second point is the assumption that standardization can only be done on data. While for some types of handoffs, data standardization might be both important and effective (for example, handoffs across organizational boundaries at logical points in care), it might be irrelevant or even harmful for others (those that occur at arbitrary points in a patient's trajectory of car). And, there are many other dimensions that might be standardized in useful ways. For example, explicitly standardizing on order among patients along a "sickest first" or "most important first" scale would help in communicating priorities among patients and also with the common problem of having to rush or truncate the handoff before it is entirely completed due to the press of or intercurrent events. Second, although the case presentation format is well known and well rehearsed, it may not be the most effective vehicle for creating shared understanding during the handoff; a journalistic format, where the "big picture view" is given first, followed by supporting detail, may have advantages over the more traditional order which begins with details and builds to a summary (Propp, 2010).

11.3.6 Residency Is Not Practice

We suspect that no one will disagree with the observation that, with rare exceptions, handoffs have been studied only in training settings. (Handoffs among nurses are an important exception to this observation). But, most inpatient healthcare is provided in community hospital settings where trainees play at most a minor role. In addition, some physicians in practice have made attempts to improve their handoff practices in ways that might usefully inform our understanding (Propp, 2010). One important difference between training and practice situations is that in practice, participants often have a long shared history of prior interactions, and thus develop strong opinions on the trustworthiness of their handoff partners; personal assessments of trust are strongly associated with assessments of the quality of the handoff (Matthews et al., 2002).

11.4 Recent Insights into Handoffs

The scientific literature on handoffs is abundant and disparate; it appears in no single location and is not indexed in a single source. In particular, healthcare-oriented databases such as Pubmed or Medline will miss a significant proportion of studies in the engineering, psychology, and communication literature. Cohen et al. at the University Michigan have developed an extremely useful resource, a web-based compendium of the handoff literature across multiple sources which they are endeavoring to keep current as new material appears (Cohen and Hilligoss, 2009). In this section, we make no attempt to summarize this extensive material but instead highlight a small number of relatively recent studies that shed new light on handoffs.

Nemeth and colleagues did extensive observational studies of handoffs in ICU settings and noted the variability in the length and depth of discussion was tailored to match variability in uncertainty about patients' conditions and/or courses (Brandwijk et al., 2003; Kowalsky et al., 2004; Nemeth et al., 2005, 2007). These analyses support the idea that at least some of the observed variability in handoffs is necessary variation required to accommodate variation in the underlying process being managed. He also noted, as did Wears and colleagues (Behara et al., 2005), that handoffs tended to be conversational, marked by back-and-forth exchanges and questions; in ED physician shift change handoffs, 60% of patient discussions involved one or more clarifying questions, emphasizing that the handoff in the setting was less a "data dump" and more a process of shared sense making (Eisenberg et al., 2005; Wears and Perry, 2010).

Horwitz and colleagues studied the ED to hospitalist handoff (Horwitz et al., 2007, 2008, 2009), and evaluated an intervention involving asynchronous communication (voice mail) in an attempt to improve the handoff process (Horwitz et al., 2009). Perceptions of the intervention were mixed—emergency physicians favored it, while hospitalists were less enthralled. In a commentary, Murphy raised an issue little noted by health professionals, that any communicative act involves two dimensions, content and relational: its subject matter, and a message about the relationship among the parties. She pointed out that changing the medium from synchronous to asynchronous preserved the content but changed the relational message and offered that as a way of explaining the observation that 30% of the voice-mails in Horwitz's study were never opened (Murphy and Wears, 2009).

Apker et al. also studied the emergency physician—hospitalist handoff (Apker et al., 2007, 2010), and developed a Handoff Communication Assessment (HCA) tool using conversational analysis in a grounded theory approach to characterize the utterances making up the handoff conversations along two dimensions, content and language form. Wears and Perry then attempted to use the HCA to characterize emergency physician shift change handoffs in order to compare them to the ED—hospitalist handoff (Wears and Perry, 2010). They found the HCA to be only partially useful in characterizing the ED shift change handoff due to missing categories and unused categories, and concluded that this represented

evidence of the situatedness of the HCA. Essentially, it was so accurately representative of the ED hospitalist interchange that it did not represent some elements of the ED shift change interchange well. Nevertheless, they were able to identify some significant differences between the two types of handoffs in both subject matter (e.g., physical findings and patient's overall condition were prominent in the ED shift change handoffs but never mentioned in the ED hospitalist handoff); and speech form (the number of information seeking exchanges, such as clarifying questions, were more prominent in the ED shift change handoffs).

11.5 Possible Framings for Handoffs and Their Implications

The multiple contending views of handoffs, particularly when unarticulated and unconsciously adopted, contribute to confusion in the handoff literature, difficulty in understanding the types of problems handoffs pose, and in the sorts of interventions that might be useful. Patterson has attempted to remedy this by organizing these views into a set of "framings" (Patterson and Wears, 2010a, b). These framings are types of theoretical frameworks, and so emphasize some aspects and ignore others for particular purposes. It is important to note that these framings are not mutually exclusive—i.e., there is no "one best" framing—but rather can be viewed as levels of activity that occur simultaneously in parallel during a handoff, playing greater or a lesser roles in the activity, its analysis and potential interventions depending on the purpose. Choosing different framings leads to different implications for measurement and for improvement. Table 11.1 summarizes seven possible framings and their implications for measurement and intervention (Patterson and Wears, 2010a, b). The information processing frame is dominant in healthcare, and implies accuracy and completeness as potential metrics. Stereotypical narratives highlight deviations from typical stories, and thus offer compactness, when the stereotypes are widely shared, and emphasize the unusual. The resilience framing focuses on cross checking or critiquing, and emphasizes the role of handoffs as means of rescue. The accountability framing is also relatively common in healthcare, and assumes improvement will follow more explicit identification of who is accountable for what. The social interaction framing emphasizes the co-construction of shared meaning during the handoff, particularly as a means of generating new insights and making the handoff a source of rescue. The distributed cognition framing recognizes that understanding is not encapsulated entirely inside a single caregiver's head,

TABLE 11.1

Handoff Framings and Their Implications

Conceptual Framing	Primary Function	Intervention Example	Quality Measures	Examples
Information processing	Transfer data through a noisy communication channel	Standardized handoff protocol	Accurate information content transferred	Information units Information omissions
Stereotypical narratives	Label by stereotypical narrative and highlight deviations	Daily goals for interdisciplinary teams	Appropriate patient narrative Insightful summary synthesis	Level of information abstraction
Resilience	Cross check assumptions from fresh perspective	Critique of care by oncoming party Two challenge rule for residents questioning attending	Collaborative cross checking	Risk-adjusted mortality and morbidity Adverse events
Accountability	Transfer of responsibility and authority	Protocol explicitly assigning tasks to team members	Task completion Inappropriate tasks transferred	Dropped tasks or patients
Social interaction	Co-construction of shared meaning	Supporting interdisciplinary communication during rounds	Respectful interactions, team climate Generation of new insights	Interprofessional communication quality
Distributed cognition	Replace a member of a network of specialized practitioners	Shared repository or artifact for aiding coordination among caregivers	Effective coordination of care	Technical errors
Cultural norms	Negotiate and share group values	Guides reflection on handoff improvements during orientation	Educational interventions Policies and procedures Changes in priorities, values, and acceptable behaviors	Comfort in doing handoff Number and quality of implemented process changes

Source: Adapted from Joint Commission on Accreditation of Healthcare Organizations, 2006 National patient safety goals, Retrieved March 18, 2006, from http://www.jointcommission.org/PatientSafety/NationalPatientSafetyGoals/06_hap_cah_npsgs.htm.

but rather is distributed across other workers and a variety of artifacts; the handoff here is viewed as a way of distributing this understanding further, across time, or across organizational boundaries. And finally, the cultural norms framing recognizes that handoffs also play a role in negotiating, maintaining, and transmitting group values over time.

11.6 Summary

Far from being simple, mundane processes that sometimes lead to harm, handoffs are complex, exquisitely situated, negotiations that, like most adaptations, produce both success and failure. Efforts to study or improve handoffs should keep the following points in mind:

- Handoffs are situated and heterogeneous. Superficial variability affords a good fit to variability in situations and work processes.

Standardization can be helpful, but requires insight and creativity in thinking about what, when, whether, and how, to standardize.

- Handoffs are not just about information, but rather about responsibility, authority, and understanding. Information transfer is a means, not an end.

- The goal of handoffs should be kept firmly in mind. It is *continuity*: to assist in preparing the receiving party to act safely and effectively in their care of the patient(s).

- The handoff is not the only modality for achieving the goal of continuity. Some aspects (such as stance toward changes in plan) are best managed via handoffs; other aspects (such as details with possible relevance, like allergies) may be best managed by other means.

- An abundance of information produces a poverty of attention. Handoffs should not focus on comprehensiveness, but rather on salience.

- Understanding handoffs requires help from disciplines not commonly found in care delivery organizations. Training for the health professions does not equip people with the theories, knowledge, or tools to understand or evaluate the complexity of behaviors in handoffs. Sustained collaborations with safety science professionals—engineers, psychologists, specialists in communication or organizational behavior, etc.—are required for real progress to be made.

References

Akman, O., Gamage, J., Jannot, J., Juliano, S., Thurman, A., and Whitman, D. (2007). A simple test for detection of length-biased sampling. *Journal of Biostatistics, 1*(2), 189–195.

Apker, J., Mallak, L. A., Applegate, E. B., Gibson, S. C., Ham, J. J., Johnson, N. A. et al. (2010). Exploring emergency physician–hospitalist handoff interactions: Development of the handoff communication assessment. *Annals of Emergency Medicine, 55*(2), 161–170.

Apker, J., Mallak, L. A., and Gibson, S. C. (2007). Communicating in the "Gray Zone": Perceptions about emergency physician hospitalist handoffs and patient safety. *Academic Emergency Medicine, 14*(10), 884–894. doi: 10.1197/j.aem.2007.06.037

Ashby, W. R. (1957). Requisite variety. *An Introduction to Cybernetics* (pp. 202–218). London, U.K.: Chapman & Hall Ltd.

Ashby, W. R. (1958). Requisite variety and its implications for the control of complex systems. *Cybernetica, 1*, 83–99.

Behara, R., Wears, R. L., Perry, S. J., Eisenberg, E., Murphy, A. G., Vanderhoef, M. et al. (2005). Conceptual framework for the safety of handovers. In K. Henriksen (Ed.), *Advances in Patient Safety* (Vol. 2, pp. 309–321). Rockville, MD: Agency for Healthcare Research and Quality/Department of Defense.

Brandwijk, M., Nemeth, C., O'Connor, M., Kahana, M., and Cook, R. I. (January 27, 2003). Distributing cognition: ICU handoffs conform to Grice's maxims, Retrieved January 27, 2003, from http://www.ctlab.org/properties/pdf%20files/SCCM%20Poster%201.27.03.pdf

Caplan, R. A., Posner, K. L., and Cheney, F. W. (1991). Effect of outcome on physician judgments of appropriateness of care. *JAMA, 265*(15), 1957–1960.

Cheung, D. S., Kelly, J. J., Beach, C., Berkeley, R. P., Bitterman, R. A., Broida, R. I. et al. (2010). Improving handoffs in the emergency department. *Annals of Emergency Medicine, 55*(2), 171–180. doi: S0196-0644(09)01261-X [pii] 10.1016/j.annemergmed.2009.07.016

Cohen, M. D., and Hilligoss, P. B. (January 2009). Handoffs in hospitals: A review of the literature on information exchange while transferring patient responsibility or control, Retrieved June 12, 2009, from http://deepblue.lib.umich.edu/handle/2027.42/61522

Cohen, M. D. and Hilligoss, P. B. (2010). The published literature on handoffs in hospitals: Deficiencies identified in an extensive review. *Quality and Safety in Health Care, 19*(6), 493–497. doi: 10.1136/qshc.2009.033480

Cooper, J. B. (1989). Do short breaks increase or decrease anesthetic risk? *Journal of Clinical Anesthesia, 1*(3), 228–231.

Cooper, J. B., Long, C. D., Newbower, R. S., and Philip, J. H. (1982). Critical incidents associated with intraoperative exchanges of anesthesia personnel. *Anesthesiology, 56*(6), 456–461.

Durso, F. T., Crutchfield, J. M., and Harvey, C. M. (2007). The cooperative shift change: An illustration using air traffic control. *Theoretical Issues in Ergonomics Science, 8*(3), 213–232.

Eisenberg, E. M., Murphy, A. G., Sutcliffe, K., Wear, R., Schenkel, S., Perry, S. J. et al. (2005). Communication in emergency medicine: Implications for patient safety. *Communication Monographs, 72*(4), 390–413.

Feldman, J. (2003). Medical errors and emergency medicine: Will the difficult questions be asked, and answered? *Academic Emergency Medicine, 10*(8), 910–911.

Fischhoff, B. (2003). Hindsight ≠ foresight: The effect of outcome knowledge on judgment under uncertainty. *Quality and Safety in Health Care, 12*(4), 304–311.

Gandhi, T. K. (2005). Fumbled handoffs: One dropped ball after another. *Annals of Internal Medicine, 142*(5), 352–358.

Grusenmeyer, C. (1995). Shared functional representation in cooperative tasks: The example of shift changeover. *International Journal of Human Factors in Manufacturing, 5*, 163–176.

Haig, K. M., Sutton, S., and Whittington, J. (2006). SBAR: A shared mental model for improving communication between clinicians. *Joint Commission Journal on Quality and Patient Safety, 32*(3), 167–175.

Henriksen, K., and Kaplan, H. (2003). Hindsight bias, outcome knowledge and adaptive learning. *Quality and Safety in Health Care, 12*(Suppl 2), ii46–ii50.

Hilligoss, P. B., and Cohen, M. D. (2010). What is the goal of patient handoff? Implications for the improvement of care transitions. Paper presented at the *Computer-Supported Cooperative Work: Handoffs Workshop*, Savannah, GA.

Horwitz, L. I., Meredith, T., Schuur, J. D., Shah, N. R., Kulkarni, R. G., and Jenq, G. Y. (2009a). Dropping the baton: A qualitative analysis of failures during the transition from emergency department to inpatient care. *Annals of Emergency Medicine, 53*(6), 701–714.

Horwitz, L., Moin, T., and Green, M. (2007). Development and implementation of an oral sign-out skills curriculum. *Journal of General Internal Medicine, 22*(10), 1470–1474.

Horwitz, L. I., Moin, T., Krumholz, H. M., Wang, L., and Bradley, E. H. (2008). Consequences of inadequate sign-out for patient care. *Archives of Internal Medicine, 168*(16), 1755–1760. doi: 10.1001/archinte.168.16.1755

Horwitz, L. I., Moin, T., Krumholz, H. M., Wang, L., and Bradley, E. H. (2009b). What are covering doctors told about their patients? Analysis of sign-out among internal medicine house staff. *Quality and Safety in Health Care, 18*(4), 248–255. doi: 10.1136/qshc.2008.028654

Horwitz, L. I., Parwani, V., Shah, N. R., Schuur, J. D., Meredith, T., Jenq, G. Y. et al. (2009c). Implementation and evaluation of an asynchronous physician sign-out for emergency department admissions. *Annals of Emergency Medicine, 54*(3), 368–378.

JCAHO. (2002). Delays in treatment. *JCAHO Sentinel Event Alert*, (26). Retrieved from http://www.jointcommission.org/sentinel_event_alert_issue_26_delays_in_treatment/(accessed March 18, 2006).

Joint Commission on Accreditation of Healthcare Organizations. (2006). 2006 National Patient Safety Goals, Retrieved March 18, 2006, from http://www.jointcommission.org/PatientSafety/NationalPatientSafetyGoals/06_hap_cah_npsgs.htm

Kitch, B. T., Cooper, J. B., Zapol, W. M., Marder, J. E., Karson, A., Hutter, M. et al. (2008). Handoffs causing patient harm: A survey of medical and surgical house staff. *Joint Commission Journal on Quality and Patient Safety, 34*(10), 563–570.

Kowalsky, J., Nemeth, C. P., Brandwijk, M., and Cook, R. I. (2004). Understanding sign outs: Conversation analysis reveals ICU handoff content and form, Retrieved November 7, 2005, from http://www.ctlab.org/documents/Sccm2005%20POSTER.pdf

Lardner, R. (1996). Effective Shift Handover, Retrieved January 2, 2004, from http://www.hse.gov.uk/research/otopdf/1996/oto96003.pdf

Leonard, M., Graham, S., and Bonacum, D. (2004). The human factor: The critical importance of effective teamwork and communication in providing safe care. *Quality and Safety in Health Care, 13*(Suppl 1), i85–i90.

March, J. G. (1991). Exploration and exploitation in organizational learning. *Organization Science, 2*(1), 71–87.

Matthews, A. L., Harvey, C. M., Schuster, R. J., and Durso, F. T. (2002). Emergency physician to admitting physician handovers: An exploratory study. Paper presented at the *Proceedings of the Human Factors and Ergonomics Society 46th Annual Meeting*, Baltimore, MD.

Murphy, A. G., and Wears, R. L. (2009). The medium is the message: Communication and power in sign-outs. *Annals of Emergency Medicine, 54*(3), 379–380. doi: doi:10.1016/j.annemergmed.2006.09.019

Nemeth, C., Kowalsky, J., Brandwijk, M., Kahana, M., Klock, P. A., and Cook, R. I. (2007). Between shifts: Healthcare communications in the PICU. In C. P. Nemeth (Ed.), *Improving Healthcare Team Communication: Building on Lessons from Aviation and Aerospace* (pp. 135–154). Aldershot, U.K.: Ashgate.

Nemeth, C. P., O'Connor, M., Nunnally, M., Klock, P. A., and Cook, R. I. (2005). Distributed cognition: How hand-off communication actually works. *Anesthesiology, 103*, A1289.

Patterson, E. S., and Wears, R. L. (2010a). Measurement approaches for transitions of authority and responsibility during handoffs. In E. S. Patterson and J. Miller (Eds.), *Macrocognition Metrics and Scenarios: Design and Evaluation for Real-World Teams* (pp. 137–159). Farnham, U.K.: Ashgate.

Patterson, E. S., and Wears, R. L. (2010b). Patient handoffs: Standardized and reliable measurement tools remain elusive. *Joint Commission Journal on Quality and Patient Safety, 36*(2), 52–61.

Petersen, L. A., Orav, E. J., Teich, J. M., O'Neil, A. C., and Brennan, T. A. (1998). Using a computerized sign-out program to improve continuity of inpatient care and prevent adverse events. *Joint Commission Journal on Quality Improvement, 24*(2), 77–87.

Philibert, I., and Leach, D. C. (2005). Re-framing continuity of care for this century. *Quality and Safety in Health Care, 14*(6), 394–396. doi: 10.1136/qshc.2004.011569

Propp, D. A. (2010). Improving handoffs in the emergency department. *Annals of Emergency Medicine, 56*(2), 204–205.

Randell, R., Wilson, S., Woodward, P., and Galliers, J. (2010). Beyond handover: Supporting awareness for continuous coverage. *Cognition, Technology & Work, 12*, 1–13. doi: 10.1007/s10111-010-0138-3

Salluzo, R. F., Mayer, T. A., Strauss, R. W., and Kidd, P. (1997). *Emergency Department Management: Principles and Applications*. St Louis, MO: Mosby.

Simon, H. A. (1971). Designing organizations for an information-rich world. In M. Greenberger (Ed.), *Computers, Communications and the Public Interest* (pp. 37–72). Baltimore, MD: The Johns Hopkins University Press.

Snow, V., Beck, D., Budnitz, T., Miller, D. C., Potter, J., Wears, R. L. et al. (2009). Transitions of care consensus policy statement: American College of Physicians–Society of General Internal Medicine–Society of Hospital Medicine–American Geriatrics Society–American College of Emergency Physicians–Society of Academic Emergency Medicine. *Journal of General Internal Medicine, 24*(8), 971–976. doi: 10.1007/s11606-009-0969-x

The Joint Commission. (2008). *The National Patient Safety Goals Handbook*. Chicago, IL: The Joint Commission.

Vidyarthi, A. (2004). Fumbled handoff, Retrieved December 21, 2005, from http://webmm.ahrq.gov/cases.aspx?caseID = 55

Wears, R. L., and Perry, S. J. (2010). Discourse and process analysis of shift change handoffs in emergency departments. Paper presented at the *54th Human Factors and Ergonomics Society*, San Francisco, CA.

Wears, R. L., Perry, S. J., Shapiro, M., Beach, C., and Behara, R. (2003). Shift changes among emergency physicians: Best of times, worst of times. Paper presented at the *Human Factors and Ergonomics Society 47th Annual Meeting*, Denver, CO.

Westbrook, J. I., Coiera, E., Dunsmuir, W. T. M., Brown, B. M., Kelk, N., Paoloni, R. et al. (2010). The impact of interruptions on clinical task completion. *Quality and Safety in Health Care, 19*(4), 284–289. doi: 10.1136/qshc.2009.039255

Woods, D. D., and Cook, R. I. (1999). Perspectives on human error: Hindsight biases and local rationality. In F. T. Durso, R. S. Nickerson, R. W. Schvaneveldt, S. T. Dumais, D. S. Lindsay, and M. T. H. Chi (Eds.), *Handbook of Applied Cognition* (1st edn., pp. 141–171). New York: John Wiley & Sons.

12

High-Reliability Organizations in Health Care

Vinit Desai, Peter M. Madsen, and Karlene H. Roberts

CONTENTS

12.1 HRO Processes 174
    12.1.1 Process Auditing 175
    12.1.2 Creating Effective Reward Systems 175
    12.1.3 Avoiding Quality Degradation 176
    12.1.4 Perceiving Risk Appropriately 176
    12.1.5 Effective Command and Control Systems 176
12.2 HRO Processes in Health Care Organizations 176
12.3 Process Auditing 177
12.4 Appropriate Rewards 177
12.5 Quality Review 178
12.6 Risk Awareness 179
12.7 Command and Control 179
    12.7.1 Migration of Decision Authority 179
    12.7.2 Redundancy 180
    12.7.3 Situational Awareness 180
    12.7.4 Adhering to Formal Rules and Procedures 180
    12.7.5 Training 181
12.8 Conclusions.. 181
References.. 182

High-reliability organizations (HROs) are organizations in which errors can have catastrophic consequences, but that exhibit nearly error-free operation. The term HRO is sometimes used to denote any organization that operates in a high-hazard domain (e.g., suggesting that all commercial airlines are HROs). But we reserve the term HROs for organizations that operate on a nearly error-free basis over very long periods of time, despite many opportunities for disaster to occur. Following Roberts (1990), we suggest identifying HROs by asking the question, "how many times could this organization have failed resulting in catastrophic consequences that it did not?" "If the answer is on the order of tens of thousands of times the organization is 'high reliable'." (Roberts, 1990, p. 160).

Modern societies employ many technologies and deal with many environments that are so hazardous that periodic catastrophes must be expected unless HROs are utilized. Massive manmade catastrophes are impossible for individuals to produce, because individuals lack the resources needed to generate catastrophe. Thus, manmade catastrophe cannot occur without organizations. Similarly, modern societies typically task organizations with mitigating natural and environmental hazards and with responding to natural disasters. Consequently, natural disasters also rarely occur without the central involvement of organizations. Increasing technological innovation contributes to the possibility of both natural and manmade catastrophe; because modern communications and production technologies are sufficiently complex that mere humans can easily mismanage them. As a result, many disasters could be avoided if more organizations were to become HROs.

The purpose of this chapter is to define HRO processes and discuss their application in the health care domain. HROs were first studied in process and transportation industries. But interest in translating HRO processes identified in these industries for use in health care has been growing over the past two decades. And recent concern over patent safety and medical error has exacerbated this trend. In this chapter, we review evidence for the effectiveness of HRO processes in health care organizations. We find that the use of HRO processes in health care seems to improve patient outcomes. But we also find that building and keeping an HRO in health care is an enormously difficult process. It requires continual attention to making sure certain HRO processes do not disappear.

12.1 HRO Processes

Research on HROs has been conducted since the mid-1980s when a group of scholars at the University of California at Berkeley coined the term "high-reliability organizations" and headed off to study three organizations in which errors were considered unacceptable. Those three organizations were a commercial nuclear power plant, the U.S. Federal Aviation Administration's operation of its air traffic control system, and the U.S. Navy's aircraft carrier aviation program.

Other scholars in other places have joined this research effort, and yet other scholars do similar research under slightly different rubrics. HRO research intersects with and complements safety-related work focusing on individuals (such as naturalistic decision making), and on groups and teams (such as crew resource management). For example, work on crew resource management coming from the airline industry (e.g., Helmreich and Foushee, 1993) is essentially a mechanism for reducing the authority gradient and improving communication and decision making in flight crews. But HROs are organizations, not individuals or groups. Consequently, HRO research focuses on systemic strategies for improving safety and reliability—such as organizational structure, assessment and incentive systems, leadership, and culture.

By this time, the concepts derived from HRO work are many, and it is difficult to categorize them. This is possibly because the scholars in the area have drawn concepts from both micro and macro organizational theory and have not been very clear about the crossover. Most recently some of them (see, for example, Grabowski and Roberts, 1999; Madsen and Desai, 2010; Roberts et al., 2005; Roberts and Tadmor, 2002) have moved away from focusing on single organizations, finding that errors often occur across boundaries between organizations or in other organizations, that influence and cause errors in a focal organization. By focusing narrowly on a single organization, the error and its consequences may be missed. For example, patients are regularly transferred in hospitals from emergency to acute and less acute care, from hospitals to long- or short-term care, and then to home care. Each of these junctures offers numerous possibilities for error.

Another example of this is a hypothetical petroleum company identified as an HRO. Its product provides fuel for a major aviation disaster. The aviation system (consisting of air traffic control personnel, airline flight crews, maintenance crews, and others) is likely not an HRO in this situation. Is the petroleum company? The errors occurred elsewhere, and the petroleum company likely bore no direct or even partial technical responsibility for the accident. It does bear a social relationship with the aviation system, and its fuel now blankets an accident site. By focusing on the innocent petroleum company, we miss the larger picture of unreliability.

The system level work is, as yet, too underdeveloped to discuss with any confidence. Consequently, we focus our review here on work done at the organizational level of analysis.

This research focuses on organizational processes and how they impact the cognitive processes or organizational participants by fostering and defining reliability enhancement. Different scholars have emphasized slightly different conceptualizations of the key characteristics of HROs that allow them to operate nearly error free. But the two most prominent conceptualizations—those of Weick et al. (1999) and Libuser (1993)—converge significantly. Weick et al. (1999) articulate five characteristics of HROs:

1. Preoccupation with failure—the tendency for members of HROs to remain consistently vigilant in detecting even very minor failures and near misses.

2. Reluctance to simplify interpretations—the maintenance of complex models and shared understandings of organizational and environmental functioning.

3. Sensitivity to operations—the continual focus (even by senior leadership) on the technical work carried out in the organization.

4. Commitment to resilience—the ability to improvise responses to unexpected events.

5. Under specification of structures—the ability for organizational structure to change rapidly (especially with regard to the level of centralization or decentralization used) to meet task demands.

As can be seen from this list, Weick et al.'s conceptualization of HRO characteristics emphasizes the role of cognition management—the way organizational members are able to build and maintain viable understandings of the activity system to which they belong—to HROs.

These focuses build on Weick's notions of mindfulness and sense making in organizations (Weick, 1995). Applying these concepts to HROs, Weick and Roberts (1993) developed the concept of collective mind to explain how organizations such as nuclear-powered aircraft carriers produce reliable performance. Collective mind is conceptualized as a capacity for heedful inter-relating. Variation in this capacity affects the ability of a system to maintain a working similarity between itself and a complex, dangerous environment. The more developed the capacity for mindful performance is, the greater the comprehension is, and the fewer errors that result. Other writers talk about distributed cognition and how cognitions are melded together in organizations to complete tasks and reach goals (see, for example, Hutchins, 1995; Klein, 1989).

In contrast to Weick et al.'s (1999) emphasis on cognitive strategies in HROs, Libuser's (1993) articulation of the key characteristics of HROs focuses more directly on the structures and processes employed by HROs that allow them to achieve nearly error-free performance. To identify these processes, Libuser compared the risk management practices of failing and successful commercial banks. These HRO processes are

1. Process auditing—the creation of systems for ongoing checks to spot unexpected system states and latent errors. These systems include safety drills, equipment testing, and near-miss reporting, as well as systems for following up on problems revealed in previous audits. This process is similar to Weick et al.'s (1999) emphasis on preoccupation with failure.

2. Creating effective reward systems—the use of incentive systems that reward organizational members for the behaviors required to maintain reliability, not simply for behaviors that enhance efficiency (see Kerr, 1975), including vigilance, error and near-miss reporting, and maintenance of requisite variety in organizational models.

3. Avoiding quality degradation—refusal to accept long periods of error-free operation as evidence that improvement is not needed. This process includes defining as organizational referents those organizations that are generally regarded as the standard for quality in the industry.

4. Perceiving risk appropriately—this includes systematic processes for (a) identifying potential risks to the organization and (b) devising strategies for mitigating the hazard associated with each risk and for responding effectively to each risk if it occurs.

5. Effective command and control systems—this process refers to maintaining appropriate authority structures and communication systems, including (a) migration of decision authority to people with appropriate information and expertise regardless of rank, (b) providing a redundancy of people or hardware,

(c) building communication systems that allow for a central actor to develop global situational awareness or "the big picture," (d) adhering to formal rules and procedures, and (e) providing constant training.

As can be seen, there are significant overlaps between Weick et al.'s (1999) cognitive strategies of HROs and Libuser's (1993) HRO processes. In fact, in several cases, the processes described by Libuser constitute the organizational structures, systems, and policies that allow members of HROs to develop the shared cognitive properties described by Weick et al. Consequently, in completing our review of HRO research and exploring the application of HRO to the health care domain, we will use Libuser's HRO processes as a guiding framework.

12.1.1 Process Auditing

A number of early, as well as more recent, HRO studies find evidence for the benefits of process auditing for organizational reliability. For example, wilderness firefighters analyze characteristics of their environments using complex mental models, and conduct their activities according to well-established formal procedural guidelines. These systems aid in spotting unexpected errors or problems, since these appear as inconsistencies or aberrations which depart from the firefighters' mental understandings or formal expectations (Weick, 1993). Similarly, recent work from an HRO perspective finds evidence for the importance of near-miss reporting in HROs. For example, Madsen (2010) finds that commercial airlines that take reporting and learning from near-miss events seriously improve their safety performance more rapidly than those that do not.

12.1.2 Creating Effective Reward Systems

In HROs, error avoidance and safety are as much a part of the bottom line as is productivity. Consequently, many HRO studies find evidence that effective reward systems are central to reliability enhancement. For example, in Roberts' work on U.S. aircraft carriers (Roberts, 1989; Rochlin et al., 1987; Weick and Roberts, 1993), she and various colleagues found that U.S. naval aviation uses nonpunitive reward systems which encourage error reporting and refrain from punishing those that report even if the reported error was made by the reporter. In fact, Landau and Chisholm (1995) report an incident in which an enlisted seaman working as an aircraft mechanic on an aircraft carrier reported having lost a tool on the deck during an exercise. Because a loose tool can be quite dangerous on an aircraft carrier deck (if it is sucked into an aircraft engine, for example), the exercise was immediately halted and all aloft aircraft were directed to land bases. The next day, the seaman who reported the lost tool was publically commended for his bravery in reporting his mistake.

12.1.3 Avoiding Quality Degradation

Research examines the need for HROs to avoid quality degradation. For example, Marcus and Nichols (1999) examined two nuclear plants with contrasting safety records, finding that the safer plant anticipated potential problems while the plant with a more questionable record often responded in a reactive and in some cases resilient manner. In general, organizations such as HROs that have long periods of error-free operation must continue to actively examine and make sense of their environments, as complacency following success could otherwise cause these organizations to overlook events or warning signals which precede large-scale catastrophes (Weick, 1995).

12.1.4 Perceiving Risk Appropriately

Again, research on HROs in action point to the need for processes to help HRO members correctly perceive risk. For instance, Weick (1993) analyzed the collapse of sensemaking in a wildland firefighting team preceding a major disaster, which ultimately resulted in the death of 13 firefighters. Weick argues that the deaths resulted, in part, from the firefighters' overt reliance on formal heuristics and rules which were unsuited for the conditions they faced, as they did not yield accurate information regarding environmental risks or appropriate actions. Instead, Weick suggests, HROs and other organizations that wish to operate successfully in hazardous environments must engage in a substantial degree of improvisation, changing their plans and interpretations of risks in their environment on an ongoing basis as new information arises.

12.1.5 Effective Command and Control Systems

A number of HRO studies examine the organizational structural and communication processes that contribute to reliability enhancement (e.g., Bigley and Roberts, 2001; Eisenhardt, 1993; Reason, 1997; Roberts, 1990). Eisenhardt (1993), for example, found that there is simultaneous centralization–decentralization in organizations in what she calls "high velocity environments." It is in the form of a pattern she calls "consensus with qualification," which is a two-step decision process in which everyone in the organization affected by a decision tries to reach consensus on it but when that cannot occur, the decision is made centrally. Roberts (1990) finds that HROs move back and forth between centralization and decentralization. When nothing important

is transpiring, they can afford hierarchy but as the tempo heats up and complex interactions are required, centralization breaks down and highly decentralized decision making is observed. Reason (1997) discussed flexibility and shifting structures in HROs.

Bigley and Roberts (2001) focused on three structural mechanisms that give the organization flexibility. These are structuring mechanisms to support flexibility, organizational support for constrained improvisation on the part of lower level people, and cognition management methods that, in combination, lead to exceptional organizational reliability under volatile environmental conditions. Structuring mechanisms include elaborating the structure, switching roles within it, migrating authority, and resetting the system as external conditions change.

In summary, research on HROs is now done at a variety of universities. It generally focuses on cognitive or organizational processes that contribute to reliability enhancement. Suggestions from this work are slowly turning researcher attention to the larger systems in which these processes occur, but this work is under developed. In the next section, we will review the health care literature and highlight studies of health care organizations that have adopted some of these organizational processes to enhance reliability.

12.2 HRO Processes in Health Care Organizations

Early HRO research defined its domain to include only organizations that operated complex technologies and processes under conditions of tight coupling and extreme time pressure, where failure could lead to large-scale catastrophe. This definition pushed HRO researchers toward research on organizations in the processes and transportation industries, and away from health care. Although health care organizations can certainly be classified under these criteria, initial research sought to establish HRO concepts in the context of highly visible, large-scale, industrial disasters. In fact, Roberts and Rousseau (1989) explicitly excluded health care organizations from the purview of HRO: "Hospital emergency rooms, for instance, are characterized by several [HRO] dimensions...Yet other dimensions (e.g., hypercomplexity and large number of decision makers) are largely irrelevant. Emergency rooms also seldom self-destruct" (p. 133). Some medical researchers share this view, arguing that HRO principles are not applicable to health care settings and that alternative processes are needed (e.g., Nemeth and Cook, 2007).

However, recent work by both HRO and health care scholars with an interest in enhancing patient safety has explored the similarities (rather than the differences) between the challenges faced by the organizations that had previously been studied by HRO scholars and the challenges faced by health care organizations, and found that, in many instances, they were the same or very similar (Amalberti et al., 2005; Barach and Small, 2000; Dixon and Schofer, 2006; Faraj and Xiao, 2006; Gaba, 2000; Gaba et al., 2003; Hugh, 2002; Luria et al., 2006; Reason, 2005; Shapiro and Jay, 2003). As this work introduced HRO terminology into the health care domain, it was quickly incorporated into the literatures on medical error and patient safety.

Our main purpose in this chapter is to review substantive work on the application of HRO processes to health care organizations. In conducting literature reviews for this purpose, we found several hundred instances of the terms "high-reliability organization" and "HRO" in health care journals. However, we quickly determined that the vast majority of these references used HRO terminology merely to point out that health care is a domain characterized by significant hazards to patient safety (similar to the faulty definition of an HRO as any organization that operates in a high hazard domain, mentioned previously). A much smaller number of health care studies examine substantive applications of HRO processes to health care. In the following sections we attempt to review this work (this is the first such attempt that we are aware of). In doing so, we will structure our discussion by following Libuser's (1993) typology of HRO processes.

12.3 Process Auditing

Process auditing refers to systematic checks and formal audits to inspect for problems in the "process." In health care organizations, the process is providing critical patient care, often in an environment of physiologic uncertainty and instability. Health care researchers interested in patient safety have recently focused a great deal of attention on process auditing, especially incident and near-miss reporting. Not all of this work cites HRO as a motivation for this focus, but several studies do. For example, Shapiro and Jay (2003) suggest that many doctors and nurses feel that hyper vigilance is part of the medical tradition, as evidenced by the admonition "first, do no harm." However, these authors argue that health care has begun to adopt HRO characteristics by instituting "anonymous medical error reporting systems, more open discussion regarding error, and new requirements for error disclosure" (p. 239).

Similarly, Hobgood et al. (2004) surveyed doctors, nurses, and EMTs from several hospital emergency departments. Less than half of their respondents had noticed and reported a clinical error during the year prior to the survey and less than 15% of their respondents had ever received any type of formal training in error reporting. Hobgood et al. argue that this evidence indicates that emergency departments need to implement error reporting systems and provide substantial training in using such systems before emergency departments can become HROs.

A variety of other research, while not specifically alluding to HRO characteristics, describes the role of process auditing in preventing or mitigating medical errors and adverse outcomes. For instance, Henriksen and Dayton (2006) describe the value of organizational factors that emphasize the need to understand system complexities and interconnections, and suggest that health care leaders are obligated to audit their activities on an ongoing basis and to proactively detect unexpected or concerning problems that may arise from these interactions. Clarke (2006) suggests that processes to detect and report threats to patient safety are increasingly required to ensure ongoing reliability and to prevent adverse health outcomes. Finally, Barach and Small (2000) argue that similar conditions and events precede both catastrophic failures and near misses alike, but that organizations which prevent outright failures are distinguished by the presence and effectiveness of processes to spot unexpected problems early on and to recover from them. The researchers extend work on process auditing by suggesting that systems should be in place in health care organizations to detect, report, investigate, and understand near-miss events, not only events involving outright failure.

12.4 Appropriate Rewards

Less work considers the rewards or sanctions associated with promoting reliability in health care organizations. Much of this work overlaps with, and forwards, research on process auditing, by suggesting that anonymous as well as nonpunitive error reporting systems are required to ensure that health care organizations receive early, proactive, and ongoing information regarding potential adversities (e.g., Barach and Small, 2000; Clarke, 2006). For instance, Lamb and Nagpal (2009) argue that HROs in other industries, such as commercial aviation, treat near-miss events similarly to catastrophic failures, since near misses can indicate deeper, unresolved, or systemic problems that may prompt failure in the future if left untreated. They suggest that health care organizations must establish error reporting systems that encourage near-miss reporting, and establish rewards or incentives that encourage adverse event reporting.

Madsen et al. (2006) report on a broader organizational effort to improve reliability and quality outcomes in a major pediatric intensive care unit. The unit's doctors aimed to reward caregiver behaviors associated with highly reliable outcomes, such as vigilance, attention toward patient safety, as well as initiative and innovation. However, to establish and incentivize these behaviors, unit leadership had to undergo extensive efforts to redesign the unit's structure and processes. For instance, increased training was required for nurses, respiratory technicians, and other caregivers prior to providing them with greater autonomy and discretion to analyze patient conditions and direct care and treatment in collaboration with the unit's doctors. While these efforts were successful for several years, the unit's approach ultimately failed as a result of external political pressure.

While some research separately considers reward systems within health care organizations that strive for high reliability, much of the work we uncovered, as reviewed earlier, combines the discussion of processes that facilitate reliable behaviors with incentives and reward systems aiming to achieve similar outcomes. This is perhaps intuitive, given that highly effective organizations tend to establish rewards that promote behaviors which are consistent with formal policies, procedures, or desired outcomes.

12.5 Quality Review

Libuser's (1993) typology also suggests that highly reliable organizations strive to continuously evaluate, review, and improve their quality, even following prolonged periods of satisfactory performance. This process includes benchmarking with referents; evaluating quality improvement practices at highly reliable peers; and engaging in other steps to identify, transfer, and improve best practices for quality and reliability improvement.

In this arena in particular, health care organizations face substantial internal and external pressures to review quality and avoid degradation. For instance, health care organizations increasingly strive to enhance quality outcomes, given the attention placed on these measures by the Centers for Medicare and Medicaid Services and the Joint Commission for Accreditation of Hospitals Organization (Resar, 2006). In addition, the federal government's Agency for Healthcare Research and Quality (AHRQ) aims to identify current patient

safety initiatives in place throughout the industry, to facilitate their transfer, and to work with organizations aiming to improve associated performance (Dixon and Shofer, 2006). Accordingly, Dixon and Shofer (2006) discuss interviews with senior leadership in multiple major health care organizations, finding a number of ongoing benchmarking and quality improvement efforts. Major challenges to these efforts include limited historical experience; uncertainty regarding how to appropriately transfer, tailor, and implement specific improvement efforts identified elsewhere to the organization's localized context; and issues related to the timing and measurement of quality improvement efforts and outcomes.

Resar's (2006) review identified additional challenges to health care organizations' historical quality improvement efforts. For example, Resar suggests that quality improvement practices are rarely specifically articulated in a concrete fashion that facilitates measurement and assessment. Even when practices are measured, benchmarking to mediocre or average outcomes provides caregivers with a false sense of reliability. In addition, due to wide caregiver autonomy and discretion, much quality depends upon caregivers' constant vigilance, dedication, and hard work; characteristics which some health care organizations are more successful at encouraging and rewarding than others.

These barriers are echoed and extended by Amalberti et al. (2005), who argue that health care organizations cannot merely benchmark and implement processes believed to promote quality elsewhere, as has long been suggested by HRO research in other industries. Instead, the authors suggest that considerable attention must be placed on tailoring the practices to an organization's environment and local needs, since inappropriately transferred practices may create unexpected outcomes or unanticipated errors within the health care industry's complex operating environment.

To summarize, health care organizations increasingly aim to identify, analyze, and implement quality improvement practices, especially given consistent stakeholder and regulatory pressures to engage in these efforts. However, health care research systematically suggests substantial challenges that must be overcome for these approaches to yield success. A notable gap in research, as well as in practitioner understanding, involves how quality improvement practices are transferred, translated, and implemented in target organizations. Research increasingly identifies this implementation process as a significant hurdle, but little research that we are aware of provides solutions to this challenge. Related research could stand to advance health care research on quality and reliability enhancement as well as the field of HRO research as it pertains to other industries.

12.6 Risk Awareness

Much early work on high reliability and risk awareness in the health care industry aimed to assess and document the extent of risks and hazards which exist in the field, and by extension to communicate the importance of risk awareness to patient safety and highly reliable organizational performance within the industry. For example, Gaba et al. (2003) used a standardized survey to compare perceptions of safety climate within hospital organizations with those shared within naval aviation, which is commonly regarded as a high-reliability enterprise. The authors found that 5.6% of surveyed naval aviators perceived the absence of a safety climate, compared with 17.5% of hospital personnel. The study suggested that workforce perceptions regarding the health care operating environment indicate the presence of substantial risks to patient safety, trust, collaboration, and a variety of quality-oriented outcomes. Similarly, Singer et al. (2003) surveyed a large sample of caregivers and administrators across hospitals, finding that perceptions of risk and safety differed considerably across hospitals, but also by clinical status and job class within each hospital. They interpreted their findings as a suggestion to continue research on risk awareness and perceptions of climate across individuals within health care settings, since differences of opinion or uncertainties regarding environmental hazards could complicate quality of care.

Beyond establishing the risks present in the health care environment and comparing these risks with those present in other industries, some research attempts to provide tools to aid risk awareness and assessment. Hugh's (2002) review, for instance, suggests that error detection and prevention techniques adapted from pilot training procedures to laparoscopic surgery helped to reduce error rates by providing surgeons with a more comprehensive and accurate understanding of risks and other factors in their surgical environments.

Collectively, the early research on risk in health care was associated with a trend toward recognition of patient safety, health care quality, and organizational reliability as important forms of performance within the industry, and major regulatory and industry associations increasingly advocate for careful attention toward related improvements (Resar, 2006). As a result, research on high reliability and risk awareness in health care has increasingly shifted toward identifying processes, characteristics, and factors that aid health care organizations to accurately assess, understand, and respond to environmental risks. For instance, Despins et al. (2010) use high-reliability theory to argue that patient safety is driven, to a large extent, by how accurately nurses, as well as other care providers, differentiate between signals and background noise in evaluating the risks and hazards faced by patients in specific environments. Their review of literature on patient risk and safety suggests that interventions and training can be designed to improve the accuracy of signal detection and risk analysis in health care.

12.7 Command and Control

As indicated earlier, HROs strive to establish effective command and control systems in order to maintain effective authority structures, communication systems, and to coordinate actions even during significant crises, during which the relationships between actions and outcomes may be unclear (Libuser, 1993). Establishing command and control systems requires several related steps; these include the migration of decision making authority to people with appropriate information and expertise regardless of their rank, providing redundancy of people or hardware, building systems that allow central actors to develop global situational awareness, adhering to formal rules and procedures (and, in some cases, recognizing the need to depart from these), and providing constant and ongoing training to organizational members. Each of these is discussed in turn in the following, in the context of health care organizations.

12.7.1 Migration of Decision Authority

In general, health care organizations have dramatically advanced the knowledge and practice of authority migration. Many organizations in the field are increasingly adopting formal policies or informal norms to ensure that teams of caregivers have the autonomy to direct patient care and treatment. This is especially prominent in emergency care, since uncertainties regarding workflow and extreme time sensitivity prevent caregivers from waiting for orders from the formal hierarchy of authority (Faraj and Xiao, 2006; Madsen et al., 2006). For instance, Faraj and Xiao (2006) report a case study of a major trauma center, where teams of surgeons, anesthesiologists, nurses, and other caregivers would split into smaller subgroups to accommodate new patient arrivals whenever demand in the unit spiked. These splits were accompanied by the migration of decision making authority; each patient care team was initially responsible for making basic care decisions to stabilize and treat their patient's injuries.

Madsen et al. (2006) report a long-term effort to establish formal authority migration policies in a pediatric intensive care unit. Through the unit's development and growth, nurses were placed in charge of making some patient care decisions in collaboration with attending doctors. This practice required changes in training and education policies, as the unit's leadership aimed to select and retain nurses who had gained the specialized skills, knowledge, and abilities required for their roles as primary decision makers. Several innovations in patient care arose as a result of this authority migration. For instance, in one situation, nurses and respiratory therapists decided to slightly change the mix of gases used for some children on ventilators. The new mix gave the children more energy and allowed for longer periods of play. Despite these benefits, the case also illustrates a primary challenge accompanying authority migration; organizations must ensure that caregivers with decision making authority have the appropriate training and background to accurately understand their environment and make sound treatment choices.

12.7.2 Redundancy

Redundancy in health care organizations implies that the organizations have established multiple procedures or technologies to independently arrive at similar outcomes, in order to double-check or ensure that routines and activities are conducted appropriately and to detect errors in their implementation. Carroll and Rudolph (2006) provide several examples. For example, they suggest that drug prescription errors can be avoided through a combination of overlapping strategies such as stronger controls over who can prescribe, more double checks of patient data, better training, color coding, and single dose technologies. They also suggest that a combination of procedures, including the placement of a physical mark on the appropriate site, could prevent wrong-site surgeries.

Similarly, Welch and Jensen (2007) argue that redundancy is a major component of and contributor to high reliability in health care organizations. They suggest that redundant processes help to double-check major care decisions so that errors can be detected before they escalate. They provide examples of redundancy across functions and departments within major health care settings, such as allowing a pharmaceutical unit to oversee medications administered within the emergency department and to query physicians regarding possible contraindications, or allowing a radiology department, lab, or other unit to similarly oversee relevant care decisions made elsewhere within the broader organization.

12.7.3 Situational Awareness

Situational awareness refers to the ability to quickly and effectively integrate relevant information from multiple sources in order to develop an accurate understanding of the environment, even under high uncertainty or rapid

change. Research on HROs suggests that highly reliable organizations establish structures and processes which facilitate the development of situational awareness (Libuser, 1993), and several researchers have extended this argument into the realm of health care organizations. For instance, Carroll and Edmondson (2002) suggest that a primary challenge facing leadership in health care organizations, in addition to developing a culture emphasizing safety, is to establish structures that facilitate the development of accurate mental models regarding the relationships between particular actions and their possible outcomes. Autrey and Moss (2006) argue that the lack of situational awareness can give rise to a variety of errors in health care organizations, ranging from the application of inappropriate mental models to incorrect predictions regarding future conditions. They suggest that hospitals and other health care organizations must encourage the development of situational awareness through training and education, as well as rigorous communication efforts aimed at ensuring decision makers receive relevant information in a timely manner.

Challenges also exist with respect to developing situational awareness within health care organizations. For example, Tamuz and Harrison (2006) relate an example of a hospital that struggled to balance the development of situational awareness with their provision of multiple forms of redundancy, another element of command and control systems within HROs. The hospital's pharmacy computer system routinely tended to produce a variety of redundant but irrelevant warnings, making it difficult for pharmacists to differentiate and attend to critical warning messages.

Despite the difficulties and challenges related to the development of situational awareness in practice, research on reliability in health care increasingly recognizes the importance of situational awareness in promoting quality and preventing errors (e.g., Autrey and Moss, 2006; Faraj and Xiao, 2006; Kerfoot, 2006). Generally, health care organizations as well as organizations in other fields that establish structures and processes to help members develop accurate mental models regarding conditions in their environments tend to facilitate more effective managerial decision making, since individuals within these organizations have more accurate knowledge regarding environmental risks and hazards as well as the relationships between actions and potential outcomes.

12.7.4 Adhering to Formal Rules and Procedures

A variety of research suggests that highly reliable health care organizations encourage their members to adhere to formal rules, and create systems of reward and sanction to facilitate consistent behavior. Carroll and Rudolph

(2006) suggest that formal rules comprise one component of a broader organizational system, also including professional culture, hierarchy, and leadership, which helps to control members' actions and ensure consistent and highly reliable organizational performance. Carroll and Edmondson (2002) argue that rules and formal standards are essential to ensuring health care quality and reliability, but that health care practitioners often resist such formalities as infringements on their professional standing and autonomy. These authors suggest that one solution to this challenge involves the nature of formalization; rather than dictating how surgeons and other caregivers should operate or conduct their practice, effective rule systems should govern work practices in ways that allow discretion but also maximize patient health and well-being.

In contrast, research also suggests that organizational reliability depends on members' abilities to deviate or depart from formal rules when they are inappropriate in a specific environment (Weick, 1993). Faraj and Xiao (2006) discuss the importance of breaking protocol when medically necessary, and provide the example of an anesthesiologist who removed a stabilizing neck brace against hospital policy because the neck brace was interfering with airway management. The decision was reached after careful consideration of the costs and benefits of various approaches. The situation underscores the fact that rule adherence is only one component of organizational reliability; when combined with situational awareness, training, and other components, it provides a comprehensive system that can help the organization's members accurately interpret various situations and evaluate the consequences of either following or departing from formal guidelines.

12.7.5 Training

Ongoing training regarding quality, communication, teamwork, and other elements of reliability is a central component of health care organizations, given that training and continuing education are relatively formalized and institutionalized aspects of the health care environment. As such, organizations within the field increasingly tend to combine other quality improvement efforts with ongoing training and education. For example, Autrey and Moss (2006) argue that redundancy and other elements of highly reliable systems have limitations, and must be combined with efforts to train members to identify errors and enhance competencies related to optimum team performance. The authors suggest that training should emphasize skills needed to cope with disasters, such as communication, preparation, and planning.

Wilson et al. (2005) argue that training in highly reliable health care organizations should emphasize several distinct competencies. First, training should familiarize members with each others' roles and responsibilities, since crises often involve shared efforts and substantial coordination across roles. Second, training should enhance situational awareness by providing members with experience noticing differences between various crisis and noncrisis situations, since the boundary between crises and routine operations can be difficult to discern in practice. Training should also provide members with substantial practice coordinating their efforts, and provide members with practice responding to and correcting errors as they unfold.

Beyond the content of training, research on training and reliability in health care organizations also provides insight regarding various training techniques and tools. For instance, Wilson et al. (2005) discuss the benefits of scenario-based training, and Autrey and Moss (2006) also argue that training should be conducted with realistic and challenging scenarios to help members practice developing situational awareness and coordinating actions to respond to crises.

Gaba (2004) discusses the development and future of patient simulation in training teams of caregivers such as anesthesiologists, surgeons, and technicians, arguing that simulation training provides cost benefits, realism, and avoids actual harm that could arise from training on live patients.

Despite growing research and the increasingly central role of training in highly reliable health care organizations, some authors suggest that more can be done to expand and revise training models within the field. For example, McKeon et al. (2006) suggest that training should emphasize patient safety, collaboration, and provide members with experience managing unexpected events. Despite efforts in this direction, they suggest that many health care organizations suffer because professional boundaries challenge cross-functional collaboration and teamwork. As a result, they argue that the distance between professions within health centers continues to grow, even though crisis response and reliability enhancement efforts require that this distance instead shrinks over time.

12.8 Conclusions

Health care providers are often faced with unusual, unknown situations in which trained intuition and previous experience with developing situational awareness and sensemaking can guide future experiences. There is a growing amount of research about how to achieve highly reliable organizational processes, most of it in industries other than health care. We do not yet know how generalizable that work is to health care settings. While some very good organizational research has been conducted in the health care domain, more of it is needed. Known reliability processes uncovered in other industries need to be assessed in health care and new ones need to be discovered.

Health care professionals want the best outcomes for their patients. We ask these people to examine the reliability enhancing processes, discussed here, because there is empirical evidence that these processes are associated with increased reliability and safety. Several questions would aid health care practitioners as they seek to accomplish these goals. First, are ongoing processes audited to assess whether they are working as they should? Are new processes considered? Are the rewards in place successful at reinforcing desired behavior? Are we rewarding behavior A while hoping for behavior B (Kerr, 1975)? Is quality of constant concern? Is the right medical staff in place? Does hospital administration perceived and respond to risk appropriately? This question is especially important, in light of empirical evidence that found higher level employees perceived their organizations as safer than lower level employees, in health care and other settings (Gaba et al., 2003). Finally, health care managers should determine whether appropriate command and control processes are in place.

We do not know which, if any, of these processes can be overlooked in organizations without decreasing reliability. We do know that organizations which include them all are more reliable than those that do not. We also know that HROs require constant nurturance and are highly susceptible to dismantling. Without constant effort, organizations will lose their resilience and ability to respond to changing and uncertain events. All of these efforts require resources and managers need to think about the potential consequences to reliability of cutbacks, mergers, and other strategies designed to save money. Business models and appropriate health care models require effort to integrate.

Finally, we urge health care professionals to think about the interdependencies of units in their own organizations and of their organization with respect to other organizations. The larger story of HROs cannot be told until we know more about how these interdependencies should operate and how they actually do operate (Heath and Staudenmayer, 2000).

References

Amalberti, R., Auroy, Y, Berwick, D., and Barach, P. (2005). Five system barriers to achieving ultrasafe health care. *Annals of Internal Medicine*, 142, 756–764.

Autrey, A.P. and Moss, J. (2006). High-reliability teams and situation awareness: Implementing a hospital emergency incident command system. *Journal of Nursing Administration*, 36, 67–72.

Barach, P. and Small, S.D. (2000). Reporting and preventing medical mishaps: Lessons from non-medical near miss. *British Medical Journal*, 320, 759–763.

Bigley, G.A. and Roberts, K.H. (2001). Structuring temporary systems for high reliability. *Academy of Management Journal*, 44, 1281–1300.

Carroll, J.S. and Edmondson, A.C. (2002). Leading organizational learning in health care. *Quality and Safety in Health Care*, 11, 51–56.

Carroll, J.S. and Rudolph, J. (2006). Design of high reliability organizations in health care. *Quality and Safety in Health Care*, 15, i4–i9.

Clarke, J.R. (2006). How a system for reporting medical errors can and cannot improve patient safety. *The American Surgeon*, 72, 1088–1091.

Despins, L.A., Scott-Cawiezell, J., and Rouder, J.N. (2010). Detection of patient risk by nurses: A theoretical framework. *Journal of Advanced Nursing*, 66, 465–474.

Dixon, N.M. and Shofer, M. (2006). Struggling to invent high-reliability organizations in health care settings: Insights from the Field. *Health Services Research*, 41, 1618–1632.

Eisenhardt, K. (1993). High reliability organizations meet high velocity environments: Common dilemmas in nuclear power plants. In K.H. Roberts (Ed.), *New Challenges to Understanding Organizations*. New York: Macmillan, pp. 33–54.

Faraj, S. and Xiao, Y. (2006). Coordination in fast-response organizations. *Management Science*, 52, 1155–1169.

Gaba, D.M. (2000). Structural and organizational issues in patient safety: A comparison of health care to other high-hazard industries. *California Management Review*, 43, 83–102.

Gaba, D. (2004). The future vision of simulation in health care. *Quality and Safety in Health Care*, 13, i2–i10.

Gaba, D.C., Singer, S.J., Sinaiko, A.D., Bowen, J.D., and Ciavarelli, A. (2003). Differences in safety climate between hospital personnel and naval aviators. *Human Factors*, 45, 173–185.

Grabowski, M. and Roberts, K.H. (1999). Risk mitigation in virtual organizations. *Organization Science*, 10, 704–721.

Heath, C. and Staudenmayer, N. (2000). Coordination neglect: How lay theories of organizing complicate coordination in organizations. In B.M. Staw and R.I. Sutton (Eds.), *Research in Organizational Behavior*, Vol. 22. New York: Elsevier, pp. 153–191.

Helmreich, R.L. and Foushee, H.C. (1993). Why crew resource management?: The history and status of human factors training programs in aviation. In I.E. Weiner, B. Kanki, and R.L. Helmreich (Eds.), *Cockpit Resource Management*. New York: Academic Press.

Henriksen, K. and Dayton, E. (2006). Organizational silence and hidden threats to patient safety. *Health Services Research*, 41, 1539–1554.

Hobgood, C, Hevia, A., and Hinchey, P. (2004). Profiles in patient safety: When an error occurs. *Academic Emergency Medicine*, 11, 766–770.

Hugh, T.B. (2002). New strategies to prevent laparoscopic bile duct injury—Surgeons can learn from pilots. *Surgery*, 132, 826–835.

Hutchins, E. (1995). *Cognition in the Wild*. Cambridge, MA: MIT Press.

Kerfoot, K. (2006). Reliability between nurse managers: The key to the high-reliability organization. *Nursing Economics*, 24, 274–275.

Kerr, S. (1975). On the folly of rewarding A while hoping for B. *Academy of Management Journal*, 18, 769–783.

Klein, G. (1989). Recognition-primed decision. *Advances in Man–Machine Systems Research*, 5, 47–92.

Lamb, B. and Nagpal, K. (2009). Patient safety: Importance of near misses. *British Medical Journal*, 339, b3032.

Landau, M. and Chisholm, D. (1995). The arrogance of failure: Notes on failure-avoidance management. *Journal of Contingencies and Crisis Management*, 2, 221–227.

Libuser, C. (1993). Organizational structure and risk mitigation. Unpublished doctoral dissertation. Los Angeles, CA: University of California.

Luria, J.W., Muething, P.J., and Schoettker, U.R. (2006). Reliability science and patient safety. *Pediatric Clinics of North America*, 53, 1121–1133.

Madsen, P.M. (2010). The strategy of learning from small and large losses: Organizational learning from near-misses and accidents in commercial aviation. *Marriott School of Management Working Paper*, Brigham Young University, Provo, Utah.

Madsen, P.M. and Desai, V. (2010). Failing to learn? The effects of failure and success on organizational learning in the global orbital launch vehicle industry. *Academy of Management Journal,* 53(3), 451–476.

Madsen, P.M., Desai, V., Roberts, K.H., and Wong, D. (2006). Designing for high reliability: The birth and evolution of a pediatric intensive care unit. *Organization Science,* 17, 239–248.

Marcus, A.A. and Nichols, M.L. (1999). On the edge: Heeding the warnings of unusual events. *Organization Science*, 10(4), 482–499.

McKeon, L.M., Oswaks, J.D., and Cunningham, P.D. (2006). Safeguarding patients—Complexity science, high reliability organizations, and implications for team training in healthcare. *Clinical Nurse Specialist*, 20, 298–304.

Nemeth, C. and Cook, R. (2007). Reliability versus resilience: What does healthcare need? In *Human Factors and Ergonomics Society Annual Meeting Proceedings*, Oct. 1–5, 2007; Baltimore, MD, pp. 621–625.

Reason, J. (1997). *Managing the Risks of Organizational Accidents*. Aldershot, U.K.: Ashgate.

Reason, J. (2005). Safety in the operating theatre—Part 2: Human error and organisational failure. *Quality and Safety in Healthcare*, 14, 56–60.

Resar, R.K. (2006). Making noncatastrophic health care processes reliable: Learning to walk before running in creating high-reliability organizations. *Health Services Research*, 41, 1677–1689.

Roberts, K.H. (1989). New challenges to organizational research: High reliability organizations. *Industrial Crisis Quarterly*, 3, 111–125.

Roberts, K.H. (1990). Some characteristics of one type of high reliability organization. *Organization Science*, 1, 160–176.

Roberts, K.H., Madsen, P.M., and Desai, V.M. (2005). The space between in space transportation: A relational analysis of the failure of STS 107. In M. Farjoun and W. Starbuck (Eds.). *Organization at the Limit: NASA and the Columbia Disaster*. Oxford, U.K.: Blackwell.

Roberts, K.H. and Rousseau, D.C. (1989). Research in nearly failure free high reliability organizations: Having the bubble. *IEEE Transactions on Engineering Management*, 36, 132–139.

Roberts, K.H. and Tadmor, C.T. (2002). Lessons learned from non-medical industries: The tragedy of the USS Greeneville. *Quality and Safety in Health Care*, 11, 355–357.

Rochlin, G.I., T.R. La Porte, and K.H. Roberts. (1987). the self-designing high-reliability organization: Aircraft carrier flight operations at sea, *Naval War College Review*, 40, 76–90.

Shapiro, M. and Jay, G. (2003). High reliability organizational change for hospitals: Translating tenets for medical professionals. *Quality and Safety in Healthcare*, 12, 238–239.

Singer, S.J., Gaba, D.M., Geppert, J.J., Sinaiko, A.D., Howard, S.K., and Park, K.C. (2003). The culture of safety: Results of an organization-wide survey in 15 California hospitals. *Quality and Safety in Healthcare*, 12, 112–118.

Tamuz, M. and Harrison, M. (2006). Improving patient safety in hospitals: Contributions of high-reliability theory and normal accident theory. *Health Services Research*, 41, 1654–1676.

Weick, K.E. (1995). *Sensemaking in Organizations*. Thousand Oaks, CA: Sage.

Weick, K.E. (1993). The collapse of sensemaking in organizations: The Mann Gulch disaster. *Administrative Science Quarterly*, 38, 628–652.

Weick, K.E. and Roberts, K.H. (1993). Collective mind and organizational reliability: The case of flight operations on an aircraft carrier deck. *Administrative Science Quarterly*, 38, 357–381.

Weick, K.E., Sutcliffe, K.M., and Obstfeld, D. (1999). Organizing for high reliability: Processes of collective mindfulness. In R. Sutton and B. Staw (Eds.), *Research in Organizational Behavior*, 81–124. Greenwich, CT: JAI.

Welch, S. and Jensen, K. (2007). The concept of reliability in emergency medicine. *American Journal of Medical Quality*, 22, 50.

Wilson, K.A., Burke, C.S., Priest, H.A., and Salas, E. (2005). Promoting health care safety through training high reliability teams. *Quality and Safety in Health Care*, 14, 303–309.

13

The Relation between Teamwork and Patient Safety

David P. Baker, Eduardo Salas, James B. Battles, and Heidi B. King

CONTENTS

13.1 Introduction 185
13.2 Background 185
13.3 Structure of This Chapter 186
13.4 Teamwork 186
    13.4.1 What Is a "Team"? 186
    13.4.2 What Is Team Performance? 186
    13.4.3 Training Teams 188
    13.4.4 Summary 188
13.5 Team Training in Health Care 189
    13.5.1 Anesthesia Crisis Resource Management Program 189
    13.5.2 MedTeams™ Purpose and Strategy 190
    13.5.3 TeamSTEPPS® 190
    13.5.4 Does Team Training Work in Health Care? 193
    13.5.5 Summary 194
13.6 Future Directions for Research 194
Acknowledgments 196
References 196

13.1 Introduction

This chapter reviews the empirical evidence concerning the relation between teamwork and patient safety. Since its original publication, there has been significant increase in understanding the relation between improved teamwork and improved care. New empirical tests have been conducted and new methodologies have been proposed. The available evidence suggests that training teams of health-care providers is a pragmatic, effective strategy for enhancing patient safety by reducing medical errors.

13.2 Background

In 1999, the Institute of Medicine (IOM) published *To Err is Human: Building a Safer Health System*, a frightening indictment of the inadequate safety that the United States medical establishment too often provides its patients (Kohn et al., 1999). Extrapolating from data gathered through the Harvard Medical Practice Study (HMPS) and the Utah–Colorado Medical Practice Study (UCMPS) (Studdert et al., 2002), the IOM report concluded that medical errors cause between 44,000 and 98,000 deaths annually (Kohn et al., 1999).

The report also noted that medical errors are financially costly. The IOM estimated that, among U.S. hospital inpatients, *medication* errors alone cost approximately 2 billion dollars annually. Besides their direct costs, errors result in opportunities lost, given that funds spent in correcting mistakes cannot be used for other purposes, as well as in higher insurance premiums and copayments. In addition, due to their effect on diminished employee productivity, decreased school attendance, and a lower state of public health, such errors exact a price from the society at large. Specifically, the IOM estimated that the total indirect cost of medical errors that result in patient harm lies between 17 and 29 billion dollars annually.

Key to the present chapter's orientation toward teamwork, the IOM noted that the majority of medical errors result from *system* failures, rather than from individual providers' substandard performance. Thus, in conjunction with its drive to implement organizational safety systems by delivering safe practices (Tier 4), the IOM recommended establishing interdisciplinary team training programs (Kohn et al., 1999).

The primary responsibility for conducting and supporting research to address the IOM's recommendations currently rests with the Agency for Healthcare Research and Quality (AHRQ). This responsibility encompasses three broad areas: (1) identifying the causes of errors and injuries in health-care delivery; (2) developing, demonstrating, and evaluating error-reduction and patient-protection strategies; and (3) distributing effective strategies throughout the U.S. health-care community (Agency for Healthcare Research and Quality, 2000).

One of AHRQ's initial efforts to address the patient safety crisis was to commission a review of the existing evidence base for different safe patient practices. Evidence Report 43 entitled *Making Health Care Safer: A Critical Analysis of Patient Safety Practices* presents existing data on practices viewed as having the potential to improve patient safety. Within this report, Pizzi et al. (2001) identified Crew Resource Management (CRM)—a team training approach—as a strategy that has tremendous potential to improve patient safety based on its success in aviation.

Since this initial report, AHRQ, in collaboration with the Department of Defense (DoD), has led the way when it comes to improving teamwork in health care. AHRQ funded a more extensive literature review of the relation of teamwork and patient safety (Baker et al., 2003a), collaborated with DoD on an evaluation of existing DoD medical team training programs (Baker et al., 2003b), funded the development of Team Strategies and Tools to Enhance Performance and Patient Safety (TeamSTEPPS®), a team training program for health professionals, and launched the national implementation of the TeamSTEPPS toolkit (King et al., 2008). The results of these efforts, as documented throughout this chapter, have led to numerous and growing efforts to improve teamwork in the delivery of care.

13.3 Structure of This Chapter

This chapter reviews the evidence concerning the extent to which team training improves patient safety outcomes. This chapter provides the most comprehensive, current review of the evidence base that supports the importance of teamwork and presents a compelling argument for its relation to patient safety. In the next section, we define the key characteristics of a team and discuss the principles that underlie successful teamwork. Next, we introduce current trends and issues in medical team training as well as review the available evidence of

its effectiveness in health care. Finally, we offer a set of conclusions and recommendations for future research.

13.4 Teamwork

13.4.1 What Is a "Team"?

It goes without saying that teamwork is critical to effective health care. Small groups of individuals work together in intensive care units (ICUs), operating rooms (ORs), labor and delivery (L&D) wards, emergency departments (EDs), and family-medicine practices. Physicians, nurses, pharmacists, technicians, and other health professionals must coordinate their activities to make safe and efficient patient care a priority. Katzenbach and Smith (1993) describe a team as a small number of people with complementary skills who are committed to a common purpose, performance goals, and approach for which they are mutually accountable.

To identify the key features of a team for the purposes of this chapter, we reviewed several oft-cited definitions, as well as other relevant literature (Dyer, 1984; Guzzo and Shea, 1992; Mohrman et al., 1995; Salas et al., 1992). The definition we adopted for this discussion comprises the following five characteristics: (1) teams consist of a minimum of two or more individuals; (2) team members are assigned specific roles, perform specific tasks, and interact or coordinate to achieve a common goal or outcome (Dyer, 1984; Salas et al., 1992); (3) teams make decisions (Orasanu and Salas, 1993); (4) teams have specialized knowledge and skills and often work under conditions of high workload (Cannon-Bowers et al., 1995); and (5) teams differ from small groups because teams embody the coordination that results from *task interdependency*; that is, teamwork characteristically requires team members to adjust to one another other, either sequentially or simultaneously, to achieve team goals (Brannick et al., 1997).

13.4.2 What Is Team Performance?

Teamwork has traditionally been described in terms of classical systems theory, which posits that team *inputs*, team *processes*, and team *outputs* are arrayed over time. Team *inputs* include the characteristics of the task to be performed, the elements of the context in which work occurs, and the attitudes team members bring to a team situation. Team *process* constitutes the interaction and coordination that are required among team members if the team is to achieve its specific goals. Team *outputs* consist of the products that result from team performance (Hackman, 1987; Ilgen, 1999; McGrath, 1984). Thus, teamwork occurs in the process phase during which team members interact and work together to produce team outputs. Teamwork does not require team members to work together permanently; it is sustained by a shared set of teamwork skills, not by permanent assignments that carry over from day-to-day. Finally, installing a team structure in an organization does not automatically result in effective teamwork.

When it comes to dissecting team process, researchers have distinguished two components: taskwork and teamwork (Morgan et al., 1986). Morgan et al. (1994) describe taskwork as the individual responsibilities that team members have that involve understanding the nature of the task, how to interact with equipment, and how to follow proper policies and procedures. Taskwork activities have also been referred to as operational or technical skills and are representative of "what" teams do (Davis et al., 1985). In health care, taskwork includes such activities as taking a patient's history, taking an x-ray, or inserting a central venous line.

Much of the research that has investigated the influence of taskwork on team performance was conducted in the 1970s. During this time, several studies found that teams comprising individuals with higher levels of task proficiency, abilities, and skills outperformed teams that were made up of individuals that were less task proficient and had lower abilities (Kabanoff and O'Brien, 1979). Specifically, researchers reported that the average skill of team members was positively related to both the speed and accuracy in which teams performed their tasks (Kabanoff and O'Brien, 1979).

While taskwork represents the specific steps and strategies associated with the requirements of a particular job, teamwork refers to the behaviors team members use to coordinate their actions or tasks (Cannon-Bowers et al., 1995; Smith-Jentsch et al., 1998a). Thus, teamwork describes "how" teams coordinate work that comprises interdependent tasks. For example, in studying emergency medicine teams, we might be interested in how priorities are established, as well as how team members monitor each other's performance to ensure that tasks are accomplished correctly.

Historically, researchers have sought to identify generic skills that define the behaviors that comprise teamwork. Cannon-Bowers et al. (1995) conducted a review of this literature and found over 130 labels used to describe different teamwork skills. Relying upon this information, these researchers defined clusters of team knowledge, skill, and attitude (KSA) competencies that were hypothesized to be essential for successful teamwork. Cannon-Bowers et al. also suggested that team competency requirements vary as a function of the task performed (i.e., taskwork) and the nature of the team performing that task. For example, a team comprising the same set of teammates performing the same tasks was viewed to require a different set of team KSAs than

a team that had variable membership and performed a wide variety of tasks. Teams that perform a wide variety of tasks, with varying teammates, like those found in health care, were viewed to require more generic team knowledge, attitudes, and skills that are transportable from one team to the next.

More recently, Salas et al. (2005) questioned whether or not there was a "Big Five" set of teamwork competencies; a smaller set of core KSAs that apply to all teams. In reviewing the literature, Salas et al. proposed that all teams require team leadership, mutual performance monitoring, backup behavior, adaptability, and team orientation. In addition, three support mechanisms were identified: mutual trust, closed-loop communication, and shared mental models. Together, these processes are proposed to yield improved team effectiveness and efficiency regardless of type of task performed or the stability of team membership.

Recently, health care has adopted a model of teamwork that is based on the earlier Salas et al. (2005) review. Alonso et al. (2006) identified four skills that they argued were trainable and important for all health-care teams: team leadership, mutual support, situation monitoring, and communication. All of these skills when learned and honed by team members result in positive performance, knowledge, and attitudinal outcomes. For example, in the Alonso et al. model, improved team knowledge in the form of a shared mental model was viewed to result from effective situation monitoring and communication. Table 13.1 presents each skill and its definition.

Amodeo et al. (2011) used confirmatory factor analysis to test the validity of the Salas et al. (2005) model in comparison to the Alonso et al. (2006) model, and a third model of teamwork comprising four core skills

TABLE 13.1

Teamwork Skill Competencies and Definitions

Competency	Definition
Situation monitoring	Tracking fellow team members' performance to ensure that the work is running as expected and that proper procedures are followed
Mutual support	Providing feedback and coaching to improve performance or when a lapse is detected; assisting teammate in performing a task; completing a task for the team member when an overload is detected
Leadership	Ability to direct/coordinate team members, assess team performance, allocate tasks, motivate subordinates, plan/organize, and maintain a positive team environment
Communication	The initiation of a message by the sender, the receipt and acknowledgment of the message by the receiver, and the verification of the message by the initial sender

proposed by Smith-Jentsch et al. (1998a). Data were collected from 152 health-care workers (physicians [15.2%], registered nurses [21.9%], and technicians [26.5%]) using the TeamSTEPPS Teamwork Perceptions Questionnaire (T-TPQ, Baker et al., 2010a). The T-TPQ was originally designed to align with the Alonso et al. framework, so items from the T-TPQ were remapped to both the Salas et al. and Smith-Jentsch et al. models of teamwork by individuals highly knowledgeable about teamwork. While all the models yielded good fit indices, the Alonso et al. model produced the best fit. This study is significant because it provides one of the few empirical tests of competing theories of teamwork.

In sum, a validated model of teamwork in health care is important because it defines what teams do, how they do it, and what factors of team performance are related to patient safety. For example, the Alonso et al. (2006) model calls out four critical skills: leadership, situation monitoring, mutual support, and communication. In most, if not all clinical settings, leadership is important because the medical team leader, usually the physician, defines the care plan and updates the team regarding progress regarding the plan. This planning and updating can take place during a formal team briefing, during clinical rounding or by updating a patient's record. Situation monitoring is important for all team members as team members have different responsibilities. For example, the L&D nurse who monitors a laboring mother's fetal heart rate will provide regular updates to the floor charge nurse and attending physician. Mutual support may or may not come into play depending on if team members need assistance—for example, the L&D nurse who has to leave a patient to participate as a scrub nurse for a C-section may be backed up by the floor charge nurse while the section is performed. Finally, communication is the glue that holds all of this together. Whether verbal or nonverbal, instructions and updates are distributed throughout the team so that a shared understanding of how to treat the patient is maintained. This is a complex process, and when breakdowns occur, patient safety can quickly be compromised.

13.4.3 Training Teams

Team training can be defined as applying a set of instructional strategies, that rely on well-tested tools (e.g., simulators, lectures, videos), to specific team competencies (Salas et al., 1999, 2000; Salas and Cannon-Bowers, 2000). Effective team training reflects general principles of learning theory, presents information about requisite team behaviors, affords team members the opportunity to practice the skills they are learning, and provides remedial feedback.

A great deal of research has been devoted to the most effective strategies and techniques for training specific

team KSA competencies. A comprehensive review of this research has presented an extensive collection of principles and guidelines concerning the design and delivery of team training. For example, guidelines exist for assertiveness training (Smith-Jentsch et al., 1996), cross-training (Volpe et al., 1996), stress management training (Driskell and Johnston, 1998), and team self-correction (Smith-Jentsch et al., 1998b).

In recent years, Salas et al. have engaged in a series of meta-analyses to identify the true efficacy of team training. In the most comprehensive analysis, Salas et al. (2008) reviewed the findings from 48 studies that yielded 84 effect sizes. Across these investigations, Salas et al. found that team training improves team process by upward of 20% and team outcomes by 14.4%. Assuming the IOM statistics are accurate (98,000 lives and 17 billion dollars per year), even a 5% performance improvement would yield approximate 5,000 lives saved and over 800 million dollars. Therefore, the potential value of team training for improving team performance in the delivery of quality care cannot be dismissed.

It is important to note that successful team training programs constitute more than developing team members' KSAs. For example, because organizational factors outside the training program itself affect the program's success, conducting a needs analysis prior to designing a training intervention is essential to determining the best delivery method or instructional strategy. In addition, training developers should take advantage of the increased practice opportunities provided by certain training tools, such as advance organizers (e.g., outlines, diagrams, graphic organizers), preparatory information, pre-practice briefs, attentional advice, goal orientation, and metacognitive strategies (Cannon-Bowers and Salas, 1998). In health care, the development and use of patient simulators has been critical in allowing health-care teams to practice and refine team skills without presenting additional risks to patients (see e.g., Beaubien and Baker [2004], and other chapters in this book). Finally, AHRQ has developed and validated a series of safety culture surveys so that organizations can assess the extent to which their culture values a team approach.

13.4.4 Summary

The preceding section discussed the elements that typify effective teamwork and effective team training. Similar, yet differing models have been proposed and research has sought to develop a common framework. In health care, Alonso et al. (2006) model that advocates four core skills has been tested and validated (Amodeo et al., 2011). This model serves as the basis for the AHRQ TeamSTEPPS program, which was released as a public domain resource in 2006 and has now been implemented at hundreds of health-care institutions across the United States as well as abroad in countries like Japan, Australia, and Taiwan.

New meta-analytic findings have confirmed the efficacy of team training and its impact on team performance. Team training works and this conclusion cannot be disputed. Nonetheless, team training does not occur in a vacuum and many organizational factors must be considered to make training successful. AHRQ has been instrumental in developing and releasing tools to address the complexity of making a team training program successful.

13.5 Team Training in Health Care

Medical team training programs began with the introduction of Anesthesia Crisis Resource Management (ACRM) training at Stanford University School of Medicine and at the Anesthesiology Service at the Palo Alto Veteran Affairs Medical Center (Howard et al., 1992). AHRQ's 2001 review of in-place patient safety practices critiqued the ACRM model, citing it as high in impact but low in evidence supporting its effectiveness (Pizzi et al., 2001). The DoD was also instrumental in using team training to improve health care. Specifically, MedTeams (Morey et al., 2003) was implemented in a number of Army and Navy hospitals in the years 2000–2006. Finally, and most significantly, AHRQ and DoD collaborated extensively, beginning in 2003, to produce and release TeamSTEPPS in 2006, which is now the national standard for training teamwork in health care.

13.5.1 Anesthesia Crisis Resource Management Program

Developed by David Gaba and his colleagues at Stanford University and the Palo Alto Veteran Affairs (VA) Medical Center, ACRM is designed to help anesthesiologists effectively manage crises by working in multidisciplinary teams that include physicians, nurses, technicians, and other medical professionals (Gaba, 1998; Gaba et al., 2001; Howard et al., 1992). To facilitate this goal, ACRM training provides trainees with critical incident case studies to review (Davies, 2001). In addition, ACRM provides training in technical skill and in team KSAs. Training in the selected teamwork skills is intended to enable trainees to learn from adverse clinical occurrences, and to work more effectively with different leadership, followership, and communication styles (Gaba et al., 2001).

ACRM training takes place in a simulated OR, after completing the reading assignments that precede each

module. The simulated OR includes actual monitoring equipment, a full-patient simulator, a video station for recording the team's performance, and a debriefing room that is equipped with a variety of audiovisual equipment. The full-patient simulator incorporates a series of complex mathematical models and pneumatic devices to simulate a patient's breathing, pulses, heart and lung sounds, exhaled CO_2, thumb twitches, and other physiological reactions (Gaba et al., 2001; Murray and Schneider, 1997).

The ACRM curriculum comprises 3 full separate days of simulation training, over 3 years of anesthesiology training. Day 1 provides an introduction to ACRM principles and skills. Day 2 provides a refresher on these skills and analyzes clinical events from the perspective of the clinician's technical and teamwork skills and from the perspective of the organization as a larger system. Day 3 emphasizes leadership training, debriefing skills, and adhering to the procedures established to deal with adverse clinical events. Each training module comprises a similar structure: preassigned readings, course introduction and review of materials, familiarization with the simulator, case study analysis and videotape reviews, and 6 hours of participating in simulator scenarios, followed by an instructor-led debriefing and a postcourse data collection. Each scripted training scenario is approximately 45 min long; each debriefing session lasts about 40 min (Gaba et al., 2001).

Several instructors are required to run the ACRM training scenarios. They might include an OR nurse who role-plays the circulating nurse and an anesthesiologist instructor who role-plays the operating surgeon. In addition, a director monitors and records the simulation from another room, communicating with the instructors via two-way radios. Throughout the simulation, trainees rotate through various roles, such as "first responder," "scrub technician," and "observer" (Gaba et al., 2001).

ACRM training, complete with yearly refresher training, is currently used at several major teaching institutions in the United States and around the world (Australia, Israel, Denmark). At some centers, ACRM training is offered for experienced practitioners as well as for trainees. Moreover, some malpractice insurers (i.e., Harvard Risk Management Foundation) have lowered their rate structure for ACRM-trained anesthesiologists (Gaba et al., 2001).

An ACRM evaluation typically assesses a variety of process-oriented criteria. *Teamwork performance* is typically assessed using behavioral markers of the 10 teamwork skills (Gaba et al., 1998). One measure of these teamwork behaviors comprises a checklist, analogous to the Line/Line Operational Simulation (LOS) Checklist used in CRM programs (Helmreich et al., 1995). Using a five-point rating scale, trained raters evaluate team performance on each dimension (Gaba et al., 1998). Measures of inter-rater agreement exhibited r_{wg} values (James et al., 1984) ranging between .60 and .93 (Gaba et al., 1998); an r_{wg} of .70 is considered sufficiently high to reflect a satisfactory degree of agreement among raters.

Most of the thousands of participants who have undergone ACRM training evaluate it favorably, even the "death scenario," which is specifically designed to assess how trainees handle losing a patient; these positive responses generally last for up to 6 months after training (Gaba et al., 2001). Furthermore, recent research suggests that participation in ACRM training also increases trainees' self-efficacy and decreases their self-reported anxiety (Tays, 2000).

13.5.2 MedTeams™ Purpose and Strategy

The primary purpose of MedTeams is to reduce medical errors through interdisciplinary teamwork. MedTeams was initially developed for the ED on the premise that most errors result from breakdowns in systems-level defenses that occur over time (Simon et al., 2000). According to the MedTeams ED curriculum, each team member has a vested interest in maintaining patient safety and is expected to take an assertive role in breaking the error chain. MedTeams defines a core ED team as a group of 3–10 (average = 6) medical personnel who work interdependently during a shift and who have been trained to use specific teamwork behaviors to coordinate their clinical interactions. Each core team includes at least one physician and one nurse. A coordinating team that assigns new patients to the core teams and provides additional resources as necessary manages several core teams.

MedTeams training was developed from an evaluation-driven course design. Based on needs analysis data, five critical dimensions were identified. Then 48 specific, observable behaviors were linked to these dimensions and Behaviorally Anchored Rating Scales (BARS) (Smith and Kendall, 1963) were constructed. Finally, to establish its content validity, the MedTeams curriculum was reviewed and refined during three 5-day expert panel sessions that included ED physicians and nurses from 12 hospitals of various sizes (Simon et al., 1998). Expert panel review and modification of the curriculum has been used to create L&D and OR versions of MedTeams.

MedTeams uses a train-the-trainer approach to implement the training. Individuals, designated by their facility, receive comprehensive training on how to teach MedTeams and are certified as MedTeams instructors. The course consists of an 8 hour block of classroom instruction that contains an introduction module, five learning modules, and an integration unit. After completing the classroom training, each team member participates in a 4 hour practicum that involves practicing teamwork behaviors and receiving feedback from a trained instructor. Coaching, mentoring, and review sessions are also provided during regular work shifts (Simon et al., 1998).

MedTeams training has been evaluated using a quasi-experimental research design (Morey et al., 2002, 2003) in which a variety of process factors (e.g., quantity of teamwork behaviors) and enabling factors (e.g., attitudes toward teamwork, staff burnout) were measured over a 1 year period. An analysis of these data indicated a positive effect of training on *outcome criteria* (e.g., medical errors, patient satisfaction) (Morey et al., 2002). However, this study suffered two significant limitations; participating hospitals self-selected into either the experimental or control groups and observers were not blind to the experimental conditions. To address this limitation, a subsequent random clinical trial of MedTeams in L&D units was conducted. Unfortunately, this study did not yield any positive effects on the clinical outcomes investigated (Nielsen et al., 2007).

13.5.3 TeamSTEPPS®

TeamSTEPPS was initiated in January 2003, when AHRQ and DoD convened a national panel of experts on human factors, human error, and medical team training. At this meeting, approximately 30 of the nation's leading experts discussed the needs, requirements, and strategies for effective teamwork in health care. Topics included competency requirements for medical teams, appropriate training strategies for teams, how to reliably measure teamwork, and what health care could learn from aviation and other disciplines. The end result was a road map guiding the research that followed.

A comprehensive review of the literature on the evidence-based relation between teamwork and patient safety was then conducted. As noted at the beginning of this chapter, an earlier review by Pizzi et al. (2001) had been published in AHRQ's *Evidence Base 43 Report*, which identified patient safety practices in other domains that should be tested in health care. In their chapter, Pizzi and colleagues argue that CRM training has a great deal of promise for addressing teamwork in health care. Because this review only focused on CRM, a broader review was needed. Therefore, Baker and his colleagues reviewed the larger discipline of teamwork and team training (Baker et al., 2003a). Because much of this work had been accomplished in the military, AHRQ and DoD felt this research could be directly extended to health care.

In addition to the literature review, DoD sought to examine their existing medical team training programs to identify any changes or updates that were required. Since 2001, the DoD Healthcare Team Coordination

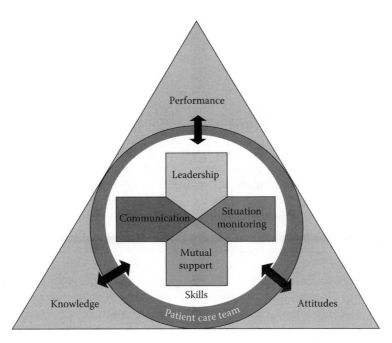

FIGURE 13.1
TeamSTEPPS instructional framework.

Program (HCTCP) has been conducting medical team training. By 2003, a number of different medical team training programs existed within DoD. To assess the strengths and weaknesses of these programs, Baker et al. (2003b) conducted a case-study analysis of three existing DoD medical team training programs (i.e., MedTeams, Medical Team Management, and Dynamic Outcomes Management©*). While results revealed that each program possessed strengths, the issued report called for the development of a medical team training specification that was evidence-based and would guide developers of medical team training programs. Baker et al. recommended that this document be constructed like a Federal Aviation Administration (FAA) Advisory Circular. Such FAA documents specify the requirements carriers must meet when implementing certain programs like CRM training (i.e., refer to FAA AC120-51E). In the end, AHRQ and DoD decided that a new and updated, evidenced-based program that was wholly owned by the Federal Government was warranted to enable wide-scale dissemination.

The teamwork competencies presented in Table 13.1 and their research basis served as the foundation for TeamSTEPPS. With that information as a starting point, the goal of AHRQ and DoD was to take this academically-oriented information and convert it to a framework that was meaningful from an instructional standpoint. As an example, theories of teamwork point to the importance of adaptability/flexibility as a central

skill (Salas et al., 2005). Yet it is difficult to directly train to the skill of adaptability/flexibility, which is required when responding to an unpredictable situation a team may encounter. Therefore, TeamSTEPPS instructs team members to monitor performance of others and provide assistance, plan and organize team roles, and communicate with one another efficiently and effectively. Combined, these skills yield a highly adaptable and flexible team.

To develop the TeamSTEPPS instructional model, teamwork competencies from the literature were classified as trainable or competencies that are the result of employing these trainable skills (i.e., outcomes). For example, shared mental models were viewed as an outcome of using monitoring and backup behaviors. The resulting TeamSTEPPS instructional framework is presented in Figure 13.1 where the core competencies include the trainable skills of leadership, situation monitoring (mutual performance monitoring), mutual support (backup behavior), and communication. These core competencies are encircled by the patient care team, which encompasses the patient. Performance, knowledge, and attitudinal outcomes are then depicted in the corners, resulting from proficiency on the central skills or core competencies.

The TeamSTEPPS curriculum contains an introductory module relating to the history of team training, a testimonial from Sue Sheridan, and the structure of teams. The introduction provides participants with insight into the importance of teamwork in health care. Four didactic-based modules discuss the core

---

* Dynamic Outcomes Management has since been renamed *Lifewings*.

competencies/skills of: (1) leadership; (2) situation monitoring; (3) mutual support; and (4) communication, respectively. Emphasis is placed on defining team skills, demonstrating the tools and strategies team members can use to gain proficiency in the competencies/skills, and identification of tools and strategies that can be used to overcome common barriers to achieve desired outcomes. Specialty case scenarios and video vignettes are used to further reinforce the learning.

The TeamSTEPPS initiative also consists of several sessions devoted to implementation, a multiphase process based upon John Kotter's model of organizational change (Kotter and Rathgeber, 2006). The process is carried out by a cadre of trainer/coaches who champion the effort within their unit, department or institution. A successful TeamSTEPPS initiative requires a carefully developed implementation and sustainment plan that is captured in Figure 13.2. It is based on lessons learned, and the literature of quality and patient safety, and culture change.

The goal of Phase I of the TeamSTEPPS implementation plan is to determine organizational readiness for undertaking a TeamSTEPPS initiative. During the pretraining assessment of Phase I, the organization or work unit will identify leaders and key champions that will comprise the organizational-level change team. The role of this organizational-level change team is to identify specific opportunities for improvement that can be realized by employing a teamwork initiative. A site assessment is conducted to determine readiness of the institution to include vital support of leadership, potential barriers to implementing change, and whether resources are in place to successfully support

the initiative. Such practice is typically referred to as a training needs analysis; a necessary first step to implementing a teamwork initiative (Salas and Cannon-Bowers, 2000).

The AHRQ Hospital Survey on Patient Safety Culture (http://www.ahrq.gov/qual/hospculture/) is a tool available to the public to conduct a site assessment. This survey can assist health-care organizations and systems in evaluating employees' perceptions and attitudes about the existing culture and issues related to patient safety. Information gathered from this assessment enables leaders to evaluate a variety of organizational factors that have an impact on patient safety to include: awareness about safety issues, evaluating specific patient safety interventions, tracking of change in patient safety over time, setting internal and external benchmarks, and fulfilling regulatory requirements or directives. To further demonstrate the need for improved teamwork and the importance of acting now, facility specific data (e.g., root cause analyses, occurrence reporting, patient and staff satisfaction questionnaires) can be used to further support the cause.

Final determination is based on whether improved team performance, to include employing a TeamSTEPPS Initiative, is the appropriate intervention necessary to impact change. A thorough needs analysis may uncover many underlying issues within the institution (e.g., systems problems, equipment problems, staffing shortages, etc.). The role of leadership is to assess the overall needs of the organization based upon the analysis, and determine the appropriate interventions.

Once organizational readiness is determined and a decision to proceed with a TeamSTEPPS initiative is

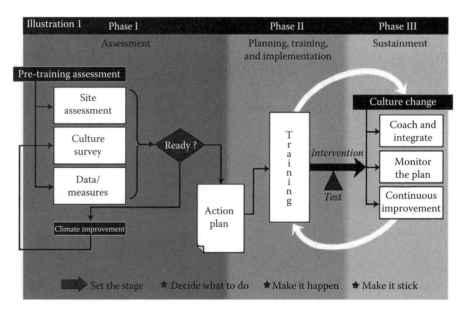

FIGURE 13.2
TeamSTEPPS implementation plan.

made, the role of the organizational-level change team is enhanced. Usually appointed by facility leadership, the change team will determine within which units or departments to deploy the initiative, develop an implementation and action plan for the organization, train the staff or other trainers, serve as the champions responsible for ongoing coaching and reinforcement of the team behaviors and skills on the unit or department.

Phase II is the planning and execution segment of the TeamSTEPPS Initiative. Typically, the change team (or specific designees) complete a 2½ day intensive TeamSTEPPS train-the-trainer session (as described later, AHRQ is developing an infrastructure to support such training). Provided in this session, is the core TeamSTEPPS curriculum to include scenarios, case studies, multimedia, and simulation. Culture change and coaching workshops that entail the provision of skills and strategies necessary for implementation, sustainment, and spread of the initiative are introduced. A 4 hour block of time is devoted to participant development of a customized TeamSTEPPS action plan. Each unit or department produces a tangible report detailing exactly how the initiative will be executed to best meet their unique circumstances. At the end of the session, participants are provided an opportunity to practice teach a module of the core curriculum using specialty-specific scenarios appropriate to their units or departments. Peer and instructor feedback serves to reinforce understanding of the content, along with refinement of presentation skills.

TeamSTEPPS was designed to be tailored to the organization in which it is being implemented. Options include implementation of all tools and strategies throughout the entire organization, a phased-in approach that targets specific units or departments, or selection of individual tools introduced at specific intervals (a dosing strategy). As long as the primary learning objectives are maintained, the TeamSTEPPS materials are extremely adaptable.

The goal of Phase III is to sustain and spread improvements in teamwork performance, clinical processes, and outcomes resulting from the TeamSTEPPS initiative. During this phase, users will (1) integrate teamwork skills and tools into daily practice, (2) monitor and measure the ongoing effectiveness of the TeamSTEPPS intervention, and (3) develop an approach for continuous improvement and spread of the intervention throughout the organization or work unit. Sustainment is managed by the designated change team through coaching and active observation of team performance. Continuing training of the core curriculum through refresher courses and newcomers' orientation, conducting continual evaluations of teams throughout the organization, and providing meaningful, ongoing feedback to staff members in the workplace where day-to-day

health care is provided. The key objective is to ensure that there are opportunities to implement the tools and strategies taught, practice and receive feedback on the trained skills, and continual reinforcement of the TeamSTEPPS principles on the unit or within the department.

Since the release of TeamSTEPPS in late 2006, AHRQ has received numerous requests for the materials, as well as guidance on implementation strategies. To address this need, AHRQ and DoD established a national support network for TeamSTEPPS through the Centers for Medicare and Medicaid Services (CMS) Quality Improvement Organizations (QIOs). Following an agricultural extension model, the QIOs serve as partners in the diffusion and adoption of TeamSTEPPS, further aiding health-care entities in improving patient outcomes through tracking of multiple performance metrics. QIOs were trained and supported via an AHRQ contract that established five TeamSTEPPS training centers; Carilion Clinic, Creighton University Medical Center, Duke University Medical Center, University of Minnesota Medical Center, and the University of Washington Medical Center.

13.5.4 Does Team Training Work in Health Care?

The fundamental standard by which ACRM, MedTeams, and TeamSTEPPS is measured is whether or not a direct link between improved teamwork and improvements in patient outcomes can be established. To date, such research has been limited but has gained more attention in recent years. For example, Weaver et al. (2010) recently published a review of existing team training programs in health care and associated research. In total, 40 articles are presented describing different team training activities and their effectiveness. In the following, we briefly review the findings to date related to the training programs we reviewed here.

Early work on this topic focused on MedTeams. As described earlier, MedTeams training was evaluated using a quasi-experimental research design (Morey et al., 2002, 2003) in which a variety of process factors (e.g., quantity of teamwork behaviors) and enabling factors (e.g., attitudes toward teamwork, staff burnout) were measured over a 1-year period. A positive effect of training on *outcome criteria* (e.g., medical errors, patient satisfaction) was reported (Morey et al., 2002). However, this study suffered significant limitations as discussed previously.

More recently, Nielsen et al. (2007) conducted a cluster-randomized controlled trial at 15 U.S. hospitals. Seven received MedTeams training and the remaining eight hospitals made up a control group. A total of 1,307 individuals were trained, and 28,536 deliveries were analyzed. The only process measure that differed significantly after team training was in the interval from deciding to perform an immediate cesarean delivery to making the incision, which decreased from 33 to 21 min. Regarding clinical outcomes, the study failed to show that MedTeams in obstetrical practices had any important clinical impact.

Regarding TeamSTEPPS, two studies have attempted to establish a link between this training and patient outcomes. First, Mann et al. (2006) tested many of the tools that comprise the TeamSTEPPS curriculum. These researchers analyzed discharge data on more than 80,000 deliveries per year from 2000 to 2004. The average Adverse Outcomes Index (AOI, a composite measure of quality in L&D) was the dependent measure. Teamwork tools to promote a culture of patient safety were implemented in 2002. As a result, the AOI score for high-risk premature births improved 47%, term deliveries 14%, and 16% overall from 2001 to 2004. Malpractice claims, suits, and observations (monies placed in reserves for potential claims) dropped by more than 50% in the period from 2002 to 2005.

Capella et al. (2010) investigated the impact of TeamSTEPPS on team behaviors in the trauma resuscitation bay. The study used a pretraining/posttraining design in which the intervention, TeamSTEPPS, was augmented by simulation. Data were collected regarding each level of Kirkpatrick's (1967) multilevel evaluation framework including reactions (posttraining only), learning, behavior, and outcomes. Learning was measured regarding participant attitudes, using the T-TPQ, posttraining only (Baker et al., 2010b); participant knowledge, using items from the TeamSTEPPS Learning Benchmarks test found in the TeamSTEPPS curriculum and items developed specifically regarding trauma resuscitation; and skill, using a human patient simulation. Behavior was measured pre- and posttraining by conducting observations in the trauma bay with trained observers using the Trauma Team Performance Observation Tool (TPOT) (Baker et al., 2011). Pretraining observations were collected from November 2008 to February 2009 (n = 33) and posttraining observations were collected from May to July 2009 (n = 40). Finally, outcomes were assessed by collecting clinical data from the trauma registry. The clinical parameters included time from arrival to computed tomography (CT) scanner, arrival to intubation, arrival to OR, arrival to Focused Assessment Sonography in Trauma (FAST) examination, time in ED, hospital length of stay (LOS), ICU LOS, complications, and mortality.

Results showed significant improvement in learning regarding performance across three simulations, and behavior, as observed in the trauma bay. Regarding clinical outcomes, the time from arrival to the CT scanner (26.4–22.1 min), endotracheal intubation (10.1–6.6 min, for those patients requiring intubation), and the OR

(130.1–94.5 min) decreased significantly after the training; however, a significant effect for training was not observed for other clinical outcomes. Similar to Nielsen et al. (2007), TeamSTEPPS was found to impact the time required for trauma teams to execute clinical tasks but not necessary more critical clinical outcomes like patient mortality, complications, or LOS.

13.5.5 Summary

This section reviewed the most prevalent team training programs in health care and research on these programs' effectiveness. Suffice it to say, team training has become widely recognized by health care as an effective strategy to improve the delivery of care and improve patient safety. AHRQ's development and launch of TeamSTEPPS has yielded a standardized approach based on solid evidence and demonstrated effectiveness across a broad spectrum of high risk occupations (Salas et al., 2008). Through AHRQ's promotion of TeamSTEPPS, more and more organizations are adopting this training annually and reporting success with this approach. Nonetheless, research on the true impact of team training in health care is only beginning to emerge. In the final section, we describe specific research needs that the medical community and health services researchers should seek to address.

13.6 Future Directions for Research

Our review of team training programs clearly shows that the medical community is striving to implement team training across a number of medical domains. We recommend that this trend be continued and we believe at this point there is no question that it will be. However, the extent to which these programs are being implemented with the help of what we know from the science of learning, of team performance and of training is less clear and more evaluation research is required. Therefore, the following recommendations are cast to support what we believe are the crucial requirements to further substantiate the relation between teamwork and patient safety.

Recommendation 1: Align training and measurement of team performance with current models of teamwork

It is only within the last 5 years that research has blossomed regarding the core competencies of health-care teams. Both the work by Salas et al. (2005) and Alonso et al. (2006) are steps in providing a uniformed set of KSA competencies that are related to most health-care teams. Nonetheless, the extent to which these competencies are generic and apply to all health-care teams needs to be tested. For example, Capella et al. (2010) contextualized TeamSTEPPS to trauma teams and added a simulation component to enable skills practice. AHRQ has also developed contextualized versions of the TeamSTEPPS program with the release of TeamSTEPPS for Rapid Response Teams and the development of TeamSTEPPS for Medical Homes and Long-term Care. Each of these programs are based on a customized version of the Alonso et al. model, but research needs to determine if this content driven approach is more effective than a generic approach advocated by the original TeamSTEPPS curriculum.

Research needs to test competing theories of teamwork to determine the relative validity of different models. Amodeo et al. (2011) present a first comparative test of the type of research we advocate for here but additional research is required.

Recommendation 2: Assess both process and outcomes when determining the impact of team training

The ultimate criterion to establish team training's true effectiveness in health care is the extent to which training reduces negative patient outcomes. However, the low base rate of serious errors such as sentinel events precludes this "ultimate criterion" from serving as a viable outcome construct. Given the vast number of medical procedures conducted each day without incident, applying this "ultimate criterion" to the team training in health care is impractical to say the least, despite the prevalence of errors cited in the IOM report *To Err is Human*.

In an effort to resolve the base rate issue, we suggest that research on the impact of team training should account for both process and outcome measures. For example, in addition to specifying the skill competencies that comprise teamwork, the Alonso et al. (2006) model specifies knowledge, attitudinal, and performance outcomes. Alonso et al. suggest that high levels of teamwork lead to team members holding more accurate shared mental models (i.e., knowledge) and more positive attitudes toward teamwork (i.e., mutual trust, belief in the importance of teamwork). Moreover, though not sequential, when these outcomes occur, performance should improve. In the Alonso et al. model performance examples might include the time taken to execute the initial decision in an emergency-medicine unit, the time it takes to intubate a patient that has stopped breathing, or the reduction of health care acquired infections in a given ICU. Research by Nielsen et al. (2007) and Capella et al. (2010) indicate that team training can improve such process-oriented measures, though tests in other clinical domains are required.

Another important outcome that health care should consider is tapping voluntary event reporting systems that capture "near misses." Such events can be viewed as a proxy criterion for error. The Patient Safety and Quality Improvement Act of 2005 (Patient Safety Act), which authorized the creation of Patient Safety Organizations (PSOs) provides a significant resource for such investigations. The Patient Safety Act encourages clinicians and health-care organizations to voluntarily report and share quality and patient safety information without fear of legal discovery.

Recommendation 3: New measures must be developed and tested

Our basic conclusion is that research on measurement and the development of measures to assess team performance needs to catch up with the deployment of interventions to improve quality in health care. The impact of team training in health care cannot be established without reliable and valid tools being available to capture its effects. Since the IOM report over 10 years ago, there has been an explosion of tools and strategies but not regarding the development of measures. For example, AHRQ is supporting the national deployment of TeamSTEPPS (teamstepps.ahrq.gov), but few tools are available to assess the impact of the TeamSTEPPS tools and strategies. Lacking these metrics, it will be difficult for organizations consuming these patient safety interventions to know if they worked and how they did work. This is critical in health care where financial resources are extremely limited. Therefore, researchers need to not only focus on how to improve quality, team performance, and safety but also concurrently develop measurement tools so that the effects of these interventions can be established.

Recommendation 4: Embed team training within professional training and development

Our final recommendation focuses on what we consider the imperative need to embed team training in professional development. By "embedding" we mean implementing and regulating team training throughout a health-care provider's career. Baker et al. (2005) outlined an approach for implementing team training throughout the professional career of physicians and we know of a number of institutions that are examining how team training and simulation can be incorporated into a clinician's professional training. New medical schools, like the Virginia Tech Carilion School of Medicine, are even including interprofessional blocks of instruction where nursing, physician assistant, and medical students take classes together. The problem-based learning (PBL) model of medical education seems to provide an effective strategy for introducing team concepts early on and reinforcing them throughout medical training. However, research again needs to examine how and when in a clinican's education the development of teamwork skills can best be accomplished.

In sum, we believe that a great deal has been accomplished since the original publication of this chapter. In 5 years, new tools have been developed and team training has been widely deployed and adopted by health care. Given the vast progress that has been made, we believe the future looks bright, but it will only be through ongoing research and further development of the evidence base in health care that this future will be realized.

Acknowledgments

This research was supported by an earlier grant from the Agency for Healthcare Research and Quality and the Department of Defense, TriCare Management Activity. The views expressed herein are those of the authors.

References

Agency for Healthcare Research and Quality. (2000). *Doing What Counts for Patient Safety: Federal Actions to Reduce Medical Errors and Their Impact*. Rockville, MD.

Alonso, A., Baker, D., Holtzman, A., Day, R., King, H., Toomey, L., and Salas, E. (2006). Reducing medical error in the military health system: Is team training the right prescription? *Human Resources Management Review*, 16, 396–415.

Amodeo, A. M., Kurtessis, J., Bhupatkar, A., Baker, D. P., and Gallo, J. (2011). Is there a Big 5 of teamwork? An empirical test. Poster to be presented at the *26th Annual Conference for the Society for Industrial and Organizational Psychology*, Chicago, IL.

Baker, D. P., Amodeo, A., and Gallo, J. (2010a). Development and validation of the TeamSTEPPS teamwork perceptions questionnaire. Poster presented at the *25th Annual Conference for the Society for Industrial and Organizational Psychology*, Atlanta, GA.

Baker, D., Amodeo, A., Krokos, K., Slonim, A., and Herrera, H. (2010b). Assessing teamwork attitudes in healthcare: Development of the TeamSTEPPS teamwork attitudes questionnaire. *Quality and Safety in Health Care*, 19(6), e49 (1–4).

Baker, D. P., Beaubien, J., and Holtzman, A. (2003b). *DoD Medical Team Training Programs: An Independent Case Study Analysis*. Washington, DC: American Institutes for Research.

Baker, D. P., Capella, J., Gallo, J., and Hawkes, C. (2011). Development of the trauma team performance observation tool. Poster to be presented at the *26th Annual Conference for the Society for Industrial and Organizational Psychology*, Chicago, IL.

Baker, D. P., Gustafson, S., Beaubien, J., Salas, E., and Barach, P. (2003a). *Medical Teamwork and Patient Safety: The Evidence-Based Relation*. Washington, DC: American Institutes for Research.

Baker, D., Salas, E., Barach, P., King, H., and Battles, J. (2005). The role of teamwork in the professional education of physicians: Current status and assessment recommendations. *Joint Commission Journal on Quality and Safety*, 31(4), 185–202.

Beaubien, J. and Baker, D. (2004). The efficacy of simulation for training teamwork skills in health care: How low can you go? *Quality and Safety in Healthcare*, 13, 51–56.

Brannick, M. T., Salas, E., and Prince, C. (1997). *Team Performance Assessment and Measurement*. Mahwah, NJ: Erlbaum.

Cannon-Bowers, J. A. and Salas, E. (1998). *Making Decisions under Stress: Implications for Individual and Team Training*. Washington, DC: American Psychological Association.

Cannon-Bowers, J. A., Tannenbaum, S. I., Salas, E., and Volpe, C. E. (1995). Defining competencies and establishing team training requirements. In R. A. Guzzo, E. Salas, and Associates (Eds.), *Team Effectiveness and Decision-Making in Organizations* (pp. 333–380). San Francisco, CA: Jossey-Bass.

Capella, J., Smith, S., Philp, A., Putnam, T., Gilbert, C., Fry, W., Harvey, E. et al. (2010). Team training improves clinical care of trauma patients. *Journal of Surgical Education*, 67, 439–443.

Davies, J. M. (2001). Medical applications of crew resource management. In E. Salas, C. A. Bowers, and E. Edens (Eds.), *Improving Teamwork in Organizations: Applications of Resource Management Training* (pp. 265–281). Mahwah, NJ: Erlbaum.

Davis, L. T., Gaddy, C. D., and Turney, J. R. (1985). *An Approach to Team Skills Training of Nuclear Power Plant Control Room Crews*, (NUREG/CR-4258GP-R-123022). Columbia, MD: General Physics Corporation.

Driskell, J. E. and Johnston, J. H. (1998). Stress exposure training. In J. A. Cannon-Bowers and E. Salas (Eds.), *Making Decisions under Stress—Implications for Individual and Team Training* (pp. 191–217). Washington, DC: American Psychological Association.

Dyer, J. L. (1984). Team research and training: A state of the art review. In F. A. Muckler (Ed.), *Human Factors Review* (pp. 285–323). Santa Monica, CA: Human Factors and Ergonomics Society.

Gaba, D. M. (1998). Research techniques in human performance using realistic simulation. In L. C. Henson and A. H. Lee (Eds.), *Simulators in Anesthesiology Education* (pp. 93–101). New York: Plenum.

Gaba, D. M., Howard, S. K., Fish, K. J., Smith, B. E., and Sowb, Y. A. (2001). Simulation-based training in anesthesia crisis resource management (ACRM): A decade of experience. *Simulation & Gaming*, 32, 175–193.

Gaba, D. M., Howard, S. K., Flanagan, B., Smith, B. E., Fish, K. J., and Botney, R. (1998). Assessment of clinical performance during simulated crises using both technical and behavioral ratings. *Anesthesiology*, 89, 8–18.

Guzzo, R. A. and Shea, G. P. (1992). Group performance and inter-group relations in organizations. In M. D. Dunnette and L. M. Hough (Eds.), *Handbook of Industrial and Organizational Psychology* (2nd edn., pp. 269–313). Palo Alto, CA: Consulting Psychologists Press.

Hackman, J. R. (1987). The design of work teams. In J. W. Lorsch (Ed.), *Handbook of Organizational Behavior* (pp. 315–342). Englewood Cliffs, NJ: Prentice Hall.

Helmreich, R. L., Butler, R. E., Taggart, W. R., and Wilhelm, J. A. (1995). *Behavioral Markers in Accidents and Incidents*. Report Commissioned by The FAA (Rep. No. NASA/UT/FAA Technical Report 95–1). Austin, TX: The University of Texas.

Howard, S. K., Gaba, D. M., Fish, K. J., Yang, G., and Sarnquist, F. H. (1992). Anesthesia crisis resource management training: Teaching anesthesiologists to handle critical incidents. *Aviation, Space, and Environmental Medicine*, 63, 763–770.

Ilgen, D. R. (1999). Teams embedded in organizations: Some implications. *American Psychologist*, 54, 129–139.

James, L. R., Demaree, R. G., and Wolf, G. (1984). Estimating within group interrater reliability with and without response bias. *Journal of Applied Psychology*, 69, 85–98.

Kabanoff, B. and O'Brien, G. E. (1979). The effects of task type and cooperation upon group products and performance. *Organizational and Human Performance*, 23, 163–181.

Katzenbach, J. R. and Smith, D. K. (1993). The wisdom of teams. *Small Business Reports*, 18(7), 68–71.

King, H., Battles, J., Baker, D. P., Alonso, A., Salas, E., Webster, J., Toomey, L., and Salisbury, M. (2008). TeamSTEPPS™: Team strategies and tools to enhance performance and patient safety. In K. Henriksen, J. Battles, M. Keyes, and M. Grady (Eds.), *Advances in Patient Safety: New Directions and Alternative Approaches, Vol. 3 Performance and Tools* (pp. 5–20). AHRQ Publication No. 08-0034-1. Rockville, MD: Agency for Healthcare Research and Quality.

Kirkpatrick, D. L. (1967). Evaluation of training. In R. L. Craig (Ed.), *Training and Development Handbook: A Guide to Human Resource Development* (pp. 18.1–18.27). New York: McGraw-Hill.

Kohn, L. T., Corrigan, J. M., and Donaldson, M. S. (1999). *To Err Is Human*. Washington, DC: National Academy Press.

Kotter, J. and Rathgeber, H. (2006). *Our Iceberg is Melting: Changing and Succeeding under Any Conditions*. New York: St. Martin's Press.

Mann, S., Pratt, S., Gluck, P., Nielsen, P., Risser, D., Greenberg, P., Marcus, R. et al. (2006). Assessing quality in obstetrical care: Development of standardized measures. *Joint Commission Journal on Quality and Patient Safety*, 32(9), 497–505.

McGrath, J. E. (1984). *Groups: Interaction and Performance*. Englewood Cliffs, NJ: Prentice Hall.

Mohrman, S. A., Cohen, S. G., and Mohrman, A. M. (1995). *Designing Team-Based Organizations: New Forms for Knowledge Work*. San Francisco, CA: Jossey-Bass.

Morey, J. C., Simon, R., Jay, G. D., and Rice, M. M. (2003). A transition from aviation crew resource management to hospital emergency departments: The MedTeams story.

In R. S. Jensen (Ed.), *Proceedings of the 12th International Symposium on Aviation Psychology* (pp. 1–7). Dayton, OH: Wright State University Press.

Morey, J. C., Simon, R., Jay, G. D., Wears, R., Salisbury, M., Dukes, K. A. et al. (2002). Error reduction and performance improvement in the emergency department through formal teamwork training: Evaluation results of the MedTeams project. *Health Services Research*, 37, 1553–1581.

Morgan, B. B., Glickman, A. S., Woodward, E. A., Blaiwes, A. S., and Salas, E. (1986). *Measurement of Team Behaviors in a Navy Environment*. NTSC Technical Report No. 86–014. Orlando, FL: Naval Training Systems Center.

Morgan, B. B., Jr., Salas, E., and Glickman, A. S. (1994). An analysis of team evolution and maturation. *The Journal of General Psychology*, 120, 277–291.

Murray, W. B. and Schneider, A. J. L. (1997). Using simulators for education and training in anesthesiology. *American Society of Anesthesiologists Newsletter*, 61, 1–2.

Nielsen, P. E., Goldman, M., Mann, S., Shapiro, D., Marcus, R., Pratt, S., Greenberg, P. et al. (2007). Effects of teamwork training on adverse outcomes and process of care in labor and delivery: A randomized controlled trial. *Obstetrical & Gynecological Survey*, 62, 294–295.

Orasanu, J. M. and Salas, E. (1993). Team decision making in complex environments. In G. Klein, J. Orsanu, and R. Calderwood (Eds.), *Decision Making in Action: Models and Methods* (pp. 327–345). Norwood, NJ: Ablex.

Pizzi, L., Goldfarb, N. I., and Nash, D. B. (2001). Crew resource management and its applications in medicine. In K. G. Shojana, B. W. Duncan, K. M. McDonald, and R. M. Wachter (Eds.), *Making Health Care Safer: A Critical Analysis of Patient Safety Practices* (pp. 501–510). Rockville, MD: Agency for Healthcare Research and Quality.

Salas, E. and Cannon-Bowers, J. A. (2000). The anatomy of team training. In N. S. Tobias and D. Fletcher (Eds.), *Training and Re-Training: A Handbook for Business, Industry and Military* (pp. 312–335). Farmington Hills, MI: MacMillan.

Salas, E., DiazGranados, D., Klein, C., Burke, C. S., Stagl, K., Goodwin, G. F., and Halpin, S. M. (2008). Does team training improve team performance? A meta-analysis. *Human Factors*, 50, 903–933.

Salas, E., Dickinson, T. L., Converse, S. A., and Tannenbaum, S. I. (1992). Toward an understanding of team performance and training. In R. W. Swezey and E. Salas (Eds.), *Teams: Their Training and Performance* (pp. 3–29). Norwood, NJ: Ablex.

Salas, E., Rhodenizer, L., and Bowers, C. A. (2000). The design and delivery of crew resource management training: Exploiting available resources. *Human Factors*, 42, 490–511.

Salas, E., Rozell, D., Mullen, B., and Driskell, J. E. (1999). The effect of team building on performance: An integration. *Small Group Research*, 30, 309–329.

Salas, E., Sims, D. E., and Burke, S. C. (2005). Is there a "Big Five" in teamwork? *Small Group Research*, 36(5), 555–599.

Simon, R., Langford, V., Locke, A., Morey, J. C., Risser, D., and Salisbury, M. (2000). A successful transfer of lessons learned in aviation psychology and flight safety to health care: The MedTeams system. In *Patient Safety Initiative 2000-Spotlighting Strategies, Sharing Solutions* (pp. 45–49). Chicago, IL: National Patient Safety Foundation.

Simon, R., Morey, J. C., Rice, M. M., Rogers, L., Jay, G. D., Salisbury, M. et al. (1998). Reducing errors in emergency medicine through team performance: The MedTeams project. In A. L. Scheffler and L. Zipperer (Eds.), *Enhancing Patient Safety and Reducing Errors in Health Care* (pp. 142–146). Chicago, IL: National Patient Safety Foundation.

Smith, P. C. and Kendall, L. M. (1963). Retranslation of expectations: An approach to the construction of unambiguous anchors for rating scales. *Journal of Applied Psychology*, 47, 149–155.

Smith-Jentsch, K. A., Johnston, J. H., and Payne, S. C. (1998a). Measuring team-related expertise in complex environments. In J. A. Cannon-Bowers and E. Salas (Eds.), *Making Decisions under Stress: Implications for Individual and Team Training* (pp. 61–87). Washington, DC: American Psychological Association.

Smith-Jentsch, K. A., Salas, E., and Baker, D. P. (1996). Training team performance-related assertiveness. *Personnel Psychology*, 49, 909–936.

Smith-Jentsch, K. A., Zeisig, R. L., Acton, B., and McPherson, J. A. (1998b). Team dimensional training. In J. A. Cannon-Bowers and E. Salas (Eds.), *Making Decisions under Stress: Implications for Individual and Team Training* (pp. 271–297). Washington, DC: APA Press.

Studdert, D. M., Brennan, T. A., and Thomas, E. J. (2002). What have we learned from the Harvard Medical Practice Study? In M. M. Rosenthal and K. M. Sutcliffe (Eds.), *Medical Error: What Do We Know? What Do We Do?* (pp. 3–33). San Francisco, CA: Jossey-Bass.

Tays, T. M. (2000). Effect of anesthesia crisis resource management training on perceived self-efficacy. Unpublished doctoral dissertation. Pacific Graduate School of Psychology.

Volpe, C. E., Cannon-Bowers, J. A., Salas, E., and Spector, P. E. (1996). The impact of cross training on team functioning: An empirical investigation. *Human Factors*, 38, 87–100.

Weaver, S. J., Lyons, R., Diazgranados, D., Rosen, M. A., Salas, E., Olgesby, J., Birnbach, D. J. et al. (2010). The anatomy of healthcare team braining and the state of practice: A critical review. *Academic Medicine*, 85, 1746–1760.

14

Health-Care Work Schedules

Ann E. Rogers

CONTENTS

14.1 Common Work Schedules 199
    14.1.1 Physician Work Schedules... 199
    14.1.2 Registered Nurses... 200
    14.1.3 Other Health-Care Providers 201
14.2 Impact of Work Schedules on Patient Safety, Fatigue, and Performance 201
14.3 Recommendations 203
14.4 Conclusions... 207
References... 208

Hospitals provide care 24 hours a day, 7 days a week, regardless of financial constraints, sometimes without the services of key specialists. Moreover, the common practice of working overtime and double shifts exacerbates the stressful environment of such demanding workplaces. Despite the documented adverse effect of subsequent fatigue experienced by professionals in such health-care settings, as well as in other industries (e.g., aviation and transportation), there is only limited regulation of the length of shifts in these and other professions. Indeed, some employees are reluctant to eliminate their option of working double shifts and overtime due to financial or personal scheduling reasons.

This chapter will examine the recent regulations issued by the Accreditation Council for Graduate Medicine Education (ACGME) aimed at attempting to limit shift hours of resident physicians. Subsequently, the chapter will explore the results of groundbreaking studies on sleep deprivation of registered nurses (RNs) who have long work schedules and the varying attempts made by health-care institutions to remedy that problem. The chapter will also briefly consider sleep deprivation in a variety of other health professionals before examining the impact of these work schedules on patient safety, fatigue, and performance. Finally, there is a list of recommendations for reducing the impact of current work schedules.

14.1 Common Work Schedules

14.1.1 Physician Work Schedules

Prior to 2003, when the ACGME issued regulations limiting shift hours, resident physicians worked an average of 83 h/week (Baldwin et al. 2003), and some reported working up to 100 h/week (Defoe et al. 2001).

Work schedules for residents are now limited to (1) an 80 h workweek averaged over a period of 4 weeks, including all in-house calls; (2) 1 day off in 7 without any clinical duties or call, averaged over 4 weeks; (3) in-house call frequency of no more than every third night, averaged over 4 weeks; (4) maximum onsite duty up to 24 h with and additional 6 h allowed for didactic education and transfer of patients; and (5) at-home or pager calls limited in frequency so that the resident can rest and have a reasonable amount of personal time (Friedman 2007). However, these calls do not count toward the every third or 24 + 6 h limit.

Although there has been a debate about the degree of compliance with work hour reform, (IOM [Institute of Medicine] 2009) there is evidence that the work hours of postgraduate year 1 (PGY-1) residents (interns) has decreased from 70.7 h/week prior to the institution of the ACGME regulations (Baldwin et al. 2003) to 66.6 h/week in 2003–2004 (Landrigan et al. 2006). In addition, most institutions were in compliance with the regulations limiting resident physicians to an average of 80h/week (IOM [Institute of Medicine] 2009). The mean length of extended duty periods for interns also declined, from 32.1h to 29.9h, although violations regarding the length of duty period, time off between duty periods, and call frequency continue to be reported (IOM [Institute of Medicine] 2009).

Further modifications of resident duty hours and work schedules were recommended by the Committee on Optimizing Graduate Medical Trainee (Resident) Hours and Work Schedules to Improve Patient Safety in December, 2008. This Institute of Medicine Committee recommended that (1) scheduled continuous duty periods not exceed 16h unless a 5h uninterrupted continuous sleep period is provided between 10 p.m. and 8 a.m.; (2) extended duty periods (e.g., 30h that include a protected sleep period) not be more frequent than every third night, with no averaging; (3) after completing duty periods residents must be allowed continuous off-duty intervals that vary according to the duration of the shift worked, whether the work period occurs during the nighttime or during the day; (4) night-float or night shift duty not exceed four consecutive nights and be followed by a minimum of 48h off duty after three or four consecutive nights; (5) at least one 24h off-duty period be provided per 7 day period without averaging; and (6) residents be permitted, but not required, in exceptional circumstances to remain on duty longer than the scheduled time to ensure patient safety or engage in critical learning experiences (IOM [Institute of Medicine] 2009).

Although the ACGME did not adopt all of these recommendations, its Task Force on Quality Care and Professionalism proposed several standards including limiting interns to working no more than 16 consecutive hours, and limiting other residents to working no more than 24 consecutive hours (with the encouragement of strategic napping after working 16h and between the hours of 10 p.m. and 8 a.m.) (ACGME 2010). These standards will be implemented on July 1, 2011.

However, there are no restrictions on the hours physicians work after they complete their training, and most continue to work long and unpredictable hours. According to the U.S Bureau of Labor Statistics, more than one-third of fully licensed physicians work 60h or more per week (U.S. Department of Labor 2007), with work hours varying by specialty. For example, surgeons with an office-based practice report working an average of 60h/week, while those in primary care report working on average 50h/week (Weiss 2005). Certain specialties are more likely than others to report working more than 80h/week. For example, of the 23 specialties surveyed in 2005, the following specialties reported that at least 15% of their office-based practice members worked more than 80h/week: urologists

(15%), obstetricians-gynecologists (16%), pulmonologists (16%), hematologists–oncologists–immunologists (17%), infectious disease specialists (17%), general surgeons (19%), cardiologists (20%), neurosurgeons (23%), and thoracic surgeons (33%) (Weiss 2005). Even less information is available about the work schedules of hospitalists, and attending physicians in academic medical centers.

14.1.2 Registered Nurses

While numerous studies have tracked resident physician hours, this group represents only one population of health-care professionals who are sleep deprived. However, it was only in 2004 that reports of significant work schedule studies of the largest group of health-care providers, RNs, in the United States were published. That year Rogers and her colleagues began publishing a pioneering series of studies about hospital staff nurse work schedules (Rogers et al. 2004a,b; Scott et al. 2006a,b) that provided much needed information about the hours worked by these professionals and the effects of these work hours on patient safety. The importance of these studies lay in the fact that over half (56.2%) of the almost 3 million RNs licensed to practice in the United States work in hospitals (Health Resources and Services Administration 2007). (There is still little known about the work schedules of nurses working in ambulatory care settings, public/community health settings, nursing homes/extended care facilities, or in roles such as nurse managers, nurse practitioners, or nurse educators.)

The majority of hospital staff nurses now work 12h shifts (Rogers et al. 2004b; Scott et al. 2006b, 2010; Trinkoff et al. 2006). The majority of these shifts are scheduled from 7 a.m. to 7:30 p.m. or 7 p.m. to 7:30 a.m., although some are scheduled from 3 a.m. to 3:30 p.m. and 3 p.m. to 3:30 a.m. (particularly in emergency departments).

Work breaks and lunch periods free of patient care responsibilities are rare (Rogers et al. 2004a; Stefancy 2009; Trinkoff et al. 2006), and working longer than scheduled is the norm rather than the exception (Kalish and Lee 2009; Lipscomb et al. 2002; Rogers et al. 2004b; Scott et al. 2010; Trinkoff et al. 2006; Tucker and Edmondson 2003). In addition, nurses in certain specialty areas such as critical care units, postoperative recovery units (postanesthesia care units (PACUs)), and operating rooms, are often required to take calls in addition to their regularly scheduled work shifts (Rogers et al. 2004b). If called into work, they are expected to work their next regularly scheduled 12h shift without additional time off (Association of Perioperative Registered Nurses 2008).

Over 16 states including Illinois, New Jersey, New York, Oregon, Texas, and Washington have passed legislation limiting and/or banning mandatory overtime (American Nurses Association 2011); however, there are no limits on how long RNs may voluntarily work during a day (24 h period) or a week (7 day period). Data collected during the Staff Nurse Fatigue and Patient Safety Study showed that 12.2% of 895 staff nurses worked more than 16 consecutive hours during the 28 day data-gathering period and that some nurses worked almost 24 consecutive hours (23 h and 40 min) (Rogers et al. 2004b; Scott et al. 2006b). Although the average number of hours worked per week by participants in the study was 40.2 ± 12.9 h, weekly work hours of up to 97.2 h/week were reported (Rogers et al. 2004b; Scott et al. 2006b). A later survey of 11,785 nurses in Texas (Texas Board of Nursing 2008), showed that 65% of the participants sometimes or routinely worked more than 12.5 consecutive hours and 47% sometimes or routinely worked more than 60 h/week. Anecdotal data suggests that neonatal nurse practitioners commonly work 24 and 48 h shifts, as this group of advanced practice nurses substitute for residents in neonatal ICUs (National Association of Neonatal Nurses and the National Association of Neonatal Nurse Practitioners 2008).

Although some hospitals have limited the number of hours worked by RNs, and others have eliminated double shifts, few have made the extensive changes that the Veterans Administration (VA) Health System made in December 2006, in response to a Congressional mandate (U.S. House and Senate 2004). Nurses employed by the VA Health System are forbidden to provide direct patient care in excess of 12 h in 24, and/or to work more than 60 h in a 7 day period.

14.1.3 Other Health-Care Providers

Despite this recent series of studies, there is still scanty data on the typical work hours or schedules of other groups of health-care providers, such as physician assistants, pharmacists, respiratory therapists, technicians, and other support personnel. It is possible, however, that some personnel such as certified medical assistants (CMAs) and nursing assistants work even longer hours than RNs. For example, interviews with staff members who worked at nursing homes studied by Louwe and Kramer (Louwe and Kramer 2001) found that in 13 of 17 facilities, at least one nursing staff member, and usually more, had worked between one and three double (16 h) shifts during the previous 7 days. In five of the facilities, at least one staff member had worked four to seven double shifts in the last 7 days. And in one facility, more than one-third of the nursing staff had worked between eight and eleven double shifts in the past 14 days. Although all direct-care nursing staff (RNs, licensed practical nurses/licensed vocational nurses

(LPNs/LVNs), CMAs, and nursing assistants) worked extra hours, nursing assistants worked the majority of double shifts.

14.2 Impact of Work Schedules on Patient Safety, Fatigue, and Performance

Although extended work hours have been associated with increased worker injuries and accidents, the impact of extended work hours on patient safety remains unclear. A series of studies by Rogers and her colleagues demonstrated that the risk of error increased as much as threefold when hospital staff nurses worked shifts of 12.5 h or longer. In addition, errors were 1.46–1.96 times more likely to occur when they worked more than 40 h/week (Rogers et al. 2006; Scott et al. 2006b). The study relied on self-reporting, and hospitals were not identified; therefore, errors were probably underreported and it was impossible to determine if the reported errors adversely affected patient outcomes. An analysis of the errors reported by study participants suggested, however, that most errors were minor and unlikely to result in harm (Balas et al. 2004, 2006).

Studies of resident physicians have also had mixed findings. Many of these studies, which tested cognitive skills, occurred prior to work hour reform in 2003. Some rarely showed deficits associated with long work hours, while others reported more errors interpreting electrocardiograms (ECGs), problems with attention and concentration, and lower performance on the American Board of Family Practice in-training examinations (Friedman et al. 1971; Jacques et al. 1990; Kahol et al. 2008). Only later did researchers conclude that residents originally classified as "rested" were actually chronically sleep deprived. These mixed findings might have occurred because the skills tested often involved familiar tasks, for example, adding numbers, or because there was little difference between the "rested" and "nonrested" conditions.

Other issues that confounded reliable comparison of results among different studies included multiple definitions of sleep deprivation, the outcome measures chosen, the lack of control groups, and the presence or absence of mitigating factors, for example, whether or not the resident was assisted in performing a procedure by an attending physician.

In order to overcome the flaws associated with these earlier studies, the Harvard Work Hours Group conducted a prospective, cross-over observational study in two critical care units (Landrigan et al. 2004; Lockley et al. 2004). Twenty interns were randomly assigned to work 3-week rotations on both schedules in two different ICUs. One group followed a traditional schedule that was extended (24 h or longer) every other shift, while the other group followed the intervention schedule (shifts no longer than 16 h). Errors were assessed using trained physician observers, voluntary staff reporting, chart review, and computerized event detection monitors. Interns on the intervention schedule had 36% fewer errors compared to those on the traditional schedule, with the greatest decrease occurring in diagnosis-related errors (a reduction from 18.6 to 3.3 per 1000 patient days) (Landrigan et al. 2004). The number of procedural errors did not decrease, nor were there statistically significant differences in the number of intern-related potential adverse events—a measure of harm that reached the patient.

In 2003, ACGME implemented regulations restricting resident physicians from working more than 80 h/week over a 4 week period or more than 24 consecutive hours plus an additional 6 h for transitional activities. The council also required that residents be permitted 24 h off per week, averaged over 4 weeks (ACGME [Accreditation Council Graduate Medical Education] 2008). Changes in resident duty hours, however, have not lowered the risks of a prolonged hospital stay, the number of potentially preventable safety-related events as measured by the Agency for Healthcare Research and Quality (AHRQ), patient safety indicators among VA or Medicare patients, mortality rates of intensive care unit patients hospitalized in teaching hospitals, or 30 day mortality rates for hospitalized Medicare beneficiaries (Prasad et al. 2009; Rosen et al. 2009; Volpp et al. 2007a,b). In fact, the only significant change noted by Volpp (2007a,b), was a decrease in 30 day mortality rates for VA patients with four common medical conditions the second year after duty hour reform.

Explanations for the absence of significant improvements associated with the decrease in hours include increased handoffs and issues of continuity of care, increased burden on the nursing staff, noncompliance with duty hour regulations, and the hypothesis that significant sleep deprivation still occurred with 30 h (26 + 4) shifts required by ACGME (Landrigan et al. 2007; Rosen et al. 2009). Moreover, residents working outside certain specialty areas (emergency departments and radiology suites) generally do not provide hands-on care. Rather, they are part of a team of health-care professionals that discusses cases daily. In this context, the responsibilities of residents mostly include diagnosis, as well as writing orders and following test results, both of which customarily receive backup review by pharmacists and RNs. Indeed, in some institutions, hospital pharmacists also make rounds with the physicians.

Although most of these studies have not demonstrated a direct association between adverse events and the long hours worked by resident physicians and hospital staff nurses, there is ample evidence that many health-care providers are not obtaining sufficient sleep. Interns studied by the Harvard Work Hours Group obtained on average 6.6 ± 0.8 h of sleep daily (measured by polysomnograph) when working the traditional schedule and 7.4 ± 0.9 h when their work shifts were shortened to 16 h/day (Lockley et al. 2004). Nurses who participated in the Staff Nurse Fatigue and Patient Safety study self-reported an average of only 6.8 h sleep on workdays. This figure must be interpreted cautiously, however, since polysomnography (PSG) is considered the most accurate measure of sleep, while self-reported times tend to underestimate actual sleep duration (Jean-Louis, 2000). Furthermore, the reported work sleep average of 6.8 h does not reflect the fact that participants reported getting less than 6 h sleep prior to working approximately one in four shifts worked (2,830/11,329 shifts), and that 80% of the participants failed to obtain at least 6 h sleep during the 28 day data-gathering period. Additionally, over two-thirds of the participants reported struggling to stay awake on duty at least once during the 28 day data-gathering period and 20% reported actually falling asleep on duty (Scott et al. 2006b).

Baseline data from a later pilot study of a Fatigue Countermeasures Program for Nurses found that at baseline the 62 nurse participants averaged only 6.81 ± 1.54 h (range 1.5–9.73) on workdays (Scott et al. 2010). The majority of participants (85%) reported poor quality sleep as measured by scores on the Pittsburg Sleep Quality Index (mean score = 8.35 ± 2.7), and 48% had scores greater than 10 on the Epworth Sleepiness Scale indicative of abnormal daytime sleepiness. Moreover, those who reported difficulties remaining awake were also those who reported shorter sleep durations and were more likely to make errors.

The association between shortened sleep durations and the increase risk of errors in these field studies of health-care providers (Dorrian et al. 2008; Lockley et al. 2004; Scott et al. 2006b) is not surprising, since several laboratory studies previously showed an association between sleep loss and prolonged periods of wakefulness and decreased performance and errors (Dawson and Reid 1997; Pilcher and Huffcutt 1996; Van Dongen et al. 2003).

Pilcher and Huffcutt's meta-analysis (Pilcher and Huffcutt 1996) showed that while sleep deprivation had little effect on the performance of motor tasks, cognitive tasks, particularly longer and more complex tasks, were quite sensitive to sleep deprivation. Paradoxically, partial sleep deprivation (<5 h sleep in 24) had a more profound adverse effect on cognitive tasks than either short-term (≤45 h) or long-term sleep deprivation (>45 h).

Laboratory studies have also shown that moderate levels of prolonged wakefulness can produce performance impairments equivalent to or greater than levels of intoxication deemed unacceptable for driving, working, and/or operating dangerous equipment (Dawson and Reid 1997; Lamond and Dawson 1999). Furthermore, performance on neurobehavioral tests remained relatively stable during the first 17 h of testing, a period the researchers called the normal working day, then decreased linearly, with the poorest performance occurring after 25–27 h of wakefulness (Lamond and Dawson 1998). Performance on the most complex task—grammatical reasoning—was impaired several hours before performance on vigilance accuracy and response latency (20.3 h versus 22.3 and 24.9 h, respectively). Although Dawson and his colleagues (Dawson and Reid 1997; Lamond and Dawson 1998) were the first to report that prolonged periods of wakefulness (20–25 h without sleep) can produce performance decrements equivalent to a blood alcohol concentration (BAC) of 0.1%, numerous other studies have shown that prolonged wakefulness significantly impairs speed and accuracy of hand-eye coordination, decision-making, and memory (Babkoff et al. 1988; Florica et al. 1968; Gillberg et al. 1994; Linde and Bergstrom 1992; Mullaney et al. 1983).

A person who is not sleep deprived performs tasks more efficiently after prolonged wakefulness and can cope better with nonstandard work hours (nights or rotating shifts) than someone with a sleep deficit (Dinges et al. 1996). Individuals working nights and rotating shifts rarely obtain such optimal amounts of quality sleep. Their sleep is shorter, lighter, more fragmented, and less restorative than sleep at night (Knauth et al. 1980; Lavie et al. 1989; Walsh et al. 1981). Older workers have more difficulty coping with shiftwork (Harma et al. 1990; Matsumoto and Morita 1987), experience more frequent sleep disruption (Harma et al. 1990; Matsumoto and Morita 1987; Parkes 1994; Pavard et al. 1982), and have more severe decreases in neurocognitive performance when tested during simulated 12 h night shifts (Reid and Dawson 2001). (The "older" subjects in the latter study (Reid and Dawson 2001) had a mean age of 43.9 ± 6.8 years (range 35–56), similar to that of the nurse workforce in the United States (Steiger et al. 2007).

Finally, studies have shown that accident rates rise after 9 h of consecutive work, double after 12 (Hanecke et al. 1998), and triple by 16 h (Akerstedt 1994). Data from National Transportation Safety Board (NTSB) aircraft accident investigations also show higher rates of error after 12 h (National Transportation Safety Board 1994). Finally, night shifts longer than 12 h and day shifts longer than 16 h have consistently been found to be associated with reduced productivity and more accidents (Rosa 1995).

While extended work hours have previously been associated with increased worker injuries and accidents in other professions, these findings cannot easily be extrapolated to RNs working extended hours and

failing to obtain restorative sleep. In addition, while studies that focused on RNs have produced highly suggestive results, further work is needed to confirm these initial data. For example, Rogers et al. showed that the risk of error increased as much as threefold when hospital staff nurses worked shifts of 12.5 h or longer; but the study relied on self-reporting, and hospitals were unidentified. Thus, errors were probably underreported and it was impossible to determine if the reported errors adversely affected patient outcomes.

Likewise, studies of resident physicians also suffered from conflicting findings and confounding issues that were not taken into account. For example, residents in one study were misclassified as rested when they were actually sleep deprived. Moreover, the use of differing definitions of sleep deprivation, outcome measures, and lack of control groups further confuses attempts at drawing broad conclusions, thus reducing the usefulness of the existing database of studies.

Thus, the result of many studies on the impact of work schedules on patient safety, fatigue, and performance are very suggestive but, ultimately, inconclusive. Currently, we cannot claim there is a direct link between a certain number of work hours and a specific calculable risk to patient safety. The logical supposition that overworked and sleep-deprived nurses are at increased risk of making significant errors in patient care requires further studies to collect unequivocal data.

14.3 Recommendations

While none of the thousands of different shift schedules that have been developed for health-care providers are ideal, some scheduling practices are less disruptive to sleep than others. Critical elements to consider when developing such schedules include time of day, rotation rate, direction of rotation, shift length, and workload (see Table 14.1). Educational programs to alert shift workers and resident physicians about the hazards associated with long work hours and fatigue have also been suggested.

Shifts scheduled at night and those that begin early in the morning are associated with a high risk of sleep deprivation. Night shift workers face not only the challenge of wakefulness promoted by circadian factors such as light and temperature rhythms (National Transportation Safety Board 1994) but also the stress of daily family and domestic responsibilities may further reduce the opportunity to sleep. In fact, a study of paper mill workers showed that they obtained 2 fewer hours of sleep per day when working night shift (10 p.m.–6 a.m.) compared to working afternoon shifts

TABLE 14.1

Recommendations for Work Schedules

Area of Interest	Recommendations
Number of hours worked	1. Minimize the number of shifts of 12.5 h or longer and restrict the number of hours worked per week to <60
	2. Resident physicians should not be allowed to work more than 80 h/week or more than 16 consecutive hours without a protected period of sleep
Night shifts	1. Because of the significant sleep debt accrued when working night shift, employees should not be allowed to work more than three to four consecutive night shifts
	2. Night shift workers should be allotted several consecutive days off (e.g., 3–4) to allow recovery from sleep deprivation
	3. Night shift workers should be allowed to take short naps during their breaks to prevent fatigue
Rotating Shifts	1. Shifts should rotate forward (e.g., days to evenings to nights; or days to nights) rather than backwards (e.g., days to nights, nights to evenings, or nights to days)
	2. Employees should receive several consecutive days off (e.g., 3–4) to recover from sleep deprivation associated with working rotating shifts
Rest breaks	1. Are necessary during a work shift to prevent fatigue
	2. Adequate staffing should be available to cover breaks and all staff should be encouraged to take a short break every 2–3 h
	3. Breaks should be formally scheduled, since employees only rarely accurately judge their level of fatigue and often delay taking breaks longer than appropriate
Duration between shifts	Although the new ACGME recommendations require at least 10 h off between scheduled or expected duty periods for resident physicians, these and other health-care workers should receive more time off between shifts so that they can obtain adequate sleep, spend time with friends and family members, and take care of household tasks
Educational programs	Employees should be educated about the necessity of obtaining sufficient sleep taking regular breaks from work, and taking measures to facilitate their adaptation to night and rotating shifts. These programs should be tailored to the needs of each group of health-care providers
Root cause analysis	When an adverse event occurs, the root cause analysis should include obtaining information about the possibility of fatigue contributing to the adverse event

(2–10 p.m.) (Torsvall et al. 1989). These shortened sleep durations are typically associated with increased sleepiness on duty. For example, 20% of the paper mill workers slept an average of 43 min while on duty at night (Torsvall et al. 1989), and half of the eight nurses working 12 h night shifts in an ICU reported shorter daytime sleep periods and took naps on duty during at least 75% of their work shifts (Daurat and Foret 2004).

Thus, because of the progressive sleep debt that accumulates during successive night shifts, employees should be scheduled for no more than three or four such consecutive shifts. In fact, Folkard and Tucker (2003) have shown that the risk for incidents rises significantly each successive night, being approximately 6% higher on the second night, 17% higher on the third night, and 36% higher on the fourth night. This progressive rise in night-shift risk could not be explained as due only to fatigue associated with successive work days since the day-shift rise was approximately 2% higher on the second day shift, 7% higher on the third day shift and 17% higher on the fourth day shift (Figure 14.1).

Because of the profound sleepiness that develops after working several consecutive night shifts, employees should have several days off between each group of night shifts to allow adequate recovery. In addition, employees who are expected to rotate from night shift to day shift require an additional day off, since they might spend part of the first day off sleeping. Although employees might require up to 3 or 4 days to recover

from night shift work (Akerstedt et al. 2000), some authorities have suggested that recovery from night shift work requires a minimum of two nocturnal sleep periods with at least 9 h spent in bed each night (Jay et al. 2007; Lamond et al. 2007).

Shifts starting between 4 a.m. and 7 a.m. are also associated with significant sleep deprivation and excessive sleepiness (Kecklund et al. 1997; Tucker et al. 1998). Nursing shifts traditionally start at 7 a.m., although data collected during the Staff Nurse Fatigue and Patient Safety study demonstrated that many nurses arrived early and began working before their shift was officially schedule to begin. Resident physicians also routinely begin their workday before 7 a.m. Evening shift workers, although often isolated from family and social activities (Drake and Wright 2010), usually obtain the most sleep, averaging 7.6 h/night compared 6.8–7.0 h/night for most day shift workers (Drake et al. 2004; Pilcher et al. 2000).

Workers on rotating shifts, however, experience nearly as much sleep restriction as permanent night shift workers (Pilcher et al. 2000). In fact, even on their days off, rotating shift workers remain sleepier than day shift workers (Lowden et al. 1998). Both the speed and direction of rotation influence sleep durations. According to a meta-analysis of shift rotations by Pilcher and her colleague (Pilcher et al. 2000), rapid shift rotations (multiple rotations within 1 week) were associated with reduced sleep durations compared to slow rotations (e.g., at least 3 weeks per shift rotation).

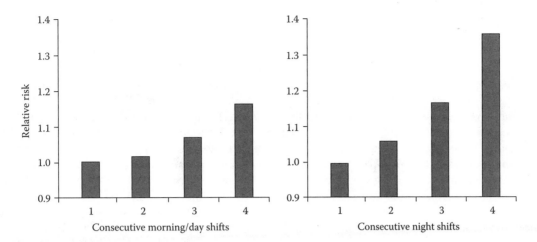

FIGURE 14.1
Risk of accidents and injuries across four consecutive morning or day shifts (left) and across four consecutive night shifts (right), expressed relative to the first shift in the sequence. Data were compiled from five published studies on 8h shift systems. (Reprinted from Dinges, D.F., Baynard, M., and Rogers, N.L., Chronic sleep restriction, in *Principles and Practice of Sleep Medicine*, M.H. Kryger, T. Roth, and W. Dement (Eds.), W.B. Saunders, Philadelphia, PA, Copyright 2005, with permission from Elsevier.)

Both clockwise and counterclockwise rotations negatively impact sleep duration and are associated with circadian misalignment (Czeisler et al. 1982). Nevertheless, most authorities recommend clockwise or forward rotation (Czeisler et al. 1999; Tepas et al. 1997) for example, days to evenings, evenings to nights, and nights to days, since delaying sleep onset is easier than going to sleep earlier. Some workers, however, prefer counterclockwise rotations since it usually provides them with a longer period of time off between rotations to evening shift from night shift (Tepas et al. 1997). The paucity of well-designed studies and the contradictory findings they provide, however, preclude making reliable recommendations. For example, at least one well-controlled study revealed no differences in sleep duration and vigilance when rapidly rotating clockwise and counterclockwise rotations were compared (Cruz et al. 2003), while a second study showed that clockwise rotation is better.

The limited data available on shift schedules of hospital staff nurses suggest that only a minority rotate shifts (9.4%), usually alternating between day and night shifts. The remainder of hospital staff nurses report being schedule to work exclusively day shifts (56.5%), night shifts (28.5%) or evening shifts (5.6%) (Rogers 2003, 2005). Data regarding shift patterns of other health-care providers is not available.

At present it is difficult to determine if working permanent night shifts is more harmful than working rotating shifts. Previously it was believed that those working permanent night shifts would eventually adapt, especially if they worked night shifts consistently for long periods (Drake and Wright 2010). However, more recent examination of melatonin secretion, generally considered one of the best known indicators of the state of the endogenous circadian clock, showed that only about 3% of permanent night shift workers show evidence of complete adjustment of their endogenous melatonin rhythm to night work (Folkard 2008). Nevertheless, night shift workers routinely choose to resume a normal schedule of being active during the day and sleep during the night on their days off (Drake and Wright 2010).

The limited data available suggest that longer shifts are associated with shorter sleep durations (Baldwin and Daugherty 2004; Hayashi et al. 1996; Kageyama et al. 1998; Nakanishi et al. 1999; Park et al. 2001; Sasaki et al. 1999). Prior to the 2003 duty hour reforms, first year residents reported sleeping an average of 5.7±0.90h per night; second year residents reported obtaining 5.98±0.98h of sleep per night (Baldwin and Daugherty 2004). However, the lack of similar national studies of resident sleep times since 2003 precludes making conclusions on the current situation.

Participants in the Staff Nurse Fatigue and Patient Safety Study reported they had an average of 6.9±1.7h sleep when working shifts of <12.5h and 6.5±1.8h sleep when working shifts ≥12.5 consecutive hours (p<0.0001). These data are particularly a concern since the majority of hospital staff nurses now work 12h shifts (Rogers et al. 2004b; Scott et al. 2006b), and neither nurses nor resident physicians get 7–8h sleep per night that three large well-designed studies (Banks and Dinges 2007; Dinges et al. 2005; Van Dongen et al. 2003) have suggested is necessary to avoid cognitive and performance deficits.

Participants averaged 6.70±1.79h sleep, including naps, on workdays and 8.12±2.13h sleep on their days off (p<0.0001), which is a mean difference of 1.42h. Nurses regularly reported to work with limited amounts of sleep (median sleep duration was 6.5h). Only 186 of the 895 participants (20.8%) reported obtaining at least 6.0h sleep prior to every shift they worked. Thirty-eight nurses (4.2%) reported obtaining no sleep in the 24 h period preceding their work shift at least on one occasion during the data-gathering period (range 1–3 shifts). The remaining participants averaged 4.67 ± 3.74 shifts (range 1–19 shifts) where they reported obtaining <6 h sleep in the 24 h period preceding their work shift.

In general, long work shifts require workers to take frequent days off to recover (Akerstedt et al. 2000). In fact, several studies have suggested that at least 24 h of sleep ad libitum is needed to minimize performance deficits when sleep is restricted by long work hours over a 6 day period (Belenky et al. 2003; Van Dongen et al. 2003).

In addition to recommending a full day off each week for recovery, a recent report from the Institute of Medicine (IOM) recommended that resident physicians have at least one 48 h period free of any clinical responsibilities each month (IOM 2009). In addition, the new ACGME regulations specify that by July 2011 (1) work hours for first year residents (PG1 or interns) must not exceed 16 h; (2), the 6 h transition period at the end of a 24 h call period for more experienced residents (PG2 and above) must be shortened to 4 h and that no patient contact may occur during that transition time; (3) there must be at least 10 h between each scheduled or expected duty period; and (4) the 80 h per week limit on duty hours include all in-house call periods and moonlighting (ACGME 2010). The IOM report also recommends strategic napping, especially after 16 h of continuous duty, especially between 10 p.m. and 8 a.m.

While it might be ideal to limit the number of hours worked by health-care providers, not all health-care providers agree with this course of action. Although medical residents have petitioned the Occupational Safety and Health Administration (OSHA) in the past regarding the length of their work hours (Public Citizen 2010), there is significant resistance to limiting the hours worked by nurses (Geiger-Brown and Trinkoff 2010; Janney and Landrigan 2008; Texas Board of Nursing 2008). For example, over 95% of the 11,785 nurses who responded to a Texas Board of Nursing Examiners survey in 2007 opposed any limitation on work hours and 85% of the respondents opposed employers limiting their work hours. In fact, some respondents indicated that "no one has the right to place any kind of limit on the amount of hours a nurse works, even if the nurse works full-time 12 h days at one job and full-time 12 h nights at another job with rotating and/or overlapping shifts on the same day"(Texas Board of Nursing 2008). Others stated that they knew when they were tired (Texas Board of Nursing 2008), an assertion contradicted by numerous studies showing that fatigued individuals regularly underestimate their level of impairment (Van Dongen et al. 2003).

Moreover, nurse managers and leaders—professionals likely aware of the hazards associated with long work hours and insufficient sleep—often feel compelled to offer 12 h shifts in order to recruit and retain sufficient numbers of nurses to provide patient care (Geiger-Brown and Trinkoff 2010a,b; Janney and Landrigan 2008). They are also aware that a 3-day work week is attractive to many nurses, because it allows those desiring greater income to work additional shifts through an agency at premium pay (Janney and Landrigan 2008).

Despite heavy workloads, nurses and other health-care providers should be encouraged and possibly required to take breaks during the workday. It is not uncommon for hospital staff nurses to report working an entire shift without a break from patient care (Rogers et al. 2004a; Tucker and Edmondson 2003). In fact, the Staff Nurse Fatigue and Patient Safety Study found that in only 42.9% of the 11,164 shifts, nurses were completely relieved of patient care responsibilities for a meal and/or break periods during their work shift. Average break periods reported were 26 min, suggesting that the only time the nurse was relieved from patient care responsibilities was for a meal. During the majority of shifts (47.1%), nurses reported being able to sit down for a meal or break, but not being relieved of patient care responsibilities; during 10.1% of the shifts however, participants reported they did not sit down for a break or meal during the entire shift.

Numerous studies however, have suggested that short, frequent breaks can improve performance and reduce fatigue (Galinsky et al. 2000; Rosa 1995; Rosekind et al. 2000). Although many industries, including health care, have adopted extended shifts (e.g., 10 and 12 h), very little information is available regarding the time, duration, and frequency of breaks required when employees work longer shifts. However, most authorities recommend that additional, more frequent breaks be allowed; Costa (Costa et al. 1995) even cautioned that extension of nurses night shifts to 10 h was acceptable only if their workload was reduced and sufficient rest breaks provided.

Workload is another important factor to consider when devising recommendations for shift hours. Laboratory and field studies have shown that task intensity as well as time on task influences work performance (Gander et al. 2010; Spencer et al. 1997). For example, a study of air traffic controllers found that under low workload conditions, their fatigue ratings remained relatively stable for continuous work periods of up to 4 h. However, when the workload was high, their fatigue ratings increased rapidly after 2 h of continuous work (Spencer et al. 1999).

Likewise, residents and RNs commonly work beyond the end of their regularly scheduled shifts to complete patient care (IOM 2009; Rogers et al. 2004b; Scott et al. 2006b; Tucker and Spear 2006).

Indeed, the recent IOM report on resident physician work hours notes that although duty hours were reduced in 2003, this reduction was not accompanied by a reduction in the caseload that residents were expected to manage (Baskies et al. 2008; Ferguson et al. 2005; IOM 2009). Furthermore, as mentioned earlier, the report recommended that residents have at least one 48 h period free of any clinical responsibilities each month (IOM [Institute of Medicine] 2009) and be allowed strategic napping, especially after 16 h of continuous duty, and especially between the hours of 10 p.m. and 8 a.m.

Although the relationship among long hours, fatigue, and workload has received limited attention in the health-care environment, studies have implicated heavy resident workloads with errors by these health-care professionals as well as delays in patient care, and possible detrimental effects on patient outcomes (Jagsi et al. 2008; Ong et al. 2007; Vidyarthi et al. 2007). Furthermore, the IOM report noted that, "Working beyond shift length because of workload contributes to violations in duty hour limits" by both residents and RNs. A variety of reports cited by IOM note that heavy workload, time pressures (e.g., patient severity and nonnursing tasks), and reduced supervision can all contribute to poorer patient care and patient outcomes (e.g., treatment errors and mortality).

Finally, shiftwork education programs have also been recommended (Jha et al. 2010; Tepas et al. 1997), but the efficacy of these programs has not been established. In fact, Tepas and his colleagues (Tepas 1993; Tepas et al. 1997) are highly critical of shiftwork education programs that do not address scheduling issues but instead try to transfer all solutions to shiftwork problems to the worker. The SAFER Program, developed by the American Academy of Sleep Medicine and used by many teaching hospitals for educating resident physicians, consists of a 60–90 min lecture on the neurobiology of sleep, the effects of sleep loss on the personal and professional lives of resident physicians, and outlines effective measures to reduce fatigue and improve performance (Arora et al. 2007). Objective measurements of sleep using actigraphy showed that residents did not obtain more sleep after attending the SAFER presentation, nor did they take prophylactic naps and/or utilize cross-coverage to obtain more sleep when on call.

In response to such problems, a recent study evaluated the feasibility of implementing a Fatigue Countermeasures Program for Nurses, modeled on the SAFER Program and the NASA-AMES Research Center's Fatigue Countermeasures Program. Staffing was adjusted to allow participants to take regular breaks during their work shift, hospital policies that prohibited sleeping during breaks were waived during the study period, and each unit provided a quiet area equipped with a recliner and alarm clock so that participants could take a nap during the night shift. These changes were associated with a significantly increased sleep duration on work nights from 6.82 ± 1.54 to 7.44 ± 1.41 h. In addition, there was a significant decrease in the number of episodes of drowsiness and unplanned sleep on duty (Scott et al. 2010). However, despite the significant increase in sleep duration, almost half the nurses (48%) continued to report abnormal levels of excessive sleepiness on the Epworth Sleepiness Scale and 92% continued to report poor sleep quality on the Pittsburg Sleep Quality Index.

14.4 Conclusions

Laboratory and field studies suggest that sleep deprivation can degrade patient care given by resident physicians and RNs. However, confounding factors in some studies and the potential for misinterpreting observations in others preclude making a definitive link between long work hours and health-care errors. In addition, no studies have shown that errors made by sleep-deprived care workers are generally serious.

Nevertheless, the potential for life-threatening errors exists whether or not health-care providers are sleep deprived; therefore, the potential negative effect of sleep deprivation on performance cannot be ignored. Furthermore, the effect of sleep deprivation on the performance of physician assistants, pharmacists, respiratory therapists, technicians, and other support personnel is still mostly unknown. Therefore, the issue of the role of sleep deprivation in the performance of health-care professionals remains to be resolved. Indeed, the pioneering series of studies about sleep deprivation among hospital staff nurses by Rogers et al. does suggest that there is much to be concerned about, yet much more to learn. In addition, the recent regulations issued by the ACGME, aimed at attempting to limit shift hours of resident physicians, the IOM's rules limiting resident work hours, and the initial attempts by some states to legally limit mandatory overtime demonstrate a broad concern that sleep deprivation degrades the performance of health-care professionals.

Thus, there is a need for further research to resolve previous, conflicting findings and to clarify and verify more recent observations. Unlike the transportation industry, which routinely considers whether fatigue contributed to a particular accident, few health-care institutions gather information about the role of fatigue in patient care errors. The recent studies of fatigue among health-care workers point to the need to follow the lead of the transportation industry by continuing and expanding these studies on behalf of patient safety.

References

ACGME (Accreditation Council Graduate Medical Education). 2008. *The ACGME's Approach to Limit Resident Duty Hours: The Common Standards and Activities to Promote Adherance*, 2003 (cited December 27, 2008). Available from http://www.acgme.org/acWebsite/GME_infor/history.GME/pdf

ACGME. 2010. *Proposed Standards Developed by the ACGME Task Force on Quality Care and Professionalism*, June 23, 2010 (cited July 22, 2010). Available from http://acgme-2010standards.org/

Akerstedt, T. 1994. Work injuries and time of day-national data. Paper read at *Work Hours, Sleepiness, and Accidents*, September 8–10, Stockholm, Sweden.

Akerstedt, T., G. Kecklund, M. Gillberg, A. Lowden, and J. Axelsson. 2000. Sleepiness and days of recovery. *Transportation Research Part F* 3:251–261.

American Nurses Association. 2011. *Mandatory Overtime*, August 5, 2010 (cited January 9, 2011). Available from http://www.nursingworld.org/MainMenuCategories/ANAPoliticalPower/State/StateLegislativeAgenda/MandatoryOvertime.aspx

Arora, V.M., E. Georitts, J.N. Woodruff, H.J. Humphrey, and D. Meltzer. 2007. Improving the sleep hygiene of medical interns: Can the sleep, alertness and fatigue education in residency programs help? *Archives of Internal Medicine* 167 (16):1738–1744.

Association of Perioperative Registered Nurses. 2008. *AORN Position Statement: Safe Work/On Call Practices*, 2005 (cited August 12, 2008). Available from http://www.aorn.org/PracticeResources/AORNPositionStatements/Position_SafeWorkOnCallPractices/

Babkoff, V., M. Mikulincer, T. Caspy, D. Kempinski, and H. Sing. 1988. The typology of performance curves during 72 hours of sleep loss. *The Quarterly Journal of Experimental Psychology* 324:734–756.

Balas, M.C., L.D. Scott, and A.E. Rogers. 2004. The prevalence and nature of errors and near errors reported by hospital staff nurses. *Applied Nursing Research* 17 (4):224–230.

Balas, M.C., L.D. Scott, and A.E. Rogers. 2006. The prevalence and nature of errors and near errors reported by critical care nurses. *Canadian Journal of Nursing Research* 38:21–41.

Baldwin, Jr., D.C. and S.R. Daugherty. 2004. Sleep deprivation and fatigue in residency training: Results of a national survey of first-and second-year residents. Review of April 2004. *Sleep* 27 (2):217–223.

Baldwin, D.W., S.R. Daugherty, R. Tsai, and M.J. Scotti. 2003. A national survey of residents' self-reported work hours: Thinking beyond speciality. *Academic Medicine* 78 (11):1154–1164.

Banks, S. and D.F. Dinges. 2007. Behavioral and physiological consequences of sleep restriction. *Journal of Clinical Sleep Medicine* 3 (5):519–528.

Baskies, M.A., D.E. Ruchelsman, C.M. Capeci, J.D. Zuckerman, and K.A. Egol. 2008. Operative experience in an orthopedic surgery residency program: The effect of work hour restrictions. *Journal of Bone and Joint Surgery—American* 90 (4):924–927.

Belenky, G., N.J. Wesensten, D.R. Thorne, M.L. Thomas, H.C. Sing, D.P. Redmond, M.B. Russo, and T.J. Balkinm. 2003. Patterns of performance degradation and restoration during sleep restriction and subsequent recovery: A sleep dose-response study. *Journal of Sleep Research* 12 (1):1–12.

Costa, G., E. Gaffuri, G. Ghirlanda, D.S. Minors, and J.M. Waterhouse. 1995. Psychophysiological conditions and hormonal secretion in nurses on a rapidly rotating shift schedule and exposed to bright light during night work. Review of February 2004. *Work & Stress* 9 (2–3):148–157.

Cruz, C., A. Boqut, C. Detwiler et al. 2003. Clockwise and counterclockwise rotating shifts: Effects on vigilance and performance. *Aviation, Space and Environmental Medicine* 74:606–614.

Czeisler, C.A., J.F. Duffy, T.L. Shanahan et al. 1999. Stability, precision and near 24-hour period of human circadian pacemaker. *Science* 284:2177–2181.

Czeisler, C.A., M.C. Moore-Ede, and R.H. Coleman. 1982. Rotating shift work schedules that disrupt sleep are improved by applying circadian principles. *Science* 217:460–463.

Daurat, A. and J. Foret. 2004. Sleep strategies of 12-hour shift nurses with emphasis on night sleep episodes. *Scandinavian Journal of Work Environment & Health* 30 (4):299–305.

Dawson, D. and K. Reid. 1997. Fatigue, alcohol, and performance impairment. *Nature* 388 (6639):235.

Defoe, D.M., M.L. Power, G.B. Holzman et al. 2001. Long hours and little sleep: Work schedules of residents in obstetrics and gynecology. *Obsetrics and Gynecology* 97:1015–1018.

Dinges, D.F., M. Baynard, and N.L. Rogers. 2005. Chronic sleep restriction. In *Principles and Practice of Sleep Medicine*, M.H. Kryger, T. Roth, and W. Dement (eds.). Philadelphia, PA: W.B. Saunders.

Dinges, D.F., R.C. Graeber, M.R. Rosekind, A. Samuel, and H.M. Wegman. 1996. *NASA Technical Memorandum 110404, Principles and Guidelines for Duty and Rest Scheduling in Commercial Aviation*. Moffett Field, CA: NASA.

Dorrian, J., C. Tolley, N. Lamond, C. van den Heuvel, J. Pincombe, A.E. Rogers, and D. Dawson. 2008. Work hours, sleep and errors in a group of Australian hospital nurses. *Applied Erogonomics* 39 (5):605–613.

Drake, C.L., R. Roehrs, G. Richardson et al. 2004. Shift work sleep disorder: Prevalence and consequences beyond that of symptomatic day workers. *Sleep* 27:1453–1462.

Drake, C.L. and K.P. Wright, Jr. 2010. Shift work, shift-work disorder, and jet lag. In *Principles and Practice of Sleep Medicine*, M.H. Kryger, T. Roth, and W.C. Dement (eds.). St. Louis, MO: Elsevier.

Ferguson, C.M., K.C. Kellogg, M.M. Hutter, and A.L. Warshaw. 2005. Effect of work-hour reforms on operative case volume of surgical residents. *Current Surgery* 62 (5):535–538.

Florica, V., E.A. Higgins, P.F. Iampietro, M.T. Lategola, and A.W. Davis. 1968. Physiological responses of man during sleep deprivation. *Journal of Applied Physiology* 24:169–175.

Folkard, S. 2008. Do permanent night workers show circadian adjustment? A review based on the endogenous melatonin rhythm. *Chronobiology International* 25:215–224.

Folkard, S. and P. Tucker. 2003. Shift work, safety, and productivity. *Occupational Medicine (Oxford)* 53 (2):95–101.

Friedmann, P. 2007. The ACGME Approach to Limiting Resident Duty Hours: Promoting Paitient Safety, Resident Education, and Resident Well-being. http://www.com.edu/workforce/ResidentDutyHours/TheACGMEApproachtoLimitingResidentDutyHoursFriedmann.pdf. December 3.

Friedman, R.C., J.T. Bigger, and D.S. Kornfeld. 1971. The intern and sleep loss. *NEJM* 285:201–203.

Galinsky, T.L., N.G. Swanson, S.L. Sauter, J.J. Hurrell, and L.M. Schleifer. 2000. A field study of supplementary rest breaks for data-entry operators. *Ergonomics* 43 (5):622–638.

Gander, P., R.C. Graeber, and G. Belenky. 2010. Fatigue risk management. In *Principles and Practice of Sleep Medicine*, M.H. Kryger, T. Roth, and W.C. Dement (eds.). St. Louis, MO: Elsevier.

Geiger-Brown, J. and A.M. Trinkoff. 2010a. Is it time to pull the plug on 12-hour shifts? Part 1. The evidence. *Journal of Nursing Administration* 40 (3):100–102.

Geiger-Brown, J. and A.M. Trinkoff. 2010b. Is it time to pull the plug on 12-hour shifts? Part 3, Harm reduction strategies if keeping 12-hour shifts. *Journal of Nursing Administration* 40 (9):357–359.

Gillberg, M., G. Kecklund, and T. Akerstedt. 1994. Relations between performance and subjective ratings of sleepiness during a night awake. *Sleep* 17:236–241.

Hanecke, K., S. Tiedemann, F. Nachreiner, and H. Grzech-Sukalo. 1998. Accident risk as a function of hour at work and time of day as determined by accident data and exposure models for the German working population. *Scandinavian Journal of Work Environment and Health* 24 (Suppl 3):43–48.

Harma, M., P. Knauth, J. Ilmarinen et al. 1990. The relation of age to the adjustment of the circadian rhythms of oral temperature and sleepiness in shift work. Review of June 2003. *Chronobiology International* 7:277–333.

Hayashi, T., Y. Kobayashi, K. Yamaoka, and E. Yano. 1996. Effect of overtime work on 24-hour ambulatory blood pressure. Review of November 2004. *Occupational and Environmental Medicine* 38 (10):1007–1011.

Health Resources and Services Administration. 2007. *Preliminary Findings: 2004 National Sample Survey of Registered Nurses*, 2005 (cited February1, 2007). Available from http://bhpr.hrsa.gov/healthworkforce/reports/rnpopulation/preliminaryfindings.htm

IOM (Institute of Medicine). 2009. *Resident Duty Hours: Enhancing Sleep, Supervision and Safety*. Washington, DC: The National Academies Press.

Jacques, C.H., J.C. Lynch, and J.S. Samkoff. 1990. The effects of sleep loss on cognitive performance of resident physicians. *Journal of Family Practice* 30 (2):223–229.

Jagsi, R., D.F. Weinstein, J. Shapiro, B.T. Kitch, D. Dorer, and J.S. Weissman. 2008. The Accreditation Council for Graduate Medical Education's limits on resident work hours and patient safety: A study of resident experiences and perceptions before and after work hour reductions. *Archives of Internal Medicine* 168 (5):493–500.

Janney, M. and C.P. Landrigan. 2008. Improving nurse working conditions: Towards safer models of hospital care. *Journal of Hospital Medicine* 3 (3):181–183.

Jay, S.M., N. Lamond, S.A. Ferguson, J. Dorrian, C.B. Jones, and D. Dawson. 2007. The characteristics of recovery sleep when recovery opportunity is restricted. *Sleep* 30 (3):353–360.

Jha, A.K., B.W. Duncan, and D.W. Bates. 2010. *Chapter 46: Fatigue, Sleepiness and Medical Errors*. AHRQ 2001 (cited December 20, 2010). Available from http://archive.ahrq.gov/clinic/ptsafety/chap46a.htm

Kageyama, T., N. Nishikido, T. Kobayashi, Y. Kurokawa, T. Kaneko, and M. Kabuto. 1998. Long commuting time, extensive overtime and sympathodominant state assessed in terms of short-term heart rate variability among male white collar workers in the Tokyo megalopolis. *Industrial Health* 36:209–217.

Kahol, K., M.J. Leyba, M. Deka, V. Deka, S. Mayes, M. Smith, J.J. Ferra, and S. Panchanatha. 2008. Effect of fatigue on psychomotor and cognitive skills. *American Journal of Surgery* 195 (2):195–204.

Kalish, B.J. and H. Lee. 2009. Nursing teamwork, staff characteristics, work schedules and staffing. *Health Care Management Review* 34 (4):323–333.

Kecklund, G., T. Akerstedt, and A. Lowden. 1997. Morning work: Effects of early rising on sleep and alertness. *Sleep* 20:215–223.

Knauth, P., K. Landau, C. Droge, M. Schwitteck, M. Widynski, and J. Rutentranz. 1980. Duration of sleep depending on the type of shift work. *International Archives of Occupational and Environmental Health* 46:167–177.

Lamond, N. and D. Dawson. 1999. Quantifying the performance impairment associated with fatigue. *Journal of Sleep Research* 8(4): 255–262.

Lamond, N.S., S.M. Jay, J. Dorrian, S.A. Ferguson, C.B. Jones, and D. Dawson. 2007. The dynamics of neurobehavioral recovery following sleep loss. *Journal of Sleep Research* 16 (1):33–41.

Landrigan, C.P., L.K. Barger, B.E. Cade, T.A. Ayas, and C.A. Czeisler. 2006. Interns' compliance with accreditation council for medical education work-hour limits. *JAMA* 296 (9):1063–1070.

Landrigan, C.P., C.A. Czeisler, L.K. Barger et al. 2007. Effective implementation of work-hour limits and systematic improvements. *Joint Commission Journal of Quality and Patient Safety* 33 (Suppl 11):19–29.

Landrigan, C.P., J.M. Rothschild, J.W. Cronin, R. Kaushal, E. Burdick, J.T. Katz, C.M. Lilly, P.H. Stone, S.W. Lockley, D.W. Bates, and C.A. Czeisler. 2004. Effect of reducing interns' work hours on serious medical errors in intensive care units. *New England Journal of Medicine* 351 (18):1838–1848.

Lavie, P., N. Chillag, and R. Epstein et al. 1989. Sleep disturbance in shift workers: A marker for maladaptation syndrome. *Work & Stress* 3:33–40.

Linde, L. and M. Bergstrom. 1992. The effect of one night without sleep on problem-solving and immediate recall. *Psychological Research* 54:127–136.

Lipscomb, J.A., A.M. Trinkoff, J. Geiger-Brown, and B. Brady. 2002. Work-schedule characteristics and reported musculoskeletal disorders of registered nurses. Review of November 2004. *Scandinavian Journal of Work Environment and Health* 28 (6):394–401.

Lockley, S.W., J.W. Cronin, E.E. Evans, B.E. Cade, C.J. Lee, C.P. Landrigan, J.M. Rothschild, J.T. Katz, C.M. Lilly, P.H. Stone, D. Aeschbach, and C.A. Czeisler. 2004. Effect of reducing interns' weekly work hours on sleep and attentional failures. *New England Journal of Medicine* 351 (18):1829–1837.

Louwe, H. and A. Kramer. 2001. Case studies of nursing care facility staffing issues and quality of care. In *Report to Congress: Appropriateness of Minimum Nursing Staffing Ratios in Nursing Homes Phase II Final Report*, A. Associates (ed.). Baltimore, MD: Centers for Medicare & Medicaid Services.

Lowden, A., G. Kecklund, J. Axelsson, and T. Akerstedt. 1998. Change from an 8-hour shift to a 12-hour shift, attitudes, sleep, sleepiness and performance. Review of March 2003. *Scandinavian Journal of Work Environment and Health* 24 (Suppl 3):69–75.

Matsumoto, K. and Y. Morita. 1987. Effects of night-time nap and age on sleep patterns of shift workers. Review of June 2003. *Sleep* 10:580–589.

Mullaney, D., D.F. Kripke, P.A. Fleck, and L.C. Johnson. 1983. Sleep loss and nap effects on sustained continuous performance. *Psychophysiology* 20:643–651.

Nakanishi, N., K. Nakamura, S. Ichikawa, K. Suzuki, and K. Tatara. 1999. Lifestyle and the development of hypertension: A 3-year follow-up study of middle-aged Japanese male office workers. *Occupational Medicine* 49:109–114.

National Association of Neonatal Nurses and the National Association of Neonatal Nurse Practitioners. 2008. *Position Statement #3043, Neonatal Advanced Practice Nurses Shift Length, Fatigue and Impact on Patient Safety*, 2007 (cited August 15, 2008). Available from www.nann.org/pdf/APNFatigue_FINAL_08.2007_rev09.07.pdf

National Transportation Safety Board. 1994. *A Review of Flightcrew-Involved Major Accidents of U.S. Air Carriers, 1978 though 1990*. Washington, DC: NTSB.

Ong, P.E., A. Bostrom, A. Vidyarthi, C. McCulloch, and A. Auerbach. 2007. House staff team workload and organization effects on patient outcomes in an academic general internal medicine service. *Archives of Internal Medicine* 167 (1):47–52.

Park, J., Y. Kim, H.K. Chung, and N. Hisanaga. 2001. Long working hours and subjective fatigue symptoms. Review of December 2002. *Industrial Health* 39:250–254.

Parkes, K. 1994. Sleep patterns, shiftwork, and individual differences: A comparison of onshore and offshore control room operators. Review of June 2003. *Ergonomics* 34:827–844.

Pavard, A., J. Vladis, J. Foret et al. 1982. Age and long term shiftwork with mental load: Their effects on sleep. Review of June 2003. *Journal of Human Ergology* 11 (Suppl):303–309.

Pilcher, J.J. and A.I. Huffcutt. 1996. Effects of sleep deprivation on performance: A meta-analysis. Review of November 2004. *Sleep* 19 (4):318–326.

Pilcher, J.J., B.J. Lambert, and A.I. Huffcutt. 2000. Differential effects of permanent and rotating shifts on self-report sleep length: A meta-analytic review. *Sleep* 23:755–762.

Prasad, M., T.J. Iwashyna, J.D. Christie, A.A. Kramer, J.H. Silber, K.G. Volpp, and J.M. Kahn. 2009. Effect of work-hours regulations on intensive care unit mortality in United States teaching hospitals. *Critical Care Medicine* 37 (9):2564–2569.

Public Citizen. 2010. *Petition to Reduce Medical Resident Work Hours*, 2010 (cited November 17, 2010). Available from http://www.citizen.org/hrg1917

Reid, K. and D. Dawson. 2001. Comparing performance on simulated 12 hour shift rotation in young and older subjects. Review of June 2003. *Occupational and Environmental Medicine* 58 (1):58–62.

Rogers, A.E. 2003. Hospital staff nurses regularly report fighting to stay awake on duty. *Sleep* 26 (Suppl):A 423.

Rogers, A.E. 2005. Staff nurse fatigue and patient safety. In *Kansas State Nurses Association Annual Meeting*. Topeka, KA.

Rogers, A.E., W.-T. Hwang, and L.D. Scott. 2004a. The effects of work breaks on staff nurse performance. *Journal of Nursing Administration* 34 (11):512–519.

Rogers, A.E., W.-T. Hwang, L.D. Scott, L.H. Aiken, and D.F. Dinges. 2004b. The working hours of hospital staff nurses and patient safety. *Health Affairs* 23 (4):202–212.

Rogers, A.E., W.-T. Hwang, L.D. Scott, and D.F. Dinges. 2006. A diary based examination of nurse sleep patterns and patient safety [abstract]. *Sleep* 29 (Suppl):A115.

Rosa, R.R. 1995. Extended workshifts and excessive fatigue. *Journal of Sleep Research* 4 (Suppl 2):51–56.

Rosekind, M.R., E.L. Co, K.B. Gregory, and D.L. Miller. 2000. Crew factors in flight operations XIII: A survey of fatigue factors in corporate/executive aviation operations. Moffitt Field, CA: NASA Ames Research Center.

Rosen, A.K., S.A. Loveland, P.S. Romano, K.M.F. Itani, J.H. Silber, O.O. Even-Shoshan, M.J. Halenar, Y. Teng, J. Zhu, and K.G. Volpp. 2009. Effects of resident duty hour reform on surgical and procedural patient safety indicators among hospitalized veterans health administration and medicare patients. *Medical Care* 47 (7):723–731.

Sasaki, T., K. Iwasaki, T. Oka, N. Hisanaga, T. Ueda, Y. Takada, and Y. Fujiki. 1999. Effect of working hours on cardiovascular-autonomic nervous functions in engineers in an electronics manufacturing company. *Industrial Health* 37 (1):55–61.

Scott, L.D., N. Hofmeister, N. Rogness, and A.E. Rogers. 2010. An interventional approach for patient and nurse safety: A fatigue countermeasures feasibility study. *Nursing Research* 59:250–258.

Scott, L.D., W.-T. Hwang, and A.E. Rogers. 2006a. The impact of multiple care giving roles on fatigue, stress, and work performance among hospital staff nurses. *Journal of Nursing Administration* 36 (2):86–95.

Scott, L.D., A.E. Rogers, W.-T. Hwang, and Y. Zhang. 2006b. The effects of critical care nurse work hours on vigilance and patient safety. *American Journal of Critical Care* 15 (4):30–37.

Spencer, M.B., A.S. Rogers, and C.L. Birch. 1999. A diary study of fatigue in air traffic controllers during a period of high workload. *Shiftwork International Newsletter* 16:88.

Spencer, M.B., A.S. Rogers, and B.M. Stone. 1997. A review of the current scheme for the regulation of air traffic controller hours (SCRATCOH). Farnborough, England: Defence Evaluation and Research Agency.

Stefancy, A.L. 2009. One-hour, off-unit meal breaks. *American Journal of Nursing* 109 (1):64–66.

Steiger, D.M., S. Bausch, B. Johnson, and A. Peterson. 2007. *The Registered Nurse Population: Findings from the March 2004 National Sample Survey of Registered Nurses.* Washington, D.C.: US Department of Health and Human Services, Health Resources and Services Administration, Health Resources and Services Administration, Bureau of Health Professions.

Tepas, D.I. 1993. Educational programs for shift workers, their families, and prospective shift workers. *Ergonomics* 36 (1–3):199–209.

Tepas, D., M. Paley, and S. Popkin. 1997. Work schedules and sustained performance. In *Handbook of Human Factors and Ergonomics*, G. Salvendy (ed.). New York: John Wiley.

Texas Board of Nursing. 2008. *Nursing Work Hours Public Hearing Summary April 2007, Nurses Offer Feedback on Proposed Nursing Work Hours Position Statement*, 2008 (cited September 23, 2008). Available from http://www.bon.state.tx.us/practice/pdfs/nwh-summary.pdf

Torsvall, L., T. Akerstedt, K. Gillander et al. 1989. Sleep on the night shift: 24 hour EEG monitoring of spontaneous sleep/wake behavior. *Psychophysiology* 26:352–358.

Trinkoff, A., J. Geiger-Brown, B. Brady, J. Lipscomb, and C. Muntaner. 2006. How long and how much are nurses now working? *American Journal of Nursing* 106 (4):60–71.

Tucker, A.L. and A.C. Edmondson. 2003. Why hospitals don't learn from failures: Organizational and psychological dynamics that inhibit system change. Review of November 2003. *California Management Review* 45 (2):55–72.

Tucker, P., L. Smith, I. Mcdonald, and S. Folkard. 1998. The impact of early and late shift changeovers on sleep, health, and wellbeing in 8- and 12-hour shift systems. Review of November 2004. *Journal of Occupational Health Psychology* 3 (3):265–275.

Tucker, A.L. and S.J. Spear. 2006. Operational failures and interruptions in hospital nursing. *Health Services Research* 41 (3):643–662.

U.S. Department of Labor, Bureau of Labor Statistics. 2010. *Occupational Outlook Handbook, 2008–2009, Physicians and Surgeons, on the Internet.* U.S. Department of Labor 2007 (cited July 22, 2010). Available from http://www.bls.gov/oco/ocos074.htm

U.S. House and Senate. 2004. *Department of Veterans Affairs Health Care Personnel Enhancement Act of 2004.* 108th Congress, S.2484, Section 7456A Nurses alternate work schedules.

Van Dongen, H.P.A., G. Maislin, J.M. Mullington, and D.F. Dinges. 2003. The cumulative cost of additional wakefulness: Dose-response effects on neurobehavioral functions and sleep physiology from chronic sleep restriction and total sleep deprivation. Review of April 2003. *Sleep* 26 (2):117–126.

Vidyarthi, A.R., A.D. Auerbach, R.M. Wachter, and P.P. Katz. 2007. The impact of duty hours on resident self reports of errors. *Journal of General Internal Medicine* 22 (2):205–209.

Volpp, K.G., A.K. Rosen, P.R. Rosenbaum, P.S. Romano, O. Even-Shoshan, A. Canamucio, L. Bellini, T. Behringer, and J.H. Silber. 2007a. Mortality among patients in VA hospitals in the first 2 years following ACGME resident duty hour reform. *JAMA* 298 (9):984–992.

Volpp, K.G., A.K. Rosen, P.R. Rosenbaum, P.S. Romano, O. Even-Shoshan, Y. Wang, L. Bellini, T. Behringer, and J.H. Silber. 2007b. Mortality among hospitalized medicare beneficiaries in the first 2 years following ACGME resident duty hour reform. *JAMA* 298 (9):975–983.

Walsh, J.K., D.I. Tepas, and P.D. Moss. 1981. The EEG sleep of night and rotating shift workers. In *Biological Rhythms, Sleep, and Shift Work*, L.C. Johnson, D.I. Tepas, W.P. Colquhon, and M.J. Colligan (eds.). New York: Spectrum.

Weiss, G.G. 2005. Exclusive survey-productivity: Work hours up, patient visits down. *Medical Economics* 83 (22):57–58, 60, 62–63.

Section IV

Physical Ergonomics

15

The Physical Environment in Health Care

Carla J. Alvarado

CONTENTS

15.1 Introduction 215
15.2 Environment of Care: Standards, Guidelines, Rules, and Reference Texts 217
    15.2.1 Environment Design for the Disabled 219
    15.2.2 Environmental Design Provisions for Disasters 219
15.3 Individual Built Environment Components 219
    15.3.1 Space and Physical Constraints 219
    15.3.2 Arrangement of Components 220
    15.3.3 Environment Components and Health-Care-Acquired Infection 222
15.4 Climate and Thermal Environments 222
    15.4.1 Effects of Clothing on Heat Exchange 223
15.5 Air Quality 224
    15.5.1 Air-Handling Systems in Health-Care Facilities 225
    15.5.2 Hospital Environments with Special Air Handling 225
15.6 Noise 226
    15.6.1 Noise and Performance 227
    15.6.2 Alarms 228
15.7 Vibration 229
15.8 Illumination 229
15.9 Conclusion 231
References 232

> It is the unqualified result of all my experience with the sick that, second only to their need of fresh air, is their need of light; that, after a close room, what hurts them most is a dark room and that it is not only light but direct sunlight they want.
>
> **Florence Nightingale (1860)**

15.1 Introduction

Florence Nightingale is mostly known for her radical innovations in nursing care. But besides being a nurse, reformer, statistician, epidemiologist, and humanitarian, Nightingale was a hospital designer. Nightingale's *Notes on Hospitals* (Nightingale, 1863) provided detailed recommendations on the proper physical environment for civilian health-care institutions. Some of these recommendations remain valid even today. Nightingale felt patients needed fresh air circulation for infection prevention and sunlight and windows with outside views to raise patient spirits and quicken healing. Just as patient safety and the patient experience were at the forefront of Nightingale's work in the design of the hospital environment, so does it continue in the design of modern health-care facilities.

Despite a dramatic economic downturn in the U.S. economy, predictions point to resumption of construction/renovation at health-care facilities in the coming future (Hrickiewicz, 2009). This will include an increasing emphasis on human factors engineering (HFE) and more proactive HFE involvement in the design of the environment of care from the architectural concept to patient occupancy. The goal of the this chapter is to identify overall human factors components and specific key environmental elements (light, sound, climate, arrangement of space, etc.) that directly affect the patient experience in the health-care system. These human factors components and elements not only influence patients' satisfaction, privacy, confidentiality, safety, stress, but also their health-care providers' safety, quality of care, and quality of working life.

This chapter discusses facilities common to communities in the United States. Facilities with unique services such as free-standing ambulatory surgical centers, long-term care and/or assisted living, hospice care, home care, mobile emergency care, psychiatric hospitals, and special care facilities such as Alzheimer's and other dementia units may require special environmental considerations. However, sections of the chapter will be useful for all care settings and the human factors principles presented applicable in most settings.

The care environment discussed in this chapter is constituted by those features in a built health-care entity that are created, structured, and maintained to support quality health care (AIA and FGI, 2001). As more patient care moves out of the traditional health-care settings and into the patient's home, patients and their families have become more involved in the course of care. Medical devices continue to increase the technology of care both in the built environment, the mobile environment, and now the home care setting. Facilities need to respond to the changing physical environment requirements for accommodations for both people and technologies,

considering human factors principles in their response to the surrounding needs for physical space.

The physical environment does not constitute the total environment of care for patient safety. Social factors and organizational factors are also highly important and contribute to the human factors/systems approach to patient safety. It is important to ensure that the physical environment enhances the dignity of the patient and complies with federal and state regulations through environmental features that permit patient privacy and confidentiality in all care settings. Other chapters in this book address the social, organizational, and systems factors related to health care and patient safety.

Human factors is generally defined as the study of human beings and their interaction with products, environments, and equipment in performing tasks and activities (Czaja, 1997). These interactions take place within a physical environment. If one revisits the late Dr. Avidis Donabedian's classic model of quality of care, which developed into the now-standard method for evaluating hospitals in the country, it incorporates three major components of care: structure, process, and outcome (Donabedian, 1988). Within Donabedian's well-known structure–process–outcome model, Carayon and colleagues embed a work system/human factors model (Figure 15.1) developed by Carayon and Smith

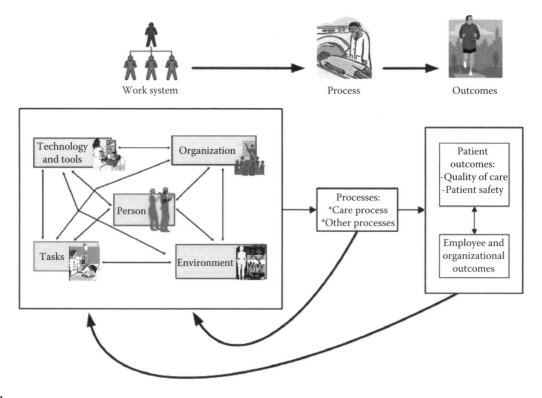

FIGURE 15.1

SEIPS model of work system and patient safety. (From Carayon, P. et al., *Qual. Saf. Health Care*, 15(suppl 1), 50, 2006; Carayon, P. and Smith, M.J., *Appl. Ergon.*, 31, 649, 2000; Smith, M.J. and Carayon-Sainfort, P., *Int. J. Ind. Ergon.*, 4, 67, 1989; Smith, M.J. and Carayon, P., Balance theory of job design, in W. Karwowski (Ed.), *International Encyclopedia of Ergonomics and Human Factors*, Taylor & Francis, London, U.K., pp. 1181–1184.)

(Smith and Carayon-Sainfort, 1989; Smith and Carayon, 2000; Carayon et al., 2006).

This Systems Engineering Initiative for Patient Safety model explicitly recognizes the interdependent nature of the five major aspects of a work system—a *care provider* performing various *tasks* using *tools and technology* in a given *environment* within an established *organization*. Thus, the model provides necessary specification to the structural elements posed by Donabedian. The *physical environment* affects the processes of care or the *task*, the *people* performing or receiving this care, and the *tools and technology* being used in the specific care *organization*. Research has now shown the effect of the physical or built environment on clinical outcomes. Some elements of the physical environment affect staff performance more directly. Health-care acquired infections (HAIs) and physiologic responses to light, noise, temperature, etc. are the most often studied (Topf, 1985, 1988, 1992, 2000; Topf and Dillon, 1988; Buemi et al., 1995; Cabrera and Lee, 2000; Topf and Thompson, 2001; Morrison et al., 2003; Sehulster and Chinn, 2003; Bartley et al., 2010). Potential patient safety hazards in the environment are often localized in the medical devices used by the worker; therefore, much of patient safety analysis in the literature focuses on hazards associated with care devices and procedures. However, as shown in Figure 15.1, patient safety hazards are created by a combined interaction of the task, equipment, organizational, and environmental conditions with the patient or the health-care provider. The physical environment can either support or impede what the patients and staff want and need to do in a health-care facility. Thus, the physical environment can add to caregiver and/or patient stress and can be a major detriment to the course of a patient's care. The physical environment should be designed to reduce patient, family, and staff stress wherever possible.

Patient and employee satisfaction is another outcome that may be influenced by the physical environment but is less studied. Although elements of the physical environment may support or detract from patients' health-care experience and staff's quality of working life, these issues will not be addressed in this chapter. Research- and evidence-based materials are available on the subject of patient satisfaction and staff's quality of working life to support these environmental design goals (Rubin et al., 1998). Additionally, medical device controls and displays, workstation design, mechanical hazards (strains, sharps injury, slips, and falls), electrical hazards (shock, fire), toxic substances (medical gases, therapy agents, radiation), and fire hazards are all associated with the physical environment of patient care. Well-established standards and guidelines exist for these areas and are referenced throughout in the chapter but shall not be expanded upon.

15.2 Environment of Care: Standards, Guidelines, Rules, and Reference Texts

Several excellent human factors texts are available for a detailed discussion of the physical environment. Much of the discussion and recommendations made in this chapter are referenced to these texts, see Table 15.1.

In addition to these human factors texts, there are many professional society guidelines, accreditation standards, codes, and federal, state, and local rules addressing the environment of care. But be aware, many manufacturers' recommendations and societal guidelines or standards are based on research documentation in reviewed scientific literature and some are not. Specific contents of many standards and guidelines are based on older technologies and may not meet current needs; others simply are not in agreement in content or recommendations and the extent of their disagreement can be quite

TABLE 15.1

Human Factors Texts

Title	Authors	Publisher
Handbook of Human Factors and Ergonomics, 3rd Edition (2006)	Salvendy G. (Ed.)	New York: John Wiley & Sons
An Introduction to Human Factors Engineering, 2nd Edition (2003)	Wickens, C. D., Lee, J. D., Liu, Y., and Gordon Becker, S. E.	Upper Saddle River, NJ: Pearson Prentice Hall
Human Factors in Engineering and Design, 7th Edition (1993)	Sanders, M. S. and McCormick, E. J.	New York: McGraw-Hill, Inc.
Kodak's Ergonomic Design for People at Work, 2nd Edition (2003)	Chengalur, S. N., Rodgers, S. J., and Bernard, T. E.	Hoboken, NJ: John Wiley & Sons.
Handbook of Human Factors in Medical Design, (2010)	Weinger, M. B., Wiklund, M. E., and Gardner-Bonneau, D. J.	Boca Raton, FL: CRC Press, Taylor & Francis Group

large. Applying external recommendations in isolation without considering the timeliness, science, and interactions of other related documents may result in less than optimum outcomes. Additionally, the interactions of the various components of the Systems Engineering Initiative for Patient Safety (SEIPS) model (Figure 15.1) should always be taken into consideration in the design of the built environment. Based on the model's elemental interactions, the physical environment and its impact on quality care can be ever changing.

The chapter will not generate yet another set of detailed recommendations for environmental design nor will it go into great detail regarding the standards, rules etc. on the various environmental elements. Therefore, it presents a summary of current U.S. standards, guidelines, and rules to consider when addressing the human factors elements and design of any health-care environment (Table 15.2).

In 2006, the American Institute of Architects (AIA) revised their edition of the *Guidelines for Design and Construction of Hospitals and Health Care Facilities* (AIA, 2006). The *Guidelines for Design and Construction of Health*

Care Facilities is referenced by architects, engineers, and health-care professionals throughout the United States and in other countries who are planning new or renovated health-care facility construction. Authorities in 42 states, the Joint Commission, and several federal agencies use the Guidelines as a reference, code, or standard when reviewing construction designs and plans and completed health-care facilities. Applicable environmental aspects of this document include the therapeutic environments discussed in this chapter and should be consulted. Perhaps the most controversial issue the AIA revised document addressed was the need for patient single versus semi-private room occupancy. Consider, for example, the human factors elements in this proposed recommendation. Sources of stress for patients are perceived lack of privacy and control, noise, and crowding (Schmaker and Pequegnat, 1989) and a single room may address these issues. HAIs may be passed between patients occupying the same room via the hands of their health-care providers and/or shared equipment. However, most falls occur in patient rooms, among elderly patients, attempting to go to the

TABLE 15.2

Originators of U.S. Codes, Guidelines, Standards, and References Applicable to the Health-Care Environment

Organization	Web Address
Agency for Healthcare Research and Quality (AHRQ) The Hospital Built Environment	http://www.ahrq.gov/qual/hospbuilt/hospenv.pdf
American Society of Healthcare Engineering of the American Hospital Association	http://fgiguidelines.org/
American Institute of Architects (AIA)	http://aia.org
American Society of Heating, Refrigerating and Air-Conditioning Engineers (ASHRAE)	http://www.ashrae.org
American National Standards Institute	www.ansi.org
American Society of Mechanical Engineers (ASME)	http://asme.org
American Society for Testing Materials (ASTM)	http://www.astm.org
Americans with Disabilities Act (ADA)	http://www.ada.gov/
Association for the Advancement of Medical Instrumentation (AAMI)	http://www.aami.org
Association for Professionals in Infection Control and Epidemiology (APIC)	http://apic.org
Centers for Disease Control and Prevention (CDC)—Hospital Infection Control Practices (HICPAC)	http://www.cdc.gov
Department of Defense (DOD)	http://www.defense.gov/
ECRI	http://ecri.org
Food and Drug Administration (FDA)	http://www.fda.gov
Human Factors and Ergonomics Society	http://www.hfes.org
Illuminating Engineering Society of North America (IESNA)	http://www.iesna.org
International Code Council (ICC)–International Building Code	http://www.iccsafe.org
The Joint Commission	http://www.jointcommission.org/
Leadership in Energy and Environmental Standards	http://www.usgbc.org/DisplayPage.aspx?CategoryID=19
National Council on Radiation Protection and Measurements (NCRP)	http://www.ncrp.com
National Fire Protection Association (all fire codes are adopted for use by CMS)	http://www.nfpa.org/codes/index/html
Nuclear Regulatory Commission (NCR)	http://www.ncr.gov
Occupational Safety and Health Administration	http://www.osha.gov

bathroom and when patients are alone (Kerzman et al., 2004). Multioccupancy patient rooms may provide more constant supervision and therefore may be more appropriate for patient safety in this scenario. Recently, more evidence suggests a single patient room is associated with reduction of risk of cross infection, greater flexibility in operation, and fewer room transfers (Chaudhury et al., 2003). This literature affirms conversion to single rooms can substantially reduce the rate at which patients acquire infectious organisms while in the intensive care units (ICU) (Bracco et al., 2007; Peleg and Hooper, 2010; Teltsch et al., 2011). The current AIA guideline states single-bed rooms as the minimum standard for medical/surgical and postpartum nursing units in general hospitals. There is no reason to believe, given the studies' outcomes, this guideline recommendation will change in the future and single patient rooms are the standard of care in new and remodeled health-care facilities. Nonetheless, it is important to remember that the ultimate interpretation of information contained in these guidelines is the responsibility of the state or federal authority having jurisdiction over the health-care facilities.

15.2.1 Environment Design for the Disabled

The Americans with Disabilities Act (ADA) became law in 1990 ("Americans with Disabilities Act of 1990," 1990). Under Titles II and III of the ADA, public, private, and public service hospitals and other health-care facilities environment will need to comply with the accessibility and usability guidelines (DOJ, 2010). It should be recognized, however, that the users of health-care facilities often have very different environmental accessibility needs from the typical adult individual with disabilities addressed by the ADA model standards and rule. Hospital patients, and especially long-term care nursing facility residents, due to their stature, reach, and strength characteristics, typically require the assistance of caregivers during transfer maneuvers or activities of daily living. Many prescriptive requirements of model accessibility standards place both older persons and caregivers at greater risk of injury. Flexibility in the requirements may be permitted for the use of assistive devices/configurations that provide safety considerations for this assistance should be considered in the human factors/ergonomics assessment of the physical environment.

15.2.2 Environmental Design Provisions for Disasters

In health-care settings where there is recognized potential for hurricanes, tornadoes, flooding, earthquakes or other regional disasters, environmental design should consider the need to protect the life safety of all health-care facility patients and workers. Additionally, the environment must address the very significant need for continuing care services following such disasters. Acute care facilities with emergency and critical care services can serve as receiving, triage, and treatment centers in the event of nuclear, biological or chemical exposures. Human factors principles are applicable to the designation and design of specific areas for these functions and the threat of terrorism initiatives (IOM, 2002). The U.S. Department of Defense (DoD) determined mass smallpox vaccinations can be conducted safely with very low rates of serious adverse events. Program implementation emphasized human factors issues, such as careful staff training, contraindication screening, recipient education, and attention to bandaging. DOD experience suggests that broad vaccination programs may be implemented with fewer serious adverse events than previously believed (Grabenstein and Winkenwerder, 2003). Additionally, health-care data security is an essential requirement in all antibioterrorism efforts. Developers of medical information systems should utilize the existing security development and human factors evaluation methods to foresee as many of the environmental and technical factors that may endanger data security as possible, and apply appropriate precautions against this threat (Niinimaki et al., 1998).

15.3 Individual Built Environment Components

15.3.1 Space and Physical Constraints

The health-care environment presents many challenging venues in space and physical constraints. The physical space in which patient care will be provided must be sufficiently large to accommodate the functions, people, and devices for which it is intended. In the case of health care, if the space is too constrained, not only could patients and staff be uncomfortable and care may be impaired, but there might also be a greater risk of errors resulting in harm to patients and caregivers. Hospitals and clinics are forever adding patient care equipment and on-site storage is often limited or not considered at the time of purchase. Equipment that cannot be placed in a convenient location within a work area is likely to be used less effectively or even less often (Carayon et al., 2010). The goal of human factors is to design systems that reduce human error, increase productivity, and enhance safety and comfort (Wickens et al., 2004). Workspace design and component placement (medical equipment, furniture, workstation, etc.) is one of the major areas that improve the fit between humans, medical devices, and the patient care physical environment. Important issues include workspace, equipment and furnishings, placement, reach dimensions, clearance dimensions, and visual dimensions in the design of the ambient environment. Ideally, we would like to place each component in an optimum location for providing safe quality care. This optimum would be predicated on the human capabilities and characteristics, including sensory capabilities, and anthropometric (data on body dimensions of various populations) and biomechanical (mechanics of muscular activity) characteristics, the interindividual differences of the users (Allread and Israelski, 2010). Unfortunately, in the crowded environment of a health-care facility this is not always possible. Only so many medical device displays can fit in the optimum viewing area; only so many patient rooms can be close to the nursing station; only so many controls can be placed in an area for rapid response. An additional factor that must be considered is that in the same individuals, these differences change over time. The average age of a nurse in the United States is now in the mid to late forties. Individuals' sensory and musculoskeletal systems change over time and the physical environment must be accommodated as the human body ages.

Sanders and McCormick (Sanders and McCormick, 1993) provide specific principles that should be applied when determining the arrangements of components within a physical space:

1. Importance principle. This principle states that important components be placed in convenient locations. "Importance" refers to the degree to which the component is vital to the achievement of the objectives of the system. As an example, the scanning device used in a medication bar coding system would need to be located in a convenient location for the nursing staff use as it is essential to the bar coding scan.

2. Frequency-of-use principle. This principle states that frequently used components be placed in convenient locations. As examples, medication carts should be accessible and convenient to nurses and pharmacy personnel alike; computer order entry equipment should be convenient to patient care venues, not hidden in a report room.

3. Function principle. The function principle of arrangement provides for the grouping of components according to their function. Thus, temperature indicators and temperature controls might well be grouped; central venous catheter placement equipment and sterile field supplies stored together on a designated "line-placement cart."

4. Sequence-of-use principle. In the use of items, sequences or patterns of relationships frequently occur in performing some task. In applying this principle, the items would be arranged so as to take advantage of such patterns. As examples, patient resuscitation equipment is located at the head of the patient bed; commercial packs for urinary drainage placement start with sterile gloves on the top layer of the pack to facilitate removal and donning prior to catheter insertion.

15.3.2 Arrangement of Components

Arranging components in a work space, be the computer workstations and monitors in a nursing station or controls and displays at an ICU bed, requires that availability of the relevant data and the use of certain methods in applying that data. The types of data relevant for arranging components in the environment generally will fall into three broad categories (Sanders and McCormick, 1993):

1. Basic data about human beings. Anthropometric and biomechanical data on sensory, cognitive, and psychomotor skills. Such data generally come from research and are published in various tables, source books, and papers (Panero and Zelnick, 1979; Gordon et al., 1989;. Kroemer, 1989; ISO, 1996; Kroemer et al., 2001).

2. Task analysis data. These are data about the work activities of the people who are involved in the specific system or work environment. Task analysis is at the heart of the human factors contributions to arranging components in a work space. Often care providers have experience performing some level of task analysis as they participate in root cause analysis (RCA), failure modes and effects analysis (FMEA), or usability testing exercises in their health-care facilities.

3. Environmental data. This category covers any relevant environmental features of the patient care situation such as lighting, noise, temperature, etc. A discussion of these topics follows in the chapter.

When modifying an existing environment, data relating to this environment is appropriate to collect. The data can be obtained in various methods. Observations, videotaping, and interviews with both new and experienced personnel are all useful methods to obtain the necessary data.

In the case of new systems, facilities information about the activities to be performed can be based on the same methods as above and inferred from tentative drawings, plans, procedures or concepts available (Reiling et al., 2004).

Relationships between components, whether they are care providers, patients, or things, are called *links*. Links fall generally into three classes:

1. Communication links:
    a. Visual (person to person or person to equipment)
    b. Auditory, voice (person to person, person to equipment, or equipment to person)
    c. Touch (person to person, person to equipment, or equipment to person)
2. Control links:
    a. Equipment controls (person to equipment)
3. Movement links from one location to another:
    a. Eye movements
    b. Manual and/or foot movements or both
    c. Body movements

The kind of information collected about links includes how often the components are linked and in what sequence the links occur. As an example, how often does the nurse leave the patient bedside to procure care supplies? Where do these needed supplies reside? How often is a patient status monitor viewed and immediately before or after viewing, the information communicated to another care provider?

Probably the most common method of arranging components by using link data is through trial and error. The designer physically arranges scale drawings of the components, trying to keep the most frequently used components in the most advantageous locations. In recent literature on designing and building a patient safe hospital, environmental failure modes and effects analysis or enhanced (E)FMEA was performed (Reiling et al., 2004). Regardless of what technique is used, it is important that the resulting arrangement be validated by using a mock-up or simulator with actual health-care providers carrying out actual or simulated tasks (Sanders and McCormick, 1997). No matter how thorough the design data, end-users' "hands on" input is vitally important.

Additionally, the environmental assessment needs to consider equipment storage requirements and reprocessing requirements and just how the equipment interacts with the available space and other critical equipment in that space. Devices designed for clinical arenas are often difficult to modify for use in patient homes or mobile care units such as ambulances and helicopters, where use and storage space may be far more limited and other

environmental factors such as noise, vibration, temperature, and moisture influence the device efficacy.

The design of the inanimate environment has influence on the care providers' personal ease, with important components for patient care being placed at convenient locations for frequent retrieval, line of sight and control, and consistency with other layouts within the system. HFE as it applies to this user-environment-control strategy can effectively reduce the patient risk within these components.

The following physical space issues should be considered in all health-care venues:

- Clearance problems are among the most often encountered health-care-space issues. They are most important with medical care devices. The space needed for the medical equipment and between and around it is critical for the operators' proper and safe use and the patients' safety. The transport care environment, such as ambulances and helicopters, are extremely limited in space and often pose a clearance challenge. Home care may pose an extreme challenge, where the use and storage of the medical equipment must compete with the family's personal belongings and no governmental safety standards or fire codes are applicable.

- Reach requirements of the medical devices and environment should not exceed the arms of the smallest user, as they operate a hand-operated device and/or activate a foot pedal (Wickens et al., 2004).

- In addition to reach requirements, critical controls or devices must be accessible and not placed in tandem with other critical controls or devices that may inadvertently be activated or deactivated. Although link analysis may suggest minimizing distances between components, such as the sequential links between controls, there is physical space required in the operation of various controls to avoid touching other controls. One must consider the combination of anthropometric factors (such as larger fingers and hands or longer and shorter arms, etc.) and the psychomotor movements made in the use of the device controls.

- A well-designed environment should also consider the cleaning and maintenance requirements of the medical equipment and environmental surfaces. What space is required for the reprocessing or maintenance? What physical hazards, that is, moisture, electrical cords, brushes, tools etc. will be added to the physical space for this work? What cleaning requirements would pose hazards to patients and care providers in daily environmental hygiene?

- Medical equipment placement should ensure that the visual displays can be easily seen and read by the care providers in the inanimate environment. This requires proper positions and clear line of sight with respect to the medical equipment design and other equipment in the area. The normal line of sight is usually considered to be about 15° below the horizon (Sanders and McCormick, 1993). The area for most convenient visual displays has generally been considered to be defined by a circle roughly 10°–15° in radius around the normal line of sight. However, when arranging the care environment, one must be mindful of patient privacy and patient-sensitive information being displayed inappropriately.

- Falls resulting in injury are relatively common in health care for both patients and workers (Morse, 2002; Herwaldt and Pottinger, 2003; Tinetti, 2003; Kerzman et al., 2004). Fall hazards may be created by a combination of medical equipment placement and environmental conditions. People can fall and sustain injuries in a number of ways associated with the placement of objects including furniture placement, proximity of bathroom or portable commode to patient bed, slipping on wet surfaces created by the patient, their care equipment, medications, placement of patients' drinking water, tripping over equipment cords or tubing associated with the equipment or tripping over the equipment placement in the care environment (Edelberg, 2001; Morse, 2002; Kerzman et al., 2004).

15.3.3 Environment Components and Health-Care-Acquired Infection

The inanimate health-care facility environment as it directly relates to the environment and patient care procedures was rarely implicated in disease transmission, except among patients who are immunocompromised (Maki et al., 1982; Sehulster and Chinn, 2003). However, the importance of the inanimate environment has undergone renewed emphasis, given the emergence of multi-drug-resistant organisms (MDROs) present in health-care settings. Environmental surfaces and patient care equipment can transfer the organisms onto the hands of care providers (Duckro et al., 2005; Vonberg and Gastmeier, 2006; Hota et al., 2009). Inadvertent exposures to environmental pathogens via medical devices (e.g., endoscopes, respiratory therapy equipment) (Rutala, 1996), improper heating, ventilating, and air conditioning (HVAC) systems and airborne pathogens (e.g., *Mycobacterium tuberculosis*

and *varicella-zoster* virus), contaminated environmental surfaces contact, and lack of provider hand hygiene (MDROs), device-related chemical exposure (heat intolerant medical devices requiring lower temperature liquid sterilants or ethylene oxide (ETO) sterilization and aeration), or medical devices creating potentially infective aerosols in the environment can result in adverse patient outcomes and cause illness among patients and health-care providers (CDC, 1994; Rutala, 1996; Sehulster and Chinn, 2003; Tablan et al., 2003).

The incidence of health-care-environment-associated adverse outcomes including patient infections can be minimized by the following:

- Accurate and clearly written manufacturers' instructions on medical device care, use, and storage in infection prevention

- Appropriate use of cleaners and disinfectants with medical devices and in the inanimate environment

- Appropriate maintenance of the HVAC system and water systems

- Adherence to water-quality standards for equipment using water from main lines (e.g., water systems for hemodialysis, ice machines, hydrotherapy equipment, dental unit water lines, and automated endoscope reprocessors)

- Ambient temperature and ventilation standards for specialized care environments (e.g., airborne infection isolation rooms, protective environments, or operating rooms)

- Following general human factors guidelines in designing individual patient room and workplace accessibility to personal protective and isolation gear such as gloves, masks, and gowns and especially hand hygiene opportunities

15.4 Climate and Thermal Environments

There are six main factors that should be considered in order to assess the health-care thermal environment:

- Air temperature
- Radiant temperature
- Air velocity and quality
- Humidity
- The activity of the occupants—both patients and their care providers
- The clothing worn by the patients (or lack thereof) and their care providers

Personal comfort varies with time of day, season, diet, health status, and clothing choices as well as job or task stress and cultural variables and expectations (Chengalur et al., 2004). When the body becomes "too hot or too cold" it reacts in a way that is consistent with maintaining core temperature at a relatively constant level. The initial reaction to heat is vasodilatation, where peripheral blood vessels dilate and transfer blood, and hence heat, to the surface of the body where it is lost to the surrounding environment and sweating occurs. When the body is exposed to cold, the initial reaction is vasoconstriction, where the peripheral blood vessels constrict and reduce the flow of blood to the body surface that reduces heat loss (Parsons, 2000).

There have been numerous studies of the effects of thermal environments on human performance (Parsons, 1993). A number of general conclusions can be made on these effects. Important individual considerations are: subjective and psychological parameters, degree of acclimatization and factors that contribute to individual differences, that is, body composition, gender. As heat stress increases, there will be effects on physical and mental performance, but decrements in performance occur not only at high environmental temperatures. Performance at vigilance tasks, frequent in health care, can be lowest in slightly warm environments that can have soporific effects (Parsons, 2000). For example, tasks such as monitor surveillance may be compromised as the technician becomes drowsy in a warm environment or tasks caregivers perform comfortably in street clothes or hospital uniforms become "warmth discomfort" when carried out in patient isolation grab (moisture resistant gowns, masks, gloves) over clothes or uniforms.

The effects of cold on health-care providers' performance are often ignored and can be very significant. While there are few effects on mental performance, cold can act as a "secondary task," that is, shivering, or physical movement to "keep warm," hence increasing workload and possibly decreasing mental and physical performance of the given task. The effects of cold on manual performance are attributed to physiological reactions to cold. Slowing of movements, due to stiffening of joints and slow muscle reaction, numbness, and a loss of strength are all associated with vasoconstriction caused by cold. These reactions cause deterioration in manual dexterity and hence performance at many manual tasks (Parsons, 1993). Tactile sensitivity is related to skin temperature. In general, cold stress is less of an occupational hazard than heat stress. However, providing emergency care out of doors often exposes the care provider to the cold and elements such as snow and ice. Under these extreme conditions, a person may not even be able to determine what they are touching without visual reference. Additionally, due to the requirements of the Occupational Safety and Health Administration

(OSHA) Bloodborne Pathogens Rule (DOL, 2001) for moisture resistance and the use of personal protective gowns during surgical procedures, many operating rooms (OR) have lowered ambient room temperature to accommodate the medical staff wearing this extra protective gear. Other OR health-care providers, not directly performing the surgical procedure are often cold in this setting, causing shivering and reduced tactile sensitivity and all patients undergoing surgical procedures, no matter how minor, are at risk, to varying degrees, of developing hypothermia.

The heat exchange process is very much affected by four environmental conditions: air temperature, humidity, air flow, and the temperature of surrounding surfaces, that is, walls, ceilings glass windows, heat producing equipment, etc. (Sanders and McCormick, 1993). The interaction of these conditions is quite complex and cannot be covered in detail here. But some familiar examples in health-care environments are the extremes in heat temperatures associated with reprocessing of medical devices and instruments, instrument washers and steam sterilizers, hospital laundry facilities, facilities' kitchen equipment such as stove tops, ovens, and automated dishwashers and sanitizers. As mentioned above modern health-care environments rarely exhibit extremes of cold, but health-care providers and patients can suffer reactions to cold in air conditioning and medical transport. However, care provider and patient safety related to extremes in temperature become more problematic in pre-hospital, emergency care, and transportation (Rossman, 1992; Helm et al., 2003). For example, drugs used in pre-hospital emergency medical service (EMS) in principle are subject to the same storage restrictions as hospital-based medications. The pre-hospital emergency environment, however, often exceeds these storage recommendations. Main stress factors are sunlight, vibration, and extreme temperature, which may lead to alteration in chemical and physical stability of stored pharmaceuticals, as well as microbiological contamination and concentration enhancement of pharmacological inserts (Helm et al., 2003). Additionally, risk of weather-related emergency vehicle accidents is associated with adverse weather conditions (Weiss et al., 2001).

15.4.1 Effects of Clothing on Heat Exchange

The clothes the care providers wear can have a profound effect on the heat exchange process. It is the insulating effect of clothing that reduces heat loss to the environment. When it is cold, such as out of doors or in extreme air conditioning, reduced heat loss is beneficial but when it is hot, clothing interferes with the heat loss and can be harmful (Sanders and McCormick, 1993). The insulation value of most clothing materials is associated with the amount of trapped air within the weave and fibers and permeability of the material to moisture. In hot environments, evaporation of sweat is vital to maintain thermal equilibrium and materials that interfere with this process can result in heat stress or even heat stroke. In a cold environment, if evaporation of sweat is impeded, a garment can become soaked with perspiration, thus reducing its insulating capacity and warmth. Consider the example of the operating room and surgical instrument decontamination and reprocessing. The surgical reprocessing technician is required to be outfitted according to Bloodborne Pathogens (BBP) precautions, as exposure to human blood and other infectious materials is highly anticipated in the job duties. The technician will most likely wear undergarments, socks, shoes and shoe covers, a tight weave cotton scrub suit or uniform, a moisture proof cover gown or apron, protective gloves, head cover, a splash/moisture resistant mask, and a face-shield or safety glasses. The work environment contains sources of high heat—hot water, steam, and hot stainless steel surgical instrument trays. Under prolonged exposures to this scenario, heat stress could result, effecting quality of working life and perhaps quality of patient care. Although most modern health-care facilities are closely climate controlled, specific areas of high heat and cold may remain an environmental hazard to patient and worker safety.

Additionally, in cold environments the amount of insulation capacity needed may interfere with the performance of patient care tasks; heavy gloves reduce manual dexterity, heavy coats or jackets may reduce range of motion, hats or hoods may reduce hearing, vision, etc. When health-care providers are in cold or hot environments, efforts should be made to optimize their comfort, protect their health, and maximize their safe performance.

15.5 Air Quality

In addition to the effects of heat and cold, there are other aspects of thermal comfort associated with the air. Draught can be a serious problem in ventilated and air conditioned buildings. Draught has been identified as one of the most annoying environmental factors in workplaces. People are most sensitive to air movement at the head region, that is, head, neck, and shoulders (Sanders and McCormick, 1993). Draught in the health-care environment may be associated with HVAC systems requiring rapid air exchange such as in operating rooms or employing high-efficiency particulate air (HEPA) filtration for patient isolation techniques.

Humidity is the amount of moisture in the air. The indoor humidity to a large extent tracks the prevailing outdoor humidity. In cool, dry winter weather, heating the air lowers the relative humidity, and in summer, because most air-conditioning systems do not humidify, dry conditions cannot be avoided. Exposure to low humidity can occur in situations such as out of doors (i.e., emergency medical technicians (EMTs), first aid stations) care in most desert regions during summer and patient care in air-conditioned buildings. Similar dry conditions can occur during winter months in heated buildings. Low humidity can result in dryness in noses and throats, dry skin and chapped lips (Sanders and McCormick, 1993). A perceivable level of eye irritation is experienced by both wearers and nonwearers of contact lens when the relative humidity is at or below 30% and the effect becomes pronounced after 4 h (Rohles and Konz, 1987). Although no peer-reviewed scientific publications associate adverse patient safety events and humidity-associated dry eyes, theoretically there could be caregiver vision irritation and impairment due to low environmental humidity. As humidity increases, discomfort will be felt at the higher end of the thermal comfort zone. If at all possible, patient care and medical devices should not increase the humidity above 55% for person performing light work (Chengalur et al., 2004).

Popular belief holds that weather influences moods. The Weather Channel© discovered that the most frequent expressions used to describe the effects of heat and humidity are as follows: lazy and tired, slow or unproductive, sick or uncomfortable, and irritable and angry. However, little support in the scientific literature for this belief has been provided by the few studies assessing impact of humidity and temperature in the workplace. With regard to associated injury illness, humans are more tolerant of high humidity than high temperatures (Sanders and McCormick, 1993). Most high humidity effects on thermal comfort are conditions associated with the high temperature of the environment. If a person cannot efficiently evaporate heat-induced sweat away from their body, due to the high moisture content in the surrounding air, their thermal comfort declines.

Humid conditions associated with patient safety can occur and may lead to error when the care providers do not consider that medical devices are now used in many various arenas of care. For example, in home use, changing temperatures or high humidity (such as a bathroom or shower) may affect medical device performance. Areas of high humidity and patient care must also consider medical device storage requirements and degradation of device components such as adhesives and chemical reagents and relative humidity requirements for optimal use. Elevated humidity above the desired range could result in condensation and/or moist supplies, damaged equipment, and microbial build up in the environment. There is no prescribed specific humidity level for health-care settings.

15.5.1 Air-Handling Systems in Health-Care Facilities

When designing an air-handling system for use in health-care facilities, use AIA construction and environmental guidelines (AIA, 2006) and American Society of Heating, Refrigerating and Air conditioning Engineers (ASHRAE) Standard 170 Ventilation of health-care facilities (ASHRAE, 2003, 2008) as minimum standards where state or local regulations are not in place for design and construction of health-care facility ventilation systems. Individuals performing the environmental human factors analysis should be familiar with these guidelines and consider all the potential interactions between the people, medical devices, and the air-handling systems in various care settings. The Joint Commission's Environment of Care standards require hospitals to install and maintain appropriate pressure relationships, air exchange rates, and filtration efficiencies for ventilation systems serving areas designed to control airborne contaminants, such as biological agents, gases, fumes and dust (JCAHO, 2005). It is important to note that care technology, medical devices or equipment, should not generate any airborne contaminants that exceed these prescribed containment standards and guidelines. Most importantly, medical devices used in this environment must not create an exhaust flow sufficient to modify the direction of the room's pressure, that is, that an infectious disease isolation room would maintain negative pressure to the general unit hallway or an operating room positive pressure to corridor air.

15.5.2 Hospital Environments with Special Air Handling

In addition to general hospital ventilation, four areas of patient care are considered to have additional or special air-handling requirements; therefore, these additional areas need to be considered when conducting a human factors and environmental assessment or design for these areas (CDC, 1994; Sehulster and Chinn, 2003; Tablan et al., 2003). Although the actual percentage of HAIs directly related to the environment is unknown, the mortality, morbidity, and costs of mitigation are considerable in hospital construction and/or remodeling (Vonberg and Gastmeier, 2006).

1. *Airborne infection isolation* (AII) refers to the isolation of patients infected with organisms spread via airborne droplet nuclei <5 μm in diameter. This isolation area receives numerous air changes per hour (ACH) (>12 ACH for new construction as of 2001; >6 ACH for construction before 2001), and is under negative pressure, such that the direction of the air flow is from the outside adjacent space (e.g., the corridor) into the room. The air in an AII room is preferably exhausted to the outside, but may be recirculated provided that the return air is filtered through a HEPA filter. The use of personal respiratory protection is also indicated for persons entering these rooms when caring for tuberculosis (TB) or smallpox patients and for staff who lack immunity to airborne viral diseases (e.g., measles or varicella zoster virus [VZV] infection). If one applies the SEIPS model to this environment, it is apparent that the medical equipment should be accessible and usable when the operator is in full personal protective equipment: mask, isolation gown, gloves, and protective eyewear. Medical equipment used in this environment must not create an exhaust flow sufficient to modify the direction of the isolation room's negative pressure. Additionally, if not disposable, the medical device often stays in the room environment; hence, power supplies, device care, and maintenance must be considered in the human factors analysis and design of the environment for patient safety, as the device often does not come out of the room until the patient leaves.

2. *Protective environment* (PE) is a specialized patient-care area, usually in a hospital, with a positive air flow relative to the corridor (i.e., air flows from the room to the outside adjacent space). The combination of HEPA filtration, high numbers of air changes per hour, and minimal leakage of air into the room creates an environment that can safely accommodate the immunocompromised patient. Immunocompromised patients are those patients whose immune mechanisms are deficient because of immunologic disorders (e.g., human immunodeficiency virus [HIV] infection or congenital immune deficiency syndrome), chronic diseases (e.g., diabetes, cancer, emphysema, or cardiac failure), or immunosuppressive therapy (e.g., radiation, cytotoxic chemotherapy, antirejection medication, or steroids). These patients who are identified as high-risk patients have the greatest risk of infection caused by airborne or waterborne microorganisms. The caregiver will often be gowned, gloved, and masked, protecting the patient from outside the room microbes. The medical device in this environment must facilitate use under such protective gear and conditions. Medical devices used in this care environment should not generate or collect dust (possible reservoir *Aspergillus spp.*), not create a wet environment or reservoirs of

water (i.e., *Pseudomonas spp.*, atypical mycobacteria associated with contaminated device reservoirs), and not produce unfiltered exhaust. If the medical equipment generates an exhaust, HEPA filtration of the exhaust must be incorporated into the environment or equipment design.

3. *Ventilation requirements for ORs* maintain positive-pressure ventilation with respect to corridors and adjacent areas. People and medical equipment for the OR use should not block the room ventilation or create an unfiltered exhaust in this environment. The equipment must be easily used by care personnel in full surgical gowns, gloves, hair coverings, and protective eyewear. Ultraviolet (UV) lights should not be used in the equipment design, as they pose a risk to both patient and operator and are without efficacy in preventing air-associated surgical-site infections. An environment that requires OR doors to remain open for extended periods of time should not be designed. OR doors should be closed except for the passage of equipment, entry to essential personnel and the patients.

4. *Procedures for infectious TB patients* who also require emergency surgery. All medical devices in this setting must be adaptable for use with the operator wearing an N95 respirator approved by the National Institute for Occupational Safety and Health without exhalation valves or Powered Air Purifying Respirators (PAPR). Because of these Personal Protective Equipment (PPE) requirements, other environmental components such as hearing, sight, and range of motion may be compromised for patient safety and should be addressed.

Other potentially infectious hazards associated with the air in the health-care setting include medical lasers. In addition to the direct hazards to the eye and skin from the laser beam itself, it is also important to address other hazards associated with the use of lasers. These nonbeam hazards, in some cases, can be life threatening, for example, electrocution, fire, and asphyxiation. Because of the diversity of these hazards, the human factors environmental analysis may wish to include input from laser safety experts and/or industrial hygienists to the hazard evaluations. Surgical lasers and dental lasers should be designed for operator use while wearing appropriate PPE, including N95 or N100 respirators, to minimize exposure to laser plumes and splatter. Most likely the operator of the laser and the assistants will be gowned, masked, gloved, and wearing laser protective eyewear. The physical environment

cannot interfere with any of these prescribed motions in the use of the patient care equipment and, in turn, the protective gear should not compromise hearing, sight, and range of motion.

15.6 Noise

Prior to the era of technology, our health-care environmental noise consisted of human voices, activities of daily living such as housekeeping, meal preparation/distribution, or the sounds of nature coming from the outdoors. The hospital atmosphere of the 1940s and 1950s was still one of austere silence, as in a library reading room. Hallways displayed a ubiquitous picture of a uniformed nurse, finger to the lips, sometimes accompanied by the words, "Quiet Please." Signs on the street read, "Hospital Zone–Quiet." The occasional overhead page for a physician signaled a true emergency. But that subdued setting has gradually been replaced by one of turbulence and frenzied activity. People now dart about in a race against time; telephones ring loudly; intercom systems blare out abrupt, high-decibel messages that startle the unsuspecting listener. These sounds are superimposed on a collection of beeps and whines from an assortment of electronic gadgets—pocket pagers, call buttons, telemetric monitoring systems, electronic intravenous machines, ventilator alarms, patient-activity monitors, and computer printers. The hospital, designed as a place of healing and tranquility for patients and of scholarly exchanges among physicians, has become a place of beeping, buzzing, banging, clanging, and shouting (Grumet, 1993). Noise has become such a pervasive aspect of health-care settings and daily life that we refer to as noise pollution, as we might reference a chemical spill or infectious agent introduced into the environment. Noise is a source of stress for patients (Schmaker and Pequegnat, 1989). Excess noise can lead to increased anxiety and pain perception, loss of sleep, and prolonged convalescence (Baker, 1984, 1993; Williams, 1988; Baker et al., 1993). Noise can increase heart rate, subjective stress, and annoyance and possibly impair task performance, concentration, and complex problem solving (Morrison et al., 2003). Noise can mask the sound of critical patient monitor alarms, interfere with communication, and can cause misinterpretation of care measures such as breath or chest sounds or blood pressure. The impact of sound in the ICU environment affects staff members as well as patients and their families, and may contribute to elevated heart rate, stress, and annoyance levels independent of the stress produced by caring for critically ill patients. These stressful conditions may contribute to "burnout" and negatively impact staff retention (Oates and Oates, 1995, 1996; Walsh-Sukys et al., 2001; Morrison et al., 2003).

But noise in certain environmental circumstances may actually be helpful. For example, low levels of continuous noise (the hum of an HVAC system fan) can mask the more disruptive effects of distracting noise (the tick of the clock or the conversation at the nursing station). Soft background music in the environment may even be soothing and stress-reducing to patients and care staff (Cabrera and Lee, 2000).

As the reader knows, the ear is the sensory organ associated with hearing. The stimulus for hearing is sound, a vibration (actually compression and rarefaction) of the air molecules. The acoustic stimulus can therefore be presented as a sine wave, with amplitude and frequency. The frequency of the stimulus more or less corresponds to its *pitch*, and the amplitude corresponds to its *loudness*. When describing the effects of sound on hearing, the amplitude is typically expressed as a *ratio* of sound pressure, measured in *decibels* (dB). In addition to amplitude and frequency, two other critical dimensions of the sound are its *temporal characteristics*, sometimes referred to as the *envelope* in which a sound occurs and its *location* (Wickens et al., 2004). For example, temporal characteristics are what distinguish the wailing of an ambulance siren from a steady horn blast of someone's car alarm, and the location (relative to the listener) is of course what distinguishes the ambulance siren pulling up behind from that of the ambulance about to cross the intersection in front (Wickens et al., 2004).

A detailed description of the ear as a sensory organ is found in many texts (Sanders and McCormick, 1993; Wickens et al., 2004) and will not be discussed here. However, it should be mentioned that a physical factor responsible for loss in sound transmission is the potential loss of hearing of the listener. Simple human aging is responsible for a large portion of hearing loss, *Prebycusis*, particularly sound in the high-frequency regions. Degeneration of the hair cells of the organ of Corti in the cochlea of ear results in aging nerve deafness. Once nerve degeneration has occurred, it rarely can be remedied, only accommodated with hearing devices. As the care provider work force continues to age, hearing critical alarms and performing hearing-reliant tasks may become compromised. Additionally, our aging patient population becomes an important consideration in alarm choice and design for warnings, especially in nursing homes.

15.6.1 Noise and Performance

The effects of noise on human physical and mental performance can be divided into effects on nonauditory task and auditory task performance (e.g., interference with speech communication). The effects of noise on nonauditory task performance in the literature have been inconclusive: different studies indicating noise has no effect on or even increases task performance (Sanders and McCormick, 1993, Parsons, 2000).

Noise can interfere with auditory communication of information (speech, warning signals, etc.) and therefore decrease task performance and possibly safety. Humans can detect signals within a background noise. It is important to know the "efficiency" of this detection within a specific type of background noise to be able to assess the effects of background noise on communication or to design a warning signal for that environment. Without going into great detail, the detection threshold of a signal within a background noise can be represented as signal-to-noise (SIN) ratio over noise frequency and criteria exists and indices established defining an articulation index (Parsons, 2000). As with other environmental stressors, noise can add to the workload of a care task and can potentially affect safe performance. Additionally, the onset of a loud noise will cause a *startle response*, often characterized by muscle contractions, blink, and head or hand jerk movement (Sanders and McCormick, 1993). These responses are relatively transient and settle back to normal or near normal levels very quickly; but, for example, consider the safety outcomes of a startle response in delicate, precise surgery or invasive bedside procedures such as central venous catheter placement. The startle response can disrupt perceptual motor function.

A variety of health-care environment noise sources have been measured and guidelines have been established for the ambient environment and auditory warning and alarm signals. As previously discussed, masking is when one aspect of the sound environment reduces the sensitivity of the ear to another component of the sound environment. The basic idea in this SIN is that if an auditory signal occurs in the presence of noise, the threshold of detection of that signal is raised, and this elevated threshold should be exceeded by the signal if it is to be detected (Sanders and McCormick, 1993). This may lead to what could be referred to as "auditory signal inflation" as the environment gets louder, so must the alarm. Perhaps the better solution is to reduce the background noise and in turn the signal. This of course may not be possible in medical transport where the competing noise is generated by the vehicle engine itself, helicopter rotors, ambulance sirens, etc. Studies found that blood pressure values obtained by medical personnel in a quiet environment were significantly more accurate than those obtained in an ambulance (Prasad et al., 1994), and that emergency care providers could correctly identify 96% of breath sounds in a quiet environment, only 54% in an ambulance (Brown et al., 1997).

The World Health Organization and the Environmental Protection Agency (EPA) recommend that hospital noise levels not exceed 40–45 dB(A) during the day and 35 dB(A) at night. Sound levels above 50 dB(A) are sufficient to cause sleep disturbance, and sustained levels above 85 dB(A) can damage hearing (Morrison et al., 2003). Although the levels of noise in hospitals do not approach industrial time weighted averages of noise (manufacturing plant lines can yield continuous noise at 80–95 dB(A), requiring PPE hearing protection), substantial levels of noise are encountered in hospitals. The average equivalent decibel level found in an operating room was in the range of 60–65 dB(A), but the sound-reflective tile surfaces and stainless-steel equipment tended to enhance harsh and reverberant noises (Grumet, 1993). A simple task like popping open an envelope of sterile surgical gloves yields a sound of 86 dB(A), dropping a stainless-steel bowl can generate an abrupt or explosive "impulse" sound at 108 dB(A), and the sound of escaping gas as anesthesia tanks are changed reaches 103 dB(A). When an impulse sound exceeds the background level by about 30 dB, a "startle" reaction can occur in the listener. Patients in the recovery room are generally exposed to 50–70 dB(A), but shouts between staff members and the shifting of bedrail positions expose them to short-term noise at 90 dB(A). Similarly, hospital laboratories generally have background hums in the 60–68 dB(A) range, whereas the noisy automated machines in laboratory workplaces are in the 65–74 dB(A) range. Computer rooms at peak activity reach 85 dB(A), and a pneumatic-tube carrier arrives with an 88 dB(A) thud (Grumet, 1993).

Sleep loss is a major contributor to "intensive care unit psychosis," especially within the surgical ICU, which is generally noisier than its medical counterpart (Falk and Woods, 1973). A patient in the ICU on a life-support system may be attached to 10 different audible warning devices (Pownall, 1987), and hospitalized patients who often feel estranged and fearful to begin with, experience hospital noises as far more sinister than staff members realize (Nolen, 1973): they often falsely construe the various alarm systems as indicating life-threatening events.

15.6.2 Alarms

In a study of alarm systems used by anesthesiologists during surgical procedures, 75% of the alarms were found to be spurious, caused by factors like the patient's movement or simple mechanical events, whereas only 3% of the sounds indicated an actual risk to the patient (Kestin et al., 1988). A survey of Canadian anesthesiologists found that 67% of them had deliberately disabled audible signals to cope with the many false alarms and "sonic overkill" during surgery (McIntyre, 1985). A brief overview of effective critical alarms strategies is addressed here with broad-based recommendations based on the human factors literature.

In 2002, The Joint Commission, known as Joint Commission for the Accreditation of Hospitals & Healthcare Organizations (JCAHO) at that time, set a 2003 national patient safety goal to improve the effectiveness of clinical alarm systems. This goal takes the form of two recommendations for hospitals [JCAHO, 2002]:

- Implement regular preventive maintenance and testing of alarm systems
- Assure that alarms are activated with appropriate settings and are sufficiently audible with respect to distances and competing noise within the [care] unit

As in other areas of human factors environment and task analysis, most guidelines and principles need to be accepted with a few grains of salt because specific circumstances unique to safe patient care may argue for their violation. With such a caveat in mind, a few guidelines for the use of auditory displays/alarms are available (Sanders and McCormick, 1993; Salvendy, 1997; ECRI, 2002; Wickens et al., 2004; Carayon et al., 2010):

1. Principles of Auditory Display (Sanders and McCormick, 1993; Wickens et al., 2004)
   - Compatibility—signal should exploit learned or natural relationships
     Wailing signals with emergency (an ambulance approaching a traffic light)
   - Approximation—two-stage signals
     Attention-demanding to attract attention and identify a general category of information (Code Blue ascending and descending alarm)
     Designation signal to follow with the exact information identified with the first signal
     Example: an auditory paging system "fire alarm"—the fire alarm sounds first and then the operator announces the exact location of the fire in the facility
   - Dissociability—signals should be discernible from any ongoing audio input
   - Parsimony—input signals to operator should not provide more information than is necessary
   - Invariance—the same signal always equals the same information at all times
2. Principles of Presentation
   - Avoid extreme dimensions—high intensity can cause a startle
   - Intensity relative to ambient noise—signal should not be masked by ambient noise
   - Use interrupted or variable signals—avoid steady-state signals, and vary the frequency

- Do not overload the auditory channel—too many signals overload the operator, for example, during the Three Mile Island accident, more than 100 alarms were activated with no means of suppressing unimportant ones (Reason, 1990)

Both the Joint Commission and Emergency Care Research Institute (ECRI) believe that hospitals should not determine audibility by measuring sound levels of alarms. Instead, users should examine whether an alarm is sufficiently audible in the environment in which it is being used (ECRI, 2002). Various factors of the SEIPS model need to be included in this examination, such as organizational staffing levels, arrangement of components in the built environment, tasks that may compete with the alarm, background noise in the environment, and the frequency of the alarm signal. Hearing, when coupled with the other senses, can offer the brain an overwhelming array of information. Be mindful that each sensory modality appears to have particular strengths and weaknesses and must not be considered independent of the others.

15.7 Vibration

Vibration may be distinguished in terms of whether it is specific to a particular limb, such as the vibration produced by a surgical drill or saw, or whether if influences the whole body, such as that from a helicopter or ambulance (Wickens et al., 2004). Vibration can significantly affect the comfort and performance of health-care providers in emergency vehicles (Bruckart et al., 1993). It is unlikely that the patients and care providers would be exposed to such vibration levels in hospitals. However, vibration to the hands of surgeons, dentists, and other care providers from vibrating dental or surgical tools or devices such as drills and saws may be significant and could cause physical damage. Poorly designed medical devices not only jeopardize performance and productivity but are a major cause of cumulative trauma disorders (CTD) such as tendonitis, neuritis, Raynaud's phenomenon, trigger finger, epicondylitis (tennis elbow; golfers elbow), and Carpal Tunnel Syndrome (CTS) (Armstrong et al., 1993; Chaffin et al., 1999).

Vibration appears to principally affect visual and motor performance. Visual performance is generally impaired most by vibration frequencies in the range of 10–25 Hz. The degradation in performance is probably due to the movement of the image on the retina, which causes the image to appear blurred. Impacts of vibration include medical device display vibration, which can reduce the ability to see fine detail in displays, and impair performance using nonmoving/isometric controls (McLeod and Griffin, 1989; Griffin, 1997). Vibration in an emergency vehicle, for example, makes medical devices with touch screens extremely unreliable as control input devices (Wickens et al., 2004). In summary, considerable research has demonstrated the effects of vibration on performance. The effects of vibration are dependent on the difficulty of the task, type of display, and type of device used, and physical attributes of the user (Chaffin et al., 1999).

Low frequency vibration, such as the regular sea swell on a ship, the rocking of a light aircraft or the environment of a closed cab in a moving ambulance, can lead to motion sickness. Quite simply, the discomfort of the sickness is sufficiently intrusive that it is hard to concentrate on anything else, including the care tasks at hand (Thornton and Vyrnwy-Jones, 1984; Sanders and McCormick, 1993). The results of motion sickness, including emesis and anxiety, not only are uncomfortable but also could adversely effect the patient's care and outcome (Thornton and Vyrnwy-Jones, 1984; Sanders and McCormick, 1993; Fleischhackl et al., 2003; Weichenthal and Soliz, 2003). The incidence of patient motion sickness during ambulance transport is reported to be between 20% and 33%; this is often associated with nausea and vomiting, and may therefore be a risk factor for aspiration (Fleischhackl et al., 2003).

15.8 Illumination

The design of artificial illumination systems does have an impact on the performance and comfort of those using the environment as well as on the effective responses of the people to the environment. It is not the intention of this chapter to make anyone an expert on illumination for illuminating engineering is both an art and a science (Sanders and McCormick, 1993). The Illuminating Engineering Society of North America (IESNA) Recommended Practice (IESNA, 1995) provides guidelines for good lighting in those areas unique to health-care facilities, and is intended for both lighting designers and health-care professionals. The aim of this chapter is to familiarize the reader with some basic characteristic of illumination—illuminance, luminance, contrast, glare, and shadow—and illustrate the importance of proper illumination in the patient care environment.

Essentially all visual stimuli that the human can perceive may be described as a wave of electromagnetic energy. The wave is represented as a point along the visual *spectrum* and referred to as its *wavelength*. While we can measure or specify the hue of a stimulus reaching the eye by its wavelength, the measurement of brightness is more complex because there are several different meanings of light *intensity*. The source of light may be characterized by its *luminous intensity* or *luminous flux*, the actual light energy of the source, the *lumen*. But the actual amount of this energy that strikes the surface of an object to be seen is described as the *illuminance* and measured in units of $1\,lm/ft^2$—a *footcandle* or $1\,lm/m^2$—a *lux*. Hence, the term illumination characterizes the lighting quality of a given working environment. How much illuminance an object receives depends on the distance of the object from the light source. *Luminance* of an object is the amount of light reflected off the object to be detected, discriminated, and recognized by the observer when these objects are not themselves the source of light (Wickens et al., 2004). Luminance is different than illuminance because of differences in the amount of light that surfaces either reflect or absorb. *Glare* is produced by brightness within the field of vision that is sufficiently greater than the luminance to which the eyes are adapted. Although we may be concerned about the illumination of light sources in direct viewing, that is, looking directly into the O.R. lights or the sun, the *direct glare* produced the light source; *reflected glare*, for example, a high intensity examination lamp reflecting off a polished steel instrument container, can also cause annoyance, discomfort, or loss in visual performance (Sanders and McCormick, 1993).

Poorly designed lighting environment, such as too little or too much light, veiling reflections and glare, can contribute to visual discomfort and eyestrain (Boyce, 1981). The effects of lighting on performance are somewhat complex and limits of the visual system influence the nature of the visual information that arrives at the brain for the more elaborate perceptual interpretation (Wickens et al., 2004). Defects occur in the person's visual system, such as color defects (color blind) and myopia, and may place additional constraints and limits on this visual information. The following Table 15.3 suggests the environment minimum average illuminance for a general hospital (Sanders and McCormick, 1993; IESNA, 1995; AIA and FGI, 2001).

Besides the measurement of the environmental illuminances recommended in Table 15.3, there are basic concepts of illumination that should be considered for any human factors environment and task analysis in a patient care setting:

- What are sources of illumination in health-care settings
- Use of color
- Visibility
- Distribution of light
- Glare
- Reflectance
- Effects of the lighting on performance
- Use of lighting and elderly workers
- Lighting and computer use

Furthermore, the human factors environmental analysis must consider that excessive exposure to light can cause direct effects on health. Medical devices such as UV lights and medical lasers can cause damage to the eye. The sun's radiation can damage and cause cancer to the unprotected skin of care providers in the outdoors. In addition to these direct effects on health, eyestrain can be caused by inadequate lighting conditions. Too little or too much light, veiling reflections, disability and discomfort glare, and flicker can cause eyestrain (Boyce, 1981). This can cause irritation in the eyes, a

TABLE 15.3

Health-Care Environment Illuminance Recommendations

Environment	Ambient Light Lux/Footcandles	Task Lighting Lux/Footcandles
Patient rooms	300/30	750/75
Nursing station (day)	300/30	500/50
Nursing station (night)	100/10	500/50
Medication preparation	300/30	1,000/100
Exam room	300/30	1,000/100
Operating rooms	300–500/30–50	10,000–20,000/1,000–2,000
Hallways (active shifts)	300/30	
Hallways (sleeping)	100/10	

Sources: The American Institute of Architects (AIA) and Facility Guidelines Institute (FGI), *Guidelines for Design and Construction of Hospital and Health Care Facilities*, The American Institute of Architects, Washington DC, 2001; Illuminating Engineering Society of North America (IESNA), *Lighting for Hospitals and Health Care Facilities ANSI Approved*, Vol. IESNA Publication RP-29-95, Illuminating Engineering Society of North America, New York, 1995; Sanders, M.S. and McCormick, E.J. *Human Factors in Engineering and Design*, 7th edn., McGraw-Hill, Inc., New York, 1993.

breakdown of vision and possible patient safety risks. There are nonvisual effects of light on the body (seasonal affective disorders (SADS); influence of various glands in the body, etc.); however, not much is known about their effects quantitatively (Parsons, 2000) and certainly not as they relate to patient care and safety issues. Light can cause discomfort as well as positive sensations in patients and care providers. Lighting conditions that produce definite discomfort can usually be identified and criteria for the concepts of illumination are applicable to resolve the discomfort (IESNA, 1995).

Visual performance is the only performance outcome that is affected directly by changing the lighting conditions. Lighting itself cannot produce work output. What it can do is make details easier to see and colors easier to discriminate without producing discomfort or distraction; the greater the contribution of vision to the task the greater the effect of lighting on task performance. Recommendations for improving the visual performance include

1. Changes in the task:
   a. Increase the size of detail—make the patient care procedures or medication instructions or device monitors in a larger font
   b. Increase luminance contrast of the detail in task; use color contrast or light and shadow to define boundaries or highlight important areas
   c. In a cluttered visual field make the object to be detected clearly different from surrounding objects—size, color, shape, and contrast
2. Changes in environment:
   a. Increase luminance
   b. Change to a lamp with better color properties
   c. Provide lighting that is free from disability glare and veiling reflections
   d. Add or change to lighting to increase apparent size or luminance contrast of the object

Lastly, special attention should be given to visual performance for the elderly care provider and elderly patient. Individual visual performance differences accompany aging. *Presbyopia* (farsightedness) decreases in visual acuity, contrast sensitivity, and color discrimination and an increase in time to adapt to sudden changes in luminance and sensitivity to glare are common attributes associated with age. Possible environmental solutions may be to increase illuminance, taking note that glare and veiling reflections can also increase with increased illuminance, and by creating transition zones in the environment, that is, going from a movie to the

theater lobby to the outdoors, giving the person more time to acclimate to the brighter conditions (Sanders and McCormick, 1993). Finally, most visual physical disability is not one of total blindness but rather one of partial sight. The most common causes for visual disability partial sight are cataracts, macular degeneration, glaucoma, and diabetes, a disease currently at epidemic levels in the United States, affecting both patients and care provider populations. Several illumination strategies are suggested for improving visual performance under these conditions (Trace Center, Engineering, and Wisconsin-Madison, 2005):

- Cataracts: Limit glare
- Macular degeneration: More light and magnification
- Glaucoma: Limit patterns, use high contrast color, and spatial separate items
- Diabetes: Increase light in early stages of retinal change and magnification.

Just as our patients' acuity increases as patients' age increase, the economics and worker shortages are keeping older patient care providers in the workforce, retiring later. The environment must be designed to accommodate the interaction of environmental conditions associated with this aging population.

15.9 Conclusion

Aspects of the health-care environment that were not addressed in this chapter and can have direct influence on patient and care provider safety include: safety and security, energy and cost-effectiveness, disaster planning, infection control, and information technology (IT) and health-care technology and communications. For an extended understanding of health-care facility environment, the reader is encouraged to consult current reference literature on these subjects as well as other chapters in this book.

Although built or physical environments of care are usually assessed in terms of the effects of their separate component parts, patients and care staff are exposed to whole, integrated environments (Parsons, 2000). This built environment is one part of the system and constantly changes in the interactions of other systems components (Carayon et al., 2006). This last point brings me back to reemphasize one final human factors/patient safety issue that this chapter has touched on repeatedly: the importance of a thorough *task analysis* (Wickens et al., 2004). The full impact of the physical environment on human performance and patient safety can never be adequately predicted without a clear understanding of what parts of the SEIPS or system model will be present and relate with the other, which person will interact with them, which person *must* interact with them, and what the cost will be to patients, care providers, and the organization if the task performance is degraded.

References

Allread, W. G. and Israelski, E. W. (2010). Anthropometry and biomechanics. In M. B. Weinger, M. E. Wiklund, and D. J. Gardner-Bonneau (Eds.), *Handbook of Human Factors in Medical Device Design*. Boca Raton, FL: CRC Press/Taylor & Francis Group.

American Society of Heating, Refrigerating, and Air-conditioning Engineers (ASHRAE). (2008). *ANSI/ASHRAE?ASHE Standard 170*. Atlanta, GA: ASHRAE Press.

American Society of Heating, Refrigerating, and Air-Conditioning Engineers (ASHRAE). (2003). *ASHRAE-HVAC Design Manual for Hospitals and Clinics*. Atlanta, GA: ASHRAE.

Armstrong, T. J., Buckle, P., Fine, L. J., Hagberg, M., Jonsson, B., Kilbom, A. et al. (1993). A conceptual model for work-related neck and upper-limp muscosceletal disorders. *Scandinavian Journal of Work and Environmental Health, 19*, 73–84.

Baker, C. F. (1984). Sensory overload and noise in the ICU: Sources of environmental stress. *CCQ Critical Care Quarterly, 6*(4), 66–80.

Baker, C. F. (1993). Annoyance to ICU noise: A model of patient discomfort. *Critical Care Nursing Quarterly, 16*(2), 83–90.

Baker, C., Garvin, B., Kennedy, C., and Polivka, B. (1993). The effect of environmental sound and communication on CCU patients' heart rate and blood pressure. *Research in Nursing & Health, 16*(6), 415–421.

Bartley, J., Olmsted, R., and Hass, J. (2010). Current views of health care design and construction: Practical implications for safe, cleaner environments. *American Journal of Infection Control, 38*(5), S1–S12.

Boyce, P. R. (1981). *Human Factors in Lighting*. New York: MacMillan.

Bracco, D., Dubois, M. J., Bouali, R., and Enggimann, P. (2007). Single rooms may help to prevent nosocomial bloodstream infection and cross-transmission of methicillin-resistant *Staphylococcus aureus* in intensive care units. *Intensive Care Medicine, 33*(5), 836–840.

Brown, L. H., Gough, J. E., Bryan-Berg, D. M., and Hunt, R. C. (1997). Assessment of breath sounds during ambulance transport. *Annals of Emergency Medicine, 29*(2), 228–231.

Bruckart, J. E., Licina, J. R., and Quattlebaum, M. (1993). Laboratory and flight tests of medical equipment for use in U.S. Army Medevac helicopters. *Air Medical Journal, 1*(3), 51–56.

Buemi, M., Allegra, A., Grasso, F., and Mondio, G. (1995). Noise pollution in an intensive care unit for nephrology and dialysis. *Nephrology Dialysis Transplantation, 10*, 2235–2239.

Cabrera, I. N. and Lee, M. H. (2000). Reducing noise pollution in the hospital setting by establishing a department of sound: A survey of recent research on the effects of noise and music in health care. *Preventive Medicine, 30*(4), 339–345.

Carayon, P. and Smith, M. J. (2000). Work organization and ergonomics. *Applied Ergonomics, 31*, 649–662.

Carayon, P., Hundt, A., Karsh, B.-T., Gurses, A., Alvarado, C., Smith, M. et al. (2006). Work system design for patient safety: The SEIPS model. *Quality and Safety in Health Care, 15*(Suppl 1), i50–i58.

Carayon, P., Karsh, B.-T., and Alvarado, C. J. (2010). Environment of use. In *Human Factors in Medical Device Design*, Weinger, M. B., Wiklund, M. E., and Gardner-Bonneau, D. J. (Eds.) (pp. 63–96). Boca Raton, FL: CRC Press/Taylor & Francis Group.

Centers for Disease Control (CDC) (1994). Guidelines for preventing the transmission of mycobacterium tuberculosis in health-care facilities. *Morbidity and Mortality Weekly Report (MMWR), 43*(RR13), 1–132.

Chaffin, D. B., Andersson, G. B. J., and Martin, B. J. (1999). *Occupational Biomechanics*. New York: John Wiley & Sons, Inc.

Chaudhury, H., Mahmood, A., and Valente, M. (2003). Coalition of healthcare environmental research. The use of single patient rooms versus multiple occupancy rooms in acute care environments. *Coalition for Health Environments Research (CHER Report)*. Retrieved from http://www.premierinc.com/quality-safety/tools-services/safety/topics/construction/

Chengalur, S. N., Rodgers, S. J., and Bernard, T. E. (2004). *Kodak's Ergonomic Design for People at Work* (Vol. 2nd edn.). Hoboken, NJ: John Wiley & Sons.

Czaja, S. J. (1997). Systems design and evaluation. In *Handbook of Human Factors and Ergonomics*. G. Salvendy (Ed.), (2nd edn., pp. 17–40). New York: John Wiley & Sons, Inc.

Department of Justice (DOJ). 2010 ADA Standards for accessible design (2010).

Donabedian, A. (1988). The quality of care. How can it be assessed? *Journal of the American Medical Association, 260*(12), 1743–1748.

Duckro, A. N., Blom, D. W., Lyle, E. A., Weinstein, R. A., and Hayden, M. K. (2005). Transfer of vancomycin-resistant enterococci via health care worker hands. *Archives of Internal Medicine, 165*, 302–307.

Edelberg, H. K. (2001). How to prevent falls and injuries in patients with impaired mobility. *Geriatrics, 56*(March), 41–45.

Emergency Care Research Institute (ECRI) (2002). Critical alarms and patient safety. *Health Devices, 31*(11), 397–417.

Falk, S. A., and Woods, N. F. (1973). Hospital noise–levels and potential health hazards. *New England Journal of Medicine, 289*(15), 774–781.

Fleischhackl, R., Dorner, C., Scheck, T., Fleischhackl, S., Hafez, J., Kober, A. et al. (2003). Reduction of motion sickness in prehospital trauma care. *Anaesthesia, 58*(4), 373–377.

Gordon, C. C., Churchill, T., Clauser, C. E., Bradmiller, B., McConville, J. T., Tebbitts, I. et al. (1989). 1988 Anthropometric survey of U.S. Army personnel: Summary statistics interim report. Natick, MA: U.S. Army Natick Research, Development and Engineering Center.

Grabenstein, J. D. and Winkenwerder, W. (2003). US military smallpox vaccination program experience. *The Journal of the American Medical Association, 289*, 3278–3282.

Griffin, M. J. (1997). Vibration and motion. In *Handbook of Human Factors and Ergonomics* G. Salvendy (Ed.) (2nd edn., pp. 828–857). New York: John Wiley & Sons.

Grumet, G. W. (1993). Pandemonium in the modern hospital. *New England Journal. of Medicine, 328*(6), 433–437.

Helm, M., Castner, T., and Lampl, L. (2003). Environmental temperature stress on drugs in prehospital emergency medical service. *Acta Anaesthesiologica Scandinavica, 47*(4), 425–429.

Herwaldt, L. and Pottinger, J. (2003). Preventing falls in the elderly. *Journal of the American Geriatrics Society, 51*(8), 1175–1177.

Hota, S., Hirji, Z., Stockton, K. et al. (2009). Outbreak of multidrug-resistant *Pseudomonas aeruginosa* colonization and infection secondary to imperfect intensive care unit room design. *Infection Control and Hospital Epidemiology, 30*, 25–33.

Hrickiewicz, M. (2009). To build or not?. *Health Facilities Management.* Retrieved from Institute of Medicine (IOM), M. P. o. B. I. (2002). *Countering Bioterrorism The Role of Science and Technology.* Washington, DC: The National Academies Press.

Illuminating Engineering Society of North America (IESNA) (1995). *Lighting for Hospitals and Health Care Facilities ANSI Approved* (Vol. IESNA Publication RP-29-95). New York: Illuminating Engineering Society of North America.

Institute of Medicine (IOM) (2002). *Countering Bioterrorism: The Role of Science and Technology.* Washington, DC: The National Academy Press.

International Organization for Standardization (ISO) (1996). Basic human body measurements for technological design (ISO 7250). Geneva: International Organization for Standardization.

Joint Commission on Accreditation of Healthcare Organizations (JCAHO) (2002), from http://www.jcaho.org/ptsafety

Joint Commission on Accreditation of Healthcare Organizations (JCAHO) (2005). *Environment of Care® Essentials for Health Care* (5th edn.). Chicago, IL: Joint Commission Resources.

Kerzman, H., Chetrit, A., Brin, L., and Toren, O. (2004). Characteristics of falls in hospitalized patients. *Journal of Advanced Nursing, 47*(2), 223–229.

Kestin, I., Miller, B., and Lockhart, C. (1988). Auditory alarms during anesthesia monitoring. *Anesthesiology, 69*, 106–109.

Kroemer, K. H. E. (1989). Engineering anthropometry. *Ergonomics, 32*, 767–784.

Kroemer, K. H. E., Kroemer, H., and Kroemer-Elbert, K. (2001). *Ergonomics-How to Design for Ease and Efficiency* (2nd edn.). Upper Saddle River, NJ: Prentice Hall.

Maki, D., Alvarado, C., Hassemer, C., and Mary, A. Z. (1982). Relationship of the inanimate environment to endemic nosocomial infection. *New England Journal of Medicine, 307*, 1562–1566.

McIntyre, J. (1985). Ergonomics: Anaesthetists' use of auditory alarms in the operating room. *International Journal of Clinical Monitoring and Computing, 2*, 47–55.

McLeod, R. W. and Griffin, M. J. (1989). Review of the effects of translational whole-body vibration on continuous manual control performance. *Journal of Sound and Vibration, 133*, 55–115.

Morrison, W. E., Haas, E. C., Shaffner, D. H., Garrett, E. S., and Fackler, J. C. (2003). Noise, stress, and annoyance in a pediatric intensive care unit. *Critical Care Medicine, 31*(1), 113–119.

Morse, J. M. (2002). Enhancing the safety of hospitalization by reducing patient falls. *American Journal of Infection Control, 30*(6), 376–380.

Nightingale, F. (1863). *Notes on Hospitals* (3rd edn.). London: Longman, Green, Longman, Roberts, & Green.

Niinimaki, J., Savolainen, M., and Forsstrom, J. J. (1998). Methodology for security development of an electronic prescription system. *Proceedings of the AMIA Symposium, 1998*, 245–249.

Nolen, W. (1973). Can you hear me? *New England Journal of Medicine, 289*, 803–804.

Oates, P. R. and Oates, R. K. (1996). Stress and work relationships in the neonatal intensive care unit: Are they worse than in the wards. *Journal of Paediatrics and Child Health, 32*, 57–59.

Oates, R. K. and Oates, P. K. (1995). Stress and mental health in neonatal intensive care units. *Archives of Disease in Childhood, 72*, F107–F110.

Panero, J. and Zelnick, M. (1979). *Human Dimensions and Interior Space*: New York: Whitney Library of Design/Watson-Guptill Publications.

Parsons, K. C. (1993). *Human Thermal Environments.* London, U.K.: Taylor & Francis.

Parsons, K. C. (2000). Environmental ergonomics: A review of principles, methods and models. *Applied Ergonomics, 31*, 581–594.

Peleg, A. Y. and Hooper, D. C. (2010). Hospital-acquired infections due to gram-negative bacteria. *New England Journal of Medicine, 362*, 1804–1813.

Pownall, M. (1987). Medical alarms: Ringing the changes. *Nursing Times, 83*(18–19).

Prasad, N. H., Brown, L. H., Ausband, S. C., Cooper-Spruill, O., Carroll, R. G., and Whitley, T. W. (1994). Prehospital blood pressures: Inaccuracies caused by ambulance noise? *American Journal of Emergency Medicine, 12*(6), 617–620.

Reason, J. (1990). *Human Error.* Cambridge, U.K.: Cambridge University Press.

Reiling, J. G., Knutzen, B. L., Wallen, T. K., McCullough, S., Miller, R. H., and Chernos, S. (2004). Enhancing the traditional design process: A focus on patient safety. *The Joint Commission Journal on Quality Improvement, 30*(3), 115–124.

Revised OSHA Bloodborne Pathogens Act, Pub. L. No. 29 CFR 1910.30 (2001).

Rohles, F. H. and Konz, S. A. (1987). *Climate* (1st edn.). New York: John Wiley & Sons.

Rossman, L. (1992). Protecting yourself. *Journal of Emergency Medical Services, 17*(11), 48–49.

Rubin, H. R., Owens, A. J., and Golden, G. (1998). *An Investigation to Determine Whether the Built Environment Affects Patients' Medical Outcomes*. Martinez, CA: The Center for Health Design, Inc.

Rutala, W. A. (1996). APIC guideline for selection and use of disinfectants. *American Journal of Infection Control, 24*, 313–342.

Salvendy, G. (Ed.) (1997). *Handbook of Human Factors and Ergonomics* (2nd edn.). New York: John Wiley & Sons.

Sanders, M. S. and McCormick, E. J. (1993). *Human Factors in Engineering and Design* (7th edn.). New York: McGraw-Hill, Inc.

Schmaker, S. A. and Pequegnat, W. (1989). Hospital design, health providers, and the delivery of effective health care. In *Advances in Environment, Behavior, and Design*, Zube E. H. and Moore G. T. (Eds.). New York: Plenum Press, p. 355.

Sehulster, L. and Chinn, R. Y. W. (2003). Guidelines for environmental infection control health-care facilities. *Centers for Disease Control and Prevention (CDC), 52*, 1–42.

Smith, M. J. and Carayon-Sainfort, P. (1989). A balance theory of job design for stress reduction. *International Journal of Industrial Ergonomics, 4*, 67–79.

Smith, M. J. and Carayon, P. (2000). Balance Theory of job design. In *International Encyclopedia of Ergonomics and Human Factors*, W. Karwowski (Ed.). London, U.K.: Taylor & Francis, pp. 1181–1184.

Tablan, O. C., Anderson, L. J., Besser, R., Bridges, C., and Hajjeh, R. (2003). Guidelines for preventing health-care–Associated Pneumonia, 2003. *Centers for Disease Control and Prevention (CDC), 53*, 1–36.

Teltsch, D. Y., Hanley, J., Loo, V., Goldberg, P., Gursahaney, A., and Buckeridge, D. (2011). Infection acquisition following intensive care unit room privatization. *Achieves of Internal Medicine, 171*(1), 32–38.

The American Institute of Architects (AIA) (2006). *Guidelines for Design and Construction of Health Care Facilities*. Washington DC: The American Institute of Architects.

The American Institute of Architects (AIA) and Facility Guidelines Institute (FGI) (2001). *Guidelines for Design and Construction of Hospital and Health Care Facilities*. Washington DC: The American Institute of Architects.

Thornton, R. and Vyrnwy-Jones, P. (1984). Environmental factors in helicopter operations. *Journal of the Royal Army Medical Corps., 130*(3), 157–161.

Tinetti, M. (2003). Clinical practice. Preventing falls in elderly persons. *New England Journal of Medicine, 348*(1), 42–49.

Topf, M. (1985). Noise-induced stress in hospital patients: Coping and nonauditory health outcomes. *Journal of Human Stress, 11*(3), 125–134.

Topf, M. (1988). Noise-induced occupational stress and health in critical care nurses. *Hospital Topics, 66*(1), 30–34.

Topf, M. (1992). Stress effects of personal control over hospital noise. *Behavioral Medicine, 18*(2), 84–94.

Topf, M. (2000). Hospital noise pollution: An environmental stress model to guide research and clinical interventions. *Journal of Advanced Nursing, 31*(3), 520–528.

Topf, M. and Dillon, E. (1988). Noise-induced stress as a predictor of burnout in critical care nurses. *Heart & Lung: Journal of Acute & Critical Care, 17*(5), 567–574.

Topf, M. and Thompson, S. (2001). Interactive relationships between hospital patients' noise-induced stress and other stress with sleep. *Heart & Lung, 30*(4), 237–243.

Trace Center, Engineering, C. o. and Wisconsin-Madison, U. o. (2005), 2005, from www.trace.wisc.edu

U.S. Department of Justice Americans with Disabilities Act (ADA), Pub. L. No. Public Law 101–336. (1990).

Vonberg, R. P. and Gastmeier, P. (2006). *Nosocomial aspergillosis* in outbreak settings. *Journal of Hospital Infection, 63*, 246–254.

Walsh-Sukys, M., Reitenbach, A., Hudson-Barr, D., and DePompei, P. (2001). Reducing light and sound in the Neonatal Intensive Care unit: An evaluation of patient safety, staff satisfaction and costs. *Journal of Perinatology, 21*, 230–235.

Weichenthal, L. and Soliz, T. (2003). The incidence and treatment of prehospital motion sickness. *Prehospital Emergency Care, 7*(4), 474–476.

Weiss, S. J., Ellis, R., Ernst, A. A., Land, R. F., and Garza, A. (2001). A comparison of rural and urban ambulance crashes. *American Journal of Emergency Medicine, 19*(1), 52–56.

Wickens, C. D., Lee, J. D., Liu, Y., and Becker, S. E. G. (2004). *An Introduction to Human Factors Engineering* (2nd edn.). Upper Saddle River, NJ: Prentice Hall.

Williams, M. A. (1988). The physical environment and patient care. *Annual Review of Nursing Research, 6*, 61–84.

16

Assessment and Evaluation Tools for Health-Care Ergonomics: Musculoskeletal Disorders and Patient Handling

Sue M. Hignett, Michael J. Fray, and Mary Matz

CONTENTS

16.1 Assessment and Evaluation Tools 235
    16.1.1 OWAS 236
    16.1.2 NIOSH 237
    16.1.3 RULA 238
    16.1.4 REBA 238
    16.1.5 QEC 239
16.2 Participatory Ergonomics in Health Care 239
    16.2.1 Patient Handling 240
16.3 Intervention Evaluation Tool 241
    16.3.1 Safer Lifting Policies 242
    16.3.2 Ergonomics in Health-Care Design 243
16.4 Conclusion 243
References 243

Physical ergonomics projects usually focus on staff (caregiver) well-being, including both static and dynamic musculoskeletal risks. This can include the design of buildings, equipment (technology), and systems. The assessment of problems, interventions (design and change projects), and subsequent evaluations are key components of any project. Many physical ergonomics tools have been used for both assessment and evaluation. In this chapter, examples are given from a range of health-care activities including operating theaters, midwifery, physiotherapy, radiography, ultrasonography, and ambulance work.

Caring for people can involve contaminated, physically demanding, and emotionally challenging work in situations where the patient can be both physically and mentally vulnerable. Nursing work is often physically heavy (involving lifting weights that would be unacceptable in other industries); physically dirty (involving tasks such as washing soiled bodies); and highly repetitive (Hignett, 2001; Lee-Treweek, 1997). The role of the patient in the interactions interface is significant and they should always be seen as an active participant. The type and level of participation will vary between both clinical specialties and individual patients (Hignett, 2003a; Hignett et al., 2005). This will be explored within the framework of Participatory

Ergonomics (PE) with examples from hospital orderlies and laboratory design.

Patient handling has been identified as the major musculoskeletal risk for nursing staff for many years. Many interventions have been implemented but evaluations have been difficult. A new evaluation tool has been produced to try to address the complexity of these multifactorial interventions; this will be used to both describe the range of interventions and validated evaluation tools. Incorporating ergonomics in health-care design is relatively new but gaining ground. The chapter concludes with a brief synopsis of research and recommendations for health-care facility building design.

16.1 Assessment and Evaluation Tools

Physical workload measures can include many outcome measures, for example, postural analysis, forces applied, electromyography, and biomechanical measures (e.g., Lumbar Motion Monitor) (Table 16.1).

Many of the known musculoskeletal risks to health-care workers are assessed by the quantity of work. Frequency counts can use real time observational data

TABLE 16.1

Physical Work Load Measures

Risk	Outcome Measure	Tool
Exposure	Injury records Self-completed logs Frequency and repetition	
Posture	Joint angles	OWAS, RULA, REBA, QEC
Biomechanical	Force measure Movement Speed or acceleration	Biomechanical models, e.g., NIOSH
Physiological	Heart rate Oxygen uptake EMG	

Source: Fray, M., A comprehensive evaluation of outcomes from patient handling interventions, PhD thesis, Loughborough University, Loughborough, U.K., 2010.

but are costly and time-consuming to achieve a satisfactory level of accuracy. Self-reporting methods have been used to good effect to report not only the effects of musculoskeletal disorders (MSDs) but also measure the exposure to hazards (Hollmann et al., 1999; Knibbe and Friele, 1999; Warming et al., 2009).

Several postural analysis and biomechanical tools have been used to assess physical stress in health-care tasks. Some are used to diagnose problems and evaluate changes. These include Ovako Working posture Analysis System (OWAS), National Institute for Occupational Safety and Health (NIOSH), Rapid Upper Limb Assessment (RULA), Rapid Entire Body Assessment (REBA), and Quick Exposure Check (QEC). Other more general assessment tools might be used to review compliance with legislation or professional recommendations (e.g., MAC, http://www.hse.gov.uk/msd/mac/index.htm).

16.1.1 OWAS

OWAS was developed as an observational technique for evaluating work postures, with a set of criteria for redesigning working methods and places based on the evaluation of experienced workers and ergonomics experts (Karhu et al., 1977). Although OWAS was developed in the steel industry, it has been widely used in other industrial sectors from forestry to health care.

OWAS has 84 basic posture types (4 back, 3 arm, 7 legs, and 3 weight/force) that are combined to give a total possible 252 posture types. Data are collected by split second observations (snap shots) of the workers' posture. Equal interval observations (time sampling) are recommended (Hignett and McAtamney, 2005) and a minimum of 80 observations is recommended for accuracy (Suurnäkki et al., 1988), although a 10% error limit is reported for 100 observations (Training

Publication No. 11, 1992). The postures are analyzed to produce an action category (AC) rating that gives an indication of the level of severity of the postural load (Louhevaara and Suurnäkki, 1992) starting from AC 1, where no action is required, through to AC 4, where action is required immediately.

OWAS has been used to evaluate working postures for many health-care activities, for example, operating room (OR) staff, nurses, physiotherapists, and ambulance staff. Kant et al. (1992) looked at the postures of OR staff (surgeons, assistant anesthetists, instrumentation nurses, and circulating nurses) using OWAS as an occupational health survey and research tool. They found that the postures adopted in this environment were predominantly static with instrumentation nurses and surgeons showing a high level of poor postures.

Hignett (1996a) used OWAS to compare the postures adopted by nurses on Care of the Elderly wards when 'carrying out' patient handling tasks (Figure 16.1) with performance of nonpatient manual handling (e.g., bed making, collecting and operating equipment). She found that the percentage of harmful postures adopted during patient handling tasks was significantly higher than during nonpatient handling tasks.

Jackson and Liles (1994) observed the working practices of physiotherapy students using OWAS. They reported that more hazardous postures were observed in neurological rehabilitation settings (AC 3–4; mostly standing transfers without using patient lifting equipment) than in outpatient physiotherapy. A pilot study looking at working postures of physiotherapists

FIGURE 16.1
Patient handling task. (From *Appl. Ergon.*, 27(3), Hignett, S., Postural analysis of nursing work, 171–176, Copyright 1996, from Elsevier.)

(Hignett, 1995) found significantly more harmful postures (higher OWAS ACs) when the physiotherapists were not able to make modifications to either their work station, system, and work, or the position of the patient due to treatment requirements.

Doormaal et al. (1995) looked at the extent to which the working postures, specific tasks, and activities constituted a physical load for ambulance staff. They found that 16%–29% of the working postures analyzed with OWAS during a shift were spent in harmful postures, including static and short duration maximal exertion lifting tasks. Ferreira and Hignett (2005) used OWAS to investigate the static and dynamic working postures of Emergency Medical Technicians (EMTs, paramedics) when delivering care and treatment. Overall, only 26% of the time-sampled postures within the patient compartment required some corrective measures (AC = 2, 3, or 4). Most of the postures were AC 1 (no corrective measures required) due to long periods of time when the EMTs were seated, talking with patients, and completing patient information forms. Although postural risks were greater when the ambulance was stationary, other hazards were observed when the ambulance was in motion, with 11% of postures recorded as nonsitting postures during transportation, indicating that paramedics were moving about the patient compartment (without a harness or seat belt).

16.1.2 NIOSH

NIOSH first developed a lifting equation in 1981 to assist safety and health practitioners evaluate lifting demands in the sagittal plane based on biomechanical demands of weight, frequency of lift, and horizontal and vertical displacement. This equation was revised and expanded in 1991 to include asymmetrical lifting tasks and complex coupling situations (Waters et al., 1993). Although it was not originally intended for use in the evaluation of lifting people it has been used to evaluate both patient and nonpatient handling tasks and as a benchmark for the recommended limit of compressive forces. Waters (2007) applied the revised lifting equation to the conditions for patient lifting and calculated a maximum manual lifting load for manual patient handling. He determined that caregivers should lift no more than 35 lb (15.9 kg) of patient weight under the best of circumstances (e.g., no tubes, contractures, combative behavior, etc.).

Steinbrecher (1994) used NIOSH to assess the task of loading hospital linen sacks into a disposal chute. She found that the recommended weight limit (RWL) of 23 kg was exceeded when the linen sack was being placed in the chute. The recommendations included the provision of linen carts to facilitate the movement of the sack. Torma-Krajewski (1987) evaluated manual

handling tasks performed by ambulance workers. She used the NIOSH guidelines to grade the biomechanical analysis and concluded that lifting tasks for paramedics were very stressful due to the high weights and large horizontal distances of the hands from the trunk.

Radiography or x-ray technology is one of the primary methods of diagnosing pathology and trauma and can be found in most hospitals, clinics, and trauma centers with varying numbers of staff and shift patterns (some providing 24 h cover). Kumar et al. (2003) reported a study looking at the activities of 7 x-ray technologists working in different areas and analyzed the biomechanical stress for 16 tasks: (1) wearing a lead apron; (2) loading a small x-ray cassette (10 N); (3) loading a large x-ray cassette (20 N); (4) pushing/pulling an x-ray tube; (5) pushing a mobile x-ray unit on the floor; (6) pulling a mobile x-ray unit on the floor; (7) pushing a patient stretcher in the hallway; (8) pushing a wheelchair with a patient in the hallway; (9) repositioning a patient horizontally in bed; (10) repositioning a patient on side in bed; (11) repositioning a patient to an upright seated position in bed; (12) repositioning a cassette under a patient; (13) slider board transfer of a patient; (14) spine board transfer of a patient; (15) pulling the spine board; (16) lifting a patient from a wheelchair. The patient handling tasks were found to produce the highest levels of biomechanical stress, but the more radiographic-specific tasks (1–6, 12) also contributed to the biomechanical load due to repetition (2–3) with the task of repositioning a cassette under a patient (12) exceeding the maximum NIOSH (1981) limit.

Zhuang et al. (1999) looked at the postural stresses from the patient handling task of transferring a sitting patient from a bed to a chair using different transfer methods:

- Manual method with two carers
- Walking belt with two carers
- Sliding board with one carer
- Stand up lift with one carer
- Overhead lift with one carer
- Basket sling lift with one carer

In the preparation stage before the transfer, the manual lift, walking belt, sliding board, and stand up lift all exceeded the maximum NIOSH recommended limit for back compressive forces. Their recommendation was to use the basket sling and overhead lift as they significantly reduced the biomechanical load on the carers back.

Marras et al. (2009) compared the use of ceiling-mounted and floor (mobile) lifts using the Lumbar Motion Monitor. They found that none of the spine compression forces approached the 3400 N threshold for endplate damage regardless of whether a ceiling-based system or a floor-based system was used. However, for many of the floor-based system patient handling maneuvers, anterior/posterior shear forces were high enough to cause possible disc damage and degeneration for the caregiver at the mid to upper levels of the lumbar spine especially during turning maneuvers and in confined spaces, such as bathrooms. They concluded that ceiling-based lifts were preferable to floor-based patient lift systems.

16.1.3 RULA

RULA (McAtamney and Corlett, 1993) is a quick survey method that can be used as part of an ergonomic workplace assessment where MSDs are reported. RULA assesses biomechanical and postural loading on the neck, shoulders, and upper limbs and was designed to assess predominantly sedentary work. It allocates scores based on the position of groups of body parts with additional scores for force/load and muscle activity. The final RULA score is a relative rather than an absolute score, and gives an indication of the risk level on a four-point AC scale from AC 1, where the posture is acceptable, to AC 4 where investigation and changes are needed immediately. RULA has been used to look at static working postures in the OR and ultrasonography.

Cowdry and Graves (1997) used hierarchical task analysis and RULA to look at surgical access to pediatric patients. They found that many surgeons were using adult operating tables with the result that the tables were much wider than the patients. This contributed to the awkward working postures observed, particularly for seated surgeons. Pediatric operating tables and chairs were recommended as well as consideration of alternative patient positioning.

Ultrasonography is used for general investigations of soft tissues as well as specific diagnostic tests in a range of clinical specialties, for example, gynecology, obstetrics, cardiology, vascular, and pediatrics. Sonographers use a hand-held transducer that is applied to the area needing investigation, linked to a scanning machine, which relays the images collected by the probe. It typically involves a static posture to maintain the arm in a fixed position (unsupported abduction) while pressing the transducer against the patient (Figure 16.2).

Crawford et al. (2002) used RULA to assess the musculoskeletal risk to obstetric sonographers. The highest scores were recorded when performing the scan; this was suggested to relate to both the complexity of the sonographer–transducer–patient interface and the lack of flexibility in design of the equipment and workstation.

16.1.4 REBA

REBA is a whole body assessment tool that was initially designed to provide a pen-and-paper postural analysis

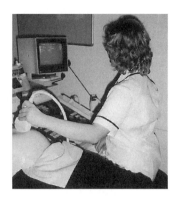

FIGURE 16.2
Obstetric sonography. (From Hignett, S., Ergonomics in health and social care, in Smith, J. (Ed.), *The Guide to the Handling of Patients*, 5th edn., Back Care/Royal College of Nursing, Teddington, Middlesex, U.K., 2005, pp. 41–48.)

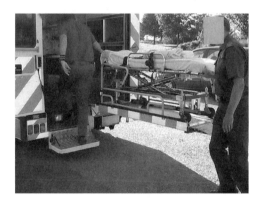

FIGURE 16.3
Ambulance loading with tail lift. (From Hignett, S. and Jones, A., *Emerg. Med. J.*, 24, 200, 2007. BMJ.)

tool to be used in the field by direct observations or with photographic stills/video (Hignett and McAtamney, 2000). It was developed to assess the type of unpredictable working postures found in health care and other service industries and was validated using examples from the electricity, health care, and manufacturing industries. Data are collected about the body posture, forces used, type of movement or action, repetition, and coupling. A final REBA score is generated giving an indication of the level of risk and urgency with which action should be taken on a five-point AC scale of 0–4 from no action required through to action necessary now.

Hignett and Jones (2007) compared three systems for loading ambulances; the easi-loader, ramp and winch, and tail lift (Figure 16.3). Postural snapshots were collected every 2 s and analyzed using REBA. Six-hundred and sixty-two postures were analyzed and the average REBA scores were calculated. They reported that the easi-loader had the greatest postural risks (AC 3), whereas the tail lift and ramp and winch were in REBA AC 2.

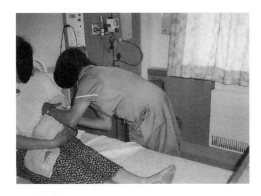

FIGURE 16.4
Assisting breastfeeding. (From Hignett, S., *Occup. Environ. Med.*, 60(9), e6 (electronic paper), 2003b, http://wwwoccenvmed.com/cgi/content/full/60/9/e6. BMJ.)

16.1.5 QEC

Midwifery is a very reactive area of health care with respect to both childbirth and breast feeding. The establishment of breast feeding is strongly encouraged in the United Kingdom (Department of Health, 1993) and that assisting breast feeding can be a lengthy (up to 1 h per feed) and frequent task with midwives reporting that they can spend most of a night shift (8 h) going from one mother to the next to assist with breast feeding (Figure 16.4; Hignett, 1996b). Steele and Stubbs (2002) used several methods to investigate midwifery involvement in breast feeding: focus groups, postural analysis, and body mapping. The physical discomfort was measured using the body mapping and the QEC (David et al., 2008). The QEC results suggested that midwives were frequently flexed (trunk) between 20° and 60°.

16.2 Participatory Ergonomics in Health Care

Researchers have discussed the challenge of describing the complexity of the hospital organizational structure in terms of a hierarchy, using a classical model. The problem of subjecting "human" services in the public sector to a Fordist model was a criticism of Fordism in the 1970s as it was difficult to increase productivity and control costs because of the labor-intensive and professionalized nature of industries like health and education (Ackroyd and Bolton, 1999). The core business of the hospital, providing the public service of health care (both public and private sector organizations) is the fundamental difference compared to other industries. Van Cott (1994) called this "people-centered and people-driven" in contrast to technology-centered, where the human role is to monitor the equipment or supervise small numbers of other staff. Although it would be

hoped that examples would be available for both staff and patient involvement, only projects involving staff have been identified within the PE framework.

The definition of participatory approaches includes interventions at macro (organizational, systems) levels as well as micro (individual) where workers are given the opportunity and power to use their knowledge to address ergonomic problems relating to their own working activities. PE has been used as an approach to tackle musculoskeletal problems (Vink and Wilson, 2003). Projects using PE have historically tended to be at a micro (individual) level, with mixed results; however, participatory interventions at macro (organizational) levels also have much to offer (Hignett et al., 2005). The first two examples look at musculoskeletal risks from static and dynamic activities; the third example describes the use of PE in a building design project.

In many clinical areas staff are required to work in sterile conditions, which may involve working in a flow cabinet. Lee et al. (2002) carried out an initial assessment of the task of working at a laminar flow hood using injury data, staff questionnaires, task analysis, and RULA. The results were used to develop design criteria for three prototypes that were subsequently evaluated by staff in a PE project. The resultant recommendations supported a design offering variability in working height and depth.

Evanoff et al. (1999) reported a PE project carried out with hospital orderlies to see if direct worker participation in problem solving would improve job satisfaction, injury rates, lost time, and musculoskeletal symptoms. The intervention was evaluated using the OSHA 200 log, workers compensation insurance records, self-administered surveys of workers at 1, 7, and 15 months. They found a decrease in risks of work injury, with a reduction in the relative risk of 50%, for both OSHA 200 log and injury rate as well as a reduction in total days lost. The survey found a large and statistically significant reduction in the proportion of employees with musculoskeletal symptoms.

Estryn-Behar et al. (2000) used a PE methodology for the conception of a new hospital laboratory. A number of different analyses (activity analysis of seven professional groups, space analysis, noise map, and lighting map) were conducted to provide information for discussion at three workshops. The workers were able to plan a more practical new space layout. This was tested with a 3D scale model to give every worker the opportunity to modify the model. The final layout produced an improved functional distribution of the available space.

Rivilis et al. (2008) reported a systematic review looking at the effectiveness of PE interventions at improving workers health. They concluded that there was partial to moderate evidence that PE interventions have a positive impact on musculoskeletal symptoms, reducing injuries and workers' compensation claims, and a reduction in lost days from work or sickness absence. Decision-making was most commonly through consultation with groups or individually. In all except one study, changes to the physical design of equipment and workplaces were implemented. Some studies included changes in work tasks, job teams, or work organization.

16.2.1 Patient Handling

Patient handling is a known cause of musculoskeletal risk for health-care staff. A range of ergonomic and other approaches have been used to reduce the effects of these tasks, for example, risk assessment and management, training, equipment provision, culture change. Buckle (1987) summarized the epidemiological findings from 1960s to the 1980s, confirming that nursing was among the highest risk occupations with respect to low back problems, with a point prevalence of approximately 17%, an annual (period) prevalence of 40%–50% and a lifetime prevalence of 35%–80%. More recently, Estryn-Behar et al. (2003) collected data from over 30,000 nurses in 10 European Union countries to investigate the physical load among nursing staff as part of the NEXT study (Hasselhorn et al., 2003). The results found that MSDs were still common with more than 25% of respondents having a medical diagnosis of MSD, suggesting that nurses are still exposed to a high risk of back injury associated with their working activities.

There have been at least four systematic reviews of patient handling interventions. Three used a Cochrane approach (Amick et al., 2006; Dawson et al., 2007; Martimo et al., 2008). Although they only reviewed a small number of studies they concluded that there is

- A moderate level of evidence for the effect of Occupational Health and Safety (OHS) interventions on musculoskeletal conditions in health-care settings (Amick et al., 2006).
- A moderate level evidence for multicomponent patient handling interventions and physical exercise interventions (Amick et al., 2006).
- Moderate evidence that training in isolation was not successful and that multidimensional interventions were effective (Dawson et al., 2007).
- No evidence that training with or without lifting equipment was effective in the prevention of back pain or consequent disability. They suggested that either the advocated techniques did not reduce the risk of back injury or that training did not lead to adequate change in lifting and handling technique (Martimo et al., 2008).

The fourth (Hignett et al., 2003) used a more modern mixed methods approach (Pluye et al., 2009) to systematic review methodology by including all study types (quantitative and qualitative). To achieve this heterogeneity, each study was defined within a study type hierarchy. A quality score was then allocated by using an appraisal/extraction tool within each category rather than comparatively between categories. Interventions were grouped as multifactorial, single factor, and technique training-based. The results were reported as summary statements with the associated evidence level (strong, moderate, limited, or poor). The findings are summarized as follows:

- Strong evidence that interventions predominantly based on technique training have no impact on working practices or injury rates.
- Multifactor interventions, based on a risk assessment program, are most likely to be successful in reducing musculoskeletal injuries related to patient handling activities.

The seven most commonly used strategies were identified (Table 16.2). It is suggested that these seven factors could form the basis of a generic program, although it is likely that an intervention strategy and program will need to be further developed and extended in order to be responsive to local organizational and cultural factors. The risk assessment process could facilitate the detailed design of the program and identification of additional appropriate strategies, with the allocation of priorities based on local negotiation with managers and staff.

Based on the findings of the systematic review, Hignett (2003c) recommended the provision of a minimum set of equipment for all clinical environments

TABLE 16.2

Intervention Strategies for Multi-Factor Interventions

Intervention Strategy	No. of Occurrences	Ave. Quality Rating of Studies (%)
Equipment provision/ purchase	18	50
Education and training (range of topics)	18	54
Risk assessment	13	55
Policies and procedures	10	50
Patient assessment system	8	43
Work environment redesign	7	58
Work organization/ practices changed	7	63

Source: Hignett, S., Occup. Environ. Med., 60(9), e6 (electronic paper), 2003b, http://wwwoccenvmed.com/cgi/content/full/60/9/e6. BMJ.

where patient handling occurs on a regular basis: lifts (mobile and ceiling), stand-aids (standing lifts), sliding sheets, lateral transfer boards, walking belts, adjustable height beds and baths.

16.3 Intervention Evaluation Tool

Fray (2010) has developed an Intervention Evaluation Tool (IET) to allow the comparison of patient handling interventions across different types of outcomes. The IET includes evaluation tools that have been designed to measure a specific type of intervention or outcome for 12 outcomes (Table 16.3). The IET was developed, and has been used and tested, in four European countries (Finland, Italy, Portugal, and United Kingdom), with translation/cross-translation of the collection tools and data (Fray and Hignett, 2009). It was found that tools to evaluate patient outcomes were poorly represented in the literature, so new measurement tools were developed (but not validated).

1. Safety culture

Safety culture is a measure of organizational behavior and how the organizational systems manage the patient handling risks. The data for safety culture in the IET is from interviews with the ward/unit manager, with additional questions about management commitment to managers, advisors, and staff. This section is based on the Patient Handling Observational Question Set (PHOQS; Hignett and Crumpton, 2007). It gives an audit of procedures rather than behaviors, for example, policy, risk assessment, records of training, etc., and should measure support for the prevention program both financially and organizationally.

2. Musculoskeletal health (MS) measure

The MSD Health Measure uses a shortened and validated version of the Nordic Questionnaire (Dickinson et al., 1992) as a self-completion questionnaire. It provides a measurement of the level of MSD in the working population including injuries, chronic conditions, fitness for work, etc.

3. Competence compliance

There is agreement in most situations on safe practice for patient transfers. This section uses the DiNO score system designed and evaluated by Johnsson et al. (2004). This observational checklist looks at staff's individual behavior to complete patient transfers, including competence, skill, compliance with safe methods and equipment use.

4. Absence or staff health

Sickness absence data are collected in the organization with a standardized form to record the time away from work or lost productivity due to patient handling related MSD, days/shifts lost, staff on reduced work capacity, staff turnover. It does not record data from self-reported systems. The IET calculation is standardized for exposure per work hour per individual, to allow comparisons between work areas and different size samples.

5. Quality of care

The quantification and assessment of quality of care has challenged researchers and there are many suggestions

TABLE 16.3

Twelve Outcomes in the Intervention Evaluation Tool

Outcome	Definition
1. Safety culture	A measure of organizational behavior and how management systems control patient handling risk
2. Musculoskeletal health measure	Level of MSD in the working population
3. Compliance, competence	Measures of the staff individual behavior to complete patient transfers
4. Absence or staff health	Measures the time away from work or lost productivity due to MSD from patient handling activities
5. Quality of care	Patient requirements for dignity, respect, safety, empathy
6. Incidents and accidents	Relating to patient handling where staff could have been injured
7. Psychological well-being	Mental health status, stress, strain, job satisfaction, etc.
8. Patient condition	Length of stay, treatment progression, level of independence
9. Patient perception	The subjective assessment of a patient for individual transfers
10. MSD exposure measures	Physical workload factors that place the staff under strain
11. Patient injuries	Records of incidents, accidents, or injuries to patients
12. Financial	The financial impact of MSD in an organization

Source: Fray, M., A comprehensive evaluation of outcomes from patient handling interventions, PhD thesis, Loughborough University, Loughborough, U.K., 2010.

for calculations. Nelson et al. (2008) related measures of quality of care to patient handling to give a complex observational tool of all aspects of care delivery (with over 30 different measures). The IET evaluates whether patient needs are being considered for dignity, respect, safety, and security when they are moved or handled during a hospital stay.

6. Accident numbers

The inclusion of incident numbers that have patient handling factors is unclear and complex. The IET scoring system includes an underreporting ratio from the unit manager and self-reports of unsafe practice by the staff as well as accidents or near misses from patient handling where staff could have been injured.

7. Psychological well-being

The measurement of the staff mental health status, psychological stress, strain, and job satisfaction, etc., is based on a three-section assessment tool used by Evanoff et al. (1999) as a single page staff questionnaire.

8. Patient condition

As with quality of care, there were few precedents of measuring patient condition (length of stay, treatment progression, level of independence) in patient handling studies. The concept of being able to improve the patient's condition due to high standards of patient handling management is unproven, but has a high level of intent among practitioners.

Meeting the clinical needs of the patient and improving care delivery is evaluated using a questionnaire given to staff and management, as it is suggested that patients may not have enough understanding of what should happen to them in a care situation.

9. Patient perception

The subjective assessment of a patient when being moved in a single transfer or mobility situation, fear, comfort, etc., is recorded as a direct assessment (comfort, security, fear) of the transfer or task using a 9-point scale (Kjellberg et al., 2004).

10. MSD exposure measures

This section evaluates the physical workload factors that place the staff under strain, for example, forces, postures, frequency of tasks, and workload measures. The question set used in the IET was developed from three studies: self-reported workload measure (Knibbe and Friele, 1996), patient parameters and workload (Cohen et al., 2004), and the Arjo Resident Gallery (Arjo Ab, 2006).

11. Patient injuries

Accident reporting systems were examined for patient harm accidents (bruises, lacerations, tissue damage, falls, etc.) and pressure ulcer prevalence scores related to the movement and positioning of patients.

12. Financial

The financial impact of MSD in an organization is due to lost staff time, lost productivity costs, compensation claims, litigation, and all direct and indirect costs against the costs of any prevention program. These organizational outcomes are recorded as part of the management interview for the cost of

- Days lost
- Reduced capacity days
- MSD claims
- Treatments for the MSD (internal or external)
- Interventions extra to the organizational set up

The data are standardized using the OSHA formula (Charney, 1997, 2008; Collins et al., 2004). The calculation can then be used as a cost benefit model (e.g., Siddarthan et al., 2005).

Each of the 12 sections calculates a section score and there is an overall performance score. All 13 scores delivered by the IET can be used to compare departments, organizations, and pre- and postinterventions. The individual section scores can be used to develop future intervention strategies to improve the weak areas.

16.3.1 Safer Lifting Policies

Nelson et al. (2007) summarized international initiatives to manage the risks associated with patient handling (American Nurses Association, 2004; Australian Nurses Federation, 1998; Smith, 2005). These initiatives have been called "No Lift," "Zero Lift," "Minimal Lift," "Safe Patient Handling and Movement," and "Safer Patient Handling." They aim to reduce patient handling through organizational management, risk assessment, equipment, and other technologies. Most have the same key elements of

- Elimination of manual lifting of patients in all but exceptional or life threatening circumstances.
- Use of mechanical lifting aids and other handling equipment whenever they can help to reduce the risk.
- Assessment of the handling needs of the patient and documentation by nursing staff, with a physiotherapist where possible.
- Maintenance of a patient's independence by encouraging mobility. Patients should be encouraged to assist in their own transfers, including bearing their own weight, as much as possible.

The European Panel on Patient Handling Ergonomics (EPPHE) was formed in 2004 to share information about research on patient handling from 13 European countries: Denmark, Finland, France, Germany, Greece, Ireland, Italy, the Netherlands, Poland, Portugal, Sweden, Switzerland, and United Kingdom. There is European legislation on manual handling to "ensure that workers are protected against the risks involved in the handling of heavy loads" (Council Directive 90/269/EEC). In 2007, Hignett et al. reported a collaboration to review the implementation of this directive for patient handling in nine European countries and discuss the residual problems (barriers) to safer patient handling. It was found that five countries implemented the directive in 1993 (France, Ireland, Portugal, Sweden, and United Kingdom), followed by Finland, Greece, and Italy in 1994, and Germany in 1996. The provision of guidance for manual handling activities in health and social care varied with only Finland, Sweden, and United Kingdom having national official manuals and the guidance for patient handling. Although the other countries reported having no official handling manual, there were various guidance publications from different individual and regional sources. Residual barriers included a lack of

- Staff and equipment
- Research evidence on patient handling methods (e.g., postural load of nursing aides and physicians) and equipment
- Standards for educational programs
- Guidance based on ergonomic standards
- Development of technical aids

16.3.2 Ergonomics in Health-Care Design

It has been acknowledged that the physical environment has a significant impact on health and safety (including confidentiality, cross infection, and travel time) and that hospitals are not designed with the explicit goal of enhancing staff and patient safety through facility design innovations (Reiling et al., 2004). It has been suggested that "there is significant correlation between the layout and design of a workplace and the risk of injury" (WorkSafe, 2007). There have been many recommendations for the spatial requirements in health-care facilities (Hignett and Lu, 2010). Some have empirical evidence (Hignett and Evans, 2006; Hignett and Lu, 2007, Hignett et al., 2008; Nuffield Provincial Hospitals Trust, 1955; Villeneuve, 2006) and have been included in national guidelines (FGI, 2010; Department of Health, 2008). Others offer professional advice (Arjo Ab, 2005).

The Facility Guidelines Institute (FGI, 2010) have introduced a requirement for new and renovation construction projects to carry out a patient handling and movement assessment (PHAMA) to determine patient handling requirements (with an accompanying cost/benefit analysis). These requirements should be used to assist designers in ensuring conditions in the patient care environment that will safely accommodate fixed, floor-based, and other patient handling equipment.

16.4 Conclusion

This chapter has explored a range of research areas looking at musculoskeletal issues for health-care workers. Previous research was found to have looked predominantly at nursing and medical workers. A number of postural analysis tools have been used to assess health-care activities and brief description for four of these (field) tools were given. Dynamic and static postural loading was explored with examples for patient handling, operating theaters, midwifery, radiography, ultrasonography, physiotherapy, and ambulance work.

There is an increasing body of research looking at patient safety issues. It will be interesting to see whether links can be drawn between risks to care givers and patients. It would seem likely that working in a posture with a high level of physical stress might be a contributory factor to patient safety incidents due to poor sight lines or difficulties in manipulating equipment, consumables, and controls.

References

Ackroyd, S., Bolton, S. (1999). Notes & issues: It is not Taylorism: Mechanisms of work intensification in the provision of gynaecological services in a NHS Hospital, *Work, Employment and Society*, 13, 2, 369–387.

The Facilty Guidelines Institute (FGI) (2010). *Guidelines for Design and Construction of Health Care Facilities*. Chicago American Society for Healthcare Engineering (ASHE). http://www.fgiguidelines.org (accessed June 26, 2011).

American Nurses Association. (2004). Handle with Care. http://cms.nursingworld.org/MainMenuCategories/OccupationalandEnvironmental/occupationalhealth/handlewithcare/hwc.aspx (accessed March 1, 2010).

Amick, B., Tullar, J., Brewer, S., Irvine, E., Mahood, Q., Pompeii, L., Wang, A., Van Eerd, D., Gimeno, D., Evanoff, B. (2006). *Interventions in Healthcare Settings to Protect Musculoskeletal Health: A Systematic Review*. Toronto, Ontario, Canada: Institute for Work and Health.

Arjo Ab. (2005). *Guidebook for Architects and Planners* (2nd edn.) http://www.arjo.com/int/Page.asp?PageNumber=268 (accessed March 29, 2010).

Arjo Ab. (2006). *The Residents Gallery*. Arjo Publication.

Australian Nurses Federation (Victoria Branch). (1998). No lift policy. http://www.anfvic.asn.au/multiversions/3555/FileName/NoLifting.pdf (accessed March 1, 2010).

Buckle, P. (1987). Epidemiological aspects of back pain within the nursing profession. *International Journal of Nursing Studies*, 24, 4, 319–324.

Charney, W. (1997). The lift team method for reducing back injuries: A 10 hospital study. *AAOHN*, 45, 6, 300–304.

Cohen, M., Village, J., Ostry, A., Ratner, P. et al. (2004). Workload as a determinant of staff injury in intermediate care. *International Journal of Occupational and Environmental Health*, 10, 4, 375–383.

Collins, J. W., Wolf, L., Bell, J., Evanoff, B. (2004). An evaluation of 'best practices' musculoskeletal injury prevention program in nursing homes. *Injury Prevention*, 10, 206–211.

Cowdry, I. M., Graves, R. J. (1997). Ergonomic issues arising from access to patients in paediatric surgery. In S. Robertson (Ed.), *Contemporary Ergonomics*. London, U.K.: Taylor & Francis, pp. 32–38.

Crawford, J. O., McHugo, J., Vaughan, R. (2002). Diagnostic ultrasound: The impact of its use on sonographers in obstetrics and gynaecology. In P. T. McCabe (Ed.), *Contemporary Ergonomics*. London, U.K.: Taylor & Francis, pp. 21–26.

David, G., Woods, V., Li, G., Buckle, P. (2008). The development of the Quick Exposure Check (QEC) for assessing exposure to risk factors for work-related musculoskeletal disorders. *Applied Ergonomics*, 39, 1, 57–69.

Dawson, A. P., McLennan, S. N., Schiller, S. D., Jull, G. A., Hodges, P. W., Stewart, S. (2007). Interventions to prevent back pain and back injury in nurses: A systematic review. *Occupational and Environmental Medicine*, 64, 642–650.

Department of Health. (1993). *Changing Childbirth*. Report of the Expert Maternity Group. London, U.K.: HMSO.

Department of Health. (2008). *HBN 09–02 Maternity Care Facilities*. London, U.K.: The Stationary Office.

Dickinson, C., Campion, K., Foster, A., Newman, S., O'Rourke, A., Thomas, P. (1992). Questionnaire development: An examination of the Nordic musculoskeletal questionnaire. *Applied Ergonomics*, 23, 3, 197–201.

Doormaal, M., Driessen, A., Landeweerd, J., Drost, M. R. (1995). Physical workload of ambulance assistants. *Ergonomics*, 38, 2, 361–376.

Estryn-Behar, M., le Nézet, O., Laine, M., Pokorski, J., Caillard, J.-F. (2003). Physical load among nursing personnel. In H.-M. Hasselhorn, P. Tackenberg, and B. H. Müller (Eds.), *Working Conditions and Intent to Leave the Profession among Nursing Staff in Europe*. Stockholm, Sweden: National Institute for Working Life, Chapter 12.

Estryn-Behar, M., Wilanini, G., Scialom, V., Rebouche, A., Fiette, H., Artigou, A. (2000). New conception of a hospital laboratory with a participatory ergonomics methodology. In *Proceedings of the 14th Triennial Congress of the International Ergonomics Association and the 44th Annual Meeting of the Human Factors and Ergonomics Society*, San Diego, CA, July 29–August 4, 2000. Santa Monica, CA: The Human Factors and Ergonomics Society.

Evanoff, B. A., Bohr, P. C., Wolf, L. D. (1999). Effects of a participatory ergonomics team among hospital orderlies. *American Journal of Industrial Medicine*, 35, 4, 358–365.

Facility Guidelines Institute. (2010). *Guidelines for Design and Construction of Health Care Facilities*. American Society for Healthcare Engineering. www.fgiguidelines.org

Ferreira, J., Hignett, S. (2005). Reviewing ambulance design for clinical efficiency and paramedic safety. *Applied Ergonomics*, 36, 97–105.

Fray, M. (2010). A comprehensive evaluation of outcomes from patient handling interventions. PhD thesis. Loughborough, U.K.: Loughborough University.

Fray, M. J., Hignett, S. (2009). Measuring the success of patient handling interventions in healthcare across the European Union. In *Proceedings of the 17th Triennial Congress of the International Ergonomics Association*, Beijing, China, August 9–14, 2009.

Hasselhorn, H.-M., Tackenberg, P., Müller, B. H. (2003). *Working Conditions and Intent to Leave the Profession among Nursing Staff in Europe*. Stockholm, Sweden: National Institute for Working Life.

Hignett, S. (1995). Fitting the work to the Physiotherapist. *Physiotherapy*, 81, 9, 549–552.

Hignett, S. (1996a). Postural analysis of nursing work. *Applied Ergonomics*, 27, 3, 171–176.

Hignett, S. (1996b). Manual handling risks in midwifery—Identification of risks factors. *British Journal of Midwifery*, 4, 11, 590–596.

Hignett, S. (2001). Embedding ergonomics in hospital culture: Top-down and bottom-up strategies. *Applied Ergonomics*, 32, 61–69.

Hignett, S. (2003a). Hospital ergonomics: A qualitative study to explore the organisational and cultural factors. *Ergonomics*, 46, 9, 882–903.

Hignett, S. (2003b). Intervention strategies to reduce musculoskeletal injuries associated with handling patients: A systematic review. *Occupational and Environmental Medicine*, 60, 9, e6 (electronic paper). http://wwwoccenvmed.com/cgi/content/full/60/9/e6

Hignett, S. (2003c). Systematic review of patient handling activities starting in lying, sitting and standing positions. *Journal of Advanced Nursing*, 41, 6, 545–552.

Hignett, S. (2007). Physical ergonomics in healthcare. In P. Carayon (Ed.), *Handbook of Human Factors and Ergonomics in Healthcare and Patient Safety*. Mahwah, NJ: Lawrence Erlbaum Associates Inc., pp. 309–321, Chapter 20.

Hignett, S., Crumpton, E. (2007). Competency-based education for patient handling. *Applied Ergonomics*, 38, 7–17.

Hignett, S., Crumpton, E., Alexander, P., Ruszala, S., Fray, M., Fletcher, B. (2003). *Evidence-Based Patient Handling: Tasks, Equipment and Interventions*. London, U.K.: Routledge.

Hignett, S., Evans, D. (2006). Spatial requirements for patient handling in hospital shower/toilet rooms. *Nursing Standard*, 21, 3, 43–48.

Hignett, S., Fray, M., Rossi, M. A., Tamminen-Peter, L., Hermann, S., Lomi, C., Dockrell, S., Cotrim, T., Cantineau, J. B., Johnsson, C. (2007). Implementation of the manual handing directive in the healthcare industry in the European Union for patient handling tasks. *International Journal of Industrial Ergonomics*, 37, 415–423.

Hignett, S., Jones, A. (2007). Safe access/egress systems for emergency ambulances. *Emergency Medicine Journal*, 24, 200–205.

Hignett, S., Lu, J. (2007). Evaluation of critical care space requirements for three frequent and high-risk tasks. *Critical Care Nursing Clinics of North America*, 19, 167–173.

Hignett, S., Lu, J. (2010). Space to care and treat safely in acute hospitals: Recommendations from 1866–2008. *Applied Ergonomics*, 41, 666–673, doi. 10.1016/j.apergo.2009.12.010

Hignett, S., Lu, J., Morgan, K. (2008). *Empirical Review of NHS Estates Ergonomic Drawings*. Department of Health Estates and Facilities Management Research Report B(02)13. London, U.K.: The Stationary Office.

Hignett, S., McAtamney, L. (2000). Rapid entire body assessment (REBA). *Applied Ergonomics*, 31, 201–205.

Hignett, S., McAtamney, L. (2005). REBA and RULA: Whole body and upper limb rapid assessment tools. In W. Marras and W. Karwowski (Eds.), *Occupational Ergonomics Handbook* (2nd edn.). Boca Raton, FL: CRC Press.

Hignett, S., Wilson, J. R., Morris, W. (2005). Finding ergonomic solutions: Participatory approaches. *Occupational Medicine*, 55, 200–207.

Hollmann, S., Klimmer, F., Schmidt, K. H., Kylian, H. (1999). Validation of a questionnaire for assessing physical work load. *Scandinavian Journal of Work, Environment & Health*, 25, 105–114.

Jackson, J., Liles, C. (1994). Working postures and physiotherapy students. *Physiotherapy*, 80, 7, 432–436.

Johnsson, C., Kjellberg, K., Kjellberg, A., Lagerström, M. (2004). A direct observation instrument for assessment of nurses' patient transfer technique (DINO). *Applied Ergonomics*, 35, 591–601.

Kant, I., de Jong, L. C. G., van Rijssen-Moll, M., Borm, P. J. A. (1992). A survey of static and dynamic work postures of operating room staff. *International Archives of Occupational and Environmental Health*, 63, 423–428.

Karhu, O., Kansi, P., Kuorinka, I. (1977). Correcting working posture in industry: A practical method for analysis. *Applied Ergonomics*, 8, 4, 199–201.

Kjellberg, K., Lagerström, M., Hagberg, M. (2004). Patient safety and comfort during transfers in relation to nurses work technique. *Journal of Advanced Nursing*, 47, 3, 251–259.

Knibbe, J. J., Friele, R. D. (1996). Prevalence of back pain and characteristics of the physical workload of community nurses. *Ergonomics*, 39, 2, 186–198.

Knibbe, J., Friele, R. (1999). The use of logs to assess exposure to manual handling of patients, illustrated in an intervention study in care home nursing. *International Journal of Industrial Ergonomics*, 24, 445–454.

Kumar, S., Moro, L., Narayan, Y. (2003). A biomechanical analysis of loads on x-ray technologists: A field study. *Ergonomics*, 46, 5, 502–517.

Lee, E. J., Lochang, J., Engst, C., Robinson, D. (2002). Applying an ergonomic design process to injury prevention in a hospital pharmacy. In *The Proceedings of the 33th Annual Conference of the Association of Canadian Ergonomists*. Banff, Alberta, Canada.

Lee-Treweek, G. (1997). Women, resistance and care: An ethnographic study of nursing auxiliary work. *Work, Employment and Society*, 11, 1, 47–63.

Louhevaara, V., Suurnäkki, T. (1992). *OWAS—A Method for the Evaluation of Postural Analysis during Work*. Helsinki, Finland: Institute of Occupational Health, Centre for Occupational Safety.

Marras, W. S., Knapik, G. G., Ferguson, S. (2009). Lumbar spine forces during manoeuvring of ceiling-based and floor-based patient transfer devices. *Ergonomics*, 52, 3, 384–397.

Martimo, K. P., Verbeek, J., Karppinen, J., Furlan, A. D., Takala, E. P., Kuijer, P., Jauhianen, M., Viikari-Juntura, E. (2008). Effect of training and lifting equipment for preventing back pain in lifting and handling: Systematic review. *BMJ*, 336, 429–431, doi: 10.1136/bmj.39463.418380.BE

McAtamney, L., Corlett, E. N. (1993). RULA: A survey method for the investigation of WRULD. *Applied Ergonomics*, 24, 2, 91–99.

Nelson, A., Baptiste, A., Matz, M., Fragala, G. (2007). Evidence-based interventions for patient care interventions. In P. Carayon (Ed.), *Handbook of Human Factors and Ergonomics in Healthcare and Patient Safety* (1st edn.), Mahwah, NJ: Lawrence Erlbaum Associates Inc., pp. 323–345, Chapter 21.

Nelson, A., Collins, J., Siddharthan, K., Matz, M., Waters, T. (2008). Link between safe patient handling and patient outcomes in long-term care. *Rehabilitation Nursing*, 33, 1, 33–43.

Pluye, P., Gagnon, M., Griffiths, F., Johnson-Lafleur, J. (2009). A scoring system for appraising mixed methods research, and concomitantly appraising qualitative, quantitative and mixed methods primary studies in mixed studies reviews. *International Journal of Nursing Studies*, 46, 4, 529–546.

Rivilis, I., Van Eerd, D., Cullen, K., Cole, D., Irvin, E., Tyson, J., Mahood, Q. (2008). Effectiveness of participatory ergonomic interventions on health outcomes: A systematic review. *Applied Ergonomics*, 39, 342–358.

Siddarthan, K., Nelson, A., Weisenborn, G. (2005). A business case for patient care ergonomic interventions. *Nursing Administration Quarterly*, 29, 1, 63–71.

Smith, J. (Ed.). (2005). *The Guide to the Handling of Patients* (5th edn.). Teddington, Middlesex, U.K.: BackCare/Royal College of Nursing.

Steele, D., Stubbs, D. (2002). Measuring working postures of midwives in the healthcare setting. In P. McCabe (Ed.), *Contemporary Ergonomics*. London, U.K.: Taylor & Francis, pp. 39–44.

Steinbrecher, S. (1994). The revised NIOSH lifting guidelines. *AAOHN Journal*, 42, 2, 62–66.

Suurnäkki, T., Louhevaara, V., Karhu, O., Kuorinka, I., Kansi, P., Peuraniemi, A. (1988). Standardised observation method of assessment of working postures: The OWAS method. In A. S. Adams, R. R. Hall, B. McPhee, and M. S. Oxenburgh (Eds.), *Ergonomics International 88. Proceedings of the 10th Triennial Congress of the International Ergonomics Association*, Sydney, Australia, pp. 281–283.

Torma-Krajewski, J. (1987). Analysis of lifting tasks in the health care industry. In Occupational Hazards to Health Care Workers. *American Conference of Governmental Industrial Hygienists*, Cincinnati, OH, pp. 51–68.

Training Publication No. 11. (1992). *OWAS: A Method for the Evaluation of Postural Load during Work*. Institute of Occupational Health, Centre for Occupational Safety, Helsinki, Finland.

Van Cott, H. (1994). Human errors: Their causes and reduction. In M. S. Bogner (Ed.), *Human Error in Medicine*, Hillsdale, NJ: Erlbaum, pp. 53–65.

Villeneuve, J. (2006). Physical environment for provision of nursing care: Design for safe patient handlin. A. Nelson (Ed.), *Safe Patient Handling and Movement: A Practical Guide for Health Care Professionals.* New York: Springer Publishing Company, Inc.

Vink, P., Wilson, J. R. (2003). Participatory ergonomics. *Proceedings of the XVth Triennial Congress of the International Ergonomics Association and the 7th Joint Conference of the Ergonomics Society of Korea/Japan Ergonomics Society. Ergonomics in the Digital Age.* August 24–29, 2003, Seoul, Korea.

Warming, S., Precht, D., Suadicani, P., Ebbehoj, N. (2009). Musculoskeletal complaints among nurses related to patient handling tasks and psychosocial factors—Based on logbook registrations. *Applied Ergonomics*, 40, 569–576.

Waters, T. R. (2007). When is it safe to manually lift a patient. *American Journal of Nursing*, 107, 8, 53–58.

Waters, T. R., Putz-Anderson, V., Garg, A., Fine, L. J. (1993). Revised NIOSH equation for the design an evaluation of manual lifting tasks. *Ergonomics*, 36, 7, 749–776.

Worksafe Victoria. (2007). *A Guide to Designing Workspaces for Safer Handling of People for Health, Aged Care, Rehabilitation and Disability Facilities* (3rd edn.) http://www.worksafe.vic.gov.au/wps/wcm/connect/d39b9b004071f551a67e-fee1fb554c40/VWA531.pdf?MOD=AJPERES (accessed March 29, 2010).

Zhuang, Z., Stobbe, T. J., Hsiao, H., Collins, J. W., Hobbs, G. R. (1999). Biomechanical evaluation of assistive devices for transferring residents. *Applied Ergonomics*, 30, 285–294.

Section V

Technology

17

Human Factors and Ergonomics of Health Information Technology Implementation

Ben-Tzion Karsh, Richard J. Holden, and Calvin K. L. Or

CONTENTS

17.1 Introduction 249
17.2 Computerized Provider Order Entry 250
    17.2.1 CPOE Human Factors and Ergonomics Issues 251
    17.2.2 CPOE Summary 252
17.3 Bar Coded Medication Administration 253
    17.3.1 BCMA Human Factors and Ergonomics 253
    17.3.2 BCMA Summary 254
17.4 Clinical Decision Support 254
    17.4.1 CDS Human Factors and Ergonomics Issues 255
    17.4.2 CDS Summary 255
17.5 Consumer Health Information Technology 256
    17.5.1 CHIT Summary 257
17.6 Summary 258
References 258

17.1 Introduction

In February 2009, the American Recovery and Reinvestment Act (ARRA) was signed by President Obama. ARRA has profound implications for clinicians—soon they will be expected to adopt and actively use electronic health records (EHRs) that incorporate some form of electronic prescribing or computerized provider order entry (CPOE) and clinical decision support (CDS), or they will be subject to financial penalties (US Department of Health and Human Services, 2009, 2010).

In ARRA there is funding for health information technology (HIT) as part of the Health Information Technology for Economic and Clinical Health (HITECH) Act (US Congress, 2009; Stark, 2010). The funding includes $27 billion in the form of incentives for physicians and hospitals to adopt electronic health records (EHRs) and $1.2 billion for 70 regional extension centers to help physicians implement the systems they have purchased (Blumenthal and Tavenner, 2010). The incentives are only realized if the physicians adopt EHRs *and* become "meaningful users" of the EHR as defined by the Office of the National Coordinator for Health IT

(ONC) (US Department of Health and Human Services, 2010). The definitions of "meaningful use" became final on July 13, 2010 and define what physicians need to demonstrate in order to receive incentive payments. Examples of the initial set of core meaningful use measures required to qualify for incentive payments include

- Over 50% of patients' demographic data recorded as structured data
- Over 80% of patients have at least one medication entry recorded as structured data
- Over 40% of prescriptions are transmitted electronically using certified EHR technology
- One clinical decision support rule implemented
- Over 10% of patients are provided patient-specific education resources

In addition, the meaningful use requirements translate into "required capabilities and related standards and implementation specifications that Certified EHR Technology will need to include to, at a minimum, support the achievement of meaningful use Stage 1 by eligible health-care providers under the Medicare and Medicaid EHR Incentive Program regulations" (US Department of Health and Human Services, 2010). The U.S. National Institute of Standards and Technology (NIST) is working on standards for EHR usability (NIST, 2010).

Interestingly, in March 2009, one month after ARRA was signed, the National Research Council (NRC) concluded that the nationwide deployment of health IT might set back twenty-first century health-care goals. That belief was based on what the NRC referred to as the "central conclusion" of their report: current health IT systems do not provide sufficient cognitive support for health-care providers (Stead and Lin, 2009). Similarly, the director of the ONC stated in an article on EHR adoption, "Many certified EHRs are neither user-friendly nor designed to meet HITECH's ambitious goal of improving quality and efficiency in the health care system" (Blumenthal, 2009). Also, on the same day as the Meaningful Use final rule was announced, NIST, the Agency for Healthcare Research and Quality (AHRQ), and the ONC sponsored a workshop on HIT usability concerns (Redish and Lowry, 2010).

The seemingly mixed signals regarding the meaningful use legislation parallel the mixed state of evidence regarding HIT generally. Evidence is mixed for the effectiveness of CDS (Shea et al., 1996; Bates et al., 2001; Kho et al., 2008), CPOE (Bates et al., 2001; Kaushal and Bates, 2001; Mekhjian et al., 2002; Kaushal et al., 2003; King et al., 2003; Kuperman and Gibson, 2003; Potts et al., 2004; Kaushal et al., 2006; Shamliyan et al., 2008), bar coded medication administration (BCMA) systems (Patterson et al., 2002; Koppel et al., 2008), and EHRs (Chaudhry et al., 2006; Linder et al., 2007; Zhou et al., 2009; Himmelstein et al., 2010; Jones et al., 2010; Poon et al., 2010b).

As mentioned, one hypothesis regarding the mixed evidence for HIT is that current HIT does not sufficiently support the cognitive work of individual or teams of clinicians and patients (Karsh et al., 2010). Similar conclusions were reported in October 2, 2009 AHRQ-funded reports on EHR usability (Armijo et al., 2009a,b) calling for AHRQ to fund research on the relationships between clinician cognitive work and EHRs.

In recognition of the fact that current HIT may not sufficiently support clinician cognitive work, the ONC released, in December 2009, a funding opportunity announcement for the new Strategic Health IT Advanced Research Projects (SHARP) that would fund four centers, each for 4 years, including one called "Patient-Centered Cognitive Support." That center's stated purpose will be to "addresses the challenge of harnessing the power of health IT so that it integrates with, enhances, and supports clinicians' reasoning and decision making, rather than forcing them into a mode of thinking that is natural to machines but not to people." That center, the SHARP-C, is now funded (http://www.uthouston.edu/nccd/), as are other projects focused on workflow and

usability issues with HIT. In this chapter, we will review known human factors and ergonomics issues related to CPOE, BCMA, CDS, and consumer HIT (CHIT). We do not cover EHRs specifically for three reasons: (a) the HITECH Act that motivated this chapter requires EHRs that include electronic prescribing/CPOE and CDS, not stand-alone EHRs per se, (b) EHRs involve much more extensive functionalities than the other health IT reviewed and designs vary widely—so drawing conclusions may be misleading, and (c) many studies of CPOE and CDS were of systems integrated into EHRs and it is difficult to disentangle results.

17.2 Computerized Provider Order Entry

CPOE refers to "a variety of computer-based systems that share the common features of automating the medication ordering process and that ensure standardized, legible, and complete orders" (Kaushal et al., 2003, p.1410) and various processes (e.g., documentation) that support ordering (Ash et al., 2007). Adoption of CPOE continues worldwide (Aarts and Koppel, 2009), but historically has been low, especially in the United States and in smaller practices and hospitals (Ash et al., 2004b; Furukawa et al., 2008; Jha et al., 2008; Jha et al., 2009). There has also been some debate about the costs and benefits of investing in these somewhat expensive systems. Although there is evidence that CPOE systems can generate a favorable return on investment (e.g., Kaushal et al. 2006), support for CPOE is generally predicated on its potential to improve medication safety and the quality of patient care (Bates and Gawande, 2003; The Leapfrog Group, 2008).

Early evidence, mainly accumulated at large academic medical centers using "home-grown" systems, suggested that CPOE could prevent medication errors (e.g., Bates et al., 1999). Findings of safety benefits are supported by more recent reviews (e.g., Ammenwerth et al., 2008; Shamliyan et al., 2008) and have been replicated in specific settings such as pediatrics and intensive care (Potts et al., 2004; van Rosse et al., 2009). At the same time, reviews show that among studies demonstrating improvement there are also studies showing no safety benefits of CPOE (Eslami et al., 2007), that mainly minor errors are reduced (Reckmann et al., 2009), that error reduction is not always associated with actual reduction of harm/adverse drug events (Wolfstadt et al., 2008; van Rosse et al., 2009), and that the greatest benefits result not from CPOE alone but from CPOE systems that are integrated with decision support functionality (Kaushal et al., 2003). There are also reports that CPOE does not reduce downstream (e.g., medication administration) errors (Fitzhenry et al., 2007), that new errors are introduced (Koppel et al., 2005; Campbell et al., 2006; Chuo and Hicks, 2008), that mortality rates increase post-CPOE (Han et al., 2005), and that CPOE systems are sometimes aborted or abandoned (Ornstein, 2003; Aarts et al., 2004).

In sum, there has been a great deal of debate about whether the effects of CPOE are "positive" or "negative" (see discussions in, e.g., Ammenwerth et al., 2006). Nevertheless, two things seem clear. (1) Ordering errors can be averted with CPOE and (2) CPOE can lead to unintended consequences on work, communication, and coordination. Human factors can help provide an understanding of *why* and *how* CPOE affects safety and work, that is, an understanding of the safety-shaping mechanisms that CPOE may trigger. A human factors approach to understanding the safety-shaping mechanisms of CPOE begins with an understanding of how CPOE transforms the structures and processes—both social and cognitive—of the sociotechnical systems where CPOE is implemented (for further reading, see Holden et al., Accepted; Karsh et al., 2006; Karsh et al., 2010; Holden, 2011a).

17.2.1 CPOE Human Factors and Ergonomics Issues

Studies of CPOE depict numerous sociotechnical impacts of CPOE (Harrison et al., 2007). CPOE can introduce new processes and work demands (Campbell et al., 2006). Some of these are seen as "extra work," particularly the requirement of physicians to do electronic data entry (Massaro, 1993; Aarts et al., 2004). Holden (2010) suggests that the work is not actually new, but rather is work previously done by others (nurses, pharmacists, radiologists) shifted to the ordering physician (see also Shu et al., 2001; Georgiou et al., 2007a; Niazkhani et al., 2009). This shift and the similarity between electronic order entry and the work of "less-educated" data entry specialists and nurses create role conflicts (Holden, 2011b). Alternatively, new processes facilitated by CPOE can be viewed as new ways of accomplishing work in a better (e.g., safer) way. For example, Callen et al. (2006) describe how CPOE was used in one hospital's emergency department to better verify test results, something that was not being systematically done pre-CPOE. As another example, physicians in one study viewed new requirements to maintain a centralized electronic problem list (depicting a patient's conditions) as extra work but many recognized that having an updated problem list made it easier to find information, keep track of data, and make decisions (Holden, 2011a).

Time studies reveal that entering orders via CPOE can take more time than using paper (Eslami et al., 2007; Niazkhani et al., 2009), a phenomenon also feared by physicians considering adopting CPOE (Ash and Bates,

2005; Viccellio, 2008) and perceived by those using it (Callen et al., 2006; Niazkhani et al., 2009; Holden, 2010, 2011a; but see Schectman et al., 2005). CPOE increases time spent on most types of orders in hospital settings (Tierney et al., 1993; Bates et al., 1994; Evans et al., 1998; Shu et al., 2001; but see Mutimer et al., 1992; Ostbyte et al., 1997). In primary care settings, both Overhage et al. (2001) and Tamblyn et al. (2006) found that although physicians' electronic prescribing time improved over time, most electronic prescribing tasks took longer compared to paper orders. In contrast, after adjusting for site differences, Hollingworth et al. (2007) found that physicians at two ambulatory care sites spent only 12 s more per medication order using electronic prescribing compared to physicians using paper at a third site. The authors implied that this was because the system was "carefully implemented" (p. 729).

There is some concern that more time spent on CPOE will take time away from patient care (Niazkhani et al., 2009). Yen et al.'s (2009) time-motion study in a pediatric emergency department (ED) found that post-CPOE physicians spent more time on the computer but no less time on direct patient care. Indeed, studies often find that extra time related to physicians' CPOE use is at least partially offset by the elimination of various intermediary or peripheral tasks (e.g., Bates et al., 1994; Niazkhani et al., 2009; Yen et al., 2009). However, Han et al. (2005) argued that in their critical care setting, the additional time needed to enter orders via CPOE caused delays in initiating patient care, one possible cause of the post-CPOE increased risk of mortality observed in that study. In another study, most hospital physicians perceived either no change or a decrease in time spent with patients post-CPOE, whereas most nurses perceived an increase (Weiner et al., 1999). Far less studied, though reported by some (Koppel et al., 2005; Holden, 2010, 2011a), is the time lost during CPOE shutdowns and crashes.

Although less is known about how CPOE redistributes nonphysician workers' time, there is evidence that post-CPOE nurses take more time documenting medication administration (Evans et al., 1998) and spend more time using the computer (Hollingworth et al., 2007), "away from the bedside, effectively reducing staff-to-patient ratios" (Han et al., 2005, p. 1510). For pharmacists, one study of two Canadian tertiary care centers found that pharmacists spent more time reviewing electronic chemotherapy orders compared to paper orders (Beer et al., 2002). Consistent with the idea that work is shifted from nonphysicians to physicians, Taylor et al. (2002) demonstrated time savings for nurses, pharmacists, and other staff (see also Niazkhani et al., 2009). Yet, timing changes for individual steps in the integrated, multistep, multiprofessional medication management process are not good indicators of the efficiency or safety of that process. For example, medication orders that are created far more quickly than they can be reviewed by pharmacists or laboratory orders that are sent (by physicians) well in advance of the specimen being sent (by nurses) result in "overproduction waste" (Ohno, 1988) and confusion (Georgiou et al., 2007a). Collapsing across professional boundaries, however, the emerging picture is that of reductions in overall order turnaround times when using CPOE (Cordero et al., 2004; Holden, 2010). Reviews by Niazkhani et al. (2009) and Georgiou et al. (2007b) find evidence for decreased turnarounds for medication, radiology, and laboratory orders related to electronic order entry.

The sequence of tasks may be altered as well. Callen et al. (2006) described how CPOE system delays during the order entry process led some hospital physicians to interleave other tasks when entering orders. In the Han et al. (2005) study, a medication process that was done in parallel to critical patients being transported to the facility was made sequential after the introduction of CPOE, such that several time-sensitive steps (e.g., order entry and medication dispensing) had to be delayed until the patient arrived. Rosenbloom et al. (2006) and Gesteland et al. (2006) suggest that these "bad" workflow changes, not CPOE in and of itself, had much to do with increased mortality in that study.

CPOE changes not only "how" but also "where" work is done. On the one hand, with multiple computer stations spread throughout the hospital or clinic, wireless connectivity, and remote access (e.g., from home), the ordering process can be completed virtually anywhere, assuming that the needed hardware, software, and connectivity are available (Holden, 2010). On the other hand, unless portable computers are being used, physicians can no longer enter orders directly at the patient's bedside and sometimes not even in the same room as the patient (Callen et al., 2006; Holden, 2010), potentially resulting in interruptions of activity (Niazkhani et al., 2009).

Changes in professional communication and coordination are the other well-documented sociotechnical consequence of CPOE (Beuscart-Zephir et al., 2005; Campbell et al., 2006). Studies have shown a reduction in the amount of talking done by physicians (Shu et al., 2001), the amount of time nurses spent talking to others about patients (Yen et al., 2009), and care related, synchronous, face-to-face communication in general (Han et al., 2005; Campbell et al., 2006). In many cases, CPOE puts in place a work structure that reinforces professional divisions, with the physician increasingly entering orders alone and the nurse receiving those orders without having had input into order creation (Dykstra, 2002; Cheng et al., 2003; Gorman et al., 2003; Ash et al., 2004a; Beuscart-Zephir et al., 2005; Han et al., 2005). This "silo" organization hinders communication

across professions (e.g., physician–nurse) and opposes "the negotiated and co-constructed nature of ordering practice" (Niazkhani et al., 2009, p. 544; see also Gorman et al. 2003). As a result, nurses may be "left out of the loop" (Carpenter and Gorman, 2001; Shabot, 2004; Beuscart-Zephir et al., 2005; Aarts et al., 2007) and both pharmacists and nurses may lose their flexibility to make adjustments based on their clinical judgment (Cheng et al., 2003; Beuscart-Zephir et al., 2005; Campbell et al., 2006; Pitre et al., 2006). Another communication problem associated with CPOE is what is called the "illusion of communication" (Dykstra, 2002): the incorrect assumption that the generation of an order is equivalent to the notification and mobilization of the order's recipient (e.g., nurse acting on that order) (Ash et al., 2004a; Campbell et al., 2006). Positive communication effects have also been documented, for example, better communication of pathology orders because "more professionals are able to access the system electronically making exchange of information easier and faster" (Georgiou et al., 2006, p. 132).

Communication between physicians—and by extension physicians' ability to work together effectively and efficiently—is also changed by CPOE. The effects are mixed. On one hand, communication between physicians using CPOE is more legible, more complete, and is transmitted quickly (Niazkhani et al., 2009; Holden, 2010, 2011a). On the other hand, information entered using pull down menus, templates, or prestructured forms can be irrelevant, incomplete or fragmented, and may suffer from the problem of "garbage in-garbage out" (Ash et al., 2004a; Niazkhani et al., 2009; Holden, 2011a).

17.2.2 CPOE Summary

As with other health IT, the sociotechnical impacts of CPOE are numerous and mixed. CPOE creates new work that can be seen as burdensome or beneficial, redistributes workers' time and responsibilities, requires additional time to use, reduces order turnaround times, alters the sequence and location of ordering work, and has both positive and negative implications for communication and collaboration. The presence of undesirable, unintended consequences of CPOE is conspicuous in light of the strong push to adopt and use CPOE (Campbell et al., 2006; Ash et al., 2007; Harrison et al., 2007); yet such consequences are the reasons why CPOE sometimes fails (Aarts et al., 2004; Han et al., 2005).

Many of the undesirable sociotechnical consequences can be linked to usability problems (Beuscart-Zephir et al., 2005; Koppel et al., 2005; Holden, 2011a) and to CPOE systems that are *ecologically invalid,* that is, whose design embodies an inaccurate model of the ordering process and related work (Goorman and Berg, 2000; Gorman et al., 2003; Wears and Berg, 2005). The new work that is required by ecologically invalid CPOE design is ill suited to the conditions of the work environment (e.g., unit complexity) and conflicts with existing roles and practices (Han et al., 2005; Wears and Berg, 2005; Aarts et al., 2007; Georgiou et al., 2007a).

It deserves noting that as undesirable consequences emerge, health-care providers, management, and others in the implementing organization (e.g., IT departments) adaptively implement countermeasures that among other things prevent CPOE-related changes from resulting in patient harm (Harrison et al., 2007; Harrison and Koppel, 2010). In the presence of CPOE, physicians invent strategies to improve their own workflow, including reverting to pre-CPOE work practices, using workarounds (e.g., entering a hard stop rather than text in a required field in the order form [Callen et al., 2006]), and supplementing CPOE use with the use of paper artifacts (Cheng et al., 2003; Campbell et al., 2006; Holden, 2011a). Institution-wide solutions are also common (Georgiou et al., 2007a). For example, in dealing with the "illusion of communication" problem mentioned earlier, some institutions have implemented systems (e.g., computerized alerts, printouts) to notify workers of the need to address a recently entered order (Niazkhani et al., 2009). Similarly, Wright et al. (2006) report on a proactive effort in an acute care clinic to improve communication between physicians, nurses, and nonclinical staff (unit secretary) in anticipation of CPOE.

17.3 Bar Coded Medication Administration

It is estimated that about 19% of medical errors are medication related (Leape et al., 1991), of which 40% occur in the administration and dispensing phases (Bates et al., 1995). The medication administration stage accounts for 26%–32% of adult-patient medication errors (Bates et al., 1995; Kopp et al., 2006) and 4%–60% of pediatric patient medication errors (Walsh et al., 2005). The Institute of Medicine (IOM) estimates that on average hospitalized patients experience one medication administration error per day (Institute of Medicine, 2007), where an error can be an omitted drug, wrong drug, unauthorized drug, wrong dose, extra dose, wrong route, wrong form, wrong technique, or wrong time (Barker et al., 1982; Barker et al., 2002). Errors in the administration stage are less likely to be intercepted and more likely to reach patients than errors in any preceding stage (Bates et al., 1995; Leape et al., 1995; Kopp et al., 2006; Shane, 2009).

The most recommended technology for controlling medication administration errors is BCMA (Bates, 2000; Wald and Shojania, 2001; Bates and Gawande, 2003; Institute of Medicine, 2007; Cescon and Etchells, 2008), and in 2010 the first rigorous scientific assessment of BCMA effectiveness was published (Poon et al., 2010a). The study demonstrated that BCMA could reduce medication administration errors by 41.4% (from 11.5% on units without BCMA to 6.8% on units with BCMA) and reduce potential adverse drug events by 50.8% (from 3.1% on units without BCMA to 1.6% on units with BCMA). The aforementioned study confirmed the effectiveness of BCMA that other, less rigorous studies, had suggested (Lawton and Shields, 2005; Cochran et al., 2007; Sakowski et al., 2008).

BCMA systems, if designed and implemented appropriately, help to ensure the five "rights" of medication administration: right medication, right patient, right dose, right route, right time (Neuenschwander et al., 2003; Cummings et al., 2005) and to ensure complete and accurate documentation of the administration process (McRoberts, 2005). BCMA systems at minimum require machine readable bar code labels that uniquely identify medications, nursing staff, and patients (Cummings et al., 2005). BCMA systems are typically integrated with an electronic medication administration record (eMAR) and electronic nurse documentation (Pedersen and Gumpper, 2008). The eMAR is an electronic version of the paper MAR and contains, for each patient, the names of medications to be administered, dose, route, time and/or schedule, and the form of the medication. eMARs can be used with or without BCMA systems.

BCMA hardware generally comes in two varieties: (1) wireless, handheld devices that both read bar codes and have a screen that displays the eMAR, or (2) a handheld scanner that is tethered to a wireless computer-on-wheels (COW) that displays the eMAR and allows access to other relevant software. Alternatively, a tethered scanner can be attached to a fixed computer station in the patient room.

17.3.1 BCMA Human Factors and Ergonomics

BCMA research has demonstrated fundamental changes to nursing work following BCMA implementation (Karsh et al., 2011). Regarding time on tasks, one study (Poon et al., 2008) found that the percent of time spent on information retrieval (e.g., looking up drug information), verifying patient ID, and waiting (e.g., for the computer to operate) increased after BCMA implementation, while the percent of time spent managing orders and delivering medications decreased. Also, post-BCMA, the percent of time that nurses spent communicating with patients or patients' families decreased. These results suggest that BCMA changed the task distributions for nurses in ways that are not clearly positive.

Another study assessing nurse self-reports about BCMA systems found more conflicting results (Holden et al., Accepted). In the study, nurses rated medication administration processes—matching the medication to the MAR, checking patient identification (ID), and documenting administration, and the overall administration process—on seven dimensions. Nurses reported more positive perceptions of the usefulness, accuracy, and consistency of checking patient ID. Nurses perceived a lower likelihood of error and a higher likelihood of error detection during the ID check process post-BCMA. Post-BCMA, nurses perceived the documentation process to be less useful, less easy to perform, and less time-efficient. These finding can be explained in terms of design of the BCMA technology. In particular, BCMA automated ID matching, as long as the nurse scanned the patient ID band, so BCMA adequately supported the checking task. On the other hand, even though documentation could also be automated—with a simple scan of the medication or push of a button indicating what was administered, when, and to whom—the disconnect between the actual administration process and the workflow assumed by the BCMA system, coupled with the system's rigidity, made documentation difficult.

Holden et al. (2008a) described a case that could further explain the aforementioned findings. A medication administration on a unit using BCMA was delayed despite the nurse having the correct dose for the patient. The problem was that the BCMA software expected one 20 mg dose, but pharmacy dispensed two 10 mg doses. The BCMA system could not handle the deviation from what it expected. In such cases, if the nurse administers anyway, the nurse must spend time manually entering the reason for what the BCMA system thinks is a discrepancy.

Others have also found BCMA to cause nurse work problems. Patterson et al. (2002) identified five "negative side effects" of BCMA including nurse confusion over "automation surprises" (Sarter et al., 1997), degraded coordination between nurses and physicians, workarounds, increased prioritization of monitored activities during goal conflicts, and decreased ability to deviate from routine sequences. Many of these findings have been confirmed by others (Carayon et al., 2007; Koppel et al., 2008; van Onzenoort et al., 2008; Vogelsmeier et al., 2008). Koppel et al. (2008) provided a comprehensive analysis of the causes of nurse BCMA workarounds and problems. They found 31 different causes and categorized them according to the Systems Engineering Initiative for Patient Safety model (Carayon et al., 2006): technology-related (e.g., difficult-to-read or navigate screens), task-related (e.g., BCMA was perceived as slowing performance), organization-related (e.g., patients or medications without bar codes), patient-related (e.g., patients sleeping), and environment-related (e.g.,

doorways and patient-room configurations that hinder bedside access of BCMA systems that use computers-on-wheels). Only one study appears to have quantitatively examined the extent of violations or workarounds during medication administration before and after BCMA implementation (Alper et al., 2008). Nurses' self-reported violations did not change for the process step of matching medications to the MAR, but decreased for checking patient ID, and increased for documentation after BCMA implementation.

On the other hand, there are ways in which BCMA has improved nursing work, for example, by providing new methods for supporting cognitive work needs. For example, Holden et al. (2008a) found that BCMA can enhance nurse ability to keep track of which medications to administer, when to do so, and when previous doses of the medication were administered, by allowing nurses to access, sort, or print a list of medications, scheduled administration times, and records of when medications were scanned in the past. BCMA systems support other types of cognitive work by, for example, alarming when a medication or patient is scanned and there is a mismatch (Koppel et al., 2008).

17.3.2 BCMA Summary

There is now compelling evidence that BCMA can reduce medication administration errors and potential adverse drug events, but at the same time there are human factors and ergonomic challenges. BCMA systems fundamentally change nursing work and not always in positive ways. Evidence shows that design that poorly supports nurses' cognitive and physical performance needs leads to workarounds and protocol violations, and these in turn may increase the risk of patient safety events. In other cases, poor design prevents nurses from doing their work at all, for example, when data on their BCMA hand-held devices cannot be read. At the same time, BCMA does support some nurse needs in ways that previously paper systems could not. Much more work is needed to design BCMA systems to adequately support the complex cognitive work of nurses.

17.4 Clinical Decision Support

(This section contains some abbreviated material from a more comprehensive review of CDS usability by Karsh [2009] combined with new human factors insights into CDS.)

CDS is supposed to provide clinicians and/or patients with computerized clinical information at the appropriate time and in an understandable format (Osheroff et al., 2005). CDS systems are typically designed to aid decision making for prevention, screening, diagnosis, treatment, drug dosing, test ordering, and/or chronic disease management, and "push" the information to the decision maker (Berlin et al., 2006). However, there is no agreement on the types of features or information technologies that constitute CDS. The broad definition given earlier would include alerts, reminders, structured order forms, pick lists, patient-specific dose checking, guideline support, medication reference information, and "any other knowledge-driven interventions that can promote safety, education, workflow improvement, communication, and improved quality of care" (Teich et al., 2005).

Certain types of CDS have been shown to be effective. Computerized alerts may decrease error rates and improve therapy (Bates et al., 2001). Computerized clinical reminders can increase compliance with guidelines (Kho et al., 2008) and preventive screening (Shea et al., 1996), improve clinical practice (Kawamoto et al., 2005), and may even save physicians time (Kho et al., 2008). At the same time, "…there are few CDS implementations to date in routine clinical use that have substantially delivered on the promise to improve healthcare processes and outcomes, though there have been an array of successes at specific sites …Yet even these successes have generally not been widely replicated" (Sittig et al., 2008).

17.4.1 CDS Human Factors and Ergonomics Issues

One of the reasons that many believe CDS has not achieved its promise is the high rate of alerts and reminders that are ignored or overridden (Weingart et al., 2003; Van der Sijs et al., 2006; Tamblyn et al., 2008). Clearly, the misuse, disuse, and abuse of automation are important human factors and ergonomics topics (Parasuraman and Riley, 1997). Evidence about why CDS is ignored or overridden is similar to what has been found studying automation in other domains—the research on CDS has identified low specificity, low sensitivity, unclear information content, unnecessary workflow disruptions, and poor usability all as contributing factors (Miller et al., 2005; Saleem et al., 2005; Gaikwad et al., 2007). In fact, in a paper titled "Grand Challenges in Clinical Decision Support" (Sittig et al., 2008) 10 grand challenges for CDS were posed, and the #1 ranked challenge was improving the human–computer interface. The authors felt that CDS automation needed to be transformed into "one that supports and does not interrupt the clinical workflow." The authors went on to say, "We need new HCIs (human computer interfaces) that will facilitate the process by which CDS is made available to clinicians to help them prevent both errors of omission and commission. Improved HCI design may include increased

sensitivity to the needs of the current clinical scenario; provide clearer information displays, with intrusiveness proportional to the importance of the information; and make it easier for the clinician to take action on the information provided." Others have reached similar conclusions (Feied et al., 2004; Teich et al., 2005).

Not surprisingly, evidence as to what leads to effective CDS also matches human factors research into other forms of automation (Parasuraman et al., 2000; Sheridan and Parasuraman, 2006; Parasuraman and Wickens, 2008). Features or designs of CDS that support its use include automatic provision of support as part of workflow, attention to usability, provision of recommendations and not just assessments, and provision of support at the time and location of decision making (Kawamoto et al., 2005; Blaser et al., 2007; Zaidi et al., 2008). All of this demonstrates that the main barriers to effective use of CDS are poor usability and lack of integration with workflow.

However, from a deeper human factors and ergonomics assessment, it is likely that the reasons CDS has not been effective are even more complicated. Karsh (2010) conducted a work system analysis of causes of poor CDS effectiveness and use and identified a more diverse set of issues. These problems included not having the appropriate data entered prior to seeing a patient—data needed to trigger CDS—and not being able to enter those data during the visit, clinicians not logging into the computer system in the first place, clinicians not being trained how to incorporate computers into patient visits such that CDS might have a chance to be triggered, brittle CDS algorithms that did not match patient realities, physical layouts that reduced the likelihood that physicians would use the computer in the room, and EHR architectures that inhibited easy integration of CDS. Clearly, the factors that seem to inhibit effective use of CDS have all been found before in human factors research of automated systems and computer technology.

17.4.2 CDS Summary

CDS has not enjoyed widespread success. Rates of ignoring or overriding CDS are very high, with some reports as high as 90% (Weingart et al., 2003; Van der Sijs et al., 2006). There are clinical reasons that explain some of this—sometimes physicians ignore alerts and reminders for legitimate reasons. These reasons might include that the alert is not relevant for a particular patient, or for that patient at that time, or that the physician and patient both know it is time for a particular preventive screening exam, but the patient has already refused it. But clearly there are also many human factors and ergonomics design and implementation reasons. Some of them relate to simple usability 101 improvements, but others are more complicated macroergonomics issues related to workflow integration, and design for appropriate reliance and compliance given highly variable and unpredictable cognitive work.

17.5 Consumer Health Information Technology

The desire to cut health-care cost and empower patients to participate more actively in their own care highlight the need for CHITs in disease management. CHITs refer to patient-focused interactive Web- or technology-mediated applications that are designed to improve information access and exchange, enhance decision making, maintain records of essential health parameters, provide social and emotional support, and facilitate behavior changes for health issues (Slack, 1997; Gustafson et al., 2005). They are frequently developed for home use by patients for self-care of chronic illnesses or other major diseases. Examples include Comprehensive Health Enhancement Support System (CHESS) (Gustafson et al., 1992), ComputerLink (Brennan et al., 1991), Health Buddy (LaFramboise et al., 2003), and the HomMed automated monitoring system (Scott et al., 2004).

CHITs can support patient care at home by extending services and resources to patients who have limited or no access to health care (Gustafson et al., 1999; Hailey et al., 2002; Martinez et al., 2006). However, the introduction of CHITs into the home can bring with it a host of human factors challenges because they are not always user-friendly or are not designed mindful of specific needs, capabilities, and limitations of patient users or their home environments. As a result, the technologies may be perceived as inefficient and the users may eventually abandon them (Rogers and Mead, 2004; Stoop et al., 2004; Liddy et al., 2008).

Changes in physical and cognitive characteristics of patient users due to chronic disabling conditions or aging is a major issue in health innovation design (Vanderheiden, 2006; Or et al., 2009). Many CHITs have been developed for chronically ill patients who are usually frailer and older than the general population and may suffer from age-related cognitive decline and physical disabilities, such as loss of vision and dexterity. These functional impairments limit users' abilities to read text and graphics on the screen, operate essential peripheral devices (e.g., mouse and keyboard), and remember the login name and password (Chaparro et al., 1999; Finkelstein et al., 2003; Kaufman et al., 2003; Tak and Hong, 2005; Lober et al., 2006). Challenges can be magnified when the patient users are required to perform complex visual search and pattern matching

while using a mouse and keyboard or when the applications are not user-friendly, such as small size of the font, irrelevant or excessive use of information and graphical representations, or low contrast screens. To address these challenges, guidelines and principles that focus on technology interface development for aging and disabilities should be applied to system designs and can be found in a number of human factors and ergonomics handbooks, standards, and research reports (Sanders and McCormick, 1993; Karwowski, 2003; Vanderheiden, 2006; Czaja and Lee, 2007; Web Content Accessibility Guidelines [WCAG] 2.0., 2008; Or et al., 2009; Or et al., 2011).

Patients' computer skills and confidence level in their ability to use technology are other important *person factors* for CHIT design and implementation. For example, Mead et al. (2003) examined factors affecting patients' interest in using the Internet as a resource for health information. They found that lack of confidence in own ability to use the technology was associated with lower interest in using it to get health information. With respect to computer skills, evidence has showed that lack of computer skills contributed to why patients did not use CHITs for disease management (Gustafson et al., 2001; Carroll et al., 2002; Zimmerman et al., 2004). User confidence and skills should be augmented when designing and implementing CHITs. This can be achieved by providing users with sufficient computer skills training and ongoing technical support (Kalichman et al., 2006). Also, giving adequate and effective instructions is essential. For instance, Evangelista et al. (2006) revealed that patients with limited computer skills can and will use CHITs when sufficient instructions on how to use the technology they were provided.

Usefulness and usability of technologies are other significant human factors issues that affect CHIT implementation success and users' willingness to use CHIT. For usefulness, for example, Stoop et al. (2004) developed a computer-based patient education system for laypersons (patients with amblyopia and family members), but the implementation failed. The authors found that the designers had made unwarranted assumptions when building the system about what the users needed and wanted. As a consequence, the users felt that the system was not useful and rejected it as it was not geared toward the patients' actual needs and expectations. Moreover, usefulness is associated with patients' CHITs acceptance. For instance, Or et al. (2011) examined factors affecting acceptance of a web-based, interactive self-management technology among home care patients with chronic cardiac disease. They found that patients were more likely to accept and use the technology if they realized that the technology was useful—using the technology would enhance their ability and effectiveness in managing their disease. The evidence report by the U.S. Agency for Healthcare Research and Quality (Jimison et al., 2008) and the systematic review by Or and Karsh (2009) also provided evidence to support that system usefulness is a significant determinant of acceptance and use of CHITs.

Usability is exceptionally important in CHIT applications development because designers should give specific attention to patient users whose needs and functional characteristics (cognitive, perceptual, and psychomotor) can be unique and different from the general population. Poor usability can lead to product abandonment or unintended adverse consequences. Rogers and Mead (2004) showed that patients who did not use CHITs reported that the applications were too complicated and difficult to use. Usability evaluation can be applied to improve the usability of CHITs and to ensure that the system design can reflect better the true needs and preferences of the users (Bates and Gawande, 2003; Gosbee and Gosbee, 2007). In fact, evidence has shown how usability evaluation can contribute to the success of CHIT design and implementation. For instance, Stinson et al. (2006) performed usability testing of an electronic pain diary prototype in adolescents with arthritis. They found that the system interface design was suboptimal and impeded users' task performance and suggested redesign recommendations. Lin et al. (2009) tested the usability of a computer-mediated health education program in older adults. Based on the test, they tailored the design of the system interface to accommodate older users to enhance use efficacy. Kaufman et al. (2003) evaluated the usability of a home-based medical information system in elderly patients. Their evaluation discovered the challenges elderly patients confronted in using the system and the interface design problems that negatively affected task performance. Many other studies—too many to list here—have demonstrated the significance and effectiveness of usability evaluation in CHIT development.

To promote usefulness and usability, CHITs must be designed such that system applications are in consonance with patients' needs. The technology should improve patients' ability to manage their disease, save them time managing their disease, and make them more effective at managing their disease (Or et al., 2011). They should also promote learnability, efficiency, memorability, error reduction and error recovery, and user satisfaction (Nielsen, 1993). Achieving these goals will likely require systematic usability evaluation and cognitive analyses, such as cognitive work and task analyses (Bisantz and Burns, 2008), followed by careful attention to human factors engineering (Karsh et al., 2006) and cognitive engineering principles for designing automation to support cognitive work (e.g., managing a disease) (Littner et al., 2002; Hollnagel and Woods, 2005). One of the principal methodologies is that technology design

and development processes should involve intended target users that allows technology designers or health informatics practitioners to solicit users' feedback and understand how they react to the system applications (Hartwick and Barki, 1994; Demiris et al., 2001; Holden et al., 2008b). Such analyses and user involvement can ensure that necessary and desired content and features are present to support disease management.

The physical environment of patients' homes will influence how CHITs are implemented and used. Considering human factors engineering principles for work environment design is a good practice for improving the interaction between patient users, the technology, and the environment. Lighting, noise, and thermal comfort are the most common environmental aspects to consider. A room that does not have an appropriate level of lighting, for example, will make using a device (e.g., monitor display) physically uncomfortable and affect task performance (Sanders and McCormick, 1993). Similarly, a patient will have a difficult time using technology such as a videophone or an interactive telemedicine device with their care providers at a remote site if the technology is placed in a room that is often noisy. Furthermore, some safety hazards that could exist in patients' homes will preclude successful and safe use of technology. Dedicated electrical outlets for power supply to the technology may be limited in some patients' homes (Gardner-Bonneau, 2001). Because of the unavailability or inadequacy of electrical outlets, it would not be unusual for electrical strips or computers' power cables to run across the room for the equipment. They are hazardous to patients as they could be tripped by the wiring when they are working or walking around the room (Brauer, 2006). Limited space in patients' homes is another issue. For example, a heavy and cumbersome computer monitor, computer chassis, and other peripheral devices may be placed on a fragile or small desk surface; this puts the users at risk because the equipment could fall from the surface. Appropriate placement of the equipment is critical to maintain safety. Detailed information and design recommendations concerning work environment design are available for reference (Sanders and McCormick, 1993; Karwowski, 2003; Brauer, 2006).

17.5.1 CHIT Summary

There are significant human factors challenges for the design and implementation of CHIT, ranging from the design of the software or device, to the unique needs of the chronically ill and elderly, to the unique physical environments of individuals' homes (Or et al., 2009). Fortunately, evidence demonstrates that applications of user-centered design approaches can increase the likelihood of CHIT acceptance and use.

17.6 Summary

The rate of adoption of all forms of health IT in outpatient and inpatient environments, nursing homes, and the community will continue to increase rapidly due to ever improving designs, and government and public pressures. An increase in adoption rates does not automatically mean that care or patient outcomes will improve, however. The evidence reviewed in this chapter has demonstrated that all types of health IT change the nature of work for the users, and whether that change helps or hinders care is related to the design of the systems to meet the needs of the user and the users' context of use. The issue of health IT usability is now recognized by various U.S. government agencies as needing significant attention. Human factors and ergonomics practitioners and scientists are in a unique position to contribute to the work on health IT usability; our hope is that many more will get involved.

References

Aarts, J., Ash, J. S., and Berg, M. (2007). Extending the understanding of computerized physician order entry: Implications for professional collaboration, workflow and quality of care. *International Journal of Medical Informatics, 76S,* S4–S13.

Aarts, J., Doorewaard, H., and Berg, M. (2004). Understanding implementation: The case of a computerized physician order entry system in a large Dutch university medical center. *Journal of the American Medical Informatics Association, 11,* 207–216.

Aarts, J., and Koppel, R. (2009). Implementation of computerized physician order entry in seven countries. *Health Affairs, 28,* 404–414.

Alper, S. J., Holden, R. J., Scanlon, M. C., Patel, N., Murkowski, K., Shalaby, T. M. et al. (2008). Violation prevalence after introduction of a bar coded medication administration system. Paper presented at the *2nd International Conference on Healthcare Systems Ergonomics and Patient Safety,* Strasbourg, France.

Ammenwerth, E., Schnell-Inderst, P., Machan, C., and Siebert, U. (2008). The effect of electronic prescribing on medication errors and adverse drug events: A systematic review. *Journal of the American Medical Informatics Association, 15,* 585–586.

Ammenwerth, E., Talmon, J., Ash, J. S., Bates, D. W., Beuscart-Zephir, M. C., Duhamel, A. et al. (2006). Impact of CPOE on mortality rates—Contradictory findings, important messages. *Methods of Information in Medicine, 45,* 586–593.

Armijo, D., McDonnell, C., and Werner, K. (2009a). *Electronic Health Record Usability: Evaluation and Use Case Framework* (No. AHRQ Publication No. 09(10)-0091-1-EF). Rockville, MD: Agency for Healthcare Research and Quality.

Armijo, D., McDonnell, C., and Werner, K. (2009b). *Electronic Health Record Usability: Interface Design Considerations* (No. AHRQ Publication No. 09(10)-0091-2-EF). Rockville, MD: Agency for Healthcare Research and Quality.

Ash, J. S., and Bates, D. W. (2005). Factors and forces affecting EHR system adoption: Report of a 2004 ACMI discussion. *Journal of the American Medical Informatics Association, 12,* 8–12.

Ash, J. S., Berg, M., and Coiera, E. (2004a). Some unintended consequences of information technology in health care: The nature of patient care information system-related errors. *Journal of the American Medical Informatics Association, 11,* 104–112.

Ash, J. S., Gorman, P. N., Seshadri, V., and Hersh, W. R. (2004b). Computerized physician order entry in U.S. hospitals: Results of a 2002 survey. *Journal of the American Medical Informatics Association, 11,* 95–99.

Ash, J. S., Sittig, D. F., Dykstra, R. H., Guappone, K., Carpenter, J. D., and Seshadri, V. (2007). Categorizing the unintended sociotechnical consequences of computerized provider order entry. *International Journal of Medical Informatics, 76S,* S21–S27.

Barker, K. N., Flynn, E. A., Pepper, G. A., Bates, D. W., and Mikeal, R. L. (2002). Medication errors observed in 36 health care facilities. *Archives of Internal Medicine, 162*(16), 1897–1903.

Barker, K. N., Mikeal, R. L., Pearson, R. E., Illig, N. A., and Morse, M. L. (1982). Medication errors in nursing-homes and small hospitals. *American Journal of Hospital Pharmacy, 39*(6), 987–991.

Bates, D. W. (2000). Using information technology to reduce rates of medication errors in hospitals. *British Medical Journal, 320,* 780–791.

Bates, D. W., Boyle, D. L., and Teich, J. M. (1994). Impact of computerized physician order entry on physician time. Paper presented at the *Proceedings of the American Medical Informatics Association Symposium.*

Bates, D. W., Cohen, M., Leape, L. L., Overhage, J. M., Shabot, M. M., and Sheridan, T. (2001). Reducing the frequency of errors in medicine using information technology.[see comment]. *Journal of the American Medical Informatics Association, 8*(4), 299–308.

Bates, D. W., Cullen, D. J., Laird, N., Petersen, L. A., Small, S. D., Servi, D. et al. (1995). Incidence of adverse drug events and potential adverse drug events. Implications for prevention. *JAMA—Journal of the American Medical Association, 274*(1), 29–34.

Bates, D. W., and Gawande, A. A. (2003). Patient safety: Improving safety with information technology. *New England Journal of Medicine, 348*(25), 2526–2534.

Bates, D. W., Teich, J. M., Lee, J., Seger, D., Kuperman, G. J., Ma'Luf, N. et al. (1999). The impact of computerized physician order entry on medication error prevention. *Journal of the American Medical Informatics Association, 6,* 313–321.

Beer, J., Dobish, R., and Chambers, C. (2002). Physician order entry: A mixed blessing to pharmacy? *Journal of Oncology Pharmacy Practice, 8,* 119.

Berlin, A., Sorani, M., and Sim, I. (2006). A taxonomic description of computer-based clinical decision support systems. *Journal of Biomedical Informatics, 39*(6), 656–667.

Beuscart-Zephir, M. C., Pelayo, S., Anceaux, F., Meaux, J.-J., Degroisse, M., and Degoulet, P. (2005). Impact of CPOE on doctor-nurse cooperation for the medication ordering and administration process. *International Journal of Medical Informatics, 74,* 629–641.

Bisantz, A. M., and Burns, C. M. (2008). *Applications of Cognitive Work Analysis.* Boca Raton, FL: CRC Press.

Blaser, R., Schnabel, M., Biber, C., Baeumlein, M., Heger, O., Beyer, M. et al. (2007). Improving pathway compliance and clinician performance by using information technology. *International Journal of Medical Informatics, 76*(2–3), 151–156.

Blumenthal, D. (2009). Stimulating the adoption of health information technology. *New England Journal of Medicine, 360*(15), 1474–1479.

Blumenthal, D., and Tavenner, M. (2010). The "meaningful use" regulation for electronic health records. *New England Journal of Medicine, 363*(6), 501–504.

Brauer, R. L. (2006). *Safety and Health for Engineers* (2nd edn.). Tolono, IL: John Wiley & Sons, Inc.

Brennan, P. F., Ripich, S., and Moore, S. M. (1991). The use of home-based computers to support persons living with AIDS/ARC. *Journal of Community Health Nursing, 8*(1), 3–14.

Callen, J. L., Westbrook, J. I., and Braithwaite, J. (2006). The effect of physicians' long-term use of CPOE on their test management work practices. *Journal of the American Medical Informatics Association, 13,* 643–652.

Campbell, E. M., Sittig, D. F., Ash, J. S., Guappone, K. P., and Dykstra, R. H. (2006). Types of unintended consequences related to computerized provider order entry. *Journal of the American Medical Informatics Association., 13,* 547–556.

Carayon, P., Hundt, A. S., Karsh, B., Gurses, A. P., Alvarado, C. J., Smith, M. et al. (2006). Work system design for patient safety: The SEIPS model. *Quality and Safety in Healthcare, 15*(Suppl I), i50–i58.

Carayon, P., Wetterneck, T. B., Hundt, A. S., Ozkaynak, M., Desilvey, J., Ludwig, B. et al. (2007). Evaluation of nurse interaction with bar code medication administration technology in the work environment. *Journal of Patient Safety, 3*(1), 34–42.

Carpenter, J. D., and Gorman, P. N. (2001). What's so special about medications: A pharmacist's observations from the POE study. Paper presented at the *Proceedings of the American Medical Informatics Association Symposium.*

Carroll, C., Marsden, P., Soden, P., Naylor, E., New, J., and Dornan, T. (2002). Involving users in the design and usability evaluation of a clinical decision support system. *Computer Methods and Programs in Biomedicine, 69*(2), 123–135.

Cescon, D. W., and Etchells, E. (2008). Barcoded medication administration—A last line of defense. *JAMA—Journal of the American Medical Association, 299*(18), 2200–2202.

Chaparro, A., Bohan, M., Fernandez, J., Kattel, B., and Choi, S. D. (1999). Is the trackball a better input device for the older computer user? *Journal of Occupational Rehabilitation, 9*(1), 33–43.

Chaudhry, B., Wang, J., Wu, S. Y., Maglione, M., Mojica, W., Roth, E. et al. (2006). Systematic review: Impact of health information technology on quality, efficiency, and costs of medical care. *Annals of Internal Medicine, 144*(10), 742–752.

Cheng, C. H., Goldstein, M. K., Geller, E., and Levitt, R. E. (2003). The effects of CPOE on ICU workflow: An observational study. *Proceedings of the Symposium of the American Medical Informatics Association,* pp. 150–154.

Chuo, J., and Hicks, R. W. (2008). Computer-related medication errors in neonatal intensive care units. *Clinics in Perinatology, 35*(1), 119.

Cochran, G. L., Jones, K. J., Brockman, J., Skinner, A., and Hicks, R. W. (2007). Errors prevented by and associated with barcoded medication administration systems. *Joint Commission Journal on Quality and Patient Safety, 33,* 293–301.

Cordero, L., Kuehn, L., Kumar, R. R., and Mekhjian, H. S. (2004). Impact of computerized physician order entry on clinical practice in a newborn intensive care unit. *Journal of Perinatology, 24,* 88–93.

Cummings, J., Bush, P., Smith, D., and Matuszewski, K. (2005). Bar-coding medication administration overview and consensus recommendations. *American Journal of Health-System Pharmacy, 62,* 2626–2629.

Czaja, S. J., and Lee, C. C. (2007). Information technology and older adults. In J. A. Jacko and A. Sears (Eds.), *Human–Computer Interaction Handbook: Fundamentals, Evolving Technologies and Emerging Applications* (2nd edn., pp. 777–793). New York: Lawrence Erlbaum Associates.

Demiris, G., Finkelstein, S. M., and Speedie, S. M. (2001). Considerations for the design of a web-based clinical monitoring and educational system for elderly patients. [Review]. *Journal of the American Medical Informatics Association, 8*(5), 468–472.

Dykstra, R. (2002). Computerized physician order entry and communication: Reciprocal impacts. *Proceedings of the American Medical Informatics Association Symposium,* pp. 230–234.

Eslami, S., Abu-Hanna, A., and de Keizer, N. F. (2007). Evaluation of outpatient computerized physician medication order entry systems: A systematic review. *Journal of the American Medical Informatics Association, 14,* 400–406.

Evangelista, L. S., Strömberg, A., Westlake, C., Ter-Galstanyan, A., Anderson, N., and Dracup, K. (2006). Developing a web-based education and counseling program for heart failure patients. *Progress in Cardiovascular Nursing, 21*(4), 196–201.

Evans, K. D., Bensham, S. W., and Garrard, C. S. (1998). A comparison of handwritten and computer-assisted prescriptions in an intensive care unit. *Critical Care, 2,* 73–78.

Feied, C. F., Handler, J. A., Smith, M. S., Gillam, M., Kanhouwa, M., Rothenhaus, T. et al. (2004). Clinical information systems: Instant ubiquitous clinical data for error reduction and improved clinical outcomes. *Academic Emergency Medicine, 11*(11), 1162–1169.

Finkelstein, J., Khare, R., and Ansell, J. (2003). Feasibility and patients' acceptance of home automated telemanagement of oral anticoagulation therapy. Paper presented at the *Annual Symposium of American Medical Informatics Association.*

Fitzhenry, F., Peterson, J. F., Arrieta, M., Waitman, L. R., Schildcrout, J. S., and Miller, R. A. (2007). Medication administration discrepancies persist despite electronic ordering. *Journal of the American Medical Informatics Association, 14,* 756–764.

Furukawa, M. F., Raghu, T. S., Spaulding, T. J., and Vinze, A. (2008). Adoption of health information technology for medication safety in U.S. hospitals, 2006. *Health Affairs, 27,* 865–875.

Gaikwad, R., Sketris, I., Shepherd, M., and Duffy, J. (2007). Evaluation of accuracy of drug interaction alerts triggered by two electronic medical record systems in primary healthcare. *Health Informatics Journal, 13*(3), 163–177.

Gardner-Bonneau, D. (2001). Designing medical devices for older adults. In W. A. Rogers and A. D. Fisk (Eds.), *Human Factors Interventions for the Health Care of Older Adults* (pp. 221–237). Mahwah, NJ: Erlbaum.

Georgiou, A., Westbrook, J. I., Braithwaite, J., and Iedema, R. (2006). Multiple perspectives on the impact of electronic ordering on hospital organisational and communication processes. *Health Information Management Journal, 34,* 130–135.

Georgiou, A., Westbrook, J. I., Braithwaite, J., Iedema, R., Ray, S., Forsyth, R. et al. (2007a). When requests become orders—A formative investigation into the impact of a computerized physician order entry system on a pathology laboratory service. *International Journal of Medical Informatics, 76,* 583–591.

Georgiou, A., Williamson, M., Westbrook, J. I., and Ray, S. (2007b). The impact of computerised physician order entry systems on pathology services: A systematic review. *International Journal of Medical Informatics, 76,* 514–529.

Gesteland, P. H., Nebeker, J. R., and Gardner, R. M. (2006). These are the technologies that try men's souls: Common-sense health information technology. *Pediatrics, 117,* 216–217.

Goorman, E., and Berg, M. (2000). Modelling nursing activities: Electronic patient records and their discontents. *Nursing Inquiry, 7,* 3–9.

Gorman, P. N., Lavelle, M. B., and Ash, J. S. (2003). Order creation and communication in healthcare. *Methods of Information in Medicine, 42,* 376–384.

Gosbee, J. W., and Gosbee, L. L. (2007). Usability evaluation in healthcare. In P. Carayon (Ed.), *Handbook of Human Factors and Ergonomics in Health Care and Patient Safety* (pp. 679–692). Mahwah, NJ: Lawrence Erlbaum Associates.

Gustafson, D. H., Bosworth, K., Hawkins, R. P., Boberg, E. W., and Bricker, E. (1992). CHESS: A computer-based system for providing information, referrals, decision support and social support to people facing medical and other health-related crises. Paper presented at the *Annual Symposium on Computer Applications in Medical Care.*

Gustafson, D. H., Hawkins, R., Boberg, E., Pingree, S., Serlin, R. E., Graziano, F. et al. (1999). Impact of a patient-centered, computer-based health information/support system. *American Journal of Preventive Medicine, 16*(1), 1–9.

Gustafson, D. H., Hawkins, R., Pingree, S., McTavish, F., Arora, N. K., Mendenhall, J. et al. (2001). Effect of computer support on younger women with breast cancer. *Journal of General Internal Medicine, 16*(7), 435–445.

Gustafson, D. H., Hawkins, R. P., Boberg, E. W., McTavish, F., Owens, B., Wise, M. et al. (2005). CHESS: 10 years of research and development in consumer health informatics for broad populations, including the underserved. In D. Lewis, H. B. Jimison, G. Eysenbach, R. Kukafka and P. Z. Stavri (Eds.), *Consumer Health Informatics: Informing Consumers and Improving Health Care* (pp. 239–247). New York: Springer-Verlag.

Hailey, D., Roine, R., and Ohinmaa, A. (2002). Systematic review of evidence for the benefits of telemedicine. *Journal of Telemedicine and Telecare, 8,* 1–30.

Han, Y. Y., Carcillo, J. A., Venkataraman, S. T., Clark, R. S. B., Watson, R. S., Nguyen, T. C. et al. (2005). Unexpected increased mortality after implementation of a commercially sold computerized physician order entry system. *Pediatrics, 116,* 1506–1512.

Harrison, M. I., and Koppel, R. (2010). Interactive sociotechnical analysis: Identifying and coping with unintended consequences of IT implementation. In K. Khoumbati, Y. K. Dwivedi, A. Srivastava and B. Lal (Eds.), *Handbook of Research on Advances in Health Informatics and Electronic Healthcare Applications: Global Adoption and Impact of Information Communication Technologies* (pp. 31–49). Hershey, PA: IGI Global.

Harrison, M. I., Koppel, R., and Bar-Lev, S. (2007). Unintended consequences of information technologies in health care—An interactive sociotechnical analysis. *Journal of the American Medical Informatics Association, 14,* 542–549.

Hartwick, J., and Barki, H. (1994). Explaining the role of user participation in information system use. *Management Science, 40*(4), 440–465.

Himmelstein, D. U., Wright, A., and Woolhandler, S. (2010). Hospital computing and the costs and quality of care: A national study. *The American Journal of Medicine, 123*(1), 40–46.

Holden, R. J. (2010). Physicians' beliefs about using EMR and CPOE: In pursuit of a contextualized understanding of health IT use behavior. *International Journal of Medical Informatics, 79,* 71–80.

Holden, R. J. (2011a). Cognitive performance-altering effects of electronic medical records: An application of the human factors paradigm for patient safety. *Cognition, Technology & Work, 13,* 11–29.

Holden, R. J. (2011b). Social and personal normative influences on healthcare professionals to use information technology: Towards a more robust social ergonomics. *Theoretical Issues in Ergonomics Science,* in press.

Holden, R. J., Alper, S. J., Scanlon, M. C., Murkowski, K., Rivera, A. J., and Karsh, B. (2008a). Challenges and problem-solving strategies during medication management: A study of a pediatric hospital before and after bar-coding. Paper presented at the *Proceedings of the 2nd International Conference on Healthcare Systems Ergonomics and Patient Safety,* Strasbourg, France.

Holden, R. J., Brown, R. L., Alper, S. J., Scanlon, M. C., Patel, N. R., and Karsh, B. (2011). That's nice, but what does IT do? Evaluating the impact of bar coded medication administration by measuring changes in the process of care. *International Journal of Industrial Ergonomics 41,* 370–379.

Holden, R. J., Or, C. K. L., Alper, S. J., Rivera, A. J., and Karsh, B. (2008b). A change management framework for macroergonomic field research. *Applied Ergonomics, 39*(4), 459–474.

Hollingworth, W., Devine, E. B., Hansen, R. N., Lawless, N. M., Comstock, B. A., Wilson-Norton, J. L. et al. (2007). The impact of e-prescribing on prescriber and staff time in ambulatory care clinics: A time-motion study. *Journal of the American Medical Informatics Association, 14,* 722–730.

Hollnagel, E., and Woods, D. D. (2005). *Joint Cognitive Systems: Foundations of Cognitive Systems Engineering.* New York: CRC Press.

Institute of Medicine. (2007). *Preventing Medication Errors.* Washington, DC: National Academy Press.

Jha, A. K., DesRoches, C. M., Campbell, E. G., Donelan, K., Rao, S. R., Ferris, T. G. et al. (2009). Use of electronic health records in U.S. hospitals. *New England Journal of Medicine, 360,* 1628–1638.

Jha, A. K., Doolan, D., Grandt, D., Scott, T., and Bates, D. W. (2008). The use of health information technology in seven nations. *International Journal of Medical Informatics, 77,* 848–854.

Jimison, H., Gorman, P., Woods, S., Nygren, P., Walker, M., Norris, S. et al. (2008). Barriers and drivers of health information technology use for the elderly, chronically ill, and underserved. Evidence Report/Technology Assessment No. 175 (Prepared by the Oregon Evidence-based Practice Center under Contract No. 290-02-0024). AHRQ Publication No. 09-E004. Rockville, MD: Agency for Healthcare Research and Quality. November 2008.

Jones, S. S., Adams, J. L., Schneider, E. C., Ringel, J. S., and McGlynn, E. A. (2010). Electronic health record adoption and quality improvement in US hospitals. *American Journal of Managed Care, 16*(12 Spec No), SP64–SP71.

Kalichman, S. C., Cherry, C., Cain, D., Pope, H., Kalichman, M., Eaton, L. et al. (2006). Internet-based health information consumer skills intervention for people living with HIV/AIDS. *Journal of Consulting and Clinical Psychology, 74*(3), 545–554.

Karsh, B. (2009). *Clinical Practice Improvement and Redesign: How Change in Workflow Can Be Supported by Clinical Decision Support* (No. AHRQ Publication No. 09-0054-EF). Rockville, Maryland: Agency for Healthcare Research and Quality.

Karsh, B. (2010, September 27–October 1). What the doctor ordered? The role of cognitive decision support systems in clinical decision-making & patient safety. Paper presented at the *Human Factors and Ergonomics Society 54th Annual Meeting,* San Francisco, CA.

Karsh, B., Alper, S. J., Holden, R. J., and Or, C. K. L. (2006). A human factors engineering paradigm for patient safety—Designing to support the performance of the health care professional. *Quality and Safety in Healthcare, 15*(Suppl I), i59–i65.

Karsh, B., Weinger, M. B., Abbott, P. A., and Wears, R. L. (2010). Health information technology: Fallacies and sober realities. *Journal of the American Medical Informatics Association, 17,* 617–623.

Karsh, B., Wetterneck, T. B., Holden, R. J., Rivera-Rodriguez, A. J., Alper, S. J., Scanlon, M. et al. (2011). Bar coding in medication administration. In Y. Yih (Ed.), *Handbook of Healthcare Delivery Systems* (pp. 47–41 to 47–17). Boca Raton, FL: CRC Press.

Karwowski, W. (2003). *Handbook of Standards and Guidelines in Ergonomics and Human Factors.* Mahwah, NJ: Lawrence Erlbaum Associates.

Kaufman, D. R., Patel, V. L., Hilliman, C., Morin, P. C., Pevzner, J., Weinstock, R. S. et al. (2003). Usability in the real world: Assessing medical information technologies in patients' homes. *Journal of Biomedical Informatics, 36*(1–2), 45–60.

Kaushal, R., and Bates, D. W. (2001). Computerized physician order entry (CPOE) with clinical decision support systems (CDSSs). In K. G. Shojania, B. W. Duncan, K. M. McDonald and R. M. Wachter (Eds.), *Making Health Care Safer: A Critical Analysis of Patient Safety Practices* (pp. 59–69). Rockville, MD: Agency for Healthcare Research and Quality.

Kaushal, R., Jha, A. K., Franz, C., Glaser, J., Shetty, K. D., Jaggi, T. et al. (2006). Return on investment for a computerized physician order entry system. *Journal of the American Medical Informatics Association, 13*(3), 261–266.

Kaushal, R., Shojania, K. G., and Bates, D. W. (2003). Effects of computerized physician order entry and clinical decision support systems on medication safety: A systematic review. *Archives of Internal Medicine, 163*(12), 1409–1416.

Kawamoto, K., Houlihan, C. A., Balas, E. A., and Lobach, D. F. (2005). Improving clinical practice using clinical decision support systems: A systematic review of trials to identify features critical to success. *British Medical Journal, 330*(7494), 765E–768E.

Kho, A. N., Dexter, P. R., Warvel, J. S., Belsito, A. W., Commiskey, M., Wilson, S. J. et al. (2008). An effective computerized reminder for contact isolation of patients colonized or infected with resistant organisms. *International Journal of Medical Informatics, 77,* 194–198.

King, W. J., Paice, N., Rangrej, J., Forestell, G. J., and Swartz, R. (2003). The effect of computerized physician order entry on medication errors and adverse drug events in pediatric inpatients. *Pediatrics, 112*(3 Pt 1), 506–509.

Kopp, B. J., Erstad, B. L., Allen, M. E., Theodorou, A. A., and Priestley, G. (2006). Medication errors and adverse drug events in an intensive care unit: Direct observation approach for detection. *Critical Care Medicine, 34,* 415–425.

Koppel, R., Metlay, J. P., Cohen, A., Abaluck, B., Localio, A. R., Kimmel, S. E. et al. (2005). Role of computerized physician order entry systems in facilitating medication errors. *JAMA—Journal of the American Medical Association, 293,* 1197–1203.

Koppel, R., Wetterneck, T. B., Telles, J. L., and Karsh, B. (2008). Workarounds to barcode medication administration systems: Occurrences, causes and threats to patient safety. *Journal of the American Medical Informatics Association, 15,* 408–428.

Kuperman, G. J., and Gibson, R. F. (2003). Computer physician order entry: Benefits, costs, and issues. *Annals of Internal Medicine, 139*(1), 31–39.

LaFramboise, L. M., Todero, C. M., Zimmerman, L., and Agrawal, S. (2003). Comparison of Health Buddy (R) with traditional approaches to heart failure management. *Family & Community Health, 26*(4), 275–288.

Lawton, G., and Shields, A. (2005). Bar-code verification of medication administration in a small hospital. *American Journal of Health-System Pharmacy, 62*(22), 2413–2415.

Leape, L. L., Bates, D. W., Cullen, D. J., Cooper, J., Demonaco, H. J., Gallivan, T. et al. (1995). Systems analysis of adverse drug events. *Journal of the American Medical Association, 274*(1), 35–43.

Leape, L. L., Brennan, T. A., Laird, N., Lawthers, A. G., Localio, A. R., Barnes, B. A. et al. (1991). The nature of adverse events in hospitalized-patients—Results of the Harvard Medical-Practice Study-II. *New England Journal of Medicine, 324*(6), 377–384.

Liddy, C., Dusseault, J. J., Dahrouge, S., Hogg, W., Lemelin, J., and Humbert, J. (2008). Telehomecare for patients with multiple chronic illnesses: Pilot study. *Canadian Family Physician, 54*(1), 58–65.

Lin, C. A., Neafsey, P. J., and Strickler, Z. (2009). Usability testing by older adults of a computer-mediated health communication program. *Journal of Health Communication, 14*(2), 102–118.

Linder, J. A., Ma, J., Bates, D. W., Middleton, B., and Stafford, R. S. (2007). Electronic health record use and the quality of ambulatory care in the United States. *Archives of Internal Medicine, 167*(13), 1400–1405.

Littner, M., Hirshkowitz, M., Davila, D., Anderson, W. M., Kushida, C. A., Woodson, T. et al. (2002). Practice parameters for the use of auto-titrating continuous positive airway pressure devices for titrating pressures and treating adult patients with obstructive sleep apnea syndrome. *Sleep, 25*(2), 143–147.

Lober, W. B., Zierler, B., Herbaugh, A., Shinstrom, S. E., Stolyar, A., Kim, E. H. et al. (2006). Barriers to the use of a personal health record by an elderly population. Paper presented at the *Annual Symposium of American Medical Informatics Association*.

Martinez, A., Everss, E., Rojo-Alvarez, J. L., Figal, D. P., and Garcia-Alberola, A. (2006). A systematic review of the literature on home monitoring for patients with heart failure. *Journal of Telemedicine and Telecare, 12*(5), 234–241.

Massaro, T. A. (1993). Introducing physician order entry at a major academic medical center: I. Impact on organizational culture and behavior. *Academic Medicine, 68,* 20–25.

McRoberts, S. (2005). The use of bar code technology in medication administration. *Clinical Nurse Specialist, 19*(2), 55–56.

Mead, N., Varnam, R., Rogers, A., and Roland, M. (2003). What predicts patients' interest in the Internet as a health resource in primary care in England? *Journal of Health Services Research and Policy, 8*(1), 33–39.

Mekhjian, H. S., Kumar, R. R., Kuehn, L., Bentley, T. D., Teater, P., Thomas, A. et al. (2002). Immediate benefits realized following implementation of physician order entry at an academic medical center. *Journal of the American Medical Informatics Association, 9*(5), 529–539.

Miller, R. A., Waitman, L. R., Chen, S. T., and Rosenbloom, S. T. (2005). The anatomy of decision support during inpatient care provider order entry (CPOE): Empirical observations from a decade of CPOE experience at Vanderbilt. *Journal of Biomedical Informatics, 38*(6), 469–485.

Mutimer, D., McCauley, B., Nightingale, P., Ryan, M., Peters, M., and Neuberger, J. (1992). Computerised protocols for laboratory investigation and their effect on use of medical time and resources. *Journal of Clinical Pathology, 45,* 572–574.

Neuenschwander, M., Cohen, M. R., Vaida, A. J., Patchett, J. A., Kelly, J., and Trohimovich, B. (2003). Practical guide to bar coding for patient medication safety. *American Journal of Health-System Pharmacy, 60*(8), 768–779.

Niazkhani, Z., Pirnejad, H., Berg, M., and Aarts, J. (2009). The impact of computerized provider order entry systems on inpatient clinical workflow: A literature review. *Journal of the American Medical Informatics Association, 16,* 539–549.

Nielsen, J. (1993). *Usability Engineering.* Boston, MA: Academic Press.

NIST. (2010). Health information technology. Retrieved December 27, 2010, from http://www.nist.gov/itl/hit/

Ohno, T. (1988). *Toyota Production System: Beyond Large-Scale Production.* New York: Productivity Press.

Or, C. K. L., and Karsh, B. (2009). A systematic review of patient acceptance of consumer health information technology. *Journal of the American Medical Informatics Association, 16*(4), 550–560.

Or, C. K. L., Karsh, B., Severtson, D. J., Burke, L. J., Brown, R. L., and Brennan, P. F. (2011). Factors affecting home care patients' acceptance of a web-based interactive self-management technology. *Journal of the American Medical Informatics Association, 18*(1), 51–59.

Or, C. K. L., Valdez, R., Casper, G., Carayon, P., Burke, L. J., Brennan, P. F. et al. (2009). Human factors and ergonomics in home care: Current concerns and future considerations for health information technology. *WORK: A Journal of Prevention, Assessment and Rehabilitation, 33*(2), 201–209.

Ornstein, C. (2003). Hospital heeds doctors, suspends use of software: Cedars-Sinai physicians entered prescriptions and other orders in it, but called it unsafe. *The Los Angeles Times.*

Osheroff, J. A., Pifer, E. A., Teich, J. M., Sittig, D. F., and Jenders, R. A. (2005). Improving outcomes with clinical decision support: An implementer's guide: Health Information Management and Systems Society.

Ostbyte, T., Moen, A., Erikssen, G., and Hurlen, P. (1997). Introducing a module for laboratory test order entry and reporting of results at a hospital ward: An evaluation study using a multi-method approach. *Journal of Medical Systems, 21,* 107–117.

Overhage, J. M., Perkins, S., Tierney, W. M., and McDonald, C. J. (2001). Controlled trial of direct physician order entry: Effects on physicians' time utilization in ambulatory primary care internal medicine practices. *Journal of the American Medical Informatics Association, 8,* 361–369.

Parasuraman, R., and Riley, V. (1997). Humans and Automation: Use, misuse, disuse, abuse. *Human Factors, 39*(2), 230–253.

Parasuraman, R., Sheridan, T., and Wickens, C. D. (2000). A model for types and levels of human interaction with automation. *IEEE Transactions on Systems Man and Cybernetics Part A-Systems and Humans, 30*(3), 286–297.

Parasuraman, R., and Wickens, C. D. (2008). Humans: Still vital after all these years of automation. *Human Factors, 50*(3), 511–520.

Patterson, E. S., Cook, R. I., and Render, M. L. (2002). Improving patient safety by identifying side effects from introducing bar coding in medication administration. *Journal of the American Medical Informatics Association, 9*(5), 540–553.

Pedersen, C. A., and Gumpper, K. F. (2008). ASHP national survey on informatics: Assessment of the adoption and use of pharmacy informatics in US hospitals-2007. *American Journal of Health-System Pharmacy, 65*(23), 2244–2264.

Pitre, M., Ong, K., Huh, J. H., and Fernandes, O. (2006). Thorough planning and full participation by pharmacists is key to MOE/MAR success. *Healthcare Quarterly, 43*, 4.

Poon, E. G., Keohane, C. A., Bane, A., Featherstone, E., Hays, B. S., Dervan, A. et al. (2008). Impact of barcode medication administration technology on how nurses spend their time providing patient care. *Journal of Nursing Administration, 38*(12), 541–549.

Poon, E. G., Keohane, C. A., Yoon, C. S., Ditmore, M., Bane, A., Levtzion-Korach, O. et al. (2010a). Effect of bar-code technology on the safety of medication administration. *New England Journal of Medicine, 362*(18), 1698–1707.

Poon, E. G., Wright, A., Simon, S. R., Jenter, C. A., Kaushal, R., Volk, L. A. et al. (2010b). Relationship between use of electronic health record features and health care quality results of a statewide survey. *Medical Care, 48*(3), 203–209.

Potts, A. L., Barr, F. E., Gregory, D. F., Wright, L., and Patel, N. R. (2004). Computerized physician order entry and medication errors in a pediatric critical care unit. *Pediatrics, 113*(1 Pt 1), 59–63.

Reckmann, M. H., Westbrook, J. I., Koh, Y., Lo, C., and Day, R. O. (2009). Does computerized provider order entry reduce prescribing errors for hospital inpatients? A systematic review. *Journal of the American Medical Informatics Association, 16*, 613–623.

Redish, J., and Lowry, S. Z. (2010). *Usability in Health IT: Technical Strategy, Research, and Implementation: Summary of Workshop and Usability in Health IT, July 13, 2002*: NIST, Department of Commerce.

Rogers, A., and Mead, N. (2004). More than technology and access: Primary care patients' views on the use and nonuse of health information in the Internet age. *Health & Social Care in the Community, 12*(2), 102–110.

Rosenbloom, S. T., Harrell, F. E., Lehmann, C. U., Schneider, J. H., Spooner, S. A., and Johnson, K. B. (2006). Perceived increase in mortality after process and policy changes implemented with computerized physician order entry. *Pediatrics, 117*, 1452–1455.

Sakowski, J., Newman, J. M., and Dozier, K. (2008). Severity of medication administration errors detected by a bar-code medication administration system. *American Journal of Health-System Pharmacy, 65*(17), 1661–1666.

Saleem, J. J., Patterson, E. S., Militello, L., Render, M. L., Orshansky, G., and Asch, S. M. (2005). Exploring barriers and facilitators to the use of computerized clinical reminders. *Journal of the American Medical Informatics Association, 12*, 438–447.

Sanders, M. S., and McCormick, E. J. (1993). *Human Factors in Engineering and Design* (7th edn.). New York: McGraw-Hill Science.

Sarter, N. B., Woods, D. D., and Billings, C. E. (1997). Automation surprises. In G. Salvendy (Ed.), *Handbook of Human Factors and Ergonomics* (2nd edn.). New York: Wiley.

Schectman, J. M., Schorling, J. B., Nadkarni, M. M., and Voss, J. D. (2005). Determinants of physician use of an ambulatory prescription expert system. *International Journal of Medical Informatics, 74*(9), 711–717.

Scott, R. L., Park, M. H., Uber, P. A., Ventura, H. O., and Mehra, M. R. (2004). Electronic home monitoring reduces medical resource utilization in chronic heart failure: A prospective investigation of the HomMed-Sentry telemonitoring system. *Journal of Cardiac Failure, 10*(4), S110–S111.

Shabot, M. M. (2004). Ten commandments for implementing clinical information systems. *Baylor University Medical Center Proceedings, 17*, 265–269.

Shamliyan, T. A., Duval, S., Du, J., and Kane, R. L. (2008). Just what the doctor ordered. Review of the evidence if the impact of computerized physician order entry system on medication errors. *Health Services Research, 43*(1), 32–53.

Shane, R. (2009). Current status of administration of medicines. *American Journal of Health System Pharmacy, 66*(Suppl 3), S42–S48.

Shea, S., DuMouchel, W., and Bahamonde, L. (1996). A meta-analysis of 16 randomized controlled trials to evaluate computer-based clinical reminder systems for preventive care in the ambulatory setting. *Journal of the American Medical Informatics Association, 3*(6), 399–409.

Sheridan, T. B., and Parasuraman, R. (2006). Human-automation interaction. In R. S. Nickerson (Ed.), *Reviews of Human Factors and Ergonomics* (Vol. 1, pp. 89–129). Santa Monica, CA: Human Factors and Ergonomics Society.

Shu, K., Boyle, D., Spurr, C., Horsky, J., Heiman, H., O'Connor, P. et al. (2001). Comparison of time spent writing orders on paper with computerized physician order entry. *MEDINFO, 84*, 1207–1211.

Sittig, D. F., Wright, A., Osheroff, J. A., Middleton, B., Teich, J. M., Ash, J. S. et al. (2008). Grand challenges in clinical decision support. *Journal of Biomedical Informatics, 41*(2), 387–392.

Slack, W. (1997). *Cybermedicine*. San Francisco, CA: Jossey-Bass.

Stark, P. (2010). Congressional intent of the HITECH Act. *American Journal of Managed Care, 16*(12 Spec No), SP24–SP28.

Stead, W. W., and Lin, H. S. (Eds.). (2009). *Computational Technology for Effective Health Care: Immediate Steps and Strategic Directions*. Washington, DC: National Academies Press.

Stinson, J. N., Petroz, G. C., Tait, G., Feldman, B. M., Streiner, D., McGrath, P. J. et al. (2006). e-Ouch: Usability testing of an electronic chronic pain diary for adolescents with arthritis. *Clinical Journal of Pain, 22*(3), 295–305.

Stoop, A. P., Van't Riet, A., and Berg, M. (2004). Using information technology for patient education: Realizing surplus value? *Patient Education and Counseling, 54*(2), 187–195.

Tak, S. H., and Hong, S. H. (2005). Use of the Internet for health information by older adults with arthritis. *Orthopaedic Nursing, 24*(2), 134–138.

Tamblyn, R., Huanc, A., Talylor, L., Kawasumi, Y., Bartlett, G., Grad, R. et al. (2008). A randomized trial of the effectiveness of on-demand versus computer triggered drug decision support in primary care. *Journal of the American Medical Informatics Association, 15*, 430–438.

Tamblyn, R., Huang, A., Kawasumi, Y., Bartlett, G., Grad, R., Jacques, A. et al. (2006). The development and evaluation of an integrated electronic prescribing and drug management system for primary care. *Journal of the American Medical Informatics Association, 13*, 148–159.

Taylor, R., Manzo, J., and Sinnett, M. (2002). Quantifying value for physician order-entry systems: A balance of cost and quality. *Healthcare Financial Management, 56*, 44–48.

Teich, J. M., Osheroff, J. A., Pifer, E. A., Sittig, D. F., Jenders, R. A., and Panel, C. D. S. E. R. (2005). Clinical decision support in electronic prescribing: Recommendations and an action plan. *Journal of the American Medical Informatics Association, 12*(4), 365–376.

The Leapfrog Group. (2008). Computerized physician order entry. Retrieved October 30, 2009, from http://www.leapfroggroup.org/media/file/Leapfrog-Computer_Physician_Order_Entry_Fact_Sheet.pdf

Tierney, W. M., Miller, M. E., Overhage, J. M., and McDonald, C. J. (1993). Physician inpatient order writing on microcomputer workstations: Effects on resource utilization. *JAMA—Journal of the American Medical Association, 269*(3), 379–383.

US Congress. (2009). *American Recovery and Reinvestment Act of 2009*. In *GovTrack.us (database of federal legislation)*. Retrieved July 1, 2011, from http://www.govtrack.us/congress/bill.xpd?bill=h111-1

US Department of Health and Human Services. (2009). Retrieved August 21, 2009, from http://healthit.hhs.gov/

US Department of Health and Human Services. (2010). Electronic Health Records and Meaningful Use. Retrieved December 1, 2010, from http://healthit.hhs.gov/portal/server.pt?open = 512&objID = 2996&mode = 2

Van der Sijs, H., Aarts, J., Vulto, A., and Berg, M. (2006). Overriding of drug safety alerts in computerized physician order entry. *Journal of the American Medical Informatics Association, 13*(2), 138–147.

van Onzenoort, H. A., van de Plas, A., Kessels, A. G., Veldhorst-Janssen, N. M., van der Kuy, P. H. M., and Neef, C. (2008). Factors influencing bar-code verification by nurses during medication administration in a Dutch hospital. *American Journal of Health-System Pharmacy, 65*(7), 644–648.

van Rosse, F., Maat, B., Rademaker, C. M. A., van Vught, A. J., Egberts, A. C. G., and Bollen, C. W. (2009). The effect of computerized physician order entry on medication prescription errors and clinical outcome in pediatric and intensive care: A systematic review. *Pediatrics, 123*, 1184–1190.

Vanderheiden, G. C. (2006). Design for people with functional limitations. In G. Salvendy (Ed.), *Handbook of Human Factors and Ergonomics* (3rd edn., pp. 1387–1417). New York: John Wiley & Sons.

Viccellio, P. (2008). Turnaround time and transaction costs. *Annals of Emergency Medicine, 51*, 186–187.

Vogelsmeier, A. A., Halbesleben, J. R. B., and Scott-Cawiezzel, J. R. (2008). Technology implementation and workarounds in the nursing home. *Journal of the American Medical Informatics Association, 15*, 114–119.

Wald, H., and Shojania, K. G. (2001). Prevention of misidentifications. In K. G. Shojania, B. W. Duncan, K. M. McDonald and R. M. Wachter (Eds.), *Making Health Care Safer: A Critical Analysis of Patient Safety Practices*. Rockville, MD: Agency for Healthcare Research and Quality.

Walsh, K. E., Kaushal, R., and Chessare, J. B. (2005). How to avoid paediatric medication errors: A user's guide to the literature. *Archives of Disease in Childhood, 90*(7), 698–702.

Wears, R. L., and Berg, M. (2005). Computer technology and clinical work: Still waiting for Godot. *JAMA—Journal of the American Medical Association, 29*, 1261–1263.

Web Content Accessibility Guidelines (WCAG) 2.0. (2008). W3C Recommendation December 11, 2008. Accessed January 2, 2011, http://www.w3.org/TR/WCAG20/

Weiner, M., Gress, T., Thiemann, D. R., Jenckes, M., Reel, S. L., Mandell, S. F. et al. (1999). Contrasting views of physicians and nurses about an inpatient computer-based provider order-entry system. *Journal of the American Medical Informatics Association, 6*, 234–244.

Weingart, S. N., Toth, M., Sands, D. Z., Aronson, M. D., Davis, R. B., and Phillips, R. S. (2003). Physicians' decisions to override computerized drug alerts in primary care. *Archives of Internal Medicine, 163*(21), 2625–2631.

Wolfstadt, J. I., Gurwitz, J. H., Field, T. S., Lee, M., Kalkar, S., Wu, W. et al. (2008). The effect of computerized physician order entry with clinical decision support on the rates of adverse drug events: A systematic review. *Journal of General Internal Medicine, 23*, 451–458.

Wright, M. J., Frey, K., Scherer, J., and Hilton, D. (2006). Maintaining excellence in physician nurse communication with CPOE: A nursing informatics team approach. *Journal of Healthcare Information Management, 20*, 65–70.

Yen, K., Shane, E. L., Pawar, S. S., Schwendel, N. D., Zimmanck, R. J., and Gorelick, M. H. (2009). Time motion study in a pediatric emergency department before and after computer physician order entry. *Annals of Emergency Medicine, 53*, 462–468.e1.

Zaidi, S. T. R., Marriott, J. L., and Nation, R. L. (2008). The role of perceptions of clinicians in their adoption of a web-based antibiotic approval system: Do perceptions translate into actions? *International Journal of Medical Informatics, 77*, 33–40.

Zhou, L., Soran, C. S., Jenter, C. A., Volk, L. A., Orav, E. J., Bates, D. W. et al. (2009). The relationship between electronic health record use and quality of care over time. *Journal of the American Medical Informatics Association, 16*(4), 457–464.

Zimmerman, L., Barnason, S., Nieveen, J., and Schmaderer, M. (2004). Symptom management intervention in elderly coronary artery bypass graft patients. *Outcomes Management, 8*(1), 5–12.

18

Human–Computer Interaction Design in Health Care

Frank A. Drews and Heidi S. Kramer

CONTENTS

18.1 Introduction 265
18.2 Types of Human–Computer Interfaces 266
    18.2.1 Devices and Interfaces 267
        18.2.1.1 Medical Monitoring 267
        18.2.1.2 Computer-Controlled Devices 268
        18.2.1.3 Clinical Data Systems 268
18.3 Basic Interface Design Principles 270
    18.3.1 The User 270
        18.3.1.1 Perception 271
        18.3.1.2 Cognition/Planning 271
        18.3.1.3 Response Processes/Execution of Plans 272
    18.3.2 Task 275
        18.3.2.1 Understanding the Task 275
        18.3.2.2 Methods to Identify Task Requirements 275
    18.3.3 Context 276
        18.3.3.1 Distractions and Interruptions 276
        18.3.3.2 Collaboration and Distributed Cognition 276
        18.3.3.3 Multidisciplinary Collaboration 277
18.4 Conclusions 277
References 277

18.1 Introduction

Good interface design contributes to safer medical equipment and improved provider and patient safety. In addition, good interfaces have an impact on a number of additional dimensions: they make software easier to learn, improve performance speed, increase user satisfaction, and reduce error (Schneiderman, 1992), while "poor user interface design greatly increases the likelihood of error in equipment operation" (Kaye and Crowley, 2000). The health care environment is particularly challenging for interface design, because it is a complex, uncertain, highly demanding, and highly stressful environment. Problems with interface design will be acerbated because the health care professional may not be able to compensate for bad interface design as is necessary in many other domains.

Poor interface design in health care is also a serious problem. The U.S. Food and Drug Administration (FDA) reports that up to 50% of device recalls from 1985 to 1989 were due to poor product design, including problems with software and user interfaces. As a result, in 1990, Congress passed the Safe Medical Devices Act (SMDA) requiring device producers to follow good manufacturing practices.

The goal of this chapter is to provide a guide on the topic of human–computer interaction in health care. This guide will use the basic assumption (as illustrated in Figure 18.1) that successful task performance is a function of interface design, user properties, and the context in which the task is performed.

In the next sections, we will discuss principles and challenges of interface design in health care. One of the reasons why human–computer interaction is such an important topic is because error in medicine is widespread

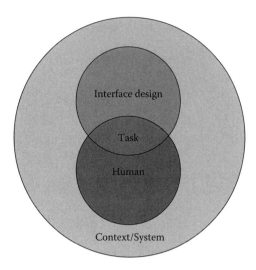

FIGURE 18.1
Factors that affect task performance.

(Institute of Medicine, 2000; see Chapter 22) and among the contributing factors to human error in health care are the growing complexity of science and technology on one hand and constraints of exploiting information technologies on the other hand. Well-developed information technology can help leverage the advances of science and technology, but badly implemented, it will handicap these advances and decrease patient safety.

18.2 Types of Human–Computer Interfaces

Computers and software operate in a highly invisible way: often, the user receives only limited information about the operational or organizational state of the system (Norman, 1993). The user interface provides the operating environment for the user to allow interaction with the system and to receive feedback about the system's status. Several types of interfaces can be distinguished. One type of user interface that still can be found in institutional health care environments are character-based interfaces (DOS and UNIX user interfaces). These legacy systems were developed to meet very specific aims of a particular organization. One challenge of these interfaces to novice and intermediate users is the high cognitive load they impose: the user has to remember the command syntax, the spelling, and the procedures of the system. With these interfaces, having knowledge in the head (Norman, 1990) is the only way to interact with computer.

A more complex variant of character-based interfaces are full-screen interfaces. The user can switch between entry fields all over the screen. Full-screen interfaces can be found, for example, in patient documentation

systems, where a form-filling dialogue is used to enter the patient information. The interaction with these displays utilizes menus and function keys. A problem associated with full-screen interfaces is that the menu structure has to be optimized for the user's needs, and the analysis to identify those needs is often not performed. Another problem with these types of interfaces is that moving through entry boxes can be time consuming, and, as a consequence, interaction with full-screen interfaces does not allow for optimal performance. In addition, sometimes, functionality like autocompletion of entries increases the likelihood of error. Because of their lack of user friendliness, character-based interfaces were abandoned in favor of graphical user interfaces (GUIs).

GUIs often consist of *w*indows, *i*cons, *m*enus, and a *p*ointing device (WIMP systems). GUIs allow the user to directly manipulate visual representations of the dialogue objects on the screen. Using a GUI supports the user in multiple ways. The interface supports recognition of information that is presented to a user while character-based interfaces require active recall of information. One of the important features of GUIs is that they can apply interface metaphors (e.g., the commonly used desktop metaphor). This metaphor use allows a user to apply existing knowledge about the real objects represented in the metaphor to directly interact with the interface (e.g., the paper basket disposes documents). Consequentially, the time to learn how to interact with such system is minimized allowing for efficient interaction.

Finally, Web applications and Internet-enabled applications are commonly associated with two characteristics: shared access to data and navigation using hyperlinks. Accessing applications through a Web browser enables access to the same tools, functionality, and data from multiple computers. This high level of flexibility can be a great benefit in the health care context considering that clinicians frequently need to enter or access patient data from multiple locations. Another major advantage of Web applications over desktop-based applications is related to software administration: it is the relative ease of deploying new versions of software. Software is installed on a single server or a few servers compared to downloading and installing the new software to multiple desktops. However, the ease of deployment may increase the likelihood of frequent updates in functionality, which can negatively affect users. Consistent with all change, the impact of modifications to Web applications should be carefully considered as unanticipated or frequent updates may confuse or frustrate users.

Navigation using hyperlinks presents unique opportunities and challenges while retaining some of the same challenges that exist when supporting user interaction in more traditional applications. Hyperlinks in

Web and Internet-enabled applications support access to the wealth of information available on the World Wide Web as well as supporting navigation within the application. The type of link that provides access to the information (internal or external) should be communicated clearly to the user. Additionally, the destination, what to expect and how to return to the point of origin should be clearly communicated. Users should not have to follow a link to know where it will lead (Card et al., 2001).

When using hyperlinks to control user navigation and interaction, the same care should be used for defining and grouping hyperlinks as when creating menu options for desktop applications. The text labels, icons, and groupings need to be meaningful to the targeted user group (Bailey, 2000; Mobrand and Spyridakis, 2007). In addition, constraints need to be enforced with regard to the placement of navigation hyperlinks (whether top, left, or right side of the screen), since only consistency of these links throughout the application will improve performance.

Unfortunately, there is a common misconception that Web applications are easier or faster to build than desktop applications (Cooper et al., 2007). The design and development process is similar for both types of applications. One facet of the development of GUI's or hypertext-based systems is that they are often developed for specific applications and environments but follow some general conventions (Microsoft, 1995), thus allowing some standardization across applications, which facilitates learning and interaction. However, there are also some limitations concerning the type of interface used for specific purposes. In the next section, we will discuss specific types of devices and interfaces in more detail.

18.2.1 Devices and Interfaces

There is more and more evidence that hazards resulting from medical device use might by far exceed hazards based on technology failure of a device. These problems can partly be attributed to professionals dealing with the interfaces of these devices and illustrates that effective human–computer interaction design needs to be an integral part of the device development process. According to Leape et al. (1998), one of the main problems in this context is that many systems are not designed for safety. One assumption that is implicit in the design process is the assumption of "error-free performance." In the case of human error when interacting with a device, enforcement is performed by punishing the operator. However, sustained error-free performance in a high stress and high-stake environment is impossible (Reason, 1990), but the occurrence of human error can be reduced to a minimum by good interface design. To allow a better understanding of some of the devices that are currently used in

health care environments, we will now describe different functions of devices that pose different requirements for the interface designer. We will focus on the most prominent devices in health care: medical monitoring devices, computer-controlled devices, and clinical data systems.

18.2.1.1 Medical Monitoring

Interfaces of devices designed to support patient monitoring provide feedback about the status of the patient. Patient monitoring can be defined as "repeated or continuous observations or measurement of the patient, his or her physiological function, and the function of life support equipment, for the purpose of guiding management decisions" (Gardner and Shabot, 2001). Patient monitoring can be found in a wide range of contexts (e.g., intensive care, perinatal care, perioperative care, postsurgical care, and coronary care) utilizing a broad variety of devices. The devices used can range from patient monitoring systems in the operating room (e.g., anesthesia monitors) to devices measuring blood oxygen saturation in perinatal care. Outside the clinical environment monitoring devices can be found in home health care contexts (blood pressure and heart rate monitors). Monitoring systems provide the care giver with information about the patient's current status, thus supporting diagnosis and planning of treatment. However, clearly monitoring devices do not perform treatment, the change of the patient's status is achieved through means ranging from noninvasive to invasive means (e.g., changing the patient position and administration of drugs). Figure 18.2 illustrates the elements and relationships in the context of patient monitoring.

Current patient monitoring interfaces frequently follow the single sensor single indicator (SSSI) approach (Drews, 2008; Gardner and Shabot, 2001; Goodstein, 1981), where for each sensor used, a single variable is displayed. Such SSSI design results in sequential, piecemeal data gathering, which makes it difficult and effortful for the clinician to develop a coherent understanding of the relationships of monitored variables and their underlying mechanisms (Vicente et al., 1995, Drews and Westenskow, 2006). An alternative to the SSSI approach are interfaces that support the clinician by providing patient information in an integrated way, thus allowing rapid detection, diagnosis, and treatment. For example, Drews et al. (2001) developed a cardiovascular display that incorporates anesthesiologists' mental model of the cardiovascular system. In this object/configural display, symmetry shows normal values of important patient variables, while asymmetry of the graphical display elements is used to show deviations from the normal status of the patient. The authors incorporated emerging features and implemented patterns that matched particular diagnoses (see also Agutter et al., 2003). One

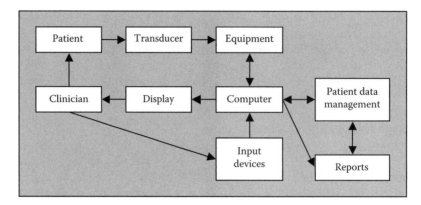

FIGURE 18.2
Elements and relationships in the context of monitoring patient variables.

of the limitations of these approaches is that they make assumptions about "normal" states of the patient to provide reference points. However, these assumptions, unlike in the context of monitoring technical systems, are not always warranted (for an alternative approach see Drews, 2008).

18.2.1.2 Computer-Controlled Devices

A second type of interface that is commonly used in heath care allows the control of computer-operated equipment. For example, an infusion pump interface provides information about the pump's current status (e.g., mode of operation) and the means to change its status to accomplish a particular goal. Equipment device interfaces aim at making interaction with the device simple, fast, and ideally error free. To accomplish these goals, equipment interfaces should provide information (e.g., menus) that reflects the structure of the task the clinician tries to accomplish, and the work flow or the procedures that represent standard or best current clinical practice. Because the clinician interacting with the device is often forced to simultaneously perform other tasks or is interrupted by tasks that may require immediate attention (Drews, under review), information about the current or last operational status of the device should be provided to allow smooth resumption of the interaction previous to the interruption. In addition, equipment interfaces should allow for easy and natural navigation, provide information about performed actions, have an option to reverse and cancel actions, and facilitate correct actions and prevent or discourage potentially hazardous actions. The recent emergence of new types of devices that provide emergency care in the home and workplace context makes these interface requirements even more relevant, since here the user is often not familiar with the device but needs to operate it under high time pressure under unfamiliar conditions. Devices of this type include,

for example, automatic defibrillators. In addition to the principles outlined earlier, there are additional challenges for interface design for this type of device (see Drews et al., 2007).

Both patient monitoring and equipment interfaces practice medicine in a technical sense and are regulated by the FDA. Because of this special status and the potential dangers associated with their use, the software of these interfaces is subjected to special verification methods, and the required safety certification ensures safe technical operation of the equipment.

18.2.1.3 Clinical Data Systems

Clinical data systems serve as a repository for patient data. In addition to the medical monitoring and computer-controlled devices previously discussed, the use of technology supports other areas of health care including digital imaging, computerized physician order entry, and medication administration. Unfortunately, the data generated and collected by these devices and systems are rarely, if ever, available to clinicians through a single system. A comprehensive electronic health record (EHR) system is needed to resolve this issue.

18.2.1.3.1 Electronic Health Records

EHR systems were first described in 1959, and dozens of systems were under development in the 1960s (Dick et al., 1997). Despite these early attempts for system development, the adoption of complete EHR systems has been a slow and difficult process (Berner et al., 2005; Gans et al., 2005; Wears and Berg, 2005). However, the HITECH Act—part of the American Recovery and Reinvestment Act (ARRA)—in an effort to integrate records and reduce health care costs seeks to promote the development and use of EHRs (U.S. Department of Health and Human Services, 2010).

There is clearly a tension between the fact that EHR systems were described half a century ago, but it took until today for a clear interest and incentives to be in place to support the implementation of such systems. This raises the questions of what are EHRs and why, after 50 years, are effective and efficient systems that contribute to the quality and safety of care still few and far between (Wears and Berg, 2005)?

The EHRs properties and capabilities include the availability of a longitudinal electronic record of patient health information generated by one or more encounters in any care delivery setting (HIMMS, 2010). This comprehensive information includes patient demographics and financial information, clinical documentation (e.g., progress notes, vital signs, past medical history, and immunizations), pathology and laboratory data, device monitoring data, radiology reports, and current and past medications. The data are collected and stored in the EHR Data storage facility and are made available to different areas of clinical care, like primary care, acute care, home health care, long-term care, and tertiary care. A conceptual overview of an EHR system as described here is provided in Figure 18.3. The ultimate goal of this approach is the creation of a system that provides a complete longitudinal record of all clinical encounters a patient had in both ambulatory (outpatient) and acute care settings to provide all patient relevant data to health care providers as needed.

In addition to data storage, the ideal EHR system would be supportive of the clinician's workflow including ordering capabilities and allow the clinician efficient and secure access to the patient's history, no matter who provided the prior clinical services. Further, the system would also provide decision support functionality to support users in dealing with the high level of complexity of managing patients, given the large body of knowledge that is available in medicine. In addition to providing information to all providers, an EHR system would facilitate communication and information flow between the providers (Committee on Data Standards for Patient Safety, 2003; Miller et al., 2003). Current systems do not support this full range of functionality and consequentially create a significant burden for health care providers.

18.2.1.3.1 Barriers to Efficient Effective EHR systems

The development of EHR systems is a complex undertaking that has seen multiple efforts since the pioneering efforts of the 1960s and 1970s. Hundreds of systems have been created to collect clinical data. The number and variability of the systems created should not be surprising considering the multiple data requirements in care settings. For example, in acute care settings, nursing charting systems provide a means for documenting patient vital signs, inputs (liquid and food) and outputs, medication administration, and other aspects of patient care. Pharmacies receive, fill, and track medication orders. Radiology and labs tests are ordered, performed, and reported. Computerized physicians order entry (CPOE) systems track orders for tests, medications, and record aspects of care. In addition, integration of data into a single system becomes highly challenging considering that hundreds of devices for monitoring and/or administering medications have been developed by independent vendors, each with proprietary interfaces to databases and reporting mechanisms.

The ideal unified EHR system requires that multiple data types such as patient identifiers, diagnoses, and

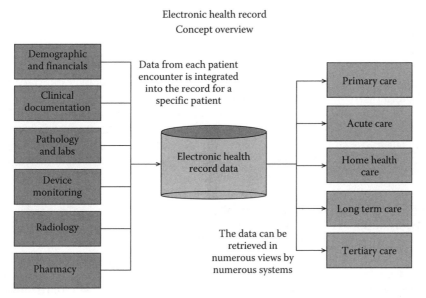

FIGURE 18.3
Conceptual overview of an electronic health record system.

procedures are stored using consistent, shared vocabularies, and data structures. The challenge is to standardize terms and codes used by multiple clinicians, vendors, and payors to identify procedures tests and diagnoses. There are several initiatives for standards, including Logical Observation Identifiers Names and Codes (LOINC), which started as a standard identifier for labs and has expanded to nursing diagnosis. The American Medical Association developed Current Procedural Terminology (CPT). Medicare uses Healthcare Common Procedure Coding System (HCPCS) that is based on CPT to identify conditions and procedures (e.g., hiccups are identified as 786.8). Health Level 7 (HL7) supports electronic interchange. Overall, these standards provide a basis for a unified EHR system but significant work remains.

Clearly, the need for EHR systems has increased over the last decades. As previously discussed, it is apparent that patient data collection is complex and highly fractured. However, even after resolving the data storage issues, health care will continue to be provided in a highly interconnected, complex system that requires active support of clinicians since the amount of data that are available and need to be integrated exceed limited human information processing capabilities.

The discipline of human factors has a unique ability to assist with the broad critical area of developing cognitive support tools that can be implemented in EHR systems to provide important information to the clinician without negatively affecting clinical performance. The National Research Council Committee on Engaging the Computer Science Research Community in Health Care Informatics (2009) concluded that emphasis on providing cognitive support to clinicians is necessary for successful deployment of health care Information Technology (IT). The committee described the current situation as, "The health care IT systems of today tend to squeeze all cognitive support for the clinician through the lens of health care transactions and the related raw data, without an underlying representation of a conceptual model for the patient showing how data fit together and which are important or unimportant. As a result, an understanding of the patient can be lost amidst all the data, all the tests, and all the monitoring equipment" (National Research Council, 2009). Therefore, cognitive support is clearly more than an add-on feature in devices and EHR systems.

To develop effective cognitive support, it is important to analyze and effectively support clinician's information processing at all levels of the human–computer interaction. Support of clinicians by clinical systems should be implemented in several ways: clinical systems need to follow the workflow, the data provided should be presented to facilitate rapid clinical problem identification, aid in formulating responses to those problems, and support documentation of actions in a systematic and comprehensive fashion (Sittig and Singh, 2009). Another important aspect of cognitive support in this context is a system's ability to facilitate the collaboration and information sharing between a number of providers, including physicians, pharmacists, nurses, social workers, and many others. This is clearly one of the most challenging aspects of interface design that goes much beyond the types of interfaces discussed in earlier sections, because these interfaces were centered on individuals and not created to facilitate collaboration, which is essential in heath care.

In the following sections, we will discuss how cognitive support that clearly needs to be available in a complex system like an EHR can be provided effectively. This type of support should pursue the goal of increasing performance in a range of clinical activities. These activities include the ability to assess the situation, comprehension of the situation, planning for interventions, specifying goals for interventions, selection of specific interventions, execution and coordination of these interventions, and providing feedback about the success of the implementation of interventions. All these activities need to be included when pursuing the goal of providing a comprehensive approach toward cognitive support. Interface design can contribute significantly to this formidable challenge. In the next section, we will discuss basic interface design principles that apply to the different device and system functions discussed earlier. The structure of this section will reflect the movement from low-level cognitive support to higher level cognitive support that needs to be implemented in interfaces to accomplish the goal of providing real cognitive support.

18.3 Basic Interface Design Principles

Often, the design of user interfaces is described as a process that incorporates principles that focus on the user and the user's task. Because of its collaborative nature, in health care, a broader systems perspective is needed. The next sections describe general principles of good interface design that center around the user, the task, and the context in which users perform their tasks (see Figure 18.1).

18.3.1 The User

Several cognitive stages of information processing are involved when a user interacts with a computer interface. Following a standard information processing framework, these stages are a perceptual stage, a cognitive stage, and a response stage. The following design recommendations loosely follow these three stages; though, in some cases, the principles affect more than one stage.

18.3.1.1 Perception

Interface design, especially for patient monitoring and device interfaces but also for clinical data systems, deals with an interesting challenge: on one hand, the design has to support rapid perception of information, because action has to be taken fast; on the other hand, action also has to be implemented without error, under stress, or with concurrent cognitive demand from other tasks. The following principles of interface design support visual perception.

Visibility and affordance help to create an interface that is highly *transparent*.

18.3.1.1.1 Visibility

Good interface design supports visibility of functionality. Visibility helps to identify the required actions to perform the intended operations (Norman, 1988). GUIs support the principle of visibility by showing visible features: the user can perceive what to do to reach a goal. In contrast to a visible feature, an invisible feature forces the user to remember the existence of a particular feature. Perception provides no guidance.

18.3.1.1.2 Affordance

The concept of affordance refers to attributes of objects that allow people to know how to use the object (e.g., a mouse button invites pushing it). Norman (1988) defines affordance as providing a clue about which interaction with the interface is needed to perform a particular operation. Graphical elements like buttons and icons afford different, specific activities.

To facilitate fast recognition, display elements should be coded redundantly. *Redundant* coding exploits the redundancy gain, that is, information coded on more than one dimension (e.g., red and highest location on traffic light) is easier to identify than when coded on one dimension only. However, when using redundancy in color, it is important to follow guidelines for color use (Rice, 1991).

18.3.1.1.3 Gestalt Principles

The usage of gestalt principles when designing an interface (Rock and Palmer, 1990) supports fast perception of information. The gestalt principles describe that humans tend to see things as belonging together, if they are located proximally, are enclosed by a box or lines, move or change together, or look alike in color, size, or shape. Information that is organized following the gestalt principles can be understood better, because the user can extract the organizational structure of the interface.

18.3.1.1.4 Search Processes

A well-designed interface makes important information salient. This can be done by having information pop out, which facilitates the search process. Elements that pop out of a display can be identified faster, because a parallel visual search process can be conducted. Without saliency, a slower serial search process has to be performed that requires inspection of each individual element until the target is found (Treisman and Gelade, 1980).

Discriminability of elements on a display is important, because similar elements increase the search time due to additional processes helping to distinguish between these elements. In addition, similar display elements increase error in performance, especially under stress and time pressure.

18.3.1.1.5 Visual Clutter

Finally, it is important to avoid visual clutter on a computer interface. Clutter can be defined as having too many elements on a display that have no utility. Good human–computer interface design reduces the use of graphical elements to a minimum, because the more irrelevant information is displayed the longer it will take the user to identify relevant information.

18.3.1.2 Cognition/Planning

Once, perceived information has to be processed and integrated into existing knowledge, either following a fast and effortless pattern matching process or a serial, slow, and cognitively demanding process (Drews and Westenskow, 2006).

18.3.1.2.1 Mental Model

A mental model (Craik, 1943) is an internal representation of some aspects of the external world, used to make predictions and inferences. An interface designer has to understand a user's mental model to understand his goals, actions, and information needs. An interface reflecting a user's mental model can increase performance, by reducing time for interaction and learning, by facilitating understanding of the interface structure, and by reducing error. Currently, interfaces often reflect the mental model of the developer and not the user, reducing the likelihood of successful and efficient interaction with a system.

The use of metaphors in graphical interface design attempts to exploit a user's existing domain-independent mental models. For example, common metaphors used in health could guide interface development. This is in contrast with home health care devices. Here, the user has usually no mental model about the domain, and metaphors implemented to support interaction with the devices have to very general. A solution is to provide common metaphors that are easily comprehensible.

An example for such general and successfully used metaphor is the desktop metaphor that helps a user to understand how a computer works by linking computer operations to the functions of a desktop. Such fit between a display and a user's conceptual knowledge facilitates interaction (Carroll et al., 1988; Wozny, 1989).

18.3.1.2.2 Proximity Compatibility Principle

Wickens and Carswell (1995) introduced the proximity compatibility principle that reflects the fact that cognitive integration of information is facilitated when related information is presented in close proximity. For example, to make an assessment of the cardiovascular status of a patient, information from several cardiovascular variables has to be cognitively integrated. If the relevant variables are in proximity on a patient monitor, the information can be integrated faster and with less cognitive demand.

18.3.1.2.3 Memory: Knowledge in the World versus Knowledge in the Head

Another important design principle is based on Norman's (1988) distinction between knowledge in the world (the interface provides relevant information) and knowledge in the head (the user has to recall how to interact with the interface). That it is easier to deal with a system that provides knowledge in the world is reflected in the popularity of GUIs. Infrequently used systems need to provide information about its use because the user likely will have forgotten instructions and procedures of how to use the system.

18.3.1.2.4 Pattern Recognition

A large number of theories on human cognition usually differentiate between two types of higher level cognitive processes. This distinction can also be found in the context of theories about diagnostic processes for clinical diagnosis: The first type is an analytical process, where inferences are made either from the clinical data available to articulate a hypothesis or where a particular hypothesis leads the clinician to evaluate data either to support or better to falsify this hypothesis. Clinicians seem to use this strategy frequently (Groen and Patel, 1985; Patel and Groen, 1986).

The second type of process is based on pattern recognition. This is a process that, for example, when dealing with images, can be called a visual problem-solving process, which is prevalent in domains where diagnoses are based on visual information such as x-rays, dermatological slides, or electrocardiographic waveforms. In other contexts, where this process is not based on visual information, clinicians are using patterns of data to make inferences about the status of a patient.

What sets the two processes apart from each other are the speed and the cognitive effort involved in each

of them. The first problem-solving process is relatively slow, laborious, and cognitively effortful but highly flexible. This can be contrasted with the pattern recognition-based process, which is relatively fast, highly automatic, and cognitively effortless.

This distinction is consistent with the hypothesis by Norman et al. (1992, 1989) that an expert's superiority in diagnosis is attributable to highly developed perceptual abilities rather than highly developed analytical skills. According to those authors, medical diagnosis has a nonanalytical basis in areas where patient information is visualized. When previous experience shares features with the current situation, case-based reasoning becomes the basis for a diagnosis (Kolodner, 1993). A physician retrieves diagnoses of similar cases from memory, integrates the various pieces of data, and applies them to the current case (Weber et al., 1993). When this data integration constitutes a familiar *pattern* or gestalt, a clinician can make an almost effortless diagnosis (Klein, 1997, 1998; Nyssen and De Keyser, 1998).

Pattern recognition processes are emphasized in the naturalistic decision-making (NDM) literature as a fundamental part of the rapid, expert decision-making process (Klein, 1997, 1998). Decision-making research has shown that people, including clinicians, do not make decisions by consuming information to generate multiple options and then carefully evaluate those options for the optimal decision. Rather, people use a process based on prior experience to rapidly categorize situations through pattern matching, assess the fit of the data with previous experience, and either proceed with the decision or seek more information (Klein, 2008).

Given the fact that based on changes in technology, the amount of information available to clinicians has increased dramatically and that clinical systems are moving toward an integration of this wealth of information, it is necessary to structure this information in such way that it is easily accessible to clinicians. This can be accomplished wither either by providing the information in such way that it facilitates automatic pattern recognition or by providing it in such way that it requires more effortful problem solving. To strike a balance between these two types of processes in the design of interfaces is a formidable task. And most recent changes that relate to the availability of new monitoring devices, digital imaging, and lab tests that provide highly detailed information on individual patients make this an even more pressing issue for interface design in health care.

18.3.1.3 Response Processes/Execution of Plans

After the cognitive process of the information provided by an interface and the generation of an intervention, the health care professional has to implement this plan.

Several design principles can support the fast and error-free implementation of an intervention. Because interfaces and patient information systems allow direct interaction and manipulation, these design principles apply more to these interfaces than to patient monitor systems.

18.3.1.3.1 Consistency

Good interface design is consistent design. Consistency refers to the idea that similar operations performed with an interface use similar elements to achieve similar tasks. If a system interface is designed consistently, then it allows the user to learn faster how to interact with this system, and the system is easier to use and less likely to lead to user error. Consistency of design applies to multiple levels: within an individual interface, within components of systems, and between interfaces of different systems or devices. Modern health care is provided in a highly interconnected, complex system. Health care professionals are not limited to working in a single setting, but move between settings (e.g., when transitioning a patient between surgery or radiology and rooms). The various settings are equipped differently, and the heterogeneity of interfaces potentially negatively impacts performance of clinicians. Even within a particular setting, often different devices from different manufacturers are present, and professionals are expected to be able to interact with these devices proficiently. Inconsistent response mapping negatively affects behavior, and some degree of interface standardization would increase performance and reduce human error. The need for "a national consensus on comprehensive standards for the definition, collection, coding, and exchange of clinical data," which includes condition, procedures, medications, and laboratory data, is described in the Institute of Medicine (2001). This need for standards extends to the design of interfaces.

18.3.1.3.2 Mapping

Mapping refers to the relationship between controls and their effects in the world. The principle of pictorial realism (Roscoe, 1968) states that the information on a display should have the appearance of the variable represented by it. The principle of moving parts (Roscoe, 1968) describes how to produce a good mapping between controls and effect. According to this principle, dynamically displayed information should reflect the user's mental model about the movement in the real system. Both principles can be used to assure consistent mapping in a display.

18.3.1.3.3 Direct Manipulation

Direct manipulation allows the user to perform actions on the visible objects on the GUI. Direct manipulation has several advantages: it can be consistent with a user's mental model of a task, thus facilitating performance, and it is easy to learn.

18.3.1.3.4 Response Alternatives

Providing shortcuts for experienced users allows them to respond faster than by direct manipulation. The issue with shortcuts is that it requires time to learn them, and some users never acquire this knowledge because they make a tradeoff between the effort of learning shortcuts and decrease in response times.

A problem related to providing response alternatives is customization of the interface. In general, customization provides the user with more flexibility, which can be beneficial. But customization can be confusing when several people interact with the same system. For example, in the case of a customized patient monitoring system, valuable time could be spend figuring out the controls diverting attention from the monitoring task.

18.3.1.3.5 Providing a Natural Dialogue

Interfaces should be as simple as possible, providing information about computer concepts to user concepts in a natural way. Therefore, it is necessary to analyze the user's task and the context in which the task is performed and to use task-related terminology for labeling when designing an interface. Related to this is the goal to minimize the effort of navigating through a menu structure, by making it easy to understand. This can be achieved by using task-based terminology and by creating menu items that reflect the task structure.

18.3.1.3.6 Feedback

The concept of feedback is related to visibility. Feedback about what action has been performed and what was accomplished by this action is important to facilitate ease of interaction. Delays in feedback or nonexisting feedback have negative consequences on behavior. Interfaces frequently use visual or auditory feedback, though verbal, tactile, or combinations of all of these are possible. Miller (1968) and Card et al. (1991) recommend specific times for interface feedback: according to these authors, 0.1 s is the limit for the user to feel that the system responds immediately, which means that no special feedback is needed. With an increase in system response time, feedback about the current computer operations has to be provided (after 10 s users often want to perform a different task while waiting for the computer to finish).

Another important element of feedback is to provide informative error messages. Cryptic error messages require further investigation that is unlikely to be done, especially if the error seems inconsequential. Well designed and informative error messages provide information about the source of the error, information about the consequences of the error, and information how to fix the error using naturalistic language that is comprehensible to the targeted user group.

Additionally, supporting the user to find out how to achieve desired goals is another important aspect of effective feedback design. Good interface design allows the user to get specific help information quickly and instructs the user, in simple steps, how to reach the desired goal.

18.3.1.3.7 Predictive Information

Finally, another way of providing feedback in interface design is to provide the clinician with modeled data that help anticipate future states, for example, the state of a patient to make predictions and to test these predictions effectively. Drews et al. (2006) demonstrated the usefulness of modeled data in the context of delivery of anesthesia. Based on pharmacokinetic and pharmacodynamic models, it is possible to predict the potential interactions of medications and to predict the effectiveness of a particular intervention. Tools that allow a clinician to preview the potential impact of an intervention based on modeled data have the potential to improve clinical performance significantly and should be included in effective interfaces.

18.3.1.3.8 Constraints

Constraints refer to determining ways of restricting the kind of user interaction that can take place at a particular moment. Deactivation of menu options while operating in a particular device mode creates such constraints. Restricting the user under certain circumstances reduces the likelihood of error. Norman (1999) distinguishes between three constraint types: physical, logical, and cultural constraints. Physical constraints refer to the way how physical objects constrain the movement of objects. Logical constraints rely on people's understanding how the world works, that is, people's common sense about actions and consequences. Cultural constraints rely on learned conventions, for example, specific colors that express warnings (red) or desaturated blood (blue). Ecological interface design (Vicente, 2002) emphasizes the importance of making constraints visible. Constraints that are currently successfully implemented are based on limiting the range of drug concentrations that can be administered to a patient. Such constraint has the potential to reduce the likelihood of overdosing a medication in a clinical context, for example, when programming an infusion pump.

18.3.1.3.9 Error Tolerant Systems and Ease of Error Recovery

To err is human; thus, interfaces should minimize the negative consequences of errors and allow fast error recovery (see also Chapter 22 in this volume). Error tolerant interfaces prevent or mitigate dangerous and disastrous consequences when human error occurs. For the interface designer, it is important to identify potential user errors during the design and evaluation process and to implement solutions for making the system error tolerant. Common approaches are providing an undo option and making verification requests in case of irreversible commands.

Systems with multiple operation modes are prone to human error. Two ways of implementing modes can be distinguished: Real modes and user-maintained modes (Sellen et al., 1992). An example for a real mode in medical devices was reported by Cook (2005), who pointed out that some glucose monitors provided glucose values in either American or European standard units (mg/dL or mmol/L) based on the setting in the device. However, this setting could change when users attempted to set the date or the time in the device. Additional situations under which the mode changed were, according to the FDA, when the glucose meter is dropped or its battery was changed. A similar type of mode error was reported by Ramundo and Larach (1995) who report a case where physicians noticed a mismatch between their patient observations and the variables on the patient monitor: the patient's blood pressure was constantly 120/70. Wondering about the constant values, the physicians realized that the monitor was in demonstration mode and not displaying data of the patient at all. User-maintained modes differ, because here the user constantly pushes a key to keep the mode activated. This creates awareness that the particular mode is activated making a mode error less likely.

18.3.1.3.10 Individual User Characteristics

An important issue regarding the development of efficient user interfaces is to address individual user characteristics. For example, given the fact that 8% of the male population has limitations related to color vision, display designers should be concerned about the high percentage of color-blind users. Another important user characteristic is age. Aging affects perceptual, cognitive, and response processes; though these changes are gradual and started in the 1920s and 1930s, they become more serious with increasing age. Other variables to be considered in interface design are user experience, level of education, work experience, and previous computer experience.

18.3.1.3.11 Stress

Another important facet of the task environment in health care is that health care professionals deal with high levels of stress, because decisions with potentially far reaching implications have to be made fast and interventions have to be implemented rapidly. Sources of stress may be high expectations and workload accompanied by time pressures and poor social support (Glasberg et al., 2007). Under stress, information processing is often suboptimal, for example, not all task relevant information is processed and included in a decision resulting in potentially suboptimal clinical decision making.

18.3.1.3.12 Sleep Deprivation

While the results are controversial and somewhat inconclusive, sleep deprivation, especially for clinicians on the night shift, has been shown to have detrimental effects (Weinger and Ancoli-Israel, 2002). These effects range from mood variations to motor and cognitive tasks. More specifically, short-term recall and reaction times can deteriorate with lack of sleep (Veasey et al., 2002).

18.3.2 Task

Behavior of experts is goal oriented and structured (Hollan, 2000). Thus, analyzing behavior in its context helps to identify important aspects and constraints of the user's task. Computer-based clinical systems fail when the designers' model of health care clashes with the nature of the clinical work (Wears and Berg, 2005). A coherent understanding of a system's function and the physiologic mechanisms is a necessary precondition for high levels of performance (Craik, 1943; Nyssen and De Keyser, 1998). What data are presented and the context, or how and where data are presented, makes a difference. For example, when developing an infusion pump interface, it is important to analyze how infusions are administered, what steps and procedures are required and actually performed before and during administration, and what current standards of best practice are.

18.3.2.1 Understanding the Task

A task analysis helps to understand the task and its environment. "Successful interaction design requires a shift from seeing the machinery to seeing the lives of the people using it" (Winograd, 1997), which is central for a comprehensive approach toward interface design in health care. Task analysis investigates the situation of the user performing a task with a focus on analyzing the rationale and the purpose of a person's actions. The results help identify the current standard of practice and the user requirements.

Several techniques can be used to perform a task analysis. Hierarchical task analysis (Annett and Duncan, 1967) involves the identification of the task and its decomposition into smaller, less complex subtasks. At the lowest level is the description of the user's physical operations. At the end of the analysis, the lowest level subtasks are grouped together as plans specifying how the task

might be performed. Thus, the hierarchical task analysis focuses on the physical and observable actions of a user.

Another well-known task analysis method is goals, operators, methods, and selection rules (GOMS) (Card et al., 1983). GOMS helps the designer to understand potential problems when a user interacts with interfaces. GOMS assumes hierarchical organization of user action: a user formulates a goal and subgoals and attempts to achieve them by using methods and operators. A method is a sequence of steps that are perceptual, cognitive, or motor operations. Selection rules guide the user in selecting the method to reach the goal. GOMS can be used to describe software functionality and interface characteristics, which can lead a systematic analysis of potential usability problems. One problem with GOMS is based in the assumption of error-free user performance (Nielsen, 1993), which is unrealistic and lead to an underestimation of problems when interacting with an interface.

To complement the aforementioned approaches, a cognitive task analysis (CTA) should be conducted to allow the identification of the cognitive characteristics of the user's task. CTA identifies a user's skills and knowledge when interacting with a system, thus allowing focusing the interface development on specific users and their background. Typical phases of CTA are knowledge elicitation, analysis, and knowledge representation. During knowledge elicitation, information about cognitive events, structures, or mental models are elicited from domain experts. The analysis phase structures the elicitated knowledge and analyses its meaning. Finally, the knowledge representation phase shows relationships between important constructs and describes the meaning of the collected data.

18.3.2.2 Methods to Identify Task Requirements

Taylor (2000) analyzed the success and failure of IT projects and found that a major contributor to failure is the lack of clear objectives and requirements. Good interface development identifies the user's needs and requirements as a first step. Requirements describe what the intended interface should do and how it should perform in a particular environment (Robertson and Robertson, 1999). Types of requirements are, for example, functional requirements (what the systems should do), data requirements (what data have to be used), environmental requirements (circumstances under which the product will be expected to operate), user requirements (characteristics of the intended user group), and usability requirements (capture the usability goals).

Gaining an understanding of the task and its requirements in the context of the clinician's work environment is often not straightforward. A significant reason for the difficulty is that work practices are not always written down or even consistent across individuals and teams. Even when there are "official" policies and guidelines a crucial question is whether or not these policies and guidelines are actually effective or followed. Expert panels, focus groups, questionnaires, and surveys are often used to extract information from the user regarding requirements. While these methods may provide information on group dynamics, concerns, and perceptions, they are conducted outside actual work and often provide little guidance for identifying the full range of functional data, environmental, user, and usability requirements. In addition, they are unlikely to reveal the actual tasks, why the process breaks down or how to fix core issues. Besides being conducted outside actual work, another reason these methods are often inadequate lies in the fact that individuals often have poor insight into their own work processes. In fact, trained researchers who conduct structured observations and task analysis of actual work flow are often better at identifying the patterns of behavior and motivations than the individuals who are immersed in the work (Norman, 2004).

Ethnographic and naturalistic studies seek to overcome these limitations. Ethnographic studies by definition include direct observation of users performing the actual tasks of interest in the relevant environment. Although arguably the best understanding of the task context and requirements comes from structured observation by trained researchers, such studies are often time and resource intensive. Therefore, given schedule and resource constraints, these studies can be balanced with other methods to gather the information about the requirements. For example, the designer can perform analysis of documents that describe procedures, standards, and regulations and use well-targeted questionnaires, structured, semistructured, or unstructured interviews to complete the process. Both task analysis and analysis of requirements are of importance for interface development, because they help to understand the user's goals as they approach the task, identify the information needs, and help to understand how the user deals with routine events, nonroutine events, or emergencies.

18.3.3 Context

In addition to the aforementioned factors of user and task, it is important to consider contextual factors in the process of interface design. This is a consequence of the fact that performance does not occur in a vacuum, and high levels of performance are often based on collaboration and distributed cognition.

18.3.3.1 Distractions and Interruptions

For a clinician, the health care environment is full of distractions and interruptions including ambient noise and

alarms. Additionally, clinicians are sought after to provide information to colleagues, patients, and/or members of a patient's family. Studies in intensive care units (ICUs) have observed that clinicians experienced one distraction or interruption approximately every 5 min (Coiera et al., 2002; Collins et al., 2006; Drews et al., under review). Study results also showed that the average time to complete the interrupted task was over 5 min (Collins et al., 2006). With the time to task completion being longer than the mean time between interruptions, it is no surprise that interruptions are associated with 37% of the errors committed in an ICU (Donchin et al., 2003). With interruptions rated highest as the reason for errors in health care settings (Conklin et al., 1990; Walters, 1992; Williams, 1996), it is important to understand how this interruption-driven environment affects the user's interaction with the electronic systems. The system must be designed to support coping with interruptions and allow easy recovery from those interruptions (i.e., allow movement to another task while maintaining the data and configuration of the first task). Thus, design that includes a context analytic approach leads to the creation of robust systems whose performance is not based on the assumption of error free performance of the user.

18.3.3.2 Collaboration and Distributed Cognition

Another important systems aspect of the health care environment is that it is a highly collaborative environment. This has implications for the design of interfaces. A more recent perspective on human cognition emphasizes the fact that cognition is not limited to individuals but the result of an interaction between individuals and resources and artifacts in their environment (Hollan et al., 2000; Hutchins, 1995). One of the important aspects of this perspective is that cognitive processes are possibly distributed over a number of interacting individuals, involve coordination between internal and external structure, and finally may be distributed through time with events that occur earlier potentially transforming later events. This leads to the interpretation of social organizations as a form or cognitive architecture, which allows the study of transmission and transformation of information. But this also implies the importance of developing interfaces that allow for shared cognition and facilitate socially distributed cognition. One of the benefits of this approach is that it allows a rich perspective that can contribute to improve the effectiveness of interface design in general and in health care in particular.

A strength of the analysis of distributed cognition and shared cognition is that it provides some recommendations for interface design that supports these functions. One of the recommendations is based on the fact that shared objects often have a use history that reveals what information is used more frequently than other types of information. A paper-based patient record that shows signs of wear for the pages describing laboratory results communicates to other providers that this information has been reviewed often. Digital objects can allow for the integration of such use history (as graphical representation) that allows different providers to not only understand what information was retrieved, but also by whom this information was used. Work that used this approach to present document interactions is presented in Hill and Hollan (1994). Another approach is to make the history of menu choices more transparent by highlighting frequently used information (Hollan et al., 2000). History enriched digital objects; exploring the history of other peoples interactions.

18.3.3.3 Multidisciplinary Collaboration

Nurse information systems. With teamwork and sharing of patient-related information being essential, information about the status of patients and the performed or planned treatment has to be distributed to all team members. For example, nurses collaborate with other nurses and other clinical colleagues. Information systems for nurses have to reflect this multidisciplinary collaboration. Nurses use information from different sources. At the bedside, they may record particular patient information that is similar or identical to the data physicians use, though, the data may have different implications and are used in differently by both groups. As a consequence, interface design has to ensure that data are presented consistent with a health worker's mental models, but also reflect the communalities between different disciplines in terms of their information requirements.

18.4 Conclusions

User interface design in health care has to overcome several challenges to successfully affect quality of patient care. One of these challenges is the variety of basic applications in the clinical context (e.g., patient monitoring, device interfaces and clinical information systems) where interface design is relevant, but where the requirements for the design differ dramatically between applications. This implies that the basic purpose of a system and its user interface has to be identified and that the interface designer has to take this purpose into account when developing the interface. In addition, the context in which the task is performed that utilizes the interface, and the larger social context that may require dissemination of information has also to be a consideration in interface development.

At this point, it seems that a simple interface solution that serves all these different purposes is not possible and attempts to develop such solutions run danger of creeping functionalism, adding unnecessary complexity, and consequentially losing a desired high level of integration.

Another challenge for interface design is related to the fact that the health care work environment is a highly stressful, high-risk, and high-stake environment. This affects the health care professional's ability to process information and to make decisions. As a consequence, information processing should not be assumed as being at the highest level and error free. Interface design has to take this into account.

Another important challenge for the interface designer is related to the fact that health care is a highly cooperative environment where communication and shared cognition are essential for success. Interface design that does not take this into account has the potential to exclude team members from access to information and therefore negatively impact efficient team work and patient outcomes.

Another important facet of the health care environment is that the professional has to interact with many different devices from many different manufacturers using very different approaches for user interfaces. One of the potential consequences of such lack of standards is that performance can be suboptimal, because time has to be spent in identifying the particular interface and the way how interaction is structured.

All these challenges can be answered, and there is significant potential to improve the quality of health care when interface designers actively face these challenges. As identified by the Institute of Medicine (2000), one of the contributing factors to error in medicine is the growing complexity of science and technology on one hand and constraints of exploiting information technologies on the other hand. Interface design that realizes the potential of information technologies can help to make the complexity of science and technology manageable and applies technology and science to improve health care and patient safety.

References

Agutter, J., Drews, F. A., Syroid, N. D., Westenskow, D. R., Albert, R. W., Strayer, D. L., Bermudez, J. C., and Weinger, M. B. (2003). Evaluation of a graphical cardiovascular display in a high fidelity simulator. *Anesthesia & Analgesia*, 97, 1403–1413.

Annett, J. and Duncan, K. D. (1967). Task analysis and training design. *Occupational Psychology*, 41 211–221.

Bailey, R. W. (2000). *Link Affordance.* Retrieved April 9, 2010, from http://webusability.com/article_link_affordance_11_2000.htm

Berner, E. S., Detmer, D. E., and Simborg, D. (2005). Will the wave finally break? A brief view of the adoption of electronic medical records in the United States. *Journal of the American Medical Informatics Association 12*(1), 3–7.

Card, S. K., Moran, T. P., and Newell, A. (1983). *The Psychology of Human-Computer Interaction.* Hillsdale, NJ: Lawrence Erlbaum Associates.

Card, S. K., Pirolli, P., Van Der Wege, M., Morrison, J. B., Reeder, R. W., Schraedley, P. K. et al. (2001). *Information scent as a driver of Web behavior graphs: Results of a protocol analysis method for Web usability.* Paper presented at the *Proceedings of the SIGCHI Conference on Human Factors in Computing Systems.*

Card, S. K., Robertson, G. G., and Mackinlay, J. D. (1991). *The information visualizer: An information workspace.* Paper presented at the Proceedings of the SIGCHI conference on human factors in computing systems: Reaching through technology, New Orleans, LA.

Carroll, J. M., Mack, R. L., and Kellogg, W. A. (1988). Interface metaphors and user interface design. In M. Helander (Ed.), *Handbook of Human-Computer Interaction* (pp. 67–85). Amsterdam, The Netherlands: Elsevier Science.

Coiera, E. W., Jayasuriya, R. A., Hardy, J., Bannan, A., and Thorpe, M. E. C. (2002). Communication loads on clinical staff in the emergency department. *The Medical Journal of Australia, 176*(9), 415–418.

Collins, S., Currie, L., Bakken, S., and Cimino, J. J. (2006). Interruptions during the use of a CPOE system for MICU rounds. *AMIA Annual Symposium Proceedings,* 2006, 895.

Committee on Data Standards for Patient Safety. (2003). Key Capabilities of an Electronic Health Record System: Letter Report (N. Academies, Trans.). Washington, DC: Institute of Medicine.

Conklin, D., MacFarland, V., Kinnie-Steeves, A., and Chenger, P. (1990). Medication errors by nurses: Contributing factors. *AARN Newsletter, 46*(1), 8–9.

Cook, R. (2005). Mode error leads to recall of medical device. Retrieved on 6/28/2011 from: http://catless.ncl.ac.uk/risks/24.10.html#subj2

Cooper, A., Reimann, R., and Cronin, D. (2007). *About Face 3: The Essentials of Interaction Design* Indianapolis, IN: Wiley Publishing, Inc.

Craik, K. J. W. (1943). *The Nature of Explanation.* Cambridge: Cambridge University Press.

Dick, R. S., Steen, E. B., and Detmer, D. E. (Eds.). (1997). *The Computer-Based Patient Record: An Essential Technology for Health Care, Revised Edition.* Washington DC: Institute of Medicine, National Academy Press.

Drews, F. A., Agutter, J., Syroid, N. S., Albert, R. W., Westenskow, D. R., and Strayer, D. L. (2001). Evaluating a graphical cardiovascular display for anesthesia. In *Proceedings of the 41st Human Factors* Meeting. Human Factors and Ergonomics Society, Santa Monica, CA.

Drews, F. A. (2008). Patient monitors in critical care: Lessons for improvement. In: *Advances in Patient Safety: From Research to Implementation,* Vol. 5, AHRQ Publication Nos. 050021 (1–5). Rockville, MD: Agency for Healthcare Research and Quality.

Drews, F. A., Picciano, P., Agutter, J., Syroid, N., Westenskow, D. R. and Strayer, D. L. (2007). Development and evaluation of a just-in-time support system. Human Factors, *49,* 543–551.

Drews, F. A. and Westenskow. D. R. (2006). The right picture is worth a thousand numbers: Data displays in Anesthesia. *Human Factors, 48,* 59–71.

Drews, F. A., Syroid, N., Agutter, J., Strayer, D. L., and Westenskow, D. R. (2006). Drug delivery as control task: Improving patient safety in anesthesia. *Human Factors, 48,* 85–94.

Donchin, Y., Gopher, D., Olin, M., Badihi, Y., Biesky, M., Sprung, C. L. et al. (2003). A look into the nature and causes of human errors in the intensive care unit. *Quality and Safety in Health Care, 12,* 143–148.

Gans, D., Kralewski, J., Hammons, T., and Dowd, B. (2005). Medical groups' adoption of electronic health records and information systems. *Health Affairs, 24*(5), 1323–1333.

Gardner, R. M. and Shabot, M. (2001). Patient monitoring systems. In L. E. P. G. W. L. F. E.H. Shortliffe (Ed.), *Medical informatics: Computer Applications in Health Care and Biomedicine* (2nd edn., pp. 443–484). New York: Springer.

Glasberg, A. L., Eriksson, S., and Norberg, A. (2007). Burnout and stress of conscience among healthcare personnel. *Journal of Advanced Nursing, 57*(4), 392–403.

Goodstein, L. P. (1981). Discriminative display support for process operators. In J. Rasmussen and W. B. Rouse (Eds.), *Human detection and diagnosis of system failure* (pp. 433–449). New York: Plenum.

Groen, G. J. and Patel, V. L. (1985). Medical problem-solving: Some questionable assumptions. *Medical Education, 19,* 95–100.

Hill, W. C. and Hollan, J. D. (1994). History-enriched digital objects: Prototypes and policy issues. *Information Society, 10,* 139–145.

HIMMS. (2010). *EHR Electronic Health Record.* Retrieved March 14, 2010, from http://www.himss.org/ASP/topics_ehr.asp

Hollan, J. D., Hutchins, E., and Kirsh, D. (2000). Distributed cognition: toward a new foundation for human-computer interaction research. *ACM Transactions on Human-Computer Interaction: Special Issue on Human-Computer Interaction in the New Millennium, 7*(2), 174–196.

Hutchins, E. (1995). *Cognition in the Wild.* Cambridge, MA: MIT Press.

Institute of Medicine. (2000). *To Err is Human: Building a Safer Health System.* Washington, DC: National Academy Press.

Institute of Medicine. (2001). *Crossing the Quality Chasm: A New Health System for the 21st Century.* Washington, DC: National Academy Press.

Kaye, R., and Crowley, J. (2000). *Medical Device Use-Safety: Incorporating Human Factors Engineering into Risk Management.* Washington, DC: Center for Devices and Radiological Health.

Klein, G. (1997). Developing expertise in decision making. *Thinking and Reasoning.* East Sussex, UK: Psychology Press.

Klein, G. (1998). *Sources of power: How people make decisions.* MIT Press: Cambridge, MA Sellen, A. J., Kurtenbach, G. P., and Buxton, W. (1992). The prevention of mode errors through sensory feedback. *Human-Computer Interaction, 7,* 141–164.

Klein, G. (2008). Naturalistic decision making. *Human Factors: The Journal of the Human Factors and Ergonomics Society, 50*(3), 456–460.

Kolodner, J. L. (1993). *Case-Based Reasoning.* San Francisco, CA: Morgan Kaufmann Publishers Inc..

Leape, L. L., Woods, D. D., Hatlie, M. J., Kizer, K. W., Schroeder, S. A., and Gd, L. (1998). Promoting patient safety by preventing medical error. *JAMA, 280*(16), 1444–1447.

Microsoft. (1995). *The Windows Interface Guidelines for Software Design: An Application Design Guide.* Redmond, WA: Microsoft Press.

Miller, R. B. (1968). Response time in man-computer conversational transactions. In *Proceedings of the AFIPS Spring Joint Conference* (33rd edn., pp. 267–277.).

Miller, R. H., Sim, I., and Newman, J. (2003). *Electronic Medical Records: Lessons from Small Physician Practices.* San Francisco, CA: California HealthCare Foundation.

Mobrand, K. A. and Spyridakis, J. H. (2007). Explicitness of local navigational links: Comprehension, perceptions of use, and browsing behavior. *Journal of Information Science, 33*(1), 41–61.

National Research Council. (2009). *Computational Technology for Effective Health Care: Immediate Steps and Strategic Directions.* Washington, DC: The National Academies Press.

Nielsen, J. (1993). *Usability Engineering.* San Francisco, CA: Morgan Kauffman.

Norman, D. (1990). *The design of everyday things* (Re-issue edition ed.). New York, NY: Doubleday/Currency.

Norman, D. (1993). *Things That Make Us Smart: Defending Human Attributes in the Age of the Machine.* New York: Addison-Wesley.

Norman, D. (1988). *The Psychology of Everyday Things.* New York: Basic Books.

Norman, D. (1999). Affordances, conventions and design. *ACM Interactions Magazine,* May/June, 38–42.

Norman, D. (2004). *Emotional Design: Why We Love (or Hate) Everyday Things.* New York: Basic Books.

Norman, G. R., Coblentz, C. L., Brooks, L. R., Babcook, C. J. (1992). Expertise in visual diagnosis: a review of the literature. *Academic Medicine, 67,* S78–S83.

Norman, G. R., Rosenthal, D., Brooks, L. R., Allen, S. W., Muzzin, L. J. (1989). The development of Expertise in Dermatology. *Archives of Dermatology, 125*(8):1063–1068.

Nyssen A.-S. and De Keyser, V. (1998). Improving training in problem solving skills: Analysis of anesthesists' performance in simulated problem situations. *Le Travail Humaine, 61,* 387–401.

Patel, V. L. and Groen, G. J. (1986). Knowledge-based solution strategies in medical reasoning. *Cognitive Science, 10,* 91–116.

Ramundo, G. B. and Larach, D. R. (1995). Monitor with a mind of its own. *Anesthesiology, 82*(1), 317–318.

Reason, J. (1990). *Human Error.* New York: Cambridge University Press.

Rice, J. F. (1991). Display color coding: 10 rules of thumb. *IEEE Software, 8*(3), 99–111.

Robertson, S. and Robertson, J. (1999). *Mastering the Requirements Process.* Boston, MA: Addison-Wesley.

Rock, I. and Palmer, S. (1990). The legacy of Gestalt psychology. *Scientific American, 263,* 48–61.

Roscoe, S. N. (1968). Airborne displays for flight and navigation. *Human Factors, 10*(4), 321–332.

Schneiderman, B. (1992). *Designing the User Interface: Strategies for Effective Human-Computer Interaction* (2nd edn.). Reading, MA: Addison Wesley.

Sellen, A. J., Kurtenbach, G. P., and Buxton, W. A. S. (1992). The prevention of mode errors through sensory feedback. *Human Computer Interaction, 7*(2), 141–164.

Sittig, D. F. and Singh, H. (2009). Eight rights of safe electronic health record use. *JAMA, 302*(10), 1111–1113.

Taylor, A. (2000). IT projects: Sink or swim. *The Computer Bulletin, 42*(1), 24–26.

Treisman, A. M., and Gelade, G. (1980). A feature-integration theory of attention. *Cognitive Psychology, 12*(1), 97–136.

U.S. Department of Health and Human Services, Centers for Medicare and Medicaid Services. (2010). *Medicare and Medicaid Programs; Electronic Health Record Incentive Program; Proposed Rule.* Retrieved March 14, 2010, from http://edocket.access.gpo.gov/2010/pdf/E9-31217.pdf

Veasey, S., Rosen, R., Barzansky, B., Rosen, I., and Owens, J. (2002). Sleep loss and fatigue in residency training: A reappraisal. *JAMA, 288*(9), 1116–1124.

Vicente, K. J. (2002). Ecological interface design: Progress and challenges. *Human Factors, 1,* 62–78.

Vicente, K. J., Christoffersen, K., and Pereklita, A. (1995). Supporting operator problem solving through ecological interface design. *IEEE Transactions on Systems, Man, and Cybernetics, 25*(4), 529–545.

Walters, J. A. (1992). Nurses' perceptions of reportable medication errors and factors that contribute to their occurrence. *Applied Nursing Research, 5*(2), 86–88.

Wears, R. L. and Berg, M. (2005). Computer technology and clinical work: Still waiting for godot. *JAMA, 293*(10), 1261–1263.

Weber, E. U., Böckenholt, U., Hilton, D. J., and Wallace, B. (1993). Determinants of diagnostic hypothesis generation: Effects of information, base rates, and experience. *Journal of Experimental Psychology: Learning, Memory, and Cognition, 19*(5), 1151–1164.

Weinger, M. B. and Ancoli-Israel, S. (2002). Sleep deprivation and clinical performance. *JAMA, 287*(8), 955–957.

Wickens, C. D. and Carswell, C. M. (1995). The proximity compatibility principle: Its psychological foundations and its relevance to display design. *Human Factors, 37,* 473–494.

Williams, A. (1996). How to avoid mistakes in medicine administration. *Nursing Times, 92*(13), 40–41.

Winograd, T. (1997). The design of interaction. In P. Denning and R. Metcalfe (Eds.), *Beyond Calculation: The Next Fifty Years of Computing* (pp. 149–162). New York, NY: Copernicus/Springer-Verlag.

Wozny, L. A. (1989). The application of metaphor, analogy, and conceptual models in computer system. *Interacting with Computers, 1*(3), 273–283.

19

Trust in Health Technologies

Enid Montague and John D. Lee

CONTENTS

19.1 Introduction 281
    19.1.1 Toward Ubiquitous Health Care Technology: Labor Shortages and Increasingly Capable Technology 282
    19.1.2 Defining Trust in a Sociotechnical System 282
19.2 The Scope of the Chapter 282
    19.2.1 Patient Trust in Care Providers 283
    19.2.2 Worker Trust (Appropriate, Over-Trust, and Distrust) in Technologies 283
19.3 Types of Trust 283
    19.3.1 Individual Trust 284
    19.3.2 Organizational Trust 284
    19.3.3 Societal Trust 284
    19.3.4 Technological Trust 285
    19.3.5 Automation 285
19.4 Relationships of the Different Types of Trust and Health Outcomes 286
    19.4.1 Patient and Care Provider 286
    19.4.2 Worker and Worker 286
    19.4.3 Provider, Technology, and Patient (Trust through Technology) 287
    19.4.4 Worker, Technology, and Worker (Trust through Technology) 288
19.5 Organization, Society, and Technology 288
19.6 Trends and Recommendations 289
References 290

19.1 Introduction

Health care depends on relationships. Doctors and nurses that establish good relationships with patients see better health care outcomes. Good relationships make patients more likely to comply with treatment regimes and more likely to benefit from a given treatment (Pearson and Raeke 2000). Patients develop a web of relationships with health care providers, health care institutions, and hospitals that strongly influence the nature of services they receive.

More broadly, relationships between care providers are also central to achieving positive patient health and safety outcomes. Emergency responders must work with doctors and nurses as a team to save lives, and surgeons must work with their surgical team to achieve good outcomes. These workers must also use tools and technologies to help them provide high-quality care. They must also have positive relationships with the hospitals and clinics they work for. The complex web of relationships among patients, caregivers, institutions, and technology mediates health outcomes.

Central to these relationships is trust. Patients must trust care providers and care providers must trust the other providers that they work with. Trust is a social emotion that guides behavior when complexity and situational dynamics make it impossible or impractical to completely understand the situation. Such dynamic complexity is common in health care, making the issue of trust central to many relationships at the interpersonal, organizational, team, and technological levels.

Patients must trust their care providers to provide them with high-quality care and sound medical advice. When patients do not trust aspects of health care institutions, they do not seek care when they need it, they request superfluous procedures, and they do not adhere to medical advice. Health care providers must trust the organizations that they work for will make good decisions about the technologies they choose to implement and that these organizations employ competent workers to manage, maintain, and calibrate the technologies. Without trust, these relationships are dysfunctional at best and will cease to exist at worst. Without trust, health care workers are inefficient, teams fail to communicate, fail to compensate for each other's mistakes, and care is compromised. Without trust in organizations, care providers leave their hospitals and clinics, therefore leading to a deficit of qualified health care providers. When trust is compromised in technologies, workers and patients might choose not to use technologies that they do not trust, limiting their ability to provide and/or receive high-quality care.

The increasing ubiquity and sophistication of health care technology makes the issue of trust more, not less, critical. Highly sophisticated technology is often difficult to understand and use, requiring patients and caregivers to respond to it on the basis of trust or distrust. Technology also complicates relationships between people, changing the information and interactions that govern trust-based relationships. As an example, relationships that begin through face-to-face interactions and are maintained through email when the parties are separated by distance are more likely to persist (Coutu 1998; Cummings et al. 2006). Trust in and through technology represents a major challenge for health care.

19.1.1 Toward Ubiquitous Health Care Technology: Labor Shortages and Increasingly Capable Technology

The increasing role and complexity of medical technology reflects both the advent of new diagnostic and treatment tools and labor shortages in hospitals. The shortage of nurses and other skilled clinical employees is projected to reach a 30% deficit of nurses worldwide by the year 2020. Similarly, there is a projected 13% deficit of physicians for 2015 (Center for Workforce Studies Association of American Medical Colleges 2009), with some areas such as primary care and obstetrics facing an even larger deficit. Sultz and Young (2006) attribute the labor shortage in health care to increased frustration with staff reductions and excessive workload that make it difficult to provide quality patient care. This shortage places a cost burden on hospitals to both find and retain skilled employees and forces some hospitals to restructure their organization to redistribute work.

One strategy in redistributing work in response to the shortage of workers is technological innovation. Technological innovation in health care is growing rapidly with the aspiration of augmenting human capabilities to create more efficient work practices and care provision. The combination of increasingly capable technology and the need to redistribute work from overburdened workers suggest that increasing technology

use in health care is inevitable. Therefore, understanding how trust is established and maintained in the web of technology, care givers, and patients is crucial to the success of current and future health care systems.

19.1.2 Defining Trust in a Sociotechnical System

Trust is a person's belief that another person, organization, or tool will not fail them. Research in organizational and interpersonal relationships has shown trust, or lack of trust, to have profound effects on individuals and the organizations they work for (Mayer et al. 2006). Scholars have considered trust as an important concept in describing how people accept, rely, and comply with technology that is introduced to augment or replace the role of the human (Lee and See 2004). People rely on technology they trust and reject technology they that they do not trust. Individuals interact with multiple elements of the work system: technology and tools, organization, work-related tasks, environment, and other individuals (Smith and Carayon-Sainfort 1989). These interactions both influence trust and are influenced by trust.

19.2 The Scope of the Chapter

This chapter describes trust relationships that exist in health care work systems and the importance of considering trust between individuals, between individuals and technologies, and between individuals and other system elements such as organizational culture. Considering health care as a sociotechnical system, individual workers interact with patients, other workers, teams, technologies, organizations, and society (see Figure 19.1). Trust

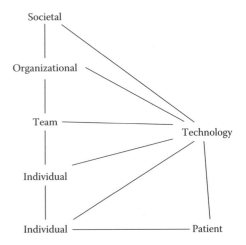

FIGURE 19.1
The web of sociotechnical relationships in health care mediated by trust.

can mediate each of these interactions. Drawing upon research in interpersonal trust, organizational trust, and technological trust, this chapter shows the importance of considering trust in health care system design. The following relationships will be the focus of the chapter:

- Trust between health care providers
- Trust between providers and patients
- Providers' trust in technologies
- Patients' trust in technologies
- The influence of technology on trust between providers
- The influence of technology on trust between patients and providers

19.2.1 Patient Trust in Care Providers

Trust between patients and care providers is integral to achieving positive health outcomes. Patient trust in care providers influences important patient outcome variables such as satisfaction, sustained enrollment in health plans, utilization of preventive services, adherence to medical advice, malpractice litigation, health status, and health service seeking behaviors (Kao et al. 1998; Thom and Campbell 1997). Patients' trust has also been linked to important organizational and economic factors such as decreases in the possibility of a patient leaving a care provider's practice and withdrawing from a health plan (Thom et al. 1999; Zheng et al. 2002). For example, Safran et al. (1998) used validated interpersonal trust instruments to evaluate the relationship between patient and health care providers in primary care environments. They examined the effect of patient trust in their provider on adherence to the physician's advice, patient satisfaction, and health status; they found no link between trust and improved health status, but they did find trust to be a strong correlate of satisfaction and adherence. Patients' trust has also been linked to important organizational and economic factors such as decreases in the possibility of a patient leaving a care provider's practice and withdrawing from a health plan (Pearson and Raeke 2000).

19.2.2 Worker Trust (Appropriate, Over-Trust, and Distrust) in Technologies

One of the goals in designing effective work systems is to facilitate appropriate trust between individuals and the technological systems (Sheridan and Ferrell 1981). However, the goal is not always to increase trust. When individuals over-trust a technology, they can overrely on the technology and become complacent (Parasuraman et al. 1993); while undertrust can influence individuals to disuse technologies that might otherwise benefit their lives (Muir and Moray 1996).

Workers' trust in technologies can be shaped from experiences like usability, the technology's fit with work tasks, or how their colleagues use technologies. An individual's trust can also be influenced by how the work system perceives the technology; does the system provide the worker with usable, well maintained, technology and appropriate technical support? Finally, an individual's trust can be influenced by society; is there pressure from the medical literature, legal, educational, and social systems to adopt or reject certain practices, technologies, and techniques?

When workers do not trust technology, they might choose not to use technologies that might otherwise be helpful. In this scenario, work systems may have wasted resources on costly tools that workers will not use. Conversely, workers may be less efficient or safe when they choose not to use technologies that can augment their capabilities or add extra safe guards. When workers over-trust technologies, they may become complacent and fail to closely monitor technologies that are not perfectly reliable. When workers trust technologies inappropriately, they may use technologies in ways that they were not designed for, which can lead to errors or reliance on inaccurate sources of information to make health decisions. If workers use technologies in ways they were not designed for, other workers may also adopt inappropriate methods, which can be problematic for care provision and for the creation of best practices and standards of care.

19.3 Types of Trust

Trust research in health care systems has focused on trust between patients and providers (Pearson and Raeke 2000) and trust between patients and the health care institution (Hall et al. 2001), but there has been little scholarship on trust in health-related technologies or the effects of technologies on other trust relationships in the health care system. Multiple types of trust exist that describe the trusting relationship between a person and another person or system element. Interpersonal trust describes trust between two individuals, while organizational trust describes a worker's trust in the organization they work in; trust in technology describes an individual's trust in a technology, and social trust describes a person's trust in a social system such as health care. This section will define the types of trust that can be present in health care systems and describe how technology influences or affects these relationships. In health care systems, multiple relationships are connected and can affect outcomes of trust (see Figure 19.1). Individual workers have relationships with other individuals, patients, technologies, and the organization they work for. These relationships are often reciprocal, as trust relationships build over time and influence each other. More globally, attitudes and practices within the health care systems are influenced by society.

19.3.1 Individual Trust

At the individual level, interpersonal trust is defined as trust between individuals, who may or may not be part of a team. When these relationships are better specified, they may be called team trust or organizational trust. For example, groups of people who work together for a common goal, whether they work for the same organization or not might be called a team. However, individuals who work together in the same organization, coworkers, may be considered part of an organization rather than part of an interpersonal relationship. This distinction has implications for measuring trust in work systems, as interpersonal trust is measured based on an individual's relationship with another person, while organizational trust includes the dynamics of the work relationship such as power structures in a worker's relationship with their supervisor (Cummings and Bromiley 1996). In health care systems, teams may differ from other organizations, where a team is typically defined as individuals who work together in the same place, at the same time. A health care team may be a distributed group of individuals working in different places for different organizations yet working toward a common goal; an individual patient's health and safety. An example of a team in health care would be a care team for a patient with diabetes. The team might include a primary doctor, an endocrinologist, a dietician, a diabetes educator, and a podiatrist; team members might be distributed across different clinics, health systems, or states. The care provision team could work together to ensure positive outcomes for one or several patients. An individual care provider could be a member of many different teams; each with varying structures and team members, depending the organizational structure of the care providers work system. Finally, patients may be considered members of the team depending on their and their care providers' preferences and organizational culture.

19.3.2 Organizational Trust

Organizational trust is defined as an individual's trust in the organization in which they are a part of, including other aspects of the organization such as employers and coworkers (Mayer et al. 1995). In health systems, organizational trust can affect how care providers work together and communicate. Organizational trust can affect how care providers work with patients (i.e., care provision culture) and how patients view their roles in the system. For example, patients who see themselves as part of the care providing team may view their of care providers as teammates. If their view their relationship as a service transaction, they may view their care providers as service providers that should consider patient preferences. The nature of trust in these two perspectives is that in the former, patients may construct based on how their providers work with them collaboratively, while in the former, they may construct trust based on how well the provider provides services (i.e., based on time, accuracy, and quality).

Several measurement tools exist to assess trust in organizations, each takes a slightly different perspective on organizations (Athos and Gabarro 1978; Cook and Wall 1980; Cummings and Bromiley 1996; Scott 1980; Shaw 1997). Athos and Gabarro (1978) developed an instrument to asses an individual's trust in their employer that explores a worker's belief that their employer has integrity, is truthful, fair, and honest, while Cook and Wall's (1980) instrument measures the belief that colleagues and managers are reliable and capable. Scott's (1980) instrument measures an individual's trust in people at different levels of the organization, such as immediate supervisor, management, and consultants. Cook and Wall's (1980) and Scott's (1980) instruments might be ideal for exploring trust between workers with different roles or organizational power, such as nurse's trust in physicians or residents' relationships with attending physicians. On the other hand, Athos and Gabarro's (1978) instrument might be better suited to evaluating an individual's trust in their organization (such as a health system or clinic) as a whole.

19.3.3 Societal Trust

The social system includes media, cultural, and legal systems that affect how health systems are organized and experienced by individuals. National surveys and demographic polls are typically used to assess citizens' social trust in the health care system. Trusting or distrusting attitudes toward health care as an institution are shaped by social influences such as attitudes about the government, regulations, media, and cultural attitudes. The resultant trusting or distrusting attitudes can also influence individual behaviors in the health systems, such as when and how patients utilize the system. Gilson (2003) has written about the role of social trust in health systems and defines social trust in health care as a macrorelationship that is influenced by and influences public policy. Therefore, social events and changes have the potential to influence an individual's trust in the system as a whole and the collective trust of citizens has the potential to influence important decisions related to the organization of health care systems. In addition to individual attitudes about the health care system, the social system can influence which types of technologies are used, how and when care is provided, and which patients are allowed access to care. The importance of considering the social system and individuals attitudes of trust to the social system will be illustrated later in this manuscript.

19.3.4 Technological Trust

Health technology includes specific instruments, equipment, medication, knowledge, and best practices used by professionals and patients (Andersen and Kristensen 2008). Health technologies can be grouped into several categories (see Table 19.1) ranging from drugs and biological preparations, to equipment, technologies and devices, to procedures, knowledge, and organizational systems (Andersen and Kristensen 2008). Health technologies can be further classified by their purpose, such as prevention, screening, diagnosis, treatment, and rehabilitation. Technologies also exist in various stages of maturity: future, experimental, tested, established, and obsolete. A broad perspective of health technologies is useful, as technologies evolve over time. For example, elements of the work system that might be considered low technologies, such as knowledge and education systems, can also be redesigned as high-tech medical devices.

19.3.5 Automation

Conversations about trust in technologies are often described as trust in automation. Determining which aspects of the system to automate is useful in understanding, the relationships between technologies and user trust in those technologies. Automation is defined as using mechanical or electrical technology to accomplish a work-related task that humans are capable or incapable of performing (Wickens and Hollands 2000). Automation is defined by (1) its intentions, (2) the human functions it substitutes, and (3) the advantages and disadvantages it demonstrates in the interaction between humans and automated apparatuses (Wickens and Hollands 2000, p. 538). Automation is not solely limited to replacing a complete task; there are four categories of automation purposes (Wickens and Hollands 2000):

The first purpose is using automation is to complete functions that the human is incapable of accomplishing due to human limitations or safety concerns. In this category, automation is necessary and required. Examples of this type of automation would be robotic arms that allow surgeons to complete surgery in locations that they could not otherwise reach with their own hands. The second purpose is using automation to carry out functions that humans would otherwise perform poorly. Examples in this category would be complex monitoring systems that provide real-time assessments of bodily functions and health status; such as heart rate and blood pressure monitors. The third purpose is supplementing or supporting performance when humans have limitations; this category is different from replacing fundamental features of the task because its goal is to aid supporting jobs that are needed in order to achieve the focal task. Examples of this type automation are functions such as digital information retrieval devices (e.g., electronic health records). The last purpose is based on economic need. Wickens and Hollands (2000) say that automation is introduced in these circumstances because it is more expensive for the organization to use humans as opposed to technology. An example of this could include technologies such as virtual care providers or telemedical systems that help clinicians provide care or consultation to locations where their expertise may be needed for a short amount of time and travel is expensive and infeasible.

TABLE 19.1

Types and Examples of Health Technology

Type of Health Technology	Examples of Health Technology
Drugs	Cytostatic drugs for chemotherapy, drugs for parkinsonism, and vitamins
Biological preparations	Vaccines, blood products, and gene therapy
Equipment, devices, and supplies	Pacemakers, CT-scanners, diagnostic test kits and elastic stockings, and AED
Medical and surgical procedures	Manual therapy, nutrition therapy, medical dementia treatment, cognitive therapy, and surgical treatment of myopia
Diagnostic procedures and techniques	Determination of functional capacity, diagnosis of depression, and palpation
Presentation of knowledge	Patient schools, preventive health interviews, and diet campaigns
Support systems	Telemedicine systems, electronic booking systems, drug selections, clinical laboratories, and blood bank
Organizational and managerial systems	Outreach psychosis teams, visiting nurse service, free choice of hospital, vaccination programs, and health-insurance reimbursement for dental treatment

Source: Adapted from Andersen, S.E. and Kristensen, F.B., The technology, in Kristensen, F.B. and Sigmund, H. (Eds.), *Health Technology Assessment Handbook*, 2 edn., National Board of Health Denmark, Copenhagen, Denmark, 2008, pp. 89–103.

19.4 Relationships of the Different Types of Trust and Health Outcomes

Relationships in health care systems include meaningful interactions between patients and care providers, providers and other providers, and humans and technologies. In the conceptual model of health system relationships (see Figure 19.2), relationships include (1) the relationship between an individual worker and a patient, (2) the relationship between multiple workers, (3) the relationship between the individual worker and the patient through or with technology, (4) the relationship between workers through or with technologies, (5) the relationship between the health care team and other entities, (6) the relationship between the health care system or organization and other entities in the system, and (7) the relationship between society and other entities in the system.

In this model, each of these relationships depends on a type of trust, which can affect which technologies are used and how the technologies are used. The following section defines these trust relationships and illustrates the importance of the trust relationship using health care examples.

19.4.1 Patient and Care Provider

Fundamental to the care provision process is the patient provider relationship (see #1 in Figure 19.2). Trust is an important factor to maintaining this relationship and contributes important system outcomes such as adherence to care plans and care seeking behavior. Kao et al. (1998) conducted a study using survey methods with 292 patients about their trust attitudes toward care providers; they found that "patients' trust in their physician is related to having a choice of physicians, having a longer relationship with their physician, and trusting their managed care organization" (p. 681). In Piette's et al.'s (2005) study of patients with diabetes, the researchers found that trust was an important factor in predicting medication adherence. Though cost was a primary barrier for adherence for patient, trust in care providers acted as a moderating variable in adherence. Their results showed that "having a low income was only associated with cost-related adherence problems in the context of low physician trust" (p. 1749).

19.4.2 Worker and Worker

Workers have relationships with other workers in health care systems (see #2 in Figure 19.2); including relationships between clinicians, nurses, and nonclinical staff. Positive relationships between care providers have been regarded as integral to achieving care and efficiency outcomes. However, interpersonal trust, communication, and conflict have been regarded as barriers to these goals.

The nurse–doctor relationship is fundamental to health system quality and clinical outcomes (Hojat et al.

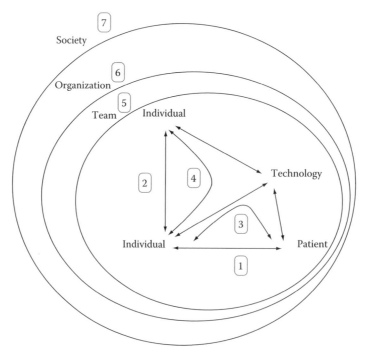

FIGURE 19.2
Trust relationships in health care systems.

2001; Pavlovich-Danis et al. 1998). For example, studies that explore nurse–doctor relationships in intensive care units found that patients who received care from doctor–nurse teams that worked collaboratively had lower mortality rates than those that were cared for by less collaborative teams (Baggs et al. 1999; Knaus et al. 1986; Larson 1999). The relationship between doctor and nurse is defined as a relationship of interdependence and collaboration, where each individual is dependent on the other, and the team must work closely together to share information and develop plans to ensure patient health, recovery, and safety. Unfortunately, the nurse–doctor relationship has been characterized as complicated by role ambiguity and conflict (Prescott and Bowen 1985; Schmalenberg and Kramer 2009). Huntington and Shores (1983) detail the history of nurse–doctor conflict in their article *From conflict to collaboration*. Conflict between workers can influence how technologies are used. When teams do not trust each other, they may use technology to provide a record of events to support them in the face of conflict. When workers trust each other, they may develop shared attitudes about the technologies they use and collaborate to identify ways that the technology can be used to improve patient care.

Communication and interpersonal trust are also important in relationships between clinicians. In a qualitative study of health care practices in pediatric intensive care units (PICU), Vivian et al. (2009) found that poor communication between staff members impacted interpersonal trust and compromised care-giving practices in the PICU. Studies have also shown that doctors often have difficulty communicating in practice, which can have implications for trust and system outcomes. For example, in a study of clinicians in academic medical center, Bell et al. (2009) surveyed 1722 primary care providers and found that only 77% were aware that their patients were admitted and of those providers, only 23% received direct communication from the inpatient clinician. In this case, technology may contribute to a reduction in traditional means of communication and thus trust. When providers can communicate information through electronic systems, they may cease to use traditional methods such as telephones and face-to-face communication.

19.4.3 Provider, Technology, and Patient (Trust through Technology)

Fundamental to the care provision process is the patient–provider relationship. However, this relationship is often mediated by technology (see #3 in Figure 19.2). Computer-supported cooperative work scholarship looks at how collaborative activities and their coordination can be supported with technologies such as computer systems (Carstensen and Schmidt 1999). In a computer-supported

TABLE 19.2

Computer-Supported Cooperative Work Place Time Matrix

	Same Time (Synchronous)	Different Times (Asynchronous)
Same place (Collocated)	Synchronous Collocated	Asynchronous Collocated
Different places (Distributed)	Synchronous Distributed	Asynchronous Distributed

Source: Adapted from Johansen, R., *Groupware: Computer Support for Business Teams*, The Free Press, New York, 1988.

cooperative work paradigm, technology-mediated interaction includes technologies used by multiple people based on time and place (see Table 19.2):

- Different time and different place (e.g., bathroom walls, room schedules, and surgical whiteboards).
- Different time and different place (e.g., e-mail, newsgroups, and listervs).
- Same time and different place (e.g., virtual mediated communication, shared documents remotely, such as physicians sharing x-rays, MRIs, and virtual communities).
- Same time and same place, such using a device or technology in the same place collaboratively (e.g., electronic health records and digital whiteboards).
- In face-to-face encounters, computers and other technologies are becoming more prevalent in interpersonal health encounters between doctors and patients.

Technologies have the potential to affect the trust relationship in positive and negative ways. In a study of electronic fetal monitors, Montague et al. (2010) found that patients and care providers developed trust in the monitoring technologies in different ways. For patients, the way the doctor used the technologies influenced the patients trust in the technology, this included whether or not the doctor was able to use the technology effectively or if the patient felt the doctor was using the technology inappropriately. The technology's trustworthiness was also related to patients' trust in the care provider. For example, if care providers are unable to use technologies effectively in front of patients, patients' trust in the technology and the care provider may be influenced. These relationships illustrate the need for usable technologies for care providers, where providers and patients can develop appropriate trust in the technologies.

Technologies such as computers can also alter nonverbal communication patterns that are associated with building trust and rapport between doctors and patients. Margalit et al. (2006) evaluated the use of computers in clinical encounters by video taping clinical visits. They found that the way physicians used the computers affected the nonverbal interactions with patients; care providers in their study spent between 25% and 42% of the visit gazing at the computer screen. They also found that 24% of the visits include heavy keyboarding. Margalit et al. (2006) suggest computer use in clinical encounters, specifically how clinicians use computers, can diminish dialog between patient and clinicians, and affect patient-centered practice.

19.4.4 Worker, Technology, and Worker (Trust through Technology)

Workers often work with other workers through technologies (see #4 in Figure 19.2); these situations range from using devices and technologies collaboratively in face-to-face encounters, to working and communicating virtually through technologies. When technologies are introduced and implemented into the health care systems, workflow and working relationships for care providers and staff may be affected. Some of the implications of the new technologies might include altered interpersonal interactions that affect the formation of interpersonal trust. Other implications might be changing attitudes about the work system or difficulty accomplishing tasks with the new technologies. In virtual health care work systems, such as virtual intensive care units, workers must work with together through virtual technologies to monitor and provide care to patients. In these work systems, virtual (or tele-ICU) nurses and doctors monitor patients remotely by viewing patient data through computerized health monitoring systems, audio, and sometimes video. In these systems, teleworkers work with doctors and nurses who care for patients on site. In virtual work places, opportunities to develop interpersonal relationships may be limited, which can have implications for the formation and maintenance of trust. McGuire et al. (2010) found that communication opportunities between tele-ICU workers and ICU workers varied, some factors included time of day and the ICU receptiveness to working with tele-ICU worker. In their work, they found that the methods teleworkers used to communicate with ICU nurses had strengths and weaknesses related to the technology mediated nature of work.

19.5 Organization, Society, and Technology

The social system can impact when and how technologies are used in health systems and other relationships.

One example of the social systems influence on technology use and work practices in the health care system would be the use of continuous use electronic fetal heart monitoring in obstetrics. Because electronic fetal monitors are often used in the legal system to support malpractice claims, they have thus become standards of care in many health care systems in childbirth (Eddy 2002; Rosenblatt et al. 1990). The use of devices as standards of care may influence the perception of the device's trustworthiness. Patients may have a higher level of trust in the technologies because they are standards of care (Beck 1980). Care providers may have lower levels of trust in the technologies if they believe that the technologies are not related to a higher quality of care and are unreliable sources of evidence in malpractice claims against physicians (Lent 1999).

A recent study showed that doctors are implanting implantable defibrillators, invasive devices implanted in the heart to regulate heart beats, to patients who do not meet guidelines for implementation and thus do not need them (Al-Khatib et al. 2011). Conversely, this analysis showed that many patients who did need the devices were not receiving them. The researchers explored the records of 111,717 people who received defibrillators at 1,227 hospital in the United States and then compared the patients' symptoms to published guidelines for defibrillators implantation. They found that 22.5 of the patients that received the defibrillators did not meet the published guidelines. Patients that receive the technologies when they do not meet the guidelines have a higher likelihood of death or complication than those who meet the guidelines. While the researchers do not know why the doctors made decisions to provide implants to patients who did not meet the guidelines, they speculate that the doctors were making the decisions that they felt were best. This study might illustrate a tendency to over-trust technologies regardless of the risk. Second, it might also illustrate the influence of patient request of technologies on the care provider's decisions and thus their trust in the defibrillators. Patient and family member request of technologies in end of life care has become a controversial topic that is influenced by the social system's perspectives on right to life, appropriate care, morality, and health care economics. These social perspectives on end of life care may influence the health care organization's decisions about technology use, care providers' decisions about technology use, and individual patients (and their family members') decisions about technology use. Studies have shown that these societal elements are related to the request and rejection of technology in end-of-life care (Prendergast et al. 1998).

In another example, the use of mammography technologies to detect breast cancer has come in to question when the risks of false positives were perceived to out way the benefits of early screening. In 2009, the

U.S. Preventative Services Task Force released a report that recommended that women aged 40–49, that do not have risk factors for breast cancer, should forgo mammography screenings (U.S. Preventive Services Task Force 2009). The report was based on scientific evidence, but the public response indicated strong opposition to the recommendation. The conversation was amplified through popular press and media and eventually resulted in a removal of the recommendation from the agencies website and an eventual senate and congressional hearing to override the recommendation (Woolf 2010).

In another example of the relationship between trust in the social system and trust in technologies, some citizens have rejected the use of vaccination technologies. There is a belief by some individuals and nonprofit organizations that there is a link between vaccination technologies and diseases like autism. Much of this belief is based on a study that was published and later retracted from the *British Medical Journal* that linked vaccinations to autism. The effect of the public's distrust in vaccination technologies has resulted in many parents choosing not to vaccinate their children. The movement has been promulgated by the media and celebrity endorsers as well as the prevalence of high fidelity "evidence" through social networking. For example, YouTube videos of national football league cheerleader Desiree Jennings alleged depilating side effects from a flu vaccination were watched by thousands of health consumers. The effects of low social trust and a decrease in the adoption of vaccinations have led to low trust in other social aspects. For example, unvaccinated children may be a contributor to the resurgence of outbreaks such as measles and mumps (Dayan et al. 2008). Unvaccinated children and adults pose a risk to children, and others who cannot be vaccinated who then may pass diseases to others who are not vaccinated. Vaccinations are a means of social trust, where individuals in a society trust that citizens are protecting themselves from diseases, so that others who cannot be protected will be able to live without fear of contracting diseases that can otherwise be prevented with vaccination. When trust is broken, it can perpetuate future distrust and lead to a culture of blame.

19.6 Trends and Recommendations

Relationships in health care are critical to system outcomes. Developing appropriate trust in these relationships is important to developing and sustaining functional relationships. Specifically, this chapter addressed:

- Patients' relationships with care providers
- Care providers' relationships with technologies
- Care providers' relationships with other care providers
- Care providers' relationships with the organizations they work for
- Health systems' relationship with society

Through a discussion of these relationships, several trends emerged. First, different types of trust are fundamental to the relationships that comprise the health care system. Interpersonal trust is important for relationships between individuals, such as the relationship between care providers and patients. Organizational trust in the health organization is important for care providers that work together. Trust in technologies is important for both care providers and patients. A second trend is that different types of trust are interrelated. For example, interpersonal trust can influence which technologies are used and how they are used. A final theme is that trust relationships are important to health system outcomes. Interpersonal trust between patients and care providers is linked to health and organizational outcomes such as patient sustained engagement with a practice. Interpersonal trust between care providers is also related to health and system outcomes such as patient health outcomes. In each relationship, trust is related to outcomes related to system effectiveness and health outcomes.

System designers should consider the different types of trust and the effects of interventions on these types of trust. System designers should be aware of the effects of interpersonal relationships on other aspects of the system, such as technology adoption. They should be cognizant of the effects of system interventions on important interpersonal relationships. For example, the way care providers use technologies can affect interpersonal relationships between doctors and patients. If care providers have trouble using technologies that are poorly designed in front of co-workers or patients, they might be assessed as less trustworthy. Conversely, attitudes about technologies can affect organizational relationships. Workers may have differing attitudes or perceptions of trustworthiness that may lead to conflicts when working together. Training and other interventions that contribute to shared understanding of technology capabilities and limitations may minimize variances in attitudes about technologies between care providers. Technologies can bring together care providers from different locations and specialties to care for patients in meaningful ways. System designers should consider the benefits that technologies might bring to the effectiveness of the health care team. Perhaps more importantly, health system stakeholders should understand that technology in health care is a necessary part of an evolving health care institution. Therefore, stakeholders should be proactive in ensuring that their needs and relationships are considered and protected as new tools are designed, implemented, and integrated into the care provision practice.

References

Al-Khatib, S. M., A. Hellkamp, J. Curtis, D. Mark, E. Peterson, G. D. Sanders, P. A. Heidenreich, A. F. Hernandez, L. H. Curtis, and S. Hammill. 2011. Non-evidence-based ICD implantations in the United States. *JAMA: The Journal of the American Medical Association* 305 (1):43–49.

Andersen, S. E. and F. B. Kristensen. 2008. The technology. In F. B. Kristensen and H. Sigmund (Eds.), *Health Technology Assessment Handbook*. Copenhagen, Denmark: National Board of Health Denmark.

Athos, A. G. and J. J. Gabarro. 1978. *Interpersonal Behavior: Communication and Understanding in Relationships*. Englewood Cliffs, NJ: Prentice Hall.

Baggs, J. G., M. H. Schmitt, A. I. Mushlin, P. H. Mitchell, D. H. Eldredge, D. Oakes, and A. D. Hutson. 1999. Association between nurse–physician collaboration and patient outcomes in three intensive care units. *Critical Care Medicine* 27 (9):1991–1998.

Beck, C. T. 1980. Patient acceptance of fetal monitoring as a helpful tool. *JOGN Nursing* (November/December) 9 (6):350–353.

Bell, C. M., J. L. Schnipper, A. D. Auerbach, P. J. Kaboli, T. B. Wetterneck, D. V. Gonzales, V. M. Arora, J. X. Zhang, and D. O. Meltzer. 2009. Association of communication between hospital-based physicians and primary care providers with patient outcomes. *Journal of General Internal Medicine* 24 (3):381–386.

Carstensen, P. H. and K. Schmidt. 1999. Computer supported cooperative work: New challenges to systems design. Paper read at In K. Itoh (Ed.), *Handbook of Human Factors*.

Center for Workforce Studies Association of American Medical Colleges. 2009. *Recent Studies and Reports on Physician Shortages in the U.S.* Association of American Medical Colleges, 2009 August.

Cook, J. and T. Wall. 1980. New work attitude measures of trust, organizational commitment and personal need non-fulfillment. *Journal of Occupational Psychology* 53 (1):39–52.

Coutu, D. L. 1998. Trust in virtual teams. *Harvard Business Review* 76:20–21.

Cummings, L. L. and P. Bromiley. 1996. The organizational trust inventory (OTI): Development and validation. In R. Kraut, M. Brynin, and S. Kiesler (Eds.), *Trust in Organizations: Frontiers of Theory and Research* 302:330–330.

Cummings, J., J. Lee, and R. Kraut. 2006. Communication technology and friendship during the transition from high school to college. In *Computers, Phones, and the Internet: Domesticating Information Technology*, New York: Oxford University Press, pp. 265–278.

Dayan, G. H., M. P. Quinlisk, A. A. Parker, A. E. Barskey, M. L. Harris, J. M. H. Schwartz, K. Hunt, C. G. Finley, D. P. Leschinsky, and A. L. O'Keefe. 2008. Recent resurgence of mumps in the United States. *New England Journal of Medicine* 358 (15):1580–1589.

Eddy, A. 2002. Consent in obstetrics: A legal view. *The Obstetrician and Gynaecologist* 4 (2):97–100.

Gilson, L. 2003. Trust and the development of health care as a social institution. *Social Science & Medicine* 56 (7):1453–1468.

Hall, M. A., E. Dugan, B. Zheng, and A. K. Mishra. 2001. Trust in physicians and medical institutions: What is it, can it be measured, and does it matter? *The Milbank Quarterly* 79 (4):613–639.

Hojat, M., T. J. Nasca, M. J. M. Cohen, S. K. Fields, S. L. Rattner, M. Griffiths, D. Ibarra, A. A. G. de Gonzalez, A. Torres-Ruiz, and G. Ibarra. 2001. Attitudes toward physician–nurse collaboration: A cross-cultural study of male and female physicians and nurses in the United States and Mexico. *Nursing Research* 50 (2):123–128.

Huntington, J. A. and L. Shores. 1983. From conflict to collaboration. *The American Journal of Nursing* 83 (8):1184–1186.

Johansen, R. (1988). *Groupware: Computer Support for Business Teams*. New York: The Free Press.

Kao, A. C., D. C. Green, N. A. Davis, J. P. Koplan, and P. D. Cleary. 1998. Patients' trust in their physician: Effects of choice, continuity, and payment method. *Journal of General Internal Medicine* 13 (10):681–686.

Knaus, W. A., E. A. Draper, D. P. Wagner, and J. E. Zimmerman. 1986. An evaluation of outcome from intensive care in major medical centers. *Annals of Internal Medicine* 104 (3):410–418.

Larson, E. 1999. The impact of physician–nurse interaction on patient care. *Holistic Nursing Practice* 13 (2):38–46.

Lee, J. D. and K. A. See. 2004. Trust in automation: Designing for appropriate reliance. *Human Factors* 46 (1):30–80.

Lent, M. 1999. The medical and legal risks of the electronic fetal monitor. *Stanford Law Review* 51 (4):807–837.

Margalit, R. S., D. Roter, M. A. Dunevant, S. Larson, and S. Reis. 2006. Electronic medical record use and physician–patient communication: An observational study of Israeli primary care encounters. *Patient Education and Counseling* 61 (1):134–141.

Mayer, R. C., J. H. Davis, and F. D. Schoorman. 1995. An integrative model of organizational trust. *The Academy of Management Review* 20 (3):709–734.

Mayer, R. C., J. H. Davis, F. D. Schoorman, and J. Pointon. 2006. An integrative model of organizational trust. *Organizational Trust: A Reader*.

McGuire, K., Khunlertkit, A., Carayon, P., Hoonakker, P. L. T., and Wiegmann, D. 2010. Communication in the tele-ICU. Paper read at *Human Factors and Ergonomics Society (HFES) 54th Conference*, San Francisco, CA, September 27–October 1, 2010.

Montague, E., Winchester, W. W., and Kleiner, B. M. 2010. Trust in medical technology by patients and health care providers in obstetric work systems. *Behaviour & Information Technology* 29 (5):541–554.

Muir, B. M. and N. Moray. 1996. Trust in automation. Part II. Experimental studies of trust and human intervention in a process control simulation. *Ergonomics* 39 (3):429–460.

Parasuraman, R., R. Molloy, and I. L. Singh. 1993. Performance consequences of automation-induced complacency. *The International Journal of Aviation Psychology* 3 (1):1–23.

Pavlovich-Danis, S., H. Forman, and P. P. Simek. 1998. The nurse–physician relationship: Can it be saved? *Journal of Nursing Administration* 28 (7/8):17–20.

Pearson, S. D. and L. H. Raeke. 2000. Patients' trust in physicians: Many theories, few measures, and little data. *Journal of general internal medicine* 15 (7):509–513.

Piette, J. D., M. Heisler, S. Krein, and E. A. Kerr. 2005. The role of patient–physician trust in moderating medication nonadherence due to cost pressures. *Archieves of Internal Medicine* 165 (15):1749–1755.

Prescott, P. A. and S. A. Bowen. 1985. Physician–nurse relationships. *Annals of Internal Medicine* 103 (1):127–133.

Prendergast, T., M. Claessens, and J. Luce. 1998. A national survey of end-of-life care for critically ill patients. *American Journal of Respiratory and Critical Care Medicine* 158 (4):1163–1167.

Rosenblatt, R. A., G. Weitkamp, M. Lloyd, B. Schafer, L. C. Winterscheid, and L. G. Hart. 1990. Why do physicians stop practicing obstetrics? The impact of malpractice claims. *Obstetrics & Gynecology* 76 (2):245–250.

Safran, D. G., D. A. Taira, W. H. Rogers, M. Kosinski, J. E. Ware, and A. R. Tarlov. 1998. Linking primary care performance to outcomes of care. *Journal of Family Practice* 47 (3):213–220.

Schmalenberg, C. and M. Kramer. 2009. Nurse–physician relationships in hospitals: 20000 nurses tell their story. *Critical Care Nurse* 29 (1):74–83.

Scott, C. L. 1980. Interpersonal trust: A comparison of attitudinal and situational factors. *Human Relations* 33 (11):805–812.

Shaw, R. 1997. *Trust in the Balance: Building Successful Organizations on Results, Integrity an Concern*. San Francisco, CA: Jossey-Bass.

Sheridan, T. B. and W. R. Ferrell. 1981. *Man–Machine Systems: Information, Control, and Decision Models of Human Performance*. Cambridge, Mass., and London, England: MIT Press.

Smith, M. J. and P. Carayon-Sainfort. 1989. A balance theory of job design for stress reduction. *International Journal of Industrial Ergonomics* 4 (1):67–79.

Sultz, H. A. and K. M. Young. 2006. *Health Care USA: Understanding its Organization and Delivery*. Gaithersburg, MD: Aspen Publishers.

Thom, D. H. and B. B. Campbell. 1997. Patient–physician trust: An exploratory study. *Journal of Family Practice* 44 (2):169(8).

Thom, D. H., K. M. Ribisl, A. L. Stewart, and D. A. Luke. 1999. Further validation and reliability testing of the Trust in Physician Scale. *Medical Care* 37 (5):510–517.

U.S. Preventive Services Task Force. 2009. *Screening for Breast Cancer: Recommendation Statement*, AHRQ Publication No. 10–05142-EF-2.

Vivian, L., A. Marais, S. McLaughlin, S. Falkenstein, and A. Argent. 2009. Relationships, trust, decision-making and quality of care in a paediatric intensive care unit. *Intensive Care Medicine* 35 (9):1593–1598.

Wickens, C. D. and J. G. Hollands. 2000. *Engineering Psychology and Human Performance*. Upper Saddle River, NJ: Prentice-Hall.

Woolf, S. H. 2010. The 2009 breast cancer screening recommendations of the US Preventive Services Task Force. *JAMA* 303 (2):162–163.

Zheng, B., M. A. Hall, E. Dugan, K. E. Kidd, and D. Levine. 2002. Development of a scale to measure patients' trust in health insurers. *Health Services Research* 37 (1):187–202.

20

Human Factors in Telemedicine

Peter Hoonakker

CONTENTS

20.1 Introduction 293
20.2 Backgrounds 294
   20.2.1 Definition 294
   20.2.2 Current Telemedicine Applications 294
      20.2.2.1 Tele-Consultation 294
      20.2.2.2 Tele-Education 295
      20.2.2.3 Tele-Monitoring 295
      20.2.2.4 Tele-Surgery 295
      20.2.2.5 Tele–Home Care 295
20.3 Effects of Telemedicine on Costs, Quality of Care, and Patient Safety 296
   20.3.1 Effectiveness of Telemedicine 296
   20.3.2 Telemedicine and Quality of Care 296
   20.3.3 Telemedicine and Patient Safety 297
20.4 Barriers and Facilitators for Telemedicine Implementation 297
   20.4.1 Advantages of Telemedicine 297
   20.4.2 Barriers against Telemedicine Implementation 298
      20.4.2.1 Costs 298
      20.4.2.2 Risks 298
      20.4.2.3 State of Technology 298
      20.4.2.4 Interconnectivity 298
      20.4.2.5 Usability 299
      20.4.2.6 Process Redesign (Changes in Workflow) 299
      20.4.2.7 Attitudes Such as Resistance to Change 299
   20.4.3 Dealing with These Barriers from a Human Factors Perspective 299
20.5 Conclusion 301
Acknowledgments 302
References 302

20.1 Introduction

The word telemedicine is a combination of the Greek word *Tele* (τελε) meaning far, at a distance, or remote, and the Latin word *Mederi*: to heal. In other words, telemedicine is medicine at a distance. There are many different sorts of telemedicine, for example, astro-telemedicine (telemedicine in space), tele-cardiology, tele-dermatology, tele-neurophysiology, tele-obstetrics, tele-oncology, tele-ophthalmology, tele-pathology, tele-psychiatry, tele-surgery, tele-care, and the tele-intensive care unit (tele-ICU).

Telemedicine or remote medicine is not a recent development. For example, hundreds of years ago, in case of serious diseases such as the bubonic plague, villagers in medieval Europe and Africa used smoke signals to warn people to stay away from their village (Wikipedia). Another example of care at a distance is the use of the postal mail to send medical data. Telemedicine took a great leap forward with the invention of modern telecommunication technologies (see Table 20.1). For example, one of the first uses of telegraphy during the Civil War (1861–1865) was to send out casualty lists and order supplies (Wootton and Craig, 1999). Already, in 1905, Einthoven, the inventor of the electrocardiogram (ECG)

TABLE 20.1

Main Phases of Telecommunication Development

Technology	Period
Telegraphy	1830s–1920s
Telephone	1870s–now
Radio	1920s–now
Television	1950s–now
Space technologies, e.g., satellite-based communications	1960s–now
Digital technologies	1990s–now

Source: Adapted from Norris, A.C., *Essentials of Telemedicine and Telecare,* John Wiley & Sons, New York, 2002.

and Nobel Prize winner, began to transmit ECGs from the hospital to his laboratory 1.5 km away via a telephone cable (Einthoven, 1906). With the invention of radio, radio-medical services became available, for example, for ships. The Italian International Radio Medicine Centre (CIRM) started in 1935 and by 1996 had assisted more than 42,000 patients, mainly seafarers (Stanberry, 1998). The invention of the television added visual images to the auditory information sent from a distance. In 1964, the Nebraska Psychiatry Institute developed a two-way link with Norfolk State Hospital 112 miles away for education and consultation (Benschoter, 1971; Norris, 2002). In the 1970s, the first commercial communication satellites became available, and several countries (Foote, 1977; House and Roberts, 1977; Watson, 1989) started telemedicine programs to improve rural healthcare, such as the Alaska Satellite Biomedical Demonstration Program (Foote, 1977). With the arrival of digital technologies and the Internet, telemedicine made a great leap forward. Digital technologies made telemedicine possible all over the world and in many countries, such as Australia, Canada, Germany, Greece, Italy, Japan, Norway, the United Kingdom, and the United States. In countries such as Algeria, Egypt, India, Kosovo, Macedonia, Montenegro, Morocco, Russia, Tunisia, and Turkey, telemedicine programs have been initiated.

There are two main drivers for telemedicine (Norris, 2002; Perednia and Allen, 1995): (1) to improve quality of care in rural areas and (2) to provide remote, low-cost specialty services, where full-time service is impractical, not available, or very expensive. For example, more than 50 million people in the United States or nearly 20% of the population live in rural areas, but only 9% of the nation's physicians practice in rural communities. Recruiting and retaining rural physicians has turned out to be difficult, for example, because of professional isolation (Zollo et al., 1999). Different telemedicine applications can possibly reduce this isolation, enhance lifelong learning opportunities for rural healthcare providers, and enhance quality of care for patients.

Healthcare is characterized by shortages of resources; shortage of qualified personnel is one of the most important ones. There is a critical shortage of qualified nurses, medical aides, laboratory personnel, pharmacists, radiologists, technicians, and qualified specialists, for example, in surgical and obstetrical care in the United States. Furthermore, this shortage will only get worse now that the baby boom generation has started to retire.

20.2 Backgrounds

20.2.1 Definition

"Modern" telemedicine is defined as follows: "Telemedicine utilizes information and telecommunications technology to transfer medical information for diagnosis, therapy and education" (American Telemedicine Association, 1999). Modern telemedicine is most beneficial for populations living in isolated communities and remote regions and is currently being applied in virtually all medical domains. Over the years, many different applications of telemedicine have been developed.

20.2.2 Current Telemedicine Applications

The following broad categories of telemedicine can be identified (Hoonakker et al., 2010; Norris, 2002):

1. Tele-consultation
2. Tele-education
3. Tele-monitoring
4. Tele-surgery
5. Tele-care

20.2.2.1 Tele-Consultation

Tele-consultation is defined as consultation via telecommunication systems, generally for the purpose of diagnosis or treatment of a patient at a site remote from the patient or primary care physician. Tele-consultation can take place between two (or more) healthcare providers or between the patient and one (or more) healthcare provider. Tele-consultation can take place in real-time or as store-and-forward. The best example of the latter is tele-radiology with the transmissions of large x-ray files (Norris, 2002). X-ray files are sometimes sent overnight to other countries such as India, where specialists examine them, make a diagnosis, and—thanks to the time difference between the different countries—are able to send their findings back to Western countries where they will be ready for the following working day. Tele-consultation is the currently most frequently used form of telemedicine. Tele-consultation services are now available all across the world, mostly to connect remote and/or rural areas to urban centers.

20.2.2.2 Tele-Education

Tele-education, also known as distant learning or tele-learning, involves training and education with the use of information or tele-communication technology (e.g., television and the Internet) on a remote basis. There are many forms of tele-education, ranging from public education via the Internet, to clinical education, access to the National Library of Medicine, Cochrane databases, Medline, etc., to computer-based training (CBT) and even academic studies via the Internet. Evidently, there are many advantages to tele-education. The need to travel is reduced; students can study when it is convenient for them instead of having to follow a fixed schedule; information is available everywhere the students are at least when they are connected to the Internet; and education is open to many other people than students. However, tele-education has also disadvantages. For example, many people with health complaints search the Internet for information about their health complaints, and the information that is provided to them does no always originate from the best sources. In other words, there can be problems with the validity and reliability of the information available through tele-education. Furthermore, most lay people are not physicians and are not able to make a differential diagnosis. This can lead to misinformation and even to panic.

Results of studies that compare tele-education with classroom teaching in healthcare show that tele-education can be very effective. For example, results of a randomized controlled trial (RCT) that compared the effects of (1) tele-education and (2) class-room training on neonatal resuscitation targeted at nurses showed that the training resulted in a significant gain in knowledge and skill scores in both groups and that the results of the two groups were comparable (Jain et al., 2010).

20.2.2.3 Tele-Monitoring

Tele-monitoring involves the use of telecommunications links to gather routine or repeated data on a patient's condition (Norris, 2002). The purpose of monitoring is to decide if and when an adjustment is needed in the patient's treatment. Examples of tele-monitoring are as follows: automatic reminders to patients at home to take their medication, automatic tele-monitoring of blood pressure, heart rate, glucose and hemoglobin levels, and tele-Intensive Care Units (tele-ICUs).

There are several advantages of tele-monitoring for both the patient and the healthcare provider. For the healthcare provider, tele-monitoring can be an efficient way to gather necessary patient information without a great time commitment. Tele-monitoring is useful to patients, because they can receive healthcare provider feedback on their vital signs and symptoms more often and more rapidly than they otherwise might receive. Furthermore, because the patient is more involved in his or her own treatment, the patient will become more aware of his or her vital signs and symptoms and can possibly gain a better sense of what affects these signs and symptoms and how the signs and symptoms in turn affect how he or she feels. Several studies have shown benefits of relatively simple tele-monitoring systems (Friedman et al., 1996; Logan et al., 2007). Other studies have shown that more advanced tele-monitoring systems, such as the tele-ICU, can improve clinical care outcomes (e.g., reduced length of stay, reduced mortality, and reduced complications) and reduce healthcare costs (Rosenfeld et al., 2000; Breslow et al., 2004; Zawada et al., 2009).

20.2.2.4 Tele-Surgery

Tele-surgery concerns surgical procedures carried out at a distance. A remotely located surgeon can participate in a surgery without being in the same operating room where the patient is located. This technology enables surgeons in different places to work together on the same patient. Typically, one surgeon, or multiple surgeons, remains at the side of the patient while the other surgeon controls a robotic device. One of the most striking examples of tele-surgery is Operation Lindbergh (Marescaux et al., 2001). A remote surgeon in New York used a Zeus surgical system, a computer-assisted device, to remove the gall bladder of a patient 4,300 miles away in Strasbourg, France.

Patients in remote and nonremote areas can benefit from tele-surgery. Some of the benefits of tele-surgery include tele-education (i.e., training of new surgeons) and tele-consulting (i.e., assisting surgeons in developing countries or underserved regions where access to surgeons is limited). Tele-surgery may reduce the cost of surgeries because it reduces the travel time of the patient or the surgeon. The technology has also been developed for NASA, which would use a portable robotic unit to perform tele-surgery in extremely remote areas such as space or on the battlefield.

20.2.2.5 Tele–Home Care

Tele–home care is a telemedicine application that does not stand on its own, as the other four applications do, but is a combination of tele-consultation, tele-monitoring, and tele-education. Tele–home care enables people to remain independent in their own home by providing person-centered technologies to support the individual and/or their caregivers. Patients are monitored electronically, and the data are sent (automatically) to the physician or the nurse. The patient can also directly contact his or her physician or nurse using interactive audio/video technology (Chumbler et al., 2004). Another variant is that instead of the physician visiting the patient at home, a specially trained community nurse visits the patients and reports to the physician (Terschuren et al., 2007).

20.3 Effects of Telemedicine on Costs, Quality of Care, and Patient Safety

20.3.1 Effectiveness of Telemedicine

Many studies on telemedicine focus on cost-effectiveness of telemedicine as the main research topic. Despite the relatively young history of telemedicine, many literature reviews have been conducted to determine the effectiveness of telemedicine (e.g. Balas et al., 1997; Bergmo, 2009; Hakansson and Gavelin, 2000; Jarvis-Selinger et al., 2008; Roine et al., 2001). Results of these studies show that telemedicine can be cost-effective, depending on the telemedicine application being examined and the method used to calculate costs and benefits. For example, Roine et al. (2001) conducted a systematic review of the telemedicine literature in the period 1966–2000 to examine the evidence for the effectiveness and economic efficiency of telemedicine. They identified a total of 1124 articles with telemedicine as the topic. They applied the following inclusion criteria to the 1124 articles: articles had to take in consideration, in a scientifically valid method, the outcomes of a form of telemedicine in terms of administrative changes, patient outcomes, or economic assessment. One of the criteria for scientific validity was that the studies had a control group. Note that their inclusion criteria did not limit the studies to only studies with a RCT. Applying the criteria resulted in 50 articles. They concluded that based on the classification system developed by Jovell and Navorro-Rubin (1995) to assess methodological strength of studies that the quality of the studies ranged in most cases from fair to poor. Only 6 of the 50 selected studies used a RCT, which is considered the strongest study design. Results of the study by Roine et al. (2001) showed that tele-radiology, tele-neurosurgery (transmission of CT images before patient transfers), tele-psychiatry, transmission of echo-cardiographic images, and the use of electronic referrals enabling e-mail consultations and video-conferencing are among the most cost-effective telemedicine applications.

The methods used to calculate the cost-effectiveness of telemedicine vary widely: evidently, it matters which costs are taken into account (costs to the patient, the physician, the clinic, the health system, the insurer, or society at large). For example, several studies when comparing the costs of a telemedicine application with a hospital visit take the hours traveled by patients (and their family) into account. Evidently, telemedicine can greatly reduce the time and costs associated with patient travel. For example, Raza et al. (2009) examined over a 6-year period the effects of the use of videoconference telemedicine for providing consultation to a remote, underserved clinic, 215 miles from the main pulmonary telemedicine clinic. Results show that in that 6-year period, a total of 364 patients (684 visits) received telemedicine consultations. Only 8% of the patients required an in-clinic visit at the main medical center. Telemedicine saved patients nearly 3,00,000 miles of travel in the 6-year period (Raza et al, 2009).

However, results of a recent systematic literature review by Bergmo (2009) on the different methods used to evaluate cost-effectiveness of telemedicine show that the majority of the evaluation studies are not in concordance with standard evaluation techniques and that especially the generalizability of the studies (being able to generalize the results of a particular study in a particular setting to other settings) is lacking.

20.3.2 Telemedicine and Quality of Care

To justify the often expensive implementation of telemedicine applications, many studies have been conducted to demonstrate that telemedicine applications provide at least equal or better quality of patient care. Results of these studies show that there is limited evidence that telemedicine can enhance quality of care. Most of the studies that find a positive impact of telemedicine on quality of care are demonstration projects or were studies that are limited in their *study design* (e.g., no control group) or in their sample size. Results of the systematic reviews make that clear.

For example, Currell et al. (2000) conducted a Cochrane literature analysis that compared telemedicine versus face-to-face patient care in the period 1960–1999. Only studies that meet the strongest inclusion criteria (RCTs and time-series analysis) are included in a Cochrane analysis. Currell and colleagues identified more than 200 studies with telemedicine as the research topic, but only 24 met the inclusion criteria of a Cochrane analysis. After examining the study designs more closely, another 16 studies had to be excluded for several reasons and only 7 studies were used in the analysis. Most of the studies had tele–home care as the topic. Results of the analysis showed that none of the seven included studies showed any detrimental effects from the interventions (telemedicine implementation), but neither did they show unequivocal clinical benefits, nor did the findings constitute evidence of the safety of telemedicine (Currell et al., 2000). We describe one example of the studies included in the analysis. Ahring et al. (1992) examined the effects of glucose self-monitoring for insulin-dependent diabetics in an RCT. Patients in both the control and experimental groups took five daily blood glucose measurements for a period of 12 weeks. The patients in the control group took their measurements to their clinic visits, and the experimental group transferred their results by telephone modem. The patients in the experimental group received telephone counseling on their diabetic management. Results of the study showed that blood glucose (HbA1c) of respondents in the experimental (telemedicine) group significantly improved but not in the control group. However, there were no significant changes in other variables such as weight, random blood glucose, or insulin (Ahring et al., 1992).

In a more recent review with less stringent inclusion criteria, Sibbald et al. (2007) reviewed the literature on the effects of shifting care from hospital to the community. Telemedicine implementation was part of the review, and 25 studies on telemedicine were included in the review. Results showed that telemedicine does indeed improve access for remote populations, but there is insufficient evidence of the impact of telemedicine on health outcomes. However, the authors suggested that some of the problems with telemedicine, such as difficulties with correctly diagnosing the problem, may improve with advances in technology.

20.3.3 Telemedicine and Patient Safety

A far as we know, no systematic review has been conducted to specifically examine the effects of telemedicine on patient safety. Currell et al. (2000), based upon their systematic review of the telemedicine literature, concluded that although none of the studies in their review showed any detrimental effects from telemedicine implementation, neither did they show unequivocal benefits, and the findings did not provide evidence of the safety of telemedicine.

Few studies have examined the impact of the different telemedicine applications on patient safety. One of the exceptions is tele-monitoring through the tele-ICU. Several tele-ICU studies have shown an impact of tele-monitoring on patient safety. Examples of increased patient safety through tele-ICU monitoring range from preventing that an ICU patient falls out of his/her bed to improving the safety climate in ICUs that are being monitored. For example, results of a study by Thomas et al. (2007) using the Safety Climate Score (SCS) from the Safety Attitudes Questionnaire showed that the mean SCS score increased significantly from 66.4 (pre-tele-ICU implementation) to 73.4 (post-tele-ICU implementation) in the ICUs that were being monitored, while the overall SCS scores for the hospitals decreased from 69.0 to 65.4.

Other examples provide indirect evidence of the impact of tele-ICUs in patient safety. For example, the literature shows that patient safety improves when ICUs have an intensivist. However, there is a huge shortage of intensivists, which is one of the main reasons why the tele-ICU was developed. One tele-intensivist can monitor more than 150 ICU-patients in several different hospital and ICUs. There is some evidence for the impact of the tele-intensivist model on patient safety (Breslow et al., 2004).

20.4 Barriers and Facilitators for Telemedicine Implementation

Implementation of information technology (IT) systems in healthcare has proven to be very difficult. Already, in 1995, in a review of medical informatics between 1950 and 1990, Collins concluded that implementing IT seems to be "a more complex task than putting a man on the moon" (Collen, 1995, p. 464). And if health IT implementation has proven to be difficult *within* healthcare organizations (HCOs), one can wonder about the success rates of telemedicine implementation where two or more HCOs are involved. Currently, telemedicine applications have not been implemented on a large scale. In this paragraph, barriers and facilitators of telemedicine implementation will be discussed.

20.4.1 Advantages of Telemedicine

Evidently, telemedicine has distinct advantages for patients living in remote areas (Nesbitt et al., 2000; Norton et al., 1997). Telemedicine offers rural patients better access to better medicine (Norris, 2002). Patients may be able to receive high-quality medical care without having to travel long distances (Jaatinen et al., 2002; Loane et al., 2000; Oakley et al., 2000). Telemedicine may also prevent referrals to specialists (Jaatinen et al., 2002; Whited et al., 2002) or reduce the time patients have to wait for specialty care (Kedar et al., 2003; Whited et al., 2002). For rural physicians, telemedicine also has advantages: through tele-consulting and tele-education so they can keep up-to-date with the latest medical developments and technologies. Some studies show that as a result of telemedicine implementation, rural healthcare providers reported more social networking with other physicians, higher job satisfaction, and reduced turnover among nurses and medical staff (Dixon et al., 2008). The use of telemedicine has the possibility to increase quality of care (Miyasaka et al., 1997; Woods et al., 2000). Finally, telemedicine can lead to reduction in costs and better resource utilization (Norris, 2002).

20.4.2 Barriers against Telemedicine Implementation

However, there are also many barriers to telemedicine implementation, and several studies have identified specific barriers. According to several authors (Doarn, 2008; Graschew et al., 2008; Norris, 2002), the costs of capital investment and operations are one of the biggest challenges to telemedicine implementation. Other authors point out the legal, ethical, and clinical risks associated with telemedicine (Attarian, 2007; Benedict, 2001; Darkins, 1996; Stanberry, 2000, 2006). Others focus on the technological challenges that telemedicine applications face (Doarn, 2008; Harnett, 2008). Telecommunication is the key to any telemedicine application, and telecommunication can be limited by technical barriers such as insufficient bandwidth or unreliable signals, and especially people in remote areas may experience delays in telecommunications or not have access to the network at all (Doarn, 2008). Broens et al. (2007) identified support, training, usability, quality, and acceptance as major determinants of the success of future telemedicine implementations. Finally, several authors (Doarn, 2008; Groves et al., 2008) point out that attitudes, such as resistance to change, are important barriers to telemedicine. The barriers to telemedicine implementation are summarized in Table 20.2.

20.4.2.1 Costs

The initial investments in telemedicine are often high. The technology and the infrastructure to transfer the data are very expensive. To provide just one example: the implementation costs of a tele-ICU (see tele-monitoring) are between 3 and 5 million dollars (Groves et al., 2008). That is just to set up the system and get it functional. To keep it running, additional tens of thousands of dollars are required each year (Groves et al.,

TABLE 20.2

Barriers to Telemedicine Implementation

1.	Costs
2.	Risks
3.	State of technology
4.	Connections between different technological systems
5.	Usability
6.	Process redesign (changes in workflow)
7.	Attitudes such as resistance to change

2008). Currently, the costs are being paid by the healthcare delivery system because there are currently no mechanisms in place to collect third reimbursement for remote physician services, neither in the United States (telemedicine costs are not reimbursed by Medicaid) nor in Europe. Current research on telemedicine is aimed at justifying these high costs by emphasizing the financial benefits and, to a lesser extent, improvements in quality of care (Carayon et al., 2010).

20.4.2.2 Risks

Several liabilities issues have been raised for telemedicine applications. Unclear delineation of roles and responsibilities and poor communication between the different parties involved can lead to mistrust that can cause medico-legal risks for all parties (Groves et al., 2008). Some significant legal issues related to operating a telemedicine network include the corporate practice of medicine, patient confidentiality and privacy, malpractice, informed consent, licensure and credentialing, intellectual property, Medicare and Medicaid payment, fraud and abuse, medical device regulation, and antitrust (Edelstein, 1999). However, most of these risks are not particular for telemedicine, and most risks can be managed with appropriate risk-management, such as asking patients for informed consent (Stanberry, 2006). However, because of the potential difficulties of communication and without clear roles and responsibilities, there are certain risks to telemedicine.

20.4.2.3 State of Technology

Evidently, telemedicine is very dependent on the technology. If the technology does not work properly, the use of telemedicine is not possible. And, in some cases, the use of telemedicine stretches the boundaries of what is technical possible. For example, telemedicine systems often need a lot of bandwidth to transmit all the data. Consequently, if people have to deal with many technological mishaps, their enthusiasm for telemedicine applications will be greatly reduced. However, there are also telemedicine applications that use relatively simple electronic devices such as telephone modems that have proven to be successful.

20.4.2.4 Interconnectivity

Currently, more and more technological systems are being implemented in healthcare settings and in the community, ranging from "smart" medical devices to advanced administration systems, such as electronic health records (EHRs), electronic medication administration records (eMARs), electronic nursing flowsheets, and all kinds of computer decision support tools

(CDSTs) such as computerized provider order entry, just to name a few. One of the problems is that these devices and the software do not communicate very well with each other. The connections are hardly standardized; therefore, data often have to be entered twice (or even more often) into the different systems. The situation can be compared to the early days of personal computers (the 1980s), in which users had to know a lot about computers, programming, etc., in order to set up computers and keep them running (attaching different devices, downloading printer and other device drivers, manually making changes in start-up files, etc.). Only after Windows came to dominate the market, standardization took place, and, as a result, it became easier to connect the different computer peripherals, install new software, etc. These developments speeded up with the arrival of the Internet, which made it much easier to both find and download the necessary information such as printer drivers, etc. Only recently has "plug-and-play" become feasible and new devices or programs can be installed by plugging in the *Universal* Serial Bus (USB) connection, and the computer itself "automatically" takes care of the necessary connections.

20.4.2.5 Usability

The problems described earlier can have as a result that users stop using the technology. User acceptance is determined by usefulness and ease of use, also called usability (Baroudi et al., 1986). If hardware or software is not useful or not easy to use, users will never start using it or will stop using it. Since the introduction of personal computers in the early 1980s, researchers have been interested in evaluating end-user satisfaction with computer systems. Both theory (Fishbein and Ajzen, 1975) and empirical studies (Baroudi et al., 1986) suggest that user satisfaction contributes to technology use rather than the reverse. Therefore, it is important to pay attention to the usability of telemedicine technologies in order to enhance the actual use of telemedicine.

20.4.2.6 Process Redesign (Changes in Workflow)

Telemedicine evidently has major consequences for workflow (Carayon et al., 2010; Currell et al., 2000). For example, instead of seeing a patient in the clinic, a tele-consultation takes place, in which the patient, the rural physician, and the specialist in the clinic take part. Apart from the clinical workflow that may or may not change a great deal (e.g., does the specialist have to perform a medical history or is that data already available?) and the administrative workflow, the technical details also have a great impact on workflow. Computers have to be started up, connections (e.g., a video-link) have to be made and tested, etc. Most of the time that involves, beside the physician(s) and the patient, an IT support person.

20.4.2.7 Attitudes Such as Resistance to Change

People resist change. Many health IT implementation projects have either been aborted or advance were very slow because of user resistance (Aarts and Koppel, 2009; Bates, 2006; Connolly, 2005). Obviously, health IT implementation has a huge impact on clinical and administrative workflows, and therefore the way the work is organized needs to change.

Often, it is physician resistance that causes most problems, and one of the causes is that physicians fear losing their autonomy (Groves et al., 2008). In the case of telemedicine, physicians have to rely on "strangers," that is, other physicians with whom they do not have a personal relationship and whom they probably have never met in person. Other causes of physician resistance are the often difficult relationship with the healthcare system (e.g., hospital) they work for, and physicians may feel that change is being forced upon them by the healthcare system (Groves et al., 2008; Navein, 1998).

Nurses' resistance is often less entrenched than physician resistance. Initially, nurses' issues are similar to those of physicians: autonomy and privacy (Groves et al., 2008; Kirkley and Stein, 2004). In addition to autonomy and privacy issues, nurses who are already overloaded have to deal with the changes in workflow. For example, as described earlier, because of bad interconnections between the different health IT systems and devices, nurses may have to enter the same data in several different health IT systems.

20.4.3 Dealing with These Barriers from a Human Factors Perspective

From a human factors perspective, telemedicine is especially challenging. Human factors and, especially, the socio-technical approach study how human beings interact with tools, equipment, machines, and technology. The human factors tradition dates back to the beginning of the twentieth century with the work of Taylor (Taylor, 1911) and the invention of mass production with the assembly-line conveyer belt at the Ford company and the problems this caused for the human–machine interaction. The socio-technical systems (STS) approach received much attention with the work of Trist and Emery in the 1950s and 1960s at the Tavistock Institute in London (Emery and Trist, 1960, 1965; Trist and Bamforth, 1951).

With the rapid technological developments described earlier (radio, television, space technologies, and digital technologies), the interaction between people and technology has become increasingly complex. Especially the development of the digital technologies (Internet, e-mail, etc.) has a profound impact not only on how people interact with technology, but also on how people interact with each other. Just think of developments such as e-mail, but also new social network applications such as MySpace, Facebook, and Twitter. The development of digital technologies has led to the development of specializations in the human factors approach, for example, usability engineering. From the usability viewpoint, we know that software programs and interfaces need to be easy to use and useful for users; otherwise, people will not use them. This has a great impact on how hardware, software, and interfaces are designed. Developments in telemedicine take this one step further. People work in virtual teams and need to collaborate, often at a great distance from each other, even though they may never have actual met face-to-face. This evidently has important consequences for communication, but also for trust among people. The virtual team organization makes it difficult for people who collaborate with other people around the world at companies and other organizations; in the case of medicine, where lives can be at stake when errors are made; this creates an altogether different impact.

Human factors can play a role in the design, implementation, and evaluation phases of telemedicine projects, including usability and heuristic evaluations, prospective risk assessments, and redesign of workflow. For example, if we examine the barriers described earlier more closely, human factors can play an important role in dealing with barriers 5–7. We can assume that barriers 1–4 will be dealt with over time. In general, costs of IT come down when they are implemented in large numbers. Furthermore, we expect that in the near future, a solution will be found to reimburse remote physician services. The risks are in most cases not particular for telemedicine, and most risks can be managed with risk management. Possible problems with the state of technology, and, related to that, problems with interconnectivity will be solved over time as well. Technological solutions will be found, because it is a problem that is not particular to telemedicine but to all health IT implementations.

Some people suggest that barriers 5–7 will be harder to deal with than the barriers 1–4.

For example, several authors suggest that *resistance to change*, and, in particular, *physician resistance to change* is one of the major barriers to successful health IT and telemedicine implementation (Campbell et al., 2001; Groves et al., 2008; Navein, 1998; Thomas et al., 2009). Several suggestions have been made to overcome physician resistance, although the problem will probably always remain an important barrier.

We know from the virtual team literature (Kirschner and Van Bruggen, 2004; Majchrzak et al., 2004) that it will help if the physicians will get to know each other personally, or if existing relationships were used and built upon. Special meetings can be organized before and during implementation in which the participating physicians get to meet each other. However, this will not always be possible, especially if large distances are involved. Therefore, time needs to be devoted to consensus building and the development of local champions, well before implementation starts (Groves et al., 2008).

Furthermore, good communication and feedback loops need to be established. Physicians need to be informed about progress, but also about problematic issues. Feedback has to be given on a regular basis about progress being made, using several quality of care indicators. Therefore, it will be helpful to start collecting data on quality of care indicators, well before implementation is begun, in order to compare pre- and postimplementation quality of care data.

Nurses' resistance to change is another important factor, but, in general, nurses seem to be less entrenched than physicians. Nurses do not seem to resist technology that much, but often the extra work it causes (Kirkley and Stein, 2004). For example, computer charting may take more time than paper charting. And sometimes, because of bad interconnectivity between systems, nurses are asked to enter the same information twice. To deal with nurses' resistance, the same recommendations as for physicians can be used.

Thus, collaboration will be improved if the different parties involved get to know each other personally. For example, in several tele-ICUs, the tele-ICU nurse manager organizes regular meetings between the tele-ICU nurses and the ICU nurses they collaborate with, or the ICU nurses are invited to see how the tele-ICU works. These meetings seem to be very effective in enhancing collaboration and reducing resistance to change. Nurses get to know each other personally, can discuss their activities, the goals of the tele-ICU, etc., which, in turn, facilitates communication and trust.

Evidently, nurses also need to be kept in the loop with regard to problematic issues as well as the progress that is made. For example, results of a longitudinal study by Fraenkel et al. (2003) showed that 7 months after implementation of a clinical information system in an ICU significant improvement in quality indicators and nurses' perception of the system had been achieved.

One of the most important barriers to telemedicine implementation is that it changes the way the work is organized: *it changes the flow of work*. Currell et al. (2000) concluded already in 2000 that the implementation of telemedicine can have a major impact on the organization of healthcare services and service delivery and administration, but that so far these factors have been largely ignored in studies on telemedicine. Now, 10 years later, the situation has not changed drastically. Results of a literature review on the impact of telemedicine on workflow have shown that relatively few studies have *focused* on work organization and workflow (Carayon et al., 2010). Most of the studies that were part of the literature review focused on cost-benefit assessment of telemedicine and clinical effectiveness. With regard to the effects of telemedicine on workflow, most of the studies show that work is transferred from hospital centers and clinics to primary care clinics of primary care physician offices. Work activities are transferred from medical specialists to primary care physicians, physician assistants, nurse practitioners, nurses, and technicians. Most of the physicians (medical specialists and primary care physicians) are satisfied with telemedicine applications implementations as well as most of the patients. Results of many studies show that from a clinical effectiveness point of view, the clinical results of telemedicine are most of the time equal or better than traditional care, although there are also several studies that report less effectiveness. For most of the patients, telemedicine means less travel time and less lost work time.

Evidently, telemedicine implementation does have a huge impact on workflow, but based on the results of the literature review, Carayon et al. (2010) concluded that (1) telemedicine has not been implemented on a large scale and therefore have relatively few impacts on (clinical) workflow and (2) that until now, there are relatively few studies that have focused on the impact of telemedicine on workflow.

Therefore, more studies should—instead of focusing on specific drugs, devices, and services—focus on the comparative effectiveness of care delivery and work processes. For example, Berenson et al. (2009) suggested that researchers examining the effects of tele-ICU implementation should focus on the impact of the tele-ICU on work-process innovations, instead of focusing on quality and cost outcomes.

Human factors specialists can play an important role in redesigning work processes in the context of telemedicine implementation. The human factors approach has developed many tools to describe and analyze the work process (e.g., flow diagrams and process mapping). These tools can be used to examine the processes in healthcare, and how the processes change as a result of health IT implementation, such as telemedicine. The results of these analyses can then be used to redesign the work processes to facilitate the implementation and use of telemedicine applications.

20.5 Conclusion

Telemedicine has much potential and in the future probably more and more telemedicine applications will be implemented, especially in rural and underdeveloped and underserved areas. Both patients and healthcare providers can benefit from telemedicine implementation. However, telemedicine implementation also faces several barriers. The technological and administrative (e.g., reimbursement of healthcare personnel) barriers will probably be solved in the near future. Other barriers are more difficult to solve. Telemedicine implementation faces many of the same barriers as other health IT implementations. In the recent past, we have seen that it is very difficult to implement health IT. Some people have joked that it is easier to put a man on the moon than to implement health IT (Collen, 1995). Some authors suggest that 75% of health IT implementations fail (Berg, 1999; Levitt, 1994; Wyatt, 1994), but that number is hard to substantiate. Still, for example, Computerized Provider Order Entry (CPOE) has only been successfully implemented in 15% of healthcare settings (Ash et al., 2004; Delbanco, 2006). And no telemedicine applications have been implemented on a large scale.

Our review of the literature has shown that telemedicine can be cost-effective, depending on the telemedicine application being examined and the method used to calculate costs and benefits. Results also show that there is limited evidence that telemedicine can enhance quality of care. Most of the studies that find a positive impact of telemedicine on quality of care are demonstration projects or studies with limited study design (e.g., no control group) or small sample size. Very few studies have focused on the impact of telemedicine on patient safety, and we recommend that future studies be conducted to this purpose.

Despite possible advantages of telemedicine, especially for rural areas, there are many barriers, and a human factors approach can probably help to overcome these barriers. One of the reasons why it is so difficult to implement health IT is that it has a profound effect on the way healthcare is organized. Health IT implementation has an effect on many administrative and organizational processes: it changes the workflow. Telemedicine implementation has an even bigger impact on administrative and organizational processes than other forms of health IT. Healthcare providers involved in telemedicine implementation projects work at a distance and have often not even met in person.

Human factors engineering can support the effective, efficient, and safe implementation of health IT and telemedicine. Based on their review of the telemedicine literature, Allen and Perednia concluded that "most failures of telemedicine programs are associated with the human aspects of implementing telemedicine" (Allen and Perednia, 1996, p. 22). Human factors specialists have over the years acquired significant experience with IT implementation. They know from experience how important aspects, such as usability, usefulness, and ease of use are for successful (health) IT implementation.

Moreover, they are specialized in studying workflow and the impact of technology on workflow and have many tools at their disposal to examine workflow, and the impact of health IT and telemedicine on workflow; these methods can help redesign healthcare processes and facilitate telemedicine implementation.

However, a final word of caution. Berg (1999) already in 1999 pointed out that IT implementation in healthcare is not an easy straightforward, top-down, technology-centered process because healthcare processes are complex, fluid, nonstandardized processes in which many different parties participate and collaborate. Therefore, "classical" engineering tools such as flow charting, process mapping, etc., maybe very helpful, but they will probably fail to grasp the complexity and flexibility of healthcare processes. Healthcare is a complex STS. Berg (1999) therefore recommends two approaches to collect the necessary knowledge to be able to implement health IT: (1) end users should be involved and (2) qualitative research methods such as interviews and observations of with end users should be used to fully understand the different activities, processes, and flow of information.

Health IT implementation, including telemedicine implementation, is not a process of simply installing and using a new technology. On the contrary, it is an iterative process because it is not possible to foresee all of the consequences, and it is sometimes difficult to distinguish different phases such as analysis, design, implementation, and evaluation because of the many feedback loops involved (Berg, 1999).

Acknowledgments

Partial support for this chapter was provided by a grant from the National Science Foundation [NSF 08-550: Virtual Organizations as Sociotechnical Systems (VOSS), PI: Pascale Carayon], and a contract with the Agency for Healthcare Research and Quality (AHRQ), Rockville, MD (Contract No. HHSA 290-2008-10036C, Pascale Carayon and Ben-Tzion Karsh, PIs).

References

Aarts, J. and Koppel, R. (2009). Implementation of computerized physician order entry in seven countries. *Health Affairs, 28*(2), 404–414.

Ahring, K. K., Ahring, J. P., Joyce, C., and Farid, N. R. (1992). Telephone modem access improves diabetes control in those with insulin-requiring diabetes. *Diabetes Care, 15,* 971–975.

Allen, A. and Perednia, D. A. (1996). Telemedicine and the healthcare executive. *Telemedicine Today, 1996, Winter (Special Issue),* 4–34.

American Telemedicine Association. (1999). *Telemedicine: A Brief Overview.* Paper presented at the Congressional Telehealth Briefing, from www.atmeda.org/news/overview.html

Ash, J. S., Gorman, P. N., Seshadri, V., and Hersh, W. R. (2004). Computerized physician order entry in U.S. hospitals: Results of a 2002 survey. *Journal of the American Medical Informatics Association, 11*(2), 95–99.

Attarian, D. E. (2007). Medical liability in cyberspace. *American Academy of Orthopaedic Surgeons (AAOS) Bulletin.* Retrieved January 13, 2011, from http://www.aaos.org/news/bulletin/oct07/managing2.asp

Balas, E. A., Jaffrey, F., Kuperman, G. J., Boren, S. A., Brown, G. D., Pinciroli, F. et al. (1997). Electronic communication with patients: Evaluation of distance medicine technology. *JAMA, 278*(2), 152–159.

Baroudi, J. J., Olson, M. H., and Ives, B. (1986). An empirical study on the impact of user involvement on system usage and information satisfaction. *Communications of the ACM, 29*(3), 232–238.

Bates, D. W. (2006). Invited commentary: The road to implementation of the electronic health record. *Proceedings (Baylor University Medical Center), 19*(4), 311–312.

Benedict, S. (2001). Legal ethical and risk issues in telemedicine. *Computer Methods and Programs in Biomedicine, 64*(3), 225–233.

Benschoter, R. (1971). CCTV-pioneering Nebraska medical centre. *Educational Broadcasting* (October), 1–3.

Berenson, R. A., Grossman, J. M., and November, E. A. (2009). Does telemonitoring of patients—The eICU—Improve intensive care? *Health Affairs Web Exclusive, 28*(5), w937–w947.

Berg, M. (1999). Patient care information systems and health care work: A sociotechnical approach. *International Journal of Medical Informatics, 55*(2), 87–101.

Bergmo, T. (2009). Can economic evaluation in telemedicine be trusted? A systematic review of the literature. *Cost Effectiveness and Resource Allocation, 7*(1), 18.

Breslow, M., Rosenfeld, B., Doerfler, M., Burke, G., Yates, G., Stone, D. et al. (2004). Effect of a multiple-site intensive care unit telemedicine program on clinical and economic outcomes: an alternative paradigm for intensivist staffing. *Critical Care Medicine, 32,* 31–38.

Broens, T. H., Huis In't Veld, R. M., Vollenbroek-Hutten, M. M., Hermens, H. J., van Halteren, A. T., and Nieuwenhuis, L. J. (2007). Determinants of successful telemedicine implementations: A literature study. *Journal of Telemedicine and Telecare, 13,* 303–309.

Campbell, J. D., Harris, K. D., and Hodge, R. (2001). Introducing telemedicine technology to rural physicians and settings. *The Journal of Family Practice, 50*(5), 419–424.

Carayon, P., Karsh, B.-T., Cartmill, R. S., Hoonakker, P. L. T., Hundt, A. S., Kruger, D. J. et al. (2010). *Incorporating Health Information Technology into Workflow Redesign. Summary Report.* (No. contract HHSA 290-2008-10036C. AHRQ Publication 10-0098-EF). Rockville, MD: Agency for Healthcare Research and Quality.

Chumbler, N., Mann, W., Wu, S., Schmid, A., and Kobb, R. (2004). The association of home-telehealth use and care coordination with improvement of functional and cognitive functioning in frail elderly men. *Telemedicine Journal & e-Health, 10*(2), 129–137.

Collen, M. F. (1995). *A History of Medical Informatics in the United States 1950 to 1990*. Bethseda, MD: American Medical Informatics Association.

Connolly, C. (2005, March 21). Cedars-Sinai doctors cling to pen and paper. *Washington Post* P. A01.

Currell, R., Urquhart, C., Wainwright, P., and Lewis, R. (2000). Telemedicine versus face to face patient care: Effects on professional practice and health care outcomes. *Cochrane Database of Systematic Reviews,* (2). Retrieved from http://www.mrw.interscience.wiley.com/cochrane/clsysrev/articles/CD002098/frame.html, doi:10.1002/14651858. CD002098

Darkins, A. (1996). The management of clinical risk in telemedicine applications. *Journal of Telemedicine and Telecare, 2*(4), 179–184.

Delbanco, S. (2006). Usage of CPOE steadily increasing, Leapfrog says: But top exec is disappointed with rate of adoption. *HealthCare Benchmarks and Quality Improvement, 13*(3), 33–34.

Dixon, B. E., Hook, J. M., and McGowan, J. J. (2008). *Using Telehealth to Improve Quality and Safety: Findings from the AHRQ Portfolio (Prepared by the AHRQ National Resource Center for Health IT under Contract No. 290-04-0016)* (No. AHRQ Publication No. 09-0012-EF). Rockville, MD: Agency for Healthcare Research and Quality.

Doarn, C. R. (2008). The last challenges and barriers of the development of telemedicine programs. In R. Latifi (Ed.), *Current Principles and Practices of Telemedicine and e-Health* (pp. 45–54). Amsterdam, the Netherlands: IOS Press.

Edelstein, S. A. (1999). Careful telemedicine planning limits costly liability exposure. *Healthcare Financial Management, 53*(12), 63–66.

Einthoven, W. (1906). Le telecardiogramme. *Archives Internationales Physiologie, 4,* 132.

Emery, F. E., and Trist, E. L. (1960). Socio-technical systems. In C. H. Chirchman and M. Verhulst (Eds.), *Management Science, Models En Techniques* (Vol. 2, pp. 83–97). New York: Pergamon.

Emery, F. E., and Trist, E. L. (1965). The causal texture of organizational environments. *Human Relations, 18,* 21–31.

Fishbein, M., and Ajzen, I. (1975). *Belief, Attitude, Intention and Behavior: An Introduction to Theory and Research*. Reading, MA: Addison-Wesley.

Foote, D. R. (1977). Satellite communication for rural health in Alaska. *Journal of Communication, 27*(4), 173–182.

Fraenkel, D. J., Cowie, M., and Daley, R. (2003). Quality benefits of an intensive care clinical information system. *Critical Care Medicine, 31*(1), 120–125.

Friedman, R. H., Kazis, L. E., Jette, A., Smith, M. B., Stollerman, J., Torgerson, J. et al. (1996). A telecommunications system for monitoring and counseling patients with hypertension. Impact on medication adherence and blood pressure control. *American Journal of Hypertension, 9*(4 Pt 1), 285.

Graschew, G., Roelofs, T. A., Rakowsky, S., and Schlag, P. M. (2008). Network design for telemedicine: E-Health using satellite technology. In R. Latifi (Ed.), *Current Principles and Practices of Telemedicine and e-Health* (pp. 67–82). Amsterdam, the Netherlands: IOS Press.

Groves, R. H., Holocomb, B. W., and Smith, M. L. (2008). Intensive care telemedicine: Evaluating a model for proactive remote monitoring and intervention in the critical care setting. In R. Latifi (Ed.), *Current Principles and Practices of telemedicine and e-Health* (pp. 131–146). Amsterdam, the Netherlands: IOS Press.

Hakansson, S., and Gavelin, C. (2000). What do we really know about the cost-effectiveness of telemedicine? *Journal of Telemedicine and Telecare, 6*(1), S133–S136.

Harnett, B. (2008). Creating telehealth networks from existing infrastructures. In R. Latifi (Ed.), *Current Principles and Practices of Telemedicine and e-Health* (pp. 55–65). Amsterdam, the Netherlands: IOS Press.

Hoonakker, P. L. T., McGuire, K., and Carayon, P. (2010). Sociotechnical issues of tele-ICU technology. In D. M. Haftor and A. Mirijamdotter (Eds.), *Information and Communication Technologies, Society and Human Beings: Theory and Framework* (pp. 225–240). Hershey, PA: IGI Global.

House, A. M., and Roberts, J. M. (1977). Telemedicine in Canada. *Canadian Medical Association Journal, 117*(4), 386–388.

Jaatinen, P. T., Aarnio, P., Remes, J., Hannukainen, J., and Koymari-Seilonen, T. (2002). Teleconsultation as a replacement for referral to an outpatient clinic. *Journal of Telemedicine and Telecare, 8*(2), 102–106.

Jain, A., Agarwal, R., Chawla, D., Paul, V., and Deorari, A. (2010). Tele-education vs classroom training of neonatal resuscitation: A randomized trial. *Journal of Perinatology, 30*(12), 773–779.

Jarvis-Selinger, S., Chan, E., Payne, R., Plohman, K., and Ho, K. (2008). Clinical telehealth across the disciplines: Lessons learned. *Telemedicine and e-Health, 14*(7), 720–725.

Jovell, A. J., and Navarro-Rubin, M. D. (1995). Evaluation de la evidencia cientifica. *Medicina Clinica (Barcelona), 105,* 740–743.

Kedar, I., Ternullo, J. L., Weinrib, C. E., Kelleher, K. M., Brandling-Bennett, H., and Kvedar, J. C. (2003). Internet based consultations to transfer knowledge for patients requiring specialised care: Retrospective case review. *BMJ, 326*(7329), 696–699.

Kirkley, D., and Stein, M. (2004). Nurses and clinical technology: Sources of resistance and strategies for acceptance. *Nursing Economics, 22*(4), 216–222.

Kirschner, P. A., and Van Bruggen, J. (2004). Learning and understanding in virtual teams. *Cyberpsychology & Behaviour, 7*(2), 135–139.

Levitt, J. I. (1994). Why physicians continue to reject the computerized medical record. *Minnesota Medicine, 77,* 17–21.

Loane, M., Bloomer, S., Corbett, R., Eedy, D., Hicks, N., Lotery, H. et al. (2000). A randomized controlled trial to assess the clinical effectiveness of both realtime and store-and-forward teledermatology compared with conventional care. *Journal of Telemedicine and Telecare, 6,* S1–S3.

Logan, A. G., McIsaac, W. J., Tisler, A., Irvine, M. J., Saunders, A., Dunai, A. et al. (2007). Mobile phone-based remote patient monitoring system for management of hypertension in diabetic patients. *American Journal of Hypertension, 20*(9), 942–948.

Majchrzak, A., Malhotra, A., Stamps, J., and Lipnack, J. (2004). Can absence make a team grow stronger? *Harvard Business Review, 82*(5), 131–137, 152.

Marescaux, J., Leroy, J., Gagner, M., Rubino, F., Mutter, D., Vix, M. et al. (2001). Transatlantic Robot-Assisted Telesurgery. *Nature, 413,* 379–380.

Miyasaka, K., Suzuki, Y., Sakai, H., and Kondo, Y. (1997). Interactive communication in high-technology home care: Videophones for pediatric ventilatory care. *Pediatrics, 99*(1), E11–E16.

Navein, J. (1998). Physician resistance to telemedicine. *Journal of Telemedicine and Telecare, 4*(1), 103–104.

Nesbitt, T. S., Hilty, D. M., Kuenneth, C. A., and Siefkin, A. (2000). Development of a telemedicine program: A review of 1,000 videoconferencing consultations. *Western Journal of Medicine, 173*(3), 169–174.

Norris, A. C. (2002). *Essentials of Telemedicine and Telecare.* New York: John Wiley & Sons.

Norton, S. A., Burdick, A. E., Phillips, C. M., and Berman, B. (1997). Teledermatology and underserved populations. *Archives of Dermatology, 133*(2), 197–200.

Oakley, A. M. M., Kerr, P., Duffill, M., Rademaker, M., Fleischl, P., and Bradford, N. (2000). Patient cost-benefits of realtime teledermatology-a comparison of data from Northern Ireland and New Zealand. *Journal of Telemedicine and Telecare, 6*(2), 97–101.

Perednia, D. A., and Allen, A. (1995). Telemedicine technology and clinical applications. *Journal of the American Medical Association, 273*(6), 483–488.

Roine, R., Ohinmaa, A., and Hailey, D. (2001). Assessing telemedicine: A systematic review of the literature. *Canadian Medical Association Journal, 165*(6), 765–771.

Rosenfeld, B., Dorman, T., Breslow, M., Pronovost, P., Jenckes, M., Zhang, N. et al. (2000). Intensive care unit telemedicine: alternative paradigm for providing continuous intensivist care. *Critical Care Medicine, 28,* 3925–3931.

Sibbald, B., McDonald, R., and Roland, M. (2007). Shifting care from hospitals to the community: A review of the evidence on quality and efficiency. *Journal of Health Services Research & Policy, 12*(2), 110–117.

Stanberry, B. A. (1998). *Legal and Ethical Aspects of telemedicine.* London, U.K.: Royal Society of Medicine.

Stanberry, B. A. (2000). Telemedicine: Barriers and opportunities in the 21st century. *Journal of Internal Medicine, 247*(6), 615–628.

Stanberry, B. A. (2006). Legal and ethical aspects of telemedicine. *Journal of Telemedicine and Telecare, 12*(4), 166–175.

Taylor, F. W. (1911). *Principles of Scientific Management.* New York: Harper & Brothers.

Terschuren, C., Fendrich, K., van den Berg, N., and Hoffmann, W. (2007). Implementing telemonitoring in the daily routine of a GP practice in a rural setting in northern Germany. *Journal of Telemedicine and Telecare, 13*(4), 197–201.

Thomas, E. J., Chu-Weininger, M. Y. L., Lucke, J., Wueste, L., Weavind, L., and Mazabob, J. (2007). The impact of a tele-ICU provider attitudes about teamwork and safety climate. *Critical Care Medicine, 35,* A145.

Thomas, E. J., Lucke, J., Wueste, L., Weavind, L., and Patel, B. (2009). Association of telemedicine for remote monitoring of intensive care patients with mortality, complications, and length of stay. *JAMA, 302*(24), 2671–2678.

Trist, E. L., and Bamforth, K. (1951). Some social and psychological consequences of the long-wall method of coal getting. *Human Relations, 4,* 3–39.

Watson, D. S. (1989). Telemedicine. *Medical Journal of Australia, 151*(2), 62–66.

Whited, J., Hall, R., Foy, M., Marbrey, L., Grambow, S., Dudley, T. et al. (2002). Teledermatology's impact on time to intervention among referrals to a dermatology consult service. *Telemedicine Journal and e-Health, 8*(3), 313–321.

Woods, K. F., Johnson, J. A., Kutlar, A., Daitch, L., and Stachura, M. E. (2000). Sickle cell disease telemedicine network for rural outreach. *Journal of Telemedicine and Telecare, 6*(5), 285–290.

Wootton, R., and Craig, J. (1999). *Introduction to Telemedicine.* London, U.K.: Royal Society of Medicine.

Wyatt, J. C. (1994). Clinical data systems, part 3: Development and evaluation. *Lancet, 344,* 1682–1688.

Zawada, E. T. J., Herr, P., Larson, D., Fromm, R., Kapaska, D., and Erickson, D. (2009). Impact of an intensive care unit telemedicine program on a rural health care system. *Postgraduate Medicine, 121*(3), 160-170.

Zollo, S. A., Kienzle, M. G., Henshaw, Z., Crist, L. G., and Wakefield, D. S. (1999). Tele-education in a telemedicine environment: Implications for rural health care and academic medical centers. *Journal of Medical Systems, 23(3),* 107–122.

21

Using Health IT to Improve Health Care and Patient Safety

James M. Walker

CONTENTS

21.1 Introduction 305
21.2 Minimizing Risk and Maximizing Benefit (Safety and Effectiveness) 307
    21.2.1 Software-Safety Principles 307
21.3 Levels of Approach 310
    21.3.1 Societal 310
    21.3.2 Health IT Manufacturers 312
    21.3.3 Care-Delivery Organizations 313
        21.3.3.1 CDO Clinicians 314
        21.3.3.2 CDO Informatics Team 314
        21.3.3.3 CDO Information Technology 316
    21.3.4 Individual 316
    21.3.5 Non-CDO HIT Service Providers 316
21.4 Conclusion 317
Acknowledgment 317
References 317

> The major difference between a thing that might go wrong and a thing that cannot possibly go wrong is that when a thing that cannot possibly go wrong goes wrong it usually turns out to be impossible to get at or repair.
>
> **Douglas Adams,** *Mostly Harmless*

21.1 Introduction

As is the case with any healthcare intervention, the safety and effectiveness of health IT (HIT) need to be designed in, confirmed before sale, implemented carefully, and improved continuously during and after implementation. According to the Institute of Medicine, care-delivery organizations (CDOs) "... should expect any new technology to introduce new sources of error..." (Donaldson 2000). The Committee on Certifiably Dependable Software Systems of the National Academies "subscribes to the view that software is 'guilty until proven innocent,'" noting that "... [*serious*] accidents have already occurred, and, without intervention, the increasingly pervasive use of

software—especially in arenas such as transportation, health care, and the broader infrastructure—may make them more frequent and more serious" (Jackson et al. 2007).

Because the use of HIT entails a particularly complex set of policy and technology innovations—affecting job definitions, working conditions, communications, workflows, and job security—a systematic approach to HIT safety is especially important (Carayon and Karsh 2000; Eason 2001; Kushniruk et al. 2006). This chapter reviews HIT safety in the context of systems safety and software safety and suggests both strategies and tactics for increasing the safety of HIT-supported care.

Definitions: Because HIT and the healthcare transformation it can support are in the early stages of development, clarity about the way this chapter uses common terms is essential:

- Accident: A term limited by the fact that it reflects a global judgment—usually difficult to substantiate—regarding the intent and performance of the many actors involved (directly and indirectly). *Events that are referred to as "accidents" are often associated with suboptimal safety practices.*

- Adverse effect: Any care-process compromise, with or without patient harm. (*The frequently used "unintended consequence" and "unanticipated consequence" introduce questions of intent and thoroughness, which are difficult or impossible to answer, rendering them too imprecise to be useful: Adverse effects of HIT use must be intended rarely if ever; "unanticipated" begs the question regarding how much skill and effort were expended in trying to anticipate and prevent the adverse effect.*) *Compromised care processes are adverse effects in that they require clinicians to identify the compromise and return the process to normal, often straining their cognitive, physical, and organizational resources. This increases the risk of patient (and clinician) harm. Another reason for treating care-process compromises as adverse events is that it is often difficult (and may be impossible) to confirm definitively that no patient harm occurred.*

- Care-process compromise: Any abnormal care-process event that has the potential to contribute to patient harm. *This construct is equivalent to Weinger's "nonroutine event," but is intended to express more clearly the "dysfunctional" and "potentially dangerous" character of nonroutine events that Weinger notes (Weinger and Slagle 2002). (One of the limitations of the frequently used phrase "near miss" is that it is commonly applied to both care-process compromise that is corrected before it harms a patient and to serious patient harm.)* (Masand 2010; PDR Secure 2010).

- Consumer: Any person who uses or anticipates using healthcare services or products.

- Electronic Health Record (EHR): HIT used by clinicians (including, e.g., decision support and bar-code-supported medication administration).

- Error (HIT-related): (1) The creation of an HIT-related hazard by an HIT manufacturer, HIT implementer, or the designer of a non-HIT healthcare system; (2) an HIT user's failure to cope adequately with an HIT-related hazard (a "forced error"); or (3) an HIT user's failure to use optimally designed and implemented HIT appropriately (an "unforced error") (Jackson et al. 2007; Vicente 1999). *In the HIT safety literature, error is attributed almost exclusively to the users of health IT—to the exclusion of HIT manufacturers and implementers (Cook et al. 2000). As Norman observes, "People do err, but primarily because they are asked to perform unnatural acts: to do detailed arithmetic calculations, to remember details of some lengthy sequence or statement, or to perform precise repetitions of actions, all the result of the artificial nature of invented artifacts"* (Norman 1998).

- HIT hazard: Any aspect of HIT or its interactions with other healthcare systems that increase the risk of care-process compromise and patient harm. *This reflects the reality that a hazard may arise due to the interface between two HIT (or other) systems, each of which is functioning appropriately (Harrison 2009). (The construct, "hazard" largely overlaps with Shappell and Wiegmann's "precondition for unsafe act"; "hazard" is used here because of the central role of hazard control in safety engineering.)* (Lehto and Clark 1990; Wiegmann and Shappell 2003).

- Health-information exchange (HIE): HIT that enables clinicians and patients to access information from other members of the healthcare team that work for multiple companies.

- HIT: Any electronic information system used by a consumer or by any member of a patient's healthcare team (including the patient herself) to support patient care (e.g., EHRs, networked and freestanding PHRs, and HIE).

- HIT manufacturer: Any organization that substantially designs and codes HIT software, either for internal use or for sale. *This definition excludes CDOs that do limited custom coding of commercial software—although such organizations do incur the responsibility to perform integrated testing that includes any custom coding.*

- HIT safety: The extent to which HIT use improves or compromises patient-care processes and patient outcomes. *This is highly context-specific, depending on the organizational context, the specific HIT applications used, and the quality of the HIT implementation.*

- Healthcare informatics: The art and science of meeting the information needs of the healthcare team.

- Healthcare team: Every person who provides care to patients. *The team includes but is not limited to patients, caregivers, physicians, nurses, case managers, clerks, pharmacists, long-term-care providers, home health nurses, consumers, and information providers (such as the National Library of Medicine).*

- Human factors engineering: The design of care processes and HIT to fit organizational, team, and individual capabilities and limitations.

- Information security: The maintenance of data integrity, data availability, and patient confidentiality (i.e., control of who accesses patient data) (Harrison 2009).

- Patient: Any individual who employs a physician-led team to support their healthcare.

- Patient harm: Physical, psychological, financial, or reputational.
- Patient healthcare record (PHR): HIT used by patients and their caregivers and other healthcare consumers.
- Process: A set of activities (more or less structured) that produce a specific service or product. *Most patient-care processes involve multiple actors and multiple workflows.*
- Quality and safety: Although safety is often referred to separately from quality to emphasize its importance, safety is a core constituent of healthcare quality (Committee on Quality of Health Care in America, 2001; McDonough 1999).
- Software dependability: A context-specific construct that combines effectiveness and safety. *"A system is dependable when it can be depended on to produce the consequences for which it was designed, and no adverse effects, in its intended environment"* (Jackson et al. 2007).
- Workflow: The work of an individual or team represented as a sequence of tasks.

21.2 Minimizing Risk and Maximizing Benefit (Safety and Effectiveness)

While the elimination of adverse effects is an appropriate goal of safety engineering, it is often not possible (Harrison 2009). Furthermore, the environment in which an intervention is used conditions the level of risk that is acceptable. For example, it is widely accepted that screening healthy populations for health risks should only be undertaken when the benefits have been shown to be substantial and the risks negligible in clinical trials (Sackett et al. 1991). On the other hand, if a patient is gravely ill in an ICU without other options, the use of a risky intervention may be life-saving. As the eminent safety engineer, Nancy Leveson notes, "the healthcare system is extremely complex, and the approach to risk is very different [than in other fields]: it is risky to treat someone with a drug, but it is also risky to not treat someone" (Leveson 2008b).

One appropriate measure of HIT safety might be that the benefits of implementing HIT are greater than the risks it introduces. For example, in the case of drug safety, Paul Seligman, Associate Director of Safety Policy and Communication in the Center for Drug Evaluation and Research (CDER) of the FDA, holds that "FDA's job is to balance a drug's benefits against the risks for the drug's intended use. The agency approves a

drug only after determining that the benefits outweigh the risks for most people, most of the time" (Seligman 2007). Expressed as the ratio of the number of patients needed to treat to achieve one positive outcome (NNT) to the number of patients needed to treat to produce one (roughly equivalent) instance of harm (NNH), the FDA's standard for drug safety can be stated as NNT/NNH < 1/1 (McQuay and Moore 1997; Thompson et al. 2003). While this might constitute a reasonable minimum when a patient risks death or permanent injury if not treated, the goal for most interventions—including HIT—must be to minimize the ratio, achieving an NNT/NNH ratio of no more than 1/10 and ideally less than 1/100, although no such goal or standard has been promulgated. This absence is due, in part, to a lack of rigorous studies that would establish the NNT and NNH of HIT, although both encouraging and disturbing preliminary studies have been published (Black et al. 2011; Bonnabry et al. 2008). In one implementation, patients were harmed overall, with a number needed to treat to cause 1 death (NNH) of 27 (Han et al. 2005). However, this implementation included so many deviations from implementation best practices that the results are hard to generalize—beyond emphasizing the potential for poorly implemented HIT to compromise patient safety (Del Beccaro et al. 2006; Longhurst et al. 2006).

It is worth noting that this absence of evidence regarding benefit and risk is not unique to HIT: "The field of software engineering suffers from a pervasive lack of evidence about the incidence and severity of software failures; about the dependability of existing software systems; about the efficacy of existing and proposed development methods; about the benefits of certification schemes; and so on..." (Jackson et al. 2007).

21.2.1 Software-Safety Principles

Despite limited data on HIT safety specifically, much is known about software safety in general (Canada Infoway et al. 2007). This section discusses principles that are important from the shared perspective of society, HIT manufacturers, CDOs, and patients and consumers. Principles that are pertinent primarily from a single perspective are discussed in the following relevant section.

1. *Safety is a characteristic of whole systems*: While it may not be accurate to say that HIT is no safer than the most unreliable of its many components and interactions with other healthcare systems, it is true that "Safety in health care depends more on dynamic harmony among actors than on reaching an optimum level of excellence at each separate organizational level" (Amalberti et al. 2005). Achieving such a dynamic harmony is particularly difficult due

to the complexity of healthcare. As management consultant Peter Drucker observed, "Even small health care institutions are complex, barely manageable places... large health care institutions may be the most complex organizations in human history" (Drucker 1989). Characteristics that healthcare shares with other complex, open systems include the large number of actors and their multiple perspectives, (Kohn et al. 2000) ambiguity, distributed decision-making, conflicting goals, time pressures, frequent interruptions (Carayon 2007; Coiera 2000), multiple connections among sub-systems, rapid change, the potential for catastrophic failure, and multiple forms of process automation (Vicente 1999). Organizing this complexity into care processes and supporting HIT that maximize safety and help the healthcare team respond effectively to unexpected demands is a daunting task (Hollnagel et al. 2006). For example, an integrated delivery network may spend 40,000 person-hours on integrated testing of its care processes, HIT (including EHR, networked PHR, and HIE) and scores of interfaced information systems before each major software upgrade (Adams 2009).

A corollary of this principle is that safety is not an *intrinsic* property of any HIT system. "A software component that may be dependable in the context of one system might not be dependable in the context of another" (Jackson et al. 2007). For this reason, specification of the work domain (or environment) in which the HIT will be used must inform the manufacture, certification, and implementation of HIT. One of the most powerful of these contextual determinants of HIT safety is a given CDO's focus on the possibility of failure and its systematic efforts to avoid failure: "They [*safety-conscious organizations*] continually rehearse familiar scenarios of failure and strive hard to imagine novel ones. Instead of isolating failures, they generalize them. Instead of making local repairs, they look for system reforms" (Reason 2000).

2. *Most adverse effects "are not the result of unknown scientific principles* but rather of a failure to apply well-known, standard engineering practices" (Leveson 1995). For example, such standard practices as isolating network routers or providing redundant power supplies are expensive and easy to exclude from already-large IT budgets. Of course, even when good safety-engineering principles have been followed, there will be some hazards that were not identified or

could not be fully controlled; they will have the potential to contribute to care-process compromise and patient harm.

3. *Proactive hazard control throughout the HIT life-cycle* is critical to safety (Battles 2007). As Leveson puts it, "Hazard analysis is accident analysis before the accident happens" (Leveson 2008a). To be effective and affordable, this hazard control must be initiated at the beginning of the HIT life-cycle (i.e., during the manufacturer's needs-assessment and design phases) (Harrison 2009; Jackson et al. 2007; McDonough 1999). As Figure 21.1 illustrates, the proactive identification and control of hazards arising from errors in HIT design and from the interaction of HIT with other healthcare systems prevent those hazards from ever stressing clinicians or harming patients. In this way, proactive hazard control achieves substantially greater safety benefits than the retrospective analysis of care-process compromises with or without patient harm (i.e., safety incidents and accidents). Of course, eliminating all hazards is probably not feasible. The remaining hazards will still have the potential to contribute to "forced" HIT-use errors, and there will also be "unforced" HIT-use errors, which are due solely (as far as can be determined) to user factors such as fatigue and lack of training. Finally, various forms of hazard control produce variable safety benefits: "In general, the accepted hierarchy of control from most to least effective is: (1) elimination of hazards, (2) containment of hazards, (3) containment of people, (4) training of people, and (5) warning of people" (Lehto and Clark 1990).

4. *Retrospective hazard analysis*: When hazards elude proactive hazard control efforts, retrospective hazard analysis (e.g., root-cause analysis) has the potential to identify previously unknown hazards, enabling their inclusion in future hazard-control efforts, as the right side of Figure 21.1 illustrates (Pronovost et al. 2009).

5. *Simplicity of systems and software*: An often-unrecognized adverse effect of the flexibility of software is "the ease with which partial success is attained, often at the expense of unmanaged complexity.... Attempting to get a poorly designed, but partially successful, program to work all of the time is usually futile; once a program's complexity has become unmanageable, each change is as likely to hurt as to help. Each new feature may interfere with several old features, and each attempt to fix an error may

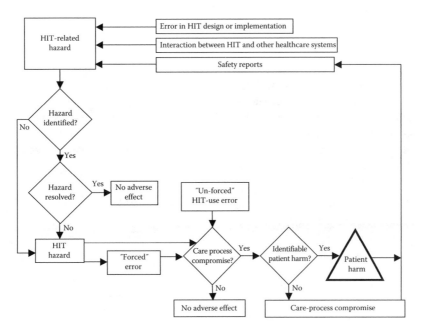

FIGURE 21.1
Proactive and retrospective hazard control.

create several more" (Leveson 1995). Hence the safety-engineering dictum, "This year's safety fix becomes next year's root cause." Because of the difficulty of making complex software systems safe, the Committee on Certifiably Dependable Software Systems of the National Academies concluded that, "one key to achieving dependability at reasonable cost is a serious and sustained commitment to simplicity, including simplicity of critical functions and simplicity in system interactions…. To achieve high levels of dependability in the foreseeable future, striving for simplicity is likely to be by far the most cost-effective of all interventions" (Jackson et al. 2007).

6. *Automate cautiously*: Because of its perceived potential to overcome perceived manager and clinician inertia, HIT-driven automation can seem (to both administrators and clinicians) to be a panacea. However, automation needs particularly careful hazard control (Strom et al. 2010). As Donaldson notes in *To Err is Human*, "Health care professionals… should adopt the custom of automating cautiously, always alert to the possibility of unintended harm…" (Donaldson 2000). An aspect of automation that deserves particular attention is the balance of automation and flexibility that is appropriate to each specific workflow in a care process (Bloomrosen et al. 2011; Walker and Carayon 2009). The following stratification of levels of automation is adapted from Parasuraman: (Parasuraman et al. 2000)

a. The automated system initiates an intervention—

   i. Without notifying a human.

   ii. And notifies a human.

   iii. Allowing a human a restricted time to veto it.

b. The automated system initiates an intervention after a human's approval.

c. The automated system suggests one action among many options for human approval.

d. The automated system narrows the options for action to a few.

e. The automated system offers a full set of reasonable options.

f. The automated system presents data clearly, but offers no other assistance.

Particularly levels a, b, and c introduce the risk that "The shift from performing tasks to monitoring automated systems can actually inhibit one's ability to detect critical signals or warning conditions" (Parasuraman and Mouloua 1996). This inhibition is problematic because "there will always be a set of conditions under which the automation will reach an incorrect decision" (Vicente 1999).

7. *Software testing is necessary, but not sufficient* to establish HIT safety. In addition to testing, "A dependability case will generally require many forms of analysis, including (1) the validation of environmental assumptions, use models, and

fault models; (2) the analysis of fault tolerance measures against fault models; (3) schedulability analysis for temporal behaviors; (4) security analysis against attack models; (5) verification of code against module specifications; and (6) checking that modules in aggregate achieve appropriate system-level effects. Indeed, the dependability case for even a relatively simple system will usually require all of these kinds of analysis, and they will need to be fitted together into a coherent whole" (Jackson et al. 2007).

8. *Certification of HIT must reflect best safety practices*: Across all sectors, existing certification schemes "have a mixed record. Some are largely ineffective, and some are counterproductive.... Few, if any, existing certification schemes encompass the combination of characteristics recommended in this report ("Software for Dependable Systems: Sufficient Evidence?")—namely, explicit dependability claims, evidence for those claims, and a rigorous argument that demonstrates that the evidence is sufficient to establish the validity of the claims" (Jackson et al. 2007).

9. *Provide task-focused information*: HIT has the widely recognized potential to overload clinicians with information. As Nobel laureate economist Herbert Simon famously observed, "What information consumes is rather obvious: it consumes the attention of its recipients. Hence a wealth of information creates a poverty of attention, and a need to allocate that attention efficiently among the overabundance of information sources that might consume it" (Simon 1971). One result of information overload is that "When people are overloaded with information, they tend to do nothing until they have figured out what is going on..." (Leveson 1995). For this reason, Simon asserts that "The proper aim of a management information system is not to bring the manager all the information he needs, but to reorganize the manager's environment of information so as to reduce the amount of time he must devote to receiving it" (Simon 1971). If HIT is to help the patient's healthcare team work more safely and efficiently, it must be designed to provide selected, task-focused information at the right time and in the right format to meet the needs of each member of the team. Of course, the differentiation of information into what should be presented initially and what should be available a click (or two) away is a demanding discipline, requiring a deep understanding of information theory and of the needs, capabilities, and workflows of each member of the

healthcare team (Vicente and Rasmussen 1990). The widespread failure to practice this discipline is likely to be an important contributor to the ineffectiveness (or outright failure) of many HIT implementations. Since much HIT continues to be conceived and developed with the goal of providing exhaustive information, this principle needs particular attention from HIT researchers, manufacturers, and implementers.

10. *Normalization of deviance*: Finally, even safe systems can contribute to care-process compromise and patient harm if they are used (by organizations and by individuals, consciously and unconsciously) to justify relaxing standard safety practices, so-called normalization of deviance (Vaughan 2005). One critical contribution of a strong organizational safety culture must be to help administrators and clinicians avoid the temptation of decreasing human vigilance in response to the increasing process-management capabilities of HIT systems (cf. Principle 6).

21.3 Levels of Approach

According to Software-Safety Principle 1, HIT safety depends critically on the concerted efforts of actors at multiple organizational levels: at least societal, HIT manufacturers, CDOs, and individuals (Carayon 2006; Moray 1994; Rasmussen 1997). Currently, the most authoritative American set of recommendations relevant to HIT safety at any of these levels is the report of the Committee on Certifiably Dependable Software Systems of the National Academies, which addresses the first three levels (Jackson et al. 2007; See also the British standards for HIT manufacture and implementation; Harrison 2010.)

21.3.1 Societal

This level includes the public and its affinity groups, policy makers, regulators, CDOs, professional societies, researchers, HIT manufacturers, educators, employers, and other payers. This is the level at which the need for safety and the cost of achieving safety are balanced—often implicitly—with those balances sometimes promulgated as explicit policies. To optimize this balancing, researchers and policy makers will need to educate themselves and the public regarding the potential benefits and risks of HIT—as manufactured and as implemented—along with methods for maximizing the benefits and minimizing the risks (Grassley 2010). Policy makers and the public will need to achieve consensus on appropriate, understandable balances between, for example, the benefits of information availability and the risks of loss of confidentiality.

Standards for the following requirements for HIT safety are probably most effectively addressed at the societal level:

1. *Data security (including patient and consumer confidentiality and patient identification across venues of care)*: Work on policies and technical standards for data security (including confidentiality) is underway (HIT Policy Committee PCAST Report Workgroup 2011; HIT Policy Committee Privacy and Security Tiger Team 2010; President's Council of Advisors on Science and Technology 2010).

2. *Accountability for HIT safety*: Allocating responsibility for HIT safety is an urgent need. The Committee on Certifiably Dependable Software Systems asserts that "it should always be explicit who is accountable for any failure to achieve it [software dependability]. Such accountability can be made explicit in the purchase contract, or as part of certification of the software, or as part of a professional licensing scheme, or in other ways…. Clearly, no software should be considered dependable if it is supplied with a disclaimer that withholds the manufacturer's commitment to provide a warranty or other remedies for software that fails to meet its dependability claims" (Goodman 2010; Jackson et al. 2007).

3. *Certification of HIT safety (as manufactured and as implemented)*: While considerable progress in testing the safety of EHRs *as implemented* has been made, HIT safety needs to be included in certification testing (Classen et al. 2010). In addition, the role of dependability analysis in the certification of HIT requires careful consideration. (See Software-Safety Principle 3, mentioned earlier, and HIT Manufacturers, in the following; Jackson et al. 2007.)

4. *Systematic reporting of HIT-associated hazards and care-process compromises*: (Jackson et al. 2007; Kohn et al. 2000) Despite preliminary studies (Ash et al. 2006, 2007a and b, 2009; Campbell et al. 2007; Campbell and Sittig 2006; Sittig and Singh 2009), the lack of a standard categorization (and supporting ontology) for HIT-related hazards (including safety incidents), the lack of a usable and useful tool for managing and reporting hazards, and the concerns of CDOs and HIT manufacturers about confidentiality have thus far hobbled the sharing and further study of the thousands of hazards that safety-conscious

CDOs have identified and mitigated (some of which have contributed to care-process compromise and patient harm; Bates 2005; Goodman et al. 2010). To meet this need, the author has developed and is testing with Andrea Hassol (under Agency for Healthcare Quality contract # HHSA2902006000011i) the HIT Hazard Manager™, a tool designed to enable HIT manufacturers and CDOs to characterize hazards consistently, manage hazards proactively (and retrospectively), and share comparable information about hazards confidentially throughout the healthcare sector—all under the protection of a Patient-Safety Organization (PSO). The Hazard Manager relates to similar tools as outlined in the Table 21.1 (Harrison 2010; PDR Secure 2010).

To benefit from full participation of manufacturers, implementers, and users, such a reporting system may need regulatory support, potentially in the form of a Joint Commission-like requirement that every HIT manufacturer and CDO has a formal plan for controlling and reporting hazards and adheres to it.

To what extent such standards should be promulgated and enforced through legislation, regulation, the purchasing policies of government agencies, governmental persuasion, private–public partnerships, and/or the play of market forces is beyond the scope of this chapter (Hoffman and Podgurski 2010).

As with standards, many HIT-related research topics can be addressed most cost-effectively at the societal level. Among the most pressing questions is to determine how HIT—in various implementations—affects patient outcomes and the quality and efficiency of care processes in everyday practice, particularly in smaller clinics and hospitals. The AHRQ and Research and the Office of the National Coordinator for HIT (ONC) are funding research that addresses these questions in the context of HIEs and of the Beacon Communities (AHRQ 2008; ONCHIT 2010).

Widespread adoption of safe HIT practices will also depend on the results of research into the relative costs and benefits of various HIT-safety practices. Using this information, policy makers, certification bodies, regulators, and professional bodies could provide guidance regarding which practices are obligatory for their constituents and which are recommended. HIT manufacturers and implementers could then be required (either by regulation or by the market) to disclose their compliance with these evidence-based HIT safety practices to their existing and potential customers (Jackson et al. 2007).

Finally, public affinity groups such as AARP and Consumers' Union can exert a powerful influence on HIT safety by requiring policy makers, HIT manufacturers, and CDOs to report on their performance of their respective roles in supporting HIT safety.

TABLE 21.1

Characteristics of Tools for HIT Management and Reporting

	HIT Hazard Manager	DSCN14–2009	EHR Event
Proactive hazard control	Yes	Yes	Yes
Retrospective hazard control	Yes	Yes	Yes
Supports CDO hazard control?	Yes	No	No
Domain-specific hazard ontology?	Yes	No	No[a]
Integration with ARHQ common formats	In testing	n/a	No
CDO confidentiality	Yes	Probable	Yes
PSO protection	Yes	n/a	Yes
CDO access to hazard reports of fellow customers	Yes	No	No
HIT manufacturer confidentiality	Yes	Probable	?
HIT manufacturer access to its aggregated hazards	Yes	No	No
Aggregated view of all HIT hazards	Yes	No	Probable
Usability and usefulness validated	In testing	No	No

[a] Many choices include multiple answers. For example, "Inability to capture, save, view, retrieve, perceive important data."

21.3.2 Health IT Manufacturers

The most safety-critical decision health IT (HIT) manufacturers make—because the "largest class of problems arises" in the requirements-definition phase (Jackson et al. 2007)—is how much resource to spend on the complex and expensive tasks of analyzing (1) the multiple work domains of healthcare, (2) the cognitive demands of healthcare tasks, and (3) the information needs of the multiple teams and individual actors who will use the HIT—in order to determine the requirements relevant to the various types of HIT systems (Lopez et al. 2010; Kushniruk 2001; Vicente 1999). These analyses are critical, because one of the features "shared by large-scale accidents,... [is] that they were prompted by situations that were both unfamiliar to workers and that had not been anticipated by system designers..." (Vicente 1999).

A second safety-critical decision—because it addresses the "second major class of problems"—is how much resource to expend on user-centered design (Jackson et al. 2007). HIT informed by user-centered design makes a critical contribution to better and more efficient care by enabling the various members of the healthcare team to work more efficiently and with fewer errors (remembering that patients and their caregivers are increasingly HIT users; Software Engineering Institute 2011). Furthermore, usability encourages HIT adoption: the Technology Adoption Model's prediction that usability (along with usefulness) is a key determinant of technology adoption has been borne out by multiple studies (Venkatesh 1999). Additionally, substantial data suggest that user-centered design can decrease software development costs and increase sales (Bias and Mayhew 1994). Finally, effective user-centered design would also position manufacturers to be effective developers of much of the training that HIT users need to use

HIT safely and effectively—an activity that strains the resources of most CDOs (Goodman et al. 2010; Harrison 2009; Jackson et al. 2007).

Despite these considerations, HIT manufacturers have been prone to neglect usability. Some reasons for this include the fact that HIT manufacturers often do not have expertise in user-centered design, that HIT purchasers rarely have usability assessment as a core competency, (Kadry et al. 2010) and that working clinicians frequently are not given meaningful input into purchasing decisions. (Since users are typically unable to bring all of their needs to consciousness unaided, such approaches as domain analysis and cognitive task analysis are needed to supplement interviews, focus groups, and surveys of users.) (Hundt et al. 2005; McDonnell and Wendel 2010; Vicente 1999).

A key contributor to usability and to safety is system simplicity (See Principle 5.). This simplicity takes two primary forms: simplicity of system interactions and simplicity of safety-critical software functions. HIT manufacturers can achieve simplicity by using modularity, component isolation, and redundancy to increase component independence, "in which particular system-level properties are guaranteed by individual components... which can preserve these properties despite failures in the rest of the system" and to decrease interactive complexity (the interaction of HIT components in ways that are difficult to predict, even with extensive testing) (Jackson et al. 2007).

Another safety-critical question for manufacturers is how much resource to expend on hazard control and whether and how to communicate hazards that may affect implementers or users. Because hazard control is most effective when begun during the earliest phases of the HIT lifecycle—needs assessment, system design, and software coding—manufacturers are ideally situated to practice this method of improving HIT safety (Jackson et al. 2007). Almost all HIT manufacturers have a hazard-communication plan, and the EHR Association is developing a set of best practices for hazard communication and management (Dvorak 2011).

The Committee on Certifiably Dependable Software Systems of the National Academies recommends the following best practices for developing dependable software:

- Safe programming languages
- Systematic configuration management
- Static analysis of software code
- Hazard analyses and threat models
- Automated regression testing (Jackson et al. 2007)

Summing up all of these activities, manufacturers should develop and publish (at least to customers) a dependability case that defines the real-world effects (safety and effectiveness) their HIT is intended to achieve and the evidence (derived from both testing and analysis) that it dependably achieves them (Jackson et al. 2007).

21.3.3 Care-Delivery Organizations

Uniquely among the various actors, CDOs have a direct, professional responsibility to their patients to acquire, implement, and continuously improve safe and effective HIT. CDOs include those who manufacture, implement, or optimize HIT themselves and those who outsource any or all of those services. While outsourcing can both simplify and complicate the quest for HIT safety, it does not alter the CDO's core responsibility to provide safe and effective HIT to its healthcare teams. The first, critical step in outsourcing is to "... consider the nature of the service required carefully *before* seeking a supplier capable of producing it, by defining specifications" (Domberger 1999).

The first safety principle applied to CDOs asserts that "Organizational structures as well as human–machine interfaces may contribute to accidents..." (Perrow 1983). To the extent that a CDO is committed—as an organization—to improving its care-process performance and patient outcomes, safe and effective HIT will be seen as a strategic necessity and for that reason be more likely to be achieved. If such a commitment is absent or inconsistent, HIT is likely to become an organizational burden and even a safety liability (Berg 2001; Harrison et al. 2007).

Granted a basic commitment to improved care and patient outcomes, several specific organizational competencies are needed to support HIT safety:

- A systematic patient-safety program
- The ability to design safe and effective, patient-centered, interdisciplinary care processes; (Walker and Carayon 2009)

- Effective support (including training) for patient-centered healthcare teams (Salas et al. 2008)
- Human-factors engineering (Carayon 2007)
- Clinically astute informatics and IT teams that focus on healthcare-team needs rather than technology for its own sake
- Project management (the disciplined application of management practices to maximize project benefits and minimize risks) (Graham and Englund 1997)
- Information security (including an information-security office independent of the IT department that will develop and enforce policies to secure patient information, oversee HIT access, and respond to patient concerns)
- Delivery of effective training, based on adult learning theory
- The ability to manage high-quality care processes efficiently (Business Process Management)
- Continuing, integrated optimization of care processes and HIT

As the preceding list implies, budgeting adequately to achieve safe and effective HIT is a stringent challenge for any CDO. In addition to the considerable purchase price of HIT systems, ongoing maintenance costs average about twice the original purchase price. And nonsoftware project costs account for about 80% of the total direct costs of a typical HIT project. Given these high costs, the budget for HIT safety—including the user-centered design and ongoing optimization that contribute to safety—is often slighted. While 9% of total operating expenses is a typical overall IT budget in other information-intensive industries, 2%–3% has been typical of CDOs. As a rule of thumb, a CDO that intends to provide safe and effective HIT will need to budget at least 4% of total operating expenses for its overall IT budget.

Finally, selecting from among currently available HIT systems is a complex process that includes inevitable compromises with the ideal. The CDOs who are most successful in supporting improved care with HIT have typically chosen commercial HIT manufacturers who are cooperative partners in the continuing, integrated optimization of care processes, and the HIT systems needed to support them (Richards 2005c). Said differently, a CDO does well to think of itself as purchasing a manufacturer's organizational culture as least as much as it purchases software products. As the Committee for Dependable Software Systems notes, "the culture of an organization in which software is produced can have a dramatic effect on its quality and dependability." While many HIT customers may not have expertise in software design and safety engineering, all can assess the "meticulous attention to detail, high aversion to risk, and realistic assessment of software, staff, and process" that characterize HIT manufacturers with robust safety cultures (Jackson et al. 2007).

After purchase, CDOs can work as partners with the manufacturer to support HIT safety in many ways:

1. By helping the manufacturer understand healthcare work domains and healthcare-team needs. (This is true even in the case of established HIT systems—because so few healthcare work domains, care processes, and team needs have been carefully analyzed or incorporated into HIT systems.)

2. By learning the safety value of simplicity and helping the manufacturer distinguish between critical and optional functions.

3. By providing a test bed for user-centered HIT design.

4. By identifying and reporting potential HIT-related hazards to the manufacturer.

21.3.3.1 CDO Clinicians

Although clinicians may feel unprepared to judge HIT systems, their understanding of the various domains of clinical work and of their organizations' efforts to improve the quality and efficiency of care enable them to make irreplaceable contributions to HIT safety (Harrison 2009b). These contributions should begin with defining the healthcare team's information needs. This needs definition will enable the care-delivery organization (CDO) to distinguish among HIT manufacturers whose products are designed to meet those needs and products that reflect little understanding of those needs. Clinicians will also need to support adequate funding for the less visible but critical aspects of HIT safety, such as a fault-tolerant hardware infrastructure (in CDO Information Technology, in the following), process redesign, usability design and testing, and competency-based user training (pre- and postimplementation; Berg 2001). Many clinicians will need to participate for years in teams that lead the continuing integrated optimization of care-process and HIT. Finally, HIT safety requires the refinement and practice of traditional elements of healthcare professionalism such as communicating promptly and clearly with patients and with other clinicians, securing patient information, and protecting patient confidentiality (Brennan et al. 2002).

21.3.3.2 CDO Informatics Team

The following recommendations for improving HIT safety are derived from the Software-Safety Principles discussed earlier and published and unpublished best practices that have been used successfully in at least one care-delivery organization (Walker et al. 2005). Baker and Harrison have developed a protocol for HIT safety that has been implemented in the British National Health Service; its usability, usefulness, and cost-effectiveness have not been assessed (Baker 2011; Harrison 2010)

1. Analyze the organization's primary work domain(s) and most important care processes (An integrated-delivery network may identify about 100 primary processes. Less vertically integrated CDOs are likely to identify far fewer) (Vicente 1999).

2. Support clinical redesign of care processes and relevant clinical policies for quality and efficiency (Adams et al. 2005). Implementing redesigned processes and policies at least 6 months before HIT go-live will decrease the likelihood that the confusion caused by new processes and policies will decrease the safety of the critical initial phase of HIT use.

3. Identify the information each member of the patient's healthcare team needs to perform their part of each process.

4. Particularly, complex processes should undergo prospective risk assessment. Bonnabry's method for comparing the risks to patient safety before and after implementation of an order-entry (CPOE) system provides a particularly compelling, widely applicable example of the potential of prospective risk assessment to measure the safety effects of HIT proactively and to identify actionable opportunities for improvement (Bonnabry et al. 2008).

5. Support the selection of HIT systems beginning with an assessment of the HIT manufacturer's understanding of the relevant healthcare work domains and information needs, their safety-engineering practices, and the usability of their products (Richards 2005b; Software Engineering Institute 2011).

6. Minimize the number of nonintegrated HIT components (and the resulting interfaces) to decrease interactive complexity. Perform integrated testing on all HIT and interfaced systems to identify new hazards and to estimate the likelihood of occurrence of those previously identified (Boyer et al. 2005; Jackson et al. 2007).

7. Configure and test the HIT system for usability (Schaffer 2004).

8. Control HIT-related hazards both proactively and retrospectively (Moizer 1994; Pronovost et al. 2009).

a. Identify hazards before implementation.

b. Require the vendor to design significant hazards out, if feasible.

c. Perform custom programming to remove significant hazards, if feasible.

d. Do not implement individual HIT functions or applications that pose significant hazards and that cannot be designed out.

e. Assign small, dedicated teams to make all HIT changes (e.g., to guarantee the integrity of evidence-based order sets).

f. Train users to avoid hazards (that cannot be otherwise controlled) using competency-based training.

g. Monitor production HIT systems for the adverse effects of hazards that could not be resolved and to identify new hazards (Boyer and soback 2005; Sittig and Classen 2010).

h. Monitor retrospective analyses of care-process compromises and patient harm (e.g., Joint Commission inspections and root-cause analyses) for previously unidentified hazards.

i. Manage all hazards actively until they are controlled as fully as feasible.

j. Report hazards, process compromises, and patient harm to the CDO's safety team and to a national clearinghouse.

9. Use adult-learning theory (i.e., just-in-time, task-based, competency-based training) to enable optimal HIT use—at go-live and on a continuing basis (Harrison 2010; Krum and Latshaw 2005). Adult-learning theory predicts that the adult members of the healthcare team will learn material that is relevant to their needs far more effectively than they learn information that is not task-relevant. This means that an effective HIT-training program will need to identify teachable moments and provide task-based, just-in-time, just-enough training to take advantage of those moments (Tang et al. 2006). For patients, a new diagnosis of diabetes may provide the motivation to take a more active role in their care by learning to use their healthcare team's networked PHR to report signs and symptoms, to update their medicine list, to renew medicines, and to receive and act on reminders for disease-monitoring tests. For clinicians, HIT upgrades and redesigned care processes are predictable motivators of learning. For both patients and clinicians, online education is an increasingly acceptable, effective, and

affordable form of education (Fordis et al. 2005; Mahoney et al. 2002).

10. Test process-automation tools (such as order sets, interaction checkers, alerts, and reminders) for safety before implementation. It is critical that only invariably appropriate, low-risk interventions (e.g., ordering peak and trough drug levels with a gentamicin order) are fully automated (i.e., set at level "a" in Software-Safety Principle 6.)

11. Use phased HIT implementation and small pilots with rapid assessment to identify hazards. Cognitive psychology, adult-learning theory, human factors engineering, and extensive practical experience suggest that phased implementation is safer than "big-bang" implementation (Culp et al. 2005; Canada Infoway et al. 2007).

12. During the critical period when HIT is first in widespread use, the informatics team can take several steps to increase safety:

a. Provide "shadow" trainers in every patient-care area around the clock until they are no longer needed. Although optimal staffing levels have not been evaluated empirically, Geisinger's experience is that one trainer for every four HIT users at go-live strikes a reasonable balance between providing a trainer wherever they are needed and cost, with decreasing numbers of trainers needed throughout the first 4 weeks of HIT use.

b. Staff a go-live command center, with every potentially needed expertise physically available around the clock until the number of calls decreases significantly (about 1 week after go-live), with decreasing coverage until almost no new issues are being identified (about 4 weeks).

c. Provide specially trained clinician "super-users" in every clinical area to provide just-in-time support to HIT users during go-live and beyond (Boyer and soback 2005).

d. After go-live, staff a dedicated, ongoing 24/7 production-support team that is available via phone or e-mail, with specialist backup on call (Boyer and soback 2005).

e. Back up the production-support team with an HIT rapid-response team (with members from clinical leadership, legal, risk management, public relations, safety, information security, medical records, informatics, and IT) that can meet on 2 hours notice to oversee the emergency management of HIT-related care-process compromises that may have caused patient harm (See "Safety first for NHS IT systems" for a societal-level method of organizing a rapid-response system; Baker 2009).

13. Support continuing, integrated optimization of care processes and HIT, beginning at 6 months postimplementation—the approximate time at which users become skilled enough as HIT users that they are ready to learn new processes and new HIT skills (Berg 2001; Culp 2005; Henry 2005). This integrated optimization of care processes and HIT is synergistic and continues to produce measurable improvements in patient outcomes for decades after the initial implementation (Gilfillan et al. 2010).

21.3.3.3 CDO Information Technology

Supporting high-availability, integrated HIT systems that serve the patient's needs across all venues of care (patient-centered care) requires technical capabilities that are much more advanced and expensive than those of the pre-HIT state. When IT departments do not meet those requirements, HIT access and the care processes that depend on that access are compromised. For this reason, a clinically astute, process-focused IT team that understands the complexity and time-sensitivity of care processes is a critical contributor to HIT safety (Richards 2005a).

Both patient safety and the high cost of coping with downtimes necessitate provision of high-reliability systems, including the following: (Anderson 2002; Campbell et al. 2007)

- Redundant power supplies with emergency backup power.
- Redundant servers, with a fail-over "shadow" version of the production server that is housed in a different building from the production server and is available for full production use within seconds of the production server's failure.
- Redundant data storage, including frequent local backup, and daily remote backup.
- Redundant, physically secured, raised-floor data facilities.
- Identification of critical software systems for which time-specified recovery plans are developed and reviewed annually.
- Development and testing (by the informatics and IT teams) of downtime contingency plans that include backup computers connected to emergency power and to a dedicated printer in every patient-care area. These computers should contain the last 48 hours of information for each

hospital patient and summary information for each ambulatory patient who is scheduled for a visit within the next 24 hours, as well as policies and procedures for ordering test and communicating their results, ordering and administering medications, and updating HIT systems after the downtime ends.

- Rapid, positive downtime notification to every clinical unit (including reminders of downtime procedures and tools).

Even for large CDOs, carrying out an effective HIT-safety program is a daunting, resource-intensive task (which appears anecdotally to be more honored in the breach than the observance). Many smaller physician practices, hospitals, and long-term, postacute care organizations will not have the resources to implement adequate levels of HIT-safety best practices (Diamond et al. 2004; Hagen and Richmond 2008). They—and their patients—may be best served by one of the developing services through which large CDOs provide access to their redesigned care processes and supporting HIT systems on a subscription basis. Such arrangements have the potential to make the integrated process redesign and HIT optimization (including safety provision) of larger CDOs available to smaller CDOs, enabling the smaller CDOs to provide measurably better care at lower total cost—thereby improving the well-being of their patients.

21.3.4 Individual

While the other societal actors share an obligation to optimize HIT safety, patients and other healthcare consumers must be able to rely on the safety of HIT regardless of their personal level of knowledge or activation. They must also have opportunities to contribute actively to HIT safety. One of their most effective options is to ask the members of their healthcare team about their HIT-safety policies and practices. They can also practice HIT safety personally by learning how to use HIT to support their care, managing their IDs and passwords carefully, and reporting problems with the HIT systems they use to their healthcare teams.

21.3.5 Non-CDO HIT Service Providers

Most clinician members of care-delivery organization (CDOs) have publicly committed themselves to the principles (the primacy of patient welfare, patient autonomy, and social justice) and commitments (among others, to honesty with patients, to patient confidentiality, and to never exploiting patients for personal financial gain) of medical professionalism (Brennan et al. 2002). While CDOs have not always lived up to them, these principles and commitments have shaped the safety culture of CDOs to a significant extent. They have also provided objective criteria according to which CDOs and their members are judged—by themselves and others. The result is that these principles and commitments exert a powerful influence on the context in which CDOs implement and use HIT. A second powerful influence on this context of use is the professional responsibility of primary-care-physician-led teams and others for all aspects of the patient's care—embodied particularly clearly in the patient-centered medical home (Gilfillan et al. 2010; Patient-Centered Primary Care Collaborative 2007). In the absence of analogous public principles, commitments, and responsibilities, the contexts in which non-CDO organizations provide HIT services and products to consumers—although various and not well characterized to date—are almost certainly significantly and systematically different. For example, it is standard practice in the networked PHRs operated by some CDOs that every incoming patient message is acted on by a professional member of the patient's healthcare team within 2 working days of receiving the message (Wendling 2011). No comparable best practice appears to have been promulgated for freestanding PHRs (Dick et al. 1997; NCVHS 2006; Tang et al. 2006). Given the importance of the context of HIT use on HIT safety, the safety impact of these different contexts deserves careful measurement and reporting.

21.4 Conclusion

Although much remains to be learned, the principles and methods of safety engineering, human factors engineering, healthcare informatics, and computer science provide actionable guidance in the shared effort to increase HIT safety and effectiveness.

Acknowledgment

The author is indebted to Bob Wears for his provocative comments on an early version of this chapter.

References

Adams, J. (2009). Integrated testing resource use.
Adams, J. (2009). Integrated testing resource use. Personal communication.
Adams, J. A., L. Culp et al. (2005). Workflow assessment and redesign. In *Implementing an Electronic Health Record System*. J. Walker, E. Bieber, and F. Richards (Eds.). London, U.K.: Springer Verlag.
AHRQ. (2008). AHRQ Portfolios of Research.
AHRQ. (2008). AHRQ Portfolios of Research. http://www.ahrq.gov/fund/portfolio.htm
Amalberti, R., Y. Auroy et al. (2005). Five system barriers to achieving ultrasafe health care. *Annals of Internal Medicine* **142**(9): 756–764.
Anderson, M. (2002). The toll of downtime. A study calculates the time and money lost when automated systems go down. *Healthcare Informatics* **19**(4): 27–30.
Ash, J., D. Sittig et al. (2006). An unintended consequence of CPOE implementation: Shifts in power, control, and autonomy. *AMIA Annual Symposium Proceedings* 11–15.
Ash, J. S., D. F. Sittig et al. (2007a). Categorizing the unintended sociotechnical consequences of computerized provider order entry. *International Journal of Medical Informatics* **76**(Suppl 1): S21–S27.
Ash, J. S., D. F. Sittig et al. (2007b). The extent and importance of unintended consequences related to computerized provider order entry. *JAMIA* **14**(4): 415–423.
Ash, J. S., D. F. Sittig et al. (2009). The unintended consequences of computerized provider order entry: Findings from a mixed methods exploration. *International Journal of Medical Informatics* **78**(Suppl 1): S69–S76.
Baker, M. (2009). Safety first for NHS IT systems, NHS Connecting for Health.
Baker, M. (2009). Safety first for NHS IT systems, NHS Connecting for Health, West Yorkshire, U.K. www.connectingforhealth.nhs.uk
Baker, M. (2011). DSCN 18/2009 testing.
Baker, M. (2011). Implementing DSCN 14/2009 and DSCN 18/2009. Personal communication.
Bates, D. (2005). Computerized physician order entry and medication errors: Finding a balance. *JBI* **38**(4): 259–261.
Battles, J. (2007). Patient safety and technology, a two-edged sword. In *Handbook of Human Factors and Ergonomics in Health Care and Patient Safety*. P. Carayon (Ed.). Mahwah, NJ: Lawrence Erlbaum Associates, Inc. Publishers, pp. 383–391, Chap. 24.
Berg, M. (2001). Implementing information systems in health care organizations: Myths and challenges. *International Journal of Medical Informatics* **64**(2–3): 143–156.
Bias, R. G. and D. J. Mayhew (Eds.). (1994). *Cost-Justifying Usability*. San Diego, CA: Morgan Kaufmann.
Black, A. D., J. Car et al. (2011). The impact of eHealth on the quality and safety of health care: A systematic overview. *PLoS Medicine* **8**(1): e1000387.
Bloomrosen, M., J. Starren et al. (2011). Anticipating and addressing the unintended consequences of health IT policy: A report from the AMIA 2009 Health Policy Meeting. *JAMIA* **18**: 82–90.
Bonnabry, P., C. Despont-Gros et al. (2008). A risk analysis method to evaluate the impact of a CPOE system on patient safety. *JAMIA* **15**: 453–460.
Boyer, E. A., J. A. Adams et al. (2005). System integration. In *Implementing an Electronic Health Record System*. J. Walker, E. Bieber, and F. Richards (Eds.). London, U.K.: Springer Verlag, pp. 89–94.

Boyer, E. A. and M. W. Soback. (2005). Production support. In *Implementing an Electronic Health Record System*. J. Walker, E. Bieber, and F. Richards (Eds.). London, U.K.: Springer Verlag, pp. 95–100.

Brennan, T., L. Blank et al. (2002). Medical professionalism in the new millennium: A physician charter. *Annals of Internal Medicine* 136(3): 243–246.

Campbell, E. and D. Sittig. (2006). Types of unintended consequences related to computerized provider order entry. *JAMIA* 13(5): 547–556.

Campbell, E. M., D. F. Sittig et al. (2007). Overdependence on technology: An unintended adverse consequence of computerized provider order entry. *AMIA Annual Symposium Proceedings* 2007: 94–98.

Carayon, P. (2006). Human factors of complex sociotechnical systems. *Applied Ergonomics* 37: 525–535.

Carayon, P. (2007). Human factors and ergonomics in health care and patient safety. In *Handbook of Human Factors and Ergonomics in Health Care and Patient Safety*. P. Carayon (Ed.). Mahwah, NJ: Lawrence Erlbaum Associates, Inc.

Carayon, P. and B. Karsh. (2000). Sociotechnical issues in the implementation of imaging technology. *Behaviour and Information Technology* 19(4): 247–262.

Classen, D., D. Bates et al. (2010). Meaningful use of computerized prescriber order entry. *Journal of Patient Safety* 6(1): 15–23.

Coiera, E. (2000). When conversation is better than computation. *JAMIA* 7: 277–286.

Committee on Quality of Health Care in America. (2001). *Crossing the Quality Chasm: A New Health System for the 21st Century*. Washington, DC: National Academy Press.

Cook, R., M. Render et al. (2000). Gaps in the continuity of care and progress on patient safety. *BMJ* 320: 791–794.

Culp, L. M. (2005). Optimizing specialty practices. In *Implementing an Electronic Health Record System*. J. Walker, E. Bieber, and F. Richards (Eds.). London, U.K.: Springer Verlag, pp. 128–133.

Culp, L. M., J. A. Adams et al. (2005). Phased implementation. In *Implementing an Electronic Health Record System*. J. Walker, E. Bieber, and F. Richards (Eds.). London, U.K.: Springer Verlag, pp. 111–119.

Del Beccaro, M., H. Jeffries et al. (2006). CPOE implementation: No association with increased mortality rates in an intensive care unit. *Pediatrics* 118(1): 290–295.

Diamond, C., D. Garrett et al. (2004). Achieving electronic connectivity in healthcare, Markle Foundation, New York. http://www.internet2.edu/health/files/cfh_roadmap_final_0714.pdf

Dick, R. S., E. B. Steen et al. (Eds.). (1997). *The Computer-Based Patient Record: An Essential Technology for Health Care*, Washington, DC: National Academies Press.

Domberger, S. (1999). *The Contracting Organization: A Strategic Guide to Outsourcing*, Oxford, U.K.: Oxford University Press.

Donaldson, M. (2000). An overview of to err is human: Re-emphasizing the message of patient safety. In *To Err is Human: Building a Safer Health System*. L. Kohn, J. Corrigan, and M. Donaldson (Eds.). Washington, DC: National Academy Press.

Drucker, P. F. (1989). *The New Realities: In Government and Politics/In Economics and Business/In Society and World View*, New York: Harper & Row.

Dvorak, C. (2011). EHR Association HIT-safety efforts.

Dvorak, C. (2011). EHR Association HIT-safety efforts. Personal communication.

Eason, K. (2001). Changing perspectives on the organizational consequences of information technology. *Behaviour and Information Technology* 20(5): 323–328.

Fordis, M., J. E. P. B. King, M. Christie, P. H. Jones, K. H. Schneider, S. J. Spann, S. B. Greenberg, and A. J. Greisinger. (2005). Comparison of the instructional efficacy of internet-based CME with live interactive CME workshops: A randomized controlled trial. *JAMA* 294(9): 1043–1051.

Gilfillan, R., J. Tomcavage et al. (2010). Value and the medical home: Effects of transformed primary care. *American Journal of Managed Care* 16(8): 607–614.

Goodman, K., E. Berner et al. (2010). Challenges in ethics, safety, best practices, and oversight regarding HIT vendors, their customers, and patients: A report of an AMIA special task force. *JAMIA* 18: 77–81.

Graham, R. J. and R. L. Englund. (1997). *Creating an Environment for Successful Projects: The Quest to Manage Project Management*, San Francisco, CA: Jossey-Bass.

Grassley, C. (2010). Grassley asks hospitals about experiences with federal health information technology program.

Grassley, C. (2010). Grassley asks hospitals about experiences with federal health information technology program. http://grassley.senate.gov/news/Article.cfm?customel_dataPageID_1502=24867

Hagen, S. and P. Richmond. (2008). *Costs and Benefits of Health Information Technology*, Congressional Budget Office, Washington, DC.

Hagen, S. and P. Richmond. (2008). Evidence on the costs and benefits of health information technology, Congressional Budget Office, Washington, DC. http://www.cbo.gov/ftpdocs/91xx/doc9168/05-20-HealthIT.pdf

Han, Y., J. Carcillo et al. (2005). Unexpected increased mortality after implementation of a commercially sold CPOE system. *Pediatrics* 116: 1506–1512.

Harrison, M., R. Koppel et al. (2007). Unintended consequence of information technologies in health care—An interactive sociotechnical analysis. *JAMIA* 14(5): 542–549.

Harrison, I., M. Baker et al. (2009). Health Informatics—Application of clinical risk management to the manufacture of health software. *DSCN14/2009*. NHS, National Health Service, 14/2009 V1.1. http://www.isb.nhs.uk/documents/dscn/dscn2009/dataset/142009v1_1.pdf

Harrison, I. (2010). Health Informatics—Guidance on the management of clinical risk relating to the deployment and use of health software. DSCN18/2009. NHS Connecting for Health, West Yorkshire, U.K. http://www.isb.nhs.uk/documents/dscn/dscn2009/dataset/182009.pdf

Henry, E. E. (2005). Optimizing primary-care practices. In *Implementing an Electronic Health Record System*. J. Walker, E. Bieber, and F. Richards (Eds.). London, U.K.: Springer Verlag, pp. 120–127.

HIT Policy Committee PCAST Report Workgroup. (2011). *Implications of the PCAST Report for ONC's HIT Strategy.* PCAST Report Workgroup meeting, Washington, DC.

HIT Policy Committee Privacy and Security Tiger Team. (2010). Consumer choice technology hearing. *HIT Policy Committee Privacy and Security Tiger Team*, Washington, DC.

Hoffman, S. and A. Podgurski. (2010). Finding a cure: The case for regulation and oversight of electronic health record systems. *Harvard Journal of Law and Technology* **22**(1): 103–165.

Hollnagel, E., D. D. Woods et al. (Eds.). (2006). *Resilience Engineering*, Padstow, Cornwall, U.K.: TJ International Ltd.

Hundt, A., P. Carayon et al. (2005). Evaluating design changes of a smart IV pump. In *Healthcare Systems Ergonomics and Patient Safety*. R. Tartaglia, S. Bagnara, T. Bellandi, and S. Albolino (Eds.). Florence, Italy: Taylor & Francis, pp 239–242.

iCare about health, Canada Infoway et al. (2007). Electronic health records and patient safety: A joint report on future directions for Canada. https://www2.infoway-inforoute.ca/Documents/EHR-Patient%20Safety%20Report.pdf

Jackson, D., M. Thomas et al. (Eds.). (2007). *Software for Dependable Systems: Sufficient Evidence?* Washington, DC: National Academies Press.

Kadry, B., I. C. Sanderson et al. (2010). Challenges that limit meaningful use of health information technology. *Current Opinion in Anaesthesiology* **23**(2): 184–192.

Kohn, L., J. Corrigan et al. (Eds.). (2000). To err is human: Building a safer health system. *A report of the Committee on Quality of Health Care in America*. Washington, DC: National Academy Press.

Krum, W. L. and J. D. Latshaw. (2005). Training. In *Implementing an Electronic Health Record System*. J. Walker, E. Bieber, and F. Richards (Eds.). London, U.K.: Springer Verlag, pp. 60–66.

Kushniruk, A. (2001). Analysis of complex decision-making processes in health care: Cognitive approaches to health informatics. *Journal of Biomedical Informatics* **34**: 365–376.

Kushniruk, A., E. Borycki et al. (2006). Predicting changes in workflow resulting from healthcare information systems: Ensuring the safety of healthcare. *HealthcQ.* **9(Sp)**: 114–118.

Lehto, M. and D. Clark. (1990). Warning signs and labels in the workplace. In *Workspace, Equipment and Tool Design*. W. Karwowski and A. Mital (Eds.). Amsterdam, the Netherlands: Elsevier, pp. 303–344.

Leveson, N. (1995). *Safeware: System Safety and Computers.* Reading, MA: Addison-Wesley Publishing Company.

Leveson, N. (2008a). Hazard analysis definition. Personal communication.

Leveson, N. (2008a). Hazard Analysis definition.

Leveson, N. (2008b). *An Engineering Systems Approach to Benefit/Risk Decision-Making*, Cambridge, MA, Center for Biomedical Innovation.

Longhurst, C., P. Sharek et al. (2006). Perceived increase in mortality after process and policy changes implemented with computerized physician order entry. *Pediatrics* **117**(4): 1450–1451.

Lopez, K., G. Gerling et al. (2010). Cognitive work analysis to evaluate the problem of patient falls in an inpatient setting. *JAMIA* **17**: 313–321.

Mahoney, D. F., B. J. Tarlow, R. N. Jones, and J. Sandaire. (2002). Effects of a multimedia project on users' knowledge about normal forgetting and serious memory loss. *JAMIA* **9**: 383–394.

Masand, M. (2010). Accidental strangulation with a Venetian blind cord—A near miss. *BMJ* **340**: C3458.

McDonnell, C. and K. W. L. Wendel. (2010). Electronic health record usability: Vendor practices and perspectives. *AHRQ* May 2010(09(10)-0091-3-EF). Rockville, MD.

McDonough, J. (1999). *Proactive Hazard Analysis and Health Care Policy*, Milbank Memorial Fund, New York.

McQuay, H. J. and R. A. Moore. (1997). Using numerical results from systematic reviews in clinical practice. *Annals of Internal Medicine* **126**(9): 712–720.

Mewhinney, M. and J. Les Dorr. (2006). NASA aviation safety reporting system turns 30 Washington, Federal Aviation Administration, Washington, DC. http://www.nasa.gov/centers/ames/news/releases/2006/06_82AR_prt.htm

Moizer, J. (1994). Safety hazard control in the workplace. *Safety Hazard Control in the Workplace*. T. t. I. C. o. t. S. D. Society. Stirling, Scotland.

Moray, N. (1994). Error reduction as a systems problem. In *Human Error in Medicine*. M. Bogden (Ed.). Hillsdale, NJ: Lawrence Erlbaum Associates.

NCVHS. (2006). *Personal Health Records and Personal Health Record Systems: A Report and Recommendations from the National Committee on Vital and Health Statistics*. Washington, DC.

Norman, D. (1998). *The Invisible Computer: Why Good Products Can Fail, the Personal Computer Is So Complex, and Information Appliances Are the Solution*. Cambridge, MA: The MIT Press.

Office of the National Coordinator for Health IT. (2010). Beacon Community Program. http://healthit.hhs.gov/portal/server.pt/community/healthit_hhs_gov__onc_beacon_community_program_improving_health_through_health_it/1805

Parasuraman, R. and M. Mouloua. (1996). Monitoring of automated systems. In *Automation and Human Performance: Theory and Application*. R. Parasuraman and M. Moulona (Eds.). Mahwah, NJ: Erlbaum, pp. 91–115.

Parasuraman, R., T. B. Sheridan et al. (2000). A model for types and levels of human interaction with automation. *IEEE Transactions on Systems, Man, and Cybernetics—Part A: Systems and Humans* **30**(3): 286–297.

Patient-Centered Primary Care Collaborative. (2007). Joint Principles of the Patient-Centered Medical Home.

Patient-Centered Primary Care Collaborative. (2007). Joint principles of the patient-centered medical home. http://www.pcpcc.net/content/joint-principles-patient-centered-medical-home

PDR Secure. (2010). EHR Event.

PDR Secure. (2010). EHR event. http://www.ehrevent.org/Report-EHR-Safety-Event/Page-1

Perrow, C. (1983). The organizational context of human factors engineering. *Administrative Science Quarterly* **28**: 521–541.

President's Council of Advisors on Science and Technology. (2010). *Report to the President Realizing the Full Potential of Health IT to Improve Healthcare for Americans: The Path Forward.*

President's Council of Advisors on Science and Technology. (2010). Report to the president realizing the full potential of health IT to improve healthcare for Americans: The path forward. http://www.whitehouse.gov/sites/default/files/microsites/ostp/pcast-health-it-report.pdf

Pronovost, P., C. Goeschel et al. (2009). Framework for patient safety research and improvement. *Circulation* **119**(1): 330–337.

Rasmussen, J. (1997). Risk management in a dynamic society: A modelling problem. *Safety Science* **27**(2/3): 183–213.

Reason, J. (2000). Human error: Models and management. *BMJ* **320**: 768–770.

Richards, F. (2005a). Infrastructure. In *Implementing an Electronic Health Record System*. J. Walker, E. Bieber, and F. Richards (Eds.). London, U.K.: Springer Verlag, pp. 21–35.

Richards, F. (2005b). Vendor selection and contract negotiation. In *Implementing an Electronic Health Record System*. J. Walker, E. Bieber, and F. Richards (Eds.). London, U.K.: Springer Verlag, pp. 15–20.

Richards, F. (2005c). Managing the client–vendor partnership. In *Implementing an Electronic Health Record System*. J. Walker, E. Bieber, and F. Richards (Eds.). London, U.K.: Springer Verlag, pp. 101–110.

Sackett, D. L., R. B. Haynes et al. (1991). *Clinical Epidemiology*. Boston, MA: Little, Brown.

Salas, E., N. J. Cooke et al. (2008). On teams, teamwork, and team performance: Discoveries and developments. *Human Factors* **50**(3): 540–547.

Schaffer, E. (2004). *Institutionalization of Usability: A Step-by-Step Guide*. Boston, MA: Addison-Wesley.

Seligman, P. (2007). Managing Drug Safety Issues: Q and A with Paul Seligman, M.D., M.P.H.

Seligman, P. (2007). Managing Drug safety issues: Q and A with Paul Seligman, M.D., M.P.H. http://www.fda.gov/consumer/updates/drugsafetyqa053107.html

Simon, H. (1971). Designing organizations for an information-rich world. In *Computers, Communications, and the Public Interest*. M. Greenberger (Ed.). Baltimore, MD: The Johns Hopkins Press.

Sittig, D. and D. Classen. (2010). Safe electronic health record use requires a comprehensive monitoring and evaluation framework. *JAMA* **303**(5): 450–451.

Sittig, D. and H. Singh. (2009). Eight rights of safe EHR use. *JAMA* **302**(10): 1107–1109.

Software Engineering Institute. (2011). CMMI for acquisition. Software Engineering Institute, Pittsburgh, PA. http://www.sei.cmu.edu/cmmi/tools/acq/

Software Engineering Institute. (2011). CMMI for Acquisition. *Software Engineering Institute*.

Strom, B. L., R. Schinnar et al. (2010). Unintended effects of a computerized physician order entry nearly hard-stop alert to prevent a drug interaction: A randomized controlled trial. *Archives of Internal Medicine* **170**(17): 1578–1583.

Tang, P., J. Ash et al. (2006). Personal health records: Definitions, benefits, and strategies for overcoming barriers to adoption. *JAMIA* **13**(2): 121–126.

Thompson, I. M., P. J. Goodman et al. (2003). The influence of finasteride on the development of prostate cancer. *NEJM* **349**: 215–224.

Vaughan, D. W. (2005). System effects: On slippery slopes, repeating negative patterns, and learning from mistakes. In *Organization at the Limit: NASA and the Columbia Accident*. W. Starbuck and M. Farjoun (Eds.). Oxford, U.K.: Blackwell.

Venkatesh, V. (1999). Creation of favorable user perceptions: Exploring the role of intrinsic motivation. *MIS Quarterly* **23**(2): 239–260.

Vicente, K. (1999). *Cognitive Work Analysis*. Mahwah, NJ: Lawrence Erlbaum.

Vicente, K. and J. Rasmussen. (1990). The ecology of human–machine systems II: Mediating "Direct perception" in complex work domains. *Ecological Psychology* **2**(3): 207–249.

Walker, J. and P. Carayon. (2009). From tasks to processes: The case for changing health information technology to improve health care. *Health Affairs* **28**(2): 467–477.

Walker, J. M., F. Richards et al. (Eds.). (2005). *Implementing an Electronic Health Record System*. Healthcare Informatics. New York: Springer.

Weinger, M. and J. Slagle. (2002). Human factors research in anesthesia patient safety. *JAMIA* **9**(6): S58–S63.

Wendling, C. (2011). MyGeisinger user message response rates. Personal communication.

Wendling, C. (2011). MyGeisinger User Message Response Rates.

Wiegmann, D. and S. Shappell. (2003). *A Human Error Approach to Aviation Accident Analysis: The Human Factors Analysis and Classification System*. Burlington, VT: Academic Press.

Section VI

Human Error

22

Human Error in Health Care

Frank A. Drews

CONTENTS

22.1 Introduction 323
22.2 A Brief History of Human Error 323
22.3 Human Error Deconstructed: Models of Human Error 325
    22.3.1 Individual Contribution to Human Error 325
    22.3.2 Error in Context 328
    22.3.3 Situational and Systemic Influences on Human Error 328
        22.3.3.1 Latent Conditions 328
        22.3.3.2 Error-Producing Conditions 328
        22.3.3.3 Violations 329
        22.3.3.4 Violations in Health Care 330
        22.3.3.5 Violation-Producing Conditions 331
        22.3.3.6 Hazards 332
        22.3.3.7 Defenses 332
        22.3.3.8 Adverse Event 332
22.4 Integrative Model of Human Error in Health Care 333
22.5 Studies of Human Error in Health Care 334
    22.5.1 Critical Care 334
    22.5.2 Surgery 335
    22.5.3 Anesthesia 337
22.6 Conclusions 338
References 338

> Human error in medicine, and the adverse events which may follow, are problems of psychology and engineering not of medicine.
>
> **John Senders (1994, p. 159)**

22.1 Introduction

This chapter provides an overview of the current literature on human error in health care with a particular focus on psychological models developed to explain fallible behavior. After reviewing the history of the study of human error, some models of human error will be described; next, contextual, situational, and systemic influences on erroneous behavior will be discussed. Then, an integrative model of human error in health care is presented. This chapter concludes with a review of work on human error in health care, with a special focus on the areas of critical care, surgery, and anesthesia.

22.2 A Brief History of Human Error

Interest in human error is an expression of an interest in explaining human behavior, or as Ernst Mach (1905) writes, "Knowledge and error flow from the same mental sources; only success can tell one from the other." This interest at least dates back to Cicero's "to err is human" in the first century BC and found another expression in Sir Francis Bacon's analysis that "the human understanding, from its peculiar nature, easily supposes a greater degree of order and equality in things than it really finds" (p. 320, *novum organum*).

More recently, a number of different perspectives evolved which were discussed in Hollnagel (2007).

Despite Hollnagel's claim that from an ontological perspective the construct error is a "category mistake" (Ryle, 1949) and that it has little to no value, from a pragmatic perspective the construct appears helpful, since it describes and potentially explains the variability of human behavior in sociotechnical or socionatural systems like health care (Durso and Drews, 2010).

At least two perspectives on human error can be distinguished that mark the transition from more traditional approaches to more modern approaches or the "new look" on human error. The first approach is characterized by a number of authors who identify human error as a cause for some kind of adverse event. This interpretation of error is consistent with attempts to identify human error as a construct associated with a number of traits of a person. Approaches that belong to this class emerged in the early twentieth century. For example, Greenwood and Woods (1919) investigated industrial accidents and in the course of these investigations explored the idea that some individuals are more accident-prone than others. Here accident-proneness is interpreted as a stable characteristic of a person. Clearly, this approach is consistent with the perspective usually entertained in the mass media and is labeled the "bad apple" theory of human error (see Dekker, 2002, 2006). The bad apple theory identifies error as the cause of adverse events, with the responsibility for the event being solely attributed to an individual. Later work on identifying individuals with accident-proneness by Reason (1974) indicates a lack of a consistent pattern that would demonstrate the existence of such "accident repeaters." More recently, there are renewed attempts to associate the individual-centered perspective on error with personality factors derived from modern personality theories. For example, the constructs of agreeableness and conscientiousness of the big five personality theory (McCrae and Costa, 1987) have been identified as being moderately negative correlated with a tendency to display elevated accident rates (Cellar et al., 2001).

There are at least three approaches of studying human error: the individual-centered perspective, the phenomenon-based perspective, and the organizational perspective. Each of these perspectives has a different focus and consequently can contribute to a better general understanding of human error in health care. What separates the perspectives is that in the context of the individual-centered perspective, the focus is on the mechanisms that as a result of our cognitive architecture are creating the conditions for the genesis of error. In the context of the phenomenon-based approach, a number of researchers identified types of error and provide specific explanations for the genesis of these types of error. Finally, the organizational perspective is interested in understanding the organizational contributors to error. As a consequence, this approach focuses less on the cognitive architecture and more on workplace and organizational conditions that contribute to the genesis of error.

A more recent and more integrative perspective on human error is to conceptualize error as an endpoint in a chain of events that as a whole leads to a negative outcome (Turner, 1976, 1978). This systemic perspective on human error is based on work of a number of authors who contributed by providing elements to this integrated perspective. For example, Norman (1981) was able to identify types of errors of action (slips, lapses, and mistakes). Perrow (1984) argued that accidents are a normal part of operation in complex sociotechnical systems and are not external to the operation of a system, as a more dogmatic engineering approach would assume; Rasmussen and Jensen (1974) and Reason (1990) analyzed levels of human performance that describe human performance based on the level of attention required to perform activities at this level and the degree of automaticity that is associated with performance. Finally, Reason (1997) organized these elements in a framework and integrated them into one of the more coherent perspectives on human error. Because of its importance for understanding human error in health care, this perspective will be discussed in more detail in the following.

Another perspective on human error originates in cognitive, and more recently in social, psychology and analyzes the common fallacies and biases that are a result of the human cognitive architecture. Those fallacies and biases have been identified in the context of perception (Eysenck and Keane, 2005), thinking (Fisk, 2004; Thompson, 2004), judgment (Reber, 2004; Tversky and Kahneman, 1974), and memory (e.g., Pohl, 2004). A good overview about different types of error is provided by Pohl (2004) in general, and by Croskerry (2003) in the context of medical diagnoses. Consistent with the approach of identifying fallacies and biases of human cognition is Reason's (1990) approach to identifying two dominant error forms that are the basis for all types (i.e., slips, lapses, and mistakes) of human error. Reason (1990) argues that these error forms are based on two factors: (1) similarity and (2) frequency of information in memory. For example, when facing a situation where sufficient information is not available to guide action, Reason suggests that such underspecification leads to the use of context-driven behavior that has been frequently successful in the past. Capture errors (Norman, 1981), that is, placing a prescription under the date of January of the previous year, are an example.

Of broader interest in the context of human error in health care is a more recent publication by Reason where he proposes one of most popular models of human error: the Swiss Cheese Model (Reason, 1997). Here we will discuss Reasons' approach in more detail and expand on this work.

22.3 Human Error Deconstructed: Models of Human Error

22.3.1 Individual Contribution to Human Error

One widely recognized definition of human error is provided by Reason (1990): "a generic term to encompass all those occasions in which a planned sequence of mental or physical activities fails to achieve its intended outcome, and when these failures cannot be attributed to the intervention of some chance agency." Another definition that focuses on the contextual influences is provided by Swain and Guttman (1983): "An error is an out of tolerance action, where the limits of tolerable performance are defined by the system." However, the most comprehensive definition of human error is provided by the revised definition of medical error based on the first IOM report and expanded by the Quality Interagency Coordination task force: "An error is defined as the failure of a planned action to be completed as intended, or the use of a wrong plan to achieve an aim. Errors can include problems in practice, products, procedures, and systems." (Kohn et al., 1999).

The purpose of models of human error is to provide insight into the psychological and organizational contributors that lead to near misses and adverse events in many contexts. Over the last two decades, a literature emerged that analyzes human error in a wide range of contexts. For example, human error has been studied in the context of aviation (Helmreich, 1997; Wiegmann and Shappell, 2003), nuclear power plant operation (Lee et al., 2004), and exploration of natural resources (Wagenaar et al., 1994). Overall, work from a range of domains leads to a better understanding of how and under which conditions error occurs. Another result of this work is the development of interventions to mitigate the impact of human error or lead to a significant reduction in error rates and improvement of safety in the operation of facilities, equipment, and personnel.

Over the last decade, there has been a growing interest in studying human error using psychological and/or human factors perspectives to improve our understanding of human error in health care. This new focus of research on human error is clearly documented by the Institute of Medicine's (1999) Report "To Err is Human."

In this chapter we focus on human error as a result of limitations of the human cognitive architecture (Reason, 1990) that, with contextual factors, can result in adverse events. According to Rasmussen and Jensen (1974), it is possible to differentiate different levels of human performance to identify different cognitive control modes of behavior. This differentiation then allows the prediction of types of human error associated with these specific levels of performance. According to Rasmussen and Jensen (1974), human performance can be described on three levels as being skill-based, rule-based, or knowledge-based.

Skill-based behavior is behavior that is highly automatic and learned. It is driven by patterns that were acquired earlier as part of the process of acquisition of expertise. A person acting at this level is not required to allocate significant amounts of attention to the execution of a task. Behavior at the skill-based level can be automated; attention can either shift occasionally to another task or be completely allocated to such additional task. The withdrawal of attention from the skill-based task may result in error when changes in the environment or changes in the tools used occur. For skill-based performance to be successful, it is important that the operator, at least on occasion, monitors the progress of the behavior to keep it on track.

Rule-based performance is present when a person deals with a familiar problem where a solution to this problem is available in form of a stored production rule (if X then Y) that was learned during the development of expertise. Here a person is required to allocate attention to the task to recognize the current conditions and to match these conditions via an association to the conditional statement of a production rule. In case of such match, a particular rule is applied and an action is executed. The cognitive demand involved at this level of performance is higher compared to the demand when performing at the skill-based level. Error in this context can be a result of mismatching the conditional elements of a rule to a situation.

Finally, knowledge-based behavior occurs when a person deals with a new and unfamiliar problem. Performance at this level requires undivided attention, and the cognitive processes involved in dealing with this type of problem are slow and sequential. Error at this level is a result of a lack of comprehension of a system or a problem. Figure 22.1 illustrates the different levels of performance and cognitive processes involved. Reason (1990) applies this classification of human behavior to develop a framework that integrates different literatures on Human Error. The result of this work is the generic error modeling system (GEMS).

The GEMS distinguishes three stages of cognitive processing (planning, storage, and execution) and three levels of control, which vary according to the intensity of cognitive effort (automatic, mixed, and effortful). A combination of stages and levels of cognitive control leads to different modes that result in behavior being skill-based, rule-based, or knowledge-based. The skill-based mode corresponds to habitual activities that are nearly effortless with respect to cognitive load; the rule-based mode depends on pattern-matching against a set of internal problem-solving rules; the knowledge-based mode applies to novel, difficult, dangerous, or critical

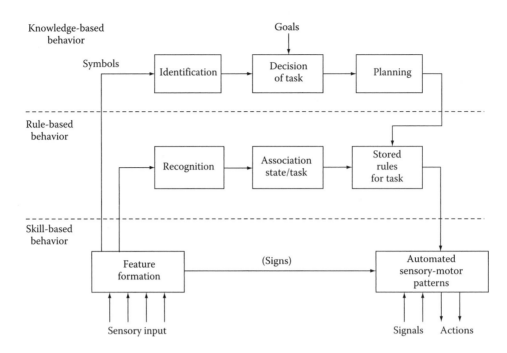

FIGURE 22.1
Levels of behavior and cognitive processes. (From Goodstein, L.P. et al., *Tasks, Errors and Mental Models*, Taylor & Francis, London, U.K., 1988.)

problems, or when automatic control has to be overridden to prevent performance of a task in the habitual way. Thus, the level of control depends on the complexity and novelty of the task. For a given task, humans typically gravitate toward the lowest mode possible to minimize cognitive effort (Fisk and Taylor, 1991). Another way of describing the context under which humans act is Hollnagel's (2004) efficiency–thoroughness trade-off (ETTO) principle, which states that any action is performed in the tension between efficiency and thoroughness, and thoroughness is often sacrificed for efficiency.

The GEMS model makes the prediction that the same cognitive processes that govern everyday activities influence the occurrence of error. Each level and stage of cognitive processing corresponds to a different type of error. Slips are the result of errors at the execution phase of a skill-based activity, whereas lapses are skill-based errors at the storage stage. An example of a slip is writing a routine medication refill for digoxin but inadvertently switching the "qod" frequency to "qd." Mistakes are categorized as errors of rule- or knowledge-based modes of control. An example for a mistake in health care would be the failure to suspect a retropharyngeal abscess in an adult who presents with painful swallowing and fever because the wrong diagnostic evaluation rule, "rule out streptococcal pharyngitis," is applied.

What is important in this context is to understand that all different types of error constitute instances of active failures. Active failures are occasions where performance at the "sharp end" breaks down. Active failures occur where human operators interact with a system.

What makes Reason's approach appealing is that not only does it allow one to distinguish between the basic error types, but these error types can be described and differentiated on a number of different dimensions that allows in turn the prediction of the occurrence of particular types of errors under specific circumstances. Among those dimensions that differentiate the error types are type of activity, the focus of attention, the control mode, the level of expertise and predictability of error types, the likelihood of error compared to successful behavior, situational influences, and the relationship to change. In the next section, we will discuss these dimensions in more detail in the context of error types.

Type of activity. This aspect is related to the activity that is performed by a person. If an activity includes routine behavior that does not require explicit problem-solving, a slip may result. For example, a nurse who has programmed an infusion pump many times and has developed automaticity in doing so may accidentally press an incorrect key, which leads to a slip that results in administration of an incorrect dose of a drug. Here a particular, highly automatic sequence of behavior is unsuccessfully executed. This situation can be contrasted with activities that are on the rule-based or at the knowledge-based level of behavior. According to Reason, this level of performance results in a different type of error. Rule- and knowledge-based behavior is often displayed in response to the perception of a problem where the nature of the problem makes it impossible to execute behavior at the skill-based level. Mistakes are a result of awareness of such a challenge and reflect a person's attempt to deal with a problem.

Focus of attention. The focus of attention also differentiates between the different error types. A slip may be the result of a person's attention being captured by an external event, that is, an interruption or another type of distraction. Another reason for this potentially dangerous allocation of attention is that a person might be preoccupied with some issue other than the task at hand, resulting in allocation of limited attention to the task. In the context of rule-based and knowledge-based behavior, it is unlikely that attention is refocused by external stimuli; however, external stimuli can potentially distract the person at this level of performance.

Control mode. For both rule-based and skill-based behavior driven by existing knowledge structure, behavior is executed based on expectations about the environment. Thus, behavior is feed-forward controlled. This situation contrasts with performance on the knowledge-based level where control is based on feedback. Information about the environment is used to drive behavior, since the person has very few expectations based on his/her lack of previous experience with the situation or problem. The cognitive effort that is associated with performance at this level of behavior is very high, limiting the sustainability of this effort.

Level of expertise. Error at the knowledge-based and the rule-based level is related to knowledge of existing rules and relationships. On the skill-based level, error is based on the fact that an existing rule becomes activated and as a result a skill-based behavior is displayed that is not appropriate in the present context. On the rule-based level, the conditions that normally specify what rule to apply and the current conditions that are triggering application of a rule are not matched correctly, leading to the execution of an inappropriate rule. A mistake results from the application of a wrong rule. On the knowledge-based level of performance, error is a result of the limited cognitive resources available and the fact that the person has to generate options and courses of action using sequential laborious processes. Finally, if a person has an inadequate mental model of processes relevant in this context, this mental model will result in the application of a strategy that is wrong because of the incorrect assumptions driving the selection of an action. Experts are unlikely to resort to this level of performance because being an expert implies that a person should be able to operate on the skill- and rule-based level of performance for most activities in the domain of expertise.

Likelihood of error. Because behavior in work contexts manifests itself mostly at the skill- and rule-based level of performance, the opportunity for error and the frequency of error encountered is highest at this level of behavior. On the other hand, the relatively rare occasions when a person has to perform at the knowledge-based level of performance allows for the commission of mistakes. This is reflected in the fact that the relative frequency of error as an expression of erroneous behavior/opportunity for error casts a different light on this relationship. Error as an outcome of performance at the skill-based and the rule-based level is rare, but it is much more frequent at the knowledge-based level of performance.

Situational influences. According to Reason (1990), situational factors are an important contributor to error. Those factors can be related to the task at hand and to the larger context in which the task is performed. Skill-based slips are a result of attention being captured by the context in which a task is performed combined with the presence of an experience (a schema) that allowed successful performance of the task in the past. However, the past experience likely will activate an action that is wrong in the current context if it is different from conditions in the past. The situation is similar for rule-based mistakes where the specific trigger conditions for the application of a rule are not correctly matched, leading to the execution of an incorrect rule. For knowledge-based mistakes, the situation is more complex since the preexisting experience and knowledge of a person matter and influence the strategy that is applied when dealing with a problem. Task structure and situational variables are important because they are potentially framing the way an individual approaches a problem and attempts to solve the problem. However, without any existing knowledge about the task, situational factors determine to a large extent how a person approaches a problem and therefore these factors have the potential to shape performance (see Miller and Swain, 1987).

Detectability. According to Reason (1989), it is easier for a person to detect a slip or a lapse, that is, errors that are associated with the execution of an action are easier to detect than those that are related to planning (mistakes). Overall, the odds of recovering from error are low, especially if the error is a mistake, where detection of an error depends on other people detecting the error.

Relationship to change. By changing the conditions under which a person acts, errors may be introduced. In the context of skill-based behavior, deviation from established procedures with changes in environmental conditions is a significant contributor to the genesis of error. Rule-based mistakes differ because changes are present that differ from the rules that were part of previous training or procedures. The trigger for this type of error is that the conditions under which modification should be performed are not clear. Consequentially, a person may apply an incorrect rule when the context indicates application of a different rule is necessary for successful behavior. The lack of knowledge or alternative plans leads to the genesis of knowledge-based mistakes. Here a person lacks the knowledge to perform a particular action and consequently is likely to fail.

Overall, the benefit of Reason's (1990) framework is that it allows prediction of the occurrence of specific types of error under specific conditions. The prediction of specific types of error allows the implementation of error-mitigating strategies that reduce the overall impact of error in a system. One of the limitations of the error framework as discussed earlier is that its main focus is the cognitive architecture that leads to certain types of error. However, cognition does not happen in isolation; it is context-driven and susceptible to contextual influences. A more complete framework of human error includes these contexts to develop a more systemic perspective on human error.

22.3.2 Error in Context

Hobbs (2001) performed an analysis of the relationship between contextual and contributing factors and prevalence of types of errors in aviation maintenance. The goal of Hobbs's work was to understand the specific contribution of contextual factors and the likelihood at which they promote certain error forms. For the purpose of data collection, Hobbs developed a safety occurrence questionnaire that assessed safety-related issues as perceived by aircraft maintenance engineers. At the core of the questionnaire, respondents were asked to report safety-related threats and to specify the chain of events that lead to the specific event. Participants were also asked to describe their understanding of why the specific safety event happened. A correspondence analysis examined the relationship between unsafe acts (types of error) and the external factors that contributed to these acts. Out of all reported events, 96% included unsafe acts. Out of all unsafe acts, 20% involved memory lapses, 17% involved violations, 13% involved slips, 10% involved rule-based errors, 12% involved knowledge-based errors, and 6% involved errors described as errors that involved failure to perceive information in the environment. The correspondence analysis plot revealed the following relationships between unsafe acts and contributing factors:

- Lapses of memory were closely associated with pressure and fatigue.
- Rule-based errors were closely related to procedural problems, difficulties in coordination, and errors that occurred earlier.
- Knowledge-based errors were mostly repeated due to lack of training.

- Slips were related to equipment problems and fatigue.
- Violations were related to pressure and equipment deficiencies.

What makes these findings relevant in the context of human error in health care is that they identify the contextual factors that lead to performance breakdowns, which are not included in a cognitive analysis of human fallibility. Other work also focused on these contributors to Human Error.

22.3.3 Situational and Systemic Influences on Human Error

22.3.3.1 Latent Conditions

Latent conditions are factors that are present in organizations for long periods of time and that are important contributors to human error. However, their contribution is less apparent than the contribution of active failures. To understand active failures better, a number of examples follow: Poor design of equipment, tools, and devices; problems in supervision; maintenance failures that are not detected over long periods of time; manufacturing defects not affecting operation under normal conditions; unworkable procedures; clumsy automation; and shortfalls in training. A good example for a device with latent conditions in health care is provided by Lin et al. (1998) for an infusion pump and by Leveson and Turner (1993) for a dual mode accelerator. Relevant in this context is that latent conditions may be present for many years before, in conjunction with active failures, a loss occurs. Latent conditions exist as a result of high-level decisions in an organization and the contributions of regulators, manufacturers, designers, and organizational managers. Latent conditions are present in all systems and organizations, and their existence is unavoidable. This is partly due to the inability of designers, managers, and others to plan for the unexpected and that changes and decisions have unanticipated consequences (Merton, 1936). Latent conditions are found at the blunt end of operations, where active failures are usually committed at the sharp end.

22.3.3.2 Error-Producing Conditions

The goal of including error-producing conditions (EPC) into a framework of human error is to help identify conditions under which the occurrence of error is more likely and by removing those conditions to implement measures that reduce the likelihood of error. This approach requires the identification of precursors or EPC that contribute to the creation of hazards. Adopting a framework based on Williams's (1988) work on the Human Error Assessment and Reduction Technique (HEART), the most important EPC are the following:

Unfamiliarity with a situation, time pressure in error detection, low noise-to-signal ratio, mismatch between an operator's mental model and that imagined by a device designer, impoverished quality of information, ambiguity in performance standards, disruption in normal work–sleep cycles, and unreliable instrumentation. The foundation of these EPC is Williams's comprehensive review of the human factors literature on performance-shaping factors.

Unfortunately, this work has yet not received its well-deserved attention in the context of health care. However, to provide some evidence for the benefit of this approach as an analytical tool to identify problems in the health care context, we will discuss one health care application of this approach.

22.3.3.2.1 EPC in Health Care

Drews et al. (2008) studied the prevalence of EPC in the ICU that are associated with specific devices. The selection of specific devices in this study was driven by their function (therapeutic or monitoring devices) and their criticality for patient safety (Samore et al., 2004).

To identify the impact of EPC, the authors developed a questionnaire that included 121 items grouped according to general EPC and device-specific EPC. The instrument was administered to 25 experienced ICU nurses who were asked to rate their level of agreement with general ("You always can tell whether something is malfunctioning or just difficult to use") and device-specific statements concerning EPC ("You feel you have received adequate training to use [device]") had to be rated. Ratings were obtained on a scale from 1 to 9 with lower scores indicating higher agreement with the statement (1 = "strongly agree"; 9 = "strongly disagree") (see Table 22.1). Next, the average ratings were aggregated for individual EPC. The results in Table 22.1 indicate that low signal-to-noise ratio, unreliable instrumentation, a mismatch between a nurse's mental model and the mental model of the designer of a device, and shortage of time rank high in their

TABLE 22.1

Mean Scores (Standard Deviations) of Ratings for Error-Producing Factors

Error-Producing Factor	Mean	SD
Unfamiliarity with situation	3.08	1.21
Shortage of time	4.42	1.74
Low signal-to-noise ratio	5.92	2.02
Operator–designer mismatch	4.63	1.31
Quality of information	3.81	1.51
Ambiguity in performance standards	1.13	0.34
Disruption of work–sleep cycles	3.81	1.55
Unreliable instrumentation	4.82	1.15

TABLE 22.2

Mean Ratings (Standard Deviations) for Device Specific Error-Producing Conditions Based on Device Criticality

| Error-Producing Condition | Criticality | | |
	High	Medium	Low
Unfamiliarity with situation	4.4 (2.9)	2.3 (1.7)	1.3 (0.5)
Shortage of time	3.8 (2.3)	2.1 (1.3)	2.5 (1.8)
Low signal-to-noise ratio	3.3 (2.1)	2.1 (1.2)	3.1 (1.8)
Operator–designer mismatch	4.4 (2.6)	2.8 (1.5)	4.0 (2.3)
Quality of information	4.0 (2.6)	2.0 (1.0)	4.0 (2.4)

importance as EPC. These higher mean rankings can be contrasted with the EPC ambiguity in performance standards that seems not to be perceived as an important contributor to conditions that are increasing the likelihood of error.

The mean scores of statements across device criticality categories (aggregated EPC ratings across device categories) are reported in Table 22.2.

Analyses of variance for each EPC identified five EPC with significant differences between devices of varying criticality. These conditions were unfamiliarity ($F(2,71) = 28.9$; $P < 0.01$), shortage of time ($F(2,71) = 5.0$; $P < 0.01$), low signal-to-noise ratio ($F(2,71) = 3.1$; $P < 0.05$), operator–designer mismatch ($F(2,71) = 7.6$; $P < 0.01$), and quality of information ($F(2,71) = 30.0$; $P < 0.01$). Post hoc tests indicated that differences were significant for all devices for the EPC unfamiliarity. In the case of time shortage, there was a significant difference between high and moderate and high and low criticality, and none between moderate and low critical devices. In cases of the EPC operator–designer mismatch and quality of information, the only significant differences were between high and moderate and moderate and low criticality devices.

These results indicate that there are differences between EPC ratings and devices that vary in terms of their criticality in such way, that nurses identify more EPC with devices that are more critical for patient care. Thus, it is quite ironic that the most critical devices for patient care and patient safety are also the ones that have the greatest potential to contribute to human error.

Overall, the results of this work indicate the importance of analyzing the context in which people perform tasks, since these circumstances are potentially facilitators to failure.

22.3.3.3 Violations

This discussion has so far centered on the concept of human error and how cognitive architecture and contextual influences affect human performance. However, the discussion of human failure is not complete without including violations of procedures. Violations are often considered the intentional or erroneous deviation from a protocol, procedure, or rule (Reason, 1997). Certainly, a person committing a violation does not intend a negative outcome, except in the case of sabotage (not be considered here). The next section will introduce types of violations that are highly relevant in the context of health care.

A violation can be defined as a deviation from a safe operating procedure, standard, or rule (Reason et al., 1994). Both organizational and individual factors contribute to the occurrence of violations. Often violations are described as behavior that is deliberate but nonmalevolent (Reason, 1990). Reason (1990) distinguished between routine, optimizing, and necessary violations. Others identified additional types of violations, for example, situational violations (Lawton, 1998). Runciman et al. (2007) distinguish between four types of violations: routine violations, corporate violations, exceptional violations, and necessary violations (see Figure 22.2).

Routine violations are violations that occur when a person perceives an alternative, more efficient way of dealing with a task than that required by a policy or protocol. As a result of this behavior, safety is sacrificed. Often, external pressures reinforce routine violations; they are then repeated on a routine basis. Corporate violations result from decisions at the administrative level that create a situation that supports the violation of procedures (e.g., excessive working hours). Exceptional violations occur in unusual or exceptional circumstances where a routine cannot be followed and an exceptional response is required. What makes these violations problematic is that they are risky because of their deviation from routine and likely to not produce the desired outcome. Violations like these occur under novel conditions. Finally, optimization violations are those violations where additional motives are involved that go beyond

task specific considerations and where these motives supersede the primary motivation to engage in a task.

Violations are an important component of a model of human behavior and breakdowns of performance, because in addition to limitations of the cognitive architecture, violations help explain which conditions foster the occurrence of error. In addition to individual, task-related, and organizational contributors, violations create conditions that make error more likely.

22.3.3.4 Violations in Health Care

In health care, the types of violations outlined earlier can be observed in many contexts. For example, a very common routine violation for health care workers is to not sanitize their hands when entering a patient room. Violations like this might be partly driven by high levels of time pressure. Optimizing violations in the health care context are violations where maximization of a personal goal may lead to the administration of a procedure that is not indicated. For example, a provider who performs a surgical procedure because the procedure is well compensated commits an optimization violation, when another more qualified provider also could provide the procedure. Nurses and doctors who omit steps in a procedure because of tight scheduling of patients and limited time available are committing necessary violations. For example, a clinician who is working under time pressure and who turns his back to a sterile field during a central line insertion is, according to protocol, required to reestablish a sterile field by redraping the patient. However, in the presence of time pressure, the clinician may not redrape the patient to reestablish sterility and therefore commits a necessary violation. An example for a corporate violation in health care is chronic understaffing of a unit so that nurses working at the unit provide care to more than the optimal number of patients with predictable negative consequences.

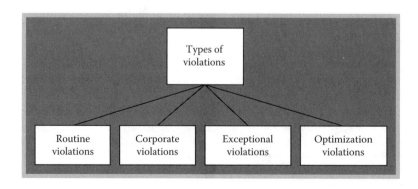

FIGURE 22.2
Four types of violations. (From Runciman, B. et al., *Safety and Ethics in Healthcare: A Guide to Getting it Right*, Ashgate Publishing, Aldershot, England, 2007. With permission.)

TABLE 22.3

Examples of VPC and EPC

EPC		VPC
Intrinsic	**Systemic**	**VPC**
High levels of diagnostic uncertainty	Poor spatial and equipment design	Gender
High decision density	High communication load	Personality
High cognitive work load	Holding admitted patients	Risk-taking behavior
Time pressure	Production pressures	Normalization of deviance
Multiple transitions of care	High noise levels	Maladaptive group pressures
Interruptions/distractions/ multitasking	Inadequate staffing	Maladaptive copying behavior
Low signal-to-noise ratio	Incompatible goals	Authority gradient effects
Surge phenomena	Poor feedback	Likelihood of detection
Resource fluctuations	Inexperience	Inconvenient safety procedures
Fatigue	Inadequate supervision	
Disruption of circadian rhythm	Physical work demands	
Sleep deprivation/debt		
Novel or infrequently occurring conditions		

Source: Adapted from Croskerry, P., When diagnoses fail. New insights, old thinking. 2003. Retrieved on June 29, 2011 from http://www.stacommunications.com/journals/pdfs/cme/cmenov2003/diagnoseserrors. pdf.

Note: Ordered based on likelihood of creating undesirable behavior. Intrinsic factors are specific to the practice of emergency medicine, while systemic factors generalize beyond emergency medicine.

22.3.3.5 *Violation-Producing Conditions*

In addition to EPC, Williams (1997) identified a number of violation-producing conditions ("VPC," see also Croskerry, 2005). In the presence of a VPC, individuals are more likely to violate procedures or protocols at the workplace. Among the conditions that Williams identified are the perceived low probability of detection, inconvenience in performing a procedure according to protocol, authority to violate a procedure, copying behavior of others who are violating procedures, the lack of a disapproving authority figure present, the perceived requirement to obey authority figure, being male, and group pressure to violate existing procedures. VPC have been identified in a number of contexts, like plant management (Rasmussen and Petersen, 1999) and traffic violations (Jason and Liotta, 1982; Van Elsande and Fouquet, 2007). Croskerry and Chisholm (2005) applied the concept of VPC to the health care context (see also Croskerry and Wears, 2003). A modified version of the factors Croskerry and Chisholm (2005) identify as EPC and VPC is presented in Table 22.3.

22.3.3.5.1 *VPC in Health Care*

Patterson et al. (2006) report the findings of a study that examined bar code medication administration in clinical settings. In their study, they observed a total of 28 nurses during the process of medication administration with the goal of identifying workaround strategies related to problems with the barcode scanning method. The workarounds they identified can be classified as routine violations since they involve disabling the technology-based defenses that were put in place to reduce the likelihood of misadministration of drugs to patients. Workarounds in the context of patient identification that were identified were entering patient information (Social Security number) by hand instead of scanning the patient's wristband and scanning surrogate wristbands not located on the patient. One of the many interesting findings in this study was that in an acute care setting only 53% of the patient identification strategies were in compliance with the protocol for drug administration. In addition to the collection of data on patient identification strategies, data on the medication administration strategy was collected. For medication administration, the procedures specify the steps of scanning the patient to identify the patient, scanning the medication, pouring the medication, and, finally, immediately administering the medication to the patient. Prepouring of the medication is strongly discouraged since it involves scanning of the medications for several patients before the medications are administered, possibly resulting in medication misadministration. The authors found that long-term care nurses were more likely to engage in a violation than nurses in acute care.

Overall, the findings of this study indicate that violations of procedures are frequent in the context of patient identification and medication administration when using bar code technology. However, the context in which these tasks are performed plays an important role. The authors express skepticism about the effectiveness of sanctions, training, or policies and procedures with the goal of increasing compliance. The main reason for this skepticism is that the observed nurses stated that high time pressure present in the unit was in the way of complying with the procedures. Similar findings on violations in the context of drug administration are reported by Koppel et al. (2008).

The previous discussion illustrates the important role that contextual factors play in contributing to erroneous performance. It is this assumption that is at the core of systemic approaches to human error, because human performance and human cognition are not isolated from the context in which they occur.

22.3.3.6 Hazards

The result of an unsafe act is a hazard. A hazard increases the likelihood of an adverse event as an outcome of an action. The difference between hazards and adverse events is that an adverse event is the realization of the potential for harm that is associated with a hazard. This transition from potential into reality can be due to several contributors. One of the contributors to this transformation is that the defenses that are normally in place to protect patients and health care workers are not effective. Another reason is that the defenses are intentionally disabled. Finally, it is possible that the conditions that lead to a particular hazard were not foreseen, and defenses were not available. Thus, the lack of effective defenses in conjunction with an unsafe act can lead to an adverse event.

22.3.3.7 Defenses

Even with potentially high rates of human error in health care, not all of the errors result in patient harm. This is due to the fact that in health care, as in other industries, defenses are in place. Defenses are protective measures put in place to reduce the likelihood of negative outcomes resulting from an unsafe act. Defenses are designed to serve one or more functions. These functions may aim at creating a better understanding and awareness of hazards present and/or to provide some guidance by having people follow protocols or checklists. The functions also provide alarms when dangers are imminent by utilizing direct feedback, although this might be difficult or impossible in some contexts. Two types of defenses can be distinguished that serve the aforementioned functions: hard and soft defenses (Reason, 1997). Hard defenses include technical devices (e.g., alarms on patient monitors, ventilators, infusion pumps), where soft defenses rely on people and

organizational factors (e.g., procedures, training, licensing). Conventional approaches from engineering make the assumption that the presence of defenses increases the safety of a system. Paradoxically, defenses do not only produce positive effects by increasing the resistance of a system, but might also increase the likelihood of error and violations. For example, unreliable bar code readers will force the identification and implementation of workarounds to avoid delays and inefficiencies, rendering this line of defense ineffective. The identification of ineffective defenses is a challenge especially in sociotechnical or socionatural systems since actors in these systems may differ in their intentions from those of the developers of the defenses, creating a potential conflict. In addition, because of the complexities of these systems, it is often challenging to evaluate the effectiveness of defenses in the context of their use. Nonetheless, when pursuing the goal of error reduction in health care, it is important to recognize ineffective defenses, to identify the context of their ineffectiveness, and to remove them, or, alternatively, to address the specific vulnerabilities that are associated with these ineffective defenses.

22.3.3.8 Adverse Event

There are a number of important differences in the use of terminology when describing and analyzing human error during a comparison between health care and other industries. These differences in terminology point toward important domain-specific differences that need to be taken into account when investigating human error in health care.

The terminology to describe the consequences of human error in industries other than health care uses frequently the term "losses" to describe the endpoint of error. In health care, "losses" are described as events that are related to patient injury, though these losses can also involve injury of health care workers or losses of resources, like waste. However, since most of the work focuses on patient-related losses, the terms "iatrogenic injury" or "adverse event" are used commonly. Often an adverse event is defined as any injury to a patient due to medical management as opposed to some underlying disease (Rothschild et al., 2005). Accordingly, a *nonpreventable adverse event* is an unavoidable injury despite appropriate medical care where a *preventable adverse event* is an injury due to a nonintercepted serious medical error, that is, a failure of defenses in conjunction with an unsafe act.

Serious medical error is medical error that causes harm or injury or has the potential to cause harm, that is, a hazard is present. The category of serious medical error includes preventable adverse events, intercepted serious errors, and nonintercepted serious errors. *Intercepted serious error* can be defined as serious medical error that, because of the effectiveness of defenses, does not reach the patient, where *nonintercepted serious error* is a serious medical error that is not caught by the defenses and reaches the patient. Additionally, there is the category of trivial medical errors that have either little or no potential for harm to the patient. This error category is rarely studied in health care since the main focus of patient safety is to reduce the number of serious injuries and fatalities. However, the frequency of trivial medical errors can serve as an indicator for deeper system-based problems. Also, in conjunction with contextual factors, a potentially serious medical error might be realized as a trivial error when there is no negative outcome. However, a more advanced theory of human error should not categorize error based on the outcome but based on the mechanisms that are involved. Thus, we will use the term "adverse event" to describe the outcome and exclude nonpreventable adverse events since they are outside the scope of human error.

22.4 Integrative Model of Human Error in Health Care

The previous sections present the elements that are included in a systemic theory of human error in health care. Such theory has, necessarily, similarities with the general models of human error; in both cases limitation of the cognitive architecture and the responsiveness of this architecture to contextual variables play an important role. However, there are also important differences between these approaches. Next, we will discuss the similarities and differences in turn.

The similarities involve an overlap of the elements that are included; for example, elements like EPC, VPC, unsafe acts, etc., are part of all models since they reflect theoretical constructs that incorporate the vulnerability of the cognitive architecture to contextual factors. Also, these models share the assumption that it is the adaptability of human behavior to change in the environment that is an essential element of any theory of human error. An additional commonality between these approaches is that theories of human error need to acknowledge the dynamic nature of actions in any system and the significant influence of performance-shaping factors. Safe acts in this context can be thought of as the result of equilibrium between attractant and repellant forces (here they are error and violation facilitating conditions) that can shape performance. A distortion of this equilibrium transforms a safe act into an unsafe act, creating a hazard, and, in conjunction with failed defenses, an adverse event.

However, there are also important differences that are related to the differences in the domains. One point that differentiates the domain for which theories of human error were developed compared to health care relates to the complexity of health care. According to Kizer (2000), health care is likely the most complex activity being undertaken by humans. What creates this high complexity are the complex technologies involved, the large number of powerful drugs available, the range of backgrounds of providers involved, the lack of clear lines of authority, the highly variable physical settings, the range of unique patients, the presence of significant communication barriers, the widely varying care processes, and the high time pressure that is often present.

Another important difference is that the health care domain differs from other domains to the extent to which there is potential for surprise (Durso and Drews, 2010). In socionatural systems like health care, interaction with the natural system creates and requires higher levels of variability, which increases the potential for erroneous behavior. Since technical systems, to a large extent, are based on the laws of physics, it is plausible to assume with Durso and Drews (2010, p. 74) that the "differences between physics and biology point to a number of safety-relevant dimensions, all of which favor aviation over health care in terms of safety."

Another difference between health care and non–health care industries relates to the issue of adaptability of behavior. In some contexts, adaptability is a requirement for unexpected conditions of operation (e.g., aviation), where in others (e.g., health care) adaptability is a requirement for successful performance even under normal conditions. For example, varying degrees of anatomical variation of the internal jugular vein can be found in up to 40% of patients (Lin et al., 1998) in need of internal jugular vein temporary angioaccess. This level of variation requires high levels of adaptability of the physician performing a catheter insertion. Thus, adaptability and flexibility are a requirement in health care and consequently are limiting the extent to which practices that were successful in non health care industries can successfully be applied to health care with the goal to reduce human error.

Figure 22.3 illustrates a theory of human error in health care that incorporates the differences outlined earlier. Behavior-shaping internal and situational influences are represented as arrows that affect an action by pulling it toward the outcome of an unsafe act or successful behavior. The overall forces that are present in professional contexts are usually asymmetrical, that is, favoring safe acts. However, since this is also a reflection of the organization learning how to structure particular acts, innovation and new types of acts required may initially, due to a lack of experience, favor safe acts less than unsafe acts. Finally, there are strong compensatory

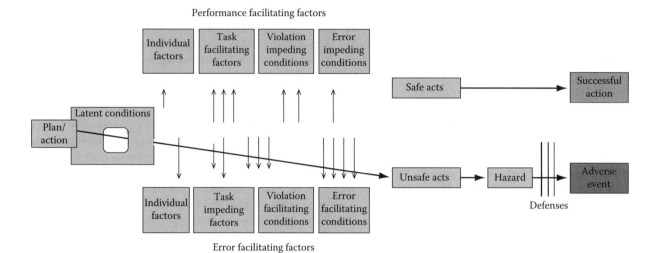

Performance facilitating factors

Error facilitating factors

FIGURE 22.3
An integrative model of human error.

mechanisms in place that support accomplishing successful behavior.

22.5 Studies of Human Error in Health Care

The IOM Report (1999) discusses complexity and technology as contributing factors to human error in health care. By focusing on the higher rates of adverse events in the highly technical surgical specialties of vascular surgery (16.1%), thoracic surgery (10.8%), and neurosurgery (9.9%) (Brennan et al., 1991), we find evidence that technological aspects are contributors to adverse events. Also, data from the Federal Drug Administration show a similar picture about device-related adverse events. For example, from 1998 to 2001, an increase from 62,000 reported adverse events to more than 100,000 adverse events in 2001 illustrates that increasing introduction of technology contributes to adverse events. However, it seems premature to draw the conclusion that it is technology that contributes to the high prevalence of human error in health care. As will be demonstrated in the next sections on error in critical care, surgery, and anesthesiology, the contributors range from organizational factors to individual contributors that is consistent with the model of human error in health care outlined earlier. Technology plays an important role as a contributor to human error, but it also carries the promise to reduce human error in health care.

22.5.1 Critical Care

Out of all hospital beds, 10% are acute care beds. In the United States, 4.4 million patients are admitted

annually to the intensive care unit (ICU) with the expectation that these numbers will increase due to the aging-related demographic changes of the US population. In addition, the acuity of illnesses of hospitalized patients will grow as a result of this demographic change (Angus et al., 2000). The average mortality rate of ICU patients is between 12% and 17%, which means that more than 500,000 ICU patients die annually (Al-Asadi et al., 1996).

Critical care medicine is a specialty in medicine that differs from others in some important ways that makes it interesting for the study of human error. Critical care medicine is an area where high-risk decisions often have to be made rapidly and are often based on incomplete patient data. In addition, the presence of comorbidities in patients creates a high level of complexity that makes decision-making difficult. Finally, patients in critical care often have lower physiologic reserves to help them cope with potential problems from suboptimal care.

One of the earliest studies that investigated the frequency of adverse events in the ICU was conducted by Donchin et al. (1995). Donchin et al. studied the ICU in a 650-bed tertiary-care hospital and collected data on error over a 4 month time period, where they identified a total of 554 human errors. Of the 464 errors that were reported by ICU staff, 46% of the errors were committed by physicians and 54% were committed by nurses. Also, the authors investigated the distribution of error over 24 h periods and report that a larger proportion of errors were committed during the day compared to the night. However, the proportionally largest number of errors was reported during the morning hours between 10 am and noon when staff activities peak. In addition, peaks for nurse error were also found during shift changes. Donchin et al. (1995) estimated that there is an occurrence of 1.7 errors per patient per day in the ICU. Further, Donchin et al. found that in the examined six-bed medical ICU a "severe or potentially detrimental error" (p. 146) occurred twice during a 24 h time period.

Andrews et al. (1997) studied adverse events in two intensive care units and one surgical unit in a large, tertiary care urban teaching hospital affiliated with a university medical school. In their prospective observational study, they found that out of the 1047 studied admissions, 480 (45.8%) patients were identified as having an adverse event. The authors developed nine categories of adverse events and analyzed the percentages of adverse events for each category. The largest category of adverse events, almost 30%, was monitoring and daily care, 19.5% of the adverse events fit in the category of complications, and 13.4% in the treatment category.

Rothschild et al. (2005) investigated the incidence rate and nature of adverse events and medical error in a Medical Intensive Care Unit (MICU) and a Coronary Care Unit (CCU). For this study, the authors collected data during nine 3 week periods, distributed over a time period of 12 months. The study occurred at a tertiary academic hospital with 720 beds. Their data analysis revealed a total of 120 adverse events in 79 patients, leading to an adverse event proportion of 20.2%. Of the 120 adverse events, 66 (55%) were nonpreventable, while 54 (45%) events were classified as preventable. In addition, they found 223 intercepted and nonintercepted serious errors, which were defined as failure of a planned action to be completed as intended (slip or lapse) or of a wrong plan to achieve the aim (mistake) that caused harm to the patient or had the potential to cause harm. Based on their findings, all adverse events occurred at a rate of 80.5 per 1000 patient days, preventable adverse events occurred at 36.2 per 1000 patient days, and serious error at 149.7 per 1000 patient days. One interesting aspect of Rothschild's et al. (2005) findings is that when analyzing all serious medical errors (intercepted and nonintercepted serious error and preventable adverse events), 148 out of 277 (53%) occurred in the categories of ordering or execution of treatments. Thus, the authors observed mostly action-based slips and memory-based lapses that contributed to the presence of medical error. Among the cases that were part of this category, wrong medication dosages were contributing with 62 errors; failure to take precautions or to follow a protocol to prevent accidental injury contributed with 22 errors. It is one of the remarkable findings of this study that error is related to the execution of plans rather than related to determining which course of action to follow (24.2% of all serious medical errors). This finding reflects Bates and Gawande's (2003) notion that in medicine, in the past, the emphasis was more on the development of plans (identifying therapy) rather

than ensuring that the plans (therapeutic interventions) were correctly executed.

Graf et al. (2005) report a study of human error in a medical ICU. They analyzed self-reported observed or committed errors or incidents that were reported on a structured incident form. During the data collection phase that included a 64 day reporting period, a total of 216 patients were admitted to the ICU. In 32 of the patients at least one medical error was reported with a total number of 50 reported errors. In 10 patients more than one error was reported. Interestingly, those patients who experienced error were more severely ill than those patients who did not experience error. Also, those patients who experienced more than one error were significantly more ill than those who experienced a single error. As contributing factors to error, the authors identified disregard of standards, rules, and orders and poor staff communication. What makes this study interesting in the present context is that the authors also analyzed the defenses that successfully averted harm from the patient. The defenses that were identified by the authors include regular rounds and supervision in 25 instances (50%), awareness in 20 instances (40%), and a monitoring alarm (2%). Luck contributed as a defense in 8% of the cases, though this might not be completely reliable. Overall, the large majority of error (92%) that was identified in this study was rated as "potentially avoidable." See Chapter 40 for additional information.

22.5.2 Surgery

De Leval et al. (2000) published one of the first studies that applied a human factors error framework to study error in the operating room (OR). In their study the authors focused on 243 thoracic and cardiovascular surgeries that involved arterial switch operations performed by a total of 21 surgeons. Overall the authors observed a patient fatality rate of close to 7% and a near-miss rate of 17.7%. Death and/or near misses occurred in 24.3% of the surgeries. Based on their observations, De Leval and colleagues (2000) established for the specific context of their study that many events that lead to surgeon error originated in latent conditions, for example, supervisory and organizational contributors. One interesting methodological finding of the study was that a self-assessment questionnaire that was used did not increase the explained variance in the multiple regression models, casting some doubt on the use of self-assessment instruments in the context of studies on human error in health care.

More recently, Rogers et al. (2006) investigated closed malpractice claims to identify factors that cause surgical error. Surgeon-reviewers examined 444 closed malpractice claims and identified surgical error resulting in patient injury in 58% of the claims. The human factors–based contributors identified in this study were separated into five categories: (1) cognitive factors, (2) lack of technical competence/knowledge, (3) communication breakdowns, (4) patient-related factors, and (5) other system factors. Of all identified cases that involved surgeon error, 23% of the cases resulted in patient death, and 65% of the cases resulted in patient disability. The majority of errors occurred during the intraoperative phase (75%), with postoperative errors ranking second (35%), and preoperative errors ranking third (25%).

The contributing human factors were analyzed by identifying subcategories within the larger five categories described earlier. However, since multiple contributors can facilitate the occurrence of error, the overall percentages exceed 100% in some cases. Within the group of cognitive factors contributing to error, the largest category were errors in judgment (169 cases; 66%), and failure of vigilance and memory (162; 63%). There were 106 (41%) occasions in the category of technical competence/knowledge. Communication breakdowns were separated into four categories with hand-off error being observed 28 (11%) times, lack of clear line of responsibility 24 (9%) times, conflict among personnel 7 (3%) times, and other 29 (11%) times. Patient-related factors were separated into anatomic/physiologic factors and contributed 91 (35%) times, complicating prior medical/surgical history contributed 41 (16%) times, abnormal or difficult anatomy was identified as contributor 24 (13%) times, and morbid obesity contributed to error in 9 (3%) cases. Among the category of patient-related factors were behavioral factors that contributed 23 (9%) times, with noncompliance being the largest subcategory contributing in 10 cases (4%), substance abuse contributing 4 times (2%), and psychiatric illness appearing as a contributor 4 (2%) times. Among systems factors that were identified, most often the lack of supervision contributed 47 times (18%), technology failure contributed 38 times (15%), workload and inadequate staffing was identified as contributor in 9 cases (3%), interruptions/distractions were contributors in 7 cases (3%), ergonomic failure contributed 5 times (2%), and fatigue was identified as a contributor in 3 cases (1%). What is remarkable in this context is that individual factors rarely occurred as a single contributor. This finding provides additional support to the systemic perspective on human error. Another important aspect of this study is that the majority of errors involved more than one contributing clinician. Clearly, it is a combination of different factors that lead to breakdowns and adverse events. Further evidence for this claim comes from the fact that about one third of all observed errors crossed phases of care and was not only present at a single phase. Overall, it is important to emphasize that two-thirds of all observed errors fall into one of the following four categories: communication breakdown, lack of supervision, technology failure, and patient-related factors, clearly indicating the

importance of contextual contributors. Finally, lack of experience or technical skills of the surgeon contributed to 41% of the cases. The importance of cognitive factors as contributors to error is highlighted by the fact that 9 out of 10 cases involved cognitive errors that acted in conjunction with other factors to the genesis of erroneous performance.

Another study by ElBardissi et al. (2007) used structured interviews to investigate the contributors to human error in the Cardiovascular Surgery Operating Room. In their study, the authors distinguished several levels of human factors contributors, consistent with the idea of systemic contributors to human error. The levels used for their analyses followed the human factors analysis classification system (HFACS; Wiegmann and Shappell, 2003) that distinguishes organizational influences, unsafe supervision, preconditions to unsafe acts, and unsafe acts. Under organizational influences, the authors assess the organizational climate, organizational processes, and resource management. Under unsafe supervision, the authors categorize inadequate supervision, problem correction, and inappropriate operations. Under preconditions to unsafe acts, they list environmental factors like the technological and the physical environment, adverse mental states, adverse physiological states, physical/mental limitations, teamwork, and personal readiness. Finally, under unsafe acts they list decision errors, skill-based errors, perceptual errors, routine violations, and exceptional violations. For data collection the authors conducted interviews with 68 members of the surgery team. They asked interviewees to rank their answers on a Likert scale according to the frequency of which factors are present during surgery and contribute to problems. Highest ratings for unsafe acts were skill-based errors, routine violations, and exceptional violations. For preconditions for unsafe acts the highest ratings were identified for crew resource management/teamwork, physical/mental limitations, and physical/mental status. Highest ratings for unsafe supervision were received by factor planning, where organizational influences resource management was ranked highest. One interesting analysis in this paper focused on the correlations between individual factors. Significant correlations were found in a number of relations that are shown in Figure 22.4.

What makes these findings interesting is that Figure 22.4 shows the relations between the individual factors at all four levels. Focusing on unsafe acts only, it becomes clear that the only unsafe act that is not highly related to other preconditions is perceptual errors, which are only correlated with teamwork and potentially specific to error in surgery. Skill-based errors, routine and exceptional violations, and even decision errors are correlated with a number of preconditions indicating that not a single precondition affects performance but that

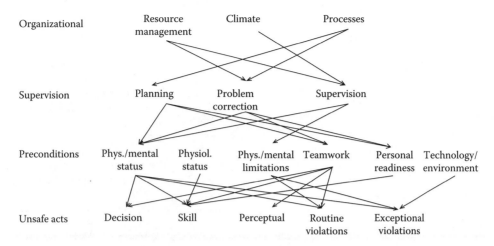

FIGURE 22.4
Significant correlations between components of HFACS for the OR. (Adapted from ElBardissi, A.W. et al., *Annals of Thoracic Surgery*, 83, 1412–1418, 2007.)

it is a number of factors that are relevant contributors. The same high level of interconnectedness applies for the relations between preconditions and unsafe supervision and unsafe supervision and organizational influences. Overall this work demonstrates the relationship between latent conditions (organizational and supervisory factors), preconditions for unsafe acts (error and violation-producing conditions and individual factors), and unsafe acts. Overall, the results of the work on human error in surgery supports the model illustrated in Figure 22.3. See Chapter 45 on human factors and ergonomics in surgery in this handbook for additional information.

22.5.3 Anesthesia

Weinger (1999) reviewed the literature on human error in anesthesia to identify the contribution of technology to adverse events. As Weinger points out, there is a certain irony in the fact that the technology that is developed to facilitate anesthesia administration also contributes to a significant extent to human error. Clearly, this irony is similar to what Bainbridge (1987) described as ironies of automation. The devices and activities that contributed to adverse events in anesthesia are misuse of monitors, ventilator missetting or malfunction, intravenous drug/fluid delivery system problems, intravenous drug dose errors, gas supply errors, breathing circuit disconnections, and others. However, it is remarkable that the number of incidents that are primarily caused by equipment failure is relatively small (estimated as being between 11% and 14%). These results are in stark contrast to the fact that in many other domains additional contributors are present and the resulting adverse event is not cause by a single failure, but by a number of breakdowns and a failure of defenses that all together lead to human error. However, others (see Gaba et al.,

1995) point out that there are also nondevice-related contributors to human error in anesthesia.

Cooper et al. (2002) analyzed preventable anesthesia mishaps with a focus on human contributors. For this purpose, they use a modified critical-incident analysis technique to retrospectively examine the contribution of human error and equipment failure to adverse events in anesthesia. In 47 interviews they identified a total of 359 preventable adverse events, which were grouped into a total of 23 categories. As in previous studies, the majority of reported problems included human error as a contributor (82%). Exclusive equipment failure contributed with only 14% to adverse events in this context. Analyzing the specific categories in terms of frequency of events, the most common events were breathing circuit disconnections, inadvertent gas flow changes, and syringe swaps, that is, confusing two syringes. The most interesting result of this study from a human error perspective is the category of factors associated with the adverse events reported. This category included predisposing circumstances that contributed to the adverse event. Overall a total of 481 observations were listed, which implies that frequently more than one of these factors was present. To identify contributors based on the framework outlined here, we recategorized the 28 factors into factors that fall into the categories of EPC or VPC. Consistent with Williams's framework, unfamiliarity was the largest category (e.g., inadequate total experience, inadequate familiarity with surgical procedure, anesthetic technique, etc.), contributing in 35% of the cases. Time shortage contributed in 7% of the cases, fatigue in 5%, and operator and design mismatch contributed in 2% of the cases. Overall, in 49% of the reported cases, EPC were present. In terms of VPC, supervisor not present contributed in 6% of the cases, and group-related factors contributed in 4% of the cases. So, in 10% of the cases VPC were present. Finally, an additional category was created that included factors that contributed to communication or interaction-related problems. This category contributed in 16% of the cases.

To summarize, an analysis of the associated factors to adverse events identified that EPC contributed in almost 50% of the cases to an adverse event. In 16% of the cases communication or interaction-related problems were a contributor and VPC contributed overall in 10% of the cases.

Overall these results of studies on human error in anesthesiology support the importance of contextual factors as contributors to human error. See the chapter on "Human Factors in Anesthesiology" in this handbook for additional information.

22.6 Conclusions

Overall, the study of human error leads to complex theories. This is a result of the fact that most errors are based on the interactions between the cognitive architecture and contextual factors that influence behavior. Only by focusing on both elements will we be able to develop theories of human error that have any explanatory and predictive value. And it is this aspect of theory development that will allow us to improve health care in general and patient safety in particular. However, without a better understanding of the underlying factors of human error, we will only scratch the surface and any improvement in patient safety will only be temporary.

References

Al-Asadi, L., Dellinger, R., Deutch, J., and Nathan, S. (1996). Clinical impact of closed versus open provider care in medical intensive care unit. *American Journal of Respiratory and Critical Care*, 153, A360.

Andrews, L.B., Stocking, C., Krizek, T. et al. (1997). An alternative strategy for studying adverse events in medical care. *Lancet*, 349, 309–313.

Angus, D.C., Kelley, M.A., Schmitz, R.J., White, A., and Popovitch, J. (2000). Current and projected workforce requirements for care of the critically ill and patients with pulmonary disease. *The Journal of the American Medical Association*, 284, 2762–2770.

Bainbridge, L. (1987). Ironies of automation. In: J. Rasmussen, K. Diuncan, and J. Leplat (Eds.) *New Technologies and Human Error*, Chichester, U.K.: Wiley, pp. 271–283.

Bates, D.W. and Gawande, A.A. (2003). Patient safety: Improving safety with information technology. *The New England Journal of Medicine*, 348, 2526–2534.

Brennan, T.A., Leape, L.L., Laird, N.M., Hebert, L., Localio, A.R., Lawthers, G., Newhouse, J.P., Weiler, P.C., and Hiatt, H.H. (1991). Incidence of adverse events and negligence in hospitalized patients. *New England Journal of Medicine*, 324, 370–376.

Cellar, D.F., Nelson, Z.C., Yorke, C.M., and Bauer, C. (2001). The five-factors model and safety in the workplace: Investigating the relationships between personality and accident involvement. *Journal of Prevention and Intervention in the Community*, 22, 1, 43–52.

Croskerry, P. (2003). The Importance of cognitive errors in diagnosis and strategies to minimize them. *Academic Medicine*, 78, 775–780.

Croskerry, P. (2005). Diagnostic failure: A cognitive and affective approach. *Advances in Patient Safety*, 2: Concepts and Methodology. Bethesda, MD: National Library of Medicine.

Croskerry, P. and Chisholm, C. (2005). What does human factors ergonomics need to know about front-line medicine? Retrieved on January 14, 2011 from: http://www.crnns.ca/documents/Pat%20Croskerry%20Article%20May2005.pdf

Croskerry, P. and Wears, R.L. (2003). *Safety Errors in Emergency Medicine*. In V.J.

de Leval, M.R., J. Carthey et al. (2000). Human factors and cardiac surgery: A multicenter study. *The Journal of Thoracic Cardiovascular Surgery*, 119, 661–672.

Dekker, S.W.A (2002). The reinvention of human error. Technical Report 2002–2001. Lund University School of Aviation, Sweden.

Dekker, S.W.A (2006). *The Field Guide to Understanding Human Error*, Burlington, MA: Ashgate.

Donchin, Y., Gopher, D., Olin, M., Badihi, Y., Biesky, M., and Sprung, C.L. (1995). A look into the nature and causes of human errors in the intensive care unit. *Critical Care Medicine*, 23, 294–300.

Drews, F.A., Musters, A., and Samore, M. (2008). Error producing conditions in the intensive care unit. In: *Advances in Patient Safety: From Research to Implementation*, Vol. 5, AHRQ Publication Nos. 050021 (1-5). Rockville, MD: Agency for Healthcare Research and Quality.

Durso, F. and Drews, F.A. (2010). Healthcare, aviation, and ecosystems—A socio-natural systems perspective. *Current Directions in Psychological Science*, 19, 71–75.

ElBardissi, A.W., Weigmann, D.A., Dearani, J.A., Daly, R.C., Sundt, T.M. III. (2007). Application of the human factors analysis and classification system methodology to the cardiovascular surgery operating room. *The Annals of Thoracic Surgery*, 83, 1412–1418.

Eysenck, M.W. and Keane, M.T. (2005). *Cognitive Psychology: A Student's Handbook*, 5th edn., Hove, England: Psychology Press.

Fisk, J.E. (2004). Conjunction fallacy. In: R. Pohl (Ed.) *Cognitive Illusions*, Hove, England: Psychology Press, pp. 23–42.

Fiske, S.T. and Taylor, S.E. (1991). *Social Cognition*, 2nd edn., New York: McGraw-Hill.

Gaba, D.M., Howard, S.K., and Small, S.D. (1995). Situation awareness in anesthesiology. *Human Factors: The Journal of the Human Factors and Ergonomics Society*, 37, 1, 20–31.

Graf, J., von den Driesch, A. Koch, K.-C., and Janssens, U. (2005). Identification and characterization of errors and incidents in a medical intensive care unit. *Acta Anaesthesiologica Scandinavica*, 49, 930–939.

Goodstein, L.P., Andersen H.B., and Olsen, SE. (1988). *Tasks, Errors and Mental Models*, London, U.K.: Taylor & Francis.

Greenwood, M. and Woods, H.M (1919). The incidence of industrial accidents upon individuals with special reference to multiple accidents. British Industrial Fatigue Research Board, Report Number 4. London, U.K.: HMS.

Helmreich, R.L. (1997). Managing human error in aviation. *Scientific American*, 276, 62–67.

Hobbs, A. (2001). The links between errors and error-producing conditions in aircraft maintenance. Presented at the *15th FAA/CAA/ Transport Canada Symposium on Human Factors in Aviation Maintenance and Inspection*, London, U.K.

Hollnagel, E. (2004). *Barriers and Accident Prevention*, Aldershot, U.K.: Ashgate.

Hollnagel, E. (2007). Human error: Trick or treat. In: F. Durso (Ed.), *Handbook of Applied Cognition*. West Sussex, U.K.: John Wiley & Sons.

Institute of Medicine. (1999). *To Err Is Human: Building a Safer Health System*. Washington, DC: National Academy Press.

Jason, L.A. and Liotta, R. (1982). Pedestrian jaywalking under facilitating and non-facilitating conditions. *Journal of Applied Behavior Analysis*, 15, 496–473.

Kizer, K. (2000). Ten steps you can take to improve patient safety in your facility. *Briefings in Patient Safety 2000*, 1, 1–4.

Kohn, L.T., Corrigan, J.M., and Donaldson, M.S. (1999), *To Err Is Human: Building a Safer Health System*, Washington, DC: National Academy Press.

Koppel, R. Wetterneck, T.B., Telles, J.L., and Karsh, B. (2008). Workarounds to barcode medication administration systems: Occurrences, causes and threats to patient safety. *JAMIA*, 15, 408–428.

Lawton, R. (1998). Not working to rule: Understanding procedural violations at work. *Safety Science*, 28, 77–95.

Lee, Y.S., Kim, Y., Kim, S.H., Kim, C., Chung, CH., and Jung, W.D. (2004). Analysis of human error and organizational deficiency in events considering risk significance. *Nuclear Engineering and Design*, 230, 11, 61–67.

Leveson, N. and Turner, C.S. (1993). An investigation of the Therac-25 accidents. *IEEE Computer*, 26, 7, 18–41.

Lin, L., Isla, R., Doniz K., Harkness H., Vicente, K.J., and Doyle D.J. (1998). Applying human factors to the design of medical equipment: Patient-controlled analgesia. *Journal of Clinical Monitoring and Computing*, 14, 253–63.

Lin, B.S., Kong, C.W., Tarng, D.C., Huang, T.P., and Tang, G.J. (1998). Anatomical variation of the internal jugular vein and its impact on temporary haemodialysis vascular access: An ultrasonographic survey in uraemic patients. *Nephrology Dialysis Transplantation*, 13, 134–138.

Mach, E. (1905). *Erkenntnis und Irrtum*. Barth: Leipzig.

McCrae, R.R. and Costa, P.T. Jr. (1987). Validation of the five-factor model of personality across instruments and observers. *Journal of Personality and Social Psychology*, 52, 81–90.

Merton, R.K. (1936). The unanticipated consequences of social action. *American Sociological Review*, 1, 894–904.

Miller, D.P. and Swain, A.D. (1987). Human error and human reliability. In: G. Salvendy (Ed.), *Handbook of Human Factors*. New York: John Wiley & Sons.

Norman, D.A. (1981). Categorization of action slips. *Psychological Review*, 88, 1–15.

Patterson, E.S., Rogers, M.L., Chapman, R.J., and Render, M.L. (2006). Compliance with intended use of bar code medication administration in acute and long-term care: An observational study. *Human Factors: The Journal of the Human Factors and Ergonomics Society*, 48, 15–22.

Perrow, C. (1984). *Normal Accidents: Living with High Risk Technologies*. New York: Basic Books.

Pohl, R. (2004). Hindsight bias. In: R. Pohl (Ed.) *Cognitive Illusions*, Hove, U.K.: Psychology Press, pp. 363–378.

Rasmussen, J. and Jensen, A. (1974). Mental procedures in real-life tasks: A case study of electronic trouble shooting. *Ergonomics*, 17, 3, 293–307.

Rasmussen, B. and Petersen, K.E. (1999). Plant functional modelling as a basis for assessing the impact of management on plant safety. *Reliability Engineering and System Safety*, 64, 2, 201–207.

Reason, J. (1974). *Man in Motion*. London, U.K.: Weidenfeld.

Reason, J. (1989). *Human Error*. Cambridge, U.K.: Cambridge University Press.

Reason, J. (1990). *Human Error*, New York: Cambridge University Press.

Reason, J. (1997) *Managing the Risks of Organizational Accidents*, Brookfield, WI: Ashgate.

Reason, J., Parker, D., and Free, R. (1994). *Bending the Rules: The Varieties, Origins and Management of Safety Violations*. Leiden, the Netherlands: Faculty of Social Sciences, University of Leiden.

Reber, R. (2004). Availablity. In: R. Pohl (Ed.) *Cognitive Illusions*, Hove, England: Psychology Press, pp. 147–164.

Rothschild, J.M., Landrigan, C.P., Cronin, J.W. et al. (2005). The critical care study: The incidence and nature of adverse events and serious medical errors in intensive care. *Critical Care Medicine*, 33, 1694–1700.

Rogers, S.O. Jr., Gawande, A.A. et al. (2006). Analysis of surgical errors in closed malpractice claims at 4 liability insurers. *Surgery*, 140, 25–33.

Runciman, B., Merry, A., and Walton, M. (2007). *Safety and Ethics in Healthcare: A Guide to Getting it Right*, Aldershot, U.K.: Ashgate Publishing,

Ryle, G. (1949). *The Concept of Mind*, Chicago, IL: New University of Chicago Press.

Samore, M., Evans, S., Lassen, A., Gould, P., Lloyd, J., Gardner, R., Abouzelof, R., Taylor, C., Woodbury, D., Mary Willy, M., and Bright, R. (2004). Surveillance of medical device-related hazards and adverse events in hospitalized patients. *The Journal of the American Medical Association*, 291, 325–334.

Senders, J.W. (1994). Medical devices, medical errors, and medical accidents. In: S. Bogner (Ed.) *Human Error in Medicine*, Hillsdale, NJ: Lawrence Erlbaum, pp. 159–177.

Swain, A.D. and Guttman, H.E. (1983). *Handbook of Human Reliability Analysis with Emphasis on Nuclear Power Plant Applications*. Albuquerque, NM: Sandia National Laboratories.

Thompson, S.C. (2004). Illusions of control. In: R. Pohl (Ed.) *Cognitive Illusions*, Hove, England: Psychology Press, pp. 115–143.

Turner, B.A. (1976). The organizational and interorganizational development of disasters. *Administrative Science Quarterly*, 21, 378–397.

Turner, B.A. (1978). *Man-Made Disaster*. London, U.K.: Wykeham.

Tversky, A. and Kahneman, D. (1974). Judgment under uncertainty: Heuristics and biases. *Science*, 185, 1124–1131.

Van Elsande, P. and Fouquet, K. (2007). Analyzing human functional failures in road accidents. TRACE report D5.1.

Wagenaar, W.A., Groeneweg J., Hudson, P.T.W., and Reason, J.T. (1994). Promoting safety in the oil industry. *Ergonomics*, 37, 12, 1999–2013.

Weinger, M.B. (1999). Anesthesia equipment and human error. *Journal of Clinical Monitoring and Computing*, 15, 319–323.

Wiegmann, D.A. and Shappell, S.A. (2003) *A Human Error Approach to Aviation Accident Analysis: The Human Factors Analysis and Classification System*, Burlington, VT: Ashgate.

Williams, J.C. (1988). A data-based method for assessing and reducing human error to improve operational performance. In: *Conference Record for 1988 IEEE Fourth Conference on Human Factors and Power Plants*, Monterey, CA, pp. 436–450.

Williams, J.C. (1997). Assessing and reducing the likelihood of violation behaviour—A preliminary investigation. *Proceedings of an International Conference on the Commercial & Operational Benefits of Probabilistic Safety Assessments*. Institute of Nuclear Engineers. Edinburgh, Scotland.

23

Medical Failure Taxonomies

Bruce R. Thomadsen

CONTENTS

23.1 Introduction 342
23.2 Definitions and Rationales for Classifying Human Performance Failures.........................342
    23.2.1 Frustration from a Database and Root Cause Analysis 342
    23.2.2 Definition of Taxonomy 342
    23.2.3 Natures of Failures 343
        23.2.3.1 Definition of Failure...343
        23.2.3.2 Distinctions between Failure, Errors, and Mistakes..................................343
23.3 Rationales for Studying the Nature of Failures..343
    23.3.1 Understanding of Human Performance Failures...344
    23.3.2 Understanding of the Effects of Environments and Other Factors on Performance Failures.............345
23.4 Error Coding versus Taxonomies..346
    23.4.1 Examples of Error Coding 346
    23.4.2 Differences between Error Coding and Classification by Taxonomy 347
23.5 Approaches to Taxonomies: Some Examples...347
    23.5.1 Psychological Viewpoint: Donald Norman 347
        23.5.1.1 Levels of Performance and Types of Failures That Occur at Each Level 348
        23.5.1.2 Complexity of Some Failures 349
    23.5.2 Industrial (Person Faces Machine) Approach: Jens Rasmussen 349
        23.5.2.1 Basis of the Approach and Underpinning Assumptions............................350
        23.5.2.2 Analyses Based on the Classifications 353
        23.5.2.3 Limitations for Use in Medical Settings 354
    23.5.3 Latent Problems: Reason...354
        23.5.3.1 General Approach..355
        23.5.3.2 Latent Errors 355
        23.5.3.3 Limitations 357
    23.5.4 Systemic Approach: The Eindhoven Taxonomy (Tjerk van der Schaaf)................358
        23.5.4.1 Basis of the Approach and Underpinning Assumptions............................358
        23.5.4.2 Labyrinth or Pinball Approaches 360
        23.5.4.3 Applications 361
        23.5.4.4 Directive Properties and Limitations..361
    23.5.5 Scope..362
        23.5.5.1 Basis of the Approach and Underpinning Assumptions............................362
        23.5.5.2 Limitations 363
    23.5.6 Some Medical Taxonomies in the Literature 363
        23.5.6.1 National Coordinating Council for Medication Error Reporting and Prevention...................363
        23.5.6.2 Edinburgh Taxonomy..367
        23.5.6.3 Nursing Practice Breakdown Research Advisory Panel.............................367
    23.5.7 Toward a More Useful Medical Taxonomy...368
        23.5.7.1 Characteristics of a Taxonomy 368
        23.5.7.2 Madison Taxonomy..369
23.6 Classifications Based on Activity and Error 373
    23.6.1 Failure Location on a Process Tree 373
    23.6.2 Failure Location on a Fault Tree 373

23.7 Remedial Actions Based on Classifications 373
    23.7.1 Preventative Actions/Remedial Actions 373
    23.7.2 Categories of Remedial Actions 374
    23.7.3 Relation between Failure Classification and Remedial Actions 374
    23.7.4 Hierarchy of Remedial Actions 374
23.8 Integrated Remedial Actions 376
    23.8.1 Consideration of the Many Dimensions of Information on Failures 376
    23.8.2 Prevention of Events Rather than Cure of Root Causes 376
23.9 Summary 377
References 383

Nearly everyone who has published in this field has devised some form of error classification.

James Reason, 1990

23.1 Introduction

Far from just being academic exercises, classification of the types of failures that contribute to events often reveals information that becomes useful in trying to rectify hazardous situations and conditions. Whether retrospectively investigating the causes of an event or prospectively analyzing potential failures, determining steps to prevent future events seldom is clear. Sometimes analyses lead to many possible actions and selecting one of them is needed. Other times, finding even a single corrective action remains elusive. In both cases, and those between, classifying the failures and potential failures by various facets of their characteristics can provide guidance. This chapter considers failure classification systems: some history of their development, how they differ in their focus and procedures, and how they can contribute to reduction of risk to patients.

23.2 Definitions and Rationales for Classifying Human Performance Failures

23.2.1 Frustration from a Database and Root Cause Analysis

Many organizations face great frustrations dealing with error prevention. Even with very complete database systems and the performance of root cause analysis (RCA) for every event, finding effective remedial actions may remain elusive. In part, the problem comes from the type of information involved in the

analyses. Most often, databases maintain information on the "hard facts" of the event, such as what happened, what conditions existed at the time, who was involved—all very important information. However, while the information can call attention to a commonality, such as a frequent step in a sequence that is prone to error, the data do not usually lead the operator toward a potential solution, nor even suggest what the solution might look like.

RCA, another very useful tool, distills the actions of an event to the earliest causes over which the facility could have any control (the progenitor causes). Compared to simply studying an event-report database, RCA at least deals in causes, instead of just data. Yet, here also, the practitioner may be left with either no idea of how to address the causes found or, worse, finds that actions that strike directly at the progenitor causes may not make any difference in the frequency of events of the same nature.

One reason these tools, and other staples in the armamentarium of risk analysis, sometimes fail is that, while they may identify the "cause," they neglect the nature of the cause. A nurse may neglect to give a patient medication, but an effective remedial action to prevent this in the future would be different if the nurse were too overworked to get to the patient or if the nurse were too involved watching a television show. Taxonomies address the nature of the causes behind events.

23.2.2 Definition of Taxonomy

A *taxonomy* is simply a classification or ordering into groups or categories. The term most often applies to placing living organisms into their classification from phyla through species, but any system with a rational structure used for classification would qualify as a taxonomy.

For applications in patient safety, the taxonomies of interest deal with classifications of failures. An important aspect of the definition should include both classification and ordering.

23.2.3 Natures of Failures

Establishing any type of classification system for failures requires an understanding of failures. That is a big order, but fortunately a great deal of progress can be made with knowledge of some of the natures of failures.

23.2.3.1 Definition of Failure

Practically, failure is simply a lack of success. But that hardly captures the essence of failure. Everyone has experienced failure, and recognizes that failure seems to have deeper impact. Even success can seem like failure if it is qualified, and even happy endings may contain many aspects of failure. Sometimes success can be failure if the goal turns out to have been less desirable than originally thought. Failure may best be thought of as performance that leaves one unsatisfied.

23.2.3.2 Distinctions between Failure, Errors, and Mistakes

Less than acceptable performance characterizes failure. Failure often can be expected. Contests among equal competitors will mean failure for all but one. Such failures often are analyzed to improve performance in the future, although sometimes success lies outside the bounds of likelihood. In other settings, failures fall outside normalcy. In health care, the assumption is that everyone involved in a case performs correctly and effectively, and while the patient may not get well, the cause is not that a member of the health-care team failed. Situations such as health care, where stakes are high and all parties are expected to perform correctly, are *high-reliability* situations. Failures in these cases form *events*.

A failure can result from not performing adequately, such as not executing a necessary function at the required time. A failure of this type goes by the name *error*. An execution error generally falls into the class of *omission or lapse*, the lack of a required action, or the class of *slip*, a botched action. Most often, errors happen without much, or any, thought.

On the other hand, planned actions may not deliver the desired end even if executed well. In addition to misguided plans, the actions may not be planned well, or the result of the planned action may not further a long-term goal. Such failures of intention constitute *mistakes*.

These definitions merely form a convention, since dictionary definitions of "error" and "mistake" share many facets, and in conversation are often used interchangeably. While it may seem silly to make such distinctions (and much finer ones are to come), what sort of failure happened gives insight into why it happened.

To be completely useful, a taxonomy must be multidimensional, reflecting the various facets of events. The dimensions refer to aspects of events that are of completely different natures. Common dimensions include a set of classifications related to

- Where in the procedure the failure occurred
- What was going on around the persons involved in the event
- The environment in which the principals worked
- The type of failure (in what way the principals failed, discussed in the following)
- What factors led the principals to fail

These dimensions form an orthogonal system because of the independence of each dimension. Where in the process a failure occurred may be independent of what was going on in the mind of an operator. On the other hand, sometimes there are patterns connecting one dimension of events with a different dimension. Such relationships become very important information since it points to a potentially hazardous interrelationship. These concepts make a large part of this chapter.

23.3 Rationales for Studying the Nature of Failures

The question of justifying study and classification of failures then turns on why does one care why a failure occurred. As noted in the first section, many different causes could motivate the same act. In the example above of the nurse not giving a patient medication on schedule, say the administration assumed that the nurse simply forgot about it so they gave the nurse a wristwatch with an alarm to set for medication times. However, the real reason for the omission was that the nurse was grossly overworked and was still giving the medications to patients from 3 h before. Obviously, the solution, which might have worked for the assumed cause, would not actually improve the situation. *Knowing the nature of failures helps address the causes that led to the failure.*

Establishing the rationale for studying the nature of failures leads to another question: Is studying the nature of failures worthwhile, compared to proactive approaches to risk reduction? Is that like shutting the barn door after the horse runs out? Obviously, the proactive approaches serve very important functions and form indispensable tools for setting up protection against failure and promoting risk reduction. However, any event carries the assumption that some of its causes may come back in another event. The very commonly referenced model for events posited by James Reason (1990) depicts many layers of procedural safeguards created during the proactive phase

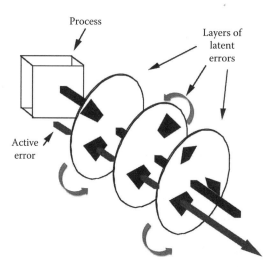

FIGURE 23.1

A model for propagation of errors through holes in the defenses. The holes constitute latent errors, discussed later in this chapter.

of risk reduction. None of the layers are perfect, however, and each has holes that could allow an unsafe action or failure to propagate. Only if holes in all the layers align would the actions propagate into an event. Figure 23.1 takes the concept a little further, picturing the holes in the safeguard layers as dynamic. While the holes in some layers might align at one time, they may not at another. Each time an event occurs, it gives clues as to the procedural locations of the holes. The holes may never again align in the manner that led to the last event, but new combinations can lead to new and different events.

23.3.1 Understanding of Human Performance Failures

Understanding human performance failures often begins by trying to understand human performance successes, that is, how humans perform functions. Modern models for human performance not only go back over a century (models in general go back more than two millennia), but cover a considerable territory. Reason (1990) presents a good summary for the interested reader, and a comprehensive review will not be presented here.

The types of performances of interest in this chapter, those that influence patient care, mostly are skilled actions that take some training. However, sometimes normal "common sense" or everyday actions also come into play. Rasmussen (1982) presents a useful model of human performance.* Most jobs first require some

---

* This discussion recognizes that models only present useful ways of looking at complex phenomena, rather than actually provide explanation for phenomena. As a result, a model may be useful when considering some aspects of the subject, but obviously completely wrong for others.

knowledge of the background related to the functions performed. An example for medical personnel would be anatomy classes or infectious disease control principles. There are also particular functions that must be mastered, such as measuring blood pressure or taking blood samples. Learning the functions begins at a level where each step must be carefully considered, possibly following notes or instruction sheets. With practice, the operation becomes routine, and eventually the operator hardly thinks about it during execution. Each operator learns many procedures and rules that dictate the appropriate situation to perform each procedure.

Rasmussen ranks actions using three groups:

1. Skill based, which when proficient, proceed automatically without intellectual intervention
2. Rule based, which considers the situation and determines what skill-based operations to apply
3. Knowledge based, which uses background information and analytic processes to address situations for which the rules do not apply

In general, humans try to handle problems using a skill if possible, searching for the appropriate rule to tell them what skill to use if that is not obvious, and only as a last resort, trying to figure out the characteristics of the situation and create an original plan to address it. Figure 23.2 illustrates a flow diagram model of the steps leading to an action, called the generic error modeling system (GEMS—Reason, 1990).

Figure 23.3 shows how inputs, operations, and outputs in normal tasks flow through the levels of actions in Rasmussen's model.

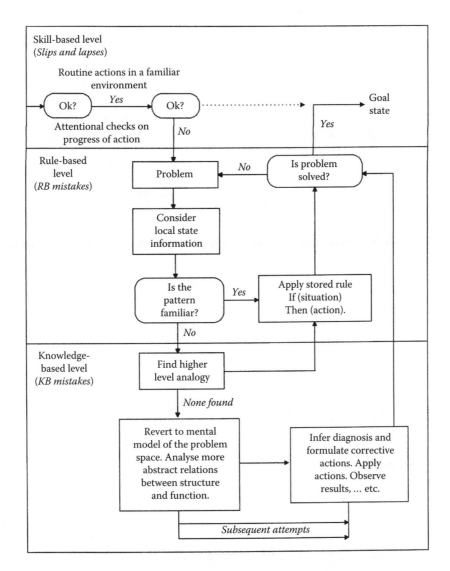

FIGURE 23.2
Flow diagram depicting the steps taken to determine actions to be taken, according to the generic error modeling system. (Reprinted from Reason, J., *Human Error,* Cambridge University Press, Cambridge, 1990. With permission of Cambridge University Press.)

23.3.2 Understanding of the Effects of Environments and Other Factors on Performance Failures

How well humans deal with situations requiring actions not only depends on their abilities and training. Surrounding conditions play a big part. While not directly responsible for how well a person performs, these other conditions influence what the person does, and thus they are called *performance shaping factors* (PSFs).

The environment, to a great extent, determines what one perceives, both how well sensory inputs are noted and how they are interpreted. For example, a room full of noise, such as from patient monitors and a paging system, may prevent a nurse from hearing the patient tell of an allergy to sulfa. Light levels, angles of computer screens, or smells from a plumbing problem all

potentially affect perception. Conditions such as temperature and humidity can add physiological stress that affects performance. See Chapter 15 for more details.

Of a different nature, but still environmental, are conditions such as the time of day or day of the week. The quality of human performance, during a normal work day, follows a pattern of highs and lows as attention waxes and wanes. The end of the work day becomes hazardous as the fatigued worker starts thinking about home. A similar pattern holds for the workweek, as attention fades on Friday afternoon, a time when many failures occur. The situation worsens for extended work days (beyond 8 h) and longer work weeks. The common practice of hospital staff working 12 h shifts places much of the working time when performance falters. Working on holidays presents a double whammy, firstly because

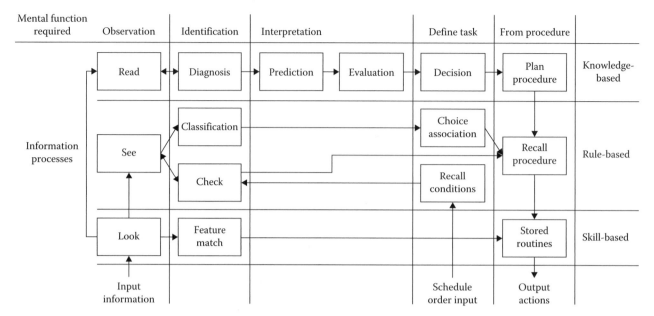

FIGURE 23.3

The flow of actions in Rasmussen's model (1982). (From *J. Occup. Accid.*, 4, Rasmussen, J., Human errors. A taxonomy for describing human malfunction in industrial installations, 311–335, Copyright 1982, with permission from Elsevier.)

the workers would rather be elsewhere (where their thoughts are) and secondly because if needed, other supporting staff members often are unavailable. See Chapter 14 for more details.

Equipment and instrumentation performance also affects human performance. Equipment malfunctions, while possibly causing a failure or event directly, can also lead to human failure. Equipment may give wrong indications or signals that are hard to understand or interpret. When equipment fails to work properly, persons that ordinarily depend on them may need to perform functions manually that they are not used to. At the least, equipment failures add stress to the work environment, increasing the likelihood of human problems.

Personal factors also affect performance. When personal problems weigh heavily on the mind, other inputs and tasks often get pushed to the back as thoughts about the problems keep creeping to the fore. The same happens with happy occasions and anything exciting. Even the morning news can affect how a person acts during the day.

Many shaping factors stem from an organization's policies. Take, for example, the overworked nurse who failed to give the patient medication on time. The basic problem in this case likely is lack of staffing, an organizational decision. Similar decisions involving equipment maintenance could likewise result in failures. The greatest organizational PSF is an administration that just does not care about the quality of care delivered. Such an attitude tends to percolate through the rest of

the organization—leading employees who care about quality to leave.

23.4 Error Coding versus Taxonomies

As the quote at the beginning of this chapter indicates, error classification systems abound. However, not all classification systems form taxonomies. The key distinction in the definition is "ordered." A structure underlies a taxonomy.

23.4.1 Examples of Error Coding

Many error classification systems, and particularly common in medicine, are error coding systems—basically lists of faults. The lists *do* usually have some organization, such as general categories of errors in medication, errors in surgery, and errors in diagnosis. Under the general categories fall subcategories, for example, the errors in medication may allow designations as wrong drug, wrong time, wrong amount, wrong patient, and the like. A good example of such an error coding system is the New York Patient Occurrence and Tracking System (NYPORTS). Tables 23.1 and 23.2 give a sample of this classification system. Error coding systems provide valuable information when assessing what errors take place, identifying procedures or actions that require concentrated resources to improve, or following trends

TABLE 23.1

The Error Codes for the New York Patient Occurrence and Tracking System

Occurrence Codes in Use 2005–2007	
Codes	Description
108[a]	Medication error that resulted in permanent patient harm
109[a]	Medication error that resulted in near death event
110[a]	Medication error that resulted in patient death
401	New pulmonary embolism (PE)
402	New deep vein thrombosis (DVT)
604	Acute myocardial infarction (AMI)
701	Burns (second or third degree burns occurring during inpatient or outpatient service encounters)
751	Falls resulting in x-ray proven fractures, subdural or epidural hematoma, cerebral contusion, traumatic subarachnoid hemorrhage and or internal trauma (e.g., Hepatic or splenic injury)
808[b]	Post-op wound infection requiring incision and drainage (I&D), intravenous (IV) antibiotics or hospital admission within 30 days

Source: New York Department of Health.
[a] Codes requiring an investigation/RCA.
[b] Code 808 Post operative wound infection was discontinued effective February 1, 2007, thus reporting is available for only 2005 and 2006.

over time. The benefits of NYPORTS can be seen at its Web site (New York State Department of Health, 2001).

23.4.2 Differences between Error Coding and Classification by Taxonomy

Taxonomies go beyond error coding systems. Error codes may form a first-order base for a taxonomy, but the taxonomy must go deeper into the classification of what happened. The order of a taxonomy places the failures in a logical or functional position in the process of performance as well as in the process performed. That is, not only does a taxonomy consider at which step the failure occurred, but why or how it occurred. The assumption behind taxonometric theory is that *understanding the full nature of failures provides clues for their prevention.* Examples of taxonomies follow.

23.5 Approaches to Taxonomies: Some Examples

Since taxonomies contain an order, they reflect the interests and perspective of their originator. There is no one taxonomy, nor is there likely to be. Taxonomies are just models—as noted before, useful for a purpose but not a true reflection of reality. Taxonomies tend to be the most successful when addressing very narrow conditions or

TABLE 23.2

Detail Codes

Codes	Description
901	Other serious occurrence warranting DOH notification
902[a]	Specific patient transfers from D&TC to hospital—for D&TC use only
911[a]	Wrong patient; wrong surgical site procedure
912[a]	Incorrect procedure or treatment invasive
913[a]	Unintentionally retained foreign body
914	Misadministration of radiation or radioactive materials
915[a]	Unexpected death (including delay in treatment, admission or omission of care, including neonate ≥28 weeks AND ≥1000 grams AND no life threatening anomalies)
916[a]	Cardiac and or respiratory arrest requiring advanced cardiac life support (ACLS) intervention (including delay in treatment, admission or omission of care)
917[a]	Loss of limb or organ (including delay in treatment, admission or omission of care)
918[a]	Impairment of limb or organ (including delay in treatment, admission or omission of care)
921	Crime resulting in death or injury
922	Suicide, and attempted suicide related to an inpatient hospitalization, with serious injury
923	Elopement from hospital resulting in death or serious injury
931	Strike by hospital staff
932	External disaster outside the control of hospital staff that which effects hospital operation
933	Termination of any services vital to the continued safe operation of the hospital or to the health and safety of its patients and staff
934	Poisoning occurring within the hospital (water, air, food)
935	Hospital fire or other internal disaster disrupting care or causing harm to patients or staff
937	Malfunction of equipment during treatment or diagnosis or a defective product which has a potential for adversely affecting patient or hospital personnel or results in a retained foreign body
938[a]	Malfunction of equipment during treatment or a defective product resulting in death or injury
961	Infant abduction
962	Infant discharged with wrong family
963	Rape of patient

Source: New York Department of Health.
[a] Codes requiring an investigation/RCA.

situations, and become less robust and useful when made more general and encompassing. The selection of taxonomies considered below only indicates that they provide particular insights into the study (and classification) of taxonomy, and, by no means, impugns excluded taxonomies.

23.5.1 Psychological Viewpoint: Donald Norman

Norman (1981) considered why and in what ways persons say the wrong thing. By considering verbal slips,

TABLE 23.3

Norman's Classifications of Slips

A Classification of Slips Based on Their Presumed Sources
Slips that result from errors in the formation of the intention
Errors that are not classified as slips: errors in the determination of goals, in decision making and problem solving, and other related aspects of the determination of an intention
Mode errors: erroneous classification of the situation
Description errors: ambiguous or incomplete specification of the intention
Slips that result from faulty activation of schemas
Unintentional activation: when schemas not part of a current action sequence become activated for extraneous reasons, then become triggered and lead to slips
Capture errors: when a sequence being performed is similar to another more frequent or better learned sequence, the latter may capture control
Data-driven activation: external events cause activation of schemas
Associative activation: currently active schemas activate others with which they are associated
Loss of activation: when schemas that have been activated lose activation, thereby losing effectiveness to control behavior
Forgetting an intention (but continuing with the action sequence)
Misordering the components of an action sequence
Skipping steps in an action sequence
Repeating steps in an action sequence
Slips that result from faulty triggering of active schemas
False triggering: a properly activated schema is triggered at an inappropriate time
Spoonerisms: reversal of event components
Blends: combinations of components from two competing schemas
Thoughts leading to actions: triggering of schemas meant only to be thought, not to govern action
Premature triggering
Failure to trigger: when an active schema never gets invoked because
The action was preempted by competing schemas
There was insufficient activation, either as a result of forgetting or because the initial level was too low
There was a failure of the trigger condition to match, either because the triggering conditions were badly specified or the match between occurring conditions and the required conditions was never sufficiently close

Source: Norman, D., *Psychol. Rev.*, 88, 1, 1981. With permission.

he developed a list of reasons behind the slips. Table 23.3 shows the list. The order and ranking of errors makes the list a taxonomy, rather than an error code. While the particular article addresses verbal slips, they serve as a model of actions in general. In his discussion, and thus in Table 23.3, Norman (1981) does not follow the convention in terminology above distinguishing errors from mistakes.

23.5.1.1 Levels of Performance and Types of Failures That Occur at Each Level

The first of three levels of errors, *errors in intention*, fall into our classification of mistakes. Norman considers three types of errors in intention. The first form high-level errors, that is, errors in cognitive processes: mistakes in thinking about the problem or situation. The

second, erroneous classification of the situation, would characterize actions that would be perfectly correct in a usual situation, but, unfortunately, the perpetrator is in a different situation. Finally, the last type of error at this level comes from not giving enough thought to what the intention should be, essentially, going off half-cocked.

The second level is entitled "results from faulty activation of schemas." A schema, a popular term at that time, refers to a well-practiced mental program. Some schemas run very automatically, for example, receptionists answering the phone with the name of a company or many of the actions in driving a car, which almost become autonomic. Others are less reflexive, but still procedures well known to the person performing them, such as the same receptionist scheduling a patient and asking for the appropriate information. Norman considers two possibilities leading to these errors, either when a schema becomes activated when it should not have been, and when, after activation, the schema shuts down prematurely. (The possibility that the schema never starts falls into the next level.) Errant schemas form an important and very common classification for errors. The situations in which schemas are likely to start occur frequently. Capture errors happen when the intention is to perform one action, but that action is closely related to a different, more common action, for example, walking into a film-processing darkroom and, instead of turning on the safelight, turning on the regular white light (and ruining the film). Data-driven activation (a somewhat misleading term) results when some stimulus initiates a schema inappropriately. With the departing salutation, "Have a good day" becoming popular, the common response is, "Same to you." This leads to the humorous, and oft-heard exchange with an airline ticket-counter agent wishing a traveler, "Have a good flight," followed by the traveler's reply, "Same to you." The response comes well before any thought process formulates it. In the last unintentional activation classification, the associative activation, a running schema starts a different schema that often is associated with the first, even though in the current context the second is not desired. In medical events, unintentional activation of schemas forms a common problem. An example could be a unit nurse preoccupied with a very unresponsive diabetic patient with a blood sugar level of 23. When the food trays arrive at dinner time, the nurse delivers a tray to a different patient who is mentally incompetent and is not supposed to eat the night before surgery (NPO), simply because the tray came on the cart with all the rest. The mostly automatic delivery of the dinner trays follows closely upon seeing the tray on the cart. With thoughts more directed toward the patient with immediate problems, grabbing the tray initiates the delivery schema without thoughts about the NPO status of the patient receiving the tray, nor of the possible, life-threatening consequences.

Loss of activation of a schema can also cause errors. Table 23.3 gives several possible ways that schema can stall once started. The last category, "Repeating steps in an action sequence," is not so much the loss of activation of a schema as its loss of effectiveness when the operator repeats steps inappropriately (assumedly forgetting that they had been completed once) and the process loops around, such as giving medication to a patient twice during the same round through the ward. The other examples lead to unfinished, or incompletely finished, procedures.

The final of the three major classifications considers schemas that either (1) become triggered at an inappropriate time—essentially by accident, or lack of forethought (or present thought)—or (2) never get triggered in situations where they ought. The three reasons given

for the latter cover a lot of territory. While not tabulated in a study, this category of omissions likely plays a significant role in medical events. While the first, preemption by another schema, is not so common, the second, the simple lack of activation due to inattention certainly plays a major role in many events. This should not be thought of as personnel daydreaming, and just forgetting a task. Rather, medical caregivers in most settings run just to stay in place. With many other activities ongoing, the omission of a task more often results from the thought of performing it not rising to immediacy. For example, a nurse in a patient ward may forget to change the saline IV bag on a fairly healthy patient while dealing with a number of very sick patients. The fact that the healthy patient has an IV may escape the nurse's consciousness. This differs from the first classification where that preemption occurs when a schema is about to begin (has been thought of and almost initiated) but a different schema starts instead. The preempting schema likely has a good reason to start in the situation, unlike the capture error, where the schema actually invoked most likely simply is performed more frequently than that intended. The final classification for failure to initiate a schema arises from poor matches between the perceived situation and that which would activate the schema. Norman (1981) gives some examples, but many more exist. The reasons may fall along the continuum between problems with the operator, such as ineffective training or perception, to unclear situations. The unclear situation could also lead to mode errors from the first major group.

23.5.1.2 Complexity of Some Failures

As the discussion in the preceding paragraph highlights, fine lines may separate many of the classifications. Indeed, a given action may carry features of several classifications. Such multiple classifications should not be considered as a failure of the taxonomy, but rather an indication of the complexity of human actions. Sometimes, the classification assigned to a performance failure reflects the perspective of either the person making the classification, or of the resultant event. In most cases with competing classifications, no one classification is correct and all bring information to the investigation.

23.5.2 Industrial (Person Faces Machine) Approach: Jens Rasmussen

Rasmussen (1982) worked with nuclear power plants, and the taxonomy he developed for analysis of events that happened in the plant reflects the work environment. Because of the incredible proactive investment in safety in the plants, they tend to run along very reliably most of the time, with little for the operator to do. Unexpected situations generally begin when some system fails, or maybe just fall outside of normal ranges. Such a situation is called an initiating event, although it may not qualify yet as an actual "event" as defined above, depending on what follows. The operator needs to respond to the abnormal conditions. Inappropriate responses result in failures. Rasmussen's system categorizes the operator's response.

23.5.2.1 Basis of the Approach and Underpinning Assumptions

Rasmussen's taxonomy focuses on the human performance, divided into categories as in Figure 23.4. The analysis of failure particularly studies three aspects of the operator's response: what failed (or better, where in the required steps the failure occurred), how it failed (what the nature of the fault was), and why the operator failed. Figures 23.5 through 23.7 show the taxonomy's pathways used to find classifications for performance failures. Beginning at the top of the what failed pathway, the investigator moves into a box with the question, "Did the operator recognize the need for activity?" Arrows labeled "yes" or "no" guide the investigator to the next box. Eventually, the path leads to a classification. The procedure is the same for the other two pathways.

The what failed pathway (Figure 23.5) parallels the steps the operator would follow to execute the appropriate response to the initiating event. The first step is

to notice that something happened and action may be required. Once noticed, the identification of what happened (or is happening) must follow. Before developing a plan of action, the operator needs to have a clear idea of the goal the action should accomplish. For the power plant, this would be the reactor state; in health care, it may be to stabilize the patient during a myocardial infarction. The goal leads to a desired target state (for our infarction patient, increasing regular flow to the heart muscle) and that state, in turn, to a task to achieve the target state (e.g., angioplasty). From the task follows a procedure and, finally, the operator executes the planned procedure. Without a failure in at least one of these steps, the response succeeds. This pathway leads through the steps linearly. Upon reaching a box at which a failure occurred, the classification is made.

The how it failed pathway (Figure 23.6) becomes more convoluted. The main pathway running down the left side progressively takes the investigation farther from the normal situation. Consideration of the classifications in this pathway gives great insights into the types of errors humans make. The first box asks if the situation was normal, one for which the operator had appropriate skill. That is, things had no particular reason for going astray, except that the operator bungle the execution, which is always a possibility since no one is perfect, or that the operator erroneously thought that something was wrong. These two possibilities are given for why the operator might fail in this situation: one is a simple blunder and the other a topographical misorientation,

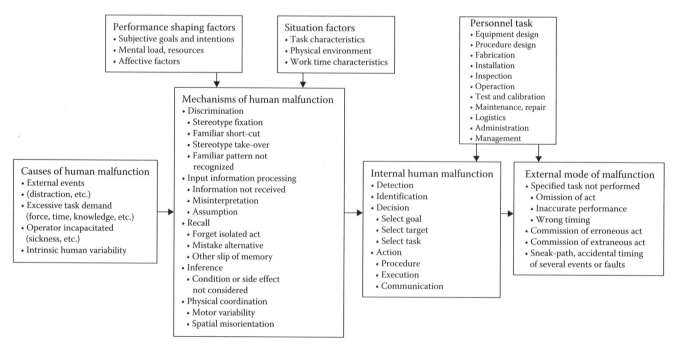

FIGURE 23.4
Rasmussen's taxonometric organization of failures in human performance. (From *J. Occup. Accid.*, 4, Rasmussen, J., Human errors. A taxonomy for describing human malfunction in industrial installations, 311–335, Copyright 1982, with permission from Elsevier.)

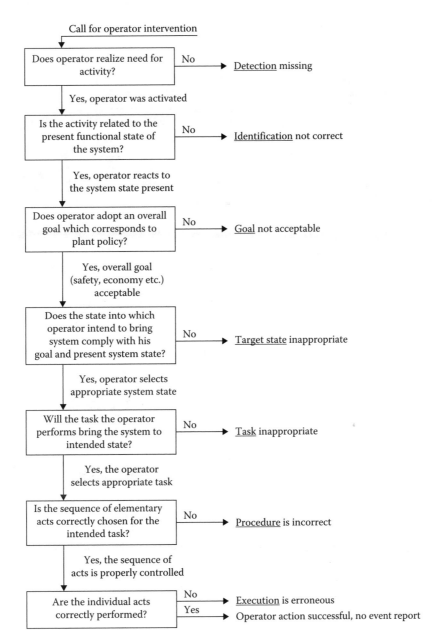

Call for operator intervention

FIGURE 23.5
Rasmussen's what failed pathway. (From *J. Occup. Accid.*, 4, Rasmussen, J., Human errors. A taxonomy for describing human malfunction in industrial installations, 311–335, Copyright 1982, with permission from Elsevier.)

that is, the operator thought the situation was different, and responded appropriately to the perceived conditions.

Answering "no" in the first box points toward the box below. While for the first box, the situation was normal but the operator goes astray, this second box in the column on the left represents a situation that deviates from normal. However, instead of acting appropriately to the abnormal situation, the operator responds as if all were normal. Many medical, and nonmedical, events contain some aspect of this, termed a stereotype fixation. Reactions to warning signals often take this form, such

as with the Omnitron event. In this case, a very small but highly radioactive source was being used to treat a cancer. For the procedure, a treatment unit passed the source, attached to a cable, into catheters that lead the source into the cancer in the patient's body. During one treatment, the source broke free from its control cable and remained in the patient while the cable returned to the shielded container. The control unit displayed a code that the operators (therapists) did not understand. When the therapists entered the room, a radiation detector on the wall was flashing because the source was not in its shielded container, where it should have

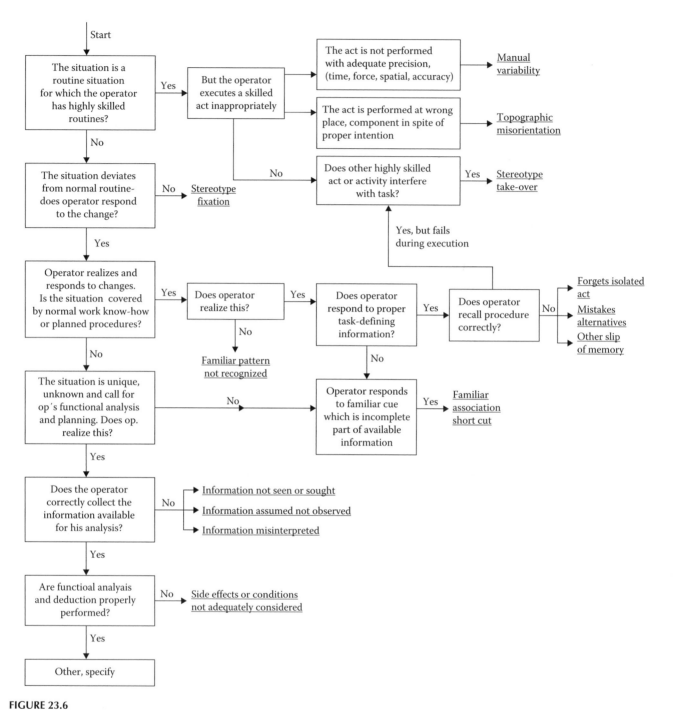

FIGURE 23.6

Rasmussen's how it failed pathway. (From *J. Occup. Accid.*, 4, Rasmussen, J., Human errors. A taxonomy for describing human malfunction in industrial installations, 311–335, Copyright 1982, with permission from Elsevier.)

been. A trainee asked the more senior therapists if the alarm indicated a problem, to which they said, no, the alarm sometimes just makes noises, and turned it off. The patient left containing the radioactive source and died as a result. Most often such failures occur without thought, such as with a worker using a computer program for which a normal value for a given parameter is so common that the program allows simply pressing a RETURN key to enter that value. If a particular patient,

however, requires a different value that the operator must calculate, after working hard to calculate the value to enter, the operator, in a moment of postsolution mental let-down just hits the RETURN key as usually is done, accepting the default value rather than entering the value just calculated.

Further along the how it failed pathway comes an abnormal situation that the operator notes, and one that the operator has procedures to handle. Failure here can result from any of several types of failures. The operator may not recognize the situation as the one covered by the procedures (familiar pattern not recognized, e.g., to clear the fault on the machine and reset the circuit breaker). Alternatively, the operator may recognize the situation but forget the seldom-used procedures, select the wrong procedure (also due to infrequent use) or just forget some aspect of what needs to take place. From the box, "Does the operator recall procedure correctly?" the arrow, "Yes, but fails during execution" leads either to the execution errors from above, or to a stereotype take-over. Similar to Norman's capture error, the stereotype take-over is where a more commonly used skill preempts a lesser-used one. An example most readers identify with would be planning on picking up milk on the way home from work. One gets into the car with the supermarket in mind, and next is aware of pulling into the driveway at home without the milk. The sequence of getting into the car, turning it on, leaving the parking lot at work etc., becomes so familiar that we do not usually think about any of the steps—they just flow as a well-learned skill: so well learned that they take over if not kept suppressed when changes are required.

Backing up to the "Does the operator respond to proper task-defining information?" box, the "no" alternative indicates that the operator jumps to an action based on premature analysis of the situation. Some aspects of the abnormal condition appear similar to a different, more common situation, and based on the quick match of those aspects, the operator performs the wrong procedures—a familiar association short cut. This differs from a stereotype fixation, where the operator did not notice the situation was abnormal. The four classifications—stereotype take-over, stereotype fixation, topographical misorientation, and a familiar association short cut—all share that actions are directed by conditions that do not exist, and that the operator acts on assumptions, that is, without full cognizance.

Information failures come next in line, with various reasons why the operator failed to have necessary information to take the proper actions. Information failures also play significant roles in medical events, particularly when personnel become so rushed they feel they have no time to gather information. The final classification, "Side effects or conditions not adequately considered" falls in the knowledge-based realm. The situation, following the boxes, is that the operator recognizes the need for action but no procedures exist to address it, and having gathered the necessary information, the operator must think out the problem. Again, Rasmussen assumes that the operator will work it out correctly unless they fail to consider all important aspects. This is hardly fair, since most unusual situations pose complex problems with interrelated

relationships. Indeed, correct conclusions at this point may actually be rare.

The last pathway attempts to identify "why" the operator failed (Figure 23.7). Here, the list seems to portray the operator as a mostly effective automaton that would perform the correct act were it not for distractions or incapacitations, or unless the required act exceeds the knowledge, training or physical capacities.* The only other category is, "spontaneous human variation."

The layout of the taxonomy's pathways assumes that the first classification encountered is *the*, unique classification. That can well be argued for the how it happened pathway. While an event may result from a failure in identifying the situation and in selecting the correct task, most often the reason the task was inappropriate stems from the misidentification. However, not uncommonly the why pathway could easily find multiple causes for the failure. Many events present with failures that seem particularly like several of the classifications in the how it failed pathway. Consideration of the number of classifications describing an event will be picked up in the discussion of the next taxonomy.

23.5.2.2 Analyses Based on the Classifications

In Figure 23.8, Rasmussen (1980) places some of the classifications for failures into the perspective of skill-rule-knowledge-based actions. In analysis of nuclear power plant events, Rasmussen (1980) maps the failures into a two-dimensional table of the what failed categories in columns and the how it failed in rows (Figure 23.9). This is an example of considering the various dimensions of the causes while looking at possible relationships between the variables. Such tables can guide remedial actions by indicating the nature of failures that occur at particular phases in the activities needed to execute a procedure.

"...It is necessary to find <u>what</u> went wrong rather than <u>why</u>," states Rasmussen. This not only explains the relative crudeness of the Why portion of his taxonomy compared to the other two parts, but implies, paradoxically, that the reasons behind the events reflect more superficial "causes" while the What failed classifications give insight into the deeper causes.

Rasmussen (1980) observes that to deal with abnormal situations, because of memory limitations, understanding a process proves more effective than training for a required skilled procedure. He also notes that most operators have a "point of no return" in making a decision, beyond which observed information becomes unlikely to change their course of action.

---

* The title of the work uses the term *human malfunction*, which reveals the author's view of the operator.

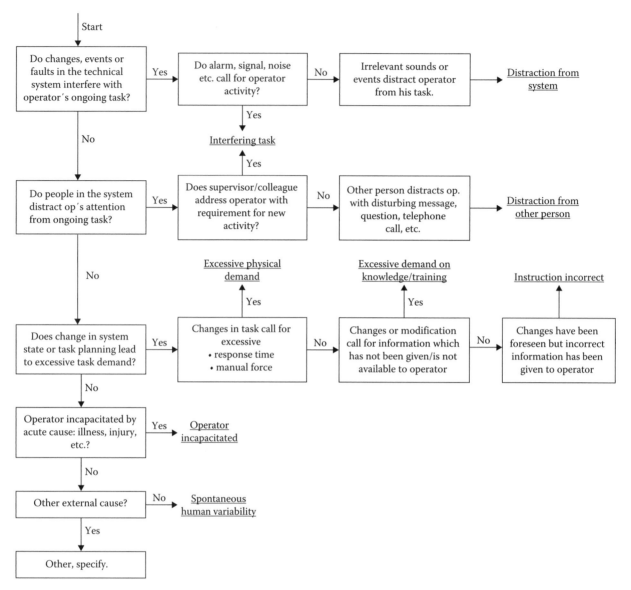

FIGURE 23.7
Rasmussen's why it failed pathway. (From *J. Occup. Accid.*, 4, Rasmussen, J., Human errors. A taxonomy for describing human malfunction in industrial installations, 311–335, Copyright 1982, with permission from Elsevier.)

23.5.2.3 *Limitations for Use in Medical Settings*

Rasmussen's taxonomy as developed over time provides valuable insight into several dimensions of the character of an event. However, the premise of the scenario that things go along humming until an initiating event, while quite appropriate for a nuclear power plant, applies only to a limited extent in medicine. Of course there are many situations where the patient will be fine until suddenly taking a turn for the worse, but often the "initiating event" is an action by the health-care team. Instead of failing in a reaction, the failure starts the chain of events. For these cases, the questions in the boxes do not always make sense, and the character of erroneous or mistaken actions

that originate the event seem unlisted. Patient care forms more of a continuum, with one action always leading to, or meshing with, another. The times are rare when the patient just goes along without the staff having to do things with the patient. In that way, the model applies poorly to medicine. The why pathway also seems hard pressed to capture the character of medical events, which offer a wider range of reasons for failure.

23.5.3 Latent Problems: Reason

James Reason, like Jens Rasmussen, has been so prolific and such a fundamental force in the field of analysis of

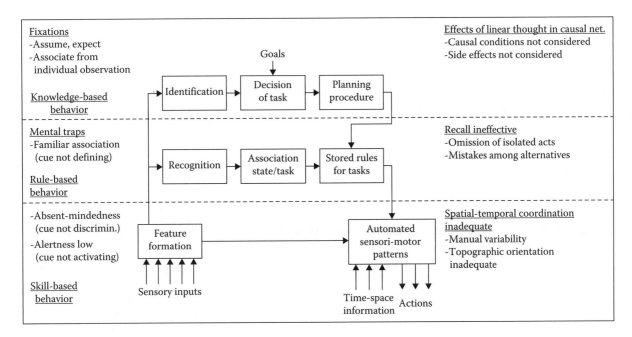

FIGURE 23.8
A correlation between some types of performance failures and levels of action. (From Rasmussen, J., What can be learned from human error reports? In K.D. Duncan, M.M. Gruneberg, and D. Wallis (Eds.), *Changes in Working Life*, 1980. Copyright Wiley-VCH Verlag GmbH & Co. KGaA. Reproduced with permission.)

human performance that summarizing his works or contributions would require more space than a chapter in this handbook. What will have to suffice is a mention of some of the indispensable concepts that find their way into much of the discussion of human performance failures. Anyone doing serious work in the field should become very familiar with Reason's (1990) *Human Error*. Unfortunately, few persons in medical organizations will have the time or inclination to study the book (as delightful reading as it is).

23.5.3.1 General Approach

Reason (1990) organizes the classification of failures around the mental level of the activity, discussed earlier as skill based, rule based, and knowledge based.

Failures, of course, can take place anywhere along the process, although the characteristics of failures at the three levels differ, as shown in Table 23.4. The causes for each type of error are discussed in detail in Reason's (1990) book, and would take too much space to cover adequately here. The interested reader (and that should be any reader) should take the time to learn the definitions. Reason discusses an important aspect of performance failures not commonly considered. Actions an operator takes occur in a continual flow of inputs and other actions. The history to the time of action strongly affects the operator's performance. Because an hour ago a malfunctioning IV pump began working after restarting, even though the manufacturer's instructions say

to retread the line through the unit, a nurse may now first try restarting a failed unit. Diagnoses also can be influenced by how recently similar symptoms were seen in a different patient. Additionally, working very hard on a problem puts all thought processes in high gear. When a solution is reached, there is a mental let down, during which vigilance and attentiveness drop temporarily.

Failures, of course, often result from unsafe acts. Unsafe acts, also of course, do not always lead to failure; if they did there would be no incentive to perform them. Much of the time, unsafe acts save effort, or at least, cost nothing. However, because of the potential harm, unsafe acts should be discouraged when possible. Figure 23.10 presents a graphical classification of unsafe acts. While the classifications related to the basic error types seem common for failure taxonomies, the bottom classification, Violations, rarely receives notice. In studies of medical events, violations play important parts. Few are sabotage, but many are intentional skipping of quality assurance step due to a rushed environment (Thomadsen et al. 2003).

23.5.3.2 Latent Errors

Figure 23.1 showed a variation of Reason's Swiss cheese model of events. Various layers of defenses against failure provide depth to the prevention mechanisms. However, none of the mechanisms is perfect. In fact, some can work in reverse and make errors more likely.

	11	12	17		5		117	36	2	200
ERROR MODE	Detection of demand	Observation - communication	Identification of system state	Goal - strategic decision	Target - tactic system state	Task - determine, select	Procedure - plan, recall	Execution	Various	Distribution across error modes
Absent-mindedness	-	1	1	-	-	-	-	1	-	3
Familiar association	-	-	5	-	1	-	-	-	-	6
Alertness low	6	2	-	-	-	-	1	1	-	10
Omission of functionally isolated act	1	-	-	-	-	-	65	2	-	68
Omissions - others	1	2	-	-	1	-	12	1	-	17
Mistakes among alternatives	-	1	-	-	-	-	1	9	-	11
Expect, assume - rather than observe	-	4	3	-	-	-	3	-	-	10
Side effect not adequately considered	-	-	1	-	1	-	13	-	-	15
Latent conditions not adequately considered	-	-	5	-	-	-	15	-	-	20
Manual variability, lack of precision	-	-	-	-	-	-	-	10	-	10
Topographic, spatial orient, weak	-	-	-	-	-	-	-	10	-	10
Various; not mentioned	3	2	2	-	2	-	7	2	2	20

Distribution across mental task phase   MENTAL TASK PHASE

FIGURE 23.9

An analysis of frequency of errors as functions of their classifications by Rasmussen's what failed and how it failed pathways. (Reprinted from Rasmussen, J., What can be learned from human error reports? In K.D. Duncan, M.M. Gruneberg, and D. Wallis (Eds.), *Changes in Working Life*, 1980. Copyright Wiley-VCH Verlag GmbH & Co. KGaA. Reproduced with permission.)

TABLE 23.4

Distinctive Features of Failures in Skill-, Rule-, and Knowledge-Based Actions.

Dimension	Skill-Based Errors	Rule-Based Errors	Knowledge-Based Errors
Type of activity	Routine actions	Problem-solving activities	
Focus of attention	On something other than the task in hand	Directed at problem-related issues	
Control mode	Mainly by automatic processors (Schemata)	(Stored rules)	Limited, conscious processes
Predictability of error types	Largely predictable "strong-but-wrong" errors (Actions)	(Rules)	Variable
Ratio of error to opportunity for error	Though absolute numbers may be high, these constitute a small proportion of the total number of opportunities for error		Absolute numbers small, but opportunity ratio high
Influence of situational factors	Low to moderate; intrinsic factors (frequency of prior use) likely to exert the dominant influence		Extrinsic factors likely to dominate
Ease of detection	Detection usually fairly rapid and effective	Difficult, and often only achieved through external intervention	
Relationship to change	Knowledge of change not accessed at proper time	When and how anticipated change will occur unknown	Changes not prepared for or anticipated

Source: Reprinted from Reason, J., *Human Error*, Cambridge University Press, Cambridge 1990. With permission of Cambridge University Press.

Decisions made by administrators who never see patients (often referred to as those who operate on the blunt end of the health-care system) can set the stage for failure, for example, by frequently depending on poorly trained temporary nursing staff for regular coverage. The environment affects performance, such as computer screens that have a narrow range of angles from which the contents can be seen. These holes in the defense that operate in the background comprise "latent errors," poor practices that weaken the defense. Most of the time, such holes in the defense against failure cause no problem. But, an unsafe act by one poorly trained person combined with the inability to read the computer monitor at the critical time can allow the unsafe act to propagate as an error, and cause an event. Despite the importance of eliminating latent errors, the taxonomy does *not* provide classifications for such. Latent errors must be found by backprojecting the human performance failure classifications over several events. Often, latent errors act as traps for humans, laying a framework that facilitates an error.

23.5.3.3 Limitations

The work of Reason forms a very comprehensive body for classifying human performance failures and

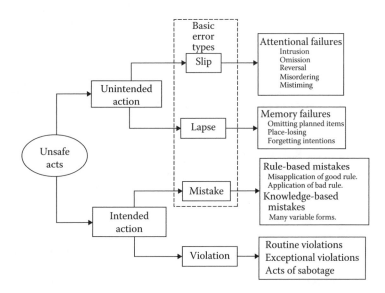

FIGURE 23.10

Unsafe act classifications. (Reprinted from Reason, J., *Human Error*, Cambridge University Press, Cambridge, 1990. With permission of Cambridge University Press.)

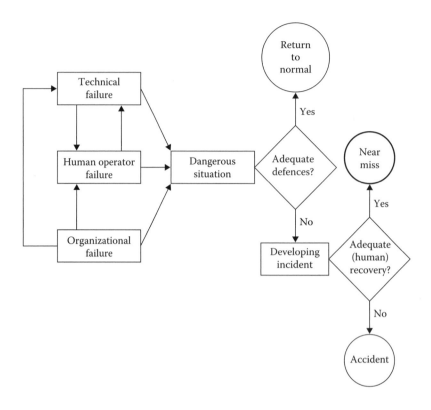

FIGURE 23.11
An overview of the Eindhoven failure taxonomy. (From Van der Schaaf, T. PRISMA medical: A brief description, Unpublished, 2005; van Vuuren, W. et al., *The Development of an Incident Analysis Tool for the Medical Field*, Eindhoven University of Technology, Eindhoven, the Netherlands, 1997. With permission.)

assessing the causes, usually the psychological or behavioral causes. The relationships to the latent errors remain more difficult to ferret out, and to that end, the lack of organizational classifications may leave the reader with little guidance. On the whole, however, the book by Reason (1990) presenting this approach to failure analysis provides an excellent discussion of the concepts.

23.5.4 Systemic Approach: The Eindhoven Taxonomy (Tjerk van der Schaaf)

Van der Schaaf developed his taxonomy while considering a chemical plant (van der Schaaf, 1992). Similar to the environment of the nuclear power plant, for the most part the plant does its job, with human intervention required at specific times or when something goes wrong. However, van der Schaaf takes a more global vantage.

23.5.4.1 Basis of the Approach and Underpinning Assumptions

This taxonomy, originally called PRISMA for Prevention and Recovery Information System for Monitoring and Analysis, was reworked for medical applications and renamed SMART (System for Monitoring and Analysis in Radiotherapy—but for wider application in medicine than just that specialty—Koppens, 1997; van Vuuran, 1997; van der Schaaf, 2005). The overall organization follows Figure 23.11. Van der Schaaf (1992) stresses that for a study of events, near misses, that is, events that include successful recovery, provide the most useful information. While studying failing events gives insights into what went wrong, the near misses contain information on what could prevent a failure following an error.

SMART contains a single pathway with three major, possible subpaths (see Figure 23.12). Beginning at "Start," the first decision is whether technical problems entered into the event, then gives selections for

- Design
- Construction
- Material problems (i.e., something broke)

The next subpath considers organizational factors:

- Knowledge transfer—inadequate training of the personnel to perform the function that failed

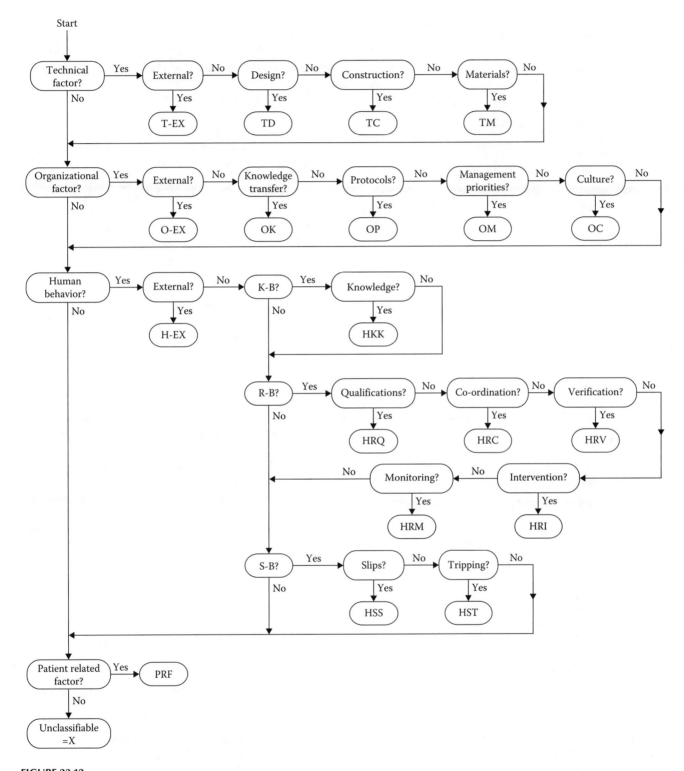

FIGURE 23.12
The classification pathway for SMART. (From Van der Schaaf, T. PRISMA medical: A brief description, Unpublished, 2005; van Vuuren, W. et al., *The Development of an Incident Analysis Tool for the Medical Field*, Eindhoven University of Technology, Eindhoven, the Netherlands, 1997. With permission.)

- Protocols—inadequate procedures or policies dealing with the situation surrounding the event ("too complicated, inaccurate, unrealistic, absent, poorly presented"; van Vuuren et al. 1997)
- Management priorities—management decisions that resulted in situations enabling failures
- Culture—a pervasive attitude in the organization that places little value on steps to insure quality performance

A large subpath for human behavior comes next, subdivided into the three bases for action:

1. Knowledge based, which only contains a self-referential category.
2. Rule based, including
   a. Qualifications—a mismatch in the operator's knowledge, training, and abilities to perform the task
   b. Co-ordination—a lack of coordinating actions between team members
   c. Verification—a failure to completely assess the situation before taking action
   d. Intervention—a failure during the action resulting from either poor planning or erroneous execution
   e. Monitoring—a failure during the execution or immediately after due to the lack of observing the progress of the task
3. Skill based, including
   a. Slips—a fine motor failure
   b. Tripping—a large motor failure

"Patient related factors" forms the last classification, defined as "failures related to patient characteristics which are beyond the control of staff and influence treatment." (van Vuuren et al. 1997)

Each of the subpaths begins with an "External" classification, each of which applies to failures beyond the control and responsibility of the organization.

23.5.4.2 Labyrinth or Pinball Approaches

For the designers of the taxonomy, the layout serves more than just a pathway to lead the investigator through the classifications. The order of the classifications also plays an important role. As designed, the first classification encountered becomes *the* classification for the event. For this to make sense, great care goes into the organization of the taxonomy. (This is some of the "ordering" that was so important in the definition.)

The rationale offered by the creators explains that any technical problems that result in events should be fixed first. In general, they are the easiest of the categories to address, and if technical problems contributed to an event, removing those problems likely would prevent future events. After technical problems come the sources of latent problems in the system. Latent problems reflect organizational weaknesses, so the organizational categories come second. Only after addressing the technical and organizational problems should the investigator look for human performance causes. This hierarchical approach could be called *labyrinth*, after the game where the player moves a steel ball through a maze with holes in the board. A player's score depends on the distance through the maze when the ball falls through a hole.

An alternative to playing labyrinth is *pinball*. In a pinball game, the ball runs the course, falling into holes and racking up points, only to pop out and continue on the course. Only after dropping in many holes (if one is good or lucky) does the ball disappear from play. Pinball players argue that following the whole pathway for every event, noting each classification that contributed to the event, yields more information than playing labyrinth, and the additional information could assist in correcting problems and preventing future events. The argument continues that, while addressing the technical problems first, and organizational problems next may be the ideal approach, in real organizations, particularly medical organizations, it may be easier to effect changes in the human behavior realm, and thus, be more effective and efficacious. In practice, both approaches have merit and should be considered in addressing events. During analysis, noting *the* classification for an event along with the other applicable classifications may provide the best set of clues for correcting problems.

In medical settings, the first technical classification, design failures, often falls outside of the facility's control. Equipment comes from vendors who control the design, and frequently, only one design exists. Thus, for many events, even the labyrinth players must dig deeper into the pathway for some classification that the facility controls.

Some would argue that the "Patient related factors" classification should never find use. Systems should be in place anticipating possible changes in, or actions by, the patient, and if the patient in some way "causes" an event, the real cause lies somewhere in the procedures used with that patient. Consider the case where a patient receives radiotherapy for a bronchial cancer using a radioactive source in a catheter passed through the nose and residing in the bronchus for a day or so. Several events have been recorded where the patient, accidentally or intentionally, removes the catheter before completion of treatment. While this could be classified as a Patient related factor (or action), in truth, the radiation oncologist probably did not fasten the catheter in place well enough to prevent such intervention on the patient's part (an Intervention classification).

23.5.4.3 Applications

The Eindhoven system has been used in radiotherapy (Koppens, 1997), anesthesiology (Anderson, 1997), and blood bank (Kaplan et al. 1998). Kaplan et al. (1998) studied 503 event reports, and verified that blood bank events usually resulted from multiple causes, and that latent causes were more important than expected. They also observed that the event classification pattern was similar to that seen in industry. The similarity of events in blood banks to industry may not predict that same finding for other medical setting because blood banks have little interactions with patients and function to a great extent as an industry. However, other studies using the Eindhoven system in patient-intensive settings have also seen this similarity (Thomadsen et al. 2003). Kaplan also noted that the system produced a high degree of agreement among those performing the analyses.

23.5.4.4 Directive Properties and Limitations

Van der Schaaf (1992) provides a table (similar to Table 23.5) that looks at potential methods of addressing problems based on the Eindhoven classification. The table emphasizes that where technical problems exist, fix them; if procedures are lacking, make them; if a human operative failed in some way, train them; and if knowledge was lacking, provide the necessary information, possibly establishing new lines, methods or modes of communications. Problems with motor skills, fine or gross, require equipment solutions to obviate the human lacking. As the table emphatically states, under no situations do motivational approaches deliver desired results. Posters and rallies seldom make any headway in promoting safety, nor does a policy of picturing the correct result before action or thinking good thoughts.

The usefulness of the suggested correction action, "training" comes under fire. Studies have shown (Rook, 1965; Harris and Chaney, 1969) that once trained, retraining serves no purpose. Quite the contrary, persons sent for retraining usually feel punished, as in the antiquated approach to events of "blame and train." Most medical personnel go through rigorous training in their jobs before becoming clinically responsible for patient care.

TABLE 23.5

Suggested Remedial Actions Based on the SMART Failure Classifications

Classification code	Technology/ Equipment	Procedures	Information and Communication	Training	Motivation	Escalation	Reflection
T-EX						x	
TD	X						
TC	X						
TM	X						
O-EX						x	
OK						x	
OP		x					
OM						x	
OC							x
H-EX						x	
HKK			x		NO		
HRQ				x			
HRC				x			
HRV				x			
HRI				x			
HRM				x			
HSS	X				NO		
HST	X				NO		
PRF[a]							
X							

Source: van der Schaaf, T., PRISMA medical: A brief description, Unpublished, 2005. With permission.

[a] If particular patient related factors (such as language problems) that cannot be prevented by the patients themselves recur, then these problems should be solved at an organizational level (i.e., escalation).

Failures come from inattention, misunderstanding, or any of the other causes discussed above. Retraining seldom has a role in preventing future occurrences. On the other hand, new procedures sometimes do begin before all personnel have been *adequately* trained. In such cases, remedial training that should have been completed previously still is a must.

The problem with Table 23.5 lies in the general nature of the suggestions. Just knowing the general heading of the appropriate type of action does not guide the investigator to the solution—it only limits the universe of solutions.

The reader should note that Table 23.5 offers no suggested solutions for Cultural problems. In general, if the organization places little value on error reduction and safety, not only will corrective actions likely be futile, but the organization would not be likely to have an investigator trying to find solutions.

23.5.5 Scope

23.5.5.1 Basis of the Approach and Underpinning Assumptions

The System for the Classification of Operator Performance Events (SCOPE) combines the descriptive information on the stage in an action sequence an event occurred with identifying the nature of the failure leading to the event (Kapp and Caldwell, 1996). The structure of SCOPE follows the sequence of actions required to react to the need for activity, similar to that described in Figure 23.3: detection, interpretation, rule selection *or* planning, and execution. SCOPE also notes failures in Linkages between tasks and Cross-unit flow of information between organizational teams. The organization and approach builds strongly upon the works of Rasmussen (1980, 1982), Reason (1990), and van der Schaaf (1992) cited earlier. As in the other taxonomies, the classification comes through following a pathway, shown in Figure 23.13a through e.

The first arena for human performance failures comes with detection of the need for action. Failure of detection may occur in one of four ways, a characteristic of SCOPE for each of its pathways. For each, the classification of the nature of the failure also serves as the designation of the cause. The classification "loss of arousal" implies that the operator is less than alert.

For the detection of the need for action to translate into correct actions, the situation must be interpreted properly. Interpretation failures likewise fall into four categories. Three of the categories relate to insufficient information to correctly interpret the detected situation: the operator failed to gather sufficient information, all the necessary information was not available, or the information was inconsistent. The final classification,

Expectation bias, plays an important part in a large proportion of medical events (e.g., Thomadsen et al. 2003). An operator exhibits an expectation bias interpreting a situation as that expected rather than the true situation, in spite of receiving information to the contrary. Usually, this appears as the operator assuming that things are normal and making models to explain away the detected indications. Expectation biases relate closely to the stereotype fixation, topographical misorientation, and familiar association short cut from the Rasmussen (1982) taxonomy.

Based on the interpretation of the situation, the operator determines a course of action. This decision is first based on selection of a rule that determines the schemas to activate (treating our operator as a robot again), if such a rule can be recalled or referenced. However, the operator may select a rule that did not completely cover the situation, an old, outdated rule, or a bad rule—one that never works. Any of those errors leads to failure, as does selecting a good rule at the wrong time, which can happen due to the loss of intention (not loss of attention), forgetting the actual situation, or confusing the situation with a different, usually more common one (Frequency/similarity bias) These last two categories share characteristics with Rasmussen's familiar association short cut, and topographical misorientation, and to some extent, familiar pattern not recognized. Determination of the correct rule leads directly to the execution phase.

Failure to find a rule leads the operator to a knowledge-based action and the decision and planning pathway. Failure to develop an appropriate plan comes from either the application of inappropriate goals (due to competing or conflicting objectives or from the lack of information about the goal) or from a failure in the planning proper. An operator may not be able to generate a good plan because the problem may be too difficult for the training and experience level of the operator (Bounded rationality) or because some necessary information may be missing. Generation of a good plan would lead to the last pathway, Execution.

If all the steps toward actions to this point were failure free, the errors must lie in the execution. Again, the pathway offers four causes for failures. The first is some external interference—the operator would have done fine except something prevented execution. Failure more often results from deviation from plan, because of entrapment by another rule or plan (faulty triggering) or forgetting what to do. Finally, the operator can simply bungle the execution.

The layout of SCOPE intended its use with the labyrinth approach, although its developers later recognized that, because many failures often contribute to events, the taxonomy also lends itself to a pinball approach.

"Lack of information" appears as a classification four times in the complete pathway, and incomplete information once. The frequent use of this name for a classification indicates the major role receiving correct and complete information plays in error reduction. But that is not news to anyone reading this book. It is interesting, however, that each of these information deficits occur in different phases of the process, showing the pervasive nature of information transmission in complex situations.

23.5.5.2 Limitations

Moving from Rasmussen's taxonomy to SMART and finally to SCOPE has shifted the focus continually to the operator as an active participant in the activities. SCOPE, as does the Rasmussen taxonomy, still tends to view the operator somewhat as reacting, and although more cognizant of the decision-making function of medical personnel (as seen by the existence of the decision and planning pathway) than Rasmussen's, the choice of failure classifications sometimes fails to reflect the richness of errors humans make of their own initiative. The taxonomy also focuses on human performance, and while considering environmental contributing factors, does not code for

organizational factors. Such organizational problems may be inherent in the model, for example, a classification of Bounded rationality automatically raises the question, "Why was the operator in the position to be faced with problems beyond their means in the first place?"

23.5.6 Some Medical Taxonomies in the Literature

23.5.6.1 National Coordinating Council for Medication Error Reporting and Prevention

The National Coordinating Council for Medication Error Reporting and Prevention (NCCMERP, 1999), an independent body sponsored by some 25 organizations with interest in medical safety, assembled a reporting system for medical events. This system falls somewhere between an error coding system and a taxonomy. (A typical taxonometric problem of trying to fit something into a category). The list of classifications spans 19 pages and events receive scores in several categories. For example, the initial category gathers information about the event, such as time of day and type of medical facility at which the event initiated as well as

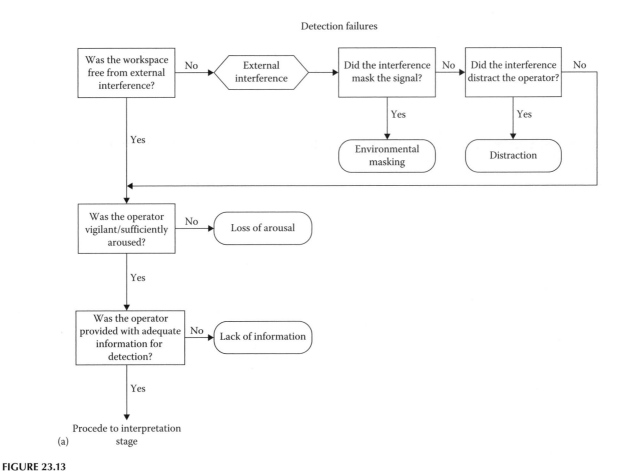

FIGURE 23.13
(a–e) SCOPE failure classification pathways. (From Kapp, E.A. and Caldwell, B., SCOPE instruction manual, Unpublished, 1996. Used with permission.)

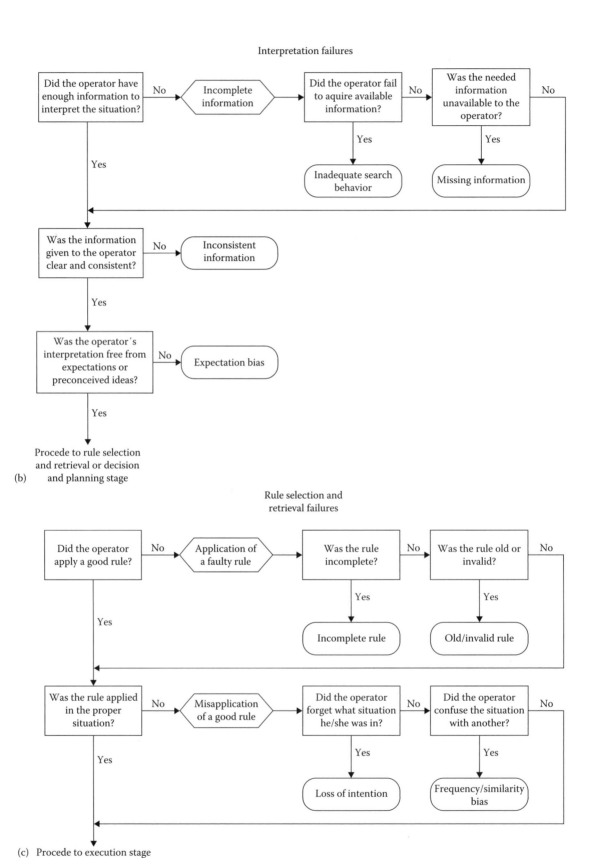

FIGURE 23.13 (continued)

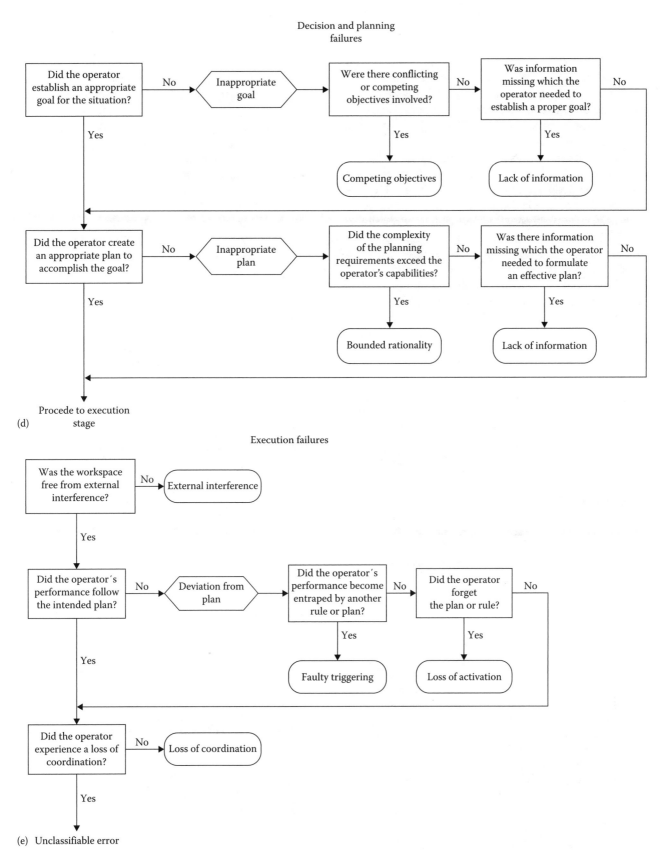

(d)

(e) Unclassifiable error

FIGURE 23.13 (continued)

where it perpetuated, followed by a free text description of the event.

The next field classifies the level of harm to the patient, followed by a very long section on the products involved. The attention to products makes sense since most errors involve medication administrations, and this taxonomy was mostly designed to address those errors. Personnel involved in the event comes next followed by what type of error occurred.

Up to this point, the system only dealt with objective, noncontroversial facts about the event. The attention then turns to deeper questions about causes. The first questions do not probe too deeply, for example, offering choices such as "NAME CONFUSION/Proprietary (Trade) Name Confusion; Suffix confusion; Prefix confusion; Sound-alike to another trade name; Sound-alike to an established (generic) name; Look-alike to another trade name; Look-alike to an established name"; etc. The next section looks at the human factors involved in the event. Table 23.6 lists the possible codes in that section. Except for the first (knowledge deficit) and the last three (stress, fatigue, and confrontation), the classifications for human errors also deal only in objective descriptions of actions (and some for very particular actions) rather than causes for, or nature of, the actions.

The final category of classification considers contributing factors, or system-related factors (Table 23.7). This last section attempts to provide information on the latent causes of the events. Some of the codes cover broad classifications, such as "Policies and Procedures," while others, for example, "pre-printed medication orders," form narrow, very specific codes. Assumedly, history plays a part in the list, where preprinted medication orders must have been a common problem during the formative period for this taxonomy. The entries in the contributing factor list mostly describe PSFs (e.g., lighting, noise), training, or organizational latent errors (e.g., staffing, policies).

The NCCMERP system orders and organizes the codes, so in that way forms a taxonomy. However, the main thrust seems to be for gathering information for databases, even in the causative sections, making it more of an error coding system. Given the number of major organizations involved with the NCCMERP, the system should find widespread use, and facilitate generation of large amounts of data, crossing institutions. Such information will be indispensable in improving patient safety, for example, by highlighting failure

TABLE 23.6

Human Factors Codes for the NCCMERP Taxonomy

87 Human Factors
87.1 Knowledge deficit
87.2 Performance deficit
87.3 Miscalculation of dosage or infusion rate
87.4 Computer error
87.4.1 Incorrect selection from a list by computer operator
87.4.2 Incorrect programming into the database.
87.4.3 Inadequate screening for allergies, interactions, etc.
87.5 Error in stocking/restocking/cart filling
87.6 Drug preparation error
87.6.1 Failure to activate delivery system
87.6.2 Wrong diluent
87.6.3 Wrong amount of diluent
87.6.4 Wrong amount of active ingredient added to the final product
87.6.5 Wrong drug added
87.7 Transcription error
87.7.1 Original to paper/carbon paper
87.7.2 Original to computer
87.7.3 Original to facsimile
87.7.4 Recopying MAR
87.8 Stress (high volume workload, etc.)
87.9 Fatigue/lack of sleep
87.10 Confrontational or intimidating behavior

Source: National Coordinating Council for Medication Error Reporting and Prevention taxonomy. With permission.

TABLE 23.7

Contributing Factors Codes for the NCCMERP Taxonomy

90 Contributing Factors (Systems Related)
[Select as many items as are applicable from this section]
90.1 Lighting
90.2 Noise level
90.3 Frequent interruptions and distractions
90.4 Training
90.5 Staffing
90.6 Lack of availability of health-care professional
90.6.1 Medical
90.6.2 Other allied health-care professional
90.6.3 Pharmacy
90.6.4 Nursing
90.6.5 Other
90.7 Assignment or placement of a health-care provider or inexperienced personnel
90.8 System for Covering Patient Care (e.g., floating personnel, agency coverage)
90.8.1 Medical
90.8.2 Other allied health-care professional
90.8.3 Pharmacy
90.8.4 Nursing
90.8.5 Other
90.9 Policies and procedures
90.10 Communication systems between health-care practitioners
90.11 Patient counseling
90.12 Floor stock
90.13 Preprinted medication orders
90.14 Other

modes common in widely different procedures or calling attention to particularly hazardous procedures.

23.5.6.2 Edinburgh Taxonomy

The Edinburgh Error Classification System attempts to combine the aspects of both the behavioral (what the operator did) and cognitive (what lead the operator to do it) approaches to classifying the types of failures (in the incarnation, Busse and Wright, 2000). The system also straddles the line separating an error coding system and a taxonomy. The authors state that the classifications focus on proximal cause rather than root cause. The investigation that leads to the classification considers *"what* happened in an incident occurrence (e.g., wrong drug administered), *how* it happened (e.g., drug confusion), and most importantly, *why* it happened (e.g., illegible handwriting on drug container...)." In other taxonomies, the examples given could just as well be called the type of event instead of what happened, what happened instead of how it happened, and how it happened instead of why. The distinctions merely emphasize that the common meaning of words leaves too much to interpretation to be used without further definition in this specialized field needing fine distinctions between concepts.

Table 23.8 presents the current classification system, expanded from an earlier system used in a study of intensive care unit (ICU) errors (Wright et al. 1991). As Busse and Wright (2000) point out, the initial classifications mostly involved PSFs while the more recent additions concentrate on "domain-specific, behavioral" categories that, instead of probing "cognitive mechanisms or organizational influences" denote "where in the task sequence a step had been omitted, or had been executed wrongly." Table 23.9 reorganizes Table 23.8 into proximal causes (these immediately leading to the event) and distal causes (those that set the stage for the event.) The developers of the Edinburgh system do not try to capture the latent causes in their classifications, but rather, draw that information out from a root-cause-analysis of the event.

The Edinburgh system has proven useful in analyses of medical events (Wright et al. 1991; Busse and Johnson, 1999; Busse and Holland, 2002), but because of the specificity of several of the categories, the system may be a little narrow for widespread application through a health-care organization. The system also leaves much of the determination of remedial actions to the experience of the investigator.

23.5.6.3 Nursing Practice Breakdown Research Advisory Panel

The Nursing Practice Breakdown Research Advisory Panel of the National Council of State Boards of Nursing,

TABLE 23.8

Categories of the Edinburgh Error Classification System

XI. Edinburgh Classification of Contributing Factors

A. Initial categories

 1. Inexperience with equipment
 2. Shortage of trained staff
 3. Night time
 4. Fatigue
 5. Poor equipment design
 6. Unit busy
 7. Agency nurse
 8. Lack of suitable equipment
 9. Failure to check equipment
10. Failure to perform hourly check
11. Poor communication
12. Thoughtlessness

B. Added categories

 1. Presence of students/teaching
 2. Too many people present
 3. Poor visibility/position of equipment
 4. Grossly obese patient
 5. Turning the patient
 6. Patient inadequately sedated
 7. Lines not properly sutured into place
 8. Intracranial pressure monitor not properly secured
 9. Endotracheal tube not properly secured
10. Chest drain tube not properly secured
11. Nasogastric tube not properly secured

Source: (A) Wright, D. et al., *Lancet*, 338, 676–678, 1991. (B) Busse, D. and Wright, D., *Topics Health Inform. Manage.*, 20, 1, 2000.

Inc. investigated 21 events to study the design of a taxonomy for use with nursing-related events (Benner et al. 2002). While the taxonomy remains under construction, the preliminary structure appears in Table 23.10.

The classification, Lack of agency, refers to failure of the nurse to act as an advocate for the patient, perhaps questioning possibly inappropriate physician orders. Lack of intervention on the patient's behalf, on the other hand, refers to the failure of a nurse to take nursing actions when such actions are required. This category differs from Lack of prevention in that the latter deals with failures involving proactive, instead of reactive, actions. The developers decided that Medication errors should be coded using the NCCMERP taxonomy.

Obviously, this taxonomy only descriptively classifies the human performance failures involved in events. Data gathered certainly will find use in following trends in errors and events, and may help point toward functions that need improvement. However, extension of the categories to causes of the failures or even the types of behavior that led to the failures probably would increase its utility.

TABLE 23.9

Edinburgh System Categories Reordered into Proximal and Distal Causes

Failure Type Categorization

A. Proximal causal factors
1. Failure to check equipment
2. Failure to perform hourly check
3. Thoughtlessness
4. Turning the patient
5. Patient inadequately sedated
6. Lines not properly sutured
7. ICP monitor not properly secured
8. En. tube not properly secured
9. Chest drain tube not properly secured
10. Nasogastric tube not properly secured

B. Distal, causal factors
1. Inexperience with equipment
2. Shortage of trained staff
3. Night time
4. Fatigue
5. Poor equipment design
6. Unit busy (refined by 9 and 10 below)
7. Agency nurse
8. Lack of suitable equipment
9. Presence of students/teaching
10. Too many people present
11, Poor visibility of equipment
12. Poor communication

Source: Based on Busse, D. and Wright, D., *Topics Health Inform. Manage.*, 20, 1, 2000.

23.5.7 Toward a More Useful Medical Taxonomy

23.5.7.1 *Characteristics of a Taxonomy*

As noted at the beginning of this chapter, to capture the many facets of events, a taxonomy needs to be multidimensional. Attempts to incorporate dimensions into failure classifications led to some of the taxonomies described above using lists that seem distinctly nonparallel. There could be many possible approaches to including the various types of

TABLE 23.10

Taxonomy of the Nursing Practice Breakdown Research Advisory Panel

Lack of attentiveness
Lack of agency/fiduciary concern
Inappropriate judgment
Medication errors
Lack of intervention on patient's behalf
Lack of prevention
Missed or mistaken physician or health-care provider orders
Documentation errors

information. One possible method would classify events by the following:

1. Where in the process the failure occurred. Such a determination would be best made with respect to a process map or chart for the procedure. This has been useful with the NCCMERP database and several databases used in radiotherapy (Ekatte et al. 2006; Mutic et al. 2010).

2. Who were the principals involved. Rarely does an event happen because of one person's actions. Ideally, activities have some checks or peer review as quality control; so several individuals may be involved at various levels.

3. Globally, in what arena did the failure take place, as depicted by the SMART taxonomy. This would guide remedial actions to the technical, organizational, or personal realm.

4. For human failures:

   a. Where along the steps necessary to perform an action did the failure occur, such as in Rasmussen's what happened pathway, and SCOPE?

   b. What was the nature of the failure, as in Rasmussen's how it happened pathway and Norman's classifications?

   c. Why the human failed. As noted earlier, Rasmussen's why it happened pathway is not very well developed, and more developmental work is needed here. However, the PSFs likely play a significant role. Also important, but more difficult to classify would be history, such as what happened to the persons involved just prior to the event.

5. Performance Shaping factors. These fall into at least two categories:

   a. Environmental (e.g., lighting, noise, temperature.)

   b. Organizational (e.g., staffing levels, supplies and maintenance, training, policies and procedures)

Taxonomies that only look at some of the dimensions can still prove useful by closing some of the holes on one of the wheels in Figure 23.1. Some remedial actions that address one aspect of a failure that lead to an event can prevent all potential future events at that step of a procedure, but that is unlikely. For example, if a radiographer failed to check if a patient had eaten in the 4 h before giving contrast (resulting in aspiration following vomiting), actions taken to train the radiographer would not likely prevent a recurrence if the cause was inattentiveness due to excess demand. Each dimension brings important information to the problem.

Van der Schaaf (1992) emphasizes the importance of capturing incidents that stopped short of becoming events, often called near misses. In these cases, someone or something intercepted the propagation of the failure—saved the day. Identifying what worked in that case might have wider application and understanding the intervention could yield substantial benefits. Taxonometric classification of near missing is just in its infancy, but is a field where investigations should proceed forthwith.

23.5.7.2 Madison Taxonomy

The Madison taxonomy (Thomadsen and Lin, 2003) evolved in the 1990s to address the difficulties in applying the existent taxonomies to medical events in radiotherapy (Thomadsen et al. 2003), and to provide a stronger correlation between the taxonometric classification and suggested remedial actions. Drawing heavily from the works of Rasmussen, van der Schaaf, and Kapp and Caldwell, this taxonomy tried to incorporate the features discussed in the last section. The failure analysis uses the location of the failure on the process and fault trees to determine where in the procedure to address concerns (see next section), and focuses on the nature of the failures.

Figure 23.14 shows the overview of the Madison taxonomy. Unlike many taxonomies that start with an initiating event, the Madison approach assumes that in medical settings, there seldom is a particular condition that upsets an equilibrium. Patients are always in flux and potentially unstable. As long as the patient is under care, there are actions, counter actions, and follow-up actions. And, there is always experimentation, testing to find problems, and further testing of possible solutions. Following events, it often becomes difficult to determine precisely when things stopped being normal, in part because no normal state may exist—each patient being unique. The beginning of the analysis starts with the last recognizable "normal" condition, or better, the last time that the situation *may* have been under control. As the situation develops, the pathway leads to the box asking if action is required. A "no" response leads to the branching indicating either action was taken when it should not have been, leading to an event, or no action being taken, and a successful outcome. Were the decision-making sequence followed according to a well-understood and practiced procedure, the outcome would be a success. If, on the other hand, the decision for action laid in the balance, with questions leading the operator both ways, the correct move, in this case inaction after considering taking action, would result in a near miss.

Following the "yes" path from the "Action required" box leads to "Action Taken," with another yes/no option. Obviously, taking no action at this point leads to failure, while taking action moves the investigation to the question of whether the action was correct. Again, the results of the correct action, success or near miss, depends on how the action came about.

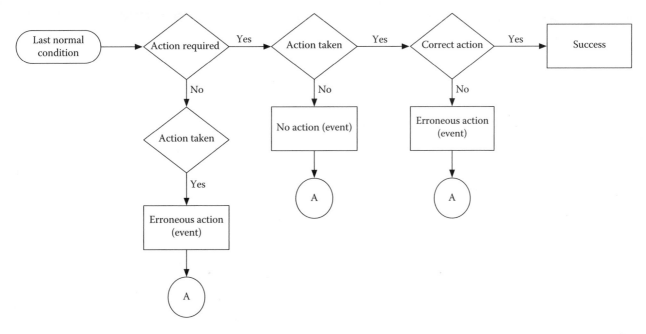

FIGURE 23.14
Overview of the Madison taxonomy. (From Thomadsen, B. and Lin, S.-W., Guidance for remedial actions based on failure classifications, Wisconsin Patient Safety Institute, November 13, 2005, 2003. With permission.)

In Figure 23.14, each of the paths leading to an event moves the analysis to the "A" in a circle and then continues on Figure 23.15. This pathway leads to the general description of where in the process the failure occurred. In this context, the "where in the process" is not the step in the procedure, but a more general view of any process that requires responses to inputs. Any of the inappropriate actions in Figure 23.14 could be described by one of the boxes in Figure 23.15. The first box, "Mistaken action," applies to an action taken not as a reaction to an immediate situation. Of course, as stated above, all actions have antecedents that provide motivation (except possibly, pathologically initiated

actions), but some rely less on response to stimuli than on initiative. For example, a nurse decides to change a patient's bed, although the sheets are not dirty, and accidentally rolls the patient out of bed. Most events, however, do occur following some need for action. The failures result from either missing that the need arose or not taking the correct action. The former is classified as a detection failure and results from not having procedures to detect the situation, or if procedures exist, they failed in their application. The latter, "Failure of reaction," results from misinterpreting the information detected, not reacting correctly (for many reasons, considered later) or not taking action when

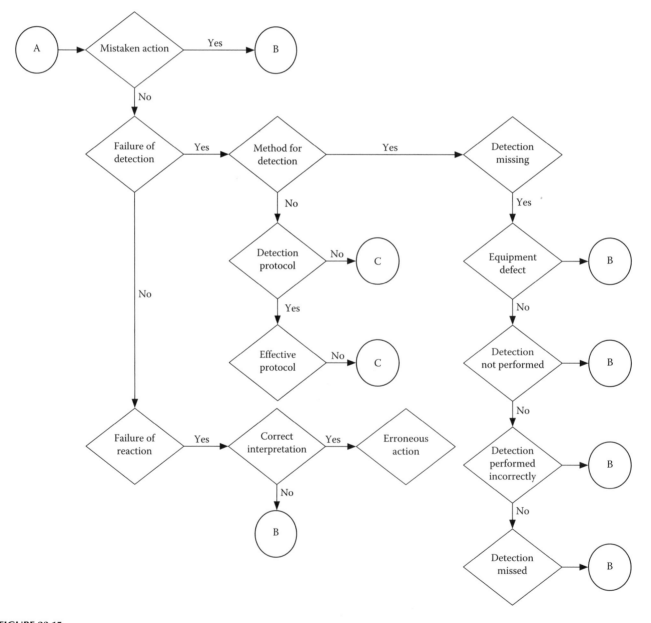

FIGURE 23.15
General characterization of the failure leading to an event. (From Thomadsen, B. and Lin, S.-W., Guidance for remedial actions based on failure classifications, Wisconsin Patient Safety Institute, November 13, 2005, 2003. With permission.)

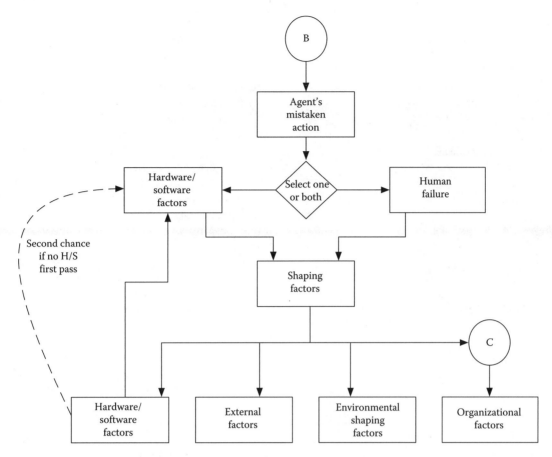

FIGURE 23.16
Action factors for the Madison taxonomy. (From Thomadsen, B. and Lin, S.-W., Guidance for remedial actions based on failure classifications, Wisconsin Patient Safety Institute, November 13, 2005, 2003. With permission.)

required. Each of the paths leads either to subsequent pathways B or C.

In Figure 23.16 at the "B," the erroneous action could be from a human performance failure (right branch) or a hardware/software failure (left branch). While purely mechanical failures remain rare in medicine, they do happen. In 1987, the very heavy head of a cobalt-60 radiotherapy unit broke from its gantry and crushed a patient, with no warning or signs of weak materials. Equipment failures fall into the classifications taken from the Eindhoven system (van der Schaaf, 1992):

1. Hardware failures
   a. Design
   b. Construction
   c. Materials (parts breaking)
   d. Maintenance (which may actually be an organizational failure)
2. Software failures
   a. Design

b. Construction
c. Maintenance (again, may be an organizational failure)

Software failures have no "materials" category since no parts break. True, programs may stop working, but the origin of such problems more directly stems from design problems than with hardware.

More commonly, medical events follow the human performance failure pathway. Classification of the nature of a human performance failure takes place at several levels along this pathway. Table 23.11 shows a sampling of classifications.

The first set of classifications simply describes the failure generally, in terms of trips, slips, omissions, blunders (conventional errors), mistakes or intentional incorrect actions (intentional violations, malfeasance or negligence).

Much of the second level of the human performance classification pathway comes from the works of Rasmussen and Norman discussed earlier. This subpathway attempts to determine the mechanisms involved in leading the operator to the action actually

TABLE 23.11

Classification of Human Performance Failures in the
Madison Taxonomy

Human error

Human error general subpath
    Tripping
    Slips
    Blunder
    Error in the intention (mistake)
    Intentional violation

Human error detailed subpath
    Physical coordination
        Manual variability (SB)
        Topographic misorientation (SB)
    Expectation bias
        Stereotype fixation (SB)
        Stereotype takeover (SB)
        Familiar association short-cut (RB)
        Familiar pattern not recognized (RB)
    Mistake choices (Recall)
    Mistakes options, mistake consequence (RB)
    Inference
    Condition or side effect not considered (KB)
    Information problem (input information processing)
        Not sought—assumed, negligent omission, not
           realized it was needed (KB)
    Vigilance
        Lack of vigilance (arousal, commitment, complacency)

SB-skill based, RB-rule based, KB-Knowledge based

executed instead of the correct action. There are similar subpathways for what phase in the performance process the failure occurred, similar to Rasmussen's what happened pathway or SCOPE.

The equipment and human pathways join again leading to "Enabling factors" or, more commonly, PSFs. Application of the term "enabling factors" intends to recall the use in psychological settlings of actions by one person that facilitate pathological or destructive actions by a second person. "Enabling factors" better project the damaging effects the actions have than the neutral term, "PSFs," which could just as well apply to performance improvement as performance degradation.* The first of the four general categories of enabling factors reprises the equipment pathway again. While equipment failures seldom directly cause medical events, such failures frequently play important roles in leading the human operators astray. The "External causes" box leads to no further classifications; being

---

* In popular psychology, the adjective "empowering" often applies to beneficial actions. It is interesting to note that just as fine distinctions must be made in developing taxonomies between words that in common usage carry equivalent meanings, here also the term "enabling" could have been applied to actions either helpful to recovery or to destruction.

outside of the facility's universe, further classification may not be very useful, and the number of classifications could become massive. Information guiding remedial actions for external causes comes from a fault tree analysis of the event, rather than from the taxonomy (see below).

The two remaining enabling categories usually contain most of the latent causes for the event under investigation. The first, "Environmental factors" covers a large territory. Table 23.12 gives a partial list of possibilities. Some of these classifications apply to the operators, others to the surrounding conditions. As noted previously, immediate attention to environmental factors improves safety immediately with relatively little cost.

The last box, "organizational factors," usually contains the preponderance of latent causes. Similar to environmental factors and most often equipment failures, the organizational factors rarely, if ever, directly cause events—they set the stage to make the event more likely, sometimes to the extent of seeming like a trap. Table 23.13 lists some organizational factors.

The semi-hierarchical organization of the taxonomy lends itself to both the labyrinth and pinball approaches, with the basic philosophy that preventing events does not necessarily require addressing the most basic, root cause nor a progenitor latent error, but that deep information may give significant clues as to which remedial actions could yield the greatest payoff. In many cases, the root causes may be the most resistant. All of the classifications give information that could be useful in determining remedial actions. Section 23.7 discusses guidance for the choice of remedial actions based on the failure.

TABLE 23.12

A Partial List of Environmental Enabling Factors

Tangible factors
    (Environmental controls) sound control—noise (non-human)
    (Environmental controls) sound control—distractions (human)
    (Environmental controls) visual control
    (Environmental controls) neatening
    (Environmental controls) cleaning
    (Environmental controls) isolation
    Environmental design

Intangible factors
    Time of day
    Day of week
    Holidays, vacations
    Personal problems
    Health

TABLE 23.13

Some Common Organizational Enabling Factors

Knowledge of leader
Management priorities
    Providing staffing
    Providing equipment
    Providing maintenance
    Providing supplies
    Safety culture (with further divisions)
Communication system
Knowledge transfer (training inadequate)
Matching abilities to job
    Tasks exceed physical ability
    Tasks exceed mental ability
    Task exceeds experience
Competition for attention
    Background
    Lack of staffing or time
    Other, competitive goals
Immediate
    Other, competitive duties
    Lack of staffing
    Other inputs into the system
Previous
    Mental fatigue due to complex work
    Concern over previous, unaddressed problems

23.6 Classifications Based on Activity and Error

The taxonomies presented classify failures in terms of types of errors, under the assumption that effective corrective actions depend on such knowledge. However, effective corrective actions also require information on *where* in the process to locate the corrective actions. While the Rasmussen Taxonomy partially provides such information (in the what failed pathway) the information there is intended to describe one aspect of the failure, not actually locate the failure in the particular procedure. This information comes from other analyses, as discussed next.

23.6.1 Failure Location on a Process Tree

The most direct location of where to place corrective actions comes from a process tree. A process tree, such as shown in Figure 23.17, charts the flow of a procedure. Process trees come in many varieties, and sometimes go by the name fishbone diagrams, process flow charts, among others. Any format that helps an investigator

understand the process would be useful. Locating the position on the chart where the failure happened very directly indicates the step requiring protection. For that, one seldom needs the chart. The chart becomes useful for two functions. The first is when the step requiring protection does not lend itself to any quality management. In such cases, the chart facilitates finding steps downstream in the procedure amenable to some verification or control procedures, or, when possible, positions upstream that could prevent the failures in the noted position.

The other situation where noting the position of a failure on the process tree becomes useful is when analyzing a large number of errors from a database. Figure 23.17 shows the result of such an analysis. The numbers on the branches of the tree give the number of failures that occurred at that step in the process. Noting where a number of events congregate suggests those steps that need particular attention.

Creation of the process tree forms an important part of proactive patient safety, but is beyond the scope of this chapter. For additional information on risk management techniques, see Chapter 29.

23.6.2 Failure Location on a Fault Tree

Similar to the information from the process tree, indicating the location of an event on a fault tree provides additional and different information yet. In this case, the tree displays *all possible failures*. Each possible failure *should* be protected by some form of quality management either at the position of the potential fault or downstream, where, for an error to propagate, there must be a concomitant failure of the quality management. While not indicating where in the procedure the additional quality management should go, the fault tree helps visualize whether all faults are covered to prevent propagation to patient injury. Figure 23.18 shows a small portion of a fault tree. As with the process tree, indicating the failures that happened on the tree when reviewing an event database shows particular faults that require attention.

23.7 Remedial Actions Based on Classifications

23.7.1 Preventative Actions/Remedial Actions

As noted several times, the goal of classifying failures that have happened is to direct corrective actions. Such analyses in no way replace the a-priori approaches for

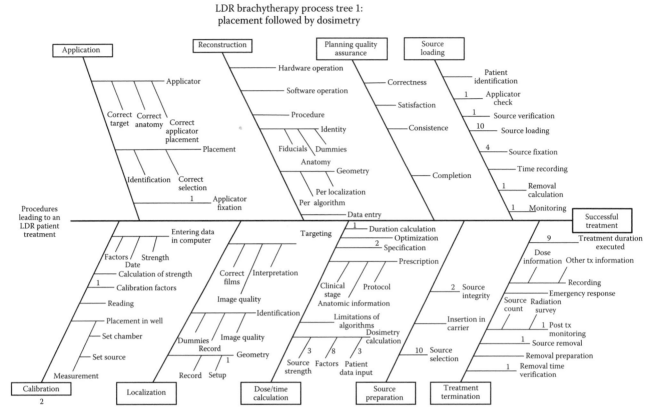

FIGURE 23.17
Process tree for low dose-rate brachytherapy showing numbers of failures occurring on the branches, based on data in a national database. (From Thomadsen, B. et al., *Int. J. Radiat. Oncol. Biol. Phys.*, 57, 1492, 2003. With permission.)

error prevention, such as studying the fault tree and making sure that all error paths are blocked by quality control, quality assurance or some other mechanism. Even with good proactive error prevention, some failures will occur and remedial actions will be necessary. A discussion of guidance for selecting remedial actions based on taxonometric classifications requires a better understanding of the nature of remedial actions themselves.

23.7.2 Categories of Remedial Actions

Van der Schaaf (1992) grouped remedial actions into several large categories (see Table 23.5). The general divisions addressed equipment, individual training, and organizational changes (creation of communication paths or improved procedures). As noted in the earlier section, these categories do not lead the investigator to very specific actions. Any systematic suggestion of remedial actions will be general and descriptive, rather than very concrete, if only because greater specificity would produce unwieldy tables, making it harder to find the appropriate classification and related action.

23.7.3 Relation between Failure Classification and Remedial Actions

The Madison taxonomy attempts to provide fairly detailed guidance to the selection of remedial action. Table 23.14 gives general categories of possible remedial action with some examples. Of course, the list cannot be exhaustive or static. Table 23.15 correlates some potential actions with the failure classification from the Madison taxonomy in this small selection from the entire matrix. The table lists some of the taxonometric classifications as headings for the rows and potential remedial actions for the columns. An "x" in the box at an intersection of a classification and an action indicates that that action may be applicable for failures with that classification. Blanks at the intersections mark those specified actions that probably would not effectively address the problems leading to the failure. Question marks indicate actions with unsure relationships to the failure classification.

23.7.4 Hierarchy of Remedial Actions

Not all remedial actions affect corrections equally. Consider the problem that syringes of the clear

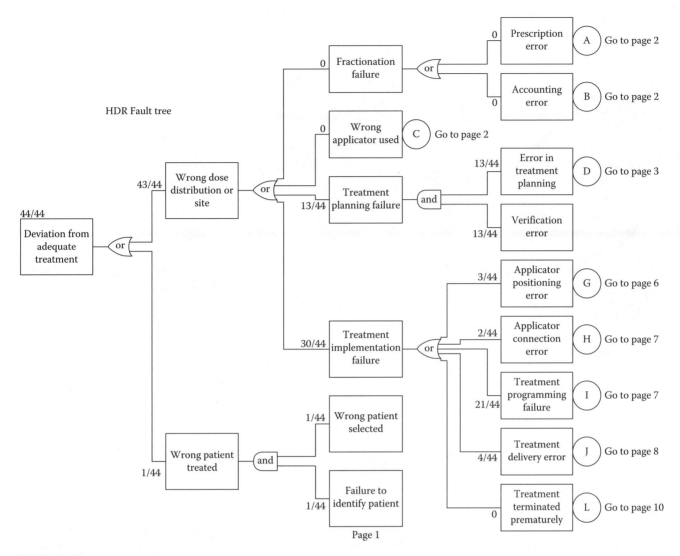

FIGURE 23.18
A fault tree for high dose-rate brachytherapy, with the number of failures along a branch over the number of total failure reported in a national database. The page numbers referred to on the far right indicate pages of the entire fault tree. (From Thomadsen, B. et al., *Int. J. Radiat. Oncol. Biol. Phys.*, 57, 1492, 2003. With permission.)

solution potassium chloride sometimes were mistaken for syringes of saline, and when injected into a patient's vein, inadvertently produced cardiac arrest. One possible remedial action would be labeling the KCl syringes with a large letters and a warning, another with color coding and still another approach would remove the syringes from the general shelves. The likely effectiveness of each of the three possible solutions differs markedly. Labels, no matter how blatant, sometimes are missed or misinterpreted, while removing the syringes eliminates simply grabbing them by mistake. The Institute for Safe Medical Practices (ISMP) lists a hierarchy for corrective action (Table 23.16).

The Madison taxonomy also ranks the likely effectiveness of possible remedial actions, as given in Table 23.17. Table 23.17 ranks actions listed in Table 23.14 by

power to significantly reduce the likelihood of failure according to Table 23.16. It is important to note that the Environmental category is not part of the ranking. Most environmental problems (or environmental conditions that facilitated human performance failures) should be corrected without regard for any other actions taken. Usually, such changes in the environment come easier than changes in practice and may improve the situation markedly with little expenditure of resources. On the other hand, some environments resist changes. Take, for example, the sounds in most ICUs. With many monitors and machines working and producing both white noise (such as from fans cooling computers) and monitor tones (such as heart beats), alarms from any given machine may go unheeded due to a combination of competition for attention and masking noise, or just from hearing

TABLE 23.14

Selected Remedial Actions Grouped by Category

Physical tools
  Barriers
  Interlocks
  Alarms
  Bar codes
  Communication devices
Informational tools
  Label
  Signs
  Reduction in similarity
Measurement tools
  Check off forms
  Operational checks
  Comparison with standards
  Increased monitoring
  Automated monitoring
  Additional status checks
  Redundant measurements
  Independent review
  Acceptance testing
Computerization
  Computerized verification
  Computerized order entry (COE)
  Computerized order entry with verification and
  feedback
Knowledge tools
  Training
  Instruction
  Experience
Administrative tools
  Mandatory pauses
  Increased staffing
  Establishing protocol
  Clarifying protocol
  Establish/clarify communication lines
  Reorder administrative priorities
  External audit
  Internal audit
Environmental actions
  Sound control
  Sight control (clearing/cleaning)
  Sight control (lighting)
  Simplification (neatening)
  Isolation (removal of competing demands
  and calls)
  Ergonometric improvements
  Safety design improvements

so many alarms (e.g., alarm fatigue, Kowalczyk, 2011.) Unyielding environments require other remedial action thrusts. See Chapter 15 for further information.

23.8 Integrated Remedial Actions

23.8.1 Consideration of the Many Dimensions of Information on Failures

Any given remedial action taken may not correct the problems that led to a failure. On this point the wisdom of the theory that the latent causes require fixing shines. Addressing the human performance problems directly may be the most expedient, and the only apparent action to take in a particular setting. However, one may find that the unaddressed latent causes lead to a different human performance problem in a completely different arena. Such a situation illustrates the value of combining a taxonometric approach with a database. The latent causes may not appear during an analysis of single, or several events, but surface over the course of time. Even multiple events that give the impression of resulting totally from human errors can point to common organizational problems or other enabling factors. Identifying the failures by their location on the process tree or fault tree, while potentially indicating procedures or potential faults that require protection with quality management, cannot direct remedial actions toward latent problems.

23.8.2 Prevention of Events Rather than Cure of Root Causes

The goal is not to prevent error but to prevent events.

Lucian Leape (2004)

While addressing the root causes may be most effective in preventing future events, any effective remedial action is likely to improve the situation. Root causes and latent problems often hide deeply in organizations, being hard to find and harder yet to root out. Often, patients improve even with placebo. Taking actions, even if not the most effective, often brings improvement in patient safety if only because the staff sees the importance of patient safety improvement. Addressing concerns also shows that reporting events brings actions and follow-up. Too often the medical staff sees nothing come of reports and become discouraged. Above all, however, remedial actions that directly address human performance problems, without going after the latent problems, frequently still provide sufficient protection against further failures that latent causes are not strong enough to bring about other events.

23.9 Summary

The following points summarize some of the important aspects of taxonomies for medical events:

1. Taxonomies assist in analyzing the nature of events.
2. To be most useful, taxonomies must classify several dimensions of a failure, such as
   a. Where it happened in the procedure
   b. What aspect of performing an action failed (such as detecting that action was needed)
   c. What was the nature of the human failures
   d. What were the contributing PSFs (environments, organizational, technical)
3. Taxomomies should have an underlying organizational structure that helps understand the failure under investigation.
4. The process of determining the failure classifications, such as using a labyrinth layout, may assist in designing remedial actions.
5. The taxonometric classifications should give guidance for selection of remedial action tools; while the strongest tools available, as based on the ISMP ranking, should be used, some the lesser-ranked tools can still be effective.
6. Classification of near-miss events gives valuable information on recovery mechanism that proved effective as the particular failure progressed, intercepting the propagation before it became an event.

Work on sculpting the classifications and the larger categories (dimensions) for taxonomies is a topic of great interest currently, as is better refining mechanisms by which the classifications guide remedial actions.

TABLE 23.15

Suggested Remedial Actions Based on the Failure Classifications in the Madison Taxonomy

	Physical Tools					Information Tools				Measurement Tools								Knowledge Tools				Administrative Tools						Equip. Related			Environmental Tools					Computerized		
	Interlocks	Bar Codes	Alarms	Barriers	Communication Devices	Reduce Similarity	Labels	Signs	Check Off Forms	Operational Checks	Comparison with Standards	Increase Monitoring	Automate Monitoring	Add Status Check	Redundant Measurement	Independent Review	Acceptance Test	Training	Experience	Instruction	Mandatory Pauses	Staffing	Establishing Protocol/Clarify Protocol	Establish/Clarify Communication Lines	Better Scheduling (Reduced Overtime)	(Administrative) Priority	Internal Audits	Repair	PMI - Preventive Maintenance Inspection	(Establish and Perform QC and QA)	(Environmental Controls) Sound Control	(Environmental Controls) Cleaning	(Environmental Controls) Neatening	(Environmental Controls) Isolation	(Environmental Controls) Visual Control	Computerized Verification	Computerized Order (Data) Entry	Computerized Order Entry with Feedback
Mistake action																												O	O	O								
Detection—no protocol																							X															
Detection—ineffective protocol																							X															
Detection—equipment defect										X							X											X	X									
Detection—not performed			X						X				X					?	?	?	?															X		
Detection—incorrect					X	X																		X														
Detection—missed	X		X		X				X			X	X	X				X	X	?											X		X			X		X
Reaction—incorrect interpretation																X																			X	X		
Reaction—erroneous action																												O	O	O								
Reaction—no action																												O	O	O								
Human error—tripping																																	X		X			
Human error—slips	X	X	X	X	X	X	X	X	X	X	X	X	X	X	X	X																				X		X
Human error—blunder	X	X	X	X	X	X	X	X	X	X	X	X	X	X	X	X																				X		X
Human error—error in the intention (mistake)	?	?	?	?											X	X																						?

Where/When: Mistake action, Detection—no protocol, Detection—ineffective protocol, Detection—equipment defect, Detection—not performed, Detection—incorrect, Detection—missed, Reaction—incorrect interpretation, Reaction—erroneous action, Reaction—no action

What: Human error—tripping, Human error—slips, Human error—blunder, Human error—error in the intention (mistake)

How and Why

	1	2	3	4	5	6	7	8	9	10	11	12	13	14	15	16	17	18	19	20	21	22	23	24	25	26	27	28
Hardware failure										X											X	X	X					
Software failure										X													X					
Enabling factor—hardware					X					X						X												
Enabling factor—software					X					X						X												
Enabling factor—external																												
Enabling factor—environmental																			X	X	X	X	X					
Enabling factor—organizational							?	?	?		?	?	?	?	X	X												
Hardware—design																?												
Hardware—construction																?												
Hardware—material								X								X												
Hardware—maintenance																X	X	X										
Software—design																?												
Software—construction																?												
Software—maintenance																X	X	X										
Manual variability (SB)																												
Topographic misorientation (SB)	X																											
Stereotype fixation (SB)	X	X	X	X	X	X	X	X	X	X	X	X	X	X	X	X					X							X
Stereotype takeover (SB)	X		X	X		X	X	X	X		X	X	X	X	?	X				X								X
Familiar association shortcut (RB)	X		X	X		X			X			X	X	?				X	X									?
Familiar pattern not recognized (RB)		X			X	X	X	X	XX		X		X		X			X	X	X	?							X
Mistakes alternatives (RB)								X					X		X	X	?		X									?
Mistake consequence (RB)							?					X		X	X	?		X										?
Forget isolated act (RB)	X	X	X	X	X			?	?	X		?		?	?	?	X			?								X
Condition or side effect no considered (KB)	X		X	X				X	X		X		X		?	?												X

(*continued*)

TABLE 23.15 (continued)

Suggested Remedial Actions Based on the Failure Classifications in the Madison Taxonomy

How and Why (continued)	Physical Tools					Information Tools				Measurement Tools								Knowledge Tools			Administrative Tools						Equip. Related				Environmental Tools					Computerized		
	Interlocks	Bar Codes	Alarms	Barriers	Communication Devices	Reduce Similarity	Labels	Signs	Check Off Forms	Operational Checks	Comparison with Standards	Increase Monitoring	Automate Monitoring	Add Status Check	Redundant Measurement	Independent Review	Acceptance Test	Training	Experience	Instruction	Mandatory Pauses	Staffing	Establishing Protocol/Clarify Protocol	Establish/Clarify Communication Lines	Better Scheduling (Reduced Overtime)	(Administrative) Priority	Internal Audits	Repair	PMI - Preventive Maintenance Inspection	(Establish and Perform QC and QA)	(Environmental Controls) Sound Control	(Environmental Controls) Cleaning	(Environmental Controls) Neatening	(Environmental Controls) Isolation	(Environmental Controls) Visual Control	Computerized Verification	Computerized Order (Data) Entry	Computerized Order Entry with Feedback
Information not sought—assumed, negligent omission, not realized it was needed	X	X	X	X	X	?	?	?	X		X			X		X						?	?	X	?											X	X	
Lack of vigilance (arousal, commitment, complacency)	X	X	X	X					?			X	X			X							?		?											X		
Organizational—knowledge of leader																X			X																			
Organizational—management priority																X			X																			
Organizational—communication system					X																			X														
Organizational—knowledge transfer																		X		X																		
Organizational—ability																	X	X	X																			
Competition for attention (background)—lack of staff or time	X		X	?					?			X	X			?						X														X		
Competition for attention (background)—other goals	X		X	?					?				X			?																				X		
Competition for attention (immediate)—other duties	?		?	?					?				X			X						X	?			X										X		

Competition for attention (immediate)—lack of staff	X	X	?			X		X		X		X
Competition for attention (immediate)—too many inputs to system	X	X	X	?		X	X		X		X	X
Competition for attention (previous)—fatigue due to complex work	X	X	X	X	X	X	X	?	X			X
Environment—(tangible) noise (non-human environmental factors)									X			
Environment—(Tangible) distraction (human related environmental factors)										X		
Environment—(intangible) environmental problem									X	X	X	X
Environment—(intangible) end of day, holiday	?	?	?	X		X	X					X
Environment—(intangible) personal problem	?	?	?	?		X	X					X
Others—exceed ability—physical												
Others—exceed ability—mental												
Others—lack of experience								? X X				

TABLE 23.16

Ranking of the Likely Effectiveness of Types of Actions to Prevent Errors, Based on the Institute for Safe Medical Practices's Toolbox (ISMP, 1999), Listed from the Most to the Least Effective

1. Forcing functions and constraints
2. Automation and computerization
3. Protocols and standard order forms
4. Independent check systems and other redundancies
5. Rules and policies
6. Education and Information

TABLE 23.17

Some Examples of Remedial Actions Ranked by Power to Affect Changes

Environment (not ranked but should be addressed first)
    (Environmental controls) sound control
    (Environmental controls) visual control
    (Environmental controls) cleaning
    (Environmental controls) neatening
    (Environmental controls) isolation
    Environmental design
Forcing functions and constraints
    Interlock
    Barriers
    Computerized order entry with feedback
Automation and computerization
    Bar codes
    Automate monitoring
    Computerized verification
    Computerized order entry
Protocols, standards, and information
    Check off forms
    Establishing protocol/clarify protocol
    Alarms
    Labels
    Signs
    Reduce similarity
Independent double check systems and other redundancies
    Redundant measurement
    Independent review
    Operational checks
    Comparison with standards
    Increase monitoring
    Add status check
    Acceptance test
Rules and policies
    External audit
    Internal audit
    Priority
    Establishing/clarify communication line
    Staffing
    Better scheduling
    Mandatory pauses
    Repair
    PMI (preventive maintenance inspection)
    Establish and perform QC and QA (hardware and software)
Education and Information
    Training
    Experience
    Instruction

Note: The environment category is not part of the ranking, since any environmental problems should be addressed independently of other actions taken.

References

Anderson, T. (1997). A case study and conceptual framework for medical event reporting systems. Masters' thesis. Madison, WI: University of Wisconsin.

Benner, P., Sheets, V., Uris, P., Malloch, K., Schwed, K., and Jamison, D. (2002). Individual, practice, and system causes of errors in nursing: A taxonomy. *Journal of Nursing Administration*, 32, 45–48.

Busse, D. and Holland, B. (2002). Implementation of critical incident reporting in a neonatal intensive care unit. *Cognition, Technology & Work*, 4, 101–106.

Busse, D. and Johnson, C.W. (1999). Human error in an intensive care unit—A cognitive analysis of critical incidents. In J. Dixon (Ed.), *17th International System Safety Conference* Orlando, FL: System Safety Society.

Busse, D. and Wright, D. (2000). Classification and analysis of incidents in complex medical environments. *Topics in Health Information Management*, 20, 1–11.

Ekaette, E.U., Lee, R.C., Cooke, D.L., Kelly, K.L., and Dunscombe, P.B. (2006). Risk analysis in radiation treatment: Application of a new taxonomic structure. *Radiotherapy & Oncology* 80, 282–287.

Harris, D. and Chaney, F. (1969). *Human Factors in Quality Assurance*. New York: John Wiley & Sons, Inc.

Institute for Safe Medical Practices. (1999). Medication error prevention "toolbox". Medication Safety Alert, June 2 http://www.ismp.org/msaarticles/toolbox.html (accessed February 15, 2011).

Kaplan, H.S., Battles, J.B., van der Schaaf, T.W., Shea, C.E., and Mercer, S.Q. (1998). Identification and classification of the causes of events in transfusion medicine. *Transfusion* 38, 1071–1081.

Kapp, E.A. and Caldwell, B. (1996). SCOPE Instruction Manual. Unpublished.

Koppens, H.A. (1997). "SMART" error management in a radiotherapy quality system. Masters' thesis. Eindhoven, the Netherlands: Eindhoven University of Technology.

Kowalczyk, L. No easy solutions for alarm fatigue. Boston Globe February 14, 2011. http://www.boston.com/lifestyle/health/articles/2011/02/14/no_easy_solutions_for_alarm_fatigue (accessed February 15, 2011).

Leape, L. (2004). Wisconsin Patient Safety Institute, Oconomowoc, WI, November 13, 2003.

Mutic, S., Brame, R.S., Oddiraju, S., Parikh, P., Westfall, M.A., Hopkins, M.L., Medina, A.D., Danieley, J.C., Michalski, J.M., El Naqa, I.M., Low, D.A., and Wu, B. (2010). Event (error and near-miss) reporting and learning system for process improvement in radiation oncology. *Medical Physics* 37, 5027–36.

National Coordinating Council for Reporting and Preventing Medication Errors. (1999). http://www.nccmerp.org/pdf/taxo2001-07-31.pdf. Accessed July 13, 2011.

New York State Department of Health. (2001). http://www.health.state.ny.us/nysdoh/hospital/nyports/annual_report/2005–2007/docs/2005–2007_nyports_annual_report.pdf

Norman, D. (1981). Categorization of action slips. *Psychological Review*, 88, 1–15.

Rasmussen, J. (1980). What can be learned from human error reports? In K.D. Duncan, M.M. Gruneberg and D. Wallis (Eds.), *Changes in Working Life*. New York: John Wiley & Sons.

Rasmussen, J. (1982). Human errors. A taxonomy for describing human malfunction in industrial installations. *Journal of Occupational Accidents*, 4, 311–335.

Reason J. (1990). *Human Error*. Cambridge: Cambridge University Press.

Rook, L. (1965). Motivation and Human Error. Report SC-TM-65–135. Albuquerque: Sandia Corp.

Thomadsen, B. and Lin, S.-W. (2003). Guidance for remedial actions based on failure classifications. Wisconsin Patient Safety Institute, Oconomowoc, WI, November 13, 2003.

Thomadsen, B., Lin, S.-W., Laemmrich, P., Waller, T., Cheng, A., Caldwell, B., Rankin, R., and Stitt, J. (2003). Analysis of treatment delivery errors in brachytherapy using formal risk analysis techniques. *International Journal of Radiation Oncology Biology Physics* 57, 1492–1508.

van der Schaaf, T.W. (1992). Near miss reporting in the chemical process industry. Doctoral thesis. Eindhoven, the Netherlands: Eindhoven University of Technology.

Van der Schaaf, T. (2005). PRISMA medical: A brief description. (Unpublished), Eindhown University of Technology, Faculty of Technology Management, Patient Safety Systems.

van Vuuren, W., Shea, C.E., and van der Schaff, T.W. (1997). The development of an incident analysis tool for the medical field. Eindhoven, the Netherlands: Eindhoven University of Technology.

Wright, D., Mackenzie, S.J., Buchan, I., Cairns, C., and Price, L.E. (1991). Critical incidents in the intensive therapy unit. *Lancet* 338, 676–678.

24

Human Error Reduction Strategies in Health Care

René Amalberti and Sylvain Hourlier

CONTENTS

24.1 Introduction 385
24.2 Health Care as a Land of Risks and Errors 386
    24.2.1 Nature and Extent of Risk 386
    24.2.2 Specific Difficulties in Coping with HE in Medicine 388
        24.2.2.1 Risks in Medicine Are Far from Being Homogeneous 388
        24.2.2.2 Topic of Human Error in Medicine Is More Difficult to Manage than in Industry 388
        24.2.2.3 In Medicine, Some Conditions of Work Are behind Usual Industry Standards and
                Facilitate Errors 389
24.3 Three Generic Solution Spaces for Error Reduction 389
    24.3.1 "Person Approach" 390
        24.3.1.1 Essential Points 390
        24.3.1.2 Recommended HE Reduction Strategies at the "Person Approach" Level 392
    24.3.2 "System Approach" 393
        24.3.2.1 Essential Points 393
        24.3.2.2 Recommended HE Reduction Strategies at the System Approach Level 394
    24.3.3 "Dynamic Approach" 395
        24.3.3.1 Essential Points 395
        24.3.3.2 Recommended Dynamic HE Reduction Strategies 395
24.4 Conclusion 397
References 398

24.1 Introduction

Since the 1980s, there has been an active attempt to increase the health care system's accountability regarding quality, safety, and efficiency. High safety standards are now mandatory for all patients. This situation represents a great challenge since medicine, for many reasons, is laden with potential risks. This chapter tries to offer a comprehensive survey of human error (HE) and related safety strategy concepts, whether derived from industry or field-dependent. The chapter is divided into three sections. This introduction gives a brief overview of the extent and specifics of risks encountered in the industry, and then in the medical field. Section 24.2 defines HE-related issues in medicine. Section 24.3 drafts three generic solution spaces for error reduction: a "person approach" (errors of individuals), a "systems approach" (errors of organizations), and a "dynamic approach" (adaptation of error reduction strategies as

safety improves). Each section suggests a set of related strategies to reduce the errors and gives directions on how to combine them.

The shock that HE can lead to disasters clearly dates back to the 1970s, with the Three Miles Island Nuclear incident. HE was not unknown at the time, but even for psychologists, it was more regarded as a mean to characterize behaviors in experiments (good vs. false responses) rather than as a topic of research of its own.

Before Three Miles Island accident, safety was merely a technical problem. The science of system reliability assessment (SRA) and probability safety assessment (PSA) had been developed in the 1960s to accompany the fast growth of technology, providing tools to more effective and more safe design (see a survey by Cox and Trait, 1991). In other words, achieving better safety was limited to the pursuit of a better and more reliable technology. Humans were hardly considered, and when taken into account, mainly perceived as instruction followers. Unfortunately, engineers suddenly found out after Three

Miles Island that humans were still in the process, and actually, more involved than ever in the control of this process and in charge of decision-making, with their emotional and social skills, and limited rationality. From that date, cognitive ergonomics and cognitive engineering became central concerns to continue in the best possible manner the extreme potential benefit of high technology to the needs and limitations of workers. Nuclear industry and aviation were pioneering fields, which gradually moved in the 1980s and 1990s to a set of "win–win" solutions, combining the extensive introduction of automation and information technology (IT) techniques with the reduction of errors (see, e.g., a review in Amalberti, 1998).

HE in medicine only started to be taken into account a decade later, with a few scientific papers and edited books (Brennan et al., 1991; Bogner, 1994; Leape, 1994; Vincent, 2001), then took another half decade for this issue to be recognized as major by medical institutions (Kohn et al., 1999, To err is human: report of the Institute of Medicine). Since the late 1990s, the number of special issues, journals, conferences, and reports dedicated to this topic has continuously grown.

Health care is now following the same pattern regarding the treatment of HE as the one taken by the safest industries a couple decades ago. The same ideas and celebrated safety models are encountered in education on HE management in medicine (with special appraisal to James Reason's model). These models and ideas tell us to know more about error, to design reporting systems as nonpunitive systems, to add defenses, to reduce contributing factors to HE, to suppress violations, to standardize protocols, to improve quality and traceability, and to create conditions for improved communications and common situation awareness with a better use of IT. Last but not least, the improvement of organizations and culture are considered key factors for long-term success.

However, common sense tells us also that medicine does not strictly behave like industry; patients are not standardized and evolution of disease is not fully predictable; moreover, for centuries, medicine was said to be closer to an art than to a science. Should these differences with industry be true, then the concept of HE in medicine will probably need to be clarified, and dedicated additional strategies will need to be developed to fully meet the challenge of error reduction.

24.2 Health Care as a Land of Risks and Errors

24.2.1 Nature and Extent of Risk

Patient safety is a growing concern. However, the risk run by patients remains hard to assess. Risks of different nature are combined: the risk posed by the disease itself, the risk entailed by access to care (waiting time, cost and privilege of individual medical insurance), the risk entailed by the medical decision made, the risk linked to implementing the therapy selected, and last but not least for managers, the managerial and occupational risks that may potentially greatly stress the social system at work (see Figure 24.1).

These risks are generally interconnected and hard to dissociate and do not generally move in the same direction; therefore, their combined effects make it hard to grasp the issue of error. For example, diseases associated with high risks may be tackled effectively only by treatments also associated with high risks; high gains for the outcome of the disease may require high risks in the treatment. Patients facing an apparently terminal illness may have a major change in their fatal prognosis, thanks to a daring medical strategy. An example of this is the treatment of hip fractures, previously fatal, which are now successfully replaced in elderly patients even when they are over 90. But, the most daring clinical strategies are also those whose levels of excellence are the least evenly distributed in the profession; they depend on craftsmanship. Inevitably, these complex approaches match those in which errors are the most frequent. Audacity is traded against safety. In short, it is almost impossible to simultaneously successfully reduce the three types of risks: (i) efficient effective disease management, (ii) probability of gaining an identical benefit through the use of an advanced treatment strategy, whatever the practitioner and medical center involved, and (iii) achieving the lowest possible rate of error (Amalberti et al., 2005; Degos et al., 2009).

The extent of HE and associated risks is also difficult to appreciate in medicine because no system was set up to monitor medical practice as is regularly carried out in the safest industries (see, e.g., systematic after flight analysis of black boxes, or line-oriented safety audits [LOSA], Helmreich, 2000); the only observatories that process epidemiological information are not very effective in finding human flaws because they focus on the patient more than on the process of care. Moreover, alerting or sentinel-event systems can only trigger alerts by monitoring large series. Insufficient, sporadic, and scattered cases, which will not pass the alert threshold (even though they eventually often add up to high numbers), are often disregarded. See Table 24.1 for a glossary of error terms used in work psychology and in medicine.

Bearing in mind these "contextual reasons" for generating human flaws and the inherent difficulty to characterize and measure HE, medicine made the choice to measure safety with the concepts of adverse event (AE) and negligence. Brennan, in the well-known pioneering Harvard Medical Practice study (1991), defines an

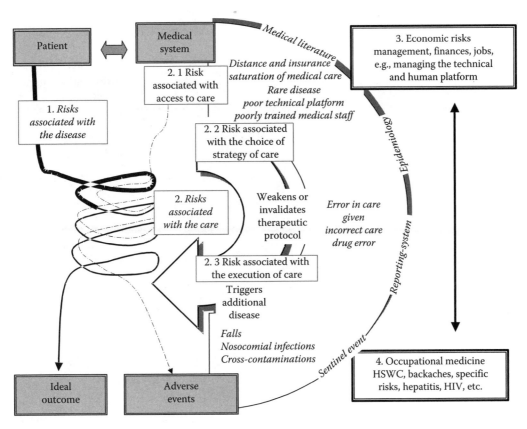

FIGURE 24.1
Conflicting sources of risks in medicine.

TABLE 24.1

Error Glossary

Generic in work psychology

- Error-HE-(Senders and Moray): [Unintentional] human action that fails to meet an implicit or explicit standard. An error occurs when a planned series of actions fails to achieve its desired outcome
- Violations (Reason): [Deliberated] deviations from safe operating practices, procedures, standards, or rules
- Active failures-AF-(Reason): unsafe acts (errors and violations) committed by those at the sharp end of the system
- Latent failures (Reason): the result of decisions taken at the higher echelons of an organization
- Near miss (Van der Shaaf): a situation where an event occurring during the course of action fails to further develop, thus not leading to consequences

Specific to medicine

- Adverse event -AE- (Brennan): injury that was caused by medical management (rather than by the underlying disease) and that prolonged hospitalization produced a disability at the time of discharge, or both
- Preventable adverse event -PAE-: all AE that can be avoided by a proper error management strategy. All PAE reflect active failures
- Negligence (Brennan): care that fell below standards expected of physicians in their community

---

"adverse event" as an injury that was caused by medical management (rather than the underlying disease) and that prolonged the hospitalization, produced a disability at the time of discharge, or both. He defines negligence as care that fell below the standard expected of physicians in their community. Preventable adverse events (PAEs) are all AEs that can be avoided by a proper error management strategy. Therefore, PAEs and

negligence are the two categories of active failures that medicine considers to be immediately linked to unsafe acts (errors and violations) committed by those at the front line of the system. However, it is important to note at this point that the two concepts (PAE and HE) should not be totally mistaken for one another. They address the same idea of human failure, but describe the failing process at a different level of abstraction and causality

TABLE 24.2

Relationships between Terms Used to Define Human Failures

	Active Failure	Human Error	Preventable AE—(PAE) Negligence
Definition	All unsafe acts committed at the sharp end, including HE and violations	Academically, should only refer to unintentional unsafe acts Common usage turns the concept into a synonym of unsafe acts, then including intentional errors (violations)	Any injury that was caused by medical management and was avoidable by a proper error management strategy
Relations between concepts	All unsafe acts are HE, but not leading to PAEs	All HE do not lead to PAEs. Most are detected and recovered far before becoming even a "near miss" or a "close call." Errors are more frequent than near misses, which occur up to 300 times more frequently than AE (Albanese, 2001)	All PAEs are related to, at least one, unsafe act Neighboring concepts "near miss" and "close-call" are related to potential PAE whose story turned out well
Usual terms used in reports for the description of the type of failure	Reported in terms of generic categories of failure using generic language of ergonomics and reliability (individual failure and organizational malfunction)	Reported in terms of error's mechanisms, error causation, and error contextual shaping factors (in the language of psychology and cognitive ergonomics)	Reported in terms of patients' (potential) consequences and professional's failure (in the language of the medical profession)

(see Table 24.2). The PAE approach is even more confusing since it mixes two theoretical backgrounds (HE for cognitive psychology vs. error and negligence for justice) that usually do not fit one another.

With these definitions, risks measured in the medical field could be well above anything experienced in safe industries (Amalberti et al., 2005).

The percentage of AEs in the various countries where studies have been conducted ranges from 3% to 12% of hospitalized patients. The proportion of PAEs is at minimum 37%, and may reach 58% in some studies[*]. Operative drug events (surgery), adverse drug events, and incorrect or delayed diagnosis are the top three areas of PAEs; this result is remarkably stable between studies. Medical errors are estimated to cause more casualties than road accidents or workplace injuries, and to represent a significant percentage of national health expenditures (Gaba, 2000; Kohn et al., 1999). Moreover, errors are per definition underreported, since hospital patients only represent a fraction of the total population exposed to medical errors. We know little on the error rate in primary care (Brami and Amalberti, 2009).

24.2.2 Specific Difficulties in Coping with HE in Medicine

Medical professions suffer from various identity crises that make the global system prone to error and the study of errors less easy than in industry.

24.2.2.1 Risks in Medicine Are Far from Being Homogeneous

- On the one hand, a number of sectors display risks of death due to AE (whether preventable or not) that exceed 10^{-2} (1 AE per 100 hospitalized patients, i.e., surgery, internal medicine), and which of course are not all related to medical errors.

- On the other hand, other sectors are extremely safe, and the risk of death due to AEs is 10–100 times lower (i.e., anesthesiology, radiology, blood transfusion). Furthermore, everything in medical practice encourages heterogeneity. This makes hospitals "sectorial environments," where different safety strategies need to be implemented according to different targets, with heterogeneous levels of regulation governing the various professionals working there, with a reduced capacity for cross-fertilization of lessons learned at local level (strongly limiting the development of a culture of safety).

24.2.2.2 Topic of Human Error in Medicine Is More Difficult to Manage than in Industry

- The pressure of insurance providers is greater than in any other industrial field. The great autonomy of physicians, which is often mentioned as a prerequisite for the profession, also makes front line actors more individually accountable for errors, with little protection from their organizations when sued and facing justice (Vincent et al., 1994). Needless to say, this

---

[*] The apparent differences among studies does not necessarily mean true differences; it may only reflect small variations in the methodology and the inclusion criteria (see Baker et al., 2004; Michel et al., 2004).

specificity does not encourage disclosure and reporting of malpractice.

- This is one of the areas where actual figures on error are hardest to collect. The figures we have are only extrapolations, and could be discussed, mainly because of the intertwined nature of contradictory dimensions of risks in medicine (see the beginning of this section), and the debatable definition and estimation of PAE (Hayward and Hofer, 2001). The dilution in medical establishments and wards, the fact that death trickles in, one patient after another without any massive single occurrence in the context of pathologies, all help cover up reality and alleviate any strong reaction of anger from the media. To make things worse, safety advances uncover safety loopholes, which, until then, were discretely hidden away. Because of this, as is the case in all fairly unsafe systems where a culture of denial and hiding is widespread, the first phase of progress is mechanically associated to an increase in known occurrences. This access to such an unbearable truth generates fears and defensive reactions from the system, hindering further safety improvements.

- Nobody can actually really pinpoint the potential gains to be expected. As Marx (2003) pointed out repeatedly in conferences, a good safety management (SM) is the result of a comparison between a safety target expected at the phase of design of the system, and the reality of occurrences observed in the field. Commercial aviation design expected 1 loss every 1,00,000,000 departures, and the reality is 1 loss every 3,500,000 departures. Space shuttle design expected 1 loss every 212 departures, and the reality is 1 loss every 60 departures. But how can safety be managed in medicine, when there is no expectation, just a reality of losses: 1 loss for every 1000 hospital visits!

24.2.2.3 In Medicine, Some Conditions of Work Are behind Usual Industry Standards and Facilitate Errors

- It is one of the few risk-prone areas where direct public pressure is so high that it considerably limits the application of wise (or common sense) safety-enhancing solutions (it is difficult to turn down anyone in an emergency ward, to regulate incoming patients to adapt patient flow to the capacity of premises, etc.).

- It is one of the few risk-prone areas where the system is extensively supported by interns who are in the process of being qualified for their jobs. First and second year resident interns, as well as nurses in training, are often brought to bear huge responsibilities.

- It is one of the few risk-prone areas where there are so many fundamental and obvious sources of HE, and where so little is done to reduce them, because they are considered as being part of the job: excess fatigue on the job because of systematic overtime, overloaded work schedules and chronic staffing shortages (Volpp, 2010), multiplication of quasisimilar products (competing number and forms of medical specialties), reassignment of tasks between the different job categories, and to top it off, staff members acutely sensitive to their own mistakes, with profound psychological consequences, and a closeness to death found in no other profession (Daugherty et al., 1998; Wu et al., 1991).

24.3 Three Generic Solution Spaces for Error Reduction

HE in medicine may be studied from various angles: how many (error rate), what type (wrong drug, delay, omission...), with what consequence for the patient (from minimal impairment to death), where (surgery, internal medicine, anesthesiology, etc.), when (week vs. weekend, nighttime vs. daytime), with what mechanism (slip and lapses vs. mistakes, errors vs. violation, active vs. latent human failure), and why (error causation, error–framing factors).

The "how many," "where," "when," and "what" questions (frequency, location, type, and consequences) are key factors to prioritize specific efforts (e.g., delay in diagnosis, surgery, and drug events are often considered as first priorities because they represent the three major sources of preventable AEs.). Conversely, the "why" questions are key factors to determine the repertoire of safety strategies to use according to the result of causal analysis.

Since this chapter focuses on error reduction strategies, the following sections give priority to the "why" questions and to the safety strategies to recommend. Reason (1990, 1995) addresses the problem by clearly opposing on the one hand "front line actors" working at the "sharp end" and making "latent errors," for example, doctors, nurses, or stretcher bearers; and on the other hand "designers, administrators, and system managers" working at the "blunt end," and making "latent errors." Design flaws, staff shortage, poor management, or growing incentive to performance are examples of

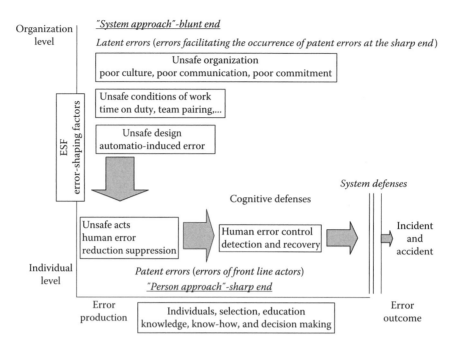

FIGURE 24.2
Summarizing James Reason's concepts: the person and system approaches.

latent errors that can potentially foster latent errors made by front line actors.

When considering possible solutions, Reason (1995) turns the aforementioned opposition into two practical approaches for error reduction: the "person approach" and the "system approach" (Figure 24.2). A third approach turns out to be complementary: the "dynamic approach." This last approach is generally missing or undersconsidered in the literature, although it is essential. Strategies adapted to error reduction should change as safety improves.

24.3.1 "Person Approach"

Chronologically, the first approach is the "person approach." It focuses on the errors and violations of individuals or groups of individuals. Most of the safety strategies relevant to this approach focus on the sharp end. Remedial efforts are directed at people at their work position.

24.3.1.1 Essential Points

At this level three principles challenge common sense and need to be revisited.

24.3.1.1.1 Total Eradication of Human Error Should Not Be an Objective

Ergonomic advances in the workplace enable discussion of the model of safety. By definition HEs are nonintentional. The literature distinguishes several types of errors: slips and lapses vs. mistakes (rule-based and

knowledge-based). Slips and lapses are assimilated to routine errors and are most frequent at expert level (Reason, 1990). Their frequency in tightly coupled systems may reach impressive figures. In aviation, numerous studies show that the minimum rate of error for professional pilots is around 1 per hour, whatever its outcome and the quality of the workplace design (Amalberti, 2001; Duffey and Saull, 1999; Helmreich, 2000). The generic top reasons for "why" errors occur are complexity of work situation and lack of familiarity with them, time pressure, lack of anticipation, and communication impairment.

24.3.1.1.2 It Is More Important to Consider Detection and Recovery than Error Itself

Allwood (1984) even considers that detection performance is the true marker of expertise, while error production is not. The rate of self-detection is then very high, above 70%, and is integrated in the natural cognitive resources that individuals have to manage. Routine errors are better detected than mistakes. Detection and recovery are sensitive to high workload, task interruptions, and system time management. In aviation, a priority principle in design is to tolerate at least one error while giving time to detect it and recover from it before loss of control (fault-tolerant system).

24.3.1.1.3 Violations Are Intentional and Should Not Be Treated as Errors

Figure 24.3 shows the different types of errors and violations. Violations happen as often as the rest of errors

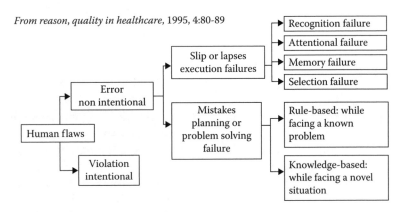

From reason, quality in healthcare, 1995, 4:80-89

FIGURE 24.3
Relations between types of errors and violations.

but generally have not been considered as having consequences on safety (Helmreich, 2000). However, additional analysis led to a different conclusion: chronic violators also have increased rates of error, specifically mistakes (Helmreich, 2000). Violations, therefore, are two-faced. On the one hand, they occur frequently but without consequences on safety. On the other hand, violations can also represent the historical source of dangerous behaviors. This duality reinforces the importance of the debate in the scientific community on the safety strategies allowing to cope with violations. To model and explain why violations occur, Amalberti extended Rasmussen's model of migrations (Amalberti et al., 2006; Rasmussen, 1997), offering to distinguish a three-phase model (see Figure 24.4).

- *The first phase* occurs during the initial design of the work situation. At that stage, systems are designed to accommodate a threefold pressure: (i) compliance with sociological rules, (ii) use of available technologies, and (iii) allowance for economic performance constraints. All the constraints describe a possible space of action, which is itself limited and enclosed by strong barriers (failsafe systems), physically blocking-off various maneuvers or accesses, as well as by virtual barriers (rules, protocols, state of the art, diverse operational limitations).

- *The second phase* occurs when the system is commissioned and must continuously adapt to new social and technical demands. The system

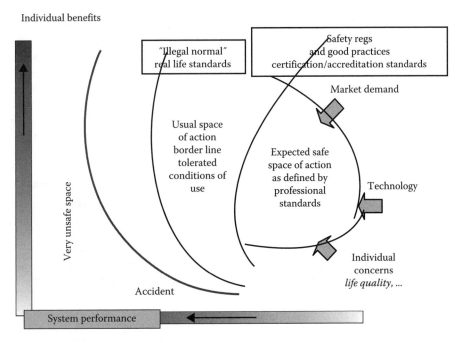

FIGURE 24.4
A model of systemic migration to boundaries and violations.

migrates toward greater performance (horizontal axis), and additional secondary benefits for the individual (vertical axis). Barriers are quickly bypassed under the pressures of real life. The first transgressions always occur at the senior management level, which, even though it is strapped for resources, is required to provide higher performance than originally expected (e.g., having the same amount of work done with less staff, missing or out-of-order equipment). Once these transgressions are achieved, the next quasi-immediate move is a second migration. This time, it is for the benefit of individuals who grant themselves secondary rights in payment for efforts made to work "officially illegally" and to routinely "officially transgress" the established rules, in view of delivering the demanded performance. The result is that the system migrates toward a "normal illegal" area of stabilized operation. At that stage, violations are better termed "border-line tolerated conditions of use" (BCTU) (Polet et al., 2003), and may be viewed as providing management and individuals with the maximum benefit for the minimum and accepted probability of harm.

- *The third and last phase* occurs after a certain amount of time. The same violations may be committed as in the second phase, but they are now "old," amplified (amplitude of deviation), and made gradually invisible to peers' and managements' inspections (part of usual collective routines). These violations also represent a major safety hazard, as there is a tendency to continue the migration in solo (less dependence with demands) until the incident or accident occurs.

It goes without saying that violation prevention must focus on the appropriate phase.

24.3.1.2 Recommended HE Reduction Strategies at the "Person Approach" Level

The best way to achieve the proper error reduction is to apply the four S recommendations: standardize, simplify, staff and share, and supervise.

24.3.1.2.1 Standardize

It requires reducing the span of medical procedures and technologies. Enlarging the span of possible solutions induces a greater risk of error. Evidence-based medicine is the main attempt to accomplish this standardization at the level of the profession, although it leads to more protocols being implemented at the ward level. Such standardization implies a significant effort in initial and recurrent training on the selected "standards"

to increase the number of "equivalent actors" (professionals sharing the same core of knowledge and know-how) and facilitate the round-the-clock coverage of patient care.

24.3.1.2.2 Simplify

It is a true challenge for medicine. It comes in addition to "standardize," with the systematic selection of the simplest solution whenever possible: technical devices, procedures, labeling, patient administration, etc. Some solutions presented as simplifying systems may have drawbacks. Extensive automation, for instance, often has a negative potential, at least when not coherently integrated. In addition to choosing the simplest solutions, the concept of "simplify" also encompasses all means to reduce opportunity to make errors and facilitate detection and recovery. The fight against confusion errors is the first priority calling for unambiguous drug labeling (Christie, 2002; Lambert et al., 2003). The facilitation of error detection relies on security redundancies, hard protections, and reminders (Reason, 2002). Medical instruments commonly have poorly designed user interfaces that promote HEs with life-threatening consequences. This is for example particularly frequent with emergency instruments: electronic syringes, patient-controlled analgesia pumps (Lin et al., 1998). The same errors that occurred long time ago in aviation (poor error analysis and techno-centered automation) are now being repeated in medicine (Woods and Cook, 2002) (see Table 24.3).

24.3.1.2.3 Staff and Share

It has long been recognized essential in aviation. Communication through briefings and debriefings is central to facilitate teamwork and avoid uncontrolled courses of action. Mortality and morbidity (M&Ms) conferences are recommended and often effective. But the concept of "staff and share" requires greater continuous interactions to synchronize, explain, and acknowledge decisions, situation awareness, and courses of

TABLE 24.3

Woods and Cook's Nine Steps to Move Forward from Error

1. Pursue second stories underneath the surface to discover multiple contributors
2. Escape the hindsight bias
3. Understand work as performed at the sharp end of the system
4. Search for systemic vulnerabilities
5. Study how practices creates safety
6. Search for underlying patterns
7. Examine how change will produce new vulnerabilities and paths to failure
8. Use new to technology to support and enhance human expertise
9. Stimulate innovation

actions of medical teams. The transfer of aviation-based teamwork skills (Crew resource management training [CRM]) to medicine is gaining ground to educate medical staff in nontechnical skills (Sexton et al., 2000). However, the generalization of these courses to medicine takes a long time and is still in its infancy, although some ideas may migrate much faster. For instance, the checklist and especially the time out (preoperative briefing) are taking a rapid central position in securing the process of communication in the operating room. The World Health Organization (WHO) has recommended the adoption of such a surgical checklist to ensure basic minimum safety standards. The use of this checklist in eight hospitals around the world was associated with a reduction in major complications (Haynes et al., 2009). Paradoxically, in addition to cutting the risk of being sued by an average 50%, sharing more information on error with patients (error disclosure) is also a mean to change attitudes in wards and is a first step toward enhancing team communication (Gallagher et al., 2007).

24.3.1.2.4 Supervise

It means designing a process to collect and analyze errors, then making ad hoc decision for improvement of priorities and actions in error reduction, and draw lessons. Reporting systems are becoming extremely popular. To be efficient, it is commonly accepted that they should remain anonymous and/or blame-free and should address near misses are well as misses. Feedback to sharp-end actors guarantees longevity and motivation to report. The technique of analysis of errors is also a challenge. Several methods have been offered (see for example the ALARM method recommended by Vincent et al. (1998), or the root cause analysis method—Battles and Shea, 2001). All of these methods need to go beyond error to evidence the latent causes. Continuous audit and practice supervision are required to control violations. Indeed, solutions to cope with violations are not trivial. On the one hand, a certain amount of flexibility with regulations and standards resulting in the presence of borderline tolerated conditions of use (BTCUs) is probably required in complex sociotechnical work to make the system efficient and adaptive. On the other hand, abuses and loss of control may represent true dangers. A compromise attitude between two extreme positions—tolerant vs. punitive approaches—is still a matter of debate. Neither mandatory nor voluntary incident reporting systems are very good tools to monitor phase II violations. These reporting systems only provide necessary information when it is too late for intelligent and smart action and control. Realistic and repetitive short periods of systematic observation of practice in wards (field audit), with a continuous dialogue about practice with peers, are probably the best

methods of elegant and intelligent control of deviations and migrations.

24.3.2 "System Approach"

The second approach is the "system approach." It traces the causal factors back into the system as a whole. Remedial efforts are directed at situations, defenses, and organization. There are schematically two major sources of error-shaping factors: bad system design implying undue work constraints (general architecture, procurement, and personnel management) and bad organizational design (governance, managerial policies, commitment, safety culture).

24.3.2.1 Essential Points

24.3.2.1.1 Changes Are Mandatory

New regulations deeply impact all structures and create instability.

24.3.2.1.2 Management Is the Key Actor for Safety Culture

It is essential to promulgate a safety culture by making sure management adopts it and then turns it into a priority.

24.3.2.1.3 Patient Safety Starts with a Good Hospital Layout

The design of the hospital and of the global care process, recruiting process of staff members, as well as the procurement and maintenance of technologies, are key latent factors that may contribute in time to the emergence of errors. For instance, an undue distance between operating room, emergency and surgery wards may cause transfer delays, transport injuries, as well as information losses between medical teams.

24.3.2.1.4 Time Management Is Crucial

The problem is twofold: first, with poor management of working hours or overtime (for example, noncoherent closing time for support services and wards), and second, with accumulation of fatigue. A great number of recent papers address the fatigue of residents, showing that eliminating interns' extended work shift may have contradictory results, good for private life and the morale of interns, but not necessarily as good for the patient outcomes, since training is affected by the reduction of time on duty (Volpp, 2010). This result clearly evidences the need for a global approach and not just patching local problems. In aviation for instance, fatigue prevention and recovery has been successfully addressed not only by controlling pilots' hours and work schedules but also by staffing the airlines appropriately.

24.3.2.2 Recommended HE Reduction Strategies at the System Approach Level

24.3.2.2.1 Use Macro-Level System Design

A series of ergonomics guidelines and recommendations were published to improve the design of medical systems at a high abstraction level (Carayon and Smith, 2000; see also a brochure from Reiling, 2003 showing an actual example of a safe hospital design). An example of this is to provide links between the structural level (architecture) and the functional properties and demands of health care processes. Task analysis at ward and hospital levels, addressing tasks, constraints, and products of the administration staff is a prerequisite for a good macroergonomics approach. Participatory ergonomics, asking the workers to participate into the design loop, namely to elaborate relevant scenarios for test the design, is a second prerequisite (see Kuutti, 1995; Noro and Imada, 1991).

24.3.2.2.2 Adapt Governance

Reason (1995) recommends assessing institutional resilience and high management support to safety by means of three concepts: commitment, cognizance, and competence.

Commitment: In the face of ever-increasing production pressures, do you have the will to make SM tools work effectively? Arbitration between economic drivers and safety issues needs to occur.

Cognizance: Do you understand the nature of "the safety war," particularly with regards to human and organizational factors?

Competence: Are your SM techniques, understood, appropriate, and properly utilized?

24.3.2.2.3 Share Responsibilities and Professionalism at the Right Level

It was recommended in the person approach section to go beyond errors at the sharp end, and to consider systems as the potential cause of errors of frontline actors. At this point, some people jump forward to conclusions and say that if *the system is the root cause of accidents*, the behavior of individual frontline operators is relatively unimportant: each of them may then feel that what he/she does has little importance because he/she is "controlled by the system." If something goes wrong, the blame is inevitably on the system or, possibly, on managers. Others take exactly the opposite view. Here comes the big word: "responsibility." So who must be blamed in case of accidents? Those involved in the system, or the system which employs them? These two positions seem impossible to reconcile and may stay that way as long as the idea remains to find a culprit. These two apparently irreconcilable views are actually reconciled in the common term of "professionalism." This term describes an entire set of qualities specific to an *individual*, including his/her determination, his/her ability to understand, his/her commitment to safety and, eventually, his/her ability to positively express his/her individual freedom. For example, flying an aircraft can never be safe unless all participants are personally totally committed to making it so. But at the same time, "professionalism" is also the product of a certain education, the personal appropriation of a surrounding culture, and the personal implementation of a collective skill. It is globally the result of social standards, which demand safety. Whatever its mode of action on individuals—fear of reprobation, a means of obtaining recognition from colleagues, corporate culture, or anything else—professionalism bridges, first and foremost, the psychological and social dimensions of individuals (Braithwaite, 2010).

24.3.2.2.4 Implement a Safety Culture

The concept of safety culture has become a gospel in all complex technical systems. A series of celebrated models were recommended to facilitate the adoption of a safe organization. The best-known model is the high reliability organization (HRO) (Weick and Sutcliffe, 2001). Weick and Sutcliffe recommend organizations to acquire "meaningfulness." By meaningfulness, they mean that these organizations are

- Preoccupied with failure: They treat any lapse as a symptom that something is wrong with the system.

- Reluctant to simplify interpretations: They take deliberate steps to create a more complete and nuanced picture.

- Sensitive to operations: They recognize that unexpected events usually originate in latent failures. Normal operations may reveal these lessons but are visible only if [they] are attentive to the frontline, where the real work gets done.

- Committed to resilience: They develop capabilities to detect, contain, and bounce back from those inevitable errors that are part of an indeterminate world.

- Deferent to expertise: They encourage decisions to be made at the frontline and migrate authority to the people with the most expertise, regardless of rank.

Most of these traits depend on how the organization will react to a danger signal, including "weak signals." In complement to the HRO approach, Ron Westrum (2004) suggested a very useful classification of typical ways in which an organization responds to information indicating danger. Westrum described three organization modes:

- The *pathological* mode: Insufficient account is taken of risks, even in a normal situation. Safety rules are regularly set aside in the interest of economy and danger signals are ignored or concealed. "Messengers are shot."

- The *bureaucratic* mode: The organization applies standard methods conscientiously and abides by the rules. It does not suppress danger signals, or only very slightly, but does nothing to detect them outside the standardized channels. It minimizes the risks in a normal situation and loses control if important unforeseen events occur.

- The *generative* mode: The organization encourages inventiveness at all levels. The pursuit of its objectives includes a large proportion of nonstandard activities that constantly reveal new risks. However, because even very low-level staff is allowed to observe and act autonomously, risks are quickly discovered and corrected.

To end this section, it is important to acknowledge the long term continuous effort needed to succeed in installing a safety culture of transparency. It took years for aviation to install such a culture, and the personnel involved were not as diverse. Medicine will probably require more time. It is vital to continue the effort and investment, even in the absence of immediate payback. Paradoxically, and that has been already said in this chapter, the very first signs that something is changing will probably be the increase in incidents reported (access to transparency). The system and the media must be educated to have a good and clear understanding of this paradox.

24.3.3 "Dynamic Approach"

The third approach, the "dynamic approach" is generally missing or underconsidered in the literature, although it is essential. The very nature of HE and the effectiveness of safety strategies change with time and systems improvements. Because of this, any safety strategy is sensitive to a specific level of risk, and may lose all effectiveness for another level of risk (Amalberti, 2006).

24.3.3.1 *Essential Points*

24.3.3.1.1 *Effective Safety Improvement Solutions Vary with the Degree of Safety Achieved*

- When the system is unsafe, the most effective safety strategies aim at increasing the constraints placed on stakeholders, while providing rapid technological enhancements: increased training, more rules, more protocols, and maybe even a strict sanction policy toward rule breakers. Technological progress is the other high-priority goal, since it is commonly accepted that it eventually contributes more to improving safety than repressive measures. At this stage, feedback on accidents and serious incidents is sufficient to foster progress. On the one hand, these accidents/incidents are numerous enough to monitor risks, and on the other hand, they quite adequately represent future risks: at that stage, yesterday's accident will be tomorrow's accident, if nothing is done.

- For better systems, safety strategies must enhance the acquisition and generalization of a safety culture. Accidents have become few and far between, and databases now collect close calls or voluntary reports. At this stage, any accident or incident experienced will not repeat itself; it will come back as a recombination of various episodes belonging to different accidents. However, since accident precursors are the only common occurrence, they need to be clearly identified. At this phase, the system is highly regulated. New rules can be developed, adapted to new contexts or new technologies, and older rules that have become obsolete must be "pruned." Beside technical innovations, progress will mainly come from generalizing and effectively complying with safety measures, making sure that all stakeholders, in all areas of practice, actually follow the same rules (acquiring and generalizing a culture of safety). In a word, at this stage, safety is not jeopardized by the lack of rules, but by inconsistencies in practice.

- At the highest level, the nature of safety changes once again. Feedback gradually loses its predictive capabilities. Focus must be placed on a more systemic approach, while preserving enough flexibility in the system to avoid being straightjacketed in a posture that would reduce adaptation to technological progress. This is typically the challenge of resilience for ultrasafe systems (Amalberti, 2006; Morel, 2008).

24.3.3.1.2 *Safety Improvements Will Change Professional and Technical System*

Eventually, any continuous safety improvement will add constraints and supervisory systems on actors, standardize performance among clinicians, increase accountability, introduce more automation, and therefore change many standards of the profession.

24.3.3.2 *Recommended Dynamic HE Reduction Strategies*

An adaptive strategy must be chosen to address the same problem as the system becomes safer (Figure 24.5).

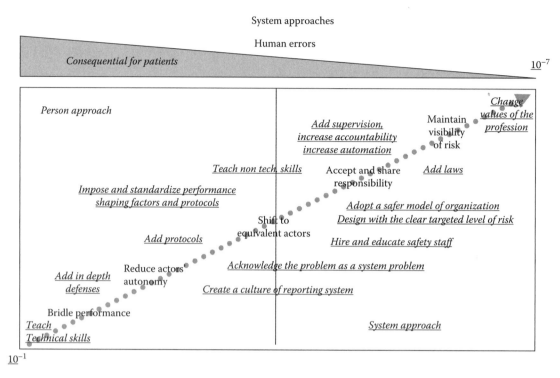

FIGURE 24.5
The dynamic approach: Variation along time of the safety strategies in the person and system approaches.

The big picture can be captured with a three-tier model, based on three dedicated actions: standardize, audit, supervise. These keywords have already been cited in the person approach, yet here they are used in a more systemic point of view. The three tiers are to be considered as a nonend cumulative model. The first tier reaches a threshold allowing for the second tier to start, and at the same time, this first tier continues to improve, moving from one set of techniques to another, as the second tier also starts improving, etc.

24.3.3.2.1 First Tier

When the system is relatively unsafe, the priority is to *standardize* people (competence), work (procedure), and technology (ergonomics). Recommended procedures in design (ergonomics) and operations (guidelines, protocols) are the main generic tools used during this phase (see person approach). The tools for standardization then move on to more official prescriptions (health care ministry policies) during the second tier, and ultimately will be turned into new federal or national laws in the last tier (patient rights, etc.). Moving from that tier to the second tier does not necessarily depend on reaching of a given threshold in safety improvement, but merely on the gradual disclosure of problems to media, which in turn, induce a shift in the way the safety level is politically managed. The governance of safety moves being from a local technical domain-dependent professional business to being submitted to new regional or national

agencies, reactive to the public's and politicians' wills and fears.

24.3.3.2.2 Second Tier

With the new governance, safety strategies benefit from new impulses. A continuous *audit* is required to address and control residual problems. Now is the time to extensively develop monitoring tools: in-service experience, reporting systems, sentinel events, and morbidity and mortality conferences. It is also time to consider safety at a systemic level, enhancing communication and the safety culture at all levels, including the management level. The teaching of nontechnical skills (CRM) becomes a priority to get people to work as a team. Macroergonomics tend to replace local ergonomics, and policies tend to replace simple guidelines (see system approach). The audit procedure shows a series of recurrent system and human loopholes that resist safety solutions based on standardization, education, and ergonomics. All solutions follow an inverted U-curve in efficiency. When the cost-benefit of a solution becomes negative (for example, ratio: time spent on recurrent education/time effective in wards), the governance of the system has to envisage transitioning to the third tier.

24.3.3.2.3 Third Tier

Here the watchword is *supervision*. Supervision means full traceability by means of IT techniques (to force standardization and personal accountability for errors) and growing automation (to progressively change the role of technical people, freeing them from repetitive, time-consuming, and error-prone techniques, and making them more focused on decisions). For instance, advances in IT and computers provided aviation with an end-to-end supervisory system using systematic analysis of onboard black boxes, and leading to the eventual deployment of a global-sky-centralized-automated-management-system called data-link. The impact on people and their jobs, whether pilots, controllers, or mechanics is far-reaching.

Most sectors of health care are today in the second tier of the model. The medical field must be aware that the conditions to enter into the third tier are not yet fully satisfied. The move is just starting with the introduction of ITs. However, it is important not to rush into the extreme solutions specific to the third tier (full monitoring and black boxes) when the top of the second tier has not even been reached. Too much haste could result in professionals severely rejecting innovations (the system being immature) and to an uncontrolled negative feedback effect in terms of payback and effectiveness of the total system. With time, this step eventually occurs, but medicine is at least 10 years behind aviation, and given the inherent specificity and complexity of the medical domain, this move might require many more decades. Possibly, a number of areas in medicine could never transition to this ultimate level, because the potential for creativity and reactivity needed to face the variety of patients may need to be preserved.

24.4 Conclusion

To paraphrase medical language, at this point, the reader should be convinced that "human error is not a disease, it is merely a symptom." To err is human and, as such, is part of natural human behavior, it is just a by-product of performance. Bearing this in mind, proper risk management should not limit itself to errors at the sharp end, for this would jeopardize global success in the long run. Because risks are numerous and diverse, and because the efficiency of risk management techniques vary according to the safety level at hand, it is best to adopt a dynamic approach of risk management. Last but not least, the first major outcome of successful risk management is a distressing increase of reported events. To avoid a counterproductive interpretation of these numbers, the media (in and outside hospitals) should be educated to the risk management process underway.

The six high level dynamic principles that need to accompany health care improvements are as follows:

- *Principle 1: Audit the safety level preliminary to the choice of a strategy.* A safety audit includes not only safety figures, but the comprehensive analysis of the source of errors. Go beyond errors.

- *Principle 2: Adopt a safety strategy consistent with your level of safety.* Safety is a nonstop incremental process. Respect the order of progression to make comprehensive changes, acceptable and affordable by workers in the horizontal and vertical axes. Do not add laws when even recommended protocols do not exist.

- *Principle 3: Show benefits of change for private interest of workers, and not benefits for the institution with only inconvenience for workers.* Respect as long as possible accountability, transfer of routines, and ecology of human strategies during transition times. Brusque transition times breaking all routines destabilize human systems and generally cost additional losses before capitalizing rooms from progresses. Take the win without the losses.

- *Principle 4: Define and orient desirable future activity.* An envelope of acceptable behaviors should be the aim, instead of setting a rigid but narrow target with Yes/No criteria of success. Humans make errors, but they are also clever managers of contradictory dimensions of risk. Give them instructions for change that have the potential to be equally powerful to cope with contradictory dimensions of the demand (save time, save money, save fatigue, do more, safer). Be careful with new tools that are intended to do—or assist to do—a much better "normal job," but fail when the job becomes nonstandard. Real situations are scarcely fully standard. This principle is also preeminent when seeking for criteria to comply with new standards, rules, or laws: do you expect 100% fit with the instruction? Less? How far can your instruction be interpreted accordingly to the context? Think about expectation before introduction.

- *Principle 5: Adopt a global strategy for change management.* Do not impose change at the sharp end without changing the blunt end.

- *Principle 6: Monitor results with relevant monitoring tools and indicators;* Be extremely cautious with indicators (they are a representation of the world, not the world): add new indicators along time to avoid considering progresses only under a few lighting spots, with the risk of disregarding the very reality of field going out of these spots. Monitor the change at the blunt end as well as the change at the sharp end.

References

Albanese, P. (2001) The cost of safety. Paper presented at the MIAS 2001, *Human Issues in Aviation System*, Toulouse, Sept 26.

Allwood, C. (1984). Error detection processes in statistical problem solving. *Cognitive Science*, 8, 413–437.

Amalberti, R. (1998). Automation in aviation: A human factors perspective. In D. Garland, J. Wise, and D. Hopkin (Eds.), *Aviation Human Factors* (pp. 173–192). Hillsdale, NJ: Lawrence Erlbaum Associates.

Amalberti, R. (2001). The paradoxes of almost totally safe transportation systems. *Safety Science*, 37, 109–126.

Amalberti, R. (2006). Optimum system safety and optimum system resilience: Agonist or antagonists concepts? In E. Hollnagel, D. Woods, and N. Levison (Eds.), *Resilience Engineering: Concepts and Precepts* (pp. 238–256). Aldershot, U.K.: Ashgate.

Amalberti, R., Auroy, Y., Berwick, D., Barach, P. (2005). Five systemic barriers keeping health care from becoming ultra safe: A conceptual framework for organizational safety. *Annals of Internal Medicine*, 142, 9, 756–764.

Amalberti, R., Vincent, C., Auroy, Y., de Saint Maurice, G. (2006). Framework models of migrations and violations: A consumer guide. *Quality and Safety in Healthcare*, 15(suppl_1), i66–i71.

Baker, R., Norton, P., Flintoft, V., Blais, R., Brown, A., Cox, J. et al. (2004). The Canadian adverse events study: The incidence of adverse events among hospital patients in Canada. *JMAC*, 170, 11, 1678–1686.

Battles, J., Shea, C. (2001). A system of analysing medical errors to improve GME curricula and programs. *Academic Medicine*, 76, 2, 124–133.

Bogner, M. (Ed.). (1994). *Human Error in Medicine*. Hillsdale, NJ: Lawrence Erlbaum Associates.

Braithwaite, J., Hyde, P., Pope, C. (eds) (2010) *Culture and Climate in Health Care Organizations*. UK: Palgrave Macmillan.

Brami, J., Amalberti, R. (2009). *La sécurité du patient en médecine générale* [*Patient Safety in Primary care*]. Paris, France: Springer Verlag.

Brennan, T., Leape, L., Laird, N., Localio, A., Lawthers, A., Newhouse, J. et al. (1991). Incidence of adverse events and negligence in hospitalized patients: Results of the Harvard medical practice survey study I. *New England Journal of Medicine*, 324, 370–376.

Carayon, P., Smith, M. J. (2000). Work organization and ergonomics. *Applied Ergonomics*, 31, 6, 649–662.

Christie, W. (2002). Standardized colour coding for syringe drug labels: A national survey. *Anaesthesia*, 57, 793–798.

Cox, S., Trait, N. (1991). *Reliability, Safety and Risk Management: An Integrated Approach*. Oxford, U.K.: Butterworth Heinemann.

Daugherty, S., deWitt, B., Beverley, R. (1998). Learning, satisfaction, and mistreatment during medical internship: A national survey of working conditions. *JAMA*, 279, 1194–1199.

Degos, L., Amalberti, R., Bacou, J., Bruneau, C., Carlet, J. (2009). The frontiers of patient safety: Breaking the traditional mold. *British Medical Journal*, 338, b2585.

Duffey, R. B., Saull, J. W. (1999). On a minimum error rate in complex technological systems. In F. S. Foundation (Ed.), *Conference on Enhancing Safety in the 21st Century* (Vol. 289–301). Rio de Janeiro, Brazil.

Gaba, D. (2000). Structural and organizational issues in patient safety. *California Management Review*, 43, 1, 83–102.

Gallagher, T. H., Studdert, D., Levinson, W. (2007). Disclosing harmful medical errors to patients. *New England Journal of Medicine*, 356, 26, 2713–2719.

Haynes, A., Weiser, T., Berry, W., Lipsitz, S., Breizat, A., Dellinger, E., Herbosa, T. et al. (2009). A surgical safety checklist to reduce morbidity and mortality in a global population. *New England Journal of Medicine*, 360, 491–499.

Hayward, R., Hofer, T. (2001). Estimating hospital deaths due to medical errors: Preventability is in the eye of reviewer. *JAMA*, 286, 4, 415–420.

Helmreich, R. (2000). On error management: Lessons from aviation. *British Medical Journal*, 320, 721–785.

Kohn, L., Corrigan, J., Donaldson, M. (1999). *To Err is Human—Building a Safer Health System*. Committee on Quality in America. Washington, DC: Institute of Medicine, National Academic Press.

Kuutti, K. (1995). Work processes: Scenarios as a preliminary vocabulary. In J. Carroll (Ed.), *Scenario-Based Design* (pp. 19–36). New-York: John Wiley & sons.

Lambert, B., Ken-Yu, C., Prahlad, G. (2003). Effects of frequency and similarity neighborhoods on pharmacist' visual perception of drug names. *Social Science & Medicine*, 57, 1939–1955.

Leape, L. (1994). Error in medicine. *JAMA*, 272, 23, 1851–1857.

Lin, L., Isal, R., Donitz, K., Harkness, H., Vicente, K., Doyle, J. (1998). Applying human factors to the design of medical equipment: Patient-controlled analgesia. *Journal of Clinical Monitoring and Computing*, 14, 253–263.

Marx, D. (September 2003). Design of safety culture. Paper presented at *The Second United States/United Kingdom Patient Safety Research Methodology Workshop: Safety in Design*, Rockville, MD.

Michel, P., Quenon, J. L., de Sarasqueta, A. M., Scemama, O. (2004). Comparison of three methods for estimating rates of adverse events and rates of preventable adverse events in acute care hospitals. *British Medical Journal*, 328, 1–5.

Mocel, G., Amalberti, R., Chauvin, C. (2008) Articulating the differences between safety and resilience: the decision-making of professional sea fishing stippers. *Human Factors*, 1, 1–16.

Noro, K., Imada, A. (Eds.). (1991). *Participatory Ergonomics*. London, U.K.: Taylor & Francis.

Polet, P., Vanderhaegen, F., Amalberti, R. (2003). Modelling the border line tolerated conditions of use. *Safety Science*, 41, 1, 111–136.

Rasmussen, J. (1997). Risk management in a dynamic society. *Safety Science*, 27, 2–3, 183–214.

Reason, J. (1990). *Human Error*. Cambridge, U.K.: Cambridge University Press.

Reason, J. (1995). Understanding adverse events: Human factors. *Quality in Health Care*, 4, 80–89.

Reason, J. (2002). Combating omissions errors though task analysis and good reminders. *Quality and Safety in Health Care*, 11, 40–44.

Reiling, J. (2003). Hospital design. *Invited Speaker US/UK Collaborative Workshop on Design for Patient Safety Meeting*. Agency for Healthcare Research and Quality (AHRQ), Rockville, Maryland, Sept 21.

Sexton, B., Thomas, E., Helmreich, R. (2000). Error, stress, and teamwork in medicine and aviation: Cross sectional surveys. *British Medical Journal*, 320, 745–749.

Vincent, C. (Ed.). (2001, 2nd edn.). *Clinical Risk Management*. London, U.K.: British Medical Journal Publications (1st edn. 1995).

Vincent, C., Adams, S., Stanhope, N. (1998). A framework for the analysis of risk and safety in medicine. *British Medical Journal*, 316, 1154–1157.

Vincent, C., Young, M., Phillips, A. (1994). Why do people sue doctors? A study of patients and relatives taking legal action. *The Lancet*, 343(June 25), 1609–1613.

Volpp, K., Friedman, W., Romano, P., Rosen, A., Silber, J. (2010). Residency training at a crossroads: Duty-hour standards. *Ann Intern Med*, 153, 826–828.

Weick, K., Sutcliffe, K. (2001). *Managing the Unexpected: Assuring High Performance in a Range of Complexity*. San Francisco, CA: Jossey-Bass.

Westrum, R. (2004). A typology of organisational cultures. *Quality and Safety in Health Care*, 13, 2, 22–27.

Woods, D., Cook, R. (2002). Nine steps to move forward from error. *Cognition, Technology, & Work*, 4, 137–144.

Wu, A., Folkman, S., Mc Phee, S. et al. (1991). Do house officers learn from their mistakes? *British Medical Journal*, 265, 2089–2094.

25

Communicating about Unexpected Outcomes, Adverse Events, and Errors

James W. Pichert, Gerald B. Hickson, Anna Pinto, and Charles Vincent

CONTENTS

25.1 General Issues Related to Error Disclosure 402
    25.1.1 Landscape of Disclosure 402
    25.1.2 Impact of Errors and Mistakes on Clinicians 403
    25.1.3 Patient/Family Responses to Medical Injury and Error Disclosure 405
    25.1.4 Injury from Medical Treatment Differs from Other Injures 405
25.2 Medical Malpractice Environment—A Barrier to Disclosure? 405
    25.2.1 Are Attorneys the Problem? 406
    25.2.2 Other Potential Barriers 406
    25.2.3 Role of Poor Communication 407
25.3 So Why Disclose? 408
25.4 Process of Error Disclosure 408
    25.4.1 Obvious Errors Resulting in Obvious Harm: Full Disclosure 408
    25.4.2 Error Disclosure in Complex Cases 410
        25.4.2.1 Error Disclosure When the Cause of an Unexpected Outcome Is Not Immediately Known 410
        25.4.2.2 When the Physician Deems the Error Unrelated to the Outcome 412
        25.4.2.3 When the Cause of an Unexpected Adverse Outcome Is a Matter of Dispute 413
        25.4.2.4 When a Subsequently Treating Physician Believes That an Error Occurred, but the Patient/Family Is Unaware 414
25.5 Disclosure Is Just the Beginning 415
    25.5.1 Ask Specific Questions about Emotional Trauma 416
    25.5.2 Continuing Care and Support 416
    25.5.3 Inform Patients of Changes 416
    25.5.4 Financial Assistance and Practical Help 416
    25.5.5 Long-Term Follow-Up and Support 416
        25.5.5.1 Denial of Communication and Apology after a Surgical Procedure Was Missed ... 417
        25.5.5.2 Explanations and Apology after Iatrogenic Cardiac Arrhythmia 417
        25.5.5.3 Anesthetic Awareness: Reducing the Fear of Future Operations 418
25.6 Conclusion 418
References 418

A Case of a Retained Sponge: "JS" is a 25 year old male brought to the emergency department with multiple gunshot wounds to his abdomen and pelvis. X-rays reveal an extensive injury pattern. He requires bowel resection and liver repair, after which he is transferred to the trauma unit. A chest x-ray taken to confirm chest tube placement reveals a retained sponge. JS needs surgery to remove the sponge. Dr. ABC has asked you for advice on whether JS should be informed about the need to go back to the operating room; after all, he has not yet awakened from the initial surgery. Dr. ABC reports that JS's family has arrived and asks whether they need to be told about the need for the second surgery. If disclosure of the retained sponge is to be discussed, Dr. ABC will want your advice about how best to explain to JS his need for a second surgery. What guidance is available?

This chapter begins with discussion of general issues related to error disclosure. These include recent disclosure-related standards and recommendations, effects of errors on health professionals, and how medical errors impact patients and families. The medical liability environment is briefly assessed, and selected research on drivers of medical malpractice is summarized in the second section. The third section presents a series of specific cases designed to highlight a range of challenges associated with discussing unexpected outcomes and adverse events when the presence or contribution of errors is uncertain or a matter of dispute. A "balance beam approach" is offered as a means for weighing the pros and cons of a range of disclosure options under various circumstances involving uncertainty. The chapter concludes with stories that highlight our strong conviction that while appropriate error disclosure is essential in health care, it is only the first step in dealing with harm-causing errors. While the authors claim no special expertise in human factors engineering, analysis, or research, we nevertheless try throughout the chapter to suggest where colleagues in those fields have opportunities to help create psychologically safe systems not only for appropriate and effective error disclosure but also for appropriate and effective management of the wider psychosocial issues relating to the aftermath of medical errors.

For purposes of this chapter, we will use the following definitions:

Adverse event: An injury that is caused by medical management rather than the patient's underlying disease.

Medical error: The failure of a planned action to be completed as intended or the use of a wrong plan to achieve an aim (Reason, 1990; Reason, 2001; Gallagher, 2009; Loren et al., 2010).

Unexpected outcome: An observed patient health outcome different from the range of outcomes—including known complications—reasonably expected and previously discussed with the patient.

25.1 General Issues Related to Error Disclosure

25.1.1 Landscape of Disclosure

The seventeenth century English poet Francis Quarles perhaps only partly exaggerated when he penned: "Physicians, of all [persons], are most happy: whatever good success soever they have, the world proclaimeth ... and what faults they commit, the earth covereth"— *Hieroglyphics of the Life of Man*.

Perhaps Quarles would be surprised, then, that disclosure of medical error is now advocated, even

expected, worldwide. In July 2001, the Joint Commission on Accreditation of Healthcare Organizations set the stage by stating, "Patients and, when appropriate, their families are informed about the outcomes of care, including unanticipated outcomes" (JCAHO, 2001). While error disclosure to patients/families is not specifically mentioned, this standard does require hospitals to tell patients when they have been harmed during treatment. The National Quality Forum Safe Practice #7 on Disclosure calls for "timely, transparent, and clear communication about" a "serious unanticipated outcome." In Australia, the Council for Safety and Quality in Healthcare has established a standard for open disclosure of errors and harm, and similar standards have been adopted throughout Canada (Levinson and Gallagher, 2007). The British National Patient Safety Agency and its National Reporting and Learning Service have promulgated a framework and a series of programs about "being open" (National Patient Safety Agency, 2009; Panesar et al., 2009). Examples of the statements and standards from professional and government organizations are shown in Table 25.1.

The standards and recommendations refer to the disclosure of harm and, in some cases, the disclosure of errors as well. Minor errors occur frequently in all walks of life, and health care is no exception. To operate a full disclosure process for any small error, such as a missed dose of a drug that had no consequences, is clearly unnecessary. In this case, open disclosure would simply imply letting the patient know with a simple, brief apology that a dose had been missed, missing the dose should have no consequences, and monitoring would continue. At the other end of the scale, when a patient has been seriously harmed, open disclosure involves not only the initial meeting to let the patient and family know what has happened, but most probably subsequent meetings and support of various kinds in the longer term. S's retained foreign body clearly falls into this latter category because the flawed counting system resulted in harm, that is, the need for repeat surgery and potential loss of trust. The nature of the disclosure and subsequent support will vary from case to case, as we discuss in the sections that follow; we also consider some complicated cases in which, for example, the cause of an unexpected outcome or adverse event is not yet known, but patients or families understandably want an explanation.

Open disclosure is sometimes presented as a defined event that is largely confined to a single, carefully, and caringly planned meeting when disclosure occurs; for relatively straightforward cases where any harm is short-lived and not serious, this may be entirely reasonable. The impact of disclosures in very serious cases needs to be considered more carefully. A clinician,

TABLE 25.1

Recommendations for Communicating about Unexpected Outcomes, Adverse Events, and Errors

Joint Commission for Accreditation in Healthcare

The "Elements of Performance" for RI.2.90 require that, at *minimum*, the practitioner or designee informs the patient (and family) about:

1. Outcomes that the patient (or family) must know about to participate in current and future decisions affecting care, treatment, or services

2. Unanticipated outcomes considered reviewable sentinel events

3. Unanticipated outcomes of care, treatment, and services

Australian Council for Safety and Quality in Health Care

"Open disclosure is the open discussion of incidents that results in harm to a patient while receiving healthcare. The elements of open disclosure are an expression of regret, a factual explanation of what happened, the potential consequences, and the steps being taken to manage the event and prevent recurrence. … While open disclosure is occurring in many elements of the health system, the Standard facilitates more consistent and effective communication following adverse events." (Australian Council for Safety and Quality in Health Care, Standard for Open Disclosure, 2003)

United Kingdom, the National Patient Safety Agency

"Open and honest communication with patients is at the heart of healthcare. Research has shown that being open when things go wrong can help patients and staff to cope better with the after effects of a patient safety incident. … *Being open* is the right thing to do, and [NPSA] encourage National Health Service boards to make a public commitment to openness, honesty and transparency." (National Patient Safety Agency, 2009)

U.S., the National Patient Safety Foundation

"When a healthcare injury occurs, the patient and the family or representatives are entitled to a prompt explanation of how the injury occurred and its short- and long-term effects. When an error contributed to the injury, the patient and the family or rep should receive a truthful and compassionate explanation about the error and the remedies available to the patient." (National Patient Safety Foundation, 2003)

U.S., The National Quality Forum (NQF)

"Following serious unanticipated outcomes, including those that are clearly caused by systems failures, the patient and, as appropriate, the family should receive timely, transparent, and clear communication concerning what is known about the event." (National Quality Forum 2009, Safe Practice #7, Disclosure)

The American Medical Association (AMA)

"Patients have a right to know their past and present medical status and to be free of any mistaken beliefs concerning their conditions … the physician is ethically required to inform the patient of all the facts necessary to ensure understanding of what has occurred. … This obligation holds even though the patient's medical treatment or therapeutic options may not be altered by the new information." (American Medical Association, 2003) (Code of Medical Ethics Opinion 8.12 on "Patient Information")

Canada, Canadian Medical Association

"Take all reasonable steps to prevent harm to patients; should harm occur, disclose it to the patient." (Canadian Medical Association, 2004) (Code of Ethics updated 2004, Item 14)

or perhaps a rapid remediation team (McDonald et al., 2010), might have an initial conversation with the patient and/or family followed by one with the attending physician in which the causes and likely consequences of a serious error are discussed. In certain cases, further follow-up meetings and a longer-term supportive relationship with a designated staff member might be required.

All intend disclosure discussions to be beneficial but, if poorly handled without thought for the long term, these conversations can make matters worse. A brisk and straightforward disclosure of the facts appropriate for minor errors might be quite distressing if the incident is serious. For example, if the patient has been seriously traumatized, some patients and families are likely to need a more gentle approach in which they will be helped to gradually face the full implications of the harm and the consequent change(s) in their lives. But

clinicians must not overgeneralize: some patients and families will want all the facts immediately. These considerations also apply to the staff involved who, in serious cases, may also be quite distressed and disturbed by what has occurred.

Therefore, before thinking about disclosure, we need to consider the reactions of those involved and how health care systems can be designed to support the kind and quality of disclosure that serves not only clinicians, but also patients who have suffered harm and their families (France et al., 2005).

25.1.2 Impact of Errors and Mistakes on Clinicians

Error disclosure is a crucial first step in helping patients (and their families) who have been harmed or distressed during their treatment. Yet, staff may also be distressed by harm-causing errors and must be supported, both for their own sake and for the patients in their care (Waterman et al., 2007; Schwappach and Boluarte, 2009). Clinicians working while distraught about an error may be so deeply affected that they simply cannot provide the kinds and amounts of support required by the affected patient and family (Bogner, 1994, 2004; Scott et al., 2009). They may also find that their distress leads them to provide poorer care, which in turn engenders more distress (Schwappach and Boluarte, 2009). Such circumstances require that the local health care system supports the acknowledgment and disclosure of the error to an appropriate institutional representative (White et al., 2008). Unfortunately, fear of the U.S. medical–legal system is often cited as a barrier to or inhibitor of early disclosure (Kachalia et al., 2003; Kaldjian et al., 2006), so health care systems must proactively educate their clinicians about how to report events to appropriate risk management or quality improvement leaders (Taheri et al., 2006) in ways that protect the report from legal discovery, yet nevertheless appropriately support the patients and the clinicians involved. Laws providing protection for apology may help, but studies of their effects are inconclusive (McDonnell and Guenther, 2008).

The typical reaction of making an error has been well expressed by Albert Wu in his aptly titled paper, "The second victim":

> Virtually every clinician knows the sickening feeling of making a bad mistake. You feel singled out and exposed - seized by the instinct to see if anyone has noticed. You agonize about what to do, whether to tell anyone, what to say. Later, the event replays itself in your mind. You question your competence but fear being discovered. You know you should confess, but dread the prospect of potential punishment and of the patient's anger (Wu, 2000).

The impact of mistakes was explored in interviews with 11 doctors by Christensen et al. (1992). Although this small study did not assess the overall importance of mistakes, a number of very important themes are discussed, in particular, the ubiquity of mistakes in clinical practice; the variability in clinicians' frequency of self-disclosure about mistakes to colleagues, friends, and family; the degree of emotional impact on the physician, so that some mistakes were remembered in great detail even after several years; and the influence of beliefs about personal responsibility and medical practice. A variety of mistakes were discussed, all with serious outcomes including four deaths. All clinicians were affected to some degree, but four described intense agony or anguish as the reality of the mistake had sunk in

- *I was really shaken. My whole feelings of self-worth and abilities were basically profoundly shaken.*

- *I was appalled and devastated that I had done this to somebody.*
- *My great fear was that I had missed something, and then there was a sense of panic.*
- *It was hard to concentrate on anything else I was doing because I was so worried about what was happening, so I guess that would be anxiety. I felt guilty, sad, had trouble sleeping, wondering what was going on* (Christensen et al., 1992).

After the initial shock, the clinicians had a variety of reactions that lasted from several days to several months. Some of the feelings of fear, guilt, anger, embarrassment, and humiliation were unresolved at the time of the interview, even a year after the mistake. A few reported symptoms of depression, including disturbances in appetite, sleep, and concentration. Many described fears related to concerns for the patient's welfare, litigation, and colleagues' discovery of their "incompetence." Perhaps as a result of these kinds of fears, Garbutt et al. (2008) found that many physicians say nothing to anyone, isolating themselves. In the authors' experience, however, many are driven to "confess" or to "rationalize" the events and their involvement in conversations with other physicians and staff, creating other kinds of problems. Whether they are talkers or not, all health care professionals need basic training in disclosure skills and safety principles both to inform informal coaching of colleagues and to promote swift event reporting so that appropriate institutional support can be provided to those involved.

The impact on clinicians of serious mistakes was highlighted in a much larger study by Wu and colleagues; they sent questionnaires to 254 interns in the United States asking the respondents to describe the most significant mistake in patient care they had made in the last year, which had serious or potentially serious consequences for the patient (Wu et al., 1991). Various types of error were reported, most frequently missed diagnoses (30%) and drug errors (29%). Almost all the errors had serious outcomes and almost a third involved a death but, at that time, less than a quarter of the mistakes were discussed with the patient or patient's family. More than a quarter of house officers feared negative repercussions from the mistake. Feelings of remorse, anger, guilt, and inadequacy were common. Nursing professionals express similar feelings (Scott et al., 2008).

Therefore, when considering how best to disclose error to patients, it is also necessary to consider how the staff involved have been affected and how this, in turn, may affect the patient and family. Most obviously, if the staff are very affected, the disclosure conversation may need to be led by a senior clinician who has not been directly involved in the patient's care. When providers show their human side, patients may derive comfort and their trust may be restored as a result of knowing that the staff have been emotionally affected by the incident. Nevertheless, these options should be assessed on a case-by-case basis depending on the severity of the incident's emotional impact on the staff as well on patient/family reactions.

25.1.3 Patient/Family Responses to Medical Injury and Error Disclosure

Errors and their disclosures can pose even greater challenges for patients and their families. Patients are vulnerable psychologically, even when diagnosis is clear and treatment goes according to plan. Even routine procedures and normal childbirth may produce posttraumatic symptoms (Clarke et al., 1997; Czarnocka and Slade, 2000). When patients experience harm or misadventure, the consequences can be significant (Dovey et al., 2003) and their reaction may be particularly severe. Their families' experiences and reactions must also be considered because families will likely be involved in error communications, may influence the injured party's intentions to pursue a claim, and, in many cases, may provide care required by the injury and/or disease process.

The full impact of some incidents only becomes apparent in the longer term. A perforated bowel, for example, may require a series of further operations and time in the hospital. The long-term consequences may include chronic pain, disability, and depression, with a deleterious effect on family relationships and ability to work (Vincent et al., 1993). Depression appears to be a more common long-term response to medical injury than posttraumatic stress disorder (Vincent and Coulter, 2002), although there is little research in this area. Whether people actually become depressed and to what degree will depend on the severity of their injury, their personality, the cognitive schemata they have formed about their condition, the support they have from family, friends and health professionals, and a variety of other factors (Kessler, 1997).

When a patient dies, the trauma is obviously more severe still, and may be particularly severe after a potentially avoidable death (Lundin, 1984). For instance, many people who have lost a spouse or child in a road accident continue to ruminate about the accident and what could have been done to prevent it for years afterward. They are often unable to accept, resolve, or find any meaning in the loss (Lehman et al., 1989). Relatives of patients whose death was sudden or unexpected may, therefore, find the loss particularly difficult to bear. If the loss was avoidable in the sense that poor treatment played a part in the death, their relatives may face an unusually traumatic and prolonged bereavement. They may ruminate endlessly on the death and find it hard

to deal with the loss. The long-term effects of serious adverse events on patients as well as the range of factors that explain patients' psychological adjustment need to be better researched and understood given the paucity of relevant empirical data.

25.1.4 Injury from Medical Treatment Differs from Other Injures

The impact of a medical injury differs from most other accidents in two important respects. First, patients have been harmed, unintentionally, by people or systems in whom they placed considerable trust, and so their reaction may be especially powerful and hard to cope with. Second, and even more important, they are often cared for by the same types of professionals, and perhaps the same people, as those involved in the original injury. As they may have been very frightened by what happened to them, and have a range of conflicting feelings about those involved, this too can be very difficult, even when staff are forthcoming, sympathetic, and supportive.

Patients and relatives may therefore suffer in three distinct ways from an injury: from the injury itself, from the way the incident is handled, and from the loss of trust in their health care system and its providers. Many people harmed by treatment suffer further trauma through the incident being insensitively and inadequately handled. Conversely, when staff come forward, acknowledge the damage, and take the necessary action, the overall impact may be reduced. Injured patients need an explanation, an apology, to know that changes will be made to prevent future incidents, and oftentimes practical and financial help. This is not to say that they necessarily need unusual treatment or that staff should be wary of talking to them—in fact, staff avoidance may result in patient fear of abandonment. Problems tend to arise when ordinary feelings are blunted by anxiety, shame, or just not knowing what to say.

Later in the chapter, we will suggest that identifying and correcting systems contributions to the error and communicating those changes to the patient and family can be very helpful as error disclosure conversations continue over time.

25.2 Medical Malpractice Environment—A Barrier to Disclosure?

Before we turn to the disclosure process itself, the malpractice environment must be addressed. Medical malpractice research reveals that clinicians frequently cite fear of litigation, and its associated consequences, as barriers to open error disclosure and seeking advice from colleagues when adverse outcomes occur. This is, at first glance, entirely understandable. Medical malpractice research also suggests what (and who) promotes lawsuits. While many medical liability disputes are resolved before trial, when they do go to trial, awards can be substantial. *Jury awards* reportedly more than doubled in the United States between 1992 and 2002 (Jury Verdict Research, 2002). Mean awards are likely greater now despite tort reforms such as caps on awards (Hyman et al., 2009), but data on this point are hard to come by because comprehensive, accessible data on malpractice claims are not available—states typically have no systematic aggregation of data from individual trial courts, and departments of insurance vary in what they collect from individual insurance companies (Mello, 2006). Widespread media reporting of high-award cases, alongside substantial and ongoing coverage of the Institute of Medicine 1999 report *"To Err is Human,"* have dramatically increased public (and juror) awareness of the potential for error (Kohn et al., 1999). Perhaps this awareness has eroded public confidence and trust in the medical profession and medical institutions generally. Physicians whose errors may have caused harm are clearly aware of the social challenges when they consider whether or how to disclose them.

25.2.1 Are Attorneys the Problem?

Attorneys make a convenient target in a legal system whose outcomes appear neither sensitive nor specific; meritorious cases are not pursued, nonmeritorious cases are pursued, and the courts seem to reward both about equal proportions of the time. It is little wonder that physicians, patients, and policy makers in the United States find the system of medical–legal dispute resolution so vexing. No one should be surprised that physician concerns about attorneys and courtrooms influence their willingness to disclose or their comfort in disclosing (Kachalia et al., 2003; Kaldjian et al., 2006). It is easy to understand—if not to pardon—physicians if they sometimes appear to advocate a Shakespearean solution to the malpractice liability climate: *"The first thing we do, let's kill all the lawyers"* (William Shakespeare, King Henry VI, Part II (1597–1598), act IV, sc. ii).

Tongue-in-cheek sentiments aside, and despite the presence of some ambulance-chasing attorneys, the idea that every error or harm will be followed by litigation is not true. First, the incidence of adverse events quoted in major studies (Kohn et al., 1999), far exceeds the extent of litigation experienced. Research based on medical record reviews suggests that adverse events occur in approximately 6% of U.S. hospital stays and that approximately one-third of those may be attributed to negligence of some sort (Brennan et al., 1991).

Interestingly, however, only about 2% of patients/families whose medical documents revealed a reason to pursue a claim did so (Localio et al., 1991). Unfortunately, at least as many patients/families made a claim whose medical records did *not* support one. Lack of social safety nets for persons with significant ongoing medical care requirements and several noneconomic factors appear to drive some families to the courtroom.

Families often find it difficult to identify an attorney who will take their case. Successful plaintiff attorneys report interviewing 50–100 prospective clients for every case they agree to take. Clayton et al. (1993) reported that families had to approach more than three lawyers before finding one who would pursue their claim. In a contingency-fee-based system, lawyers' prospects for "return on investment" mean that some cases with merit do not get filed. On the other hand, questionable cases that have a sympathetic plaintiff and potential for a large award may go forward. Mediation and other methods of alternative dispute resolution may help promote fairness (Liebman and Hyman, 2004). In the meantime, lawyers will continue to play a role in deciding which cases go forward. The crucial point is that the overall rate of claims is not high compared with the overall level of harm.

25.2.2 Other Potential Barriers

The legal landscape is hardly the only commonly cited barrier to disclosure. Disclosures when errors have affected large numbers of patients (e.g., equipment failures, revised interpretations of diagnostic results) are especially challenging—and important (Chafe et al., 2009). As health care providers and health care organizations prepare to disclose a clear error, they may reasonably fear the loss of patient trust, injury to reputation in the medical community, and potential for litigation (Kraman and Hamm, 1999; Wu, 1999). Some may even have been taught (directly or via observation of mentors) *not* to disclose errors nor to apologize. Others have learned from painful personal or vicarious experience not to discuss errors lest they be publicly humiliated in the press or in morbidity and mortality conferences designed more to attribute blame than to consider and remediate all potential contributors to an unexpected adverse outcome. It is no wonder then that several studies report that disclosures are far rarer than errors (Wu et al., 1991; Blendon et al., 2001).

Other barriers besides the intense feelings and concerns summarized in Sections 25.1.2 and 25.1.3 can inhibit provider–patient/family error communications (Kachalia et al., 2003; Kaldjian et al., 2006). These include personal and interpersonal barriers such as threats to ongoing provider–patient relationships, lack of established doctor–patient relationships on which to base disclosure conversations, uncertainty about the cause(s) of errors and their actual impacts on patient outcomes, and lack of role models/confidants. Perceived institutional barriers include absence of supportive forums for disclosure, failures to disseminate information about how to report errors, lack of guidelines for when to disclose, and lack of confidence that disclosing errors and near misses will help promote system improvements.

We are not naive; these barriers are real. Nevertheless, we take as a given that disclosures are the right thing to do as a matter of principle. In addition, as will be discussed in Section 25.3, disclosures can have salutary effects (e.g., improved safety, honest and trusting relationships, intact professional integrity and accountability, and sometimes ameliorated risk) (Kachalia et al., 2003; Kaldjian et al., 2006), but only insofar as they are supported by widely disseminated disclosure training, consistent institutional commitment, and, as Section 25.4 describes, thoughtful communication strategies. Poor communication, on the other hand, will compound all the other barriers to disclosure. As we briefly discuss next, poor communication prior to and during treatment is what often results in errors and risk management claims in the first place.

25.2.3 Role of Poor Communication

Analysis of risk management files shows that communication failures between clinicians, and between patients and clinicians, set the stage for clinician errors and patient/family anger, and play important roles in malpractice claims (Pichert et al., 1997; Pichert et al., 1998; Morris et al., 2003; White et al., 2004; White et al., 2005; Greenberg et al., 2007; Hain et al., 2007). While some health care leaders guard risk management data as proprietary and confidential, appropriate dissemination of aggregate data can focus organizations on opportunities for improving quality and safety and reducing risk (Taheri et al., 2006).

Patients and families who file malpractice claims are often seeking communication or action to prevent recurrences, not necessarily compensation. Patients and families have reported that they filed claims because they were angry and wanted explanations, apologies, and other information (Table 25.2; Beckman et al., 1994; Hickson et al., 1994; Vincent et al., 1994). Others reported filing a claim when they came to believe that a medical professional was attempting to mislead them; that is not to say that the patients' physicians intentionally misled or were not forthcoming with information, but that was the families' perception.

Research also shows that a small proportion of physicians attract disproportionate shares of malpractice claims (Sloan et al., 1989; Bovbjerg and Petronis, 1994). Several studies suggest that physicians who stand out

TABLE 25.2

Reasons[a] Parents Sued after an Adverse Event Involving Their Newborn

Advised to sue by influential other	32%
Needed money	24%
Believed there was a cover-up	24%
Child would have no future	23%
Needed information	20%
Wanted revenge, license	19%

Source: Hickson, G.B. et al., *J. Amer. Med. Assoc.*, 267, 1359, 1992.

[a] Families could offer more than one reason.

from a malpractice standpoint do so because they have difficulty establishing and maintaining rapport with patients and families. For instance, in one study, interviews were conducted with women who had recently delivered healthy children and who were not involved in malpractice claims (Hickson et al., 1992). Families who saw obstetricians who had, unknown to them, a high rate of previous malpractice claims were more likely to complain about communication failures: "Dr. X offered no information. I felt he was hiding something. He never even tried to talk to my husband." Families seeing high malpractice risk physicians were also more likely to complain about problems with access, and feeling that they were not respected as human beings: "Dr. Y was rude. She was nasty that I started labor on the [national holiday] … she gave me snappy and smart [-alec] answers." Subsequent studies have linked patients' unsolicited complaints and physicians' malpractice histories (Hickson et al., 2002; Stelfox et al., 2005; Hickson et al., 2007). Specifically, the 8%–10% of physicians in large medical groups accounted for 50% of the groups' unsolicited patient complaints—many related to poor communication and lack of apparent concern for the patient—also accounted for a disproportionate share of malpractice claims. More generally, extensive research on doctor–patient communication suggests that doctors' information giving and expression of affect are positively associated with patients' satisfaction with the health care provider and health care in general (Ong et al., 1995; Williams et al., 1998). Consequently, research suggests that a pre-outcome relationship characterized by respect, openness, and trust—or the converse—affects how receptive patients and families may be to disclosure of medical errors after the consequences are appreciated.

To summarize the malpractice literature, we know that adverse outcomes and errors occur. When patients are injured, good communication and concern are important. Poor communication prompts some patients to sue, compounding the potential impact of any error (Schwappach and Koeck, 2004; Mazor et al., 2006;

Straumanis, 2007; Van Vorst et al., 2007). Small numbers of physicians attract disproportionate shares of suits, and more patients complain about these physicians' communications, evidence of caring, accessibility, and treatment skills. So when adverse outcomes and errors occur, and the consequences of an error are appreciated, the poor prior experiences may make families less willing to hear and accept apologies. Therefore, good communication from the start sets the stage for making appropriate disclosure of unexpected outcomes, adverse events, and errors.

25.3 So Why Disclose?

So, if disclosures present difficult challenges and create psychological and/or legal risks, why disclose? There are a number of reasons, some ethical, some humanitarian, and some practical (Kachalia et al., 2003; Kaldjian et al., 2006). First, of course, disclosing an error shows the provider to be honest, forthright, a patient advocate, and a responsible professional with nothing to hide. Second, patients generally appreciate and expect appropriate disclosures when errors have occurred (Hobgood et al., 2002). Third, nondisclosure and other communication failures threaten patient–provider relationships, threaten the public's view of health care providers generally, and may promote patient nonadherence to medical regimens. Fourth, if patients perceive nondisclosures as willful, they may be driven to plaintiff attorneys, ironically the very thing that influences some physician *not* to disclose errors (Cantor, 2002; Liang, 2002). Finally, some reports have suggested that appropriate disclosure may reduce the financial consequences of errors for those who disclose (Kraman and Hamm, 1999; Boothman, 2006; Boothman et al., 2009). The overall evidence for this assertion is, however, disputed or inconclusive (Kachalia et al., 2003; Studdert et al., 2007; McDonnell and Guenther, 2008). Once again, we simply assert that appropriate disclosures are the right and professional thing to do.

Once a commitment to disclose errors is achieved by a person or an entire organization, the issue becomes not whether, but how and when. Most attention has been directed toward disclosing clear-cut errors resulting in obvious harm—but many times things are not so obvious. Under those conditions, how do health care professionals meet family (and our own) needs? In the sections that follow, we first discuss the general principles of error disclosure applicable when the error is clearly recognized as the cause of harm. We then turn to consider some more complex cases in which the relationship between error and harm is uncertain or disputed, where

additional issues arise. In each case, the aims are *not* to present scripts, but, at least initially, to share a strategy for critically examining the how and when of appropriate disclosure-related strategies.

25.4 Process of Error Disclosure

Error disclosure means communicating bad—or at least unsettling—news, about which much has already been written. Generally accepted principles for delivering bad news provide a starting point for error disclosure, and also offer clinicians a foundation and model for the actual discussion and likely reactions. Clinicians who, for instance, have had to inform patients that they have potentially serious conditions know to provide as much information as the patients/families need and desire, to expect a variety of reactions such as initial denial or anger, to pace discussions carefully, and to make sure that patients are clear about what happens next (Table 25.3). All these basic principles apply to the special case of error disclosure as well.

Fortunately, while errors occur daily in medical institutions, the need to disclose a harm-causing error will be a relatively rare event for most health care providers. Therefore, most professionals need training with practice and follow-up coaching to learn disclosure skills. Such training is usually best accomplished when cases are based on real, local incidents; involve personal experiences of clinicians who participate in the training programs: and derive from systematically studied observations of patients and professionals in real circumstances (Pichert et al., 1998; Liebman and Hyman, 2004; Gunderson et al., 2009). Human factors researchers, therefore, have obvious opportunities to contribute to our understanding of how best to disclose errors under various conditions (O'Connell and Reifsteck; 2004; Chan et al., 2005; Gallagher et al., 2007).

25.4.1 Obvious Errors Resulting in Obvious Harm: Full Disclosure

In addition to employing the general guidelines for delivering bad news listed in Table 25.3, the specific content of an error disclosure must also be carefully considered (Banja, 2001). The most *straightforward* physician–patient conversations about errors occur in cases when the errors are *evident* and *clearly caused injury*. Examples include a retained sponge, wrong-sided surgeries, and failures to follow up on clearly abnormal test/scan or procedure in cases where timeliness is important. Before reading on, pause to ask yourself what *you* want discussed if you or a loved one is the patient who

TABLE 25.3

General Guidelines for Breaking Bad News

- Choose a private area; set the stage
- Provide a brief review of the course of care (e.g., to the family of S: "As you know, we operated to repair the damage by the gunshots. We transferred him to the trauma unit for continued care. We always take an x-ray to check on things after the surgery.")
- Signal what is coming with a "warning shot" ("I'm sorry to have to tell you that we've found something on the x-ray that will require another surgery very soon.")
- Be frank but kind in the delivery of the news, pause, pause a little longer, and only then respond ("Specifically, the film I ordered has clearly revealed that a surgical sponge—a gauze pad we use to soak up blood, so we can see where we're cutting—was inadvertently left inside. That should not have happened, and I'm very sorry that the sponge was not discovered before we closed up. As I said, the main thing now is that we remove the sponge, but of course that will require another surgery, and for that all of us involved in your care also apologize.")
- Empathize by using statements that signal partnership, empathy, apology, respect, legitimization, and support (aka PEARLS) as appropriate
- Comfort with silence
- Gauge patient/family readiness for information by asking a few questions. Then respond with just enough information to answer, invite a few more questions, and respond iteratively until the patient/family appears satisfied
- Invite patient/family to ask any questions at any future time as well
- Assure family of physician availability (dispel abandonment fears)

has been harmed. Even brief reflection reveals a host of patient/family information needs. A little more reflection suggests a variety of challenges associated with meeting those needs. After all, following the disclosure, patients must deal with the knowledge that their pain or injury might have been avoided, concerns about longer-term consequences of the error, and that the party(ies) responsible for the injury may remain involved in their care (Vincent, 2001).

Table 25.4 lists error disclosure elements considered important by many patients (Gallagher et al., 2003; Gallagher et al., 2007; Iedema et al., 2008). Its contents

also *suggest* a few words that might apply to S's family's desire for information related to the retained foreign body. Not all eight elements may apply to every case, or additional considerations may apply in specific instances. And, of course, care providers will choose words and phrasing suited to *their* best communication style and to the patient/family needs, but always aiming to provide full disclosure. To be clear, these elements apply when full disclosure is appropriate, that is, when an error and resulting harm are both obvious, or at least when a consensus of opinion exists based upon review of the data. In many instances, the relationship

TABLE 25.4

The Disclosure Conversation

Disclosure Element	How It Might Be Conveyed
1. Apology (be precise); nature of error, harm	Mr./Ms. ____, I'm sorry to report that [nature of error and specific outcome]. On behalf of us all, I apologize for the [specific error and outcome]
2. When and where error occurred	Here's what happened … [explanation appropriate for patient/family understanding of medical facts]
3. Causes, results of harm, actions taken to reduce gravity of harm, actions to reduce or prevent reoccurrence	The error occurred because [explain briefly in nontechnical language]. As a result, what happened to you is [explain briefly]. Once we realized the error, we [explain actions taken to address that patient's needs, if applicable]. We have a team that reviews errors and recommends how to prevent recurrences
4. Who will manage ongoing care	If you allow me to continue, I will work [together with other team members and administrators] on your care
5. Describe error review process, reports to regulatory agencies, how systems issues are identified	We take mistakes seriously. Everything will be reviewed by experts. The results will be reported to me [and, if applicable, to the ____ Agency]. If the review reveals ways we can better care for our patients, we will work to make those changes
6. Provide contact info for ongoing communications	I [or Dr. ____] will communicate about your continuing care. When questions arise, please have me paged. Or call my assistant, [name and contact information] who can help or find me
7. Offer counseling, support if needed	People in the ___ office can talk with you and connect you with support services. Here's their card. May I call them or anyone else for you?
8. Address bills for additional care	I hope we'll agree that the review I requested fairly addresses the charges resulting from the error. The focus now is [returning you to health, restoring function, assist with grieving process …]

between an unexpected outcome and error, if any, will not be obvious, the data will not be available, and/or consensus has not emerged. In those cases, we must meet patient/family and professionals' needs for information by conducting appropriate case evaluations and disclosing the results of the assessment.

Patient/family reactions to case reviews, bad news, and error disclosures cannot be predicted. Many will react with grace, courage, and forgiveness; others with stoic acceptance. Some, who have perhaps suspected that a problem has occurred, will be relieved—at least initially—if their problem is remediable. Others will be distressed and tearful. Still others will appear stunned and numb, unable to respond otherwise at that moment. A few, though, will lash out in anger in their distress (McCord et al., 2002). Violent anger, while understandable, poses a real threat of harm to the care provider and others in the clinical environment, and the potential for anger escalating into violence must be anticipated. Generally speaking, most anger can be managed if the providers respond in ways that tend to defuse rather than inflame (Table 25.5).

Disclosure conversations are complicated by the clinician's emotions and the uncertainty of patient/family reactions and questions. Clinicians should take care to focus on the patient's/family's reactions, not their own, during error disclosures. This is not the time to share how the adverse outcome has affected you or the team. While such statements may be appropriate later, your attention needs to be on the patient's needs now and in the future, not your own. If/when patients ask who was responsible for the error or harm, it is especially helpful to have rehearsed responses that appropriately convey how individuals and systems may have failed, but that the circumstances will be carefully reviewed (if that is

true). Planning ahead, anticipating potential reactions and questions, and having alternatives at hand when the plan goes awry simply increase the odds that the discussion will go as well as possible. Remember that the goal of an error discussion is *not* to make you feel better, but to inform the affected parties and fulfill a professional and ethical obligation. Rarely, if ever, are these conversations entirely satisfactory. That is no reason to avoid them, but it is important to have realistic expectations about their outcome. Then, following the discussion, it is useful to debrief afterward over what went well, what might have gone better, and how you might do better next time.

25.4.2 Error Disclosure in Complex Cases

The general principles outlined in the previous section provide a basis for error disclosure and support in relatively straightforward cases where, even if the outcome is poor, the clinical facts are known and agreed. We now consider some more complex cases where the cause is not yet known, there is a dispute about what occurred, and, finally, where an error has occurred, but the family is unaware of the error. These types of cases are experienced daily throughout large medical centers, but how best to handle them are not well researched by the health care and human factors communities (Surbone et al., 2007). The cases are, therefore, offered in part as challenges to thoughtful clinical and human factors collaborators to help health care systems cope well when adverse outcomes are associated with real or perceived errors. Since the publication of the first edition of this book, others have also contributed to the case-based literature (e.g., Gallagher et al., 2007; Gallagher et al., 2009; Rodriguez et al., 2009; Smetzer et al., 2010; Spandorfer et al., 2010).

25.4.2.1 Error Disclosure When the Cause of an Unexpected Outcome Is Not Immediately Known

You are the general internist for J, a 74 year old man with a history of atrial fibrillation admitted for pneumonia. He was on warfarin (2.5 mg/day) at home to reduce risk of clot formation. His INRs, a measure of anticoagulation, measured 2–3 (checked monthly and just over a week ago). Admission orders included an antibiotic (ceftriaxone) and warfarin (2.5 mg/day). You have just been called at home (early on the third day of hospitalization) because J is no longer able to use the left side of his body. You order stat coagulation studies, a head CT scan, and proceed to the hospital. When you arrive, you discover his INR is 6. Review of hospital lab reports reveal that you had not ordered any coagulation studies since he arrived at the hospital. The omission concerned you because you generally obtain coagulation studies in any patient on both warfarin and a broad-spectrum antibiotic. In addition, review of his

TABLE 25.5

If Patient/Family Reacts Angrily

- Avoid defensiveness or challenging the patient's anger
- Consider the fears or affronts that underlie the anger
- Acknowledge the patient's anger directly, don't ignore it
- Use empathetic statements
- Avoid making excuses, blaming
- Use a calm, soft voice; as the angry person dials it up, dial yours down
- If an apology is appropriate, make one, but be specific; if one has already been offered, repeat it
- If an apology is *not* appropriate, avoid blaming others, the system, and so on
- If the anger appears to stem from some point of dispute, reassure the patient/family that it's okay for the MD/patient to disagree (agreeably) on issues of care

  May have to call time out and reconvene for discussion at a later time when emotions have settled

bedside drug sheet (see Figure 25.1) revealed that he had been getting 12.5 mg of warfarin per day instead of the 2.5 mg that you thought you ordered.

You go to J's room. When you enter his son asks, "Why do you think dad had a stroke?" How would you recommend the physician respond? Which of the following would you choose?

1. *No disclosure/safe facts*: "Pneumonia is a serious condition, your father has a bad heart condition … he has several risk factors for a stroke. Under these circumstances, strokes just happen sometimes, and I 'm sad to see that J has had one. Now, we need to focus on managing …"

2. *Facts, more later*: "I don't know yet. J's INR (a measure of blood thinning) was high, which might have contributed … we need to review his condition, the events leading up to the stroke, and the results of his most recent tests … to see whether we can find out the cause of the stroke."

3. *Disclose error*: "INR was high … it looks like J got too much warfarin (± "I didn't check his INR since admission …"), which may or may not have contributed, we need to review his condition, the events leading up to the stroke, and the results of his most recent tests … to see whether we can find out the cause of the stroke."

4. *Disclose error, assign responsibility*: "INR high, J got too much warfarin. The nurses misread my order (or bad handwriting); I am so sorry that this probably caused J's stroke …"

All four of these responses and their variants have merit. That is what makes the "what, when, and how" of disclosure so challenging under normal conditions of uncertainty, and why disclosure scripts, while sometimes helpful, will be neither prudent nor effective in all circumstances (Pichert et al., 1998; Hickson and Pichert, 2008). How clinicians respond in the

short-term will depend on several factors, including the type of stroke (hemorrhagic or embolic), the medical team's level of certainty that the stroke was caused by the error, and their assessment of whether this is the time and place to impart information about a possible serious error. No one response or type of response will suit all situations. Medical professionals are better served weighing the pros and cons of each strategy, factoring in knowledge about themselves and their patients/families, and the particular circumstances associated with the (perceived) adverse outcome and the medical care provided to that point. The strategy we recommend, therefore, may be characterized as a "balance beam" approach to considering what, how, and when to disclose (Figure 25.2).

To the left of the fulcrum, no disclosure of any error, real or perceived, is offered. To the right of the fulcrum, the apparent error is disclosed and linked to the adverse outcome as though no uncertainty exists. In our experience, medical professionals asked to choose what to disclose, in a hypothetical case such as J's, distribute themselves along every point of the continuum and assert strong arguments in favor of their positions. Each selection from "no disclosure" to "disclose the apparent error and assign responsibility," therefore, must have positive features. But a few moments of thought will reveal that each alternative also has potentially negative features. Before reviewing Table 25.6, see what pros and cons you would assign to each point along the continuum. In addition, ask yourself what follow-up questions might reasonably be anticipated from some families in response to each position along the balance beam.

The questions that appear in the far-right column of Table 25.6 are *not* meant to suggest that one alternative will be better or somehow easier to use than another. The important issue for medical professionals is to consider such questions in advance and, rather than being caught off-guard, be ready to provide answers to those questions as well. So, for example, if the professional postpones disclosure by promising a review, upon the subsequent disclosure some families will ask why the apparent error had not been discussed previously. The health professionals should anticipate the

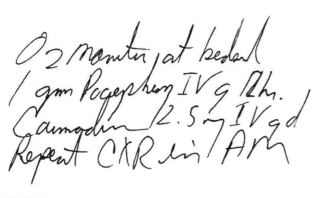

FIGURE 25.1
J's bedside drug sheet.

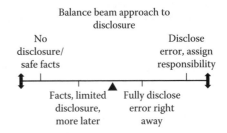

FIGURE 25.2
A balance beam approach to disclosure.

TABLE 25.6

Pros and Cons Associated with Disclosures When Not All the Facts Are Known or May Never Be Known

Alternative Response	Pros	Cons	Questions Patients/Families Might Ask If Given This Response
No disclosure/safe facts (at this time, or ever)	Some families will not press you; some just do not want to know or do not need more information; may buy some time to finish an evaluation; may be kind to not share too much at once; others?	It is short of honest; looks BAD when discovered; news gets out various ways; many people have a need to know all the information as soon as possible; may signal a "code of silence"; others?	Do you have any ideas about what may have caused this outcome? Who else can I ask? I just don't understand why this happened, what else can you tell me?
Facts, limited disclosure, more later	It is more honest; buys time for impartial review; may preserve relations; others?	Still may turn out to be short of honest; you know more, may look bad later; news still gets out; premature speculations may create unrealistic expectations or wrong understandings; others?	How long will the review take? Why will it take so long? [Later, if the error was determined to be related to the adverse outcome: Didn't you know about the error when we first spoke? Why did Dr./Nurse ___ say something different? How can such a thing happen here?]
Fully disclose apparent error right away	Shows honesty; many are willing to forgive with apology; *may* cost less later? others?	You might be wrong; it will be challenging to take it back later; empowers legal action? may negatively impact patient confidence in the institution; others?	Who is responsible for this error? Who is going to pay my bills? Why didn't Dr. ___ or Nurse ___ mention the error? How can such a thing happen?
Disclose error, assign responsibility	Shows utter honesty, no "code of silence"; may make you feel better; shows that you are on patients side; others?	Colleagues do not like jousters; you cannot take it back later; jousts drive torts, may cost more money; may be unrelated to the adverse outcome; others?	Can you arrange for the responsible party to apologize to us? We should expect the institution to compensate us, right? Can the responsible parties be disciplined (fired)?

question, so may be prepared to respond. One alternative might be to simply assert, "I was aware of the possibility, but I wasn't sure whether the error caused the stroke. That's why I wanted a review of all the circumstances, and why I promised to report the results to you, which is what I'm doing now. I just didn't want to speculate ..." The preexisting relationship sets the stage for trust in the midst of challenging disclosures. With a good prior relationship, we believe that most (but not all) families will accept this explanation.

25.4.2.2 When the Physician Deems the Error Unrelated to the Outcome

To highlight another, probably more common problem, consider the case where a physician chooses *not* to disclose the medication error immediately, but the son asks a nurse who subsequently visits the room, "Why do you think dad had a stroke?" In this case, the physician has judged that the case is to the left of the fulcrum in the balance beam figure and that a first-encounter disclosure of the Coumadin error serves no purpose. A useful exercise is to consider the pros and cons if nursing professionals respond to this challenging question by saying the following:

1. "I don't know ... did you ask your physician? (what did your physician say)?"

2. "Your physician is sad that a coagulation study was not ordered a little sooner ... doing so *might* have provided us with important information that might have prevented the stroke ..."

3. "I'm sad to report that the doctor should have but failed to order coagulation studies ... probably would have prevented this stroke."

4. "Unfortunately, your father was given too much of his blood thinner medicine, which probably caused the stroke. Let me show you the doctor's drug order ..."

The stroke poses challenges for all of J's care providers because the medication error may have played a role in causing a bleeding event, but the patient's atrial fibrillation may have caused a thrombotic event. Consequently, even though the drug error is clear, the *stroke's* cause at the time the family asks about it is not. Whether the drug error contributed or not will be unclear at least until CT scan results are read. The delay gives the physician, nurses, and others who care for the patient many opportunities to deliver mixed or even contrary messages. For example, a nurse might choose #4 out of a sense of guilt, perhaps fear of being fired for making the error or perhaps taking an opportunity to criticize a physician whose previous behavior toward the nurses had been unkind or disruptive.

25.4.2.3 When the Cause of an Unexpected Adverse Outcome Is a Matter of Dispute

HP is a 62 year old female admitted with congestive heart failure. The routinely administered admission-screening test for methicillin-resistant Staphylococcus aureus (MRSA) reports out as "negative." A central line is placed on Hospital Day #1 with full documentation of insertion bundle practices. HP shows improvement. Unfortunately, on Hospital Day #4, HP develops fever, blood cultures are drawn. Two days later, the cultures grow MRSA, so HP is placed on vancomycin, and the medical team shares the diagnosis with HP and her husband. HP's subsequent course is complicated by adult respiratory distress syndrome (ARDS). After HP recovers and has been transferred out of the intensive care unit, she asks her nurse various questions about isolation. After conversing with patient and husband for a few minutes, HP's husband asks the nurse, "How did she get MRSA ... did she get it here?"

Using the same "balance beam" approach taken in the previous case, please pause to consider the advantages and disadvantages that might be associated with each of the following strategies. How would you, as HP's nurse, respond to her husband?

1. *"No. While it's not impossible, it's highly unlikely ... Probably got MRSA from the community ... could have had it on your skin."*
2. *"It's possible, but very remote ... masks and gloves provide protection ... line infections happen, but I just don't think that's what happened."*
3. *"I don't know at present ... may never know ... let me ask the medical team to gather more information, then one of us will come back and share with you whatever is learned."*
4. *"It's certainly possible ... but there are several potential factors/causes for the infection ... will ask for a review ... be back to you."*
5. *"It's probable ... precautions aren't 100% effective ... but there are several potential factors/causes ... will ask for a review ... be back to you."*

Let us say the nurse chose number 2 (possible, very remote). HP's husband responds, "The patient next door is on isolation ... is it for MRSA?" The nurse knows it is and responds affirmatively. HP's husband then says, "You and some other nurses have been complaining about having to wear 'those gloves' and I don't see the

doctors using the alcohol gel. Are you *sure* she didn't get it here?" (You also know that hand hygiene adherence last month was 77%). What are the pros and cons for the following potential responses?

1. *"No. I am sure she didn't get while here ..."*
2. *"Possible, but probably not ... our infection control statistics are good."*
3. *"There's only a small uncertainty, but ... we just can never be 100% sure ... we may never know."*
4. *"I suppose it's possible ... I apologize for any lapses ... but we may never know."*
5. *"Probably ... I've been worried about doctor adherence, etc. ... so sorry for the lapses you observed."*

What if the review reveals that the routine MRSA admission screening test was actually incorrect and HP had MRSA when she came to the hospital? Or, what if HP were MRSA negative on admission and the MRSA strain in HP's blood culture proved identical to that of the patient next door versus not identical to that of the patient next door?

No matter how forthcoming or skilled a medical professional or team of professionals may be in sharing with patients/families concerning adverse outcomes and their potential causes, not all families will be satisfied with the explanations that are provided. Sources for patient/family nonacceptance are numerous, but must include the nature of medical probability. In the last case, the nurse's judgment may have been reasonable: based on what the nurse knew at the time, the patient might not have gotten MRSA during the admission, but the nurse cannot be 100% certain. Families may cling to a belief that an error caused harm, even if the probability is remote. Reasons for that tenacity might include denial about cause, suspicion of the medical profession in general, or the particular health care team (especially with staff who complain about procedures and family observations of apparent protocol deviations), overheard jousting by medical professionals, or the fact that the patient/family is actually correct. Because families ultimately have the final choice to accept or not, it is critical that all the health care team members seek to minimize the "inflammation" that may negatively impact disclosure. In addition, inflammation/anger/dissatisfaction with care has the potential to impact subsequent medical adherence, practice drop-out, complaining to others, and of course, in the face of a prolonged hospital stay, permanent disability or death, and decisions to file suit (Ong et al., 1995).

There are numerous reasons why medical professionals have difficulty in walking away from an encounter by simply agreeing to disagree agreeably. Professional training has a way of creating the need in physicians, nurses, and others to resolve conflict and always reach an agreed-upon conclusion. Yet consensus is not always possible, and when some families continue to disagree (as in the last case) some nurses and physicians will be angry themselves.

Also the focus in the courtroom unfortunately moves to dispute and denial versus quality improvement and moving to eliminate infectious disease risk.

25.4.2.4 When a Subsequently Treating Physician Believes That an Error Occurred, but the Patient/Family Is Unaware

H is a 58 year old male with persistent low back pain. He reports that about 1 year ago he "underwent a surgical procedure at L2-L3 by Orthopedic surgeon #1." H has come to you (Orthopedic surgeon #2) for a second opinion. You order a MRI. The radiologist reports, and you confirm, that the MRI reveals disc disease at L2-L3 and evidence of surgery at L3-L4. H is eager to know the findings.

Before considering how Orthopedist #2 might respond to H, who would you have share the new finding with H (if anyone)?

1. Orthopedist #1
2. Orthopedist #2
3. Representative of the first medical center: for example, an administrator, risk manager, or lawyer
4. Someone else

Beliefs about who should present the findings affect how the second orthopedist would respond to H's question. Consider the first alternative. What are the pros and cons of engaging the original surgeon (and colleagues) to disclose? One "pro" is that involving them provides the second team an opportunity to provide direct feedback about both the previous and the proposed procedures, and it may cause the first surgeon and team to rethink their protocols. Second, Orthopedist #1 and colleagues likely would want to know the correct data, and some members of the first team might feel professionally motivated to share their findings. If the family wants to hear an apology from those who erred, Orthopedist #1 and team might as well become involved right away. Finally, Orthopedist #1 would not want to first learn about the error from H or, worse, a plaintiff attorney.

One "con" is that providing the data to the first team and asking them to discuss the results with the family takes time. And the first team may dispute the findings ("When we operated, we found significant disease at L3-L4, so we operated there ... our op note discussing a procedure at L2-L3 was a simple oversight.") In

addition, Orthopedist #1 and team may resent any perceived "high-handedness" and go out of their way to look for payback opportunities. Or some families might not wish to return to Orthopedist #1. The pros for the second alternative (disclosure by Orthopedist #2) include Orthopedist #2 being sure that H had been told the facts, which is essential for truly informed consent. Otherwise, the pros and cons are largely opposite those of the first alternative. Finally, involving others can be helpful if those persons have outstanding interpersonal skills, experience as bearers of bad news, and knowledge of the institution's policies and procedures regarding disclosure and, if necessary, ability to authorize compensation. Obvious risks, however, include the potential for making H suspicious or feeling ganged up upon, or taking the chance that the group disputes the facts with one another during the discussion.

How, then, might H's second orthopedist respond?

1. *No disclosure*: "... we found clear evidence of disc disease (without declaring where) ... you need another surgery to get relief ... let's start working on authorization."
2. *Principal facts*: "... disc disease ... you need another surgery ... not certain today about all the findings ... I want to review records ... let's start working on authorization."
3. *Set stage for review*: "... disc disease ... findings on MRI I don't understand ... I need to review all the records ... with your permission, I'll contact Dr. Orthopedist #1 and learn her/his take on it ... then share the findings."
4. *Disclose error*: "I don't know why you were told your surgery was at L2-L3 because surgery was performed at L3-L4 and disease remains at L2-L3."

All four choices have both pros and cons. Consider a few (and add your own). The first two alternatives, no disclosure and presentation of only the principal facts, may take less time, at least initially, especially if the family either never learns the truth or chooses not to pursue any action against the first team. In addition, both strategies may be intended kindly; after all, the diagnosis is tough enough for the patient and the error likely will have no meaningful impact on the eventual outcome, so disclosure at this time might even seem cruel. On the other hand, these strategies may result in the patient/family feeling misled if/when the truth comes out later, and those affected may come to believe that all their health professionals are engaged in a conspiracy of silence. In such cases, the time saved initially may be brief compared with the time and effort required to address such concerns. Cons for the second medical team are that they will not be doing their ethical duty to provide real informed consent, and those who believe the first team provided genuinely bad care may feel guilty thinking that the patient needs to be aware in order to make decisions going forward.

The third option, setting the stage for a review, allows Orthopedist #1 and associates to decide whether Orthopedist #2 will appropriately discharge his professional duties and inform Orthopedist #1 how they wish to be involved in the discussion about the care they provided. In some cases, the first team will solicit the second team's assistance, which requires time, but may assure full disclosure. Unfortunately, if the first team wants to handle the disclosure, the strategy may backfire if the family perceives that the second team is simply passing the buck. Perhaps most problematic, however, is that the first team may not disclose the facts or in the fashion the second team believes should occur. Consider what the responses might be when you report the findings and your need to talk with H? Orthopedist #1 may confirm your suspicion and may wish to personally disclose the error to H, or Orthopedist #1 may ask your help with disclosure. But Orthopedist #1 may NOT confirm your suspicion. In fact, Orthopedist #1 may disagree with you ("changed my mind when I got in there"); refuse to call it an error; or even ask you to conceal the error); or, suspecting trouble, may avoid you and refuse to communicate. One way for Orthopedist #2 to reduce the risk that Orthopedist #1 says "no" or fails to make a promised disclosure is to provide a time limit (e.g., "H will return to my office on [date], when I will repeat the findings you've already shared).

If the response turns out to be "NO," what does Orthopedist #2 do then? What alternatives are there? Orthopedist #2 might consider reporting the clinical concerns to an appropriate peer review authority; attempt to report the experience in a "nonjudgmental discussion" with H, and/or tell H about the error and Orthopedist #1's choice not to discuss it.

The pros and cons of the fourth option—involving immediate disclosure without discussion with Orthopedist #1—are essentially the opposites of the first two alternatives. First, disclosure presents the truth and shows the second orthopedist to be an advocate for the truth. Second, by heading off patient/family speculation, disclosure may help the patient move through the grief process. In addition, if the second team discloses the error directly to H, Orthopedist #2 may avoid an uncomfortable conversation with the first. The cons are that collegial relationships might suffer as a result of these choices, and if Orthopedist #2 and colleagues happen to be wrong about the first surgery site, patient/family anger about the situation may promote litigation.

As we have suggested elsewhere in this chapter, some medical professionals do not like this "balance beam"

approach of weighing pros and cons for several alternatives. They prefer disclosure scripts or a structured list of guidelines. Such scripts can be useful when circumstances are clear-cut: where all the data are known, agreed upon, and everyone involved is a skilled communicator. Unfortunately, experience suggests that the majority of circumstances involving adverse outcomes and apparent errors simply do not meet these tests for implementing scripts. Fortunately, health care professionals have substantial experience balancing pros and cons of treatment alternatives in the face of medical uncertainty. Deciding how best to discuss adverse outcomes associated with an apparent or potential error is just another circumstance in which the skill of weighing pros and cons applies.

Finally, this case highlights the disclosure-related duties of subsequently treating physicians who recognize that patient harm resulted from an error. Once again, the balance beam encourages consideration of alternatives ranging from no disclosure to returning the patient to the initially treating physician, to declaring the error, and assigning responsibility for it. The AMA's Council on Ethical and Judicial Affairs states, *"…even if a physician is not responsible for the harm, that physician still has the ethical obligation of protecting patient welfare in general by disclosing incompetence and promoting operational improvements that enhance patient safety."* (American Medical Association, 2003) This admonition assumes that the medical professionals who believe they have identified a harm-causing error indeed possess complete data, have the expertise to evaluate it, and have humbly considered that they might be wrong. Let us be clear that no "code of silence" is being advocated. Professionals must simply recognize that declarations about the care delivered by a colleague may have serious consequences. These consequences must, therefore, be considered alongside professional duty. The complicating factor is appreciating the competing tugs to serve the best interests of patients, oneself, one's medical group, one's institution, and/or even one's liability insurer. The professional should understand these competing priorities and act honestly and with integrity.

25.5 Disclosure Is Just the Beginning

When serious harm has occurred, acknowledging, disclosing, and discussing the incident is just the first stage. The longer-term needs of patients, families, and staff need to be considered. Injured patients have their own particular problems and needs. Some will require a great deal of professional help. Others will prefer to rely on family and friends. Some will primarily require remedial medical treatment, others primarily psychological assistance. We cannot cover every eventuality, but a few basics bear keeping in mind.

25.5.1 Ask Specific Questions about Emotional Trauma

A common theme in interviews with injured patients is their perception that the professionals involved in their care failed to sufficiently appreciate the depth of their distress (Vincent, 2001). In many cases, outright psychiatric disorders were missed. Clinicians, risk managers, and others involved with these patients can ask basic questions without fear of "making things worse." Other crucial areas of inquiry are feelings of anger, humiliation, betrayal, and loss of trust—all frequently experienced by injured patients.

A proportion of patients who suffer an adverse event will become sufficiently anxious or depressed to warrant formal psychological or psychiatric treatment. While it is important that a senior clinician is involved in giving explanations and monitoring remedial treatment, the staff, of say a surgical unit, cannot be expected to shoulder the burden of formal counseling. They have neither the time nor the necessary training to deal with the more serious reactions. Patient advocates can assist with fact finding, identifying needs, and supporting (Hayden et al., 2010). When a referral to a psychologist or psychiatrist is indicated, it must be handled carefully. Injured patients are understandably very wary of their problems being seen as "psychological" or "all in the mind." Therefore, one approach might be to normalize psychological assessment and intervention by suggesting, "When patients are injured as a result of an error, not only do we address the injury and the event, but we also know that the situation can impact you emotionally, so we always recommend that patients see a mental health expert for evaluation ..."

25.5.2 Continuing Care and Support

Injured patients may receive support, comfort, and practical help—usually constructive but sometimes not—from many sources. It may come from their spouse, family, friends, colleagues, doctors, or community organizations. Especially important sources of support will be the doctors and other health professionals who are involved in their treatment—but principally if these professionals have support for their own roles in the event. It is vital that the duty of care is paramount.

After an initial mistake, most patients find it reassuring to be overseen by a single senior doctor who undertakes to monitor all facets of treatment, even if it involves several specialties (cf. Table 25.3). Where care has been substandard, the patient must if at all possible

be offered a referral elsewhere if that is what they wish, but if the incident is dealt with openly and honestly, then trust may even be strengthened (Clements, 1994).

25.5.3 Inform Patients of Changes

Studies of closed claims suggest that patients' and relatives' wishes to prevent future incidents can be seen both as a genuine desire to safeguard others and as an attempt to find some way of coping with their own pain or loss (Hickson et al., 1992). The pain may be ameliorated if they feel that, because changes were made, then at least some good came of their experiences. Relatives of patients who have died may express their motives for litigation in terms of an obligation to the dead person to make sure that a similar accident never happens again, so that some good comes of their death. The implication of this is that if changes have been made as a result of the adverse outcome, it is very important to inform the patients concerned. While some may regret that the changes were made too late for them, there will be those who appreciate the fact that their experience was understood and acted upon.

25.5.4 Financial Assistance and Practical Help

Injured patients often need immediate, practical help. They may need medical treatment, counseling, and explanations, but they may need money as well. They may need to support their family while they are recovering, pay for specialist treatment, facilities to cope with disability, and so on. In less serious cases, relatively modest sums of money to provide private therapy, alterations to the home, or additional nursing may make an enormous difference to patients both practically and in their attitude toward the medical group or medical center (Colorado Physicians Insurance Company, 2005; Mello and Gallagher, 2010). Protracted and adversarial medicolegal negotiations can be very damaging, frustrating, and above all incomprehensible to the patient and their family. One only has to imagine oneself in a similar position to appreciate this. If you were injured in a rail or aviation accident, you would expect the organization concerned to help you, not say that "you will be hearing from our lawyers in due course."

25.5.5 Long-Term Follow-Up and Support

Supporting patients in the aftermath of medical errors should not be restricted to their immediate needs after serious incidents. Patients and their families may have a host of psychological and other needs that may continue over a considerable period of time. The emotional trauma experienced in the first instance may be prolonged due to a wide range of factors such as patient personality, coping mechanisms, and lack of adequate support. Research suggests that availability and activation of social support during a life crisis is a major moderator in successfully dealing with stress (Schwarzer et al., 2004). Apart from obvious sources of support such as family, friends, and other communities in which a patient belongs, in the event of a serious medical error patient perceptions of the support they receive from the health care organization in which the incident took place may affect their longer-term adjustment and recovery. The continuing offer of support and care after patients are discharged from the hospital may be of great importance for their return to normal life.

In more practical terms, it is absolutely essential to reassure patients and their families that they will have further opportunities for support and communication with the health care organization after they are discharged from the hospital. For instance, providing patients with the contact details of people within the organization who will be available to answer questions seems a logical thing to do. Also, organizing follow-up meetings where patients will have the opportunity to ask questions, share how they get on with the incident-related challenges of their lives, or be offered an update on the investigation of the incident may be welcomed by many patients and their families.

To close this chapter, we will consider one case of a very ineffective approach to error disclosure and two examples of the benefits of a positive and proactive long-term approach to it. In the first example, even though there was no serious clinical harm, the psychological outcomes of the way the incident was handled were significant for the patient. Neither of the other two had long-term consequences, but both were very frightening to the patients concerned and, if handled badly, might well have affected their recovery and willingness to have future treatment. The examples show that ineffective communication with the patient in the event of a medical error can lead to severe patient frustration, anger, loss of trust in the health care providers, and distrust of health care in general; second, they highlight that even potentially disastrous events, like awareness under anesthesia, can be handled in a sensitive and innovative way with great benefits to staff and patients alike.

25.5.5.1 Denial of Communication and Apology after a Surgical Procedure Was Missed

A was scheduled to receive two gynecological surgical procedures. During the informed consent process, she noticed that one of them was not listed on the consent form. The patient tried a few times to raise the issue with her care team and also asked a nurse to add the procedure on the consent form. Preoperatively she checked with the nurse to make sure that *the surgeon had signed the informed consent form with the corrected procedures. The surgeon had indeed signed the corrected consent form, so the patient was prepared for surgery. However, when the patient enquired about the operations the day after the surgery, she found out that the procedure she was concerned about had not been carried out (MITSS, 2009).*

When the surgeon came to see the patient, the communication between them was handled rather poorly. Unwilling to admit that a mistake had been made, he left the patient's room obviously frustrated. The patient's expectations that the surgeon would return with more information and suggestions about the procedure that was missed were not met. The surgeon never approached the patient to express his regret for what happened or enquire about the patient's well-being, and neither did anyone else in the organization. The patient decided not to attend her postoperative appointment out of frustration and anger that no one from the hospital had followed up her case. It was only 3 months later after the patient called the hospital's complaint line to report her experience that the CEO called her to apologize and the surgeon actively asked to talk to her. Yet, once again the communication was managed rather inadequately as the surgeon insisted on assigning blame mostly to others and at one point even blamed the patient. This case shows how poor provider communication skills can exacerbate patients' frustration and emotional trauma after such incidents.

The following two examples illustrate how a proactive and sensitive approach to error disclosure can benefit all parties involved.

25.5.5.2 Explanations and Apology after Iatrogenic Cardiac Arrhythmia

A was admitted for minor day case surgery, expecting to return home later that day. A surgeon requested a weak solution of adrenaline to induce a blood-free field, but was given a stronger solution than requested. As soon as the liquid was applied the patient developed a serious cardiac arrhythmia, the operation was terminated, and she was transferred to the intensive therapy unit, where she gradually recovered.

The clinical risk manager was alerted immediately and assessed the likely consequences for the patient and her family. The first task was clearly to apologize and provide a full explanation. However, with both the patient and family in a state of shock, this had to be carried out in stages. The consultant and risk manager had a series of short meetings over a few days to explain what had happened and keep the family informed about ongoing remedial treatment. Each time the family was given the opportunity to reflect on what they had been told and come back with further questions. A small package of compensation was also arranged, primarily aimed at providing the necessary clinical and psychological support. The whole incident was resolved within 6 months and the patient expressed her thanks to the hospital for the way in which the incident had been handled, particularly the openness about the causes of the incident. Continuing follow-up support was well organized by risk management and clearly contributed to the positive reaction of the patient. Most, *but probably not all*, families would respond similarly; professionals who attempt to follow this model should be prepared for the instance when the patient and family do not respond well from the very first.

25.5.5.3 Anesthetic Awareness: Reducing the Fear of Future Operations

A woman was admitted for an elbow replacement. During the operation she awoke, paralyzed and able to hear the discussions amongst the surgical team. She was terrified, in great pain, and absolutely helpless. The lack of anesthetic was fortunately noticed, and she was next aware of waking in recovery screaming.

The risk manager visited the patient at home as soon as practicable, maintained contact, offered psychological treatment for trauma, and advised her on procedures for compensation, including an offer to pay for an independent legal assessment of the eventual offer of compensation. Emotional trauma was the principal long-term concern. In this case, a fear of future operations was a major factor, very important in a woman suffering chronic conditions requiring further treatment, so this problem required additional, imaginative measures. Specifically, when the patient felt ready, she was given a tour of the operating theater, and the anesthetic failure was explained in great detail, as were the procedural changes that had been made subsequent to the incident. This was clearly immensely important in reducing her understandable fear of future operations and minimizing the long-term impact of the incident.

25.6 Conclusion

This chapter began by presenting recent disclosure-related standards and recommendations, and discussing effects of errors on health professionals and patients. Selected research on medical malpractice suggested the importance of appropriate error disclosures. A series of specific cases was then used to highlight pros and cons of different approaches to disclosure under various challenging circumstances. Human factors abound, but have not much been analyzed or researched. As a result, experts in human factors engineering, analysis, and research have important opportunities to help assess and create

psychologically safe systems for effective error disclosure. We concluded that appropriate error disclosures are essential in health care, but they are merely the beginning of dealing with harm-causing errors and their sequelae.

References

American Medical Association (October 3, 2003). *Council on ethical and judicial affairs*. In: *Code of medical ethics*. http://www.ama-assn.org/ama1/pub/upload/mm/369/2a03.pdf (last accessed May 30, 2010).

Australian Council for Safety and Quality in Health Care (2003). Open disclosure standard: A national standard for open communication in public and private hospitals, following and adverse event in health care. http://www.safetyandquality.health.wa.gov.au/docs/open_disclosure/ACSQHC_Open_Disclosure_Standard.pdf (last accessed June 29, 2011).

Banja, J. (2001). Moral courage in medicine—Disclosing medical error. *Bioethics Forum*, 17(2), 7–11.

Beckman, H.B., Markakis, K.M., Suchman, A.I., and Frankel, R.M. (1994). The doctor–patient relationship and malpractice: Lessons from plaintiff depositions. *Archives of Internal Medicine*, 154(12), 1365–1370.

Blendon, R.J., Schoen, C., Donelan, K., Osborn, R., DesRoches, C.M., Scdoles, K., Davis, K., Binns, K., and Zapert, K. (2001). Physicians' views on quality of care: A five-country comparison. *Health Affairs*, 20(3), 233–243.

Bogner, M.S. (1994). *Human error in medicine: Misadventures in health care*. Hillsdale, NJ: L. Erlbaum Associates.

Bogner, M.S. (2004). *Misadventures in health care: Inside stories*. Mahwah, NJ: Lawrence Erlbaum.

Boothman, R.C. (2006). Apologies and a strong defense at the University of Michigan Health System. *Physician Executive*, 32(2), 7–10.

Boothman, R.C., Blackwell, A.C, Campbell, D.A., Jr., Commiskey, E., and Anderson, S. (2009). A better approach to medical malpractice claims? The University of Michigan experience. *Journal of Health & Life Sciences Law*, 2(2), 125–159.

Bovbjerg, R.R., and Petronis, K.R. (1994). The relationship between physicians' malpractice claims history and later claims: Does the past predict the future? *The Journal of the American Medical Association*, 272(18), 1421–1426.

Brennan, T.A., Leape, L.L., Laird, N., Hebert, L., Localio, A.R., Lawthers, A.G., Newhouse, J.P., Weiler, P.C., and Hiatt, H.H. (1991). Incidence of adverse events and negligence in hospitalized patients: Results of the Harvard medical practice study I. *The New England Journal of Medicine*, 324(6), 370–376.

Canadian Medical Association (2004). *Code of ethics*. Ottawa: The Canadian Medical Association. http://policybase.cma.ca/PolicyPF/PD04-06.pdf (last accessed May 31, 2010).

Cantor, M.D. (2002). Telling patients the truth: A systems approach to disclosing adverse events. *Quality and Safety in Health Care*, 11, 7–8.

Chafe, R., Levinson, W., and Sullivan, T. (2009). Disclosing errors that affect multiple patients. *Canadian Medical Association Journal, 180*(11), 1125–1127.

Chan, D.K., Gallagher, T.H., Reznick, R., and Levinson, W. (2005). How surgeons disclose medical errors to patients: A study using standardized patients. *Surgery, 138*(5), 851–858.

Christensen, J.F., Levinson, W., and Dunn, P.M. (1992). The heart of darkness: The impact of perceived mistakes on physicians. *Journal of General Internal Medicine, 7*(4), 424–431.

Clarke, D.M., Russell, P.A., Polglase, A.L., and McKenzie, D.P. (1997). Psychiatric disturbance and acute stress responses in surgical patients. *The Australian and New Zealand Journal of Surgery, 67*(2–3), 115–118.

Clayton, E.W., Hickson, G.B., Wright, P.B., and Sloan, F.A. (1993). Doctor–patient relationships. In: F.A. Sloan, P.B. Githens, E.W. Clayton, G.B. Hickson, D.A. Gentile, and D.A. Partlett, eds. *Suing for medical malpractice* (pp. 50–71). Chicago, IL: University of Chicago Press.

Clements, R. (1994). The continuing care of the injured patient. In: R. Clements and P. Huntingford, eds. *Safe practice in obstetrics and gynecology.* London, U.K.: Churchill Livingstone.

Colorado Physicians Insurance Company (2005). *COPIC's 3Rs program: Lessons learned. 2(1).* www.callcopic.com (last accessed June 29, 2011).

Czarnocka, J., and Slade, P. (2000). Prevalence and predictors of post-traumatic stress symptoms following childbirth. *British Journal of Clinical Psychology, 39*(Pt 1), 35–51.

Dovey, S.M., Phillips, R.L., Green, L.A., and Fryer, G.E. (2003). Consequences of medical errors observed by family physicians. *American Family Physician, 67*(5), 915.

France, D.J., Throop, P., Walczyk, B., Allen, L., Parekh, A.D., Parsons, A., Rickard, D., and Deshpande, J.K. (2005). Does patient centered design guarantee patient safety? Using human factors engineering to find a balance between provider and patient needs. *Journal of Patient Safety, 1*(3), 145–153.

Gallagher, T.H. (2009). A 62-year-old woman with skin cancer who experienced wrong-site surgery: Review of medical error. *The Journal of American Medical Association, 302*(6), 669–677.

Gallagher, T.H., Bell, S.K., Smith, K.M., Mello, M.M., and McDonald, T.B. (2009). Disclosing harmful medical errors to patients: Tackling three tough cases. *Chest, 136*(3), 897–903.

Gallagher, T.H., Denham, C.R., Leape, L.L., Amori, G., and Levinson, W. (2007). Disclosing unanticipated outcomes to patients: The art of practice. *Journal of Patient Safety, 3*(3), 158–165.

Gallagher, T.H., Studdert, D., and Levinson, W. (2007). Disclosing harmful medical errors to patients. *The New England Journal of Medicine, 356*(26), 2713–2719.

Gallagher, T.H., Waterman, A.D., Ebers, A.G., Fraser, V.J., and Levinson, W. (2003). Patients' and physicians' attitudes regarding the disclosure of medical errors. *The Journal of the American Medical Association, 289*(8), 1001–1007.

Garbutt, J., Waterman, A.D., Kapp, J.M., Dunagan, W.C., Levinson, W., Fraser, V., and Gallagher, T.H. (2008). Lost opportunities: How physicians communicate about medical errors. *Health Affairs, 27*(1), 246–255.

Greenberg, C.C., Regenbogen, S.E., Studdert, D.M., Lipsitz, S.R., Rogers, S.O., Zinner, M.J., and Gawande, A.A. (2007). Patterns of communication breakdowns resulting in injury to surgical patients. *Journal of the American College of Surgeons, 204*(4), 533–540.

Gunderson, A.J., Smith, K.M., Mayer, D.B., McDonald, T., and Centomani, N. (2009). Teaching medical students the art of medical error full disclosure: Evaluation of a new curriculum. *Teaching and Learning in Medicine, 21*(3), 229–232.

Hain, P.D., Pichert, J.W., Hickson, G.B., Bledsoe, S.H., Irwin, C., Hamming, D., Hathaway, J., and Nguyen, C. (2007). Using risk management files to identify and address causative factors associated with adverse events in pediatrics. *Therapeutics and Clinical Risk Management, 3*(4), 625–631.

Hayden, A.C., Pichert, J.W., Fawcett, J., Moore, I.N., and Hickson, G.B. (2010). Best practices for basic and advanced skills in health care service recovery: A case study of a re-admitted patient. *Joint Commission Journal on Quality and Patient Safety, 36*(7), 310–319.

Hickson, G.B., Clayton, E.W., Entman, S.S., Miller, C.S., Githens, P.B., Whetten-Goldstein, K., and Sloan, F.A. (1994). Obstetricians' prior malpractice experience and patients' satisfaction with care. *Journal of the American Medical Association, 272,* 1583–1587.

Hickson, G.B., Clayton, E.W., Githens, P.B., and Sloan, F.A. (1992). Factors that prompted families to file medical malpractice claims following perinatal injuries. *The Journal of the American Medical Association, 267*(10), 1359–1363.

Hickson, G.B., Federspiel, C.F., Blackford, J.U., Pichert, J.P., Gaska, W., Merrigan, M.W., and Miller, C.S. (2007). Patient complaints and malpractice risk in a regional healthcare center. *Southern Medical Journal, 100*(8), 791–796.

Hickson, G.B., Federspiel, C.F., Pichert, J.W., Miller, C.S., Gauld-Jaeger, J., and Bost, P. (2002). Patient complaints and malpractice risk. *The Journal of the American Medical Association, 287*(22), 2951–2957.

Hickson, G.B., and Pichert, J.W. (2008). Disclosure and apology. In: *National patient safety foundation stand up for patient safety resource guide.* North Adams, MA: National Patient Safety Foundation.

Hobgood, C., Peck, C.R., Gilbert, B., and Chappell, K. (2002). Medical errors—What and when: What do patients want to know? *Academic Emergency Medicine, 9*(11), 1156–1161.

Hyman, D.A., Black, B., Silver, C., and Sage, W.M. (2009). Estimating the effect of damages caps in medical malpractice cases: Evidence from Texas. *Journal of Legal Analysis, 1*(1), 355–409.

Iedema, R., Sorensen, R., Manias, E., Tuckett, A., Piper, D., Mallock, N., Williams, A., and Jorm, C. (2008). Patients' and family members' experiences of open disclosure following adverse events. *International Journal for Quality in Health Care, 20*(6), 421–432.

Joint Commission on Accreditation of Healthcare Organizations (JCAHO) (2001). *Patients, and, when appropriate, their families are informed about the outcomes of care, including unanticipated outcomes.* Standard R1.2.9. http://www.jointcommission.org/ (last accessed June 29, 2011).

Jury Verdict Research®. (2002). *Medical malpractice: Verdicts, settlements and statistical analysis.* Palm Beach Gardens, FL: LRP Publications.

Kachalia, A., Shojania, K.G., Hofer, T.P., Piotrowski, M., and Saint, S. (2003). Does full disclosure of medical errors affect malpractice liability? The jury is still out. *Joint Commission Journal on Quality and Patient Safety*, 29(10), 503–511.

Kaldjian, L., Jones, E.W., and Rosenthal, G. (2006). Facilitating and impeding factors for physicians' error disclosure: A structured literature review. *Joint Commission Journal on Quality and Patient Safety*, 32(4), 188–198.

Kessler, R.C. (1997). The effects of stressful life events on depression. *Annual Review of Psychology*, 48, 191–214.

Kohn, L.T., Corrigan, J.M., and Donaldson, M.S. (1999). *To err is human: Building a safer health care system*. Washington, DC: Institute of Medicine, National Academy Press.

Kraman, S.S., and Hamm, G. (1999). Risk management: Extreme honesty may be the best policy. *Annals of Internal Medicine*, 131(12), 963–967.

Lehman, D.R., Lang, E.L., Wortman, C.B., and Sorenson, S.B. (1989). Long-term effects of sudden bereavement: Marital and parent-child relationships and children's reactions. *Journal of Family Psychology*, 2(3), 344–367.

Levinson, W., and Gallagher, T.H. (2007). Disclosing medical errors to patients: A status report in 2007. *Canadian Medical Association Journal*, 177(3), 265–267.

Liang, B.A. (2002). A system of medical error disclosure. *Quality and Safety in Health Care*, 11, 64–68.

Liebman, C., and Hyman, C.S. (2004). A mediation skills model to manage disclosure of errors and adverse events to patients. *Health Affairs*, 23(4), 22–32.

Localio, A.R., Lawthers, A.G., Brennan, T.A., Laird, N.M., Hebert, L.E., Peterson, L.M., Newhouse, J.P., Weiler, P.C., and Hiatt, H.H. (1991). Relation between malpractice claims and adverse events due to negligence. Results of the Harvard Medical Practice Study III. *The New England Journal of Medicine*, 325(4), 245–251.

Loren, D.J., Garbutt, J., Dunagan, W.C., Bommarito, K.M., Ebers, A.G., Levinson, W., Waterman, A.D., Fraser, V.J., Summy, E.A., and Gallagher, T.H. (2010). Risk managers, physicians, and disclosure of harmful medical errors. *Joint Commission Journal on Quality and Patient Safety*, 36(3), 101–108.

Lundin, T. (1984). Morbidity following sudden and unexpected bereavement. *British Journal of Psychiatry*, 144, 84–88.

Mazor, K.M., Reed, G.W., Yood, R.A., Fischer, M.A., Baril, J., and Gurwitz, J.H. (2006). Disclosure of medical errors: What factors influence how patients respond? *Journal of General Internal Medicine*, 21(7), 704–710.

McCord, R.S., Floyd, M.R., Lang, F., and Young, V.K. (2002). Responding effectively to patient anger directed at the physician. *Family Medicine*, 34(5), 331–336.

McDonald, T.B., Helmchen, L.A., Smith, K.M., Centomani, N., Gunderson, A., Mayer, D., and Chamberlin, W. (March 1, 2010). Responding to patient safety incidents: The "seven pillars." *Quality and Safety in Health Care*, doi: 10.1136/qshc.2008.031633

McDonnell, W.M., and Guenther, E. (2008). Narrative review: Do state laws make it easier to say "I'm sorry?" *Annals of Internal Medicine*, 149, 811–815.

Mello, M.M. (2006). Medical malpractice: Impact of the crisis and effect of state tort reforms. In: *The research synthesis report 10*. Princeton, NJ: The Robert Wood Johnson Foundation.

Mello, M.M. and Gallagher, T.H. (2010). Malpractice reforms-opportunities for leadership by health care institutions and liability insurers. *Obstetrics and Gynecology*, 116(5), 1254–1256.

MITSS (2009). Adapted from: *Allison's story: A communication breakdown results in a surgery not being done*. http://www.mitss.org/patients_families_stories.html (last accessed June 24, 2010).

Morris, J.A., Carrillo, Y., Jenkins, J.M., Smith, P.W., Bledsoe, S., Pichert, J., and White, A. (2003). Surgical adverse events, risk management and malpractice outcome: Morbidity and mortality review is not enough. *Annals of Surgery*, 237(6), 844–851.

National Patient Safety Agency (2009). *Being open: Communicating patient safety incidents with patients, their families and carers*. http://www.nrls.npsa.nhs.UK/resources/?entryid45=65077. (last accessed June 29, 2011).

National Patient Safety Foundation (2003). *Statement of principle*. http://www.npsf.org/download/Statement_of_Principle.pdf (last accessed October 3, 2003).

National Quality Forum (2009). *Safe practices for better healthcare—2009 update*. http://www.qualityforum.org/Search.aspx?keyword=safe+practices+and+disclosure (last accessed June 29, 2011).

O'Connell, D., and Reifsteck, S.W. (2004). Disclosing unexpected outcomes and medical error. *The Journal of Medical Practice Management*, 19, 317–323.

Ong, L.M., De Haes, J.C., Hoos A.M., and Lammes F.B. (1995). Doctor–patient communication: A review of the literature. *Social Sciences in Medicine*, 40(7), 903–918.

Panesar, S.S., Cleary, K., and Sheikh, A. (2009). Reflections on the National Patient Safety Agency's database of medical errors. *Journal of the Royal Society of Medicine*, 102(7), 256–258.

Pichert, J.W., Hickson, G.B., Bledsoe, S., Trotter, T., and Quinn, D. (1997). Understanding the etiology of serious medical events involving children: Implications for pediatricians and their risk managers. *Pediatric Annals*, 26(3), 160–172.

Pichert, J.W., Hickson, G.B., and Trotter, T.S. (1998). Malpractice and communication skills for difficult situations: Ambulatory child health. *The Journal of General and Community Pediatrics*, 4(2), 213–221.

Reason, J. (1990). *Human error*. New York: Cambridge University Press.

Reason, J. (2001). Understanding adverse events: The human factor. In C. Vincent (Ed.), *Clinical risk management: Enhancing patient safety* (pp. 9–30). London, U.K.: BMJ Publications.

Rodriguez, M.A., Storm, C.D., and Burris, H.A. III (2009). Medical errors: Physician and institutional responsibilities. *Journal of Oncology Practice*, 5(1), 24–26.

Schwappach, D.L., and Boluarte, T.A. (2009). The emotional impact of medical error involvement on physicians: A call for leadership and organizational accountability. *Swiss Medical Weekly*, 139(1–2), 9–15.

Schwappach, D.L.B., and Koeck, C.M. (2004). What makes an error unacceptable? A factorial survey on the disclosure of medical errors. *International Journal for Quality in Health Care, 16*(4), 317–326.

Schwarzer, R., Knoll, N., and Rieckmann, N. (2004). Social support. In: A. Kaptein and J. Weinman, eds. *Health psychology.* Oxford, U.K.: British Psychological Society, Blackwell Publishing.

Scott, S.D., Hirschinger, L.E., and Cox, K.R. (2008). Sharing the load of a nurse "second victim": Rescuing the healer after trauma. *Modern Medicine, 71*(12), 38–43.

Scott, S.D., Hirschinger, L.E., Cox, K.R., McCoig, M., Brandt, J., and Hall, L.W. (2009). The nature history of recovery for the healthcare provider after adverse patient events. *Quality and Safety in Health Care, 18*(5), 325–330.

Sloan, F.A., Mergenhagen, P.M., Burfield, W.B., Bovbjerg, R.R., and Hassan, M. (1989). Medical malpractice experience of physicians. Predictable or haphazard? *The Journal of the American Medical Association, 262*(23), 3291–3297.

Smetzer, J., Baker, C., Byrne, F.D., and Cohen, M.R. (2010). Shaping systems for better behavioral choices: Lessons learned from a fatal medication error. *Joint Commission Journal on Quality and Patient Safety, 36*(4), 152–163.

Spandorfer, J., Pohl, C.A., Rattner, S.L., and Nasca, T.J. (2010). *Professionalism in medicine: A case-based guide for medical students.* Cambridge, NY: Cambridge University Press.

Stelfox, H.T., Gandhi, T.K., Orav, E.J., and Gustafson, M.L. (2005). The relation of patient satisfaction with complaints against physicians and malpractice lawsuits. *American Journal of Medicine, 118*(10), 1126–1133.

Straumanis, J.P. (2007). Disclosure of medical error: Is it worth the risk? *Pediatric Critical Care Medicine, 8*(2 Suppl), S38–S43.

Studdert, D.M., Mello, M.M., Gawande, A.A., Brennan, T.A., and Wang, Y.C. (2007). Disclosure of medical injury to patients: An improbable risk management strategy. *Health Affairs, 26*(1), 215–226.

Surbone, A., Rowe, M., and Gallagher, T.H. (2007). Confronting medical errors in oncology and disclosing them to cancer patients. *Journal of Clinical Oncology, 25*(12), 1463–1467.

Taheri, P.A., Butz, D.A., Anderson, S., Boothman, R., Blanco, O.L., Greenfield, L.J., and Mulholland, M.M. (2006). Medical liability—The crisis, the reality, and the data: The University of Michigan story. *Journal of the American College of Surgeons, 203*(3), 290–296.

Van Vorst, R.F., Araya-Guerra, R., Felzien, M., Fernald, D., Elder, N., Duclos, C., and Westfall, J.M. (2007). Rural community members' perceptions of harm from medical mistakes: A high plans research network (HPRN) study. *Journal of the American Board of Family Medicine, 20*(2), 135–143.

Vincent, C.A. (2001). Caring for patients harmed by treatment. In: C.A. Vincent, ed. *Clinical risk management. Enhancing patient safety* (pp. 461–479). London, U.K.: BMJ Publications.

Vincent, C.A., and Coulter, A. (2002). Patient safety. What about the patient? *Quality and Safety in Health Care, 11,* 76–80.

Vincent, C.A., Pincus, T., and Scurr, J.H. (1993). Patients' experience of surgical accidents. *Quality in Health Care, 2*(2), 77–82.

Vincent, C., Young, M., and Phillips, A. (1994). Why do people sue doctors? A study of patients and relatives taking legal action. *Lancet, 343*(8913), 1609–1613.

Waterman, A.D., Garbutt, J., Hazel, E., Dunagan, W.C., Levinson, W., Fraser, V.J., and Gallagher, T.H. (2007). The emotional impact of medical errors on practicing physicians in the United States and Canada. *Joint Commission Journal on Quality and Patient Safety, 33*(8), 467–476.

White, A.A., Pichert, J.W., Bledsoe, S.H., Irwin, C., and Entman, S.S. (2005). Cause-and-effect analysis of closed claims in obstetrics and gynecology. *Obstetrics & Gynecology, 105*(5 Pt 1), 1031–1038.

White, A.A., Waterman, A.D., McCotter, P., Boyle, D.J., and Gallagher, T.H. (2008). Supporting health care workers after medical error: Considerations for health care leaders. *Journal of Clinical Outcomes Management, 15*(5), 240–247.

White, A.A., Wright, S., Blanco, R., Lemonds, B., Sisco, J., Bledsoe, S., Irwin, C., Isenhour, J., and Pichert, J.W. (2004). Cause-and-effect analysis of risk management files to assess patient care in the emergency department. *Academic Emergency Medicine, 11*(10), 1035–1041.

Williams, S., Weinman, J., and Dale, J. (1998). Doctor–patient communication and patient satisfaction: A review. *Family Practice, 15,* 480–492.

Wu, A.W. (1999). Handling hospital errors: Is disclosure the best defense? *Annals of Internal Medicine, 131*(12), 970–972.

Wu, A. (2000). Medical error: The second victim. *British Medical Journal, 320,* 726–727.

Wu, A.W., Folkman, S., McPhee, S.J., and Lo, B. (1991). Do house officers learn from their mistakes? *The Journal of the American Medical Association, 265*(16), 2089–2094.

26

Human Factors Engineering of Health Care Reporting Systems

Christopher W. Johnson

CONTENTS

26.1 Introduction to the Human Factors of Adverse Events..424
    26.1.1 Estimating the Costs of Adverse Health Care Events 424
26.2 Medical Device Reporting Systems to Identify Human Factors Problems............................424
26.3 Human Factors Problems with Patient Safety Reporting Systems..426
26.4 Human Factors of Incident Reporting..427
26.5 Underreporting..428
26.6 Elicitation and Form Design...434
26.7 Form Content and Delivery Mechanisms...437
    26.7.1 Delivery Mechanisms 437
    26.7.2 Preamble and Definitions of an Incident...438
    26.7.3 Identification Information 438
    26.7.4 Time and Place Information..439
    26.7.5 Detection Factors and Key Events 440
    26.7.6 Consequences and Mitigating Factors...442
    26.7.7 Causes and Prevention..443
26.8 Explanations of Feedback and Analysis...445
26.9 Conclusions..446
References...446

It is notoriously difficult to elicit information about adverse events in health care. This is partly due to the continuing pressures on health care professionals. In air traffic management, if a controller is involved in a near-miss incident he/she will be immediately removed from their position. This is intended both to reduce the stress on the individual and also to mitigate the risk from the controller becoming involved in further incidents as they come to terms with a previous adverse event. The watch supervisor will then, typically, guide them through the reporting process while their colleagues ensure the continued safety of the airspace (Johnson et al., 2009). In contrast, clinicians are typically expected to continue working. The priority is to help mitigate any consequences for the patient. This often involves urgent interaction with colleagues and coworkers. Relatives and friends of the individuals involved must also be informed. These tasks must be integrated with the continuing demands from other patients within their care.

Typically, it is not until the end of the shift that clinicians have any opportunity to access online- or paper-based reporting systems. It is difficult to underestimate the fatigue and stress that can be compounded by self-blame and the fear of both professional and legal action. These factors exacerbate any underlying usability problems with the reporting forms. It can be difficult to complete many dozens of fields that are required in some reporting systems, especially when any individual is only responsible for a portion of the care that the patient receives. In other cases, it can be difficult for clinicians to determine what they should write in the "free-text" sections of incident reports. In consequence, many accounts of health care incidents provide partial or contradictory information. This chapter addresses the human factors engineering of health care reporting systems; the aim is to help clinicians provide the information that is needed if we are to avoid the recurrence of adverse events.

26.1 Introduction to the Human Factors of Adverse Events

Previous chapters have shown how human factors problems trigger accidents and incidents in health care. For example, a physician might make a slip if they write down 10 mg of an appropriate medication when the intention was to prescribe 1 mg. Alternatively, they might make a mistake by giving a medication that was not intended as part of the patient's treatment. They could also lapse by forgetting to deliver an intended drug. Finally, clinicians can commit violations by deliberately ignoring recommended practice. All of these different forms of human "error" have been noted and studied in a range of hospital and primary care settings (Johnson, 2003a; Snijders et al., 2009; Waring et al., 2010).

At the same time, there is a growing body of research into "resilience engineering" in health care (Vincent et al., 2010). This has identified many different ways in which clinicians intervene to avoid or mitigate many potential incidents. We rely upon individual and team decision making to guide most aspects of diagnosis and treatment. We rely on the skill and judgment of clinicians to decide when and when not to intervene. We depend upon their vigilance to determine when mistakes have been made or, ideally, to intervene before colleagues make a mistake.

We also depend upon the commitment and professionalism of clinicians and support staff in ensuring that we have accurate information about adverse events when they do occur. It is for this reason that the U.K. National Patient Safety Agency (NPSA) has recently developed standards for the reporting of adverse events through the National Reporting and Learning Service (RLS):

> Improving the quality of data submitted to the RLS is critical to creating opportunities for national and local learning. More accurate data, particularly in reporting the degree of harm to the patient, means that (we) can be more effective in identifying critical risks and driving learning (NPSA, 2009a, 2009b).

This creates a significant challenge, especially when many health care professionals are reluctant to report adverse events to administrators.

26.1.1 Estimating the Costs of Adverse Health Care Events

It is difficult to underestimate the significance of human error in health care. We are surrounded by newspaper items and television broadcasts that reinforce concern over a succession of incidents and accidents. The products of research in this area inform much of this media interest. For example, a series of studies have argued that almost 100,000 patients die from preventable causes in U.S. hospitals. This annual toll exceeds the combined number of deaths and injuries from motor and air crashes, suicides, falls, poisonings, and drowning (Barach and Small, 2000). It has been estimated that there are 850,000 adverse incidents every year in the U.K. National Health Service. Such statistics are, however, very difficult to validate. National figures rely on interpolation from relatively small samples. The biases within these samples further confound interpretation. For instance, some hospitals routinely report a high number of minor events, such as the misapplication of a bandage, while others reported virtually nothing. The underreporting of adverse events to national monitoring organizations is estimated to range from 50% to 96% annually (Institute of Medicine, 1999; Hirose et al., 2007).

26.2 Medical Device Reporting Systems to Identify Human Factors Problems

A number of initiatives have established "lessons learned" and incident reporting applications across both local and national health care systems. These can be used to ensure that information about previous failures and near misses informs future practice. There are further advantages if these schemes capture near-miss information as well as reports of adverse occurrences. These near misses can be used to find out why accidents DON'T occur; this argument builds on the concepts of resilience engineering that were introduced in previous sections. Not only do we seek to avoid those situations that lead to human error or system failure in health care, but we should also promote those interventions that help to mitigate the adverse consequences of potential incidents.

Incident reports also provide means of monitoring potential problems as they recur during the lifetime of an application. They can be used to elicit feedback that keeps staff "in the loop." The data (and lessons) from incident reporting schemes can be shared. Incident reporting systems provide the raw data for comparisons both within and between industries. If common causes of incidents can be observed, then, it is argued that common solutions can be found. Incident reporting schemes are cheaper than the costs of an accident. A further argument in favor of incident reporting schemes is that organizations may be required to exploit them by regulatory agencies.

There are many different types of reporting systems in health care. One class of applications has been developed for reporting problems with medical devices. For instance, the U.S. Center for Devices and Radiological Health operates a range of schemes that feed into the Manufacturer and User Facility Device Experience Database (MAUDE). For example, the following report describes how the drug calculator of a medication assistant in a patient monitoring application would occasionally round up values to a second decimal place. The users complained that this could easily result in a medication error and that the manufacturer was failing to acknowledge the problem. The manufacturer initially responded that vigilant nursing staff ought to notice any potential problems when calculating the medication. The clinicians countered this by arguing that they had explicitly taught nursing staff to trust the calculation function as a means of *reducing human error*. Subsequent reports from the device manufacturer stressed that clinicians can configure the resolution of medication measurements through a unit manager menu:

> THIS IS BEST METHOD FOR CLINICAL STAFF, IT PRE-CONFIGURES DRUG CALCULATIONS AND ALLOWS SETTINGS TO REFLECT HOW DRUGS ARE PREPARED BY THE PHARMACY. CUSTOMER WAS TOLD, DRUG CONCENTRATION ROUNDING TO NEAREST HUNDREDTHS, COULD BE EASILY ADDRESSED IN UNIT MANAGER SETUP, TO REFLECT HIGHER RESOLUTION. THEREBY, ADDRESSING ANY CONCERN OF A ROUNDING ISSUE. manufacturer HAS REVIEWED CUSTOMER'S CONCERN AND HAVE DETERMINED THAT "DRUG CALCULATIONS" FEATURE IS FUNCTIONING AS DESIGN. ADDITIONALLY, manufacturer HAS REVIEWED WITH CUSTOMER, THE USER'S ABILITY TO CHANGE UNITS OF MEASURE, TO ACHIEVE DESIRED RESOLUTION. THE DEVICE IS PERFORMING AS DESIGNED (**MDR TEXT KEY: 1601404**)

The U.K. Medicines and Healthcare Products Regulatory Agency (MHRA) also, provides several mechanisms for reporting adverse health care events including the Manufacturers' On-line Reporting Environment (MORE). These applications help to implement a series of different national and international requirements. In Europe, the Medical Devices Directive (Directive 2007/47/EC) requires that "2.3.5. All serious adverse events must be fully recorded and immediately notified to all competent authorities of the Member States in which the clinical investigation is being performed." In addition, manufacturers must "notify the competent authorities of the following incidents immediately on learning of them and the relevant corrective actions: (i) any malfunction or deterioration in the characteristics and/or performance of a device, as well as any inadequacy in the labeling or the instructions for use that might lead to or might have led to the death of a patient or user or to a serious deterioration in his state of health; (ii) any technical or medical reason connected with the characteristics or performance of a device for the reasons referred to in subparagraph (i) leading to systematic recall of devices of the same type by the manufacturer." These provisions include adverse events that stem from usability or human factors issues during the operation of the device. Each member state within the European Union enacts national legislation to ensure that they conform to the requirements in these directives. For instance, the U.K. regulatory framework is based around the Statutory Instruments 2002 No. 618 (Consolidated legislation), 2003 No. 1697 (amendments to cover the reclassification of breast implants and additional requirements covering devices utilizing materials from TSE-susceptible animal species), Medical Devices Regulations 2007 No. 400 (amendment to cover the reclassification of total hip, knee, and shoulder joints), and the Medical Devices (Amendment) Regulations 2008 No 2936, which transpose Directive 2007/47/EC into U.K. law. These were passed by Parliament in December 2008 and fully came into force in March 2010.

The net effect of all of this is to ensure that incident reports are one of several events that will trigger regulatory intervention and inspection by the MHRA. In addition, they will intervene to inspect a sample of manufacturers who market their devices in the United Kingdom, whether those companies have had any adverse events.

The U.S. Safe Medical Devices Act of 1990 (SMDA) guides the reporting of adverse events involving health care technology. Under the provisions of this act, end users must report device-related deaths to the FDA and the manufacturer. Serious injuries must also be reported to the manufacturer or to the FDA if they do not know how to contact the manufacturer. The FDA established a number of schemes to meet the requirements of the SMDA. These were confirmed under the Medical Devices Amendments of 1992 (Public Law 102-300; the Amendments of 1992) to Section 519 of the Food, Drug, and Cosmetic Act relating to the reporting of adverse events. This established a single reporting standard for device user facilities, manufacturers, importers, and distributors The Medical Devices Reporting Regulation implements the reporting requirements contained in the SMDA of 1990 and the Medical Device Amendments of 1992. More recently, the 1998 Food and Drug Administration Modernization Act (FDAMA) reduced some of the regulatory burden on manufacturers by removing an obligation to provide annual reports on adverse events. End users could file an annual report instead of semiannual reports to summarize adverse event reports.

The Canadian reporting system is governed by the Medical Devices Regulations. Australian practice is guided by the Therapeutic Goods Act. Japanese regulations are informed by the Ministry of Health and Welfare. The key point here is to recognize the diversity of different national reporting systems. This can create vulnerabilities if information about adverse events in one country cannot easily be used to inform practice in another. For instance, U.S. reporting requirements covers a broad range of usability issues including poor labeling and instruction as well as design flaws. In contrast, European regulations are perceived not to address usability issues except where they stem from manufacturing problems or inadequate labeling. It is also important to acknowledge the wider limitations of these device-related reporting systems. The focus on particular items of equipment implies that many human-related incidents will fall beyond the scope of these national and international schemes. The 1999 U.S. Institute of Medicine report "To Err is Human" identified a broad range of adverse health care events including "transfusion errors and adverse drug events; wrong-site surgery and surgical injuries; preventable suicides; restraint-related injuries or death; hospital-acquired or other treatment-related infections; and falls, burns, pressure ulcers, and mistaken identity." Many of these preventable incidents fall outside of the scope of the existing device-related reporting systems. In consequence, it was argued that a new national reporting framework should be established to ensure that as much information as possible is gathered about adverse events in health care. The proposal was to establish a wide-ranging mandatory system for more serious occurrences and a voluntary scheme to elicit information about less serious incidents and near misses. This multitiered approach was intended to ensure that lessons were learned both from those adverse events that did occur but also more proactively to learn from those that were narrowly avoided in the past but which might occur in the future.

Shortly after the Institute of Medicine Report, the U.K. NHS (2000) Expert group on learning from adverse events in health care issued a document entitled "Organization with a Memory." This argued that reporting systems are "vital in providing a core of sound, representative information on which to base analysis and recommendations." It was critical of current reporting practice in the national health care system and made four key recommendations. First, a "unified" mechanism should be developed for reporting and analysis when things go wrong. Second, a more open culture should be established to ensure that errors or service failures can be reported and discussed. Third, techniques should be developed for ensuring that necessary changes are put into practice. Finally, there should be a wider appreciation of the value of the system approach in preventing, analyzing, and learning from errors. These requirements not only encourage greater reporting of adverse events involving human factors issues. Requirements to improve the reporting "culture" also crucially depend upon an appreciation of human factors issues in order to encourage reporting in the first place.

26.3 Human Factors Problems with Patient Safety Reporting Systems

The publication of "To Err is Human" and "Organization with a Memory" served to increase the prominence of voluntary reporting systems that were already in existence at a local or regional level in several different countries. For instance, the New York state NYPORTS program was established in 1985. These early state-based schemes tended to focus on more severe accidents that resulted in patient injuries or on facility issues, including structural problems and fire hazards. They elicited reports from large secondary health care providers, such as regional hospitals and nursing homes. In Connecticut, 14,000 of the 15,000 reports received in 1996 came from these homes. Success was very mixed (Institute of Medicine, 1999). For example, the Colorado program initially received less than eight reports per year. However, with a concerted campaign to increase awareness over the benefits of reporting this increased over a 10 year period to more than 1000 reports per annum.

These pioneering U.S. systems can be contrasted with a number of more local reporting initiatives that grew up across the United Kingdom. The Edinburgh incident reporting scheme was set up in an adult intensive care unit (ICU) in 1989 (Busse and Wright, 2000). It served a ward with eight beds, three medical staff, one consultant, and up to eight nurses per shift. A study of the incidents reported over the first 10 years of this scheme found that most fell into four task domains: relating to ventilation, vascular lines, drug administration, and a miscellaneous group. The scheme encouraged staff to describe adverse events in narrative form, as well as noting contributing factors, detection factors, and the grade of staff involved in the event. Approximately, one-third of the reporters had been involved in the incident that was being reported. Fewer than 10% of the reports were made by medical as opposed to nursing staff. Waring (2004) analyzed the success of local reporting systems and found that medical doctors were more inclined to report incidents where the process of reporting was localized and integrated within medical rather than managerial systems of quality improvement. Underlying these variations, it is suggested that medical reporting is more likely when physicians have greater control or ownership of incident reporting, as this fosters confidence in the purpose of reporting, in particular its capacity to make meaningful service improvements whilst maintaining a sense of collegiality and professionalism.

However, local systems make it difficult to aggregate data, so that we can determine whether specific incidents form part of a wider pattern. It was for this reason that the Australian Patient Safety Foundation's system was established in 1989. The work of Runciman (2002) and his colleagues at the Australian Patient Safety Foundation had a profound impact on many health care professionals because it helped to establish a framework for what was arguably the first national, voluntary reporting system with a specific remit to elicit information about human factors in adverse health care events. The Federal Agency for Health Care Research and Quality (AHRQ) and the National Patient Safety Foundation (NPSF) have helped to coordinate similar initiatives in the United States. The NPSA fulfills this role in the United Kingdom. These organizations promote a range of initiatives that are intended to reduce "human error" in health care. They promote voluntary incident reporting systems as a means of detecting and then addressing common features in adverse events. The U.K. NPSA has developed the National Reporting and Learning System (NRLS). Since 2003, this has complemented local reporting arrangements. The intention is that health care staff will be able to submit anonymous patient safety reports. These will then be analyzed to identify national patterns, to identify patient safety priorities, and to promote consistent solutions.

The NPSA's NRLS was initially intended to help the NHS meet a series of targets. By 2005, the aim was to

> reduce by 25% the number of instances of negligent harm in the field of obstetrics and gynecology which result in litigation (currently these account for over 50% of the annual NHS litigation bill); by 2005, reduce by 40% the number of serious errors in the use of prescribed drugs (currently these account for 20% of all clinical negligence litigation); by 2005, reduce to zero the number of suicides by mental health inpatients as a result of hanging from non-collapsible bed or shower curtain rails on wards (currently hanging from these structures is the commonest method of suicide on mental health inpatient wards) (NHS, 2000).

Many of these targets remain to be achieved. Such high objectives must be balanced against a number of prosaic problems that limit the effectiveness of incident reporting systems. For instance, there is a danger that they will act as repositories of information without inspiring direct intervention to correct particular problems. This can lead to incident starvation if potential contributors feel that their reports are being ignored. Further problems stem from the observation that reporting systems often elicit information about known issues, such as maladministered spinal injections or communication problems between particular hospital departments. The collection of such information does little to suggest possible interventions that might be used to address these long-term and deep-seated issues.

Many of the problems that undermine incident reporting can be directly related to human factors issues. These include the problems of underreporting. The Royal College of Anesthetist's pilot reporting system found that self-reporting retrieves only about 30% of incidents that can be detected by independent audit. Other issues relate more narrowly to the biases that can affect the analysis of incident reports once they have been achieved. Finally, the human factors of incident reporting can also complicate the monitoring that must be used to determine whether local and national systems are having any measurable impact on patient safety. If clinician's do not report adverse events then it can be difficult to obtain reliable data to show that particular interventions are having any effect in reducing the frequency or mitigating the consequences of those events (Kingston et al., 2004).

26.4 Human Factors of Incident Reporting

Some of the problems that limit the effectiveness of clinical incident reporting systems are largely technical, for instance, automated support may be necessary to identify common patterns across the thousands of documents that can be submitted to national schemes (Johnson, 2003a, 2009). However, most barriers to the successful application of incident reporting stem from human factors issues. This creates a recursive problem in which we must first address a series of human factors issues in order to elicit reports about the underlying causes of, for example, human error in medicine. The following sections focus on several aspects of this problem. These include the difficulty of eliciting reports in the first place. This issue can be divided into two subproblems, first, how to persuade potential contributors of the benefits of their involvement and second, how to ensure that they then provide all necessary information

(Kingston et al., 2004). We also briefly examine the problems of causal analysis; it can be difficult to avoid blaming individuals so that systemic failures can be examined. Equally, there are some incidents in which personal responsibility should not be ignored if external bodies are to believe in the probity of the system. Finally, we consider the human factors issues that arise in the development and implementation of recommendations that are intended to ensure previous events do not recur.

26.5 Underreporting

A number of attempts have been made to estimate the scale of underreporting in health care systems. For instance, Barach and Small (2000) state that the "underreporting of adverse events is estimated to range from 50% to 96% annually." The U.K. Royal College of Anaesthetist's concluded that only about 30% of the total number of incidents detected by independent audit will be contributed by voluntary reporting systems. These caveats also affect device-related reporting systems. A more recent study of pediatric reporting used 140 survey responses to show that 34.8% of respondents indicated that they had reported <20% of their perceived medical errors in the previous 12 months and 32.6% had reported <40% of perceived errors committed by colleagues (Taylor et al., 2004). The U.S. General Accounting Office (1997) conducted a study into submission frequencies 2 years after the requirement was introduced for manufacturers and importers to report all device-related deaths, serious injuries, and certain malfunctions to the FDA. They concluded that less than 1% of device problems occurring in hospitals were reported to the FDA. The more serious the problem, the less likely it was to be reported. A GAO follow-up study concluded that the subsequent implementation of the Medical Device Reporting (MDR) regulation, introduced in previous sections, had not corrected the problems of underreporting (FDA, 2002).

Mackenzie et al. (1996) compared "deficiencies" in the management of patient airways using self-reporting and through exhaustive video analysis. The self-reporting fell into three different categories: anesthesia records constructed during the treatment, retrospective anesthesia quality assurance reports, and a posttrauma treatment questionnaire that was filled in immediately after each case. Video analysis of 48 patient "encounters" identified 28 deficiencies in 11 cases. These included the omission of necessary tasks and practices that "lessened the margin of patient safety." In comparison, anesthesia quality assurance reports identified

none of these incidents. Anesthesia records identified two and the posttrauma treatment questionnaire suggested contributory factors and corrective measures for five deficiencies. Similarly, Jha et al.'s (1998) work on adverse drug events compared the efficacy of three different detection techniques: voluntary incident reporting, the computer-based analysis of patient records, and exhaustive manual comparisons of the same data. In one study, they focused on patients admitted to nine medical and surgical units in an 8 month period. Both the automated system and the chart review strategies were independent and blind. The computer monitoring strategy identified 2620 incidents. Only 275 were determined to be adverse drug events. The manual review found 398 adverse drug events. Voluntary reporting only detected 23.

The wide variation in reporting rates within voluntary systems stems, in part, from the obvious methodological problems that arise when assessing the total number of adverse events that might have occurred but were not reported. The work of Mackenzie et al. (1996) shows that the retrospective use of patient records will yield different observations of the baseline error rate than the use of more detailed contemporary video analysis. Similarly, the study by Jha et al. (1998) shows that further differences in the baseline rate can be obtained if manual inspections are supported by computer-based search techniques within medical records. It is important not to underestimate the practical consequences of inaccuracies in these baseline rates. For instance, a number of agencies have sought to establish reporting quotas based on the estimates of underlying error rates. The ability to meet this quota is then interpreted as a measure of the quality of the reporting system. This then provides an indirect measure of the safety culture in the host organization. In 1998, there was considerable controversy when the U.S. Health Care Financing Administration attempted to place a cap of 2% on the medication error rate in the Medicare Conditions of Participation. This implied that it was "acceptable" if there were errors in up to 2% of medications (Shaw Phillips, 2001). The subsequent controversy also pointed to the difficulty of establishing this 2% figure as a benchmark for adverse medication events. Many different factors could introduce local variations on the underlying error rate. These include differences in the size of health care institutions, their funding profile and equipment provision, the nature and extent of the demands on their services, the profile, and characteristics of the population they serve etc. More recent studies have avoided the use of quotas but instead have observed that incident reporting rates from acute hospitals increase with time from connection to national reporting systems, such as the NPSA's NRLS, and are positively correlated with independently defined measures of safety culture, "higher reporting rates being associated with a more positive safety culture" (Hutchinson et al., 2009).

Having raised caveats about the difficulty of assessing baseline figures for adverse events, it is still possible to assess changes in the contribution rate over time. However, this is more complex than it might at first appear. For instance, the introduction of a reporting systems can encourage a "confessional" phase in which the rate of submissions is temporarily increased by the publicity and availability of a new scheme. It can also be difficult to interpret the cause of any longer-term changes in submission rates. For instance, an increase in the number of contributions might reflect a rise in the error rate and this, in turn, can be the result of changes in the activities of the reporting organization (Johnson, 2003a). Alternatively, any increase may be due to an increase in the willingness to report adverse events. This ambiguity can have unfortunate implications if risk managers are forced to explain why the reported number of adverse events appears to increase over time. Conversely, any fall in the number of submissions might be either due to specific safety improvements or due to a lack of interest in the benefits of contributing to a reporting system. Several other factors can influence submission rates. Most notably, participation can depend upon confidence in the individuals who run the system (Waring, 2004).

Having recognized the difficulty of accurately assessing the scale of underreporting, a number of authors have sought to identify the reasons why health care professionals fail to contribute to incident reporting systems. For instance, Lawton and Parker (2002) issue a series of questionnaires to 315 doctors, nurses, and midwives who volunteered to take part in the study from three English NHS trusts. These questionnaires included nine short scenarios describing either a violation of a protocol, compliance with a protocol, or improvisation, where no protocol exists. Different versions of the questionnaire were presented to different volunteers so that each scenario was presented with a good, poor, or bad outcome for the patient. Participants were asked to indicate how likely they were to report the incident described in the scenario to a senior member of staff. The study showed that doctors were particularly reluctant to report adverse events to a superior. The participants were more likely to report incidents with an adverse outcome than those that might be described as "near misses." They were also more likely to report to a senior member of staff, irrespective of outcome, if the incident involved the violation of a protocol rather than incidents in which a protocol was followed or the clinicians improvised in the absence of such guidelines. The results of this study are, however, difficult to apply across many reporting systems because the questionnaires and the associated scenarios were drafted to identify the likelihood of report to a "senior colleague"

rather than through a confidential or anonymous reporting system.

Van Geest and Cummins (2003) build on this work when they assess the reasons why physicians fail to report or even detect adverse events. Their work formed part of an NPSF project that was established in 2001 to better understand the physicians' and nurses' experience of health care errors. The intention was then to identify the needs of each group in order to help them "combat" these adverse events. The needs assessment was conducted in two phases. First, the NPSF convened a series of focus groups to determine the origins of, and ways to reduce, health care error. These groups considered the cultural and systemic barriers to identifying, reporting, and analyzing errors in health care. The second phase of this requirement elicitation was conducted through a self-administered mail survey of physicians and nurses. The physician survey utilized a random sample of 1084 physicians from the American Medical Association's database of all physicians practicing in the United States. The nurse survey used a sample of 1148 nurses from the American Nursing Association.

The focus group discussions with the physicians helped to reveal a common concern over the growing complexity of many health care systems. This complexity increases the likelihood of adverse events because clinicians may not have a full understanding of the technology that they are expected to operate. Similarly, increasing complexity also stems from the close interaction between varied groups of coworkers. Communication and coordination issues also increase the likelihood of misunderstandings and other forms of adverse events. These problems result in "inefficient therapeutic approaches, lack of follow-up on ordered tests, and failure to monitor medications." The physician's focus groups went on to argue that this complexity can prevent clinicians from identifying and, therefore, reporting adverse health care events. This observation has recently been confirmed by an independent study of telemedical incidents (Johnson, 2003b).

Participation can also be undermined by the feeling that reporting adverse events will not generate the funds or political support necessary to make sustained improvements. There is also a reluctance to question the professional competence of their colleagues, especially if there is a tendency to blame individuals rather than seek more appropriate safeguards. These insights illustrate some of the ambiguities that characterize attitudes toward incident reporting. The physicians in Van Geest and Cummins (2003) study felt that certain forms of error were tolerated, while others would elicit a punitive response. In the questionnaire survey, 69% of respondents had identified errors in patient care. However, only 50% reported "working with" nonpunitive systems for error reporting and examination. The physicians stated they knew the proper channels to report safety concerns (61%).

The focus groups involving nurses identified safety more as a "systems issue" rather than an issue that might be associated with particular individual erroneous actions. This was again explained in terms of the growing integration and complexity of many health care applications. It also reflected the nurses' perception that their individual work was embedded within that of their team of coworkers. However, only 49% of survey respondents agreed that safety was best addressed at the patient level. They argued that safety was better focused at the level of adverse effects on individual patients. The authors of the NPSF study argued that the nurses in the focus group felt communications failures were one of the most important barriers to the reporting of medical errors. The adverse reporting "culture" was also identified. The nurses explained this tolerance of error in terms of a historical focus on efficiency in health care provision rather than on safety. Nurses also identified a "code of silence" that permeates much of the health care system. They felt this to be particularly problematic for nurses who often are the first to identify the consequences of an adverse event but are not "empowered" within the medical hierarchy. The focus groups described the sense of isolation that nurses can feel when they either commit or observe an adverse event. Both of these events can alienate them from their coworkers. More than 80% of the survey respondents stated that they had identified error in health care. Thirty-five percent indicated that they had worked with nonpunitive systems for error reporting or examination. Eighty-seven percent of respondent nurses indicated that they knew the proper channels to report safety concerns. Ninety-three percent reported discussing patient safety concerns with colleagues and/or supervisors. Only 30% stated that they had read one of the Institute of Medicine's reports on patient safety. Seventy-two percent of nurses were actively engaged in practices to identify medication errors.

Similar evidence for the causes of underreporting can be obtained from Cohen et al.'s (2003) survey of 775 nurses across the United States. Although this study focuses on attitudes toward the reporting of medication errors, it reveals a number of more general attitudes and opinions. The survey seemed to provide a consensus in favor of the benefits of incident reporting. Fifty-eight percent of respondents agreed that error reporting is a valuable tool to measure a nurse's medication competency while 42% disagreed with this statement. Ninety-one percent concurred that "A good way to understand why errors occur is through a thorough analysis of information obtained from incident reports." While only 36% agreed with the statement that "during my nursing career, I failed to report one or more medication errors

because I thought reporting an error might be personally or professionally damaging" and 64% disagreed. These positive statements in support of incident reporting cannot easily help to explain the problems of underreporting. However, greater insights are provided by the 51% of respondents who observed that incident reports of my medication errors are placed in my personnel file. Individuals may be reluctant to submit reports about their colleagues if it is felt that those reports will adversely affect the career prospects of coworkers. Further insights are provided by the results for the question "I initiate an incident report when I catch":

Mistake done by:	Always	Sometimes	Never
Another nurse's mistake	37%	54%	9%
A pharmacist's mistake	45%	42%	14%
A physician's mistake	42%	39%	19%

As can be seen, nurses reveal a slightly greater ambivalence when reporting another nurse's mistake. Cohen et al. (2003) then went on to analyze these responses in terms of their respondents' experience and work setting. Nurses working in a hospital were less likely to report another nurses "error." Those in intensive care (23%) and orthopedic settings (29%) were least likely to report another nurse's mistake compared to other hospital settings. The proportion of nurses stating that they would always report varied from 32% to 53% in these areas. Nurses working in "home health care" were least likely to report a physician's mistake (32%). Sixty-seven percent of student nurses admitted being prepared to initiate a report for a nurse's mistake compared with 32% of licensed practical nurses and 50% of registered nurses. Nurses with less than 1 year (54%) or more than 15 years of experience (50%) are more likely to report a pharmacist's or physician's mistake than nurses with 1–15 years of experience. The proportion stating that they would report such incidents in this group varied from 21% to 45%. Finally, this survey also probed some of the ethical issues involved in terms of admitting adverse events to patients and their relatives. Only 18% agreed that they would always tell a patient or their relative if they had made a mistake. Fifty-two percent would sometime take this action and 31% admitted that they would never disclose these details.

Taylor et al.'s (2004) pediatric study listed reasons for underreporting including a lack of certainty about what is considered an error (indicated by 40.7% of respondents, from 140 questionnaires) and concern about implicating others (37%). There were also important differences between the types of scenarios that would be reported by the participants in this qualitative survey. Almost all respondents agreed that they would report a 10-fold overdose of morphine leading to respiratory depression in a child. However, only 31.7% would report an event in which a supply of breast milk is inadvertently connected to a venous catheter but is discovered before any breast milk went into the catheter. Respondents felt that the problems of underreporting might be addressed by increased education about which errors should be reported (stated by 65.4% of respondents), "feedback on a regular basis about the errors reported (63.8%) and about individual events (51.2%), evidence of system changes because of reports of errors (55.4%), and an electronic format for reports (44.9%)."

The NPSA (2009a, 2009b) have summarized the barriers to incident reporting from more than a decade of studies into the promotion of "lessons learned" applications across health care:

- *Sense of failure:* There is a professional culture that expects perfection. In consequence, some health care professionals find it difficult to acknowledge that things have not gone as well as expected.

- *Fear of blame:* Even when health care institutions have developed a strong "safety culture" that supports a "no blame" approach to incident reporting, the public and media have a radically different attitude toward the accountability of health care professionals.

- *Reports being used out of context:* There is a concern that safety-related information will be misused or misinterpreted, for instance, by the public or the media. It may also be cited in hearings about professional conduct or in subsequent litigation.

- *Benefits of reporting are unclear:* A lack of feedback about the interventions in response to reported incidents can lead to disillusionment. Changes may take too long for staff to see the benefits, especially if they change duties. Reporting something that has not affected the patient can be seen as pointless and time consuming.

- *Lack of resources:* Staff often lack the time or computational resources to provide incident information during a busy shift. Afterward, they may be too tired to be involved in other professional commitments to complete the necessary forms.

- *Not my job:* Incident reporting can be seen as the responsibility for other groups in the chain of care. Nurses may feel that they have little influence of particular procedures. Technicians may be unsure of the clinical outcome of the incidents that they observe and so on.

- *Difficulty in reporting:* Incident reporting can be seen as complicated and time consuming. It is often unclear what should and what should not be reported within particular systems.

- *Organizational barriers:* The NPSA has recognized the enormous diversity of organizations that contribute to health care. These include pharmacy or optometry companies and other independent contractors that function across numerous sites. These companies also have their own requirements for "corporate governance" that include incident reporting. In such cases, a single reporting format is ideal if the burden of reporting is to be minimized.

Many of these findings were supported by Robinson et al.'s (2002) study of physician and public opinion on the quality of health care and medical "error." They compared a mail survey of 1000 Colorado physicians (n = 594) and 1000 national physicians (n = 304) with a telephone survey of 500 Colorado households. The main aim was to assess their differing attitudes toward some of the main findings in the Institute of Medicine report "To Err is Human," mentioned in previous sections. They found that 70% of Colorado physicians believed the reduction of medical errors should be a national priority. Only 29% of physicians believed that "quality of care was a problem," compared to 68% of the wider population in this sample. Similarly 24% of physicians believed that a national agency is needed to address the problem of medical errors, while 60% of the public agreed with this statement. All of the physicians believed that the fear of medical malpractice was a barrier to reporting of errors and that greater legal safeguards are necessary for a reporting system to be successful. Sixty percent of the physicians agreed that it is difficult to differentiate errors due to negligence from unintended errors.

The NPSA's review of the problems that prevent the reporting of health care incidents led to an enumeration of the steps that can be taken to encourage participation, many of these relate to the human factors of reporting systems:

- Make it simple to report, and communicate this widely.
- Ensure timely and valuable feedback—one of the biggest challenges in the patient safety agenda but vital to ensuring continued reporting.
- Provide ongoing training sessions to explain the process and the benefits and demonstrate the importance of reporting. Use stories of change to show how changes can be made.
- Inform all new staff on orientation, including all grades of staff and professions.
- Disseminate safety information through newsletters, local intranet sites,

presentations, safety focus meetings, safety briefings, executive walkabouts/drive-abouts, etc.

- Disseminate success stories, good practice, and improvement tips.
- Ensure clinical and managerial leadership support.
- Provide a 'reporting pack' of background information, key contacts, roles and responsibilities, for example, feedback reports, patient safety definitions, etc.
- Undertake surveys and audits of reporting levels, evaluate the percentage of staff who report (typically between 70% and 80% of reporting is from nursing staff, with between 2% and 3% of reporting by doctors) and focus presentations and training to those that underreport. Find out what would make it easier for them to report and demonstrate what particular benefits they can gain (NPSA, 2009a, 2009b).

Mandatory reporting systems provide alternative means of encouraging participation. The Joint Commission on Accreditation of Health Care Organizations ran a voluntary scheme between 1995 and 2000. Only 798 adverse events were submitted. Two-thirds of these came from self-disclosure, however, one-third were notified as a result of media involvement (MDH, 2000). This level of participation can be contrasted with a mandatory system operating across New York state where 1200 mishaps were reported by hospitals in a single year with approximately 20,000 total submissions. Mandatory systems provide the opportunity to offer a "carrot and stick" approach; the incentive of "no blame" reporting can be combined with legal sanctions for the failure to participate. However, this raises important ethical questions, especially for near-miss or low-criticality events. It can be difficult to determine whether or not a clinician had the opportunity to observe an incident or even whether an incident was reportable in the first place. There may, therefore, be a tendency for clinicians to "over report"; high reporting frequencies can hide problems that arise when lessons learned systems are "cluttered" with many minor events.

Cohen's (2000) argues that the U.S. SMDA of 1990 led to mandatory reporting requirements that have been "unsuccessful in gaining compliance with reporting requirements for user error." As we have seen, the intention behind this federal bill was that health care facilities and manufacturers must report adverse events related to the failure or misuse of specific medical devices. However, Cohen argues that little action is taken unless the system receives reports about a large number of serious events. He also argues that the state-based mandatory systems are used "almost exclusively to punish individual practitioners or health care organizations." In consequence, mandatory systems often fail to provide insights into the deeper causes of adverse events, which Cohen argues are largely "systemic" rather than "individual."

In April 2000, the U.S. National Academy for State Health Policy conducted an investigation into the State Reporting of Medical Errors and Adverse Events. They found that 15 states (Colorado, Florida, Kansas, Massachusetts, Nebraska, New Jersey, New York, Ohio, Pennsylvania, Rhode Island, South Carolina, South Dakota, Tennessee, Texas, and Washington) required the mandatory reporting of adverse events from general and acute care hospitals. The levels of participation and the scope of these schemes were very different. The types of events to be reported included: unexpected deaths, wrong site surgery, major loss of function, and errors in medication The diverse practices identified for these mandatory systems motivated a not-for-profit group, known as the U.S. National Quality Forum, to propose a national strategy for health care quality measurement and reporting.

The proponents of mandatory systems argue that some adverse events are so serious that they must be reported in order to reassure the public and ensure that appropriate action is taken. The proponents of voluntary systems, in contrast, point to the problems of underreporting in mandatory systems and to the difficulty in "policing" reporting requirements. They also point to the success that some voluntary systems have had in encouraging participation when health care professionals are offered protection against legal sanction. For example, United States Pharmacopeia and Institute for Safe Medication Practices have established the Medication Errors Reporting Program. This confidential, voluntary medication error-reporting scheme has received around 1000 error reports each year. The quality of reports made to this voluntary system is just as significant as the number of submissions. These submissions have informed a number of significant interventions. After a series of accidents, the Institute for Safe Medication Practices persuaded manufacturers to include the maximum dose for cisplatin on phial caps and seals.

Few of the proponents on either side of this debate advocate exclusively mandatory or voluntary schemes as a solution to the problems of underreporting. In contrast, controversy surrounds the extent to which health care professionals should have the discretion to determine what is reportable under each of the various schemes. As mentioned, the Institute of Medicine advocates a national mandatory system for more serious mishaps and a local voluntary system feeding information up through regional schemes in the case of less serious adverse events. This architecture is intended to ensure that a national voluntary system is not inundated by a mass of low-risk incidents; local managers help to filter the passage of information up through state schemes to national systems whereas the more serious events merit a more immediate focus at a higher level. This mixed approach of mandatory and voluntary reporting will only successfully tackle the problems of underreporting if the schemes are supported by legal protection for individual participants. Any breach of confidentiality in general and the (ab)use of voluntary reports in any consequent litigation would undermine confidence in the scheme. Partly as a result of these concerns, a number of initiatives have attempted to reduce underreporting by ensuring that voluntary incident reports are subject to the same legal protection offered by similar schemes in other domains, in particular by NASA and the FAA's Aviation Safety Reporting System (ASRS).

The ASRS operates an elaborate mechanism whereby reports are initially passed to NASA. They then screen each submission to ensure that information relating to a criminal offense is passed to the Department of Justice and the FAA. Information about accidents rather than incidents is passed to the US National Transportation Safety Board (NTSB) and the FAA. All remaining reports fall within the scope of the ASRS and are, therefore, protected under the following provisions. Section 91.25 of the Federal Aviation Regulations prohibits the use of any reports submitted to NASA in any disciplinary action. However, appropriate action can be taken if information about an incident is derived from a source other than the ASRS submission. In addition to the provisions that protect contributors, the action of filing a report is considered to be "indicative of a constructive attitude." Accordingly, the FAA will not seek to impose a civil penalty or suspend a license if the individuals involved submit a report within 10 days of the incident and the violation was inadvertent, if it did not involve a criminal offense or accident. These exemptions apply providing that the person has not committed a violation for a 5 year period prior to the date of the incident.

These guidelines within the field of aviation are worth citing because they have provided a blueprint for similar protection, which is being offered under health care reporting systems. For example, the Veteran's Health Administration's (VA) National Center for Patient Safety has established two systems. The first is a mandatory reporting scheme for more serious adverse events. The second, known as the Patient Safety Reporting System (PSRS), was developed from a joint initiative between the VA and NASA in May 2000. Unlike the ASRS, the VA's PSRS is intended to collect information on adverse events, as well as near misses. NASA collects the reports and maintains the confidentiality of the system. Under the agreement, the VA may not review any report or data

until it has been de-identified. Concern over the inadvertent disclosure of contributor information has led to the decision that the initial report will not be held once the event has undergone an initial analysis. The contributors' identity is also protected under Privacy Act and recognized exemptions to the Freedom of Information Act. Records created for the VA as part of a medical quality assurance program, such as patient safety reports, have additional protections beyond those of other government agencies. United States Code (USC) 5705 with certain exceptions provides that records and documents created by the VA "as part of a medical quality-assurance program" are "confidential and privileged and may not be disclosed to any person or entity." Raymond and Crane's (2001) review of these confidentiality measures raises the caveat that "although federal law appears to provide the VA considerable protection against the discovery and disclosure of data, these unique legal shields are not afforded to non-VA hospitals."

Previous paragraphs have described how a range of human factors issues, including a fear of retribution and concern over the efficacy of any contribution, help to create the problems associated with underreporting. We have also reviewed a wide range of initiatives to address these problems including the development of mandatory and voluntary schemes as well as the provision of legal protection against disclosure in confidential systems. There are other approaches that help to address the problem of underreporting. In particular, "sentinel" focus resources more narrowly on a small number of representative institutions or work teams. These groups are given additional training and resources to both encourage and support any reporting. Monitoring systems and exhaustive reviews of patient records may also be used to catch any incidents that are missed. The results from these investigations can then be extrapolated to provide additional insights into the potential scale of any problems at a regional or national level.

The FDA (1999) was amongst the first to realize that sentinel systems can be used to address the problem of underreporting in national systems. They found it difficult to ensure active participation from more than 60,000 "end user" organizations for health care–related devices. They, therefore, decided to conduct a trial in which a small number of organizations were provided with additional support to explicitly encourage participation in a voluntary reporting system. Seventeen hospitals and six nursing homes participated in the 12 month "DEVICENET" study. Coordinators were identified in each institution; these individuals were typically clinical risk managers. They were either offered a 1 day group training in Washington, DC or a slightly shorter course in their own organizations. Videos were also prepared for each of the participant institutions. These were intended for use during in-house staff orientation and in-service training sessions. The video encouraged individuals to follow their facility's internal procedures for reporting of adverse events. After viewing the video, each staff member in the participating institution was given a one-page sheet summarizing the local provision. These sheets also provided information about the confidentiality safeguards offered to participants. Each report had any individual identification information removed as soon as possible after it had been received. After 30 days, the facility ID was removed, so that it was no longer possible to link the report to the facility. This period enabled the study team to link the original report with any follow-up reports and provided an opportunity to discuss any questions about the report with the study coordinator. The sentinel trial also enabled participants to contribute anonymous reports. At the end of the year's study, the coordinators had gathered 315 reports of which 14 were anonymous. They argued that this level of activity was "far above" the average for reporting device-related incidents. This study also illustrated some of the limitations of sentinel reporting. A continuing problem for the FDA is that many nursing homes fail to contribute any reports of adverse events even though they operate many of the devices and procedures that give rise to problems in other health care settings. In spite of all of the additional support offered in this trial, none of the 315 reports came from any of the six participating nursing homes.

Witham et al. (2006) provide a more recent example of the sentinel approach. Their work stemmed from a recognition of the problems associated with the incident reporting (IR1) system used in the National Health Service, including the lack of anonymity, failure of medical staff to fill in reports, bias toward certain types of events (e.g., falls), and lack of feedback toward staff. They, therefore, focused resources to increase awareness about a local adverse incident reporting system in a Medicine for the Elderly ward in Ninewells Hospital, Tayside, Scotland. Before its introduction, an orientation session took place at the weekly ward multidisciplinary meeting. Reports were collated at trimonthly intervals by the middle grade doctor on the ward (MDW) and presented to the multidisciplinary team in aggregate form. These aggregate reports were then used to select a small number of areas on which to focus ward-level quality improvement activities: 32 of 72 (44%) of sentinel audit incidents were reported by nursing staff and 37 of 72 (51%) by medical staff, with three reports not attributable. No incidents were reported by allied health professionals. There was no overlap in reports between the sentinel audit and the national IR1 system.

Sentinel schemes reduce the problems of underreporting by focusing resources on a number of "representative" institutions. A limitation with this is that sentinel schemes may lack the resources to ensure that focused support is provided across all procedures and departments even within one of these favored organizations. In consequence, patient safety organizations have also funded centers to focus on different aspects of patient safety. The VA established four of these institutions, each with an annual budget of approximately $500,000. One looked at patient safety in the operating room, another at elderly patients (Weeks and Bagian, 2000). Although these units were established to support research and development, their work addressed many of the adverse events elicited through national and local reporting systems.

26.6 Elicitation and Form Design

The previous section has focused on the human factors issues that lead to underreporting. In contrast, the following pages look more at the problems of ensuring that adequate information is obtained once a health care professional has decided to submit an incident report. This is not as simple as it might seem. In particular, it may not be possible to interview staff in order to elicit additional details in anonymous schemes. In confidential systems, there is also the danger that any subsequent contacts with managers may inadvertently disclose the identity of the contributor in the process of providing further information. In such circumstances, it is imperative that human factors and human–computer interaction expertise be used to ensure that reporting forms are designed to support the skills and expectations of potential contributors.

The design of incident reporting forms has remained a focus of debate amongst the handful of research groups that are active in this area (Johnson, 2000; Wu et al., 2008). Meanwhile, hundreds of local, national, and international systems are using ad hoc, trial and error techniques to arrive at appropriate forms. It is important to stress that there are several different approaches to the presentation and dissemination of incident reporting forms. For example, some organizations provide printed forms that are readily at hand for the individuals that work within particular environments. This approach clearly relies upon the active monitoring of staff who must replenish the forms and who must collect completed reports. Other organizations rely on computer-based forms. These can either exist in formats such as Adobe's PDF, which must be printed and completed by hand, or in electronic form so that they can be completed online. In either case, there is an assumption that staff will have access to appropriate hardware and software resources. This is not always the case in many health care domains. Many of these machines may also be located in public areas where colleagues and coworkers can observe the submission of an incident report. Each of these different approaches may also be supplemented by, for instance, telephone-based reporting for situations in which forms are unavailable.

This plethora of submission techniques is further complicated by the observation that personnel are increasingly expected to file reports through multiple systems. For instance, local voluntary systems such as those proposed by the Institute of Medicine currently operate alongside several mandatory state-based schemes at the same time as federal agencies, including the US Centre for Devices and Radiological Health (CDRH), also operate national systems. There are also often different parallel schemes for reporting incidents that injure employees rather than patients.

Given this diversity, it can be very difficult to establish which system to file a report under. For instance, many of these schemes define the severity of an incident that should be reported to them. In many cases, however, health care workers may not know what the ultimate outcome of a mishap will be. For instance, medication errors often have uncertain, long-term effects. Should an individual begin by reporting to a local system and then file successive reports to regional and national systems and the results of the incident become more certain? Alternatively, some hospitals have established "one-stop-shops," where all reports are filed via a risk manager who ensures that local information is fed into regional and national schemes depending on the nature of the incident.

The design and layout of reporting forms remains a critical issue irrespective of whether individuals report directly to external agencies or via a local safety manager. If potential contributors cannot use the fields of these documents to accurately provide necessary information then there is little likelihood that incident reporting systems will provide an effective tool for "organizational learning." Form design is, therefore, a critical area for human factors input in the development of most reporting systems. It is surprising; therefore, that many systems are implemented without even the most cursory forms of user testing (Johnson, 2003a). In consequence, it can be difficult to determine whether underreporting stems from a widespread rejection of the system or from acute frustration with the electronic and paper-based forms that are intended to elicit feedback about past failures. User testing is important because a vast range of different approaches have been used to elicit information about adverse events. For example, Busse and Wright (2000) describe a paper-based reporting form that was developed for a local system within a U.K. Neonatal Intensive Care Unit. This used open "free-text" fields for individuals to describe the incident that they have witnessed. Such open-ended questions are appropriate in systems where it is possible

for analysts to go back and ask additional questions to clarify any information that is either missing or only partially understood. The benefit of the approach is that it makes only minimal assumptions about the information that the contributor wishes to report. They are not forced to select particular items from a predefined list that may unduly constrain their selections.

Problems arise when analysts must translate the information provided by these "open" forms into the format that is required by regional or national agencies. It can be difficult to identify common patterns or trends across the colloquial terms and natural language accounts that characterize submissions to many local reporting systems. In contrast, national and regional systems rely on taxonomies that use a number of key terms to classify many different adverse events. Clinical risk managers must examine these free-text accounts and then reclassify them using the agreed descriptions that can be incorporated into the regional or national systems. NPSA (2010) guidance can be used to assess the outcome of incidents in health care. The classification process can be difficult because it is not always possible to define an ideal match between the details of a particular adverse event and national or regional categories, particularly, if, for example, an incident involved a mixture of events:

- *No harm*: Impact prevented—any patient safety incident that had the potential to cause harm but was prevented, resulting in no harm to people receiving NHS-funded care. Impact not prevented—any patient safety incident that ran to completion but no harm occurred to people receiving NHS-funded care.

- *Low*: Any patient safety incident that required extra observation or minor treatment and caused minimal harm, to one or more persons receiving NHS-funded care.

- *Moderate*: Any patient safety incident that resulted in a moderate increase in treatment and which caused significant but not permanent harm, to one or more persons receiving NHS-funded care.

- *Severe*: Any patient safety incident that appears to have resulted in permanent harm to one or more persons receiving NHS-funded care.

- *Death*: Any patient safety incident that directly resulted in the death of one or more persons receiving NHS-funded care.

Computer-based tools can help to guide potential contributors through the classification process. Participants need never see the hundreds of individual fields in the full taxonomy. Instead, they are only shown those options that are relevant to the incident they are reporting. This relevance is partially determined by the contributor's previous responses to questions about the adverse event. In this limited sense, these computer-based tools are context sensitive. They tailor the elicitation to match the incident that is being described. A limitation, however, is that it can be difficult to ensure that any two contributors will assign the same keywords to similar incidents. This is, arguably, more likely to happen when risk managers are trained to classify the free-text accounts of their coworkers. It is difficult to ensure that every potential contributor receives a similar level of training. Similarly, problems can arise when potential participants cannot find the keywords to match the incident that they have witnessed. Incident taxonomies address this problem in a number of ways. First, the detailed categories usually include the value "other" as a catch-all. There is, of course, a danger that contributors will too readily use the "other" classification if they do not understand what is meant by the more detailed terms. Second, the many national systems also include a question at the end of their classification, which asks contributors to state, "Please tell us how you think this form could be improved."

The problem created by an incomplete taxonomy can be overcome by extending the list of terms to ensure that it is broad enough to cover every likely eventuality. However, this creates further problems if users have to navigate hundreds and even thousands of descriptions to find the one that matches the event they are looking at. Computer-based reporting systems can help to overcome these problems by guiding the users, so that the answers to previous questions can help to filter the options that they are presented with. For example, if the user indicates that they have witnessed a patient-related accident then they are not usually presented with menu options or check boxes that relate to an error in diagnosis. Of course, there may be some unusual incidents that stem from precisely this combination of issues and so extensive user testing is required to ensure that users can exploit tool support without becoming so frustrated that they will abandon a submission.

User testing can also help to reveal other biases. For instance, there is a tendency for users always to select items from the top of a scrolling list or menu. Few users will scroll to the bottom of long and complex widgets. This can have an unfortunate influence on the findings that may be derived from a reporting system where the position of an item on the display can determine whether or not users recognize it as an attribute of a particular incident. The NPSA has acted to address these problems by conducting a series of field studies into the application of the NRLS between January and May 2003. Thirty-nine organizations from a range of health care settings worked with the NPSA to test reporting methods. Most of this work focused on the completeness and consistency of the taxonomy rather than an evaluation of the computer-based systems being developed by the NPSA's commercial partners.

The local, paper-based reporting system from an ICU forms a strong contrast with the demands for national reporting, illustrated by the NPSA's initiatives. There are, however, a number of other reporting systems that do not fit into either of these different stereotypes. For instance, Staender et al. (1999) within Swiss Departments of Anaesthesia have pioneered an online system for incident reporting.* Anaesthesia Critical Incident Reporting System (CIRS) embodies a number of assumptions about the individuals who are likely to use the form. Perhaps the most obvious is that they must be computer literate. This is significant because CIRS exploits a diverse range of dialogue styles or interface widgets. These include check boxes and pull-down menus as well as free-text fields. This system is different from the one proposed by the NPSA because it was established as the result of a self-help initiative from a number of motivated clinicians. It was not set up as part of a government system, although it subsequently attracted this support. Equally, it differs from the local system because it developed beyond a single hospital and hence could not easily be sustained using limited resources and a paper-based approach. CIRS also exploits a number of predetermined categories to characterize each incident. Users must select from one of 16 different types of surgical procedure that are recognized by the system. They must also characterize human performance along a number of numeric Likert scales. These are used to assess lack of sleep, amount of work-related stress, amount of nonwork-related stress, effects of ill or healthy staff, adequate or inadequate knowledge of the situation, and appropriate skills and experience. For example, if the individuals involved in the incident had no sleep in the last 24 h, then the score should be 1. If they had more than 7 h sleep, then the scope should be 5. Scores between these two extremes should be allocated in proportion to the amount of sleep that had been obtained by the participants. This approach is relatively straightforward when referring to objective amounts of sleep. However, the CIRS workload scale is more difficult to interpret in the same range from 5 (unusually heavy) to 1 (unusually light). The introspective ability to independently assess such factors and provide reliable self-reports again illustrates how many incidents reporting forms reflect the designers' assumptions about the knowledge, training, and expertise of the target workforce.

One of the most innovative features of the CIRS system is that it is possible for health care professionals to use the Internet as a means of reviewing information

---

* http://www.medana.unibas.ch/cirs/, last accessed February 2010.

about previous incidents. The anonymous cases can be read online and comments can be appended to create a dialogue between individuals who either request additional (anonymous) information or who have experienced similar incidents in other organizations. For example, the following report describes a drug misadministration:

> *Incident Description:* Female patient 11 y/o was scheduled for tonsillectomy. She was NORMAL as regards the physical examination and lab values. The operation was done as usual without any abnormal events in anesthesia or the recovery. She was discharged awake from PACU to the ward. Shortly after her arrival, the ward's nurse injected her by what she thought was antibiotic. But soon she discovered that it was BROFEN (Ketoprofen) suspension. The poor child developed convulsions and cyanosis at once. She was transmitted quickly again to the OT. The patient was hypotensive (70/40) and tachycardic (180), O2 saturation was 75%, and the end tidal CO_2 was 70.

Another clinician accessing this report left the following observation and request for additional information:

> *Sorry about the sad case. Side question: why was an antibiotic given and why afterwards? It seems that some accidents occur as a consequence of an action that wasn't unnecessary in the first place. For example, I once heard of an appendicectomy case that got an epidural injection where a mix-up also occurred with fatal consequences.*

A key point here is that the initial report acts as a focus for further discussion about common factors in previous incidents. From a technical standpoint, this type of facility also requires that the reporting system be extended beyond the forms that elicit information about the initial adverse event to include some mechanism for further dialogue.

26.7 Form Content and Delivery Mechanisms

Irrespective of whether a reporting system is intended to collect information about local incidents or national statistics about adverse events, it is important that managers consider the range of information that must be elicited in the aftermath of an accident or a near miss. This includes factual data about what precisely happened. Reporting forms can also prompt potential contributors for more analytical information about what they consider to be potential causes for an adverse event. This section reviews some of the human factors issues that

must be considered during the development of reporting forms. It also considers the usability issues that affect the different delivery mechanisms, which enable potential contributors to submit information about near misses and adverse events.

26.7.1 Delivery Mechanisms

Most of the early reporting systems relied upon simple paper-based forms. Increasingly, however, as NPSA and CDRH initiatives show, more schemes are relying upon Internet technologies. This approach has numerous benefits. For example, managers do not need to continually check the supply of paper-based forms nor do they have to monitor drop-boxes to check for new submissions. Electronic forms can be revised and then made accessible across many different health care organizations without the overheads associated with conventional distribution networks. The use of appropriate interface design techniques can support users by providing default values for common fields in electronic forms. Inferences can be made to populate these online documents. For example, the date of the report can be set to the day on which the form is accessed unless the end user decides to change it. Similarly, the organization in which an incident occurs might default to the one in which the reporting system is accessed.

A range of problems also complicates the use of online reporting systems. In particular, many organizations have significant concerns about security of applications that can be vulnerable to attack from people both inside and outside the reporting organization. These considerations are particularly important given the sensitive nature of confidential and anonymous reporting. There are technical solutions for many of these issues. For instance, digital signatures can be used to authenticate the sender of particular information. Electronic watermarks can be used to ensure that reports are not overwritten or unnecessarily altered after submission. However, many of these technical solutions increase the burdens of system operators (Johnson, 2009). For example, they may be excluded from the system if they do not authenticate their access through the use of an appropriate password (Nakajima et al., 2005).

Further limitations also affect the use of computer-based forms to elicit incident reports. In particular, it can be difficult to ensure that all potential contributors have easy access to the necessary technical resources. Many health care professionals only have work-time access to shared computers in public spaces. They can easily be interrupted or observed as they fill out a confidential or anonymous reporting form online. It can also be possible for other users to access information that their colleagues have entered either by accessing system logs or cached information that has inadvertently been left on disk after a session has ended.

The difficulties in ensuring access to computer-based reporting forms have led many national and regional systems to operate hybrid approaches. Online systems are provided alongside paper-based forms or telephone numbers that can be used to leave verbal accounts of adverse events on answering machines. A small number of pilots have also been conducted into various forms of reporting using mobile devices ranging from SMS text messages through to full-blown reporting applications. It seems likely that this approach will become increasingly popular even though there are significant challenges to be addressed in completing online incident reporting forms using the limited input techniques and display resolution associated with mobile devices.

26.7.2 Preamble and Definitions of an Incident

It is important to provide users with a clear idea of when they should consider making a submission to the system. For example, the local scheme mentioned in previous paragraphs explicitly states that an incident must fulfill the following criteria: (1) It was caused by an error made by a member of staff, or by a failure of equipment. (2) A person who was involved in or who observed the incident can describe it in detail. (3) It occurred while the patient was under our care. (4) It was clearly preventable. Complications that occur despite normal management are not critical incidents. But if in doubt, fill in a form.

It can be surprising that incidents, which occur in spite of normal management, do not fall within the scope of the system. This effectively prevents the system from targeting problems within the existing management system. However, such criticisms neglect the focused nature of this local system, which is specifically intended to "target the doable" rather than capture all possible incidents. CIRS exploits a wider definition of an adverse incident:

> Defining critical incidents unfortunately is not straightforward. Nevertheless we want to invite you to report your critical incidents if they match with this definition: an event under anaesthetic care, which had the potential to lead to an undesirable outcome if left to progress. Please also consider any team performance critical incidents, regardless of how minimal they seem.

This could potentially cover a vast range of adverse events. Such a definition would stretch the limited resources of many local or national systems. It also illustrates the way in which the definition of an incident both determines and is determined by the reporting system

that is intended to record it. The definition must be broad enough to capture necessary information about adverse events. However, if the definition is drawn too widely then the system may be swamped by a mass of low-risk mishaps and near misses, so that it can be difficult to identify critical events in time to take corrective actions.

Definitions of what should be reported partly determine whether or not a health care professional will actually submit a report. For example, Lawton and Parker (2002) show that adverse events are more likely to be reported when staff deviate from written protocols. They argue that professionals are unwilling to challenge a fellow professional without strong grounds. They are also reluctant to report behavior that has negative consequences for the patient when the behavior reflects compliance with a protocol or improvisation where no protocol is in place. The key issue here is that such observations about reporting behavior are often orthogonal to the abstract definitions of adverse events that form the basis of many reporting systems (Kaplan et al., 1998). Hutchinson et al. (2009) also show how higher reporting rates can be correlated with a number of wider safety culture metrics derived from hospital staff surveys and with better risk-management ratings from external audits. They did not find any apparent association between reporting rates and standardized mortality ratios, data from other safety-related reporting systems, hospital size, average patient age, or length of stay.

26.7.3 Identification Information

Many health care institutions operate a "gatekeeper" approach in which members of staff will review potential submissions to determine whether or not particular incident reports should be forwarded for more detailed analysis. These teams have a responsibility to determine whether reports should be handled locally or whether they should be passed to external agencies. This is important when health care professionals may not always understand which reporting scheme they should be using. For instance, the NPSA's frameworks must be distinguished from the Serious Untoward Incident (SUI) reporting system that informs Strategic Health Authorities and the Department of Health about incidents that require urgent attention; "they may not necessarily be patient safety incidents, and will often include identifiable information to enable action at a local level. For this reason, it is not appropriate to combine the two systems." Similarly, the NPSA are keen to point out that any incidents involving the use of a medical device will be shared with the MHRA. Gatekeepers can help to direct reports to the appropriate scheme. However, in order to do this, it is often necessary for employees to provide identification information, so that the "gatekeepers" can gather further data if a mishap requires subsequent investigation. These systems, therefore, provide confidentiality but not anonymity.

In contrast, the local system described in previous sections allows anonymous reporting. This is intended to encourage participation by reducing the fear of retribution. However, anonymity creates problems during any subsequent causal analysis. It can be difficult to identify the circumstances leading to an incident if analysts cannot interview the person making the report. Instead, health care professionals are often required to provide considerable additional details to ensure subsequent investigations have sufficient information to identify any potential lessons without the need to contact the individual making the original submission.

The anonymity of a reporting system can be compromised in local reporting systems. Inferences can be made about the identity of a contributor based on shift patterns, on clinical procedures, and on the limited number of personnel who have the opportunity to observe an incident. Clearly, there is a strong conflict between the desire to prevent future incidents by breaking anonymity to ask supplementary questions and the desire to safeguard the long-term participation of staff within the system. The move from paper-based schemes to electronic systems raises a host of complex sociotechnical issues surrounding the anonymity of respondents. For instance, each client computer connecting to a website will potentially disclose location information through its Internet Protocol (IP) address. This address is not linked to a particular user, but it can be used to trace a report back to a particular machine. If logs are kept about user activity then it will be possible to identify the contributor. Alternatively, many health care organizations routinely log users' keyboard activity; hence, there may also be more direct means of identifying the person who contributes an electronic incident report. Balanced against this concern for anonymity and confidentiality, there can also be problems if groups or individuals deliberately seek to distort the findings of a system by generating spurious reports. These could, potentially, implicate third parties. The problems of malicious reporting together with the technical difficulties of providing anonymous reports, therefore, makes it likely that future electronic systems will follow the ASRS approach of confidential submissions.

26.7.4 Time and Place Information

There is a tension between the need to learn as much as possible about the context in which a mishap occurred and the need to preserve the confidentiality or the anonymity of the person contributing a report. A frequent criticism

is that in order to protect the identity of those involved in an adverse event, reporting systems also remove information that is vital if other managers and operators are to avoid future mishaps (Johnson, 2003a; Evans et al., 2004). As mentioned in the previous paragraph, this is a particular problem in local systems where it may be possible to infer who was on duty if respondents provide information about the time at which an incident occurred. However, if this data is not provided then it may not be possible to determine whether or not night-staffing patterns played a role in the incident or whether an adverse event was effected by particular hand-over procedures between different teams of coworkers. Similarly, if location information is not provided then it can be difficult to determine whether ergonomic issues and the configuration of particular devices contributed to a mishap. If location information is provided, for instance, within an ICU, it is then often possible to name a small number of professionals to be identified as responsible for health care provision within that area. The difficulties created by the omission of location information can be seen by the frequent requests for additional unit information in the dialogues that emerge after the contribution of an incident to the CIRS application.

Location information falls into several different types. Geographical information may be important in national and regional systems to detect common patterns within specified areas. These can emerge if local groups of hospitals adopt similar working practices that may contribute to adverse events. Similarly, geographical information can be important to identify "hot spots" that can be created by a batch of medication or other common supply problems. These details need not be explicitly requested from the individuals who contribute a report. They can often be inferred from the delivery mechanism that is used to collect the report. These inferences can be relatively straightforward, for example, if reports from a particular hospital regularly arrive on a specific day of the week. They can also be based on more complex information, such as the IP address of a contributing machine. These assumptions can, however, be unfounded. For instance, problems will arise if contributors work in one location and submit a report from another. This scenario is likely to occur because many health care workers benefit from increasing job flexibility; especially in the delivery of specialist care across a relatively wide geographical area.

As mentioned, the location information requested from contributors can take several different forms. Not only do analysts often need to identify geographical patterns within a series of incident reports. They may also need to locate functional similarities within the different areas of a health care system. For instance, the FDA offers 30 or more locations in their medical devices reporting system. This taxonomy includes "612 mobile health unit" and "002 home" as well as "830 public venue" and "831 outdoors." The diversity in the classification reflects the diverse locations in which mass market health care devices might fail.

In confidential systems, location information must be obtained so that analysts can contact reporters in order to follow up any necessary additional details. Even in anonymous systems, it can be necessary to provide location information. For example, if a device has failed or if a problem involves a subcontractor then it may be necessary for the reporter to provide information about the location and identity of the supplier who was involved in the adverse event. This creates considerable opportunity for error in reporting system software. Arguably, the most frequent problems center on the misspelling of names. For example, Siemens has been entered into the FDA system under Seimens and Simens. Incidents have also been recorded under *Siemens Medical, Siemens Medical Solutions, Siemens Medical Systems*, etc. Any analysis and retrieval software must cope with such alternative spellings if potentially relevant information is not to be overlooked. An alternative approach is for the system to prevent users from typing in this information. Instead, they are compelled to select a supplier identifier from an enumerated list or menu. This is liable to have several hundred items. Such widgets pose a considerable challenge for human–computer interface design. They can also introduce considerable frustration in users who must scroll through the names of dozens of medical suppliers before they reach the one that they are looking for. This frustration is likely to increase if the supplier identity information is missing from the enumerated list. However, one benefit from this approach is that most software reporting systems can use a supplier list to automatically update address information so that users need not type in their location details. This is a significant benefit given that many end users will not have this information to hand as they begin to file a report in the aftermath of an adverse event.

Most reporting forms also prompt contributors for the time when an incident occurred. As with location information, the elicitation of this information is not as simple as it might appear. As mentioned earlier, temporal information can be used with geographical data to support inferences about the identity of a potential contributor or about the work group who are implicated in a report. A number of other human factors issues also complicate the elicitation of this information. For example, CIRS prompts the reader "at what time of the day did the incident happen (1–24)?" Other systems enable the contributor to specify that the date and time of an occurrence is "unknown." This is important when a report is filed by someone who observes the consequences of an incident without being present at the time when a failure first occurred.

There are additional complexities. For example, some incidents only emerge slowly over a prolonged period. For instance, an infusion device may administer the incorrect medication over several of these intervals. Other mishaps can take place over an even longer timescale, even extending to months and years. Similarly, the same adverse event might occur several times before it is detected or might occur at different times to several patients. It is unclear how these different circumstances might be coded within many reporting schemes. For example, contributors might be required to complete a separate form for each instance of the adverse event even though they were strongly connected by sharing the same causes and consequences for the patient. Further complexity arises because, as mentioned, the time at which an incident occurs can be different from the time at which an adverse event is detected or reported. These additional details are often elicited so that safety managers can review the monitoring mechanisms that are intended to preserve patient safety. If an adverse event is only detected many months after it has taken place, for example, if a patient returns to report the adverse consequences of a mishap, then many internal quality control and monitoring systems can be argued to have failed.

26.7.5 Detection Factors and Key Events

It is important to determine how adverse events are detected. Many incidents come to light through a combination of luck and vigilance. Unless analysts understand how contributors identified the mishap then it can be difficult to determine whether there have been other similar incidents. CIRS provides an itemized list of detection factors. These include direct clinical observation, laboratory values, airway pressure alarm, and so on. The respondent can identify the first and second options that gave them the best indication of a potential adverse event. The local ICU scheme described by Busse and Wright (2000) simply asks for the "grade of staff discovering the incident." Even though it explicitly asks for factors contributing to and mitigating the incident, it does not explicitly request detection factors.

The reporting of detection factors raises a number of problems. For example, clinical training often emphasizes the importance of "making errors visible." Nolan (2000) identifies the "double checking," of physician medication orders (prescriptions) by the pharmacist and the checking of a nurse's dose calculations by another nurse or by a computer are examples of making errors visible. Similarly, if patients are educated about their treatment then they can also play an important role in identifying errors. However, as these checks and balances become more widely integrated into health care practice, it is less and less likely that they will trigger incident reports. There is a paradox that the most effective detection factors are likely to be those that are mentioned least often in incident reports because they are accepted as part of the standard practice. Unfortunately, things are seldom this straightforward. For example, a recent report to the FDA described how the drug calculator of a medication assistant in a patient monitoring application would occasionally round up values to a second decimal place. The users complained that this could easily result in a medication error and that the manufacturer was failing to acknowledge the problem. The manufacturer responded that vigilant nursing staff ought to notice any potential problems when calculating the medication. The clinicians countered this by arguing that they had explicitly taught nursing staff to trust the calculation function as a means of *reducing human error* (Johnson, 2004). This illustrates the recursive nature of incident detection. Manufacturers assume that health care professionals will crosscheck any advice to detect potential errors. Conversely, health care professionals increasingly assume that automated systems will provide the correct advice that is necessary to help them detect adverse events.

Many detection factors are focused on the proximal or immediate events that can lead to an adverse outcome (Itoh et al., 2009). For example, a nurse observing a patient's adverse reaction to a particular medication can trigger a report. It is rare for reports to be filed when health care professionals detect the latent conditions that may eventually contribute to an incident or accident. Many nurses and doctors accept a culture of coping with limited resources and high demands on their attention. Lawton and Parker (2002), therefore, argue that the U.K. National Health Service should take a more proactive approach to incident reporting. Individuals and teams must be sensitized so that they are more likely to detect and report the conditions that will lead to error before an error actually occurs:

> Proactive systems work in part by asking people to judge how frequently each of a number of factors such as staffing, supervision, procedures, and communication impact adversely on a specific aspect of their work. So, for example, if nurses in intensive care are experiencing problems with the design of a particular piece of equipment, this will be recorded and action taken to improve the design. This kind of proactive approach allows the identification of latent failures before they give rise to errors that compromise patient safety.

Vincent et al. (1998) build on this analysis when they identify those latent conditions that should be monitored and detected prior to an adverse event. Their enumeration includes items such as heavy workload, inadequate knowledge or experience, inadequate supervision, a stressful environment, rapid change within an organization, incompatible goals (e.g., conflict between finance and clinical need), inadequate systems of communication, and inadequate maintenance of equipment and buildings. They observe that these latent factors will affect staff performance, can make errors more likely, and will impact on patient outcomes. However, few existing reporting systems or research studies have found concrete means of encouraging respondents to detect and report these latent conditions within health care institutions. Part of the explanation for this may lie in the observation that many of the latent conditions for adverse events are almost characteristic of many modern health care organizations. These include rapid change within an organization and conflict between clinical and financial needs.

Most reporting forms prompt the contributor to explain what happened. In many systems, this field is left open so that individuals and teams can describe the critical events in their own words. Additional prompts are often used to help ensure that contributors provide sufficient detail for subsequent analysis. For instance, the local ICU system breaks down the "what happened" information into a number of different categories. Respondents are first prompted to elicit information about "the incident." This information is broken down into a "description of what happened" and "what factors contributed to the incident?" They are also asked for mitigation factors; however, a more detailed discussion about this part of the form is postponed to the next section of this chapter. The local system also prompts for other information about what happened. A further section of questions address the "circumstances" of the incident. This includes temporal information, mentioned previously as well as "what procedure was being carried out," "what monitoring was being used?" and "if equipment failure, give details of equipment." A final section about what happened is intended to elicit general information about the personnel involved. Respondents are asked for the "grade of relevant responsible staff" and the "grade of staff discovering the incident." In keeping with the rest of this form, contributors can complete the form using their own terms.

This use of free text helps to avoid the irritation and fatigue that can be caused by reporting forms that ask contributors to select a small number of "tick-box" responses from hundreds of irrelevant options. Conversely, these "tick-box" enumerations can reduce the amount of time needed to complete lengthy natural language descriptions that must then be recoded into the taxonomies that were mentioned in previous sections.

The problems associated with hundreds of irrelevant "tick-boxes" can be addressed by context-sensitive reporting systems. For example, if a contributor indicated that the incident did not involve a medication error then they would not need to select options such as adverse drug reaction (when used as intended), contraindication to the use of the medicine in relation, mismatching between patient and medicine, omitted medicine/ingredient, patient allergic to treatment, wrong/omitted/passed expiry date, and so on. Similarly, the software only presents questions about what happened to a particular device once the user confirms that the incident involved a device failure. The dynamic nature of the form content helps to filter out irrelevant questions. However, the use of context-sensitive software can also create problems if users have to report combinations of, for example, adverse drug reactions and device failures. A different form of frustration can arise when health care professionals cannot find boxes to match the different aspects of the incident they want to report.

The CIRS online system also adopts a mixed approach. Initially, the web-based form prompts participants to enter information about what has happened by checking boxes associated with the various team members who were involved in an adverse event. They must then enter the number of hours "on duty without sufficient rest (if known)." This is intended to provide an insight into the workload on the provider of anesthetic care. The subjective nature of this question makes it very difficult to interpret any results. Different individuals can have very different ideas about what represents "sufficient rest" (Johnson, 2007). This again illustrates the importance of conducting usability studies and of considering the human factors issues when constructing the questions that will be asked as part of a reporting form. Respondents are then asked to provide information about the patients involved. In this case, a radio button widget is used to indicate the sex of the patient. Respondents can type numeric values into an age field. Radio buttons are also used to indicate whether the patient is undergoing elective or emergency procedure and their American Society of Anesthesiologists Physical Status classification system (Classes I–V). Class I refers to a normal and healthy patient and Class V refers to a patient who is unlikely to survive without an operation. This classification provides a crude approximation to the a priori risk involved in anesthesia. The CIRS form assumes that each incident only affects a single patient. This is an appropriate assumption within anesthesia; however, it would not be warranted if a common "error" is replicated in a number of similar procedures. The key point is that significant testing should be conducted to determine whether such assumptions can be justified within a particular domain. This testing can be performed in several different ways. For example, a

sentinel scheme can be used in the manner described in previous paragraphs. Representative institutions pilot the system before it is made more widely available. Similarly, potential contributors might be asked to use a prototype system to report information about an incident that they have observed in the past. They can also be given information about stereotypical mishaps and then asked to enter relevant information into the system "as if" they had witnessed such an adverse event.

The CIRS system elicits further information about "what happened" by asking contributors to select the "overall anesthetic technique" during which the incident occurred. A pull-down menu offers nine broad categories of activity that range from general anesthesia to regional anesthesia and care of a multiple trauma patient. These constrained fields are then followed by a number of more open questions. Contributors are asked to describe the incident in their own words. They are warned to be "careful not to present data here that could identify the patient, the team, or the institution." This section elicits information about the events leading to an incident. A second question asks respondents to "describe the management of the situation in your own words" from the moment of occurrence on. In passing, it is worth noting that CIRS warns users that "if you wish to print out this report, please stay in between the margins of the text field." This stems from a problem in the formatting of the online form that does not have a dedicated print function. Such issues are less significant for voluntary reporting systems where great pains are taken to preserve the anonymity of individual contributors. The need to keep a printed record of particular reports assumes a greater priority within mandatory and confidential schemes.

26.7.6 Consequences and Mitigating Factors

Vincent and Coulter (2002) argue that it is vital to consider the patient's perspective when assessing the consequences of an adverse health care event. They refer to the psychological trauma both as a result of an adverse outcome and through the way that an incident is managed. They urge that

> if a medical injury occurs it is important to listen to the patient and/or the family, acknowledge the damage, give an honest and open explanation and an apology, ask about emotional trauma and anxieties about future treatment, and provide practical and financial help quickly.

They also argue that patients are often best placed to report the consequences of an adverse event.

Most existing reporting systems rely upon clinicians to assess the outcomes of adverse events when they submit information about an incident or near miss. This raises a number of complex issues for health care reporting systems. It is often difficult to determine whether or not a mishap had any appreciable impact upon the patient outcome. CIRS asks contributors to select an outcome from an enumeration that includes: outcome independent from the event; patient dissatisfaction, prolongation of hospitalization; unplanned hospitalization of an outpatient; unplanned admission to an ICU; minor morbidity; major morbidity and death. This contrasts with the local system that simply provides a free-text area for the respondent to provide information about "what happened to the patient?" An incident might have no immediate effect. Hence, the distinction between immediate and long-term outcomes is an important issue. Similarly, the administration of an incorrect medication may include side effects that increase the probability of adverse consequences in the future. In such circumstances, it may only be possible to consider the likelihood of an effect rather than commit to a certain outcome. Further problems arise because the individuals who witness an incident may only be able to provide information about the consequences of that event. The lack of clinical audit and of agreed outcome measures in some areas of health care creates additional complexity. Finally, health care professionals can inadvertently compromise the confidentiality of a report by carefully monitoring the progress of particular patients involved in an incident.

Further complexity is created by the need to assess the potential consequences of near-miss incidents. Few health care systems explicitly address this issue. However, it is a common concern in aviation and maritime systems (Johnson, 2003a). Given that no adverse event has actually taken place, it can be argued that the incident had negligible consequences. Alternatively, risk managers might assume the "worst plausible consequences" when assessing the severity of a near miss. The interpretation of "plausible" consequences is subjective and varies from system to system. For example, some air traffic management systems will treat a report of an air proximity violation "as if" a collision had actually occurred if the crew rather than the controllers were forced to initiate an avoiding action.

Voluntary reporting systems are often intended to elicit information about the low consequence and near-miss incidents that are not covered by mandatory schemes. These reporting systems provide as much information about how to mitigate failure as they do about the causes of adverse events. The local ICU system, introduced in previous paragraphs simply asks what "minimized" the incident. The CIRS system adopts a mixed approach to the elicitation of mitigating factors. Like the local system, respondents are prompted to describe the management of the situation in their own words. CIRS then provides a number of explicit prompts. The

online form asks "what led you to successfully manage the event (recoveries)?" Respondents must select the most important factor using radio buttons that are grouped into a number of categories. Personal factors include knowledge, skill, experience, situation awareness, and use of appropriate algorithms. A further category of mitigation factors focuses on team intervention described in terms of extraordinary briefings, extraordinary teambuilding, extraordinary communication within the anesthetic team, extraordinary communication in the surgical team, and extraordinary communication between the teams. The form also prompts for system factors including additional monitoring or material, replacement of monitoring or material, additional personnel, and replacement of personnel. Finally, there is an "other" category. As before, this detailed enumeration can help to guide users who may not be used to thinking in terms of "mitigation factors." There is a danger that schemes that ask more open questions may fail to elicit critical information about the ways in which managerial and team factors helped to mitigate the consequences of an incident.

In some incidents, it can be relatively easy to interpret information about the mitigation of adverse events. For instance, one study identified that there were 6.5 adverse drug events for every 100 admissions in a U.S. hospital (Bates et al., 1995). Of those, it was argued that 28% could have been detected and avoided mainly by changing the systems used to order and administer drugs. Similarly, another study showed that computerized monitoring systems were significantly more likely to identify and prevent severe adverse drug events than those identified by chart review (51% vs. 42%, $p = 0.04$) (Jha et al., 1998). However, it can be far harder to interpret incident reports where claims are made about human intervention in the mitigation of adverse events. It can be difficult to identify what precisely protected patient safety if another member of staff intervenes to prevent an adverse event. At one level, a safety manager might praise the vigilance of that individual. At another level, they might use this as an example of the success of the monitoring systems within a team of coworkers. Further investigation is required to determine whether such confidence is warranted. Individuals often identify potential incidents in ways that are not directly linked to official monitoring procedures. Conversely, well-developed routines can successfully detect potential incidents even when individuals are tired or operating under extreme workload (Itoh et al., 2009).

26.7.7 Causes and Prevention

It can be difficult to agree upon the causes of adverse health care events. In particular, it is often difficult to distinguish between the contribution made by technical failure, human error, and management responsibility. Many issues center on the problems of counterfactual reasoning. Counterfactual arguments lie at the heart of most forms of causal analysis. We can say that some factor A caused an accident if the accident would have been avoided if A had not occurred. This is counterfactual in the aftermath of an adverse event because we know that A did happen and so also did the mishap. The local reporting system introduced in previous sections asks respondents to suggest, "how such incidents might be avoided?" This open question is, in part, a consequence of the definition of an incident in this scheme, which included occurrences "that might have led (if not discovered in time) or did lead, to an undesirable outcome." It also provides a further illustration of this counterfactual approach to causal information.

Byrne and Handley (1997) have conducted a number of studies into human reasoning with counterfactuals. They have shown that deductions from counterfactual conditionals differ systematically from factual conditionals. For example, the statement "either the medication was prescribed too late or the disease was more advanced than we had thought" is a factual disjunction. Studies of causal reasoning suggest that readers will think about these possible events and decide which is the most likely. It is often assumed that at least one of them took place. The statement that "had the medication been prescribed sooner or the disease been less advanced then the patient would have recovered" is a counterfactual disjunction. This use of the subjunctive mood not only communicates information about the possible outcome of the incident but also a presumption that neither of these events actually occurred. This theoretical work has pragmatic implications for incident investigation (Salter, 2003). If factual disjunctions are used then care must be taken to ensure that one of the disjunctions has occurred. If counterfactual disjunctions are used then readers may assume that neither disjunction has occurred. The key point here is that most reporting systems rely upon counterfactual definitions of causation. Human factors studies of counterfactual reasoning have identified systematic biases that make it critical for risk managers to carefully analyze the causal arguments that they receive in response to adverse events.

The close association between causation and counterfactual arguments can also be seen in reporting systems that ask respondents about potential means of preventing an accident. The CIRS system provides an enumeration that is intended to guide the contributor in their analysis; "what would you suggest for prevention?" and respondents must select the most important item from a varied list. Potential preventative measures include additional monitoring or material, improved monitoring or material, better maintenance of existing monitoring/equipment, improved management of drugs, and improved arrangement of monitoring/equipment. The CIRS enumeration also provides items for improved training/education, better working conditions, better organization, better supervision, more personnel, better communication, more discipline with existing checklists, better quality assurance, development of algorithms/guidelines, and abandonment of old routine. Finally, there is an opportunity to include other preventative measures but this time using free-text descriptions.

Reporting systems also probe for information about factors that did not directly cause an adverse event but that contributed to the course of an incident or accident. These can include communication factors, education and training factors, equipment and resources factors, medication factors, organization and strategic factors, patient factors, task factors, team and social factors, and work and environment factors. They can also include more specific issues including the failure to refer for hospital follow-up, poor transfer/transcription of information between paper and/or electronic forms, poor communication between care providers, use of abbreviation(s) of drug name/strength/dose/directions, handwritten prescription/chart difficult to read, omitted signature of health care practitioner, patient/carer failure to follow instructions, failure of compliance aid/monitored dosage system, failure of adequate medicines security (e.g., missing CD), substance misuse (including alcohol), medicines with similar looking or sounding names, poor labeling and packaging from a commercial manufacturer, etc.

The CIRS reporting form mirrors this use of an enumeration rather than a counterfactual approach to causal information. Contributors must select whether the most "important field" to identify "what led to the incident (cause)?" These fields include personal factors such as diminished attention without lack of sleep, diminished attention with lack of sleep, insufficient knowledge, etc. They also include team factors such as insufficient communication or briefing. System factors include lack of personnel and unfamiliar surroundings. It is important to stress again that the answers to causal questions should be interpreted with care. Although the individuals who directly witness an incident can provide valuable information about how future adverse events might be avoided, they may also express views that are influenced by remorse, guilt, or culpability. Subjective recommendations can also be biased by the individual's interpretation of the performance of their colleagues and their management or of particular technical subsystems. Even if these factors did not obscure their judgment, they may simply have been unaware of critical information about the causes of an incident.

26.8 Explanations of Feedback and Analysis

The human factors issues involved in incident reporting do not end with the submission of a report. Potential contributors must be confident that the information that they submit will be taken seriously. This does not imply that every report must initiate change within the host organization. It is, however, important that contributors know that their reports have been successfully received and attended to. In other domains, electronic tracking systems have been introduced so that contributors can monitor who has responsibility for handling their submission from the moment that it is logged (Johnson, 2003a). Such techniques have not been widely introduced within health care applications. In confidential systems, it is more usual for contributors to receive an acknowledgement slip in return for their contribution. Such feedback is obviously difficult to provide in anonymous schemes. Most reporting forms provide participants with information about how their contributions will be processed. For example, the local system described by Busse and Wright (2000) includes the promise that "information is collected from incident reporting forms (see overleaf) and will be analyzed. The results of the analysis and the lessons learnt from the reported incidents will be presented to staff in due course." This informal process is again typical of systems in which the lessons from previous incidents can be fed back through ad hoc notices, reminders, and periodic training sessions. The CIRS web-based system is slightly different. It is not intended to directly support intervention within particular working environments. Instead, the purpose is to record incidents so that anesthetists from different health care organizations can share experiences and lessons learned:

> Based on the experiences from the Australian-Incident-Monitoring-Study, we would like to create an international forum where we collect and distribute critical incidents that happened in daily anaesthetic practice. This program not only allows the submission of critical incidents that happened at your place but also serves as a teaching instrument: share your experiences with us and have a look at the experiences of others by browsing through the cases. CIRS© is anonymous.

NRLSs have a far wider set of responsibilities than either CIRS or the local scheme mentioned earlier. It can be far more difficult to predetermine all of the techniques that might be used to address the vast range of different adverse events reported to these large-scale systems. However, it is arguably more important to provide potential contributors to national and regional schemes with feedback on progress made from previous reports. In local systems, health care improvements can be presented in staff meetings, newsletters, and other information sources. This is not so easy for schemes that cover many thousands of diverse health care organizations. The provision of effective feedback is particularly important if national reporting systems are not to be seen as expensive "talking shops" rather than "agents for change."

As mentioned, the local scheme referred to in this chapter provided feedback to staff through periodic newsletters. CIRS provides feedback in the form of an online dialogue or forum through which professionals can add comments to the various reports that are received. National organizations such as the CDRH and NPSA use the web as a primary means of providing feedback to potential contributors. This approach is justified by the relatively low cost of Internet site development. However, it relies upon a form of information "pull." Health care professionals have to keep going back to the site to download, or pull, updated information about patient safety initiatives. In contrast, e-mail dissemination provides a form of information "push" to ensure that lessons learned are sent out to health care institutions in a timely fashion. Unfortunately, this approach raises a host of additional human factors issues. For example, the increasing problem of spam mail has increased the likelihood that many individuals will overlook or automatically delete messages that have such a mass distribution. Similarly, not everyone has convenient access to the free time to access e-mail or web in their working environment. Increasing attention is focusing on the use of mobile devices both for the submission of incident reports and also for the "information push" of safety alerts that can then be linked to more sustained documents to be accessed over the web.

Most incident reporting systems continue to use paper-based dissemination techniques. This situation is gradually changing as a result of financial and administrative pressures. For example, in 1997, the decision was taken to stop printing the *FDA's User Facility Reporting Bulletin*:

> Ten years ago, our computer capability allowed us to communicate only within FDA. Now, with advanced computer technology we can globally communicate through the Internet and through Fax machines. As you would expect, Congressional budget cuts have affected all parts of government. FDA did not escape these cuts. In the search for ways to reduce our expenses, printing and mailing costs for distribution of publications in traditional paper form have come to be viewed as an extravagant expenditure ..." (Wollerton, 1997).

26.9 Conclusions

This chapter has provided a high-level survey of the human factors issues involved in the reporting of adverse health care events. It began by reviewing recent initiatives from the U.S. Institute of Medicine and a range of national patient safety agencies that have encouraged the development of voluntary and mandatory reporting schemes. Later sections went on to example the problem of underreporting. Potential contributors are often concerned that they will be blamed for any involvement in an incident or near miss. Several schemes have arranged for limited legal protection to support participants in voluntary reporting systems. These arrangements were described and the exceptions to this protection were identified.

The middle sections of this chapter introduced a range of different architectures for incident reporting. These included local systems that are designed and operated by individual health care professionals within single units. We also described different regional and national systems. For instance, the FDA has pioneered the use of sentinel reporting to reduce the reporting biases that effect large-scale schemes. This approach focuses training resources and support onto a number of representative institutions so that all staff are sensitized to the importance of incident reporting and hence may be more likely to participate in the scheme. It is, typically, not possible to provide similar levels of resourcing across all of the thousands of organizations who contribute to less focused national systems. The increasing diversity of mandatory and voluntary reporting systems has made it difficult for many staff to know which scheme they should use after a particular adverse event, particularly when they may not be certain of the ultimate impact on any patients who were involved. In consequence, an increasing number of hospitals have introduced "gatekeeper" systems where all reports are first submitted to a local safety manager who then assumes responsibility for passing them to the relevant schemes.

Form design and distribution have a significant impact on the human factors of reporting systems. It can be difficult for individual to access online systems even once they are persuaded to share information about an adverse event. Conversely, it can be difficult to sustain the levels of funding and management interest necessary to replenish and monitor supplies of paper-based forms. Usability problems can affect online systems if users do not have access to displays with adequate resolution to present increasingly complex forms. Similarly, it can be difficult to ensure that all potential contributors have access to the software that is required for many Internet-based systems. Even if paper-based forms are used, careful consideration must be given to the design of reporting systems. Even apparently simple information requirements, such as the date when an incident occurred, can lead to problems. For instance, many forms provide no means of specifying that the same adverse event recurred on several occasions. This means that several different forms may be submitted if, for instance, an incorrect medication was administered to the same patient over a course of several days. Many forms include questions about why and incident occurred and how it might have been avoided. The closing sections of this chapter have examined recent human factors work that has pointed to the biases that influence causal analysis.

References

Barach, P. and S.D. Small (2000). Reporting and preventing medical mishaps: Lessons from non-medical near miss reporting systems. *British Medical Journal*, 320, 759–763.

Bates, D.W., D.J. Cullen, N. Laird, L.A. Petersen, S.D. Small, and D. Servi (1995). Incidence of adverse drug events and potential adverse drug events: Implications for prevention. *Journal of the American Medical Informatics Association*, 274, 29–34.

Busse, D.K. and D.J. Wright (2000). Classification and analysis of incidents in complex, medical environments. *Topics in Healthcare Information Management*, 20, 4.

Byrne, R.M.J. and S.J. Handley (1997). Reasoning strategies for suppositional deductions. *Cognition*, 62, 1–49.

Cohen, M.R. (2000). Why error reporting systems should be voluntary: They provide better information for reducing errors. *British Medical Journal*, 320, 728–729.

Cohen, H., E.S. Robinson, and M. Mandrack (2003). Getting to the root of medication errors: Survey results. *Nursing*, 33(9), 36–45.

Evans, S.M., J.G. Berry, B.J. Smith, and A.J. Esterman (2004). Anonymity or transparency in reporting of medical error: A community-based survey in South Australia. *Medical Journal of Australia*, 180 (11), 577–580.

FDA (1999). Final Report of a Study to Evaluate the Feasibility and Effectiveness of a Sentinel Reporting System for Adverse event Reporting of Medical Device Use in User Facilities, Contract # 223-96-6052, JUNE 16. http://www.fda.gov/cdrh/postsurv/medsunappendixa.html

FDA (2002). Medical Device Reporting—General Information, Centre for Devices and Radiological Health, Food and Drugs Administration, Washington, DC. http://www.fda.gov/cdrh/mdr/mdr-general.html

Hirose, M., E. Scott, I. Regenbogen, S. Lipsitz, Y. Imanaka, T. Ishizaki, M. Sekimoto, Eun-Hwan Oh, and A.A. Gawande (2007). Lag time in an incident reporting system at a university hospital in Japan. *Quality and Safety in Health Care*, 16, 101–104.

Hutchinson, A., T.A. Young, K.L. Cooper, A. McIntosh, J.D. Karnon, S. Scobie, and R.G. Thomson (2009). Trends in healthcare incident reporting and relationship to safety and quality data in acute hospitals: Results from the national reporting and learning system. *Quality and Safety in Health Care*, 18, 5–10.

Institute of Medicine (1999). *To err is human: Building a safety health system*. Washington, DC: National Academy Press.

Itoh, K., N. Omata, and H.B. Andersen (2009). A human error taxonomy for analysing healthcare incident reports: Assessing reporting culture and its effects on safety performance. *Journal of Risk Research*, 12 (3 & 4), 485–511.

Jha, A.K., G.J. Kuperman, J.M. Teich, L. Leape, B. Shea, E. Rittenberg, E. Burdick, D.L. Seger, M. Vander Vliet, and D.W. Bates (1998). Identifying adverse drug events: Development of a computer-based monitor and comparison with chart review and stimulated voluntary report. *Journal of the American Medical Informatics Association*, 5 (3), 305–314.

Johnson, C.W. (2000). Designing forms to support the elicitation of information about accidents involving human error. In P.C. Cacciabue (ed.), *Proceedings of the 19th European Annual Conference on Human Decision Making and Manual Control*, January 2000, EC Ispra, Research Center, pp. 127–134.

Johnson, C.W. (2003a). *A handbook of accident and incident reporting*. Glasgow: Glasgow University Press. http://www.dcs.gla.ac.uk/~johnson/book

Johnson, C.W. (2003b). The interaction between safety culture and uncertainty over device behaviour: The limitations and hazards of telemedicine. In G. Einarsson and B. Fletcher (eds.), *Proceedings of the International Systems Safety Conference 2003, International Systems Safety Society*, August 4–8th, 2003, Unionville, VA, pp. 273–283.

Johnson, C.W. (2004). Communication breakdown between the supplier and the users of clinical devices, biomedical instrumentation and technology. *Journal of the US Association for the Advancement of Medical Instrumentation*, 38 (1), 54–78.

Johnson, C.W. (2007). The systemic effects of fatigue on military operations. In *2nd IET Systems Safety Conference*, October 22–24, 2007. The IET, Savoy Place, London, U.K., pp. 1–6.

Johnson, C.W. (2009). Politics and patient safety don't mix: Understanding the failure of large-scale software procurement for healthcare systems. In P. Casely and C.W. Johnson (eds.), *Fourth IET Systems Safety Conference*, October 26–28, 2009. IET Conference Publications, Savoy Place, London, U.K.

Johnson, C.W., B. Kirwan, T. Licu, and P. Statsny (2009). Recognition primed decision making and the organisational response to accidents: Ueberlingen and the challenges of safety improvement in European air traffic management. *Safety Science*, 47, 853–872.

Kaplan, H.S., J.B. Battles, T.W. Van der Schaaf, C.E. Shea, and S.Q. Mercer (1998). Identification and classification of the causes of events in transfusion medicine. *Transfusion*, 38 (11–12), 1071–1081.

Kingston, M.J., S.M. Evans, B.J. Smith, and J.G. Berry (2004). Attitudes of doctors and nurses towards incident reporting: A qualitative analysis. *Medical Journal of Australia*, 181 (1), 36–39

Lawton, R. and D. Parker (2002). Barriers to incident reporting in a healthcare system. *Quality and Safety in Healthcare*, 11, 15–18.

Mackenzie, C.F., N.J. Jefferies, A. Hunter, W. Bernhard, and Y. Xiao (1996). Comparison of self reporting of deficiencies in airway management with video analyses of actual performance. *Human Factors*, 38, 623–635.

MDH (2000). Medical Errors and Patient Safety: Key Issues, Minnesota Department of Health, Health Economics Program Issue Paper, December.

Nakajima, K., Y. Kurata, and H. Takeda (2005). A web-based incident reporting system and multidisciplinary collaborative projects for patient safety in a Japanese hospital. *Quality and Safety in Health Care*, 14, 123–129.

NHS Expert Group on Learning from Adverse Events in the NHS (2000). An organisation with a memory. Technical report. National Health Service, London, U.K.

NPSA (2004a). National Reporting Goes Live, National Patient Safety Agency, London, U.K., March. http://www.npsa.nhs.uk/dataset/dataset.asp

NPSA (2004b). Healthcare Professionals: Frequently Asked Questions. National Patient Safety Agency, London, March. http://www.npsa.nhs.uk/faq/h_profFAQsView.asp

NPSA (2004c). About the NPSA. National Patient Safety Agency, London, U.K., March 2004c. http://www.npsa.nhs.uk/static/About.asp

NPSA (2009a). Seven Steps to Patient Safety for Primary Care. London, U.K., February 2009.

NPSA (2009b). Learning from Reporting: Improving the Quality of Data, Quarterly Data Summary Issue 14, Report Number 1137F1, November 2009.

Nolan, T.W. (2000). System changes to improve patient safety, *British Medical Journal*, 320, 771–773.

Raymond, B. and R.M. Crane (2001). Design Considerations for a Patient Safety Improvement Reporting System, Institute for Health Policy, Kaiser Permanente, Oakland, April.

Robinson, A.R., K.B. Hohmann, J.I. Rifkin, D. Topp, C.M. Gilroy, J.A. Pickard, and R.J. Anderson (2002). Physician and public opinions on quality of health care and the problem of medical errors. *Archives of Internal Medicine*, 162, 2186–2190.

Runciman, W.B. (2002). Lessons from the Australian patient safety foundation: Setting up a national patient safety surveillance system—Is this the right model? *Quality and Safety in Health Care*, 11 (3), 246–251.

Salter, M. (2003). Serious incident inquiries: A survival kit for psychiatrists. *Psychiatric Bulletin*, 27, 245–247.

Shaw Phillips, M.A. (2001). National program for medication error reporting and benchmarking: Experience with *MedMARx*. *Hospital Pharmacy*, 36 (5), 509–513.

Snijders, C., T.W. van der Schaaf, H. Klip, W.P.F. van Lingen, R.A. Fetter, and A. Molendijk (2009). Feasibility and reliability of PRISMA-Medical for specialty-based incident analysis. *Quality and Safety in Health Care*, 18, 486–491.

Staender, S., M. Kaufman, and D. Scheidegger (1999). Critical incident reporting in anaesthesiology in Switzerland using standard Internet technology. In C.W. Johnson (ed.), *Proceedings of the 1st Workshop on Human Error and Clinical Systems*. Glasgow Accident Analysis Group, University of Glasgow.

Taylor, J.A., D. Brownstein, D.A. Christakis, S. Blackburn, T.P. Strandjord, E.J. Klein, and J. Shafii (2004). Use of incident reports by physicians and nurses to document medical errors in pediatric patients. *Pediatrics*, 114 (3), 729–735.

U.S. General Accounting Office (1997). GAO Report to Congressional Committees Medical Device Reporting Improvements Needed in FDA's System for Monitoring Problems with Approved Devices, GAO-HEHS-97-21, Washington, DC.

Van Geest, J.B. and D.S. Cummins (2003). An Educational Needs Assessment for Improving Patient Safety Results of a National Study of Physicians and Nurses, National Patient Safety Foundation, White Paper Report Number 3. http://www.npsf.org/download/EdNeedsAssess.pdf

Vincent, C. and A. Coulter (2002). Patient safety: What about the patient? *Quality and Safety in Health Care*, 11, 76–80.

Vincent, C., S. Taylor-Adams, and N. Stanhope (1998). A framework for analysing risk and safety in clinical medicine. *British Medical Journal*, 316, 1154–1157.

Vincent, C., J. Venn, and G. Hanna (2010). High reliability in health care. *British Medical Journal*, 340, 84.

Waring, J. (2004). A qualitative study of the intra-hospital variations in incident reporting. *International Journal for Quality in Health Care*, 16 (5), 347–352.

Waring, J., E. Rowley, R. Dingwall, C. Palmer, and T. Murcott (2010). Narrative review of the UK patient safety research portfolio. *Journal of Health Services Research and Policy*, 15, 26–32.

Weeks, W.B. and J.P. Bagian (2000). Developing a Culture of Safety in the Veterans Health Administration, Effective Clinical Practice, American College of Physicians, November/December.   http://www.acponline.org/journals/ecp/novdec00/weeks.htm

Witham, M.D., P.M. Jenkins, and M.E.T. McMurdo (2006). Using a sentinel adverse incident audit on a medicine for the elderly ward. *Quality and Safety in Health Care*, 15, 446–447.

Wollerton, M.A. (1997). Last printing of user facility reporting bulletin. *FDA User Facility Reporting Bulletin*.

Wu, J.-H., W.-S. Shen, L.-M. Lin, R.A. Greenes, and D.W. Bates (2008). Testing the technology acceptance model for evaluating healthcare professionals' intention to use an adverse event reporting system. *International Journal for Quality in Health Care*, 20 (2), 123–129.

27

Error Recovery in Health Care

Tosha B. Wetterneck

CONTENTS

27.1 Introduction 449
27.2 Importance of Error Recovery in Health Care 449
27.3 What Is Error Recovery? 450
    27.3.1 Conceptualizing Error Recovery 450
    27.3.2 Conceptualizing Error Recovery in Health Care 451
27.4 Error Recovery Defined 452
    27.4.1 Error Detection 452
    27.4.2 Error Explanation 453
    27.4.3 Error Correction 454
    27.4.4 When Errors Are Detected but Not Corrected 455
27.5 Work System Redesign to Improve Error Recovery 455
27.6 Examples of Error Recovery in Health Care 456
    27.6.1 Error Recovery in Critical Care and Emergency Department Environments and the Nurse's Role ... 456
    27.6.2 Physician's Role in Error Recovery 458
    27.6.3 Pharmacist's Role in Error Recovery 458
    27.6.4 Health Information Technology and Error Recovery 459
27.7 Conclusion 459
References 460

27.1 Introduction

Since the landmark IOM report in 2000 (Kohn et al., 2000) that reported the large numbers of deaths and serious harm occurring from errors in health care delivery, the health care community has focused many resources to improving the quality and safety of care provided to patients. In doing so, health care organizations have embraced many concepts and techniques from other industries to make health care safer. Much of the patient safety efforts have been focused on error prevention by changing the health care system and processes of care to stop errors from occurring in the first place. Thus, if errors do not occur, then there is no possibility of errors reaching the patient and causing harm. However, the attempts to make health care systems function as high reliability organizations (HROs) by improving the reliability of health care processes and decreasing errors to very low levels as seen in other industry HROs are hampered because the nature of health care requires humans in most processes. Because it is not realistic to think that errors in (most) health care systems and

processes can be reduced to levels seen in HROs, other strategies must be in place to deal with errors that do occur and prevent or minimize the harm to patients from these errors. This strategy is error management through error recovery. Error recovery involves a strategy that identifies errors before they reach the patient and remedies them to avoid patient harm and would complement error prevention strategies. Research has ensued over the past 25 years to better understand error recovery and provide a framework in which to study the recovery process. This chapter gives an overview of the concept of error recovery and why it is an important strategy to improve safety in health care.

27.2 Importance of Error Recovery in Health Care

To improve patient safety and care quality, it is important for health care organizations to focus on error recovery, in addition to error prevention. Error recovery

is important for many reasons. *First*, the importance of being aware of the potential for errors in the work environment, recognizing them, and recovering from them is an important characteristic of HROs (Weick and Sutcliffe, 2001). *Second*, is the nature of providing health care. Attempts to standardize and automate health care processes are constrained by the need for health care providers (humans) to deliver the care that patients require. Humans are necessary because of the social and humanistic aspects of providing health care. Also, the complexity of patient care, including the rapidity at which patient conditions change, requires continued human involvement in monitoring, data analysis, and decision making processes. While humans allow for flexibility, creativeness, and adaptability in work processes, the occurrence of workarounds or violations, in which workers act in ways that are in contradiction with policies and procedures, may lead to errors (Alper and Karsh, 2009). Workers experience high time pressure, high workload, and long work hours, which leaves them at risk for fatigue and stress, and therefore at risk of making errors. In addition, patients, as inputs into health care processes, vary greatly on disease and health states, and sociodemographic and economic factors. Patients behave in difficult-to-predict and irrational manners. All of these patient factors make the standardization of health care delivery and its outcomes a challenge. *Third*, making errors and learning from them is an important part of the learning process. Humans need to experience errors and to develop contingency plans for what to do when things go wrong. So, error recovery efforts are important to help humans recover from their own and other people's errors. In health care, the focus on avoiding or minimizing negative consequences with recovery translates into not allowing the results of errors "reach the patient" and if errors do occur, then putting into place measures to monitor for harm and minimize any harm that may occur. *Fourth*, there is a tendency for health care workers to "normalize" deviations in health care. Some errors and failures occur at such high frequency that workers just "live with errors" and attempt to build systems to recover from the errors when they occur. While this strategy allows for care to continue to be given, it does not allow the organization to learn from the errors that are occurring to redesign systems and formalize error recovery mechanisms and still leaves patients at risk of negative consequences of errors. *Last*, there is increasing implementation of technology in health care in an effort to improve patient safety and care quality, for example, electronic health records, computer provider order entry (CPOE), clinical decision support (CDS) systems, barcoded medication administration (BCMA) technologies, and smart intravenous (IV) pumps. However, the implementation of these technologies has not necessarily decreased error

rates and may lead to new errors occurring. So the need for error recovery is great.

27.3 What Is Error Recovery?

Error recovery is the process by which errors that occur are identified and actions are taken to stop the error from having untoward consequences (van der Schaff and Kanse, 2000). The main goal of error recovery is to prevent negative consequences from occurring as a result of the error. Error recovery is considered *planned* if it is built into the system proactively as a strategy to manage errors (Kanse et al., 2006). To initiate error recovery, there must be a failure which could include either an error or a violation that then produces a deviation that is somehow possible to be detected. This chapter will focus on errors rather than violations in discussing error recovery. Error recovery is distinct from the occurrence of the error itself (Allwood, 1984) and includes three stages: detection, explanation (diagnosis), and correction of the error to avoid negative consequences (Kontogiannis, 1999). Error recovery may occur during all performance stages: the planning of an action, its execution, or while monitoring the outcomes of the action (Kontogiannis, 1997).

27.3.1 Conceptualizing Error Recovery

Several frameworks have been proposed for error recovery to guide the evaluation of the error recovery process. The Eindhoven model of incident causation was developed by van der Schaaf (1992) and used to investigate near miss chemical industry accident investigations. It states that if system defenses are not adequate to deal with systems failures and an incident develops, that if there is adequate human recovery, the incident is termed a near miss and if not, it is an accident (van der Schaaf, 1992). From this model has grown a common framework to conceptualize error recovery, as shown in Figure 27.1. The error recovery process starts with failure(s) in the work system that serves as conditions under which an error occurs. For error recovery to occur, the error must first be *detected* and then actions must be taken to *correct* the error or prevent or repair any negative *outcomes* that occur as the result of the error. At times after error detection, the operator will go through a process of determining why the error occurred called *diagnosis* or *explanation*. This may occur before, during, or after error correction, or not at all.

Recently, Patel and colleagues proposed a cognitive model of error recovery based on Norman's seven stages of action model (Norman, 1988) and used this model to

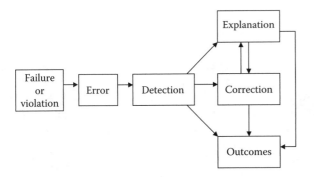

FIGURE 27.1
Conceptual model of error recovery. (From Kanse, L. et al., *Ergonomics*, 49, 505, 2006. With permission.)

evaluate error recovery in health care settings (Patel et al., in press). The model expands the three-step process (detection, explanation, and correction steps) to seven steps: (1) error perception (detection); (2) error interpretation (explanation); (3) error evaluation, evaluating the consequences of the error; (4) setting a goal for response; (5) deciding to take action; (6) specifying corrective actions; and (7) executing corrective actions. This model makes explicit the different action steps required for error correction and ultimately recovery to occur. It also proposes error evaluation separately from detection and explanation, which will be discussed in Section 27.4.4.

27.3.2 Conceptualizing Error Recovery in Health Care

While the concept of error recovery may still be new to many health care organizations, the term "near miss" is not. There has been much focus in health care on the value of near misses for organizational learning and safety improvement (Pathiraja et al., 2011). During this time, the conceptualization of near misses and even the term used for near misses in health care have evolved, led by efforts of The Joint Commission, a health care accreditation organization and leader in the patient safety improvement, and other thought leaders in patient safety. In health care, we know that a small percentage of errors that occur actually reach patients and go on to cause harm. Near misses were originally conceptualized as errors that occur but that either do not reach patients or do reach patients but do not cause harm. For this to occur, the error may be caught and fixed before reaching the patient or the process does not evolve in a manner that the error proceeds to reach the patient. For example, a medication order is placed in which there is a wrong dose ordered that could harm the patient but the medication is an "as needed" medication and is never administered. If the error reaches the patient, harm may not occur for many reasons. One is pure "luck," in other words, the situation would be

reasonably expected to cause harm but the harm does not occur, for example, a patient allergic to a medication receives this medication in error, and the rash he previously manifested as the allergy happens to not occur. Second, the error itself may not cause a situation which is inherently risky, for example, a nurse gives the wrong medication to a hospitalized patient but the medication is acetaminophen, a medication that is usually well tolerated by most people at typical dosing regimens. Last, error recovery may have occurred and measures put into place to monitor the patient and preventive treatment put into place to avoid harm. There then is a differentiation between (1) errors that occur and do not reach the patient, (2) errors that reach the patient but do not cause harm, and (3) errors that reach the patient and cause harm but the harm is mitigated through recovery efforts (Wetterneck and Karsh, 2011; Wu, 2011). Error recovery may be involved in the first two scenarios leading to harm avoidance and is the source of harm mitigation in the third scenario. It is important to distinguish between errors that do not reach patients and errors that reach patients and do not cause harm as the processes in place, which lead to these differing outcomes in health care, may be different (Kessels-Habraken et al., 2010).

From a definition standpoint in health care, the term "close call" is proposed to replace the use of near miss for events/incidents that do not reach the patient but could have caused harm as the words more closely describe what happened from a health care standpoint (Wu, 2011). "No harm event" is proposed to replace the use of near miss for incidents that reach the patient and do not cause harm and "adverse event" as an incident that causes patient harm. However, none of these terms imply that error recovery has occurred so there would need to be additional characterization of the events to identify this characteristic. Other medical error taxonomies may be useful to identify error recovery in reported errors. For example, the Agency for Healthcare Research and Quality's (2010) Patient Safety Organization Generic Common Formats for error reporting (version 1.1) has two descriptors that relate to error recovery. One is an event circumstance—the "reason the near miss did not reach the patient," and the second is considered patient information—"rescue intervention made within 24 h after discovery of an incident to reduce or halt the progression of harm to the patient." The National Coordinating Council for Medication Error Reporting and Prevention's (NCC MERP, 1998) taxonomy for medication errors also acknowledges errors and the patient outcome of the errors with consideration of error recovery. Category B errors are errors that did not reach the patient, Category C errors reach the patient but did not cause harm, and Category D errors reach the patient and require monitoring or treatment to prevent harm from occurring. Categories E–H errors represent errors with increasing levels of harm. While this taxonomy allows the categorization of errors that do and do not reach the patient as well as errors that do not lead to harm related to recovery efforts, errors that lead to some harm with recovery efforts having reduced the harm cannot be elucidated.

Error recovery in health care, then, works by stopping errors from reaching the patient or by minimizing the negative consequences that could occur if and when the error reaches the patient. Health care organizations function to serve patients; therefore, the patient is an important factor in all patient safety and quality efforts. Inserting the patient into the error recovery pathway in health care is important for health care organizations' understanding of error recovery.

27.4 Error Recovery Defined

This section reviews the elements of error recovery using the approach developed by van der Schaff and colleagues with considerations from the work of Patel and colleagues (van der Schaaf, 1992; Kanse et al., 2006; Patel et al., in press). Error recovery is comprised of three distinct elements: error detection, error explanation (diagnosis), and error correction. The end result of the recovery is avoidance or mitigation of negative consequences (Kontogiannis, 1999). Error detection and correction are necessary elements of error recovery.

27.4.1 Error Detection

Error detection is defined as the realization that an error or deviation has occurred or suspicion of such that leads to the determination that an error occurred (Allwood, 1984; Kontogiannis, 1999). The error detection step is necessary for error recovery; therefore, much of the error recovery research has focused on the mechanisms of and triggers for error detection.

There are multiple mechanisms by which errors are detected. These mechanisms are related to the type of error that has occurred, that is, slip or mistake, based on Reason's generic error-modeling system (Reason, 1990). Error detection may occur by the person who has committed the error or by another person. The following studies investigated self-detection of errors and their relationship to the type of error. Allwood (1984) studied self error detection by persons performing problem solving and found three mechanisms by which errors were detected. The first mechanism is direct error hypothesis, that is, the immediate detection and correction of an error. Second is error suspicion in which the proposed solution was unusual or unexpected and therefore provoked further investigation. Third is standard checking

in which persons checked their answers without regard to whether or not an error occurred. Eighty-seven percent of errors due to failures in execution or slips were detected versus 52% of the errors associated with choosing the wrong method (failures in planning or mistakes) (Allwood, 1984). Rizzo et al. (1987) used Allwood's detection categories to study student performance of tasks and errors types committed while using a computer database system (Rizzo et al., 1987). Most of the slips were detected by direct error hypothesis whereas knowledge-based mistakes were detected by error suspicion, and rule-based mistakes were mainly detected by direct error hypothesis and error suspicion.

Sellen (1994) developed an error detection mode taxonomy based on diary studies of 75 persons' self-detection of 527 errors. She found three major modes of slip and mistake self-detection. The first mode is *action based*, in which a person identifies a mismatch between the executed action and either their conscious intentions or planned action. For example, a physician intends to order levofloxacin for pneumonia treatment but instead selected levodopa from the computer pick list. The second mode is *outcome based* in which the error is caught based on a mismatch between the actual outcome and the expected outcome or a match between known error forms and the outcome. For example, in the aforementioned example, the physician did not initially realize that they ordered levodopa (a drug for Parkinson's disease) instead of levofloxacin (an antibiotic) for pneumonia; however, the patient condition unexpectedly worsens. Last, the limiting function mode represents error detection based on environmental constraints which prevent further action after an error has occurred. For example, a physician trying to order levofloxacin for pneumonia receives a computerized alert stating the patient is allergic to the antibiotic which prevents ordering of the drug.

Sellen also noted a fourth mode of error detection that was not a self-detection mode but rather was the detection of slips and lapses by another person (Sellen, 1994). These detections occurred during monitoring of actions by another person (e.g., supervisor) or recognition by a person who is more familiar with the plans required to obtain the expected outcome. Frequently, the person committing the error does not have the assessment capability to detect the error despite an outcome that may not match the expected outcome (Sellen, 1994; Kontogiannis, 1999; Blavier et al., 2005; Kanse et al., 2006). This is commonly seen in trainees or less experienced workers as compared to expert workers. Additionally or alternatively, the person may suffer from "cognitive tunnel vision" in which they disregard available evidence that an error has occurred for other evidence that substantiates an outcome closer to their expectations (Sellen, 1994; Kontogiannis, 1999; Blavier et al., 2005; Kanse et al., 2006). Error detection may also be more likely to occur by persons who have expertise in the task (Blavier et al., 2005). For example, pharmacists may detect and remedy more medication discrepancies when performing medication reconciliation of a patient's medication list on hospital discharge than the ordering physician as the pharmacist is an expert in medications and solely deals with patient medications (Schnipper et al., 2006).

In general, the research on error detection has found that there is a relationship between the error detection mechanism and the error type or human performance stage (van der Schaaf and Kanse, 2000). Failures in execution, namely slips, are often detectable by the person committing them and then recovered from before any negative consequences can occur. Failures in planning, namely rule-based and knowledge-based mistakes, are more difficult to detect than failures in execution (Zapf et al., 1994). Failures in planning may be detected by the person committing them but often a second person will be necessary to detect the error and recovery from it. Table 27.1 presents examples of systems defenses that promote error detection in health care.

27.4.2 Error Explanation

Error explanation deals with investigation into what caused the error or deviation that is detected. It involves determining whether the error occurred during the interpretation of the situation, the planning stage, or the execution of the tasks. Less attention has been given to this error recovery step than detection or correction. This may be because this step is not always necessary for recovery to occur; that is, correction of some errors may immediately follow detection without any explanation of why the error occurred (Kontogiannis, 1999; Kanse et al., 2006). van der Schaaf (1992) analyzed near miss reports from a steel plant and found that few reports explained why the error occurred (van der Schaaf, 1992). But just because error explanation does not always need to occur to correct an error does not mean it is not important. To ultimately prevent errors from occurring, investigation into the cause of the error must occur to identify preventive solutions for system redesign.

There are multiple mechanisms by which the reasons for errors are explained. The error detection method of standard checking has a "built-in" error explanation step. By going back through the actions just performed and making sure they are correct, the reason for an error often becomes obvious. Another mechanism is having the system present information to the user about an error that occurred along with the detection method. For example, barriers can be placed in the workplace designed to provide information to the user about the origin of the error. Computer interfaces can be designed

TABLE 27.1

System Defenses That Promote Error Detection and Health Care Examples

Warnings and alarms
- Patient monitoring equipment in intensive care units and operating rooms

Physical barriers
- Universal precautions: use of eye shields, gloves, and gowns, etc., to prevent disease transmission

Procedures
- Time-out protocols before surgical procedures

Hardware and software lockouts
- Designing electronic health records to only allow one patient record to be open at a time (to avoid wrong patient errors)

Reminders
- Clinical decision support in electronic health records to promote needed preventive screening

Human–machine redundancy
- Use of bar coding technology redundant to required visual medication check

Self-detection and correction of errors
- Double calculation of all chemotherapy doses

Forcing functions
- Hard-stop limits on Smart IV pump programming

Sources: Wetterneck, T.B. and Karsh, B.T., Human factors applications to understanding and using close calls to improve health care. In A.W. Wu (Ed.), *The Value of Close Calls in Improving Patient Safety: Learning How to Avoid and Mitigate Harm,* Joint Commission Resources, Chicago, IL, pp. 39–54, 2011. With permission. Adapted from Salvendy, G. (Ed.), *Handbook of Human Factors and Ergonomics,* 3rd edn., John Wiley & Sons, Inc., Hoboken, NJ, p. 711, 2006.

to make previous erroneous actions obvious to the user through messages sent to the user or other screen displays, thus increasing both the likelihood of error detection and explanation. For example, some word processing software uses underlining of text and color to directly indicate incorrect spelling or grammar usage on the screen display. The spelling and grammar function can be used to obtain further information about the error and how to correct it. In health care, computerized alerts sent to providers in the ordering process provide information on the error detected, for example, a drug allergy, and information about the error, for example, the drug levofloxacin was ordered and the patient has a severe reaction to the drug. At times, the source of the error is not obvious and the person will have to actively search for information about the error. This occurs commonly in situations where a second person has detected an error and is attempting to recover from it as the origin of the error itself may not be as obvious if the actions and intentions of the original operator are not transparent or known to the person. Consider the following error example: a provider chooses levodopa from a computer pick list of medications instead of levofloxacin. However, the provider does not detect the error. A pharmacist, who reviews all medication orders in the hospital, reviews the order for levodopa and realizes the dose chosen for the levodopa is incorrect but does not initially realize that it is actually a wrong medication error. The pharmacist needs to call the physician for a clarification of the dose. If the pharmacist calls the same physician who ordered the medication, that physician may recognize the wrong medication error. However, if the ordering physician is not available, another physician may not recognize the wrong medication error. The pharmacist could also review the patient's medical conditions on hospital admission and recognize that the patient does not have an indication for levodopa and that the patient has pneumonia, and an antibiotic has not been prescribed. So the pharmacist would have more information to share with the physician when calling about the incorrect medication order. So error explanation by a second person may have the extra challenges of needing to find out the actions and intentions of another person, and unless this is necessary for error recovery, this step may not occur or not occur frequently or correctly. Therefore, when possible, systems should be designed to present information to the user at the time of error detection to aid in error recovery. The need for teamwork and closed-loop communication between team members and during hand-offs becomes obvious for error explanation to proceed well (Baker et al., 2006).

There are multiple reasons why error explanation may not occur in the process of error recovery. First, error explanation may not be necessary to correct the error

and therefore not performed. For example, if a provider chooses levodopa from a computer pick list of medications instead of levofloxacin, the provider does not need to understand that there was a selection error made that was possibly related to look-alike, sound-alike medication names or a hand–eye coordination issue during the physical selection of the medication name using the computer mouse. Instead, the provider cancels the levodopa order and orders levofloxacin. In fact, health care workers frequently focus on correcting errors that occur to continue on with patient care without understanding the underlying problems (Tucker et al., 2002). This is known as first-order problem solving. However, organizational learning and improvement relies upon second-order problem solving, or the learning that occurs from investigating the causes of errors such that the system can be redesigned to prevent the error from occurring the next time (Tucker et al., 2002). Systems issues which contribute to the lack of error explanation include time pressure, high workload, lack of motivation to understand the problem, poor safety culture, onerous error reporting systems, and inability to communicate directly with persons who could help solve the error (Kontogiannis, 1999; Tucker et al., 2002).

27.4.3 Error Correction

Error correction is the last step in error recovery and along with error detection is necessary for error recovery to occur. It consists of a reassessment of the situation, modification of the plan or development of a new plan to correct or compensate for the error, and implementation of this plan (Kontogiannis, 1999). When determining the plan of action for error correction, the operator needs to consider three avenues to solve the problem. They may either bring the system state back to the same state as before the error (backward recovery), bring the system to an intermediate state to buy time to problem solve (forward recovery), or bring the system to the intended goal state using redundant system mechanisms (compensatory recovery). The ultimate goal is to prevent or reduce the negative consequences of the error. An example of backward recovery is a physician who orders an antibiotic for pneumonia to which the patient is allergic subsequently canceling the order after receiving a drug allergy alert from the computerized order entry system and ordering a new antibiotic. An example of forward recovery is when a pharmacist reviews a medication order for an antibiotic and determines the patient is allergic to the medication but the patient also has multiple other allergies making antibiotic selection difficult, so the order for antibiotics is canceled until an infectious disease specialist can consult and recommend antibiotics. Last, an example of compensatory recovery is a pharmacist reviewing a medication order and realizing that the dose is too high for the patient due to the patient's poorly functioning kidneys. The pharmacist has a protocol that allows her to change the dose without calling the physician who wrote the order and the pharmacist makes the change. The process of error correction can be a stressful time for those involved as the context of work may have changed after the error occurred resulting in increased time pressure or fewer available resources (Kontogiannis, 1999).

27.4.4 When Errors Are Detected but Not Corrected

For error recovery to occur, an error must be both detected and corrected in a manner that results in mitigation or prevention of negative consequences. However, there are circumstances when an error may be detected but not corrected, and thus not recovered. There are many reasons this may occur. Recall the Patel et al. (2011) model of error recovery and the proposed step of evaluating the error as a task separate from the act of error detection and explanation. The person detecting the error may evaluate it and ultimately judge that the consequences of the error are not enough to warrant actions to fix the error at that point in time. For example, a patient erroneously receives a dose of morphine for pain relief meant for a different patient, that is, wrong patient error, but the patient who received it also has an active prescription for morphine at the same dose and was due to receive it soon. Or the error may be judged to have no potential for harm, and therefore no actions are necessary to avoid negative consequences. There are also circumstances when there is not an opportunity to recover from the error, often due to the

occurrence of harm in a short period of time that cannot be reversed. For example, a patient may receive another patient's dose of morphine and stop breathing and die before anything can be done to recover. Lastly, there are systems issues like communication across authority gradients, poor safety culture, and inadequate expertise and resources, which may hamper error recovery efforts. Patterson et al. (2003) describe an example of an incorrect chemotherapy medication (wrong medication error) that was detected by both the nurse and the pharmacist, however, the physician that they contacted with their concerns insisted that the chemotherapy medication was correct. The physician was not the staff physician, and the concerns were not escalated up the authority gradient, rather, the patient received the incorrect medication. This speaks to the need to identify human performance issues in health care and to redesign health care systems to support error recovery.

27.5 Work System Redesign to Improve Error Recovery

Based on the known science of error detection and work systems, strategies have been suggested to support the likelihood of error detection and error recovery in industry and in health care (Table 27.2). Many of Weick and Sutcliffe's (2001) characteristics for becoming an HRO have to do with error detection and recovery. The reporting of near misses, errors, their causes, and

TABLE 27.2

Error Recovery Supporting Strategies in Health Care

Embrace HRO characteristics
- Encourage reporting of near misses and errors, their causes and recovery efforts and make it easy for staff to report
- Share reported near misses, errors, error recovery strategies, and proactive risk assessment results across the organization
- Use near misses and errors to redesign the work system to mitigate potential for patient harm
- Avoid the punitive approach when dealing with errors
- Encourage vigilance for errors that may occur, especially after changes to the work system, for example, technology implementation
- Develop a culture of safety

Train and prepare employees and trainees to be resilient by learning about the principles of error recovery and ways they can detect and correct their own errors and others errors
- Develop contingency plans for high risk and error prone situations, for example, loss of electronic health record service
- Use simulation for training to allow employees and trainees to commit errors and recover from errors in a setting that cannot harm patients

Design work systems to maximize error recovery as well as error prevention
- Adequate supervision and training of staff and trainees
- Error recovery strategies planned into job descriptions of employees in high risk and error prone situations, for example, double checking of chemotherapy dose calculations
- Implement decision support tools to assist workers in identifying and correcting errors
- Teamwork and communication across disciplines
- Improve the observability, traceability, and reversibility of errors

recovery efforts by front-line staff to the organization can promote organizational learning. Organizations can share these errors and strategies with front-line workers to educate staff and encourage vigilance for similar errors and opportunities for recovery. A non-punitive approach to errors is essential for reporting and developing a patient safety culture. In addition to the HRO principles, organizations can promote error recovery by training employees and trainees to be resilient when errors do occur. Education about medical error and error recovery strategies as well as developing and practicing contingency plans for high risk and error prone situations are critical. Also, the use of simulation as a training tool in health care is rapidly growing (Gaba, 2007; Henneman et al., 2010). Simulation provides a safe environment for employees and trainees to practice care delivery, commit errors, and learn to recover from those errors without the risk of harm to patients.

System (re)design is another important means of influencing error recovery (Table 27.2). Organizations can plan error detection methods into the processes of care, for example, standard checking of work by self or others (Kanse et al., 2006), or standard monitoring of actions and outcomes to detect deviations (Sellen, 1994). Planned error recovery steps are part of a person's job or a process and are usually formalized by the organization through policy and procedure or job description. For example, in health care, it is standard practice that pharmacists review all patient medication orders in hospitals for appropriateness and correctness to the patient condition. Second, enhancing team communication and supervision can help detection of knowledge- and rule-based errors by a second person (Kontogiannis, 1997). Third, the addition of decision support tools can provide feedback to users on actions they have taken and alert them to potential errors (Kontogiannis, 1997). Last are three essential characteristics of work systems and process design, which can aid front-line workers in error recovery: observability, traceability, and reversibility (Leplat, 1989; Kontogiannis, 1999). Observability involves making an error detectable to the person who commits the error or someone else so it can be fixed. The system can be designed to give feedback to the workers, for example, alarms on patient monitoring equipment or drug alerts in CPOE. Traceability involves being able to understand what occurred that led to the error. Reversibility involves the ability to cancel the effects of the error, thereby preventing harm and correcting the error. An example of these three features at work in health care is the design of medication ordering processes in hospitals and the use of CPOE. The ordering physician places an order for a medication that the patient is allergic to and receives an allergy alert from the computer (observability). The allergy alert shows the physician which order triggered the alert and the

patient's allergy information (traceability). By responding to the alert, the physician can choose to cancel the previous order before it has been processed and select a new medication (reversibility).

Kontogiannis (1999) outlines a taxonomy of strategies that operators use to recover from error, much of it based on the previous error detection research described earlier. The five strategies include (1) using inner feedback (e.g., cues from memory and action-based detection); (2) external communication through interactions with work colleagues, team members, supervisors, or others involved in the process; (3) exploring system feedback like using feedback from the interface to understand errors occurring; (4) planning behaviors, for example, standard checking, error suspicion, or contingency plans; and (5) error informed strategies which involve matching deviations in the current situation to errors previously recognized and coping with the frustration from error occurrence. So these strategies depend upon the operator but can also be supported by the work environment, like development of smart computer interfaces as seen in word processing programs or in health care as CDS (Kuperman et al., 2007), team training or proximity to team members, and appropriate barriers, for example, as shown in Table 27.1.

27.6 Examples of Error Recovery in Health Care

There is a growing body of literature that has examined error recovery in health care. Most commonly, error recovery has been examined by professions type (i.e., nurses, physicians, pharmacists), health care setting (e.g., critical care units or operating rooms), or during particular health care processes, for example, the medication use process or anesthesia delivery. Multiple methods have been used to study error recovery, including direct observation of staff, chart review, computerized monitoring of clinical data, and administrative databases that may indicate an adverse event or error, error reporting systems, prospective risk assessment, interviews, focus groups, and surveys (Montesi and Lechi, 2009). The following section provides examples of health care error recovery based on these categories.

27.6.1 Error Recovery in Critical Care and Emergency Department Environments and the Nurse's Role

The critical care environment, for example, intensive care units and emergency departments (EDs) in hospitals, is a complex environment with the sickest patients requiring high-intensity and often high-risk patient care from providers and staff. Therefore, the environment is prone to adverse events and errors, and recovery is an important mechanism to avoid patient harm (Patel and Cohen, 2008). Efforts to evaluate error recovery in critical care environments have documented the frequency of occurrence of medical errors and error recovery and described error recovery processes that take place. Rothschild et al. (2005a) performed the Critical Care Safety Study to evaluate errors and adverse events in two academic intensive care units (one medical unit and one coronary care unit). They defined an intercepted error as an error that was recovered from before it reached the patient. Using direct observation of intern physicians (first year of training after medical school), solicited reports of errors, computerized adverse drug event detection monitoring, and chart abstraction, they identified 223 serious errors (errors with potential to cause harm or actually caused harm) over 189 observation days. Sixty percent of serious errors were intercepted. The nurse, physician, and pharmacist were all involved in error interception (42%, 23%, and 17% of total serious errors, respectively). Seventy-eight percent of serious errors were medication related.

In a separate study, Rothschild et al. (2006) described errors that were committed and intercepted before reaching the patient in an academic coronary care unit with a specific focus on nurse interception of errors. Nurses are very important in error recovery as they provide direct patient care in hospitals around the clock and are knowledgeable about and involved with most patient care decisions and health care delivery that occurs. Using direct observation of nurses, chart abstraction, and review of incident reports, the researchers identified 142 errors over 147 days, most of which (69%) were intercepted before reaching the patient and were medication related (Rothschild et al., 2006). The errors that were recovered from where most commonly related to the execution of a task and failing to follow protocol. Nurse recovery examples included the nurses catching a medication order written on the wrong patient, the nurse identifying for the physician that the source of a patient's respiratory distress was being disconnected from the ventilator, and a nurse prompting a physician to discontinue an unnecessary urinary catheter.

Rogers and colleagues have evaluated the role of nurses and specifically critical care nurses in error discovery and correction (Balas et al., 2004; Rogers et al., 2008). In the first study, a random sample of U.S. nurses was given logbooks to record over a 28-day time period if they made an error during their work shift or if they caught themselves about to make an error (labeled a "near error") and to describe the events. Thirty-two percent of nurses (127 out of 393) recorded making at least one near error, for a total of 213 near errors. Fifty-six percent of the described near errors were medication related. Many nurses cited interruptions and team communication issues as factors relating to the near errors. In a second study, 502 U.S. critical care nurses used logbooks to record information at the end of their shift about errors they made or discovered during their work shift over a 28-day time period (Rogers et al., 2008). Of the 502 nurses, 184 reported at least one error. Discovered error was not specifically defined except that it was an error that someone else had committed. A total of 367 errors were discovered, and 12% of these were discovered before they reached the patient. Nurses most commonly discovered nurses' errors (41% of discovered errors) and physician ordering errors (9%). The most frequent types of discovered errors include medication administration errors (44%), procedural errors (31%), and charting errors (15%).

Henneman et al. (2005, 2006) have evaluated medical errors and error recovery in the ED. In an observational study of voluntarily reported medical errors, 203 medical errors were reported (Henneman et al., 2005). ED providers or staff recovered from 47% of the errors. Recovery was defined as detecting and correcting the error such that no harm occurred. A majority of these errors were corrected by nurses (60%). Henneman and colleagues went on to explore the mechanisms and strategies by which nurses recover from errors in the ED (Henneman et al., 2006). Data from four focus groups of ED nurses at a U.S. academic medical center were analyzed using content analysis to identify strategies for identifying, interrupting, and correcting errors. Nurses used five different strategies to identify errors: surveillance, anticipation, double checking, awareness of big picture, and experiential knowing. Nurse strategies for interrupting errors included patient advocacy, offer of assistance, clarification, verbal interruption, and creation of a delay. Nurses corrected errors by assembling the team to communicate and involving leadership when needed.

Carayon and colleagues used a proactive risk assessment approach to identify nursing error recovery processes in intensive care units centered on medication management (Faye et al., 2010). First, they used observation and interviews of critical care nurses in an intensive care unit at a community hospital to define the medication management process. They then used a modified failure modes and effects analysis process to identify failures in the process and recovery processes. Twenty-one recovery processes were identified for the five most critical failure modes. These recovery processes ranged from calling another person for assistance (physician, biomedical engineer, another nurse, etc.), to getting new equipment, supplies, or IV lines to use for medication administration.

27.6.2 Physician's Role in Error Recovery

The physician's role in error recovery, separate from the use of CDS technology that is discussed in the following, has been infrequently evaluated in health care. Anesthesiologist physicians and nurse anesthetists have been most studied. Cooper, one of the pioneers of patient safety studies in anesthesia, found that 93% of major errors and equipment failures in anesthesia delivery were recovered from without significant negative outcomes (Cooper et al., 1984). Gaba and colleagues outline seven steps needed for an anesthesiologist to recover from an anesthesia incident. These include detection of the incident, verification of the incident, recognition of the incident's potential or actual harm, life sustaining function maintenance, implementation of generic strategies to diagnose, correct, or compensate for the error, diagnose and treat the underlying cause of the error, and finally, monitor and follow-up to ensure error recovery efforts were successful (Gaba et al., 1987). Anesthesia recovery efforts also require the presence of adequate supplies, medications, and backup equipment.

Attending physicians caring for patients on a general medical inpatient service has also been evaluated as a means of error recovery. Chaudhry and colleagues had two hospitalist attending physicians at one institution record medical errors, near misses, and adverse events that occurred during the care of their patients over a 5½ month time period (Chaudhry et al., 2003). Near misses were defined as errors that did not cause harm to patients. A total of 63 errors were identified in the care of 528 patients. Sixty-two percent of errors ($n = 39$) were near misses. Half of the near misses were medication related and 20% were diagnostic errors. The errors were initially discovered by the attending physician (36%), the pharmacist, the resident physicians, a consultant physician, a nurse, or a laboratory technician.

Family physicians have also participated in error-reporting studies of observed or committed errors that have evaluated error recovery. One study evaluated 194 medication-related error reports using the NCC-MERP harm severity categories to code errors into no harm and harm categories (Kuo et al., 2008). The majority of these errors were recovered (84%; category B: 41%, C: 35%, and D: 8%). Pharmacists recovered many of these errors (40%), physicians 19%, patients 17%, and 7% by nurses.

Patel et al. (in press) have evaluated error recovery in surgery attending physicians and resident physicians using clinical cases with embedded errors. The physicians read two cases and were instructed to evaluate the management of the case, and their responses were analyzed. More errors were detected by the surgery attendings compared to the residents. Attendings tended to detect errors as they were reading through the case and correct the errors immediately while residents detected errors at the end while summarizing the case and were slower in decision making about error correction. Residents tended to detect errors with impact on patient outcome while attendings detected more complex errors and knowledge-based errors, consistent with previous studies showing persons with expertise make less knowledge-based errors.

27.6.3 Pharmacist's Role in Error Recovery

Many error and error recovery studies have evaluated the role of pharmacists in error recovery in the medication use process, especially in hospital settings. Leape and colleagues were among the first investigators to systematically evaluate errors in hospital medical/surgical and intensive care units and recognize the important role of the pharmacist in intercepting errors (Leape et al., 1999). They performed a pre-post study evaluating the occurrence of preventable adverse drug events (harmful medication errors) before and after the implementation of the participation of a senior pharmacist on physician rounds and being available on call during the day in an academic tertiary care center medical intensive care unit. They found that the preventable adverse drug events decreased by 66% after the pharmacist intervention. The pharmacists intervened 398 times and, in at least 63% of these interventions, recovered from errors already committed; 45% of interventions clarified or corrected medication orders, 12% recommended alternative therapy, 4% identified drug–drug interactions, and 2% identified drug allergies. Pharmacist interventions have also been found to reduce preventable adverse drug events in academic hospital general medical units with the large majority of interventions recovering from errors rather than preventing them (Kucukarsian et al., 2003). And as studies mentioned earlier have shown, pharmacists also recover from medication errors in the outpatient setting (Kuo et al., 2008).

Kanse et al. (2006) evaluated error recovery in a Dutch hospital pharmacy to better understand the near misses that occur and the factors that contributed to recovery (Kanse et al., 2006). Thirty-one near misses were analyzed, and it was found that based on the typical medication use process, there were 30 planned recovery opportunities that were missed before the eventual recovery that occurred. This means that the actual recovery occurred further downstream in the process—closer to reaching the patient—than need be. The major reason for the missed recoveries was due to organizational culture—the planned double checks in the process were not important and therefore suboptimally performed. Few of the recoveries that occurred included the error explanation step; most recoveries were detected and corrected without gaining an understanding of the causes of the problem.

27.6.4 Health Information Technology and Error Recovery

Health information technology (IT) is being increasingly implemented in health care to improve error prevention and assist health care professionals in detecting and recovering from errors that have occurred. Four health IT applications that have been studied for their role in error recovery are Smart IV pumps, BCMA technology, CPOE, and CDS systems.

Smart IV pumps are IV medication delivery devices with a built-in drug library that is programmed with common IV medications, their concentrations, and the upper and lower dosing limits for infusion and bolus therapy (Wetterneck et al., 2006). When a user programs the pump using the drug library, if the programmed dose is outside of the preset limits, the user receives an alert informing them of the out-of-limit programming. Thus, the alert detects potential errors in pump programming, and the user must decide if the alert is true or false and take action. The preset limits can be hard or soft limits. Hard limits cannot be overridden; they must be canceled or reprogrammed. Soft limits can be overridden. The pump keeps a log of users' responses to alerts, so organizations can see programming errors that were recovered from and also monitor alerts that are overridden. Unfortunately, users receive many false-positive alerts that must be overridden due to many factors, for example, limits that do not match actual practice, nursing practice for drug administration, and patients that require drug dosing outside of the typical ranges (Wetterneck et al., 2005). Two controlled trials of Smart IV pumps performed to evaluate whether Smart IV pumps decrease adverse drug events both showed nonsignificant decreases; however, there were issues with the study's power to detect a difference in adverse drug events as the numbers may be small and whether the drug library was being used in the studies (Rothschild et al., 2005b; Nuckols et al., 2008). Despite this, Smart IV pumps are becoming mainstream in hospital settings (Pedersen et al., 2009).

BCMA technology assists with error recovery during the medication administration process. BCMA has the patient schedule of ordered medication administration and is used to scan and verify the correct patient and the correct medication and dose as well as whether it is being administered at the correct time (Carayon et al., 2007). If there is a mismatch between the BCMA and the patient scan, medication scan or medication schedule, an alert is sent to the user. The user can either override the alert and document the reason for this or respond to the alert to correct the problem. Only recently have studies been performed to show that BCMA decreases medication administration errors, specifically decreasing wrong medication and wrong medication dose, which are likely directly related to BCMA alerts (Poon et al., 2010). Uptake of this technology in hospitals has been much slower than other technologies (Pedersen et al., 2009), likely related to the challenges of implementation and BCMA's enormous impact on nursing and pharmacy workflow.

CPOE technology is designed for ordering providers to directly input their orders into the computer system. The design of the CPOE system and its functionalities can impact the orders that providers enter such that safer care is provided through avoidance of ordering and transcription errors (Wetterneck et al., 2008). Much of the error recovery-based safety benefit of CPOE is gained through built-in decision support tools that alert providers to potential ordering errors (Kuperman et al., 2007). In general, CDS functionalities can improve medication ordering through generated alerts, for example, from drug order-allergy checking, dose checking, and duplicate order checking. CDS has been shown to improve provider performance if the user is automatically prompted to the presence of an alert or reminder while using the system (Garg et al., 2005).

There are many examples of error recovery evaluation in health care. Most health care studies have not evaluated the different steps involved in error recovery, that is, detection, explanation, and diagnosis. Rather, they study whether or not recovery occurred, usually referring to detection and correction of the error. Comparing rates of error recovery over time across studies is limited by the varying terms used, for example, near miss, intercepted error, discovered error, near error, etc., and the varying definitions and measures used for these terms related to whether or not the error reached the patient, and if it did, whether or not harm occurred and the impact of recovery efforts on patient harm. In fact, many studies to date have not considered error recovery to include errors that reach the patient and cause harm but the harm was mitigated through recovery efforts.

27.7 Conclusion

In summary, error recovery is an important error management strategy for health care organizations to be aware of and strive toward in health care's quest to produce HROs. The science behind error recovery has been well developed in other industries and is only recently being applied and evaluated in health care. The studies of error recovery in health care are limited by many factors. There is a lack of integration of error recovery science into the study design and measurement of recovery, likely related to much of this literature being found outside of the typical health care sources and a lack of education on human performance and patient safety. Also, health care organizations would benefit from a medical error taxonomy that includes error recovery principles such that both error prevention and error recovery efforts can be monitored locally and second-order learning can occur from error recovery cases. This single, inclusive medical error taxonomy, if used by researchers, would allow for consistent measurement of error recovery across studies to compare data over time and across institutions. Last, most studies of error recovery in health care focus on the recovery efforts of a certain person type, that is, nurses or physicians, or a specific technology. While these studies are an important first step, they do not fully discuss the context/system of work and/or the processes of care being performed, thus making it difficult to understand what the results mean and to generalize the results to other organizations. These issues should be addressed in future studies of error recovery in health care. The expertise of human factors professionals in understanding work system design, human performance, and how to redesign systems for both error prevention and error recovery is critical for the future of patient safety efforts in health care.

References

Agency for Healthcare Research and Quality (2010). Patient Safety Organization. Concepts Underlying AHRQ'S Common Formats. http://www.pso.ahrq.gov/formats/eventdesc.htm. Accessed July 30, 2011.

Allwood, C.M. (1984). Error detection processes in statistical problem solving. *Cogn Sci*, 8, 413–437.

Alper, S.J., and Karsh, B.T. (2009). A systematic review of safety violations in industry. *Accid Anal Prev*, 41(4), 739–754.

Baker, D.P., Day, R., and Salas, E. (2006). Teamwork as an essential component of high-reliability organizations. *Health Serv Res*, 41(4, Pt 2), 1576–1598.

Balas, M.C., Scott, L.D., and Rogers, A. (2004). The prevalence and nature of errors and near errors reported by hospital staff nurses. *Appl Nurs Res*, 17, 224–230.

Blavier, A., Rouy, E., Nyssen, A.-S., and De Keyser, V. (2005). Prospective issues for error detection. *Ergonomics*, 48, 758–781.

Carayon, P., Wetterneck, T.B., Hundt, A.S., Ozkaynak, M., DeSilvey, J., Ludwig, B. et al. (2007). Evaluation of nurse interaction with bar code medication administration technology in the work environment. *J Patient Saf*, 3(1), 34–42.

Chaudhry, S.I., Olofinboba, K.A., and Krumholz, H.M. (2003). Detection of errors by attending physicians on a general medicine service. *J Gen Intern Med*, 18, 1595–1600.

Cooper, J.B., Newbower, R.S., and Kitz, R.J. (1984). An analysis of major errors and equipment failures in anesthesia management: Considerations for prevention and detection. *Anesthesiology*, 60, 34–42.

Faye, H., Rivera, A.J., Karsh, B.-T., Hundt, A.S., Baker, C., and Carayon, P. (2010). Involving intensive care unit nurses in a proactive risk assessment of the medication management process. *Jt Comm J Qual Saf*, 36(8), 376–385.

Gaba, D.M. (2007). The future vision of simulation in healthcare. *Simul Healthc*, 2(2), 126–135.

Gaba, D.M., Maxwell, M., and DeAnda, A. (1987). Anesthetic mishaps: Breaking the chain of accident evolution. *Anesthesiology*, 66, 670–676.

Garg, A.X., Adhikari, N.K., McDonald, H., Rosas-Arellano, M.P., Devereaux, P.J., Beyene, J. et al. (2005). Effects of computerized clinical decision support systems on practitioner performance and patient outcomes: A systematic review. *JAMA*, 293(10), 1223–1238.

Henneman, E.A., Blank, F.S.J., Gawlinski, A., and Henneman, P.L. (2006). Strategies used by nurses to recover medical errors in an academic emergency department setting. *Appl Nurs Res*, 19, 70–77.

Henneman, E.A., Blank, F.S., Smithline, H.A., Li, H., Santoro, J.S., Schmidt, J. et al. (2005). Voluntarily reported emergency department errors. *J Patient Saf*, 1, 126–132.

Henneman, E.A., Roche, J.P., Fisher, D.L., Cunningham, H., Reilly, C.R., Nathanson, B.H. et al. (2010). Error identification and recovery by student nurses using human patient simulation: Opportunity to improve patient safety. *Appl Nurs Res*, 23, 11–21.

Hines, S., Luna, K., Lofthus, J., and Stelmokas, D. (2008). Becoming a high reliability organization: Operational advice for hospital leaders (prepared by the Lewin Group under Contract No. 290-04-0011). AHRQ Publication No. 08-0022. Rockville, MD: Agency for Healthcare Research and Quality.

Kanse, L., Van der Schaaf, T.W., Vrijland, N.D., and Van Mierlo, H. (2006). Error recovery in a hospital pharmacy. *Ergonomics*, 49, 505–516.

Kessels-Habraken, M., Van der Schaaf, T., De Jonge, J., and Rutte, C. (2010). Defining near misses: Towards a sharpened definition based on empirical data about error handling processes. *Soc Sci Med*, doi: 10.1016/j.socscimed.2010.01.006.

Kohn, L.T., Corrigan, J.M., and Donaldson, M.S. (2000). *To err is human: Building a safer health system*. Washington, DC: National Academy Press.

Kontogiannis, T. (1997). A framework for the analysis of cognitive reliability in complex systems: A recovery centered approach. *Reliab Eng Syst Saf*, 58, 233–248.

Kontogiannis, T. (1999). User strategies in recovering from error in man–machine systems. *Saf Sci*, 32, 49–68.

Kucukarsian, S.N., Peters, M., Mlynarek, M., and Nafziger, D.A. (2003). Pharmacists on rounding teams reduce preventable adverse drug events in hospital general medicine units. *Arch Intern Med*, 163, 2014–2018.

Kuo, G.M., Phillips, R.L., Graham, D., and Hickner, J.M. (2008). Medication errors reported by US family physicians and their office staff. *Qual Saf Health Care*, 17, 286–290.

Kuperman, G.J., Bobb, A., Payne, T.H., Avery, A.J., Gandhi, T.K., Burns, G. et al. (2007). Medication-related clinical decision support in computerized provider order entry systems: A review. *J Am Med Inform Assoc*, 14, 29–40.

Leape, L.L., Cullen, D.J., Clapp, M.D., Burdick, E., Demonaco, H.J., Erickson, J.I. et al. (1999). Pharmacist participation on physician rounds and adverse drug events in the intensive care unit. *JAMA*, 282(3), 267–270.

Leplat, J. (1989). Error analysis, instrument and object of task analysis. *Ergonomics*, 32, 813–822.

Montesi, G., and Lechi, A. (2009). Prevention of medication errors: Detection and audit. *Br J Clin Pharmacol*, 67(6), 651–655.

National Coordinating Council for Medication Error Reporting and Prevention (1998). NCC MERP taxonomy of medication errors. www.nccmerp.org//pdf//taxo2001-07-31.pdf. Accessed February 7, 2011.

Norman, D.A. (1988). *The psychology of everyday things*. New York: Basic Books.

Nuckols, T.K., Bower, A.G., Paddock, S.M., Hilborne, L.H., Wallace, P., Rothschild, J.M. et al. (2008). Programmable infusion pumps in ICUs: An analysis of corresponding adverse drug events. *J Gen Intern Med*, 23(Suppl 1), 41–45.

Patel, V.L., and Cohen, T. (2008). New perspectives on error in critical care. *Curr Opin Crit Care*, 14(4), 456–459.

Patel, V.L., Cohen, T., Murarka, T., Olsen, J., Kagita, S., Myneni, S. et al. (2011). Recovery at the edge of error: Debunking the myth of the infallible expert. *J Biomed Inform*, 44(3), 413–423.

Pathiraja, F., Wilson, M.C., and Pronovost, P.J. (2011). Close calls in health care. In A.W. Wu (Ed.), *The value of close calls in improving patient safety: Learning how to avoid and mitigate harm* (pp. 3–12). Chicago, IL: Joint Commission Resources.

Patterson, E.S., Render, M.L., and Ebright, P.L. (2003). Repeating human performance themes in five health care adverse events. In *Oral presentation & proceedings of the human factors and ergonomics society 46th annual meeting* (pp. 1418–1422). Baltimore, MD, September 30–October 4.

Pedersen, C.A., Schneider, P.J., and Scheckelhoff, D.J. (2009). ASHP national survey of pharmacy practice in hospital settings: Dispensing and administration—2008. *Am J Health Syst Pharm*, 66(10), 926–946.

Poon, E.G., Keohane, C.A., Yoon, C.S., Ditmore, M., Bane, A., Levtzion-Korach, O. et al. (2010). Effect of bar-code technology on the safety of medication administration. *N Engl J Med*, 362(18), 1698–1707.

Reason, J. (1990). *Human error*. Cambridge, U.K.: Cambridge University Press.

Rizzo, A., Bagnara, S., and Visciola, M. (1987). Human error detection processes. *Int J Man Mach Stud*, 27, 555–570.

Rogers, A.E., Dean, G.E., Hwang, W.-T., and Scott, L.D. (2008). Role of registered nurses in error prevention, discovery and correction. *Qual Saf Health Care*, 17,117–121.

Rothschild, J.M., Hurley, A.C., Landrigan, C.P., Cronin, J.W., Martell-Waldrop, K., Foskett, C. et al. (2006). Recovery from medical errors: The critical care nursing safety net. *Jt Comm J Qual Patient Saf*, 32(2), 63–72.

Rothschild, J.M., Keohane, C.A., Cook, E.F., Orav, E.J., Burdick, E., Thompson, S. et al. (2005b). A controlled trial of smart infusion pumps to improve medication safety in critically ill patients. *Crit Care Med*, 33(3), 533–540.

Rothschild, J.M., Landrigan, C.P., Cronin, J.W., Kaushal, R., Lockley, S.W., Burdick, E. et al. (2005a). The Critical Care Safety Study: The incidence and nature of adverse events and serious medical errors in intensive care. *Crit Care Med*, 33(8), 1694–1700.

Schnipper, J.L., Kirwin, J.L., Cotugno, M.C., Wahlstrom, S.A., Brown, B.A., Tarvin, E. et al. (2006). Role of pharmacist counseling in preventing adverse drug events after hospitalization. *Arch Intern Med*, 166(5), 565–571.

Sellen, A. (1994). Detection of everyday errors. *Appl Psychol Int Rev*, 43, 475–498.

Tucker, A.L., Edmondson, A.C., and Spear, S. (2002). When problem solving prevents organizational learning. *J Organ Change Manag*, 15(2), 122–137.

van der Schaaf, T.W. (1992). Near Miss Reporting in the Chemical Process Industry. PhD Thesis, University of Technology, Eindhoven, the Netherlands.

van der Schaaf, T.W. and Kanse, L. (2000). Errors and error recovery. *Lect Notes Contr Inform Sci*, 253, 27–38.

Weick, K.E., and Sutcliffe, K.M. (2001). *Managing the unexpected: Assuring high performance in an age of complexity*. San Francisco, CA: Jossey-Bass.

Wetterneck, T.B., Brown, R.L., Carayon, P., Kleppin, S.M., Schoofs Hundt, A., and Ozkaynak, M. (2005). End-user response to intravenous infusion pump medication dosing alerts. An analysis of user-interface events. In R. Tartalgia, S. Bagnara, T. Ballandi, and S. Albolino (Eds.), *Healthcare systems ergonomics and patient safety. Proceedings of the international conference—HEPS 2005 Florence, Italy* (pp. 235–238). London, U.K.: Taylor & Francis.

Wetterneck, T.B., and Karsh, B.T. (2011). Human factors applications to understanding and using close calls to improve health care. In A.W. Wu (Ed.), *The value of close calls in improving patient safety: Learning how to avoid and mitigate harm* (pp. 39–54). Chicago, IL: Joint Commission Resources.

Wetterneck, T.B., Paris, B., Walker, J., and Carayon, P. (2008). CPOE functionalities and medication ordering errors in the ICU. In L.I. Sznelwar, F.L. Mascia, and U.B. Montedo (Eds.), *Human factors in organizational design and management—IX* (pp. 369–376). Santa Monica, CA: IEA Press.

Wetterneck, T.B., Skibinski, K.A., Roberts, T.L., Kleppin, S.M., Schroeder, M.E., Enloe, M. et al. (2006). Using failure mode and effects analysis to plan implementation of Smart I.V. pump technology. *Am J Health Syst Pharm*, 63, 1528–1538.

Wu, A.W. (2011). Introduction. In A.W. Wu (Ed.), *The value of close calls in improving patient safety: Learning how to avoid and mitigate harm* (pp. 3–12). Chicago, IL: Joint Commission Resources.

Zapf, D., Maier, G.W., Rappensperger, G., and Irmer, C. (1994). Error detection, task characteristics, and some consequences for software design. *Appl Psychol*, 43(4), 499–520.

Section VII

Human Factors and Ergonomics Methodologies

28

Cognitive Work Analysis in Health Care

Michelle L. Rogers, Emily S. Patterson, and Marta L. Render

CONTENTS

28.1 Introduction..465
28.2 Human Performance in Work...466
28.3 What Is Cognitive Work Analysis..466
28.4 Overview of Conceptual Frameworks to Understand Work...466
28.5 Knowledge Elicitation Methods 467
    28.5.1 Ethnographic Observations...467
    28.5.2 Critical Decision Method...468
    28.5.3 Artifact Analysis...468
    28.5.4 Usability Testing...468
    28.5.5 Triangulation...468
28.6 Performing a Cognitive Work Analysis...469
28.7 Descriptions of the Phases...469
    28.7.1 Phase One: Work Domain Analysis..469
    28.7.2 Phase Two: Control Task Analysis 470
    28.7.3 Phase Three: Strategies Analysis...471
    28.7.4 Phase Four: Social Organizational Analysis..472
    28.7.5 Phase Five: Worker Competencies Analysis 472
28.8 Conclusion 473
Acknowledgments...473
References..473

28.1 Introduction

Adoption of technology has been proposed as a means to reduce medical error (Kohn et al., 1999). This reduction might be derived through automation of tasks, monitoring, and improved flexibility of data handling directed toward better situation assessment (Puckett, 1995, Sarter et al., 1997; Potter et al., 2000). Studies of accidents, however, consistently demonstrate that new computerized systems predictably affect human problem-solving ability in ways that often contribute to accidents. Experience with technology's contribution to new failures is widespread, ranging from personal catastrophes by unintentional, but unrecoverable, keystrokes that wipe out entire files (Norman, 1983) to software problems that crashed NASA's Mars exploration mission (Young et al., 2000). As health care moves toward increasing dependence on computerized tools; order entry, electronic medical records, and medication administration to name just a few, their design of these computerized tools to support human work (e.g., make it easier, faster, safer, and more accurate) also increases in importance. Recent reports, including the failure of computerized order entry at Cedars Sinai (Chin, 2003), as well as unexpected consequences with implementation of patient care information systems reported by Ash et al. (2004) and bar coded medication administration by Patterson et al. (2002), lend credence to this view. Cognitive work analysis (CWA) is a method that models how environmental, organizational, individual, and technical constraints contribute to work in order to design tools that support work. This methodology, applied to health care systems, could improve safety, efficacy, efficiency, and the acceptance of computerized tools by health care providers.

28.2 Human Performance in Work

High reliability is needed in industries that operate in an unforgiving social and political climate, have risk of serious adverse consequences, have limited opportunity for learning through experimentation, use complex processes and technology, and have potential for unexpected events and surprise (Harper et al., 1991; Weick and Sutcliffe, 2001). Health care is such a system with inherent hazard and vulnerability to failure. Workers perceive these hazards or risks for adverse outcomes and adapt individually, in groups, and organizationally to avoid or guard against them in the pursuit of their goals. These continuous adaptive efforts "make safety." Feedback on how these adaptations are working or how the environment is changing is a critical factor in preserving safety. Recognition of the boundaries or the vulnerabilities of a system show workers where failures may occur and can guide investment to cope with these paths toward failure. Change within the system may reduce the ability to identify those boundaries and vulnerabilities. The ability of the system to support technical work determines the reliability of a complex system. Such systems, in addition to facilitating completion of the task or goal, must manage access to information, facilitate coordination among team members, and support workload, management of goal conflicts by the workers, deviation from "normal," and change (or unexpected events) (Woods and Roth, 1988; Woods et al., 1994; Woods, 2003). The design of tools for a complex environment must consider operative elements of complexity including those from the work domain and those stemming from humans and organizations (Hvey and Wickens, 1993; Patterson and Woods, 2001; Woods et al., 2002).

28.3 What Is Cognitive Work Analysis

A CWA describes the goals of a human agent and the means by which they are accomplished considering the entire work system, which includes the individual, organizational, task-related, environmental, and technology factors (Smith and Sainfort, 1989). It is a conceptual framework and methodology developed by Jens Rasmussen and Annelise Mark Petersen and advanced by Vicente (1999). The CWA provides insight into what tasks are cognitively challenging and are therefore candidates for system redesign or tool development.

The method examines how many elements (environment [regulatory, work space], social [organization], individual, and technical constraints) contribute to safe, effective, and efficient work in order to identify the characteristics of these systems that are important for success. The purpose of CWA is to create tools and systems that effectively support human work. The framework and methodology grew out of the failure of the "simple" explanation—that competence in the technical core alone was sufficient for safety and productivity.

28.4 Overview of Conceptual Frameworks to Understand Work

The theoretical framework and methods of CWA presumes that a sophisticated, deep understanding of human work could facilitate the development and evaluation of tools, training, and systems to improve work and outcomes. In no industry is this more important than health care, where there is high hazard (Kohn et al., 1999), skyrocketing costs, and best practices are inconsistently applied (Patel et al., 1998; Corrigan, 2001). Three models have been used to examine and improve work. The first, a normative analysis, uses an idealized description of work in isolation, promoting formal stepwise procedures dictating movement through a process to achieve a goal (e.g., policies and procedures). The usefulness of normative approaches in complex industries is limited because of the inevitable discrepancy between how work is actually performed compared to the list of prescribed correct steps. Such a prescriptive approach cannot ever address all the variation in circumstances, goal trade-offs, and management of unexpected events found in work. For instance, a nurse in the in-patient medical unit has written procedures that dictate timely administration of the 9 o'clock medications, review of discharge instructions to a patient, support of a family and patient who have received the worst possible news, resuscitation of an arresting patient, and prevention of a fall by a frail confused patient. Obviously, all of these tasks cannot be achieved in one 30 min time interval at the present rate of staffing. Clinicians use their expertise routinely to choose sequentially how to achieve the most important goals with available resources and time. Rigid adherence to written procedure is not desirable, since all possible combinations of events cannot be articulated or even imagined.

The second approach, descriptive, examines work in a naturalistic setting (observing real people as they perform their work). This approach has the advantage of capturing varied and complex demands imposed on workers, the coordination and communication required between workers to complete a task, how workers use expertise, tools, information, and cues to problem solve, and the role that contextual and/or social factors play in promoting or inhibiting problem resolution. Devising systems, training, or tools based solely on this approach, however, unnecessarily limits the design. Some of the strategies and practices discovered from descriptive analysis of work are dependent on the design of existing tools or systems. For example, in the absence of computerized order entry, physicians rely largely on memory for drug–drug interactions, and side effect profiles. Given the time constraints and effort, it is not possible or even necessary for each physician to look up each ordered drug each time. However, the addition of a computerized textbook or automated checks of drug–drug interaction, if properly designed, might add safety without compromising other goals.

CWA is the third method used to understand technical work. The goal of CWA is to identify the technical and organizational requirements necessary in order for a tool to support work effectively. This approach builds on the normative and descriptive models, adding the advantage of adapting to worker practices while allowing innovation that improves support of technical work.

28.5 Knowledge Elicitation Methods

CWA provides a way to model the worker adaptation to unique situations (allowing users to capitalize on their expertise) while providing control of the technical system (Vicente, 1999). In order to arrive at these models, data are collected for the domain analyses through the following techniques:

- Ethnographic observations of workers in naturalistic or simulated settings
- Critical decision method (CDM) interviews
- Artifact analysis
- Usability testing

The data are then abstracted into two principal domains: technical and social. These categories are subdivided, the technical into three separate conceptual phases—work domain, control tasks, and strategies; and the social into two conceptual domains—social organization and worker competencies. This abstraction makes visible both the intrinsic tasks as well as the constraints of the domains. These constraints are additive, each conferring a separate and unique set of constraints. Through examination of constraints, a tool can be designed with the greatest degree of freedom.

28.5.1 Ethnographic Observations

Ethnographic observation is a methodology derived from anthropology that emphasizes detailed data collection of observable human behavior and interviews. Data collection in ethnographic observational studies involves trained observers sequentially capturing in detail both (1) observable activities and verbalizations, and (2) self-report data about how artifacts (tools) support performance (Hutchins, 1995). One method of analysis of observational data is the process tracing method, developed for the study of problem solving outside of the experimental laboratory, which externalizes internal processes, in order to be able to support inferences about internal cognition or problem solving by applying behavioral interaction protocols (Woods, 1993). Behavioral interaction protocols (Jordan and Henderson, 1995) follow in minute detail the sequence of observed behaviors in order to demonstrate a plan or goal. For example, when a nurse brings the medication cup into the patient, checks his armband, rechecks his armband, goes out to the med cart and gets the medication administration record, rechecks the armband against the written record, and then leaves with the medications in her hand, one can assume that she planned to give that patient medications, which on scrutiny proved to be for some other individual.

The process tracing method broadens the data sources beyond verbalizations to include a number of techniques: (1) direct observation of participant behavior, (2) "think aloud" and explanatory verbalizations while performing the task in question, (3) written records of the task performed (print-outs of medication records observed), (4) writing detailed timed notes of actions taken, (5) records of verbal communication among team members or via formal communication media, and (6) interruptions. To further increase reliability and generalizability of observation (Hutchins, 1995), practitioners are observed in their work setting over time, so that the effects of being observed in their activities are lessened. In addition, reliance solely on interview data is avoided since it is suspect from memory limitations and bias. Instead, the same types of data are captured from multiple practitioners in the same fashion to enable objective comparison and analyses. Moreover, the focus of the observation is explicitly characterized in order to target the data collection and avoid getting lost in the complexity and missing important details. Finally, data are abstracted in ways that allow comparisons across sites, hospitals, and with other domains such as aviation to increase the generalizability of the findings (Jordan and Henderson, 1995).

Traditional human factors simulation studies are developed to mimic the complex interactions of a cognitive system (a human and a machine) in a naturalistic setting (Gaba, 1992). These complex simulations that closely resemble a real-world setting, such as an operating room, reduce concerns about the extreme generalizability of findings to a real-world setting (Gaba and DeAnda, 1988; Patterson et al., 2004). In designing a simulation study, the focus is normally on stimulus sampling rather than subject sampling because the performance variability is often greater based on the problem to be solved than the subject selection. During data collection, study participants perform challenging, face valid tasks without interruption. Following the simulation run, there is a debriefing session to elicit data about the cognitive processes that were employed. A popular technique is "cued retrospective playback elicitation" (Jordan and Henderson, 1995). This technique aids memory and enforces detailed elicitation around observed actions by playing and pausing video of the simulation performance during an interview immediately following the simulation run.

28.5.2 Critical Decision Method

Interviews to assay knowledge are structured to improve validity and reliability of self-reported information. One approach, the CDM interview (Klein et al., 1989), is a seven-step process that uses a timeline of an event as a forcing function (in the form of an interview) to make it more difficult to stray from personal experience with a specific situation. An interview using the CDM follows this format:

- *Preparation*: Where the interviewer, familiar with domain, operationally defines goals for knowledge elicitation (i.e., the interviewer might focus on a specific decision involving confusion, change of a plan, interesting use of an artifact).

- *Incident selection*: Query the interviewee about what makes an incident interesting.

- *Incident recall*: The interviewee recounts the episode in its entirety.

- *Incident retelling*: The interviewer tells the story back, asking for details and clarifications.

- *Timeline verification and decision point identification*: Interviewer reviews incident a second time.

- *Deepening*: Interviewer, reviewing the incident a third time, employs probe questions that focus attention on aspects of each decision making event within the incident, usually starting with informational cues and eliciting the meaning of those cues and expectations goals and actions they engender.

- *"What if" queries*: A fourth sweep through the incident posing hypothetical changes, asking the participant to speculate on what might have happened if an element had been different.

28.5.3 Artifact Analysis

Cognitive artifacts are information tools that are part of the distributed cognition that takes place in technical work (Norman, 1990). They are objects that represent the collective intentions, current state, and past accomplishments across a period of time (Hutchins, 1995). The cognitive artifacts are present to recreate the intention originally made when the document or tool is used. They provide relief for memory and other cognitive functions that can get overwhelmed with high workload situations. These artifacts are analyzed for their use in the system and the roles they have taken on in the organization. For example, during the medication administration process, depending on the setting, the nurse may use several different paper forms to document the care of patients (e.g., a flow sheet to document inputs and outputs, laboratory communication form, medication administration record, and education record). Each of these forms was developed for a specific part of the work process and to meet an information need. As needs change and evolve, seldom are the forms removed but more often than not, additional forms become part of the process.

28.5.4 Usability Testing

Usability testing is iteratively conducted on prototypes of a software product prior to release in order to assess and improve the ability of the target user population to accomplish tasks easily and efficiently for which the software was designed (Rubin, 1994). Test participants are users who meet prespecified criteria. Three to six participants are traditionally used in a single usability test in nonrandomized order. Five test participants are reported to identify 85% of usability problems on average (Nielsen, 1993). During the usability test session, participants interact with a prototype version of the software in a specific testing scenario while thinking out loud. The scenarios are rooted in the context of the work environment. The verbalizations and actions of the study participants, captured on videotape, are analyzed for confusion, difficulty meeting task goals, time on task, and scenario-based performance metrics. Conclusions are drawn from erroneous assumption or actions, statements indicating surprise or confusion, tasks taking longer than anticipated. Test participants can complete a detailed survey to provide secondary data about the software's usability (Lewis, 1995). Following the scenario execution, there is usually a debriefing session to elicit data about the cognitive processes that were employed.

28.5.5 Triangulation

Using multiple techniques to provide converging evidence attains the rigor necessary for accurate analysis.

In order to consider a phenomenon to have sufficient authenticity or warrant, patterns need to be discovered across multiple methodologies, sites, and settings. Miles and Huberman (1984) define triangulation as a method to check a new finding against another by using other internal indices that should provide convergent evidence. For example, we use multiple health care sites and settings in our observational studies and compare our findings from the observations with our findings from the critical incident interviews to increase the reliability, validity, and generalizability. The summaries are compared and contrasted with each other to reach final conclusions in terms of their similarities, differences, and observed patterns. The goal is that the inferences drawn from one source of data would be confirmed by another source of data, that is, one theme is identified and the different data sources are compared to determine what they predict about that theme. Patton (1999) further describes the actual process of triangulation as having four different types: (1) methods, (2) sources, (3) analyst, and (4) theory/perspective. Different kinds of data may yield varying results since the methodologies are sensitive to distinctions of the setting.

28.6 Performing a Cognitive Work Analysis

The knowledge elicited as part of a CWA is inspected iteratively and analyzed in five (5) phases: work domain analysis, control task analysis, strategies analysis, social organization analysis, and worker competencies. We briefly describe them here, but recommend reviewing Vicente (1999) for a comprehensive discussion of each of the phases in CWA. The domain constructs in CWA conceptually move from environmental to cognitive constraints in the work. The analysis moves from the specific descriptive data through modeling tools to characterize intrinsic work constraints. Identification of intrinsic work constraints naturally permits identification of a series of critical elements in the system design and interventions each derived from the five domains. The interventions could result in new socio-technical systems and direct future work practices. Because constraints are fixed, they form a basis or boundary for worker adaptation while permitting the maximal flexibility for workers to respond to unexpected demands and adapt to the inevitable continuous changing environment, a critical benefit of CWA (Vicente, 1999). Following review of these phases and the modeling process, we will demonstrate the CWA in a real world example in order to clarify the approach.

28.7 Descriptions of the Phases

The five phases of a CWA include (1) work domain analysis, (2) control task analysis, (3) strategies analysis, (4) social organization analysis, and (5) worker competencies. The first phase, work domain analysis, describes the work domain of the system being examined; independent of workers, tools, events, or goals. It descriptively should "map" the work or display the "lay of the land," often most easily identified as the physical space. For example, the aircraft carrier might represent the work domain of controlling flight. Control task analysis, the second phase, concentrates on the products or goals that are to be achieved from the system, that is, what needs to be done independently of "how" or "by whom." The outcome of the analysis can be used to develop policies and procedures for workers to reach those goals. The third phase, strategies analysis, includes the processes used in the work domain to achieve the product. The social organizational phase (phase four) covers the relationship between agents, either workers or automation, and how responsibility for the processes is distributed across the actors of the system. Finally, the fifth phase, worker competencies, recognizes the constraints associated with the worker themselves—the knowledge, physical conditions, and training needed for effective function. Analysis of the constraints from each of the five phases defines the boundaries of action for the workers. These are considered behavior-shaping constraints (Rasmussen and Petersen, 1995) and lead directly to implications for systems design.

28.7.1 Phase One: Work Domain Analysis

Abstraction of data into the work domain builds a map of the system being acted on or the functional structure. The work domain characterizes two different aspects of work: the parts of the systems and the functional elements that describe the work. To aid in the systematic capture of both elements, Rasmussen (1986) developed a tool called the "Abstraction–Decomposition Space" (Figure 28.1) to break down that map into a hierarchical structure. Decomposition breaks work down into parts of the whole. This category is organized so that the top level is composed of elements from all the levels below. The abstraction hierarchy creates a structural means—end hierarchy. This hierarchy is organized so that each cell in the space represents the same work domain differently. Another way to understand how to build an abstraction hierarchy is that the top level answers the question "why" and the level below that answers the question "what," and the level below that the question "how." For instance, performance in the ICU (the top "why" level) is measured by risk adjusted mortality and length of stay, staff and patient satisfaction (the middle "what" level). These measures are collected by health information databases, laboratory databases, and surveys (the "how" level). The work domain analysis would then provide information for the control tasks to act on.

To clarify this process, we detail an abstraction–decomposition model for use of an electronic whiteboard. This map shows only the lay of the land, not the work, the processes, the people, or the systems in place. Patterson et al. (2010) conducted ethnographic observations of use of an electronic whiteboard in two separate locations. As a result, the following classification scheme can be structured in the form of an abstraction hierarchy. The representation that resulted (Figure 28.1) identified the information content and structure of the interface. The hierarchy then informs the control task analysis.

In the decomposition hierarchy, the part-whole dimension of the space consists of three levels: components (interface, etc.), subsystems (disposition, etc.), and the whole system (the electronic status board system). The highest functional analysis recognizes that the goal of the whole system is to improve the symptoms and patient outcomes through coordinating communication with accurate, timely care to people who are ill. The abstract subsystems permit receipt, triage, treatment, and disposition across systems and people. Notice that the table is filled with nouns, and not verbs. Within this abstraction model, links can be made within the categories to identify requirements and constraints.

28.7.2 Phase Two: Control Task Analysis

Control task analysis allows identification of the requirements associated with known recurring types of events. The analysis identifies what needs to be done (but not by whom), or the goals of the system, including the variety of ways that each task may be accomplished. Methods adapted from human performance studies can be used to capture how work goals are achieved (Rasmussen, 1986; Klein et al., 1989).

Structured linear reasoning reflects the processes that novices must navigate to make decisions in the workplace. Experts, however, often shortcut the process using their experiences and by matching prior actions to the present situation until one fits. These deviations are hallmarks of expertise and function to economize both on cognition and on time. In the control task analysis, a decision ladder (Figure 28.2) represented a folded linear sequence of action. It is used since arrows can represent the short cuts directly from one stage to another bypassing other stages. Methods of task analysis that have proven more helpful include the decision ladder, which is a template identifying requirements associated with information processing activities. These methods decouple what need to be done from "who" does it.

Figure 28.2 diagrams only the initiation process of an emergency department visit where the worker begins the task by examining the tools. An expert worker likely follows the bolded script where, beginning with the task, a glance of the tools informs him or her about the readiness of the system (e.g., the status of beds/locations where the patient can be placed, the list including patient identifiers and location), permitting the user

	Aggregation-decomposition		
	Whole system (Electronic Status Board)	Subsystem	Components
Functional Purpose	Reduce symptoms, Improve outcomes Determine care location		
Abstract Function		Receive Triage Coordinate disposition Support resource planning	Systems to communicate transport
Generalized Function		Triage Disposition	
Physical Function		Coordination Maintain status Housing Track duration of visit	Document urgency Transfer to correct care location Document results
Physical Form		Laboratory Hardware Plant Patient	Medications Hardware, software Beds, rooms, halls, etc. Location, identifiers

(left margin label: Physical functional)

FIGURE 28.1
Map of status board use in an emergency department.

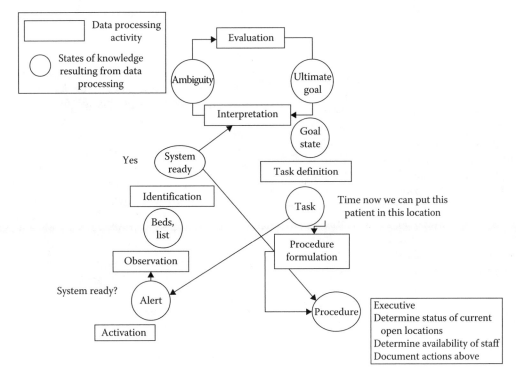

FIGURE 28.2
Control task analysis decision ladder.

to begin the assignment process. In contrast, a novice practitioner might need to complete the entire control task sequence, first activating processing, observing the interface specifically for each element, checking them off on a mental checklist, deciding that there is a location ready, evaluating the other tasks waiting for attention against the goals that needed to be achieved, and then beginning the receiving process. Workers track each element of the list of execution through the control task diagram, bypassing those elements unnecessary to the process at hand based on expert judgment.

Table 28.1 summarizes the control tasks identified from the field studies. Since the emergency department status board is a computer system, the information is communicated by entering information into a desktop workstation.

28.7.3 Phase Three: Strategies Analysis

Strategies analysis addresses how the task in question can be completed. The analyses are important since the design of a tool is often based on a single strategy

TABLE 28.1

Summary of Control Task Analysis of Emergency Department Whiteboard

Step	Description	Link	Ladder Code	Abstraction Level	Decomposition Level
1	Determine system readiness (list of patients, locations, nurse, physician, tests, etc.)	Means—ends, part whole	Goal state	Functional purpose	System
2	Select/enter patient	Topological	Define task	Generalized function	Subsystem
3	Identify room	Topological	Formulate procedure	Physical form	Subsystem
4	Match patient and room	Means—ends, part whole	Procedure	Physical function	Subsystem
5	Determine status of available rooms		Procedure	Physical form	Component
6	Evaluate for discrepancies (name, nurse, physician, etc.)		Alert	Functional purpose	Component
7	Locate patient into room		Execute	Physical form	Component
8	Record actions taken		Task	Generalized function	Subsystem

(a normative approach), but if multiple strategies are in use, that tool may not work in all environments or conditions of work. The outcome of this analysis will be a description of how the task in question can be done.

Strategies in most industries routinely vary based on situational and workload issues as it does in the emergency department. For instance, some emergency departments serve a patient population that can be cared for with the facilities available within the department. On the other hand, trauma level emergency departments may need a large number of laboratory tests and/or x-rays, thus increasing the time they are in the hospital. Moreover, emergency care is anything but a linear activity. Some patients use emergency rooms for primary care issues while trauma cases can tax the resources used by the hospital in order to give quality care. The strategies can be found in careful analysis of the observations by finding consistent patterns of behavior across practitioners. The analysis of strategies also takes into account how workers use information about the process under scrutiny to problem solve.

28.7.4 Phase Four: Social Organizational Analysis

The social organizational analysis describes how the work can be distributed across human workers and machine automation and how such actors communicate and cooperate. It answers the question "who might perform the tasks." Any such structure will need to be able to manage change rapidly, with autonomy given to the workers closest to the information source. The redundancy in the system should allow the moving of people and roles as demanded by the situation. Information related to the task must also track with the task, so that workers have the data required available as the task is completed. Rasmussen et al. (1994) identified six criteria for dividing work demands: actor competency, access to information, or action means, facilitating the communication needed for coordination, workload sharing, safety and reliability, and regulation compliance.

The social organizational evaluation of work can use the modeling tools discussed in the previous sections, the abstraction–decomposition hierarchy, and the decision ladder. The system design interventions would deal with role allocation and organizational structure. The abstraction–decomposition space identifies the information set that would be applicable to the entire organization. If the workers are distributed across the space, we can identify what subset of the status board information would be useful for each group. Data collection techniques that can be used include ethnographic observations and the process trace method as they address cooperation and communication issues.

28.7.5 Phase Five: Worker Competencies Analysis

The ultimate objective of this phase is to identify the characteristics an ideal worker would have. To understand the worker competencies, which can be used then to develop training, experts in CWA apply the skills, rules, and knowledge taxonomy (Rasmussen, 1987). Skill-based behavior is automated actions performed without conscious attention—like walking. Rule-based behavior consists of actions based on stored rules derived from prior performance, procedures, or instruction. Rule-based behavior reacts to cues in the environment. In contrast, knowledge-based behavior searches for the correct action to achieve an explicit goal with a clear mental model of the process. Skill-based behavior is cognitively the most economical, while rule-based actions require more time and effort, with knowledge-based behavior the slowest and cognitively the most resource intensive. The behavior used is a function of the skill level of the worker.

Worker competency recognizes that many tasks require varying expertise, making it practical to divide the work along lines of expertise. For instance, in the outpatient clinic, physicians, nurse educators, clerical personnel, and billing clerks all work to provide care to the diabetic patient. Access to information implies that the worker who can see the information that identifies a problem will be able to act on that information, or conversely workers responsible for a task are able to access the information necessary for the task.

To explore what this evaluation means practically, the following text provides an example from using the electronic status board:

- *Skill-based knowledge*: The worker ideally recognizes names of patients and lab results, the organization of the facility, and room numbers; is able to interact with the tool and the surrounding environment.

- *Rule-based knowledge*: Worker competencies at this level include icons or signs that support rule-based behavior in managing the status of care in an emergency department and the awareness of the strategies used to shortcut the process.

- *Knowledge-based behavior*: This competency measures what will be needed to manage the unexpected changes in patient status and access to appropriate and rapid information is important in this arena. For instance, one part of determining the disposition of a patient includes the ability to check lab results and the status of other important outcomes. The need for knowledge-based behavior stems from cues that stimulate the worker to question actions from rule-based behavior.

28.8 Conclusion

CWA is the framework of choice because it studies the actual work and the system in which it is done. While it is necessary to understand the work of clinicians and practitioners, it is not sufficient because of human performance and limitations in current practice. Similarly, it is not adequate to study only the technical systems because they are not optimized. CWA provides a more complete understanding of the work environment.

CWA brings a powerful tool in health care to design and redesign work; particularly as clinical information systems are implemented. This framework involves identification of the knowledge, mental processes, and decisions that are required to perform technical work. It is commonly used in technical work where the nature of the task, the methods used by practitioners to accomplish the task, and the factors that complicate task performance are less well understood. It is the technical work of physicians, nurses, pharmacists, and technologists that confronts inherent hazards and threats to safety. Practitioners and the system that supports their work in context actually are the means by which patient safety is created and sustained. They are the ones who face the dilemmas and conflicts of practice; they are the ones who receive the demands for production and cope with the complexity of the real world; they are the ones who bridge the gaps that modern health care products.

Acknowledgments

The authors would like to offer their sincere thanks and acknowledge David D. Woods, PhD, Department of Integrated Systems Engineering, The Ohio State University, for his contributions to a previous version of this manuscript.

References

Ash, J.S., Berg, M., and Coiera, E. (2004). Some unintended consequences of information technology in health care: The nature of patient care information system-related errors. *Journal of the American Medical Informatics Association, 11*(2), 104–112.

Chin, T. (2003). Doctors pull plug on paperless system. American Medical Association [Online] Retrieved from http://www.ama-assn.org/amednews/2003/02/17/bil20217.htm

Corrigan, J.M. (2001). *Crossing the Quality Chasm*, Washington, DC: National Academy Press.

Gaba, D.M. (1992). Improving anesthesiologists' performance by simulating reality. *Anesthesiology, 76*(4), 491–494.

Gaba, D.M. and DeAnda, A. (1988). A comprehensive anesthesia simulation environment: Re-creating the operating room for research and training. *Anesthesiology, 69*(3), 387–394.

Harper, R.R., Hughes, J.A., and Shapiro, D.Z. (1991). Harmonious working and CSCW: Computer technology and air traffic control. In J.M. Bowers and S.D. Benford (Eds.), *Studies in Computer Supported Cooperative Work. Theory, Practice and Design* (pp. 225–34), Amsterdam, the Netherlands: North-Holland.

Huey, B.M. and Wickens, C.D. (1993). *Workload Transition: Implications for Individual and Team Performance*, Washington, DC: National Academy Press.

Hutchins, E. (1995). *Cognitive in the Wild*, Cambridge, MA: MIT Press.

Jordan, B. and Henderson, A. (1995). Interaction analysis: Foundations and practice. *The Journal of the Learning Sciences, 4*, 39–103.

Klein, G., Calderwood, B., and MacGregor, D. (1989). Critical decision method for eliciting knowledge. *IEEE Transactions on Systems, Man and Cybernetics, 19*, 462–472.

Kohn, L.T., Corrigan, J.M., and Donaldson, M.S. (1999). *To Err Is Human: Building a Safer Health System*, Washington, DC: National Academy Press.

Lewis, J.R. (1995). IBM computer usability satisfaction questionnaires: Psychometric evaluation and instructions for use. *International Journal of Human–Computer Interaction, 7*(1), 57–78.

Miles, M. and Huberman, A.M. (1984). *Qualitative Data Analysis: A Sourcebook of New Methods*, Newbury Park, NJ: Sage Publications.

Nielsen, J. (1993). A mathematical model of the finding of usability problems. In *Proceedings of ACM/IFIP INTERCHI'93 Conference*, April 24–29, 1993, Amsterdam, the Netherlands.

Norman, D. (1983). Design rules based on analysis of human error. *Communication of the ACM, 26*, 254–258.

Norman, D. (1990). The "problem" with automation: Inappropriate feedback and interaction, not "over-automation." *Philosophical Transactions of the Royal Society of London, 327*, 585–593.

Patel, V.L., Allen, V.G., Arocha, J.F., and Shortliffe, E.H. (1998). Representing clinical guidelines in GLIF: Individual and collaborative expertise. *Journal of the American Medical Informatics Association, 5*, 467–483.

Patterson, E.S., Cook, R.I., and Render, M.L. (2002). Improving patient safety by identifying side effects from introducing bar coding in medication administration. *Journal of the American Medical Informatics Association, 9*(5), 540–553.

Patterson, E., Rogers, M.L., and Render, M.L. (2004). Simulation-based embedded probe technique for human–computer interaction evaluation. *Cognition, Technology, and Work, 6*(2), 197–205.

Patterson, E., Rogers, M.L., Tomolo, A.M., Wears, R.L., and Tsevat, J.M. (2010). Comparison of extent of use, information accuracy, and functions for manual and electronic patient status boards. *International Journal of Medical Informatics, 79*(12), 817–823.

Patterson, E.S. and Woods, D.D. (2001). Shift changes, updates, and the on-call architecture in space shuttle mission control. *Computer Supported Cooperative Work, 10*(3–4), 317–346.

Patton, M.Q. (1999). Enhancing the quality and credibility of qualitative analysis. *Health Services Research, 34*(5, Part II), 1189–1208.

Potter, S.S., Roth, E.M., Woods, D.D., and Elm, W. (2000). Bootstrapping multiple converging cognitive task analysis techniques for system design. In J.M.C. Schraagen, S.F. Chipman, and V.L. Shalin (Eds.), *Cognitive Task Analysis* (pp. 317–340), Mahwah, NJ: Lawrence Erlbaum Associates.

Puckett, F. (1995). Medication-management component of a point-of-care information system. *American Journal of Health-System Pharmacy, 52*, 1305–1309.

Rasmussen, J. (1986). A cognitive engineering approach to the modeling of decision-making and its organization. In *Technical Report No. RIso-M-2589*, Roskilde, Denmark: RIso Laboratory.

Rasmussen, J. (1987). Cognitive control and human error mechanisms. In: Rasmussen, J., Duncan, K. and Leplat, J. (Eds). *New Technology and Human Error* (pp. 53–61), New York: Wiley.

Rasmussen, J. and Petersen, A.M. (1995). Virtual ecology of work. In J. Flach, P. Hancock, J. Caird, and K.J. Vicente (Eds.), *Global Perspectives on the Ecology of the Human Machine System* (pp. 121–156), Hillsdale, NJ: Lawrence Elrbaum Associates.

Rasmussen, J., Petersen, A., and Goldstein, L. (1994). *Cognitive Systems Engineering*, New York: John Wiley & Sons, Inc.

Rubin, J. (1994). *Handbook of Usability Testing: How to Plan, Design, and Conduct Effective Tests*, New York: Wiley & Sons.

Sarter, N., Woods, D., and Billings, C. (1997). Automation surprise. In G. Salvendy (Ed.), *Handbook of Human Factors and Ergonomics* (pp. 1926–1943), New York: Wiley.

Smith, M.J. and Sainfort, P. (1989). A balance theory of job design for stress reduction. *International Journal of Industrial Ergonomics, 4*, 67–79.

Vicente, K. (1999). *Cognitive Work Analysis: Toward Safe Productive and Healthy Computer Based Work*, Mahwah, NJ: Lawrence Erlbaum Associates.

Weick, K.E. and Sutcliffe, K.M. (2001). *Managing the Unexpected: Assuring High Performance in an Age of Complexity* (1st ed.). University of Michigan Business School management series. San Francisco, CA: Jossey-Bass.

Woods, D.D. (1993). Process tracing methods for the study of cognition outside of the experimental psychology laboratory. In G. Klein, J. Orasanu, and R. Calderwood (Eds.), *Decision Making in Action: Models and Methods* (pp. 228–251), Norwood, NJ: Ablex Publishing Corporation.

Woods, D.D. (2003). Discovering how distributed cognitive systems work. In E. Hollnagel (Ed.), *Handbook of Cognitive Task Design* (pp. 37–54), Mahwaj, NJ: Erlbaum.

Woods, D.D., Johannesen, L., Cook, R.I., and Sarter, N. (1994). Behind human error: Cognitive systems, computer and hindsight. In *Crew Systems Ergonomic Information and Analysis Center*, Dayton, OH: Wright Patterson Airforce Base.

Woods, D.D. and Roth, E.M. (1988). Cognitive systems engineering. In M. Helander (Ed.), *Handbook of Human–Computer Interaction* (1st ed.) (pp. 3–43), New York: North-Holland (reprinted in N. Moray, editor, Ergonomics: Major Writings, Routledge/Taylor & Francis, 2005).

Woods, D.D., Tinapple, D., Roesler, A., and Feil, M. (2002). *Studying Cognitive Work in Context: Facilitating Insight at the Intersection of People, Technology and Work*, Columbus, OH: Cognitive Systems Engineering Laboratory, Institute for Ergonomics, The Ohio State University. Accessed online: http://csel.eng.ohio-state.edu/woodscta

Young, T. et al. (2000). Mars Program Independent Assessment Team Report. NASA, March 2000.

29

Human Factors Risk Management for Medical Products

Edmond W. Israelski and William H. Muto

CONTENTS

29.1 Introduction to Risk Management...476
29.2 Human Reliability and Use-Error...477
    29.2.1 Background on Human Reliability..477
    29.2.2 Defining Use-Error...477
29.3 Identification of Hazards When Risk Management Is Conducted......................................478
    29.3.1 Hazard Analysis..478
    29.3.2 Where Risk Management Fits in the Development Process....................................478
29.4 Failure Modes and Effects Analysis..479
    29.4.1 Introduction...480
    29.4.2 Basic Steps in Performing Human Factors FMEA...481
        29.4.2.1 Form a Team..481
        29.4.2.2 Perform a Task Analysis...481
        29.4.2.3 Start a Worksheet...482
        29.4.2.4 Brainstorm Potential Use-Errors (Failure Modes)....................................482
        29.4.2.5 List Potential Effects of Each Failure Mode/Operator Error................483
        29.4.2.6 Assign Severity Ratings...484
        29.4.2.7 Assign Occurrence Ratings...484
        29.4.2.8 Derive a Risk Index...484
        29.4.2.9 Prioritize the Risks..484
        29.4.2.10 Take Actions to Eliminate or Reduce the High-Priority Failure Modes............485
        29.4.2.11 Assign Effectiveness Ratings...485
        29.4.2.12 Revise the Risk Priorities...485
    29.4.3 FMEA Variations..486
        29.4.3.1 Numeric versus Descriptive Ratings...486
        29.4.3.2 Range of Numeric Values..486
        29.4.3.3 Detectability...486
29.5 Fault Tree Analysis and Other Technique Variations..486
    29.5.1 Introduction...486
    29.5.2 Basic Steps in Conducting an FTA...487
        29.5.2.1 Identify Top-Level Hazards..487
        29.5.2.2 Identify Fault Tree Events...487
        29.5.2.3 Identify the Conditions under Which the Events Can Lead to Failure............487
        29.5.2.4 Combine the Aforementioned Events into a Fault Tree.........................487
        29.5.2.5 Assign Probabilities to Each Event..487
        29.5.2.6 Calculate the Probability of Each of the Branches Leading to the Top-Level Hazard.......487
    29.5.3 Uses for Fault Tree Probability Estimates..487
    29.5.4 Advantages and Disadvantages of FMEAs and FTA...488
    29.5.5 Other Risk Management Methods (Health Care FMEA, Health Hazard Assessment)......489
29.6 Examples of Use-Error FMEAs and FTAs...490
    29.6.1 Qualitative Use-Error FMEA: Therac-25 Radiation Therapy System.................490
    29.6.2 Quantitative Use-Error FMEA: Automatic External Defibrillator.......................491
    29.6.3 Nonquantitative Fault Tree for an AED..491

29.6.4 Quantitative Fault Tree for an AED 491
29.6.5 Comparison of Quantitative versus Qualitative Risk Management 496
29.7 Criteria for Establishing Acceptable Risk 496
29.8 Common Mistakes in Performing Use-Error Risk Analysis 497
29.8.1 Blaming the User 497
29.8.2 Treating Risk Management as a Necessary Evil Rather than a Development Tool 497
29.8.3 Underestimating Risk Probability and Severity 497
29.8.4 Overlooking Critical Sources of Failure 498
29.8.5 Failure to Document Low-Level Risks 498
29.8.6 Overreliance on Training or Labeling 498
29.8.7 Failing to Validate Modes of Control 498
29.9 Standards and Regulations 498
29.9.1 Standards for Risk Management 498
29.9.2 Regulations 498
29.10 Implementation Considerations 499
29.10.1 Team Membership 499
29.10.2 Estimation Techniques 499
29.10.2.1 Brainstorming 499
29.10.2.2 Achieving Consensus 499
29.10.2.3 Delphi Techniques 500
29.11 Uses for Risk Management 500
29.11.1 Trade-Offs Regarding Design Safety 500
29.11.2 Decision Point for Usability Testing 501
29.11.3 Labeling and Training 501
29.11.4 Input to Corrective and Preventative Action (CAPA) Systems 501
29.11.5 Expanded Applications for Use-Error Risk Analysis 501
29.12 Future Directions for Use-Error Risk Management 501
29.12.1 Increased Awareness of Use-Errors Related to Patient Safety 501
29.12.2 Increased Focus on the Use Environment 501
29.12.3 Improved Medical Error Data 502
29.12.4 Improved Tools 502
Appendix A 503
References 504

Risk management and its elements of risk analysis, hazard analysis, risk evaluation, and risk control have been used as engineering tools for many years in industries such as nuclear power, weapons systems, and transportation to identify risks and to control system modes of failure. Recently, the health care industry has seen increased attention to risk management tools as a means to improve patient safety. These tools and related methods have been applied to understanding "use-errors" made with medical devices. Use-errors are defined as predictable patterns of human errors that can be attributable to inadequate or improper design. Use-errors can be predicted through analytical task walkthrough techniques and via empirically based usability testing. This chapter explains and discusses the special methodology of use-error-focused risk analysis and some of its history. Case studies are presented that illustrate the methods of use-error risk analysis such as fault tree analysis (FTA) and failure mode effects analysis (FMEA) and

some pitfalls to be avoided. The case studies include the Therac-25™, computer-controlled radiation therapy system, and an automatic external defibrillator (AED).

29.1 Introduction to Risk Management

Risk analysis in the context of use-errors in medical products and processes has received increasing attention in recent years. Although risk analysis techniques have been used for decades to assess the effect of human behavior on critical systems such as in aerospace, defense systems, and nuclear power applications, the use of these techniques in medical applications has received relatively little attention in the published literature. Because human error has been identified as a major contributing cause to patient injury and death, risk analysis techniques have seen increased attention. Use-errors are defined as a pattern of predictable human errors that can be attributable to inadequate or improper design. In Section 29.6, there is a more detailed discussion of use-errors including examples. Definitions of relevant risk management terms are included in Appendix A.

29.2 Human Reliability and Use-Error

29.2.1 Background on Human Reliability

Human reliability and related human errors have been analyzed and studied a great deal in the area of applied psychology, including a great deal of theorizing about error causes and classifications (Bailey, 1983; Miller and Swain, 1987; Reason, 1990, 1997) among others. Human error may be due to many causes including limitations in

- Attention
- Perception
- Memory
- Cognitive processing
- Response execution

Fortunately, a deep understanding of the underlying causes and theories of human error is not critical in applying use-error risk analysis. It is important, however, to be able to systematically identify/analyze potential human errors in the use of a medical product using analytical tools such as FTA and FMEA, which are more fully described in Sections 29.4 and 29.5. The use of these tools leads to a systematic effort to understand and eventually control these errors and their resulting hazards and harms to patients and other product users. The theory behind human errors plays less of a role compared to the effort to carefully catalog and control human errors. It is true that human error theories that would assist in the prediction of error would be beneficial in incorporating effective design mitigations to reduce or eliminate the risks. Again, designers of medical products can use other human factors best practice tools, such as usability testing that can be used to measure the effectiveness of their design mitigations.

29.2.2 Defining Use-Error

Use-error is characterized by a repetitive pattern of failure that indicates a failure mode that is likely to occur with use and thus has a reasonable possibility of predictability of occurrence. Use-error can be addressed and minimized by the device designer and proactively identified through the use of techniques such as usability testing and hazard analysis. An important point is that in the area of medical products, regulator and standards bodies make a clear distinction between "use-error" and the common terms "human error" and "user error." The term use-error attempts to remove the blame from the user and open up the analyst to consider other causes including

- Poor user interface design, for example, poor usability
- Organizational elements, for example, inadequate training or support structure
- Use environment not properly anticipated in the design
- Not understanding the users tasks and task flow
- Not understanding the user profile in terms of training, experience, task performance incentives, and motivation

There are several taxonomies that attempt to categorize the elements of a use-error. For more information, see Chapter 23. Figure 29.1 shows one such taxonomy from the IEC standard for medical devices and in particular the part that focuses on usability (IEC, 2007). This diagram is a formal part of an international standard that focuses exclusively on human factors for medical devices. Human factors professionals are currently debating the details of this model, which is based on Reason's models of human error (1997). An important implication of the model is that it highlights many sources of error that should be considered and analyzed by manufacturers. The designation of human error does not automatically absolve the manufacturer from responsibility. The IEC standard states that "abnormal use" is outside the scope of the standard. A controversy surrounding Figure 29.1 is that some interpret the standard as implying that manufacturers do not need to consider categories such as off-label use or inadequate training. But, it is common for manufacturers to consider these error categories in risk analysis and offer design mitigations that address them. Figure 29.2 shows one such example, illustrating a plastic lockbox that guards against narcotic substance abuse for a portable ambulatory patient-controlled analgesia (PCA) infusion pump. In the design of this product it became apparent from medical device reports (MDRs) that there could be illegal attempts by narcotic addicts to access vials of pain medication intended to manage chronic patient pain. These actions would fall under the category of "abnormal use," but prudent manufacturers of portable PCA pumps have taken proactive steps using a lockbox to reduce the risk of device abuse.

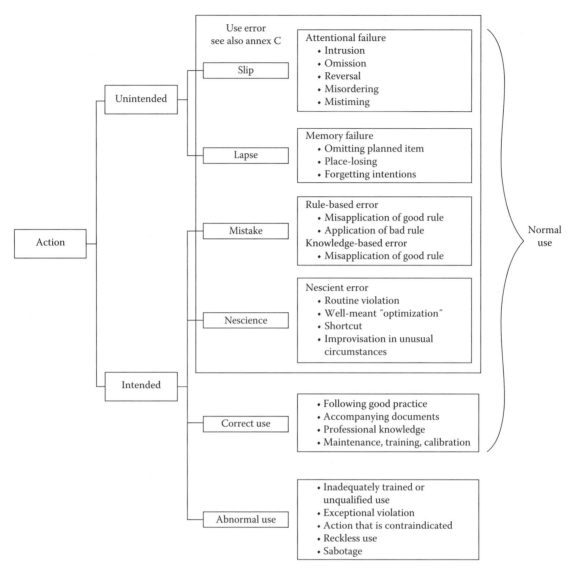

FIGURE 29.1

A possible taxonomy of use actions and use-errors in operation a medical device. (From IEC 62366, Ed 1, Medical devices—Application of usability engineering to medical devices, 2007. With permission.)

29.3 Identification of Hazards When Risk Management Is Conducted

29.3.1 Hazard Analysis

An important first step in risk management is to understand and catalog the hazards and possible resulting harms that might be caused by a medical device. Some call this hazard analysis. Unfortunately, others use the term in a more general way and use the term hazard analysis as a synonym for risk management. Hazard analysis is often done as the first part of an iterative process with early drafts being updated and expanded as additional risk management methods (e.g., FMEA and FTA) are used. Experts from medical, quality, and product

development, among other disciplines can brainstorm on harms and hazards. Technically, hazards are the potential for harms. Harms are defined as physical injury or damage to the health of people, or damage to property or the environment. Table 29.1 shows examples of harms from hazards for a pen-like automatic needle injector device. Table 29.2 shows similar harms and hazards from an AED.

29.3.2 Where Risk Management Fits in the Development Process

Use-error-focused risk analyses including FMEA and FTA are particular methods in the user-centered or human factors design process. It is the analytical complement to empirical usability assessment commonly

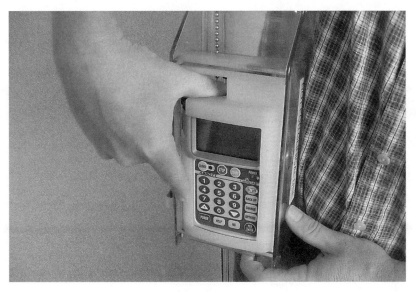

FIGURE 29.2
A protective plastic lockbox for the prevention of abuse of narcotics used in an ambulatory PCA infusion pump.

called usability testing. Figure 29.3 shows where use-error risk analysis fits into the overall human factors process for medical devices. The figure includes inputs and outputs for the various human factors process steps and also shows the iterative nature of the process as indicated by feedback loops, mainly emanating from the usability testing and postmarket analysis steps. Figure 29.4 depicts another view of where risk management fits into the development process as commonly described by design controls. The Code of Federal Regulations (CFR, 2004) regulates the development process in the United States and in these regulations the process is called design controls with distinct development stages such as concept, design input, design output, verification, and validation. In Figure 29.4, risk

management begins early in the design process and continues through the end of the process where design validation occurs.

Sections 29.6 and 29.7 describe the most used tools involved in user-error risk analysis: FMEA, and FTA. Case studies are offered for a radiation therapy system and an AED in later sections, including examples of both FMEAs and FTAs.

29.4 Failure Modes and Effects Analysis

The recommended steps for conducting a use-error risk analysis are the same as traditional risk analysis with one significant addition, namely the need to perform a task analysis. Possible use-errors are then deduced from identifying how task execution can fail. Each of the use-errors is rated in terms of severity of its effects and the probability of occurrence that comprise a risk

TABLE 29.1

Possible Harms and Hazards from Use of an Automatic Needle Injection Device

• Bleeding, bruising, or tearing of skin, leading to a possible infection	• Pain on injection
	• Increased bleeding, due to the presence of alcohol
• Incomplete injection that may lead to giving another injection, leading to over medication	• Nondelivery, wasted dose
• Under medication	• Delivery intramuscularly instead of subcutaneously
• Delay in therapy	• Possible infection from microorganisms present on skin
• Failed therapy due to unsuccessful injection	

TABLE 29.2

Possible Harms from Use of an AED

• Nondelivery of defibrillating shock.	• Set victim on fire
• Delay in delivery of defibrillating shock.	• Delivery of weak noneffective shock
• Administration of shock when not needed.	• Ignoring subsequent second episode of cardiac fibrillation
• Bystander shocked when touching patient during delivery	• Burns caused by delivery of electrodes touching each other

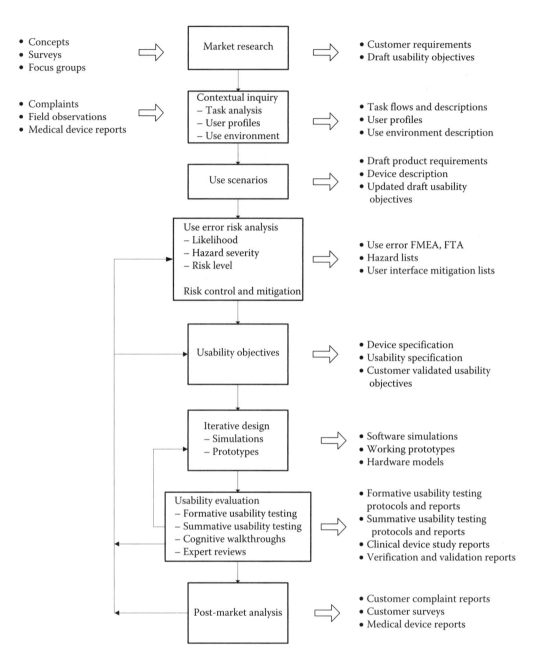

FIGURE 29.3
Human factors engineering process for medical products.

index used for prioritization. For each of the high priority items, modes (or methods) of control are assumed for the system or subsystem and reassessed in terms of risk. The process is iterated until all higher-level risks are eliminated.

29.4.1 Introduction

Among the most widely used of the risk analysis tools is FMEA and its close relative, failure modes effects

and criticality analysis (FMECA).* FMEA is a "bottom-up" design evaluation technique used to define, identify, and eliminate known and/or potential failures, problems, and errors from the system. The basic approach of an FMEA from an engineering perspective

---

* FMECA is an extension of FMEA that starts with FMEA elements and further considers ratings of criticality and probability of occurrence. Because of their common basis, FMEA and FMECA are commonly referred to as FMEA. Likewise, in this chapter FMEA and FMECA will be referred to as FMEA.

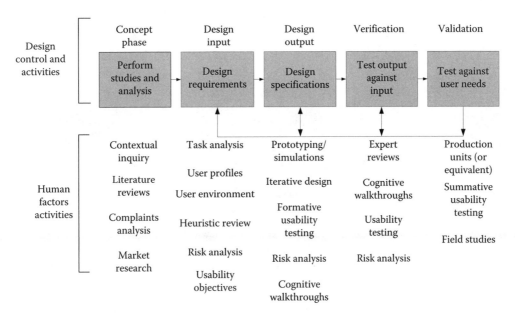

FIGURE 29.4
Human factors in the design control process.

is to answer the question: If a system component fails, what is the effect on system performance or safety? Similarly, from a human factors perspective, FMEA addresses the question, "if a user commits an error, what is the affect on system performance or safety?" A human factors risk analysis has several components that help define and prioritize such faults: (1) the identified fault or use-error, (2) occurrence (frequency of failure), (3) severity (seriousness of the hazard and harm resulting from the failure), (4) selection of controls to mitigate the failure before it has an adverse effect, and (5) an assessment of the risk after controls are applied.

A use-error risk analysis is not substantially different from a conventional design FMEA. The main difference is that rather than focusing in on component or system level faults, it focuses on user actions that deviate from expected or ideal user performance.

29.4.2 Basic Steps in Performing Human Factors FMEA

29.4.2.1 Form a Team

Although risk analyses are occasionally performed by lone individuals, the most effective risk analyses are conducted by a team of critical organizational stakeholders. In a medical device company, the critical stakeholders could include: human factors, medical affairs, development (hardware and software engineering where appropriate), and quality assurance. Other stakeholders can include marketing, regulatory,

documentation, and training. The inclusion of a variety of stakeholders maximizes the chance of performing as unbiased an analysis as possible. The objective of the first meeting is to get everyone to understand and agree to the objectives and the process of conducting the FMEA.

29.4.2.2 Perform a Task Analysis

A task analysis is a detailed sequential description (in graphical, tabular, or narrative form) of the tasks performed while operating a device or system. The analysis should cover the major task flows performed by users of the product and would usually include the following (Stamatis, 1995):

- The person performing the task (in a system that requires multiple operators)
- The task stimulus—stimulus that initiates the task
- The equipment or component used to perform the task
- The task feedback
- Required human response (including decision making)
- The characteristics of the task output, including performance requirements

There are many ways to document a task analysis. Table 29.3 shows a task analysis in the form of a task

TABLE 29.3

Example of a Hierarchical Task Description Table

Activity	Task	Subtask	Notes
Device setup	Calibration of unit	Connect sensor to load	
		Measure resistance	
		Adjust sensitivity control	If sensitivity cannot be adjusted go to troubleshooting activity tasks
	Check battery charge state	Read battery meter	
		Verify that reading is greater than 12 V but less than 15 V	If reading is not in range go to battery charging activity
	Check mounting security	Tighten clamp until knob starts to slip	
		Gently pull down on unit to test for no movement	If movement detected then readjust clamp
Start device	Initial start-up	Verify power is on	
		Check that sensor gauge reading is in starting range	
	Monitor status of reading	Check both sensor level and sensitivity readings for in range readings	
Stop device	Press stop button	Check for readings to return to zero	

description table. Figure 29.5 is an example of a task flow diagram. Other forms of task analysis can be found in Kirwan and Ainsworth (1992).

The most commonly recommended methodology for gathering task analysis data is contextual inquiry, an information gathering process with three components: task analysis, user profiling, and use environment analysis. This method may include the following techniques for data gathering:

- Observations
- Follow around studies (shadowing)
- Behavioral checklists
- Time slice sampling studies
- Time and motion studies
- Usability study repurposed to understand task behaviors
- Interviews and surveys (1:1 and focus groups)
- Review of job and task documentation. (A draft user instructions manual is one source of user task descriptions.)
- Review of training materials, job descriptions, and job evaluations
- User kept diaries and activity logs

29.4.2.3 Start a Worksheet

A use-error focused FMEA is documented using an FMEA worksheet such as the ones shown in Tables 29.13 and 29.14. Note that there are many variations of FMEA worksheets. Your company or team may choose variations to better suit your project needs. The worksheet can be developed using large sheets of paper (e.g., 11″ × 17″

or larger). Developing FMEAs directly on a computer using dedicated FMEA software tools or generic spreadsheets or word processing (using a "table" function) can be very effective for group interactions, especially when using a computer projector.

29.4.2.4 Brainstorm Potential Use-Errors (Failure Modes)

For each of the tasks identified in the task analysis, possible operator errors, actions that deviate from expected or optimal behavior, are identified. (These errors, in risk analysis parlance would be referred to as "faults" or "failure modes.") Team interactions for this phase should follow common brainstorming rules wherein inputs are recorded and the merits of the individual items are not debated. All conceivable errors that might be committed by an operator should be analyzed. Brainstorming for each task and subtask step may generate the following:

- Use-errors that may be design induced or would be anticipated to follow predictable trends or patterns.
- Common human mistakes, lapses of memory, and attention that have commonly been called user or human errors. (See, e.g., Figure 29.1)
- Each distinct use environment should be considered as different environments may lead to error causing conditions.
- Use-error data from predicate devices. Use-errors may be cataloged in customer complaint data or MDRs, which are required in the United States by the FDA.

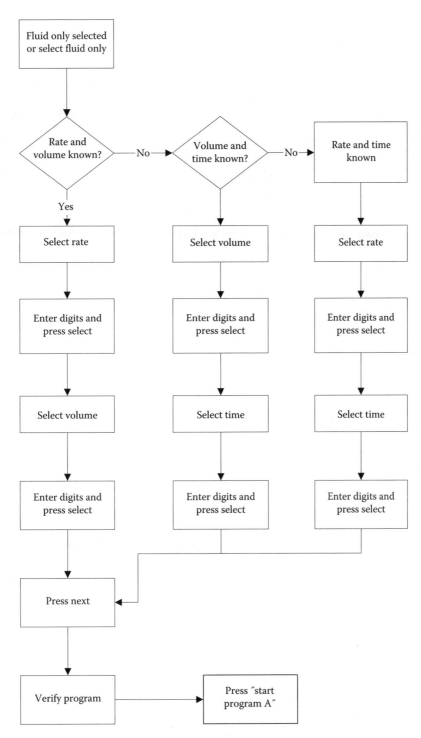

FIGURE 29.5
Example of a task flow diagram for programming an infusion pump.

29.4.2.5 List Potential Effects of Each Failure Mode/Operator Error

For each of the failure modes, the team identifies the potential effects (harm) of the failure mode if it happens. This step is important for the subsequent determination of risk ratings. Each effect can be thought of as an if-then process (McDermott et al., 1996). For most medical devices, the harm will have implications for the patient or the user of the device (e.g., causing death, reversible or nonreversible injury). When there are multiple harms, the convention is to consider the worst-case scenarios.

TABLE 29.4

Example of Numeric Ratings for Severity in FMEA

Severity Category	Severity Rating	Consequences to Patients, Users, or Environment
Catastrophic	5	May cause death, total loss of operation, or severe impact on the environment that cause death or total loss of operation
Critical	4	May cause permanent injury, permanent occupational illness, major and recoverable damage to operation, or major and recoverable impact on the environment
Significant	3	May cause recoverable injury, recoverable occupational illness, major and recoverable damage to operation, or major and recoverable impact on the environment
Marginal	2	May cause minor injury, minor occupational illness, or minor impact on operation or the environment
Negligible	1	Will not result in significant injury, occupational illness, or a significant impact on the environment. Negligible effect on operation

29.4.2.6 Assign Severity Ratings

A severity rating is a rating of how serious the effects or harm of a given fault would be if it occurs. Depending on the method used, severity can be assigned a numeric value (typically 1–5 or 1–10, with 1 being the lowest severity and 5 or 10 being the highest) or a qualitative descriptive rating. Table 29.4 shows an example list of categories used to assign a numeric rating. Table 29.5 shows an alternative rating using qualitative descriptive ratings. With medical devices, the highest severity ratings are typically reserved for faults that have severe consequences for the patient or user, usually involving severe injury or death. The lower ratings would be reserved for minor reversible injuries, such as a scrape or noninjury effect (e.g., customer inconvenience). For many medical devices, ratings of severity should be provided and/or approved by medical affairs or equivalent.

29.4.2.7 Assign Occurrence Ratings

Occurrence ratings are estimates of the predicted frequency or likelihood of occurrence of a fault. It is best to base occurrence ratings on existing data such as: customer complaint data, telephone support center call data, usability test results, etc. Where such data are not available, occurrence ratings must come from the collective judgment of a team. As with severity ratings, occurrence ratings can be assigned a numeric value (typically 1–5 or 1–10, with 1 being the lowest frequency of occurrence and 5 or 10 being the highest). Table 29.6 shows a representative example of a numeric occurrence rating scale. An alternative rating scheme, shown in Table 29.7 uses qualitative terms to define the frequency of occurrence where frequency of occurrence is rated as probable, occasional, rare, and improbable.

During this phase, the team may decide to eliminate those items that are agreed to be so improbable that their presence in the analysis is judged to have no value.

29.4.2.8 Derive a Risk Index

Using the numeric approach, a "risk index" or "risk level" is calculated by multiplying the severity rating times the occurrence rating:

$$\text{Risk index} = \text{Severity} \times \text{Occurrence}$$

When using qualitative ratings, criteria are developed for how risk levels will be defined based on combinations of severity and occurrence. Table 29.8 is an example of how risk priorities could be defined using qualitative descriptors for severity and occurrence.

29.4.2.9 Prioritize the Risks

Using the risk index (either numeric or qualitative ratings), the failure modes are sorted according to risk index from the highest to lowest. With the prioritized list, the team will determine:

- Failure modes that can be addressed immediately
- Failure modes that will be addressed at a later date or a later version of the product
- Failure modes that are determined to require no action

The criterion used by the team to decide a "cutoff" line (the threshold below which problems will be designated as needing no action) should be predefined; ideally by standardized procedures. This will ensure that the different project teams will have a uniform means to address risk levels.

Prioritization may include special rules. For example, a team might decide to apply corrective actions to all failure modes with a high severity rating, regardless of the occurrence rating. Such an approach would help eliminate incongruities where a failure mode with a low severity and high occurrence rating is ranked equal to or higher than a failure mode with high severity but with a lower occurrence rating.

TABLE 29.5

Example of Qualitative (Descriptive) Ratings for Severity in FMEA

Severity Rating	Consequences
Major	The hazard could directly result in death or serious injury of the patient or operator, or indirectly affect the patient such that delayed or incorrect information could result in death or serious injury to the patient
Moderate	The hazard could directly result in moderate injury to the patient or operator, or indirectly affect the patient such that delayed or incorrect information could result in moderate injury to the patient
Minimal	The hazard is not expected to result in negative medical consequences or any complication

29.4.2.10 Take Actions to Eliminate or Reduce the High-Priority Failure Modes

Using an organized problem-solving approach, the team identifies "modes of control" (or "methods of control" or "mitigations") for each of the higher priority failure modes. Modes of control are design or other elements that are intended to reduce the probability of occurrence and/or the severity of the consequences of the event. Although modes of control can also include warnings or training (often referred to under the broad term, "labeling"), such mitigations should be considered less effective and secondary to design modes of control. Where labeling modes of control are applied, care should be taken to validate such measures to ensure that the instructions and training materials are effective and accessible by users.

29.4.2.11 Assign Effectiveness Ratings

For each mode of control, a rating is given for the likelihood that the mode of control will be effective. If the analysis is done early in the design cycle, then the

TABLE 29.6

Example of Numeric Ratings for Probability of Occurrence in FMEA

Frequency Category	Numeric Rating	Frequency of Occurrence
Extremely likely	5	Very high probability of occurrence
Likely	4	High probability of occurrence
Possible	3	Moderate probability of occurrence
Unlikely	2	Low probability of occurrence
Extremely unlikely	1	Very low or remote probability of occurrence

TABLE 29.7

Example of Qualitative Ratings for Probability of Occurrence in FMEA

Probability of Occurrence	Description
Probable	Likely to occur regularly or many times during the life of the product under specified operating conditions
Occasional	Likely to occur infrequently or several times during the life of the product under specified operating conditions
Rare	Will rarely occur or is very unlikely to occur during the life of the product under specified operating conditions
Improbable	Not expected to occur under specified operating conditions

effectiveness of potential mode of control will be more difficult to estimate with confidence (formative usability testing and usability inspection methods can aid in these estimates). If done later in the development cycle, then data from summative usability testing will provide better estimates for the effects of the mitigating factors. Using qualitative descriptors, the effectiveness of the mitigation could be expressed as

- Does not change risk
- Reduces (risk) to medium
- Reduces (risk) to low

In a quantitative form, mitigation effectiveness is sometimes given a numeric rating. In the example in Table 29.9, the values are expressed in descending order; wherein the most effective mode of control is given a 1 and least effective, a rating of 5.

29.4.2.12 Revise the Risk Priorities

With the assumed modes of control in place, the numerical or qualitative risk indices are revised or recomputed. Any failure modes with risk indices that remain above the cutoff line should be considered for additional

TABLE 29.8

Example of a Qualitative Risk Index, Based on Previous Qualitative Occurrence and Severity Ratings

Probability of Occurrence	Hazard Severity		
	Major	Moderate	Minimal
Probable	High	High	Medium
Occasional	High	Medium	Low
Rare	Medium	Low	Low
Improbable	Low	Low	Low

TABLE 29.9

Example of Mitigation Effectiveness Ratings

Effectiveness of Mitigation	Rating
Extremely unlikely	5
Unlikely	4
Possible	3
Unlikely	2
Extremely likely	1

modes of control to further reduce the risk level. The process should be iterated until all use-errors have an acceptable level of risk.

29.4.3 FMEA Variations

29.4.3.1 Numeric versus Descriptive Ratings

Ratings for severity and occurrence can be given either a numeric or qualitative (descriptive) ratings. As can be seen in the tables shown, whether numeric or descriptive, such ratings are almost always categorical rather than precise values for occurrence or severity.

Occurrence ratings are often the most troublesome category for groups to assess. Stamatis (1995) offers probabilistic benchmarks for how occurrence ratings could be categorized. Although such categories can be useful for mechanistic or component failures, where failure data are often available; such benchmarks are seldom available when dealing with human error, mainly because there is a paucity of data related to human reliability, especially when attempting to estimate failures in parts per million or even parts per thousand. For this reason, it is incumbent on groups to provide these estimates based on device histories, knowledge of the user, the device, and the use environment.

29.4.3.2 Range of Numeric Values

When using numeric values, an analysis will incorporate a range of values for estimating severity and occurrence, traditionally 1–10. The ratings for severity, occurrence (and "detectability") are multiplied together to form a risk priority number or "RPN." A common notion is that ratings are more precise with the larger range of numeric values. However, practically speaking, numbers that extend beyond this range will likely not provide benefit. Furthermore, the extended range of numbers will almost certainly require more time and effort to debate and manipulate the resultant values.

29.4.3.3 Detectability

In addition to ratings for severity and occurrence, it is common practice, especially in manufacturing process

settings to include a third rating for "detectability." Detectability is a rating to characterize the ability of the process or user to detect a problem before it occurs. In contrast to severity and occurrence, the values for detection using numeric ratings are rated in reverse order; where high detectability is given a low numeric value (i.e., 1) and lowest detectability a 5 (or 10). Although in common use, the authors offer the following reasons why detectability may not be appropriate for use in human factors FMEAs or perhaps in any design FMEA.

- According to Schmidt (2004), detection of a hazard during use of a device may not assure that harm will be avoided. He delineates three conditions that can determine if detectability would serve to prevent harm: (1) Is there enough time to react after detection? (2) Is information provided to the user to indicate specific actions to avoid the harm? (3) Will the user have the presence of mind to remember what is done and take action? In many cases, even if the user can detect impending harm, he or she may not be capable of preventing it.

- In situations where a user can detect harm and prevent it, it is the authors' position that this should be factored into the probability of occurrence. If detectability is enhanced or enabled by the system in a way that will reduce the overall probability of occurrence or the severity, this constitutes a mode of control and therefore should be reflected by a reduction in the probability of occurrence. With both detection and occurrence reductions of the RPN value, it is the authors' contention that the risk index is being reduced artificially by overweighting the effectiveness of the detection mode of control.

29.5 Fault Tree Analysis and Other Technique Variations

29.5.1 Introduction

Another commonly used tool for analyzing and predicting failure is fault tree analysis, or FTA. FTA is a "top-down" deductive method used to determine overall system reliability and safety (Stamatis, 1995). A fault tree, depicted graphically, starts with a single undesired event (failure) at the top of an inverted tree, and the "branches" show the faults that can lead to the undesired event—the root causes are shown at the bottom of the tree. For human factors applications, FTA can be a useful tool for visualizing the effects of human error combined with device faults or normal conditions on the overall system. Further, by assigning probability estimates to the faults, combinatorial probabilistic rules can be used to calculate an estimated probability of the top-level event or hazard. As with FMEAs, fault trees can be developed by teams or by individuals with team review. The following describes basic steps in developing fault trees. For more information, refer to the literature in reliability engineering or systems safety engineering (e.g., Veseley, 1981).

Fault trees have often been avoided because the drawings and computations are seen as being labor intensive. In recent years, graphical software programs have been made available for personal computers that enable users to rapidly assemble fault trees by "dragging and dropping" standard logic symbols onto a drawing area and connections are made (and maintained automatically. These tools automatically can calculate branch and top-level probabilities based on event estimated event probabilities entered. Such tools make FTAs much more accessible and much less labor intensive.

29.5.2 Basic Steps in Conducting an FTA

29.5.2.1 Identify Top-Level Hazards

The team will brainstorm to identify the top-level hazards (undesired event) to be addressed. A fault tree will be developed for each of these hazards.

29.5.2.2 Identify Fault Tree Events

Identify faults and other events (including normal events) that could result in the top-level undesired event. These can be documented in a list or on notes posted on a wall.

29.5.2.3 Identify the Conditions under Which the Events Can Lead to Failure

For example,

- Events that may lead directly to harm (single-point failure) or cause another fault without other events occurring
- Events that must happen in conjunction with other events to cause failure
- Events that must happen in sequence to cause a failure

29.5.2.4 Combine the Aforementioned Events into a Fault Tree

Table 29.10 shows the symbols (and their usage) that are employed in fault trees. Of these, the more likely to be used are those in the left half of the table that represent individual events ("basic events," "undeveloped events," or "house events") and the most significant logic symbols, the "OR gate" and the "AND gate." Most of the remaining symbols are likely to be used in design-specific analyses or in systems with a high degree of complexity.

29.5.2.5 Assign Probabilities to Each Event

At this point, the investigative team may decide that the fault tree sufficiently characterizes the system and human interactions and requires no further development. Other situations, however, may require further quantitative analysis. If so, then the team will assign probabilities to each of the events based on quantitative data or estimates based on expert judgment.

29.5.2.6 Calculate the Probability of Each of the Branches Leading to the Top-Level Hazard

Fault tree probabilities propagate upward from the individual events. The probability of the individual gates and the overall fault tree probability are computed by using numerical combinatorial rules for various logic gates. Figure 29.6 illustrates two logic gates, OR and AND gates with two events each along with the formulas for their probabilities.

The figure also illustrates the effect of two different logic gates with two identical input events. In computing, the probability of failure with the same input probabilities of .01, the OR gate shows a resultant probability of failure of approximately .02 (or 2 in 100), whereas the AND gate shows a probability of .0001 (or 1 in 10,000).

29.5.3 Uses for Fault Tree Probability Estimates

Although fault trees can provide utility without probability estimates, fault tree probabilities can provide benefit in the following ways.

- *Hazard probability.* Probabilities derived from fault trees can be used as a decision tool to assist design teams in determining if further modes of control are needed for further risk reduction. After new modes of control are incorporated into the fault tree, revised probability estimates can be used to determine the effect on overall probabilities. A fault tree may provide a useful supplement to FMEAs that typically do not depict the combinatorial aspect of events.
- *Critical path analysis.* One of the most useful purposes of a fault tree can be realized by tracing high probability pathways through the fault tree. Such a technique, even using qualitative probability ratings may uncover high risk "pathways"

TABLE 29.10

Fault Tree Symbols

Basic event	A basic component failure or use-error
Undeveloped event	A component or use-error that has not been fully developed due to the lack of information or significance
House event	A normally occurring event in the system
OR gate	Output occurs if any of the input events occur
AND gate	Output occurs if all of the inputs occur
Transfer gate	Symbol used to link elements in a fault tree with other pages or other segments of the tree
Voting gate	Output occurs if a specified number of events occur. The input events need not occur simultaneously
Inhibit gate	A certain condition of the system must exist before one failure produces another. The inhibit condition may be either normal to the system or be the result of failure
Exclusive OR	Output occurs if any input occurs; specified inputs are mutually exclusive
NOR gate	Output occurs when all the input events are absent
NOT gate	The output occurs when the input event does not occur
NAND gate	Output occurs when at least one of the input events is absent
Priority AND gate	Output occurs when all inputs occur in the sequence specified

that require design attention to reduce risks that may otherwise have been overlooked.

- *Sensitivity analysis.* To determine the effect of a given failure on the overall system, one can substitute different probability estimates for individual events or faults and assess the effect on the top-level probability. Such an analysis may be useful for teams performing FMEAs where there is disagreement among team members on the need to mitigate certain failure modes. After constructing a fault tree consisting of the FMEA elements, substituting a range of probabilities (e.g., 0.1–1.0) for selected faults can show to what extent a given failure affects the top-level event probability. A large change in the top-level probability when increasing a single or multiple fault probabilities may indicate where sufficient safeguards are lacking and further modes of control should be considered.

29.5.4 Advantages and Disadvantages of FMEAs and FTA

Because of its tabular format, FMEA is a straightforward method for documenting potential failures in terms of a "risk index," enabling explicit documentation of the modes of control or "mitigations" for each failure mode. For these reasons, FMEA is probably the most commonly used method of documenting risk analyses.

However, because of its bottom-up nature, wherein faults are dealt with one item at a time, it is often difficult to assess how a fault or combination of faults can lead to an undesired event. Further, in complex systems, it is not uncommon for FMEAs to consist of several hundred pages. In such cases, the management of a risk analysis becomes a daunting task; the elimination of inconsistency and redundancy can be challenging. Table 29.11 summarizes advantages and disadvantages in using FMEAs.

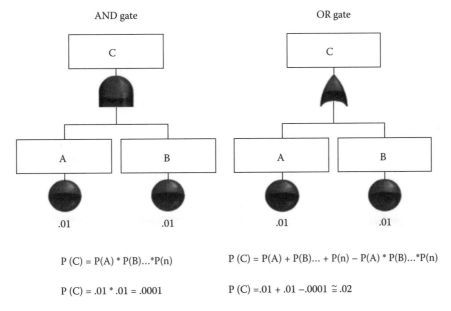

FIGURE 29.6
Illustration of "AND gate" and "OR gates" with two events and associated probability calculations.

The graphical representation of events with associated logic gates of fault trees can provide unique insight into how individual events can lead to or prevent an undesired event.

In large systems, fault trees can become very large and unwieldy. In such cases, it may also be difficult to visualize complex system interactions. The use of fault trees to compute the probability of failure enables teams to compute the probability of selected hazards as well as overall system failure. Lack of data or time, however, will often prevent teams from fully developing fault trees except for the most critical of systems. Table 29.12 summarizes the advantages and disadvantages of fault trees.

Both FMEA and FTA offer distinct advantages and disadvantages. For these reasons, we believe that both techniques are complementary.

29.5.5 Other Risk Management Methods (Health Care FMEA, Health Hazard Assessment)

There are many variations on these techniques and they appear under various names such as

- *Healthcare Failure Modes and Effects Analysis, HFMEA*™ (DeRosier et al., 2002). A process-oriented FMEA focused on health care processes, such as medication ordering in a hospital that was developed by the Veterans Affairs National Center for Patient Safety. HFMEA is defined by the Veterans Affairs (VA) as
  - A prospective assessment that identifies and improves steps in a process thereby reasonably ensuring a safe and clinically desirable outcome

TABLE 29.11

Advantages and Disadvantages of FMEA

Advantages	Disadvantages
• Risk index/RPN enables prioritization of faults	• Difficult to assess combination of events/complex interactions (unless explicitly documented)
• Explicitly documents modes of control/mitigation	• Large documents can be difficult to manage: minimize inconsistencies and redundant items
• Format useful for tracking action items	• Severity and occurrence ratings are often difficult for individuals or teams to estimate. Much time can be spent in discussions

• Easily constructed using hand-written spreadsheets or computer-based software tools:
  • Spreadsheets/word processing tables
  • Specialized FMEA tools
	• Sometimes can be overly conservative. With each fault isolated, failure to consider combinatorial events (as do fault trees) may lead to the false conclusion that every item requires explicit mitigation

TABLE 29.12

Advantages and Disadvantages of FTA

Advantages	Disadvantages
• Graphical format enables visualization of combination of events	• Drawings can become large and unwieldy in complex systems
• Enables estimation of overall probability of failure based on estimates of root causes	• Modes of control are not always explicit
• Small fault trees can be developed using common flowchart drawing tools	• Requires more training than FMEA
	• Special software required for rapid development of fault trees

- A systematic approach to identify and prevent product and process problems before they occur

- *Hazard Analysis and Critical Control Points, HACCP.* This is a process endorsed by the FDA and focuses on hazard analysis of food safety by examining critical control points in the food processing flow for fault potential.

- *Hazard and Operability Studies, HAZOPS.* A systematic analysis of industrial processing flows. It has been used for identifying potential hazards and operability problems caused by deviations from the design intent of industrial processing plants.

- *Health Hazard Assessment, HHA.* A systematic analysis to anticipate, identify, quantify, and recommend controls for health hazards associated with the occupational environment

- *Root Cause Analysis, RCA.* This is a series of analysis techniques that are typically done after an adverse or sentinel event. It is therefore a look back or retrospective analysis in contrast to FMEA, HFMEA, FTA, and the other forward-looking prospective techniques. RCA is required by JCAHO, Joint Commission on Accreditation of Healthcare Organizations, guidelines for health care facility accreditation and includes the following:

  - Affinity diagrams: They show clustering of possible causes of problems arranged in a hierarchy of root cause groups that resemble an organizational chart.

  - Fishbone diagrams: Sometimes they may be referred to as a cause-and-effect diagram. The diagram looks like the skeleton of a fish, and, hence, it is often referred to as the fishbone diagram.

The steps in the retrospective creation of a fishbone diagram are basically the same as the basic steps of RCA and include (North Carolina Department of Environmental and Natural Resources, Office of Organizational Excellence, 2002) the following:

1. Draw the fishbone diagram.

2. List the problem/issue to be studied in the "head of the fish."

3. Label each "bone" of the "fish." The major categories typically utilized are as follows:

   - The four Ms: Methods, machines, materials, and manpower

   - The four Ps: Place, procedure, people, and policies

   - The four Ss: Surroundings, suppliers, systems, and skills

     Note: You may use one of the four categories suggested, combine them in any fashion or make up your own. The categories are to help you organize your ideas.

4. Use an idea-generating technique (e.g., brainstorming) to identify the factors within each category that may be affecting the problem/issue and/or effect being studied. The team should ask "What are the machine issues affecting/causing ..."

5. Repeat this procedure with each factor under the category to produce subfactors. Continue asking, "Why is this happening?" and put additional segments under each factor and subsequently under each subfactor.

6. Continue until you no longer get useful information as you ask, "Why is that happening?"

7. Analyze the results of the fishbone after team members agree that an adequate amount of detail has been provided under each major category. Do this by looking for those items that appear in more than one category. These become the "most likely causes."

8. For those items identified as the "most likely causes," the team should reach consensus on listing those items in priority order with the first item being the most probable cause.

29.6 Examples of Use-Error FMEAs and FTAs

29.6.1 Qualitative Use-Error FMEA: Therac-25 Radiation Therapy System

The Therac-25 was a radiation therapy machine (a medical linear accelerator) designed for the treatment of tumors. The minicomputer-controlled system was activated through commands entered via a keyboard. Typing an "E" command selected a lower energy electron beam, and an "X" activated a higher energy "x-ray mode." The x-ray beam with an output of 25 MeV (million electron volts) was attenuated to proper dose levels by positioning a turntable-mounted beam-flattening filter into the radiation beam path. The filter was designed to be put into the beam path by computer control only when the system was in the high-energy x-ray mode. When using the electron beam mode, the attenuating filter was not used and was rotated away from the beam outlet. The placement of this filter is a key factor in this case (Leveson, 1995).

During this particular event, as the operator mistakenly entered the command "X" instead of "E," she quickly realized her error and pressed the up arrow key to move the cursor to the beam selection field and then typed the correct command "E." One of the causal problems was that an 8 s timer was activated in the software by this irregular command sequence and that any edits made before the 8 s elapsed were ignored. Although the screen displayed the E beam mode, the device beam selection state remained on X. Another, even more serious problem was that the device failed to command the turntable filter to move in place. After the operator commanded "B" for beam on, the system delivered the high-power x-ray energy mode without the beam-flattening attenuating filter in place. To further compound the tragedy, the system presented a cryptic error message "Malfunction 54" after the first dose, which was unrecognizable to the operator. (This is a classic case of poor system error messages that have no meaning to typical users, but do have meaning to the software designers.) Thinking that no dose was delivered, the operator pressed "P" to proceed. The same cryptic error message appeared and the screen displayed no unusual amounts of delivered radiation, so the operator entered the command to activate the beam again, activating a third dose. The repeated doses of the unattenuated beam resulted in the patient receiving over 16,000 rad instead of the intended 180 rad. The patient died about 5 months later from the massive radiation overdose (Casey, 1998).

This incident, as well as several others involved a combination of technical failures (involving software and hardware) combined with human behavior resulting in catastrophic radiation overdoses. According to Leveson (1995), AECL, the manufacturer of the Therac-25 had conducted hazard analyses in the form of FTAs. If so, what went wrong? Why were these hazards not anticipated and eliminated through design? According to Leveson, AECL did not consider possible software failures in their consideration of hazards. AECL's reasoning for this was that the software had been extensively tested and that software was judged not to be subject to risk analysis because software does not wear, fatigue, or fail from reproduction processes. We also know that use-error-focused risk

analysis was not common in 1985. If such a system were being developed today, current best practices would probably guide manufacturers to consider all of the system related considerations in the assessment of risk, including the user and environment as well as system hardware and software. Table 29.13 shows an abbreviated, hypothetical FMEA of the Therac-25 showing potential faults/use-errors. The ratings are qualitative. We propose that if this kind of analysis were performed at the time, the identification of high risk items would have lead the design team to consider redesign of certain elements to mitigate the high level risks (Israelski and Muto, 2003).

29.6.2 Quantitative Use-Error FMEA: Automatic External Defibrillator

An AED is a device used to treat sudden cardiac arrest due to ventricular fibrillation. The typical AED will detect ventricular fibrillation through chest electrodes applied by a caregiver, and automatically administer a defibrillatory shock (after the caregiver pushes a button) through the same electrodes. In a study by Andre et al., (2004), four devices were tested for usability among a group of untrained participants. Although the authors concluded that the AEDs in the test were sufficiently easy to use so that nonmedical people could effectively use them without prior training, the investigators had observed several use-error related problems. Table 29.14 depicts an example of an abbreviated FMEA using quantitative numeric ratings for hazard severity (from Table 29.4), probability of use-error (Table 29.6), and effectiveness of the method of control (Table 29.9).

29.6.3 Nonquantitative Fault Tree for an AED

One of the features of fault trees is that they can illustrate multiple "pathways" to an undesired event. Figure 29.7 depicts an example fault tree for an AED in which the top-level undesired event is nondelivery of defibrillatory shock. From this example, it appears that there are a number of potential sources of failure listed in the fault tree, some of which are technical failures and others are use-error related. The fact that all of these items are attached to "OR gates" implies that any one of these events could lead to the top-level event, the nondelivery of defibrillatory shock. Such a structure generally implies higher risk. This FTA example has no quantitative probability estimates.

29.6.4 Quantitative Fault Tree for an AED

Figure 29.8 depicts an example of a fault tree with quantitative probability estimates for an AED. This example portrays the top-level undesired event as the caregiver or bystander receiving a high voltage shock.

TABLE 29.13

Example Qualitative FMEA for Therac-25

Task	Hazard	Failure Mode/Use-Error	Use-Error Prob	Hazard Severity	Risk Level	Method of Control	Effectiveness of Control on Risk Level	Risk Acceptability
Generic Tasks								
Turn on device	Delay in therapy	Fail to hold power button for at least 2 s	Rare	Minimum	Low	Training and instructions note that power button requires holding for 2 s	Little to no impact	Acceptable
Programming Beam Modes								
Enter dose	Overdose, single activation	Typed X (high-power x-ray mode) instead of E (electron mode). Administer dose	Occasional	Major	High	Review screen shows selected dose mode Enable editing of mode selection	Reduces to medium	Acceptable with justification
	Insufficient dose, not effective	Typed E instead of X administer dose	Occasional	Moderate	Medium	Review screen shows dose mode	Reduces to medium	Acceptable with justification
	Insufficient dose; not effective	Typed B (beam) command, shutter did not open	Rare	Moderate	Low	Software robust Dosimeter failure	Little or no impact	Acceptable
Activate beam	Overdose, single activation	Computer selects wrong mode	Improbable	Major	Low	Software extensively tested	No further controls needed	Acceptable
	Overdose, single activation	Computer selects wrong energy level	Improbable	Major	Low	Software extensively tested	No further controls needed	Acceptable
	Overdose, single activation	Selected x-ray mode Tungsten filter not properly positioned (software control only—no closed loop feedback) Dosimeter failure	Occassional	Major	High	High quality stepper motor. Note to engineering Can filter position be guaranteed? Filter position sensor could assure that beam will not activate unless in place	Little or no impact Proper dosimeter operation would indicate overdose, after the fact	Unacceptable
	Overdose due to multiple activations	Typed B, beam activated User unclear about success User activates repeatedly	Occssional	Major	High	Software robust Note to engineering: how does the user know beam was actually activated, and at what power level? Could a message (and printout) confirm actual output?	Proper dosimeter operation would indicate total dosage to warn against multiple exposures (worst-case exposure?)	Acceptable with justification

TABLE 29.14

Example Use-Error FMEA for a Hypothetical AED; Product: AED; Activity: Setup

Task	Hazard	Use-Errors	Use-Error Probability	Hazard Severity	Risk Level	Method of Control	Effectiveness of Control	Risk Level with Control	Risk Acceptability
Open case	Delay in therapy	Difficulty/unable to open case	3	5	15	Use fabric case with hook-and-loop closures	1	15	Acceptable with review
	Broken/torn fingernail	Use fingernail to open latch	1	2	2	No latch	1	2	Acceptable
Tear open electrode package	No therapy delivered	Package missing as result of not being replaced from previous use	3	5	15	Design case with slot positions for accessories. Missing item obvious. Recommend admin procedures using seals	2	30	Acceptable with review
	No therapy delivered	Tear electrode when attempting to open package	3	5	15	Provide "zipper" closure that allows easy opening of sealed package. Construct electrode with nontear backing	1	15	Acceptable with review
	Delay in therapy	Difficult/unable to open package	3	5	15	Provide "zipper" closure that allows easy opening of sealed package	1	15	Acceptable with review
Expose upper chest of patient	Nondelivery of shock	Clothing not adequately removed	2	5	10	Provide scissors. Provide pictorial and auditory instructions	2	20	Acceptable with review
	Burn caused by metallic object in clothing	Wire in undergarment or metal fastener left in place	2	3	6	Provide pictorial and auditory instructions	3	18	Acceptable with review
Peel backing from electrodes	Delay in therapy	Difficulty removing backing	3	5	15	Provide extended tab that allows easy removal of backing	1	15	Acceptable with review
	Nondelivery of shock	Used without moving backing	2	5	10	Detection circuit will alarm because EKG signal will not be detected with insulated electrodes	1	10	Acceptable with review
Apply electrodes to chest	Shock not delivered properly	Improper positioning	3	5	15	Provide pictorial and auditory instructions	3	45	Acceptable with review
	Local burn	Electrodes placed too close together	2	3	6	Provide pictorial and auditory instructions	3	18	Acceptable with review

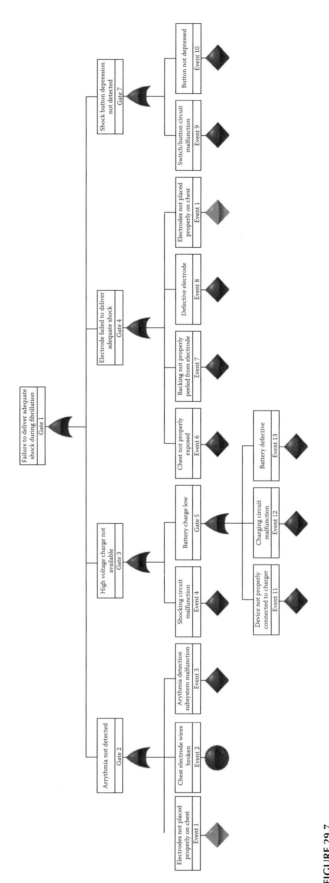

FIGURE 29.7
Example FTA for a hypothetical AED.

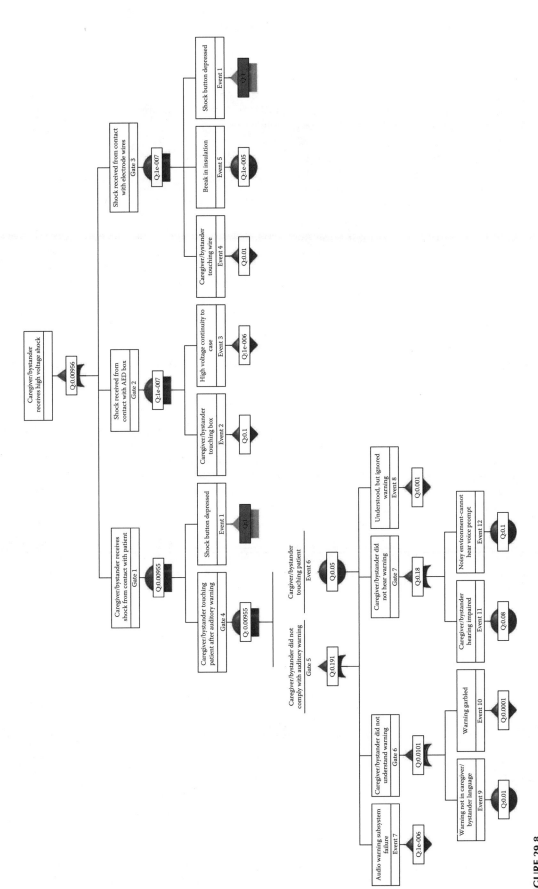

FIGURE 29.8
Example quantitative FTA for a fictitious AED.

The probability of occurrence of almost.01 is a value that would probably be considered unacceptably high in perhaps most medical devices. Most teams (after checking for errors) would probably begin looking for the cause or causes of the high failure probability. One of the techniques for identifying vulnerable areas is to perform a "critical path analysis." This is done by starting with the top-level event and identifying the gate (or event) with the highest probability. That gate will be traced further to determine which events or gates have the highest probability. This process is repeated until reaching the bottom of the fault tree. The highest probability branches through the tree will constitute a "critical path" in the fault tree, signifying a combination (or sequence) of events that are most likely to lead to the top-level event.

29.6.5 Comparison of Quantitative versus Qualitative Risk Management

Qualitative and semiquantitative methods for FMEAs are very similar. Table 29.15 shows a comparison of the two methods from which it is easier to see how similar they are. When using RPN, an explicit risk index is usually not calculated. Commonly RPN is the result of multiplying three variables, fault frequency × hazard severity × effectiveness of mitigation or detection. Of course, it is feasible to obtain a risk index or risk level quantitative value from the product of the first two variables, that is, fault frequency and hazard severity. It is the authors' recommendation that a risk index be calculated both before and after risk mitigations are considered for both methods. The outcome or residual risk is practically the same comparing the two methods. As noted earlier, many analysts do not endorse the use of a detection rating for medical device FMEAs.

29.7 Criteria for Establishing Acceptable Risk

The end point for FMEA and other risk management techniques is whether residual risks remaining after mitigation are acceptable. Three regions of residual risk are discussed

- Acceptable
- Acceptable with review or justification
- Unacceptable or intolerable

The second category of "acceptable with review" is a gray area called "as low as reasonably practicable" (ALARP) that, as shown in Figure 29.9, a manufacturer can further divide into two subregions. A criterion for defining this upper ALARP region is when a risk versus benefit trade-off is positive. What is reasonably "practicable" requires the risk versus benefit trade-off be explicitly stated and estimated. Lives saved or lives improved versus cost and potential harm must be balanced in favor of the more positive outcomes. An important consideration in making this trade-off is whether the benefit is proportionate to the cost. An example is heart bypass surgery. The risk of death during surgery and early recovery has been estimated around 3%, which is offset by the fact that death from a heart attack is almost certain without the surgery.

According to ISO 14971:2007 practicability has two areas of consideration:

- Technical practicability, which refers to the ability to reduce the risk regardless of cost, such as having so many warnings and cautions in the labeling that normal operation is hampered.

TABLE 29.15

Comparison between Qualitative and Semiquantitative (RPN) FMEA Methods

Method	Risk Severity	Fault Probability	Risk Level	Detect/ Mitigate Risk	Outcome (Residual Risk)
Qualitative	Major	Probable	High	No change	Unacceptable
	Moderate	Occasional	Medium	Lower	Acceptable with justification
	Minimal	Rare	Low	Eliminate	
		Improbable			Acceptable
Semiquantitative RPN	5	5		1	>45, unacceptable
	4	4		2	
	3	3		3	9<45, acceptable with review
	2	2		4	
	1	1		5	
					1<9, acceptable

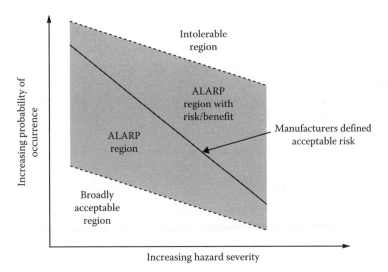

FIGURE 29.9
Illustration of risk acceptability and the ALARP region.

- Economic practicability, which refers to the ability to reduce the risk without making the provision of the medical device an unsound economic proposition. Cost and availability implications are considered in deciding what is practicable to the extent that these impact upon the preservation, promotion, or improvement of human health. Major risks should normally be reduced even at considerable cost (ISO 14971:2007).

ALARP is a difficult concept to incorporate in designing a medical device. Of course, no one wants to put a price on safety or human lives, which is something a true cost–benefit analysis sometimes requires. The ISO 14971 standard does introduce the difficult concept of costs, that is, a device cannot be made so expensive that it becomes unaffordable for medical product consumers and clearly manufacturers can only exist if they make a profit for their shareholders and employees.

29.8 Common Mistakes in Performing Use-Error Risk Analysis

For those focusing on use-error risk analysis for the first time, care should be taken to avoid the mistakes often observed among teams.

29.8.1 Blaming the User

Although there will always be errors that are unpredictable, human factors literature can provide many examples where the design of products can be shown to induce or increase the probability of error. If a user commits an error that is caused by the design, where a statistical pattern of errors is exhibited among users, the manufacturer will likely be held accountable despite the manufacturer's notion that "it's the user's fault, not ours."

29.8.2 Treating Risk Management as a Necessary Evil Rather than a Development Tool

Although risk analysis has been used to analyze and prevent failure in a myriad of systems, it is not uncommon for development teams to resist performing them. Rather than employing risk analysis as a development tool, it is often treated as a perfunctory exercise at the end of the project, with little or no chance for impact on the design. If significant use-error related defects are found in later development stages, timelines and cost become formidable obstacles to any design changes. Instead, risk analysis (including use-error analyses) should be considered to be an essential tool used early and throughout the development lifecycle.

29.8.3 Underestimating Risk Probability and Severity

When there are no objective data, practitioners must depend on estimates of fault probability and severities to estimate risk. Individuals and groups are subject to many sources of bias, especially when schedule and cost pressures are evident. Consequently, groups may unintentionally underestimate the probability or severity of various events. The first line of defense for this kind of bias is to assemble teams that include people who have had experience dealing with users of the same or similar devices, and who can represent the users' perspective (e.g., human factors, customer support, customer training, and marketing). Second, organizational independence from the development organization could be important for minimizing biases in risk estimation. Management support of the group and their output and recommendations is essential. Third, to minimize the effects of group dynamics, techniques for gathering unbiased group estimates (e.g., Delphi method) should be considered.

29.8.4 Overlooking Critical Sources of Failure

According to Leveson (1995), the developers of the Therac-25 failed to include software in their considerations of risk. By doing so, they overlooked key fault sources that should have been considered. It is clear from the literature that typical human factors considerations and use-errors were not well considered in the design. These failures allowed critical design flaws in the design of the device that had tragic consequences. One way to minimize the chance of overlooking potentially critical failures is to assemble teams that represent a wide array of relevant stakeholders that would include personnel from human factors, engineering, manufacturing, training, field service, medical, clinical research, etc. These teams should have the autonomy to make independent assessments of the overall design. These analytical methods should always be supplemented with empirically based usability testing.

29.8.5 Failure to Document Low-Level Risks

When dealing with user behavior and use-errors, it is not uncommon for individuals in a group to declare: "users would *never* do that" or "users are not *that* stupid." Attitudes associated with such statements may lead groups to omit such faults from the analysis documents. Another problem is for groups to assume that certain risks will be (or always have been) well controlled and therefore not the subject of concern. The problem with both of these scenarios is that if a situation should arise where the development team is queried as to whether a certain risk was considered (as might occur during audits or as the result of litigation) there would be no evidence that the team ever considered the problem. Although common sense should be applied here, it is recommended that teams document all conceivable hazards, and rate those that would have been omitted as having low probability.

29.8.6 Overreliance on Training or Labeling

When considering operator errors, a frequently proposed solution is to "put a label on it" (or "put a warning in the instruction manual") or "train users not to do that."

Although there are situations where training or labeling may indeed be the solution, the human factors literature is replete with examples of cases where well-trained operators (with volumes of training documents) commit errors. A main consideration should be that if a design is not consistent with good design practices or conflicts with the users' expectations, training, and labeling are not a reliable, sustainable methods for eliminating errors.

29.8.7 Failing to Validate Modes of Control

Whether the mitigations for an identified use-error are elements of design, labeling, or training, it is incumbent on the manufacturer to show evidence that the mitigations are effective. The common method for validating a user interface is *usability testing*, which involves testing representative users of the device performing representative tasks on the actual device or reasonable facsimile. Validation will depend on meeting selected usability objectives. For more information, refer to sources on human factors processes (e.g., ANSI/AAMI, 2001) and usability testing. For example, see Chapter 32.

29.9 Standards and Regulations

29.9.1 Standards for Risk Management

There are ISO standards on the use of risk analysis in the development of medical devices. The current standard is ISO 14971:2007 Medical Devices—Application of Risk Management to Medical Devices. This standard describes the FMEA and FTA processes in general. The concept of human factors or use-error focused risk analysis is also endorsed in ANSI/AAMI HE 74:2001—Human Factors Design Process for Medical Devices, which is a human factors best practices process standard. IEC 60601-1-6:2010 Collateral Standard on Usability and the more comprehensive IEC 62366:2007 Medical Devices—Application of Usability Engineering to Medical Devices include HE 74 as an attachment and also endorse the use of human factors risk analysis, evaluation, and control. The newest and most comprehensive HF standard for medical devices ANSI/AAMI HE 75:2009 Human Factors Engineering—Design of Medical Devices includes a section on managing use-error that covers much detail on the subject.

29.9.2 Regulations

In the United States, FDA guidance on human factors is very clear on the importance of performing use-error focused risk analysis (FDA, 2000). The guidance urges manufacturers to perform both analytical (e.g., risk analysis) and empirical analyses (e.g., usability testing) during product development to identify and control for use-errors.

International regulators commonly endorse international standards and look very favorably on medical device submissions for approval that conform to recognized standards from IEC and ISO. For human factors risk analysis that would include IEC 62366:2007 on usability and ISO 14971:2007 for risk management.

29.10 Implementation Considerations

29.10.1 Team Membership

It is recommended that a group perform the use-error risk analysis consisting of individuals from the product development team with detailed knowledge of the product's user interface design. This group should include

- Human factor's specialist (internal or an approved outside consultant)
- Quality assurance
- Product development/engineering
- Medical/clinical affairs
- Marketing

29.10.2 Estimation Techniques

29.10.2.1 Brainstorming

Brainstorming is the recommended method of examining far ranging, but feasible use-errors that could be committed for each task and meaningful subtask. Other investigation methods for generating possible use-errors are as follows:

- Examination of adverse events reported to regulators such as the FDA Medwatch program
- Analysis of customer complaints with a focus on use-errors attributed by customers to inadequate product design

Brainstorming needs to follow a few rules to be effective in creatively generating reasonably feasible use-errors. To conduct a productive brainstorming session:

- Verify that all team participants understand and are satisfied with the main question before you open up for ideas.
- Give everyone a few minutes to personally record a few ideas before getting started (give them a presession assignment).

- Start with going around the table, giving everyone a chance to voice their ideas or pass. After a few rounds, open the floor to everyone.
- More ideas are better. Encourage radical ideas and piggybacking and some limited tangents.
- Avoid the temptation to evaluate and critique ideas.
- Board and record exactly what is said. Avoid paraphrasing ideas. Clarify only after everyone is out of ideas.
- Do not stop until ideas become sparse. Allow for late-coming ideas.
- Eliminate duplicates and ideas that are not relevant to the topic.
- Combine ideas that fit together.

A typical brainstorming session should start with a simple question, and ends with an unedited list of ideas. It gives you a raw list of ideas. The quality will vary and some ideas will be good, and many will not be. You will spoil the session if you try to judge and evaluate ideas during the session. Just wait. At a later time, you can analyze the results of a brainstorm with other quality improvement tools, such as affinity diagrams or fishbone diagrams, as described in Section 29.8.

29.10.2.2 Achieving Consensus

The main goal in consensus building in risk management is to achieve agreement on such things as ratings for fault frequency or occurrence, hazard and harm severity, risk index, effectiveness of mitigations, and level of risk acceptability.

One software tool for facilitating group consensus is the PathMaker's Consensus Builder tool (Skymark, 2004). According to PathMaker, this tool uses two main methods to support consensual decision making:

- One is structured discussion, in which decisions are carefully framed, alternatives systematically discussed, and notes taken.
- The second method involves the effective use of voting—rating systems and multivoting—to reduce lists and quantify opinions. Multivoting allows group members to assign numerical ranks to all alternatives. The ratings are averaged and the values are used to produce a final ranking based on the averages of individual rankings.

The recommended Skymark PathMaker™ consensus building steps are as follows:

1. Carefully document by writing out the issues and alternatives.
2. Suggest many alternative candidate answers.
3. Reduce a long list (10+ items) using multivoting.
4. Carefully discuss the remaining candidates. Take notes on each.
5. Decide which criteria will be used to evaluate the alternatives.
6. Perform a rating vote.
7. Look at areas of disagreement, and discuss them further.
8. Vote again, if necessary.
9. Discuss the outcome of the vote. Has everyone been heard?
10. Has consensus been achieved? If not, iterate the process.

29.10.2.3 Delphi Techniques

The Delphi technique is a group forecasting methodology generally used for future events such as technological developments. It uses estimates from experts and feedback summaries of these estimates that allow for additional iterative estimates by these experts until reasonable consensus occurs. The authors posit its use for achieving consensus on risk management-related ratings and decisions. Details of the Delphi technique are described by the Carolla group (Carolla, 2004).

According to Carolla (2004),

> the Delphi technique was developed by the RAND Corporation in the late 1960s as a forecasting methodology. Later, the U.S. government enhanced it as a group decision-making tool with the results of Project HINDSIGHT, which established a factual basis for the workability of Delphi. That project produced a tool in which a group of experts could come to some consensus of opinion when the decisive factors were subjective, and not knowledge-based.
>
> The general steps in performing a Delphi technique are as follows:
>
> 1. *Pick a facilitation leader.* Select a person that can facilitate, is an expert in research data collection, and is not a stakeholder. An outsider is often the common choice.
> 2. *Select a panel of experts.* The panelists should have an intimate knowledge of the projects, or be familiar with experiential criteria that would allow them to prioritize the projects effectively. In this case, the department managers or project leaders, even though stakeholders, are appropriate. See Section 15.1 for

> recommendations on stakeholders for risk management Delphi panelists.
> 3. *Identify a strawman criteria list from the panel.* In a brainstorming session, build a list of criteria that all think appropriate to the projects at hand. Input from nonpanelists are welcome. At this point, there are no "correct" criteria. However, technical merit and cost are two primary criteria; secondary criteria may be project-specific.
> 4. *The panel ranks the criteria.* For each criterion (e.g., risk management–related ratings) the panel ranks it as 1 (very important), 2 (somewhat important), or 3 (not important). Each panelist ranks the list individually and anonymously if the environment is charged politically or emotionally. (This could be done via e-mail.)
> 5. *Calculate the mean and standard deviation.* For each item in the list, find the mean value and remove all items with a mean greater than or equal to 20. Place the criteria in rank order and show the (anonymous) results to the panel. Discuss reasons for items with high standard deviations. The panel may insert removed items back into the list after discussion.
> 6. *Rerank the criteria.* Repeat the ranking process among the panelists until the results stabilize. The ranking results do not have to have complete agreement, but a consensus such that all can live with the outcome. Two passes are often enough, but four are frequently performed for maximum benefit. In one variation, general input is allowed after the second ranking in hopes that more information from outsiders will introduce new ideas or new criteria, or improve the list.

The authors are aware of a recent use-error FMEA being used to analyze the risk control measures for the process of prescribing a new drug and to understand how prescribing errors might occur and possibly be mitigated. Three experienced physicians participated in a Delphi exercise in making likelihood and hazard severity estimates. They achieved consensus in three rounds.

29.11 Uses for Risk Management

29.11.1 Trade-Offs Regarding Design Safety

The goal of every manufacturer is to design and manufacture a quality product that is safe. In so doing, the manufacturer hopes to have high product sales with resultant profits. The question arises, how much quality and safety are enough? A product that has little attention paid to safety will likely fail if word-of-mouth reports the same, or if the product is subject to recalls due to safety. A product that is the subject of extreme over-design in terms of safety may not succeed because it costs significantly more than competitive products, or because it took too long to develop and missed its optimum marketing time window. The risk management tools in this chapter are useful to help identify areas of risk, and provide the means to assess risk in terms of probability and consequences. Explicit identification of these issues will allow companies to assess these risks in terms of cost-to-benefit analyses.

29.11.2 Decision Point for Usability Testing

Because usability test sessions can be time consuming, especially when dealing with complex systems, usability tests are seldom comprehensive assessments of the entire user interface. Among the tasks that should be assessed in a usability test would be those that are determined to have a high risk in terms of possible use-errors.

29.11.3 Labeling and Training

As stated earlier, it is common to apply labeling and training as a means to mitigate use-errors. Whether or not labeling/training is appropriate should depend on risk levels found in risk analysis. If a risk level is high, labels and training should probably not be used as the sole means to reduce these risks.

29.11.4 Input to Corrective and Preventative Action (CAPA) Systems

When customer complaints are reported for a given product, the manufacturer typically must respond in terms of identified causes of the problem and the solutions. Statistical trend analyses are typically done to determine how common the problem reported. However, when dealing with a small number of occurrences, risk analysis tools can also be applied to situations involving CAPA, to better understand assess the risks of occurrence (and reoccurrence). Those items involving frequent or even infrequent customer complaints, but have been identified as having high risk in risk analyses are issues that should be considered for additional modes of control.

29.11.5 Expanded Applications for Use-Error Risk Analysis

Recently, regulators in the United States and globally have requested use-error FMEA for a broader scope of applications including: packaging for controlled substance drugs, potential drug prescribing errors by physicians, mechanical autoinjectors for biologic drugs, and labeling of medical products.

29.12 Future Directions for Use-Error Risk Management

The focus on patient safety and use-errors will likely continue until accident and near-miss rates decline significantly. The following are predicted elements in that focus.

29.12.1 Increased Awareness of Use-Errors Related to Patient Safety

The recent focus on human error in medicine has increased the awareness among manufacturers and health care providers of their responsibilities in this area. A change of attitude toward the following notions will have an impact on reducing error and improving patient safety:

- Manufacturers may be held responsible for adverse events if designs reflect little or no consideration of possible use-errors.
- Human errors can be predicted and reduced or eliminated by design.
- Risk management tools used for traditional engineering/safety analysis are useful or essential for human factors related problems.

29.12.2 Increased Focus on the Use Environment

Some of the problems and risks with devices reside not in the devices, per se, but in the context of other equipment and the environment. Some examples include the following:

- Many devices in a hospital room (e.g., patient monitors, infusion pumps, ventilators) have auditory alarms designed to alert caregivers that immediate attention is needed. However, one of the common complaints regarding device alarms is that one device's alarms are not distinguishable from another. The possible confusion of alarms has been an area of concern for several human factors standards bodies (IEC, 2003).
- Electrical connections for various devices are easily confused and have allowed unintended connections to other devices or sources of power (e.g., inadvertent connection of EKG leads to 120 VAC).

- User interfaces among similar devices are most often not consistent, leading to increased risk of errors, increased time to administer treatment because of confusion, or increased training costs.

Various standards bodies (e.g., AAMI, IEC, ISO) have been attempting to develop user interface standards aimed at harmonizing various user interfaces elements (such as alarms, symbols) among diverse devices. Attempts to design comprehensive common user interface standards so far have failed for at least two reasons:

- Many manufacturers believe their user interfaces have a competitive advantage over the competitors, and are reluctant to share their ideas as part of a standard.
- Software user interface technologies have been changing continually over the past several years. Software user interfaces developed during such times may provide standards that are considered obsolete.

29.12.3 Improved Medical Error Data

One of the biggest challenges in dealing with use-errors is the lack of data to make informed decisions regarding risk levels. Consequently, teams are forced to use their own judgments (and imagination) to uncover potential failures and assess designs. In recent years, the FDA, the U.S. Veterans Administration, and others have initiated computerized event recording systems that enable caregivers and health care facilities to anonymously report accidents and near misses to a database. One of the anticipated benefits of this information is that these sources will eventually become a means for better estimating human error on a wide array of medical devices.

29.12.4 Improved Tools

The emergence of software tools to enable the more rapid construction of FMEAs, fault trees, and RCA documents have greatly improved the ability of small teams to perform risk analysis procedures in much less time than using manual methods. With similar advances in the next few years, it is likely that future tools will provide even greater ease-of-use and extended functionality that will assist users in identifying faults (in similar devices), and can automatically monitor for redundancy, automatic tracing of requirements/ actions, and so on.

Appendix A

Definitions and Acronyms for Risk Management

Term	Definition
Abnormal use	Sabotage, reckless, unqualified users, off-label use or unforeseeable use
AER	Adverse Experience Report—Description of an medical adverse event that results in death or serious injury
ALARP	As Low As Reasonably Practical—Region of risk acceptability with justification
BAT	Best Available Technology—The most current technology for mitigating design risks
Contextual inquiry	Process of observing and working with users in their normal environment to better understand the tasks they do and their workflow
Detection	The estimation of whether a particular hazard or defect will be discovered and mitigated prior to finished product release/distribution/use
FMEA	Failure Mode and Effects Analysis—A systematic approach that identifies potential failure modes in a system caused by either design or process deficiencies. It also identifies critical or significant design or process characteristics that require special controls to mitigate or detect failure modes. The procedure involves making estimates for the likelihood of each potential failure mode and to classify the severity of each resulting hazard on user or patient safety
FMECA	Failure Mode and Effects Criticality Analysis—Same as FMEA with an emphasis on analyzing criticality
Foreseeable misuse	Recognized potential misuse or misapplication of product
Formative usability testing	Usability testing that is performed early with simulations and first working prototypes and explores if usability objectives are attainable, but does not have strict acceptance criteria
Frequency	The estimation of the likelihood that a particular hazard or defect will occur
FTA	Fault Tree Analysis—A deductive, top-down method of analyzing system design and performance. It involves specifying a top event to analyze, followed by identifying all of the associated elements in the system that could cause that top event to occur. Fault trees provide a convenient symbolic representation of the combination of events resulting in the occurrence of the top event. FTAs are generally performed graphically using a logical structure of AND and OR gates
Harm	Physical injury or damage to the health of people, or damage to property or the environment
Hazard	A potential source of harm
HHA	Health Hazard Assessment—A systematic analysis to anticipate, identify, quantify, and recommend controls for health hazards associated with the occupational environment
HA	Hazard Analysis—Analysis of sources of harm to patients and other users of a medical devices
HACCP	Hazard Analysis and Critical Control Points—Used to analyze critical process control points in food processing
HAZOPS	Hazard and Operability Studies—Systematic analysis of industrial processing flows for deviations from design intent
Intended use/ intended purpose	The objective intent of the manufacturer as described in labeling, advertising materials, or oral or written representations
Prior identified criteria	Documented standards or regulations related to acceptable risk limits
Prototyping and iterative design	Iterative design involves the rapid turnaround of user interface prototypes or simulations that are usability tested and improved in an iterative cycle until usability objectives are attained
Residual risk	Risk remaining after protective measures has been taken.
Risk (risk level or risk index, RI)	Combination of the probability of occurrence of harm and the severity of that harm
Risk acceptability	An analysis of effectiveness of design controls and mitigations on the acceptability of remaining product risk leading to further design actions or the conclusion that a product is safe
Risk analysis	Systematic use of available information to identify hazards and to estimate the risk
Risk control	Process through which decisions are reached and protective measures are implemented for reducing risks to, or maintaining risks within, specified levels
Risk evaluation	Judgment, on the basis of risk analysis, of whether a risk that is acceptable has been achieved in a given context based on the current values of society
Risk management	The systematic application of management policies, procedures, and practices to the tasks of analyzing, evaluating, and controlling risk
RPN	Risk Priority Number—The result of multiplying quantitative estimates of three parameters: fault frequency, hazard severity, and detection
Severity	Measure of the possible consequence of a hazard

APPENDIX A (continued)

Definitions and Acronyms for Risk Management

Summative usability testing	Usability testing that is performed in the late stages of design. These tests include verification and validation and it is a recommended best practice to have formal acceptance criteria (e.g., usability objectives for human performance and satisfaction ratings)
Task analysis	It is a family of systematic methods that produce detailed descriptions of the sequential and simultaneous manual and intellectual activities of personnel who are operating, maintaining, or controlling devices or systems

Term	Definition
Usability inspection methods	Inspection methods involve analytical reviews and systematic walkthroughs of user interactions with simulated or working user interface designs looking to uncover usability problems
Usability objectives	Usability objectives (goals) are a desired quality of a user device interaction that may be expressed in written form, stipulating a particular usability attribute (e.g., task speed) and performance criteria (e.g., number of seconds)
UT	Usability Testing—Procedure for determining whether the usability goals have been achieved. Usability tests can be performed in a laboratory setting, in a simulated environment, or in the actual environment of intended use
Use environment analysis	The actual conditions and settings in which users interact with the device or system
Use-error	Use-error is characterized by a repetitive pattern of failure that indicates a failure mode that is likely to occur with use and thus has a reasonable possibility of predictability of occurrence. It can be addressed and minimized by the device designer, proactively identified through the use of techniques such as usability testing and hazard analysis
Use-error risk analysis	Analysis focused on the use-error component of fault and hazard analysis for medical devices
User	The person who interacts with the product
User error	User error is characterized by an isolated pattern of failure that indicates a failure mode that is due to fundamental errors by humans and has no reasonable possibility of being predicted. User error is nonpreventable and nonaddressable by the device designer
User group	Subset of intended users who are differentiated from other intended users by factors such as age, culture, or expertise that are likely to influence usability
User interface	The hardware and software aspects of a device that can be seen (or heard or otherwise perceived) by the human user, and the commands and mechanisms the user uses to control its operation and input data
User profiles	Summary of the mental, physical, and demographic traits of the end-user population as well as any special characteristics such as occupational skills and job requirements that may have a bearing on design decisions

References

Andre, A.D., Jorgenson, D.B., Froman, J.A., Snyder, D.E., and Poole, J.E. (2004). Automated external defibrillator use by untrained bystanders: Can the public-use model work? *Prehospital Emergency Care*, July/September, 8(3), 284–291.

ANSI/AAMI HE 74:2001. *Human Factors Design Process for Medical Devices* Arlington, VA.

ANSI/AAMI HE 75:2009. *Human Factors Engineering—Design of Medical Devices* Arlington, VA.

Bailey, R.W. (1983). *Human Error in Computer Systems*, Prentice Hall: Upper Saddle River, NJ.

Carolla (2004). Internet site, www.carolla.com//wp-delph.htm

Casey, S.M. (1998). *Set Phasers on Stun: And Other True Tales of Design, Technology, and Human Error*, Aegean Publishers: Santa Barbara, CA.

Code of Federal Regulations (CFR) (2004). *Quality System Regulations, Medical Devices*. 21CFR820, Title 21, Volume 8, Prentice Hall: Upper Saddle River, NJ.

DeRosier, J., Stalhandske, E., Bagian, J.P., and Nudell, T. (2002). Using health care failure mode and effect analysis: The VA national center for patient safety's prospective risk analysis system. *The Joint Commission Journal on Quality Improvement*, 28, 248–267.

FDA (2000). Medical Device Use Safety—Incorporating Human Factors Engineering into Risk Management, FDA 1497, July 18, 2000, Washington, DC.

IEC 60601-1-8:2003. Medical Electrical Equipment—Part 1–8 General Requirements for Safety: Alarm Systems—Requirements, Tests and Guidelines—General Requirements and Guidelines for Alarm Systems in Medical Electrical Equipment and in Medical Electrical Systems Geneva, Switzerland.

IEC 62366:2007. Ed 1, Medical Devices—Application of Usability Engineering to Medical Devices Geneva, Switzerland.

IEC 60601-1-6:2010. Medical Electrical Equipment—Part 1—General Requirements for Safety: Collateral standard: Usability Geneva, Switzerland.

ISO 14971:2007. Medical Devices—Application of Risk Management to Medical Devices.

Israelski, E.W. and Muto, W.H. (2003). Use-error focused risk analysis for medical devices: A case study of the Therac-25 radiation therapy system. In *Proceedings of the 47th Annual Meeting of the Human Factors and Ergonomics Society 2003*, Human Factors and Ergonomics Society: San Diego, CA.

Kirwan, B. and Ainsworth, L.K. (Eds.) (1992). *A Guide to Task Analysis*. Taylor & Francis: New York.

Leveson, N. (1995). *Safeware: System Safety and Computers*, Addison-Wesley: Reading, MA.

McDermott, R.E., Mikulak, R.J., and Beauregard, M.R. (1996). *The Basics of FMEA*, Productivity, Inc: Portland, OR.

Miller, D.P. and Swain, A.D. (1987). Human error and human reliability. In G. Salvendy (Ed.), *Handbook of Human Factors*, pp. 214–250, Wiley: New York.

North Carolina Department of Environmental and Natural Resources, Office of Organizational Excellence (2002). Internet site, http://quality.enr.state.nc.us/tools/fishbone.htm

Reason, J. (1990). *Human Error*, Cambridge University Press: Cambridge, U.K.

Reason, J.T. (1997). *Managing the Risks of Organizational Accidents*, Ashgate Pub Co.: Hampshire, U.K.

Schmidt, M.W. (2004). The use and misuse of FMEA in risk analysis. *Medical Device and Diagnostics Industry*, 26(3), 56–61.

Skymark (2004). Internet site, www.skymark.com

Stamatis, D.H. (1995). *Failure Mode and Effect Analysis*, ASQ Quality Press: Milwaukee, WI.

Vesely, W.E., Goldberg, F.F., Roberts, N.H., and Haasl, D.F. (1981). *Fault Tree Handbook. NUREG-0492*. United States Nuclear Regulatory Commission.

30

Human Factors Analysis of Workflow in Health Information Technology Implementation

Pascale Carayon, Randi Cartmill, Peter Hoonakker, Ann Schoofs Hundt, Ben-Tzion Karsh, Daniel Krueger, Molly L. Snellman, Teresa N. Thuemling, and Tosha B. Wetterneck

CONTENTS

30.1 Introduction...507
30.2 What Is the Need for Workflow Analysis?...508
30.3 Overall Approach to Workflow 508
30.4 Categories of Workflow Analysis Tools..510
30.5 When to Analyze Workflow...510
30.6 Health IT as a Tool for Workflow Analysis and Redesign..515
30.7 Conclusion 516
Acknowledgments 519
References...519

30.1 Introduction

Health information technology (IT) applications, which provide computerized clinical information to health care providers and/or patients, have been viewed as one mean of improving health care quality, enhancing patient safety, and streamlining administrative health care functions. The pace of health IT adoption in U.S. health care organizations will likely increase, owing in part to government incentive programs and pressures from purchasing groups and consumers (Institute of Medicine, 2000, 2001, 2007; Shojania et al., 2001; The Leapfrog Group, 2004; U.S. Department of Health and Human Services, 2005, 2010; The White House National Economic Council, 2006). Evaluations of the impact of health IT on quality and safety show mixed results, however. One main reason is the failure of health IT to conform to and support the cognitive work of the clinician; another is the failure to integrate health IT into clinical workflow in a way that supports workflows among organizations (e.g., between a clinic and community pharmacy), within a clinic and within a patient visit. If health IT is to provide optimum performance, it must be designed to fit the context in which it will be used, including the type of practice, patient mix, and workflows of that practice. Achieving a good fit requires the use of human factors tools to analyze and redesign those workflows. In this chapter, the primary focus is on analysis of workflows in ambulatory

practices implementing health IT; however, many of the conclusions and lessons learned can also be applied to other health care settings.

Health IT is typically designed to aid the acquisition, analysis, and display of information, and to improve decision making related to charting, prevention and screening, diagnosis, treatment, drug dosing, drug administration, test ordering, and/or chronic disease management. The technology typically pushes the information to the decision maker. Health IT applications include electronic health records (EHR); electronic medical records (EMR); computerized provider order entry (CPOE); electronic medication administration records (eMAR); computerized pharmacy systems; electronic prescribing (eRx); bar-coded medication administration (BCMA); and stand-alone and integrated computerized clinical decision support (CDS) systems, which include alerts, reminders, structured order forms, pick lists, patient-specific dose checking, guideline support, medication reference information related to prevention and screening, diagnosis, treatment, drug dosing (e.g., smart infusion pump technology), test ordering, and chronic disease management. EHRs and EMRs are differentiated by whether the information is aggregated across more than one health care organization (EHR) or a single organization (EMR).

In this chapter, we first review the need for workflow analysis in the context of health IT implementation. We then discuss the concept of workflow and describe a range of human factors tools for analyzing and redesigning workflows. Finally, we explain the potential use of health IT as a tool to analyze and redesign workflow.

30.2 What Is the Need for Workflow Analysis?

The following two case studies of EHR implementation in small practices highlight the critical need for workflow analysis before, during, and after health IT implementation.

Greenhouse Internists is a four-physician general internal medicine practice located in Philadelphia. This practice serves a diverse urban and suburban population and handles more than 16,000 patient encounters yearly. In 2004, they went live with an EHR technology (Baron et al., 2005). The transition from paper records to electronic records occurred with multiple challenges, in particular regarding workflow. The four physicians describe this transition as follows: "The process of radically redesigning 15 years of accumulated workflow in a short interval was extremely stressful" (Baron et al., 2005, p. 223). Because of the lack of attention to workflow analysis and redesign during the pre-EHR implementation phase, much of the workflow redesign occurred "on the fly." This happened in conjunction with the need to care for patients and therefore created significant increases in workload, particularly for physicians. This small practice was able to successfully implement EHR but experienced major challenges during the transition.

Pediatric Partners is a three-physician pediatric practice located in Temecula, California. Following a previous attempt to implement an EMR, the group decided to try again in 2005 and implemented a new EMR (Mohr, 2008). This second EMR implementation was successful. Significant time and resources were invested in the selection of a vendor, training, workflow analysis and redesign, and technical support. For instance, physicians were asked to work with the EMR system during off hours in order to become more familiar with the technology and the new patient workflow. A major consequence of the EMR on workflow was its impact on the integration of scheduling, billing, coding, and clinical task management. In addition, paper documents were scanned, indexed, and incorporated into the EMR for easy access and processing. The EMR was also connected to outside labs and pharmacies, which significantly reduced the number of calls to the practice (e.g., reduced need to clarify medication prescriptions).

The first case study describing the Greenhouse Internists' implementation shows that, when a practice rushes to implement an EHR system, not enough time is spent properly analyzing and redesigning workflows.

As a result, work-arounds are implemented and these tend to then be permanently integrated into the workflow. These work-arounds can undermine the usefulness of the system. In addition, integrating health IT with a less than ideal workflow does not allow the practice to achieve the expected benefits of the technology. The implementation of health IT on top of inefficient or "broken" processes only magnifies problems. Another result of the rush to implement EHR is that decisions about how to redesign workflow are generally made quickly, like "redesigning an airplane in flight" (Baron et al., 2005, p. 224). Instead, practices need to identify, evaluate, redesign, and test their workflows before implementing health IT; this can mitigate problems during and after going live with the technology (Lorenzi et al., 2009). This process may require significant time, effort, and resources as suggested by the second case study described earlier (Mohr, 2008), but will at least reduce the likelihood of major workflow-related problems related to the implementation.

As demonstrated in both case studies, very often practices do not realize that workflow analysis is critical to health IT implementation. The second case study shows how the failed attempt to implement EHR by Pediatric Partners provided an opportunity to understand the need to focus on workflow analysis and redesign (Mohr, 2008). Their second attempt led to a successful EHR implementation, probably because of the heightened attention to workflow and other related issues, such as training. In order to help practices understand the importance of analyzing workflow, it may be necessary to have coaches that encourage them to conduct these analyses. In addition, they need to have access to a range of tools and examples of workflow analysis.

30.3 Overall Approach to Workflow

Our conceptualization of workflow is grounded in the University of Wisconsin Systems Engineering Initiative for Patient Safety (SEIPS) Model of Work System and Patient Safety (Carayon et al., 2006), which has three main parts (see Figure 30.1). The *work system* describes how a person at work performs a range of tasks using specific technologies and tools, in a physical environment and under certain organizational conditions. The work system influences *processes*, or workflows, that often involve several workers and patients. These care processes create *outcomes* for the patient, the clinicians, and the organization (Carayon et al., 2006). In terms of the SEIPS model, we can examine how health IT applications affect work processes or workflows that are products of the work system of health care organizations. Most research on

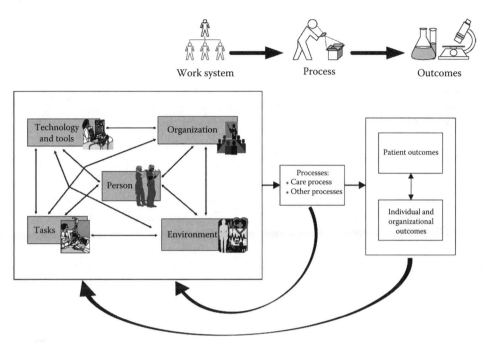

FIGURE 30.1
The SEIPS model of work system and patient safety. (From Carayon, P. et al., *Quality & Safety in Health Care*, 15(Supplement I), i50, 2006.)

health IT, however, does not focus on workflow in health care organizations. Typically, a study aims to discover the effect of a health IT application on adherence to care guidelines or on patient or organizational outcomes. For example, a researcher may examine whether the introduction of a CDS application affects the rate of screening for cancer in a specific patient population. Such research clearly implies that process changes have occurred. If, in this case, a reminder message "pops up" on the screen during a patient visit, the workflow of the provider is changed regardless of whether the provider responds to the reminder by counseling the patient about the need for cancer screening, whether the patient is screened or whether he has cancer. Unfortunately, studies vary substantially in terms of how much information they provide on workflow changes.

Traditionally, workflow has been defined and conceptualized fairly simply, as a linear sequence of activities carried out to achieve a particular organizational goal. From a human factors perspective, workflows are not necessarily simple or linear. Recently, Unertl et al. (2010) described a complex representation of workflow known as the Workflow Elements Model. This comprehensive model contains a dynamic structure with two levels, the "pervasive" and the "specific" (Unertl et al., 2010). The pervasive level contains three sublevels that represent the space, time, and organization that support workflow. Space constitutes the workspaces, both physical and virtual, that either constrain or enable workflow. The temporal element defines the coordination and scheduling of the components of flow, and the

organization identifies the relationship between people and their tasks. The specific level consists of five components: the people, tools, and tasks that flow, the characteristics that describe the tasks, and the outcomes that the tasks produce. This more complex definition and model of workflow is similar to the structure–process–outcomes models that have been proposed by others to explain how systems function (Carayon et al., 2006).

We offer an extended definition to complement that of Unertl et al. (2010). Workflow is the flow of people, equipment (including machines and tools), information, and tasks, in different places, at different levels, at different timescales continuously and discontinuously, that are used or required to support the goals of the work domain. It is also important to note that workflow in health care is not limited to clinical systems, but must also incorporate ancillary systems (such as administration, billing, and insurance) to be complete. This definition differs from prior definitions in several important ways.

First, because work is performed by people who use equipment and information, it is important to understand that what flows are people, products, information, and tasks. Second, all of the things that can flow do so through different types of space. Things can flow inter-organizationally, intra-organizationally, and even intra-visit. Third, workflow occurs at different levels. Flow can occur at an individual level, which makes an individual the unit of analysis. Flow can occur at the dyad or group level, which is flow among or by more than one person. Flow can occur at the organizational level, such as unit flow or organization flow. Fourth, our definition makes clear that workflow occurs at different timescales. The flow of a primary care visit takes place within 15–30 min blocks, but the flow of documentation may be interrupted multiple times (discontinuous) and not be completed for many hours.

Workflow in complex domains such as health care also has one key characteristic: it is *both planned and emergent*. Planned workflow refers to what *should happen*. However, in all areas of health care delivery, workflows typically *emerge* depending on patient status, time pressure, and many other factors. Therefore, planned actions may not actually happen. That workflow in health care is typically emergent and unplanned has special consequences for health IT and the notion of "integration" with workflow.

From a human factors perspective, workflow also includes communication, coordination, searching for information, interacting with information, problem solving, and planning. These activities are not necessarily unique steps in a flow diagram, but happen throughout a clinician's day and certainly affect whether individual steps in a workflow diagram can be executed. These cognitive aspects of workflow are more important than higher-level steps in a sequence diagram like "patient visit" or "check in patient" because they reveal in detail the necessary work to complete a visit or check a patient in. Those who hope to improve care, with or without health IT, must consider the detailed workflow.

30.4 Categories of Workflow Analysis Tools

We have identified approximately 100 tools for analyzing workflow. To educate users and assist them in their search for appropriate tools to meet their needs, we have created a structured organization of tools similar to the U.S. Federal Aviation Administration human factors toolkit (https://www2.hf.faa.gov/workbench-tools/). To do so, we identified categories of tools that represented common uses, based upon the categorization schemes in human factors reference guides, such as *Human Factors Methods: A Practical Guide for Engineering and Design* (Stanton et al., 2005). Because the categories are not mutually exclusive, several tools are assigned to more than one category.

The 12 tool categories are as follows:

1. Data collection tools, which provide a means of gathering information related to a task or issue
2. Data display/organization tools, which provide standard and readily comprehensible means of visually presenting data
3. Idea creation tools, which offer varying formats for identifying new or different ideas
4. Problem solving tools, which provide team members with organized, established methods for better understanding and solving problems
5. Process improvement tools, which offers a means of scrutinizing and improving processes to enhance output and outcomes
6. Process mapping tools, which offer visual means of conveying the flow and interaction of information, work, and processes
7. Project planning/management tools, which furnish participants, project managers and upper management with a means of understanding tasks associated with a project as well as progress on the project's timeline and objectives
8. Risk assessment tools, which are used for identifying and/or analyzing known or anticipated problems associated with specific processes
9. Statistical tools, which can provide meaning to data by summarizing or conveying relationships found in the data
10. Usability evaluations, which are conducted to obtain user input and/or identify design issues related to aspects of a system (e.g., a specific health IT application), such as appearance, function, and navigation
11. Task analysis tools, which permit the user to better understand tasks, generally those associated with work processes
12. Health IT applications, which collect and store information that can be used to better understand known or potential workflow issues and can also be used for performance measurement and quality reporting, and to assist in providing population-based care through the use of registries

The list of tools and their respective category(ies) are displayed in Table 30.1, except for health IT applications that are described in a following section.

30.5 When to Analyze Workflow

The adoption, implementation, and use of health IT can be described as a process that involves the following steps: (1) assessment, (2) selection of health IT, (3) implementation, and (4) continuous improvement. Throughout this process, workflows should be analyzed and reanalyzed. In the assessment phase, the actual work done in

TABLE 30.1

Tools for Analyzing and Redesigning Workflow in the Context of Health IT Implementation

Tool	Data Collection	Data Display/Organization	Idea Creation	Problem Solving	Process Improvement	Process Mapping	Project Planning/Management	Risk Assessment	Statistical Analysis	Usability	Task Analysis
5S											
5W2H				X	X						
Affinity diagrams		X									
Allocation of function analysis							X				
AΔT					X						
Balanced scorecard	X	X					X				
Benchmarking			X		X						
Benefits and barriers exercise					X		X				
Box and whisker plot		X									
Brainwriting			X								
Cause-and-effect diagram		X		X				X			
Checklist	X	X									
Cognitive task analysis	X										X
Cognitive walkthrough	X									X	
Comms usage diagram (CUD)	X		X	X	X		X				
Contingency diagram				X	X			X			
Cost-of-poor-quality analysis					X						
Critical decision method (CDM)	X										
Critical incident				X							
Critical incident technique (CIT)	X										
Critical path method				X			X				
Critical-to-quality analysis				X							
Cross-functional flowchart		X				X					
Cycle time chart		X				X					
Decision action diagrams (DADs)		X				X					
Decision matrix		X									
Decision tree		X				X					
Event tree analysis (ETA)		X						X			
Failure modes and effects analysis (FMEA)								X			
Fault tree analysis (FTA)		X						X			
Flowchart		X				X					
Focus group	X		X	X							
Force-field analysis		X		X							
Gantt charts		X					X				

(continued)

TABLE 30.1 (continued)

Tools for Analyzing and Redesigning Workflow in the Context of Health IT Implementation

Tool	Data Collection	Data Display/ Organization	Idea Creation	Problem Solving	Process Improvement	Process Mapping	Project Planning/ Management	Risk Assessment	Statistical Analysis	Usability	Task Analysis
Gap analysis					x						x
Goals, operators, methods, and selection rules (GOMS)					x						x
Groupware task analysis (GTA)				x	x						x
Heuristic evaluation										x	
Hierarchical task analysis (HTA)					x	x					x
Histogram		x									
Interview	x										
Kano analysis		x									
Kepner-Tregoe matrix	x			x	x			x			
Lean					x						
Lean six sigma					x						
List reduction				x							
Log	x										
Matrix diagram		x					x				
Metrics evaluation											
Multi-vari chart		x							x		
Multivoting				x							
Murphy diagrams				x				x			
NASA task load index (NASA TLX)	x										x
Need assessment			x	x							
Nominal group technique (NGT)			x								
Observation	x										
Operation sequence diagrams (OSD)		x						x			
Pareto chart		x									
Plan–do–check–act (PDCA) cycle					x						
Political, economic, social, and technological forces (PEST) analysis				x				x			
Potential problem analysis (PPA)								x			
Process decision program chart (PDPC)								x			
Process scorecard	x										x

Program evaluation and review technique (PERT) charts

Questionnaire for user interface satisfaction (QUIS)

Radar chart

Regression analysis

Relations diagram

Requirements and measures tree

Requirements table

Root cause analysis

Scatter diagram

Simulation

Simulation modeling

SIPOC (supplier, inputs, process, outputs, customer)

Six sigma

SMART matrix

Social network analysis (SNA)

Statistical process control (SPC)

Strategic planning

Stratification

Strengths, weaknesses, opportunities and threats (SWOT) analysis

Survey

Tabular task analysis (TTA)

Task decomposition

Time and motion study

Time value map

Top-down flowchart

Tree diagram

Trend analysis

Usability evaluation

Use case

Value stream mapping

Value-added analysis

Verbal protocol analysis (VPA)

Workflow diagram

Workflow editor/engine

Workload profile technique

Source: Carayon, P. et al., *Incorporating Health Information Technology into Workflow Redesign—Summary Report.* Prepared by the Center for Quality and Productivity Improvement, University of Wisconsin-Madison, under Contract No. HHSA 290-2008-10036C, AHRQ Publication No. 10-0098-EF, Agency for Healthcare Research and Quality, Rockville, MD, October 2010.

a clinic must be described. This step will likely involve multiple data collection and analysis methods, such as observation of individuals as they work and interviews with process stakeholders, followed by flowcharting of the process. Completing this phase will allow a practice to have a better understanding of the functions and features a specific health IT application must have in order to optimize and support these workflows after implementation.

Because practices tend to underestimate the complexity of their workflows, it is important to fully analyze the work individuals perform as members of a team. For example, the flowcharts created after observing and interviewing individuals provide a greater understanding of who performs each task, the sequence of the tasks, the information flow between tasks, and the interactions between the tasks and people. Creating a detailed flowchart depicting these interactions is critical.

Flowcharts that convey temporal progress as well as the roles involved in a process are also referred to as process maps. For example, after gaining an understanding of the clinic-based prescription writing process in the "paper world," one could create a flowchart such as that found in Figure 30.2. This flowchart contains sufficient detail to accurately depict the process, which will be essential as one identifies and assesses the appropriateness and value of each task in the process. The flowchart should then be used during vendor selection to determine whether the IT sufficiently supports the workflow.

When choosing a health IT application, it is also important to evaluate the usability of the health IT interface. Various usability evaluation methods are available for this purpose (Bastien, 2010). Li et al. (2006) conducted a heuristic evaluation of a CPOE system and identified a range of usability problems. Likewise, Hyun et al. (2009) describe a user-centered process in the design and evaluation of nursing screens in an EHR.

At the implementation stage, a clinic may use information about both the current workflows and the future workflows to identify training needs. Once again referring to the prescription writing process, a flowchart should be created depicting the anticipated or ideal process after technology implementation (Figure 30.3). Changes to workflow should be included in the pre-implementation training programs provided to staff. By explaining the changes to the workflow that will occur after the technology implementation, the transition from paper to electronic formats will likely be much easier, and clinicians and staff will not have to develop workflows "on the fly."

After the technology has been implemented, the clinic will need to engage in a continuous effort to analyze and redesign workflows. Unintended consequences may occur after health IT implementation (Ash et al., 2004, 2009). An analysis of nurses' use of BCMA technology 1–3 years after implementation identified several issues related to the design and use of the technology (Carayon et al., 2007). This analysis involved various

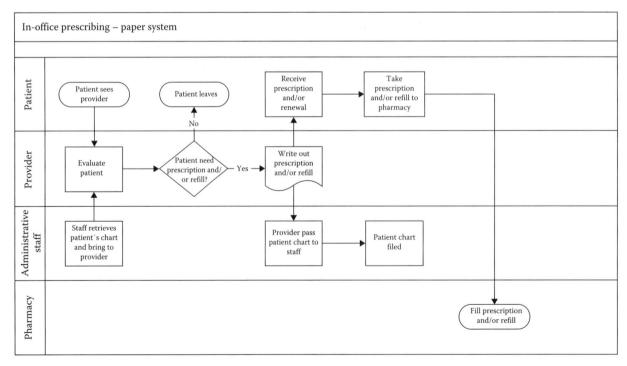

FIGURE 30.2
Example of flowchart for in-office prescribing in a paper system.

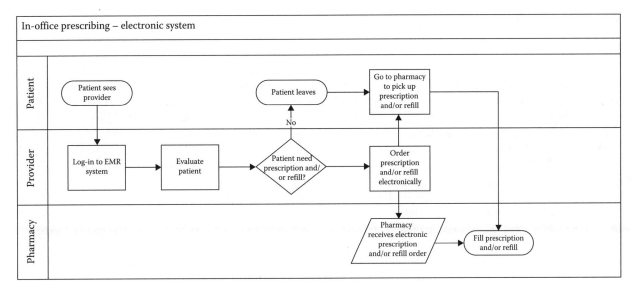

FIGURE 30.3
Example of flowchart for in-office prescribing in an electronic system.

workflow assessment tools, including direct observation of nurses' use of BCMA while administering medications, interviews with nurses, and flowcharting of the medication administration process. The observations and interviews were conducted by a human factors engineer and a pharmacist; this combination of expertise was valuable to identify all of the work system elements involved in the medication administration process. The flowcharting of the medication administration process led to the identification of 18 different ways that the BCMA-supported medication administration process could be performed. In addition, this analysis produced important information on potentially unsafe sequences in which work-arounds occurred (e.g., documentation of administration before the administration of medication is confirmed) or steps that were omitted (e.g., the patient ID band not being scanned, thereby defeating a safety feature of BCMA).

30.6 Health IT as a Tool for Workflow Analysis and Redesign

Health IT applications have the potential to be a powerful tool for analyzing and redesigning workflows. Using health IT, practices can redesign work processes to make them safer, improve their quality, and maximize efficiency. One advantage of using health IT is that data are often stored in discrete fields and therefore can be retrieved and even analyzed in an automated fashion. This obviates the need to manually review the patient record, find the needed information, and record the information in a separate database for analysis, thereby decreasing the likelihood of errors of omission and data reentry. It also reduces the human resources required for data collection. Health IT also collects and stores information about actions taken within the system. For example, EHRs log users' actions, such as order entry, with date and time stamps and the person performing the action. Thus, valuable data about the timing of events and who performed them is frequently available from health IT. This information is less reliably available from non-electronic data. There are many examples of the use of health IT as a tool to measure workflow in ambulatory clinics in the literature. The types of health IT that have been used as a workflow analysis tool include the following: (1) CDS systems (either within an EHR or freestanding), (2) telemedicine, (3) EHRs, (4) electronic informational resources, (5) electronic prescribing, (6) electronic search functionality in an EHR, and (7) an electronic prescription renewal system.

Measures of the way health IT is used are common in studying workflow. Measuring usage may be related to understanding the use of a voluntary type of health IT, one that is employed in addition to typical care delivery processes. Examples of these health IT include an informational resource (Barnett et al., 2004), a blood pressure (Goldstein et al., 2004) or diabetes (Meigs et al., 2003) CDS system, and health IT that is used to replace a process that was previously completed on paper, such as an electronic prescribing tool (Schectman et al., 2005) or laboratory decision support and ordering tool (van Wijk et al., 2001). Measured usage may include the frequency of the user accessing the health IT system or accessing specific functionalities or screens (Eccles et al., 2002; Barnett et al., 2004; Goldstein et al., 2004). For example,

Eccles et al. (2002) evaluated the number of times a physician viewed more than the first screen of an asthma decision support guideline that was incorporated into the EHR. Some studies documented the use of telemedicine by evaluating the frequency and type of information uploaded into the system. Examples of this include physicians sending and receiving referrals (Harno et al., 2000), patients entering blood pressure readings (Logan et al., 2007), or asthma medication use videos and symptom diaries (Chan et al., 2003).

Another example of evaluating workflow using health IT is found in a study by Zheng et al. (2009). The purpose of their study was to evaluate the impact of a "homegrown" EHR with CDS that was completely redesigned in response to qualitative data captured from users about poor navigational guidance. Usage logs from the EHR system were analyzed to determine the frequency of use for 17 different EHR features and typical use patterns during patient encounters (Zheng et al., 2009). The authors used sequential pattern analysis and a first-order Markov chain model to reveal recurring user-interface pathways of navigation by physicians. They found major differences between the anticipated ideal workflow using the EHR and the actual observed workflow, including rarely used functionalities that were thought to be critical features of the EHR, and a workflow that indicated suboptimal use of structured data entry.

CDS systems can be used to analyze workflow because most CDS systems capture data on the number of times the decision support is presented to a user, how the user responded to the decision support recommendation (Weingart et al., 2001) and any comments from the user about why they chose to override the CDS (van Wijk et al., 2001). Adherence to the CDS recommendations may also be evaluated, such as the rates of ordering a recommended laboratory, medication, or test within a certain time period after encountering the CDS (Litzelman et al., 1993). Because the system records the use of individual CDS functionalities within a patient's record and the identity of the user, health IT use, functioning, and adherence can be evaluated at the level of CDS type, patient type, physician (Litzelman et al., 1993), clinic/practice (Eccles et al., 2002), or type of provider, for example, attending or mid-level provider. While it may be easy to capture CDS (or other health IT) use, it is important to realize that use metrics can be misleading or meaningless and need to be carefully thought through. CDS may not be used for very good reasons, such as the CDS not being relevant to a given patient at a given time. As such, a lack of use does not indicate that clinicians are uncooperative or that the CDS has failed, but instead indicates that the physician is making a context-sensitive decision to not use the CDS. In addition, system use data often lack information about the context of use, therefore making these data difficult to interpret. What actually matters when it comes to use is "appropriate" use, which may be difficult to define.

Health IT can also be used to evaluate aspects of time related to health IT use. These include the time required to use a CDS tool to review patient information and present recommendations (Steele et al., 2005; Kuilboer et al., 2006) or the time for a digital radiology image to become available for physician interpretation (Andriole, 2002). Physician/provider and staff time have also been measured. Examples are (1) the amount of physician time spent reviewing informational resources (O'Brien & Cambouropoulos, 2000) or CDS recommendations (Kuilboer et al., 2006), (2) the time elapsed from when the CDS alert is presented until the physician/provider responds (Rollman et al., 2002) or takes action on the recommendation (Rollman et al., 2002), (3) the amount of physician time spent taking digital photographs and entering data into the computer for a telemedicine consultation (Du Moulin et al., 2003), and (4) the amount of staff time spent to assist a patient using a CDS tool (Apkon et al., 2005). In addition, patient time and distance traveled have been measured for telemedicine consultations and compared to those of conventional clinic visits (Raza et al., 2009). Lastly, the time for completion of a process or task in its entirety has been measured. For telemedicine consultations, these measures include the time until an intervention was performed (Whited et al., 2002), an appointment was scheduled (Raza et al., 2009) or a study was received, interpreted, and the results sent back to the referring provider (Mehta et al., 2001). For CDS, these measures were the duration of a task using the CDS system compared to not using it (Ong et al., 2008), the time elapsed between a CDS alert about ordering lab tests and the lab result appearing (Hoch et al., 2003), and the time elapsed from an alert to the ordering of a guideline recommended medication (Hicks et al., 2008).

30.7 Conclusion

In this chapter, we have described 12 categories of workflow analysis tools with about 100 tools. Even though a large number of tools are available, very few of them have actually been used to improve workflows in the design, implementation, and use of health IT. Therefore, we lack information about the comparative strengths and weaknesses of various tools for the different phases of health IT implementation, that is, assessment, selection of health IT, implementation, and continuous improvement. Further research is needed in these areas. The human factors community should be involved in those research and development efforts.

Human Factors Analysis of Workflow in Health Information Technology Implementation

517

APPENDIX A

List of Web Sites with Information on Workflow Analysis and Tools

Category	Organization	Mission/Goals of the Organization	Web Site URL
Federal government agencies	Agency for Healthcare Research and Quality (AHRQ)—Health IT Initiative	The mission of the AHRQ is to improve the quality, safety, efficiency, and effectiveness of health care for all Americans. The agency has focused its health IT activities on the following three goals: (1) improve health care decision making; (2) support patient-centered care; and (3) improve the quality and safety of medication management	http://healthit.ahrq.gov
	Health Resources and Services Administration (HRSA)—Office of Health Information Technology and Quality (OHITQ)	The HRSA is the primary federal agency for improving access to health care services for people who are uninsured, isolated, or medically vulnerable. HRSA's OHITQ seeks to improve the quality of health care for safety net populations and strengthen the health workforce that serves these populations	http://www.hrsa.gov/publichealth/business/healthit/
	Office of the National Coordinator for Health Information Technology (ONC)	The ONC is at the forefront of the administration's health IT efforts and is a resource to the entire health system to support the adoption of health IT and the promotion of nationwide health information exchange to improve health care	http://healthit.hhs.gov
	Centers for Medicare and Medicaid Services (CMS)	The goal of CMS is to ensure effective, up-to-date health care coverage and to promote quality care for beneficiaries	http://www.cms.hhs.gov/
National organizations and associations	American Academy of Family Physicians (AAFP)'s Center for Health Information Technology	The AAFP was founded in 1947 to preserve and promote family medicine and to ensure high-quality, cost-effective health care for patients	http://www.aafp.org
	American Academy of Pediatrics (AAP)	The AAP is committed to optimal physical, mental, and social health and well-being of infants through young adults	http://www.aap.org/
	American College of Physicians (ACP)	The ACP is a national organization whose members include internists, internal medicine subspecialists, medical students, residents, and fellows. ACP has several major initiatives involving the medical home, medical informatics, and workflow analysis	http://www.acponline.org/
	American Medical Association (AMA)	The AMA was founded in 1847 by Dr. Nathan Smith Davis. Its mission is to promote medicine and the improvement of public health	http://www.ama-assn.org/
	American Medical Informatics Association (AMIA)	AMIA promotes organization, analysis, management, and use of information to support health care. Members of AMIA promote health IT in clinical care and clinical research, personal health management, public health/population health, and translational science to improve health	https://www.amia.org/
	Association of Medical Directors of Information Systems (AMDIS)	The AMDIS was formed to advance the field of applied medical informatics. AMDIS is the professional organization for physicians interested and involved in health IT. AMDIS members are the leaders and decision-makers in their field	http://www.amdis.org
	The Center for Improving Medication Management	The Center for Improving Medication Management provides a collaborative forum to establish priorities for projects that demonstrate the value of pharmacy interoperability to improve medication management. Founding groups of the center include AAFP, Humana, Intel Corporation, MGMA, and Surescripts	http://www.thecimm.org/
	Certification Commission for Healthcare Information Technology (CCHIT®)	The CCHIT strives to improve the quality, safety, efficiency, and access of health IT with the goal to accelerate its adoption	http://www.cchit.org/
	eHealth Initiative	The mission of the eHealth Initiative is "to drive improvement in the quality, safety, and efficiency of health care through information and IT." The organization focuses on engaging stakeholders to address health care system challenges through the use of IT. The eHealth Initiative is involved in information therapy, e-prescribing, drug safety, care coordination, and comparative effectiveness	http://www.ehealthinitiative.org/

(continued)

APPENDIX A (continued)

List of Web Sites with Information on Workflow Analysis and Tools

Category	Organization	Mission/Goals of the Organization	Web Site URL
	Healthcare Information and Management Systems Society (HIMSS)	The HIMSS provides leadership on the optimal use of IT and management systems for improving health care	http://www.himss.org
	Institute for Healthcare Improvement (IHI)	The IHI is an organization dedicated to the improvement of health care throughout the world. IHI is improving health care by "building the will for change, cultivating promising concepts for improving patient care, and helping health care systems put those ideas into action"	http://www.ihi.org
	Medical Group Management Association (MGMA)	The MGMA is dedicated to improving the performance of medical group practice professionals and the organizations they represent	http://www.mgma.com/
Quality improvement organizations (QIOs)	Colorado Foundation for Medical Care (CFMC)	The CFMC is the QIO of Colorado. CFMC works to improve the quality of health care by collaborating with government programs, health providers, and managed care companies. The CFMC was funded by the CMS to examine workflow in the context of EHR adoption	http://www.cfmc.org/
	Illinois Foundation for Quality Health Care (IFQHC)	The IFQHC is the QIO of Illinois and provides assistance to Medicare consumers and health care providers who participate in the Medicare program	http://www.ifqhc.org/
	MetaStar	MetaStar is the QIO of Wisconsin that works with health care providers to improve the quality of care. Metastar believes health care should be patient-centered, safe, effective, timely, efficient, and equitable. They bring providers together to collaborate and learn from one another	http://www.metastar.com
State-level organizations and associations	Massachusetts eHealth Collaborative	The Massachusetts eHealth Collaborative was formed by the physician community "to bring together the state's major health care stakeholders for the purpose of establishing an EHR system that would enhance the quality, efficiency, and safety of care in Massachusetts"	http://www.maehc.org/
	Michigan Improving Performance In Practice (IPIP)	The IPIP is funded by a Michigan State public health grant to help primary care practices in process improvement	http://ipip-aiag.org/
	New York Primary Care Information Project (PCIP)	The PCIP works to improve health care through health IT and data exchange. The program supports the adoption and use of EHRs among primary care providers in the underserved communities of New York City	http://www.nyc.gov/html/doh/html/pcip/pcip.shtml
Nonhealthcare organizations and associations	American Society for Quality (ASQ)	ASQ is a community of experts that "advances professional development, credentials, knowledge and information services, membership community, and advocacy on behalf of its more than 85,000 members worldwide. As champion of the quality movement, ASQ members are driven by a sense of responsibility to enrich their lives, to improve their workplaces and communities, and to make the world a better place by applying quality tools, techniques, and systems"	http://www.asq.org/
	Carnegie Mellon Center for Computational Analysis of Social and Organizational Systems (CASOS)	CASOS combines computer science, dynamic network analysis, and the empirical study of complex sociotechnical systems	http://www.casos.cs.cmu.edu
	Institute of Industrial Engineers (IIE) and the Society for Health Systems (SHS)	The IIE is a professional society dedicated to the advancement of technical and managerial excellence of industrial engineers. The SHS is a society of IIE that enhances career development and continuing education of industrial engineering professionals working in the health care industry	http://www.iienet.org

Source: Carayon, P. et al., *Incorporating Health Information Technology into Workflow Redesign—Summary Report.* Prepared by the Center for Quality and Productivity Improvement, University of Wisconsin–Madison, under Contract No. HHSA 290-2008-10036C, AHRQ Publication No. 10-0098-EF, Agency for Healthcare Research and Quality, Rockville, MD, October 2010.

Health care organizations need to understand how health IT data can be used to as a tool to study workflow. As an increasing number of health care organizations implement health IT, the need to develop and refine workflow measures based on health IT data will increase. We have outlined some of the strengths and weaknesses of health IT-based methods for analyzing workflows; further research is also needed in this area. See Appendix A for a list of organizations that provide additional information about health IT implementation and workflow analysis.

The implementation of health IT will inevitably change work and workflows (Carayon et al., 2006). Analyzing and redesigning workflows before, during, and after health IT implementation is one means to understand anticipated changes and avoid the negative unintended consequences as much as possible. The workflow analysis tools that were described in this chapter offer a chance to proactively determine the kinds of workflows a health care organization desires and to then select a vendor product that best meets the needs of the workflows. Health IT can also be part of the workflow analysis and continuous improvement cycle by providing measures or indicators of workflows. However, at the current time, there is little evidence that health care organizations are studying workflows in anticipation of health IT implementations (Carayon et al., 2010). Health care organizations need help and support in analyzing and redesigning their workflows as the analytical methods require expertise that most will not have. The recently funded Regional Extension Centers (http://healthit.hhs.gov/portal/server.pt/community/hit_extension_program_-_regional_centers_cooperative_agreement_program/1335/home/16374) will provide one important source of such help, as will the online toolkit called "Workflow Assessment for Health IT" (http://healthit.ahrq.gov/workflow).

Acknowledgments

This publication was supported by contract HHSA290-2008-10036C from the Agency for Healthcare Research and Quality, U.S. Department of Health and Human Services [PIs: P. Carayon and B. Karsh], grant 1UL1RR025011 from the Clinical & Translational Science Award (CTSA) program of the National Center for Research Resources National Institutes of Health [PI: M. Drezner], grants R18-HS017899 and R01-LM008923 from the Agency for Healthcare Research and Quality, U.S. Department of Health and Human Services [PI: B. Karsh] and grant K08 HS17014 from the Agency for Healthcare Research and Quality, U.S. Department of Health and Human Services [PI: T. Wetterneck]. The opinions expressed in this document are those of the authors and do not reflect the official position of AHRQ, NIH, or the U.S. Department of Health and Human Services.

References

Andriole, K. P. (2002). Productivity and cost assessment of computed radiography, digital radiography, and screen-film for outpatient chest examinations. *Journal of Digital Imaging, 15*(3), 161–169.

Apkon, M., Mattera, J. A., Lin, Z. Q., Herrin, J., Bradley, E. H., Carbone, M. et al. (2005). A randomized outpatient trial of a decision-support information technology tool. *Archives of Internal Medicine, 165*(20), 2388–2394.

Ash, J. S., Berg, M., & Coiera, E. (2004). Some unintended consequences of information technology in health care: The nature of patient care information system-related errors. *Journal of the American Informatics Association, 11*(2), 104–112.

Ash, J. S., Sittig, D. F., Dykstra, R., Campbell, E., & Guappone, K. (2009). The unintended consequences of computerized provider order entry: Findings from a mixed methods exploration. *International Journal of Medical Informatics, 78*(Supplement 1), S69–S76.

Barnett, G. O., Barry, M. J., Robb-Nicholson, C., & Morgan, M. (2004). Overcoming information overload: An information system for the primary care physician. *Studies in Health Technology and Informatics, 107*(Pt 1), 273–276.

Baron, R. J., Fabens, E. L., Schiffman, M., & Wolf, E. (2005). Electronic health records: Just around the corner? Or over the cliff? *Annals of Internal Medicine, 143*(3), 222–226.

Bastien, J. M. C. (2010). Usability testing: A review of some methodological and technical aspects of the method. *International Journal of Medical Informatics, 79*(4), e18–e23.

Carayon, P., Hundt, A. S., Karsh, B.-T., Gurses, A. P., Alvarado, C. J., Smith, M. et al. (2006). Work system design for patient safety: The SEIPS model. *Quality & Safety in Health Care, 15*(Supplement I), i50–i58.

Carayon, P., Karsh, B.-T., Cartmill, R., Hoonakker, P., Hundt, A. S., Krueger, D. et al. (2010). *Incorporating Health Information Technology into Workflow Redesign—Summary Report*. Rockville, MD: Agency for Healthcare Research and Quality.

Carayon, P., Wetterneck, T. B., Hundt, A. S., Ozkaynak, M., DeSilvey, J., Ludwig, B. et al. (2007). Evaluation of nurse interaction with bar code medication administration technology in the work environment. *Journal of Patient Safety, 3*(1), 34–42.

Chan, D., Callahan, C., Sheets, S., Moreno, C., & Malone, F. (2003). An Internet-based store-and-forward video home telehealth system for improving asthma outcomes in children. *American Journal of Health System Pharmacy, 60*(19), 1976–1981.

Du Moulin, M. F. M. T., Bullens-Goessens, Y. I. J. M., Henquet, C. J. M., Brunenberg, D. E. M., de Bruyn-Geraerds, D. P., Winkens, R. A. G. et al. (2003). The reliability of diagnosis using store-and-forward teledermatology. *Journal of Telemedicine and Telecare*, 9(5), 249–252.

Eccles, M., McColl, E., Steen, N., Rousseau, N., Grimshaw, J., Parkin, D. et al. (2002). Effect of computerised evidence based guidelines on management of asthma and angina in adults in primary care: Cluster randomised controlled trial. *BMJ: British Medical Journal*, 325(7370), 941–944.

Goldstein, M. K., Coleman, R. W., Tu, S. W., Shankar, R. D., O'Connor, M. J., Musen, M. A. et al. (2004). Translating research into practice: Organizational issues in implementing automated decision support for hypertension in three medical centers. *Journal of the American Medical Informatics Association*, 11(5), 368–376.

Harno, K., Paavola, T., Carlson, C., & Viikinkoski, P. (2000). Patient referral by telemedicine: Effectiveness and cost analysis of an intranet system. *Journal of Telemedicine and Telecare*, 6(6), 320–329.

Hicks, L. S., Sequist, T. D., Ayanian, J. Z., Shaykevich, S., Fairchild, D. G., Orav, E. J. et al. (2008). Impact of computerized decision support on blood pressure management and control: A randomized controlled trial. *Journal of General Internal Medicine*, 23(4), 429–441.

Hoch, I., Heymann, A. D., Kurman, I., Valinsky, L. J., Chodick, G., & Shalev, V. (2003). Countrywide computer alerts to community physicians improve potassium testing in patients receiving diuretics. *Journal of the American Medical Informatics Association*, 10(6), 541–546.

Hyun, S., Johnson, S. B., Stetson, P. D., & Bakken, S. (2009). Development and evaluation of nursing user interface screens using multiple methods. *Journal of Biomedical Informatics*, 42, 1004–1012.

Institute of Medicine (Ed.) (2000). *To Err Is Human: Building a Safer Health System*. Washington, DC: National Academy Press.

Institute of Medicine (Ed.) (2001). *Crossing the Quality Chasm: A New Health System for the 21st Century*. Washington, DC: National Academy Press.

Institute of Medicine (2007). *Preventing Medication Errors*. Washington, DC: National Academy Press.

Kuilboer, M. M., van Wijk, M. A. M., Mosseveld, M., van der Does, E., de Jongste, J. C., Overbeek, S. E. et al. (2006). Computed critiquing integrated into daily clinical practice affects physicians' behavior—A randomized clinical trial with AsthmaCritic. *Methods of Information in Medicine*, 45(4), 447–454.

Li, Q., Douglas, S., Hundt, A. S., & Carayon, P. (2006). A heuristic evaluation of a computerized provider order entry (CPOE) technology. In R. N. Pikaar, E. A. P. Koningsveld, & P. J. M. Settels (Eds.), *Proceedings of the IEA2006 Congress*. Maastricht, the Netherlands: Elsevier.

Litzelman, D. K., Dittus, R. S., Miller, M. E., & Tierney, W. M. (1993). Requiring physicians to respond to computerized reminders improves their compliance with preventative care protocols. *Journal of General Internal Medicine*, 8(6), 311–317.

Logan, A. G., McIsaac, W. J., Tisler, A., Irvine, M. J., Saunders, A., Dunai, A. et al. (2007). Mobile phone-based remote patient monitoring system for management of hypertension in diabetic patients. *American Journal of Hypertension*, 20(9), 942–948.

Lorenzi, N. M., Kouroubali, A., Detmer, D. E., & Bloomrosen, M. (2009). How to successfully select and implement electronic health records (EHR) in small ambulatory practice settings. *BMC Medical Informatics and Decision Making*, 9, 15.

Mehta, A. R., Wakefield, D. S., Kienzle, M. G., & Scholz, T. D. (2001). Pediatric tele-echocardiography: Evaluation of transmission modalities. *Telemedicine Journal and e-Health*, 7(1), 17–25.

Meigs, J. B., Cagliero, E., Dubey, A., Murphy-Sheehy, P., Gildesgame, C., Chueh, H. et al. (2003). A controlled trial of web-based diabetes disease management—The MGH diabetes primary care improvement project. *Diabetes Care*, 26(3), 750–757.

Mohr, T. (2008). The second time around. *Health Management Technology*, 29(9), 22–25.

O'Brien, C. & Cambouropoulos, P. (2000). Combating information overload: A six-month pilot evaluation of a knowledge management system in general practice. *British Journal of General Practice*, 50(455), 489–490.

Ong, R. S. G., Post, J., van Rooij, H., & de Haan, J. (2008). Call-duration and triage decisions in out of hours cooperatives with and without the use of an expert system. *BMC Family Practice*, 9, 11.

Raza, T., Joshi, M., Schapira, R. M., & Agha, Z. (2009). Pulmonary telemedicine—A model to access the subspecialist services in underserved rural areas. *International Journal of Medical Informatics*, 78(1), 53–59.

Rollman, B. L., Hanusa, B. H., Lowe, H. J., Gilbert, T., Kapoor, W. N., & Schulberg, H. C. (2002). A randomized trial using computerized decision support to improve treatment of major depression in primary care. *Journal of General Internal Medicine*, 17(7), 493–503.

Schectman, J. M., Schorling, J. B., Nadkarni, M. M., & Voss, J. D. (2005). Determinants of physician use of an ambulatory prescription expert system. *International Journal of Medical Informatics*, 74(9), 711–717.

Shojania, K. G., Duncan, B., McDonald, K., & Watcher, R. M. (2001). *Making Healthcare Safer: A Critical Analysis of Patient Safety Practices*. Rockville, MD: Agency for Healthcare Research and Quality. Evidence Report/Technology Assessment No. 43; AHRQ Publication 01-E058.

Stanton, N., Salmon, P., Walker, G., Baber, C., & Jenkins, D. (2005). *Human Factors Methods: A Practical Guide for Engineering and Design*. Great Britain, U.K.: Ashgate.

Steele, A. W., Eisert, S., Witter, J., Lyons, P., Jones, M. H. A., Gabow, P. et al. (2005). The effect of automated alerts on provider ordering behavior in an outpatient setting. *PLoS Medicine*, 2(9), 864–870.

The Leapfrog Group (2004). The Leapfrog Group for Patient Safety: Rewarding Higher Standards. Retrieved March 6, 2004, from http://www.leapfroggroup.org

The White House National Economic Council (2006). Reforming Healthcare for the 21st Century. Retrieved June 25, 2007, from http://www.whitehouse.gov/stateoftheunion/2006/healthcare/healthcare_booklet.pdf

Unertl, K. M., Novak, L. L., Johnson, K. B., & Lorenzi, N. M. (2010). Traversing the many paths of workflow research: Developing a conceptual framework of workflow terminology through a systematic literature review. *Journal of the American Medical Informatics Association*, 17(3), 265–273.

U.S. Department of Health and Human Services (2005). Office of the National Coordinator for Health Information Technology: President's Vision for Health IT. Retrieved June 23, 2007, from http://www.hhs.gov/healthit/presvision.html

U.S. Department of Health and Human Services (2010). CMS: Centers for Medicare & Medicaid Services. EHR Incentive Programs. Retrieved July 1, 2010, from https://www.cms.gov/EHRIncentivePrograms/

van Wijk, M. A. M., van der Lei, J., Mosseveld, M., Bohnen, A. M., & van Bemmel, J. H. (2001). Assessment of decision support for blood test ordering in primary care—A randomized trial. *Annals of Internal Medicine, 134*(4), 274–281.

Weingart, S. N., Toth, M., Sands, D. Z., Bohnen, A. M., & van Bemmel, J. H. (2001). Assessment of decision support for blood test ordering in primary care. *Annals of Internal Medicine, 134*(4), 274–281.

Whited, J., Hall, R., Foy, M., Marbrey, L., Grambow, S., Dudley, T. et al. (2002). Teledermatology's impact on time to intervention among referrals to a dermatology consult service. *Telemedicine Journal and e-Health, 8*(3), 313–321.

Zheng, K., Padman, R., Johnson, M. P., & Diamond, H. S. (2009). An interface-driven analysis of user interactions with an electronic health records system. *Journal of the American Medical Informatics Association, 16*(2), 228–237

31

Video Analysis: An Approach for Use in Health Care

Colin F. Mackenzie and Yan Xiao

CONTENTS

31.1 Introduction..523
   31.1.1 Direct Observation 524
      31.1.1.1 Observational Teamwork Assessment for Surgery or OTAS Tool..............524
      31.1.1.2 Anesthetists' Nontechnical Skills or ANTS Behavioral Marker System 525
31.2 Use of Video Data 526
31.3 Analysis of Video Data..526
   31.3.1 Template-Based Analysis..527
   31.3.2 Critical Incident Analysis 527
   31.3.3 Extraction of Qualitative Data on Team Organizational Coordination from Video Records 527
   31.3.4 Media for Communication in the Clinical Domain...528
   31.3.5 Access to Macrocognition from Video Review 529
   31.3.6 Video Analysis Tools...530
   31.3.7 Strengths and Weaknesses of Video Analysis...531
      31.3.7.1 Strengths 531
      31.3.7.2 Weaknesses 531
31.4 Consent, Sociolegal, Privacy, and Confidentiality of Video Recording 532
   31.4.1 Consent for Video Recording...532
   31.4.2 Sociolegal Issues of Video Recording in Real Environments..................................532
   31.4.3 Privacy and Confidentiality 533
   31.4.4 Logistics of Acquisition and Storage of Video..533
      31.4.4.1 Video Acquisition System Network (VAASNET®)......................................533
      31.4.4.2 Camera Systems in VAASNET..534
      31.4.4.3 Software..536
   31.4.5 Core Support Group 536
   31.4.6 Storage of Video 537
   31.4.7 Video Buffer, Mobile Video, and Video Whiteboard..538
31.5 Conclusion 538
References...538

31.1 Introduction

Health care practices in the twenty-first century are constantly undergoing adaptions and changes in order that the medical institutions can remain competitive. Clinical personnel are trained to perform these new practices, but there is little information on whether the necessary training is reflected in adequate clinical performance. Understanding of the strengths and weaknesses of human performance has direct implications for strategies to improve quality of health care and patient safety. However, our knowledge of human performance in real, complex, and dynamic environments, such as those found in clinical care settings, is limited. Studies of health care providers in their natural settings could provide insight into how teams work under time pressure, with constant interruptions, and in suboptimal workplaces. This trend of increasing complexity of the medical workplace is not expected to diminish. A strategy used by medical organizations to meet these increasing demands for adaption involves the shift from individual to team-based work arrangements (Burke et al., 2010). Therefore, it has become important for researchers and the clinicians themselves to better understand and measure clinical performance.

Primary strengths of ethnographic observations as a tool to develop systems for measurement of performance are that they support a discovery process, draw attention to significant phenomena, and suggest new theories whose validity and generality can then be evaluated through additional studies. In comparison, studies carried out in laboratory settings do not recreate many of the real-life variables, such as risk, uncertainty, composition of teams, and the workplace domain. These are significant factors in determining performance during real-world dynamic and stressful events. Richer understanding of teams providing health care can benefit patient safety by assisting in team training development and can boost individual and team competencies in nontechnical aspects of care, such as prioritization, leadership, and decision making.

The clinic itself is the ultimate test bed for creating teamwork measures. Such measures enable assessment if and how team behaviors change after medical team training is carried out. The questions to answer with regard to team training effectiveness is what is the best way to train people in order that they will behave differently—that is, more efficiently, consistently, and safely—within the real clinical environment and how would this be measured? Examination of real-world behavior in the clinical domain can help us identify problem performance areas that can be recreated in a simulated environment and help us learn more about the real world.

Studies of communication and teamwork failures in health care indicate the influence of divergent perceptions between clinician groups (e.g., nurses, surgeons, and anesthesiologists) (Thomas et al., 2003; Makary et al., 2006) and the impact of hierarchical and cultural factors upon professionals' behavior, particularly under emergent conditions (Reader et al., 2007). There is strength in gathering intimate knowledge of the attitudes of different disciplines toward team performance, the barriers, and facilitators of effective teamwork. Team measurement within the operating room or patient resuscitation environments is a complex study because activities and decision making are time critical and must be coordinated between anesthesia, surgery, and nursing professionals, who all have their own group cultures and team dynamics.

Tools are available for measurement and assessment of the interactions that occur during interdisciplinary teamwork in complex clinical environments according to clinician's self-reported ratings. These clinician survey tools provide benefit by including ratings of interventions that focus on the strengths of the team and target opportunities for improvement (Leonard et al., 2004). These psychometrically tested tools enable rigorous assessment of safety attitudes (Sexton et al., 2006),

teamwork and patient safety (Kaissi et al., 2003), structural and organizational characteristics (Zimmerman, 1994), interpersonal and communication skills (Shortell et al.,1994), and team and safety climate measures (Hutchinson et al., 2006).

More research is needed to determine why health care professionals differ in their perceptions of teams and the contribution of teamwork skills to successful communication and performance (Murray and Enarson, 2007). It appears also that the bulk of this work may lie beyond these clinician surveys and video analysis may provide additional insights (Jeffcoat and Mackenzie, 2008).

31.1.1 Direct Observation

The structure and culture of health care organizations can act as a barrier to effective teamwork and direct observation studies identify the individual, team, and organizational precursors to breakdowns in team communication and coordination (Carthey, 2003). Understanding the relationships between the processes of care, including environmental and interpersonal interactions and outcomes, is thus essential to developing teamwork measures. Research that adopts direct observation methods can be most useful (and indirectly achieved through review of video records) (Shapiro et al., 2004).

Observational research has been particularly helpful for identifying recurrent and interrelated factors in communication during care delivery in high-risk and interdisciplinary specialties like anesthesia (Lingard et al., 2002). Additionally, lack of psychological support, cultural norms, and uncertainty as to the shared plan of action further exacerbate the situation in which anesthesiologists and other high-risk medical specialists work (Leonard et al., 2004).

For instance, studies in the emergency department show that staff do not speak up because there is a culture of not questioning superiors (Morey et al., 2002). This is particularly vital if such communications potentially can avoid or alleviate adverse impacts. Direct observation to measure aspects of team performance has been carried out at a number of different levels of analysis. An example of a highly specific assessment of an individual's team skills, excluding the performance of other team members is called observational teamwork assessment for surgery (OTAS) (Healey et al., 2005; Undre et al., 2007). An assessment that attempts to account for the whole team is called the anesthetists' nontechnical skills (ANTS) behavioral marker system (Fletcher et al., 2002, 2003).

31.1.1.1 Observational Teamwork Assessment for Surgery or OTAS Tool

Undre et al. (2007) developed OTAS to simultaneously assess two facets of the surgical process. Observer 1 monitors specific tasks carried out by team members, under the categories patient, environment, equipment, provisions, and communications. At the same time, Observer 2 uses a behavioral observation scale to rate clinician behaviors for the three surgical phases (preoperative, operative, and postoperative) against five established components of teamwork: (1) cooperation, (2) leadership, (3) coordination, (4) awareness, and, (5) communication. Two independent observers enable recording of detailed information both on what the team members do and how they do it. Potential constraints can be identified, on the effectiveness of interdisciplinary communication that might impact patient outcome.

31.1.1.2 Anesthetists' Nontechnical Skills or ANTS Behavioral Marker System

ANTS comprises the following four skill categories: (1) task management, (2) teamwork, (3) situation awareness, and (4) decision making. These skills are divided into 15 elements, each with representative behaviors as examples. The internal consistency and usability of this system was tested experimentally with a sample of 50 consultant anesthetists in 2003. The findings indicated that ANTS has a satisfactory level of validity, reliability, and usability and can be developed, alongside careful guidelines, within anesthetic training curricula. It is the first tool for nontechnical skills training in anesthesia and supports the need for objective measures of teamwork to appropriately inform real and simulated training initiatives (Fletcher et al., 2003).

There is a role for both structured and unstructured observations to help capture and measure the complexity of team interactions. The rigor of these observations has been addressed by employing strategies such as validating tools against best practice standards, using two independent observers to simultaneously collect data, and triangulation of data using multiple data collection methods (Healey and Undre, 2006). Additionally, further research is needed to ensure observation tools are robust and standardized, and also importantly, remain consistent with current best practices (Undre et al., 2006).

Carthey (2003) points out that direct observation, especially of the structured variety, may not be suited to all health care environments. Those environments that are defined by emergent rather than elective procedures—where there is unpredictable diverse case mix, larger size, and the greater movement of staff around a wider area while treating patients—can create difficulties for observers. It may be too great an attentional task to accurately record all teamwork communications due to the involvement of multiple actors and activities that can occur simultaneously (Wears, 2000).

While observational studies illustrate types of errors and incidents in health care, they are not always effective at describing and analyzing clinical conduct at the level of interactions within real complex clinical environments, such as intensive care and trauma settings. It may be that the use of video is more appropriate as it overcomes many of the disadvantages that direct observation has in complex medical settings. Video recordings can also be used in conjunction with direct observation to validate and supplement information collected by trained observers and other structured analysis (Jeffcoat and Mackenzie, 2008).

Video is perceived as the richest medium to capture the minutest and briefest particulars of human interaction while retaining the context of the event and making it available for analyses by multiple or independent subject matter experts (SMEs). Video recording in the medical environment makes it possible for clinicians to review their own activities and for analysts to extract qualitative and quantitative data. Data collection is a major challenge, particularly in studies of emergency medical care because of the unexpected nature and timing of care events. Real-time data collection is needed to counteract hindsight biases in retrospective construction of past events and to capture dynamically evolving emergency situations. An influential tool that may assist such data collection is audio–video recording.

Studies of decision making and team performance in the past decade or so have highlighted the importance of understanding human activities in real, complex environments (Klein et al., 1993, 2003). Many significant variables, such as expertise, risk, uncertainty, and composition of teams, are often difficult to replicate in usual laboratory settings. Studies in real environments and in sophisticated simulation environments with experienced practitioners are required.

A major challenge in carrying out studies in emergency medical settings, either real or simulated, is data collection. Although indirect data such as recalled past incidents can be utilized in one's modeling efforts (e.g., Klein, 1989), direct collection of behavioral data is needed to overcome potential biases in retrospective construction of past events. Tools for collecting behavioral data have become increasingly sophisticated. The most influential among these new tools is probably video recording (Dowrick and Biggs, 1983). With video recording, it is potentially possible for analysts to repeatedly examine activities and to extract detailed quantitative data. The person who was recorded can provide comments on his or her covert mental processes cued by video records. Such cognitive approaches to examination of real medical events are a powerful tool to examine performance and identify patient and practitioners safety issues.

Video recording captures rich data that can be reviewed in great detail by different reviewers; the inherent, subsecond timing data in audio and video lends itself to sophisticated time study (Laws, 1989). Video was used with simulation for medical education and in the analysis of crisis resource management trauma assessment training debriefing after patient simulation (Gaba and DeAnda, 1988; Lee et al., 2003). Video data are used as a tool to complement observation. Video by its nature is a powerful tool for behavioral researchers and its value was recognized soon after its initial consumer availability (e.g., Dowrick and Biggs, 1983). The potential utilities of video recording for studying performance in high-risk health care settings are difficult to overstate. The advances in hardware and software have made video technology a routine tool for research in individual and collaborative performance. An increasing number of research projects include video recording as a key data collection method. How this tool should be exploited methodologically and theoretically is thus a key question for researchers (Xiao and Mackenzie, 2004). Driven by advantages of video-based data collection techniques, a research program was initiated 15 years ago based on collection and analysis of video recordings made in the fast-paced, highly dynamic health care domain of trauma patient resuscitation.

An important application of video analysis that has been promoted recently is in assessment of simulated and real-life clinical performance in different medical domains including surgery, emergency departments, and during task specific procedures that have high risk as well as life saving benefits. The revision of this chapter includes updating of the current studies on the use of video in health care to include new studies and uses of video to improve health care quality, reduce errors and assess clinical performance. In addition technical aspects of video data collection, analysis, retrieval, and storage of video are updated to reflect current and potential future advances in video technology.

events that previously could not be observed due to limited space can now be analyzed through viewing of the video recorded events. Information extracted from the video data can be used for system design, identification of training needs, workload analyses, and performance evaluation and modeling. Due to the recent availability of inexpensive high-quality digital video cameras and recorders, there has been a huge increase in the use of video recording in many different settings (e.g., Gurusamy et al., 2009; Manser et al., 2009; Volandes et al., 2009). Digital video data (DVD) storage and digital cameras have simplified collection analysis and storage (Weinger et al., 2003a).

Unlike audio data, which can be readily transcribed (transcription of which is usually adequate), video data contain mostly nonverbal information. The work by deGroot (1965) and Ericsson and Simon (1984) lays the foundation of verbal protocol analysis, but, in the analysis of video data, researchers usually focus on nonverbal data and have relatively fewer previous methodological guidelines. Sanderson and Fisher (1994), based on the ideas presented by Tukey (1977), attempted to build a general framework for analyzing data that are indexed by time. They emphasized the usual exploratory nature in analyzing behavioral data, especially those collected in the real world. In addition, time-indexed data cover not only transcriptions of verbalization, but also events, transactions, nonverbal communications (e.g., gestures and facial expressions), and other observations that one can make in the review of video data. Sanderson and Fisher (1994) describe the steps in analyzing time-indexed, sequential data by eight "C"s: chunking, commenting, coding, connecting, comparing, constraining, converting, and computing. This framework provides a guide to data analysis, yet tools are still needed to increase the efficiency in dealing with video data.

31.2 Use of Video Data

The advantage of a video record is that it provides a permanent comprehensive source document. Video recording is suggested as a way to provide quality control of medical therapies and surgical intervention (Koninckx, 2008). Such video recording can be reviewed repeatedly by many different analysts. The events in complex environments such as the operating room and emergency department, where time pressure and uncertainty affect performance can now, with the powerful tool of a video record, undergo expanded analyses. Since the audio–video equipment for recording is inexpensive and small,

31.3 Analysis of Video Data

Video analysis is defined as a process by which video data are reviewed and statements (either qualitative or quantitative) on activities and performance are derived. In many aspects video analysis shares similarities with exploratory data analysis, as video recordings contain a rich amount of data on events and activities and there are no simple ways to analyze all relevant aspects. As a consequence, video data analysis tends to be very labor intensive (high sequence time to analysis time ratio; Sanderson and Fisher, 1994).

Video records of clinical care are data with multiple applications, including performance feedback, quality improvement, clinical training, as educational tools, and for human factors and ergonomic research.

31.3.1 Template-Based Analysis

A unique characteristic of real-life performance is variations in terms of tasks, duration, workload, and personnel involved. When examined across cases, aggregation methods need to be devised. Another consideration is the process of video analysis. Given that there are many facets in any given case in which one may be interested, it can be impossible to efficiently analyze all or even a significant portion of recorded cases.

One approach that enables aggregation and at the same time increases efficiency in video analysis is the development of a task template. A task template is a prototypical task sequences with known landmarks. Based on a task template, performance metrics are often possible, such as timing in reaching task landmarks and subjective performance judgment at these landmarks. Development of task templates can be achieved through task analysis by decomposing a task into a sequence of steps. This task template is used to develop review questionnaires, which speedup video reviewing processes and increased consistency of video reviews across cases. These task templates can be examined from the perspective of contrasting elective and emergency events, or trainees and experience clinicians during trauma patient resuscitation management by any members of the multidisciplinary (surgeons, anesthesiologists, nurses, technicians) team of health care providers.

Through our own experience with video-based performance modeling, it greatly helps if one can define a prototypical task sequence with a defined and clear starting and ending point. Many tasks can be analyzed in this manner. Medical procedures such as airway management, establishing intravenous lines, and placing chest tubes, are examples. Even in the highly changing field of initial trauma patient resuscitation, it is possible to model the team tasks in advanced trauma life support (ATLS) protocols, even though ATLS is formulated to guide a single person's performance (Townsend et al., 1993).

In our own work, extensive video review questionnaires were developed based on the task templates. SMEs were asked to fill in the questionnaires while reviewing video records. The questionnaires contained timing for major landmarks, subjective ratings of performance in tactic maneuvers and decisions, and judgment of the patient conditions. Results from the completed questionnaires produced the basis for further, often quantitative analysis. As an example, using this video task analysis methodology in trauma patient resuscitation identified that in emergency tasks there is compressed and increased workload, increased psychomotor activity, and an increased number of high-priority steps omitted in emergencies compared with elective task performance (Mackenzie et al., 2004).

31.3.2 Critical Incident Analysis

Over the course of recording numerous cases, some stand out with unique or interesting characteristics (Xiao and Mackenzie, 1999). The template-based analysis may not capture these characteristics. Critical incident analysis, although a technique originally developed by Flanagan (1954), represents a data analysis methodology that focus on individual episodes. These episodes often reflect interesting strategies, team performance patterns, and cognitive errors.

With comprehensive data gathering (e.g., including audio–video recordings, system logs, retrospective reviews by participant SMEs), one can potentially reconstruct these episodes in great detail (e.g., Mackenzie et al., 1996b). Examples from our critical incident analysis include those on fixation errors (Xiao et al., 1995), communication errors (Mackenzie et al., 1994), and team coordination (Xiao et al., 1998). As an illustrated example, Figure 31.1 represents the results of an analysis of flow patterns of communications between the resuscitation crew. In this example, team leader is seen to be the communication hub and perform an important role in coordinating work Xiao et al. (2002).

31.3.3 Extraction of Qualitative Data on Team Organizational Coordination from Video Records

As an example of more detailed video extraction, the qualitative data obtained from video recordings of real-life trauma patient resuscitations show that team coordination breakdowns occurred due to (1) conflicting plans, (2) inadequate support in crisis situations, (3) inadequate verbalization of problems, and (4) lack of task delegation (Xiao et al., 1998). As suggested by Hutchins in his book *Cognition in the Wild* (1995), high-performance teams have very flexible structures and are highly resistant to performance failures under the condition of varying and uncertain incoming workload. Initial analysis of data collected on trauma teams by observation and video records, suggests some dimensions that are essential for effective coordination.

First, a number of unique structuring processes are used to ensure quick response under almost any circumstance. These include the following: the usage of detailed medical protocols (e.g., ABC, airway breathing circulation for resuscitation) to ensure medical procedure homogeneity; the reliance of extensive planning and coverage plans to ensure that personnel would be available when and if needed; redundant communication

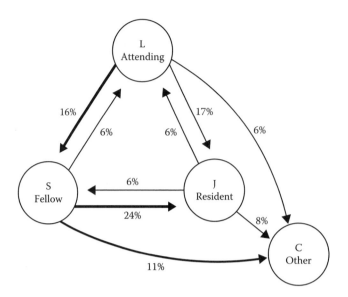

FIGURE 31.1
Overall trauma team communication pattern obtained from video analysis during patient resuscitation. The numbers beside the arrows are the average percentages of communication episodes flowing along the corresponding arrows in proportion to the total number of episodes of a specific case. All numbers in the diagram add up to 100%. Line widths indicate different frequencies of communications. L: leader; S: the senior member; J: the junior members; C: collaborating members. (From Xiao Y et al., Coordination Processes and Awareness Support in Dynamic Work Environments (National Science Foundation Award IIS-9900406) Final Report. November 2002.)

actions to ensure that needed personnel are always connected; and finally, the active management of coordination artifacts (such as boards) for information sharing and the creation of shared awareness.

A second finding relates to the surprisingly important role played by communities of practice (Brown and Duguid, 1991) for organizing trauma teams and to the specific nature of their interactions. In the trauma center setting, contrary to an expectation that team integration would occur around the patient with little functional separation, the action team that forms immediately around newly admitted patients is stylized and primarily role based. Most participants view their primary allegiance to their domain expertise (e.g., nursing, anesthesia, surgery) rather than to the temporary team forming around the patient. These tightly knit communities of practice reduce structure needs and facilitate knowledge sharing among the functional groups and ease the burdens of coordinating patient care.

A third finding is evidence of well-developed expertise coordination processes not just around individual patients but also around the whole trauma unit. Expertise coordination refers to processes that manage knowledge and skill interdependencies (Faraj and Sproull, 2000). These processes are similar and build upon team research findings about transactive memory, shared cognition, shared awareness, and distributed cognition.

Finally, a fourth finding relates to some unique cultural elements that separate this setting from other organizations as well as the more military-type higher

reliability organizations studied in the literature. Since people choose to be part of the trauma center and can leave at any time, the culture has strong elements of testing newcomers over a long period of time to see if they can be relied on to "pitch in" or be able to handle the stress. Patient safety is recognized as superceding hierarchy.

31.3.4 Media for Communication in the Clinical Domain

Verbal communications (obtained from audio–video recordings) can be viewed as one of many media that the trauma team used to communicate and coordinate. Other types of media include, in addition to utterances and explicit gestures, (1) activities, (2) workspace, (3) events, and (4) foci of attention. The use of these media was possible because trauma team members worked in closed physical workspaces. From analysis of video records of real clinical activities, several forms of noncommunication team coordination activities were noted by video analysis (Xiao and Mackenzie, 1998). These noncommunication activities were as follows:

- *Following the protocols.* Established practices (sometimes codified as protocols, such as the advanced cardiac life support or ACLS protocol), specify task distributions and priorities, immediate goals, and problems to be treated. The tasks to be done by each team member are clear. Without much communication, the surgical, anesthesia, and nursing crews commence their activities after the patient arrived. Clear task distributions were observed among the crews in resuscitation teams at the beginning of each patient admission, despite the uncertainty about the patient's status.

- *Following the leader.* Team or crew members determined what they should do by watching the crew leader. The activities of the crew leader can be viewed in some sense as the "medium" through which the team leader passed information (such as instructions) to the rest of the team. If not occupied, team members tended to follow the attention foci of team leaders. Needed materials or help were provided often without explicit solicitation.

- *Anticipation.* The team members were also found to provide unsolicited assistance through the anticipation of the team leader's response to the patient's physiological events. A gagging sound, in one case, led an assistant to offer a suctioning catheter in anticipation that the patient would vomit soon and the anesthesia crew member would have to use suction to clear the patient's airway. Thus the shared physical event space became a medium of communication for the team. The prerequisite, of course, was the ability to understand the significance of patient events. The workspace itself is also a medium through which the teams coordinate. Team members, while not under instruction to perform specific tasks, scanned the workspace and perceived tasks that needed to be carried out.

- *Activity monitoring.* The interdependencies of tasks shared by a team mean that one member's tasks could sometimes only commence after the success of another member's tasks. (e.g., surgeons can only begin certain procedures of resuscitation after the patient is anesthetized.) Thus, monitoring the progress of another member's tasks not only made it possible to compensate for a teammate's performance, but also gave lead information to prepare for the next step.

In many cases, the surgical crew did not announce their plans; however, the anesthesia crew inferred what needed to be done from the activities of the other crew. For example, during the review of the video records of a case, one participant in that case revealed that the conversation between two surgical crew members provided cues of what the surgical crew would do next, even though the conversation was not directed at the anesthesia crew.

These strategies improved task coordination, without the use of explicit, verbal or gestural, communications and enabled the resuscitation teams to perform smoothly in most situations. Information flow is an interesting aspect of team coordination including the explicit, verbal communications regarding situational assessment, and future plans, occurring usually when the team was at a decision point. Strategies of team coordination were extracted from video records of trauma patient resuscitation. Breakdowns in team coordination were observed in the following three types of crisis situations:

- *Pressure to seek alternative solutions.* In this type of situation, extreme difficulties or unexpected patient responses were encountered that prevented the implementation of routine procedures. When the patient condition was deteriorating rapidly, the team was under pressure to find an alternative solution and to act immediately.

- *Initiation of unexpected, nonroutine procedures.* When materials or expertise that had not been anticipated in advance by the supporting members of the team were unavailable, coordination breakdowns occurred, such as when no announcement was made about the need to adopt the nonroutine approach. As a result, the ability of the supporting members of the team to provide assistance was compromised. Coordination breakdowns in this type of incident were marked by the lack of anticipatory help from the team members, delays in preparing materials, and unnecessary pauses in the team leader's activities to obtain assistance.

- *Diffusion in responsibility.* In critical circumstances during patient resuscitation, a diagnostic procedure or a treatment plan may have to be abandoned if the patient condition is too unstable. Such changes in plans by crew members within the team occur during crises and under great time pressure. The rest of the team may have difficulties in adjusting itself from a diagnostic mode to action mode. The inability of the rest of the team to anticipate this sudden change in plan by crew members prevents them from adjusting their responsibilities accordingly, and results in the omission of critical supporting steps.

31.3.5 Access to Macrocognition from Video Review

Team process and performance measurement for real-world teams in complex settings is essential to predict which technical and training innovations will improve team performance. Macrocognition is "the way we think in complex situations" (Klein et al., 2003). In their book *Macrocognition Metrics and Scenarios*, Patterson and Miller (2010) view the scope of macrocognition to include the following: joint activity distributed over time and space; coordinated to meet complex, dynamic demands; occurring in uncertain, event-driven environment with conflicting goals and high consequence of failure; made possible by effective expertise in roles; shaped by organizational constraints; and producing emergent phenomena. Such a scope perfectly describes the setting of trauma patient resuscitation. Insights into macrocognition in this domain can be obtained from video records (Xiao and Mackenzie, 1997). Many studies of cognitive activities are based upon verbal data, but in the real environment, many communications are nonverbal. As illustrated in this chapter, macrocognitive analyses can be greatly assisted by video recording of individual cases that are unusual. In the trauma center, multiple video recordings with detailed retrospective accounts and substantiating records have enabled us to assess the usefulness and potential generalizability of this methodology for accessing cognitive aspects of work. Covert cognitive function can be identified by clinicians when reviewing their own video recorded care. Visual scanning patterns were used to assess information gathering (Xiao et al., 1998) as a source of access to cognition. The process of identifying task performance factors using a task template is efficient (typically 3:1 ratio of analysis to real video time), and it allows aggregation of multiple occurrences of the tasks resulting in improved validity and potential for replication across other populations and settings. It has been found to be particularly revealing to compare data from aggregated tasks at two levels of task urgency as a means for identifying frequently recurring performance omissions. Exemplary macrocognitive analysis enables the use of individual cases to illustrate how changes in practice can avoid nonoptimal performance factors and mitigate error evolution. Using a time-based scoring system for critical treatment steps (Manser et al., 2009) in simulated management of a complex anesthetic scenario, qualitative video analysis showed a significant increase in time spent on clinical and coordination activities, while higher-performing crews spent less time on task distribution and more on situation assessment. Adaptive coordination in medical teams may be informed to develop specific training for improved coordination and performance using such qualitative video analysis techniques.

31.3.6 Video Analysis Tools

Tools are needed to effectively analyze video data. Desirably, these tools should allow basic annotation and event marking/searching. Additionally, tools capable of random access to data, concurrent viewing of multiple timeline documents, touch coding, assisting iterative/recursive categorization would certainly help. Document management (in our case, patient records, questionnaires, review annotations, coding results, transcripts, patient physiological data, case logs) made video analysis more productive since data could be obtained from many different sources. A number of computer systems have been developed in assisting the manipulation and coding of video data.

- MacSHAPA (described in Sanderson et al., 1989). This tool has integrated VCR (videocassette recorder) control, annotation, and coding, with postcoding analysis functionality. Currently, it is only available in Macintosh platform. It is able to control several major models of VCRs and interface with the time code on videotapes.

- A.C.T. (described in Segal, 1994) provides touch coding (i.e., one keystroke input) and can be used both in reviewing videotapes and in real-time observation. However, it does not provide VCR control. A.C.T. only runs on a Macintosh platform.

- OCS tools (Observational Coding Systems of Tools by Triangle Research Collaborative Inc., NC) is a set of tools that enable VCR control, time code reading, and input of annotation and coding.

- VANNA (described in Harrison, 1991) is a Macintosh platform utility that allows the analysis of current multiple video sources (e.g., situations in which multiple cameras are used in recording). VANNA can display these video sources along with other time-stamped information on a single computer monitor, thus simplifying video review process.

- VINA is a tool developed during the course of our program. The VINA analysis software runs on a UNIX platform and has the following key features: (1) manual and scripted VCR control, (2) VCR control by pointing click-and-drag, (3) automatic highlighting of records in multiple timeline documents by synchronization with the time code on the videotape, (4) touch coding of events and activities, (5) temporal graphic representation of coding, and (6) digital and graphical display of recorded vital signs data synchronization with VCR.

The Observer, by Noldus (the Netherlands), is a commercial package available for the Microsoft Windows® platform. It can interface with analog media stored on video records, as well as with digital media encoded in MPEG. The package provides facilities to annotate and log video data, to analyze timeline data, and to manage coded data. The package also allows structured coding to capture multiple aspects of video data, such as activities, postures, movements, positions, facial expressions, social interactions, or any other aspect of human or animal behavior.

Several researchers developed their own video analysis tools to meet their unique demands. Guerlain et al. (2004) developed RAVE, a system to view multiple video sources from different cameras in a synchronous manner for the investigation of performance during laparoscopic procedures. In addition, Guerlain et al. (2005) developed a "black box" recorder for use in the operating room environment. Weinger et al. (2003a,b) also developed their own video capturing and analysis tool for studying nonroutine events during anesthesia.

None of these video collection and analysis tools are ideal for use in all facets of health care. Weigner's approach is anesthesia-specific. Guerlain's software is useful for multiple image review with feeds from laparoscopic images. The black box developed by Guerlain may help develop adverse event analysis and intervention research. MacSHAPA is used still by many video analysts.

Our own current software preference is called JVIDEO, and it is simple and is used by others working in the medical domain (Guzzo et al., 2006; Kilbourne et al., 2009; Sutton et al., 2010). Using a video review controller display of the video can be achieved at ½, 2×, 4×, and 8× forward speed, and backward speeds together with "jump back" (e.g., go back to the start of the last 5 s). This approach allows great flexibility for any type of video and we use it to complete a template needed for a particular video analysis. An open-source toolkit for real-time video acquisition and analysis (Motmot) is available free at http://code.astraw.com/projects/motmot and was developed by bioengineers at the California Institute of Technology for use by neuroscientists and biologists to track whole animal or cellular imaging. Several applications may be useful for human factors researchers as illustrated by a plugin ("FlyTrax") that can track the movements of fruit flies in real time (Straw and Dickinson, 2009).

31.3.7 Strengths and Weaknesses of Video Analysis

There are both strengths and weaknesses in analysis of the video record of real patient management for education, quality assurance, human factors, and ergonomic research.

31.3.7.1 Strengths

The major advantages are that the participant health care providers are not dependent as much on memory, because the video image and audio recording recreate the event, including comments by the team members and alarms that might not have been heard or noted at the time. In our experience, discussions among the participants, care providers, and SMEs, provided additional information that would not necessarily have been revealed without the video image and audio recording. It would have been more difficult to target points where covert events could be revealed by participant SMEs reviewing their own care or where performance could have been changed, without the video image and audio channel. The participants, whose care was video recorded, found video analysis useful and they noted that it allowed them to reflect on their performance in greater detail than is possible without it. They often found it revealing to discuss the video with the nonparticipant SME who could identify that the management was not ideal patient care. It was only on reviewing the video records that it became apparent how their performance could be improved. Ergonomic analyses performed from video records of complex medical domains can be used to develop improved workspace layout (Harper et al., 1995), and instrument pack redesign (Seagull et al., 2006), as a method to achieve cognitive ergonomic perspective (Laws, 1989), and for task analysis of laparoscopic surgical tool use (Mehta et al., 2002).

31.3.7.2 Weaknesses

The weakness of video analysis is that it is tedious and time-consuming. It is difficult to accurately estimate how much time was required to analyze each of these video records because much work was spent in development of the analysis system and database necessary to facilitate video analysis of these cases and others (Mackenzie and Xiao, 2003). As an example, about 10 h were spent in discussion, viewing and transcription of the 8½ min video of prolonged uncorrected esophageal intubation, reported in great detail (Mackenzie et al., 1996b), whereas, aggregated data obtained from the template analysis took about 30 min for 10 min of video record.

During our program, a considerable amount of time and effort was spent in extracting quantitative data from video data. Some of the efforts were judged successful (e.g., timing data of task landmarks template data extraction), others less so. One example of an effort to obtain quantitative data was collection of subjective stress ratings. In the video analysis of many of recorded cases, SMEs were asked to provide ratings on the perceived stress at 1 min intervals. Although valuable data were obtained, it was judged in retrospect that the efforts in collecting such ratings could have been reduced and more exploratory studies could have been accomplished.

Other weaknesses of video analysis are that even with a two-microphone system, it is difficult to pick up all the utterances and the audio record could be improved. However, it would be more obtrusive to equip care providers with microphones. The video images do not include the entire field of view and cannot identify events occurring off the screen, though the audio channel can be helpful. In analyzing a video image with the physiological data overlaid, there is a tendency to think that the participant care providers were aware of these data, when in reality, this is unlikely because of selective attention to other aspects of clinical care. Because analysis occurs after-the-fact, the care providers have time to rationalize the decisions made, as they are aware of the outcome. Thinking aloud and interviews conducted in the middle of case management have been used in simulated cases to overcome this problem (Gaba and DeAnda, 1988). Video analysis of such real events will be important in development of the database necessary for simulation and to improve clinical practice and equipment in the future (Mackenzie et al., 1995). We believe it is by systematic study of such real critical incidents that the mechanisms involved in their genesis will be understood and from these analyses preventive measures or particular approaches to training may be devised. This level of effort to collect and analyze video records may only be warranted for important research questions. The successful use of this technique is dependent on careful and extensive study design before the data are collected to insure the research questions are adequately addressed. However, in spite of the difficulties associated with this data collection strategy, it is a powerful technique that allows for exploration not possible with other techniques when applied in the real dynamic and stressful workplace.

31.4 Consent, Sociolegal, Privacy, and Confidentiality of Video Recording

A major concern of those included in the video records in health care is privacy and the confidentiality of these records. The ubiquitous use of video surveillance in the community has made health care workers and patients well aware of the detailed information that can be extracted from video. In order to enable the potential benefits of video analysis to be realized, while protecting the rights and privacy of individuals creates a dilemma for which there is no single answer.

31.4.1 Consent for Video Recording

The research subjects were the clinical care providers in a trauma center. The subjects gave consent to be video recorded. Protection of human subjects (both research subjects and patients) was secured through a formal approval process by the Institutional Review Board (IRB). For studies analyzing human performance carried out in the real workplace, the IRB agreed to allow video recording without patient consent because it was not thought feasible to obtain consent consistently in the emergency circumstances in which the video records were made, and precautions were made to ensure patient privacy by procedures such as masking patient face in video images (Mackenzie and Xiao, 2003). The original video records were retained until video analyses were completed and then erased. Video images were acquired from a ceiling-mounted camera and directed so as to minimize patient identifying features (Mackenzie et al., 2003). The research subjects knew when events were being video recorded and they consented to publication of details of video analyses and to the use of video clips (30 s–3 min duration) for presentations or in training materials.

However, video is a very powerful medium. A concern is that even when researchers obtain a subject's consent, it is not always clear that the subject understands the implications of that consent. We have used the process described in Table 31.1 as an extension of fulfilling the obligatory need to have a signed consent form. The consent process explains the implications to as many as possible of those affected by or likely to be included in video recording. It is time-consuming, but rewarding, because subjects understand what is happening. It should be noted that fewer research subject participants are needed than might be thought to obtain video data. In our first project (Mackenzie et al., 1994), the majority of the 120 video recordings were made by just six research subjects. No consent forms were signed by other colleagues in the clinical workplace or the patient. Instead the consent process described in Table 31.1 for nonresearch subject clinicians was followed. During the first 3 years only two research subjects would not consent to participate or be included in video recordings. One additional subject agreed to participate but only if their image was blurred in any retained video abstracts.

31.4.2 Sociolegal Issues of Video Recording in Real Environments

Despite advantages of video recording, challenges abound, such as gaining support from those being recorded, securing patient confidentiality, overcoming medicolegal obstacles, and effectively using the medium. A major challenge with video recording is social acceptance and potential impact on the recorded behavior. The primary concern for many people being recorded is possible legal implications of video recording. Formal reports and informal communications have

TABLE 31.1

Research Hypothesis and Plan Implementation

↓	
Research subjects	
↓	
Input into development of research protocol	
↓	
Submit to research committee and IRB	
↓	
APPROVAL →	Educational process for nonresearch subject clinicians includes:
↓	
Subjects sign consent form for video recording research	(a) Availability of research protocol
	(b) Discussion at specialty staff and one-on-one meetings
	(c) Institution wide meeting to answer repeatedly asked questions
↓	
Subject reviews raw data on video recorded project	(d) Showing of video abstract and identification of area included in video image work
↓	
Subject reviews "masked" video abstracts and signs	(e) Efforts to protect confidentiality and privacy of care provider and patients
Consent for retention	(f) Identification of possible role as SME
	(g) How the research results will be disseminated

documented the difficulties in resolving the legal concern of those who are video recorded. Due to the nature of consent, the awareness of video recording can change the very behavior being recorded. So far, the impact of video on behavior has been deemed insignificant (e.g., Pringle and Stewart-Evang, 1990).

31.4.3 Privacy and Confidentiality

Every effort was made to preserve privacy by using camera angles and tight image border control to avoid recognition of individuals. Patient identifiers were removed from paperwork associated with video records. To preserve confidentiality, only care providers and researchers were given access to the video records, which were kept secured under two sets of locks.

Technical approaches used to preserve confidentiality and maintain privacy of the video recorded individuals included video masking with blurring of the face or other distinguishing features of subjects or patients. Voices can be disguised, but these digital manipulations can impair video data analysis if qualities of speech or gaze are being analyzed. The key to the consent process and confidentiality is, in our opinion, the development of trust by those who are video recorded, that the investigators will not abuse the privilege of being allowed to acquire video data for research purposes.

Generally, the original video records were destroyed by degaussing within 4–6 weeks of collection. A sign at the entrance to the operating rooms and inside the trauma resuscitation unit (TRU) was posted to identify image recording was occurring. The wording "Be aware, filming is underway" complies with Joint Commission of Accreditation of Hospital Organizations (JCAHO) regulations for video recording in hospitals. Our experience with 11 years of video recording is that there have been no medicolegal subpoenas and no employment related or liability issues have resulted.

31.4.4 Logistics of Acquisition and Storage of Video

Acquisition of video can be as simple as the use of a home-video camera with only minimal expense to highly sophisticated multiple integrated camera systems. There are advantages to having a robust "industrial" video acquisition system to gather video data information reliably in a systematic manner. The system we have used for the past 15 years (with interval updates and expansion) is described in the sections that follow.

31.4.4.1 Video Acquisition System Network (VAASNET®)

Important features of the design and function of equipment used in audio–video acquisition in the clinical domain include unobtrusiveness, "turnkey" operation, and 24 h/day, 7 days/week availability. The equipment must also be rugged and reliable to withstand use in emergencies, and by multiple users. Because of patient occupancy of locations where the equipment is used, servicing is frequently difficult and industrial grade equipment should be used to withstand the constant wear and tear (Mackenzie et al., 2003).

In our Trauma Center a Telecontrol Center "command and control" location was built to integrate multiple sources of signal input into one location and to allow the use of the system for human factors, ergonomics and patient safety research, and as a telemedicine test bed. An infrared wireless communication system was installed for mobile audio communication among trauma team members. Each resuscitation bay (among 10) and each operating room (of 6) was cabled with 4 duplex audio, 5 video, 4 fast speed data, and 4 multimode fiber connections (Figure 31.2).

In addition, similar cabling was installed in the post-anesthesia care unit (PACU), all of which were located on the second floor of the trauma center. The Telecontrol Center was also linked with three ISDN lines and a web-based and cell phone imaging systems that provided multiple inputs from moving emergency vehicles, home-based physician consultants, other hospitals, distance education centers, and prehospital care providers (Mackenzie et al., 2003).

31.4.4.2 Camera Systems in VAASNET

An environmental video camera was positioned on the ceiling of each resuscitation bay in the 10 bays of the TRU, to provide an overview and allow remote monitoring of general activities (e.g., patient arrival, team readiness, staff present, etc.) and video recording in each location.

A second camera had pan-tilt-zoom capability (PTZ camera) and allowed remote viewing, video recording, and PTZ control of detailed activities in each bay, for example, a specific area on which the trauma team was working such as the chest (Figure 31.3).

The resolution was such that settings used on a mechanical ventilator, O_2 flow meters, intravenous infusion drip rates, etc., could be determined. The PTZ camera nearest the TRU whiteboard (recorded information about incoming patients) was also used to distribute the white board information to other locations, when not in use for monitoring resuscitation bay activities. The benefits, for example, to the anesthesiology team who covered both resuscitation and operating room activities, was that this white board image saved a 200–300 ft walk to the TRU to obtain details of impending trauma patient admissions.

The third camera system was a wireless head-mounted video camera (HMVC). This was worn by an operator during resuscitation and performance of invasive procedures (e.g., chest tube insertion to relieve blood or air trapped inside the rib cage compressing the lung). Four video transmitters placed around the TRU sent wireless video images from the HMVC to the Telecontrol Center. The HMVC images were shown in a "quad" display as a single screen that combined the environmental camera view, the PTZ image, the HMVC image, and the patient vital signs waveforms (Figure 31.4). This "quad" display

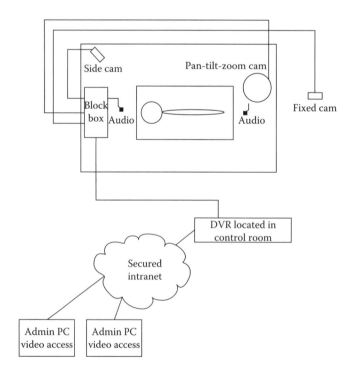

FIGURE 31.2
Schematic of VAASNET showing the most recent additions including administrative personal computer (PC) mobile remote video access (either cell phone, iPad, Tablet, or Screen display) through a secure intranet connection; digital video recorder (DVR) to replace previous analog system; and streaming through the intranet of images to the video whiteboard to coordinate OR status. This system has been implemented in over 40 locations within the University of Maryland, Medical Center, and the Shock Trauma Centers.

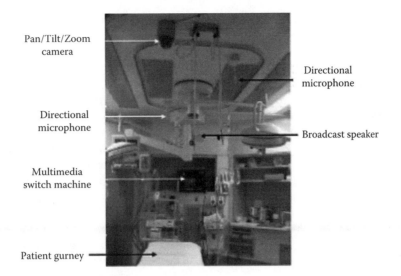

FIGURE 31.3
Shows the VAASNET installation in a TRU bay. PTZ camera provides detailed information, directional microphones at the front and back of the bay gather audio communication from team members. The multimedia switchbox sits on a shelf at the back of the bay beneath the broadcast speaker. There are four duplex audio, video, and data cable connections to and from the Telecontrol Center to each of the 10 resuscitation bays and 6 operating rooms in the trauma center.

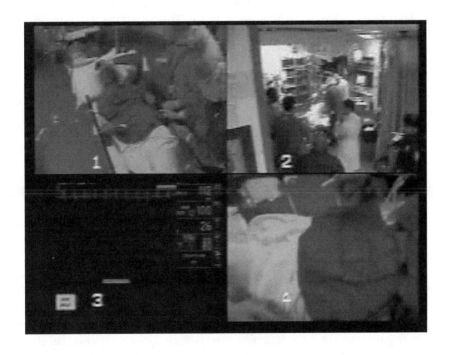

FIGURE 31.4
Screen view of four camera (Quad) acquisition display. (top right, clockwise) Image 1 = PTZ. Image 2 = overview of resuscitation bay. Image 4 = head camera worn by operator (black and white image). Image 3 = vital signs display screen dump in real time. Any one of these images could be expanded to full screen to allow detailed examination, if required.

could be video recorded and synchronized with the high-resolution PTZ image that was video recorded as a full screen display. Such a combination gave comprehensive imaging capabilities for recording events during which multiple personnel were performing simultaneous resuscitation tasks, some of which obstructed a single ceiling-mounted camera view.

In addition video streams from accessory cameras can be incorporated into the analysis to provide essential information, as an example Sutton et al. (2010) measured disruptions to surgical workflow, have been correlated with an increase in surgical errors and suboptimal outcomes in patient safety measures. They used time motion analysis techniques to measure surgeon attention during laparoscopic cholecystectomy recorded with both intra- and extracorporeal cameras. The views were synchronized to produce a video that was subsequently analyzed by a single independent observer. Each time the surgeon's gaze was diverted from the operation's video display, the event was recorded via time-stamp and the reason for looking away was also recorded and categorized. The most frequent reasons for gaze disruptions involved instrument exchange (38%) and downward gaze for extracorporeal work (28%). Improvements aimed at reducing such disruptions—and thus potentially surgical error—should center on better instrument design and realigning the axis between surgeon's eye and visual display.

31.4.4.3 Software

The first version of VAASNET was completed in 1991 (Mackenzie et al., 1995). The main data acquisition system runs under a Microsoft Windows platform. Data acquisition was in five distinct modules: (1) system control logic, (2) interface software for vital signs monitor handshake (including Marquette, Mennen, CritiCare, Nellcor, SpaceLabs, Propaq), (3) VCR communication software, (4) definition of the physiological data file sequence from multiple simultaneous sites, and (5) software to control the real-time network communication with the campus-wide fiber optic system and connections with secondary and auxiliary local area network.

31.4.5 Core Support Group

A key advantage of studying real-life performance is the access to SMEs. There four basic elements in this access: (1) data gathering, (2) data analysis, (3) development of measurement tools, and (4) interpretation of data. One should anticipate an enormous amount of effort from SMEs. To achieve these four elements, a core group of SMEs is essential. In conjunction with the equipment shown in Figures 31.2 and 31.4, a living laboratory for human factors research was created in a real clinical environment.

The VAASNET system became a test bed for assessing remote teleconsultation skills and information gathering practices of the shock trauma clinicians (Xiao et al., 1995). Using an eye tracker to measure gaze-patterns and dwell times, we documented differences in video display scanning of identical images between anesthesiologists, surgeons, nurses, and untrained observers. We concluded that training for telemedicine will require prompts and decision aids to ensure that all events are noted. High-performing trauma clinicians are not necessarily going to perform well when using audio–video-data links as clinical information without such assistance.

Quality management measures of safety performance deficiencies were compared with video analysis of actual performance to identify the strengths of video recording (Xiao and Mackenzie, 1998). The anesthesia record, the anesthesia quality management report, and the other self-reports did not detect 23 of 28 systems and care provider performance deficiencies that were identified from video records. Video recording, therefore, revealed gaps in current information collection about the factors contributing to systems problems as well as their impact on care provider performance.

An ergonomic analysis of the resuscitation workplace simplified the layout (Harper et al., 1995) by repositioning equipment such as the O_2 flow meter, noninvasive blood pressure monitor, and suction system. A more user-friendly resuscitation bay layout was achieved. With the redesign there was space to walk around the patient without the need to step over cables and device connections with the walls. With the revised layout all the patient monitoring, O_2 supply, suction, and equipment necessary for airway management were positioned so that they interfaced with the patient in one place. This interface was beneath the patient vital signs monitor and included vital signs monitor cabling, the ventilator tubing connections, suction, and O_2 supply (now brought from the roof rather than from connections with the walls).

An analysis of chest tube insertion instrument content and tray position showed the problems with existing practices (Seagull et al., 2002). Video analysis of chest tube insertion allowed us to identify the difficulties and the time trauma surgeons underwent searching for the few correct instruments among a tray of 50 or more. In addition, instrument tray position and the number of instrument trays used during insertion of chest tubes were found to be important factors in contamination of the surgical site and of operator injury due to needle sticks or knife blade cuts. A new instrument tray was produced for chest tube insertion and its use is currently being evaluated by further video recording. Abstraction of video clips identified best practices for chest tube insertion that could be contrasted with nonoptimal performance. This was powerful for training material and in supporting the quality management process.

We found that VAASNET allowed all of the uses reported by others (Dowrick, 1991). In addition, because we used the findings from video analysis to assist the clinical care providers rather than as a punitive surveillance tool, we had "buy in" for participation. Common technical errors in central line placement during trauma patient resuscitation were documented by video analysis as a means to show how mentors can improve novice operator proficiency by teaching them to avoid six common technical errors (Kilbourne et al., 2009). Some of the disadvantages of mentors, however, are that they decrease compliance with best sterile practices during central venous catheter placement. Using video capture, the number of trainees and mentors participating in central line placement and their conformity to maximum barrier precautions were identified. Even in elective placement of central lines conformity to sterile practices by mentors was inadequate. Gelbart et al. (2010) used a template-based analysis to assess performance of neonatal resuscitation. Video recording identified that some aspects of neonatal resuscitation were performed flawlessly (invasive ventilation and administration of surfactant), while others including preparation, initial steps, and assessment of communication of heart rate were not. Such analyses can facilitate targeted education and quality assurance programs.

Another use of video in health care is for development of role training. With 24 h video registration during trauma patient resuscitation, evaluation of team functioning and deviations from standardized ATLS protocols were reviewed (Lubbert et al., 2009). Errors in team organization (omission of prehospital report, no evident leadership, unorganized resuscitation, not working according to protocol, and no continued supervision of the patient) lead to significantly more deviations in the treatment than when team organization was uncomplicated. Video registration of diagnostic and therapeutic procedures by a multidisciplinary trauma team facilitates accurate analysis of possible deviations from protocol. In addition to identifying technical errors, the role of the team leader can clearly be analyzed and related to team actions. The results from this study were used to develop a training program for trauma teams that specifically focused on the team leader's functioning. Key messages about the use of video in health care are shown in Table 31.2.

31.4.6 Storage of Video

Initially video were stored on analog VHS tapes. A copy was made to be used for video analysis, annotation and communication. Other documents accompanying video included questionnaires completed by participants at the time of video acquisition and audio records of comments made by participant SMEs and nonparticipant SMEs while viewing the video record. Discharge summaries, operating room and resuscitation area records, laboratory value data, and a description of the role of each person around the patient gurney in the video record were also available.

For the purposes of obtaining inter-rater reliability data, the video records were digitized and clips (short sequences of 30 s–5 min duration) were copied onto CDs for distribution to SMEs. These SMEs rated their assessment of clinical performance shown on video clips. As many as 70 video clips could be stored on a single CD, which was then reviewed by SMEs at their own convenience at work or on their home computers. A video library was developed after the video records were digitized and any patient facial/body identifiers blurred and identification numbers removed.

Weinger et al. (2003a,b) identified the design requirement for a clinical video data analysis system. They describe a video capture system that can be moved from one clinical location to another, requiring some set-up time. Weinger et al. (2003a,b) archive video on DVD discs directly on capture, using stand-alone dedicated DVD recorders that produce an MPEG-2 video stream on a standard DVD video format disc ($1.88 per hour for DVD vs. $11.67 for digital videotape). The DVD video standard permits multiple channel surround sound, audio, choice of screen format (wide, letterbox, and pan and scan), eight tracks of separate synchronized audio, menus and random access for user interactivity, up to nine camera angles, and digital and audio copy protection. To synchronize all these data streams, a time code generator with a drop frame longitudinal SMPTE time code was sent to the video recorder.

Guerlain et al. (2005) developed a multitrack, synchronized, digital audio–visual recording system (RATE tool) to archive the complete operative environment along with the assessment tools for analysis of these data, allowing prospective studies of operative performance, intraoperative errors, team performance, and communication.

TABLE 31.2

Key Messages about Video Recording in Health Care

- It is important to develop trust with video recorded subjects
- Clinician feedback should be obtained on introduction of a new protocol or line of investigation
- Aggregated video recorded data should be reported and clinician reviews used for feedback
- Task analysis at two levels of task urgency is powerful methodology for brief and risky but beneficial tasks
- Multidisciplinary experts in surgery, anesthesiology, and nursing should be involved
- Audio records of participants should be used to explain cognitive aspects of events or covert processes
- Where events are uncertain or verbal interactions unclear, participant input is needed for clarification
- Single critical events may reveal underlying systems failures
- Video records detect quality assurance occurrences, not identified by self-reports
- Video provides powerful feedback and video clips are important training tools

Software synchronized data and allowed assignment of independent observational scores. Laparoscopic cholecystectomy cases were scored for technical performance, participants' situational awareness (knowledge of critical information), and their comfort and satisfaction with the conduct of the procedure. The RATE tool allowed real-time, multitrack data collection of all aspects of the operative environment, while permitting digital recording of the objective assessment data in a time-synchronized and annotated fashion during the procedure. The mean technical performance score was 73% ± 28% of maximum (perfect) performance. Situational awareness varied widely among team members, with the attending surgeon typically the only team member having comprehensive knowledge of critical case information. The RATE tool allows prospective analysis of performance measures such as technical judgments, team performance, and communication patterns, and offered the opportunity to conduct prospective intraoperative studies of human performance. Future uses of this system may aid teaching, failure or adverse event analysis, and intervention research.

31.4.7 Video Buffer, Mobile Video, and Video Whiteboard

We have improved our earlier VAASNET system by several additional enhancements including a buffer to allow capture and retention of video data collection from 20 sites in shock trauma, 24 h by 7 days a week, by 365 days per year. Retained video is automatically erased within 1 week of collection. The retention allows the use of the video data for quality management purposes and for IRB approved research protocols, while protecting the confidentiality of this material and minimizing the likelihood of use for medicolegal purposes. A further enhancement of the video has been used as a mobile platform for coordinating the operating rooms (OR) at the University of Maryland Medical Center. Live video feeds to a mobile device (both Apple iPad and cells phones are in use) allow instant information availability to anesthesiologists, surgeons, and nurses about OR occupancy. A hybrid whiteboard (containing magnetic strips to allow simple changes to insert emergency cases) video images and patient vital signs is in use to coordinate the activities of the OR team in the Shock Trauma Center.

risks. This chapter documents some of the benefits of a phased scale-up strategy of video recording and the outcomes of 15 years of experience. While observation is the traditional approach to identifying problems and implementing improvements, video has many advantages for data collection in complex domains. Video has application in examination of human performance in real and simulated workspace, with development of training and educational tools identified as a result of systematic analysis (Mackenzie et al., 1996a).

This review of video data extraction and analysis techniques used in a clinical environment may help other researchers in formulating their research plans and in data analysis when video recording is considered and used. This belief is based on published reports on video analysis (Goldman et al., 1972; Laughery, 1984; Mackenzie et al., 1996a,b; Mackenzie and Lippert, 1999; Mackenzie et al., 2002; Mackenzie and Xiao, 2003) as well as comments, feedback, and personal communications received from a wide variety of sources. In addition, video provides useful data to access macrocognition, not available through other sources. This chapter describing a video data collection strategy, supplements other information including the excellent Special Issue in October 1989, edited by Mackay and Tartar (Mackenzie and Xiao, 2002) on video as a research and design tool and a book chapter by Nardi et al. (1997) as well as more recent articles on the use of video for feedback in real and simulated events (Laughery, 1984; Mackenzie et al., 2002).

With video recording, events or tasks not associated with the original analysis may be detected, or information may be discarded. Whereas with observation, the data parsing occurs at the time of observation, not later, so only observed and recorded events can be analyzed. A great strength of video recording over observation was the ability to document and understand fleeting events, simultaneous interventions, or brief communications, since video allows expanded analysis of time critical, but brief or uncertain events, whereas such episodes are very difficult to observe accurately in context. Finally, video was found to be a powerful feedback and training tool that identified problems with existing practices in safety performance and ensured "buy in" to participate in change from even the most experienced shock trauma care provider.

31.5 Conclusion

Video recording in the clinical arena is an underutilized data collection tool because of medicolegal confidentiality, privacy, and employment performance–related

References

Brown JS, Duguid P (1991). Organizational learning and communities-of-practice: Toward a unified view of working, learning and innovation. *Organ Sci*, 2:40–57.

Burke CS, Salas E, Smith-Jentsch K, Rosen M (2010). Chapter 3. In Patterson ES, Miller JE (Eds.) *Macrocognition, Metrics and Scenarios*. Burlington, VT: Ashgate Publishing Company, pp. 29–43.

Carthey J (2003). The role of structured observational research in health care. *Qual Saf Health Care*, 12(Suppl II):i13–i16.

deGroot AD (1965). *Thought and Choice in Chess*. Hague, the Netherlands: Morton.

Dowrick PW (1991). *Equipment Fundamentals. Chapter 1, Practical Guide to Using Video in Behavioral Sciences*. New York: John Wiley, pp. 17–29.

Dowrick PW, Biggs SJ (1983). *Using Video Psychological and Social Applications*. Chichester, U.K.: Wiley.

Ericsson KA, Simon HA (1984). *Protocol Analysis: Verbal Reports as Data*. Cambridge, MA: MIT Press.

Faraj S, Sproull L (2000). Coordination expertise a software development teams. *Manag Sci*, 46:1554–1568.

Flanagan JC (1954). The critical incident technique. *Psychol Bull*, 51:327–358.

Fletcher G, Flin R, McGeorge P (2003). Review of Human Factors Research in Anaesthesia Report, University of Aberdeen, 10 July.

Fletcher G, Flin R, McGeorge P et al. (2003). Anaesthetists' non-technical skills (ANTS): Evaluation of a behavioural marker system. *Br J Anaesth*, 90(5):580–588.

Fletcher G, McGeorge P, Flin R et al. (2002). The role of non-technical skills in anaesthesia: A review of current literature. *Br J Anaesth*, 88(3):418–429.

Gaba DM, DeAnda A (1988). A comprehensive anesthesia simulation environment: Recreating the operating room for research and training. *Anesthesiology*, 69:387–394.

Gelbart B, Hiscock R, Barfield C (2010). Assessment of neonatal resuscitation performance using video recording in a perinatal centre. *J Paediatr Child Health*, 46:378–383.

Guerlain S, Adams RB, Turrentine FB et al. (2005). Assessing team performance in the operating room: Development and use of a "black-box" recorder and other tools for the intraoperative environment. *J Am Coll Surg*, 200(1):29–37.

Guerlain S, Turrentine B, Calland JF, Adams R (2004). Using video data for analysis and training of medical personnel. *Cog Technol Work*, 6:131–138.

Gurusamy KS, Aggarwal R, Palanivelu L, Davidson BR (2009). Virtual reality training for surgical trainees in laparoscopic surgery. *Cochrane Database Syst Rev*, 21:CD006575E.

Guzzo JL, Seagull FJ, Bochicchio GV et al. (2006). Mentors decrease compliance with best sterile practices during central venous catheter placement in the trauma resuscitation unit. *Surg Infect (Larchmt)*, 7(1):15–20.

Harper BD, Mackenzie CF, Norman KL (1995). Quantitative measures in the ergonomic examination of the trauma resuscitation unit anesthesia workplace. In *Proceedings of Human Factor and Ergonomics Society 39th Annual Meeting*, Oct 9–13, San Diego, CA, pp. 723–727.

Harrison BL (1991). Video annotation and multimedia interfaces: From theory to practice. In *Proceedings of Human Factor and Ergonomics Society 38th Annual Meeting*, Sept 2–6, Nashville, TN, pp. 319–323.

Healey AN, Undre S, Vincent C (2005). Defining the technical skills of teamwork in surgery. *Qual Saf Health Care*, 15:231–234.

Hutchins E (1995). *Cognition in the wild*. Cambridge, MA: MIT Press.

Hutchinson A, Cooper KL, Dean JE et al. (2006). Use of a safety climate questionnaire in UK health care: Factor structure, reliability and usability. *Qual Saf Health Care*, 15(5):347–353.

Jeffcoat SA, Mackenzie CF (2008). Measuring team performance in healthcare: Review of research and implications for patient safety. *J Crit Care*, 23:188–196.

Kaissi A, Johnson T, Kirschbaum MS (2003). Measuring teamwork and patient safety attitudes of high-risk areas. *Nurs Econ*, 21(5):211–218.

Kilbourne MJ, Bochicchio GV, Scalea T, Xiao Y (2009). Avoiding common technical errors in subclavian central venous catheter placement. *J Am Coll Surg*, 208(1):104–109.

Klein GA (1989). Recognition—Primed decisions. In Rouse WB (Ed.) *Advanced in Man-Machine Systems Research*, Vol. 5, Greenwich, CT: JAI Press, pp. 47–92.

Klein GA, Orasanu J, Calderwood R, Zsambok CE (Eds.) (1993). *Decision Making in Action: Models and Methods*. Norwood, NJ: Ablex.

Klein G, Ross KG, Moon BN et al. (2003). Macrocognition. *IEEE Intell Syst*, 3:81–85.

Koninckx PR (2008). Videoregistration of surgery should be used as a quality control. *J Minim Invasive Gynecol*, 12:248–253.

Laughery KR (1984). Computer modeling of human performance on microcomputers (microSAINT). *Proc Hum Fact Soc*, 12:884–888.

Laws JV (1989). Video analysis in cognitive ergonomics methodological perspective. *Ergonomics*, 32:1303–1318.

Lee SK, Pardo M, Gaba D et al. (2003). Trauma assessment training with a patient simulator: A prospective randomized study. *J Trauma*, 55:651–657.

Leonard M, Graham S, Bonacum D (2004). The human factor: The critical importance of effective teamwork and communication in providing safe care. *Qual Saf Health Care*, 13(suppl_1):i85–i90.

Lingard L, Reznick R, Espin S et al. (2002). Team communications in the operating room: Talk patterns, sites of tension, and implications for novices. *Acad Med*, 77(3):232–237.

Lubbert PH, Kaasschieter EG, Hoorntje LE, Leenen LP (2009). Video registration of trauma team performance in the emergency department: The results of a 2-year analysis in a level 1 trauma center. *J Trauma*, 67(6):1412–1420.

Mackay WE, Tatar DE (1989). Introduction to special issue on video as a research and design tool. *Spec Interes Group Comput Human Inter Bull*, 21:48–129.

Mackenzie CF, Craig GR, Parr MJ, Horst R, the LOTAS Group (1994). Video analysis of two emergency tracheal intubations identifies flawed decision making. *Anesthesiology*, 81:911–919.

Mackenzie CF, Hu PFM, Horst RC, LOTAS Group (1995). An audio–video acquisition system for automated remote monitoring in the clinical environment. *J Clin Monit*, 11:335–341.

Mackenzie CF, Hu PFM, Xiao Y, Seagull FJ (2003). Video acquisition and audio system network (VAASSNET®) for analysis of workplace safety performance. *Biomed Instrum Technol*, 221–227.

Mackenzie CF, Jefferies NJ, Hunter A et al. (1996a). Comparison of self reporting of deficiencies in airway management with video analyses of actual performance. *Hum Fact*, 38:623–635.

Mackenzie CF, Lippert FK (1999). Emergency department management of trauma. *Anesth Clin North Am*, 17:45–61.

Mackenzie CF, Martin P, Xiao Y, LOTAS Group (1996b). Video analysis of prolonged uncorrected esophageal intubation. *Anesthesiology*, 84:1494–1503.

Mackenzie CF, Xiao Y (2003). Video techniques and data compared with observation in emergency trauma care. *Qual Saf Health Care*, 12:i51–i56.

Mackenzie CF, Xiao Y, Horst R (2004). Video task analysis in high performance teams. *Cog Technol Work*, 6:139–147.

Makary MA, Sexton JB, Freischlag JA (2006). Operating room teamwork among critical care physicians and nurses: Teamwork in the eye of the beholder. *J Am Coll Surg*, 202:746–752.

Manser T, Harrison TK, Gaba D, Howard SK (2009). Coordination patterns related to high clinical performance in a simulated anesthetic crisis. *Anesth Analg*, 108:1606–1615.

Mehta NY, Haluck RS, Frecker MI, Snyder AJ (2002). Sequence and task analysis of instruments used in common laparoscopic procedures. *Surg Endosc*, 16:280–285.

Morey JC, Simon R, Jay GD (2002). Error reduction and performance improvement in the emergency department through formal teaching training: Evaluation results of the MedTeams project. *Health Serv Res*, 37:1553–1581.

Murray D, Enarson C (2007). Communication and teamwork: Essential to learn but difficult to measure. *Anesthesiology*, 106:895–896.

Nardi BA, Kuchinsky A, Whittaker S (1997). Video-as-data: Technical and social aspects of a collaborative multimedia application. Chapter 23:487–517, In Finn KE, Sellen AJ, Wilbur SB (Eds.) *Video Medicated Communication*.

Patterson ES, Miller JE (Eds.) (2010). *In Macrocognition, Metrics and Scenarios*. Burlington, VT: Ashgate Publishing Company.

Pringle M, Stewart-Evang C (1990). Does awareness of being video recorded affect doctors' consultation behavior. *Br J Gen Pract*, 40:455–458.

Reader TW, Flin R, Mearns K, Cuthbertson BH (2007). Interdisciplinary communication in the intensive care unit. *Br J Anaesth*, 98(3):347–352.

Sanderson PM, Fisher C (1994). Exploratory sequential data analysis: Foundations. *Hum Comput Interact*, 9:251–317.

Sanderson PM, James JM, Seidler KS (1989). SHAPA. An interactive software environment for protocol analysis. *Ergonomics*, 32:1271–1302.

Seagull FJ, Mackenzie CF, Bogner SM (2002). Combining experts and video clips: Ergonomic analysis for safer medical instrument trays. In *Proceedings of Human Factor and Ergonomics Society 46th Annual Meeting*, Sept 30–Oct 4, Baltimore, MD.

Seagull FJ, Mackenzie CF, Xiao Y, Bochicchio GV (2006). Video-based ergonomic analysis to evaluate thoracostomy tube placement techniques. *J Trauma*, 60(1):227–232.

Segal LD (1994). Action speak louder than words: How pilots use non verbal information for crew communications. In *Proceedings of Human Factor and Ergonomics Society 38th Annual Meeting*, Oct 24–28, Nashville, TN, pp. 21–25.

Sexton JB, Helmreich RL, Neilands TB (2006). The safety attitudes questionnaire: Psychometric properties, benchmarking data and emerging research. *Health Serv Res*, 6:44.

Shapiro MJ, Morey JC, Small SD et al. (2004). Simulation based teamwork training for emergency department staff: Does it improve clinical team performance when added to an existing didactic teamwork curriculum? *Qual Saf Health Care*, 13(6):417–421.

Shortell SM, Zimmerman JE, Rousseau DM (1994). The performance of intensive care units: Does good management make a difference? *Med Care*, 32(5):508–525.

Straw AD, Dickinson MH (2009). Motmot, an open-source toolkit for realtime video acquisition and analysis. *Source Code Biol Med*, 4:5.

Sutton E, Youssef Y, Meenaghan N et al. (2010). Gaze disruptions experienced by the laparoscopic operating surgeon. *Surg Endosc*, 24(6):1240–1244.

Thomas EJ, Sexton JB, Helmreich RL (2003). Discrepant attitudes about teamwork among critical care nurses and physicians. *Crit Care Med*, 31:956–969.

Townsend RN, Clark R, Ramenofsky ML, Diamond DL (1993). ATLS-based videotape trauma resuscitation review: Education and outcome. *J Trauma*, 34:133–138.

Tukey JW (1977). *Exploratory Data Analysis*. Reading, MA: Addison-Wesley.

Undre S, Sevdalis N, Healey AN (2006). Teamwork in the operating theatre: Cohesion or confusion? *J Eval Clin Pract*, 12:182–189.

Undre S, Sevdalis N, Healey AN et al. (2007). Observational Teamwork Assessment for Surgery (OTAS): Refinement and application in urological surgery. *World J Surg*, 31:1373–1381.

Volandes AE, Paasche-Orlow MK, Barry MJ et al. (2009). Video decision support tool for advanced care planning in dementia: Randomized controlled trial. *Br Med J*, 338:b2519.

Wears RL (2000). Beyond error. *Acad Emerg Med*, 7:1175–1176.

Weinger MB, Gonzales DC, Slagle J, Syeed M (2003a). Video capture of clinical care to enhance patient safety. *Qual Saf Health Care*, 13:136–144.

Weinger MB, Slagle J, Jain S, Ordonez N (2003b). Retrospective data collection and analytical techniques for patient safety studies. *J Biomed Inform*, 36:106–119.

Xiao Y, Faraj S, Mackenzie C et al. (2002). Coordination Processes and Awareness. Support in Dynamic Work Environments (National Science Foundation Award IIS-9900406) Final Report, November.

Xiao Y, Mackenzie CF (1997). Uncertainty in trauma patient resuscitation. In *Proceedings of Human Factors and Ergonomics Society 41st Annual Meeting*, Sept 22–26, Albuquerque, NM, pp. 168–172.

Xiao Y, Mackenzie CF (1998). Collaboration in complex medical systems. In *Collaborative Crew Performance in Complex Operational Systems. NATO HFM Symposium*, 20–22 April, 1998, Edinburgh, U.K.

Xiao Y, Mackenzie CF (1999). Micro-theory methodology in critical incident analysis. In *Proceedings of the Conference on Systems, Man and Cybernetics*, Tokyo, Japan, October 1999.

Xiao Y, Mackenzie CF (2004). Introduction to the special issue on video-based research in high risk settings: Methodology and experience. *Cogn Technol Work*, 6:127–130.

Xiao Y, Mackenzie CF, Orasanu J (1998). Visual scanning patterns during remote diagnosis. In *Proceedings of Human Factors Ergonomics Society 42nd Annual Meeting*, Oct 5–9, Chicago, IL, pp. 850–854.

Xiao Y, Mackenzie CF, the LOTAS Group (1995). Decision making in dynamic environments: Fixation errors and their causes. In *Proceedings of Human Factors and Ergonomics Society 39th Annual Meeting*, Oct 9–13, Santa Monica, CA, pp. 469–473.

Xiao Y, Mackenzie CF, Patey R, the LOTAS Group (1998). Team coordination and breakdowns in a real-life stressful environment. In *Proceedings of the Human Factors and Ergonomics Society 42nd Annual Meeting*, Oct 5–9, Chicago, IL, pp. 186–190.

Zimmerman JE (1994). Intensive care at two teaching hospitals: An organizational case study. *Am J Crit Care*, 3(2):129–138.

32

Usability Evaluation in Health Care

John Gosbee and Laura Lin Gosbee

CONTENTS

32.1 Introduction 543
32.2 Overview of Methodologies 545
    32.2.1 Usability Testing 546
        32.2.1.1 System 546
        32.2.1.2 Setting 546
        32.2.1.3 End Users 546
        32.2.1.4 Set of Tasks 546
        32.2.1.5 Data Collection 546
    32.2.2 Heuristic Evaluation 546
        32.2.2.1 Use Simple and Natural Dialog 547
        32.2.2.2 Minimize User Memory Load 547
        32.2.2.3 Maintain Consistency 547
        32.2.2.4 Provide Useful and Meaningful Feedback 547
    32.2.3 Human Factors and Health Care Standards Documents 547
32.3 Practical Guidelines: Application to Health Care 547
    32.3.1 Root Cause Analysis 548
    32.3.2 Failure Modes and Effects Analysis 548
    32.3.3 Procurement 549
    32.3.4 In-House Software Design and Development 549
    32.3.5 Policies and Protocols 550
    32.3.6 Training and Education Development 550
    32.3.7 Facility Design 551
32.4 Case Studies 551
    32.4.1 Code Cart 551
    32.4.2 IV Pumps 551
    32.4.3 Patient-Controlled Analgesia Pump 552
32.5 Human Factors Engineering as Teaching and Selling Tool 553
32.6 Summary 553
References 553

32.1 Introduction

As in many domains, the lack of human factors engineering (HFE) process can lead to some very unfortunate consequences. A plug for pediatric electrocardiograph (EKG) leads looked like a standard electrical plug and was accidentally plugged into an electrical socket. A label for a medication vial had 5 mg/mL in large bold type and 5 mL as total volume in small type at bottom of the vial. A nurse, who was asked to draw up and give 5 mg, gave the whole vial. Instructions for defibrillator leads on an automated external defibrillator (AED) were located on the peel-away paper covers, which were then blown away by the wind as a police officer attempted to use the AED outdoors.

Consider the following quiz:

1. If it is too hard to assemble life-saving equipment in a timely fashion, whom should you call and let know?

   a. The procurement committee

   b. The engineer from the company who designed and market the equipment

   c. No one. Changing the design of anything is nearly impossible

   d. a and b

2. If someone moved the leads on the defibrillator to the opposite side

   a. A few people would notice, but it would not increase accidents

   b. It would have no effect

   c. It would have a measurable effect with an increased accident rate

   d. Caution labels would solve any confusion that arose

3. If someone suggests using a computer to solve a systems issue in your hospital

   a. Go to the training courses, computerization is inevitable

   b. Ask the developers or vendors to show the usability testing data

   c. Find out if the people implementing the system have done a simulation to see if there are unintended consequences

   d. b and c

(Answers: 1. d., 2. c., and 3. d)

There is much advice about the ways that the medical device industry should embrace HFE (Wiklund, 2005). However, it has been recommended that human factors evaluation should also play a key role in health care delivery organizations (Gosbee and Gosbee, 2010). It is not entirely clear how a health care system could incorporate human factors methods, such as usability evaluation. To address this, the following chapter is filled with practical information about the role of usability evaluation for health care providers.

Table 32.1 provides an overview of patient safety activities that can incorporate HFE. For each of the two broad categories of activities (development [e.g., designing new work areas] and evaluation [e.g., root cause analysis (RCA)]), the table lists human factors concepts and methods that are applicable as well as the purpose or rationale for applying HFE.

To incorporate HFE into any organization, one of the first steps is awareness and training. In health care, there has been an increase in awareness of errors and the role that HFE could play (Gawron et al., 2006). There is much less understanding and agreement on the best way to educate and implement HFE principles and tools.

Recent literature suggests that there has been some degree of success in incorporating HFE and usability testing into health care organizations. Welch (1998) proposed avoiding troublesome devices and software by both performing usability testing of products before purchase and demanding usability testing data from manufacturers. Sunnybrook Hospital in Toronto, Canada, applied usability testing to evaluate intravenous pumps to determine possible redesign of work products and in-service training (Etchells et al., 2006). Some in the patient safety improvement arena have cited the importance of usability testing before purchase or during the implementation of a new product (Gosbee and Lin, 2001). It is also recommended that usability testing be integrated into "in-house" design and development of software and complex work areas (Espinoza, 2001).

TABLE 32.1

Summary of Patient Safety Activities That Can Incorporate HFE

Candidate Activities for HFE	HF Concepts/Methods That Should Be Applicable	Purpose of Applying HF Methods
Development activities		
Designing new work areas	User-centered design	Output from applying HF: provide functional
Policy or guideline design and development	Human factors evaluation	requirements, guide design concepts, validate
Training and education curriculum design and development	Human factors analyses (see Table 32.2). Overview of different HFE methods	design concepts, and promote user acceptance
Paper forms (e.g., labels, order forms, charts, and instruction sheets)		Overall goal: ensure usability (efficient, functional, easy to learn, easy to use, low mental workload)
Evaluation activities		
Procurement of health care systems (e.g., devices and software)	Usability evaluation	Output from applying HF: usability data for
Adverse event investigation (e.g., RCAs, FMEAs, and reporting systems)	Human factors analyses (see Table 32.2). Overview of different HFE methods	comparing products from competing vendors, identification of HF issues, or user-system interaction problems that may be causing errors
		Overall goal: identify usability issues with prospective or existing equipment, software or training programs, or existing policies and procedures that may lead to errors

Source: Adapted from Gosbee, L.L., *Jt. Comm. J.Qual. Saf.*, 30(4), 220, 2004.

There are many practical examples of how health care organizations have incorporated usability testing activities into their health care organization, which includes the following: (1) providing crucial information to support procurement decisions; (2) understanding where to focus training and other implementation efforts; and (3) creating teachable moments for human factors concepts and framework. These efforts have also helped organizational culture issues in health care that occasionally stand in the way of implementing patient safety overall.

32.2 Overview of Methodologies

There are many human factors evaluation techniques and many derivations of each type (Table 32.2). Handbooks have been developed for all types of industries, who are struggling to learn and use these helpful techniques (Rubin, 2008). The reader is referred to the standards document, *Human Factors Design Process* *for Medical Devices* (American National Standards Institute/Association for the Advancement of Medical Instrumentation, 2001). This comprehensive document provides an overview of these techniques, key references, and many medical device examples.

Table 32.2 contains a short overview of the various HFE methods one might use in a health care setting. Each row lists the name of the method (analysis activity), a short description, and the type of product one would expect from that analysis activity. The analysis activities listed in the first column give the reader a sense of the scope of human factors. While the last column, analysis products, gives specific products that can be inputs or adjuncts to patient safety or health care analyses.

To provide illustrative samples and a background to subsequent case studies for this chapter, "usability testing" and "heuristic evaluation" techniques will be described in more detail. Applicable standards documents will be briefly described, since most health care industry and government will seek HFE guidance from them.

TABLE 32.2

Overview of Different HFE Methods

HF Analysis Activity	General Description	Analysis Products
Field observations	Unobtrusively observe actual users in the typical work environment carrying out typical tasks. Taking note of how work is carried out, who carries it out, what they use to carry it out, who they interact with, environmental factors (light levels, noise, crowding from equipment, or people, etc.)	Characterizes typical work environment, identifies factors that might impact how clinicians perform (e.g., limitations of equipment, low light levels, high risk or time pressure, and frequent distractions)
Simulation or bench tests	Simulate a process or operation of a device, using different scenarios (e.g., different tasks, time pressure, lighting, and errors that one must recover from). Simulation involves end users, whereas bench tests can be performed by the analyst	Mapping of the system structure (e.g., where do all the menus in a software program lead to? Where does the pharmacist have to go to retrieve XYZ?)
Information requirements or functional needs assessment	Information requirements analysis identifies what information a user needs to carry out specific tasks or activities (from micro to macro—how can a user tell he must push that button next; how does a user know he must perform that task next; how does he know who to contact to relay information?). Similarly, a functional needs assessment identifies what tools or information a user requires in order to accomplish a task	Information and functional needs of the user. Identifies task-related activities that depend on short- or long-term memory, identifies where information should be supplied, how it should be supplied, identifies what tools a user needs to accomplish a task
Heuristic evaluation	Evaluates equipment or a process against a set of human factors principles. These principles prompt such questions as: does the software provide functionality needed by the user? Are buttons grouped in a logical fashion? Is there sufficient feedback to tell the user he has completed a task correctly? Is it obvious what a user must do next?	Identifies areas where human factors principles are violated, which may lead to unwanted consequences such as: frequent user errors, slips, high mental workload, user frustration, inefficient or inaccurate task completion, misunderstanding of policies, and deviation from prescribed guidelines or procedures
Cognitive walkthrough	A user is asked to demonstrate or walk through a device or process, thinking out loud or providing commentary on what he is doing and thinking at each step of the task or process	Characterizes where human decision making is involved in a task, factors that influence decision making, including expertise a user might rely upon, where information is retrieved from, strategies adopted, work-arounds invented to circumvent a deficiency, etc.

Source: Adapted from Gosbee, L.L., *Jt. Comm. J.Qual. Saf.*, 30(4), 220, 2004.

32.2.1 Usability Testing

Usability testing involves putting a system to the test with actual end users. The key components include the system itself, the setting, the end user, the set of tasks, and data collection approaches. Typically, several end users are brought into a testing situation one at a time and asked to accomplish certain tasks while various data are gathered. The more difficult parts of a usability test often include determining the end-user population(s), defining realistic and representative tasks, and identifying performance measures that will inform designers or decision makers. Researchers have determined that, if performed correctly, up to 90% of all design problems can be uncovered with as few as four end users (Rubin, 2008).

32.2.1.1 System

The system includes many things beyond the physical device, including accessories, labeling, instruction manuals, and related protocols. The focus of usability testing may span from the micro-level through the macro-level. At the micro-level, tests evaluate very specific characteristics or elements of an interface (e.g., readability of text messages and ambiguity of icons), whereas, at the macro-level, the focus of a test is broader and encompasses the integrated system (e.g., interactions with other equipment and performance under noisy and cluttered conditions).

32.2.1.2 Setting

Testing situations vary in realism (i.e., fidelity). For instance, low fidelity can include nonfunctional paper prototypes, and high fidelity may involve full-scale simulations that involve many people, devices, and rooms. Each of these test situations provides performance data from direct observation, video recording, or electronic capture of device data.

32.2.1.3 End Users

The importance of involving the various people who will actually be using the product or carrying out the process cannot be overstated. One or more end-user populations will have various sets of characteristics: age, gender, education, training, knowledge, experience, and end goals. The users might be new to the product or be highly experienced with it. They may work individually or rely on teamwork. These characteristics influence how the users interact with a system, and, thus, it is crucial to recruit participants who are representative of the end users.

32.2.1.4 Set of Tasks

The tasks for usability testing might include entering, updating, or retrieving information on a patient, setting up a system for use, or troubleshooting a system. The tasks chosen for testing may be individual tasks or a collection of tasks embedded in a longer realistic scenario. For example, infusion pump tasks might include changing the pump's infusion rate, attaching the infusion pump to a pole, and changing the alarm volume. There is also an opportunity to select tasks that can be completed by an individual or by a team. One may also choose to add a second task or provide realistic interruptions to observe a more representative scenario where users are dividing their time between multiple goals.

32.2.1.5 Data Collection

Data collection techniques may involve a range of activities, from an observer taking notes and using a stopwatch to video recording. Quantitative data include time on task, error type and rate, recovery type and speed, and mental workload (e.g., NASA-TLX is an example of a well-accepted measure of workload; see Wickens [1992], for other methods of assessing mental workload). Qualitative data include verbal or nonverbal reactions expressed by the participant during the course of the task that might indicate confusion, frustration, or enjoyment. Surveys and interviews also have a role and can be conducted as an additional means for understanding usability, learnability, and acceptability.

The data collection should match the goal or subgoals of the usability test. If avoidance of crucial error patterns is key, direct observation might suffice. If serial sets of tasks have to be accomplished in a timely manner (e.g., resuscitation situations), then time on task and speed of error recovery would be central to the usability test plan.

32.2.2 Heuristic Evaluation

Heuristic evaluation (i.e., expert evaluation) is the systematic inspection of a system for its conformity to HFE principles or guidelines (Kaufmann et al., 2003; Nielsen and Mack, 1995). As with usability testing, the system being evaluated can be simple paper forms, single devices, or computer screens, or a complex set of work areas. Typically, the HFE expert evaluates all aspects of a system and determines how these features deviate from a list of principles. More than one expert is used, but the research on effectiveness varies from 90% of design issues discovered by two experts to 50% found by four experts (Jacko and Sears, 2003). The inspection can be visual or interactive in nature. There are many variants of heuristic evaluation, some of which can be found in Nielsen and Mack (1995).

Next is a partial list of HFE principles from Nielsen (1993), which provides some relevant examples and definitions.

32.2.2.1 Use Simple and Natural Dialog

This refers to the display or provision of visual, auditory, or tactile item information. The dialog should be as natural as possible. The terms, icons, alarms, and other system features should be familiar to the user and easily and correctly interpreted. The dialogue should also reinforce how the system works and be in line with user expectations.

32.2.2.2 Minimize User Memory Load

Minimizing user memory load can be accomplished with such design features as the following: (1) making relevant information apparent to the user, (2) making it obvious how to navigate to a desired location or page in a paper or electronic manual, and (3) automatically prefilling information in fields where user must otherwise retype memorized information.

32.2.2.3 Maintain Consistency

Keeping elements of a system consistent refers to obvious and tangible features like visual displays, and deeper features like functionality or navigation scheme of the system. Elements of a system should behave in a consistent and predictable manner from one component or display to another.

32.2.2.4 Provide Useful and Meaningful Feedback

Feedback is the method by which a user is informed about what a system or device is doing, whether the user's intended commands were executed correctly and successfully, whether an error has occurred, or how to recover from errors.

32.2.3 Human Factors and Health Care Standards Documents

HFE and medical device standards documents are important to know about, since most health care industry and government will seek HFE guidance from them. If your hospital is discussing HFE procurement issues with industry representatives, it will help you to know the basic content and scope of AAMI HE-75, *Human Factors Engineering—Design of Medical Devices* (Association for the Advancement of Medical Instrumentation, 2009). This set of guidelines offers specific HFE design specifications for many device and system features (e.g., knob and button specifications). If your health care organization is overhauling your in-house product development process, you can gain key insights from ANSI/AAMI HE-74 *Human Factors Design Process for Medical Devices*, which provides details about

how to bring together risk assessment and control and HFE methodologies during development cycles (U.S. Food and Drug Administration, 1998).

For more information about national and international HFE standards, see Ward and Clarkson (2007).

32.3 Practical Guidelines: Application to Health Care

There is no shortage of candidate products or processes to which human factors can be applied. In this section, we outline various hospital activities that can benefit from HFE principles and methodologies. The same HFE methods can also be used in a similar fashion in the outpatient setting.

In general, the activities will have one or more of the following purposes:

1. Determine why and how errors occur (e.g., close call investigation, adverse event, investigation, and incident investigation).
2. Determine the usability of a device, process, or work area (e.g., compare two devices, information systems, or policies/protocols, prospective risk assessment on existing equipment, prospective risk assessment on policies/protocols, evaluate efficiency of work place layout, evaluate effectiveness of signs, warnings, and labels).
3. Develop a product (e.g., education program, training material, policy or protocol, software or information system, labels warnings or alerts, device or tool, and facility or work area).
4. Improve a product, process, or work area (e.g., improve a training program, revise a policy or protocol, introduce a new policy or protocol, develop learning aids, develop cognitive aids, such as reminders, cheat sheets, guides, reorganize storage areas, redesign a work area, improve labeling, or warning signs).

Hospital activities, whether they be incident investigations or development of a new protocol, focus on a variety of systems within the hospital or outpatient setting, including devices (or tools), work areas, and even facility design or architecture. *Devices* (or tools) range from simple products such as syringes or paper order forms to room-sized robotic surgery systems. *Work areas* range from familiar office style ward clerk stations to complex and changing anesthesia stations. *Architecture* or *facility design* ranges from mundane aspects of bathroom door placement to the colliding needs of privacy and visibility in intensive care units.

Activities that have any of the explicit or embedded purposes defined earlier are candidates for incorporating HFE. We outline seven activities that fall into these categories, which offer leverage for making patient safety improvements in health care organizations:

- RCA
- Failure modes and effects analysis (FMEA)
- Procurement
- In-house software development
- Policy and protocol development
- Training and education development
- Facility design

32.3.1 Root Cause Analysis

When analyzing the root cause of an accident or adverse event, an understanding of HFE can provide insight into the factors that may have contributed to it. Because health care workers must contend with a work environment that is rife with interruptions, shift work, complex equipment, a multitude of policies and protocols, time pressure, and other stressors, errors are bound to occur in these imperfect conditions. A poorly designed device, for instance, represents a latent failure, or an accident waiting to happen in the right conditions. The same can be said about a poorly designed workspace or a complex policy that is reinforced with difficult to decipher written material or an information system whose functions make it tedious or time consuming to follow the policy. Incorporating HFE methods and tools into the analysis of actual adverse events or close calls provides a means to identify such factors, directing corrective measures and interventions to the underlying systems issues and not the user, who is usually only the final trigger in a string of events or failures (Gosbee and Andersen, 2003; Schneider, 2002).

RCA teams are in an excellent position to incorporate human factors methods into their process. Heuristic evaluation discussed in Section 32.2.2, and other analysis methods can be conducted to identify vulnerabilities that contributed to an adverse event or close call. This might involve a human factors engineer trained in these methods working closely with frontline personnel. Also, usability testing can be used to identify the types of errors that might occur as well as the conditions under which their likelihood of occurring increases.

A heuristic evaluation can have a very narrow focus or a broad one. A narrow focus might include an evaluation of specific products or equipment. A broader focus would look at all the related products or processes that surround an activity, for instance, all the equipment that is used in the process of following a protocol, along with the interpersonal communication that occurs, the paperwork that is filled out, the workspace layout, the storage and retrieval of products, and the organizational factors. A broader focus may examine a wide variety of issues that may have an impact on the activity. For example:

- What were the working conditions at the time of the event?
- What aspects of the user's tasks or working environment may have interfered or influenced the event?
- What aspects of the product or equipment are counterintuitive or are incongruent with the user's knowledge, training, education, or experience?
- How could the device or paper forms be redesigned to better accommodate the user's task?
- What information was missing for the user to make an informed decision?
- Does the training on one device hinder the learning of another?
- Do the functions of the software support the user's task for following a procedure?
- Are there too many functions on the device or software, adding too much complexity to typical tasks?
- What tools should be available to the user to help them follow a policy?

32.3.2 Failure Modes and Effects Analysis

Another formal method in patient safety where human factors evaluation should be used is called FMEA. The details of this proactive risk assessment technique are beyond the scope of this chapter. The reader is referred to Stalhandske et al. (2003) for a discussion of this process and Chapter 29. One activity in the process of FMEA that can benefit from HFE is that of identifying failure modes, or things that can go wrong with a system of interest. The system could include a device or a process such as a collection of activities. Usability evaluation can be used to make the FMEA robust. A heuristic evaluation can be used to identify failure modes of a device that may cause user errors. It can also identify failure modes of a process, which arise from human factors issues, such as lack of information for decision making, poor labeling, insufficient lighting, interruptions, and so on.

Usability testing can also be used as a means for identifying various failure modes by creating test conditions (task, environment, etc.) that reflect the conditions and situations under which a process is carried out or device is used. For instance, a FMEA team might not brainstorm or identify the failure mode of "broken infusion pump after being dropped." Such a failure mode could either occur or nearly occur when five nurses grasp the new infusion pump with unwieldy handles and try to attach the pump to a pole.

32.3.3 Procurement

Procurement decisions often consider a multitude of factors, one of which is usability. Usability itself can be defined as having multiple criteria. These can be generalized into the following categories:

1. Functionality—Does it meet functional needs of end users, does it create new needs or tasks based on limitations or excessiveness in its functionality, does it present functionality rarely used, is it used for different functions depending on the user? Can it be used in the work environment that it is intended for (e.g., portability)?

2. Interface design—Does it present high risk for user errors? Is the interface or method of operating it intuitive or confusing? Are labels and warnings easily perceived and understood? Does it require daily use to become a competent user? Is it easy to tell what the device is doing? Are there annoying features that in some situations become a risk for misuse?

3. Training and learning—Does it pose training challenges for novice users, users with extensive experience on another system (transfer of training issues), users who only occasionally use the device/software (and who might experience decay of learning after prolonged periods of nonuse), different user groups who may not have the same goals, and thus use it differently?

The HFE methods covered in Section 32.2 (also in Table 32.2) are useful for ascertaining level of usability of devices or software prior to its purchase (Uzawa et al., 2008; Vignaux et al., 2009). There are a growing number of health care organizations adopting this practice. Allina Health System in Minneapolis, MN for instance incorporated human factors in their patient safety program by using HFE concepts in their workplace analysis, accident investigation, and incident reporting program. Staff training on how to conduct usability testing has led to the integration of user testing in their product selection process.

University Health Network in Toronto Canada has built a usability lab for helping procurement decisions (Cassano-Piché, 2007). One example is the application of usability testing to deciding which patient-controlled analgesia pump to purchase (Ladak et al., 2007). User testing can uncover problems ranging from inappropriate machine default settings, obscurity of critical functions, ineffective or nuisance alarms (e.g., see Wogalter and Frantz [2005] for human factors issues related to alarms), poor labeling, to ambiguous machine dialog.

Beyond a usability evaluation in the intended work environment, HFE can also be used to examine usability issues related to maintenance during the life cycle of the device. This relates to the usability of the device from the perspective of a biomedical engineer who would interact and manipulate the device differently than a nurse would. For instance, a biomedical engineer who is responsible for setting defaults for various pumps may encounter several usability issues. The default values may vary depending on whether the pump is used in the recovery room versus on the floors. The sequence of steps for doing this may be complex and prone to error. The preferred default values may also be unknown to the engineer maintaining the pump. When there are software updates, this may impact how the default settings are set up or changed. These issues should be evaluated when assessing device usability. A well-rounded usability evaluation is dependent on identifying all the different groups of end users.

32.3.4 In-House Software Design and Development

When making product selections, HFE can help identify potential risks for error. In developing in-house products, developers should be no less vigilant about usability issues (Gosbee and Gardner-Bonneau, 1998; Lopez et al., 2009). HFE can provide a framework for the development cycle (user-centered design and iterative design and testing). Such a framework provides opportunity for periodic evaluation, so that usability issues that have the potential to become latent errors are resolved during development and not after the product is rolled out.

In a user-centered design framework, the development cycle becomes iterative in nature. That is, design and testing are conducted iteratively (repeated at various intervals throughout the development cycle of the product), so that user testing and design refinements start early in the design process and continue throughout the design cycle. Design concepts and early prototypes should undergo user testing for validation and to identify any aspects of design that do not adhere to HFE design principles. What is learned from this early user testing guides any improvements to the design. User testing and design refinement are then repeated throughout the design and development.

In addition to a framework for the development cycle, HFE evaluation can be used to determine functional requirements and information requirements. These requirements influence a variety of aspects of the products design including the organization and layout (of information, controls, etc.), structure and sequence of tasks, level of customization, level of detail of information, organization and layout of the menu structure (for software), appropriate graphics, online training features or help functions, etc.

There are numerous examples of in-house software products. They include computerized provider order entry (CPOE) systems, bedside bar code medication administration (BCMA) systems, pharmacy information systems, clinical decision support systems, clinical data repositories, patient registration, and medical record systems. Although different end users of these information systems may have similar information needs for their decision making, users may differ vastly in the ways in which they search for information, in the context of their decisions, in their needs for specific representations or organization of that information, in their need for detail, and in their end goals that drive the way in which they interact with the system. These subtle but crucial differences determine whether a software product is functionally appropriate and easy to use for any user group. An HFE analysis can help to identify the functional and information needs of the various user groups, and usability testing can help to verify that the software being developed meets these needs in an efficient and usable manner.

32.3.5 Policies and Protocols

Up to this point, the discussion has centered around conducting usability evaluations on commercial or in-house products. It is important not to overlook the fact that policies and protocols are products that can be evaluated as well. HFE can be used to help assess or develop policies, protocols, or guidelines. An HFE evaluation can help to identify any special training needs with respect to specific policies/protocols that might be challenging to follow due to poor design (of equipment/software used to help carry it out), complexity, or requisite knowledge or expertise in an area to carry it out.

The task of developing policy/protocol material is a two-tiered challenge. The first is to address the problem or issue for which the policy/protocol will alleviate, and the second is to avoid creating additional problems or issues that might arise from the introduction of a new policy or protocol. To tackle these two challenges, it is helpful to first gain an understanding of the nature of the activities involved in carrying out the policy or protocol and the information, material and machines that

the user must come in contact with during the process of carrying out the policy or protocol.

To gain this understanding, the designer/developer of policies/protocols should carry out human factors evaluations. Heuristic evaluation and testing of the policy/protocol can be conducted to evaluate its readability, ease of learning, prominence of critical steps or information, complexity, mental workload, etc. One example was a group developing a more usable protocol and paper document to improve physician orders in an emergency room setting (Reingold and Kulstad, 2007). This team found that the percent of physicians drawn into using the HFE-redesigned congestive heart failure order set form increased from 9% to 72%.

When a policy or protocol is perceived to be complex or requires in specialized knowledge to carry it out, HFE can also be used to help identify portions of the process where supplemental material might be needed. Supplemental material includes any aids such as cognitive aids (e.g., resuscitation algorithms), that either help learning or to help guide users (in decision making, device operation, following steps) of a policy or protocol. Aids may also come in the form of risk communication (e.g., warning signs), whose design and evaluation can be aided by the human factors research in the area of warnings and risk communication (Isaacson et al., 2001; Lehto, 2001; Wogalter et al., 1999).

The issues that are dealt with in a heuristic evaluation or usability testing encompass how a policy or procedure is conveyed (e.g., readability, availability, and complexity), how a policy or procedure may interact or interfere with other work processes, what portions of the process should and can be standardized or automated, impact on mental or physical workload or timeliness of other time-critical activities (added paperwork, tedious, or repetitive procedures).

32.3.6 Training and Education Development

Training and education programs are also products that should be developed with the aid of HFE evaluation methods. Doing so can help developers identify specific training needs with respect to specific devices, software, and policies that might be challenging due to poor design or complexity. This in turn would provide focus areas for training material being developed and a means to evaluate it for effectiveness.

The approach is the same as that used for any other in-house product discussed in Sections 32.3.4 and 32.3.5. The goal of the end product is to teach users and provide knowledge or guidance, which in turn will help them carry out a physical or cognitive task. Developers of training and education curriculum can adopt the same methods and tools (user testing and heuristic evaluation or other methods in Table 32.2) in order to determine the functional requirements and information requirements for the training material. The product may be paper based (e.g., user manual or guide) or electronic based (educational/training software). A heuristic evaluation can help identify issues or requirements for training such as the need for supplemental material such as cognitive aids (e.g., "cheat sheets") that either helps learning or helps guide users (in decision making, device operation, following protocols, etc.); determine the frequency of refresher training to maintain competency with a product; and identify transfer of training issues, for example, clinicians who have extensive experience using a specific software program may find it more much more difficult to gain competency in a new prospective system or software upgrade. These issues apply to trainees who are either health care workers or patients. Patient education programs can be designed based on what is learned in an HFE evaluation. For the outpatient setting, this is especially important if a home medical device poses usability issues (e.g., Gosbee, 2004).

32.3.7 Facility Design

When applying HFE to the remodeling or construction of a new facility or work area, or to the purchase of office furniture and storage equipment, attention must be paid to the degree to which it conforms to or supports the work which is carried out there. HFE can help determine the requirements for layout of a new facility based on an analysis of the workflow in different areas, functional needs of the task and the users, and ergonomic specifications. HFE can also help to inform requirements for furniture or storage equipment based on the nature of the task and user characteristics or functional needs. Finally, user testing can help evaluate the efficiency that the workspace affords and the adherence to the functional needs uncovered in a human factors evaluation (e.g., distance and accessibility between related work areas, width of pathways to accommodate equipment and personnel, level of lighting required for specific tasks, and height of shelves, so that 10th percentile female can read the labels and reach stored products). This topic is discussed in more detail by Reiling and Chernos (2007).

32.4 Case Studies

32.4.1 Code Cart

At a Salt Lake City hospital, "code" teams encountered many human factors issues, including confusion and delay with medication retrieval. When a patient had a cardiac or respiratory arrest, the "code" or emergency team is assembled in a room with a code cart containing medications and devices to perform advanced life support. All systems and personnel have to fit together and work together well, since seconds count.

McLaughlin (2003) and his colleagues focused on the HFE aspects of the code cart drawer containing lifesaving medications. With upper management support, they developed a test plan with an aim to redesign the drawer for efficient and accurate retrieval by nurses given verbal orders. The scope of redesign was limited to the drawer that already fit within the several code carts deployed around the hospital.

The test plan included the following: (1) 9–11 nurses (who respond to codes) as the end users for each of five design iterations (versions); (2) the actual drawer with real medications and packaging; (3) each of ten medications was called out one at a time for retrieval (not embedded in a full resuscitation scenario); and (4) performance measures included time on task with accurate retrieval of all ten medications.

The initial "baseline drawer" resembled a laundry hamper with several loose items placed in various places. Boxed injection systems were aligned together, as were the vials and other tubes of medications. There were also irregularly shaped items stacked to the side or on top of other items. Retrieval times with this drawer (total time) ranged from 2:43 to 3:58 min, and an average of 3:07 min. The general usability/design goal was to retrieve them all within about 1 min.

Future versions of drawers included various inserts, cutouts, and labeling (or not). Some versions had adequate average time on task (1:09), but the range went as long as 1:52 min. One version had labels for each cut out area within a sheet of blue foam, but the retrieval time actually slowed, since nurses did not trust that slots were filled correctly and looked at the labels on the vial or box as well. The final version, which was subsequently introduced into all code carts in the hospital, provided a retrieval time range of 0:55–1:25 min, and an average of 1:08 min.

One bonus to this process of redesign was the interest level it generated for change. Change in health care settings is often hard to encourage or maintain in dynamic settings with a lot of traditional thinking. McLaughlin found that the usability testing with so many thought leaders helped decrease the time and effort for implementation planning and change management to almost zero.

32.4.2 IV Pumps

The U.S. Department of Veterans Affairs National Center for Patient Safety (2002) worked with American Institutes for Research (AIR) to complete a comparative usability test of intravenous infusion pumps. The purposes were the following: (1) to assess the value of usability testing as part of an operational patient safety program, (2) to generally determine whether data regarding the comparative usability of infusion pumps (i.e., IV pumps) had the potential to be useful for procurement, and (3) to make a rudimentary comparison of the value of usability testing in an ICU simulator versus a usability testing laboratory.

Seventeen registered nurses participated in the usability test. Four nurses participated in a standard usability testing lab. Thirteen participated in an ICU simulator, which was equipped with two patient manikin simulators, patient monitors, a ventilator, and many other examples of ICU equipment and furnishings.

Three, single-channel IV infusion pumps that offered comparable functionality were used. Test participants used the devices for the first time following a brief, in-service style training session. The test participants' lack of prior experience using the pumps enabled us to focus the test on initial ease of use, a strong indicator of long-term ease of use.

Each participant underwent two scenarios with two of the pumps for 30–40 min each. Performance measures included an assessment of task correctness, task times, ease-of-use ratings, confidence ratings, overall preference, and scoring according to a 10-question usability rating form. Participants performed eight major tasks, some of which included several subtasks that fit into an overall patient care scenario.

An analysis of both the objective and subjective data suggested a relatively distinct ordinal ranking among the three pumps with regard to their initial usability. Averaging all of the simulator-based test results, the highest ranked pump demonstrated superior performance in all major performance dimensions. Not surprisingly, the subjective impression of the participants was often in discordance with objective performance measures. That is, occasionally the task was not completed or completed incorrectly, but the nurse was confident the task was completed without issue.

32.4.3 Patient-Controlled Analgesia Pump

A University of Toronto study looked at improving medical device design through the application of HFE (Lin et al., 1998). Human factors engineers focused on a commonly used device as the test bed, the patient-controlled analgesia (PCA) infusion pump. According to FDA incident reports reviewed in the study, human error was found to be responsible for a majority (68%) of fatalities and serious injuries associated with the PCA in a randomly chosen year. This is similar to what was found in other studies (Callan, 1990) reviewed at that time. Of the reports where human error was involved,

nurse programming errors were found to be the most common type of human error in PCA use. Furthermore, the majority of programming errors involved setting an incorrect drug concentration, all of which led to an over-delivery of medication. Taken together, there was strong evidence that the design of the PCA Infuser had many latent errors and could be improved with HFE.

As the first step toward demonstrating the utility of HFE, the engineers develop a redesigned PCA pump based on human factors analysis techniques and design principles, including cognitive task analysis. In the field studies, nurses were observed in the recovery room doing many tasks at once and constantly being interrupted while programming the PCA. During bench tests, many difficulties were encountered even while operating the equipment under ideal conditions with the help of a user's manual. In the final step of their analysis, the study authors used human factors design principles as a checklist and identified aspects of the device that violated design principles, hence posing usability issues.

The pump's was redesigned by following the same design principles. The redesigned PCA pump included such features as logical grouping and labeling of controls, simplified and more natural language in the displayed messages, and improved status display and feedback.

The second step was to empirically evaluate the design with users. Two user groups participated: novice users (nursing students) and experienced PCA users (recovery room nurses). Participants were given a PCA order form (prescription) and their task was to program the pumps accordingly using each interface. The evaluations included performance metrics such as number of errors, time to complete the task of programming, and subjective workload measures (i.e., mental demand and frustration).

In both groups, there were marked improvements with the redesigned PCA pump. First, there was a reduction in number of errors recorded, 50% and 55% reduction for nursing students and nurses, respectively. Furthermore, there were no errors in setting the drug concentration with the redesigned system, demonstrating a degree of resistance to the most culpable error found in the Medical Device Reports[27] (Lin et al., 1998).

Accompanying the reduction in programming errors with the redesigned system was a statistically significant improvement in task completion time. Nursing students were able to complete programming tasks with the redesigned system 15% faster. The nurses showed 18% faster completion times despite having no prior experience with the new system, compared to several years of experience with the existing Abbott pump. This improvement can be attributed to, among other things, the fact that significantly fewer programming errors were made, and thus, less time was wasted recovering from errors.

The subjective workload associated with the redesigned interface was found to be lower for both user groups using the redesigned pump compared to the existing pump: 53% and 14% lower for nursing students and nurses, respectively. Finally, post-experiment interviews with the nurses and nursing students showed that an overwhelming majority (100% of nursing students and 90% of nurses) preferred the redesigned system over the current system.

32.5 Human Factors Engineering as Teaching and Selling Tool

HFE might provide a less-than-obvious benefit beyond those listed already. During the training for participation in safety activities, usability testing might be an effective tool to train a HFE mind-set for clinicians and management (Gosbee, 1999).

The HFE mind-set (turning people brains around 180°) is rarely accomplished with didactic approaches (Shapiro and Fox, 2001). This is mainly true because it is hard to teach the concept of *learned intuition*. Briefly, *learned intuition* is the phenomena whereby you cannot recall ever not knowing how to use something, and you cannot imagine someone else would not know how it should be used. Educators have created "teachable moments" about learned intuition by having students grapple with human factors design problems and having them develop and test remedies (Sojourner et al., 1993). Also, incorporating usability testing into resident, medical, pharmacy, and nursing student training has been used with success (Gosbee and Gardner-Bonneau, 1998).

One practical exercise that the authors have used with considerable success is the *usability testing small group exercise*. In this exercise, groups of 3–4 people sit around a common table or work surface, and work together to formulate a test plan and then carry it out. One person in each small group is assigned to be the "Director," who reminds the end user to think aloud, and leads the team in discussion about issues and redesign *after* usability test is complete. Another group member is assigned to be the "End User," who will then carry out a series of tasks with the system (without interference or help). The remaining group members are "Observers," who document end users' actions, facial expressions, body language, comments, cursing, and so on. Once the person has completed the task of using the system (without guidance), the team summarizes issues related to observed behavior, comments, and other observations. Observations are then used by each group to guide their recommendations for redesign. By the end of the usability testing exercise, the groups are fully engaged in discussions about the link between design deficiencies (HFE flaws) and the observations made during the usability testing exercise, hence turning their brains around 180° to be able to appreciate and adopt an HFE mind-set.

In practice, many kinds of items and levels of complexity can be used for this exercise. Simple everyday items with design features that lend themselves to teaching points can be very useful. A generic list includes various types of juice pouches or travel package size items. Some home medical devices found in pharmacies or portable medical devices in hospitals can also be used with significantly more preparation and time allotment.

32.6 Summary

This chapter focused on HFE methodologies that can be adopted in the health care setting to improve patient safety. Two methods were discussed as examples: heuristic evaluation and usability testing. These evaluation methodologies can be applied to a number of activities in a hospital setting or outpatient setting including RCA, FMEA, procurement, in-house software design and development, policies and protocols, training and education, and facility design. In all these activities, HFE can be used to gain insight into human–system interaction issues that we also call usability issues, or latent failures that are "accidents waiting to happen." The intent of this chapter was to provide guidance to the practitioner wishing to employ human factors in their patient safety program or to teach and demonstrate the concept or human factors mind-set to clinicians and management.

References

American National Standards Institute, Association for the Advancement of Medical Instrumentation (2001). *Human Factors Design Process for Medical Devices (ANSI/ AAMI HE74:2001)*. Arlington, VA: Association for the Advancement of Medical Instrumentation.

Association for the Advancement of Medical Instrumentation (2009). *Human Factors Engineering—Design of Medical Devices (AAMI HE-75)*. Arlington, VA: Association for the Advancement of Medical Instrumentation.

Callan, C.M. (1990). Analysis of complaints and complications with patient-controlled analgesia. In F.M. Ferrante, G.W. Ostheimer, and B.G. Covino (Eds.) *Patient-Controlled Analgesia*. Boston, MA: Blackwell Scientific Publications, 139–150.

Cassano-Piché, A. (2007, January). Human factors—A new lesson in patient safety. *Hospital News*, 11.

Espinoza, J. (2001). Human factors in the engineered environment. In *Proceedings of Enhancing Working Conditions and Patient Safety: Best Practices*. Pittsburgh, PA and Washington, DC: Agency for Healthcare Research and Quality. Available online at www.quic.gov/workforce/Enhance/index.htm

Etchells, E., Bailey, C., Biason, R., DeSousa, S., Fowler, L., Johnson, K., Morash, C., and O'Neill, C. (2006). Human factors in action: Getting "pumped" at a nursing usability laboratory. *Healthcare Quarterly*, 9, 69–74.

Gawron, V.J., Drury, C.G., Fairbanks, R.J., and Berger, R.C. (2006). Medical error and human factors engineering: Where are we now? *American Journal of Medical Quality*, 21(1), 57–67.

Gosbee, J.W. (1999). Human factors engineering is the basis for a practical error-in-medicine curriculum. In *Proceedings of the First Workshop on Human Error in Clinical Systems*. Glasgow, Scotland, U.K.: University of Glasgow.

Gosbee, J.W. (2004). Human factors engineering and patient safety: Starting a new series. *The Joint Commission Journal on Quality and Safety*, 30(4), 215–219.

Gosbee, J. and Anderson, T. (2003). Human factors engineering design demonstrations can enlighten your RCA team. *Qual Saf Health Care*, 12(2), 119–121.

Gosbee, J.W. and Gardner-Bonneau, D. (1998). The human factor: Systems work better when designed for the people who use them. *Health Care Informatics*, 15(2), 141–144.

Gosbee, J.W. and Gosbee, L.L. (2010). *Using Human Factors Engineering to Improve Patient Safety: Problem Solving on the Frontline* (2nd ed.). Oakbrook Terrace, IL: Joint Commission Resources.

Gosbee, J.W. and Lin, L. (2001). The role of human factors engineering in medical device and medical system errors. In C. Vincent (Ed.) *Clinical Risk Management: Enhancing Patient Safety* (2nd ed.). London, U.K.: BMJ Press.

Gosbee, L.L. (2004). Nuts! I can't figure out how to use my lifesaving epinephrine auto-injector. *The Joint Commission Journal on Quality and Safety*, 30(4), 220–223.

Isaacson, J.J., Klein, H.A., and Muldoon, R.V. (2001). Prescription medication information: Improving usability through human factors design. In M.S. Wogalter, S.L. Young, and K.R. Laughery (Eds.) *Human Factors Perspectives on Warnings* (Vol. 2, pp. 104–108). Santa Monica, CA: Human Factors and Ergonomics Society.

Jacko, J.A. and Sears, A. (2003). *The Human–Computer Interaction Handbook: Fundamentals, Evolving Technologies, and Emerging Applications*. Mahwah, NJ: Lawrence Erlbaum Associates, Inc.

Kaufman, D.R., Thompson, T.R., Patel, V.L, Page, D.L., and Kubose, T. (2003). Using usability heuristics to evaluate patient safety of medical devices. *Journal of Biomedical Informatics*, 36, 23–30.

Ladak, S.S., Chan, V.W., Easty, T., and Chagpar, A. (2007). Right medication, right dose, right patient, right time, and right route: How do we select the right patient-controlled analgesia (PCA) device? *Pain Management Nursing*, 8(4), 140–145.

Lehto, M.R. (2001). Determining warning label content and format using EMEA. In M.S. Wogalter, S.L. Young, and K.R. Laughery (Eds.) *Human Factors Perspectives on Warnings* (Vol. 2, pp. 143–146). Santa Monica, CA: Human Factors and Ergonomics Society.

Lin, L., Isla, R., Harkness, H., Doniz, D., Vicente, K.J., and Doyle, D.J. (1998). Applying human factors to the design of medical equipment: Patient-controlled analgesia. *Journal of Clinical Monitoring and Computing*, 14, 253–263.

Lopez, R., Chagpar, A., White, R., Hamill, M., Trudel, M., Cafazzo, J.A., and Logan, A. (2009). Usability of a diabetes telemanagement system. *Journal of Clinical Engineering*, 34(3), 137–151.

McLaughlin, R.C. (2003). Redesigning the crash cart: Usability testing improves one facility's medication drawers. *American Journal of Nursing*, 103(4), 64A–64F.

Nielsen, J. (1993). *Usability Engineering*. Cambridge, MA: AP Professional.

Nielsen, J. and Mack, R.L. (1995). *Usability Inspection Methods*. New York: Wiley.

Reiling, J. and Chernos, S. (2007). Human factors in hospital safety design. In P. Carayon (Ed.) *Handbook of Human Factors and Ergonomics in Health Care and Patient Safety* (pp. 275–286). Mahwah, NJ: Lawrence Erlbaum Associates.

Reingold, S. and Kulstad, E. (2007). Impact of human factor design on the use of order sets in the treatment of congestive heart failure. *Academic Emergency Medicine*, 14(11), 1097–1105.

Rubin, J.J. and Chisnell, D. (2008). *Handbook of Usability Testing: How to Plan, Design, and Conduct Effective Tests*. Indianapolis, IN: Wiley.

Schneider, P.J. (2002). Applying human factors in improving medication-use safety. *American Journal of Health-System Pharmacy*, 59(12), 1155–1159.

Shapiro, R.G. and Fox, J.E. (2001). Games to explain human factors. In *Proceedings of the Human Factors and Ergonomics Society 45th Annual Meeting*. Santa Monica, CA: Human Factors and Ergonomics Society.

Sojourner, R.J., Aretz, A.J., and Vance, K.M. (1993). Teaching an introductory course in human factors engineering: A successful learning experience. In *Proceedings of the Human Factors and Ergonomics Society 33rd Annual Meeting*. Santa Monica, CA: Human Factors and Ergonomics Society, pp. 456–460.

Stalhandske, E., DeRosier, J., Patail, B., and Gosbee, J.W. (2003). How to make the most of failure mode and effect analysis. *Biomedical Instrumentation & Technology*, 37(2), 96–102.

U.S. Department of Veterans Affairs (2002). Pilot Comparison Usability Test of Intravenous Infusion Pumps. Technical Report. Ann Arbor, MI: VA National Center for Patient Safety.

U.S. Food and Drug Administration (1998). Human factors implications of the new GMP rule: New quality system regulation that apply to human factors. In *Selections of Center for Devices and Radiologic Health Guidance Documents*. Washington, DC: U.S. Food and Drug Administration (online only).

Uzawa, Y., Yamada, Y., and Suzukawa, M. (2008). Evaluation of the user interface simplicity in the modern generation of mechanical ventilators. *Respiratory Care, 53*(3), 329–337.

Vignaux, L., Tassaux, D., and Jolliet, P. (2009). Evaluation of the user-friendliness of seven new generation intensive care ventilators. *Journal Intensive Care Medicine, 35*(10), 1687–1691.

Ward, J. and Clarkson, J. (2007). Human factors engineering and the design of medical devices. In P. Carayon (Ed.) *Handbook of Human Factors and Ergonomics in Health Care and Patient Safety* (pp. 367–382). Mahwah, NJ: Lawrence Erlbaum Associates.

Welch, D.L. (1998). Human factors in the health care facility. *Biomedical Instrumentation and Technology*, May-June, 32(3), 311–316.

Wickens, C.D. (1992). *Engineering Psychology and Human Performance* (2nd ed.). New York: Harper Collins.

Wiklund, M. (2005). *Designing Usability into Medical Products*. Boca Rotan, FL: CRC Press.

Wogalter, M.S., DeJoy, D.M., and Laughery, K.R. (1999). *Warnings and Risk Communication*. London, U.K.: Taylor & Francis.

Wogalter, M. & Frantz, J.P. (2005). *Handbook of Warnings*. London, U.K.: Taylor & Francis.

33

Medical Simulation

Mark W. Scerbo and Brittany L. Anderson

CONTENTS

33.1 Benefits of Simulation Training 557
33.2 Simulation Training in Health Care.. 558
33.3 History of Medical Simulators 559
    33.3.1 Full Body Mannequins 559
    33.3.2 Virtual Reality 560
33.4 Other Forms of Simulation 561
    33.4.1 Hybrid Simulators 561
    33.4.2 Anatomical Models/Systems 561
    33.4.3 Immersive VR Collaborative Training Systems 562
    33.4.4 Standardized Patients 563
33.5 Uses of Medical Simulation... 563
33.6 Human Factors Contributions 564
    33.6.1 Team Training 564
    33.6.2 Fidelity.. 564
    33.6.3 Haptics and Multimodal Perception.. 565
    33.6.4 Workload and Fatigue 566
33.7 Conclusion 567
References.. 568

Over the last 15 years, there has been tremendous interest in simulation within the medical and allied health care communities. Historically, physicians and other health care providers learned their procedures and skills at the bedside, by practicing on patients. However, publication of To Err is Human by the Institute of Medicine (Kohn et al., 1999) unveiled the significant fatality risk faced by patients in U.S. hospitals and set into motion unprecedented efforts by law makers, federal agencies, and medical associations to increase patient safety, increase transparency, and eliminate wasteful spending. Similar efforts have been initiated outside the United States in Australia, Canada, and countries in Europe. Consequently, there has been increased interest among educators and practitioners in the need to train medical students, residents, and other health care providers prior to practicing on real patients and they have begun to embrace simulation-based training as an important means toward that end (Dawson, 2006; Issenberg et al., 1999; Reznick and MacRae, 2006).

This chapter presents an overview of medical simulation. Much of it is devoted to medical simulators, that is, the technology, but it also covers issues surrounding the use of simulation, that is, the process. Also, the term "medical simulation" is used throughout much of the chapter because most of the technology addresses medical training; however, the technology is not limited to medicine. It is used across a wide variety of health care specialties and new simulation technology is now being developed specifically for the allied health professions.

The chapter begins with a discussion of the need for medical simulation. This is followed by a brief development history of different forms of medical simulation technology. The chapter ends with a presentation of some examples where human factors issues are impacting the evolution of this technology.

33.1 Benefits of Simulation Training

The advantages of simulation-based training are well known in other high-risk domains including aviation, the military, and nuclear plant operations (Blackburn and Sadler, 2003; Swezey and Andrews, 2001). Simulators provide a safe environment for learning and practicing risky procedures and the safety benefits extend beyond the trainee to instructors and other participants. Simulators can also provide a cost-effective means for training. For example, in the military the costs associated with live exercises using actual ground vehicles, aircraft, and artillery far exceed those of developing and utilizing simulators for training (Hayes and Singer, 1989). Simulators also expand opportunities for training: they can be available 24 h a day, 7 days a week.

Simulators also offer several training advantages that are difficult to achieve in operational settings (Farmer et al., 2003; Flexman and Stark, 1987). They can provide trainees with timely and objective performance feedback and increase opportunities for the type and pace of training to be matched to individual needs. Simulators can compress or expand the time frame needed to acquire skills. They also provide opportunities to expose trainees to rare or unusual events and circumstances. Simulators can be used by novices to acquire fundamental skills, by those at intermediate levels to maintain skills, or by experts to rehearse difficult procedures and missions. Simulators can facilitate hypothesis testing and allow users to engage in "what if" scenarios and evaluate alternative options. Most important, simulators can be designed to augment the learning experience by providing critical analyses of performance, addressing training requirements of both individuals and teams.

33.2 Simulation Training in Health Care

There are some advantages of simulators that are unique to medicine (Dawson, 2006; Reznick and MacRae, 2006). The most obvious advantage is that simulators facilitate learning and training without putting patients at risk. In fact, Ziv et al. (2003) argue that there is an ethical imperative to utilize simulation-based training whenever possible. Simulators permit trainees to make mistakes and learn from them, again without any consequences for the patients (Fritz et al., 2008; Satava, 2008).

Although there are many advantages to medical simulation, there are also some drawbacks that continue to impede its acceptance and adoption (Fritz et al., 2008). Depending on the type of simulator and the equipment necessary to run the simulation scenario, the cost can be high. Incorporating simulation-based training into a medical curriculum is a complex process. It takes time to learn, operate, and maintain the simulation technology, to write curricula, and to manage the resources and flow of trainees which places additional burdens on the teaching staff (Katz et al., 2010; Nehring and Lashley,

2004). Negative transfer of training can occur when a student learns a procedure incorrectly on the simulator and then transfers that knowledge to the real-world environment (Fritz et al., 2008). This is not a trivial concern, because negative transfer of skills acquired on the simulator could cause greater harm than if those skills had been acquired without simulation. In addition, the simulator systems that are currently available address just a small fraction of the training needs within any medical curriculum. Thus, the availability of simulators and acceptance of the technology varies across medical specialties. Simulation-based training is quickly becoming the norm within anesthesiology, trauma, and surgery, but is still rare in many of the high-risk specialties (e.g., neurosurgery). It is also true that the medical simulator industry is new and highly volatile. Thus, the lifespan of new systems can be short and one cannot always count on continued support for systems once they are purchased.

In spite of these disadvantages, medical simulation technology is currently creating a paradigmatic shift in health care education (Gallagher et al., 2005). Professional societies focused solely on medical simulation have emerged. The Society in Europe for Simulation Applied to Medicine (SESAM) was established in 1994. Ten years later, the Society for Medical Simulation (later renamed the Society for Simulation in Healthcare), a multidisciplinary professional organization for educators, practitioners, and researchers concerned with simulation was established in the United States. In 2006, the Society published the first peer-reviewed journal dedicated solely to simulation, *Simulation in Healthcare*, with David Gaba, one of the pioneers of medical simulation, as its editor-in-chief. The Advanced Initiatives in Medical Simulation organization was established in 2003 to raise awareness among U.S. law makers about the need for medical simulation to improve patient safety, reduce medical errors, and reduce health care costs (see http://www.medsim.org/index.php). The Medical Modeling and Simulation Database (MMSD; http://www.medicalmodsim.com) was established in 2008 by Eastern Virginia Medical School in collaboration with the American College of Surgeons (ACS) to provide a comprehensive database of all aspects of simulation in medicine for educators, practitioners, and researchers (Combs et al., 2006).

Moreover, simulation training is moving from novelty to necessity as more professional organizations have started to address it in their policies. For instance, the American Board of Anesthesiology presently requires candidates working on their Maintenance of Certification in Anesthesiology to complete a course in simulation as part of the program (see http://www.theaba.org/Home/anesthesiology_maintenance). In 2006, the ACS established a program to accredit simulation centers (Sachdeva et al., 2008). In the United States, the Accreditation Council for Graduate Medical Education (ACGME; http://www.acgme.org/acWebsite/home/home.asp) includes 26 specialty programs, three of which mention simulation in their requirements for graduate medical education. Specifically, the internal medicine and surgery programs require that the residents have access to simulation training, while the radiation oncology program requires the residents to perform cases using simulation.

33.3 History of Medical Simulators

The use of models in medical training actually has a long history (Bradley, 2006). Macedonia et al. (2003) reviewed records indicating ancestors of Siberian Mansai people utilized leather models to aid in teaching birthing techniques. In the fourteenth century, wax anatomical models were created from cadavers by injecting them with liquid wax (Haviland and Parish, 1970). The use of cadavers for dissection started about 500 years ago (Dyer and Thorndike, 2000). In the late eighteenth century, Felice Fontana carved a life-size wooden anatomical model of a man with 3000 moving parts (Olry, 2000). The parts could also be changed out to create a female version. By the twentieth century, manufacturers began producing anatomical models made of plastic, rubber, and latex. Aside from these physical models, surgeons have historically used a different form of simulation. Surgeons have routinely used living animals to practice their skills (Satava, 2008). Although animal models differ from humans anatomically, they permit surgeons to practice cutting and suturing on tissue in a living organism.

33.3.1 Full Body Mannequins

Mannequin simulators are life-sized physical representations of the human body and portions of its anatomy (see Figure 33.1) and are typically used for training procedures in anesthesiology, trauma, and emergency medicine (Cooper and Taqueti, 2004; Fritz et al., 2008; Lind, 2007; Satava, 2008). One of the first mannequins was *Resusci®-Anne* developed by Åsmund Lærdal in the 1950s for training life-saving procedures. *Resusci-Anne* is often recognized as the first modern day simulator and the catalyst for the development of medical simulation (Bradley, 2006; Cooper and Taqueti, 2004; Hunt et al., 2006). *Resusci-Anne* was created to teach the ABCs (airway, breathing, and circulation) of cardiopulmonary resuscitation (CPR) developed by Peter Safar (Cooper and Taqueti, 2004; Finn and Soar, 2005; Fritz et al., 2008; Grenvik and Schaefer, 2004; Lind, 2007). The

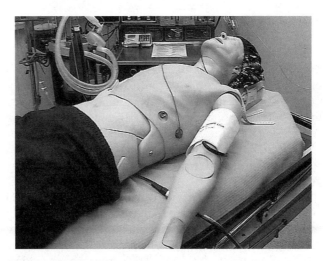

FIGURE 33.1
The SimMan full body mannequin by Laerdal, Inc.

original model allowed a trainee to practice mouth-to-mouth breathing. Later, a spring was added to the chest to allow trainees to practice all steps of CPR (Hunt et al., 2006). Modern versions of *Resusci-Anne* are still in use today.

Although *Resusci-Anne* was a successful training aid for CPR, it had very limited functionality. Developers sought to harness the power of emerging computers to create systems that would simulate more complex procedures. Stephen Abrahamson and Judson Denson created Sim One along with Sierra Engineering and the Aerojet General Corporation to teach anesthesia residents how to properly place an endotracheal tube (Abrahamson et al., 2004; Bradley, 2006; Cooper and Taqueti, 2004; Fontaine et al., 2009; Fritz et al., 2008). Sim One was the first computer-controlled full-scale mannequin simulator. It produced chest movements, eye blinks, and pupil dilation, responded to different gases and drugs, and opened and closed its jaw. Unfortunately, Sim One never became a commercial success because it was extremely expensive to build and operate and the medical community at that time did not perceive a need for an alternative form of training (Bradley, 2006; Cooper and Taqueti, 2004; Fontaine et al., 2009).

Shortly after Sim One was developed, Michael Gordon introduced the Harvey Cardiology Patient Simulator (Gordon, 1974). The Harvey system was designed to simulate a variety of physiological parameters such as heart and lung functions, venous and arterial pulses, as well as deviations from normal readings representing cardiac disease (Cooper and Taqueti, 2004). Harvey was one of the first mannequin simulators to be evaluated for its training benefits and it was shown to be comparable to the standard form of training for that time period. Harvey has continued to evolve and presently can exhibit different blood pressures, murmurs, and heart sounds. Harvey is commercially available today from Laerdal, Inc.

Significant improvements in the functionality of mannequin-based simulators required better physiological models. Philip (1986) created a computer program for instruction on the effects of anesthetic agents in the body. Fukui and Smith (1981) created SLEEPER, a computer-based tool for teaching basic physiology and pharmacology. The original version of this program required significant computing power; however, the version called BODY™ can be run on a standard PC.

At about the same time, Gaba and DeAnda (1988) created a mannequin-based simulator system to study training and performance issues in anesthesiology (although their mannequin was not a full body, just the head and thorax; Gaba, 2006). The comprehensive anesthesia simulation environment (CASE) incorporated simulated and genuine anesthesiology equipment and systems set up in a real operating room. Their system also utilized a computer-based physiological program so that the simulated patient's vital signs could be monitored in real time. The CASE system was licensed to CAE-Link, an aviation simulation company, and then MedSim, Ltd, but is no longer in production (Bradley, 2006; Cooper and Taqueti, 2004; Fritz et al., 2008). In the late 1980s, Good and Gravenstein also developed a mannequin called the Gainesville anesthesia simulator (GAS), which eventually became commercialized as the Human Patient Simulator by Medical Education Technologies (METI) Inc. Ironically, Lærdal, Inc., the company, which introduced *Resusci-Anne,* would not offer a computer-based mannequin until 2000 when they began marketing SimMan®. Unlike the other mannequin systems that incorporated models of underlying physiology, the SimMan system is largely driven by instructor scripts.

The CASE system developed by Gaba and his colleagues represents a unique departure from their predecessors. The CASE system was not developed as a technical demonstration, but instead was created to gain a better understanding of how individuals responded to critical events. Gaba and his colleagues used their system to apply and evaluate the efficacy of crew resource management concepts from aviation in anesthesiology. In this respect, it is one of the earliest examples of using simulation to evaluate human factors concepts in medicine.

33.3.2 Virtual Reality

An entirely different approach to medical simulation began to emerge in the 1990s. The development of virtual reality (VR) systems ushered in new ways to visualize information. In 1990, Delp et al. created an interactive model of the lower leg designed as a visual aid for orthopedic surgeons. Shortly thereafter, Satava (1993) described a VR simulation of internal abdominal organs that could be explored from any angle. In the mid-1990s, the National Library of Medicine sponsored the Visible Human project (Ackerman, 1998) in which a searchable database of two entire human bodies (male and female) was created from CT, MRI, and phototomography processes. The visible human is currently available to anyone on the Internet at http://www.nlm.nih.gov/research/visible/visible_human.html. Later, Delp et al. used a portion of the Visible Human to build a visual simulation of trauma from a gunshot to the upper leg (Satava and Jones, 2002). Satava and Jones (2002) describe the Visible Human project as a pivotal step in moving medicine into the information age by laying the foundation for a patient-specific holomor. In fact, by the late 1990s, two VR simulator systems had been created for surgical planning based on a patient's own CT scans (Levy, 1996; Marescaux et al., 1998).

A major benefit of VR systems is that procedures can be practiced repeatedly without cutting or altering physical components, that is, unlike mannequins, they do not require replaceable parts. VR simulators also allow students to interact with the simulator and receive objective feedback in real time (Satava, 2001). Thus, the virtual approach has been widely adopted and there are now systems that address a variety of procedures including endoscopic surgery, vascular and radiology procedures, and even sonography.

Interest in VR simulators for surgery was fueled in part by the widespread adoption of the laparoscopic or minimally invasive method. Unlike traditional surgical procedures where the surgeon operates through a single large incision made in the body, laparoscopic procedures are performed from outside the body through several small incisions in which a mini video camera and surgical instruments are inserted. The surgeon monitors his/her actions from the images on a video display.

Minimally invasive surgery was quickly embraced during the late 1980s as a safer way to operate. Patients enjoyed quicker recoveries and better cosmetic outcomes. Unfortunately, these benefits came at a cost to the surgeon. This type of surgery is more challenging for the surgeon because it creates new perceptual, spatial, and psychomotor demands (Gallagher et al., 2003). For instance, the surgeon must translate information presented on a 2D monitor into a 3D operational space and acclimate to the fulcrum effect that occurs when an instrument is moved in one direction but its image on the display appears to move in the opposite direction. In addition, laparoscopic surgery also imposes physical ergonomic stresses on the surgeon due to the shape of the instruments, the positions of the video monitor and foot controls, and even the height of the operating table (Berguer, 1999; Matern, 2004). Research also began to show that inexperienced laparoscopic surgeons were more likely to injure patients (Conner and Garden, 2006; Deziel et al., 1993; Moore and Bennett, 1995).

Consequently, there was a critical need to provide training on something other than patients. Initially, this need was served by laparoscopic box trainers consisting of a box-like structure and camera/video system that displays images from inside the box on a monitor. Standard laparoscopic instruments are used to manipulate various objects within the box. However, VR represented a promising alternative to box trainers. Laparoscopic procedures are amenable to VR simulation because the surgeon no longer views the patient's actual anatomy. Instead, the patient's anatomy is represented on a video display. Consequently, this enabled developers to create computer-generated representations of anatomy, tissue, and laparoscopic instruments.

The first VR system for laparoscopic surgery was the MIST VR (Sutton et al., 1997), which combined a laparoscopic interface with a standard PC running a simple graphical 3D representation of laparoscopic instruments interacting with geometric shapes. Users performed basic psychomotor tasks akin to tasks required in laparoscopic procedures, for example, target acquisition and placement, target transfer, traversal, target diathermy, etc. The MIST VR system (available from Mentice AB) became the first commercially successful laparoscopic VR simulator and today, is probably the most extensively studied simulator of its kind (Seymour, 2008; Stone, 1999).

Many other VR systems for laparoscopy are available today and are considered either part-task or whole-task trainers. A part-task trainer allows individuals to practice parts of a task or abstract representations of surgical tasks, rather than the whole task. For example, the MIST VR system does not provide training on a specific laparoscopic procedure. Instead, as noted earlier, trainees acquire the fundamental psychomotor skills necessary for laparoscopic surgery by performing a variety of representative task components. The advantage of a part-task trainer is that individuals can focus on basic skills or challenging subtasks that are not easily isolated within a full task sequence.

On the other hand, a whole-task trainer allows users to practice an entire procedure or major portions of the procedure. The primary advantage of a whole-task trainer is that individuals can learn the entire sequence of steps needed to perform a procedure. The LapSim® system is an example of a laparoscopic trainer that provides both part-task training on fundamental skills and whole-task training on procedures such as cholecystectomy (see Figure 33.2).

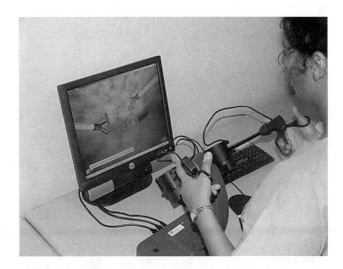

FIGURE 33.2
The LapSim laparoscopic simulator by Surgical Science. (From Schmidt, E.A. et al., Task sequencing effects for open and closed loop laproscopic skills. ISO Press, copyright retained by authors [Scerbo].)

33.4 Other Forms of Simulation

33.4.1 Hybrid Simulators

Hybrid simulators are those that incorporate aspects of VR computer-based systems into mannequin simulators (Satava, 2001). One example is the VIST system (Mentice, Inc.), a simulator designed for interventional cardiology procedures developed by Dawson et al. (2000). The system combines a physical representation of the body with synthetic fluoroscopy and a real-time 3D interactive anatomical display. The system also allows users to choose among different catheters and guide wires and insert the physical devices into the simulator. A representation of devices appears in the virtual images a radiologist or cardiologist would use for an interventional procedure. This hybrid system allows trainees to perform right- or left-sided coronary catheterization and angiography procedures using actual catheters.

The VIST system represents another major milestone in the evolution of medical simulation. In 2004, the U.S. Food and Drug Administration (FDA) recognized the necessity of simulation training for at least one complex procedure, deploying a carotid stent device. They took an unprecedented action and mandated that surgeons meet specific training criteria on the VIST system before they would be reimbursed for performing the procedure on patients (Gallagher and Cates, 2004).

33.4.2 Anatomical Models/Systems

Health care workers have used models made of rubber or silicone for many years to learn how to palpate

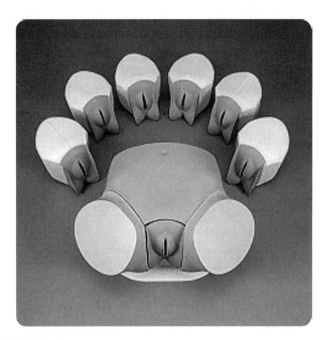

FIGURE 33.3
A female pelvic trainer with interchangeable uterine modules by Limbs and Things.

the body for abnormal structures in underlying tissue (e.g., breast, pelvic, or prostate examinations). Many of these models have interchangeable parts to convey abnormalities of different size or stages of disease (see Figure 33.3). Although these models can help learners make gross distinctions between normal and abnormal conditions, they provide very little guidance about the quality of the one's examination.

A more sophisticated model is the TraumaMan® mannequin simulator by Simulab, Inc (see Figure 33.4). The TraumaMan system is a realistic anatomical model of

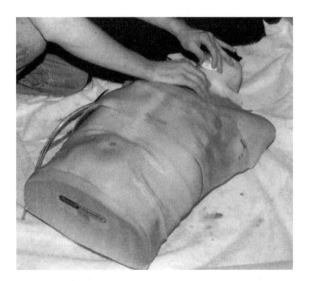

FIGURE 33.4
Resident performing a cricothyroidotomy on the TraumaMan by Simulab, Inc.

the neck, chest, and abdomen that has replaceable tissue and fluid components. It is often used for Advance Trauma Life Support (ATLS®) training and allows students to perform procedures on a realistic representation of human anatomy. However, like anatomical models, this type of mannequin simulator does not provide any means of objective performance assessment. Thus, an experienced physician must assess trainee performance by means of subjective assessments.

Pugh and Youngblood (2002) addressed this limitation for the female pelvic examination with the creation of the E-Pelvis. They outfitted a pelvic mannequin with force sensors so that they could measure which locations trainees touched, how often different areas were touched, and how much pressure they exerted. Consequently, their system could provide objective information about the accuracy and thoroughness of trainee examinations.

Gerling et al. (2009) and Wang et al. (2010) extended this approach for prostate examinations by adding a critical element of control. They embedded four to six water-filled bladders into the prostate models. A computer was used to inflate and deflate the bladders so that the type, size, and location of an abnormality could be specified precisely. Force sensors were used to record the location, duration, and amount of pressure applied during the exam. The investigators measured the performance of a sample of health care providers on several prostate configurations and were able to identify the type of finger movements examiners used to detect the most abnormalities and the most challenging configurations.

33.4.3 Immersive VR Collaborative Training Systems

Several VR-based simulations have been developed for team-based problem solving and decision making (Johnston and Whatley, 2006). Youngblood et al. (2008) developed a desktop online 3D virtual emergency department trauma bay. Users log on and are represented by an avatar. They use a mouse to navigate within the room and to select actions to perform, but communicate with other team members with headsets using an Internet voiceover protocol. The investigators compared their VR system to training with a mannequin and found that both methods produced similar levels of learning.

One benefit to this approach is that trainees do not need to be physically colocated to engage in a team exercise. Alverson et al. (2008) conducted an evaluation of interactive problem solving using text-based and VR-integrated exercises with students in New Mexico and Hawaii. They found that the VR component could be successfully incorporated into the live training sessions over these distances.

Scerbo et al. (2007) have developed a fully immersive virtual operating room modeled after a standard OR and rendered in a 10 × 10 ft cave automatic virtual environment (CAVE; see Figure 33.5). Their system differs from

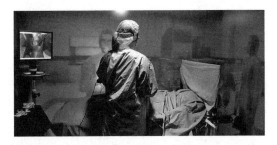

FIGURE 33.5
Student interacting with a virtual agent in a virtual operating room.

the other VR environments in that participants perform procedures on a standard mannequin and VR simulators placed within the CAVE. In addition to the virtual OR environment, virtual agents were created for the attending surgeon, anesthetist, and circulating nurse. The virtual surgical team members are distinguished by their knowledge and personality. Participants interact with the virtual agents using speech recognition software. Thus, the system augments existing medical simulator technology by providing an OR environment and a dynamic social context.

Others have begun to explore the use of Second Life for health care applications (Boulos et al. 2007). At present, much of the content on Second Life is aimed at patient education and support (Beard et al., 2009), but applications for medical education are appearing. Weicha et al. (2010) used Second Life to provide instruction on insulin therapy for continuing medical education (CME) credit.

33.4.4 Standardized Patients

Aside from manufactured forms of simulator systems, the other major type of simulation used extensively in health care is the standardized patient (SP). An SP is a normal, healthy individual trained to portray a patient. In addition, SPs can perform two other important functions. They are trained to assess physician–patient interactions according to standardized criteria and can provide immediate feedback to students about the quality of their interactions addressing their history taking, communication, and even physical examination skills (Miller, 1990). SPs are also used to assess nurses and other health professionals.

SPs were first used by Barrows in the 1960s to provide a more efficient means to evaluate each medical student's ability to take a patient history and perform a physical examination (Barrows and Abrahamson, 1964). Barrows soon began hiring additional people to not only portray a wider range of cases for his students, but also to provide him with feedback about their performance (Barrows, 1993). Although Barrows continued to develop and expand the role of SPs in his work, the rest of the medical community was slow to adopt this form

of training and assessment. It was not until the 1980s that medical educators accepted SPs as a viable method for evaluating the clinical competency of students in their formal clinical practical examination (Boulet et al., 2009). Since then, SPs have become a required component of undergraduate medical training. In 1991, the Liaison Committee on Medical Education (1991) recognized the importance of SPs for assessing clinical skills. At present, they are used in exams to certify and license physicians such as the Objective Structured Clinical Examination for medical students and in the National Board of Medical Examiners United States Medical Licensing Examination Step 2 Clinical Skills for assessing clinical competency with or without supervision and the Medical Council of Canada Qualifying Examination Part II (Boulet et al., 2009).

One problem with SPs is that they are usually healthy individuals. Thus, they are severely limited in the range of underlying pathologies or conditions they can convey during a physical examination. Some attempts have been made to extend their capabilities through moulage. McKenzie et al. (2006) used an alternative method to extend SP capabilities. They created an auditory form of augmentation for SPs, by modifying an electronic stethoscope so that it could play prerecorded abnormal heart and lung sounds. The sounds were not substitutes, but instead were combined with those of the SP's physiology. Thus, the stethoscope could be used to perform a normal exam or augmented with the prerecorded audio files for specific learning scenarios.

33.5 Uses of Medical Simulation

Today, simulation is used primarily for training and education. However, simulation also serves other purposes. It can be used for the assessment of procedural and communication skills (Gaba, 2004). Simulation can also be used to rehearse an actual procedure prior to performing it on a real patient (Bradley, 2006). This can be especially beneficial for complex and rare procedures. Hunt et al. (2006) also advocate the use of simulation for forensic testing. In the event that an actual medical procedure resulted in a negative outcome, the scenario could be recreated through simulation to determine what went wrong and how to improve performance in the future. Further, Bradley suggested that simulation can also be utilized to test new equipment and as a form of refresher training for staff.

Gaba (2004) described the breadth of health care simulation applications in 11 dimensions. Those dimensions can be organized into three higher-level areas (see Table 33.1). The first concerns the *goals* for using simulation.

TABLE 33.1

Dimensions of Simulation Applications

	Simulation Applications
Goals	Purpose, domain, skills, and patient age
User characteristics	Unit, experience, discipline, and profession
Method of implementation	Technology, site, participation, and feedback

Source: Gaba, D.M., *Quality & Safety in Health Care*, 13(suppl. 1), i2–i10, 2004.

The main objective or purpose could be education, training, assessment, rehearsal, or research. The application can address different domains such as surgery, primary care, or emergency medicine. Further, the specific aim could be focused on knowledge, technical skills, problem solving, decision making, or attitudes. Last, the age of the patient is important because training for some ages (e.g., neonates, children, the elderly) may be aimed at different specialties or require that the users have some unique knowledge. The second area concerns trainee or *user characteristics*. The simulation may be targeted to individuals, teams, or even organizations. Also, the training may be designed for novices, medical students, residents, or experts. Further, the application may target physicians, nurses, aids, or even management. The third area concerns the *method of implementation*. The simulation might use mannequins, VR, SPs, or combinations of simulation systems. The training might take place in a stand-alone facility at a medical school or hospital, on site in a clinical area, out in the field, in the office, or at home. The level of participation can vary from passive observation, to verbal interaction, to hands-on participation, to full immersion. Last, the feedback can be given in real time or delayed, by means of instructor critique, in a posttraining debriefing session, or withheld altogether. Obviously, these 11 dimensions do not fully capture all of the current simulation applications and would need to be modified as newer forms of technology impact and augment training capabilities. They do, however, reflect the broad scope of simulation training in health care that has emerged over a relatively short period of time.

33.6 Human Factors Contributions

In 2005, Scerbo described the current trends in medical simulation and argued that there were many areas where human factors professionals could make important contributions while this technology was fairly new and still evolving (Scerbo, 2005). Since then, there has indeed been a growing relationship between the human factors and medical simulation communities. Space

restrictions do not permit a thorough review of these efforts, but a few examples are noted in the following sections.

33.6.1 Team Training

Team training is an important component in health care because much of the work that is performed is done in a team setting. In the 1980s, the aviation industry recognized the need to train pilots as a crew rather than training them solely as individuals (Gaba et al., 2001). A form of team training called crew resource management (CRM) was developed and is presently required as part of the pilot training process. The medical community began to borrow principles of CRM and apply them in medicine. David Gaba first applied the principles of CRM to training anesthesia teams in 1989 and has continued to study team dynamics in anesthesia based on CRM principles (Gaba, 2010; Holzman et al., 1995). Soon after, physicians in other specialties such as emergency medicine and intensive care also began to apply CRM principles in their teams (Lighthall et al., 2003; Reznek et al., 2003).

Today, there are formal training programs designed specifically for the health care community. One example is TeamSTEPPS developed by the Department of Defense (DoD) and the Agency for Healthcare Research and Quality (AHRQ) to train military and civilian medical teams (see http://health.mil/dodpatientsafety/ProductsandServices/TeamSTEPPS.aspx). TeamSTEPPS provides training in communication, leadership, mutual support, and situation monitoring based on evidence from team research across a variety of domains.

Human factors professionals have had a significant impact in this area of health care training, largely due to successes in other team contexts. Researchers have studied team performance using medical simulation in emergency medicine and labor and delivery environments (Birnbach and Salas, 2008; Fernandez et al., 2008). Recently, Rosen et al. (2008) described a SMARTER-Team, in which they adapted TeamSTEPPS to specifically include simulation for team training.

It is not possible to address the breadth of research related to health care teams in this chapter (good reviews can be found in Baker et al., 2006; Salas et al., 2006). However, there is clearly a need for more research on the impact of team training on patient safety outcome measures across a variety of health care settings.

33.6.2 Fidelity

A major issue facing developers of medical simulators concerns fidelity. Hayes and Singer (1989) described simulation fidelity with respect to the similarity between the training situation and the operational situation. Fidelity can also be thought of in physical terms (e.g., the similarity between the equipment, materials, and the operational environment and those represented in the simulation) and functional terms (e.g., how objects behave in the simulation). Dieckmann et al. (2007) recently described fidelity and realism in physical, semantical, and phenomenal terms. Specifically, the physical mode concerns the objective characteristics (e.g., color, texture, duration, etc.). The semantical mode concerns simulation on a conceptual level and transcends physical characteristics (e.g., conveying elevated heart rate via display device or direct contact with a mannequin). The phenomenal mode concerns the unique emotional and metacognitive experiences of individuals engaged in simulation.

Designers and educators in the medical simulation community continue to debate over how much fidelity is necessary. Scerbo and Dawson (2007) argued that high fidelity in simulation does not always lead to better performance and can even interfere with performance. In other disciplines (e.g., aviation, military, etc.) data have existed for many years showing that the benefits of training with simulation have less to do with fidelity than fundamental principles of learning that address issues such as feedback, practice, and instructional guidance (Farmer et al., 2003). Recently, Issenberg et al. (2005) examined the medical simulation literature for features of high fidelity that lead to effective learning and their findings corroborate those from other disciplines. They found that the most frequently mentioned sources of efficacy were: educational feedback, repetitive practice, and curriculum integration, but not realism. Consequently, simulation-based training systems should be developed to maximize their effectiveness, not their fidelity. As Kneebone (2006) suggested, one should strike a balance between the needs of the learners, teachers, and resources available. Further, different levels within the physical, semantical, and phenomenal modes of fidelity may be needed depending upon the goals and experience levels of the trainees.

33.6.3 Haptics and Multimodal Perception

The development of VR surgical simulators has been facilitated by the integration of haptic force feedback systems, particularly for laparoscopic and endoscopic procedures because the surgeon does not touch the patient's anatomy directly. All contact with the patient's organs and tissue is felt indirectly through the laparoscopic instruments and haptic force feedback systems can simulate those sensations.

One example is the Phantom Omni® by SensAble, Inc. (see Figure 33.6), which is essentially a robotic arm that the user can move freely within a restricted 3D area. Movement of the arm can be constrained by

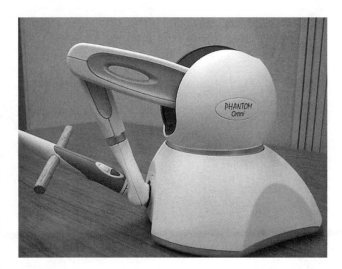

FIGURE 33.6
The Phantom Omni haptic force feedback system by SensAble Technologies.

instructions from software applications using collision detection routines that inform the device of the location, geometry, and density of objects within the operational space. When the end point of a virtual tool makes contact with a virtual object, the collision detection algorithms apply forces to the arm of the haptic device to constrain its movement according to the properties of the virtual object. Thus, a dense object will permit little movement of the arm while an elastic object will render the arm progressively more difficult to move the further the user pushes it. The collision algorithms are fairly easy to render in a point-based haptic system and therefore lend themselves to laparoscopic procedures because much of the activity occurs through single points of contact. Thus, the user can experience the haptic sensation of probing or pulling on virtual tissue with simulated laparoscopic instruments.

A variant of this type of haptic interface is used in simulators in which users penetrate tissue with a needle or feed a flexible cable through an orifice. In these systems, the range of movement is restricted to a linear direction and the force feedback mechanism modulates the force required to push the device forward. The Endoscopy AccuTouch® simulator from Immersion Medical, Inc. is an example of a whole-task trainer for colonoscopy. Users can insert a flexible scope, visualize the full length of the colon, photograph suspicious areas, and even biopsy growths. They receive haptic force feedback throughout the procedure giving them cues as to the difficulty of various maneuvers (e.g., when turns and loops of the colon must be navigated).

Scerbo et al. (2009) recently evaluated a simulator for orthopedic bone pinning (SimQuest, LLC) that incorporated two types of haptic feedback (see Figure 33.7). A force feedback system provided haptic information

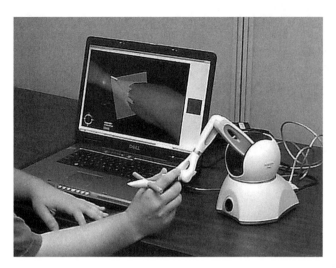

FIGURE 33.7
Student using a haptic force feedback device on an orthopedic bone pinning simulator developed by SimQuest, LLC.

about bone density and resistance as the user pushed the virtual pin through the bone. However, if users exerted excessive force on the pin, they were given a haptic warning in the form of vibrotactile feedback. The researchers found that coupling the two forms of haptic information was detrimental to performance. Instead of providing corrective information about the magnitude of force applied to the pin, the vibrotactile warning frequently caused users to stop pushing the pin altogether and introduced more inaccurate movements when the users started pushing on the pin again.

At present, the haptic models needed to faithfully reproduce the response of soft tissue are still fairly crude. Moreover, some developers and even physicians have questioned the need for high-fidelity haptics or any haptics at all (Montgomery, 2005). From a research perspective, haptic force feedback systems expand the opportunity to study characteristics of haptic and tactile perception because they provide a means to specify and reliably produce stimulus parameters that until recently could only be done with the visual and auditory systems (for example, see Klatzky et al., 2003; Lawrence et al., 2007).

The control over sensations afforded by haptic force feedback systems have opened up new opportunities to study multimodal aspects of perception, for example, how individuals integrate information from multiple perceptual sources or resolve conflicting information across sensory systems. As an example, Biocca et al. (2001) used a virtual cadaver to study cross-modal illusions. Participants were asked to remove organs from a virtual representation of a cadaver. The organs were "attached" to one another and the user's task was to grasp and pull them from the body one at a time. A graphic representation of a spring was shown to the

user to convey the degree of resistance. Participants used a pinch glove to manipulate objects in the virtual scene and when they exerted enough force to overcome a critical value of resistance, the virtual organ would snap into their hand. No haptic force feedback or auditory information was provided to the users, but on completion of the study they were asked if they perceived any sounds or felt any forces of resistance. The results showed that a proportion of users reported experiencing auditory or haptic sensations when they manipulated the objects, but the number differed for modality. Thirty-nine percent of users reported hearing sounds on some trials. By contrast, 73% of the users reported feeling resistance on some trials and 4% reported feeling resistance on every trial. These findings show that a visual representation of haptic resistance can create a compelling illusion of haptic sensation in the absence of genuine haptic stimuli and may provide a reasonable alternative to incorporating haptic force feedback systems in VR surgical systems.

33.6.4 Workload and Fatigue

Many clinicians, and particularly residents, work very long hours, with moderate to severe sleep deprivation, while managing multiple patients that vary in condition and severity. However, there has been little systematic research on workload in the OR because many of the techniques used to study it could put patients at greater risk. Medical simulators offer a reasonable compromise because workload can be manipulated in a safe, controlled environment.

Toward this end, Howard et al. (2003) used simulation to examine how sleep deprivation affects anesthesiologists. They had residents perform a 4h simulated case that included measures of vigilance. Residents were either in a sleep extended condition, where they arrived at work well rested, or in a sleep-deprived condition where they had not been allowed to sleep for 25h. The investigators found that the sleep-deprived residents had reduced psychomotor performance on the vigilance task and they reported increased feelings of sleepiness compared to the extended sleep condition, but there were no significant differences in clinical performance on the simulated task.

Klein et al. (2005) used simulation to assess mental workload in laparoscopy. Specifically, they assessed subjective mental workload using the NASA–TLX (Hart and Staveland, 1988) on a laparoscopic simulator. The NASA–TLX provides an overall index of mental workload as well as the relative contributions of six workload subscales: mental, physical, and temporal task demands; and effort, frustration, and perceived performance. The researchers had participants move foam stars between pegs and manipulated three viewing conditions: direct viewing, viewing on a display, and viewing on a display with 90° rotation. They found that performance was best for direct viewing followed by viewing on a display, which in turn was better than viewing on a display with 90° rotation. The performance effects were partly corroborated by the NASA–TLX scores. Mental workload was rated highest for the rotated viewing condition and rated as moderate for the other two conditions. Of the six subscales, effort and mental demand contributed the most to the workload scores.

More recently, Yurko et al. (2010) reported the results of several studies using the NASA–TLX to examine workload in a complex laparoscopic task (intracorporeal suturing and knot tying). Novices were trained and assessed on the fundamentals of laparoscopic surgery (FLS) simulator curriculum and were later tested on an animal model. Subjective mental workload was measured at baseline, during training, and after performance proficiency levels had been reached. They found that the NASA–TLX scores reflected performance: workload ratings declined with increasing proficiency during training. More important, the workload scores increased with increasing task difficulty (during transfer from the simulator to the operating room) and were positively related to the errors in the OR procedures.

There have been several attempts to assess laparoscopic workload using secondary task measures. Hsu et al. (2008) studied differences between novices and experts on a peg transfer task from the FLS system. The secondary task required participants to perform two-digit addition problems. They found that both groups of participants performed similarly on the peg transfer task, but the novices completed fewer math problems. More recently, Grant et al. (2009) used a time estimation/production task as a secondary measure and found variability for the interval production task was sensitive to workload, particularly at intervals around 11 s.

Stefanidis et al. (2007) also used a secondary task based on Wickens' multiple resource theory (1984). They argued that a secondary task requiring visual-spatial ability should compete for the spatial attentional resources needed for laparoscopic surgery. Their task required participants to monitor a nearby display for a temporal pattern of squares that appeared at random on either the left or right side of the display and to press a foot pedal whenever they detected a specific sequence. They studied four groups with different levels of surgical and simulator experience and had them perform a laparoscopic suturing task along with the spatial secondary task. Their results showed that the more experienced participants achieved higher scores on both the suturing and secondary tasks. However, these measures did not distinguish between experts and residents with extensive experience on the laparoscopic trainer. Instead, a difference between these two groups only emerged on the secondary task.

In a subsequent study, Stefanidis et al. (2008) investigated whether the same secondary task could reveal improvements among novices as a function of practice. They had novices practice laparoscopic suturing in weekly 1 h sessions over the course of 4 months and then perform the suturing task along with the secondary square detection task for 10 min each session. Their results showed significant improvements in suturing skills and secondary task performance for all participants. Although almost all participants achieved predetermined proficiency levels on the suturing task, no one achieved secondary task proficiency. However, longer training times were correlated significantly with higher secondary task scores, showing that extended practice resulted in an improved ability to multitask.

33.7 Conclusion

In his 2004 paper, Gaba offered alternative predictions for the future of medical simulation. In his optimistic outlook, he described a society where simulation training in health care was a requirement, where simulation was a driving force behind changes to health care curricula, where the public demanded levels of safety in health care comparable to those in aviation, and where regulatory agencies required simulation-based standards for training and evidence of efficacy for devices gathered in trials using simulation. Gaba also offered a pessimistic view where interest in simulation within the medical community waned, where practicing clinicians opposed the establishment of simulation-based training standards, and where concerns about health care costs curtailed investment in new simulation technology and research to validate simulation systems.

Six years later, it appears that we are closer to Gaba's optimistic view. Although one could argue that progress has not moved quickly enough, there is no doubt that simulation has already begun transforming health care training and education. From a pedagogical perspective, simulation is providing educators more objective criteria for assessing the competency of their trainees even if it is still not required. Simulation is providing the current cohort of health care trainees many more opportunities to practice and demonstrate competency on devices instead of on patients. Further, simulation still offers patients the possibility that all providers, irrespective of where they were educated and where they practice, meet minimal standards of competency.

However, this is just the beginning. There are tremendous opportunities to expand simulation for training to address underrepresented specialties, the skills of more advanced trainees, and even to rehearse prior to performing genuine procedures. However, there are other areas where simulation can have an important impact in health care. Simulation can be used in rehabilitation and therapy, to study disease at the cellular level or in populations, and can be applied to hospital administration and management activities. In other high-risk domains where simulation is used extensively, it changes how work is performed. Simulation is used to solve problems, to evaluate alternative hypotheses, and to create the next generation of technology that will be used to perform the work.

Often, human factors professionals are heavily involved in the development and evaluation of technology in many of these areas. As an example, Anderson et al. (2008, 2009) have used a simulator to study the ability to detect critical signals (late decelerations) in maternal-fetal heart rate tracings. Specifically, they examined detection performance for signals of different magnitudes embedded in fetal heart rates of different variability. They also looked at differences in critical signal onset times. They have shown that detection performance does indeed drop as the signal-to-noise ratio declines and that performance reaches chance level when signal onset times fall between 8 and 12 s. Moreover, they have shown that it is possible to use simulated tracings to assess one's skill at detecting these critical signals. In another application, Liu et al. (2009) have used simulation to examine how head-mounted display (HMD) technology affects monitoring performance in anesthesiology. They examined the impact of differences in focal point (near or far) on the ability of anesthesiologists to detect events while wearing an HMD. Their findings showed that the HMD did not necessarily improve detection of critical events for either focal point; however, wearing the HMD did change the anesthesiologists' monitoring behavior. The anesthesiologists spent more time looking at the patient than the anesthesia machine.

These research programs are just two examples of the value that a human factors perspective brings to medical simulation research, and the impact is not trivial. The knowledge and methods that human factors professionals have successfully used in other high-risk domains coupled with the availability of simulation technology are beginning to lift the blue curtain from the practice of medicine, by making it more transparent and amenable to scientific study.

References

Abrahamson, S., Denson, J. S., and Wolf, R. M. 2004. Effectiveness of a simulator in training anesthesiology residents. *Quality and Safety in Healthcare* 13:395–397. (Original work published 1969.)

Ackerman, J. M. 1998. The visible human project. *Proceedings of the IEEE* 86:504–511.

Alverson, D. C., Saiki, S. M., Kalishman, S. et al. 2008. Medical students learn over distance using virtual reality simulation. *Simulation in Healthcare* 3:10–15.

Anderson, B. L., Scerbo, M. W., Belfore, L. A., and Abuhamad. A. 2008. Detecting critical patterns in maternal-fetal heart rate signals. *Proceedings of the Human Factors & Ergonomics Society 52nd Annual Meeting*, pp. 1478–1482. Santa Monica, CA: Human Factors & Ergonomics Society.

Anderson, B. L., Scerbo, M. W., Belfore, L. A., and Abuhamad. A. 2009. When is a deceleration perceived as a late deceleration? *Simulation in Healthcare* 4:311.

Baker, D. P., Day, R., and Salas, E. 2006. Teamwork as an essential component of high-reliability organizations. *Health Services Research* 41:1576–1598.

Barrows, H. S. 1993. An overview of the uses of standardized patients for teaching and evaluating clinical skills. *Academic Medicine* 68:443–453.

Barrows, H. S. and Abrahamson, S. 1964. The programmed patient: A technique for appraising student performance in clinical neurology. *Journal of Medical Education* 39:802–805.

Beard, L., Wilson, K., Morra, D., and Keelan, J. 2009. A survey of health-related activities on second life. *Journal of Medical Internet Research* 11:e17.

Berguer, R. 1999. Surgery and ergonomics. *Archives of Surgery* 134:1011–1016.

Biocca, F., Kim, J., and Choi, Y. 2001. Visual touch in virtual environments: An exploratory study of presence, multimodal interfaces, and cross-modal sensory illusions. *Presence* 10:247–265.

Birnbach, D. J. and Salas, E. 2008. Can medical simulation and team training reduce errors in labor and delivery? *Anesthesiology Clinics* 26:159–168.

Blackburn, T. and Sadler, C. 2003. The role of human patient simulators in health-care training. *Hospital Medicine* 64:677–681.

Boulet, J. R., Smee, S. M., Dillon, G. F., and Gimpel, J. R. 2009. The use of standardized patient assessment for certification and licensure decisions. *Simulation in Healthcare* 4:35–42.

Boulos, M. N, Hetherington, L., and Wheeler, S. 2007. Second Life: An overview of the potential of 3-D virtual worlds in medical and health education. *Health Information and Libraries Journal* 24:233–45.

Bradley, P. 2006. The history of simulation in medical education and possible future directions. *Medical Education* 40:254–262.

Combs, C. D., Friend, K., Mannion, M., and Alpino, R. J. 2006. Simulating the domain of medical modeling and simulation: The medical modeling and simulation database. In *Medicine Meets Virtual Reality*, eds. J. D. Westwood et al., pp. 105–107. Amsterdam, the Netherlands: IOS Press.

Conner, S. and Garden, O. J. 2006. Bile duct injury in the era of laparoscopic cholecystectomy. *British Journal of Surgery* 93:158–168.

Cooper, J. B. and Taqueti, V. R. 2004. A brief history of the development of mannequin simulators for clinical education and training. *Quality and Safety in Health Care* 13:i11–i18.

Dawson, S. 2006. Procedural simulation: A primer. *Journal of Vascular and Interventional Radiology* 17:205–213.

Dawson, S. L., Cotoin, S., Meglan, D., Shaffer, D., and Ferrell, M. 2000. Designing a computer-based simulator for interventional cardiology training (with editorial comment). *Catheterization and Cardiovascular Interventions* 51:522–528.

Delp, S. L., Loan, J. P., Hoy, M. G., Zajac, F. E., Topp, E. L., and Rosen, J. M. 1990. An interactive graphics-based model of the lower extremity to study orthopedic procedures. *IEEE Transactions on Biomedical Engineering* 37:757–767.

Deziel, D. J., Millikan, K. W., Economou, S. G., Doolas, A., Ko, S. T., and Airan, M. C. 1993. Complications of laparoscopic cholecystectomy: A national survey of 4,292 hospitals and an analysis of 77,604 cases. *The American Journal of Surgery* 165:9–14.

Dieckmann, P., Gaba, D., and Rall, M. 2007. Deepening the theoretical foundations of patient simulation as social practice. *Simulation in Healthcare* 2:183–193.

Dyer, G. S. M. and Thorndike, M. E. L. 2000. Quidne mortui vivos docent? The evolving purpose of human dissection in medical education. *Academic Medicine* 775:969–979.

Farmer, E., van Rooij, J., Riemersma, J., Jorna, P., and Moraal, J. 2003. *Handbook of Simulation-Based Training.* Burlington, VT: Ashgate.

Fernandez, R., Kozlowski, S. W. J., Shapiro, M. J., and Salas, E. 2008. Toward a definition of teamwork in emergency medicine. *Academic Emergency Medicine* 15:1104–1112.

Finn, R. and Soar, J. 2005. The story of Anne. *Resuscitation* 67:5–6.

Flexman, R. E. and Stark, E. A. 1987. Training simulators. In *Handbook of Human Factors*, ed. G. Salvendy, pp. 1012–1038. New York: Wiley.

Fontaine, M. D., Cook, D. P., Combs, C. D., Sokolowski, J. A., and Banks, C. M. 2009. Modeling and simulation: Real-world examples. In *Principles of Modeling and Simulation: A Multidisciplinary Approach*, eds. J. A. Sokolowski and C. M. Banks, pp. 181–245. Hoboken, NJ: Wiley.

Fritz, P. Z., Gray, T., and Flanagan, B. 2008. Review of mannequin-based high-fidelity simulation in emergency medicine. *Emergency Medicine Australasia* 20:1–9.

Fukui, Y. and Smith, N. T. 1981. Interaction among ventilation, the circulation, and the uptake and distribution of halothane. Use of a hybrid computer. Model I. The basic model. *Anesthesiology* 54:107–118.

Gaba, D. M. 2004. The future vision of simulation in health care. *Quality & Safety in Health Care* 13(suppl. 1):i2–i10.

Gaba, D. M. 2006. What's in a name? A mannequin by any other name would work as well. *Simulation in Healthcare* 1:64–65.

Gaba, D. M. 2010. Crisis resource management and teamwork training in anaesthesia. *British Journal of Anesthesia* 105:3–6.

Gaba, D. M. and DeAnda, A. 1988. A comprehensive anesthesia simulation environment: Re-creating the operating room for research and training. *Anesthesiology* 69:387–394.

Gaba, D. M., Howard, S. K., Fish, K. J., Smith, B. E., and Sowb, Y. A. 2001. Simulation-based training in anesthesia crisis resource management (ACRM): A decade of experience. *Simulation and Gaming* 32:175–193.

Gallagher, A. G. and Cates, C. U. 2004. Approval of virtual reality training for carotid stenting. *JAMA* 292:3024–3026.

Gallagher, A. G., Cowie, R., Crothers, I., Jordan-Black, J. A., and Satava, R. M. 2003. An objective test of perceptual skill that predicts laparoscopic technical skill in three initial studies of laparoscopic performance. *Surgical Endoscopy* 17:1468–1471.

Gallagher, A. G., Ritter, E. M., Champion, H. et al. 2005. Virtual reality simulation for the operating room: Proficiency-based training as a paradigm shift in surgical skills training. *Annals of Surgery* 241:1–9.

Gerling, G. J., Rigsbee, S., Childress, R. M., and Martin, M. L. 2009. The design and evaluation of a computerized and physical simulator for training clinical prostate exams. *IEEE Transactions on Systems, Man, and Cybernetics: Part A: Systems and Humans* 39:388–403.

Gordon, M. S. 1974. Cardiology patient simulator: Development of an automated manikin to teach cardiovascular disease. *American Journal of Cardiology* 43:350–355.

Grant, R. C., Carswell, C. M., Lio, C. H., Seales, B., and Clarke, D. 2009. Verbal time production as a secondary task: Which metrics and target intervals are most sensitive to workload for fine motor laparoscopic training tasks? *Proceedings of the Human Factors & Ergonomics Society 53rd Annual Meeting*, pp. 1191–1195. Santa Monica, CA: Human Factors & Ergonomics Society.

Grenvik, A. and Schaefer, J. 2004. From resusci-anne to sim-man: The evolution of simulators in medicine. *Critical Care Medicine* 32:S56–S57.

Hart, S. G. and Staveland, L. E. 1988. Development of NASA–TLX (task load index): Results of empirical and theoretical research. In *Human Mental Workload*, eds. P. A. Hancock and N. Meshkati, pp. 139–183. Amsterdam, the Netherlands: North Holland.

Haviland, T. N. and Parish, L. C. 1970. A brief account of the use of wax models in the study of medicine. *Journal of the History of Medicine and Allied Sciences* 1:52–75.

Hayes, R. T. and Singer, M. J. 1989. *Simulation Fidelity in Training System Design: Bridging the Gap between Reality and Training.* New York: Springer-Verlag.

Holzman, R. S., Cooper, J. B., Gaba, D. M., Philip, J. H., Small, S. D., and Feinstein, D. 1995. Anesthesia crisis resource management: Real-life simulation training in operating room crises. *Journal of Clinical Anesthesia* 7:675–687.

Howard, S. K., Gaba, D. M., Smith, B. E. et al. 2003. Simulation study of rested versus sleep-deprived anesthesiologists. *Anesthesiology* 98:1345–1355.

Hsu, K. E., Man, F. -Y., Gizicki, R. A., Feldman, L. S., and Fried, G. M. 2008. Experienced surgeons can do more than one thing at a time: Effect of distraction on performance of a simple laparoscopic and cognitive task by experienced and novice surgeons. *Surgical Endoscopy* 22:196–201.

Hunt, E. A., Nelson, K. L., and Shilkofski, N. A. 2006. Simulation in medicine: Addressing patient safety and improving the interface between healthcare providers and medical technology. *Biomedical Instrumentation and Technology* 40:399–404.

Issenberg, S. B., McGaghie, W. C., Hart, I. R. et al. 1999. Simulation technology for healthcare professional skills training and assessment. *JAMA* 282:861–866.

Issenberg, S. B., McGaghie, W. C., Petrusa, E. R. et al. 2005. Features and uses of high-fidelity medical simulations that lead to effective learning: A BEME systematic review. *Medical Teacher* 27:10–28.

Johnston, C. L. and Whatley, D. 2006. Pulse!!—A virtual learning space project. In *Medicine Meets Virtual Reality*, eds. J. D. Westwood et al., pp. 240–243. Amsterdam, the Netherlands: IOS Press.

Katz, G. B., Peifer, K. L., and Armstrong, G. 2010. Assessment of patient simulation use in selected baccalaureate nursing programs in the United States. *Simulation in Healthcare* 5:46–51.

Klatzky, R. L., Lederman, S. J., Hamilton, C., Grindley, M., and Swendsen, R. H. 2003. Feeling textures through a probe: Effects of probe and surface geometry and exploratory factors. *Perception & Psychophysics* 65:613–631.

Klein, M. I., Riley, M. A., Warm, J. S., and Matthews, G. 2005. Perceived mental workload in an endoscopic surgery simulator. *Proceedings of the Human Factors and Ergonomics Society 49th Annual Meeting*, pp. 1014–1018. Santa Monica, CA: HFES.

Kohn, L. T., Corrigan, J. M., and Donaldson, M. S. (eds.) 1999. *To Err Is Human: Building a Safer Health System*. Washington, DC: National Academy Press.

Kneebone, R. L. 2006. Crossing the line: Simulation and boundary areas. *Simulation in Healthcare* 1:160–163.

Lawrence, M. A., Kitada, R., Klatzky, R. L., and Lederman, S. J. 2007. Haptic roughness perception of linear gratings via bare finger or rigid probe. *Perception* 36:547–557.

Levy, J. S. 1996. Virtual reality hysteroscopy. *Journal of the American Association of Gynecological Laparoscopy* (sup 4):S25–S26.

Liaison Committee on Medical Education. 1991. Functions and structure of a medical school: Accreditation and the liaison committee on medical education; Standards for Accreditation of Medical Education Programs Leading to the M. D. degree. Washington, DC and Chicago, IL.

Lighthall, G. K., Barr, J., Howard, S. K. et al. 2003. Use of a fully simulated intensive care unit environment for critical event management training for internal medicine residents. *Critical Care Medicine* 31:2437–2442.

Lind, B. 2007. The birth of the resuscitation mannequin, resusci anne, and the teaching of mouth-to-mouth ventilation. *Acta Anaesthesiologica Scandinavica* 51:1051–1053.

Liu, D., Jenkins, S. A., Sanderson, P. M. et al. 2009. Monitoring with head-mounted displays: Performance and safety in a full-scale simulator and part-task trainer. *Anesthesia and Analgesia* 109:1135–1146.

Macedonia, C. R., Gherman, R. B., and satin, A. J. 2003. Simulation laboratories for training in obstetrics and gynecology. *Obstetrics and Gynecology* 102:388–392.

Marescaux, J., Clement, J. M., Tassetti, V. et al. 1998. Virtual reality applied to hepatic surgery simulation: The next revolution. *Annals of Surgery* 228:627–634.

Matern, U. 2004. The laparoscopic surgeon's posture. In *Misadventures in Health Care: Inside Stories*, ed. M. S. Bogner, pp. 75–88. Mahwah, NJ: Erlbaum.

McKenzie, F. D., Hubbard, T. W., Ullian, J. A., Garcia, H. M., Castelino, R. J., and Gliva, G. A. 2006. Medical student evaluation using augmented standardized patients: Preliminary results. In *Medicine Meets Virtual Reality*, eds. J. D. Westwood et al., pp. 379–384. Amsterdam, the Netherlands: IOS Press.

Miller, G. E. 1990. The assessment of clinical skills/competence/performance. *Academic Medicine* 65:S63–S67.

Montgomery, K. 2005. Are Haptics really necessary? *TATRC's 5th Annual Advanced Medical Technology Review*. Long Beach, CA.

Moore, M. J. and Bennett, C. L. 1995. The learning curve for laparoscopic cholecystectomy. *The American Journal of Surgery* 170:55–59.

Nehring, W. M. and Lashley, F. R. 2004. Human patient simulators in nursing education: An international survey. *Nursing Education Perspectives* 25:244–248.

Olry, R. 2000. Wax, wooden, ivory, cardboard, bronze, fabric, plaster, rubber and plastic anatomical models: Praiseworthy precursors of plastinated specimens. *Journal of the International Society for Plastination* 15:30–35.

Philip, J. 1986. Gas man—An example of goal oriented computer-assisted teaching which results in learning. *International Journal of Clinical Monitoring and Computing* 3:165–173.

Pugh, C. and Youngblood, P. 2002. Development and validation of assessment measures for a newly developed physical examination simulator. *American Journal of Medical Informatics Association* 9:448–460.

Reznek, M. Smith-Coggins, R., Howard, S. et al. 2003. Emergency medicine crisis resource management (EMCRM): Pilot study of a simulation-based crisis management course for emergency medicine. *Academic Emergency Medicine* 10:386–389.

Reznick, R. K. and MacRae, H. 2006. Teaching surgical skills—Changes in the wind. *The New England Journal of Medicine* 355:2664–2669.

Rosen, M. A., Lazzara, E. H., Lyons, R. et al. 2008. SMARTER-Team: Adapting event-based tools for simulation-based training in healthcare. In *Proceedings of the Human Factors and Ergonomics Society 52nd Annual Meeting*, pp. 793–797. Santa Monica, CA: Human Factors and Ergonomics Society.

Sachdeva, A. K., Pellegrini, C. A., and Johnson, K. A. 2008. Support for simulation-based surgical education through American College of Surgeons—Accredited education institutions. *World Journal of Surgery* 32:196–207.

Salas, E., Wilson, K. A., Burke, C. S., and Wightman, D. C. 2006. Does crew resource management training work? An update, an extension, and some critical needs. *Human Factors* 48:392–412.

Satava, R. M. 1993. Virtual reality surgical simulator. *Surgical Endoscopy* 7:203–205.

Satava, R. M. 2001. Accomplishments and challenges of surgical simulation. *Surgical Endoscopy,* 15:232–241.

Satava, R. M. 2008. Historical review of surgical simulation—A personal perspective. *World Journal of Surgery* 32:141–148.

Satava, R. M. and Jones, S. 2002. Medical applications of virtual environments. In *Handbook of Virtual Environments: Design, Implementation, and Applications*, ed. K. M. Stanney, pp. 937–957. Mahwah, NJ: Erlbaum.

Scerbo, M. W. 2005. The future of medical training and the need for human factors. *Proceedings of the Human Factors & Ergonomics Society 49th Annual Meeting*, pp. 969–973. Santa Monica, CA: Human Factors & Ergonomics Society.

Scerbo, M. W., Belfore, L. A., Garcia, H. M. et al. 2007. A virtual operating room for context-relevant training. *Proceedings of the Human Factors & Ergonomics Society 51st Annual Meeting*, pp. 507–511. Santa Monica, CA: Human Factors & Ergonomics Society.

Scerbo, M. W. and Dawson, S. 2007. High fidelity, high performance? *Simulation in Healthcare* 2:224–230.

Scerbo, M. W., Turner, T. R., Meglan, D. A., and Waddington, R. 2009. Evaluating visual and haptic feedback on a virtual reality simulator for orthopedic bone pinning. *Proceedings of the Human Factors & Ergonomics Society 53rd Annual Meeting*, pp. 1116–1120. Santa Monica, CA: Human Factors & Ergonomics Society.

Seymour, N. E. 2008. VR to OR: A review of the evidence that virtual reality simulation improves operating room performance. *World Journal of Surgery* 32:182–188.

Stefanidis, D., Scerbo, M. W., Korndorffer, J. R., Jr., and Scott, D. J. 2007. Redefining simulator proficiency using automaticity theory. *The American Journal of Surgery* 193:502–506.

Stefanidis, D., Scerbo, M. W., Sechrist, C., Mostafavi, A., and Heniford, B. T. 2008. Do novices display automaticity during simulator training? *The American Journal of Surgery* 195:210–213.

Stone, R. J. 1999. The opportunities for virtual reality and simulation in the training and assessment of technical surgical skills. In *Proceedings of Surgical Competence: Challenges of Assessment in Training and Practice*, pp. 109–125. London, U.K.: Royal College of Surgeons of England.

Sutton, C., McCloy, R., Middlebrook, A., Chater, P., Wilson, M., and Stone, R. 1997. MIST VR: A laparoscopic surgery procedures trainer and evaluator. In *Medicine Meets Virtual Reality*, eds. K. S. Morgan et al., pp. 598–607. Amsterdam, the Netherlands: IOS Press.

Swezey, R. W. and Andrews, D. H. 2001. *Readings in Training and Simulation: A 30-Year Perspective*. Santa Monica, CA: Human Factors & Ergonomics Society.

Wang, N., Gerling, G. J., Krupski, T. L., Childress, R. M., and Martin, M. L. 2010. Using a prostate exam simulator to decipher palpation techniques that facilitate the detection of abnormalities near clinical limits. *Simulation in Healthcare* 5:152–160.

Weicha, J., Heyden, R., Sternthal, E., and Merialdi, M. 2010. Learning in a virtual world: Experience with using second life for medical education. *Journal of Medical Internet Research* 12:e1.

Wickens, C. D. 1984. Processing resources in attention. In *Varieties of Attention*, eds. R. Parasuraman and D. R. Davies, pp. 63–102. New York: Academic Press.

Youngblood, P., Harter, P. M., Srivastava, S., Moffett, S., Heinrichs, W. L., and Dev, P. 2008. Design, development, and evaluation of an online virtual emergency department for training trauma teams. *Simulation in Healthcare* 3:146–153.

Yurko, Y. Y., Scerbo, M. W., Prabhu, A. S., Acker, C. E., and Stefanidis, D. 2010. Increased task workload is associated with inferior performance and more errors: The value of the NASA–TLX tool. *Simulation in Healthcare* 5:267–271.

Ziv, A. Wolpe, P. R., Small, S. D., and Glick, S. 2003. Simulation-based medical education: An ethical imperative. *Academic Medicine* 78:783–788.

34

Simulation-Based Training for Teams in Health Care: Designing Scenarios, Measuring Performance, and Providing Feedback

Michael A. Rosen, Eduardo Salas, Scott I. Tannenbaum, Peter J. Pronovost, and Heidi B. King

CONTENTS

34.1 Introduction 574
34.2 SBTT in Health Care: An Overview 574
    34.2.1 Key Trends in SBTT in Health Care 574
    34.2.2 Systems Approach to SBTT 575
34.3 Scenario Design for Team Training in Health Care 575
    34.3.1 Instructional Design for Practice-Based Learning: Event-Based Methods ...575
    34.3.2 Developing Event Sets for Teamwork Training Scenarios 577
        34.3.2.1 Define a Teamwork Competency Model 577
        34.3.2.2 Identify the Teamwork Learning Objectives 580
        34.3.2.3 Identify Incidents or Situations That Will Create Opportunities to Meet the Learning Objectives 580
        34.3.2.4 Develop an Event Set and Associated Teamwork Responses 581
        34.3.2.5 Pilot Test, Refine, Implement, and Continuously Evaluate the Scenario ...581
    34.3.3 Scenario Development Summary..581
34.4 Team Performance Measurement in SBT 582
    34.4.1 Components of a Team Performance Measurement System 582
        34.4.1.1 Purpose: Why Measure 583
        34.4.1.2 Content: What to Measure 583
        34.4.1.3 Location: Where to Measure 585
        34.4.1.4 Who: Selecting, Training, and Supporting Raters 585
        34.4.1.5 Method: How to Measure 585
    34.4.2 Developing Event-Based Measures 585
34.5 Debrief Facilitation 588
    34.5.1 Phases of a Team Debrief 588
    34.5.2 Best Practices in Team Debriefing 589
        34.5.2.1 Establish a Learning Climate 589
        34.5.2.2 Team Debriefs Should Be Diagnostic 590
        34.5.2.3 Focus on a Few Key Performance Issues 590
        34.5.2.4 Maintain a Focus on Teamwork Processes 590
        34.5.2.5 Focus on the Processes of Performance 590
        34.5.2.6 Support Feedback with Objective Indicators of Performance 590
        34.5.2.7 Provide Both Individual and Team Feedback and Know When Each Is Appropriate 590
        34.5.2.8 Record the Outcomes of the Debrief 590
        34.5.2.9 Educate Trainers and Team Members on the Art and Science of Facilitation ...590
34.6 Concluding Remarks 591
References 591

34.1 Introduction

The patient safety movement has spurred the development and adoption of a great variety of methods and approaches for guarding against and learning from error and incidents of patient harm (Frankel et al., 2009; Vincent, 2010). Two of these approaches are (1) teamwork training and (2) the use of simulation as a tool for individual, team, and organizational learning and development.

First, the initial rationale for broad advocacy of teamwork training as a patient safety solution was rooted in the best practices from other high-stakes industries as well as the realization that communication-related issues are a primary contributing factor to incidents of patient harm (Kohn et al., 2000). However, in recent years, prospective studies linking both team training to improved teamwork in clinical areas (e.g., Sax et al., 2009; Weaver et al., 2010) and levels of teamwork to a variety of clinical, quality, and safety measures (e.g., Draycott and Crofts, 2006; Wolf et al., 2010) have provided stronger evidence of the value of this approach in health care (for a detailed review, see Salas et al., this volume).

Second, there are several major connections between the use of simulation and patient safety. This includes providing a solution to practical limitations and ethical concerns with having real patients serve as the primary learning opportunities for care providers to develop basic clinical skills (Ziv et al., 2006). Additionally, the development of new technologies has increased the range of skills that can be trained and assessed with simulation (for a detailed review, see Chapter 33). Simulation is also being used prospectively to identify patient safety issues within health care systems, and subsequently to develop and evaluate solutions.

This chapter deals with the intersection of these two approaches to improving safety and quality in health care. Specifically, we present a systems-based perspective on developing and implementing simulation-based team training (SBTT) in the health care domain. This chapter is intended to serve as a practical guide to SBTT rooted in the best evidence and methods available.

To that end, we address three main goals. First, we provide a brief overview of key issues in both team training and simulation-based training (SBT) in health care. Second, we review the current state of the science and practice of three interrelated activities comprising the fundamentals of developing and implementing SBTT: scenario design, performance measurement, and debrief facilitation. Third, future directions for SBTT in health care are discussed.

34.2 SBTT in Health Care: An Overview

This section describes some of the trends in SBTT in health care as well as the driving forces behind those trends. Subsequently, a systems approach to SBTT in health care is outlined.

34.2.1 Key Trends in SBTT in Health Care

There are at least four general, and interrelated, applications of simulation in health care: providing practice-based learning opportunities in *training and education* (e.g., Issenberg et al., 2005; McGaghie et al., 2010), *evaluation and assessment* of individual and team competence (e.g., Boulet et al., 2008; Singleton et al., 1999) as well as equipment and system safety, *performance support* to facilitate the application of previously acquired performance competencies (e.g., preprocedure walkthroughs and warm ups in surgery; Calatayd et al., 2010; Do et al., 2006; Kahol et al., 2008), and *innovation and exploration* (e.g., simulation for prospective risk analysis and work/procedure redesign; Davis et al., 2008). To date, applications of simulation in training and education are the most widespread; however, in SBTT, methods are being blended to address multiple goals with increasing frequency. For example, many safety improvement and culture change initiatives such as the Comprehensive Unit-based Safety Program (CUSP; Pronovost et al., 2005) and the Team Strategies to Enhance Performance and Patient Safety (TeamSTEPPS®; Alonso et al., 2006) take a change management approach where exploratory in situ simulations can be applied early in the needs analysis processes and later simulation can be used as a tool for implementing and evaluating the targeted changes. The focus of this chapter is primarily on training and education applications; however, many of the techniques discussed here are applicable for other purposes.

While a relative newcomer to the modern use of simulation, the health care domain has been making large and rapid advances in the development and implementation of simulation, both in the sophistication of the *simulators* (i.e., the technology used to simulate aspects of the health care system, most usually patients) and the soundness of *methods* employed (e.g., feedback, measurement, and curriculum design). Historically, the ability to simulate biological materials and processes necessary to train many procedural tasks has proven technologically challenging as it involves representing complex organic systems and not engineered systems with technical specifications already in existence (e.g., aircraft). However, a long line of research demonstrates the difference between *physical fidelity* (i.e., the degree to which the simulator replicates the look and feel of what it is simulating) and *functional or cognitive fidelity* (i.e., the degree to which the required performance processes of the learners are the same in the simulated and real task environments). The effectiveness of a simulation as a learning opportunity depends upon the functional fidelity and not the physical fidelity in isolation (for a detailed review of this topic, see Hays and Singer, 1989). In SBTT, much of the functional fidelity lies in the interactions between team members in the scenario, and consequently patient simulators with low levels of physical fidelity can be highly effective team training devices (Beaubein and Baker, 2004; Salas et al., 1998). Consequently, some of the earliest widespread uses of simulation in health care involved teamwork training (e.g., Gaba et al., 2001).

Related to fidelity, the location of simulation-based learning activities is an issue intensely debated in the simulation in health care community. The major distinction drawn is between center-based simulation (i.e., a dedicated facility for simulation-based learning) and in situ or mobile simulation (i.e., simulation that occurs in the job environment); however, there is a rapidly evolving continuum of simulations between these two end points: completely separate from work activities, and blended with work (Gaba, 2004; Maran and Glavin, 2003). While a complete picture of the merits and limitations of different approaches is still emerging, it appears as if complete strategy will involve multiple approaches balancing the strengths and weakness of each based on needs and practical considerations. A centralized center-based simulation approach allows for greater control over and standardization of the learning environment and scenarios, consistency over faculty or trainer skill, and limited distractions for the learner. However, it is logistically challenging to bring health care workers off-site to a dedicated simulation center. An in situ approach solves the logistical issue, affords an ability to uncover systems issues on the units (e.g., identifying unsafe equipment, work process/structure, and policy issues), and makes close connections between the behaviors practiced in simulation and the work environment; however, the control and consistency of scenarios is likely affected and staff may be pulled into work activities prematurely, compromising the value of the learning activity.

34.2.2 Systems Approach to SBTT

A systems approach to training and education involves attending to a variety of factors that occur both within and outside of the actual learning activities as well as before, during, and after a learning session. Recently, Salas et al. (2009) have developed a set of overarching principles for SBT for patient safety. These principles outline the major components of a systems-based

approach to SBTT in health care. Factors of the simulation program itself related to effectiveness include the focus on teamwork competencies, the application of sound instructional design practices, the use of team performance measures to assess performance changes over time and to provide feedback, and creating appropriate prepractice conditions. Factors outside of the actual training or educational environment can greatly impact the ultimate effectiveness of the program as well (e.g., Salas et al., 2009a) and include broader systems issues in training such as preparing the organization and transfer environments and ensuring learner motivation. Evaluating the effectiveness of the simulation program is a key method in bridging the simulated and target transfer environment. These principles are summarized in Table 34.1 and addressed throughout the following sections.

The remainder of this chapter details three core components of a systems-based approach to SBTT in health care: scenario design, performance measurement, and team debriefing. These three components can help to ensure continuity between targeted teamwork behaviors and changes in performance in the work environment.

34.3 Scenario Design for Team Training in Health Care

In SBTT, *the scenario is the curriculum* (Salas et al., 2006)—it constrains what people can and cannot do within a practice activity and consequently sets the boundaries of what can potentially be learned from the experience. Therefore, the quality of the scenario is critical to the effectiveness of the simulation program. However, unlike designing educational and training content for other delivery methods (e.g., Kern et al., 2009; Mayer, 2005), there are few clear specifications of systematic methods for creating simulation scenarios. This section provides a review of the existing scenario development methods as they relate to team training in health care as well as providing a step-by-step detail of the event-based method, a long-standing methodology for developing team training scenarios.

34.3.1 Instructional Design for Practice-Based Learning: Event-Based Methods

The existing *systematic* methods of scenario design can be categorized under the broad heading of event-based methods (Fowlkes et al., 1998; Oser et al., 1999)—*an instructional design method for creating practice-based learning scenarios with embedded critical events linked to specific learning objectives and accompanying measurement*

TABLE 34.1

Principles and Tips for SBTT in Health Care

Principle	Tips
Activities occurring in the facility	
Take a systems approach	• Think about "before," "during," and "after" influences on SBT
Prepare the organization for SBT	• Pay attention to organizational factors
	• Make sure you have top-level support
	• Send positive messages about simulation
	• Help employees see value of simulation
	• Ensure key players are on board
Prepare the transfer environment	• Create a continuous learning culture
	• Show management support
	• Create opportunities to practice what is learned
	• Provide incentives
	• Reinforce desired behaviors
	• Keep sending positive signals
Ensure trainee motivation	• Show value of SBT
	• Send positive messages about SBT
Activities occurring in the simulation environment	
SBT should focus on reinforcing and promoting the needed competencies	• Train competencies, not tasks
	• Conduct a training needs analysis
	• Conduct a job/task analysis
	• Conduct a cognitive task analysis
	• Develop learning outcomes
Apply sound instructional principles to the design of SBT	• Tailor instructional strategy to learning objectives
	• Present relevant information
	• Demonstrate skills to be learned
	• Match fidelity to task (if possible)
	• Provide guided hands-on practice
	• Provide constructive and diagnostic feedback
Develop and use performance measures	• Create scenarios to elicit the desired KSAs
Set up appropriate presimulation conditions	• Prepare trainees for training through preparatory information
	• Set an appropriate presimulation climate
Set up the simulation environment	• Provide the appropriate training setting
	• Provide resources
	• Adequately train the instructors and observers
Bridging simulated and real environments	
Evaluate the effectiveness of the SBT or education program	• Examine training objectives and link to evaluation criteria
	• Measure at multiple levels (reactions, learning, behavior, and results)

Source: Adapted from Salas et al., *Joint Commission Journal on Quality and Patient Safety, 34,* 518–527, 2008.

tools. This general approach has been applied for several decades in military domains (Cambell et al., 1997; Dwyer et al., 1999; Fowlkes et al., 1994), as well as in aviation [Federal Aviation Administration (FAA)]. In fact, it is the industry standard for developing crew resource management scenarios in U.S. aviation. More recently, variations of event-based methods have been described in the context of training individual level technical competencies as well as and team level nontechnical skills in health care (Holland et al., 2008; Lazzara et al., 2010; Rosen et al., 2008a,b).

As with most methods of instructional design, event-based methods begin with the end in mind; that is, the attributes of effective learning activities are defined and processes are then constructed to reach those ends. For SBTT, desired characteristics of a simulation scenario include: explicit opportunities to practice targeted *teamwork* behaviors, a match between learner level of proficiency and the difficulty of both teamwork and technical demands placed on learners by the scenario, and the ability to support critical and improvement-focused discussions after the practice episode (i.e., feedback in the form of facilitated debriefs). Consequently, event-based methods focus on developing and embedding critical or trigger events that (1) are linked to specific teamwork learning objectives and (2) provide clear opportunities to perform (and measure) a specific targeted teamwork behavior. These engineered event sets provide a scaffold for the post-practice debrief discussion, helping to guide the team members' learning.

34.3.2 Developing Event Sets for Teamwork Training Scenarios

This section provides a synthesis of event-based methods as they apply to SBTT in health care (Fowlkes et al., 1998; Holland et al., 2008; Lazzara et al., 2010; Rosen et al., 2008b). Specifically, the following five steps will be detailed: (1) define a teamwork competency model, (2) identify and clearly state the teamwork learning objectives, (3) identify an incident that can provide the opportunity to meet the learning objectives (i.e., a situation that requires the targeted teamwork behavior to manage effectively), (4) develop an event set and list of associated expected teamwork responses, and (5) pilot test, refine, implement, and continuously evaluate the scenario.

While these steps are presented in a linear manner, an iterative process is key for the design of all types, and scenario design is no different. Choices at one step influence choices at another and consequently, as the process moves forward, decisions made at previous steps frequently need to be revisited. Additionally, this process should be executed by a scenario or curriculum design *team* representing multiple perspectives including expertise in the clinical domain targeted, the teamwork skills being trained, and instructional design (Salas et al., 2007) as well as representation from all of the roles that will participate in the scenarios being developed (e.g., nurses, physicians, technicians, pharmacists, etc.) to help ensure that the scenario is relevant, believable, and appropriate for all parties involved.

34.3.2.1 *Define a Teamwork Competency Model*

A teamwork competency model is a specification of the knowledge, skills, and attitudes (KSAs) underlying effective team performance (Cannon-Bowers et al., 1995). This serves as the basis for the design of the training or educational program in a global sense in that it defines the desired learning outcomes—the aspects of teamwork to be developed. Table 34.2 provides a brief summary of several general teamwork KSAs commonly used in health care. For a more detailed description of teamwork competencies, see Salas and colleagues (this volume; also, Salas et al., 2009b). In addition to teamwork competencies, SBTT will frequently target key clinical skills as well, especially for more advanced learners.

34.3.2.1.1 *Sources for Teamwork Competency Models*

There are multiple sources that can inform the development of a teamwork competency model for SBTT in health care, including: preexisting consensus models of teamwork competencies for a clinical specialty or task domain; theory and empirical evidence from the science of teams and team training (e.g., Salas et al., 2005); studies conducted within a specific clinical context that identify aspects of effective teamwork (e.g., Burtscher et al., 2010); and employing methods for eliciting and defining teamwork skills in a specific clinical context (e.g., Flin and Maran, 2008) as well as traditional needs analysis techniques.

The first two approaches (consensus models and the science of teams) are top-down methods, and the third and fourth approaches (studies within a specific clinical context and elicitation techniques) are bottom-up in nature. Currently, there are few consensus models of teamwork within clinical domains. However, models of teamwork are emerging for different areas (e.g., Fernandez et al., 2008). Consequently, this will not be a major input for defining teamwork competency models for training or educational programs at this time, but as these contextualized models evolve and become validated, this will likely become the main source.

Most frequently, a mix of top-down and bottom-up approaches will be used in concert where a general competency model rooted in the science of teams is operationalized or contextualized for a given clinical domain. For example, generalizable models of teamwork competencies have been adapted from the science of teams for application in a broad array of health care domains (e.g., leadership, situation monitoring, mutual support, closed-loop communication; Salas et al., 2005; http://teamstepps.ahrq.gov/). A specific subset of these competencies can be selected and contextualized for a given clinical context using methods of needs analysis employed in traditional educational and training design (Goldstein and Ford, 2002; Lazzara et al., 2010). These methods include team-focused versions of cognitive task analysis such as the critical incident technique or critical decision method (Crandall et al., 2006), as well as person analysis, organizational analysis, and team task analysis (Burke, 2005).

34.3.2.1.2 *Transportable versus Team- and Task-Specific Competencies*

Teamwork competencies can be classified based on the degree to which they apply broadly across contexts (i.e., specific clinical situations or tasks) and teams (i.e., specific groups of individual members). Team competencies underlying effective team performance across different contexts and teams are referred to as transportable or generic competencies (Cannon-Bowers et al., 1995). Context- or team-specific competencies underlie effective team performance within a narrow range of situations or teams. The appropriateness of these different types of competencies will depend on the purpose of the SBTT program and the characteristics of the learners. For example, SBTT programs conducted during new hire orientations at facilities are used to provide all staff members with a common set of teamwork skills.

TABLE 34.2

Summary of Typical Teamwork Competencies Used in SBTT and Behavioral Markers

KSAs	Description	Example Behavioral Markers	Example Citations
Knowledge			
Accurate and shared mental models (knowledge about team members and role structure, team task, and environment)	Organized knowledge structures of the relationships between task and team members	• Team members can recognize when other team member's need information they have • Team members anticipate and predict the needs of others • Team members have compatible explanations of task information	Cannon-Bowers and Salas (1997), Klimoski and Mohammed (1994), Artman (2000), and Stout, Cannon-Bowers, and Salas (1996)
Team mission, objective, norms, and resources	An understanding of the purpose, vision, and available means to meet team goals	• Team members make compatible task prioritizations • Team members agree on the methods adopted to reach their shared goals	Cannon-Bowers et al. (1995) and Marks et al. (2001)
Skills			
Closed-loop communication	A pattern of information exchange characterized by three steps: a sender initiates a message, the receiver acknowledges the message, and the sender follows up to confirm it was appropriately interpreted	• Team members cross-check information with one another • Team members give "big picture" updates to one another • Team members proactively pass critical information to those that need it in a timely fashion	Bowers, Jentsch, Salas, and Bruan (1998), McIntyre and Salas (1995), and Smith-Jentsch, Johnston, and Payne (1998)
Mutual performance monitoring	Team members' ability to track what others on the team are doing while continuing to carry out their own tasks	• Team members recognize errors in their teammates' performance • Team members have an accurate understanding of their teammates' workload	McIntyre and Salas (1995), Dickinson and McIntyre (1997), and Marks and Panzer (2004)
Backup/supportive behavior	The ability to shift and balance workload among team members during high-workload or high-pressure periods	• Team members promptly offer and accept task assistance • Team members communicate the need for task assistance • Team members redistribute workload to members who are being underutilized	Marks et al. (2001), McIntyre and Salas (1995), and Porter et al. (2003)
Adaptability	The team's ability to adjust strategies to changing conditions	• Team members replace or modify routine strategies when the task changes • Team members detect important changes in their environment quickly • Team members accurately assess the causes of important changes	Burke, Stagl, Salas, Pierce, and Kendall (2006b), Entin and Serfaty (1999), and Kozlowski et al. (1999)
Conflict management/resolution	Preemptively setting up conditions to prevent or control team conflict or reactively working through interpersonal disagreements between members	• Team members seek solutions to conflict wherein all members gain • Team members discuss task related conflict openly • Team members find it acceptable to change positions and express doubts	De Dreu and Weingart (2003), Jordan and Troth (2004), and Simons and Peterson (2000)
Team leadership	Dynamic process of social problem solving involving information search and structuring, information use in problem solving, managing personnel resources, and managing material resources	• Leaders develop plans and communicate them to the team • Leaders organize and delegate effectively based on the plan and changing work demands • Leaders invite input from team members • Leaders proactively facilitate the resolution controversy and conflict • Team members accurately identify the person with the most appropriate skill set for leadership in a specific situation • Team members shift leadership roles in response to task demands	Burke, Stagl, Klein, Goodwin, Salas, and Halpin (2006a) and Day, Gronn, and Salas (2004)

Attitudes

Mutual trust	The shared belief among team members that everyone will perform their roles and protect the interests of their fellow team members	• Team members share information openly without fear of reprisals • Team members are willing to admit mistakes • Team members share a belief that team members will perform their tasks and roles	Alavi and McCormick (2004), and Jackson et al. (2006)
Team/collective efficacy	The team members' sense of collective competence and their ability to achieve their goals	• Team members share positive evaluations about the team's capacity to perform its tasks and meet its goals	Bandura (1986), Gibson, (2003), Katz-Navon, and Erez, (2005), and Zaccaro et al. (1995)
Team/collective orientation	Team members' preference for working with others as opposed to working in isolation	• Team members accept input from others and it is evaluated on its quality, not its source • Team member value team goals over individual goals • Team members have high levels of task involvement and participatory goal setting	Alavi and McCormick (2004), Eby and Dobbins (1997), and Mohammed and Angell (2004)
Psychological safety	The team members' shared belief that it is safe to take interpersonal risks	• Team members are interested in each other as people • Team members are not rejected for being themselves • Team members believe others on the team have positive intentions	Edmondson (2003)

Source: Adapted from Rosen, M.A. et al., How can team performance be measured, assessed, and diagnosed? In K. Frush and E. Salas (Eds.), *Improving Patient Safety through Teamwork and Team Training*, Oxford University Press, Oxford, U.K., in press.

Consequently, transportable competencies will usually be the focus as staff members will be from a mix of specialty areas and the objective is to build basic teamwork skills. However, when designing SBTT for a specific unit (e.g., part of a local improvement initiative) or clinical context (e.g., a rapid response team), more team- and context-specific competencies can be used in conjunction with transportable aspects of teamwork.

34.3.2.2 Identify the Teamwork Learning Objectives

SBTT scenarios are most effective when the development process is guided by a clear purpose for the activity—a specification of teamwork learning objectives. Unstructured or unguided practice activities are generally not as effective as those designed to meet explicit learning goals (Ehrenstein et al., 1997; Kirschner et al., 2006). As described earlier, a teamwork competency model provides the "big picture" for an entire training or educational program—the total set of teamwork KSAs targeted for development. Learning objectives are linked to this competency model, but describe a more granular or specific form of the targeted teamwork competency within a specific learning activity, in this case a scenario. This parallels the distinction in medical education between broad goals and specific objectives (Kern et al., 2009). As such, learning objectives fulfill two primary and interrelated roles: setting standards by which individuals can gauge their performance and setting criteria for the SBTT program to be evaluated.

Learning objectives clearly articulate expectations for team performance in a given scenario. This is done using three basic components: a behavioral description of the performance targeted for acquisition, a criterion by which to judge effectiveness, and the conditions under which the targeted aspects of teamwork should be performed (Goldstein and Ford, 2001). Kern et al. (2009) provide the following heuristic in the form of a question that every learning objective should be able to answer: *Who will do how much (how well) of what by when?*

Learning objectives drive the scenario design process, but are also critical in the development of measurement and evaluation tools, both for assessment of learner competence and validation of the SBTT program. Consequently, learning objectives should be articulated in manner specific enough to afford measurement.

34.3.2.3 Identify Incidents or Situations That Will Create Opportunities to Meet the Learning Objectives

Once a set of learning objectives have been decided upon, it is necessary to identify a situation—a clinical context—that requires team members to perform the targeted teamwork competencies (Rosen et al., 2008).

Not all clinical situations will be useful for training teamwork competencies as teamwork is more central to some situations than others. The choice of clinical context draws the boundaries around what can be trained in many ways. Consequently, careful consideration on this step is warranted. Two critical issues to attend to in this step are (1) the criticality and frequency of the clinical context and (2) the difficulty of teamwork and taskwork demands the scenario will place on learners.

First, SBTT often focuses on clinically infrequent yet high-stakes situations. The earliest applications of SBTT concepts to anesthesia (Gaba et al., 2001) and training for emergencies in Labor and Delivery (L&D) are prime examples. Emergencies in L&D (e.g., eclamptic seizure and breach delivery) are rare, but require a rapid and coordinated response to avoid severe negative outcomes for the child and mother. Consequently, there have been numerous training programs developed that target both the technical responses required as well as the role of teamwork in these situations (e.g., Deering et al., 2009; Draycott et al., 2005). While no doubt important, training teamwork exclusively in the context of these low-frequency high criticality events does not necessarily address the types routine daily situations where teamwork plays an important role in safety and quality. Consequently, in recent years, SBTT has been developed for more frequent and less emergent contexts such as practicing briefings, handoffs, and structured communication protocols (e.g., Chen et al., 2010; McQueen-Shadfar and Taekman, 2010).

Second, choosing the clinical context will also dictate the required technical (or clinical) competencies required for effectively managing the scenario. The appropriate mix of difficulty in teamwork and taskwork demands will vary based on the experience level of the team members.

For example, when developing an initial teamwork training for relatively novice learners in formal teamwork skills (e.g., in the early phases of health care education; Rosen et al., 2010), the focus will likely be exclusively on the teamwork behaviors; consequently, using clinical situations with clearly defined clinical responses (e.g., Advanced Trauma or Cardiac Life Support protocols) as opposed to clinical situations with high degrees of "stylistic" variation in clinical solutions is likely appropriate as it can help to focus the team debrief on teamwork behaviors. With increasingly competent learners, more difficult clinical contexts can be used.

Approaches for finding candidate clinical contexts include interviews and focus groups with subject matter experts, root cause analyses, event-reporting systems, prospective risk analysis techniques (e.g., FMEA), and case studies, preexisting scenarios (e.g., Hale and Ahlslager, 2011; Hetzel-Campbell and Daley, 2009), and closed cases from legal proceedings. Many of these potential sources draw upon real world examples of situations that have occurred. In general, the credibility of the scenario (i.e., is it clinically relevant and believable?) is easier to establish if the scenario events are based upon incidents that actually happened (Holland et al., 2008).

34.3.2.4 Develop an Event Set and Associated Teamwork Responses

Once the learning objectives and clinical context have been chosen, the raw materials for developing the event set are in place. An event set consists of trigger (or critical) events that happen during the course of the scenario that require a targeted teamwork behavior to manage effectively (Fowlkes et al., 1998). These events form the skeleton of the scenario and provide the linkages between learning objectives and opportunities to perform and measure targeted behaviors (discussed in the following section).

Examples of types of scripted events useful for SBTT include the following: (1) changes in simulated patient physiology or symptoms; (2) changes in other information critical to understanding the patient's condition (e.g., lab results are received, new patient history provided by family member); (3) scripted actions or inactions of confederates (e.g., a team member is an "actor" creating opportunities for other team members to perform); and (4) changes to workload, department flow and function, or environmental conditions (e.g., a team member is removed; increased admissions, increased wait times for beds or treatments, delayed discharges or gridlock, or environmental stressors are added such as a loud family member).

Each trigger or critical event is associated with an expected teamwork response. For example, in a scenario dealing with interactions between intensive care unit (ICU) staff and the pharmacy department, the identification of a missing dose of medication on the unit could be a trigger event for task or interpersonal conflict management behaviors between the parties involved. Expected teamwork responses can be defined using a behavioral markers approach used extensively in SBTT (Salas et al., 2007), particularly in aviation (Flin and Martin, 2001) and increasingly in health care (e.g., Fletcher et al., 2003). General examples of behavioral markers for teamwork competencies are shown in Table 34.2. Behavioral markers can serve as the basis of measurement tools as discussed in following sections.

In addition to these scripted events, the performance of trainees in the scenario can serve as an opportunity to perform teamwork behaviors. For example, if a team member is struggling with technical performance on their task, another team member may be able to provide task assistance. From an instructional perspective, the problem with these unstructured or unplanned events is that it may or may not be appropriate for them to occur within a given scenario. If they do occur, they are useful to discuss in the debrief; however, up-front planning is required to ensure they correctly occur.

34.3.2.5 Pilot Test, Refine, Implement, and Continuously Evaluate the Scenario

Scenarios rarely work exactly as intended the first time, and vetting them through several iterations of pilot testing and refinements is recommended. During this process logistical issues should be considered (e.g., can all of the planned events be executed as scheduled) as well as the trainee responses (e.g., do the events actually provide an opportunity for coordinated behavior?). After the scenario has been pilot tested and modified as needed, it is ready for initial implementation in an educational or training program. As a part of implementation, all staff involved in running the scenario must be trained. This includes the fundamental objectives, events, and critical teamwork and technical responses. The documented event set, critical responses, and scenario script produced in earlier steps of this process help to formalize and communicate what needs to happen when as well as the expectations for what the learners will be doing. Additionally, observers will need to be trained on the use of any measurement tools and their reliability established and monitored (discussed as follows). While the scenario is being used in training, data should be collected to ensure the scenario is effective. This includes eliciting learner reactions about the scenario including any suggestions for improving the realism of the scenario. But, more importantly, this involves tracking learner performance over time. A good scenario will provide both opportunities for success as well as opportunities to learn from less than perfect performance.

34.3.3 Scenario Development Summary

Table 34.3 illustrates the linkages between teamwork competencies, learning objectives, critical events, and expected behaviors described in this section with an example of a teamwork training scenario within an operating room (OR) emergency. Through this up-front investment and extensive documentation of the intent and structure of the scenario through explicit learning objectives, critical events, and behaviors, the remaining steps of performance measurement and team debrief are greatly facilitated. Section 34.4 will discuss how performance measurement can capitalize on the scenario design efforts and contribute to an effective program.

TABLE 34.3

Example Scenario Illustrating Linkages between Competencies, Learning Objectives, Critical Events, and Expected Teamwork Responses

OR Scenario: Arrest during Induction

Overview: A 79-year-old man is in the OR for an inguinal hernia repair. He will need to be anesthetized. Almost immediately after induction, the patient will suffer cardiac arrest and will need resuscitation. The patient has a history of coronary artery disease, so the arrest will not be extraordinary. The patient also has type II diabetes, which is poorly managed, and is obese

Targeted Teamwork Competencies	Learning Objectives
• Conducting effective team briefing • Communication (Callouts and Check-backs) • Assertiveness (two-challenge rule and red flag statements) • Leadership • Providing and receiving task assistance	1. Conduct an effective preprocedure briefing that includes introductions and role assignments for team members, patient and procedure information, and concerns and contingencies 2. Use call outs and check-backs for all critical information and task, medication, blood product, and fluid orders 3. Effectively request and give task assistance as needed 4. Use assertive communication when patient safety issues are encountered by applying the two-challenge rule and CUS script 5. Leaders (a) provide "big picture" situation updates when there are significant changes in patient status or plan of care, (b) delegate tasks by name or role, (c) clearly communicates plan of care

Learning Objectives	Critical Events	Expected Teamwork Responses
1. Briefs	Scenario begins	Leader conducts a prebrief addressing: • Team member roles • Patient and procedure information • Actively encourages assertiveness of other team members during procedure
2. Callouts and check-backs 3. Task-assistance 5. Leadership	Patient vitals show cardiac arrest during induction	Call-out of patient status change Leader clearly identifies him/herself Verbalization of plan by leader Clear delegation of tasks by team member role or name Check-backs on all information transfers and task, medication, blood, and fluid orders Back-up behavior provided as needed in CPR and other tasks
4. Assertiveness (two-challenge rule and CUS)	Surgeon verbalizes unsafe plan	Team member(s) assertively express safety concerns with proposed plan
2. Callouts and check-backs 3. Task-assistance	New plan verbalized based on successful two-challenge rule	Check-backs on all information transfers and medication, task, blood product, and fluid orders Back-up behavior requested and provided as needed
2. Callouts and check-backs 5. Leadership	Patient vitals stabilize	Call-out of patient status change Verbalization of new plan based on change in patient status

34.4 Team Performance Measurement in SBT

Performance measurement is a vital component of learning systems as it underlies the capacity to provide *systematic* corrective feedback, ascertain the learner's progress, and evaluate the effectiveness of the learning activities. Therefore, the purpose of this section is to provide an overview of the issues involved in developing team performance measurement systems for SBTT. Specifically, the major components of team performance measurement systems are presented within a decision point framework for the design of such systems (Rosen, Schiebel, Salas, Wu, Silvestri, and King, in press). Subsequently, a process for developing event-based measures for SBTT scenarios

that builds on the scenario design process outlined in Section 34.3 will be discussed.

34.4.1 Components of a Team Performance Measurement System

Table 34.4 summarizes a set of best practices for team performance measurement in SBTT (Rosen et al., 2008a). These best practices will be discussed in reviewing key decisions to be made when developing a team performance measurement system along with potential options and issues to consider at each decision point (Rosen et al., in press). Each of these core decisions is briefly reviewed in the following sections as they relate to SBTT.

TABLE 34.4

Best Practices for Team Performance Measurement in Health Care

1. *Ground measures in theory*
   - Use theory to answer questions about what to measure
   - Capture aspects of IPO models of team performance
2. *Design measures to meet specific learning outcomes*
   - Clearly articulate the purpose (i.e., specific learning outcomes) of measurement at the beginning of the development process
   - Design the measurement system to capture information necessary for making decisions about the learning outcomes (e.g., have they been met?)
3. *Capture competencies*
   - Explicitly link performance measures to the individual and team competencies targeted for training
4. *Measure multiple levels of performance*
   - Performance measures must be sensitive to differences in individual and team performance
5. *Link measures to scenario events*
   - Link performance measures to opportunities to perform in the simulation
   - Event-based measurement affords greater structure and standardization for observers
6. *Focus on observable behaviors*
   - Focusing on behaviors decreases the "drift" of observer's ratings over time (observers are less likely to develop idiosyncratic scoring)
   - Measuring observable behavior provides a fine level of granularity, which is necessary for providing corrective feedback
7. *Incorporate multiple measures from different sources*
   - Different measurement sources provide unique information
   - Generate a plan for rapidly integrating multiple sources of measurement
8. *Capture performance processes in addition to outcomes*
   - Team performance measurement should provide information not only about the end result of performance, but about how the team reached that performance outcome
   - Team performance measurement should provide information on how to correct team processes
9. *Create diagnostic power*
   - Team performance measurement should provide information about the causes of effective and ineffective performance
10. *Train observers and structure observation protocols*
    - Establish and implement a program of training and evaluation to ensure observers are accurately and reliably rating performance
    - Provide structured protocols to ease the information burden of observation
11. *Facilitate post-training debriefs and training remediation*
    - Team performance measurement should be quickly translated into feedback and decisions about required future training

Source: Adapted from Rosen, M.A. et al., *Simulation in Healthcare*, 3(1), 33–41, 2008.

34.4.1.1 Purpose: Why Measure

Performance measurement is an investment that should pay dividends in the form of increased capacity to make decisions. The nature of the decisions to be made with the information—the purpose of measurement—should drive the development of the measurement system. While measures can usually serve multiple purposes, there are inherent trade-offs between different approaches with some methods better suited for some purposes than others (e.g., Swing, 2002). For example, increasing the reliability of a measure at a certain threshold or criterion level of performance (an advantageous characteristic for assessment purposes) will decrease sensitivity of the measure across the full range of performance (an advantageous characteristic for diagnosing performance in learning environments). The purpose of measurement will vary with the overall purpose of the simulation activities. However, the present focus will be on two main types of purpose for measurement: scaffolding processes of learning and feedback provision, and evaluating the effectiveness of the learning opportunities or the overall impact of an SBTT implementation. Decisions in the following sections will be discussed relative to these two general aims.

34.4.1.2 Content: What to Measure

In SBTT, the core content of the measurement system will be rooted in the teamwork competency model and specific learning objectives for scenarios. This comprises the KSAs that underlie effective teamwork (Table 34.2; see also Salas et al., this volume). However, depending on the purpose and location of the SBTT activities, other aspects of performance may need to be included, such as multilevel evaluation frameworks and systems models in health care.

34.4.1.2.1 Team Science Frameworks

At least two core ideas from the science of teams can be used to develop specification of measurement content for SBTT: multilevel and input–process–output (IPO) frameworks. Figure 34.1 illustrates how these concepts can be used to develop a model of content for measurement.

First, team performance is a multilevel phenomenon (Kozlowski et al., 1999). It happens concurrently with individual level task performance and, when conducting in situ simulations, it is embedded within the context of the larger system. A measurement system needs to distinguish between individual, team, and multiteam system levels in order to diagnose performance—to understand the reasons why a certain task outcome level was reached (Cannon-Bowers and Salas, 1997; Rosen et al., 2008; Salas et al., 2007). Team performance diagnosis facilitates the provision of feedback (e.g., providing individual level feedback for technical competence deficiencies, providing team level feedback for team performance deficiencies) and decisions about remediation.

Second, team science frameworks distinguish between team inputs, processes and emergent states, and outcomes (i.e., the IPO framework of team performance; Ilgen et al., 2005). *Team inputs* include a broad variety of relatively stable features of the team, its members, the task, and the environmental context. For the purposes of SBTT, the most salient team inputs are team member attitudes and knowledge competencies (e.g., shared mental models; Cannon-Bowers et al., 1993). *Team processes* are the interdependent actions taken by team members during a performance episode and correspond to behavioral competencies. Team processes can be categorized as action, transition, or interpersonal in nature (Marks et al., 2001). *Team outputs* are the result of the team's collective efforts and include task outcomes (e.g., efficiency, effectiveness), team viability (i.e., the team members' ability to work together in the future), and team level learning.

34.4.1.2.2 Training, Education, and Program Evaluation Frameworks

When the purpose of measurement is to provide an evaluation or validation of the educational or training program, multilevel evaluation frameworks can be used to expand the content suggested by team science frameworks. Several such frameworks exist including Kirkpatrick's training evaluation framework (Kirkpatrick and Kirkpatrick, 2006; Kraiger et al., 1993), Bloom's educational taxonomy (Anderson et al., 2001), and the context input process product (CIPP; Farley and Battles, 2008) model of program evaluation.

In SBTT, multilevel training evaluation is likely the most commonly used approach. A standard multilevel training evaluation includes up to five levels: reactions (i.e., affective reactions and utility judgments; Alliger et al., 1997), learning (i.e., acquisition of targeted competencies), behavior (i.e., transfer of learned competencies), results (i.e., the degree to which the SBTT is impacting important organizational variables such as patient safety and quality of care, patient or staff satisfaction, and culture change), and return on investment.

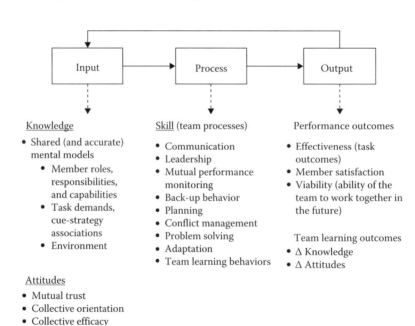

FIGURE 34.1
Examples of *teamwork* measurement content indicated by IPO frameworks.

34.4.1.2.3 Systems Models

When SBTT occurs in situ, the sphere of measurement content should expand beyond the actual team to include features of the environment and broader care delivery system as these factors are no longer controlled aspects of the scenario as they are in a simulation center. For example, a model of sociotechnical systems in health care (e.g., Flin, Winter, Sarac, and Raduma, 2009; Reason 2000) can be used to generate rough categories of broader systems issues to capture during in situ simulation. This includes the patient, work environment and equipment, individual, team, and organizational and management issues.

34.4.1.3 Location: Where to Measure

As discussed previously, simulation in health care occurs in two general locations: dedicated simulation centers and in situ, on the actual unit. This raises two main issues for measurement. First, if highly reliable and diagnostic assessment of team member or team level performance is desired (e.g., for certifications), standardized scenario content is critical. Scenario events are analogous to test items, and for a test to be valid, the items need to be consistent across respondents (i.e., teams in SBTT). Consequently, center-based simulations will likely provide more reliable and valid measurement; however, no formal comparative studies have been conducted in this regard. Second, in terms of purpose and content, if the focus is on understanding teams and broader systems issues, in situ approaches are critical.

34.4.1.4 Who: Selecting, Training, and Supporting Raters

While it is common to discuss the psychometric properties of a measurement tool, the ultimate reliability of the tool is a property of the entire measurement system including the observational protocol itself *and* the observers using it. Consequently, selecting raters with background knowledge in the clinical domain and the targeted teamwork skills is a good first step. However, developing this knowledge is possible by training raters to a criterion (e.g., using expertly scored ratings from videotaped scenarios) and monitoring their scoring performance over time for rater drift is the standard approach to achieving reliability. This process can be facilitated by providing performance support tools such as scoring guides (e.g., examples, written or videotaped, of clearly identified types and levels of performance), job aides, and refresher training.

34.4.1.5 Method: How to Measure

Measurement method refers to how data are collected and how scores are generated. There are two broad categories of measurement methods germane to SBTT in health care: self-report and observation. Several generic team performance measurement tools have been developed specifically for use in simulation, including both self-report (e.g., Malec et al., 2007) and observational methods (e.g., Guise et al., 2008). Table 34.5 provides a summary of the primary collection and scoring methods along with their associated strengths and weaknesses. For a thorough discussion of these methods, see Rosen et al. (in press).

34.4.2 Developing Event-Based Measures

In event-based methods, the basic ingredients of performance measurement tools are generated by clearly defining learning objectives, event sets, and expected teamwork responses.

Building on the example scenario discussed previously (Table 34.3), an event-based checklist is provided in Figure 34.2, and steps for generating this type of tool are discussed in the following section.

First, the content of the measurement tool created during the scenario design process needs to be organized. Specifically, the event set and targeted responses create the structure and content of the tool. The trigger events can be organized sequentially to provide a temporal structure for the scenario. Subsequently, the targeted responses can be grouped around these trigger events. This facilitates observation by focusing the attention on the observers on the behaviors to rate at any given time during the scenario.

Second, a scoring method must be chosen for the content. There are multiple ways to score or rate the same content. The primary methods have been discussed in Section 34.4.1 and are summarized in Table 34.5. In addition to these methods, simple dichotomous scoring has been widely applied to event-based checklists. Here, the targeted behaviors are rated as either a hit (i.e., the behavior was performed) or a miss (i.e., the behavior was not performed). This is a generally effective method; however, it does not capture nuances in the quality with which "hits" were performed. To address this, more response categories can be added. For example, instead of two categories, a third can be added to distinguish between "hits" performed well and "hits" performed but in need of improvement (e.g., variation in quality; Frankel et al., 2007). Whichever scoring methodology is chosen, raters need to be trained and their agreement established and monitored over time.

Third, information in addition to the scoring of behaviors may be captured. This includes whether or not instructor guidance was used to prompt or coach the targeted behavior. A learner performing a targeted behavior after intervention by an instructor is different from a learner performing the behavior independently.

TABLE 34.5

Summary of Major Methods of Rating Team Performance and Associated Strengths and Challenges

Measurement Tool Type	Description	Strengths/Challenges and Limitations	Example Citations
Self-report	Questionnaires administered to each team member individually and then aggregated in some manner to represent a team score	*Strengths* • Self-report methods are the best technique for capturing *affective teamwork competencies* (e.g., collective efficacy, mutual trust, etc.) as well as *knowledge teamwork competencies* (e.g., knowledge about role structure) *Challenges and limitations* • Self-report of the perceptions of teamwork can be subject to a number biases (e.g., referent shift bias) • Aggregating individual responses to the team level can be challenging	Malec et al. (2007) and Smith-Jentsch et al. (2008)
Global ratings scale	Observational protocol that asks raters to use a Likert scale to rate the quality of a given teamwork dimension	*Strengths* • This type of rating is familiar to people and on the surface is very simple to use *Challenges and limitations* • Can be difficult to achieve inter-rater reliability because the rating scales tend to be generic and abstract (e.g., rating "communication" as a whole) • Usually one score is given for an entire performance episode; however, performance fluctuates (i.e., sometimes things are done well, other times they are not) and raters must summate across time, which can create challenges for maintaining inter-rater reliability	Hohenhaus, Powell, and Haskins (2008)
Behaviorally anchored ratings scale (BARS)	Observer rating protocol that provides brief descriptions of teamwork behaviors as anchors for each rating dimension	*Strengths* • The ratings scales are more concrete and descriptive than general Likert anchors (e.g., highly effective to highly ineffective), which helps to facilitate a common frame of reference for different raters *Challenges and limitations* • BARS require raters to summate over time; that is, they provide one rating of a team performance dimension (e.g., communication) for a specified period of time (e.g., an entire performance episode or phase of performance) even though performance may vary within that time period	Kendall and Salas (2003) and Murphy and Pardaffy (1989)
Behavior observation scale	Observer rating protocol using a Likert type scale to rate the frequency of teamwork behaviors	*Strengths* • Produces a rating of a team's typical level of performance over time *Challenges and limitations* • These tools require raters to estimate the frequency of teamwork behaviors and therefore may be susceptible to primacy and recency effects	Salas et al. (2008) and Tziner, Joanis, and Murphy (2000)

Frequency counts	Observational protocol that requires the rater to count the number of times a specific team behavior occurs	*Strengths* • If the behaviors being looked for are specific enough, it is usually relatively easy to train observers to detect the targeted teamwork behaviors *Challenges and limitations* • It is difficult to develop and maintain inter-rater reliability of "misses" (i.e., instances where a specific teamwork behavior should have occurred, but did not). The miss count is important for understanding the big picture of a team's performance	Frankel et al. (2007) and Weaver et al. (2010)
Event-based tools	Observational protocol consisting of behavioral checklists linked to simulation scenario "trigger events," learning objectives, and teamwork competencies	*Strengths* • Maintains linkages between opportunities to measure (i.e., scenario events), targeted teamwork behaviors, and learning objectives • Focuses the observer's attention on predefined events and concrete behaviors *Challenges and limitations* • This method almost exclusively limited to use in simulation as it requires that observers be able to predict events in the environment • Measurement tools are specific to each simulation scenario	Fowlkes et al. (1998), Rosen et al. (2008), Lazzara et al. (in press)

Source: Adapted from Rosen, M.A. et al., How can team performance be measured, assessed, and diagnosed? In K. Frush and E. Salas (Eds.), *Improving Patient Safety through Teamwork and Team Training,* Oxford University Press, Oxford, U.K., in press.

Critical Events	Expected Teamwork Responses	Hit	ViQ	Miss	IG	NA
Scenario begins	Leader conducts a pre-brief addressing:					
	Member introductions and roles					
	Patient and procedure information					
	Actively encourages assertiveness of team members during procedure					
Patient vitals show cardiac arrest during induction	Call-out of patient status change (audible to all team members)					
	Leader clearly identifies him/herself					
	Verbalization of plan by leader					
	Clear delegation of tasks by team member role or name					
	Check-backs on all information transfers and task, medication, blood, and fluid orders					
	Back-up behavior provided as needed in CPR and other tasks					
Surgeon verbalizes unsafe plan	Team member(s) assertively express safety concerns with proposed plan					
New and appropriate plan verbalized	Check-backs on all information transfers and medication, task, blood product, and fluid orders					
	Back-up behavior requested and provided as needed					
Patient vitals stabilize	Call-out of patient status change					
	Verbalization of new plan based on change in patient status					

FIGURE 34.2

Example of event-based checklist for scenario described in Table 34.4. Hit = the behavior was performed meeting expectations; ViQ = variation in quality; a form of the behavior was performed, but improvement is needed; Miss = the targeted behavior was not performed; IG = the targeted behavior was performed, but only after instructor guidance; NA = not applicable, there was no opportunity to perform the behavior.

34.5 Debrief Facilitation

Whether conducted post-simulation scenario or as a part of routine practice, team debriefs are a powerful tool for building effective and adaptive teams (e.g., Ellis and Davidi, 2005; Smith-Jentsch et al., 2008). Even simple team scenarios can be perceived as complex and without a feedback system to guide learning, learners may walk away from the experience confused, or having learned the incorrect lessons. For these reasons, debriefing has been identified as a critical aspect of SBT in general, and specifically for teams (Dismukes and Smith, 2000; Fanning and Gaba, 2007).

For most scenarios involving teamwork, a facilitated debrief is the preferred method of feedback. This involves active participation on the part of the learners. The role of the facilitator is to guide the discussion and prompt the team to self-assess their performance. Performance measurement tools can help to structure this process, but ultimately the skills of the debrief facilitator remain critical. Therefore, an SBTT system must include a method for developing the debrief facilitation skills of its faculty.

This section provides an introduction to basic issues in debrief facilitation. Specifically, a short three-part structure for conducting a team debrief is provided. Subsequently, a set of best practices for conducing team debriefs are discussed.

34.5.1 Phases of a Team Debrief

The nature of effective debriefing is an active topic of debate and research in the simulation in health care community. While there may be no one single correct approach (Dismukes et al., 2006), there is a consensus among researchers and practitioners from a variety of domains on key elements of an effective debrief (Lederman, 1992; Morrison and Meliza, 1999; Rudolph et al., 2008). Specifically, debriefs are frequently described in terms of four main phases: an introduction to the process, a simple description of events, an analysis of events, and a discussion of how lessons learned can be applied or generalized to future performance.

First, as simulation is still a relatively new experience for many health care providers, an introduction to the process can help to set expectations of the role of facilitator and the learners (i.e., active participation with professionalism). Additionally, an articulation of the goals (e.g., learning and improvement, not to identify weaknesses for punitive purposes) of the debrief can help to establish the tone of the debrief and a learning climate (Rudolph et al., 2007). This has been noted as a critical yet frequently overlooked aspect of team debriefings (Dismukes et al., 2000).

Second, the purpose of the description or recap phase of a debrief is to bring all team members to the same understanding of the basic facts of the scenario that just occurred—what happened when. Because there are multiple roles in a scenario, and team members

TABLE 34.6

General Framework for the Four Phases of a Team Debrief and Practical Tips

Debrief Phase	Practical Tips
Introduction to the debrief Provide overview of the debrief process and expectations for trainee participation	• The goal of the debrief is for the team members to give their own report on what happened and to provide an in-depth analysis and critique of their own performance • The role of the debrief facilitator is to guide this process, not to lecture • The rationale for this approach is to maximize team member involvement in the process and learning from the scenario • The debrief is about *learning and improvement*, not about *evaluation*
Description Guide the team's discussion of what happened during the scenario	• Organize your personal understanding of what happened; use your notes and performance ratings to prioritize the key events and learning points • Focus discussion on critical performance aspects during the scenario; many things happened, some much more important than others • Do not let the team linger on overly lengthy descriptions of what happened (this can be a way of avoiding talking about why it happened) • Make sure that everyone has a chance to contribute: draw out quiet team members and pull attention away from more vocal team members
Analysis Guide the team's discussion of *why* things occurred as they did and the assessment of quality of performance	• Avoid finger pointing • Focus on the teamwork competencies being trained • Use appropriate questioning: open-ended questions, advocacy/inquiry • Turn team member questions and comments back on them for further discussion • Use active listening • Use silence appropriately
Application/generalization Guide the team to develop lessons learned and plans for improvement	• Prompt team members to discuss actual events on the ward that were similar to the training scenario • Prompt team members to discuss how what the team did right can be applied on the ward • Prompt team members to discuss what they would do differently • Prompt team members to discuss how what they do on the ward will be different based on what they have learned from the scenario

are usually engaged in different individual tasks, different team members will usually have differing and incomplete understandings of everything that occurred during the scenario. The description phase of the debrief provides the basic ingredients for learning.

Third, while necessary, a description of what happened is insufficient to maximize learning. Team members should be guided through a self-diagnosis of performance (Salas et al., 2007); that is, they should seek to evaluate the quality of their performance and understand the causes of their performance (e.g., why did events unfold as they did, which aspects of performance were good, and which need improvement).

Fourth, after the team members have identified opportunities to improve their performance, they need to develop specific lessons learned. Ideally, these are in the form of plans for how the team will improve upon their performance for the future.

The debrief facilitator's role through each of these steps is to guide and coach self-reflection, not to lecture. It is to ensure that all team members are able to voice their perspectives and that the team's conclusions about their performance match with the reality of their performance. Practical tips for each of these phases are provided in Table 34.6.

34.5.2 Best Practices in Team Debriefing

Recently, an evidence-based set of best practices have been produced for conducting team debriefs (Salas et al., 2008). These best practices are adapted for the context of SBT in the following sections. These best practices are rooted in a systems-based perspective as well as they deal with issues before, during, and after the actual debrief.

34.5.2.1 *Establish a Learning Climate*

Team members need to feel comfortable during a debrief as they are being asked to share their perspectives about the team's performance (Wilson et al., 2005). This involves mutual trust and respect on the part of the learners and facilitator. This will not happen if people perceive the situation to be punitive in nature (e.g., people are "pointing fingers" or "playing the blame game") or that people will lose face for admitting errors or showing vulnerabilities.

34.5.2.2 Team Debriefs Should Be Diagnostic

The debrief process should focus on understanding team's strengths and weaknesses (Morrison and Meliza, 1999). Just as an underlying cause for a patient's condition is inferred from the symptoms, a team's needs for further development (i.e., competencies or behavioral skill areas needing improvement) should be assessed on the basis of performance in the scenario. It is one of the facilitator's primary responsibilities to ensure that the discussion moves deeper than a discussion of what happened and focuses on the underlying causes of the team's performance.

34.5.2.3 Focus on a Few Key Performance Issues

In most cases, learners will only leave an SBTT episode with a few core ideas or learning points. The debrief then should focus as much as possible on exploring and reinforcing a limited set of teamwork skills. This strategy can be more effective than touching on many issues, but addressing each in a very shallow manner.

34.5.2.4 Maintain a Focus on Teamwork Processes

Most clinicians are not accustom to discussing teamwork issues and consequently team debriefs can easily gravitate toward discussion of clinical content. This is natural, but if the purpose of the scenario is to develop teamwork skills, the debrief discussion needs to focus on teamwork behaviors that occurred during the scenario that contributed to the team's overall performance (Smith-Jentsch et al., 2008). It is the facilitator's role to guide the discussion toward teamwork, but this can be greatly aided by designing scenarios with concretely correct clinical solutions for novice learners.

34.5.2.5 Focus on the Processes of Performance

In many situations, health care included, it is possible to "do everything right" and still have a bad outcome. Consequently, the outcomes of the team's performance are less important than the processes they engaged in during the scenario (Cooper et al., 1984; Savoldelli et al., 2006). Outcomes cannot be trained, only the processes that lead to desired outcomes.

34.5.2.6 Support Feedback with Objective Indicators of Performance

Tools can be used to help structure the debrief process, including checklists and the use of video recordings of simulation scenarios to clarify the details of what did and did not happen during a scenario as well as performance ratings made by observers/trainers and self-report ratings by the team members. These different sources of information can be valuable additions to the debrief process by providing more context for the discussion and enabling team members to draw more concrete connections between the discussion points and their performance; however, they are not always effective, and more research is needed to provide specific guidance on when video is most useful (Savoldelli, Naik, Park, Joo, Chow, and Hamstra, 2006).

34.5.2.7 Provide Both Individual and Team Feedback and Know When Each Is Appropriate

Ultimately a team's effectiveness is determined by how individual members perform their own tasks and how they work together. In SBT, teamwork is the focus, but individual performance issues will frequently need to be addressed. Feedback needs to be provided to each of these levels as appropriate. If there is an issue with team's performance as a whole, it is a topic that should be addressed in the team debrief. If there are issues with an individual's technical performance, this feedback should most likely be given to the individual alone. This avoids occupying the team's time with information not relevant to everyone as well as putting the individual in the awkward position of receiving feedback on poor performance in front of the group.

34.5.2.8 Record the Outcomes of the Debrief

The conclusions, lessons learned, and plans for improving team performance should be recorded. These can be used to assess performance over time as well as points of reference in future debriefs. Team members can identify what they are improving and what remains a chronic issue with their performance.

34.5.2.9 Educate Trainers and Team Members on the Art and Science of Facilitation

Facilitating a debrief is not a natural skill for most people. In SBT, trainers need to understand how to run an effective debrief. Additionally, team members can use debriefing skills on the unit as a part of their continuous improvement. Therefore, SBT can be an opportunity not just to practice team performance behaviors, but of effective processes for learning from experience as a team. The overview information provided here is a start, but developing facilitation skills requires a long-term commitment to training, education, and assessment.

34.6 Concluding Remarks

SBTT has proven to be a foundational approach to improving quality and safety in health care. However, SBTT applications are diversifying and evolving to best meet the needs of the health care community. This includes growing past a purely training-based approach seeking to improve learner competence to methods that include process improvement approaches for broader systems issues. SBTT is also being applied at earlier phases of the development of health care professionals. Additionally, clinical domains that have historically been underserved with SBTT such as ambulatory care clinics are making efforts to manage extensive culture and change processes through, in part, the application of teamwork interventions. Even with this growth in the area, the fundamental components of an SBTT program discussed in this chapter will remain critical to effectiveness.

References

Alavi, S. B. and McCormick, J. (2004). Theoretical and Measurement Issues for Studies of Collective Orientation in Team Contexts. *Small Group Research, 35*(2), 111–127.

Alliger, G. M., Tannenbaum, S. I., Bennett, W., Traver, H., and Shotland, A. (1997). A meta-analysis of the relations among training criteria. *Personnel Psychology, 50*, 341–358.

Alonso, A., Baker, D., Holtzman, A., Day, R., King, H., Toomey, L., and Salas, E. (2006). Reducing medical error in the military health system: How can team training help? *Human Resource Management Review, 16*(3), 396–415.

Anderson, L. W., Krathwohl, D. R., Airasian, P. W., Cruikshank, K. A., Mayer, R. E., Pintrich, P. R. et al. (2001). *A Taxonomy for Learning, Teaching, and Assessning: Revision of Boom's Taxonomy of Educational Objectives*. New York: Longman.

Artman, H. (2000). Team situation assessment and information distribution. *Ergonomics, 43*(8), 1111–1129.

Bandura, A. (1986). *Social foundations of thought and action: A social cognitive theory*. Rockville, MD: National Institutes of Mental Health.

Beaubein, J. M. and Baker, D. P. (2004). The use of simulation for training teamwork skills in health care: How low can you go? *Quality and Safety in Health Care, 13*, i51–i56.

Boulet, J. R., Murray, D., Kras, J., and Woodhouse, J. (2008). Setting performance standards for mannequin-based acute-care scenarios: An examinee-centered approach. *Simulation in Healthcare, 3*(2), 72–81.

Bowers, C. A., Jentsch, F., Salas, E., and Braun, C.C. (1998). Analyzing communication sequences for team training needs assessment. *Human Factors, 40*(4), 672–679.

Burke, C. S., Stagl, K. C., Klein, C., Goodwin, G. F., Salas, E., and Halpin, S. M. (2006a). What type of leadership behaviors are functional in teams? A meta-analysis. *The Leadership Quarterly, 17*, 288–307.

Burke, C. S., Stagl, K. C., Salas, E., Pierce, L., and Kendall, D. (2006b). Understanding team adaptation: A conceptual analysis & model. *Journal of Applied Psychology, 91*(6), 1189–1207.

Burke, C. S. (2005). Team task analysis. In N. Stanton, A. Hedge, K. Brookhuis, E. Salas, and H. Hendrick (Eds.), *Handbook of Human Factors and Ergonomics Methods* (pp. 51–58). Boca Raton, FL: CRC Press.

Burtscher, M. J., Wacker, J., Grote, G., and Manser, T. (2010). Managing nonroutine events in anesthesia: The role of adaptive coordination. *Human Factors, 282*–294.

Calatayd, D., Arora, S., Aggarwal, R., Kruglikova, I., Schulze, S., Funch-Jensen, P. et al. (2010). Warm-up in a virtual reality environment improves performance in the operating room. *Annals of Surgery, 251*(6), 1181–1185.

Cambell, C. H., Deter, D. E., and Quinkert, K. A. (1997). *Report on the Expanded Methodology for Development of Structured Simulation-Based Training Programs*. Alexandria, VA: U.S. Army Research Institute for the Behavioral and Social Sciences.

Cannon-Bowers, J. A. and Salas, E. (1997). A framework for developing team performance measures in training. In M. T. Brannick, E. Salas, and C. Prince (Eds.), *Team Performance and Measurement: Theory, Methods, and Applications* (pp. 45–62). Mahwah, NJ: Erlbaum.

Cannon-Bowers, J. A., Salas, E., and Converse, S. (1993). Shared mental models in expert team decision making. In N. J. J. Castellan (Ed.), *Individual and Group Decision Making* (pp. 221–246). Hillsdale, NJ: Erlbaum.

Cannon-Bowers, J. A., Tannenbaum, S. I., Salas, E., and Volpe, C. E. (1995). Defining competencies and establishing tea training requirements. In R. Guzzo and E. Salas (Eds.), *Team Effectiveness and Decision Making in Organizations*. San Francisco, CA: Jossey-Bass.

Chen, J. G., Mistry, K. P., Wright, M. C., and Turner, D. A. (2010). Postoperative handoff communication: A simulation-based training method. *Simulation in Healthcare, 5*(4), 242–247.

Cooper, J. B., Newbower, R. S., and Kitz, R. J. (1984). An analysis of errors and equipment failures in anesthesia management: Considerations for detection and prevention. *Anesthesiology, 60*, 34–42.

Crandall, B., Klein, G., and Hoffman, R. R. (2006). *Working Minds: A Practitioner's Guide to Cognitive Task Analysis*. Cambridge, MA: MIT Press.

Davis, S., Riley, R. H., Gürses, A. P., Miller, K., and Hansen, H. (2008). Failure modes and effects analysis based on in situ simulations: A methodology to improve understanding of risks and failures.

Day, D. V., Gronn, P., and Salas, E. (2004). Leadership capacity in teams. *Leadership Quarterly, 15*(6), 857–880.

Dickinson, T. L. and McIntyre, R.M. (1997). A Conceptual framework for team measurement. In M. T. Brannick, Salas, E., & Prince, C. (Ed.), *Team Performance Measurement: Theory, Methods, and Applications* (pp. 19–43). Mahwah, NJ: Erlbaum.

Deering, S., Rosen, M. A., Salas, E., and King, H. B. (2009). Building team and technical competency for obstetric emergencies: The mobile obstetric emergency simulator (MOES) system. *Simulation in Healthcare, 4*(3), 166.

Dismukes, R. K., Gaba, D. M., and Howard, S. K. (2006). So many roads: Facilitated debriefing in healthcare. *Simulation in Healthcare, 1*(1), 23–25.

Dismukes, R. K., McDonnell, L. K., and Jobe, K. K. (2000). Facilitating LOFT debriefings: Instructor techniques and crew participation. *The International Journal of Aviation Psychology, 10*(1), 35–57.

Dismukes, R. K. and Smith, G. M. (Eds.) (2000). *Facilitation and Debriefing in Aviation Training Operations.* Aldershot, U.K.: Ashgate.

De Dreu, C. K. and Weingart, L. R. (2003). Task versus relationship conflict, team performance, and team member satisfaction: a meta-analysis. *Journal of Applied Psychology, 88*(4), 741–749.

Do, A. T. Cabbad, M. F., Kerr, A., Serur, E., Robertazzi, R. R., and Stankovic, M. R. (2006). A warm-up laparoscopic exercises improves the subsequent laparoscopic performance of Ob-Gyn residents: A low-cost laparoscopic trainer. *Journal of the Society of Laparoendoscopic Surgeons, 10*(3), 297–301.

Draycott, T. and Crofts, J. (2006). Structured team training in obstetrics and its impact on outcome. *Fetal and Maternal Medicine Review, 17*(3), 229–237.

Draycott, T., Sibanda, T., Owen, L., Akande, V., Winter, C., Reading, S. et al. (2005). Does training in obstetric emergencies improve neonatal outcome? *BJOG: An International Journal of Obstetrics and Gynaecology, 113,* 177–182.

Dwyer, D. J., Oser, R. L., Salas, E., and Fowlkes, J. E. (1999). Performance measurement in distributed environments: Initial results and implications for training. *Military Psychology, 11*(2), 189–215.

Eby, L. T. and Dobbins, G.H. (1997). Collectivistic orientation in teams: an individual and group-level analysis. *Journal of Organizational Behavior, 18,* 275–295.

Edmondson, A. C. (2003). Speaking up in the operating room: How team leaders promote learning in interdisciplinary action teams. *Journal of Management Studies, 40*(6), 1419–1452.

Ehrenstein, A., Walker, B. N., Czerwinski, M., and Feldman, E. M. (1997). Some fundamentals of training and transfer: Practice benefits are not automatic. In M. A. Quinones and A. Ehrenstein (Eds.), *Training for a Rapidly Changing Workplace: Applications of Psychological Research* (pp. 119–147). Washington, DC: American Psychological Association.

Entin, E. E. and Serfaty, D. (1999). Adaptive team coordination. *Hum Factors, 41*(2), 312–325.

Ellis, S. and Davidi, I. (2005). After-even reviews: Drawing lessons from successful and failed experience. *Journal of Applied Psychology, 90*(5), 857–871

Fanning, R. M. and Gaba, D. M. (2007). The role of debriefing in simulation-based learning. *Simulation in Healthcare, 2*(1), 115–125.

Farley, D. O. and Battles, J. B. (2008). Evaluation of the AHRQ patient safety initiative: Framework and approaches. *Health Services Research, 44*(2), 628–645.

Fernandez, R., Kozlowski, S. W. J., Shapiro, M. J., and Salas, E. (2008). Toward a definition of teamwork in emergency medicine. *Academic Emergency Medicine, 15*(11), 1104–1112.

Fletcher, G., Flin, R., McGeorge, P., Glavin, R. J., Maran, N., and Patey, R. (2003). Anaesthetists' non-technical skills (ANTS): Evaluation of a behavioural marker system. *British Journal of Anaesthesia, 90*(5), 580–588.

Flin, R. and Maran, N. (2008). Non-technical skills: Identifying, training, and assessing safe behaviors. In R. H. Riley (Ed.), *Manual of Simulation in Healthcare* (pp. 303–319). Oxford, U.K.: Oxford University Press.

Flin, R. and Martin, L. (2001). Behavioral markers for crew resource management: A review of current practice. *The International Journal of Aviation Psychology, 11*(1), 95–118.

Flin, R., Winter, J., Sarac, C., and Raduma, M. A. (2009). *Human factors in patient safety: Review of topics and tools*: World Health Organization.

Fowlkes, J. E., Dwyer, D. J., Oser, R. L., and Salas, E. (1998). Event-based approach to training (EBAT). *The International Journal of Aviation Psychology, 8*(3), 209–221.

Fowlkes, J. E., Lane, N. E., Salas, E., Franz, T., and Oser, R. (1994). Improving the measurement of team performance: The TARGETs methodology. *Military Psychology, 6,* 47–61.

Frankel, A., Gardner, R., Maynard, L., and Kelly, A. (2007). Using the communication and teamwork skills (CATS) assessment to measure health care team performance. *Joint Commission Journal on Quality and Patient Safety, 33,* 549–558.

Frankel, A., Leonard, M., Simmonds, T., Haraden, C., and Vega, K. B. (Eds.) (2009). *The Essential Guide for Patient Safety Officers.* Oakbrook Terrace, IL: Joint Commission Resources and the Institute for Healthcare Improvement.

Gaba, D. M. (2004). The future vision of simulation in health care. *Quality and Safety in Health Care, 13*(Suppl. 1), i2–i10.

Gaba, D. M., Howard, S. K., Fish, K. J., Smith, B. E., and Sowb, Y. A. (2001). Simulation-based training in anesthesia crisis resource management (ACRM): A decade of experience. *Simulation and Gaming, 32*(2), 175–193.

Gibson, C. B. (2003). The Efficacy Advantage: Factors Related to the Formation of Group Efficacy. *Journal of Applied Social Psychology, 33*(10), 2153–2186.

Goldstein, I. L. and Ford, K. (2001). *Training in Organizations: Needs Assessment, Development, and Evaluation* (4th ed.). New York: Wadsworth Publishing.

Goldstein, I. L. and Ford, J. K. (2002). *Training in Organizations* (4th ed.). Belmont, CA: Wadsworth.

Guise, J., Deering, S. H., Kanki, B. G., Osterweil, P., Li, H., Mori, M. et al. (2008). Validation of a tool to measure and promote clinical teamwork. *Simulation in Healthcare, 3*(4), 217–223.

Hale, T. J. and Ahlsclager, P. M. (2011). *Simulation Scenarios for Nursing Education.* Clifton Park, NY: Delmar Cengage Learning.

Hays, R. T. and Singer, M. J. (1989). *Simulation Fidelity in Training System Design.* New York: Springer-Verlag.

Hetzel-Campbell, S. and Daley, K. M. (Eds.) (2009). *Simulation Scenarios for Nurse Educators: Making It Real.* New York: Springer.

Hohenhaus, S. M., Powell, S., and Haskins, R. (2008). A practical approach to observation of the emergency care setting. *Journal of Emergency Nursing, 34*(2), 142–144.

Holland, C., Sadler, C., and Nunn, A. (2008). Scenario design: Theory to delivery. In R. H. Riley (Ed.), *Manual of Simulation in Healthcare* (pp. 139–149). Oxford, U.K.: Oxford University Press.

Ilgen, D. R., Hollenbeck, J. R., Johnson, M., and Jundt, D. (2005). Teams in organizations: From input-process-output models to IMOI models. *Annual Review of Psychology, 56,* 517–543.

Issenberg, S. B., McGaghie, W. C., Petrusa, E. R., Gordon, J. A., and Scalese, R. J. (2005). Features and uses of high-fidelity medical simulations that lead to effective learning: A BEME systematic review. *Medical Teacher, 27*(1), 10–28.

Jackson, C. L., Colquitt, J.A., Wesson, M.J., and Zapata-Phelan, C.P. (2006). Psychological Collectivism: A Measurement Validation and Linkage to Group Member Performance. *Journal of Applied Psychology, 91*(4), 884–899.

Jordan, P. J. and Troth, A.C. (2004). Managing emotions during team problem solving: Emotional intelligence and conflict resolution. *Human Performance, 17*(2), 195–218.

Kahol, K., Satava, R. M., Ferrara, J., and Smith, M. L. (2008). Effect of short-term pretrial practice on surgical proficiency in simulated environments: A randomized trial of the "preoperative warm-up" effect. *Journal of the American College of Surgeons, 208*(2), 255–268.

Katz-Navon, T. Y. and Erez, M. (2005). When collective- and self-efficacy affect team performance: The role of task interdependence. *Small Group Research, 36*(4), 437–465.

Kern, D. E., Thomas, P. A., and Hughes, M. T. (2009). *Curriculum Development for Medical Education* (2nd ed.). Baltimore, MD: The Johns Hopkins University Press.

Kendall, D. L. and Salas, E. (2004). Measuring Team Performance: Review of Current Methods and Consideration of Future Needs. In T. J.W. Ness, V., & Ritzer, D. (Ed.), *The Science and Simulation of Human Performance* (pp. 307–326). Boston: Elsevier.

Kirkpatrick, D. L. and Kirkpatrick, J. D. (2006). *Evaluating Training Programs*. San Francisco, CA: Berrett-Koehler Publishers.

Kirschner, P. A., Sweller, J., and Clark, R. E. (2006). Why minimal guidance during instruction does not work: An analysis of the failure of constructivist, discovery, problem-based, experiential, and inquiry-based training. *Educational Psychologist, 41*(2), 75–86.

Klimoski, R. and Mohammed, S. (1994). Team mental model: Construct or metaphor? *Journal of Management, 20*(2), 403–437.

Kohn, L. T., Corrigan, J. M., and Donaldson, M. S. (2000). *To Err Is Human*. Washington, DC: Institute of Medicine.

Kozlowski, S. W. J., Gully, S. M., Nason, E. R., and Smith, E. M. (1999). Developing adaptive teams: A theory of compilation and performance across levels and time. In D. R. Ilgen and E. D. Pulakos (Eds.), *The Changing Nature of Work and Performance: Implications for Staffing, Personnel Actions, and Development*. San Francisco, CA: Jossey-Bass.

Kraiger, K., Ford, J. K., and Salas, E. (1993). Application of cognitive, skill-based, and affective theories of learning outcomes to new methods of training evaluation. *Journal of Applied Psychology, 78*(2), 311–328.

Lazzara, E. H., Weaver, S. L., DiazGranados, D., Rosen, M. A., Salas, E., Wu, T. S. et al. (2010). TEAM MEDSS: A tool for designing medical simulation scenarios. *Ergonomics in Design, 18*(1), 11–17.

Lederman, L. C. (1992). Debriefing: Toward a systematic assessment of theory and practice. *Simulation & Gaming, 23*(2), 145–160.

Malec, J. F., Torsher, L. C., Dunn, W. F., Wiegmann, D. A., Arnold, J. J., Brown, D. A. et al. (2007). The mayo high performance teamwork scale: Reliability and validity for evaluating key crew resource management skills. *Simulation in Healthcare, 2*(1), 4–10.

Maran, N. J. and Glavin, R. J. (2003). Low to high fidelity simulation: A continuum of medical education? *Medical Education, 37*, 22–28.

Marks, M. A., Mathieu, J. E., and Zaccaro, S. J. (2001). A temporally based framework and taxonomy of team processes. *Academy of Management Review, 26*, 356–376.

Marks, M. A. and Panzer, F.J. (2004). The influence of team monitoring on team processes and performance. *Human Performance, 17*(1), 25-41.

Mayer, R. E. (Ed.) (2005). *The Cambridge Handbook of Multimedia Learning*. Cambridge, U.K.: Cambridge University Press.

McGaghie, W. C., Issenberg, S. B., Petrusa, E. R., and Scalese, R. J. (2010). A critical review of simulation-based medical education research: 2003–2009. *Medical Teacher, 44*, 50–63.

McIntyre, R. M. and Salas, E. (1995). Measuring and managing for team performance: Emerging principles from complex environments. In R. A. Guzzo & E. Salas (Eds.), *Team Effectiveness and Decision Making in Organizations* (pp. 9–45). San Francisco, CA: Jossey Bass.

McQueen-Shadfar, L., and Taekman, J. (2010). Say what you mean to say: Improving patient handoffs in the operating room and beyond. *Simulation in Healthcare, 5*(4), 248–253.

Morrison, J. E. and Meliza, L. L. (1999). *Foundations of the after Action Review Process*. Alexandria, VA: U.S. Army Research Institute for the Behavioral and Social Sciences.

Oser, R. L., Cannon-Bowers, J. A., Salas, E., and Dwyer, D. J. (1999). Enhancing human performance in technology-rich environments: Guidelines for scenario-based training. In E. Salas (Ed.), *Human/Technology Interaction in complex systems* (Vol. 9, pp. 175–202). Stamford, CT: JAI Press.

Porter, C. O., Hollenbeck, J. R., Ilgen, D. R., Ellis, A. P., West, B. J., and Moon, H. (2003). Backing up behaviors in teams: the role of personality and legitimacy of need. *Journal of Applied Psychology, 88*(3), 391–403.

Pronovost, P., Weast, B., Rosenstein, B., Sexton, J. B., Holzmueller, C. G., Paine, L. et al. (2005). Implementing and validating a comprehensive unit-based safety program. *Journal of Patient Safety, 1*(1), 33–40.

Reason, J. (2000). Human error: Models and management. *BMJ, 320*, 768–770.

Rosen, M. A., Feldman, M., Salas, E., King, H. B., and Lopreiato, J. (2010). Challenges and opportunities for applying human factors methods to the development of non-technical skills in healthcare education. In V. Duffy (Ed.), *Advances in Human Factors and Ergonomics in Healthcare* (pp. 1–30). Boca Raton, FL: CRC Press.

Rosen, M. A., Salas, E., Wilson, K. A., King, H. B., Salisbury, M., Augenstein, J. S. et al. (2008a). Measuring team performance for simulation-based training: Adopting best practices for healthcare. *Simulation in Healthcare, 3*(1), 33–41.

Rosen, M. A., Salas, E., Silvestri, S., Wu, T. S., and Lazzara, E. H. (2008). A measurement tool for simulation-based training in Emergency Medicine: The Simulation Module for Assessment of Resident Targeted Event Responses (SMARTER) Approach. *Simulation in Healthcare, 3*(3), 170–179.

Rosen, M. A., Salas, E., Wu, T. S., Silvestri, S., Lazzara, E. H., Lyons, R. et al. (2008b). Promoting teamwork: An event-based approach to simulation-based teamwork training for emergency medicine residents. *Academic Emergency Medicine, 15*(1–9).

Rosen, M. A., Schiebel, N., Salas, E., Wu, T. S., and Silvestri, S. (in press). How can team performance be measured, assessed, and diagnosed? In K. Frush and E. Salas (Eds.), *Improving Patient Safety through Teamwork and Team Training*. Oxford, U.K.: Oxford University Press.

Rudolph, J., Simon, R., Raemer, D., and Eppich, W. J. (2008). Debriefing as formative assessment: Closing performance gaps in medical education. *Academic Emergency Medicine, 15*(11), 1010–1016.

Rudolph, J., Simon, R., Rivard, P., Dufresne, R., and Raemer, D. (2007). Debriefing with good judgment: Combining rigorous feedback with genuine inquiry. *Anesthesiology Clinics, 25*(2), 361–376.

Salas, E., Almeida, S. A., Salisbury, M., King, H. B., Lazzara, E. H., Lyons, R. et al. (2009a). What are the critical success factors for team training in healthcare? *The Joint Commission Journal on Quality and Patient Safety, 35*(8), 398–405.

Salas, E., Bowers, C. A. and Rhodenizer, L. (1998). It is not how much you have but how you use it: Toward a rational use of simulation to support aviation training. *The International Journal of Aviation Psychology, 8*(3), 197–208.

Salas, E., Priest, H. A., Wilson, K. A., and Burke, C. S. (2006). Scenario-based training: Improving military mission performance and adaptability. In A. B. Adler, C. A. Castro, and T. W. Britt (Eds.), *Military Life: The Psychology of Serving in Peace and Combat* (Vol. 2: Operational Stress, pp. 32–53). Westport, CT: Praeger Security International.

Salas, E., Rosen, M. A., Burke, C. S., and Goodwin, G. F. (2009b). The wisdom of collectives in organizations: An update of the teamwork competencies. In E. Salas, G. F. Goodwin, and C. S. Burke (Eds.), *Team Effectiveness in Complex Organizations: Cross-Disciplinary Perspectives and Approaches* (pp. 39–79). New York: Routledge.

Salas, E., Rosen, M. A., Burke, C. S., Nicholson, D., and Howse, W. R. (2007). Markers for enhancing team cognition in complex environments: The power of team performance diagnosis. *Aviation, Space, and Environmental Medicine Special Supplement on Operational Applications of Cognitive Performance Enhancement Technologies, 78*(5), B77–B85.

Salas, E., Klein, C., King, H. B., Salisbury, M., Augenstein, J. S., Birnbach, D. J. et al. (2008). Debriefing medical teams: 12 evidence-based best practices and tips. *Joint Commission Journal on Quality and Patient Safety, 34*, 518–527.

Salas, E., Sims, D. and Burke, C. S. (2005). Is there a big five in teamwork? *Small Group Research, 36*(5), 555–599.

Savoldelli, Naik, Park, Joo, Chow, and Hamstra, 2006; Savoldelli, G. L., Naik, V. N., Park, J., Hoo, H. S., Chow, R., and Hamstra, S. J. (2006). Value of debriefing during simulated crisis management. *Anesthesiology, 105*(2), 279–285.

Sax, H. C., Browne, P., Mayewski, R. J., Panzer, R. J., Hittner, K. C., Burke, R. L. et al. (2009). Can aviation-based team training elicit sustainable behavioral change? *Archives of Surgery, 114*(12), 1133–1137.

Simons, T. L. and Peterson, R. S. (2000). Task conflict and relationship conflict in top management teams: The pivotal role of intragroup trust. *Journal of Applied Psychology, 85*(1), 102–111.

Singleton, A., Smith, F., Harris, T., Ross-Harper, R., and Hilton, S. (1999). An evaluation of the team objective structured clinical examination (TOSCE). *Medical Education, 33*(1), 34–41.

Smith-Jentsch, K. A., Cannon-Bowers, J. A., Tannenbaum, S. I., and Salas, E. (2008). Guided team self-correction. *Small Group Research, 39*(3), 303–327.

Smith-Jentsch, K. A., Johnston, J. A., and Payne, S. C. (1998). Measuring team-related expertise in complex environments. In J. A. Cannon-Bowers & E. Salas (Eds.), *Making decisions under stress: Implications for individual and team training* (pp. 61–87). Washington, DC: American Psychological Association.

Stout, R. J., Cannon-Bowers, J.A., and Salas, E. (1996). The role of shared mental models in developing team situational awareness: Implications for training. *Training Research Journal, 2*, 85–116.

Swing, S. R. (2002). Assessing the ACGME general competencies: General considerations and assessment methods. *Academic Emergency Medicine, 9*(11), 1278–1288.

Tziner, A., Joanis, C., and Murphy, K. R. (2000). A comparison of three methods of performance appraisal with regard to goal properties, goal perception, and ratee satisfaction. *Group & Organization Management, 25*(2), 175–190.

Vincent, C. (2010). *Patient Safety* (2nd ed.). Hoboken, NJ: Wiley-Blackwell.

Weaver, S. L., Rosen, M. A., DiazGranados, D., Lazzara, E. H., Lyons, R., Salas, E. et al. (2010). Does teamwork improve performance in the operating room? A multi-level evaluation. *Joint Commission Journal on Quality and Patient Safety, 36*(3), 133–142.

Wilson, K. A., Burke, C. S., Priest, H. A., and Salas, E. (2005). Promoting health care safety through training high reliability teams. *Qaul Saf Health Care, 14*, 303–309.

Wolf, F. A., Way, L. W., and Stewart, L. (2010). The efficacy of medical team training: Improved team performance and decreased operating room delays: A detailed analysis of 4863 cases. *Annals of Surgery, 252*(3), 477–483.

Wright, B. G., Phillips-Bute, B. G., Petrusa, E. R., Griffin, K. L., Hobbs, G. W., and Taekman, J. M. (2009). Assessing teamwork in medical education and practice: Relating behavioural teamwork ratings and clinical performance. *Medical Teacher, 31*(1), 30–38.

Zaccaro, S. J., Blair, V., Peterson, C. and Zazanis, M. (1995). Collective Efficacy. In J. E. Maddux (Ed.), *Self-Efficacy, Adaptation, and Adjustment: Theory, Research, and Application*. New York, NY: Plenum.

Ziv, A., Wolpe, P. R., Small, S. D., and Glick, S. (2006). Simulation-based medical education: An ethical imperative. *Simulation in Healthcare, 1*(4), 252–256.

Section VIII

Human Factors and Ergonomics Interventions

35

Ergonomics Programs and Effective Implementation

Michael J. Smith

CONTENTS

35.1 Introduction 597
    35.1.1 What Is Ergonomics? 597
    35.1.2 Why Is Ergonomics Important? 597
    35.1.3 Purpose and Content of This Chapter 598
35.2 Brief Overview of Workplace Musculoskeletal Injuries 598
    35.2.1 Recognized Work-Related Ergonomics Risk Factors 599
    35.2.2 Some General Resources for Understanding and Dealing with WMSDs 599
35.3 Need for a "Holistic" Approach to Ergonomics 600
    35.3.1 Balance Model as a "Holistic" Approach 601
        35.3.1.1 Person 601
        35.3.1.2 Machinery, Technology, and Materials 601
        35.3.1.3 Task Factors 602
        35.3.1.4 Work Environment 602
        35.3.1.5 Organizational Structure (Macroergonomics) 602
35.4 General Approach to an Ergonomics Program 603
    35.4.1 Macroergonomics Guidelines for an Occupational Ergonomics/Safety Program 604
    35.4.2 Defining How Well a Company Is Doing Using Injury Data 605
    35.4.3 Evaluating Ergonomics Risk Factors (Hazards) as a Basis for Improvement 606
35.5 Summary of Recommendations for an Ergonomics Program 607
    35.5.1 Organizational Design 607
    35.5.2 Ergonomics Processes 608
    35.5.3 Evaluate Your Progress 608
    35.5.4 Make Changes When and Where Needed 608
35.6 Conclusions 608
References 609

35.1 Introduction

35.1.1 What Is Ergonomics?

Ergonomics is the science that matches the capabilities of people and the work they perform. The primary interest is an understanding of how the design of work affects people's safety, health, performance, and productivity. The purpose is to match the capabilities of employees with the requirements of work by designing work processes and tasks to provide the best fit for each employee. Ergonomics takes knowledge from many fields including engineering, physiology, medicine, psychology, anthropology, sociology, and business management to develop an understanding of how to make effective and healthy working conditions. Of specific interest is the prevention or reduction of musculoskeletal injuries produced by working.

35.1.2 Why Is Ergonomics Important?

Ergonomics is important because research and experience have shown that employees can be injured if their work exceeds their capacities. Ergonomics provides a way to evaluate an employee's capacities as well as the demands that work activities put on her or him. Ergonomics helps employers determine if working conditions are well designed to protect employees' health and enhance their performance. It provides tools to assess if there is a misfit between employee capabilities and the work demands. Ergonomics also provides guidance for improving the fit between the capabilities of employees and the demands of work. This leads to fewer worker injuries, better worker performance and productivity, and lower worker's compensation and hospitalization costs. It is also possible that this will bring about better patient care and safety since the staff will be performing at a higher level with less stress and strain to detract from their performance.

35.1.3 Purpose and Content of This Chapter

The purpose of this chapter is to provide general guidance in how to establish and implement an ergonomics program to reduce the risk of musculoskeletal injuries in health care workers. In terms of health care operations and patient safety, an ergonomics program is important because health care workers who are injured, experiencing discomfort, or fatigued have a higher probability of making errors that can lead to patient injuries. Other chapters in this book provide specific information about the nature of the ergonomics risks and safety risks that occur in health care settings, so this chapter will only discuss such risks as examples in need of ergonomics solutions. Thus, this chapter will not cover the entire range of ergonomics risks or the entirety of specific ergonomics solutions for particular risks. Rather it will provide an approach for establishing an ergonomics program, for defining ergonomics risks (hazards), for evaluating the seriousness of the risk, and general guidance for a process for reducing adverse ergonomics exposures (hazards).

35.2 Brief Overview of Workplace Musculoskeletal Injuries

The scientific literature has established that there is a relationship between working in certain occupations or particular types of job exposures and the occurrence of work-related musculoskeletal disorders (WMSDs; see Hagberg et al., 1995; NIOSH, 1997a; NRC/IOM, 2001). For example, warehouse workers have a higher prevalence of low-back injury, meat cutters have a higher prevalence of carpal tunnel syndrome (CTS), painters have a higher prevalence of shoulder pain, and nurses have a higher prevalence of low-back injuries.

The research evidence necessary to define the specific levels of workplace risk factors that cause each of these WMSDs is incomplete, and this makes it hard to establish a risk threshold for any particular exposure. In addition, the research evidence examining the effectiveness of specific solutions to control the ergonomics risk

factors and to reduce injuries has limited and mixed findings. Thus, the control of WMSDs using ergonomics solutions is best done using a process similar to quality improvement.

In this approach, a broad program establishes methods of analysis and definition of problems, incremental changes are made to address identified problems, and the results of the changes are regularly evaluated. If the incremental changes lead to a stable reduction in injuries, then occasional continued monitoring is carried out to assure continued stability. If the solutions are not effective in reducing injuries, then additional solutions are undertaken until the results show success.

Many WMSDs are unique among occupational injuries in the complexity of their causation and in the longtime course of their development. Some WMSDs only occur after a long period of exposure to ergonomics risk factors, while others occur due to acute events of overexertion. In addition, many of the hypothesized "pathomechanisms" for the cumulative causation of WMSDs are not completely understood. There are several plausible and reasonable theories that provide the basis for defining possible ergonomics hazards that are likely to increase the risk for WMSDs such as microtrauma to soft tissues that accumulates over time.

What appears clear from the scientific literature is that WMSDs are more likely to occur when there are multiple ergonomics and personal risk factors present. There is insufficient evidence in the scientific literature to establish dose–response relationships among particular ergonomics risk exposures (for instance, the number of repetitions of motions or the weight of the load being handled) and the specific medical conditions that may occur such as low-back pain or CTS.

In addition, it currently is not possible to define the exact duration of exposures that will produce specific injury outcomes (hours per day, weeks, months, years) for a particular person. Therefore, there is no definitive basis to establish specific levels of a particular ergonomics risk exposure that is safe. For this reason, ergonomics solutions for preventing WMSDs may determine that it is best to use conservative ergonomics risk criteria to establish the permissible limits of exposure. It is not unreasonable to customize these criteria for specific companies, operations, or a particular workforce or a particular employee to account for specific circumstances of exposure and employee susceptibility to provide reduced ergonomics risk.

Jobs are an integration of many work demands, which taken together put loads on the person's body (and the mind). These loads in themselves are not bad for health as it is important for good health to exercise your body and mind. The problem occurs when the load is too great for the person to handle, which leads to physical and mental strain. It is the combined totality of all work and life demands that produce the combination of ergonomics exposures that can increase the risk of developing WMSDs. It is unlikely that a single job exposure such as the frequency of back or hand movements, the weight of objects being lifted, and the posture of the back or the wrists can define the entirety of the risk for developing a WMSD. Most often employees are exposed to multiple ergonomics risk factors simultaneously.

In addition to the physical workplace risk factors, there are personal susceptibilities, personal work habits, and individual work methods that also contribute to the development of WMSDs. This means that a "holistic" approach to an ergonomics program is necessary to account for the entirety of exposures and risk (Smith and Carayon, 1996; Carayon et al., 1999). The ergonomics program must look for ways to improve all of these considerations if the best results are to be attained.

35.2.1 Recognized Work-Related Ergonomics Risk Factors

The traditional workplace ergonomics risk factors that have been correlated with the development of WMSDs of the upper extremities in industrial, food processing, warehousing, and service settings are (1) a high frequency of sustained repetitive motion, combined with (2) heavy exertion due to loads from the weight of materials being handled or tools held, and/or combined with (3) extreme postures of the joints, limbs, shoulders, and neck from neutral positions. In addition, mechanical vibration may also increase the risk, especially in combination with any of the aforementioned factors (primarily high repetition; Silverstein et al., 1987; Hagberg et al., 1995; Nordstrom et al., 1997; NIOSH, 1997a; NRC/IOM, 2001).

The traditional workplace ergonomics risk factors for low-back injury that have been identified in a wide variety of work settings include (1) handling a very heavy weight that puts too much load on the spine or back muscles and leads to acute strain; (2) moving heavy loads a long distance in poor postures; (3) the proximity of the load far away from the body's center of gravity, which overloads the spine; (4) overexertion in handling the load caused by the load being too heavy or by the inability to grip the load securely; (5) a very high frequency of lifts per work shift for prolonged periods of time; (6) the improper posture of the back, legs, and shoulders during the lift, which loads the spine, shoulders, and legs; (7) certain situations of prolonged whole body exposure; and (8) certain situations of prolonged static seated posture (NIOSH, 1997a; Marras, 2000, 2008; NRC/IOM, 2001; Waddell and Burton, 2001).

There is less compelling evidence regarding the relationship between work activities and lower extremity WMSDs. The scientific literature on work-related cumulative trauma knee injuries has generally found moderate evidence that repeated acute impacts to the knees, prolonged kneeling, and heavy work as probable workplace risk factors for degenerative knee injuries (Felson et al., 1991; Jensen and Eenberg, 1996). In the evaluation of several studies Lievense et al. (2001) stated that there is moderate evidence for a positive association between heavy physical work and hip osteoarthritis, but the highest quality studies showed the opposite result. There were also moderate relationships between lifting weights of 55 lb or greater and hip osteoarthritis, and being a farmer for more than 10 years and hip osteoarthritis.

The etiology of WMSDs is very complex, and factors independent of workplace exposures are also substantive risk factors for WMSDs. For instance, for lower back pain, disc rupture, or CTS, some of the following nonoccupational factors have been shown to be contributing factors. These include personal constitution, health status, heredity, age, gender, weight, and psychological status, as well as activities that are not occupational (deKrom et al., 1990; Hanrahan et al., 1993; Nordstrom et al., 1997; NIOSH, 1997a; NRC/IOM, 2001).

The nonoccupationally related cases of low-back pain, CTS, and other WMSDs are as prevalent in the general population as occupationally related cases. In other words, working is not the only cause of low-back pain, CTS, shoulder tendinitis, elbow epicondylitis, shoulder pain, or neck pain. The nonoccupational causes of WMSDs affect the effectiveness that a workplace ergonomics program can have in reducing the WMSDs that your employees incur. Ergonomics programs need to recognize that nonoccupational risk factors are important contributors to employee musculoskeletal health. Educational programs that make employees aware of these nonoccupational risks should be considered when developing a holistic ergonomics program. Providing incentives to encourage employees to engage in healthy lifestyle activities such as exercise, healthy eating habits, smoking cessation, and positive mental health (stress reduction) can be an important adjunct to an occupational ergonomics program.

35.2.2 Some General Resources for Understanding and Dealing with WMSDs

Table 35.1 identifies general resource documents that can be helpful in recognizing and dealing with WMSDs and for developing an ergonomics program. The table provides the authors of the document, the source of the document, and a short statement about the information in the document.

TABLE 35.1

Ergonomics Program Resource Documents

Author/Year	Title/Source of Publication	General Material
Chaffin (1987)	Manual materials handling and the biomechanical basis for prevention of low-back pain in industry—an overview. *American Industrial Hygiene Association Journal, 48,* 989–996	Advice on how to control back injuries
Kilbom (1994a, b)	Repetitive work of the upper extremity: Part I—Guidelines for the practitioner. *International Journal of Industrial Ergonomics, 14,* 51–57	Advice on how to control upper extremity injury risk
Konz and Johnson (2007)	*Work Design: Industrial Ergonomics.* Scottsdale, AZ: Holcomb Hathaway Publishers	General ergonomics advice on workplace evaluation and improvement
Moore and Garg (1995)	The strain index: A proposed method to analyze jobs for the risk of distal upper extremity disorders. *American Industrial Hygiene Association Journal, 56,* 443–458	A way to measure the risk for upper extremity disorders
NRC/IOM (2001)	*Musculoskeletal Disorders and the Workplace.* Washington, DC: National Academy Press	Review of scientific literature on the causes and cures of work-related musculoskeletal disorders
NIOSH (Putz-Anderson, V.) (1988)	*Cumulative Trauma Disorders.* London, U.K.: Taylor & Francis	Overview of the causes and ways to control workplace musculoskeletal disorders
NIOSH (1981)	*Work Practices Guide for Manual Lifting* (NIOSH Publication 81–122). Cincinnati, OH: NIOSH	Guidance for lifting
NIOSH (Bernard, B.) (1997a)	*Musculoskeletal Disorders and Workplace Factors.* DHSS (NIOSH) Publication No. 97–141. Cincinnati, OH: NIOSH	Review of the scientific literature on causes of workplace musculoskeletal disorders
NIOSH (1997b)	*Elements of Ergonomics Programs.* DHSS (NIOSH) Publication, No. 97–117. Cincinnati, OH: NIOSH	Provides guidance for establishing an ergonomics program
OSHA (2010)	http://www.osha.gov/SLTC/ergonomics/index.html. Washington, DC: OSHA	Provides guidance on ergonomics programs
Salvendy (1997, 2006, 2011)	*Handbook of Human Factors and Ergonomics.* New York: John Wiley & Sons, Inc.	Provides expert advice on ergonomics topics
Snook (1988)	Approaches to the control of back pain in industry: Job design, job placement and education/training. *Professional Safety, 33,* 23–31	Suggestions for reducing back injuries at work
Snook (1991)	The design of manual handling tasks: Revised tables of maximum acceptable weights and forces. *Ergonomics, 34,* 1197–1213	Advance on design of work tasks and lifting loads
Waters, Putz-Anderson, and Garg (1993)	Revised NIOSH equation for the design and evaluation of manual lifting tasks. *Ergonomics, 36,* 749–776	Revised NIOSH guidance on lifting tasks and weight limits
U.S. DoA (2003)	*Ergonomics Program (Medical Services).* Washington, DC: Department of the Army	Provides guidance for establishing an ergonomics program
WorkSafeBC (2009)	*Preventing Musculoskeletal Injury (MSI): A Guide for Employers and Joint Committees.* Contact: http://www.worksafebc.com	Provides guidance for establishing an ergonomics program
USMIL (2008)	*Military Standard 1472F Human Engineering Design Criteria for Military Systems, Equipment and Facilities.* Washington, DC: U.S. Department of Defense. Contact: http://www.redstone.army.mil/amrdec	Provided guidance and data for designing human-technology systems
U.S. DoD (2009)	*Directory of Design Support Methods.* San Diego: DTIC-A, U.S. Department of Defense	Provided a listing and description of military design handbooks, data bases and standards

35.3 Need for a "Holistic" Approach to Ergonomics

Ergonomics and human factors have been around for approximately 70 years. During the World War II, there was a need to deal with employee and soldier/flyer fatigue, stress, injuries, and poor performance due to mismatches between people and technology, unusual work schedules, and demanding and threatening working conditions. Over the ensuing 70 years, ergonomics researchers and practitioners have learned that the best way to address ergonomics concerns is to take a broad view of the problems encountered combined with focused solutions and proper management (macroergonomics). The broad view of a work process and improvements addresses the need to make sure that improvements in one area do not lead to problems in another area. It accounts for the issues that often occur elsewhere in the work system when just a focused approach is used in just one area. This has led to the understanding that a bigger view allows the ergonomics practitioner who is trying to solve a problem both to address the specific problem (the "microergonomics") and to deal with the ramifications in the entire work process (the macroergonomics).

Microergonomics factors deal with designing the proper characteristics of the tasks, tools/technology, environmental conditions, and/or capacity/knowledge of each individual or groups of employees to accomplish work effectively and safely. It addresses the specific ergonomics risk factors in a particular task, job, or operation. Macroergonomics factors deal with larger issues such as the organization of the work process; the coordination of tasks and activities among employees and groups; the supervision of processes and employees; and how people, technology, tasks, and environmental features are integrated. Ergonomics improvements and solutions must account for both levels and their effects on the work system to be successful.

35.3.1 Balance Model as a "Holistic" Approach

As indicated earlier, it is important for an employer to recognize that there are many personal and workplace factors that interact together to produce situations that lead to musculoskeletal injuries. Therefore, any strategy to control these situations should consider a broad range of factors and their influences on each other. One holistic model of human workplace interaction is the "balance theory," which is presented in Figure 35.1 (Smith and Sainfort, 1989; Smith and Carayon, 1995; Carayon et al., 1999; Smith et al., 1999; Carayon and Smith, 2000). Each element of this model can produce demands or loads on the worker that can lead to musculoskeletal injury. In combination, their effects can produce even greater demands. Each one of these elements of the model has specific features that can lead to stress and strain in the person, and this stress and strain may produce injury. In addition, each element of the model can be improved to reduce the risk of WMSDs, or a combination of elements may need to be improved to obtain the desired benefits.

35.3.1.1 Person

There is a wide range of individual attributes that can affect injury potential. These include perceptual/motor abilities, physical capabilities, and capacity such as strength and endurance, body mass index, current health status, susceptibilities to disease, personality, intelligence, heredity, and behavior (work practices, task technique). Perceptual/motor skills level can affect the ease and quality with which a task is carried out, physiological energy expenditure, posture, balance, and fatigue. Personal physiological considerations such as strength, endurance, stress tolerance, conditioning, and health status affect susceptibility to injury. Intelligence affects the ability for hazard recognition and training in hazard recognition and elimination. It is critical that a proper "fit" is achieved between employees and other characteristics of the model to achieve proper balance. Each one of the aforementioned personal attributes and characteristics can be enhanced to reduce the risk of a WMSD. For example, a person can become stronger or better conditioned physically. A person can be trained to improve the work practices they use to handle loads. A person who has recently had a strain can be given light duty tasks or excused from work until the strain heals. A person who is injured can be given proper treatment to provide the maximum possible healing.

35.3.1.2 Machinery, Technology, and Materials

There are characteristics of machinery, tools, technology, and materials used by the employee that can influence the potential for injury. One consideration is the extent to which machinery and tools influence the use

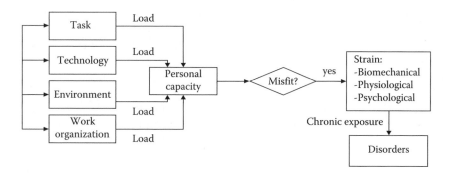

FIGURE 35.1
Balance model (reprinted from *Int. J. Industr. Ergonom.*, 4, Smith, M. J. and Sainfort, P., A balance theory of job design for stress reduction, 67, Copyright 1989, with permission from Elsevier) and stressor effects (reprinted from *Work-Related Musculoskeletal Disorders*, Smith, M. J., Karsh, B.-T., and Moro, F. B., A review of research on interventions to control musculoskeletal disorders, 200, Copyright 1999, with permission from Elsevier).

of the most appropriate and effective perceptual/motor skills and energy resources. The relationship between the controls of a machine and the action of that machine dictates the level of perceptual/motor skill necessary to perform a task, the repetition level of motions, the postures of the joints and back, and force necessary for activation. The proper design considerations for the technology to accommodate the human factors issues is well established and identified in human factors and ergonomics books such as the book by Konz and Johnson (2007) and Salvendy (2006, 2011) and by the U.S. Military Standard as defined in the document *Human Engineering Design Criteria for Military Systems, Equipment and Facilities* (MIL-STD-1472; USMIL, 2008). In addition, technology can be used to reduce loads on the employees by providing leverage, physical power, and mental power (decision-making tools). For example, lifting aids can be used to reduce back strain, and sliding boards can be used to aid in patient transfer from bed to wheelchair.

35.3.1.3 Task Factors

The demands of a work activity and the way in which tasks are conducted can influence employee strain and subsequent injury. Work task demands can be broken into the physical requirements, the mental requirements, and psychological considerations. The physical requirements influence the amount of energy expenditure necessary to carry out a task, and the forces imposed on the body during the task. Excessive physical requirements can lead to fatigue and/or strain that may lead to employee injury. Employees can only tolerate high physical demands for short time periods. Longer exposure to heavy workloads or multiple exposures to shorter duration heavy workloads can reduce an employee's capacity to respond properly. This can lead to strain and injury. Other physical task demands include the pace or rate of work, the amount of repetition in task activities, and the work pressure due to high workload. Task activities that are highly repetitive and where the pace of the work is controlled by the technology rather than employee paced tend to be more stressful.

Psychological task demands are tied to the amount of control an employee has over the work process, the level of participation in decision making, the ability to use knowledge and skills, the amount of esteem associated with the job, and product identity. These can influence an employee's job satisfaction, motivation, and attention. They also can cause psychological stress that can affect the employee's sensitivity to discomfort and pain and produce increased muscle tension.

Ergonomics improvements can be achieved by setting the appropriate workload for each task using industrial engineering work methods analysis procedures and by having adequate staffing to meet production requirements. In addition, establishing the proper work methods provides the basis for proper work practices to be used by the employees. With adequate employee training, the improved work practices will replace the bad methods that employees previously used and will lead to reduced strain.

35.3.1.4 Work Environment

The work environment exposes employees to climates, materials, chemicals, and physical agents that can cause harm or injury if the exposure exceeds safe limits. Such exposures vary widely from job to job and from task to task. Hazard exposures in the work environment influence the capacity of employees to perform tasks and the probability for an injury or illness. Climatic conditions can hamper the ability of employees to use their senses properly and can affect endurance (poor lighting, excessive noise, too hot/cold). Cold temperatures make muscles more susceptible to strain, and hot temperatures affect the extent of energy expenditure and body fluid balance. The environment should be compatible with worker perceptual/motor skill requirements, energy expenditure, and motivational needs to encourage proper work methods.

35.3.1.5 Organizational Structure (Macroergonomics)

Many aspects of organizational management can affect health and safety performance. These include management policies and procedures, the organization of tasks, the style of employee supervision, the motivational culture in the facility, the amount of socialization and interaction between employees, the amount of social support employees receive, and management attitudes toward safety and ergonomics. The latter point, management attitude, has often been cited as the most critical element in a successful safety program (Cohen, 1977; Smith et al., 1978; Cleveland et al., 1979). If the individuals that manage an organization are not interested in good ergonomics, then employees tend to be less motivated to be interested in ergonomics and safety. Conversely, if the management attitude is one in which safety and ergonomics considerations are important, then managers, supervisors, and employees will have interest in ergonomics and put energy into positive ergonomics efforts.

A consistent factor in injury causation is management pressure for higher production, or faster output, or a quick response to production problems. Technology malfunctions, insufficient staffing, and improper workload standards can exacerbate work pressure caused by management's insistence on greater production. Management emphasis on reducing costs and enhancing profits may stretch the limits of the capabilities of the workforce and the technology. When breakdowns occur or operations are not running normally, employees tend to work harder to keep production up to speed. It is during these heavy load situations that many ergonomics related musculoskeletal injuries occur.

Management must provide adequate resources to meet production goals and to accommodate heavy demand operations. This means adequate staffing, effective technology, proper process design, and scientifically determined work standards (see Konz and Johnson, 2007). Management must also establish policies to ensure that employees and supervisors accept the importance of proper ergonomics.

35.4 General Approach to an Ergonomics Program

An ergonomics program standard proposed by Occupational Safety and Health Administration (OSHA) in 1999 provides guidance that companies can follow to develop an effective ergonomics program (see http://www.osha.gov). In addition, comprehensive guidance for developing and implementing ergonomics programs has been provided by NIOSH (1997b), the Workers' Compensation Board of British Columbia, Canada (WorkSafeBC, 2009), and the Medical Services of the U.S. Department of the Army (DoA, 2003). All of these sources have been consistent in their advice for ergonomics programs.

The first step in establishing an ergonomics program is to have a corporate policy statement that defines a commitment and approach to ergonomics improvements. This is followed by defining a structured process for evaluating ergonomics risks, developing solutions, implementing solutions, and assessing the effectiveness of the solutions. There is no one best way for establishing the structured process or for the management of the process. It is often best to integrate the ergonomics program within existing processes that have proven to be effective to capitalize on the established methods and procedures that everyone is familiar with. When undertaking an ergonomics program, it is essential to the success of the program that management is committed to and active in the program. In addition, ergonomics programs are more effective when employees have an active role in defining ergonomics risks and potential solutions, and when they can participate in the improvement process.

OSHA has provided guidance on means to prevent musculoskeletal injuries since 1990 (OSHA, 1990, 2000, 2002, 2003, 2009, 2010; Federal Register, 1999). More recently, OSHA has developed guidelines for ergonomics actions to prevent musculoskeletal disorders in nursing homes (OSHA, 2009), and much in these guidelines

has application to other health care settings. Since 1990, OSHA has proposed central elements of an ergonomics program. These are (1) management leadership and support, (2) employee involvement, (3) hazard identification and reporting, (4) job hazard analysis and control, (5) implementing solutions, (6) employee training, (7) WMSD management, and (8) program evaluation. These same program elements are also aspects of the recommendations made by NIOSH (1997b), WorkSafeBC (2009), and U.S. Army medical services guidance (DoA, 2003). The consistency across independent experts for including these elements in an ergonomics program indicates their general acceptance and importance.

Management leadership calls for employers to demonstrate commitment and direction for the ergonomics program. This recommendation draws on the research on safety programs by Cohen (1977) and Smith et al. (1978) that showed the importance of management committing time and resources to the program, and active management involvement in the program through "face time" in facilities talking with employees about safety. In addition, the OSHA ergonomics program directive specified that employers must provide mechanisms for employees and unions to report ergonomics hazards and WMSDs symptoms and signs. The company policies must not discourage employees from reporting injuries and hazards or from participating in the ergonomics program.

To achieve these goals, the company can assign the responsibility for promoting the ergonomics program to managers and supervisors. The program administration and coordination can be placed in a logical operating unit such as the safety department, employee health services, or the human resources department. Ergonomics activities such as hazards evaluation or solution development can be centralized as a corporate function or decentralized to operating divisions and departments depending on whatever fits best with the corporate culture and management structure. It is important that top management gives ergonomics responsibilities to specific people for managing and carrying out the ergonomics activities, and that all supervisors and employees are encouraged to participate in ergonomics activities. Both research evidence and anecdotal evidence suggest that ergonomics and safety programs are more effective when employees are involved in the development, implementation, and evaluation of the program and when the company regularly communicates with employees about the program activities.

In terms of hazard information and reporting, the company needs to provide ways for employees to (1) be aware of WMSDs hazards and the etiology of WMSDs, (2) timely report WMSDs and symptoms of WMSDs to the company, and (3) get prompt response about their reporting. This requires a central reporting location that will aggregate symptoms and injuries and to report the findings back to employees periodically.

The hazard analysis requires the use of ergonomics methods to define hazards for WMSDs. This calls for a job hazard analysis for any person or occupational group in which employees are reporting WMSDs and/ or WMSDs symptoms. Some areas of evaluation in the job hazard analysis that OSHA recommends be examined are (1) high exertion activities, (2) activities with highly repetitive motions, (3) the absence of breaks in work when there is repetition, (4) activities with long reaches, (5) improper working surface heights, (6) static postures, (7) sitting too long, (8) the use of hand tools and powered tools, (9) contact with vibrating machines or surfaces, (10) compression of tissue on surfaces, (11) contact stresses of the hand, (12) gripping/holding objects with the hands or body while working, (13) using gloves that fit poorly, (14) moving heavy objects or people, (15) long horizontal reaches away from the body, (16) vertical reaches below the knees or above the shoulders, (17) moving objects a long distance, (18) bending or twisting the back, (19) slippery objects or objects without grips/ handles, and (20) uneven, slippery, or sloped floor surfaces. These of course do not define all of the possible ergonomics hazards but provide a good starting point. There are checklists available to identify ergonomics hazards (NIOSH, 1997b; WorkSafeBC, 2009).

Employees are a good source of information and inspiration for developing ergonomics hazard controls. Companies will want to ask affected employees about recommendations for improvements. After changes are made, companies can get good input from employees about the progress of the controls in eliminating or reducing the ergonomics hazards. Good ideas often come from the people who are closest to the ergonomics problems. Including them in the process of improvement will increase their motivation to cooperate and work to make the improvements a success. Participation is a powerful motivator.

OSHA proposed the use of any combination of engineering solutions, administrative changes, or work practice improvements that eliminate or reduce the ergonomics hazards. Engineering controls require physical changes to the work environment or equipment to reduce the hazard level. For example, providing a lifting device for handling any load over 45 lb. Administrative changes deal with modifying the work process to reduce the hazard level. For example, limiting the amount of weight that one person can lift to 45 lb. Another set of administrative changes could be limiting the number of continuous hours that can be worked to a maximum of 12 h per 24 h period, and the number of consecutive days worked to no more than 6 days per week. New work practices change the way a task is carried out. For example, by forbidding one

employee from lifting a patient and requiring that two or more employees must handle each patient who is lifted. Work practice improvements often reduce hazards by eliminating unsafe behaviors such as trying to pick up a fallen patient without help.

OSHA recommends that employees in higher ergonomics risk jobs, their supervisors, and those employees who are responsible for the ergonomics program must be trained. OSHA believes that retraining should occur at least every 3 years. The training must include information about the company's ergonomics program, WMSDs hazards, and how to reduce WMSDs risks.

OSHA believed that the company was responsible for promptly making available WMSD medical management for any injured employee at no cost. This is usually covered by worker's compensation insurance in most states. But there has been controversy about this OSHA recommendation since not all state worker's compensation boards recognize all musculoskeletal injuries as work related. OSHA indicates that an injured employee should have work restrictions imposed to protect her or him from exposure to work conditions that could exacerbate the injury until she or he has recovered from the injury. The basic concept in this OSHA advice is early intervention in the injury process to limit damage and speed up recovery.

Finally, OSHA recommended periodic evaluation of the effectiveness of the ergonomics program at least every 3 years. This can involve a compilation of the ergonomics hazards identified, ergonomics hazards eliminated, and a determination of any reduction in WMSDs frequency and severity.

35.4.1 Macroergonomics Guidelines for an Occupational Ergonomics/Safety Program

There are a number of other elements to consider in developing an ergonomics program or upgrading your current program. These include organizational policies, managing various elements of the program, motivational practices, hazard control procedures, dealing with employees' concerns, injury investigations, and injury recording and surveillance. There has been considerable research into the necessary elements for a successful safety program (Cohen, 1977, Smith et al., 1978; Cleveland et al., 1979) and how these elements should be applied. These can be used as a base for organizing an ergonomics program.

One primary factor that emerges from every study of a successful company safety program and a major aspect of the recommended OSHA ergonomics program is the program will not be successful unless there is a commitment to the program by top management. This indicates that there needs to be a written organizational policy statement on the importance of safety and ergonomics, and the policy should specify the general approach and procedures the organization intends to use. Having such a policy is the first step toward effective management commitment and a successful ergonomics program.

Smith et al. (1978) have shown that it takes more than just a written policy to ensure a successful safety program and good safety performance. It also takes involvement in the program by all levels of management. For the top managers, it means they must periodically get out to where employees are working to show their concern, to talk to employees about working conditions, and to view safety and ergonomics problems. This has been shown to be more effective when done on an informal basis. For middle managers and supervisors, there is a need to participate in safety and ergonomics program activities with employees. These safety and ergonomics activities put middle managers in touch with employee concerns about potential hazards and educate them about operational problems. Management involvement demonstrates to the employees that management cares about their safety and health and that ergonomics is important. This reinforces employee commitment to safe work practices.

Another aspect of management commitment is the level of resources that are made available for safety and ergonomics programming. Cohen (1977) found that organizational investment in full-time safety staff was a key feature to good plant safety performance. The effectiveness of safety and health staff was greater the higher they were positioned in the management structure. However, Smith et al. (1978) found that safety staff was not as critical to program effectiveness as management and employee active involvement in the safety program efforts. A full-time ergonomics staff may not be as necessary when the supervisors and employees are actively involved in ergonomics efforts.

An organization's motivational practices will influence employee safety behavior and compliance with safe work practices. Research has demonstrated that organizations that exercise participative management approaches have better safety performance. These approaches encourage employee involvement in defining the risks and in developing solutions such as improved work practices. Employee involvement leads to a greater awareness of ergonomics and leads to higher motivation levels conducive to proper employee behavior and work practices. Research has shown that organizations that use punitive motivational techniques for influencing safety behavior have poorer safety records than those using positive approaches (Cohen, 1977; Smith et al., 1978).

Another important motivational factor is encouraging communication between various levels of the organization (employees, supervisors, managers). Such communication increases participation in ergonomics issues and builds employee and management commitment to safety and ergonomics goals and objectives. Often informal communication between supervisors and employees is a potent motivator and provides more meaningful information for ergonomics hazard control.

All safety and ergonomics programs should have a formalized approach to hazard detection and control. This often includes an inspection system to define ergonomics workplace hazards, injury investigations to identify critical causes, record keeping to monitor progress, the review of new purchases to ensure compliance with safety and ergonomics guidelines, and good housekeeping for a comfortable environment. All of these elements contribute to a positive "safety climate" that encourages employee compliance with proper safety and ergonomics procedures. The importance of formalized hazard control programs is that they establish the groundwork for other programs such as employee training for work practices improvements and equipment purchasing requirements (Smith et al., 2002; Smith and Carayon, 2011).

35.4.2 Defining How Well a Company Is Doing Using Injury Data

The OSHA proposed ergonomics standard provides some general guidance for hazard analysis and specific areas to examine. The following details an approach for assessing WMSDs experience and for conducting hazard analysis. The first step in assessing ergonomics risks is to define where WMSDs are occurring in your organization. This requires an injury surveillance system. The first place to start such a surveillance system is by using the records required by OSHA for injury reporting. Occupational injuries are as recorded on the OSHA 300 Injury Log. More detailed information about an injury is recorded on the supplemental injury form (OSHA 301 reports or worker's compensation reports). In addition, there may be medical department or infirmary reports can be used. These injury reports can serve as the basis for looking at the number, type, and seriousness of injuries in specific departments or jobs.

There are four main uses of injury statistics: (1) to identify high-risk jobs or work areas, (2) to evaluate company health and safety performance, (3) to evaluate the effectiveness of hazard abatement approaches, and (4) to identify factors related to illness and injury causation. An illness and injury reporting and analysis system requires that detailed information must be collected about the characteristics of illness and injuries and their frequency and severity.

The 1970 OSH Act requirements specify that any illness or injury to an employee that causes time lost from the job, treatment beyond first aid, transfer to another job, loss of consciousness, or an occupational illness must be recorded on a daily log of injuries (OSHA 300 form). This log identifies the injured person, the date and time of the injury, the department or plant location where the injury occurred, and a brief description about the occurrence of the injury highlighting salient facts such as the chemical, physical agent or machinery involved and the nature of the injury. The number of days that the person is absent from the job is also recorded upon the employee's return to work. In addition to the daily log, a more detailed form (OSHA 301) is filled out for each injury that occurs. This form provides a more detailed description of the nature of the injury, the extent of damage to the employee, the factors that could be related to the cause of the injury, such as the source or agent that produced the injury, and events surrounding the injury occurrence. A worker's compensation form can be substituted for the OSHA 301 form, as equivalent information is gathered on these forms.

The OSH Act injury and illness system specifies reporting a procedure for calculating the frequency of occurrence of occupational injuries and illnesses, and an index of their severity. These can be used by companies to monitor their health and safety performance over time and to compare their experience with those companies in the same industry category. National data by major industrial categories are compiled by the U.S. Bureau of Labor Statistics annually and can serve as a basis of comparison of individual company performance within an industry sector in the United States. Thus, a company can benchmark its injury rate to see how it compares other companies in its industry. This industry-wide injury information is available on the OSHA Web site (http://www.osha.gov).

The OSHA system uses the following formula in determining a company's annual injury and illness incidence. The total number of recordable injuries are multiplied by 200,000 and then divided by the number of hours worked by the company employees. This gives an injury frequency per 100 person-hours of work (injury incidence). These measures can be compared with an industry average.

$$\text{Incidence rate (IR)}$$
$$= \frac{\text{Number of WMSDs} \times 200,000}{\begin{array}{c}\text{The number of hours worked}\\\text{by company employees}\end{array}}.$$

The equation for calculating the IR is as follows:

$$\text{IR} = \frac{\#\,\text{WMSDs} \times 200,000}{\#\,\text{person-hours}},$$

where

1. The number of recordable injuries and illnesses is taken from the OSHA 300 daily log of injuries.
2. The number of hours worked by employees is taken from payroll records and reports prepared for the government.

It is also possible to determine the severity of a company's injuries. Two methods are typically used. In the first, the total number of days lost due to injuries is compiled from the OSHA 300 daily log and is divided by the total number of injuries recorded on the OSHA 300 daily log. This gives an average number of days lost per injury. In the second approach, the total number of days lost is multiplied by 200,000 and then divided by the number of person-hours worked by the company employees. This gives a severity rate per 100 person-hours of work. These measures can also be compared to an industry average, or across departments within a facility, or among jobs in a facility to identify high-risk areas.

35.4.3 Evaluating Ergonomics Risk Factors (Hazards) as a Basis for Improvement

A good understanding of the demands of specific jobs and tasks provides the base for developing ways to better match and fit tasks and activities to employees. Knowledge of the loads and forces imposed on employees, the energy requirements of work, the mental demands, and the psychological stress leads to an evaluation of the potential risks for employee injury or reduced performance. In order to successfully control occupational hazards that lead to illness and injuries, it is necessary to define their nature, and predict when and where they will occur. This requires a hazard detection process that can define the frequency of the hazards, their seriousness, and their amenability to control.

Traditionally, two parallel systems of information have been used to define occupational hazards. One system is hazard identification, such as plant inspections, fault-free analysis, and employee hazard reporting programs, which are used to define the nature and frequency of the hazards. With these approaches, preventive action can be taken before an injury or illness occurs. The second system is after the fact because it uses employee injury and company loss control information to define company problem spots based on the number of injuries and the costs to the organization. When pre- and post-injury systems are integrated, they can be used to define high-risk jobs, tasks, plant areas, or working conditions where remedial programs can be established for hazard control.

Hazard identification prior to the occurrence of an occupational injury is a major goal of a hazard inspection program. In the United States, such programs have been formalized in terms of federal and state regulations that require employers to monitor and abate recognized occupational health and safety hazards defined by the standards. These standards of unsafe exposures define the benchmarks that must be achieved by companies. Unfortunately, there are no U.S. federal or state standards for ergonomics hazards that define the upper limits of specific exposures that are dangerous.

Research has shown that formal inspections are most effective in identifying permanent, fixed physical and environmental hazards that do not vary over time. Inspections are not very effective in identifying transient physical and environmental hazards, or improper employee behaviors, as these hazards may not be present when an inspection is taking place (Smith et al, 1971). A major benefit of conducting an inspection, beyond the definition of serious hazards, is the positive motivational influence on supervisors and employees. Inspections demonstrate that management is interested in the health and safety of employees and to a safe working environment. To capitalize on this positive motivational influence, an inspection should not be a punitive process. Indicating the good aspects of a work area and not just the hazards is important in this respect. It is also important to have employees participate in hazard inspections as this increases hazard recognition skills and increases motivation for safe and proper behavior.

The first step in an inspection program is to develop a checklist that identifies all potential hazards. A good starting point for an ergonomics checklist is to examine the references cited earlier (NIOSH, 1997b; OSHA, 2000, 2009, 2010; WorkSafeBC, 2009). In addition, the topical areas defined in the proposed 1999 OSHA ergonomics program standard (OSHA, 2000)* delineate important ergonomics aspects to evaluate. Many insurance companies have developed general checklists that can be tailored to a particular company. Once the checklist is completed, a process for conducting inspections can be established.

A systematic inspection procedure is preferred. This requires that the inspectors know what to look for, where to look for it, and have the proper tools to conduct an effective assessment. Aspects of the inspection that need to be determined are (1) the factors to be examined, (2) the frequency of inspection necessary to detect and control hazards, (3) the individuals who should conduct and/or participate in the inspection, and (4) the instrumentation needed to make measurements of the hazard(s). The factors to be inspected include (1) the technology, machinery, tools, and materials; (2) the environmental conditions; (3) task risk factors; and (4) employee work practices.

The frequency of inspections should be based on the nature of the hazards being evaluated. Intermittent hazards may need to be evaluated more often than fixed physical hazards. Random spot-checking is a useful method that "catches" transient hazards that may be due to employee behaviors or a particular patient (very overweight, mental problems). These intermittent hazards require more frequent inspection. In many cases, monthly inspections are warranted, and in some cases, daily inspections are reasonable. Sometimes the transient hazards are poor work practices that occur due to specific circumstances, for example, handling a very heavy bedridden patient alone rather than getting help. When such work practices are observed, they need to be called to the attention of the employee and instructions given on proper procedures.

In contrast to the periodic, formal inspection process, Smith et al. (2002) presented a hazard survey approach that is based on continuous reporting of potential hazards by employees as they encounter a hazard. This approach has formal methods for reporting and evaluating the hazards to bring about a quick response to the problem. Employees in each work area report any ergonomics hazards that they observe to a supervisor or lead employee. Imminent serious ergonomics hazards are reported to the designated ergonomics authority and dealt with as quickly as possible. Less imminent serious ergonomics hazards are presented at weekly meetings of all of the lead employees from each area. The ergonomics hazards are discussed, and significant ergonomic hazards are passed on the ergonomics authority.

One weakness of the approach is that it is unlikely that employees will report poor work practices and/or behaviors of their peers. An approach for effective identification of these behavioral ergonomics hazards is to have periodic observations by supervisors. This assumes that supervisors have knowledge of which behaviors are hazardous, that they will report/record the behaviors, and that they will take appropriate action to improve employee behaviors.

35.5 Summary of Recommendations for an Ergonomics Program

35.5.1 Organizational Design

1. Provide a written company policy that defines management commitment to the ergonomics program, describes the formal structure and processes, and established roles for all employees.

2. Ensure that everyone in the company has a chance to participate in ergonomics activities.

---

* This standard was rescinded by the U.S. Congress in 2001. Copies available from the Federal Register, 2000.

3. Include managers and employees in the implementation of the ergonomics program and in ongoing ergonomics activities.

4. Have a formal process for conducting ergonomics activities.

35.5.2 Ergonomics Processes

1. Define responsibilities for various ergonomics activities such as management of the ergonomics program, defining ergonomics hazards, developing solutions, implementing the solutions, evaluating progress and success of solutions, longer term monitoring progress and the need for additional solutions, benchmarking progress (hazard reduction, injury rate, injury severity score), and determining the need for further management intervention.

2. Use well-accepted methods for each aspect of the ergonomics program that are available from NIOSH, OSHA, and other recognized sources identified in this document.

3. Adapt checklists from OSHA, NIOSH, and WorkSafeBC for identifying potential ergonomics hazards. Add to the checklists as needed to address your unique ergonomics hazards.

4. Use the expertise you have available from among your workforce to develop ergonomics solutions. Their intimate knowledge of the tasks and macroergonomics issues is very helpful to success. Call in outside experts when you do not have the analytical skills or solution skills among your employees.

35.5.3 Evaluate Your Progress

1. Use your ergonomics hazard analysis process to evaluate and monitor the success of your solutions. Have you eliminated and/or reduced the ergonomics hazards in your facilities? Have you found new ergonomics hazards and dealt with them?

2. Monitor your musculoskeletal injuries on a regular basis. Is the number and rate of injuries dropping? Is the severity of injuries dropping? What departments and jobs are having high rates of musculoskeletal injuries? How does your injury rate compare with your industry?

35.5.4 Make Changes When and Where Needed

1. Ergonomics solutions are similar to quality improvement solutions. They require continuous monitoring to evaluate the results and for quickly making changes when the results are unsatisfactory.

2. Some employees will not participate in the ergonomics program and in the solutions to control ergonomics hazards. When this happens, you can try greater encouragement, additional training, and/or replacing them if they threaten the success of solutions.

3. Be flexible in how you carry out your ergonomics program. There is a variety of good advice on how to do ergonomics activities. If your initial approach to an issue is not working be willing to try another approach. The success of an ergonomics program is partly due to "technical" skills, partly due to acceptance by your employees, partly due to your unique situation, and partly due to your "art" in carrying out the program.

35.6 Conclusions

A "holistic" ergonomics program is necessary to be able to develop good specific ergonomic solutions to reduce the risk of WMSDs of your employees. This is based on the belief that a holistic approach is needed to deal with occupational and nonoccupational exposures. This program should include appropriate management commitment and involvement and employee involvement. There should be mechanisms established for defining ergonomics hazards that can cause WMSDs, to prioritize the importance of the ergonomics hazards, and to define solutions to remove or control the risks due to serious ergonomics hazards. Solutions might require engineering redesigns or new technology or might use administrative procedures and/or work practice improvements to resolve the risk due to the ergonomics hazards. All of these aspects require a structured approach for managing the ergonomics program, hazard identification and control, and assessment of success.

There are many resources that can provide guidance in developing specific ergonomics solutions that have been identified in this chapter. Of particular significance is the *Handbook of Human Factors and Ergonomics* (Salvendy, 1997, 2006, 2011), WorkPlaceBC (2009), and several ergonomic guidelines for specific industries published by OSHA such as the nursing home guidelines (see http://www.osha.gov).

Participative approaches for ergonomics hazard definition and ergonomics hazard control are more effective. These approaches provide critical information about the sources of hazards, motivation to employees to be actively involved in ergonomics improvements, and motivation for developing improved work practices. In addition, ergonomics approaches that employ multiple solutions for ergonomics hazard reduction have been the most successful (Smith et al., 1999; Karsh et al., 2001). A holistic approach is the best approach.

References

Carayon, P. and Smith, M. J. (2000). Work organization and ergonomics. *Applied Ergonomics*, 31, 649–662.

Carayon, P., Smith, M. J., and Haims, M. (1999). Psychosocial aspects of work-related musculoskeletal disorders. *Human Factors*, 41(6), 644–663.

Chaffin, D. B. (1987). Manual materials handling and the biomechanical basis for prevention of low-back pain in industry—An overview. *American Industrial Hygiene Association Journal*, 48, 989–996.

Cleveland, R. J., Cohen, H. H., Smith, M. J., and Cohen, A. (1979). *Safety Program Practices in Record Holding Companies* (DHEW (NIOSH) Publication No. 79-136). Washington, DC: United States Government Printing Office.

Cohen, A. (1977). Factors in successful occupational safety programs. *Journal of Safety Research*, 9, 168–178.

deKrom, M. C. T. F. M., Kester, A. D. M., Knipschild, P. G., and Spaans, F. (1990). Risk factors for carpal tunnel syndrome. *American Journal of Epidemiology*, 132, 1102–1110.

Federal Register (1999, November 23). Part 1910, subpart Y—Ergonomic program standard. *Federal Register*, 64(225), 66067–66078.

Felson, D. T., Hannan, M. T., Naimark, A., Berkeley, J., Gordon, G., Wilson, P. W. F., and Anderson, J. (1991). Occupational physical demands, knee bending, and knee osteoarthritis: Results from the Framingham Study. *The Journal of Rheumatology*, 18(10), 1587–1592.

Hagberg, M., Silverstein, B., Wells, R., Smith, M. J., Hendrick, H. W., Carayon, P., Perusse, M., Kuorinka, I., and Forcier, L. (1995). *Work-Related Musculoskeletal Disorders (WMSDs): A Reference Book for Prevention*. London, U.K.: Taylor & Francis.

Hanrahan, L. P. Higgins, D., Anderson, H., and Smith, M. (1993). Wisconsin occupational carpal tunnel syndrome surveillance: The incidence of surgically treated cases. *Wisconsin Medical Journal*, December, 685–689.

Jensen, L. K. and Eenberg, W. (1996). Occupation as a risk factor for knee disorders. *Scandinavian Journal of Work, Environment & Health*, 22, 165–175.

Karsh, B.-T., Moro, F. B. P., and Smith, M. J. (2001). The efficacy of workplace ergonomic interventions to control musculoskeletal disorders: A critical analysis of the peer-reviewed literature. *Theoretical Issues in Ergonomics Science*, 2, 23–96.

Kilbom, A. (1994a). Repetitive work of the upper extremity: Part I—Guidelines for the practitioner. *International Journal of Industrial Ergonomics*, 14, 51–57.

Kilbom, A. (1994b). Repetitive work of the upper extremity: Part II—The scientific basis (knowledge base) for the guide. *International Journal of Industrial Ergonomics*, 14, 59–86.

Konz, S. and Johnson, S. (2007). *Work Design: Industrial Ergonomics* (7th ed.). Scottsdale, AZ: Holcomb Hathaway Publishers, Inc.

Lievense, A., Bierma-Zeinstra, S., Verhagen, A., Verhaar, J., and Koes, B. (2001). Influence of work on the development of osteoarthritis of the hip: A systematic review. *The Journal of Rheumatology*, 28(11), 2520–2528.

Marras, W. S. (2000). Occupational low back disorder causation and control. *Ergonomics*, 43, 880–902.

Marras, W. S. (2008). *The Working Back: A Systems View*. Hoboken, NJ: John Wiley & Sons, Inc.

Moore, J. S. and Garg, A. (1995). The strain index: A proposed method to analyze jobs for the risk of distal upper extremity disorders. *American Industrial Hygiene Association Journal*, 56, 443–458.

NIOSH (1981). *Work Practices Guide for Manual Lifting* (NIOSH Publication 81-122). Cincinnati, OH: NIOSH.

NIOSH (1988). *Cumulative Trauma Disorders—A Manual for Musculoskeletal Diseases of the Upper Limbs*. London, U.K.: Taylor & Francis.

NIOSH (1997a). *Musculoskeletal Disorders and Workplace Factors* (DHSS (NIOSH) Publication No. 97-141). Cincinnati, OH: NIOSH.

NIOSH (1997b). *Elements of Ergonomics Programs: A Primer Based on Workplace Evaluations of Musculoskeletal Disorders (No. 97-117)*. Cincinnati, OH: NIOSH.

Nordstrom, D. L., Vierkant, R. A., DeStefano, F., and Layde, P. (1997). Risk factors for carpal tunnel syndrome in a general population. *Occupational and Environmental Medicine*, 54, 734–740.

NRC/IOM (2001). *Musculoskeletal Disorders and the Workplace*. Washington, DC: National Academy Press.

OSHA (1990). *Ergonomics Program Management Guidelines for Meatpacking Plants* (OSHA Publication No. 3121). Washington, DC: OSHA.

OSHA (2000). *Final Ergonomics Program Standard—1910.900*, USDOL: OSHA, taken from the internet site for OSHA on November, 2000; http://www.osha.gov

OSHA (2002). *Ergonomic Guideline for Nursing Homes*. See http://www.osha.gov

OSHA (2003). *Draft Ergonomic Guideline for Retail Grocery Stores*. See http://www.osha.gov

OSHA (2009). *Guidelines for Nursing Homes: Ergonomics for the Prevention of Musculoskeletal Disorders (OSHA 3182-3R)*. Washington, DC: Occupational Safety and Health Administration, U.S. Department of Labor.

OSHA (2010). http://www.osha.gov/SLTC/ergonomics/index.html

Putz-Anderson, V. (Ed.) (1988). *Cumulative Trauma Disorders—A Manual for Musculoskeletal Diseases of the Upper Limbs*. London, U.K.: Taylor & Francis.

Salvendy, G. (Ed.) (1997). *Handbook of Human Factors and Ergonomics*. New York: John Wiley & Sons, Inc.

Salvendy, G. (Ed.) (2006). *Handbook of Human Factors and Ergonomics*. New York: John Wiley & Sons, Inc.

Salvendy, G. (Ed.) (2011). *Handbook of Human Factors and Ergonomics*. New York: John Wiley & Sons, Inc.

Silverstein, B. A., Fine, L. J., and Armstrong, T. J. (1987). Occupational factors and carpal tunnel syndrome. *American Journal of Industrial Medicine*, 11, 343–358.

Smith, M. J. et al. (1971). *Wisconsin Inspection Effectiveness Project* (Contract Report L71-171). Washington, DC: U.S. Department of Labor.

Smith, M. J. and Carayon, P. C. (1995). New technology, automation and work organization: Stress problems and improved technology implementation strategies. *International Journal of Human Factors in Manufacturing*, 5, 99–116.

Smith, M. J. and Carayon, P. (1996). Work organization, stress and cumulative trauma disorders. In S. Moon and S. Sauter (Eds.), *Beyond Biomechanics: Psychosocial Aspects of Cumulative Trauma Disorders*. London, U.K.: Taylor & Francis, pp. 23–42.

Smith, M. J. and Carayon, P. (2011). Controlling occupational safety and health hazards. In L. E. Tetrick and J. C. Quick (Eds.) *Handbook of Occupational Health Psychology*. Washington, DC: American Psychological Association, pp. 75–93.

Smith, M. J., Cohen, H., Cohen, A., and Cleveland, R. (1978). Characteristics of successful safety programs, *Journal of Safety Research*, 10(2), 5–15.

Smith, M. J., Karsh, B.-T., Carayon, P., and Conway, F. T. (2002). Controlling occupational safety and health hazards. In J. C. Quick and L. E. Tetrick (Eds.) *Handbook of Occupational Health Psychology*. Washington, DC: American Psychological Association, pp. 35–68.

Smith, M. J., Karsh, B.-T., and Moro, F. B. (1999). A review of research on interventions to control musculoskeletal disorders. In *Work-Related Musculoskeletal Disorders*. Washington, DC: National Research Council, pp. 200–229.

Smith, M. J. and Sainfort, P. C. (1989). A balance theory of job design for stress reduction. *International Journal of Industrial Ergonomics*, 4, 67–79.

Snook, S. H. (1988). Approaches to the control of back pain in industry: Job design, job placement and education/training. *Professional Safety*, 33, 23–31.

Snook, S. H. (1991). The design of manual handling tasks: Revised tables of maximum acceptable weights and forces. *Ergonomics*, 34(9), 1197–1213.

U.S. Department of the Army (U.S. DoA) (2003). *Medical Services Ergonomics Program. Pamphlet 40-21*. Washington, DC: Department of the Army.

U.S. Department of the Defense (U.S. DoD) (2009). *Directory of Design Support Methods*. San Diego: DTIC-A, U.S. Department of Defense. Contact: http://www.dtic,mil/dticasd/ddsm

USMIL (2008). *Military Standard 1472F Human Engineering Design Criteria for Military Systems, Equipment and Facilities*. Washington, DC: U.S. Department of Defense. Contact: http://www.redstone.army.mil/amrdec

Waddell, G. and Burton, A. K. (2001). Occupational health guidelines for the management of low back pain at work: Evidence review. *Occupational Medicine*, 51, 124–135.

Waters, T. R., Putz-Anderson, V., and Garg, A. (1993). Revised NIOSH equation for the design and evaluation of manual lifting tasks. *Ergonomics*, 36, 749–776.

WorkSafeBC (2009). *Preventing Musculoskeletal Injury (MSI): A Guide for Employers and Joint Committees*. Vancouver, BC: Workers' Compensation Board of British Columbia. Available on the Web site http://www.worksafebc.com

36

Work Organization Interventions in Health Care

Kari Lindström, Gustaf Molander, and Jürgen Glaser

CONTENTS

36.1 Introduction 611
36.2 What Are the Stages of Prevention and the Goals and Target Groups of Organizational Interventions? 612
    36.2.1 Stages of Prevention in Organizational Interventions 612
    36.2.2 Goals of Organizational Interventions 612
    36.2.3 Target Groups in Organizational Interventions 612
36.3 Job Stressors and Lowered Well-Being as Targets of Health Care Interventions 612
36.4 Planned Organizational Interventions and Job Redesign as Developmental Processes 613
    36.4.1 Phases of the Participatory Intervention Process 613
    36.4.2 Survey Feedback Method Used in Health Care Organization 615
    36.4.3 Conference Method for Maximizing Employee Participation 616
    36.4.4 Organizational Participatory Intervention to Reduce Stress at Work 616
    36.4.5 Managing Structural Changes and Mergers in Health Care 617
    36.4.6 Multilevel Organizational Health Promotion 617
    36.4.7 Stress Management and Handling of Negative Emotions at Work 618
    36.4.8 Total Quality Management and Employer–Employee Discussion 619
    36.4.9 Reorganizing Nursing Work 620
36.5 Evaluation of Interventions and Critical Factors 621
    36.5.1 Evaluation of Interventions 621
    36.5.2 Critical Factors in Planning and Implementing Interventions 622
36.6 Conclusions 623
References 623

36.1 Introduction

Good interpersonal skills and competence to deliver high-quality care are important characteristics of health care workers and of an entire health care organization. During the last decade, health care systems in many countries have experienced a crisis caused by the rising cost of care, the long patient waits, and the rigidity and slowness of health care organizations to respond to changing external demands and challenges. For example, Ramanujam and Rousseau (2006) identified organizational learning and implementation of effective management practices as critical issues. "Challenges are organizational, not just clinical" they argued. Reasons for this are multiple and conflicting missions of health care organizations, complex external environments with multiple stakeholders, and ambiguous and dynamic tasks of multiple professions. Thus,

interventions targeting both organizational structures and functioning are needed.

Organizational interventions must be contextualized with respect to the nature of the work and the type of health care organization in which they are being carried out. Hospitals, nursing homes, and other health care settings are quite distinct with respect to clients, tasks, structures, and occupations. Hence, work organization interventions have to be carefully tailored to the specific domains in health care. They can focus on the jobs, the apparent organizational stressors, and the effects of organizational stressors on the well-being and health of personnel. Therefore, a short description of the psychological, social, and organizational factors that should be improved or changed and the well-being and health problems of health care personnel that should be targeted is given at the beginning of this chapter.

The main focus of the chapter is the various kinds of organizational intervention processes and methods applied, as well as their results. Emphasis is not only placed on participatory intervention processes such as the survey feedback method and participatory action research, but also on interventions involving health promotion, job stress, emotions at work, total quality management (TQM), and various nursing modes. The aim is to describe organizational interventions that have been successful in health care and to illustrate the main factors contributing to their success.

Issues dealing with the organizational culture of health care organizations are, however, not described here.

36.2 What Are the Stages of Prevention and the Goals and Target Groups of Organizational Interventions?

36.2.1 Stages of Prevention in Organizational Interventions

The prevention of occupational stress at work can be considered as primary prevention. It includes changing work conditions, such as providing well-defined job descriptions, redesigning work content and its organization, applying good ergonomic practices, using joint employee–employer action to improve the work environment and work life balance, and offering training to workers to help develop their competencies. Often, however, interventions involve forms of secondary prevention (e.g., to help employees control their reactions to stressful work factors). These interventions focus on individuals and their symptoms. Tertiary interventions, in turn, concentrate on disease and disability management. Organizational interventions are seldom tertiary forms of prevention (Murphy and Sauter, 2004).

36.2.2 Goals of Organizational Interventions

Organizational interventions and the management of change processes involve the following four general aspects of job perception that should be considered targets when the content and process of an intervention are planned and evaluated: changing job characteristics, role characteristics, leadership behavior, and work group characteristics. Of course, the direct effects on well-being, health, and productivity, and the effects on these factors via changes in the aforementioned aspects of job characteristics are also intervention goals.

The targets of organizational interventions can be conceptualized on the basis of either the work climate

(e.g., James and James, 1989) or open system theories of organizational behavior. For example, the causal model of organizational performance and change by Burke and Litwin (1992) describes organizational behavior at the transformational (organizational culture) and transactional (group and individual) levels. The process variables in organizational interventions are important, such as how actions are prepared and carried out (Nytrø et al., 2000).

The nature and development of work organizations have been explored using, for example, the concept of organizational healthiness. According to this concept, the psychosocial subsystem of the organization can be described as representing perceived internal functioning with regard to task completion, problem solving, and staff development (Cox and Leiter, 1992). These elements are also relevant with respect to health care organizations.

36.2.3 Target Groups in Organizational Interventions

Most studies and organizational interventions in health care have focused on the work of nurses and physicians. Segregation between various professional groups and their tasks is common in health care, especially in hospital settings. Such an approach easily leads to intervention practices that do not target the whole work unit. In primary health care, the individual-based working model has been the object of change, and team building among personnel has been the goal. In hospitals, nurses have often tried to enhance their mutual collaboration and the organization of daily patient care routines. One key issue in hospitals has been the tension between the managerial and medical subcultures. Professional autonomy and relationships between colleagues have been emphasized (Loan-Clarke and Preston, 1999). Interprofessional education has also been shown to improve collaboration as well as patient care (Reeves et al., 2009).

36.3 Job Stressors and Lowered Well-Being as Targets of Health Care Interventions

Much research data are available on job stressors and their possible effects on the well-being of health care personnel. Job and organizational stressors are often the starting points for organizational interventions. In addition to the traditional job and organizational stressors, functional and structural change processes are also sources of stress in health care organizations (Roald and Edgren, 2001).

According to a comprehensive British literature review (Michie and Williams, 2003), long work hours, work overload and pressure, and their effects on personal life are associated with both ill health and absenteeism among health care professionals. For example, a longitudinal study on hospital ward overcrowding showed that an exposure of over 6 months to an average bed occupancy rate over 10% in excess of the recommended limit was causally associated with antidepressant use of nurses and physicians (Virtanen et al., 2008). In addition, lack of control over work, lack of participation in decision making, poor social support, and ambiguous management and work roles are related to ill health and absenteeism. Successful organizational interventions have been able to affect and change these factors.

Among nurses, more than in other comparable professional groups, the level of psychological disorders, from mild emotional exhaustion to suicide, is higher according to a British report (Williams et al., 1998). The job-related sources of these psychological problems among nurses are high workload and its effect on personal life, as well as staff shortages, unpredictable staffing and scheduling, lack of time for providing emotional support to patients, role conflicts and ambiguity, poor leadership, lack of reward, workplace aggression, and lack of supervision and resources (Edwards et al., 2001; Lim et al., 2010; McVicar, 2003; Williams et al., 1998).

Physicians form another professional group that has been studied in health care to a great extent. Their main job stressors have also been high workload and pressure at work (Williams et al., 1998). Poor teamwork seems to contribute to the sick leave absences of hospital physicians even more than do traditional psychosocial risks, such as workload and work pressure (Kivimäki et al., 2001). Long work hours are also typical for physicians, and they often lead to problems in the balance between work and private life. In addition, other job factors specific to the field of specialization and the type of organization can be stressors or sources of job satisfaction for physicians. For example, among both British and Swedish psychiatrists, high job satisfaction was explained by a lower workload, a positive view of leadership, low work-related exhaustion, and a sense of participation in the organization. Also noteworthy were the few differences between British and Swedish psychiatrists (Thomsen et al., 1998). Low organizational justice has been found to increase the risk of psychological distress among male physicians but not among female physicians (Sutinen et al., 2002). In addition, management and decision-making practices, especially in large hospitals, are often problematic. Finnish studies have shown that fair leadership and fair decision making are important factors for the well-being of health care workers in general (Elovainio et al., 2002). In a review of studies

on physicians' well-being, it has been shown that work stressors not only impair physicians' health but also the quality of care of patients (Wallace et al., 2009). Those physicians who frequently work shifts longer than 24h make more serious medical errors than do those working shifts shorter than 24h (Landrigan et al., 2004). Depressed and burnt out residents have higher rates of medication errors (Fahrenkopf et al., 2008).

For other occupational groups in health care, the main stressors are generally at the same level as those for nurses and physicians. High psychosocial workload is expected because health care work requires much social interaction, both with other personnel and with patients and their relatives.

Health problems that have been found to be related to the aforementioned job stressors include reduced well-being (usually measured with the General Health Questionnaire), psychological disturbances such as depression and anxiety, and cardiovascular illnesses. In addition, back and joint pain have been strongly associated with sickness absenteeism (Williams et al., 1998).

Even though the stressors in health care work are well known, in order for interventions to be successful, the organizational structure, organizational culture, type of patients, and, finally, the financing model used for activities must be taken into account.

36.4 Planned Organizational Interventions and Job Redesign as Developmental Processes

Most of the methodologically acceptable interventions among health care workers have aimed at improving general physical and psychosocial problems. The general health-improving interventions are systemic organizational programs, which also include various kinds of personnel and management training.

36.4.1 Phases of the Participatory Intervention Process

We propose that organizational intervention processes in health care should be participatory, meaning that all stakeholders should be involved in the whole intervention process, from the beginning to the end. The key factors of the intervention process are good planning and clear responsibilities for process management. The impact of the intervention may depend even more on the process factors than on the actual goal of the intervention (Nytrø et al., 2000).

A planned organizational intervention needs a project group that is responsible for the intervention. It should be comprised of representatives of the employees and the employer, as this joint representation guarantees participation and strengthens the commitment of both groups. The process usually needs an internal or external consultant who insures that the process proceeds smoothly. The employer should be aware of the fact that the intervention needs time and this should be taken into account when planning the project.

Organizational interventions usually follow a specific series of steps (Figure 36.1): (1) The intervention process starts with a preliminary problem definition and the expectation of both the employees and the employer. (2) The second phase is the commitment of all the stakeholders to the intervention, so that a shared vision is formed of the goals, the possible content, and the process of the intervention. (3) Once a shared vision is attained, the first step is to organize a project group responsible for carrying out the program, possibly along with external consultative support. (4) It is then possible to start the intervention process with an organizational diagnosis, for example, a survey of psychosocial stressors at work. (5) Using a participatory planning process makes it possible to choose the method to be applied, the targets, and the time schedule. (6) The implementation of the planned actions requires the commitment of both the employers and employees and usually consultative support from an external consultant. (7) The last phase is to evaluate the process, which, at the very least, should include a second measurement with the same survey instrument used in the beginning of the intervention. However, data should be also collected as the intervention process proceeds. These data provide

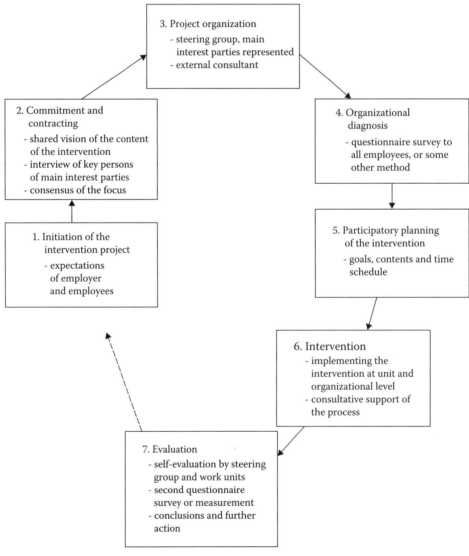

FIGURE 36.1
Phases of a participatory organizational intervention.

information about the successes and failures of the intervention.

This general process model can include several kinds of group work activities based on methods that promote the active participation of employees. One of them is the conference method, which is based on a democratic dialog between the participants (Gustavsen and Engelstad, 1986).

The participative interventions described in this section are the survey feedback method, the conference method, and multilevel organizational interventions. In addition, job stress interventions, management of structural changes, organizational health promotion, and TQM including employee–employer confidential discussions and various nursing model implementations are described. The target groups of these forms of intervention have varied from small work units to several hospitals or organizations. The length of the intervention varies from several days to 2 or even 3 years.

36.4.2 Survey Feedback Method Used in Health Care Organization

The survey feedback method follows the general steps of an organizational intervention process (Figure 36.1). It stems from the sociotechnological tradition of organizational psychology (Harrison, 1994). Many workplaces, especially in the health care sector, annually carry out work climate surveys that involve self-administered questionnaires. These questionnaires generally cover the main psychological, social, and ergonomic aspects of work and also include measures of the well-being and subjective health of personnel. Questionnaire tools that are useful for these purposes are, for example, the NIOSH Job Stress Questionnaire (Hurrell and McLaney, 1988) or the Nordic Questionnaire for Psychological and Social Factors at Work (QPSNordic; Lindström et al., 2000a). A context-specific instrument for work analysis in health care settings has been developed and validated in Germany, which has been widely applied in different German speaking countries (for an overview, Glaser, 2005).

The results of the questionnaire survey can be used by management and employees of work units for the so-called survey feedback process that is often accompanied by practical organizational interventions and their evaluation. At least, the evaluation should be based on a comparison of survey results before and after the intervention.

An example from a municipal health care organization illustrates how the survey feedback method that formed the main structure of the intervention, can aim at managing job and organizational stress problems (Lindström and Kivimäki, 1999). The target organization employed about 900 people and consisted of the

following five departments: primary health care, general hospital care, dental care, hygiene control, and the supplies and maintenance department.

In the initiation phase, the organization shared the need to manage psychosocial stress and workload and to improve the organizational climate, but the intervention plans of the employer and employees were contradictory. The situation with respect to psychosocial stress and organizational climate had worsened significantly as the organization had to cut personnel costs at a time when the staff already considered their workload to be too high. The employer was interested in saving personnel costs and reorganizing the work. The city mayor, together with representatives of the local trade union of nurses, contacted a consultant in order to implement an organizational intervention program in the entire organization.

A shared vision of the content of the intervention program was created after interviews and joint discussions of the main interested parties. The project was organized and implemented by a steering group that included representatives of the main stakeholders. The problems of the organization were diagnosed on the basis of questionnaire results. The participatory planning of intervention actions was defined in survey feedback meetings. The intervention plan was implemented with the support of an outside consultant. Actions were taken to improve work organization and to reduce stress and strain. These actions included reducing the work overload and underload peaks by revising the timing of shifts in the hospital wards, establishing work counseling groups to promote professional competence among the health care personnel, and enhancing cooperation between the physicians and nurses in primary health care.

In addition, ergonomic interventions were initiated. For instance, interdepartmental exercise groups were set up to help those suffering from neck and shoulder problems, lifting aids were acquired to facilitate the handling of patients, and training in patient handling and the use of lifting aids was provided. The evaluation included a questionnaire survey similar to the questionnaire used in the diagnosis phase, as well as documentation of the intervention process. All units produced a report on how they had succeeded in their own projects, and the results were discussed in the steering group. The overall effect was that collaboration within the workers' own work group and professional competence were improved, but the climate at the organizational level was still poor.

Controlled studies on employees' high sickness absence (as in the case of health care workers) have shown that the survey feedback method might also be useful for secondary prevention. For example, in a randomized trial study subjects in an intervention group received personal feedback of their survey results and an invitation to a consultation at their local occupational health service. The main purpose of the consultation was the construction of an action plan, and if appropriate, referral to a further consultation by a specialist or psychologist. The intervention turned out to be cost saving and more effective than usual occupational health care (Taimela et al., 2008).

36.4.3 Conference Method for Maximizing Employee Participation

The conference method, or search seminars, used in organizational interventions is an example of the participatory approach in that the setting of goals and the intervention process are based on a democratic dialog between the participants in the organization. This method is used especially in Scandinavian countries, in both health care and in other types of organizations (Gustavsen and Engelstad, 1986). The method can be used as such or in combination with other approaches. The conference method facilitates joint discussion in organizational interventions and aids the participatory intervention process.

When this method was used in an organizational merger in a municipal health care organization, the aim was to create and transfer the common goals of the merger to everyday practice in the new organization and to guarantee the availability of good services for all residents of the municipality. Several small projects were also launched to improve work practices (Lindström and Kivimäki, 1999).

A process based on the conference method usually starts with a 1 day search seminar with persons from different units, tasks, and hierarchical levels. At the beginning of the seminar, the chair or leader presents the future plans and vision of the organization. Thereafter, discussions are held in small parallel groups. The same topics (i.e., the vision of the organization, potential problems in the realization of the vision, and actions needed to realize these visions and reach the goals) are discussed by the groups. In order to strengthen people's commitment to the project, the dialog in the working conferences increases mutual understanding and helps generate new ideas. The project group then utilizes these ideas to produce main ideas and goals for action. During the 1 or 2 year project, the management and personnel meet at least once more to exchange experiences in order to check on the progress of the program and to generate new ideas on how to continue. This exchange of experiences and common learning improves the process and helps people commit to it. At the end of the project, a third seminar is organized to evaluate what has been learned during the process. These seminars are forums for exchanging experiences and increasing commitment.

36.4.4 Organizational Participatory Intervention to Reduce Stress at Work

One Norwegian example may be typical of how participatory organizational stress intervention is usually carried out. A short-term participatory intervention to reduce stress at work was implemented in a health care institution. The participants were randomly allocated to an intervention group and a control group (Mikkelsen et al., 2000).

The intervention proceeded as follows. Two weeks before starting the intervention, the researchers informed all the employees and supervisors about the project. An external organizational development (OD) consultant was responsible for carrying out the intervention together with a group comprised of the supervisor, trade union representative, and employees' safety representative. The aim was to start a learning process to identify and solve work problems in order to continuously improve worker health and organizational performance on a long-term basis. The process started with a 6 h seminar in the form of a participative search conference (Emery and Purser, 1996). The seminar was held to collectively create a plan that the members themselves would implement. The key questions asked were: (1) what are the key factors in the work unit that insure a good work environment and (2) what actions do you see as necessary for reducing the gap between the desired outcome and existing reality? The planned actions were based on the main problems reported by the participants, which were lack of information, poor communication, and insufficient respect between the different professions. There was also a need for professional and personal development. After the initial orientation seminar, the work unit held nine individual group meetings, each lasting 2 h.

The evaluation of the process was based on the logbooks and written reports of the supervisor and the work groups. The measurement instruments included a questionnaire scale of work-related stress, a health inventory of subjective health, job demands and control, social support, learning climate, and leadership style.

Because the intervention was carried out during work hours, it was difficult to find meeting times that were suitable for everyone (e.g., because of shift work). Initially, the intervention did not have a high priority among the employees. The supervisors of the two intervention units held opposite views, one was skeptical and the other one was enthusiastic. As a result, the intervention had a positive, but limited effect on work-related stress, and psychological work demands improved in the intervention group but not in the control group.

The intervention seemed to have begun a positive change process. In spite of the short-term changes, it did not give a consistent and convincing picture of the role of organizational intervention in reducing work stress (Mikkelsen and Saksvik, 1998). The intervention period was too short, and the work methods were not sufficiently established to encourage the participants to continue the process after the project period. The employees reported that the introduced methods were very useful for problem solving, but that the methods were not used in the normal staff meetings until the end of the project. A more focused strategy concentrating on some of the problems might have worked better. The participation process itself had a positive impact on well-being, and it reduced stress at work.

Health circles are another form of organizational participatory intervention to reduce stress at work. This health promotion approach has been developed in Germany. It emphasizes organizational and psychosocial factors while actively involving employees in the process. A review on health circles (Aust and Ducki, 2003) identified 11 studies presenting the results of 81 health circles. The results of the review suggest that health circles are an effective tool for the improvement of physical and psychosocial working conditions and have a favorable effect on workers' health, well-being, and sickness absence.

A trial-control study was performed in a nursing home for the elderly using the health circle approach (Glaser et al., 2007). The intervention aimed on the reduction of work stressors. Work stressors on a "model" and a "control" ward were assessed by shift observations and self-report surveys. Within several group sessions on the model ward, solutions to overcome work stressors were developed. Results of repeated shift observations and self-report survey showed that the participatory intervention was successful.

In a longitudinal study of 37 medical clinics, it has been shown that integrative methods of open communications and participation in implementing organizational change moderate the relationship between cognitive diversity among employees and clinic revenue, productivity, and patient satisfaction (Van de Ven et al., 2008).

36.4.5 Managing Structural Changes and Mergers in Health Care

Hospital mergers are typical structural changes that have both positive and negative effects on employee work conditions and well-being. For example, in the 1980s, the goal of most hospital mergers in North America was to strengthen their financial situation (Bazzoli et al., 2002). Overlapping services were discontinued, wards were combined, the number of full-time and especially auxiliary personnel decreased, and the number of part-time employees increased. Later, in the 1990s, the main trend was to downsize the number of personnel in general in order to cut costs. Many of these mergers and downsizing procedures were carried out without the participation of the personnel (Bazzoli et al., 2002). The negative consequences of the mergers and downsizing have been found to lead to an elevated workload for the remaining employees and, therefore, reduced well-being and job dissatisfaction and increased stress-related sickness absenteeism—and, in the long run, even increased cardiovascular diseases and elevated mortality (Vahtera et al., 2004).

An important focus of organizational intervention in hospitals is how to facilitate structural and cultural changes when wards and hospitals with different cultures are merged (Cavanagh, 1996). Employees' reactions to mergers can vary from neutral to strong resistance.

The management process in a merger in a Finnish hospital with 20,000 employees included the following phases:

- Informing personnel about the coming structural change
- Meetings with supervisors to help them prepare for the change process
- Questionnaire survey of all personnel immediately after the merger
- Feedback discussion about the questionnaire survey results and the implementation of supportive actions
- Intensified information communication within the organization
- Special consultation support for units with major changes

In a prospective study of this merger, negatively perceived changes were explained by the extent of the change, poor management of the change, and poor leadership in general. Low employee participation in the organizational change process was also related to negative attitudes toward the change (Lindström et al., 2003). It is therefore crucial to prepare the merger carefully and to inform the personnel (Roald and Edgren, 2001; Schweiger and DeNisi, 1991). A corporation of Swedish hospitals showed that the restructuring proceeded in a neutral manner due to its careful implementation by management and resulted in a realistic preview of the change (Sverke et al., 1999).

36.4.6 Multilevel Organizational Health Promotion

Intervention in the form of health promotion has usually been individual oriented, focused on the lifestyle, well-being, and health of the individual (e.g., Hurrell and Murphy, 1996). The Finnish extension of such a health promotion strategy is called the maintenance of work ability (MWA). It refers to participatory workplace health promotion interventions targeted toward work and organizational factors, the physical work environment, and the health and professional competence of the employees. The interventions are carried out by the health promotion team with the help of occupational health personnel or an outside organizational consultant (Elo and Leppänen, 1999). Organizational interventions represent one part of this health promotion strategy. Some of the interventions have been evaluated and found to be successful when planned and carried out systematically (Lindström et al., 2000b). A large-scale organizational health promotion program improved both individual well-being and productivity at the company level. Continuous improvement practices and leadership practices brought about the most positive changes. The most effective actions were those focused on social interaction and leadership.

In a multilevel and multicomponent intervention process of workplace health promotion in a nursing home, the conference method was used to structure the intervention process. The intervention focused on the individual/organization interface at both the group and organization level. All 600 employees of the nursing home participated in the study (Lindström, 2001). The 2½ year process was led by a project group that included the head of the organization, the occupational health physician, a personnel development expert, a union representative, and an outside consultant. The project started by diagnosing the situation on the basis of the results of a questionnaire survey and a search seminar. The intervention was comprised of lectures on patient care and work organization (e.g., learning organization principles), followed by participatory group work in intensive case projects at specific wards. The group work dealt with topics like team building, customer-oriented work, utilizing the competence of employees of various age, and developing the supervisor's role. Every 6 months, a feedback and evaluation meeting was organized at the organizational level, and, at the end of the project, a final evaluation meeting took place. In addition, a questionnaire survey was carried out at the end of the project. The intervention was evaluated from the following four perspectives:

- How it influenced the actual problems in the nursing home.
- How effective it was in improving organizational functioning.
- How it contributed to organizational and personnel learning.
- How it influenced the future perspectives of the nursing home.

The improvements with respect to psychosocial stressors and professional competence were related to persons' frequencies of participation in the various intervention events (Lindström, 2001).

The success factors and experiences from this case of organizational health promotion were the extensive participation of both the employers and the employees, the external consultant's facilitation of the process structuring, the motivation of the project group, the established joint-learning process, and the documentation of the whole process. In some units of the nursing home, however, the simultaneous structural and functional changes in the organization had a negative effect on the results of the intervention.

A systematic review on the effectiveness of different work stress interventions in health care has shown that person-directed interventions as well as organizational interventions reduce stress symptoms (Ruotsalainen et al., 2008). However, work organization interventions have a more diverse effect than interventions at the individual level. Improvements should not only focus on design issues, but also on thorough documentation and subgroup analyses. The multilevel combination of person-oriented and organization-oriented approaches is the most promising (Semmer, 2006).

36.4.7 Stress Management and Handling of Negative Emotions at Work

Several stress management and burnout interventions have been implemented for health care personnel (Schaufeli and Enzmann, 1998). Stress management interventions can be directed toward the individual, the individual/organization interface, or the levels of organization. Stress and burnout interventions are, however, dealt with in this chapter only with respect to innovative interventions and the handling of negative emotions at work.

Stress management programs have traditionally focused on individuals and their possibilities to cope with stressors at work. The systematic actions taken have included recognition of the job stressors and means of coping with the strain.

Stress management intervention at the individual level can focus on didactical stress management, the promotion of a healthy lifestyle, relaxation, and cognitive therapy techniques. Interventions focusing on the individual/organization interface deal with time management, balancing work and private life, improving peer support, or some type of counseling. At the organizational level, the methods used include stress audits, changes to improve job content and the work environment, conflict management, and various types of organizational interventions.

An innovative climate at work has been found to contribute to satisfaction with leadership among hospital ward personnel (Kivimäki et al., 1994). A British study compared innovative interventions and stress management interventions in health care (Bunce and West, 1996). A traditional stress management intervention and an intervention promoting innovative responses to stressors were carried out among mixed community-based and hospital-based staff. The traditional cognitive-behavioral stress management intervention was comprised of interactive sessions/workshops. The innovation program was based on the idea of innovative coping, and the participants were encouraged to develop innovative responses to stressors through group discussions and individual action planning. At first, the general stress level decreased in the stress management group, but this effect did not last very long. The innovative stress management group displayed more long-lasting effects in the level of innovation, but the result was not evident immediately after the intervention sessions. The general conclusion was that, in a stress management program, the process itself influences the results.

Positive effects were found among health care personnel when short- and long-term stress management interventions were carried out. The long-term intervention included a stress management session that was repeated three times. Exhaustion symptoms decreased and stayed at a lower level among those who attended the repeated stress management sessions. In addition, cynicism was lower in the group with repeated sessions, and professional competence was higher (Rowe, 2000).

The effectiveness of workplace stress management for nurses was assessed through a systematic review of randomized controlled trials and prospective cohort studies (Mimura and Griffiths, 2003). There was more evidence for the effectiveness of programs based on providing personal support than for environmental management to reduce stressors. However, the number of high-quality studies was low, and the authors concluded that further research is needed.

Situations involving negative emotions like jealousy and envy have been found to cause strain in nursing. Both individual and work-related issues have been found to provoke these emotions (Vecchio, 1999).

Among the nurses, the emotional workload was related to situations like nurse–patient or nurse–nurse interactions with a lack of social support and fair treatment, high responsibility for the patients, difficult ethical issues, dissatisfaction on the part of patients, and balancing one's work and private life (Molander, 2003). These situations produced feelings of shame or guilt in the nurses. In order to solve and handle emotions in joint discussion groups, a systematic problem-solving procedure is needed.

Problem-solving procedures were developed for these emotionally burdensome situations and their usefulness was tested in an intervention study in a nursing home (Molander, 2003). Two negatively loaded situations that had actually occurred were discussed in a counseling group. They included situations of anger in the team when one of the coworkers became the supervisor and feelings of guilt when an employee had forgotten something that was important for the patient. These feelings of anger and guilt were discussed in joint meetings in which an outside consultant was present as a facilitator. The group situation proceeded through the following steps:

- Presenting the anger- or guilt-provoking event for joint discussion in the group
- Negotiating and trying to find a joint solution
- Agreeing about the ways in which to solve the problem

The group helped the person feeling guilt or anger to define the problem, to put it into a broader context, and to find ways of handling the negative emotions provoked by the incident. The developed process models were then tested by another group. These step-by-step models helped people to cope with emotionally difficult situations, which are hard to avoid in patient work (Molander, 2003).

36.4.8 Total Quality Management and Employer–Employee Discussion

TQM can potentially improve the quality of products, the services provided, and different aspects of work design and quality of work life. Its principles are customer focus, continuous improvement, and teamwork. Some aspects are also dealt with at the individual level using annual confidential employer–employee discussions.

TQM has been found to be positive or negative or to produce no effects on work design. In public sector organizations in the United States, the main positive impact of TQM concerned job content, job control, participation, and social relations. At the same time, negative changes were reported for workload and the uncertainty/lack of clarity of duties. The results, however, varied between and within departments (Carayon et al., 1999). TQM may also have detrimental effects on health care workers. For example, it has been shown that the ongoing standardization of care and the growing bureaucracy restrict physicians' decision making and autonomy and, thus, increase job dissatisfaction and stress of physicians (Sundquist and Johansson, 2008).

Implementing or improving TQM systems in health care has been one goal of organizational interventions. TQM applied in health care organizations produces an OD process that aims to improve the quality of care processes, the work flow and collaboration between various professionals involved in the same care process. There are numerous studies on the relationship between TQM and the realignment of critical process flow, cost effectiveness, client satisfaction, and level of quality (Sommer and Merrit, 1994). Results show that some professional groups have a better collective work orientation than others, and they also better know the impact of their own activity.

TQM has mainly been applied and studied by consultants, and little research data are available. In addition, there are contradictory opinions about the effects of TQM. One element of TQM is the use of quality circles. People have thought that quality circles increase the innovativeness and participation of personnel, but they have had only a limited effect in changing employees' attitudes and organizational culture (e.g., Adam, 1991). Thus far, results on the relationship between personnel well-being and TQM are inconclusive and contradictory.

A longitudinal case-control study (Kivimäki et al., 1997) was conducted in a surgical clinic, which had yielded an international prize for successful TQM implementation, and two other surgical clinics that served as controls. The support for the potential economic impact of TQM was strong, but the evidence concerning the effects of TQM in promoting the well-being and work-related attitudes of the personnel was weak. Although the proponents of TQM have argued that it creates a cultural change that enhances the well-being of personnel, this assumption was not supported by the study. Commitment to the new system was high among the employees, except for the physicians. Among the physicians, the implementation of TQM led to a loss of autonomy due to increased teamwork, and there was no perception of their future being connected to the success of the whole organization. TQM systems, therefore, need continuous monitoring regarding their benefits and drawbacks in the organization.

Annual confidential conversations between a supervisor and an employee are often recommended in order to increase an awareness of goals, to provide feedback, and to enhance innovativeness (Katzell and Thompson, 1990). Typically, these conversations last 1 or 2 h and cover topics such as the participatory setting and a clarification of work goals and consideration of what might help employees to attain them,

feedback on past performance, and a discussion of innovative ideas. The aim of these discussions is to enhance good leadership and increase employee job satisfaction. In practice, they usually include planning of the personal objectives of the employee for the next year and receiving feedback concerning the past year's accomplishments.

Although the concept for these confidential discussions stems from management by objectives (e.g., Drucker, 1976), the feedback can help develop the employee's professional skills. Research has been conducted to examine whether these conversations improve the perception of work goal clarity, the sufficiency of feedback and innovativeness, and satisfaction with leadership. Perceived work characteristics and supervisory practices of a group that performed such discussions were compared to those of a control group by means of measurements before and after discussion implementation. The sufficiency of feedback was significantly better in the group with confidential discussions, but no change was found in goal clarity, innovation, or satisfaction with leadership (Kivimäki, 1996). Therefore, the benefits were rather narrow compared to theoretical assumptions of confidential discussions.

A TQM approach in nursing has been developed by Büssing and Glaser (2003) according to their concept of employee and client orientation. The basic assumption behind that approach was joint optimization of the quality of employee's working life and the quality of care. Empirical studies have shown that the improvement of the work situation of nurses—in terms of reduced stressors and strengthened resources—is an enabler of a high quality care from the perspective of patients.

36.4.9 Reorganizing Nursing Work

There are at least three different theoretical models for organizing nursing work (i.e., functional nursing, team nursing, and primary nursing; Thomas and Bond, 1990). Functional nursing is a task-oriented mechanistic model in which the work is divided into separate tasks and allocated to nurses according to their skills. This model has been further developed, and transitions to team nursing and primary nursing have taken place. In team nursing, every nurse carries out a more comprehensive set of tasks than in individual nursing. Primary nursing, on the other hand, is characterized by task completeness with respect to certain patients and horizontal colleague support. In practice, many wards have a work organization that is a mixture of these three models. The amount of stress among nurses has been studied in comparing these models. Patient-focused work allocation, opportunity to write nursing notes, and accountability for patient care contributed to the nurses' satisfaction with supervision and opportunities for personal growth (Mäkinen et al., 2003).

Interventions in which new models have been implemented have shown some differences in job stressors, such as work overload and interpersonal conflicts. The transition from team and task-oriented nursing to primary nursing has been found to lead to increased autonomy and decreased difficulty with work and work pressure (e.g., Büssing and Glaser, 1999; Manley et al., 1997).

However, the results of implementing different nursing models have shown that changes in perceived job stressors and well-being are not merely dependent on the nursing model. In addition, ward characteristics are important, such as the number of beds, the nurse/patient ratio, and the level of education and skills of the nursing staff. The adoption of a new model may take several years and may therefore lead to additional stress due to change. A recent study has shown that the ways used to organize nursing and nursing staff increased the work motivation and job satisfaction of nurses, but their role in job stress was limited (Kivimäki et al., 1994; Mäkinen et al., 2003).

The process of redesign and OD from a functional to a holistic nursing system was evaluated with regard to its influence on the work stressors and its implications for burnout and interactional stress (Büssing and Glaser, 1999). The longitudinal study design included a comparison between one model hospital and two control hospitals. Both survey data of various work stressors that were predictors of burnout and qualitative data from group discussions were collected. Work stressors were substantially reduced during the process, whereas emotional exhaustion and depersonalization increased. According to the qualitative data, the implementation of a holistic nursing system was accompanied by an intensification of interactional stress and emotional work.

36.5 Evaluation of Interventions and Critical Factors

36.5.1 Evaluation of Interventions

Work organization interventions are increasingly used to improve the occupational safety and health of health care organizations and their personnel. Usually a research-based evaluation of the impact of interventions is difficult because traditional experimental designs, by themselves, are not sufficient and relevant. Therefore, it has been difficult to identify effective intervention methods and encourage people to use them (e.g., Goldenhar and Schulte, 1994). The evaluation of an organizational intervention should both subsume a process evaluation and an evaluation comparing the pre- and postmeasurements of expected outcomes. The weakest evidence of an intervention is based on the description and reporting of the process and possible changes. The strongest evidence is based on randomized control studies (Murphy, 1996). Planned study designs in organizational interventions are usually the bases for evaluation, but natural changes inside and outside organizations make the evaluation challenging.

Organizational interventions have different targets. Some aim to change job design and content, some are directed more at social interaction or leadership, whereas others are rather focused on the symptoms of psychological distress or health complaints. The evaluation method varies depending on these goals of intervention.

Because organizational interventions are often long lasting processes, their evaluation requires the following:

- Documentation of the whole intervention process
- Analysis of the changes at micro- and macrolevels (i.e., the dynamics of the change)
- Contextualization of the intervention process with regard to the target organization and simultaneous changes occurring in society
- Analysis of the roles of various actors during the process

In the evaluation, the nature of behavioral change and organizational dynamics should be taken into account. Organizational factors and process issues, and people's conflicting preferences in changing or intervening should be considered (Griffiths, 1999; Mikkelsen et al., 2000). Intervention not only consists of the "treatment" or planned intervention, but also of everything related to the target organization and the process of administering the "treatment."

The following two main reasons for clearly describing the intervention process have been recommended for health care organizations:

- To strengthen the evidence that an observed effect was really due to the actions implemented by identifying links in the chain of causes and effects that could have led to the observed outcome
- To recommend which factors should be given special attention in future attempts to arrive at a similar or better result of the intervention (Wickström et al., 2000)

Although qualitative data from the process are important in organizational interventions, quantitative data should not be overlooked.

A Norwegian participatory organization intervention on job stress in community health care institutions is a good example for the importance of documenting the process for evaluation (Mikkelsen et al., 2000). This short-term intervention targeted toward the problems perceived by employees. Positive but small effects were found in relation to work-related stress, job characteristics, learning climate, and management style. A beneficial change process was also initiated. The intervention process took time to be established as an integrated process and as a learning process. The intervention period was, however, too short for any conclusions to be drawn about long-term effects. The intervention was not integrated into the normal daily routines of the work units. The participative process itself, however, had a positive effect (Mikkelsen et al., 2000). If important qualitative process elements are not documented, the conclusions about the pre- and postresults may be very superficial, even misleading.

36.5.2 Critical Factors in Planning and Implementing Interventions

There are general success factors of organizational interventions that also fit for health care organizations. Because of their nature as service organizations dealing with people's lives and their deaths, they have many specific requirements that either promote or prevent the planned improvements or positive changes.

According to the results of European case studies on organizational interventions, several success and failure factors can be identified (Kompier and Cooper, 1999). The national legal framework for preventing occupational stress facilitated OD activities at the workplace level. The main prerequisite of a successful organizational intervention was to apply a step-by-step model starting from careful planning and ending in proper evaluation. Also very important was the commitment of supervisors and employees from the very beginning (Murphy, 1999). Especially, top management should provide the support needed, both initially and later, in order for the intervention process to survive.

In organizational interventions and structural changes, as in mergers, the patient care itself and the competence of the organization, teams, or individuals need to be taken into account, along with job stressors and well-being. This requirement was clearly seen in a postmerger situation of a surgical clinic at which the patient waiting lists grew and old work routines and practices did not work. In this case, innovators from inside the organization took the initiative and started to reorganize patient care with the participation

of personnel. The problems were solved because this group had management support and personnel participation (Lindström and Turpeinen, 2004).

The combination of a bottom-up (participation) approach and a top-down (top management support) approach promotes the goals of intervention. This combination is especially critical in health care, where organizations are big and hierarchically multilevel.

It is important to realize that, in health care, there are separate professional subcultures that need to be taken into account when intervention processes are planned. Therefore, the intervention should have a sufficient number of cross-professional goals and subprojects. Involving as many employees as possible in subprojects of the intervention increases motivation.

The main success factor has usually been the application of a participatory approach in which employees have been able to commit and contribute actively to the intervention. Especially in the Swedish hospital corporation project, the positive results were interpreted as being related to the careful preparation and information that resulted in a realistic change process among the employees (Sverke et al., 1999). In this case, however, the change did not have a major influence on employees.

Well-defined responsibilities are needed for people involved in intervention processes. It is important that the project has an employer who is committed to the plans and actions. The process also needs external facilitators or a process consultant because there are obstacles along the way that must be dealt with.

Factors obstructing stress prevention have been listed in earlier studies (Kompier and Cooper, 1999). Most of these factors are also relevant for organizational intervention in health care. One of the main problems has usually been too tight a time allocation. In health care work, time must be reserved for personnel to participate in the intervention—otherwise participation stays low. Head ward nurses or middle managers used to be key persons both in motivating personnel and in organizing the work schedules and allocating work so that personnel has time to participate in joint training and intervention sessions.

Making a decision about the focus of an intervention is critical because questionnaire surveys used as diagnostic tools reveal the frequency of problems, but not necessarily their real importance. Practical and scientific aims of intervention compete because evidence and documentation for practical purposes is not so crucial as for scientific purposes.

The success factors for organizational interventions in health care are, for the most part, similar to those in other sectors or organizations, but the type of work, especially patient work, and the care culture have specific effects. Primary prevention should be combined with an approach to strengthen individual coping resources.

36.6 Conclusions

Well-established evaluation studies on work organization interventions in health care as well as in other lines of business are rather scarce. A meta-analysis of 36 studies with 55 interventions was conducted to determine the effectiveness of stress management interventions in occupational settings (Richardson and Rothstein, 2008). Cognitive-behavioral programs consistently produced larger effects than other types of interventions. Relaxation interventions were most frequently used, and organizational interventions continued to be scarce.

Organizational interventions in health care have traditionally dealt mainly with individual job stressors and strain. In addition, various professional groups have their own work culture, and job redesign intervention usually focuses either on implementing various nursing models or, at a higher level, a new management system that does not include all occupational groups at the same time. Implementing a new management system does not necessarily influence the entire care process and take into account parallel goals of well-being. Competing with organizational intervention aimed at improved well-being and a higher quality of work life in a health care organization are the restructuring and mergers taking place to cut costs. These structural changes will continue in the future because of rising health care costs. Therefore, multilevel and multitargeted organizational interventions are needed in which well-being and productivity goals can be balanced simultaneously. This balance requires much more management commitment to planned interventions. The participatory approach, however, and employee involvement can be seen as the most important factors when patient work, which is demanding emotionally, socially, and ethically, is being dealt with. Organizational culture and values should receive much more attention.

References

Adam, E. E. (1991). Quality circle performance. *Journal of Management, 17,* 25–39.

Aust, B. and Ducki, A. (2003). Comprehensive health promotion interventions at the workplace: Experiences with health circles in Germany. *Journal of Occupational Health Psychology, 9,* 258–270.

Bazzoli, G., LoSasso, A., Arnould, R., and Shalowitz, M. (2002). Hospital reorganization and restructuring achieved through merger. *Health Care Management Review, 27,* 7–20.

Bunce, D. and West, M. A. (1996). Stress management and innovation interventions at work. *Human Relations, 49,* 209–232.

Burke, W. W. and Litwin, G. H. (1992). A causal model of organizational performance and change. *Journal of Management, 18,* 523–545.

Büssing, A. and Glaser, J. (1999). Work stressors in nursing in the course of redesign: Implications for burnout and interactional stress. *European Journal of Work and Organizational Psychology, 8,* 401–426.

Büssing, A. and Glaser, J. (2003). Employee and client orientation: Concept and evaluation of quality in health care. In J. Hellgren, K. Näswall, M. Sverke, and M. Söderfeldt (Eds.), *New Organizational Challenges for Human Service Work* (pp. 115–135). München, Germany: Hampp.

Carayon, P., Sainfort, F., and Smith, M. J. (1999). Macroergonomics and total quality management: How to improve quality of working life? *International Journal of Occupational Safety and Ergonomics, 5,* 303–334.

Cavanagh, S. J. (1996). Mergers and acquisition: Some implications of cultural change. *Journal of Nursing Management, 4,* 45–51.

Cox, T. and Leiter, M. (1992). The health of health care organizations. *Work & Stress, 6,* 219–227.

Drucker, P. (1976). What results should you expect? A users' guide for MBO. *Public Administration Review, 36,* 12–19.

Edwards, D., Burnard, P., Coyle, D., Fothergill, A., and Hannigan, B. (2001). Stress and burnout in community mental health nursing: A review of the literature. *Journal of Psychiatric and Mental Health Nursing, 7,* 7–14.

Elo, A.-L. and Leppänen, A. (1999). Efforts of health promotion teams to improve the psychosocial work environment. *Journal of Occupational Health Psychology, 4,* 87–94.

Elovainio, M., Kivimäki, M., and Vahtera, J. (2002). Organizational justice: Evidence of a new psychosocial predictor of health. *American Journal of Public Health, 92,* 105–108.

Emery, M. and Purser, R. E. (1996). *The Search Conference. A Powerful Method for Planning Organizational Change and Community Action.* San Francisco, CA: Jossey-Bass.

Fahrenkopf, A. M., Sectish, T. C., Barger, L. K., Sharek, P. J., Lewin, D., Chiang, V. W., Edwards, S., Wiedermann, B. L., and Landrigan, C. P. (2008). Rates of medication errors among depressed and burnt out residents: Prospective cohort study. *British Medical Journal, 336,* 488–491.

Glaser, J. (2005). Analysis and design of nursing work—A decade of research in different fields of health care. In C. Korunka and P. Hoffmann (Eds.), *Change and Quality in Human Service Work* (pp. 13–31). München, Germany: Hampp.

Glaser, J., Lampert, B., and Weigl, M. (2007). Interaction, workload, health and work design in nursing for the elderly. In P. Richter, J. M. Peiró, and W. B. Schaufeli (Eds.), *Psychosocial Resources in Health Care Systems* (pp. 13–25). München, Germany: Hampp.

Goldenhar, L. M. and Schulte, P. A. (1994). Intervention research in occupational health and safety. *Journal of Occupational Medicine, 36,* 763–773.

Griffiths, A. (1999). Organizational interventions: Facing the limits of the natural science paradigm. *Scandinavian Journal of Work Environment and Health, 25,* 589–596.

Gustavsen, B. and Engelstad, P. H. (1986). The design of conferences and the evolving role of democratic dialogue in changing work life. *Human Relations, 39*, 101–116.

Harrison, M. I. (1994). *Diagnosing Organizations. Methods, Models, and Processes*. Thousand Oaks, CA: Sage.

Hurrell, J. J. Jr., and McLaney, A. M. (1988). Exposure to job stress: A new psychometric instrument. *Scandinavian Journal of Work Environment and Health, 14*, 27–28.

Hurrell, J. J. Jr., and Murphy, L. R. (1996). Occupational stress intervention. *American Journal of Industrial Medicine, 29*, 338–341.

James, L. A. and James, L. R. (1989). Integrating work environment perceptions: Explorations into the measurement of meaning. *Journal of Applied Psychology, 74*, 739–751.

Katzell, R. A. and Thompson, D. E. (1990). Work motivation: Theory and practice. *American Psychologist, 45*, 144–153.

Kivimäki, M. (1996). Confidential conversations between supervisor and employee as a means for improving leadership: A quasi-experimental study in hospital wards. *Journal of Nursing Management, 4*, 325–335.

Kivimäki, M., Kalimo, R., and Lindström, K. (1994). Contributors to satisfaction with management in hospital bed wards. *Journal of Nursing Management, 2*, 229–234.

Kivimäki, M., Mäki, E., Lindström, K., Alanko, A., Seitsonen, S., and Järvinen, K. (1997). Does the implementation of total quality management (TQM) change the well-being and work-related attitudes of health care personnel? *Journal of Organizational Change Management, 10*, 456–470.

Kivimäki, M., Sutinen, R., Elovainio, M., Vahtera, J., Räsänen, K., Töyry, S., Ferrie, J. E., and Firth-Cozens, J. (2001). Sickness absence in hospital physicians: 2 year follow up study on determinants. *Occupational & Environmental Medicine, 58*, 361–366.

Kompier, M. and Cooper, C. (1999). Stress prevention: European countries and European cases compared. In M. Kompier and C. Cooper (Eds.), *Preventing Stress, Improving Productivity. European Case Studies in the Workplace* (pp. 312–336). London, U.K.: Routledge.

Landrigan, C. P., Rothschild, J. M., Cronin, J. W., Kaushal, R., Burdick, E., Katz, J. T., Lilly, C. M., Stone, P. H., Lockley, S. W., Bates, D. W., and Czeisler, C. A. (2004). Effect of reducing interns' work hours on serious medical errors in intensive care units. *New England Journal of Medicine, 351*, 1838–1848.

Lim, J., Bogossian, F., and Ahern, K. (2010). Stress and coping in Australian nurses: A systematic review. *International Nursing Review, 57*, 22–31.

Lindström, K. (2001). Promoting organizational health as a part of the maintenance of work ability (MWA) activity in Finland. In C. Weikert, E. Torkelson, and J. Pryce (Eds.), *Occupational Health Psychology in Europe 2001. European Academy of Occupational Health Psychology Conference Proceedings Series* (pp. 129–133). Nottingham, U.K.: I-WHO Publications.

Lindström, K., Elo, A.-L., Skogstad, A., Dallner, M., Gamberale, F., Hottinen, V., Knardahl, S., and Ørhede, E. (2000a). *User's Guide for the QPSNordic. General Nordic Questionnaire for Psychological and Social Factors at Work* (TemaNord 2000:603). Copenhagen, Denmark: Nordic Council of Ministers.

Lindström, K. and Kivimäki, M. (1999). Organizational interventions and employee well-being in health care settings. In P. M. Le Blanc, M. C. W. Peters, A. Büssing, and W. B. Schaufeli (Eds.), *Organizational Psychology and Health Care* (pp. 135–151). München, Germany: Hampp.

Lindström, K., Schrey, K., Ahonen, G., and Kaleva, S. (2000b). The effects of promoting organizational health on worker well-being and organizational effectiveness in small and medium-sized enterprises. In L. R. Murphy and C. L. Cooper (Eds.), *Healthy and Productive Work* (pp. 83–104). New York: Taylor & Francis.

Lindström, K. and Turpeinen, M. (2004). How to manage patient overflow after a hospital merger. In C. R. Johansson, A. Frevel, B. Geibler-Gruber, and G. Strina (Eds.), *Applied Participation and Empowerment at Work* (pp. 151–161). Lund, Sweden: Studentlitteratur.

Lindström, K., Turpeinen, M., and Kinnunen, J. (2003). Effects of data system changes in job characteristics and well-being of hospital personnel. A longitudinal study. In D. Harris, V. Duffy, M. Smith, and C. Stephanidis (Eds.), *Human-Centered Computing. Cognitive, Social and Ergonomic Aspects: Proceedings of HCI International 2003* (pp. 88–92). Mahwah, NJ: Lawrence Erlbaum Associates.

Loan-Clarke, J. and Preston, D. (1999). Organizational climate and culture issues amongst nurses within a community healthcare trust in England. In P. M. Le Blanc, M. C. W. Peters, A. Büssing, and W. B. Schaufeli (Eds.), *Organizational Psychology and Health Care* (pp. 205–219). München, Germany: Hampp.

Mäkinen, A., Kivimäki, M., Elovainio, M., Virtanen, M., and Bond, S. (2003). Organization of nursing care as a determinant of job satisfaction among hospital nurses. *Journal of Nursing Management, 11*, 299–306.

Manley, K., Hamill, J. M., and Hanlon, M. (1997). Nursing staff's perceptions and experiences of primary nursing practice in intensive care 4 years on. *Journal of Clinical Nursing, 6*, 277–287.

McVicar, A. (2003). Workplace stress in nursing: A literature review. *Journal of Advanced Nursing, 44*, 633–642.

Michie, S. and Williams, S. (2003). Reducing work related psychological ill health and sickness absence: A systematic literature review. *Occupational Environment Medicine, 60*, 3–9.

Mikkelsen, A. and Saksvik, P. Ø. (1998). An evaluation of the implementation process of an organizational intervention to improve work environment and productivity. *Review of Public Personnel Administration, 2*, 5–22.

Mikkelsen, A., Saksvik, P. Ø., and Landsbergis, P. (2000). The impact of a participatory organizational intervention on job stress in community health care institutions. *Work & Stress, 14*, 156–170.

Mimura, C. and Griffiths, P. (2003). The effectiveness of current approaches to workplace stress management in the nursing profession: An evidence based literature review. *Occupational and Environmental Medicine, 60*, 10–15.

Molander, G. (2003). The emotional burden of work in the care of elderly people, and coping with it. In *Proceedings of XIth European Congress on Work and Organizational Psychology* (pp. 51). Lisboa, Portugal.

Murphy, L. R. (1996). Stress management in work settings: A critical review of the health effects. *American Journal of Health Promotion, 11*, 112–135.

Murphy, L. R. (1999). Organisational interventions to reduce stress in health care professionals. In J. Firth-Cozens and R. L. Payne (Eds.), *Stress in Health Professionals* (pp. 149–162). Chichester, U.K.: Wiley.

Murphy, L. R. and Sauter, S. L. (2004). Work organization interventions: State of knowledge and future directions. *International Journal of Public Health, 49*, 79–86.

Nytrø, K., Saksvik, P. Ø., Mikkelsen, A., Bohle, P., and Quinlan, M. (2000). An appraisal of key factors in the implementation of occupational stress interventions. *Work & Stress, 14*, 213–225.

Ramanujam, R. and Rousseau, D. M. (2006). The challenges are organizational not just clinical. *Journal of Organizational Behavior, 27*, 811–827.

Reeves, S., Zwarenstein, M., Goldman, J., Barr, H., Freeth, D., Hammick, M., and Koppel, I. (2009). Interprofessional education: Effects on professional practice and health care outcomes. *Cochrane Database of Systematic Reviews, 1*, CD002213.

Richardson, K. M. and Rothstein, H. R. (2008). Effects of occupational stress management intervention programs: A meta-analysis. *Journal of Occupational Health Psychology, 13*, 69–93.

Roald, J. and Edgren, L. (2001). Employee experience of structural change in two Norwegian hospitals. *International Journal of Health Planning Management, 16*, 311–324.

Rowe, M. M. (2000). Skills training in the long-term management of stress and occupational burnout. *Current Psychology, 19*, 215–228.

Ruotsalainen, J., Serra, C., Marine, A., and Verbeek, J. (2008). Systematic review of interventions for reducing occupational stress in health care workers. *Scandinavian Journal of Work Environment & Health, 34*, 169–178.

Schaufeli, W. and Enzmann, D. (1998). *The burnout companion to study and practice: A critical analysis (issues in occupational health series)*. London, U.K.: Taylor & Francis.

Schweiger, D. L. and DeNisi, A. S. (1991). Communication with employees following the merger: A longitudinal field experiment. *Academy of Management Journal, 34*, 110–135.

Semmer, N. (2006). Job stress interventions and the organization of work. *Scandinavian Journal of Work Environment & Health, 32*, 515–527.

Sommer, S. M. and Merrit, D. E. (1994). The impact of a TQM intervention on workplace attitudes in a health care organization. *Journal of Organizational Change Management, 7*, 53–62.

Sundquist, J. and Johansson, S. E. (2008). High demand, low control, and impaired general health: Working conditions in a sample of Swedish general practitioners. *Scandinavian Journal of Public Health, 28*, 123–131.

Sutinen, R., Kivimäki, M., Elovainio, M., and Virtanen, M. (2002). Organizational fairness and psychological distress in hospital physicians. *Scandinavian Journal of Public Health, 30*, 209–215.

Sverke, M., Hellgren, J., and Öhrming, J. (1999). Organizational restructuring and health care work: A quasi-experimental study. In P. M. Le Blanc, M. C. W. Peters, A. Büssing, and W. B. Schaufeli (Eds.), *Organizational Psychology and Health Care* (pp. 15–32). München, Germany: Hampp.

Taimela, S., Justén, S., Aronen, P., Sintonen, H., Läärä, E., Malmivaara, A., Tiekso, J., and Aro, T. (2008). An occupational health intervention programme for workers at high risk for sickness absence. Cost effectiveness analysis based on a randomised controlled trial. *Occupational and Environmental Medicine, 65*, 242–248.

Thomas, L. H. and Bond, S. (1990). Towards defining the organization of nursing care in hospital wards: An empirical study. *Journal of Advanced Nursing, 15*, 1106–1112.

Thomsen, S., Dallender, J., Soares, J., Nolan, P., and Arnetz, B. (1998). Predictors of a health workplace for Swedish and English psychiatrists. *British Journal of Psychiatry, 173*, 80–84.

Vahtera, J., Kivimäki, M., Pentti, J., Linna, A., Virtanen, M., and Ferrie, J. E. (2004). Organisational downsizing, sickness absence and mortality: The 10-town prospective cohort study. *British Medical Journal, 328*, 555–557.

Van de Ven, A. H., Rogers, R. W., Bechara, J. P., and Sun, K. (2008). Organizational diversity, integration and performance. *Journal of Organizational Behavior, 29*, 335–354.

Vecchio, R. P. (1999). Jealousy and envy among health care professionals. In P. M. Le Blanc, M. C. W. Peters, A. Büssing, and W. B. Schaufeli (Eds.), *Organizational Psychology and Health Care* (pp. 121–132). München, Germany: Hampp.

Virtanen, M., Pentti, J., Vahtera, J., Ferrie, J. E., Stansfeld, S. A., Helenius, H., Elovainio, M., Honkonen, T., Terho, K., Oksanen, T., and Kivimäki, M. (2008). Overcrowding in hospital wards as a predictor of antidepressant treatment among hospital staff. *American Journal of Psychiatry, 165*, 1482–1486.

Wallace, J. E., Lemaire, J. B., and Ghali, W. A. (2009). Physician wellness: A missing quality indicator. *Lancet, 374*, 1714–1721.

Wickström, G., Joki, M., and Lindström, K. (2000). Description of the intervention process. In G. Wickström (Ed.), *Intervention Studies in the Health Care Work Environment* (pp. 103–111). (Arbete och Hälsa nr 2000:10.) Stockholm, Sweden: National Institute for Working Life.

Williams, S., Michie, S., and Pattani, S. (1998). *Report of the Partnership on the Health of the NHS Workforce*. London, U.K.: Nuffield Trust.

37

Team Training for Patient Safety*

Eduardo Salas, Sallie J. Weaver, Michael A. Rosen, and Megan E. Gregory

CONTENTS

37.1 Team Training for Patient Safety 627
    37.1.1 Purpose and Structure of This Chapter 628
37.2 Teams and Teamwork: Definitions and Competencies 628
    37.2.1 What Is a Team? 628
    37.2.2 What Is Teamwork? 629
    37.2.3 What Are Teamwork Competencies? 629
37.3 Systems View of Teams and Team Training 629
37.4 Before Training 631
    37.4.1 Team Task Analysis 632
    37.4.2 Person Analysis 632
    37.4.3 Organizational Analysis 633
37.5 During Training: Strategies and Methods for Optimizing Teamwork in Health Care 633
    37.5.1 Team Training Strategies for Health Care 634
        37.5.1.1 Team Adaptation and Coordination Training 634
        37.5.1.2 Cross-Training 634
        37.5.1.3 Assertiveness Training 637
        37.5.1.4 Metacognition Training 637
        37.5.1.5 Guided Team Self-Correction 637
        37.5.1.6 Error Management Training 638
    37.5.2 Methods for Team Training in Health Care 638
37.6 After Training: Transfer and Sustainment 638
37.7 The Road Ahead 639
37.8 In Closing 641
References 641

37.1 Team Training for Patient Safety

In the time period since the original publication of this book, the call to develop health care as the practice of expert teams, not just teams of experts, has risen to a near mandate. While communication is the only teamwork process explicitly outlined in the National Patient Safety Goals (NPSGs, Joint Commission, 2011), elements of teamwork underlie nearly every other goal as well. For example, accurate and complete medication reconciliation (Goal 8) requires communication, coordination, and cooperation among multiple providers, the patient, and often their family. Achieving clinical benchmarks for care quality, such as administering prophylactic antibiotics prior to surgery, are also often inherently the product of team processes.

As reviewed by Baker and colleagues in Chapter 13, the base of evidence linking teamwork to quality of care, patient outcomes, provider outcomes (e.g., burnout, stress), and organizational outcomes (e.g., efficiency) has emphasized the need to ensure that care providers, administrators, and support staff are equipped with the knowledge, skills, and attitudes (KSAs) necessary to work as effective team members. The question is then how to develop groups of individuals from diverse educational, professional, and demographic backgrounds working in what are often complex, dynamic environments under high-workload demands into high-performing teams? Furthermore, how can the patient and

---

* Revision and update of Salas et al. (2007).

their family be incorporated as active members of this high-performing team and how can teams of teams work together across the continuum of care to support patient safety and care quality?

Team training—the focus of this chapter—is the primary mechanism through which expert teams can be developed. In the time since this chapter was originally published, a significant body of work has been dedicated to developing, implementing, and evaluating team training in health care. For example, findings from a 2008 meta-analysis investigating the effectiveness of team training across multiple high-risk environments included an analysis of 80 medical teams (Salas et al., 2008a). While interpretation of results must be tempered given sample size limitations, the findings suggested that team training positively impacts health care team performance outcomes ($\rho = 0.23$; 80% CV 0.11–0.35). Simultaneously, these results emphasize that team training for patient safety must be approached from a systems perspective; that is, with nearly 80% of team performance being influenced by nontraining-related factors, a comprehensive approach to developing and supporting expert team performance is vital.

Team training is defined as a set of theoretically derived strategies and methods for developing the cognitive (e.g., shared mental models, transactive memory), behavioral (e.g., communication), and attitudinal (e.g., collective efficacy, trust) competencies that underlie effective teamwork (Cannon-Bowers et al., 1995). As such, training is defined as more than simply a single program, intervention, or place. Application of a systems perspective (Laszlo, 1972; Midgley, 2002) furthers this definition by defining training as one component nested within a constellation of elements that influence team performance. Systems thinking suggests a holistic orientation to considering how to optimize team performance rather than a reductionist view. From this perspective, effective team training requires consideration of not only the strategies and methods utilized during the training event itself but also elements of the trainees, task, organization, structural environment, social environment, and other factors that collectively influence the degree to which the KSAs developed during training are transferred, generalized, and instantiated into the actual practice environment.

37.1.1 Purpose and Structure of This Chapter

The purpose of this chapter is to summarize the current state of the science regarding: (1) the core competencies of effective teamwork that form the foundation for team training content, (2) training strategies and methods for promoting effective teamwork, and (3) to highlight future research needs surrounding the question of how to develop and support effective teamwork within

health care. Specifically, these issues are addressed from a systems perspective. This perspective emphasizes that developing expert teams in health care involves much more than just training. Systems thinking advocates that training is but a single component in a broader network of organizational, task, and individual factors that affect team performance. As such, the scope and definition of team training presented here encompass factors that can influence the effectiveness of team training before, during, and after the actual training event itself. Our goal is to provide a summary of what we know so far about developing expert teams in health care and where we need to go in terms of both science and practice to continue advancing the field.

To this end, we first define teamwork, outline the core competencies that comprise the concept of teamwork, and summarize the KSAs underlying effective team performance. In the second section, we summarize the current state of the science regarding strategies and methods for optimizing teamwork through team training. Here we distill a short list of principles for developing expert teams before, during, and after training. In the third and final section, we highlight future research needs surrounding teamwork within health care and practical implications of these, as of yet, unanswered questions.

37.2 Teams and Teamwork: Definitions and Competencies

This section defines key concepts related to team training and discusses their application in health care domains.

37.2.1 What Is a Team?

A team is more than a collection of individuals. It is a set of two or more people working in an interdependent manner toward shared and valued goals (Dyer, 1984; Salas et al., 1992). Teams involve the dynamic and often adaptive interaction of individuals, each with specified roles and responsibilities. Two features of this definition are particularly important: shared goals and interdependence. First, a team's members must share meaningful goals. This does not mean that all goals will be identical for all team members, but there must be some level of overlap in the aims members are seeking to achieve (e.g., safe, timely, effective, efficient, equitable, and patient-centered care). Second, team members must have meaningful connections between their individual tasks (Saavedra et al., 1993). Their ability to complete their work and reach their goals must be dependent upon or influenced by the performance of other team members. This parsimonious definition of a team is complicated by several factors in health care organizations including frequently fluid membership (e.g., consulting specialists on a care team, multiple shifting roles on a trauma team) and boundary definitions due to the highly interdependent nature of health care delivery systems. For example, the core of a surgical team is usually taken to be the members in the room at the time of the operation: the physician and physician's assistant, anesthesia provider(s), scrub and circulator nurses, as well as technicians. However, staff outside of this core group (e.g., administrators, allied health professionals, supply) work interdependently with the core team to achieve safe and optimal use of the operating theaters. While sometimes difficult, clear definitions of membership and boundaries should be made when designing and implementing a team training program.

37.2.2 What Is Teamwork?

In order to define teamwork, it is useful to distinguish it from taskwork, the components of an individual team member's task that can be are completed in isolation without input from other members of the team (Morgan et al., 1986). Teamwork then consists of the processes used to manage interdependent aspects of team members' work (Salas et al., 2007b). It is the actions taken by team members while communicating, cooperating, and collaborating (Salas et al., 2008b), and not the outcomes. While many frameworks and theories of team process have been proposed (Ilgen et al., 2005; Salas et al., 2007b), at a high level, teamwork processes can be categorized as *action, transition,* or *interpersonal* (LePine et al., 2008; Marks et al., 2001). Action processes, such as coordination, monitoring, and backup behavior, occur during the periods of time when teams conduct activities leading directly to goal accomplishment. Transition processes, such as mission analysis, goal specification, and strategy formulation, occur during planning activities. Interpersonal processes—conflict management and affect management—occur in both transition and action phases. Just as with individual taskwork, there are a set of competencies underlying effective teamwork.

37.2.3 What Are Teamwork Competencies?

Teamwork competencies are the KSAs underlying effective teamwork (Cannon-Bowers et al., 1995). They are what team members must know, do, and feel in order to interact effectively in dynamic environments and manage their resources (material resources and expertise of team members) in order to achieve their goals. A growing list of teamwork competencies has been identified in the scientific literature on teamwork (Salas

et al., 2009b); however, communication, situational awareness, leadership, and role clarity are currently the most frequently targeted teamwork competencies in health care team training programs (Weaver et al., 2010a). The Big Five Model of teamwork (Salas et al., 2005b) has served as the basis for competency models in health care team training programs including the Team Strategies and Tools to Enhance Performance and Patient Safety (TeamSTEPPS®), jointly developed by the Department of Defense and the Agency for Healthcare Research and Quality. This program targets leadership, situation monitoring, communication, mutual support, and team structure (e.g., role clarity) teamwork competencies. This broad competency framework has been applied successfully in a variety of clinical areas including surgery (Weaver et al., 2010b), trauma (Capella et al., 2010), and labor and delivery (Nielsen et al., 2007). However, different clinical domains impose different types of teamwork demands on health care workers. Consequently, consensus teamwork competency models contextualized for the needs of specific clinical domains are beginning to emerge (e.g., Fernandez et al., 2008). Specifying teamwork competencies is crucial in developing systematic teamwork training programs, as they drive learning objectives and the team training design process (Salas et al., 2006a). Table 37.1 provides examples of common teamwork competencies targeted for training in health care.

37.3 Systems View of Teams and Team Training

As Section 37.2 underscores, the definition of teams has evolved such that they are now described as: "…complex and dynamic systems that affect, and are affected by, a host of individual, task, situational, environmental, and organizational factors that exist both internal and external to the team" (Cannon-Bowers & Bowers, 2010, p. 604).

As such, developing expert teams in health care involves much more than just training, in the commonly defined sense. Systems thinking advocates that training is but a single component in a broader network of organizational, task, and individual factors that affect team performance. As such, the scope and definition of team training presented here encompass factors that can influence the effectiveness of team training before, during, and after the actual training event itself. Specifically, training is defined as: "…the systematic acquisition of knowledge (i.e., what we need to know), skills (i.e., what we need to do), and attitudes (i.e., what we need to feel) (KSAs) that together lead to improved performance in a particular environment" (Salas et al., 2006c, p. 473).

TABLE 37.1

Example Teamwork Competencies

KSAs	Definition	Example Citations
Knowledge		
Accurate and shared mental models (knowledge about team members and role structure, team task, and environment)	Organized knowledge structures of the relationships between task and team members	Cannon-Bowers and Salas (1997), Klimoski and Mohammed (1994), Artman (2000), and Stout et al. (1996)
Cue-strategy associations	Compatible repertoire of performance strategies or courses of action associated with frequently occurring situations or problems shared among team members	Cannon-Bowers et al. (1995) and Marks et al. (2001)
Skills		
Closed-loop communication	A pattern of information exchange characterized by three steps: a sender initiates a message, the receiver acknowledge the message, and the sender follows up to confirm was appropriately interpreted	Bowers et al. (1998), McIntyre and Salas (1995), and Smith-Jentsch et al. (1998)
Team leadership	Dynamic process of social problem solving involving information search and structuring, information use in problem solving, managing personnel resources, and managing material resources	Burke et al. (2006) and Day et al. (2004)
Mutual performance monitoring	Team members' ability to track what others on the team are doing while continuing to carry out their own tasks	McIntyre and Salas (1995), Dickinson and McIntyre (1997), and Marks and Panzer (2004)
Backup/supportive behavior	The ability to shift and balance workload among team members during high-workload or high-pressure periods	Marks et al. (2001), McIntyre and Salas (1995), and Porter et al. (2003)
Conflict management	Preemptively setting up conditions to prevent or control team conflict or reactively working through interpersonal disagreements between members	De Dreu and Weingart (2003), Jordan and Troth (2004), and Simons and Peterson (2000)
Mission analysis	Formalizing an understanding of the team's core tasks, goals, and the conditions under which members will function as well as the resources available to the team	Marks et al. (2001) and Mathieu and Schulze (2006)
Team adaptation	The team's ability to adjust strategies to changing conditions and learn from past performance episodes	Burke et al. (2006), Entin and Serfaty (1999), and Kozlowski et al. (1999)
Attitudes		
Mutual trust	The shared belief among team members that everyone will perform their roles and protect the interests of their fellow team members	Alavi and McCormick (2004), Driskell and Salas (1992), and Jackson et al. (2006)
Team/collective efficacy	The team members' sense of collective competence and their ability to achieve their goals	Bandura (1986), Gibson (2003), Katz-Navon and Erez (2005), and Zaccaro et al. (1995)
Team/collective orientation	Team members' preference for working with others as opposed to working in isolation	Alavi and McCormick (2004), Eby and Dobbins (1997), and Mohammed and Angell (2004)
Psychological safety	The team members' shared belief that it is safe to take interpersonal risks	Edmondson (1999)

Source: Adapted from Salas, E. et al., The wisdom of collectives in organizations: An update of the teamwork competencies, In E. Salas, G.F. Goodwin, & C.S. Burke (Eds.), *Team Effectiveness in Complex Organizations: Cross-Disciplinary Perspectives and Approaches*, pp. 39–79, Routledge, New York, 2009.

In this sense, a systems view of team training is focused on transfer of the KSAs developed in training to the actual working environment, as well as maintenance and sustainment of their use during daily care processes over time. For example, other developmental interactions, such as coaching and mentoring, may be integral components of a systematic approach to developing expert teams in health care that may be considered in a systems approach to team training. This perspective aligns with definitions of learning as a relatively permanent change in behavior that results from a particular set of experiences or practice. While others define learning as a change in the *range of potential behaviors* (Huber, 1991), the systems view of team training emphasizes consideration of person-oriented factors (e.g., individual difference characteristics), organizational-oriented factors (e.g., supervisory and peer support for use of team KSAs on the job), and task-oriented factors (e.g., level of interdependence) that affect transfer of training and actual use of effective teamwork behaviors on the job. Borrowing from training need analysis, a method for identifying training objectives, developing training programs, and examining the generalizability of training programs (Goldstein & Ford, 2002; McGehee & Thayer, 1961), the remainder of this chapter is discussed in terms of these task-related factors, person-related factors, and organizational factors that can influence team training effectiveness before, during, and after a given training event. Most importantly, these issues are relevant to consider across the continuum of provider development and education. The priniciples presented here outline a systematic framework for the planning, implementation, and maintenance of team training that can be applied across the educational spectrum: from undergraduate and graduate coursework to continuing education for established clinicians. Table 37.2 summarizes several best practices underlying a systems approach to team training for patient safety.

37.4 Before Training

To develop effective team training that addresses organizational objectives in a strategic way, research on training, adult learning, and implementation science have established that conducting a team training needs analysis is a vital first step in the training process. As noted earlier, the purpose of such an analysis is to understand the constellation of factors that will influence learning and transfer before trainees step into a team training event or environment. The overarching purpose of such an analysis is threefold: (1) to define core learning objectives, (2) identify factors that will influence the training strategy and methods of delivery, and (3) identify and/or develop indicators of training effectiveness (Salas et al., 2006c) In this section, we discuss each component of a team training needs analysis, summarizing the current state of the science regarding individual, team, and

TABLE 37.2

A Systems Approach to Team Training for Patient Safety

Before training	1. Conduct a team training needs analysis
	1a. Define team tasks
	• Define the target job, procedure, or care process
	a. What are the core tasks/duties performed? Under what conditions?
	b. Who are the team members involved in each task and in linking each component task (i.e., step) together?
	• Have front line providers, administrators, and others with expert knowledge of the job/procedure/care process rate each component task on criticality, frequency, difficulty, etc.
	• Categorize most critical, frequent, and difficult component tasks as primarily taskwork related or teamwork related
	a. Which tasks require team members to communicate, coordinate their activities, or cooperate?
	• Translate tasks into KSAs
	a. What knowledge, skills, or attitudes must one hold to effectively perform relevant tasks?
	1b. Identify individual difference characteristics likely to impact training effectiveness
	• Trainee motivation: to participate, to learn
	• Attitudes toward training
	• Goal orientation
	• Previous experience
	1c. Identify organizational characteristics likely to impact training effectiveness
	• Organizational culture/climate
	• Policies and procedures
	• Trainee selection and notification
During training	2. Choose a team training strategy or a combination of strategies that suits KSAs identified in team training need analysis
	3. Utilize a combination of team training methods that ensure trainees have the opportunity to practice targeted team competencies with feedback
	4. Integrate formative assessment methods to facilitate diagnostic feedback during training
After training	5. Ensure implementation objectives remain aligned with organizational goals for patient safety and care quality
	6. Demonstrate organizational support in the form of visible executive leader participation and policy changes
	7. Demonstrate active buy-in and advocacy from frontline clinical leaders for transfer of trained teamwork competencies into daily practice
	8. Facilitate maintenance and sustainment through on-going with practice opportunities and by establishing a positive climate for teamwork
	9. Establish a system to measure and evaluate the effectiveness of the implementation and to reinforce sustainment of effective teamwork in daily practice

organizational factors that have been found to influence team training effectiveness.

37.4.1 Team Task Analysis

The purpose of a team task analysis (TTA) is to (1) identify if teamwork subprocesses are actually core components of a given task or care process and (2) if they are, to identify the specific aspects of teamwork underlying effective performance. While the first step may sound intuitive, it is critical to begin by first identifying whether a particular job or care process is indeed the product of interdependent work that requires communication, coordination, and cooperation among a group of individuals with shared goals (i.e., a team). This is necessary in order to ensure that team training is the appropriate "solution" to an identified opportunity for improvement. As outlined in Section 37.2, successful team performance is the product of both taskwork (i.e., the declarative and procedural knowledge and skill necessary to complete a task) and teamwork competencies (i.e., the KSAs necessary to communicate, cooperate, and coordinate with others to complete a task) (Glickman et al., 1987). Identifying both taskwork and teamwork requirements for a given job, procedure, or care process and then mapping the underlying KSAs necessary to effectively fulfill these requirements are the core objectives of a TTA.

While a detailed discussion of the individual steps involved in TTA is beyond the scope of this chapter (see Burke, 2005; Gandall et al., 2006; Morgeson & Dierdorff, 2011), there are three primary steps: (1) delineating the specific tasks, responsibilities, and processes involved in a particular team task or care process into a list of task statements, (2) organizing the list of task statements in clusters/categories, and (3) identifying the competencies (i.e., KSAs) that are associated with each cluster and specific task.

TTA is similar to traditional Socio-Technical Risk Assessment strategies (ST-PRA: see Mohaghegh et al., 2009, for a detailed discussion), given that both are dedicated to mapping out specific task/event processes. However, TTA goes further in identifying which components of a given process require communication, coordination, and/or cooperation among team members and additionally mapping the KSAs that team members need to engage in these teamwork processes effectively. A critical feature of a systems-oriented approach to TTA is consideration of both internal and external teamwork requirements; that is, identification of communication, cooperation, and coordination requirements both within and between teams.

37.4.2 Person Analysis

The individual characteristics of trainees can influence team training effectiveness both before and after a given team training experience. The evidence to date suggests that trainee attitudes toward training and teamwork, previous experiences related to training content, and goal orientation help to shape their motivation related to the training experience (i.e., to attend, to participate, to transfer) and how they apply targeted KSAs on the job (Baldwin et al., 1991; Chernyshenko et al., 2011). For example, individual attitudes regarding teamwork (i.e., attitudes regarding its value and relevance) and attitudes regarding training as a mechanism for development can influence their motivation to attend and participate in training, as well as their motivation to apply trained skills on the job. For example, one's use of context-appropriate assertiveness and other teamwork behaviors has been related to their beliefs regarding the degree that assertiveness is appropriate in the team context (Jentsch & Smith-Jentsch, 2001; Pearsall & Ellis, 2006; Smith-Jentsch et al., 1996). The general training literature has also demonstrated that organizational commitment, defined as an individual's commitment to their specific employer (Meyer et al., 1993), and job involvement, the degree to which an individual considers their job central to their personal identity (Blau, 1985), are related to an individual's motivation to learn during training (Tannenbaum et al., 1991) and training transfer (Tesluk et al., 1995).

Previous experience is one of the key mechanisms affecting trainee attitudes toward training and, more specifically, their attitudes regarding training content such as teamwork. Previous experiences color the lens through which adult trainees enter, interpret, and apply the concepts presented in team training. However, the existing evidence underscores that it is not merely the quantity of experience, but rather the type of experiences the trainee has had that can affect training outcomes. For example, studies that have operationalized experience in terms of organizational or professional tenure have tended to find little evidence for a relationship with training outcomes (Gordon et al., 1986). Conversely, studies that have examined specific previous experience related explicitly to training content have found evidence of a relationship with training transfer and post-training performance. For example, Smith-Jentsch et al. (1996) studied assertiveness training in a sample of pilots and found that pilots who had a great number of negative pretraining cockpit experiences (i.e., life threatening events involving flying with a captain using unsafe procedures or events where they felt pressured to take flight despite mechanical or environmental concerns) performed better in a behavioral exercise carried out 1 week after assertiveness training. Similar studies suggest that negative pretraining experiences may enhance training effectiveness by increasing trainee motivation (Holt & Crocker, 2000). This suggests that pretraining activities that engage trainees to think and reflect upon their own previous experiences related to training objectives and content may enhance learning.

Another important person-oriented consideration is goal orientation. Goal orientation refers to an individual's goal preferences in achievement and development settings, such as training (Dweck, 1986). Conceptually, individuals tend to assume either a learning goal orientation (LGO) or a performance goal orientation (PGO). An LGO focuses on a desire to gain competence and mastery, whereas a PGO focuses upon either proving one's existing competence (performance prove goal orientation) or on avoiding negative judgments about one's existing level of knowledge or skill (performance avoid orientation) (VandeWalle, 1997). Additionally, goal orientation (GO) has been conceptualized as existing as both an individual trait and also as a state construct— meaning that both individual and situational factors can help shape an individual's goal orientation toward a particular training or development opportunity.

Overall, the existing evidence suggests that LGO is positively associated with indicators of knowledge gain and transfer of trained skills given that it facilitates critical learning behaviors like feedback seeking and the use of varied learning strategies such as experimentation (Bunderson & Sutcliffe, 2003; Payne et al., 2007). Additionally, learning orientation has been associated with the affective components of effective learning such as self-efficacy (Bell & Kozlowski, 2002). Most important for developing team training in health care, experimental studies have demonstrated that goal orientation can be manipulated prior to a given learning event (e.g., Kozlowski et al., 2001). This suggests developing a pre-training environment that frames training as a learning-oriented event designed to further optimize mastery of the KSAs underlying effective team processes may enhance training effectiveness.

37.4.3 Organizational Analysis

Organization analysis is dedicated to determining the environmental factors and the availability of resources that may affect training objectives, delivery methods, and transfer of training into the actual practice environment (Goldstein & Ford, 2002). Specifically, organizational goals, structure, culture, climate, policies, practices, norms, history, and resources can influence trainee motivation to engage in team training, as well as their ability and motivation to utilize KSAs targeted in training on the job (Cannon-Bowers et al., 1995). Organizational characteristics are critical to consider given that they represent the environment into which providers will attempt to apply teamwork KSAs developed and reinforced in training. Additionally, organizational characteristics can help shape some of the person characteristics described earlier, such as goal orientation.

Organizational climate and organizational culture are two core characteristics identified as influential to training transfer and effectiveness. Climate refers to shared employee perceptions of organizational policies, procedures, and norms, whereas culture refers to the deep-seated beliefs and values that drive overt organizational goals, policies, and procedures (Schneider et al., 2010). For example, organizational climate for transfer of training is defined as shared perceptions regarding the priority of transferring trained teamwork skills into daily care processes relative to other organizational goals, such as efficiency (Rouiller & Goldstein, 1993). Specifically, the research on transfer climate suggests that organizations should ensure that the workplace environment provides both *situational cues* and *consequence cues* that support training objectives. Situational cues provide information regarding opportunities to use teamwork skills on the job and are conceptualized as including task cues, goal cues, self-control cues, and social cues. Consequence cues supplement situational cues by providing information regarding the consequences of using trained skills on the job in the form of feedback (either positive or negative), lack of feedback, and/or punishment.

For example, organizational processes regarding trainee selection and notification have been found to influence training effectiveness. Trainee motivation has been linked with perceptions that training selection processes are fair, that participation is voluntary and that trainees feel a sense of control over participation, and that the training experience is framed positively (i.e., framed as an opportunity to advance one's career rather than as punishment for poor performance) (Baldwin & Ford, 1988; Quiñones & Ehrenstein, 1997). Furthermore, voluntary participation has been linked with higher pretraining self-efficacy (i.e., perceptions that one will succeed in training and be able to learn from training), more positive reactions to training, and also with greater gains in learning outcomes (Baldwin & Magjuka, 1997; Colquitt et al., 2000). Providing realistic information to trainees regarding the rigor and time commitment required for training has also been linked with higher levels of trainee motivation (Russ-Eft, 2002; Webster & Martocchio, 1995).

37.5 During Training: Strategies and Methods for Optimizing Teamwork in Health Care

Approaches to optimizing teamwork tend to fall into two categories: team building and team training. Both aim to improve team performance and effectiveness; however, these approaches differ in both their focus and application. Team training focuses on developing the affective, cognitive, and behavioral competencies underlying effective teamwork (Cannon-Bowers & Bowers, 2010). The goal of team training is often to develop generalizable competencies that can be transported across multiple teams, whereas team building refers to a family of interventions specifically targeting intact work teams (i.e., teams with a history and future together) designed to clarify member roles and responsibilities (Salas et al., 2005a). Team building also focuses more specifically on fostering positive interpersonal relationships among team members in order to build affective team performance factors such as cohesion and trust. Meta-analytic findings have found both team building and team training strategies to be positively linked to team outcomes (team training $\rho=0.34$ [Salas et al., 2008a]; team building $\rho=0.31$ [Klein et al., 2009]). The critical difference is that team training strategies can develop team-specific and team-generic competencies that can be applied across multiple teams, while team building tends to focus on building team-specific competencies. Given that health care is characterized by dynamic team membership, membership on multiple teams, and fairly high turnover, team training strategies that develop transportable teamwork competencies applicable across multiple teams, tasks, and care environments are the focus of the current chapter. Thus, the subsequent section reviews several team training strategies relevant for health care. We are not advocating that this is an exhaustive review of all strategies that have been utilized to develop team performance; however, these strategies have been chosen as a sample given that they represent the greatest proportion of evidence to date. Table 37.3 provides a summary of each strategy, best practices for implementation, and examples of where each has been utilized in health care. It is notable that many examples could have been categorized across multiple training strategies given their comprehensive approaches to training.

37.5.1 Team Training Strategies for Health Care

37.5.1.1 Team Adaptation and Coordination Training

Team adaptation and coordination training (TACT) has been defined as, "a team training intervention in which team members are trained to alter their coordination strategy and to develop more structured, efficient patterns of communication" (Salas et al., 2007a, p. 475). This strategy is designed on the premise that team coordination (i.e., how teams integrate individual efforts toward mutual goals) underlies team effectiveness. Meta-analytic findings have found TACT strategies to be some of the most effective team training strategies across a variety of team types and environments ($\rho=0.47$–0.57; Salas et al., 2008a). By helping team members develop

shared accurate mental models, these strategies help members to anticipate both how a situation will evolve, as well as their team members' needs (Serfaty et al., 1998), thus improving performance. TACT also teaches teams how to adjust their coordination strategies under dynamic conditions, including high-stress situations (Entin & Serfaty, 1999).

Crew resource management (CRM), one form of team coordination and adaptation training, has helped this strategy become one of the most prominently studied forms of team training to date. Originating in aviation, CRM has spread to other fields and is now one of the most well-developed team-training strategies utilized in health care (Weaver et al., 2010a). CRM attempts to train team members to use all available resources (e.g., people, information, and equipment), as well as the competencies necessary for developing and maintaining shared situational awareness and effective coordination. A full checklist for designing, implementing, and evaluating CRM training is available in Salas et al. (2006b).

37.5.1.2 Cross-Training

Cross-training involves team members learning about the roles and responsibilities of their teammates. This strategy may involve providing information about the tasks, duties, and responsibilities of each team role (positional clarification); team members discussing, modeling, and practicing the jobs of their teammates (positional modeling); or team members actively assuming another role on the team for a designated period of time (positional rotation; Cannon-Bowers et al., 1998). Specifically, cross-training addresses shared task knowledge and interpositional knowledge; that is, the knowledge a team member has about teammates' tasks, roles, and appropriate behaviors (Cannon-Bowers & Bowers, 2010). Thus, cross-training is advantageous in that it can lead to shared knowledge structures among team members (Salas et al., 2007c) and increased understanding of teammates' needs (Cannon-Bowers & Bowers, 2010) and even empathy (Cannon-Bowers & Bowers, 2010). In turn, this can support team communication processes, help team members to engage in more accurate mutual performance monitoring, and provide more effective and efficient backup behavior. For example, a meta-analysis by Salas et al. (2008a) found moderate positive effects for cross-training on team performance ($\rho=0.40$), based on 14 effect sizes and 432 teams.

However, the effectiveness of each strategy has been related to the interdependency of the task at hand (Cannon-Bowers et al., 1998). Low-interdependency tasks may benefit from just positional clarification; however, high-interdependency tasks may require more indepth learning through positional rotation in order to

TABLE 37.3

Team Training Strategies Relevant for Health Care

Team Training Strategy	Definition	Primary Teamwork Competencies Targeted	Best Practices	General Sources	Examples of Uses of Strategy or Variations of Strategy in Health Care
Assertiveness training	Dedicated to developing communication strategies that support task-relevant and team performance relevant assertiveness	• Backup behavior • Closed-loop communication • Conflict management • Mutual trust • Psychological safety • Team leadership	• Clearly define training objectives around task-relevant and team performance assertiveness rather than general assertive behaviors and differentiate from aggressive behaviors • Compare and contrast behavioral models of both effective- and ineffective-assertive behaviors • Engage learners with opportunities to practice appropriate assertiveness behaviors that include feedback. Practice should also strive to include realistic time pressures or other stressors to allow practice using and reacting to appropriate assertiveness under such conditions	Jentsch and Smith-Jentsch (2001) and Smith-Jentsch et al. (1996)	Buback (2004), Cohen (1983), Freeman and Adams (1999), Kirkpatrick and Forchuk (1992), McCabe and Timmins (2003), McVanel and Morris (2010), and Milstead (1996)
Cross-training	Strategy in which team members learn the roles that comprise the team and the tasks, duties, and responsibilities fulfilled by fellow team members	• Accurate and shared mental models of team roles and responsibilities	• The level of interdependency should drive the type of cross-training you choose • Include information about the roles and responsibilities of other team members and how they operate to achieve these • Include explanations of why they must operate in this way—including communication patterns that clarify who they depend on for information • Provide opportunities for team members to walk in the shoes of another role or shadow a team member in another role if possible • Provide feedback during cross-training, which facilitates the formation of reasonable expectations of one another	Cannon-Bowers et al. (1998) and Volpe et al. (1996)	DeVita et al. (2005), Inman et al. (2005), Masson and Fain (1997), and Frommater et al. (1995)
Error management training	Training strategy based in active learning in which participants are encouraged to make errors during training scenarios, analyze these errors, and practice error recognition and management skills	• Collective efficacy • Cue-strategy associations • Shared mental models • Team adaptation	• Ensure trainees understand the purpose of this training strategy, which is to encounter errors and to have the opportunity to practice managing them in a safe environment • Frame errors as positive opportunities for learning • Embed the opportunity to make errors into training scenarios by providing minimal guidance during the scenario • Follow the scenario with immediate feedback and discussion to facilitate learning	Keith and Frese (2008)	Institute for Healthcare Improvement (2010) and Moorthy et al. (2006)

(continued)

TABLE 37.3 (continued)

Team Training Strategies Relevant for Health Care

Team Training Strategy	Definition	Primary Teamwork Competencies Targeted	Best Practices	General Sources	Examples of Uses of Strategy or Variations of Strategy in Health Care
Guided team self-correction	Team training strategy designed around a cycle of facilitated briefings and debriefings that occur around a training scenario or live event	• Backup behavior • Collective orientation • Closed-loop communication • Cue-strategy associations • Mission analysis • Mutual trust • Shared mental models • Team adaptation • Team leadership	• Define the team self-correction skills to be trained prior to team self-correction training • Record positive and negative examples of teamwork dimensions during team performance episode • Classify and prioritize observations, diagnose strengths and weaknesses, and identify goals for improvement before beginning debrief • Set the stage for team participation and solicit examples of teamwork behavior during debrief	Smith-Jentsch et al. (2008); Blickensderfer et al. (1997) Rasker et al. (2000), and Tannenbaum et al. (1998)	Berenholtz et al. (2009), McGreevy and Otten (2007), Neily et al. (2010), and Percarpio et al. (2010)
Metacognition training	Focuses on developing cognitive aspects of team performance by teaching strategies dedicated to analyzing, updating, and aligning team mental models of the team's task, coordination strategy, and contingency plans	• Cue-strategy associations • Mission analysis • Shared mental models • Team adaptation	• Develop training objectives around cognitive processes such as planning, monitoring, and reanalysis • Structure metacognitive practice tasks • Around a subject that trainees have preexisting knowledge about	Cohen et al. (1996), Flavell (1979), and Jentsch (1997)	Banning (2006) and Kuiper and Pesut (2004)
Team adaptation and coordination training	Focuses on how to effectively use all available resources (i.e., people, information, etc.) through the use of effective team communication, coordination, and cooperation. Training objectives often focus on developing transportable teamwork competencies and tools that can support effective team processes, such as checklists. CRM is a form of TACT	• Backup behavior • Closed-loop communication • Cue-strategy associations • Mission analysis • Mutual performance monitoring • Leadership • Shared mental models	• Develop training objectives that address around transportable teamwork competencies for ad hoc teams (no history or future) • Training team-specific competencies can also be incorporated for intact teams • Train intact teams together if possible • Create opportunities for both guided and unguided practice • Develop feedback mechanisms that engage self-reflection and team self-correction following practice opportunities • Develop tools that support effective teamwork, but recognize that tools alone (e.g., checklists) cannot optimize team performance (and alone, may negatively impact performance)	Salas et al. (2009c), Serfaty and Entin (1998), Wiener et al. (1993), and Salas et al. (1999)	Gaba (2010), Gaba et al. (2001), Helmreich and Schaefer (1994), Flin and Maran (2004), Musson and Helmreich (2004), Neily et al. (2010), Reznek et al. (2003), Thomas et al. (2007), and Wolf et al. (2010)

Source: Adapted from Weaver, S.J. et al., How to build expert teams: Best practices. In R.J. Burke & C.L. Cooper (Eds.), *The Peak Performing Organization*, pp. 129–156, Routledge, New York, 2009.

affect team performance (Cannon-Bowers et al., 1998; Marks et al., 2002).

37.5.1.3 Assertiveness Training

Ensuring that all team members—regardless of rank, status, or tenure with the team—are both willing to communicate their unique knowledge and views and equipped with the communication strategies to do so effectively is a critical aspect of successful team performance (Smith-Jentsch et al., 1996). Assertiveness is a critical behavioral marker of psychological safety; that is, perceptions that the team environment is safe for personal risk taking such as raising concerns regarding potential (or existing) problems or making suggestions (Edmondson, 1999). It is also an important part of participative decision making. Furthermore, assertiveness has been found to be positively related to both team performance outcomes and team satisfaction (Pearsall & Ellis, 2006).

Assertiveness training focuses on developing communication strategies that facilitate team members speaking up to request help, to convey information or concerns, to suggest possible solutions or action plans, and to initiate actions related to these observations. In this sense, the purpose of assertiveness training is to ensure that critical information does not remain siloed and unshared (Salas et al., 2007b). The critical element of assertiveness training is its focus on task-relevant and team performance–relevant assertiveness. Training objectives focus on teaching communication strategies through which team members can effectively convey their contributions in a way that is constructive and focused on the task and processes the team is using to achieve their goals. Thus, assertiveness training can address behavioral competencies such as information exchange, intrateam feedback, and backup behavior, as well as attitudinal (i.e., affective) components of effective team performance such as trust and cohesion.

37.5.1.4 Metacognition Training

Defined as "an individual's knowledge of and control over his or her cognitions" (Ford et al., 1998, p. 218), the concept of metacognition has roots in cognitive psychology. Specifically, metacognition includes the cognitive processes such as planning, monitoring, reflection, and reanalysis that individuals use to inform their inductive and deductive reasoning and to adapt behavior (Jentsch, 1997). As such, metacognition is conceptualized often as critical thinking skills (Cohen et al., 1997, Cohen and Freeman, 1997). The early study of metacognition focused on how individuals engage in these cognitive activities in the context of learning-oriented activities such as training and formal education (Brown, 1978; Flavell, 1979). For example, at the individual level, metacognition has been positively related to knowledge acquisition during training, performance upon the completion of training, and transfer of trained skills to novel environments, as well as self-efficacy (Ford et al., 1998). Additionally, self-evaluation has been identified as a critical component of the development of expertise (Ericsson & Charness, 1994).

These metacognitive processes of planning, reflection, and updating existing mental models are also critical for team performance (Kozlowski et al., 1996) and training strategies to develop them within the team context have been developed. For example, Cohen and Thompson (2001) developed a metacognitive training strategy that focuses on teaching learners how to think critically about their mental models regarding the team's purpose, their mental models regarding how efforts among team members will be coordinated, and their mental models of how the team will manage uncertainty or changes. This strategy focuses on teaching how to critically assess these mental models in order to ensure that they are both accurate and shared among team members.

Metacognitive training is a viable strategy for developing expert health care teams because it focuses on competencies that can help teams more effectively manage uncertainty, novelty, and time-pressured decision making and it also requires preexisting knowledge regarding a particular subject area (Cohen et al., 1996; Jentsch, 1997). Therefore, it is one team training strategy that lends itself for integration with clinical skills practice as well.

37.5.1.5 Guided Team Self-Correction

Closely related to metacognitive training strategies are guided team self-correction strategies. Guided team self-correction focuses on helping teams to develop more effective teamwork processes such as information exchange, communication delivery, supporting behaviors, and leadership using a cycle of facilitated pre-briefing and debriefing around a given training event or a live performance event (Smith-Jentsch et al., 2008). Using team self-correction strategies, team members focus discussion around the processes through which they achieved a given outcome, rather than only focusing on the final outcome itself. Furthermore, team self-correction emphasizes that debriefings can be facilitated by the team leader or by any team member trained to in guided self-correction. The goal of this cycle is to directly foster shared knowledge and knowledge organizational structures among team members and to also improve team communication processes that, in turn, support future backup behavior, mutual performance monitoring, and affective components of effective team performance such as mutual trust.

37.5.1.6 *Error Management Training*

Error management training is a scenario-based strategy distinguished by two core characteristics: (1) it is an active learning strategy in which minimal guidance is provided to participants during training events, and (2) the training environment is specifically developed such that trainees know they will encounter errors, errors are framed as positive opportunities for learning, and opportunities to make errors are specifically embedded into the training scenario (Heimbeck et al., 2003; Keith & Frese, 2011). Error management training has been shown to positively impact individual training outcomes, especially in adaptive transfer scenarios (i.e., novel situations; Keith & Frese, 2008). At the team level, teamwork breakdowns are critical antecedents of errors in interdependent tasks (Wilson et al., 2007). Thus, error management training offers the opportunity to learn the cues that signal when an error or near miss has the potential to occur in teams, to develop coordination and communication strategies to manage and recover from errors, and to develop adaptive processes that can support reliably safe outcomes (Bell & Kozlowski, 2011; Weaver et al., 2011). However, most of the evidence to date has examined the effects of error management training at the individual level of analysis however. Therefore, there is a definite need to evaluate the effects of error management as a team training strategy.

37.5.2 Methods for Team Training in Health Care

A variety of training delivery methods or strategies have been applied in health care settings including information-, demonstration-, and practice-based methods. Each is reviewed in the following separately; however, in practice, team training interventions usually include multiple delivery methods.

Information-based training methods include traditional classroom-based techniques such as didactics and reading materials provided to learners. A recent review indicates that information-based methods of delivery in team training programs in health care are effective in terms of positive staff reactions to training, learning, behavior change, and in some cases positive influence on clinical processes (Rabøl et al., 2010). Due in large part to the logistical challenges involved in scheduling health care workers for training sessions, distributed learning approaches for teamwork have been implemented and shown to be effective at changing care delivery (Deering et al., in press). These systems use information-based methods delivered in a computer-based format such that staff members can complete training when time is available.

Demonstration-based training is a "strategy of training development and delivery involving the systematic design and use of observational stimuli intended to develop specific KSAs in the learner" (Rosen et al., 2010b, p. 597). In this approach, examples of effective or ineffective task performance are presented to learners along with additional preparatory, concurrent, retrospective, and prospective instructional features (Rosen et al., 2010b). This can be either standardized and pre-recorded demonstrations as is the case with video recorded examples of effective and ineffective teamwork behaviors, or through live role modeling. While little research has been conducted on the relative effects of demonstrations for health care team training, they are commonly used and have shown to be effective in a variety of other domains (Taylor et al., 2005).

Practice-based methods of training delivery involve the use of systematically engineered opportunities to practice targeted behaviors (Salas & Cannon-Bowers, 2001). While information-based methods for teamwork training can be effective, consistent with research and best practices in team training from other high-risk domains, simulation-based (or practice-based) methods have shown to add value to didactics (Shapiro et al., 2004). Both center-based and in situ simulation delivery have been shown to be effective for building teamwork skills (Merien et al., 2010), and ultimate outcomes are more dependent upon the features of design and delivery than the specific technological capabilities of the simulators (Beaubein & Baker, 2004; Salas et al., 2005c). To that end, guiding principles and methods of designing scenarios, measuring performance, and providing feedback in simulation-based teamwork training have been developed and are discussed in detail elsewhere in this volume (Chapter 34).

37.6 After Training: Transfer and Sustainment

Team training has been advocated as a foundational component of cultural transformation in health care (Corrigan et al., 2001). This is no small task and goes well beyond the typical application of training in that many of the teamwork behaviors targeted for acquisition (e.g., assertiveness in communication) are counter-normative. They are in direct opposition to the current hierarchical nature of the health care culture (Wilson et al., 2005). This creates a difficult transfer of training problem, as the goal is not just to develop new knowledge and skill, but replace old forms of deeply ingrained interaction rooted in the disciplinary traditions and educational systems of the industry.

Like most organizational change interventions in health care (Wensing et al., 2006), behavior change on the job and the ultimate results achieved with team training programs can be variable (Buljac-Smardzic et al., 2010).

To address this, recent research has focused on the organizational factors that contribute to or inhibit the transfer and sustainment of teamwork behaviors. Based on the science of learning, transfer of training, organizational change literatures, and experiences implementing the TeamSTEPPS® program in the Military Healthcare System, a set of organizational factors and implementation characteristics linked to effective team training programs has been proposed (Salas et al., 2009a). First, *the implementation objectives should be aligned with organizational goals*. Specific learning objectives should be tied to the safety values and goals of the organization in a meaningful and visible way (Brannick & Levine, 2002). Second, *organizational support in the form of visible executive leader participation and policy changes* is a recurrent finding across safety and quality improvement research in health care (Ovretveit, 2010; Parand et al., 2010). One clear example comes from the implementation of the Veteran's Affairs Medical Team Training program, where leadership engagement during the training sessions strongly predicted sustainment of teamwork improvements over time (Paull et al., 2009). Third, *buy-in and advocacy from frontline clinical leaders* is critical for providing immediate reinforcement of training objectives on the job as well as promoting application of new behaviors in clinical practice (Rouiller & Goldstein, 1993). Fourth, *the availability of required resources such as staff time to attend training sessions* should be ensured through commitments from leadership. Fifth, *the transfer of teamwork behaviors to the clinical environment should be facilitated with practice opportunities* and by establishing a positive climate for teamwork (Lim & Johnson, 2002). Sixth, *the use of a system to measure and evaluate the effectiveness of the implementation* ensures that the aims are met and provides information on how to improve results (Salas et al., 2006a).

37.7 The Road Ahead

Overall, a significant body of work has emerged regarding team training in health care since publication of the first volume of this book. Important questions remain on the horizon however. Thus, we offer several areas for continued development as food for thought for both researchers and practitioners dedicated to continuing to build the base of evidence regarding methods for developing expert teams in health care.

> A systems oriented approach to team training and development strategies must continue to be further adapted to address specific needs of healthcare teams.

Health care shares common features with other high-risk industries, but is also characterized by a unique set of needs and constraints. The content and methods continue to evolve and be adapted to the uniqueness of the health care domain. Recent studies highlight the need for an increased focus on long-term perspectives on teamwork improvement initiatives. One study demonstrated an overall significant decrease in mortality rates due to team training, but in a dose–response relationship such that mortality rates significantly decreased for each quarter the program was implemented at a facility (Neily et al., 2010). Additionally, the implementation of rapid response teams is usually associated with significant improvements in clinical process measures (i.e., decreases in the rates of cardiac arrests outside of intensive care units), but not with significant changes in mortality (Chan et al., 2010). However, this common finding changes when a longer term perspective is taken. Significant decreases in mortality rates have been found, but only after 4 years of implementation (Santamaria et al., 2010). These findings reinforce the idea that teamwork training and organizational interventions are long-term cultural transformation efforts, and not "one-shot" training programs. The methods of training development, implementation, and evaluation need to continue to evolve to address the longer term issues of building and sustaining teamwork improvements.

> A richer understanding regarding contextual influences that moderate team training effectiveness in healthcare is needed.

As the systems view of team training underscores, training effectiveness is the product of a constellation of factors above and beyond the actual strategies and methods used to develop teamwork skills. Thus, training strategies and methods that were developed and validated in one context may or may not demonstrate similar effects when applied in other settings, in other organizations, or even among different groups within the same organization. The training literature describes that intra-organizational validity is concerned with whether a training approach will work in the same way in other units within the same organization, while inter-organizational validity focuses on understanding if a given approach will work in the same way in a different organization (Goldstein & Ford, 2002). The team training needs analysis framework that describes analyses of person, organizational, and task factors that can influence effectiveness of a given training program is one lens through which these questions can be examined.

Improvement science and implementation science also offer a lens through which to address these questions. For example, in describing the hurdles in understanding contradictory findings regarding performance improvement and quality improvement interventions such as rapid response systems, Berwick (2008) articulates that they are often: "a complex, multicomponent intervention—essentially, a process of social change. The effectiveness of these systems is sensitive to an array of influences: Leadership, changing environments, details of implementation, organizational history, and much more" (p. 1183).

We offer that effective team training similarly represents a complex social change that, at its core, is asking providers to develop a shared mindfulness regarding the processes through which they communicate, cooperate, and coordinate. To better understand and identify the multitude of influences that have the potential to either help or hinder team training, efforts could build on existing meta-analyses to investigate the unique aspects of study design, implementation processes, characteristics of the organization, trainee population, and other factors. To achieve this, authors reporting on team training should strive to follow reporting guidelines such as the Standards for Quality Improvement Reporting Excellence (SQUIRE) (Davidoff et al., 2008). These guidelines provide a roadmap for describing intervention planning, the context and setting, potential confounds, and other possible reasons for observed outcomes.

> A better understanding of multiteam systems and "team-like" structures in healthcare is needed.

Though it may appear rudimentary, one of the most difficult questions to answer when developing team training for health care teams is how to define "the team" and identify who is part of the team. Teams can be defined and classified based on the type of task they perform, their structure and composition, the coordination processes they use to complete their tasks, typical products and outputs they produce, lifespan, and level of authority within an organization (e.g., Klimoski & Jones, 1995; Sundstrom et al., 1990, 2000). Within health care, administrative teams and project teams have been differentiated from care delivery teams, who have been further classified by patient type (e.g., geriatric, pediatric), disease process (e.g., cardiovascular, neurological), and delivery setting (e.g., acute, long-term care, ambulatory) (Lemieux-Charles & McGuire, 2006). As such, providers are often members of multiple teams simultaneously and team membership is often fluid and distributed, with interdependent activities cycling dynamically with independent activities, and team members working asynchronously toward shared goals. These evolving forms of collective effort often stretch the boundaries of traditional definitions of teams and teamwork (Mortensen et al., 2007). Thus, an important

conversation for the future theory and empirical study of teams is the degree to which existing team training strategies and methods remain applicable to these complex, dynamic forms of collaboration that characterize integrated, interdisciplinary, patient-centered care today.

For example, the recent focus on the concept of a Patient Centered Medical Home (PCMH) model of care (also known as a patient-aligned care team model) emphasizes a team-based delivery model focused on whole-person health and coordination of care across both settings and specialties (AAFP, 2007). Comprising both a core care team and an extended team, the PCMH model is a team of teams, known as a multiteam system (MTS). Such MTSs are defined as a: "tightly coupled constellation of teams offering specialized skills, capabilities, and functions aimed at attaining goals too large to be performed by a single team" (DeChurch & Marks, 2006, p. 311).

Teams in these systems are collectively working toward a shared goal (i.e., patient health), but also have more proximal, independent team goals that impact their behavior. Additionally, MTSs require boundary spanners—individuals who communicate, coordinate, and cooperate across teams in addition to functioning within their own component team. At present, however, insight regarding how these systems function and how to optimize MTS performance is only beginning to be developed (e.g., DeChurch & Marks, 2006; DeChurch & Mathieu, 2009; Marks et al., 2001, 2005). Thus, insight regarding ways to best develop, implement, and evaluate team training for such increasingly complex forms of collaborative behavior must continue to grow.

> The institutionalization of medical team training across the continuum of provider development and education is needed to support continuous learning and to sustain effective teamwork processes over time.

In order to achieve the goal of cultural transformation based in part on teamwork training programs, the principles and content of teamwork training need to be integrated into the educational and training systems currently in place. To date, most teamwork training programs are implemented in the clinical environment, with staff members who have already completed most or all of their formal education and training. However, efforts have been mounting to integrate teamwork training concepts throughout the continuum of health care education and career-long learning (Lerner et al., 2009; Rosen et al., 2010a; Weaver et al., 2010c). Table 37.4 summarizes challenges with the content and delivery of teamwork training across the spectrum of undergraduate, graduate, and continuing medical education.

TABLE 37.4

Potential Approaches and Challenges to Implementing Team Training Content and Delivery Methods across the Continuum of Medical Education

	Undergraduate Medical Education	Graduate Medical Education	Continuing Medical Education
Content	Develop awareness of the importance of teamwork competencies to patient safety and quality care Develop general and transferable teamwork skills	Build specialty specific teamwork behaviors Develop interpositional knowledge about the roles of other health care professions	Develop mentoring and coaching skills to help develop teamwork skills in others Build facilitation skills to enhance learning from experience on the job
Delivery	Use blended learning approaches for developing awareness Use role-play activities to demonstrate behaviors Use team-based learning exercises to assess teamwork skills Use low-fidelity practice for generic teamwork skills	Use high-fidelity practice activities emphasizing teamwork Use on the job coaching to reinforce concepts developed in simulation exercises	Use high-fidelity practice activities that blend teamwork and technical performance in complex ways
Challenges	Learners have not yet acquired technical competence Logistical challenges for meaningful interprofessional interaction	Resident work hour limitations restrict the amount of time that can be spent in training activities	Balancing the amount of Continuing Medical Education dedicated to nontechnical vs. technical topics

Source: Adapted from Rosen, M.A. et al., Demonstration-based training: A review and typology of instructional features, *Hum. Factors*, 52, 596, 2010.

37.8 In Closing

Overall, this chapter has provided a brief summary of the growing body of work dedicated to optimizing team performance through a systems view of team training. We hope this framing echoes other calls for focusing not only the interventions designed to support patient safety but also the planning, implementation, and evaluation processes that critically impact their effectiveness.

References

Alavi, S. B., & McCormick, J. (2004). A cross-cultural analysis of the effectiveness of the learning organizational model in school contexts. *The International Journal of Educational Management, 18*(7), 408–416.

American Academy of Family Physicians (AAFP), American Academy of Pediatrics (AAP), American College of Physicians (ACP), American Osteopathic Association (AOA). (2007, February). Joint principles of the patient-centered medical home. Retrieved February 11, 2011, from http://www.aafp.org/online/etc/medialib/aafp_org/documents/policy/fed/jointprinciplespcmh0207.Par.0001.File.tmp/022107medicalhome.pdf

Artman, H. (2000). Team situation assessment and information distribution. *Ergonomics, 43*(8), 1111–1128.

Baldwin, T. T., & Ford, J. K. (1988). Transfer of training: A review and directions for future research. *Personnel Psychology, 41*(1), 63–105.

Baldwin, T. T., Magjuka, R. J., & Loher, B. T. (1991). The perils of participation: Effects of choice on trainee motivation and learning. *Personnel Psychology, 44*, 51–65.

Baldwin, T. T., & Magjuka, R. J. (1997). Training as an organizational episode: Pretraining influences on trainee motivation. In J. K. Ford, S. W. J. Kozlowski, K. Kraiger, E. Salas, & M. S. Teachout (Eds.), Improving training effectiveness in organizations (pp. 99—127). Hillsdale, NJ: Erlbaum.

Bandura, A. (1986). *Social Foundations of Thought and Action: A Social Cognitive Theory.* Engelwood Cliffs, NY: Prentice-Hall.

Banning, M. (2006). Nursing research: Perspectives on critical thinking. *British Journal of Nursing, 15*(8), 458–461.

Beaubein, J. M., & Baker, D. P. (2004). The use of simulation for training teamwork skills in health care: How low can you go? *Quality and Safety in Health Care, 13*, i51–i56.

Bell, B. S., & Kozlowski, S. W. J. (2002). Goal orientation and ability: Interactive effects on self-efficacy, performance, and knowledge. *Journal of Applied Psychology, 87*(3), 497–505.

Bell, B. S., & Kozlowski, S. W. J. (2011). Collective failure: The emergency, consequences, and management of errors in teams. In D. A. Hoffman & M. Frese (Eds.), Errors in Organizations. New York, NY: Routledge.

Berenholtz, S. M., Schumacher, K., Hayanga, A. J., Simon, M., Goeschel, C., Pronovost P.J., Shanley, C.J., & Welsh, R.J. (2009). Implementing standardized operating room briefings and debriefings at a large regional medical center. *Joint Commission Journal on Quality and Patient Safety, 35*(8), 391–397.

Berwick, D. M. (2008). The science of improvement. *JAMA, 299*(10), 1182–1184.

Blau, G. J. (1985). A multiple study investigation of the dimensionality of job involvement. *Journal of Vocational Behavior, 27*, 19–36.

Blickensderfer, E., Cannon-Bowers, J. A., & Salas, E. (1997, April). Training teams to self-correct: An empirical investigation. Paper presented at the *12th Annual Meeting of the Society for Industrial and Organizational Psychology*, St. Louis, MO.

Bowers, C. A., Jentsch, F., Salas, E., & Braun, C. C. (1998). Analyzing communication sequences for team training needs assessment. *Human Factors, 40*(4), 672–679.

Brannick, M. T., & Levine, E. L. (2002). *Job Analysis: Methods, Research, and Applications for Human Resources*. Thousand Oaks, CA: Sage.

Brown, A. L. (1978). Knowing when, where and how to remember: A problem of metacognition. In R. Glaser (Ed.), *Advances in Instructional Psychology* (Vol. 1, pp. 77–165). Hillsdale, NJ: Erlbaum.

Buback, D. (2004). Assertiveness training to prevent verbal abuse in the OR. *AORN Journal, 79*(1), 147–164.

Buljac-Samardzic, M., Dekker-van Doorn, C. M., van Wijngaarden, J. D. H., & van Wijk, K. P. (2010). Interventions to improve team effectiveness: A systematic review. *Health Policy, 94*, 183–195.

Bunderson, J. S., & Sutcliffe, K. M. (2003). Management team learning orientation and business unit performance. *Journal of Applied Psychology, 88*(3), 552–560.

Burke, C. S. (2005). Team task analysis. In N. Stanton, E. Salas, H. Hendrick, S. Konz, & K. Parsons (Eds.), *Handbook of Human Factors and Ergonomics Methods* (pp. 56.1–56.8). London, U.K.: Taylor & Francis.

Burke, C. S., Stagl, K. C., Klein, C., Goodwin, G. F., Salas, E., & Halpin, S. M. (2006). What type of leadership behaviors are functional in teams? A meta-analysis. *The Leadership Quarterly, 17*, 288–307.

Burke, C. S., Stagl, K. C., Salas, E., Pierce, L., & Kendall, D. (2006). Understanding team adaptation: A conceptual analysis and model. *Journal of Applied Psychology, 91*(6), 1189–1207.

Cannon-Bowers, J. A., & Bowers, C. (2010). Team development and functioning. In S. Zedeck (Ed.), *APA Handbook of Industrial and Organizational Psychology, Vol. 1, Building and Developing the Organization* (pp. 597–650). Washington, DC: American Psychological Association.

Cannon-Bowers, J. A., & Salas, E. (1997). Teamwork competencies: The interaction of team member knowledge, skills, and attitudes. In H. F. O'Neil, Jr. (Ed.), *Workforce Readiness: Competencies and Assessment* (pp. 151–174). Mahwah, NJ: Erlbaum.

Cannon-Bowers, J. A., Salas, E., Blickensderfer, E., & Bowers, C. A. (1998). The impact of cross-training and workload on team functioning: A replication and extension of the initial findings. *Human Factors, 40*, 92–101.

Cannon-Bowers, J. A., Tannenbaum, S. I., Salas, E., & Volpe, E. (1995). Defining team competencies and establishing team training requirements. In R. Guzzo & E. Salas (Eds.), *Team Effectiveness and Decision Making in Organizations* (pp. 117–151). San Francisco, CA: Jossey-Bass.

Capella, J., Smith, S., Philp, A., Putnam, T., Glibert, C., Fry, W., Harvey, E. et al. (2010). Teamwork training improves the clinical care of trauma patients. *Journal of Surgical Education, 6*, 439–443.

Chan, P. S., Jain, R., Nallmothu, B. K., Berg, R. A., & Sasson, C. (2010). Rapid response teams: A systematic review and meta-analysis. *Archives of Internal Medicine, 170*, 18–26.

Chernyshenko, O. S., Stark, S., & Drasgow, F. (2011). Individual differences: Their measurement and validity. In S. Zedeck (Ed.), *APA Handbook of Industrial and Organizational Psychology, Vol. 2, Selecting and Developing Members for the Organization* (pp. 3–41). Washington, DC: American Psychological Association.

Cohen, S. (1983). Programmed instruction: Assertiveness in nursing: Part two. *The American Journal of Nursing, 83*(6), 911–928.

Cohen, M. S., & Freeman, J. T. (1997). Improving critical thinking. In R. Flin & L. Martin (Eds.), *Decision Making under Stress: Emerging Themes and Applications* (pp. 155–190). Hampshire, England: Ashgate.

Cohen, M. S., Freeman, J. T., & Wolf, S. (1996). Meta-recognition in time-stressed decision making: Recognizing, critiquing, and correcting. *Journal of the Human Factors and Ergonomics Society, 38*(2), 206–219.

Cohen M. S., Freeman, J. T., & Thompson, B. T. (1997). Training the naturalistic decision maker. In C. E. Zsambok & G. Klein (Eds.), *Naturalistic decision making*, (pp. 257–268). Mahwah, NJ: Erlbaum.

Cohen, M. S., & Thompson, B. B. (2001). Training teams to take initiative: Critical thinking in novel situations. In E. Salas (Ed.), *Advances in Cognitive Engineering and Human Performance Research* (Vol. 1). pp. 251–291. Oxford, U.K.: JAI/Elsevier.

Colquitt, J. A., LePine, J. A., & Noe, R. A. (2000). Toward an integrative theory of training motivation: A meta-analytic path analysis of 20 years of research. *Journal of Applied Psychology, 85*(3), 679–707.

Corrigan, J. M., Donaldson, M. S., & Kohn, L. T. (Eds.). (2001). *Crossing the Quality Chasm*. Washington, DC: National Academy Press.

Crandall, B., Klein, G., & Hoffman, R. R. (2006). *Working Minds: A Practitioner's Guide to Cognitive Task Analysis*. Cambridge, MA: Bradford Books/MIT Press.

Davidoff, F., Batalden, P., Stevens, D., Ogrinc, G., & Mooney, S. (2008). Publication guidelines for quality improvement in health care: Evolution of the SQUIRE project. *Quality and Safety in Health Care, 17*(Supplement 1), i3–i9.

Day, D. V., Gronn, P., & Salas, E. (2004). Leadership capacity in teams. *Leadership Quarterly, 15*(6), 857–880.

DeChurch, L. A., & Marks, M. A. (2006). Leadership in multiteam systems. *Journal of Applied Psychology, 91*, 311–326.

DeChurch, L. A., & Mathieu, J. E. (2009). Thinking in terms of multiteam systems. In E. Salas, G. F. Goodwin, & C. S. Burke (Eds.), *Team Effectiveness in Complex Organizations: Cross-Disciplinary Perspectives and Approaches* (pp. 267–292). New York: Taylor & Francis.

De Dreu, C. K., & Weingart, L. R. (2003). Task versus relationship conflict, team performance, and team member satisfaction: A meta-analysis. *Journal of Applied Psychology, 88*(4), 741–749.

Deering, S., Rosen, M. A., Ludi, V., Munroe, M., Pocrnich, A., Laky, C., & Napolitano, P. (in press). On the front lines of patient safety: Implementation and evaluation of team training in Iraq. *Joint Commission Journal on Quality and Patient Safety*.

DeVita, M. A., Schaefer, J., Lutz, J., Wang H., & Dongilli, T. (2005). Improving medical emergency team (MET) performance using a novel curriculum and a computerized human patient simulator. *Quality and Safety in Health Care, 14*, 326–331.

Dickinson, T. L., & McIntyre, R. M. (1997). A conceptual framework for team measurement. In M. T. Brannick, E. Salas, & C. Prince (Eds.), *Team Performance Measurement: Theory, Methods, and Applications* (pp. 19–43). Mahwah, NJ: Erlbaum.

Driskell, J. E., & Salas, E. (1992). Collective behavior and team performance. *Human Factors, 34*, 277–288.

Dweck, C. S. (1986). Motivational processes affecting learning. *American Psychologist, 41*, 1040–1048.

Dyer, J. L. (1984). Team research and team training: A state of the art review. In F. A. Muckler (Ed.), *Human Factors Review* (pp. 285–323). Santa Monica CA: Human Factors Society.

Eby, L., & Dobbins, G. (1997). Collectivistic orientation in teams: An individual and group-level analysis. *Journal of Organizational Behaviour, 18*, 275–295.

Edmondson, A. C. (1999). Psychological safety and learning behavior in work teams. *Administrative Science Quarterly, 44*(2), 350–383.

Entin, E. E., & Serfaty, D. (1999). Adaptive team coordination. *Human Factors, 41*(2), 312–325.

Ericsson, K. A., & Charness, N. (1994). Expert performance: Its structure and acquisition. *American Psychologist, 49*(8), 725–747.

Fernandez, R., Kozlowski, S. W., Shapiro, M. J., & Salas, E. (2008). Toward a definition of teamwork in emergency medicine. *Academic Emergency Medicine, 15*, 1104–1112.

Flavell, J. H. (1979). Metacognition and cognitive monitoring: A new area of cognitive–developmental inquiry. *American Psychologist, 34*(1), 906–911.

Flin, R., & Maran, N. (2004). Identifying and training non-technical skills for teams in acute medicine. *Quality and Safety in Health Care, 13*(supplement 1), i80–i84.

Ford, K. J., Gully, S. M., Salas, E., Smith, E. M., & Weissbein D. A. (1998). Relationships of goal orientation, metacognitive activity, and practice strategies with learning outcomes and transfer. *Journal of Applied Psychology, 83*, 218–233.

Freeman, L. H., & Adams, P. F. (1999). Comparative effectiveness of two training programmes on assertive behaviour. *Nursing Standard, 13*(38), 32–35.

Frommater, D., Marshall, D., Halford, G., Rimmasch, H., & Coons, M. C. (1995). How a three-campus heart service line improves clinical processes and outcomes. *Joint Commission Journal on Quality and Patient Safety, 21*(6), 263–276.

Gaba, D. M. (2010). Crisis resource management and teamwork training in anaesthesia. *British Journal of Anesthesia, 105*(1), 3–6.

Gaba, D. M., Howard, S. K., Fish, K. J., Smith, B. E., & Sowb, Y. A. (2001). Simulation-based training in anesthesia crisis resource management (ACRM): A decade of experience. *Simulation & Gaming, 32*(2), 175–193.

Gibson, C. B. (2003). The efficacy advantage: Factors related to the formation of group efficacy. *Journal of Applied Social Psychology, 33*(10), 2153–2186.

Glickman, A. S., Zimmer, S., Montero, R. C., Guerette, P. J., Morgan, B. B., & Salas, E. (1987). The evolution of teamwork skills: An empirical assessment with implications for training. Technical report for the Naval Training Systems Center. Orlando, FL: University of Central Florida.

Goldstein, I. L., & Ford, K. (2002). *Training in Organizations: Needs Assessment, Development, and Evaluation* (4th ed.). Belmont, CA: Wadsworth.

Gordon, M. E., Cofer, J. L., & McCullough, P. M. (1986). Relationships among seniority, past performance, interjob similarity, and trainability. *Journal of Applied Psychology, 71*, 517–521.

Heimbeck, D., Frese, M., Sonnentag, S., & Keith, N. (2003). Integrating errors into the training process: The function of error management instructions and the role of error management instructions. *Personnel Psychology, 56*(2), 333–361.

Helmreich, R. L., & Schaefer, H.-G. (1994). Team performance in the operating room. In M. S. Bogner (Ed.), *Human Error in Medicine* (pp. 225–253). Hillsdale, NJ: Lawrence Erlbaum Associates.

Holt, D. T., & Crocker, M. (2000). Prior negative experiences: Their impact on computer training outcomes. *Computers & Education, 35*(4), 295–308.

Huber, G. P. (1991). Organizational learning: The contributing processes and the literature. *Organization Science, 2*, 88–115.

Ilgen, D. R., Hollenbeck, J. R., Johnson, M., & Jundt, D. (2005). Teams in organizations: From input-process-output models to IMOI models. *Annual Review of Psychology, 56*, 517–543.

Inman, R. R., Blumenfeld, D. E., & Ko, A. (2005). Cross-training hospital nurses to reduce staffing costs. *Health Care Management Review, 30*(2), 116–125.

Institute for Healthcare Improvement. (2010). Web&ACTION: Effective crisis management of serious clinical events. Retrieved February 22, 2010, from http://www.ihi.org/IHI/Programs/AudioAndWebPrograms/WebACTIONEffectiveCrisisMgmtSeriousClinicalEvents.htm

Jackson, C. L., Colquitt, J. A., Wesson, M. J., & Zapata-Phelan, C. P. (2006). Psychological collectivism: A measurement validation and linkage to group member performance. *Journal of Applied Psychology, 91*(4), 884–899.

Jentsch, F. (1997). Metacognitive training for junior team members: Solving the co-pilot's catch-22. Unpublished dissertation. Orlando, FL: University of Central Florida.

Jentsch, F., & Smith-Jentsch, K. (2001). Assertiveness and team performance: More than "just say no." In E. Salas, C. Bowers, & E. Edens (Eds.), *Improving Teamwork in Organizations* (pp. 73–94). Mahwah, NJ: Lawrence Erlbaum Associates, Inc.

Joint Commission. (2011). National patient safety goals. Retrieved February 23, 2011, from http://www.jointcommission.org/hap_2011_npsgs/

Jordan, P. J., & Troth, A. C. (2004). Managing emotions during team problem solving: Emotional intelligence and conflict resolution. *Human Performance, 17*(2), 195–218.

Katz-Navon, T. Y., & Erez, M. (2005). When collective- and self-efficacy affect team performance: The role of task interdependence. *Small Group Research, 36*(4), 437–465.

Keith, N., & Frese, M. (2008). Performance effects of error management training: A meta-analysis. *Journal of Applied Psychology, 93*, 59–69.

Keith, N., & Frese, M. (2011). Learning through errors in training. In D. A. Hoffman & M. Frese (Eds.), *Errors in Organizations*. New York: Routledge Academic.

Kirkpatrick, H., & Forchuk, C. (1992). Assertiveness training: Does it make a difference? *Journal of Nursing Staffing Development, 8*(2), 60–65.

Klein, C., DiazGranados, D., Salas, E., Le, H., Burke, C. S., Lyons, R., & Goodwin, G. F. (2009). Does team building work? *Small Group Research, 40*, 181–222.

Klimoski, R., & Jones, R. G. (1995). Staffing for effective group decision making: Key issues in matching people and teams. In R. A. Guzzo & E. Salas (Eds.), *Team Effectiveness and Decision Making in Organizations* (pp. 291–332). San Francisco, CA: Jossey-Bass.

Klimoski, R., & Mohammed, S. (1994). Team mental model: Construct or metaphor? *Journal of Management, 20*(2), 403–437.

Kozlowski, S. W. J., Gully, S. M., Brown, K. G., Salas, E., Smith, E. A., & Nason, E. R. (2001). Effects of training goals and goal orientation traits on multi-dimensional training outcomes and performance adaptability. *Organizational Behavior and Human Decision Processes, 85*, 1–31.

Kozlowski, S. W. J., Gully, S. M., Nason, E. R., & Smith, E. M. (1999). Developing adaptive teams: A theory of compilation and performance across levels and time. In D. R. Ilgen & E. D. Pulakos (Eds.), *The Changing Nature of Work and Performance: Implications for Staffing, Personnel Actions, and Development*. pp. 240–292. San Francisco, CA: Jossey-Bass

Kozlowski, S. W. J., Gully, S. M., Salas, E., & Cannon-Bowers, J. A. (1996). Team leadership and development: Theory, principles, and guidelines for training leaders and teams. In M. Beyerlein, S. Beyerlein, & D. Johnson (Eds.), *Advances in Interdisciplinary Studies of Work Teams: Team Leadership* (Vol. 3, pp. 253–292). Greenwich, CT: JAI Press.

Kuiper, R., & Pesut, D. (2004). Promoting cognitive and meta-cognitive reflective reasoning skills in nursing practice: Self-regulated learning theory. *Journal of Advanced Nursing, 45*(4), 381–391.

Laszlo, E. (1972). *Introduction to Systems Philosophy: Toward a New Paradigm of Contemporary Thought*. New York: Harper & Row.

Lemieux-Charles, L., & McGuire, W. (2006). What do we know about health care team effectiveness? A review of literature. *Medical Care Research and Review, 63*(3), 1–38.

LePine, J. A., Piccolo, R. F., Jackson, C. L., Mathieu, J. E., & Saul, J. R. (2008). A meta-analysis of teamwork processes: Tests of a multidimensional model and relationships with team effectiveness criteria. *Personnel Psychology, 61*(2), 273–307.

Lerner, S., Magrane, D., & Friedman, E. (2009). Teaching teamwork in medical education. *Mount Sinai Journal of Medicine, 76*, 318–329.

Lim, D. H., & Johnson, S. D. (2002). Trainee perceptions of factors that influence learning transfer. *International Journal of Training and Development, 6*, 36–48.

Marks, M. A., DeChurch, L. A., Mathieu, J. E., Panzer, F. J., & Alonso, A. (2005). Teamwork in multiteam systems. *Journal of Applied Psychology, 90*, 964–971.

Marks, M. A., Mathieu, J. E., & Zaccaro, S. J. (2001). A temporally based framework and taxonomy of team processes. *Academy of Management Review, 26*, 355–376.

Marks, M. A., & Panzer, F. J. (2004). The influence of team monitoring on team processes and performance. *Human Performance, 17*(1), 25–41.

Marks, M. A., Sabella, M. J., Burke, C. S., & Zaccaro, S. J. (2002). The impact of cross-training on team effectiveness. *Journal of Applied Psychology, 87*(1), 3–13.

Masson, L., & Fain, J. (1997). Competency validation for cross-training in surgical services. *AORN Journal, 66*(4), 651–659.

Mathieu, J. E., & Schulze, W. (2006). The influence of team knowledge and formal plans on episodic team process-performance relationships. *Academy of Management Journal, 49*(3), 605–619.

McCabe, C., & Timmins, F. (2003). Teaching assertiveness to undergraduate nursing students. *Nurse Education in Practice, 3*(1), 30–42.

McGehee, W., & Thayer, P. (1961). *Training in Business and Industry*. Oxford, England: Wiley.

McGreevy, J. M., & Otten, T. D. (2007). Briefing and debriefing in the operating room using fighter pilot crew resource management. *Journal of the American College of Surgeons, 205*, 169–176.

McIntyre, R. M., & Salas, E. (1995). Measuring and managing for team performance: Emerging principles from complex environments. In R. A. Guzzo & E. Salas (Eds.), *Team Effectiveness and Decision Making in Organizations* (pp. 9–45). San Francisco, CA: Jossey-Bass.

McVanel, S., & Morris, B. (2010). Staff's perceptions of voluntary assertiveness skills training. *Journal for Nurses in Staff Development, 26*(6), 256–259.

Merien, A. E. R., van de Ven, J., Mol, B. W., Houterman, S., & Oei, S. G. (2010). Multidisciplinary team training in a simulation setting for acute obstetric emergencies: A systematic review. *Obstetrics and Gynecology, 115*, 1021–1031.

Meyer, J. P., Allen, N. J., & Smith, C. A. (1993). Commitment to organizations and occupations: Extension and test of a three-component conceptualization. *Journal of Applied Psychology, 78*(4), 538–552.

Midgley, G. (Ed.) (2002). *Systems Thinking* (Vol. 1–4). London, U.K.: Sage Publications.

Milstead, J. A. (1996). Basic tools for the orthopaedic staff nurse. Part I: Assertiveness. *Orthopedic Nursing, 15*(1), 23–30.

Mohaghegh, Z., Kazemi, R., & Mosleh, A., 2009. Incorporating organizational factors into probabilistic risk assessment (PRA) of complex socio-technical systems: A hybrid technique formalization. *Reliability Engineering & System Safety, 94*, 1000–1018.

Mohammed, S., & Angell, L. C. (2004). Surface- and deep-level diversity in workgroups: Examining the moderating effects of team orientation and team process on relationship conflict. *Journal of Organizational Behavior, 25*(8), 1015–1039.

Moorthy, K., Munz, Y., Forrest, D., Pandey, V., Undre, S., Vincent, C., & Darzi, A. (2006). Surgical crisis management skills training and assessment: A stimulation-based approach to enhancing operating room performance. *Annals of Surgery, 244*, 139–147.

Morgan, B. B., Jr., Glickman, A. S., Woodward, E. A., Blaiwes, A. S., & Salas, E. (1986). *Measurement of Team Behaviors in a Navy Environment* (No. 86-014). Orlando, FL: Naval Training Systems Center.

Morgeson, F. P., & Dierdorff, E. C. (2011). Work analysis: From technique to theory. In S. Zedeck (Ed.), *APA Handbook of Industrial and Organizational Psychology, Vol. 2, Selecting and Developing Members for the Organization* (pp. 3–41). Washington, DC: American Psychological Association.

Mortensen, M., Woolley, A., W., & O'Leary, M. (2007). Conditions enabling effective multiple team membership. In K. Crowston, S. Sieber, & E. Wynn (Eds.), *Virtuality and Virtualization* (Vol. 236, pp. 215–228). Boston, MA: IFIP International Federation for Information Processing.

Musson, D. M., & Helmreich, R. L. (2004). Team training and resource management in healthcare: Current issues and future directions. *Harvard Health Policy Review, 5*(1), 25–35.

Neily, J., Mills, P. D., Young-Xu, Y., Carney, B. T., West, P., Berger, D.H., Mazzia, L.M., Paull, D.E., & Bagian, J.P. (2010). Association between implementation of a medical team training program and surgical mortality. *JAMA, 204*(15), 1693–1700.

Nielsen, P. E., Goldman, M. B., Mann, S., Shapiro, D. E., Marcus, R. G., Pratt, S. D., Greenberg, P. et al. (2007). Effects of teamwork training on adverse outcomes and process of care in labor and delivery: A randomized controlled trial. *Obstetrics and Gynecology, 109*, 48–55.

Ovretveit, J. (2010). Improvement leaders: What do they and should they do? A summary of a review of research. *Quality and Safety in Health Care, 19*, 490–492.

Parand, A., Burnett, S., Benn, J., Iskander, S., Pinto, A., & Vincent, C. (2010). Medical engagement in organization-wide safety and quality-improvement programmes: Experience in the UK Safer Patients Initiative. *Quality and Safety in Health Care, 19*, 1–5.

Paull, D. E., Mazzia, L. M., Izu, B. S., Neily, J., Mills, P. D., & Bagian, J. P. (2009). Predictors of successful implementation of preoperative briefings and postoperative debriefings after medical team training. *The American Journal of Surgery, 198*, 675–678.

Payne, S. C., Youngcourt, S. S., & Beaubien, J. M. (2007). A meta-analytic examination of the goal orientation nomological net. *Journal of Applied Psychology, 92*, 128–150.

Pearsall, M. J., & Ellis, A. P. J. (2006). The effects of critical team member assertiveness on team performance and satisfaction. *Journal of Management, 32*(4), 575–594.

Percarpio, K. B., Harris, F. S., Hatfield, B. A., Dunlap, B., Diekroger, W. E., Nichols, P.D., Mazzia, L.M., Millls, P.D., & Neily, J.B. (2010). Code debriefing from the department of veterans affairs (VA) medical team training program improves the cardiopulmonary resuscitation code process. *Joint Commission Journal on Quality and Patient Safety, 39*(9), 424–429.

Porter, C. O., Hollenbeck, J. R., Ilgen, D. R., Ellis, A. P., West, B. J., & Moon, H. (2003). Backing up behaviors in teams: The role of personality and legitimacy of need. *Journal of Applied Psychology, 88*(3), 391–403.

Quiñones, M. A., & Ehrenstein, A. (1997). Psychological perspectives on training in organizations. In M. A. Quiñones & A. Ehrenstein (Eds.), *Training for a Rapidly Changing Workplace: Applications of Psychological Research*. pp. 201–222. Washington, DC: American Psychological Association.

Rabøl, L. I., Østergaard, D., & Mogensen, T. (2010). Outcomes of classroom-based team training interventions for multiprofessional hospital staff: A systematic review. *Quality and Safety in Health Care, 19*, 1–11.

Rasker, P. C., Post, W. M., & Schraagen, J. M. (2000). Effects of two types of intra-team feedback on developing a shared mental model in command & control teams. *Ergonomics, 43*(8), 1167–1189.

Reznek, M., Smith-Coggins, R., Howard, S., Kiran, K., Harter, P., Sowb, Y., Gaba, D., & Krummel, T. (2003). Emergency medicine crisis resource management (EMCRM): Pilot study of a simulation-based crisis management course for emergency medicine. *Academic Emergency Medicine, 10*(4), 386–389.

Rosen, M. A., Feldman, M., Salas, E., King, H. B., & Lopreiato, J. (2010a). Challenges and opportunities for applying human factors methods to the development of non-technical skills in healthcare education. In V. Duffy (Ed.), *Advances in Human Factors and Ergonomics in Healthcare* (pp. 1–30). Boca Raton, FL: CRC Press.

Rosen, M. A., Salas, E., Pavlas, D., Jensen, R., Fu, D., & Lampton, D. (2010b). Demonstration-based training: A review and typology of instructional features. *Human Factors, 52*, 596–609.

Rouiller, J. Z., & Goldstein, I. L. (1993). The relationship between organizational transfer climate and positive transfer of training. *Human Resources Development Quarterly, 4*, 377–390.

Russ-Eft, D. (2002). A typology of training design and work environment factors affecting workplace learning and transfer. *Human Resource Development Review, 1*, 45–65.

Saavedra, R., Earley, P. C., & Van Dyne, L. (1993). Complex interdependence in task-performing groups. *Journal of Applied Psychology, 78*(1), 61–72.

Salas, E., Almeida, S. A., Salisbury, M., King, H., Lazzara, E. H., Lyons, R., Wilson, K.A., Almeida, P.A. & McQuillan, R. (2009a). What are the critical success factors for team training in health care? *Joint Commission Journal on Quality and Patient Safety, 35*(8), 398–405.

Salas, E., & Cannon-Bowers, J. A. (2001). The science of training: A decade of progress. *Annual Review of Psychology, 52*, 471–499.

Salas, E., DiazGranados, D., Klein, C., Burke, C. S., Stagl, K. C., Goodwin, G. F., & Halpin, S. M. (2008a). Does team training improve team performance? A meta-analysis. *Human Factors, 50*(6), 903–933.

Salas, E., Dickinson, T., Converse, S., & Tannenbaum, S. (1992). Toward and understanding of team performance and training. In R. Swezey & E. Salas (Eds.), *Teams: Their Training and Performance*. pp. 3–29. Norwood, NJ: Ablex Publishing.

Salas, E., Nichols, D. R., & Driskell, J. E. (2007a). Testing three team training strategies in intact teams. *Small Group Research, 38*(4), 471–488.

Salas, E., Priest, H. A., & DeRouin, R. E. (2005a). Team building. In N. Stanton, H. Hendrick, S. Konz, K. Parsons, & E. Salas (Eds.), *Handbook of Human Factors and Ergonomics Methods* (pp. 48-1–48-5). London, U.K.: Taylor & Francis.

Salas, E., Priest, H. A., Wilson, K. A., & Burke, C. S. (2006a). Scenario-based training: Improving military mission performance and adaptability. In A. B. Adler, C. A. Castro, & T. W. Britt (Eds.), *Military Life: The Psychology of Serving in Peace and Combat* (Vol. 2: Operational Stress, pp. 32–53). Westport, CT: Praeger Security International.

Salas, E., Prince, C., Bowers, C. A., Stout, R. J., Oser, R. L., & Cannon-Bowers, J. A. (1999). A methodology for enhancing crew resource management training. *Human Factors, 41*, 161–172.

Salas, E., Rosen, M. A., Burke, C. S., & Goodwin, G. F. (2009b). The wisdom of collectives in organizations: An update of the teamwork competencies. In E. Salas, G. F. Goodwin, & C. S. Burke (Eds.), *Team Effectiveness in Complex Organizations: Cross-Disciplinary Perspectives and Approaches* (pp. 39–79). New York: Routledge.

Salas, E., Sims, D. E., & Burke, C. S. (2005b). Is there a big five in teamwork? *Small Group Research, 36*(5), 555–599.

Salas, E., Stagl, K. C., Burke, C. S., & Goodwin, G. F. (2007b). Fostering team effectiveness in organizations: Toward an integrative theoretical framework of team performance. In R. A. Dienstbier, J. W. Shuart, W. Spaulding, & J. Poland (Eds.), *Modeling Complex Systems: Motivation, Cognition and Social Processes, Nebraska Symposium on Motivation* (Vol. 51, pp. 185–243). Lincoln, NE: University of Nebraska Press.

Salas, E., Wilson, K. A., Burke, C. S., & Priest, H. A. (2005c). Using Simulation-based training to improve patient safety: What does it take? *Journal on Quality and Patient Safety, 31*(7), 363–371.

Salas, E., Wilson, K. A., Burke, C. S., Wightman, D. C., & Howse, W. R. (2006b). A checklist for crew resource management training. *Ergonomics in Design, 14*(2), 6–15.

Salas, E., Wilson, K. A., & Edens, E. (Eds.) (2009c). *Critical Essays on Human Factors in Aviation: Crew Resource Management.* Hampshire, U.K.: Ashgate.

Salas, E., Wilson, K. A., Murphy, C., King, H. B., & Salisbury, M. (2008b). Communicating, coordinating, and cooperating when lives depend on it: Tips for teamwork. *Joint Commission Journal on Quality and Patient Safety, 6*, 333–341.

Salas, E., Wilson, K. A., Priest, H. A., & Guthrie, J. W. (2006c). Design, delivery, and evaluation of training systems. In G. Salvendy (Ed.), *Handbook of Human Factors and Ergonomics* (3rd ed., pp. 472–512). New York: John Wiley & Sons.

Salas, E., Wilson-Donnelly, K. A., Sims, D. E., Burke, C. S., & Priest, H. A. (2007c). Teamwork training for patient safety: Best practices and guiding principles. In P. Carayon (Ed.), *Handbook of Human Factors and Ergonomics in Health Care and Patient Safety* (pp. 803–822). Mahwah, NJ: Lawrence Erlbaum Associates.

Santamaria, J., Tobin, A., & Holmes, J. (2010). Changing cardiac arrest and hospital mortality rates through a medical emergency team takes time and constant review. *Critical Care Medicine, 38*, 445–450.

Schneider, B., Ehrhart, M. G., & Macey, W. H. (2010). Perspectives on organizational climate and culture. In S. Zedeck (Ed.), *APA Handbook of Industrial and Organizational Psychology* (Vol. II, pp. 373–414). Washington, DC: American Psychological Association.

Serfaty, D., & Entin, E. E. (1998). Team adaptation and coordination training. In R. Flin, E. Salas, M. Strab, & L. Martin (Eds.), *Decision Making under Stress: Emerging Themes and Application* (pp. 170–184). Aldershot, U.K.: Ashgate.

Serfaty, D., Entin, E. E., & Johnston, J. H. (1998). Team coordination training. In J. A. Cannon-Bowers & E. Salas (Eds.), *Making Decisions under Stress: Implications for Individual and Team Training* (pp. 221–245). Washington, DC: American Psychological Association.

Shapiro, M. J., Morey, J. C., Small, S. D., Langford, V., Kaylor, C. J., Jagminas, L., Suner, S., Salisbury, M.L., Simon, R., & Jay, G.D. (2004). Simulation based teamwork training for emergency department staff: Does it improve clinical team performance when added to an existing didactic teamwork curriculum? *Quality and Safety in Health Care, 13*, 417–421.

Simons, T. L., & Peterson, R. S. (2000). Task conflict and relationship conflict in top management teams: The pivotal role of intragroup trust. *Journal of Applied Psychology, 85*(1), 102–111.

Smith-Jentsch, K. A., Cannon-Bowers, J. A., Tannenbaum, S. I., & Salas, E. (2008). Guided team self-correction: Impacts on team mental models, processes, and effectiveness. *Small Group Research, 39*(3), 303–327.

Smith-Jentsch, K. A., Johnston, J. A., & Payne, S. C. (1998). Measuring team-related expertise in complex environments. In J. A. Cannon-Bowers & E. Salas (Eds.), *Making Decisions under Stress: Implications for Individual and Team Training* (pp. 61–87). Washington, DC: American Psychological Association.

Smith-Jentsch, K. A., Salas, E., & Baker, D. P. (1996). Training team performance-related assertiveness. *Personnel Psychology, 49*, 909–936.

Stout, R. J., Cannon-Bowers, J. A., & Salas, E. (1996). The role of shared mental models in developing team situational awareness: Implications for training. *Training Research Journal, 2*, 85–116.

Sundstrom, E., De Muse, K. P., & Futell, D. (1990). Work teams: Applications and effectiveness. *American Psychologist, 45*(2), 120–133.

Sundstrom, E., McIntyre, M., Halfhill, T., & Richards, H. (2000). Work groups: From the Hawthorne studies to the work teams of the 1990s. *Group Dynamics, 4*, 44–67.

Tannenbaum, S. I., Mathieu, J. E., Salas, E., & Cannon-Bowers, J. A. (1991). Meeting trainees' expectations: The influence of training fulfillment on the development of commitment, self-efficacy, and motivation. *Journal of Applied Psychology, 76*, 759–769.

Tannenbaum, S., Smith-Jentsch, K., & Behson, S. (1998). Training team leaders to facilitate team learning and performance. In J. A. Cannon-Bowers & E. Salas (Eds.), *Making Decisions under Stress: Implications for Individual and Team Training* (pp. 247–270). Washington, DC: American Psychological Association.

Taylor, P. J., Russ-Eft, D. F., & Chan, D. W. L. (2005). A meta-analytic review of behavior modeling training. *Journal of Applied Psychology, 90*(4), 692–709.

Tesluk, P. E., Farr, J. L., Matheiu, J. E., & Vance, R. J. (1995). Generalization of employee involvement training to the job setting: Individual and situational effects. *Personnel Psychology, 48,* 607–632.

Thomas, E. J., Taggart, B., Crandell, S., Lasky, R. E., Williams, A. L., Love, L.J., Socton, J.B., Tyson, J.E., & Helmreich, R.L. (2007). Teaching teamwork during the neonatal resuscitation program: A randomized trial. *Journal of Perinatology, 27*(7), 409–414.

VandeWalle, D. (1997). Development and validation of a work domain goal orientation instrument. *Educational and Psychological Measurement, 57,* 995–1015.

Volpe, C. E., Cannon-Bowers, J. A., Salas, E., & Spector, P. E. (1996). The impact of cross training on team functioning: An empirical investigation. *Human Factors, 38,* 87–100.

Weaver, S. J., Bedwell, W. L., & Salas, E. (2011). Training teams to cope with errors: A multi-level framework for instructional strategies & transfer. In D. A. Hoffman & M. Frese (Eds.), *Errors in Organizations.* New York: Routledge Academic.

Weaver, S. J., Lyons, R., DiazGranados, D., Rosen, M. A., Salas, E., Oglesby, J., Birnbach, D. J., Augenstein, J. S., Robinson, D. W., & King, H. B. (2010a). The anatomy of health care team training and the state of practice: A critical review. *Academic Medicine, 85*(11), 1746–1760.

Weaver, S. L., Rosen, M. A., DiazGranados, D., Lazzara, E. H., Lyons, R., Salas, E., Knych, S.A., McKeever, M., Alder, L., Barker, M., & King, H.B. (2010b). Does teamwork improve performance in the operating room? A multi-level evaluation. *Joint Commission Journal on Quality and Patient Safety, 36*(3), 133–142.

Weaver, S. L., Rosen, M. A., Salas, E., Baum, K. D., & King, H. B. (2010c). Integrating the science of team training: Guidelines for continuing medical education. *Journal of Continuing Education in the Health Professions, 30,* 208–220.

Weaver, S. J., Wildman, J. L., & Salas, E. (2009). How to build expert teams: Best practices. In R. J. Burke & C. L. Cooper (Eds.), *The Peak Performing Organization* (pp. 129–156). New York: Routledge.

Webster, J., & Martocchio, J. J. (1995). The differential effects of software training previews on training outcomes. *Journal of Management, 21,* 757–787.

Wensing, M., Wollersheim, H., & Grol, R. (2006). Organizational interventions to implement improvements in patient care: A structured review of reviews. *Implementation Science, 22,* 1–9.

Wiener, E., Kanki, B., & Helmreich, R. (Eds.). (1993). *Cockpit Resource Management.* San Diego, CA: Academic Press.

Wilson, K. A., Burke, C. S., Priest, H. A., & Salas, E. (2005). Promoting health care safety through training high reliability teams. *Quality and Safety in Health Care, 14,* 303–309.

Wilson, K. A., Salas, E., Priest, H. A., & Andrews, D. (2007). Errors in the heat of battle: Taking a closer look at shared cognition breakdowns through teamwork. *Human Factors, 49,* 243–256.

Wolf, F. A., Way, L. W., & Stewart, L. (2010). The efficacy of medical team training: Improved team performance and decreased operating room delays: A detailed analysis of 4863 cases. *Annals of Surgery, 252*(3), 477–483.

Zaccaro, S. J., Blair, V., Peterson, C., & Zazanis, M. (1995). Collective efficacy. In J. E. Maddux (Ed.), *Self-Efficacy, Adaptation, and Adjustment: Theory, Research, and Application.* New York: Plenum.

38

Human Factors Considerations in Health IT Design and Development

Marie-Catherine Beuscart-Zéphir, Sylvia Pelayo, Elizabeth Borycki, and Andre Kushniruk

CONTENTS

38.1 Introduction 650
38.2 Characteristics of Health IT and Health Care Work Situations 651
    38.2.1 What Is Health IT? 651
        38.2.1.1 Wide Range of Applications and Systems 651
        38.2.1.2 Implemented at Multiple Levels 651
        38.2.1.3 With Mandatory Interoperability 651
    38.2.2 Main Characteristics of Health Care Work Situations 651
        38.2.2.1 Change in Paradigm in Health Care 651
        38.2.2.2 Collaborative Nature of Health Care Work 651
        38.2.2.3 Heterogeneous Organizations 652
        38.2.2.4 Uncertainty and Criticality 653
        38.2.2.5 Interruptive Tasks and Time Constraints 653
        38.2.2.6 Safety Critical Process 653
38.3 Resulting Requirements for Health IT Design and Development: Challenges and Recommendations 654
    38.3.1 Support of Mandatory Interoperability 654
        38.3.1.1 Challenge 654
        38.3.1.2 Recommendations 654
    38.3.2 Make the System a Team Player 654
        38.3.2.1 Challenge 654
        38.3.2.2 Recommendations 655
    38.3.3 Make the System a Clinician's Partner 655
        38.3.3.1 Challenge 655
        38.3.3.2 Recommendations 656
    38.3.4 Provide Safe and Easy-to-Use Parameterization/Customization Functions 656
        38.3.4.1 Challenge 656
        38.3.4.2 Recommendations 657
    38.3.5 Incorporate User-Centered Design Methods along with Iterative Safety-Oriented Usability
        Evaluations during the Design and Development Life Cycle 657
        38.3.5.1 Challenge 657
        38.3.5.2 Recommendations 657
    38.3.6 Support Safety Critical Features of the Health Care System 658
        38.3.6.1 Challenge 658
        38.3.6.2 Recommendations 658
38.4 Design Problems in Health IT 658
38.5 Integrating HF Methods and Models in the Design and Development Life Cycle of Health IT 659
    38.5.1 Design of Health IT as an Iterative Process 659
    38.5.2 Incorporating HF Tasks in the Design Process 663
        38.5.2.1 Usability and Health IT 663
        38.5.2.2 Usability Engineering for Design of Usable and Safe Health IT 663
    38.5.3 Which Initiatives to Encourage Better Design and Development Methods Aiming at More
        Usable and Safer Health IT 665

38.5.3.1 Integration of Usability Criteria in the Certification of Health IT 665
38.5.3.2 Encourage and Verify the Application of User-Centered Process during the Design and
         Development of Health IT 666
38.5.3.3 Usability Standards to Guide the Design of Health IT.. 666
38.6 Discussion and Conclusion 667
References.. 668

38.1 Introduction

Health information technology (Health IT) is being increasingly deployed and is transforming our health care. Health IT includes applications such as electronic patient record (EPR) systems. They are currently being implemented worldwide in order to streamline and modernize health care. Other major health information systems being implemented include decision support systems, imaging systems, educational systems, and patient information systems. The move toward using Health IT is associated with a number of benefits. These include improving communication of health data to be used in decision making by patients, health professionals, and health administrators while at the same time improving the efficiency and effectiveness of health care resources and reducing the rate of occurrence of medical errors. For example, the electronic health record (EHR) is increasing the speed with which information can be made available for health care decision making electronically. The empirical literature suggests that the EHR and related decision support technologies will greatly improve the quality of clinical decision making through features and functions such as electronic alerts and reminders about patient health care needs (e.g., the need for immunizations and the presence of an allergy to a particular medication). As a result, research has demonstrated that Health IT can lead to cost savings as well as improved health care services for patients (Shortliffe & Cimino, 2006).

Health IT has profoundly modified the work environment of the health care professional. Moreover, the role of Health IT systems is to handle, transform, and transfer medical information and to provide varying types of decision support to health care professionals. Health IT systems have become important *actors* in the health care process. As such, in an ideal world, these applications should act as reliable *clinician partners and health care team players*. In addition to this, Health IT should be designed to be easy to learn and easy to use (i.e., highly usable), in order to be well accepted by all health care professionals and smoothly integrated in their work environment.

The design of such health care IT is certainly a complex endeavor and has met with success as well as considerable failure. Indeed, many Health IT implementations have failed to be delivered or meet user needs and expectations (Gorman & Berg, 2002; Koppel et al., 2008; Peute et al., 2010), and some Health IT systems have even been accused of facilitating or generating errors, therefore endangering patient safety (Han et al., 2005; Koppel et al., 2005b; Kushniruk et al., 2005). Issues related to human–computer interaction (HCI), human factors (HF), and usability of systems have been cited as major problems in the design of Health IT (Beuscart-Zephir et al., 2007; Graham et al., 2004; Khajouei & Jaspers, 2008, 2010; Zhang et al., 2003). In this chapter, we will present the complex process of Health IT design and development from an HF perspective and will discuss the essential ergonomics dimensions that need to be taken into account for the development of safe and effective Health IT applications.

The chapter starts with a presentation of the characteristics of Health IT systems and health care work situations (Section 38.2) and then lists the corresponding requirements and challenges of system design (Section 38.3). It then presents some typical examples of HF/usability problems that have been reported for Health IT applications and identifies the corresponding design flaws (Section 38.4). The last section (Section 38.5) presents the design process of Health IT and illustrates how to integrate HF tasks and methods in this design process, in line with national and international initiatives to ensure Health IT usability through certification and alike procedures (e.g., CE marking). The usability of Health IT is discussed as a critical factor for ensuring the appropriate and safe design of systems that will be adopted by end users. Usability engineering methods are detailed within the context of how they can be applied within the systems development life cycle (SDLC) of Health IT to produce safe and usable Health IT. The conclusion of the chapter summarizes the progress made to date and the remaining challenges to ensure proper consideration of HF in the design of safe Health IT. Throughout the chapter, two types of Health IT applications are used as illustrative examples, that is, medication-related computerized physician order entry (CPOE) and computerized decisions support systems (CDSS).

38.2 Characteristics of Health IT and Health Care Work Situations

38.2.1 What Is Health IT?

38.2.1.1 Wide Range of Applications and Systems

There is a wide and ever increasing range of Health IT as evidenced by the many types of health information systems that are currently in use. These include the EHR, which is meant to be a life-long electronic repository of a person's health information. The EHR is currently revolutionizing and modernizing health care practice by allowing for features such as electronic entry and retrieval of patient data, decision support, and linkage to other health care systems. The EHR incorporates information from other applications including the following: (1) electronic medical records (EMR), which exist in a physician's office; (2) EPRs, which exist in health care organizations such as hospitals; and (3) personal health records (PHRs), which contain health data accessed and managed by the patient by him/herself (Shortliffe & Cimino, 2006). There are also many other types of supporting and ancillary health information applications, such as imaging systems, identity-management systems, laboratory systems, pharmacy systems, medication systems, and hospital administration systems.

38.2.1.2 Implemented at Multiple Levels

To add to this complexity, health information systems and the information flow they support can be considered at multiple levels. For example, we can focus on health information systems at the level of primary care (i.e., in the health professional's individual office) or at the level of the hospital (which is often referred to as a hospital information system [HIS] accessed by multiple health professionals), at the regional level to address the issue of continuity of care (in linking primary care health data with hospital data), and even at the national level to take advantage of the availability of health data to support public health policy.

38.2.1.3 With Mandatory Interoperability

One of the major challenges in designing Health IT is that individual systems cannot typically be considered in isolation, since systems are increasingly being interconnected to become part of a complex whole. Challenges along these lines include the need for interoperability and structuration of data. Interoperability refers to the ability to transfer data across separate heterogeneous systems and databases, which may range from transfer of data across local areas to entire nations. Structuration of data (i.e., use of specific formats and standards for organizing health data) serves as a tool or method for supporting interoperability and exchange of data. For example, it is important for health care professionals to readily identify from the patient record essential patient information such as diagnoses (e.g., chronic or acute diagnoses), past history, treatments, and other patient information. If this information is not properly structured or coded to ensure consistent transfer and interpretation across health professionals and organizations, gaps in information, misinterpretation of data, and medical errors will result. In the next section, we will discuss the main characteristics of health care work situations and describe how a new paradigm in health care is emerging based on the presence of electronically available and increasingly structured health information.

38.2.2 Main Characteristics of Health Care Work Situations

38.2.2.1 Change in Paradigm in Health Care

In the last few decades, health care work situations have evolved profoundly, partly due (but not only) to the introduction and continuous evolution of new technologies. These evolutions have progressively led to a major change of paradigm in health care, moving the focus of health care from isolated patient management (and limited one-to-one lines of communication of information) to global integrated approaches, involving highly collaborative exchange of information, with many communication and information flow paths. For example, in Figure 38.1, we see that in the left-hand side of the figure, personalized, one-to-one intentional communication is shown between a doctor and another doctor (e.g., in the form of a referral to a specialist) as well as one-to-one communication between a doctor and a patient. Communications were mostly oral ones and supported by exchange of paper-based documents.

In the right-hand side of the figure, which illustrates one out of the many possible combinations of electronic and oral exchanges, we see that in the new paradigm, the doctor is now essentially documenting for and exchanging data with many other people for multiple purposes. In some cases, the clinician documenting the EHR does not even know how the information entered will be used and the future intentions of others who may now access the patient's information.

38.2.2.2 Collaborative Nature of Health Care Work

As a consequence of the aforementioned change in paradigm, but also due to the ever increasing

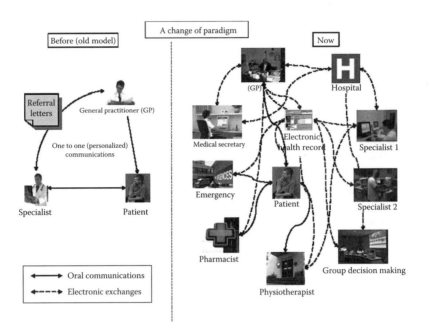

FIGURE 38.1
Schematic representation of the change of paradigm in information exchanges in health care. (Adapted from Hix, D. and Hartson, H.: *Developing User Interfaces: Ensuring Usability through Product & Process.* 1993. Copyright Wiley-VCH Verlag GmbH & Co. KGaA.)

specialization of health care professionals in complex specialty domains such as medicine, health care work is becoming more and more collaborative in nature. A great deal of health care professionals' activities comes down to medical, clinical, physiological, biological, and psycho-sociological information management, that is, information gathering, processing, selection, interpretation, documentation, and transmission. This information management in turn is the foundation for health professional decision making and for the planning and execution of health care processes. In health care professionals' daily activities, the necessary medical information is distributed across different technical artifacts (documents, computers, and medical devices [MDs]), in different people's minds (colleagues and patients), and in different locations, therefore characterizing health care work system as a distributed cognition situation (Hazlehurst et al., 2008; Horsky et al., 2003; Hutchins, 1995; Zhang & Norman, 1994). In this work environment, health care workers are constantly in the process of sharing patients' medical information or handing this information over to their colleagues, so that they can accomplish their part of the care process or take over a patient's care. Collaborative activities characterize the care process at all levels: primary care, hospital care, homecare, and concern all actors who are participating. The work system involves multiple players with various levels of expertise, all of them interacting with a wide variety of simple to complex technologies and IT applications. For example, a

physician may enter a medication order for a patient into CPOE system after consulting with the nurse who is knowledgeable about the patient's health status. The medication is then reviewed by the pharmacist and administered by a nurse using a medication administration system.

38.2.2.3 Heterogeneous Organizations

Health care systems vary widely across continents and countries, and as a result so does the organization of health care work. At the international level, there are differences in (1) the way physicians are employed by the health care system or by hospitals, (2) the distribution of health care between the private and public sectors, (3) physician/organizational differences in reimbursement for patient care by various insurance systems, (4) regulations that govern technologies, (5) the distribution of tasks between various health care professionals, and so on. On a national level, there is also a large variety of possible organizations in a health care system. For example, primary care may rely on general practitioners installed in single practice, or on small groups of physicians installed in collective practice and sharing office services (secretary), or on primary care clinics, etc. Finally, within an institution, for example, a hospital, an incredible number of different procedures and work routines may be identified in different medical departments, partly independent of type of medical specialty.

Even when similar technologies are installed such as barcode medication administration (BCMA) systems, the variety of workarounds observed (Koppel et al., 2008) are related to differences in organizations that may affect work. At all levels, this heterogeneity in the organization of work has a major impact upon the transmission and management of patients' medical information.

38.2.2.4 Uncertainty and Criticality

Health care primarily deals with sick patients, and sickness is a dynamic process. From a cognitive point of view, the patient represents a dynamic system or process whose variables are functions of time and whose physiological status inevitably evolves, partly in reaction to the treatment applied by the health care team (Hoc & Amalberti, 1995). Therefore, the care process and the corresponding work situation are dynamic, and the health care professionals' cognitive activities focus on the management and control of dynamic environments (Hoc & Amalberti, 1995; Xiao, 1994). All actors (i.e., health professionals) have to make decisions in an uncertain environment. In this dynamic environment, where the process evolves continuously, both on its own and under the impact of the health care worker's actions, uncertainty and criticality are essential dimensions of patients' medical information, and those essential dimensions vary across time.

In such a dynamic environment, each encounter involving the patient and other actors (i.e., health professionals) requires a display of relevant updated information, tagged according to its current criticality (in order to support and update the health professionals' current representation of the situation).

38.2.2.5 Interruptive Tasks and Time Constraints

Health care work is highly interruptive. The evolution of a patient's health status remains partly unpredictable. In hospital environments, where there are many patients who require attention at the same time, it is difficult for health professionals to plan all tasks and subtasks and it is even more difficult to carry them out according to a preset plan. Clinicians and especially nurses are constantly rescheduling and adjusting their work plans in response to patient needs. Interruptions in health care (Grundgeiger & Sanderson, 2009) have been repeatedly documented as disruptive to nurses' work in hospital settings (Brixey et al., 2008; Kalisch & Aebersold, 2010). High rates of interruption may contribute to errors and consequently hinder patient safety. Most clinicians use cues to remember to complete tasks that may have been interrupted (Grundgeiger & Sanderson, 2009).

Moreover, the fundamental collaborative nature of work creates tight dependencies between tasks and activities carried out by the different members of the health care team. Therefore, it is not possible to stretch tasks indefinitely or to take too much time to resume interrupted tasks because activities unperformed by a member of the team would rapidly block the execution of other activities by colleagues. As a consequence, time constraints and interruptions are tightly coupled in most health care work environments, and these constraints increase with the criticality of the patients' illness. Time constraints are more prevalent in ICU or emergency departments than in standard medical or surgical units and less important in geriatric long-stay units. These major characteristics of health care work greatly impact the collective management of medical information, and therefore constitute important HF and ergonomics challenges.

38.2.2.6 Safety Critical Process

Health care is now rightly considered as a safety critical dynamic process. As passengers may be harmed in the process of transportation by plane, train, or car, patients may be harmed by the care process they have to go through (Kohn et al., 1999). The care process has been interestingly compared to aviation (Amalberti & Hourlier, 2007) and is similar to process control activities in dynamic environments (Hoc & Amalberti, 1995; Xiao, 1994). The EHR is similar to a control panel displaying all the patient's physiological, medical, and clinical parameters, necessary to support collective as well as individual decision making about the treatment and the entire care process for a patient. Safety models and procedures are imported from the aviation domain and adapted to health care (Cacciabue & Vella, 2010). This parallel holds for the development of technologies designed to support aviation and health care work. In the aviation industry, engineers would not release a new control panel or dashboard for an airplane unless the design and development of this panel complied with user-centered design recommendations and was first evaluated (inspected and tested) for its usability, ergonomics, and safety characteristics. Why there has been very little regulation to date that has imposed this process for the development of health care technologies remains an unanswered question.

38.3 Resulting Requirements for Health IT Design and Development: Challenges and Recommendations

38.3.1 Support of Mandatory Interoperability

38.3.1.1 Challenge

Interoperability relies on structured data, and a huge amount of effort is being undertaken internationally to establish acceptable standards and codes. Unfortunately, for the time being, the requirement for structured medical data engenders difficult to solve HF problems. Structured and coded data certainly represent an ideal in the computerized world, but they estrange clinicians from the EHR. Clinicians reason in natural language and document their reasoning and orders in free text.

ILLUSTRATION WITH MEDICATION NAMES

Doctors and nurses already have to remember trade names (e.g., "Augmentin®"), generic names that depend on the country of use (e.g., British Approved Name (BAN) is "Co-amoxiclav" while U.S. Approved Name (USAN) and recommended

International Nonproprietary Name (rINN) is "amoxycillin–clavulanate") and active ingredients of branded drugs (e.g., "amoxicillin trihydrate + potassium clavulanate," which in the French Dénomination Commune Internationale (DCI) would be translated into "amoxicillin trihydratée exprimée en amoxicillin + clavulanate de potassium exprimé en acide clavulanique"). There is no way they can remember or use the corresponding international Anatomical Therapeutic Chemical (ATC) Classification System code J01CR02 in addition to this.

The same statement holds for the coding of diagnoses under the International Classification of Diseases–version 10 (ICD-10) coding system although most institutions make such coding mandatory for billing purposes. ICD-10 coding is always carried out as a post hoc, disliked activity that requires looking up corresponding codes for free text documented diagnoses.

38.3.1.2 Recommendations

In terms of usability, the main recommendation would then be to *help and provide support to health professionals entering, working with, and documenting structured and interoperable patient data.* Efficient solutions would require functions for (semi-)automatic coding. For instance, efficient drug databases incorporated into CPOEs could have all the drugs already coded in rINN + national recommended names + ATC codes in a way that allows physicians to work only with text-based prescriptions. Order entry is usually supported by pick lists recognizing the targeted drug from the initial letters that are typed in the display of the patient's current treatment and list of medications. For diagnoses, there are emerging tools for automatic recognition of "diagnosis" items in free text discharge or consultation letters that may propose ICD-10 codes for validation by the physician.

38.3.2 Make the System a Team Player

38.3.2.1 Challenge

Typically, the usual design of Health IT applications divides complex workflows and work processes into sequences of tasks and activities to be performed by specified roles/actors in a linear way (Ash et al., 2004; Wears & Berg, 2005). For example, most medication CPOE systems assume that the medication use process starts with a physician prescribing drugs, then continues with pharmacists controlling the prescription and delivering the meds to the medical unit, and then ends with the nurse controlling the preparation of the drugs, administering them to the patient and documenting this administration. But this "ideal" model of work is rarely observed in the field.

> **ILLUSTRATION WITH ORGANIZATION OF DOCTOR–NURSE COMMUNICATION**
>
> In many places, nurses collaborate with doctors in the medication decision-making phase. They discuss with the physician the patient's medical, clinical, and nursing status. Nurses may anticipate problems (that could occur on administration) and may negotiate with the physician about the patent's treatment. This cooperation may take place during common medical rounds or during doctor–nurse meetings or discussions (Pelayo & Beuscart-Zephir, 2010).

To our knowledge, there is no system able to support such collaborative decision making. Often, the collaborative aspects of the work disappear in the Health IT application (i.e., they become a simplified, linear, sequence of tasks). The implementation of such Health IT systems often has a negative impact on the collaborative aspects of health care work.

38.3.2.2 *Recommendations*

The high-level recommendation for the design of Health IT is simple: *make the system a team player*. The system should match and support collaborative decision-making processes and enhance the appropriate sharing of information among health care professionals. Implementing such high-level recommendations requires that collaborative processes are described and modeled in a way that makes them easy to understand by designers. It is within this context that the authors recommend the following:

- Make the decisions of each health professional transparent to others including the computer system, for example, DSS should include functions to track and provide information to other health providers on the team, whether the computer-provided recommendation has been responded to by the clinician.
- Consider collaborative design metaphors and technologies to support collaborative health care activities (e.g., group medical decision making) such as web-based meeting tools, and group decision support tools and approaches

from the field of computer supported collaborative work (CSCW) that would allow for several different team participants to collaborate in planning and arriving at a treatment decision.

- Design for flexible access to patient information that needs to be shared among a team of health professionals in a way that supports team care (e.g., do not "lock" or prevent certain users on the team from viewing key information they would need for effective collaborative work).
- Make sure you include all user groups when designing Health IT applications (this requires a preliminary HF-based analysis of the work system). Start design only after a complete user–task matrix has been worked out and checked for the exhaustiveness of the users listed in the table with scenarios.
- Take into account, when designing and developing health information systems, that the information in the system must be usable, accessible, and reused by multiple types of users and organizations, and at multiple levels (from a clinician's office, to a hospital, to a regional health authority or health system) for multiple different purposes. Organize the information, so that it is usable by different categories of users and different specialties.

38.3.3 Make the System a Clinician's Partner

38.3.3.1 *Challenge*

The usual motto in the medical and health informatics domain is that Health IT should deliver to clinicians relevant patient and medical information, when needed and where needed, tailored to support the procedure, task, or decision at hand. The fact is that the required information is often available "somewhere" in the information system because Health IT is good at storing huge amounts of medical information and at making it available in multiple places at any time. But it requires additional knowledge or intelligence to properly identify the context of use (user, time, task at hand, etc.) and to reconstruct adapted and usable (over)-views of the relevant information. Designers of Health IT indeed implement knowledge in the systems, usually in deep layers such as the data model, or in the way they design roles/actors/tasks and communication protocols. The problem is that the implemented knowledge may not match clinicians' actual work and thought processes (Wears & Berg, 2005). Resulting usability problems are very difficult to fix because they require intensive redesign and reengineering of the deep layers of the system, which is of course also costly.

ILLUSTRATION WITH CPOE DISPLAY OF PATIENT TREATMENTS AND CDSS FUNCTIONS

1. Most clinicians want to have the possibility of filtering or sorting the patient's list of medications by therapeutic indication. The physician's intention is rarely documented at the time of the prescription, but the sorting of medications by ATC code (or equivalent) may provide an acceptable substitute. This implies that ATC codes have been implemented in the data model describing the drugs in the system.

2. Most CPOE systems incorporate DSS functions that may suggest action when a dangerous situation is identified, for example, "prescription of vitamin K antagonist (VKA) requires monitoring of INR." It is not in favor of usability and acceptance of the system to deliver this alert to the physician while

   a. INR is already properly monitored and the results are in normal range.
   b. The INR test is already ordered for the upcoming hours.
   c. The clinician is about to open the lab tests order entry screen for INR test ordering.

(See also Miller et al., 2005, for similar concerns.) As a matter of fact, doctors resent that type of situation.

Then the challenge for the system is to incorporate a proper knowledge model (e.g., optimal frequency of monitoring) and a proper model of work supporting ordering of medications while at the same time identifying the laboratory of diagnosis tests taking into consideration the context of the patient's health management (e.g., relevant actions already carried out).

38.3.3.2 Recommendations

In order for *Health IT applications* to be *a clinician's partner*, designers and developers have to understand that usability- and HF-related issues are not restricted to the design of the HCI/graphic user interface (GUI). Models of work and models of knowledge are encapsulated in the *deep layers* of the applications. All design teams should incorporate (from the start) HF experts specializing in the health care domain along with health care professional representatives to ensure the accuracy of these models. At this early stage of the design process, a review of the literature about HF and usability problems and recommendations for the targeted type of application is mandatory.

38.3.4 Provide Safe and Easy-to-Use Parameterization/Customization Functions

38.3.4.1 Challenge

Health IT experts are well aware of the incredible diversity of work at differing levels of the health care organization and system. Most designers and vendors address this problem through parameterization or customization.

ILLUSTRATION WITH HOSPITAL ORGANIZATION OF PROCEDURES FOR DRUG DISPENSING

Most of the medication CPOE systems developed by European Companies are adapted to a common hospital organization that uses ward stocks of drugs. Differing dispensing procedures coexist in many hospitals depending on the toxicity and cost of the drugs. Dangerous or toxic drugs (e.g., morphine) would be delivered on a unit dose dispensing basis, serious or expensive drugs (e.g., quinolone antibiotics) would be delivered on nominative dispensing basis, and common supposedly not dangerous drugs (e.g., proton pump inhibitor and paracetamol) would be delivered on a "global dispensing" procedure (i.e., here nurses are in charge of "dispensing" and preparing the medication before administration). The list of drugs on unit dose, nominative or global dispensing, vary from hospital to hospital, and from region to region.

Most Health IT applications are delivered with a default configuration along with a set of complex advanced functions for parameterization and fine tuning to accommodate a variety of work systems. The usability problem here is that only experts or very advanced users are usually able to master and efficiently set these parameterization functions. As a consequence, many installations end up with maladjusted and unsafe default configurations and remain so for a long time before they are corrected. The flip side of this problem is that not all organizations that existed prior to the Health IT installation are safe and efficient. Installing health IT may also be the opportunity to fix poor organizational processes and approaches to work. Therefore, it is necessary to identify which features of the work system have to be maintained and which might benefit from a change when Health IT is implemented.

38.3.4.2 Recommendations

Health IT has to be designed for heterogeneous organizations and users, allowing for adaptation and customization to fit specific organizational requirements. Designers of health information systems should consult with HF and safety experts to determine those elements of the work system (including the computer system to be deployed) that should be designed to be flexible. The usability of parameterization functions is too often overlooked when designing systems. These functions should be properly evaluated for their usability and safety like all other more "visible" functions of the system.

38.3.5 Incorporate User-Centered Design Methods along with Iterative Safety-Oriented Usability Evaluations during the Design and Development Life Cycle

38.3.5.1 Challenge

Safety concerns are relatively recent in health care (Kohn et al., 1999) and even more so are Health IT-related safety issues (Koppel et al., 2005b; Kushniruk et al., 2005). In comparison to other safety critical domains like aviation, where the designers of new technologies are well aware of potential technology-induced errors, Health IT designers and vendors tend to minimize their responsibility for those types of errors (Koppel & Kreda, 2009). In the transportation domain, any new technology to be used by a pilot or a driver undergoes iterative, safety-oriented simulation and user tests. In health care, safety concerns are slowly penetrating the Health IT industry, with companies becoming more and more receptive to safety-related users' feedback. However, vendors remain quite reluctant to move toward user-centered methodologies and (iterative) formal usability evaluations and testing.

ILLUSTRATION WITH VENDOR PERSPECTIVES ON EHR USABILITY

The U.S. Agency for Healthcare Research and Quality (AHRQ) reported on vendor practices and perspectives regarding EHR usability (Agency for Healthcare Research and Quality, 2010). The conclusion section of the report emphasizes the importance of usability as a key feature of health IT products. "Vendors believe that [...] industry leaders will begin to differentiate themselves from the rest of the market based on usability" (p.10). But at the same time, vendors tend to reject (iterative) formal usability evaluation. "Formal usability assessments, such as task-centered user tests, heuristic evaluations, cognitive walkthroughs, and card sorts, are not commonly undertaken during the design and development process by the majority of vendors. Lack of time, personnel and budget resources were cited as reasons for this absence [...]. There was a common perception among the vendors that usability assessments are expensive and time consuming to implement during design and development phases."

These findings are perfectly in line with the authors' many years of experience in the field of HF and usability for Health IT, which involved repeated collaborations with differing companies and various products. In fact, most of the vendors see usability as a kind of key (magic) feature for gaining market share, but they do not consider it within the context of a proper usability engineering process because they have never applied it or they do not even know about it. The "usability" vendors talk about is not the same as the "usability" HF experts try to achieve.

In sum, designers, developers, and vendors of Health IT

- Are starting to become concerned about safety issues, but they usually doubt that their product per se could generate safety problems related to usability flaws.
- Want the best "usability" for their product but doubt that good usability directly results from the application of a proper usability engineering methodology during the design and development of systems.

38.3.5.2 Recommendations

Safety critical health IT tools should undergo systematic, safety-oriented simulation and user tests, following standard procedures and methods described in the HF domain (International Standards Organisation, 1999) that have already been adapted to MDs (International Electrotechnical Commission, 2008). Compliance with such evaluation/optimization procedures should be made visible to customers and purchasers, for example, through certification procedures.

38.3.6 Support Safety Critical Features of the Health Care System

38.3.6.1 Challenge

Health IT applications are typically not context aware. They are unable to detect and respond to time constraints or interruptions. They have no awareness of the uncertainties associated with differing types of medical data nor do they have the ability to understand the criticality of a patient's disease processes. In information systems, data may be well structured and organized, but they are fundamentally undifferentiated. Additional knowledge is required to characterize subsets of data as more important than others. This knowledge becomes rapidly complex, especially in the absence of local/national/international agreement on standards. When it comes to managing the importance of data over time, the required knowledge becomes even more complex.

ILLUSTRATION: DISPLAY OF PATIENT'S ANTECEDENTS AND HISTORY

It is probably not interesting for a specialist in respiratory diseases to know that her new 40 year old patient has undergone appendectomy surgery at the age of 10 years. But an early mention of an episode of asthma due to allergy at the same age might be relevant.

If the same patient was to consult an anesthesiologist before surgery, the early antecedent of appendectomy would be interesting if it was the only surgical antecedent, but it would be of little interest if there were other, more recent surgical operations.

Additionally, current systems are not good at representing the context of systems use, at identifying which user is at work or what the system is being used for. Again models of work are necessary to identify (1) the essential part of the initiated task that is missing and (2) the task that is interrupted (rather than abandoned or handed over to another health care professional). Likewise, systems are not sensitive to the presence of urgency or criticality of an emergent patient situation, while clinicians are very good at identifying an upcoming emergency.

38.3.6.2 Recommendations

It is unreasonable to think that the complex knowledge mentioned earlier will be soon available for implementation. In the meantime, we should trust the clinicians' expertise and therefore design flexible systems that clinicians can adapt to patient situations. This high-level recommendation could include the following:

- Time constraints:
  - Allow for override of lengthy sequences of tasks under certain types of emergency situations.
  - Adjust the system to the actual time available in the clinical workflow.
- Criticality:
  - Ensure a high level of quality assurance (correctness) for life-threatening situations.
  - Allow for highlighting of critical information (i.e., allow health professional users to highlight, tag, or annotate critical information they have entered into the system for a given period of time).
  - Manage (or support the management of) critical information over time appropriately.
- Uncertainty:
  - Incorporate support for risk assessment in the system.
  - Incorporate probabilities with data where applicable.
- Interruptions:
  - Design the system so that if the user is interrupted, she or he may easily return to the task (e.g., screen and context) she or he was working on before being interrupted; eventually display reminders for resuming the interrupted task.

38.4 Design Problems in Health IT

In the first part of this chapter, we have described the key features of Health IT and of health information systems from the HF point of view. We have also listed the corresponding HF and ergonomic requirements for the design of safe health information systems and applications. It is also interesting to look at the "lessons learned." A number of scientific papers have reported on usage problems with health IT. For example, poorly designed Health IT may cause (i.e., "induce" or "facilitate") users to make errors in entering health data, such as medications and their associated dosages (Koppel et al., 2005b; Kushniruk et al., 2005). In the case of severe safety issues (Han et al., 2005) resulting from system use, controversies may arise over the interpretation and understandings of the usage problems (as described in problem reports, that is, is the problem a design flaw? Does it result from faulty implementation? Was the system used in the wrong way?). It is difficult to describe the current "state of affairs" of health IT design from an HF and safety perspective (this topic could encompass an entire chapter on its own). For illustration purposes, we revisit one of the most famous papers reporting on usage problems and technology-induced errors with medication CPOE from an ergonomics and usability perspective. The paper entitled "Role of Computerized Physician Order Entry Systems in Facilitating Medication Errors" was published in *JAMA* in 2005 by Ross Koppel et al. Koppel and colleagues documented 18 usage problems with a CPOE that had been in use for 7 years in a major tertiary-care teaching hospital with 750 beds, most of these problems being frequent and having the potential to induce medication errors. This paper provoked quite a shock in the Health IT domain and has been repeatedly commented on (Bates, 2005; Horsky et al., 2005; Koppel et al., 2005a; Nemeth & Cook, 2005). One very interesting comment was posted by Nielsen (2005) in the Alertbox of his Web site "useit. com". Nielsen reinterpreted six of Koppel's observations in terms of usability problems and design flaws. In Table 38.1, we extend this short analysis to all the usage problems documented by Koppel et al. The aim is to identify whether the design (and design process) is part of the problem, and what kind of design problem an HF expert might identify as well as what solutions could fix or prevent future problems in similar applications. The same exercise can be repeated with most of the papers reporting technology-induced errors in Health IT.

In the Koppel et al.'s paper, about 12 out of the 16 problems mentioned may be considered as usability problems, which in turn are symptoms of design flaws. These problems would have been prevented by the application of a proper user-centered process and usability evaluations in the form of usability inspections and usability tests. This belief is shared by other authors who have commented on the same paper: "The authors (*Jan Horsky, Jiaje Zhang, and Vimla Patel*) are conveying their belief that it is the flawed design and poor integration with clinical work rather than the technology itself that is at the root of its suboptimal performance" (Horsky et al., 2005).

38.5 Integrating HF Methods and Models in the Design and Development Life Cycle of Health IT

38.5.1 Design of Health IT as an Iterative Process

The design of Health IT can be conceived of as a process that consists of multiple interrelated and interleaved activities. Health IT design is a dynamic activity and requires an integration of ideas and concepts from several phases, beginning with initial system design concepts and continuing with the overall design of system components and the design of specific system features (e.g., computer screens and user interfaces). An organizing framework from which we can consider Health IT design is that of the System Development Life Cycle—the SDLC, which provides guidance and structure for helping in the design and development of information systems. The "traditional" SDLC consists of a set of stages in system design and development. The first stage is project planning, which is concerned with the initial definition of the problem needing to be solved by new Health IT. After the project planning stage, the traditional SDLC involves a phase for collecting and refining the requirements for new Health IT (i.e., defining "what" the technology should be able to do, and what the usability goals are). The requirements phase is followed by the design phase, where the design is worked out (i.e., "how" the system will work), including the design of the main system components, user interfaces, and network interconnections. After the design is complete, the implementation phase follows, where the design is translated into working computer software (including computer programming and initial system testing and installation). Once the implementation phase is complete, the system is now deployed in a real-world work context (e.g., a hospital or doctor's office), which is referred to as the maintenance phase. The process is cyclic in that in the traditional model after the system is released there may need to be changes made and new systems created to take into account changes in the environment and emerging user needs. In the design of Health IT, considerable flexibility is needed to include feedback and input to design from end users and HF experts throughout the entire SDLC. Furthermore, in thinking about how to apply the recommendations discussed in this chapter, we must consider the type of design process that has been adopted and the stage at which an application is at in its SDLC (McConnell, 1996). It has been noted by many that health care is an extremely complex domain and that this must be taken into account in the design of Health IT. In particular, adoption of a traditional "waterfall" approach to the SDLC, where the stages of the SDLC are fixed and take place in a rigid order (with little allowance of reworking previous stages), has been criticized as being too inflexible for use in the design of many health care applications (Kushniruk, 2002). In the domain of health care, it is very difficult to define "once-and-for all" (at the beginning of design and development phases) all the user requirements, which makes the "traditional approach" to the SDLC problematic in Health IT. In applying the "traditional approach," user requirements are obtained at the "front end" of the SDLC, and user feedback and input do not come into play until far later in the SDLC (at the stage of implementation when it may be too late or costly to make significant changes to

TABLE 38.1

List of Usage Problems with a Medication CPOE

Problems Described (Original Labels from Koppel et al.)	Consequence (from Koppel et al.)	Reinterpretation: Usability/Design Problem	Possible HF-Based Solution and Corresponding HF-Required Tasks
PB 1: "Assumed Dose Information" "The dosages listed in the CPOE display are based on the pharmacy's warehousing and purchasing decisions, not clinical guidelines"	Dosing errors for unusual prescriptions	(Nielsen) Violation of common usability heuristics (Nielsen, 1993) Misleading default values	(N) "Each screen should list the typical prescription as a guidance" HF tasks = cognitive engineering tasks: • Cognitive analysis of clinicians decision making • Elicitation and incorporation of proper pharmaceutical/ clinical knowledge in the system
PB 2: "Medication Discontinuation Failures" Medication canceling ambiguities: Ordering new or modifying existing medications is not differentiated from canceling ("discontinuing") an existing medication	Duplicative medication (doubloon) or conflicting medication	(N) Violation of common usability heuristics: New commands not checked against previous ones	(N) "If users are doing something they have already done, the system should ask whether both operations should remain in effect or whether the new command should overrule the old one" HF task: usability inspection (a usability inspection would have caught this violation of a usability heuristic)
PB 3: "Procedure-Linked Medication Discontinuation Faults" When procedures accompanied by medications are canceled, no software link automatically cancels medications	Unintended administration of useless medication	(Authors) Lack of protocols or order sets integrating procedures with accompanying medication (A) Root cause: The model of work incorporated in the system is inadequate	In case of cancelation of a procedure or test involving the administration of medication, the system should ask the user whether the corresponding drugs have to be canceled too HF tasks: detailed analysis and modeling of workflows and dependencies between clinical activities Usability inspections and tests
PB 4: "Immediate Orders and Give-as-Needed" "Medication Discontinuation Faults" NOW (immediate) and PRN (give as needed) orders may not enter the usual medication schedule and are seldom discussed at handoffs	These orders may not be charted or canceled as directed	(A) The system is too rigid (inflexible) and does not meet users' needs (A) Violation of common usability heuristics: compatibility with users' activities (Bastien & Scapin, 1993; Scapin & Bastien, 1997) (A) The model of work incorporated in the system is inadequate	The HF-based analysis of the work system would have identified these needs for specific orders and recommended that corresponding functions be developed, that is, the system should propose NOW and PRN (and eventually OTHER) medication schedules HF tasks: detailed analysis and modeling of clinical and ordering activities Usability inspections and tests
PB 5: "Antibiotic Renewal Failure" Unintended delays in the reapproval procedure: Doctors forget to ask for continuation of approval every 3 days, nurses remind them on paper stickers, and doctors enter their orders on computers	Gaps in antibiotics therapy	(A) Insufficient or weak customization/ parameterization (A) Violation of common usability heuristics: compatibility with users' activities (A) The model of work incorporated in the system is inadequate (A) Disruption of doctor–nurse communication	Parameterization/customization: The system should automatically propose reapproval continuation every 3 days to accommodate local regulations The system should authorize the nurses to post comments for the doctors on the CPOE HF tasks: detailed analysis and modeling of clinical/ordering activities and collaboration processes Usability inspections and tests
PB 6: "Diluent Options and Errors" House staff are required to specify diluents for administering antibiotics but they are unaware of impermissible combinations	Prescription errors	(A) Not necessarily a usability/design problem: a new local regulation is imposed through the CPOE, which is not in accordance with current staff knowledge	

Problem	Consequence	Classification	Recommendations
PB 7: "Allergy Information Delay" Mixed problems: • Inadequate (post hoc) timing of alert • Scrolling problems • Low quality of allergy documentation, etc.	Shift of responsibility of checking allergy information to pharmacists	(A) Violation of common usability heuristics: memory overload (A) The model of work incorporated in the system is inadequate (A) The low quality of allergy documentation is a well-known difficulty (Kuperman et al., 2007) and is not a design problem	Support a good quality documentation of allergies by proposing ad hoc forms to clinicians Comply with usability heuristics during the design process HF tasks: detailed analysis and modeling of clinical activities; iterative usability inspections
PB 8: "Conflicting or Duplicative Medications" The CPOE system does not display information available on other hospital systems	Necessary control of house staff orders by pharmacists to prevent errors	(A) Not necessarily a usability/design problem: lack of integration of different IT systems	
PB 9: "Human Machine Interface Flaws" Small fonts, patient name not on all screen pages, different colors for the same information in different pages, etc.	Wrong patient selection	(N) Violation of common usability heuristics: • Poor readability • Lack of consistency	Comply with usability heuristics during the design process HF tasks: iterative usability inspections
PB 10: "Wrong Medication Selection" Information on the patient's medication is scattered over many pages	Wrong medication selection and ordering	(N) Violation of common usability heuristics: memory overload (A) Lack of overview of the patient treatment (A) Wrong model of the cognitive activities for medication ordering;	Allow shifting rapidly between different views of the patient's treatment depending on the step of the medication ordering (overview/detailed view) (Pelayo et al., 2005) Comply with usability heuristics, for example, "minimize users' memory load" HF tasks: detailed analysis and modeling of clinicians cognitive activities during the medication prescription process Usability inspection and usability tests
PB 11: "Unclear Log On/Log Off" Physicians can order medications at computer terminals not yet "logged out" by the previous physician	Medication errors (wrong medication or wrong patient)	(A) Not necessarily a usability/design problem: unsafe security local regulations/procedures with computers	
PB 12: "Failure to Provide Medications after Surgery" Difficulties to "activate" renewed orders when they have been suspended for surgery	Unintended interruptions or delays	(A) Violation of common usability heuristics: compatibility with users' activities (A) The model of work incorporated in the system is inadequate	Most CPOE have usability problems with "suspended" drugs (Beuscart-Zephir et al., 2010): these functions have to be carefully designed following a user-centered process. For example, the system should support the documentation of the duration of the suspension and then automatically ask the clinician whether she or he wants to resume the medication at the end of the suspension time HF tasks: detailed analysis and modeling of clinical workflows, specific recommendations for the design of functions dealing with modifications of existing medication orders Usability inspection and usability tests
PB 13: "Post Surgery Suspended Medication" Problems related to the fact that actual patient location has not been properly documented (e.g., not logged out of post anesthesia care)	Unintended interruptions or delays	(A) Not necessarily a usability/design problem: organizational problems with documentation of patients' location	

(continued)

TABLE 38.1 (continued)
List of Usage Problems with a Medication CPOE

Problems Described (Original Labels from Koppel et al.)	Consequence (from Koppel et al.)	Reinterpretation: Usability/Design Problem	Possible HF-Based Solution and Corresponding HF-Required Tasks
PB 15: "Sending Medication to Wrong Rooms When the Computer System Has Shut Down" When the system is down, patients' moves are not documented/tracked and pharmacy may send medications to wrong rooms/patients	Increased risk of wrong patient/wrong medication	(A) Not necessarily a usability/design problem: technical/management problems with maintenance and backup procedures	
PB 16: "Late in Day Orders Lost for 24 hours" Clinicians can enter orders late in day for "tomorrow—7a.m." Orders are processed only the next day, and drugs sent and administered the day after	Delay in medication administration	(N) Violation of common usability heuristics: "Date description errors"	(N) The interface should not allow prescription for "tomorrow" Violation of usability rule HF tasks: Usability inspection and usability tests
PB 17: "Inflexible Ordering Screens, Incorrect Medications" Because of CPOE inflexibility, nonstandard orders or specifications are often impossible to enter *(problem similar to problem 4)*	Medication errors due to uncertain canceling an reordering	(A) The system is too rigid (inflexible) and does not meet users' needs (A) Violation of common usability heuristics: compatibility with users' activities (A) The model of work incorporated in the system is inadequate	(A) The HF-based analysis of the work system would have identified these needs for specific orders and recommended that corresponding functions be developed HF tasks: detailed analysis and modeling of ordering activities, specific recommendations for the design of functions dealing with specific medication orders Usability inspection and usability tests
PB 18: "Role of Charting Difficulties in Inaccurate and Delayed Medication Administration" "To chart drug administrations, nurses must stop administering medications, find a terminal, log on, locate that patient's record, and individually enter each medication's administration time. If medications are not administered (e.g., patient was out of the room), nurses must scroll through several additional screens to record the reason(s) for nonadministration"	Delayed medication administration	(N) Overly complicated workflow (A) The model of work incorporated in the system is inadequate	(N) "Whenever you see users resorting to sticky notes or other paper-based workarounds, you know you have a failed user interface" HF tasks: set clear usability goals based on analysis and modeling of nurses activities Usability inspection and usability tests

Source: According to Koppel, R. et al., *JAMA*, 293, 10, 1197–1203, 2005.

Note: Columns 1 and 2 list the usage problems described by Koppel et al. along with their consequences in terms of risk of medication errors. Column 3 proposes a reinterpretation of the problem in terms of HF and usability expertise (source = Nielsen's comments, indicated with a (N) or authors comments, indicated by a (A)). Column 4 provides HF/usability-based possible solutions and indicates which HF/usability tasks would have permitted preventing the problem should they have been executed. Nielsen's or authors' solutions are again identified by a (N) or a (A).

the system). In contrast, newer design methodologies, such as rapid prototyping (McConnell, 1996), involving continual input from users in terms of feedback and testing of systems (right through requirements gathering to evaluation of design prototypes), are recommended for design of complex and highly interactive systems such as Health IT (Kushniruk, 2002).

38.5.2 Incorporating HF Tasks in the Design Process

38.5.2.1 Usability and Health IT

According to the International Standards Organization (ISO), "usability" can be defined as "the effectiveness, efficiency, and satisfaction with which a specified set of users can achieve a specified set of tasks in a particular environment" (International Standards Organisation, 1998). Here, effectiveness refers to the accuracy and completeness with which users achieve specific goals. Efficiency refers to the resources expended in relation to effectiveness. Satisfaction refers to the comfort and acceptability of use as subjectively experienced by the end users.

Along similar lines, according to Nielsen (1993), usability can be defined as a measure of ease of use of system in terms of the following dimensions: learnability, efficiency, memorability, errors, and satisfaction. Learnability refers to how easy the system is to learn how to use and master. Efficiency should lead to users being able to develop a high level of productivity with the system. Memorability refers to how easy it is to remember the system, so that if the user returns to the system at a later date they remember how to use it. Errors refer to the need for systems to have low error rates (and are therefore related to system safety). Finally, satisfaction refers to the need for the system to be pleasant to user (others have also indicated that the related concept of "enjoyability" is a major aspect of usability [Preece et al., 1994]).

In addition, with Health IT in particular, we need to distinguish between "ease of use" and "usefulness." A system may be easy to use (i.e., be easy to navigate, to learn how to use) but if the information presented by the system to the user (e.g., presentation of clinical guidelines to a doctor or nurse) is old or out of date, the system will not be "useful" and would (and should) not be adopted by users. On the other hand, Health IT that may be considered very useful in terms of either its functions or content, may be so difficult to use, that it would be abandoned by all potential users quickly.

Usability is critical in Health IT as there are many examples of violations of the aforementioned usability principles. For example, if a system is not designed with learnability in mind, and the resultant system is known to be difficult to learn, this may become a negative factor

in the marketing of the Health IT product. Furthermore, in hospitals with large turnaround (i.e., many staff leaving and others coming in), the issue of repeatedly training new staff on difficult-to-learn systems becomes a major, costly, and difficult problem for the organization. With regard to efficiency, systems that require too many steps to achieve a goal (e.g., too many steps to enter a medication) will lead to inefficiencies and problems in completing tasks within a reasonable time. Memorability has become an issue as health professionals are now expected to learn and use a variety of different systems, therefore requiring their operations to be memorable. There are now many documented examples of design of systems that have led to safety issues and have either directly or indirectly led to medical errors (e.g., see Table 38.1). Finally, systems that do not lead to user satisfaction will be difficult to implement and adopt.

38.5.2.2 Usability Engineering for Design of Usable and Safe Health IT

It is essential to test and evaluate emerging Health IT systems throughout the SDLC repeatedly in order to ensure the design will lead to a system that will meet user needs and usability goals. In testing the emerging design of the Health IT system (and also for the evaluation of more completed versions of the system to ensure user needs are met), application of methods from usability engineering is essential. Usability engineering methods are scientific approaches that can be used to assess the usability of an IT design or product, and can be applied throughout the SDLC. The two major types of usability evaluation methods that have been employed are usability inspection methods and usability testing methods (Nielsen, 1993).

Usability inspection involves several analysts systematically "stepping through" or "inspecting" a user interface or system to identify potential usability problems and issues. A popular form of usability inspection is known as heuristic evaluation. Heuristic evaluation involves one or more analysts systematically examining a user interface to determine if there are any violations of good principles (i.e., "heuristics") of design and HCI. A widely used set of heuristics defined by Nielsen (1993) are the following: (1) "visibility of system status"—for example, ensure that the status of the system is clearly indicated to the user; (2) "match of the system to the real world"—for example, in the user interface use terminology that is part of the users' natural vocabulary; (3) "user control and freedom"—for example, allow users to be able to redo or undo operations where appropriate; (4) "consistency and standards"—for example, provide consistent user interface layout and operations to simplify learning and use; (5) "error prevention"—for example, design the interface to prevent user errors;

(6) "minimize memory load"—for example, limit the amount of information in a menu; (7) "flexibility and efficiency of use"—for example, allow for "shortcuts" for experienced users; (8) "aesthetic and minimalist design"—for example, do not clutter the screen; (9) "help users recognize, diagnose, and recover from errors"—for example, provide informative error messages that will help the user recover from problems; and (10) "availability of help and documentation"—for example, provide effective online help on system use. Results from heuristic evaluation include reports of frequency of violation of these heuristics. In addition, each violation of a heuristic can be rated as to its potential impact (i.e., severity) on a scale from 0 to 4—where 0 indicates no problem, 1 indicates a cosmetic problem, 2 indicates a minor problem, 3 indicates a major problem, and 4 indicates a "usability catastrophe," which cannot be left unaddressed. Usability inspection methods such as heuristic evaluation have been successfully used to improve the usability of a wide range of Health IT, including EHR systems, decision support systems, as well as devices such as infusion pumps.

In contrast to usability inspection that involves a trained analyst examining a system or interface, usability testing involves the observation (typically video recording) of representative users of a system carrying out representative tasks while interacting with the system. For example, a usability test of a new prototype decision support tool might involve video recording users (e.g., physicians) and computer screens as they carry out decision-making tasks using the electronic tool (Kushniruk & Patel, 2004). In many studies, this has involved having subjects (e.g., physicians or nurses) "think aloud" or verbalize their thoughts as they use the system under study. Using methods from cognitive psychology for analyzing verbal reports from subjects, such as protocol analysis (Ericsson & Simon, 1984), in conjunction with quantitative and qualitative approaches to analyzing video recordings of users interacting with systems, major user problems can then be identified, coded, and classified (e.g., number of usability problems within specific parts of a system, i.e., problems encountered by users in navigating through a user interface, problems resulting from lack of consistency in the design of the interface) and communicated to system designers. In Health IT, a wide range of evaluations involving usability testing have been reported, including analysis of EHR systems, CPOE, CDSS, clinical guidelines, and patient clinical information systems.

Usability engineering methods have been used throughout the SDLC of a variety of different types of Health IT, and experience has shown that the earlier such evaluation is carried out to detect problems (and user information needs and preferences), the more

impact such studies can potentially have on improving system usability. Indeed, user-centered approaches (i.e., where improved understanding of user information needs, processing requirements, and cognitive limitations can lead to improved system design and implementation) have been applied to examine not only completed Health IT but can also be used for evaluating early emerging design possibilities, including examination of Health IT "design metaphors," which can be defined as organizing frameworks for helping users conceptualize the organization and operation of a system (e.g., the well-known "desktop" metaphor of windows-based interfaces, where objects on the screen are arranged and organized as desktops with file folders, calendars, etc.). The design metaphor adopted during Health IT design has been shown to have an important impact on user satisfaction and adoption of systems, in particular for EHR systems, where a standardized user interface design approach has not yet been achieved (despite some major efforts along these lines including the United Kingdom's National Health Service (NHS) common user interface (CUI) project (National Health Service, 2010), which has led to user interface design specifications for EHR systems).

To illustrate how evaluation (using methods such as those emerging from usability engineering) is central to design of complex and highly interactive IT systems, the star model was developed (Hix & Hartson, 1993). The model can be used to drive the integration of evaluation within design processes by considering where user-centered evaluation can take place within the SDLC. As can be seen from Figure 38.2, evaluation is seen as being central in the software development life cycle for information systems and should be applied at each major phase in the process of developing IT, including during requirements gathering, prototyping, and implementation. Thus, evaluation involving user input is seen as being essential at all stages from early design through to implementation of Health IT.

While user-centered design focuses on the early involvement of users, iterative design processes, and continual user evaluation of emerging systems (Preece et al., 1994), participatory design takes the role of the user further by including users in the design itself, as participants, or "consultants" in the process itself. Pioneered in Scandinavia, participatory design can be considered a form of sociotechnical design whereby explicit consideration is given to both the social and technical requirements of information system during design. In health care, sociotechnical approaches to Health IT design have emerged as an important trend in order to ensure systems developed ultimately meet not only individual information processing needs, but also fit within the complex social and organizational milieu of health care organizations (Aarts & Nøhr, 2010).

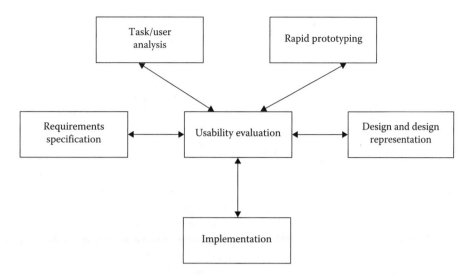

FIGURE 38.2
The star life cycle model. (Adapted from Hix, D. and Hartson, H.: *Developing User Interfaces: Ensuring Usability through Product & Process.* 1993. Copyright Wiley-VCH Verlag GmbH & Co. KGaA.)

38.5.3 Which Initiatives to Encourage Better Design and Development Methods Aiming at More Usable and Safer Health IT

In Health IT, usability is closely intertwined with safety as a number of human errors compromising patient safety occur because of a product's design and usability flaws (see Section 38.4). Moreover, usability is seen by vendors, purchasers, and users as one of the most desirable features of a Health IT application (Agency for Healthcare Research and Quality, 2010). At the same time, many do not acknowledge the relationship between a high level of usability and safety in the final product and the importance of implementing user-centered and usability engineering methods across the design and development life cycle of a product (Kushniruk, 2002). As a consequence, health authorities and administrators at various levels (e.g., National level, European level) have undertaken initiatives that

- Verify the final usability of Health IT products reaching the market (e.g., certification)
- Encourage the application of a complete user-centered process for design of Health IT products (see example of CE marking for MDs) (International Electrotechnical Commission, 2008)
- Propose that usability standards be used to guide the design of safety critical applications, such as e-prescribing systems (e.g., NHS recommendations) (NHS National Patient Safety Agency National Reporting and Learning Service, 2010)

38.5.3.1 Integration of Usability Criteria in the Certification of Health IT

In the past 5 years, the United States has led the way in the certification of Health IT, especially the EHR. The Certification Commission for Health IT (CCHIT) has been given a prominent role in developing scenarios and procedures for certification, and a number of EHRs have been certified in this manner. At the end of 2009, CCHIT integrated usability certification for ambulatory EHRs (Certification Commission for Health Information Technology, 2010). EHR certification applicants are required to participate, but the usability rating does not affect the certification outcome (i.e., vendors, should they wish, are free to disclose the results to their customers). However, a close examination of this document raises some concerns and questions. The CCHIT certification of usability relies on a "quick and dirty" procedure (no more than 40 min and no formal usability tests). While the vendor executes the certification clinical scenarios, a panel of jurors performs a rapid inspection of the application and then rates the usability of the system on three different questionnaires. It is not clear whether the jurors are HF experts; two questionnaires (i.e., After Scenario Questionnaire [ASQ] and System Usability Scale [SUS]) seem to call for end-users rating while the third one (i.e., Perceived Usability [PERUSE]) seems to call for HF experts' judgments. There is no actual usability inspection as the jurors do not have the opportunity to use the application themselves. The CCHIT usability certification is better than no attempt to assess usability, but it is still not sufficiently rigorous nor is it satisfactory from an HF perspective.

Additionally, U.S. EHR certification processes have gone through turmoil in 2010. The Office of the National

Coordinator for Health Information Technology (ONC) issued a final ruling (U.S. Department of Health and Human Services, 2010) establishing a temporary certification program to test and certify Health IT. This document does not mention usability. As well, the final ruling does not designate CCHIT as a temporary certification entity, meaning that CCHIT will need to apply to become an ONC Authorized Testing and Certification Body (ONC-ATCB). In this context, the future of usability certification is uncertain. The May 2010 AHRQ publication on EHR usability calls for usability certification programs to be carefully designed and validated, acknowledging the complexity of usability certification for Health IT.

38.5.3.2 Encourage and Verify the Application of User-Centered Process during the Design and Development of Health IT

While national and international regulation over Health IT usability remains yet to be determined, the example of new European regulation for MDs usability is of interest. Since March 2010, a usability file is mandatory to get CE marking for all new MDs or new versions of existing MDs. CE marking is mandatory for MD vendors to be authorized to sell their product in the European Economic Area. Vendors may rely on two (similar) international standards to establish their usability file: "Medical Electric Equipment—Part 1–6: General Requirements for Safety—Collateral Standard: Usability" (EN 60601-1-6) and the most recent one, which is more developed: "Medical Devices: Application of Usability Engineering to Medical Devices" (EN 62366). The main objective of the EN 62366 standard is to address "use error caused by inadequate usability," which in turn is due to the fact that the product was "developed without applying usability engineering process." IEC 62366 claims that "the usability engineering process is intended to achieve reasonable usability, which in turn is intended to minimize use errors and use-associated risks." The standard describes how to apply the usability engineering process to the design and development of MDs and explains the relationship between the usability engineering process and the risk management process (ISO standard 14971) (International Standards Organisation, 1998) that also applies to CE marking. The standard encourages the application and documentation of a user-centered design (ISO standard 13407) of MDs, including iterative formal usability assessments, aiming at eliminating (as far as possible) all usability-related risks. If some risks related to difficult usability problems remain at the end of the process, these risks have to be documented in the risk management process, as illustrated in Figure 38.3.

38.5.3.3 Usability Standards to Guide the Design of Health IT

One of the biggest issues in design of usable health care IT, in particular the design of user interfaces for EHRs and related Health IT, is the current lack of a common or de facto standard for user interface design, layout, and operation. As a consequence, users of one system (e.g., physicians or nurses) may need to learn the details of user interfaces of each system they encounter in their daily practice (this is particularly problematic if the health professional works at multiple institutions, each with their own separate and different Health IT). This has important implications both for the time taken to learn how to use a system (with each system requiring its own learning curve) and for the potential for systems to lead to medication and other errors due to user lack of knowledge of details of systems operations and user interface idiosyncrasies. Indeed, user interface consistency has been pointed out by Nielsen (1993) as being a key to usability. To address these issues, projects have appeared with the goal of developing usability standards to guide the design of Health IT. One of the largest projects along these lines is the United Kingdom's CUI design guidelines and toolkit. The guidelines and toolkit attempt to provide improved patient safety by setting guidelines for a common look and feel for systems throughout the National Health System in the United Kingdom (these guidelines are also available internationally [see http://www.mscui.net]). This work, conducted in conjunction with Microsoft Corporation, has led to a "Clinical Documentation Solution Accelerator," which allows clinicians and system designers to create documents such as discharge summaries. In addition, a series of guidelines for the safe on-screen display of medication information has also appeared from this work, with specific recommendation about displaying drug names, text and symbols, numbers and units of measure as well as general information (NHS National Patient Safety Agency National Reporting and Learning Service, 2010). For example, regarding the issue of potential user confusion caused by inconsistent use of abbreviations, the recommendations in the NHS document suggest to avoid abbreviating medication routes, sites, and frequencies (except for standard units of measure). As another example is based on research that found that omitting leading zeros in front of values (e.g., volume .6 mL) introduces a high possibility of misreading errors if the decimal point in front of the number is not noticed by the user—hence, the recommendation for this issue is to use leading zeros when a decimal point appears. In general, future advances in improving Health IT will depend on development of such evidence-based heuristics or guidelines. In the area of developing web-based user interfaces, a set of general

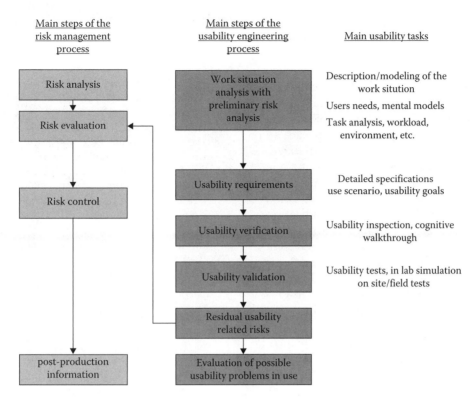

Main steps of the risk management process

Main steps of the usability engineering process

Main usability tasks

Risk analysis

Risk evaluation

Risk control

post-production information

Work situation analysis with preliminary risk analysis

Description/modeling of the work sitution

Users needs, mental models

Task analysis, workload, environment, etc.

Usability requirements

Detailed specifications use scenario, usability goals

Usability verification

Usability inspection, cognitive walkthrough

Usability validation

Usability tests, in lab simulation on site/field tests

Residual usability related risks

Evaluation of possible usability problems in use

FIGURE 38.3
Relationship between the usability engineering process and the risk management process for the design of MDs.

guidelines, based on published literature, have been developed by the Department of Health and Human Resources in the United States (see http://www.usability.gov/guidelines/).

38.6 Discussion and Conclusion

In this chapter, we have described the complexity of health care work and the corresponding difficulties and challenges in designing safe and usable Health IT. Indeed, Health IT has already proven useful and beneficial for the health care system and for the care process (Shortliffe & Cimino, 2006), but there is certainly room for improvement in terms of usability and safety for a number of applications intertwined with complex clinical work. Such significant improvement in the usability and safety of Health IT will be obtained through better design (or redesign) of the products, and more specifically through better integration of usability engineering tasks in the design life cycle. In the last 10 years, significant progress has been achieved. All stakeholders in the Health IT domain are now aware of the possibility of technology-induced errors and of the importance of the "usability" dimension, which is now considered

as a very desirable feature of all Health IT systems. In conclusion, we need to move forward and to support initiatives to speed up this improvement process in the following ways:

1. We need to improve our knowledge of the "state of affairs" of user difficulties and problems in Health IT. Review papers and large studies collecting all known usability problems with a given type of Health IT application are extremely useful but are too rare. Reporting systems for medical errors and incidents are implemented in a number of countries (e.g., in Denmark, in United Kingdom), and this is a promising trend. In a similar manner, development of specific, research oriented, repositories for technology-related problems would help us to better understand and prevent dangerous usage problems with Health IT.

2. We also need to increase our efforts to develop usability guidelines and standards for Health IT. In this respect, initiatives such as the NHS design publications are promising. It would be interesting to determine whether their guidelines for safe on-screen display of medication information (NHS National Patient Safety

Agency National Reporting and Learning Service, 2010) are acceptable and usable in other countries and at an international level. Moreover, design guidelines should also address the deep layers of the system that are essential to achieving a good usability at the user interface level. For example, if the data model describing drugs and dosages in the application is incomplete or ill organized, then it may be impossible to implement some essential recommendations for the display of medication information on the GUI. Research toward a common data model for medication CPOE and CDSS is well on its way in Europe (Chazard et al., 2009) (see also the European Project epSoS at [http://www.epsos.eu/] for information on "possibilities of reaching an agreement on a common structure of e-Prescriptions, and thus about the minimum and maximum common data sets to be shared between Member States"). It would be helpful to check the compatibility of these emerging recommendations for a common data model with usability-oriented recommendations such as those published under the CUI project.

3. To date, in the Health IT industry, there has been limited investment in usability and the application of proper user-centered design methods. As a consequence, there are few "success stories" based on efforts around usability. Contrary to other domains (e.g., telecommunications, computer games, and transportation), which invest heavily in HF methods and usability, the Health IT industry has limited awareness of the rewards of good usability and there is little knowledge on how to achieve it. National and international health authorities should encourage the Health IT industry to adopt design life cycle strategies compliant with the user-centered methodology (e.g., International Standard ISO 13407). Efforts have been underway to adapt the ISO 13407 for the design of MDs and to propose it as a contributing criterion for the CE marking of those MDs. Similar efforts could be initiated to adapt the user-centered methodology to the design of Health IT and eventually to include it in the certification process of these applications.

All these points require some coordination at the international level, which is not easy to achieve. Hopefully, research in HF and ergonomics aimed at achieving safe and usable Health IT will lead the way toward such international cooperation.

References

Agency for Healthcare Research and Quality. 2010. Electronic health record usability—Vendor practices and perspectives (Rep. No. 09(10)-0091-3-EF). Maryland.

Amalberti, R. and Hourlier, S. 2007. Human error reduction strategies in health care. In *Handbook of Human Factors and Ergonomics in Health Care and Patient Safety*, ed. P. Carayon, pp. 561–579. Hillsdale, NJ: Lawrence Erlbaum Associates.

Ash, J. S., Berg, M., and Coiera, E. 2004. Some unintended consequences of information technology in health care: The nature of patient care information system-related errors. *J. Am. Med. Inform. Assoc.,* 11: 104–112.

Bastien, C. and Scapin, D. L. 1993. Ergonomic criteria for the evaluation of human computer interface (Rep. No. 156). Roquencourt, France: INRIA.

Bates, D. W. 2005. Computerized physician order entry and medication errors: Finding a balance. *J. Biomed. Inform.,* 38: 259–261.

Beuscart-Zephir, M. C., Elkin, P., Pelayo, S., and Beuscart, R. 2007. The human factors engineering approach to biomedical informatics projects: State of the art, results, benefits and challenges. *Yearb. Med. Inform.,* 1: 109–127.

Beuscart-Zephir, M. C., Pelayo, S., and Bernonville, S. 2010. Example of a human factors engineering approach to a medication administration work system: Potential impact on patient safety. *Int. J. Med. Inform.,* 79: e43–e57.

Brixey, J. J., Tang, Z., Robinson, D. J., Johnson, C. W., Johnson, T. R., Turley, J. P., Patel, V. L., Zhang, J. 2008. Interruptions in a level one trauma center: A case study. *Int. J. Med. Inform.,* 77: 235–241.

Cacciabue, P. C. and Vella, G. 2010. Human factors engineering in healthcare systems: The problem of human error and accident management. *Int. J. Med. Inform.,* 79: e1–e17.

Certification Commission for Health Information Technology. 2010. Usability testing guide for ambulatory EHR's. http://www.cchit.org/certify/2011/cchit-certified-2011-ambulatory-ehr. (Last accessed 07/23/2001).

Chazard, E., Merlin, B., Ficheur, G., Sarfati, J. C., and Beuscart, R. 2009. Detection of adverse drug events: Proposal of a data model. *Stud. Health Technol. Inform.,* 148: 63–74.

Ericsson, K. A. and Simon, H. A. 1984. *Protocol Analysis: Verbal Reports as Data.* Cambridge, MA: MIT Press.

Gorman, P. N. and Berg, M. 2002. Modeling nursing activities: Electronic records and their discontents. In *IMIA Yearbook of Medical Informatics*, eds. R. Haux and C. Kulikowski, pp. 358–364. Stuttgart, Germany: Schattauer Publishers.

Graham, M. J., Kubose, T. K., Jordan, D., Zhang, J., Johnson, T. R., and Patel, V. L. 2004. Heuristic evaluation of infusion pumps: Implications for patient safety in intensive care units. *Int. J. Med. Inform.,* 73: 771–779.

Grundgeiger, T. and Sanderson, P. 2009. Interruptions in healthcare: Theoretical views. *Int. J. Med. Inform.,* 78: 293–307.

Han, Y. Y., Carcillo, J. A., Venkataraman, S. T., Clark, R. S., Watson, R. S., Nguyen, T. C., Bayir, H., Orr, R. A. 2005. Unexpected increased mortality after implementation of a commercially sold computerized physician order entry system. *Pediatrics,* 11 (6): 1506–1512.

Hazlehurst, B., Gorman, P. N., and McMullen, C. K. 2008. Distributed cognition: An alternative model of cognition for medical informatics. *Int. J. Med. Inform.*, 77: 226–234.

Hix, D. and Hartson, H. 1993. *Developing User Interfaces: Ensuring Usability through Product & Process.* New York: John Wiley & Sons.

Hoc, J. M. and Amalberti, R. 1995. Diagnosis: Some theoretical questions raised by applied research. *Curr. Psychol. Cogn.*, 1: 73–101.

Horsky, J., Kaufman, D. R., Oppenheim, M. I., and Patel, V. L. 2003. A framework for analyzing the cognitive complexity of computer-assisted clinical ordering. *J. Biomed. Inform.*, 36: 4–22.

Horsky, J., Zhang, J., and Patel, V. L. 2005. To err is not entirely human: Complex technology and user cognition. *J. Biomed. Inform.*, 38: 264–266.

Hutchins, E. (1995). *Cognition in the Wild.* Cambridge, MA: MIT Press.

International Electrotechnical Commission. 2008. Medical devices—Application of usability engineering to medical devices (Rep. No. EN 62366). Geneva, Switzerland: IEC.

International Standards Organisation. 1998. Ergonomic requirements for office work with visual display terminals (VDTs)—Part 11: Guidance on usability (Rep. No. ISO 9241-11). Geneva, Switzerland: ISO.

International Standards Organisation. 1999. Human centered design processes for interactive systems (Rep. No. ISO 13407). Geneva, Switzerland: ISO.

Kalisch, B. J. and Aebersold, M. 2010. Interruptions and multitasking in nursing care. *Jt. Comm. J. Qual. Patient Saf.*, 36: 126–132.

Khajouei, R. and Jaspers, M. W. 2008. CPOE system design aspects and their qualitative effect on usability. *Stud. Health Technol. Inform.*, 136: 309–314.

Khajouei, R. and Jaspers, M. W. 2010. The impact of CPOE medication systems' design aspects on usability, workflow and medication orders: A systematic review. *Methods Inf. Med.*, 49: 3–19.

Kohn, L. T., Corrigan, J. M., and Donaldson, M. S. 1999. *To Err is Human.* Washington, DC: National Academy Press.

Koppel, R., and Kreda, D. 2009. Health care information technology vendors' "hold harmless" clause: Implications for patients and clinicians. *JAMA*, 301: 1276–1278.

Koppel, R. Localio, A. R., Cohen, A., and Strom, B. L. 2005a. Neither panacea nor black box: Responding to three journal of biomedical informatics papers on computerized physician order entry systems. *J. Biomed. Inform.*, 38: 267–269.

Koppel, R., Metlay, J. P., Cohen, A., Abaluck, B., Localio, A. R., Kimmel, S. E., Strom, B. L. 2005b. Role of computerized physician order entry systems in facilitating medication errors. *JAMA*, 293 (10): 1197–1203.

Koppel, R., Wetterneck, T., Telles, J. L., and Karsh, B. T. 2008. Workarounds to barcode medication administration systems: Their occurrences, causes, and threats to patient safety. *J. Am. Med. Inform. Assoc.*, 15: 408–423.

Kuperman, G. J., Bobb, A., Payne, T. H., Avery, A. J., Gandhi, T. K., Burns, G., Classen, D. C., Bates D. W. 2007. Medication-related clinical decision support in computerized provider order entry systems: A review. *J. Am. Med. Inform. Assoc.*, 14: 29–40.

Kushniruk, A. 2002. Evaluation in the design of health information systems: Application of approaches emerging from usability engineering. *Comput. Biol. Med.*, 32: 141–149.

Kushniruk, A. W. and Patel, V. L. 2004. Cognitive and usability engineering methods for the evaluation of clinical information systems. *J. Biomed. Inform.*, 37: 56–76.

Kushniruk, A. W., Triola, M. M., Borycki, E. M., Stein, B., and Kannry, J. L. 2005. Technology induced error and usability: The relationship between usability problems and prescription errors when using a handheld application. *Int. J. Med. Inform.*, 74: 519–526.

McConnell, S. 1996. *Rapid Development: Taming Wild Software Schedules.* Redmond, WA: Microsoft Press.

Miller, R. A., Waitman, L. R., Chen, S., and Rosenbloom, S. T. 2005. The anatomy of decision support during inpatient care provider order entry (CPOE): Empirical observations from a decade of CPOE experience at Vanderbilt. *J. Biomed. Inform.*, 38: 469–485.

National Health Service. 2010. Common user interface. http://www.cui.nhs.uk/Pages/NHSCommonUserInterface.aapx. (Last accessed 07/23/2001).

Nemeth, C. and Cook, R. 2005. Hiding in plain sight: What Koppel et al. tell us about healthcare IT. *J. Biomed. Inform.*, 38: 262–263.

NHS National Patient Safety Agency National Reporting and Learning Service. 2010. Design for patient safety: Guidelines for safe on-screen display of medication information. http://www.nrls.npsa.nhs.uk/home/. (Last accessed 07/23/2001).

Nielsen, J. 1993. *Usability Engineering.* New York: Academic Press.

Nielsen, J. 2005. Medical Usability: How to kill patients through bad design. 2010. http://www.useit.com/alertbox/20050411.html. (Last accessed 07/23/2001)

Pelayo, S. and Beuscart-Zephir, M. C. 2010. Organizational considerations for the implementation of a computerized physician order entry. *Stud. Health Technol. Inform.*, 157: 112–117.

Pelayo, S. Leroy, N., Guerlinger, S., Degoulet, P., Meaux, J. J., and Beuscart-Zephir, M. C. 2005. Cognitive analysis of physicians' medication ordering activity. *Stud. Health Technol. Inform.*, 116: 929–934.

Peute, L. W., Aarts, J., Bakker, P. J., and Jaspers, M. W. 2010. Anatomy of a failure: A sociotechnical evaluation of a laboratory physician order entry system implementation. *Int. J. Med. Inform.*, 79: e58–e70.

Preece, J., Rogers, H., Benyon, D., Holland, S., and Carey, T. 1994. *Human-Computer Interaction.* New York: John Wiley & Sons.

Scapin, D. L. and Bastien, C. 1997. Ergonomic criteria for evaluating the ergonomic quality of interactive systems. *Behav. Inf. Technol.*, 16: 220–231.

Shortliffe, E. H. and Cimino, J. J. 2006. *Biomedical Informatics: Computer Applications in Health Care and Biomedicine (Health Informatics).* New York: Springer.

U.S. Department of Health and Human Services. 2010. Establishment of the Temporary Certification Program for Health Information Technology: Final Rule. http://www.hhs.gov/news/press/2010pres/06/20100618d.html.(Last accessed 07/23/2001)

Wears, R. L. and Berg, M. 2005. Computer technology and clinical work: Still waiting for Godot. *JAMA*, 293: 1261–1263.

Xiao, Y. 1994. *Interacting with Complex Work Environments: A Field study and a planning model*. Toronto, ON: University of Toronto.

Zhang, J., Johnson, T. R., Patel, V. L., Paige, D. L., and Kubose, T. 2003. Using usability heuristics to evaluate patient safety of medical devices. J. *Biomed. Inform.*, 36: 23–30.

Zhang, J. and Norman, D. A. 1994. Representation in distributed cognitive tasks. *Cogn. Sci.*, 18: 87–122.

39

Human Factors and Ergonomics in Patient Safety Management

Tommaso Bellandi, Sara Albolino, Riccardo Tartaglia, and Sebastiano Bagnara

CONTENTS

39.1 Background: Peculiarities of Health Care Services and Challenges for Human Factors and Ergonomics..... 671
39.2 Introduction..672
39.3 Making a System out of Individual Methods and Tools..672
    39.3.1 Risk Identification and Measurement..673
    39.3.2 Risk Analysis and Assessment..677
        39.3.2.1 Experience of Risk Analysis and Assessment on a Regional Scale: The Case of Tuscany...... 678
    39.3.3 Risk Anticipation and Control...681
        39.3.3.1 Risk Anticipation and Control in Tuscany..682
39.4 Macroergonomics Aspects of PSM..682
    39.4.1 Governance for PSM...682
        39.4.1.1 Relationship between Quality and PSM..683
        39.4.1.2 Centralization versus Distribution of Responsibilities in PSM....................684
        39.4.1.3 Information and Communication Technology as an Enabler of PSM...........684
    39.4.2 Organizational Assets for PSM..684
        39.4.2.1 Complexity and PSM..684
        39.4.2.2 Formalization and Standardization of the Organization for PSM...............686
        39.4.2.3 Centralization of Decision Making and PSM...686
39.5 Conclusions..687
References...687

39.1 Background: Peculiarities of Health Care Services and Challenges for Human Factors and Ergonomics

Health care organizations have unique characteristics that should to be taken into account when studying activities performed by health care professionals and others and their risks, and when designing effective solutions to improve safety for patients and workers. Indeed, Bagnara et al. (2010) described the unique features of health care systems as compared to other high-reliability organizations (HROs) and challenged the suggestion that hospitals may become HROs. The results of the comparison between hospitals and HROs by Bagnara et al. (2009) are summarized in Table 39.1. In particular, we describe the characteristics of hospitals that are crucial for the design of a system for patient safety management (PSM), such as the frequency and severity of incidents, the designated victim, the nature of the interactions between workers and the process, as well as characteristics of the organization and the tasks.

Moreover, Amalberti et al. (2005) described cultural barriers that are unique to health care services: a wide range of risk among medical specialties, difficulty in defining medical error, and various structural constraints (such as public demand, teaching role, and chronic shortage of staff). Altogether, the evidence shows that health care organizations are complex systems with a large variety of activities carried out in numerous settings within an ever-changing environment: for example, on one side, the traditional homecare for childbirth, and on the other side, the high-tech laparoscopic surgery to conduct mini-invasive liver transplants. Furthermore, the success of any intervention to improve quality and safety is often difficult to measure because of the different criteria and contrasting points of view that should be considered. For instance, the participation of users in the design and production

TABLE 39.1

Health Care Organizations versus HROs

Hospitals	High-Risk Organizations
Small but frequent accidents	Few accidents
Epidemics	Catastrophes
Designated victim: patient	Designated victim: operator
Double human being systems	Human-artifact systems
Emotional, negotiation-based decision making	Rational decision making
Ever-changing organizations	Stable organizations
Diverse interactions	Defined interactions
Experimentation-based practice	Procedure-based practice

Source: Reprinted from *Appl Ergon.*, 41(5), Bagnara, S. et al., Are hospitals becoming high reliability organizations?, 713, Copyright 2010, with permission from Elsevier.

of health care services and in the assessment of their results is a well-recognized priority of health care policies at the international level; therefore, the development of patient-centered health care services is largely influenced by this priority.

In such a situation, the discipline of human factors and ergonomics is called to play a crucial role in the analysis of user experience and in the design of services, but also in the development of methods for identifying how patients may act as coactors of the health care services rather than as passive users. After all, health care organizations are important social institutions that regulate and contribute to human life, how human beings come to life, grow up, age, and die. Moreover, the health care system is likely to be the industry with the highest worldwide economic and human resources deployment. All this calls for the application of human factors and ergonomics theoretical knowledge and methods. Although well-established human factors and ergonomics science can be successfully applied to health care systems, for example, for assessing and designing safe services, much research is needed to further develop methods and tools in order to cope with the gap between the needs of health care (Department of Health Expert Group 2000; Committee on Quality of Health Care in America and Institute of Medicine 2001) and the specificities of health care systems and the actual capabilities of the human factors and ergonomics discipline to provide effective solutions.

The International Ergonomics Association has been working on that by endorsing and organizing the Healthcare Systems Ergonomics and Patient Safety (HEPS) Conferences. Moreover, the World Health Organization Alliance for Patient Safety has shown a high consideration for the human factors discipline as witnessed by its decision to include human factors and ergonomics among the core competencies in their proposed patient safety curriculum (WHO 2010).

39.2 Introduction

The themes of interactions between humans, technology, and organization of the user-centered approach and productivity represent core elements of the human factors and ergonomics discipline and are the issues that are discussed in relation to the design and development of a PSM system in this chapter. We begin with the challenge to integrate the specific components of the PSM in a system and then address the macroergonomics aspects related to the governance and the organizational model of the PSM.

The point of view adopted in this chapter is embedded in the European political and scientific environment where health care services are mostly publicly ran and managed. Our framework refers to European regulations and national and regional institutions that typically plan and control the activities in the health care sector. Therefore, our experience as researchers, organizational designers, and policy makers is deeply rooted in this context and should be taken into account by the reader.

39.3 Making a System out of Individual Methods and Tools

Safety management requires a systematic approach to be successful and sustainable. Field studies and research of successful organizations have demonstrated that safety management is a key function of high-risk industries (Reason 1997; Weick and Sutcliffe 2001; Hines et al. 2008), where an accident may become a disaster with severe consequences both for the community and the organization itself. The 2010 British Petroleum oil spill in the Gulf of Mexico is a good example of a disaster with tragic immediate effects because it caused the deaths of 11 workers, tremendous effects on the environment and wildlife in the mid- and long-term, as well as a major economic loss that may undermine the future existence of the company itself.

Even though each accident in health care seldom has the magnitude of a disaster, the large volume of work of health care services and the well-known incidence of adverse events (de Vries et al. 2008) represent a daily disaster distributed worldwide rather than in a single location; an average of 9% of patients admitted to hospitals suffer of an adverse event but not in the same place and at the same time like the fatalities happening, for example, in transportation disasters. Therefore, risk in health care services can be considered a global threat and as such, needs to be a priority at both the local and international levels.

It is also important to remember that patient safety is a young field of study and intervention. The available evidence on solutions to prevent risks is quite limited and there is a need for both an ongoing commitment to safety and subsequent actions to appraise and control risks at the front line as well as at the management level; this is particularly important because no "prepackaged" set of rules can prevent all the risks in an environment so complex and specific as health care (Bagnara et al. 2009).

Some have argued that the so-called glocal approach invented by the ecologist movement can be an inspiration for patient safety initiatives (Lilja Pedersen 2007). The ecologist movement has succeeded, in particular, in certain countries. They have influenced politicians to consider the consequences to the environment and to change behaviors of individuals even when they are rooted in poor attitudes. Just as an example, the "zero waste" campaign involves both international and national institutions, which are responsible to require a reduction of waste in production processes and to encourage reuse and recycling. Additionally, individual citizens are encouraged to separate waste and realize that they can substitute plastic with reusable glass bottles.

From the point of view of ergonomics, these arguments are very enlightening because macroergonomic approaches and methods can help to influence the policy makers and management level of health care institutions, while cognitive and physical ergonomics methods and tools can be successfully applied at the microsystem level (Mohr et al. 2004). The case of patient falls can be a useful example of this multilevel approach: updated guidelines are available at the institutional level, participatory interventions can facilitate patient risk assessment at the front line, and flexible furniture and room design can effectively reduce the risk of falls (Bellandi et al. 2008b).

In order to be systematic, PSM needs a governance and an organizational model, as discussed later in this chapter, as well as a set of effective methods and tools to identify, assess, and prevent risks at the frontline level. Given the international attention to patient safety, one of the biggest problems has been the classification of health care accidents along with the definition of a shared vocabulary between the emerging community of professionals dedicated to patient safety, health care workers, the patients, and the public in general. Recently, the WHO (2009) released the conceptual framework for an International Classification for Patient Safety (ICPS) with the aim to share lessons learned from investigations of accidents and the actions taken to reduce the risks.

Also, the ISO 31000:2009 "Risk Management: Principles and Guidelines" and the related ISO guide 73:2009 "Risk Management—Vocabulary" may provide useful guidance to the design and development of a system for PSM even though these standards are not specifically intended for health care.

Since the early 2000, the concept of "PSM" has progressively replaced the concept of "clinical risk management" to better focus on the end point of the efforts put in place at the institutional and operational levels rather than on the problems. Even though different terms are used, we can describe the components of PSM according to available literature and case studies. We can also appraise the contribution of ergonomics to the design, development, and assessment of the PSM itself. The three key components of PSM are as follows:

1. *Risk identification and measurement*
2. *Risk analysis and assessment*
3. *Risk anticipation and control*

Figure 39.1 shows how each PSM activity is either handled by frontline clinicians or managers. In a mature organizational form of PSM, as discussed later in the chapter, the two lines of activities can be associated with the functions of a clinical risk manager and those of a patient safety manager: the clinical risk manager supervises risk identification, analysis, and control, and the patient safety manager is accountable for risk measurement, assessment, and control.

39.3.1 Risk Identification and Measurement

The first basic component of PSM brings together risk identification and risk measurement. This includes the way workers and organizations figure out and eventually quantify patient safety incidents. We use the term risk "identification" when we refer to methods able to

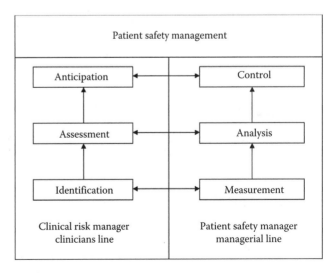

FIGURE 39.1
PSM model.

qualify incidents and we use risk "measurement" for methods that can quantify the risks. The different methods may be further classified according to the nature of the data. Methods are subjective when based on worker incident reports (IRs) or patient claims, while they are relatively objective when extracted through the use of standardized procedures using data from discharge records or patient records.

For the daily practice of risk identification, it is crucial to foster frontline clinicians' awareness of error-prone situations and their capacity to report an incident (Reason 2004). On the other hand, risk measurement is crucial for second-level problem solvers (Pronovost et al. 2008b) to help them prioritize the problems and then the actions necessary to address the problems effectively (Table 39.2).

In any high-risk industry such as aviation, incident reporting, automatic process monitoring, and chart reviews are well-known and consolidated methods that are routinely applied; but in health care, it is still challenging to disseminate these practices (Leape and Berwick 2005) for numerous reasons:

- Health care workers, in particular doctors, have been brought up with the myth of infallibility (Vincent 2006); therefore, they fear to report incidents or receive claims because in both cases their reputation is challenged.

- Health care data infrastructure and information systems are still outdated; therefore, sometimes the records are not sufficiently reliable to measure risks (Committee on Data Standards for Patient Safety 2004).

- Focus on mistakes rather than on general performance has contributed to the survival of the blame culture despite better intents (Cook and Woods 1994).

- The cultural framework of medicine favored a view where risk identification is a matter of surveillance, such as the incidence of a disease in a population, whereas a patient safety incident

is more a symptom of a pathology of the organization and not of the individual; therefore, each IR should be considered a contribution to the understanding of the underlying conditions in a framework where more reports mean more potential control of risks and not the other way round (Department of Health 2006).

The ergonomics approach is particularly suitable to overcome these barriers because it provides knowledge of the tools developed in other industries and a methodology for the implementation of the tools. We briefly discuss incident reporting and record review to show how these two methods can be designed and implemented effectively in patient safety and ergonomics.

Recent studies have demonstrated that no single method to identify or measure risks can provide a complete and thorough picture of the patient safety risks in a health care organization (Sari et al. 2006; Olsen et al. 2007; Naessens et al. 2009). Despite the expectations and investments in incident reporting, it is now clear that policy makers and health care managers probably misunderstood the aim of incident reporting, thinking it could be the only way to collect data on risks and to almost automatically provide the necessary analysis and priorities for actions. Other authors have discussed in detail (Barach and Small 2000; Johnson 2007) the contribution of human factors to the design of reporting systems in health care; in this section, we want to point out the strengths and limitations of this method in connection with the overall development of PSM.

The case of the National Reporting and Learning System (NRLS) in the United Kingdom is a very good and well-known example (Department of Health 2006) of the undue expectations of incident reporting, as well as the capacity to later review the scope and the way the reporting system is managed, following an external audit. Initially, the NRLS and the National Patient Safety Agency (NPSA) in charge of the NRLS were organized in a centralistic and quite bureaucratic fashion; they then became more integrated horizontally with other agencies at the national level and vertically with health care providers. Today, the NRLS is a module of a most extensive system that includes incident analysis conducted systematically in cooperation with experts and local teams, as well as alert reports (ARs), safety advices, and notices developed in connection with other agencies such as the National Institute for Clinical Excellence (NICE).

Another, less well-known though very successful, experience of a national reporting system has been taking place in Denmark where a dedicated law regulates incident reporting and, in general, PSM (Danish Act on Patient Safety 2004). According to this law, every health care worker is responsible to report any patient safety

TABLE 39.2

Risk Identification versus Risk Measurement

	Risk Identification	Risk Measurement
Type of task	Qualification of incidents	Quantification of risks
Nature of the data	Subjective	Objective
Source of the data	IRs, patients claims, and complaints	Patient records and discharge records
Actors involved	Frontline clinicians	Second-level problem solver
Goal	Learning from incidents	Prioritize IAs

incident and each trust is responsible to respond to the report, analyze the risks, and act to improve safety. This law is the only example, to our knowledge, of complete protection of workers who report a patient safety incident: it states that a health care worker cannot be sanctioned or prosecuted as a result of the reporting. Actually, this means that the information included in the report and in the analysis later carried out cannot be used either by hospital management or by a judge to punish the worker, even in the event of active failures that resulted in severe adverse events. Obviously, a judge can still investigate professional responsibilities, in case of a claim, but he/she cannot access information included in the PSM system. Besides the legal protection to workers who report an incident, the Danish system has a very well-designed information flow and organizational structure to support local, regional, and national activities for patient safety.

Figure 39.2 shows how a patient safety report is sent into the Danish PSM system and then analyzed at the local and regional levels. A dedicated software supports the actions taken by all the actors involved in the system: health care workers who report and get feedback, safety management staff at the local and regional levels who run the analysis and prompt the action plan, and hospital management who receives the report and pull together resources to implement actions. This system proved to be effective according to the many process and outcomes measures published by Danish health care personnel, even with important differences between regions (Lilja Pedersen 2007). This success can be attributed to the law that created the conditions to encourage reporting; other

factors may have also contributed to the success of this Danish system (Nuti et al. 2007):

- The strong leadership of the Danish Foundation for Patient Safety: a national organization representing all the stakeholders of the health care services (clinicians, patients, hospital management, institutions, medical devices, and drugs producers) with the aim to sustain PSM and promote campaigns for patient safety
- The well-designed information flow coherent with the organizational model of PSM
- The continuous feedback given to hospital management and clinicians on lessons learned from incident analysis

The WHO (2005) draft guidelines for reporting and learning systems (RLSs) strongly advice to set up incident reporting only if the organization is able to analyze and respond to reports at the local and institutional levels. In other words, incident reporting can be effective at the organizational level when integrated in an RLS (Bellandi et al. 2007), and used to analyze risks in connection with other methods (Sari et al. 2006; Olsen et al. 2007; Naessens et al. 2009). Incident reporting can be effective at the local level when it is used to record the experience of an incident of low or moderate severity (NPSA 2004a) related to active skill-based errors that the operators can detect (Thomas and Petersen 2003), when it helps clinicians cope with the stress generated by eventual individual errors (Kaldjian et al. 2006), and

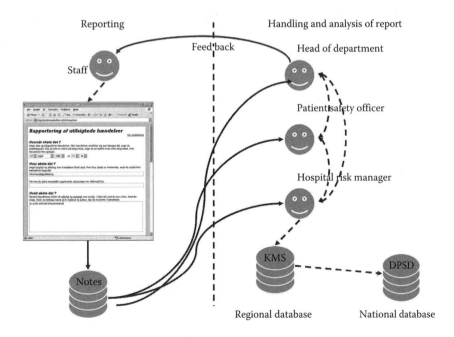

FIGURE 39.2
The RLS in Denmark.

when it sustains the awareness of potential mishaps among teams (Albolino et al. 2010).

The design of incident reporting systems is therefore crucial for its success and should consider the characteristics of the information flow and the tools used to make the report. Usually, the information flow is designed to provide a central collection of data on all the IRs coming from the front line; the reports are produced in a paper-based or electronic form. According to the Tuscan patient safety system experience (Bellandi et al. 2005, 2007, 2008a; Albolino et al. 2010), from the user point of view, the form used for reporting patient safety events has to be simple and easy to use, while the information flow must be transparent and confidential to overcome the fear of being punished as a consequence of the reporting. It is important to keep in mind that the report should be valuable to the user and acknowledged by leaders as a contribution to improvement rather than as an occasion to blame.

Most incident reporting systems take inspiration from the design of the Australian Incident Monitoring System or AIMS (Runciman et al. 1993; Runciman and Moller 2000), where a structured electronic form is sent to a central database managed by a group of experts who periodically deliver statistical analysis and in some case in-depth analysis of single events according to a robust framework. The report contains the description of the event, the causes and contributory factors, possible moderating factors, and immediate corrective actions undertaken at the local level. AIMS, therefore, includes both the identification and analysis of the event. AIMS is considered the "parent" of all incident reporting systems in health care, and has been quite successful, in particular in anesthesia, to address some risks and subsequent IAs; however, it remains a centralized tool.

In this chapter, we argue that ergonomic incident reporting shall be managed at the frontline level. Reports should go from the operator experiencing the incident to a facilitator/patient safety officer who may be a doctor or, more often a nurse, in any case a member of the professional community. This person then dedicates time to collect and analyze the reports, and decides when to trigger an in-depth analysis or when to simply record the reports, hopefully in an electronic database. We also argue that the report can be made through a paper-based or electronic form and even informally and later recorded into the system by the facilitator who analyzes the event along with colleagues who experienced the incident.

The design sketched earlier is consistent with a network-based organizational model of PSM with distributed power (Butera 2001; Scott and Davies 2006) and with the specificities of health care services: reporting becomes embedded in daily practices. This calls for a proactive management of any individual incident at the local level, and provides to the managerial level structured data both on incident occurrence and the capacity of workers to identify and control the related risk factors.

Quite obviously, the process for so-called sentinel events or never events should be slightly different: when a patient dies or suffers major disabilities as a result of the incident, then the hospital management needs to be involved in the immediate analysis of the event, especially because the event can represent a real crisis for the organization. Therefore, the information flow shall be direct and the communication instantaneous from the ward to the risk manager. The different approaches for the analysis of incidents and sentinel events are discussed later in this chapter.

Besides the Australian experience, there is little evidence for the effectiveness of incident reporting in identifying actual risks and eventually improving patient safety outcomes (Øvretveit 2005), while it has been demonstrated that PSM can be successful when it promotes and integrates a set of methods for the identification and analysis of adverse events (Sari et al. 2006). Furthermore, according to the proposed distinction between subjective and objective methods to respectively qualify and quantify risks, the traditional issue of underreporting is not so important because the measure of success for incident reporting is more the capacity to learn at the individual and organizational levels from any single event rather than to have precise estimates for the incidence of adverse events. Then, incident reporting must be integrated with the analysis and prevention of adverse events; the measure of its effectiveness will depend on the capacity of the organization to tackle IAs while we acknowledge that only some of the real risks can be controlled with this process.

The review of patient records has been found more sensitive and specific to measure the incidence of adverse events than any other method used either in research or in practice (Olsen et al. 2007; Sari et al. 2006). In a pilot study conducted in Tuscany in 2006 (Tartaglia et al. 2008), the application of the record review method allowed the detection of many important adverse events that had not been previously identified by the units involved in the study.

Record review is a method where one or more professionals systematically review a set of patient records, usually using a standard form with a list of criteria and structured questions as a guide that helps the reviewer to identify a patient safety incident. It is the method used in most of the published research studies on the incidence of adverse events; it is assumed to be objective when used retrospectively to measure incidents on the basis of the complete history of one or more hospitalizations. The most famous report of studies using the Retrospective Record Review (RRR) is the IOM report on "To Err Is Human" published in 2000. The authors of the IOM report used data from the Harvard Medical Practice Study to calculate the number of patients dying as a consequence of an adverse event, simply by transforming the incidence rate of deaths from the study sample to the population of patients discharged from hospitals in 1 year in the United States. Nevertheless, this projection has often been questioned because it produced an estimate that is considered too high, or too low, or methodologically biased, though the approximate number of 100,000 deaths in U.S. hospitals is probably the news that moved policy makers in health care to put patient safety at the top of their agendas. The RRR method is considered the major method to measure patient safety risks and has been used in many countries in the past decade (Vincent 2010). At the minimum, it provides a baseline of adverse events that can be used to set up actions and campaigns for safety improvement.

Early in the chapter, we listed methods to measure risks and, besides record review, we considered the application of a "patient safety indicator" (PSI) to discharge records. PSIs are quite promising because they can be applied automatically to screen for potential incidents in a large number of hospital admissions, but they seem to be more valuable when applied at a minimum on a regional scale, where eventual variations between hospitals on one or more indicators can be evaluated and further investigated. For example, the ratio of post-op deep vein thrombosis (DVT) needs to be compared, on one side, with the expected DVT complication ratio associated with patient comorbidities, and, on the other side, with benchmark averages so that significant variations might be investigated and eventually attributed to failures in the delivery of health care services, given that hospitals have similar characteristics.

At the local level, patient record review has been found effective in daily practice (Neale et al. 2003; Naessens et al. 2009), and new tools and procedures are available for prospective rather than retrospective review.

Once again, from the point of view of ergonomics, these tools and procedures can be designed and implemented in a manner that considers user and system requirements to facilitate the application of otherwise complex tools and the integration within the system for PSM.

According to recent work on record reviews (Thomas et al. 2002; ASSR 2007; Zegers et al. 2007; Hutchinson et al. 2010), also referred to as case-note reviews, the limits as well as the basic requirements for an effective implementation of the method in PSM can be summarized as follows (Table 39.3).

When applying the aforementioned requirements (see Table 39.3), record review can be very effective in a research study to measure the incidence of adverse events through a retrospective review (Aranaz-Andrés et al. 2009), in the daily practice of PSM in a simpler prevalence

study design (Requena et al. 2010), or in a prospective procedure (Neale et al. 2003; Naessens et al. 2009).

Even though the use of electronic health records is still quite limited, the next challenge will be the integration of the criteria used in the structured record review to automatically alert the clinicians about patient safety events and report to the PSM emerging patient safety risks while recording and connecting patient data (Committee on Data Standards for Patient Safety 2004; Kilbridge and Classen 2008; Remine consortium 2010).

39.3.2 Risk Analysis and Assessment

Human factors and ergonomics professionals and researchers have historically contributed significantly to the analysis and assessment of risks in many industries (Reason 1997), often convincing policy makers and safety managers to change their points of view on the nature of accidents (Perrow 1999) and on the actions needed to improve safety. The so-called new vision on incidents, human errors, and safety asks for an interpretation of individual incidents and a dataset on risks that takes into account the interactions between the components of a system and the dynamic nature of safety and risks (Cook and Rasmussen 2005; Amalberti and Hourlier 2007).

Many health care experts have embraced and promoted the systemic approach to patient safety, specifically referring to human factors as a key competence area that can help to develop and apply this new vision, while recognizing that a deep cultural change is needed in a sector where the myth of doctors infallibility and omnipotence of medicine (Vincent 2006) is still a widespread value and where if an incident is ever investigated, it is often to blame an individual.

Two very well-known reports, "To Err is Human" (Kohn et al. 1999) and "An Organization With a Memory" (Department of Health Expert Group 2000), have embraced the new vision and called for the development of PSM on the basis of the experience of other sectors such as aviation. As a consequence, the methods used for incident investigation and risk analysis have been transferred to health care (Woloshynowych et al. 2005), often without adequate translation and adaptation to the specific characteristics of health care organizations.

Moreover, the scientific literature is still quite poor on the description and application of the methods used to analyze and assess risks in health care (Vincent 2010). Usually, they are classified into proactive and reactive methods. Proactive methods are applied to a critical process before an accident happens whereas reactive methods collect data on incidents that have happened in the past. This distinction is questionable because the selection of a process or of an incident to be analyzed

TABLE 39.3

Record Review Limits and Requirements

Limits	Requirements
Low concordance between reviewers (Thomas et al. 2002)	1. The reviewers receive a basic training program in patient safety, including the presentation of the record review method and some examples of real cases, with the aim to share a common language and a common approach at the identification of patient safety
	2. A procedure is in place to guide the reviewer and to accurately define the end point (i.e., patient safety incidents, adverse events, and near misses), inclusion and exclusion criteria for each question/indicator included in the review form
Subjective judgments (Hutchinson et al. 2010)	3. One or more review forms are used to record the review and later analyze the single case or the aggregate data depending on the modality of the review (perspective or retrospective, for a research purpose or in daily practice)
Time-consuming (Tartaglia et al. 2008)	4. A dedicated software includes the review forms and provides contextual guidance to the reviewer, if an electronic patient record is in place, it can contain a module for review
Unreliable generalization and comparisons (Zegers et al. 2007)	5. A clear protocol describes the definitions used in the research/application of the method, the eventual review tools, the sampling procedures according to the general and specific objectives, the statistical elaborations of the data, and the reviewers profile and training program

has to start from the recognition of the actual risks at the front line and of the safety goals at the management level. Therefore, using time as a criterion, process and incident analyses are both reactive because they begin after risk identification and proactive because their aim is safety improvement.

According to the organizational model for PSM, risk analysis and assessment should take place at a different level in health care organizations because they can promote a complete interpretation and help to develop a common view of risks and safety at the front line and at the management level (Cook and Woods 1994).

At the local frontline level, risk analysis helps clinicians to detect and anticipate risks, primarily because of the understanding they gain if they are involved in the analysis process (Reason 1997, 2004, 2009). At the central management level, the knowledge created through process and incident investigation may provide evidence on risk factors to the decision makers that can define the boundaries of acceptable and unacceptable performance and help to allocate resources for safety and quality improvement.

Risk analysis and assessment are, therefore, a function of PSM crucially interconnected to identification and measurement as well as anticipation and control, and need to be designed according to the systemic approach to patient safety and to the specificities of health care. We used the terms of analysis and assessment because the first one is more general and particularly suitable when the goal is the appraisal of risk factors through a qualitative investigation on either a single or a set of incidents, while the latter refers to the classification and quantification of the type and nature of incidents, failure modes, and their causes.

In Section 39.2.3.1, the Tuscan experience of implementing the risk analysis and assessment function on a regional scale is presented and compared with other

experiences described in the literature or through personal contacts.

39.3.2.1 Experience of Risk Analysis and Assessment on a Regional Scale: The Case of Tuscany

In 2004, through the approval of a regional law, the Regional Government of Tuscany founded the Centre for Clinical Risk Management and Patient Safety. The main objective of the Centre is to promote a patient safety system inside each local health care trust of the Tuscany region, based on a culture of safety that is a no-blame culture that promotes learning from errors at the organizational level.

This is basically an experience limited to the regional scale, though it can be compared with other international experiences because methods and tools have been developed according to scientific evidence in ergonomics and patient safety, and benchmarks with PSM activities carried out especially in Europe.

With the coordination of the Regional Centre, a system based on a lean network model has been developed (Bellandi et al. 2005). The main activities of the Centre are organized along four axes: (1) organization for patient safety, (2) training, (3) PSM, and (4) campaigns for patient safety (Bellandi et al. 2008a). For each activity, a set of minimum requirements has been defined and has become part of the quality accreditation standards through a specific regional law.

The PSM system started in 2004, and has been providing regional guidelines for methods and tools based on available literature and existing information sources (i.e., claims and complaints). At the beginning, trusts asked the Centre for Clinical Risk Management and Patient Safety to organize the PSM, initiate the activities, and demonstrate that they were able to "close the circle," in other words, to complete the process from risk identification to analysis and control. A basic set of methods to analyze and assess risks were selected. Significant event audits (SEA) (Pringle et al. 1995) and mortality and morbidity (M&M) reviews (AHRQ 2010) are used at the ward level. Root cause analysis (RCA; NPSA 2004b) and failure modes and effects analysis (FMEA, Kirwan 1994) are used at the trust or regional level. Training programs were designed and delivered throughout the Regional Healthcare Service (RHS) to support the application of the methods (Bellandi et al. 2008).

All of the methods have been adopted so that, despite procedural differences, they share the systems analysis conceptual framework (Vincent et al. 1998). The criteria used to select and redesign the methods include severity of the risks, complexity of the investigation process, legitimate participants, and goal. Table 39.4 shows the preferred methods according to these criteria. SEA can be used for analyzing near misses on the basis of the judgment of the patient safety officer, that is, the facilitator, who can decide to conduct a more detailed analysis than the quite informal M&M review. SEA can also be used for adverse events when the risk manager decides to conduct an investigation into a single incident, in particular, when it involves more than one department and may have significant risk factors.

The Regional Centre verifies how many IRs are formally submitted and managed, the number of single IAs promoted as a result of risks analysis and the delivery of ARs; those reports contain the analysis of the event or the critical process and the action plan to prevent it from happening again in the future according to regional guidelines.

The PSM system started up in most trusts during the second semester of 2006. Fourteen out of 16 trusts have adopted the regional preliminary, paper-based, incident reporting system resulting in 443 reports, with a mean of 28 reports per trust in 6 months. A total of 116 ARs and 251 IAs have been recorded. All trusts have met the requirement to have at least one improvement action resulting from a systematic analysis of an adverse event or a critical process; this occurred even for the two trusts that did not have IRs but conducted investigations on the basis of patients' claims and complaints or informal reports.

At the end of the start-up period, we decided to review the PSM through a comparison with other experiences (Nuti et al. 2007) to improve the effectiveness of the patient safety initiatives. The already mentioned Danish system for patient safety has been described as a good standard of comparison because of the similarity in health care services between Tuscany and Denmark: both systems are more than 90% public owned and organized in three levels of services, where local trusts manage the first and second levels care with community wards and city hospitals, while university hospitals are the third level of care for high-level specialties and consultancies. The population in Tuscany is 3.7 million and Denmark has 5.5 million inhabitants.

In Denmark (Lilja Pedersen 2007), at the end of 2004, that is, the start-up year, the national system for incident reporting collected about 800 reports in 1 month. Between 2001 and 2005, 90 RCAs were conducted, each of them including an improvement plan. There was a slow but constant increase in analysis and a considerable reduction in time needed to run a single RCA. In 2006, the frequency of root causes and strength of IAs were analyzed. Results showed that the first two root causes are communication and procedure/guidelines, while only half of the improvement plans were effective to reduce avoidable risks.

Comparing the Tuscan data with those of the Danish at the start-up stage, the first evident gap is about quantities; for instance, the quantity of IRs in Tuscany has been very low. At the same time, Tuscan trusts are documenting many more investigations conducted with different methods (SEAs, M&M reviews, FMEA, and RCAs), and there are more single IAs included in a plan called "alert report." In Tuscany and more globally in Italy, the low number of IRs is often attributed to the lack of legislation that can protect clinicians from liability as a result of the report, unlike in the Danish Patient Safety Act (2004). This is probably a barrier to reporting, but we must notice that in Tuscany all the reports are confidentially managed at the ward level and deidentified once they leave the ward. Moreover, clinicians are invited to mainly report unsafe acts, near misses, and low-severity incidents because incidents with harm

TABLE 39.4

Methods for Risk Analysis and Assessment

Severity of the Risks	Near Miss	Adverse Event	Critical Process	Sentinel Event
Complexity of the investigation process	Low	Moderate	High	High
Legitimate participants	Peer review	Internal analysis coordinated by the risk manager	Internal or external analysis coordinated by the risk manager	External analysis supported by an expert group
Goal	Risk analysis	Risk analysis	Risk assessment	Risk assessment
Preferred method	M&M review and SEA	SEA and FMEA	FMEA	RCA

to patients are appraised through claims, complaints, and patients record review. A cultural more than legal barrier is probably the main cause of underreporting. Therefore, we are currently implementing a survey to evaluate safety culture as previously done in a national project (Tartaglia et al. 2008; Albolino et al. 2010).

The second gap is more qualitative than quantitative. Indeed, in Tuscany, the Regional Centre did not systematically assess risk factors and the effectiveness of IAs at the regional level until 2008 because the trusts did not provide ARs to the Regional Centre for Patient Safety. During the start-up stage of the PSM system, only the information provided in the annual patient safety reports submitted by each trust according to accreditation requirements was collected and analyzed. The annual reports included only summaries of the events analyzed, the main risk factors assessed, and the actions undertaken as a result of the investigations. The reasons for lack of access to the ARs by the Center were confidentiality and the eventual economic burden of litigation associated with the disclosure of sentinel events, even though the ARs did not contain personal information on patients or clinicians involved in the case.

In 2007, the Ministry of Health launched the national reporting system of sentinel events and asked Italian regions to set up dedicated policies and procedures. Therefore, in Tuscany, a regional act was issued to regulate reporting and analysis of adverse events and sentinel events. This act stated that sentinel events must be reported within 48 h, and an RCA must be conducted and sent to the Regional Centre within 15 days, while the national deadlines are 5 days for reporting and 45 days for the RCA.

Moreover, an expert group was formed and became active. This group is comprised of clinicians and risk managers dedicated to the analysis of sentinel events, and can be called by the Regional Centre to provide external support to the trusts in running the RCA and managing the consequences of the event. Since then, thanks to the structured information flow and to the expert group, the Regional Centre is able to access detailed information on sentinel events and perform aggregate analysis, as well as assess risk factors and improvement plans. The Tuscan data for 2007–2008 (Table 39.5) are consistent with findings of the Italian national reporting system of sentinel events (Ministero della Salute 2009), as well as with those of the Joint Commission even if the American classification of the types of events is slightly different (The Joint Commission 2010).

At the same time, the analysis and assessment of adverse events remain at the local level, but the ARs have been standardized and trusts have been asked to present them as a mandatory attachment to their annual report. Also, SEA and M&M reviews have become part of the balance scorecard used to assess the performance

TABLE 39.5

Sentinel Events in Tuscany 2007–2008

Type of Event	Occurrence
In-patient suicide	15
Death of newborn (weight >2500 g) not related to genetic illness	14
Any other events resulting in unexpected death or severe injury	10
Unexpected death or severe injury related to a surgical procedure	9
Death, coma, or severe injury related to medication errors	2
Retained materials in the surgical site	2
Maternal death or severe injury related to the delivery	2
Patient falls resulting in death or severe injury	2
ABO reaction due to wrong blood transfusion	2
Invasive procedure on the wrong patient	1
Wrong procedure on the correct patient	1
Death or severe injury related to the wrong attribution of the triage code at the accident and emergency department	1
Violence or abuse on patient during the hospital stay	1
Wrong surgical site	0
Violence or abuse on workers	0
Death or severe injury related to breakdowns of ambulance services	0
TOTAL	62

of trusts: each department should perform at least three SEAs and six M&M reviews to get a positive score on the RLS indicator. The motivation behind this goal is to enhance the organizational capability to learn from adverse events and critical processes. Table 39.6 shows the ongoing increase of the RLS throughout the region.

Even though the number of IRs is still quite low when compared to Denmark (Bjørn et al. 2009), between 2006 and 2008 we saw increases in the learning ratio and the improvement ratio that are the indicators used in Tuscany to assess the PSM. The learning ratio is calculated by dividing the number of ARs, produced as a result of the analysis conducted, that is, the numerator, with the number of IRs, that is, the denominator; it is the indicator used to assess the effective use of IRs for the analysis of risks. The improvement ratio measures the capacity to translate learning into practice; it is calculated by dividing the number of IAs with the number of IRs. We need to monitor these indicators over time and possibly refine their definition, but actually they have already proven to be useful. In 2007, we had a slight decrease of both learning and improvement ratios. We analyzed the problem in all 16 trusts and realized that there was a lack of integration between clinical risk management, quality management, and human resources. Then we provided a new regional regulation to foster the recognition of SEAs and M&M reviews as learning

TABLE 39.6

RLS Data 2006–2008

		IRs	ARs	IAs	Learning Ratio (AR/IR)	Improvement Ratio (IA/IR)
2006	Total	443	116	251	0.26	0.57
	Trust average	27.69	7.25	15.69		
2007	Total	587	151	288	0.26	0.49
	Trust average	36.69	9.44	18		
2008	Total	1137	453	876	0.40	0.77
	Trust average	71.06	28.31	54.75		

activities in the continuous education framework, and a new organizational model and procedures to link the results of the investigations with the quality improvement programs.

Yet, the current challenge is the integration of other methods in the PSM and the evaluation of the effectiveness of IAs over time.

39.3.3 Risk Anticipation and Control

Risk anticipation is the capacity of health care workers and organizations to use tools and procedures that can reduce, mitigate, or eliminate actual and potential risk factors. On the other hand, risk control includes activities that are currently in place to evaluate risks and the effectiveness of actions taken to improve safety over time.

The design and implementation of safety solutions is a field where ergonomics is already providing a clear contribution, for example, in the development of checklists (Haynes et al. 2009), the use of tools to improve teamwork (Flin et al. 2008; Mishra et al. 2009), the organization of work (Carayon et al. 2007), the design and assessment of biomedical devices (Ward and Clarkson 2006), and information systems (Bates and Gawande 2003). Research has demonstrated the effectiveness of these activities; it is now time to dedicate attention and effort to the hard task of actually integrating these activities into a coherent PSM system. In other words, the transfer of knowledge in practice requires methods and managerial tools so that safety solutions can become safe practices that are integrated in daily patient care processes (Bervick 2008).

Some institutions, such as the World Alliance for Patient Safety of the WHO or the Institute for Healthcare Improvement (IHI), are working specifically to promote the application of safe practices. Nowadays, it is broadly recognized that safety improvement requires a cultural change among both health care managers and clinicians. The issue of cultural change has been largely studied in the field of occupational safety, and ergonomics has provided an effective approach to reduce injuries through systemic intervention and cultural change.

Institutions such as WHO and IHI are mainly working to develop patient safety campaigns targeted at policy makers and hospital boards; they also develop bundles of tools and operational procedures to support frontline workers. We recognize the importance of the design of the campaigns and the human factor issues related to the characteristics of the communication strategy and artifacts; in this chapter, we focus our discussion on the design, implementation, and assessment of the bundles as proposed by institutions such as WHO and IHI.

One of the most successful patient safety initiatives is the campaign to eliminate hospital-acquired infections associated with the use of central venous catheters (CVC). This initiative is based on a bundle of procedures and tools that were developed based on knowledge about the human, technological, and organizational factors contributing to the risks of CVC-associated infections, that is, a well-known complication that used to be managed ineffectively without a systems approach. The impressive results reported by Pronovost et al. (2008a) were obtained with the implementation of multifaceted interventions targeted at clinician behavior (hand washing before CVC insertion and use of sterile barriers to prevent contamination), internal procedures (avoid the femoral site for insertion and remove unnecessary CVC), and technologies (availability of chlorhexidine to disinfect skin before CVC insertion and of sterile mask, cap, gown, gloves, and large drape). In several papers, Pronovost et al. (2005, 2008b, 2010) have described the methodological approach used to implement, sustain, and spread changes aimed at reducing CVC-associated infections. At the macrolevel, the focus is on top management commitment to patient safety and the assessment of safety culture, and at the microlevel, key actions include specific and evidence-based interventions related to safety concerns, including training, documentation of improvements and trade-offs, and sharing of results. Clearly, these actions can be implemented in the PSM. Activities carried out at the ward level can be sustained over time by integrating clinicians' effort to anticipate risks and the managerial responsibility to control risks; this will close the gap between the sharp end and the blunt end (Cook and Woods 1994), recognizing that only through a shared commitment patient safety can be improved.

In the United Kingdom, the National Health Service Litigation Authority (NHSLA) developed a scheme to assess the performance of trusts in terms of their capacity to anticipate and control risks over time (NHSLA 2010). The scheme contains a set of standards, requirements, and checklists to apply and evaluate safety solutions. Using this scheme to control the level of risks, NHSLA defines the costs of the insurance each trust must pay to cover their potential economic liability.

39.3.3.1 Risk Anticipation and Control in Tuscany

In Tuscany, the Regional Centre for Clinical Risk Management and Patient Safety combined the NHSLA scheme into a system to anticipate and control risks through the application and certification of good practices (Bellandi et al. 2008). Each good practice is described in a technical document with an explanation for its rationale, requirements for the practice's application and internal assessment, a list of references, and a toolbox that can be used and customized to local needs. The safety issues covered by the good practices are selected on the basis of available evidence and prioritized according to aggregate risk assessment performed at the regional level.

Before the release of the technical document, good practices are pilot tested in a sample of hospitals and departments. A regional team composed of ergonomists, clinicians, and patient advocates (in some cases) coordinate the pilot study and then promote the diffusion and external assessment of the good practices. Clinicians included in this process often represent their scientific societies, while patients are selected from a list of accredited associations that have participated in a training program on PSM available for citizens.

Once the good practice is approved and released by the Regional Health Council (RHC), the trust's board may decide to ask for an external assessment of the good practice, which is conducted by expert auditors appointed by the RHC; the good practice may eventually receive the certification if the requirements are met. Late in 2010, most of the good practices for patient safety have become part of the redesigned Regional Accreditation System; therefore, they are mandatory requisites of the patient pathways, and they are formally assessed every 2 years by the Regional Quality and Safety Commission.

The Regional Centre for Patient Safety organizes annual Patient Safety Walk rounds (Frankel et al. 2008) that involve medical directors, expert clinicians, and patient advocates who visit other RHS trusts; this helps to enhance the commitment of the leadership and contributes to benchmarking between various trusts. During the walk rounds, good practices are observed; therefore, frontline workers can perceive the commitment and informal social control of external actors, and visitors may evaluate the local implementation of the good practice and acknowledge eventual barriers to the implementation.

Good practices and patient safety walk rounds prove to be useful methods to an effective and widespread anticipation and control of risks, in particular when these activities are integrated in a PSM along with other control tools, such as the quantitative indicators related to the identification and analysis of risks.

39.4 Macroergonomics Aspects of PSM

The development of the PSM and its strategic activities worldwide show the need for developing a systemic, macroergonomic approach to improve patient safety.

Before the publication of the Institute of Medicine (IOM) report "To Err is Human," the issue of clinical risk was already known among clinicians, especially in those specialties where risk is high due to the complexity and uncertainty of the work (Cooper et al. 1978; Gaba et al. 1987; Morris 2000). But, as mentioned before, the publication of the IOM report provided evidence about the extent of the patient safety problem, but also introduced a new approach by focusing on the systemic point of view and the human factors perspectives. As the description of the PSM methods and tools in this chapter shows, this new approach introduces ergonomics into patient safety and brings up the challenge of bridging disciplines and professions aimed at improving patient safety. Since the publication of the IOM report, there has been a worldwide effort to develop strategies and programs on PSM based on principles of human factors and ergonomics.

39.4.1 Governance for PSM

All of the international experiences discussed previously demonstrate that, in order to be effective, PSM needs a governance and an organizational model that follow a systemic macroergonomic approach. Although the majority of improvement initiatives have involved the microlevel, there is now growing recognition that patient safety improvement programs need to focus on the level of the whole organization or care system (Benn et al. 2009). According to Øvretveit and Gustafson (2002), little research evidence exists about the effectiveness of quality improvement programs that target whole organizations or health care systems.

When considering the long-term, sustained impact of an improvement program, researchers must consider how to define and measure the capability for continuous safe and reliable care as a property of the entire health care system. This requires a sociotechnical approach rather than focusing on the microsystem, a unique disciplinary perspective or a single level of the system (Benn et al. 2009).

The macroergonomics approach, which focuses on the design of the overall work system (Carayon 2006; Hendrick 2007) and related methods, can help policy makers and management to develop more effective and sustainable patient safety systems. International experiences with the promotion and development of patient safety systems show that strategic decisions need to be made about the governance of the system, organizational assets, the cultural approach, and the focus of the interventions. These decisions strongly affect the way activities are actually carried out.

39.4.1.1 Relationship between Quality and PSM

The first strategic aspect in developing PSM in a health care organization is the *governance of the system*, that is, the way the system is governed by policy makers and institutions in a systematic approach to maintain and improve quality of patient care within the health system.

With regard to governance, patient safety experiences have been characterized by different styles and approaches, depending on different political and social contexts and the specificities of the health care systems. There are key elements in the decision-making process about governance of the patient safety system that can have a major impact.

The first decision is about the possibility to *consider patient safety as a strategic dimension of quality that needs to be managed inside existing quality programs or to consider patient safety as an independent variable* that significantly affects the quality of the system but needs specific strategies and initiatives that are implemented and developed over time. The adoption of one of the two options has relevant consequences for the governance of the patient safety system and its objectives.

Evidence shows that quality improvement programs are seldom effective to improve the performance of health care services (Øvretveit and Gustafson 2002); therefore, some experts discourage the integration of safety in quality management (Vincent 2010). One possible reason for the shortcomings of quality programs is the bureaucratic burden associated with those programs, in particular when the initiatives are based on scientific management principles that have been used and applied in the manufacturing industry; in such a system, the organization of work, the nature of the activities, and worker-process interactions are very different. On the other hand, others argue that safety can benefit from integration with quality (Committee on Quality of Health Care in America and Institute of Medicine

2001) because it can gain relevance in the broad health care agenda and be embedded in patient pathways and information systems.

In France, since 2004, the patient safety issue has been integrated in quality improvement policies. The Haute Autorité de Santé, which is an independent public scientific institution set up by law on August 13, 2004, promotes the quality of health care at both individual and collective levels and has created an integrated program to implement campaigns aimed at improving health care quality and safety. A specific department for the improvement of quality and safety of health care (DAQSS) has also been created in the Haute Autorité de Santé. Safety has always been considered as a strategic dimension of quality and one of the most important aspects of the accreditation process of the health care organizations (Nuti et al 2007). There is also a very good monitoring of claims and complaints and integration of the different data sources to define a coherent strategy for improving quality and safety. But the centralized management of strategic processes and the data makes the national coordination of the strategies and policies for patient safety at the local level very challenging and complex.

In the United Kingdom, the government set up a dedicated institutional body that is the health authority for patient safety activities, that is, the NPSA. NPSA contributes to improved and safe patient care by informing, supporting, and influencing organizations and people working in the health care sector. Following the systemic approach, the agency has led many different initiatives aimed at improving patient safety through three main lines of business: the NRLS, the National Clinical Assessment Service, and the National Research Ethics Service.

The NPSA is structured in sectors that reproduce the main phases of a PSM system. This organizational design is consistent with the macroergonomic approach of designing systems according to task and activities that need to be performed. This helped the English health care system to implement a very structured and systematic approach to patient safety.

The decision of the policy makers to establish an institutional and organizational effort to the specific issue of patient safety had a significant impact on the development of programs and initiatives to improve health care. From the ergonomic point of view, as underlined by Vincent (2010), considering patient safety independently from quality programs has contributed to a new focus on ergonomic and psychological issues related to health care system performance and on new tools and techniques (similar to the human factors discipline and used in other fields such as aviation and power plant companies) for the analysis of health care processes and systems.

39.4.1.2 Centralization versus Distribution of Responsibilities in PSM

Another key element in the governance of patient safety is *the modality of controlling and coordinating activities and roles involved in patient safety inside a system*. There are two main trends: either to centralize the coordination of activities in a single actor such as a central national government, a dedicated institutional body, or a private organization, as is the case for the English or the Danish system; or to decentralize and distribute the control and coordination to the different actors, as has happened in the United States or Spain. The first organization involves hierarchical networks with centralized power, while the second case involves networks with distributed power with different levels of coordination and control (Butera 2001; Scott and Davies 2006).

The centralization of activities and data is typical of the English patient safety system, but also of the Danish system, whose extension in terms of number of persons and hospitals included is very limited, therefore facilitating such a decision. The creation of hierarchical networks can help to diffuse policies and actions in a fast and homogeneous way, and to integrate local information and data. But the concentration of power and knowledge in a few nodes of the network is not consistent with the systemic and collaborative approach, and over time it becomes difficult to manage and address safety improvement (Department of Health 2006).

At the national level in Italy, starting in 2003, the Ministry of Health constituted some technical groups with the aim to deliver recommendations to prevent specific sentinel events; later on, in 2007, a national reporting system of sentinel events was set up that was based on the U.S. Joint Commission process. From the very beginning, the national initiatives have been integrated in the quality and clinical governance framework, even though the National Health Service with its federative structure where the regions are in charge of organizing health services weakened the national initiative. In 2008, an agreement between the national government and the regions defined a set of requirements including the mandatory reporting of sentinel events.

39.4.1.3 Information and Communication Technology as an Enabler of PSM

The third strategic element for governance is *the development of information and communication technology (ICT) as a tool for supporting patient safety systems*. From the ergonomic point of view, ICT needs to be a supportive element that, as described before in this chapter, should be designed according to the system's needs. Information systems are very important tools but their design and implementation have to be driven by local patient safety

policies and programs. In the development of patient safety systems, especially at the beginning, the technology was used as a driver element, especially in the development of strategic activities such as the implementation of reporting systems.

39.4.2 Organizational Assets for PSM

The second strategic aspect at the macrolevel for PSM is the organizational assets, that is, the models used to organize the patient safety systems; this should consider the three core dimensions of the macroergonomics point of view: complexity, formalization, and centralization (Hendrick 2007).

Before considering how these three dimensions can be developed through the analysis of some examples, it is necessary to note that recent studies (Burnett et al. 2010; Parand et al. 2010) show the importance of assessing the degree of organizational readiness to change and taking the time necessary to create an organizational infrastructure, process, and culture; these are premises to the start of organization-wide quality and safety improvement initiatives. This analysis would allow policy makers to set realistic expectations about the outcomes of patient safety programs. From the human factors and ergonomics viewpoint, this step represents the analysis of users' needs and is strategic to address subsequent actions in promoting patient safety inside a health care system.

According to recent studies by Burnett et al. (2010), organizational readiness is based on the organizational culture and its members' attitudes toward quality and safety, the characteristics of the operational systems and infrastructure, and the availability of resources. Burnett et al. (2010) maintains that culture is the most critical element for organizational readiness to change. Later on in this section, we discuss patient safety culture as a strategic element of the organizational asset.

From the ergonomics point of view, the analysis of the organization's initial readiness is critical to build organizational-level improvement programs that are tailored to local variations in context and to improve the work environment (Benn et al. 2009). The analysis of preexisting organizational conditions and patient safety culture is one of the most meaningful elements in the Danish patient safety program. The Danish Society for Patient Safety founded in 2005 is the governmental body in charge of promoting and managing patient safety in the Danish health care system; one of its permanent initiatives is the evaluation of patient safety culture in Danish hospitals.

39.4.2.1 Complexity and PSM

The first dimension of organizational assets to consider for the development of the PSM is *complexity*, that is, the degree of differentiation and integration that exists within a work system (Hendrick 2007). In the development of a patient safety system, differentiation relates to the definition of roles and responsibilities. A key role in PSM has been recognized as fundamental worldwide and is the clinical risk manager together with the patient safety manager. These two roles represent the coordinating roles for patient safety systems and, in most international systems for clinical risk management, they both need to have human factors and ergonomics skills.

WHO (2010) has defined the topics that need to be included in the patient safety curriculum at medical schools, and human factors and ergonomics is a key discipline to understand human performance under difficult circumstances and the interactions between people and other components of the system (technology, organization, and other people) that can produce system breakdowns. As described previously in the chapter, the knowledge and application of an ergonomic approach for the design of an incident reporting form or for the development of an information system can help to improve the human–system interface by producing better-designed systems and processes.

Although in all the countries where a PSM system has been developed, there are specific training programs for people responsible for patient safety activities, it is difficult to find well-defined lists of characteristics of the different professional profiles dedicated to clinical risk management and patient safety.

The Tuscan Patient Safety System adopted a training model that is strongly focused on human factors and ergonomics skills, which has strongly influenced the definition of skills and activities of the different roles dedicated to patient safety (Bellandi et al. 2008). The clinical risk manager is defined as an expert in ergonomics and quality who is responsible to coordinate the local safety management system, to supervise the facilitators, and to manage adverse events and clinical risks with regard to event monitor and analysis of related information.

Professionals that can become clinical risk managers are senior physicians, nurses, psychologists, or experts in social sciences. The training program is a postgraduate academic course and a valid path to become a certified European Ergonomist in a manner consistent with CREE standards. The goal is to create a new professional role for health care professionals who are able to manage, coordinate, and supervise local patient safety initiatives.

The patient safety manager needs to have the same skills but she or he is responsible for promoting actions for improvement according to the risk analysis and to international and national best practices and recommendations for patient safety. She/he has the task of monitoring, over a long period of time, the implementation of improvements and preventive actions that emerged from SEAs and M&M reviews.

Another strategic role is the facilitator who is in charge of the incident reporting system at the ward level. She/he promotes reports of significant events and the analysis of processes at the front line. This role, sometimes called "patient safety officer" in other countries, can be performed by physicians or nurses who are perceived as reliable, competent, and trustworthy at the ward level. The training program for this role is a 40 h postgraduate professional course that enables professionals to organize and promote incident reporting and SEAs.

Also the Tuscan process for selecting health care professionals who will play those two roles inside the organization follows what has been recommended in the literature: risk management should be performed by charismatic leadership in order to fight the idea that patient safety is a split off, tedious, and reactive concept (Cozens 2001).

It is also important to give key patient safety roles to people at the front line. In the literature, a great deal of evidence suggests that strategies that involve clinicians and patients are more effective that the ones involving only managers (Cook and Rasmussen 2005; Vincent 2006; Flin et al. 2008; Schouten et al. 2008). The participative approach of human factors and ergonomics, as described earlier in this chapter, involves methods and tools that are particularly useful and adequate to support the development of the bottom-up approach to the development of PSM.

For instance, recent studies by Burnett et al. (2010) found that in the implementation of patient safety initiatives, a bottom-up frontline-led approach was associated with high ratings of organizational readiness. Other research in this area has found that organizational change is more likely to succeed if staff feels empowered where formal leaders are not afraid to let others lead the change. Again, Burnett et al. (2010) found that clinicians viewed medical engagement as essential for patient safety programs' success, in accordance with other research in this area. Research has also highlighted that engaging clinicians in quality improvement and fostering staff-driven process improvement that engages the front line are commonly reported factors for success (Buchanan et al 2005; Vaughan 2007).

The spreading of responsibility throughout all levels of the organization will reduce the level of emotional response that takes place at the sharp end when an error occurs, and this should allow learning to occur more readily and more appropriately (Cozens 2001).

The roles of clinical risk manager and patient safety manager are integrated in a complex system through *coordination and communication systems* designed for the organization. In the PSM, an important integrating factor is the cooperation that takes place within and between teams. In fact, clinicians need to integrate two different needs: the need to work autonomously with the need of harmonizing their contribution to the objective of the organization they work for. Teams are very important for developing patient safety programs, and they have been shown to be a very important organizational resource in many experiences (Cozens 2001; Albolino et al. 2007; Flin et al. 2008). For instance, in the Tuscan patient safety system, each hospital has a patient safety team, according to the ergonomic approach. The team is interdisciplinary and includes representatives of all different specialties and disciplines. A multidisciplinary approach to decision making creates a broader knowledge base and as a result increases the ability of the team to address its tasks well (Cozens 2001). Also, the use of teams as a symbol for promoting patient safety at all organizational levels is recognized as particularly effective for the large and complex health care context. To achieve successful PSM activities, teams need to involve people at the sharp end and pay attention to frontline activities (Cozens 2001). Additionally, there is a specific need for good team leadership.

39.4.2.2 Formalization and Standardization of the Organization for PSM

The second organizational dimension important for the development of the PSM is *formalization*, which refers to the degree of standardization of jobs inside the work system (Hendrick 2007).

For PSM, it is important to establish standardized procedures that are implemented for some specific events, such as sentinel events, near misses, the promotion of best practices, and recommendations for patient safety inside the health care organizations. In HROs, such as hospitals, it is fundamental to standardize behaviors as much as possible to better concentrate efforts in finding innovative attitudes and solutions to manage the unexpected, which is a critical component of daily activities in these kinds of organizations (Weick 1995). Furthermore, the adoption of a standardized procedure helps the organization to keep track of what happens and to build its own memory, which are fundamental to a system that learns from its own mistakes. Ten years after the publication of the IOM report, many standardized best practices for patient safety are recognized and promoted worldwide (i.e., WHO, AHRQ, NPSA). There are many well-known and established methods and techniques for analyzing risks and improving patient safety, such as the SEA (NPSA), the M&M review (AHRQ), and the RCA (The Joint Commission). All of them are based on the systemic approach and use the basic principles of macroergonomics and human factors such as focus on understanding complex multifaceted

systems, analysis of the interaction between the human factor and the technological environment, and organization and the building of solutions with people directly involved in the process.

In the Tuscan health care system, the application of these methods has become part of the system for the evaluation of CEO performance. In order to be successful in patient safety, the health care trusts need to perform three SEAs and six M&M reviews a year. Boards of directors for the trusts are responsible for the implementation of international best practices for patient safety that are promoted at the regional level and for a system for monitoring and preventing patient falls. The achievement of these goals is connected to economic incentives so that part of the CEO's income depends on how she or he promotes patient safety in her or his organization.

39.4.2.3 Centralization of Decision Making and PSM

The third important organizational dimension to consider for the development of the PSM is *centralization*, that is, the degree in which decision making is concentrated in few people within a work system (Hendrick 2007).

As already discussed in Section 39.3.1, the level of decision-making process concentration strongly affects the development of the PSM. From an ergonomics point of view, it is important to promote a distributed decision-making process aimed at building sense making among clinicians about the level of risks and safety. This decision-making model is easier to develop in an organizational model that is based on a network with distributed power, as the one developed in Tuscany or in Spain, where patient safety is organized regionally. This decision-making process needs to promote safety and naturalistic cognitive processes that characterize people who work in environments of high uncertainty and time constraints (Klein et al. 2010; Weick 1995).

Sense making is a useful concept for understanding coping mechanisms under stress. In a narrow sense, the need for sense making arises when a group's current experience needs to be understood in connection to a past event (Albolino et al. 2007). More broadly, the need for sense making arises from the recognition that current conditions have in them signals that are in conflict or can lead to a future conflict and that choosing how to proceed is likely to have important consequences for the group.

From the macro system point of view, organizational culture is the dimension that enables all the strategic dimensions for the PSM (Reason 1997, 2000) and the real enabler of the whole process of building a patient safety system. As Amalberti et al. (2005) stated, the most important difference among industries lies not so much in methods and techniques used to improve safety, but in an industry's willingness [or "readiness" as suggested by Benn et al. (2009) and Vincent et al. (2010)] to change its cultural precedents and beliefs toward a culture of safety. Thus, as previously described, in Tuscany the main stated objective of the Centre for Clinical Risk Management and Patient safety is to promote a culture of safety based on a no-blame attitude.

Organizational culture and the practices that underpin it are essential targets for change toward greater patient safety. In particular, the necessary cultural change toward openness and accountability is fundamental as a premise for developing a patient safety system based on a systemic approach. The goal is to have a culture where reporting mistakes, including near misses, is routine, and where learning from mistakes is a behavior that is clearly valued and rewarded.

Training and other initiatives, for example, meetings, communication campaigns, and promotion of best practices, promote a patient safety culture (Cozens 2001). These activities, as described earlier in this chapter, have usually addressed two main objectives: the improvement of the system and the diffusion of a culture of safety through the direct involvement of clinicians at the sharp end.

39.5 Conclusions

PSM may become a top function of health care organizations, but needs to be better described and assessed using currently available experience of success in other sectors, while at the same time taking into account the specificities of health care. The integration of methods and tools in a PSM system and the development of policies and organizational assets to sustain the PSM activities are fundamental requirements to effectively improve patient safety and reduce risks for patients, clinicians, and the organization itself.

Yet, the effectiveness of PSM has to be measured against a classical structure-process-outcomes set of criteria, including the viewpoints of the different stakeholders: patients, clinicians, managers, and policy makers who may have different values and hence expectations. Ergonomics, in particular its cognitive and organizational domains, can make an original contribution on the basis of the evidence coming from other industrial sectors; it can provide a theoretical and methodological approach where the strategic goal of safety can face the challenges of efficiency and productivity of health care services.

The economic as well as the human costs of adverse events are well known worldwide. It is now time to demonstrate the benefit of PSM for the different stakeholders.

The application of individual methods and tools is not sufficient to improve patient safety and address human and economic costs over time. As discussed in the chapter, with the implementation of PSM in the Tuscan region, a set of indicators has been included to assess health care organization, and the litigation ratio is going down while learning and improvement indicators are slowly going up.

Since January 2010, a patient claim for an adverse event may be directly managed and eventually compensated by a single trust following the regional policy on disclosure and compensation. The shift to an internal management of claims was made possible because the PSM provided data demonstrating the capacity to anticipate and control risks at the frontline and management levels. As a consequence, the trusts gained at least one third of the budget for claims compared to previous contracts with external brokers and insurance companies; this occurred because adverse events can be reduced when local solutions or regional good practices are adopted. Moreover, patients and their families receive compensation more quickly (if it is due) and without a litigation process that undermines trust toward caregivers, especially considering that those caregivers are often involved in providing the care needed after the adverse event. Health care workers can be protected against legal prosecution because the patient safety manager initiates a negotiation process after each claim. Finally, clinicians and managers are more aware of risks and become more accountable for taking actions to improve patient safety, as the economic burden of adverse events is not transferred to external actors.

This experience shows that the implementation of a PSM system based on ergonomics principles can improve patient safety by providing concrete results. In the future, the comparison of PSM between health care services and other high-risk industries will be a fruitful research topic for ergonomics.

References

Agency for Healthcare Research and Quality (AHRQ). 2010. Web M&M. http://webmm.ahrq.gov/ (accessed June 30, 2011).

Albolino, S., Cook, R., and O'Connor. 2007. Making sense of sense making: What are physicians doing on "rounds" in the intensive care unit? *Cogn Technol Work* 9(3):131–137.

Albolino, S., Tartaglia, R., Bellandi, T., Amicosante, E., Bianchini, E., and Biggeri, A. 2010. Patient safety and incident reporting: The point of view of the Italian healthcare workers. *Qual Saf Health Care* 19:8–12.

Amalberti, R., Auroy, Y., Berwick, D., and Barach, P. 2005. Five system barriers to achieving ultrasafe health care. *Ann Intern Med* 142(9):756–764.

Amalberti, R. and Hourlier, S. 2007. Human error reduction strategies in healthcare. In *Handbook of Human Factors and Ergonomics in Healthcare and Patient Safety*, ed. P. Carayon, pp. 525–577. Mahwah, NJ: Lawrence Erlbaum Associates.

Aranaz-Andrés, J.M., Aibar-Remón, C., Vitaller-Burillo, J., Requena-Puche, J., Terol-García, E., Kelley, E. et al. 2009. Impact and preventability of adverse events in Spanish public hospitals: Results of the Spanish National Study of Adverse Events (ENEAS). *Int J Qual Health Care* 21(6):408–414.

Bagnara, S., Parlangeli, O., and Tartaglia, R. 2010. Are hospitals becoming high reliability organizations? *Appl Ergon* 41(5):713–718.Epub January 27, 2010.

Barach, P. and Small, S.D. 2000. Reporting and preventing medical mishaps: Lessons from non-medical near miss reporting systems. *BMJ* 320(7237):759–763.

Bates, D.W. and Gawande, A.A. 2003. Improving safety with information technology. *N Engl J Med* 348:2526–2534.

Bellandi, T., Albolino, S., Fiorani, M., Ranzani, F., and Tartaglia R. 2008a. The real practice of the Tuscany's regional system for patient safety management: The results of the first two years 2005–2006. Strasbourg: Proceedings of the International Conference HEPS2008.

Bellandi, T., Albolino, S., and Tomassini, C. 2007. How to create a safety culture in the healthcare system: The experience of the Tuscany region. *Theor Issues Ergon Sci* 8(5):495–507.

Bellandi, T., Forgeschi, G., and Fiorani, M. 2008b. Patient falls in the Tuscany Regional Healthcare Service: An evaluation of the hospital environmental hazards. *Proceedings of the 40th Annual Conference of the Nordic Ergonomics Society*, August 11–13, 2008, Reykjavik, Iceland.

Bellandi, T., Tartaglia, R., and Albolino, S. 2005. The Tuscany's model for clinical risk management. In *Healthcare Systems Ergonomics and Patient Safety*, ed. R. Tartaglia, S. Bagnara, T. Bellandi, and S. Albolino, pp. 94–98. London, U.K.: Taylor & Francis.

Berwick, D. 2008. The science of improvement. *JAMA* 299(10):1182–1184. doi:10.1001/jama.299.10.1182.

Bjørn, B., Anhøj, J., and Lilja, B. 2009. Reporting of patient safety incidents: Experience from five years with a national reporting system [in Danish]. *Ugeskr Laeger* 171(20):1677–1680.

Butera, F., ed. 2001. *Il campanile e la rete: e-Business e piccole e medie imprese*. Milano, Italy: Il Sole 24 Ore.

Carayon, P., Alvarado, C.J., and Hundt, A.S. 2007. Work system design in healthcare. In *Handbook of Human Factors and Ergonomics in Healthcare and Patient Safety*, ed. P. Carayon, pp. 61–79. Mahwah, NJ: Lawrence Erlbaum Associates.

Committee on Data Standards for Patient Safety. 2004. *Patient Safety: Achieving a New Standard for Care*. Washington, DC: National Academies Press.

Committee on Quality of Health Care in America and Institute of Medicine. 2001. *Crossing the Quality Chasm*. Washington, DC: National Academies Press.

Cook, R. and Rasmussen, J. 2005. "Going solid": A model of system dynamics and consequences for patient safety. *Qual Saf Health Care* 14:130–134.

Cooper, J.B., Newbower, R.S., Long, C.D., and McPeek, B. 1978. Preventable anesthesia mishaps: A study of human factors. *Anesthesiology* 49(6):399–406.

Cozens, J.F. 2001. Cultures for improving patient safety through learning: The role of teamwork. *Qual Health Care* 10:ii26–ii31.

Danish Act on Patient Safety. 2004. http://patientsikkerhed. dk/en/about_the_danish_society_for_patient_safety/ act_on_patient_safety/ (accessed July 12, 2010).

Department of Health. 2006. Safety first—A report for patients, clinicians and healthcare managers. http://www. dh.gov.uk/en/Publicationsandstatistics/Publications/ PublicationsPolicyAndGuidance/DH_062848 (accessed July 12, 2010).

Department of Health Expert Group. 2000. An organization with a memory. http://www.dh.gov.uk/prod_consum_ dh/groups/dh_digitalassets/@dh/@en/documents/ digitalasset/dh_4065086.pdf (accessed July 12, 2010).

de Vries, E.N., Ramrattan, M.A., Smorenburg, S.M., Gouma, D.J., and Boermeester, M.A. 2008. The incidence and nature of in-hospital adverse events: A systematic review. *Qual Saf Health Care* 17:216–223.

Flin, R., Crichton, M., and O'Connor, P. 2008. *Safety at the Sharp End. A Guide to the Non-Technical Skills*. Burlington, VT: Ashgate.

Frankel, A., Grillo, S.P., Pittman, M., Thomas, E.J., Horowitz, L., and Page, M. 2008. Revealing and resolving patient safety defects: The impact of leadership WalkRounds on frontline caregiver assessments of patient safety. *Health Serv Res* 43(6):2050–2066.

Gaba, D., Maxwell, M., and DeAnda, A. 1987. Anesthetic mishaps: Breaking the chain of accident evolution. *Anesthesiology* 66:670–676.

Haynes, A.B., Weiser, T.G., Berry, W.R., Lipsitz, S.R., Breizat, A.H., Dellinger, E.P. et al. 2009. A surgical safety checklist to reduce morbidity and mortality in a global population. *N Engl J Med* 360(5):491–499.

Hendrick, H.W. 2007. A historical perspective and overview of macroergonomics. In *Handbook of Human Factors and Ergonomics in Healthcare and Patient Safety*, ed. P. Carayon, pp. 41–61. Mahwah, NJ: Lawrence Erlbaum Associates.

Hines, S., Luna, K., Lofthus, J., Marquardt, M., and Stelmokas, D. 2008. *Becoming a High Reliability Organization: Operational Advice for Hospital Leaders*. (Prepared by the Lewin Group under Contract No. 290-04-0011.) AHRQ Publication No. 08-0022. Rockville, MD: Agency for Healthcare Research and Quality.

Hutchinson, A., Coster, J.E., Cooper, K.L., McIntosh, A., Walters, S.J., Bath, P.A. et al. 2010. Comparison of case note review methods for evaluating quality and safety in health care. *Health Technol Assess* 14(10):iii–iv, ix–x, 1–144.

Johnson, C. 2007. Human factors of health care reporting systems. In *Handbook of Human Factors and Ergonomics in Healthcare and Patient Safety*, ed. P. Carayon, pp. 525–560. Mahwah, NJ: Lawrence Erlbaum Associates.

Kaldjian, L.C., Jones, E.W., and Rosenthal, G. 2006. Facilitating and impeding factors for physicians' error disclosure: A structured literature review. *Jt Comm J Qual Patient Saf* 32(4):188–198.

Kilbridge, P.M. and Classen, D.C. 2008. The informatics opportunities at the intersection of patient safety and clinical informatics. *J Am Med Inform Assoc* 15:397–407.

Kirwan, B. 1994. *A Guide to Practical Human Reliability Assessment*. London, U.K.: Taylor & Francis.

Kohn, L.T., Corrigan, J.M., and Donaldson, M.S., eds. 1999. *To Err Is Human, Building a Safer Health System*. Washington, DC: Institute of Medicine National Academy Press.

Leape, L.L. and Berwick, D.M. 2005. Five years after To Err Is Human: What have we learned? *JAMA* 293(19):2384–2390.

Lilja Pedersen, B. 2007. Incident Reporting and Patient Safety in the Danish Healthcare System. Lecture at the Advanced Course on Clinical Risk Management, Sant'Anna University, Pisa, Italy.

Ministero della Salute. 2009. Protocollo di Monitoraggio degli eventi sentinella 2°Rapporto. http://www.salute.gov.it/imgs/C_17_pubblicazioni_1129_allegato.pdf (accessed July 12, 2010).

Mishra, A., Catchpole, K., and McCulloch, P. 2009. The Oxford NOTECHS system: Reliability and validity of a tool for measuring teamwork behaviour in the operating theatre. *Qual Saf Health Care* 18(2):104–108.

Mohr, J.J., Batalden, P., and Barach, P. 2004. Integrating patient safety into the clinical microsystem. *Qual Saf Health Care* 13(Suppl II):ii34–ii38.

Morris, G.P. 2000. Anaesthesia and fatigue: An analysis of the first 10 years of the Australian Incident Monitoring Study 1987–1997. *Anaesth Intensive Care* 28(3):300–304.

Naessens, J.M., Campbell, C.R., Huddleston, J.M., Berg, B.P., Lefante, J.J., Williams, A.R. et al. 2009. A comparison of hospital adverse events identified by three widely used detection methods. *Int J Qual Health Care* 21(4):301–307.

National Health Service Litigation Authority (NHSLA). 2010. Risk management. http://www.nhsla.com/RiskManagement/ (accessed July 12, 2010).

National Patient Safety Agency (NPSA). 2004a. Seven steps to patient safety: Full reference guide. http://www.nrls.npsa.nhs.uk/resources/collections/seven-steps-to-patient-safety/?entryid45=59787 (accessed July 12, 2010).

National Patient Safety Agency (NPSA). 2004b. Root cause analysis toolkit. http://www.nrls.npsa.nhs.uk/resources/collections/root-cause-analysis/?entryid45=59901 (accessed July 12, 2010).

Neale, G., Woloshynowych, M., and Vincent, C. 2003. Retrospective case record review: A blunt instrument that needs sharpening. *Qual Saf Health Care* 12:2–3.

Nuti, S., Tartaglia R., and Niccolai F. 2007. *Rischio Clinico E Sicurezza Del Paziente, Modelli E Soluzioni Nel Contesto Internazionale*. Bologna, Italy: il Mulino.

Olsen, S., Neale, G., Schwab, K., Psaila, B., Patel, T., and Chapman, E.J. 2007. Hospital staff should use more than one method to detect adverse events and potential adverse events: Incident reporting, pharmacist surveillance, and local real time record review may all have place. *Qual Saf Health Care* 16:40–44.

Øvretveit, J. 2005. Which interventions are effective for improving patient safety?—A review of research evidence. http://www.lime.ki.se/uploads/images/996/SafetyreviewOvretveit.pdf (accessed July 12, 2010).

Øvretveit, J. and Gustafson, D. 2002. Evaluation of quality improvement programmes. *Qual Saf Health Care* 11:270–275.

Parand, A., Burnett, S., Benn, J., Iskander, S., Pinto, A., and Vincent, C. 2010. Medical engagement in organisation-wide safety and quality-improvement programmes: Experience in the UK Safer Patients Initiative. *Qual Saf Health Care* 19: 1–5.

Perrow, C. 1999. *Normal Accidents*. Princeton, NJ: Princeton University Press.

Pringle, M., Bradley, C.P., Carmichael, C.M., Wallis, H., and Moore, A. 1995. Significant event auditing. A study of the feasibility and potential of case-based auditing in primary medical care. *Occas Pap R Coll Gen Pract* (70):i–viii, 1–71.

Pronovost, P., Berenholtz, S.M., Goeschel, C., Thom, I., Watson, S.R., Holzmueller, C.G. et al. 2008a. Improving patient safety in intensive care units in Michigan. *J Crit Care* 23(2):207–221.

Pronovost, P., Goeschel, C.A., Colantuoni, E., Watson, S., Lubomski, L.H., and Berenholtz, S.M. 2010. Sustaining reductions in catheter related bloodstream infections in Michigan intensive care units: Observational study. *BMJ* 340:c309.

Pronovost, P.J., Rosenstein, B.J., Paine, L., Miller, M.R., Haller, K., Davis, R. et al. 2008b. Paying the piper: Investing in infrastructure for patient safety. *Jt Comm J Qual Patient Saf* 34(6):342–348.

Pronovost, P., Weast, B., Rosenstein, B.J., Seymour, L., Holzmueller, C.G., Paine, L. et al. 2005. Implementing and validating a comprehensive unit based safety program. *J Patient Saf* 1:33–40.

Reason, J. 1997. *Managing the Risks of Organizational Accidents*. London, U.K.: Ashgate Publishing Company.

Reason, J. 2000. Human error: Models and management. *BMJ* 320:768–770.

Reason, J. 2004. Beyond the organisational accident: The need for "error wisdom" on the frontline. *Qual Saf Health Care* 13(Suppl II):ii28–ii33.

Reason, J. 2009. *The Human Contribution*. London, U.K.: Ashgate Publishing Company.

Remine consortium. 2010. Remine project. http://www.remine-project.eu/ (accessed July 12, 2010).

Requena, J., Aranaz, J.M., Gea, M.T., Limón, R., Mirallesa, J.J., and Vitaller, J. 2010. Evolución de la prevalencia de eventos adversos relacionados con la asistencia en hospitales de la Comunidad Valenciana. *Rev Calid Asist* 25(5): 244–249.

Runciman, W.B. and Moller, J. 2000. *Iatrogenic Injury in Australia*. Adelaide, Australia: Australian Patient Safety Foundation Inc.

Runciman, W.B., Sellen, A., Webb, R.K., Williamson, J.A., Currie, M., Morgan, C. et al. 1993. The Australian Incident Monitoring Study. Errors, incidents and accidents in anaesthetic practice. *Anaesth Intensive Care* 21(5):506–519.

Sari, A., Sheldon, T.A., Cracknell, A., and Turnbull, A. 2006. Sensitivity of routine system for reporting patient safety incidents in an NHS hospital: Retrospective patient case note review. *BMJ* 334:79.

Schouten, L.M., Hulscher, M.E., van Everdingen, J.J., Huijsman, R., and Grol, R.P. 2008. Evidence for the impact of quality improvement collaboratives: Systematic review. *BMJ* 336(7659):1491–1494.

Scott, W.R. and Davies, G.F. 2006. *Organizations and Organizing: Rational, Natural and Open Systems Perspectives*. Upper Saddle Riner, NJ: Prentice Hall.

Tartaglia, R., Albolino, S., and Bellandi, T. 2008. Gestione del rischio clinico e sicurezza del paziente: Le esperienze delle Regioni. *Monitor* 19:31–41.

The Joint Commission. 2010. Sentinel events statistics. http://www.jointcommission.org/SentinelEvents/Statistics/ (accessed July 12, 2010).

Thomas, E.J., Lipsitz, S.R., Studdert, D.M., and Brennan, T.A. 2002. The reliability of medical record review for estimating adverse event rates. *Ann Intern Med* 136(11):812–816.

Thomas, E.J. and Petersen, L.A. 2003. Measuring errors and adverse events in health care. *J Gen Intern Med* 18(1):61–67.

Vincent, C. 2006. *Patient Safety*. London, U.K.: Churchill Livingstone.

Vincent, C. 2010. *Patient Safety*. 2nd edn. Chichester, U.K.: Wiley Blackwell.

Vincent, C., Taylor-Adams, S., and Stanhope, N. 1998. Framework for analysing risk and safety in clinical medicine. *BMJ* 316:1154–1157.

Weick, K. 1995. *Sensemaking in Organizations*. London, U.K.: Sage Publications.

Weick, K.E. and Sutcliffe, K.M. 2001. *Managing the Unexpected*. San Francisco, CA: Jossey-Bass.

Woloshynowych, M., Rogers, S., Taylor-Adams, S., and Vincent, C. 2005. The investigation and analysis of critical incidents and adverse events in healthcare. *Health Technol Assess* 9(19):1–143, iii.

World Health Organization (WHO). 2005. WHO draft guidelines for adverse event reporting and learning systems. http://www.who.int/patientsafety/events/05/Reporting_Guidelines.pdf (accessed July 12, 2010).

World Health Organization (WHO). 2009. The conceptual framework for the International Classification for Patient Safety. http://www.who.int/patientsafety/taxonomy/icps_full_report.pdf (accessed July 12, 2010).

World Health Organization (WHO). 2010. Patient safety curriculum guide for medical schools. http://www.who.int/patientsafety/education/curriculum/download/en/index.html (accessed July 12, 2010).

Zegers, M., de Bruijne, M.C., Wagner, C., Groenewegen, P.P., Waaijman, R., and van der Wal, G. 2007. Design of a retrospective patient record study on the occurrence of adverse events among patients in Dutch hospitals. *BMC Health Serv Res* 7:27.

Section IX

Specific Applications

40

Human Factors and Ergonomics in Intensive Care Units

Ayse P. Gurses, Bradford D. Winters, Priyadarshini R. Pennathur, Pascale Carayon, and Peter J. Pronovost

CONTENTS

40.1 Intensive Care Units and Patient Safety 693
40.2 ICU as a Work System 694
40.3 Individual Characteristics 694
    40.3.1 Patients 694
    40.3.2 Providers 695
40.4 Organizational Characteristics 696
    40.4.1 Teamwork and Communication 696
    40.4.2 Multidisciplinary Rounds 697
    40.4.3 Handoffs in ICUs 697
40.5 Tools and Technologies 697
    40.5.1 Medical Devices and Alerting Systems 697
    40.5.2 Health Information Technologies 699
    40.5.3 Low-Tech Tools 700
40.6 Physical Environment 700
40.7 Tasks 701
40.8 Processes in ICUs 702
40.9 Human Factors Interventions in ICU 702
40.10 Conclusions 703
Acknowledgments 704
References 704

40.1 Intensive Care Units and Patient Safety

Intensive care units (ICUs) are one of the largest and most expensive components of American health care. Approximately 5890 hospitals in the United States provide critical care medicine and ICUs account for approximately 15% of inpatient acute care beds. In 2005, the total ICU patient days in the United States were 152.5 million. Up to one-half of people in the United States are admitted to an ICU in the last year of their life (Barnato et al. 2004; Garland 2005), and one-fifth die while being treated in an ICU (Angus et al. 2004; Garland 2005). The cumulative effect of this costly resource results in an annual cost of approximately $82 billion, which accounts for 0.66% of the U.S. gross national product (Halpern and Pastores 2010).

ICUs often provide lifesaving care, yet they also pose risks to patients. Typically, patients with very high acuity who require intensive monitoring and care are treated in ICUs. Due to the nature of clinical and other tasks conducted there and the tremendous information burden and high workload experienced by ICU care providers, this high-paced environment is prone to medical errors (Donchin et al. 1995; Graf et al. 2005; Osmon et al. 2004). For example, a study conducted in a medical–surgical ICU used both self-reports and direct observations and found 1.7 errors per patient per day (Donchin et al. 1995). Of these errors, 29% had the potential to cause significant harm including death. Given an average ICU length of stay (LOS) of 3 days, most patients will suffer a significant error. Patients in the ICU received an average of 178 activities per day, of which approximately 1% were judged to involve errors (Wu et al. 2002). A more recent observational study in a medical ICU found that the risk of experiencing a medical error or incident increases by 8% per ICU day (Graf et al. 2005).

Several studies have further examined various types of errors in ICUs. For example, a prospective, observational study found that human errors were involved in 67% of the major iatrogenic complications including severe hypotension, respiratory distress, pneumothorax, and cardiac arrest. The risk of ICU mortality was about twofold higher for patients with iatrogenic complications (Giraud et al. 1993). Tissot et al. (1999) detected 132 medication preparation and administration errors in a total of 2009 observed events (6.6%). Cullen et al. (1997) compared the frequency of preventable adverse drug events and potential adverse drug events in ICUs and non-ICUs. The combined rate of preventable adverse drug events and potential adverse drug events in ICUs was 19 events per 1000 patient days. Incidents involving invasive line, tube, and drain (LTD) placement, maintenance, and removal accounted for 14% of all the incidents identified in 18 U.S. ICUs based on a voluntary, anonymous web-based patient safety reporting system. More than 60% of these LTD-related incidents were judged to be preventable. Fifty-six percent of the patients who had an LTD incident experienced a physical injury and 23% had an anticipated increase in their hospital LOS (Needham et al. 2005).

Overall, the research in patient safety and medical errors in ICUs shows that errors are frequent in ICUs and result in poor clinical outcomes. Hence, there is an urgent need to improve patient safety in ICUs. Human factors and systems design issues have been identified as major contributors to medical errors in ICUs. This chapter describes how the ICU work system design affects patients and clinicians, provides an overview of human factors and systems design issues in ICUs, and discusses directions for future research and quality improvement efforts in ICUs.

40.2 ICU as a Work System

There are various theoretical and conceptual human factors and ergonomics (HFE) models that can be used to describe and explain how ICUs operate. In this chapter, we use one of these models, the Systems Engineering Initiative for Patient Safety (SEIPS) Model (Carayon et al. 2006), to provide an overview of HFE issues in ICUs. The SEIPS model was developed based on the well-known work system model, the Balance Theory Model (Carayon and Smith 2000; Smith and Carayon-Sainfort 1989). According to the Balance Theory Model, a work system such as an ICU can be characterized by five elements: at the center of the model is an *individual* in an ICU (e.g., intensivist, nurse, respiratory therapist, and patient) with their physical characteristics, psychological characteristics, education, skill level, and motivation affecting their performance. This individual performs various *tasks* that have different physical, mental, temporal demands on the individual (e.g., inserting a central line, administering medication, managing airway). The

individual uses different *tools and technologies* (e.g., infusion pump, computerized provider order entry (CPOE) system, a paper-based checklist) to perform these tasks. These tasks are carried out in the ICU's *physical environment* that has certain characteristics (e.g., open bay design versus private rooms, noise level) under specific *organizational conditions* (e.g., safety culture, level of teamwork, leadership support).

According to the SEIPS model, the characteristics of the five elements in the ICU work system and their interactions with each other determine the performance of processes (e.g., compliance with evidence-based guidelines, effectiveness of supply chain management process), which in turn impact patient (e.g., medical errors, mortality), employee (e.g., stress, job satisfaction), and unit/organizational outcomes (e.g., clinician turnover rates, profitability). Depending on how a change is designed and implemented, changing a characteristic of a particular element in the ICU work system can have negative or positive effects on other elements and processes and may vary across outcomes, reducing risks for one outcome while increasing the risks for another.

Based on the SEIPS model (Figure 40.1), many HFE issues can be identified in an ICU work system. In this chapter, we describe some of these important issues using the SEIPS model as a guide. These include patient and provider characteristics (individual); workload (tasks); health information technologies (HITs), medical devices, and paper-based information tools (tools and technologies); physical environment; teamwork and communication, multidisciplinary rounds (MDRs), and handoffs (organizational characteristics); and compliance with evidence-based guidelines (processes). In the final section of this chapter, we describe some of the HFE interventions that have been shown to improve ICU work systems, processes, and outcomes.

40.3 Individual Characteristics

40.3.1 Patients

Newly admitted ICU patients generally fall into two major management categories. The first category is patients who are already unstable and in a life-threatening state on arrival and require immediate and aggressive care to stabilize their physiology. They may require airway control and mechanical invasive or noninvasive ventilation for respiratory failure, fluid resuscitation and vasopressor support to maintain blood pressure, and end-organ perfusion and/or renal replacement therapy. Placement of invasive arterial and central venous access is common in these situations. The other category is patients who are admitted to the

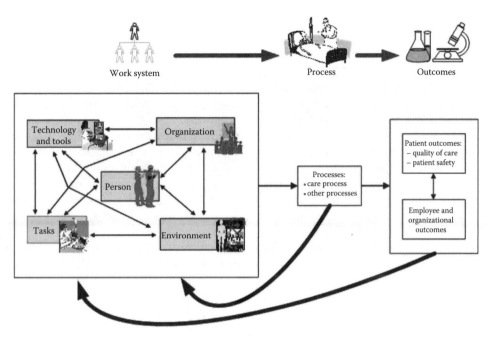

FIGURE 40.1
SEIPS model of work system and patient safety. (From Carayon, P. et al., *Qual. Saf. Health Care*, 15(Suppl. 1), i50, 2006.)

ICU for expectant management. These patients arrive in a mostly stable physiological state but their underlying disease process (medical or surgical) puts them at high risk of rapid deterioration into an unstable life-threatening condition. The role of the ICU team in these latter patients is proactive and preventative, seeking to avoid deterioration, while in the former category, their role is reactive trying to restore stability. Despite the best proactive efforts, some of the expectantly managed patients may still deteriorate into an unstable state and patients who have prolonged ICU stays frequently go through cycles of physiological stability and instability. These cycles may reflect their underlying illness but are often secondary to iatrogenic causes such as central line associated bloodstream infections (CLABSI), ventilator-associated pneumonia, adverse drug reactions, and other medically induced injuries. Therefore, the varied, dynamic, rapidly changing conditions of patients significantly affect the tasks performed by ICU providers and require them to quickly adapt to those changing conditions.

40.3.2 Providers

ICUs also differ structurally in their care models (Angus et al. 2006; Barie et al. 2000; Chang et al. 2005; Lustbader and Fein 2000; Milstein et al. 2000; Pronovost et al. 2002, 2007) and this results in significant variability in adaptive work (i.e., teamwork, communication, safety culture) and even in technical work (e.g., practice variations) that the human factors engineers must take into

account. In some ICUs, the ICU care providers manage the patient along with the patient's primary care physician or surgeon. In other units, ICU physicians are not involved in the care or are involved only as consults, and the primary care physician or surgeon provides care in those cases. Even in those units with ICU physicians, some patients may be cared for by the ICU physicians while others are not. In addition, while some ICU physicians have admission and discharge authority, in other instances, authority rests with the primary care physician, often resulting in power struggles for using the precious resource of an ICU bed. Despite this variation, the presence of an ICU physician results in a 30% reduction in mortality and LOS (Pronovost et al. 2002). As a result of this evidence, the Leapfrog Group, a consortium of large employers, created safety standards that ICUs should be staffed with ICU physicians. Yet only 25% of ICUs have this lifesaving intervention. There is a major shortage of critical care physicians in the United States that is predicted to only worsen, leaving many ICUs without even partial intensivist coverage (Krell 2008). Many services have physician assistants and nurse practitioners on the physician team to help extend coverage. Other units rely on the tele-ICU technology to provide access to critical care expertise.

This lack of ability to adhere to any national standard for critical care staffing or care models has significance for ICU culture, particularly for the goal of using HFE to help create a culture of safety and quality. When no central physician ICU service exists or if its presence and activities are limited, engaging physicians on the other widely disparate services that may admit to the ICU and adhering to the principles of safe design may be more difficult. There is a need for a central role to manage the expensive and limited resource of ICU beds. HFE evaluations can help determine which patients may need an ICU bed based on robust decision-making processes.

When intensivists do play a role in the ICU, they come from a variety of primary specialties (pediatrics, surgery, internal medicine, anesthesiology, or emergency medicine) prior to a fellowship in critical care, which adds further variability to practice patterns. Human factors engineers will find that this variability in the physician team coverage influences interactions with and among other providers including nursing, respiratory therapy, and occupational/physical therapy, resulting in locally developed communication structures and practice patterns. ICU nurses and other providers may have also come from other care areas that are vastly different from the ICU adding further influences.

In summary, human factors engineers designing systems solutions in the ICU environment need to be cognizant of the diverse types of ICUs, highly variable staffing patterns and the effects these variations may have on adaptive work as well as the technical work performed. The large number of permutations created by all of this variability in practice combined with the high-risk, high-stress environment has implications for the applicability of HFE solutions across ICU settings.

40.4 Organizational Characteristics

40.4.1 Teamwork and Communication

High diversity in the skills and responsibilities of care providers and the need for complex interactions among them for patient care demand teamwork and communication as major organizational factors in determining the safety of care in ICUs. Teamwork behaviors that prevent medical errors in ICUs include conducting briefings, managing and distributing workload among team members, communicating plans and sharing a mental model, developing contingency plans, monitoring/cross-checking the care that is provided continuously, being vigilant, asking questions to clarify and confirm plans (inquiry), being assertive, evaluating and adapting plans when necessary, and resolving conflicts effectively (Helmreich and Merritt 1998). Nurses and physicians often have discrepant attitudes about teamwork in ICUs. Compared to physicians, nurses generally perceive more difficulties in speaking up, report that conflicts are not resolved appropriately, and input from nurses is not adequately received (Thomas et al. 2003).

Studies have shown that there is a link between teamwork and patient outcomes in ICUs (Manser 2009; Wheelan et al. 2003). For example, a survey study conducted among 394 staff members of 17 ICUs indicated that mortality rates were lower in ICUs that had higher stages of group development, more structured and organized teams, and less dependent and more trusting team members (Wheelan et al. 2003). Units that have a dedicated ICU care team including an intensivist, an ICU nurse, respiratory therapist, and a pharmacist have the lowest mortality rates and costs (Brilli et al. 2001). Furthermore, when teamwork and communication improve in ICUs, employee outcomes also improve (Manser 2009). For example, implementing multidisciplinary structured work shift evaluations, during which care providers come together and evaluate how the working day went in the last 30 min of each shift, was shown to improve team communication and decrease emotional exhaustion among care providers in a pediatric intensive care unit (PICU) (Sluiter et al. 2005). Human factors research indicates that employee outcomes such as stress, fatigue, burnout, and emotional exhaustion can have a negative effect on clinicians' abilities to provide safe and quality care (Carayon et al. 2006).

One of the requirements of good teamwork is effective communication. Effective communication among ICU team members has been identified as a critical factor for patient safety and efficiency of care delivery (Baggs et al. 1999; Donchin et al. 1995; Knaus et al. 1986; Shortell et al. 1994). Problems of communication between physicians and nurses were identified as the major source of errors, contributing to 37% of errors (Donchin et al. 1995). Studies conducted in ICUs indicated that as the communication openness among team members improved, the degree to which individual care providers reported understanding patient care goals (situational awareness) also increased (Reader et al., 2007a, 2007b). Furthermore, a study in 42 ICUs identified effective communication and collaboration as an organizational characteristic leading to better patient outcomes (Shortell et al. 1994).

Both explicit (e.g., verbal discourse) and implicit communication is used in health care systems such as ICUs, for information sharing and coordination of care. Implicit communication is the transmission of information through the environment: for example, through the use of information tools and physical objects in the work environment. Human factors research in high-risk industries indicate that under stress, time pressure, and relatively routine situations, team members use implicit communication to complete the task in a safe and efficient manner (Entin and Serfaty 1999). Therefore, it is natural to expect that the care providers working in the busy, chaotic, and stressful environment of ICUs create and use information tools extensively to represent and distribute information in ways more efficient and reliable than verbal means. Two of the most important communication and information transfer mechanisms, that is, MDRs and handoffs, are discussed in the following section to provide insights on their impact on patient care and to identify opportunity areas for interventions.

40.4.2 Multidisciplinary Rounds

MDRs are one of the major communication mechanisms in ICUs, which typically occur on a daily basis. During MDR, care providers from different disciplines convene to manage care in the complex ICU system. MDR serve multiple purposes: communication of critical information among clinicians from different disciplines, assessment of patient's current condition, developing/clarifying the plan of care and goals for patient, and assignment of tasks and responsibility (Gurses and Xiao 2006). Implementing MDR has been shown to improve outcomes such as LOS (Dutton et al. 2003; Young et al. 1998), ventilator-associated pneumonia (Johnson et al. 2009), and costs (Young et al. 1998) in ICUs.

A human factors analysis using the SEIPS model can reveal various barriers to an effective and efficient MDR in ICUs: low staffing levels and high workload (task), high noise levels and interruptions in the ICU (physical environment) (Gurses and Xiao 2006), inadequate knowledge and skills of the intensivist leading the MDR (individual), use of information tools that cannot support clinicians' need of rapid access to information and rapid annotation during MDR (technology and tools) (Gurses and Xiao 2006), and ICU culture specifically as related to active involvement of families (Manias and Street 2001) in MDR (organization). For example, in an observational study of Surgical Intensive Care Unit (SICU) rounds, two interruptions (defined as disruptions for reasons unrelated to the patient being discussed during MDR) per daily MDR were observed (Friesdorf et al. 1994). To reduce paging interruptions during MDR, a medical–surgical ICU designated a consulting resident every day to be responsible for handling all telephone calls and consultations that arose during MDR (Dodek and Raboud 2003). Further research using HFE principles and methods is needed to improve effectiveness and efficiency of MDR.

40.4.3 Handoffs in ICUs

Handoffs, which can be defined as the transfer of role and responsibility for patient care, occur in ICUs multiple times in multiple ways and forms every day. Interdepartmental handoffs occur when patients are admitted to the ICU from another unit/service and when they are transferred from the ICU to a step-down unit or discharged home. Within the ICU, various handoffs occur between care providers from same or different

disciplines at the end of their shift such as nurse to nurse or nurse practitioner to resident or one attending physician to another. Handoffs are recognized as a point of vulnerability in the care of ICU patients. Poor and inadequate handoffs often threaten patient safety and lead to adverse events (Dracup and Morris 2008). Some of the HFE issues that can negatively impact the quality of handoffs include inadequately designed information tools to support the handoff process (tools/technologies), high workload (Watts et al. 2006) (task), distractions in work environment or not having adequate and quiet physical space in the ICU to give handoff report (Gurses and Carayon 2009) (physical environment), inadequate communication (Watts et al. 2006; Whittaker and Ball 2000) (organizational structure), and care providers having no formal training on how to give a handoff report (Solet et al. 2004) (individual).

Several interventions aimed at improving different types of handoffs in ICUs have been reported in the literature. For example, effective but simple changes such as simplifying and standardizing a partially automated handoff form used for resident handoffs improved accuracy, completeness, and clarity as to the exact time of transfer of care between residents (Wayne et al. 2008). Another research study demonstrated that reconciling medications with a paper-based discharge survey nearly eliminated medication errors in ICU discharge orders (Pronovost et al. 2003a,b). Introduction of a discharge information brochure was found to be useful for families to answer questions related to transfer of their child from PICU to the general floor (Linton et al. 2008).

Although various interventions have been designed and implemented to improve handoffs in ICUs, poor handoffs continue to be a major threat to patient safety. A major reason for this is that improving the safety of handoffs is a very complex problem, often involving personnel, team, and organizational barriers. Handoffs have very localized structure with practice variations even within a unit, with information content that needs to be interpreted by the receiver, and therefore poses severe constraints on time and demand. Further research that uses human factors engineering principles and methods is needed to identify patient safety hazards related to handoffs and to develop practical and sound interventions.

40.5 Tools and Technologies

40.5.1 Medical Devices and Alerting Systems

The ICU is a work environment filled with various medical devices and equipment, and various forms of HIT. This high-technology environment is constantly evolving and new technologies are introduced to improve care for the critically ill patients (Clemmer 2004). Many of these technologies are implemented to improve patient safety, in particular medication safety (Hassan et al. 2010). However, these technologies may create a range of unintended consequences because of their poor HFE design and implementation (Ash et al. 2009) or because of their impact on other work processes. If not appropriately designed and implemented, these technologies may increase rather than decrease risks and contribute to the "hostile environment of the intensive care unit" described by Donchin and Seagull (2002), adding to the many physical, cognitive, and psychosocial workplace stressors of the ICU (Alameddine et al. 2009).

For instance, smart IV pump technology has been proposed as a technological solution to prevent medication errors at the administration stage. By providing decision support at the point of care when IV medications are administered, this technology has the potential to reduce errors and adverse drug events (Malashock et al. 2004). However, a prospective randomized time-series trial of the implementation of smart infusion pumps in a cardiac surgical ICU and step-down units did not show any impact of the technology on serious medication errors (Rothschild et al. 2005). The researchers explain that this negative finding may have been due to nurses' bypass of the drug library and overriding of alerts. In an editorial accompanying this paper, Leape (2005) argues that these behaviors were probably related to the lack of usability of the technology, in particular, in a high-pressure environment such as the ICU: under clinical pressure to care for patients, nurses may not have taken the time to go through the extra steps for using the library and did override alerts. Another study of nurses' acceptance of smart IV pump technology confirms the importance of usability of the technology for acceptance and use by ICU nurses (Carayon et al. 2010).

Along with smart IV pump technology, bar coding medication administration is another technology that has been proposed to deal with the many medication errors and adverse drug events that occur at the administration stage. The study by Poon et al. (2010) shows significantly lower nontiming error rate and potential adverse drug events, and the elimination of transcription errors in hospital units, including ICUs using bar coding medication administration technology in comparison to units that did not use the technology. Even though bar coding medication administration technology may produce patient safety benefits, it may still create unintended consequences, work-arounds, and other problems related to the design of the technology resulting in a lack of fit between the technology and the work system in which it is used (Carayon et al. 2007; Koppel et al. 2008).

A major challenge of medical devices relates to the design of alerts appropriate for a hectic environment such as the ICU. In a Joint Commission review of death or injury events due to mechanical ventilation, from a total of 23 events, 65% were related to alarm issues (ECRI 2002; Korniewicz et al. 2008). Although each alert might be important on its own, an ICU almost never has only a single alert activated at a time. The numerous alerts that are activated in ICUs simultaneously cause an information overload and annoyance for the care providers (Chambrin et al. 1999). Furthermore, how many of these alarms truly have the potential to warn the provider of patient condition remains a significant question, as majority (about 95%) of the alarms or alerts in the ICU environment tend to be false alarms (Chambrin et al. 1999; Gorges et al. 2009; Imhoff and Kuhls 2006; Lawless 1994). For example, in a prospective and observational study to assess the functionality of monitoring alarms as warning systems in an adult ICU, Chambrin et al. (1999) reported that only 6% of the alarms in an ICU needed physician action, while the majority of the alarms were "false-positive" requiring no action from the care provider. Similarly, in another observational study aimed at evaluating the effectiveness of alerting mechanism of infusion pumps, only 12% of infusion pump alerts in an ICU were found to result in physician changing the original input. Eighty-eight percent of the infusion pump alerts were manually overridden (confirming original input) and 39% of all alerts were overridden but readjusted by the physicians. In addition, 56% of the alerts were not readjusted to be within recommended limits. This is another important issue that needs to be addressed by designing constraints (e.g., blocking medication delivery on alert until a change to the value or a confirmation occurs) in safety mechanisms like the alerts, so that they are certainly attended to (Rayo et al. 2007). Differentiating criticality and providing an integrated system of alerts are the two areas that can significantly benefit from HFE principles and methods.

Bria and Shabot (2005) describe their successful design and implementation of various integrated alerts for critical lab values and medication issues. Unlike alerts that become active only when significant thresholds are reached, these integrated alerts process and synthesize real-time clinical values and send directly to the physician's pagers. Alerts also need to be customizable to serve the individual physician with a better patient care scenario (Catley et al. 2003). For example, Reddy et al. (2005) found that not following a hierarchy in transmitting pager alerts resulted in loss of control for the residents, as the physician learns of the critical value before the residents could act on them. Similarly, physicians were overloaded with alerts, thus removing the distinction between critical alerts that needed their attention and decision making, and alerts that could have been easily handled by the residents. Thus, the design of alerts needs to focus on reducing disruptions to patient care (Korniewicz et al. 2008) and provider workflow, balancing frequency of alarms (Korniewicz et al. 2008) with severity of situation, and including safety constraints or interlocks as part of the design process.

40.5.2 Health Information Technologies

A range of HITs have been proposed to improve safety and quality of care in the ICU (Clemmer 2004). For instance, CPOE has been shown to reduce prescription errors in ICUs (Colpaert et al. 2006; Shulman et al. 2005). However, CPOE implementation in ICUs has also led to the emergence of new errors such as duplicate prescriptions, drug monitoring errors (Colpaert et al. 2006), and incorrect medication doses because of selection errors (Thompson et al. 2005). In addition, the implementation of CPOE in the ICU can engender major workflow problems, especially if the CPOE technology is not designed to fit the complex ICU environment (Cheng et al. 2003). Cheng et al. (2003) have identified three major workflow issues in the implementation of CPOE in ICU. The first issue concerns the increasing need for coordination between clinicians; this involves clinicians verifying orders entered in the CPOE with each other. This additional workload may add to an already high workload of ICU clinicians. The second workflow issue of CPOE is related to the poor ergonomic design of the CPOE interface. Because of inconveniences or ambiguities with the interface and other usability problems, clinicians may actually work around the safety features of CPOE by, for instance, entering orders as free text or using another physician's login. Thirdly, because physicians can enter orders from any location where they can access a computer, there is no longer a visual cue to nurses. These visual cues usually are physical flags on paper charts to alert the nurses or ward clerks to a new order. CPOE systems tend to lack such strong safe design strategies. Also the layout of electronic record interfaces often does not conform to the way physicians and nurses typically transfer information (e.g., organ system–based clinical rounds). These may negatively affect the coordination between physicians and nurses. For example, prior to CPOE, physicians would walk to the patient's bedside to write an order, and commonly have a discussion with the nurse about the patient and the treatment plan. With CPOE, this discussion may not occur. This phenomenon has also been documented by Beuscart-Zephir et al. (2005) in a study of CPOE implementation in a French hospital.

Problems in the design of various HIT in ICUs have been documented, even though many of them produce improvements in outcomes such as significant reduction of antibiotic prescribing and significant reduction

in patient LOS with the use of a handheld computer-based decision support system (Sintchenko et al. 2005). Lapinsky et al. (2004) designed and implemented a handheld computer-based knowledge access system linking a central academic ICU to multiple community-based ICUs. The objective of the system was to improve access to medical knowledge by ICU physicians. The users of the technology identified several benefits, such as small size and portability, and access to pharmaceutical information and literature updates. They also identified a range of barriers: small text fonts for reading, technical problems (e.g., battery discharge), inability to access information rapidly, and errors during text entry using handwriting recognition. Only half of the physicians regularly used the handheld device to access information on a regular basis. The authors speculate that this low level of acceptance and use was probably due to the design of the technology. Ramsay et al. (1997) found several usability problems in an evaluation of an information system for the support of cardiac intensive care, including problems in the prescribing and recording of medication dosage and administration.

Some limited research has examined the impact of HIT on ICU nurses. Kossman and Scheidenhelm (2008) gathered multiple sources of data (i.e., questionnaire, interview, and observation) from 46 nurses in medical–surgical and ICUs at two community hospitals. Nurses reported that they preferred electronic health records (EHR) to paper charts and that their use of EHR enhanced nursing work through increased information access, improved organization and efficiency, and helpful alert screens. On the negative side, nurses perceived that the EHR technology hindered nursing work through impaired critical thinking, decreased interdisciplinary communication, and a high demand on work time (73% reported spending at least half their shift using the records). The issue of the impact of EHR and other HIT on time spent on work activities and workload is receiving increasing attention. The literature review conducted by Mador and Shaw (2009) on the impact of "critical care information systems" on time spent charting and direct patient care showed discrepant findings. Some authors have suggested that studies need to go beyond calculating percentages of time spent on various tasks and analyze the workflow characteristics (Zheng et al. 2010).

Other studies have identified sociotechnical benefits of electronic medical record (EMR)/EHR implementation in ICU. For instance, Reddy et al. (2008) conducted a sociotechnical analysis of an EMR implementation in a SICU with a focus on collaborative activities using observations and interviews. They found that during morning rounds and medication administration, the EMR provided awareness because it was designed to incorporate data concerning others' work activities and an individual's own work activities.

A newer HIT in ICUs is the tele-ICU, that is, a form of telemedicine that allows critical care nurses and intensivists to monitor ICU patients remotely (Breslow et al. 2004). Research has shown mixed results for the impact of tele-ICU on various patient outcomes. For instance, the original study by Breslow et al. (2004) shows major improvement in LOS, mortality, and cost after the implementation of tele-ICU. More recent studies failed to find improvement in patient outcomes after the implementation of tele-ICU (Morrison et al. 2010; Thomas et al. 2009; Zawada et al. 2009). Very limited research has examined the impact of the tele-ICU on the work of ICU clinicians or on tele-ICU staff. Results from an observational study of nurses and physicians in a tele-ICU show that physicians and nurses interact with ICU staff about seven and six times per hour, respectively. This study also highlighted the importance of capturing data on interruptions and sources of interruptions, as well as data on the use of various information sources (Tang et al. 2007). Anders et al. (2007) performed 40 h of observation of eight nurses and one physician in one tele-ICU. Results showed that the tele-ICU fulfills three functions: (1) anomaly response: virtual ICU nurses processed information related to alerts and alarms and contacted other staff in the virtual ICU or the ICU if they perceived the need for follow-up or action; (2) access to specialized expertise: experienced virtual ICU nurses were observed to mentor junior ICU nurses, in other words ICU nurses had access to expertise of the virtual ICU nurses; and (3) sensemaking: virtual ICU nurses can make sense of what is happening with patients because they have access to many sources of data and have the resources (time, expertise) to synthesize the data. The question remains whether the tele-ICU is the "panacea" for the shortage of intensivists and critical care nurses and will produce improvements in patient outcomes (Sapirstein et al. 2009).

40.5.3 Low-Tech Tools

In addition to electronic tools, ICU care providers also develop and use low-tech tools such as paper forms (Clarke et al. 2004), checklists (Winters et al. 2009), bulletin boards, and whiteboard (Wikstrom and Larsson 2004) to support their work. Care providers create and tailor information objects in situated manners to facilitate collaborative work in ICUs. These information objects are placed and continually reconfigured in such a way that they facilitate information transfer in various physical locations under appropriate contexts. Handoff of responsibilities, for example, is sometimes accomplished by handing over a list of patients to be taken care of, while collaborators stand in front of a whiteboard. How clinical information is presented affects

physicians' and nurses' decision making in ICUs (Miller et al. 2009). If designed well and used appropriately, these "low-tech," mostly paper-based information tools can improve shared situation awareness of ICU team members by providing a shared representation of critical information, facilitate compliance with best practices, and improve patient and provider outcomes. For example, researchers have studied the impact of using a "daily goals" sheet (information tool) during rounds (Pronovost et al. 2003a,b). After implementing the daily goals sheet, more than 95% of the residents and nurses reported understanding the goals of care for the day as opposed to less than 10% before the intervention, which was an indication of effective communication and improved shared situation awareness. In addition, the ICU LOS decreased from a mean of 2.2–1.1 days after the intervention.

In summary, technologies have been constantly evolving to provide sophisticated functionalities to the ICU care team. Although some of the technological solutions have been successful in reducing errors and improving patient safety, there is still a gap in developing technologies that rank high on usability and has a good fit between the functions that the technology offer and the needs of the care providers. Consequently, workarounds and unintended consequences from technology implementation are abundant. Understanding the workflow characteristics of the ICU system and iteratively designing and implementing technologies with user-centered perspectives will help alleviate the current gap in technological support.

40.6 Physical Environment

Physical environmental design of an ICU can have significant impact on patients, families, and care providers. An HFE evaluation regarding transition from an open bay to a private room design in a neonatal intensive care unit (NICU) based on a survey study revealed that staff viewed the new private room design better in terms of overall physical environment, quality of patient care, job quality, interaction with NICU technology, and quality of life off the job. However, the perceived quality of patient care team interaction was significantly reduced after transition to the private room design mainly due to the more disseminated nature of this layout compared to the open bay design (Smith et al. 2009). Furthermore, the single-patient rooms and use of High Efficiency Particulate Air (HEPA) filter when indicated have been shown to significantly reduce infection rates (Ulrich et al. 2004).

The spaghetti syndrome of tangled cords, leads, and lines is a constant problem in ICUs that may cause adverse events such as accidental disconnection of breathing tubes or injecting wrong drug via wrong line. The electric cables, breathing tubes, lines, and cords also act as a physical barrier and may hinder access to patient and important equipment during emergency situations. Wireless systems may be able to address these problems but bandwidth may be limited given the number of devices required.

High noise levels, inadequate workspace, and crowdedness are other physical environmental issues common in ICUs that affect both patient care and care providers' quality of working life. Noise level in today's ICUs is far beyond recommended standards, which is not surprising given the various monitoring alarms, complex and noisy equipment such as ventilators, and constantly ringing phones in these units. A survey study found a significant correlation between noise-induced stress and emotional exhaustion among ICU nurses (Topf and Dillon 1988). Using a structural equation modeling approach, Gurses et al. (2009) showed that ICU nurses who reported working in a poor physical work environment (noisy, hectic, crowded) also reported higher perceived nursing workload, which in turn has a negative impact on nurses' quality of working life and perceived quality and safety of care.

The quality and safety of care and the quality of working life of care providers in ICUs can be significantly improved if HFE principles are used in the (re) design and (re)configuration of physical environment. However, compared to other characteristics of ICUs, ICU physical design is probably the least studied area from HFE point of view. HFE researchers and funding sources should focus more of their attention and resources on studying the impact of physical environment design on ICUs.

40.7 Tasks

There are many task-related HFE issues in ICUs. ICU care providers perform a high variety of tasks, most of which are critical for patient's care and require high clinical skills. Yet many of these tasks are poorly specified and most lack explicit training or competencies. The nature of tasks that need to be completed in an ICU, coupled with the shortage of staffed ICU beds, makes the jobs of ICU care providers challenging. High workload, time pressure, and need for continuous vigilance and attention are norm rather than an exception in today's ICUs. In the following paragraphs, we will focus on one of the most prominent task-related issues in ICUs: care provider workload.

Workload is a pervasive concept throughout the HFE literature, a concept important to both researchers and practitioners. Excessively high levels of workload can lead to errors and system failure, whereas underload can lead to inefficiency, complacency, and eventually errors (Braby et al. 1993). Since today's working ICU work environment imposes high demands on care providers, understanding how workload impinges on care providers' performance and ICU outcomes is critical.

High workload is a major problem in ICUs, negatively affecting providers' quality of work life (i.e., job satisfaction, stress, and fatigue) and the safety and quality of care. For example, a substantial number of studies investigated the impact of nursing workload on ICU outcomes and found that high ICU nursing workload was associated with higher infection rates, increased risk for respiratory failure and reintubation, and higher mortality rates (Penoyer 2010). Furthermore, research shows that workload is one of the most important job stressors among ICU nurses (Crickmore 1987; Malacrida et al. 1991; Oates and Oates 1995, 1996). Although limited in number, studies investigating physician workload in ICUs indicate similar findings (Dara and Afessa 2005; Landrigan et al. 2004; Pronovost et al. 2002). For example, a systematic review revealed that high-intensity physician staffing in ICUs was associated with a lower ICU mortality and reduced hospital and ICU LOS (Pronovost et al. 2002). Furthermore, a prospective randomized study evaluated the impact of replacing the first year residents' traditional extended (24 h or more) work shift with one that limits the consecutive hours of work to approximately 16 h and the maximum number of hours per week to 63 in two ICUs. Results indicated that reducing the number of work hours significantly reduced the serious medical errors (a medical error that causes harm or has substantial potential to cause harm) made by the first year residents (Landrigan et al. 2004).

Human factors research indicates that workload is a complex and multidimensional construct. Hence, there are multiple ways to conceptualize and measure workload. This is true for workload experiences in ICUs as well. For example, a literature review by Carayon and Gurses (2005) identified four different categories of measures for nursing workload in ICUs:

1. ICU-level measures: Measures workload at a macro-level such as the nurse/patient ratio.

2. Job-level measures: Measures that characterize workload as a stable characteristic of the job.

3. Patient-level measures: Measures that estimate workload based on the condition of the patient.

4. Situation-level measures: Measures how the design characteristics of an ICU work system affect demands on individual care providers.

Each type of workload measure has strengths and weaknesses and can be appropriate to use in certain situations. For example, nurse/patient ratio measures the overall workload in an ICU at a macro-level and can be used to compare ICUs and clinical outcomes. However, this measure does not provide much information about how the various design characteristics of an ICU work system affect workload. Workload measures at the situation level should be used to be able to evaluate the impact of ICU work design characteristics on workload. For example, Gurses et al. (2009) used the quantitative workload scale (Caplan et al. 1975), a well-known situation-level measure of workload, to study the impact of performance obstacles on nursing workload in ICUs. Results indicated that performance obstacles including poor physical work environment, having to deal with many family issues, disorganized supplies area, spending much time searching for patients' charts, delay in getting medications from pharmacy, and inadequate stocking of patient rooms significantly increased nursing workload, which in turn increased nurses' stress and fatigue levels and reduced quality and safety of care as perceived by the ICU nurses.

In summary, high workload is a major issue in ICUs that have adverse effects on both the quality and safety of care and care providers' quality of work life. There is a significant knowledge base within the human factors engineering discipline on how to measure and improve workload based on many years of research. We suggest that not only ICUs but also any health care system in general to use this knowledge base in their efforts to improve workload issues.

40.8 Processes in ICUs

One of the major care process-related issues in ICUs is care providers' consistent compliance with evidence-based guidelines. There is high variation in compliance with evidence-based guidelines and improving compliance can significantly improve patient safety and quality of care in ICUs (Chassin 1990; Institute of Medicine 1990; McGlynn et al. 2003). Compliance with evidence-based guidelines is very much influenced by the work system design (Gurses et al. forthcoming). For example, ambiguities in the work system design such as responsibility ambiguity (e.g., who is responsible for completing a particular step of the guideline?) and method ambiguity (e.g., who to call for help with following a guideline?) have been identified as barriers to guideline compliance (Gurses et al. 2008). Clinician characteristics such as care providers' knowledge and attitudes about complying with a particular guideline, subjective norm (perceived

social pressure on clinicians to comply with a guideline), and perceived behavioral control (clinicians' perceptions of their ability to comply with a guideline by overcoming constraints and difficulties) can also impact compliance rates significantly (Cabana et al. 1999).

In addition to care processes, there are numerous other processes that support the delivery of care in an ICU. Supply chain management, maintenance, and housekeeping are examples of supporting processes that play a critical role in the safe and timely completion of care processes. For example, if central line carts are restocked only once per day (at night) and full drapes are missing from the central line insertion cart for an afternoon procedure in the ICU, the physician will find it harder to comply with the central line insertion guidelines that may negatively affect the safety of care. A recent study indicates that poor design/management of supporting processes including inadequate restocking of patient rooms, delays in delivery of medications from pharmacy, and management/maintenance of equipment negatively affect the quality and safety of care as perceived by ICU nurses and the ICU nurses' stress and fatigue levels (Gurses et al. 2009).

Overall, the safety and timely completion of care processes and supporting processes have a significant impact on ICU patient and provider outcomes. HFE can contribute significantly in improving processes by (re) designing the ICU work environment.

40.9 Human Factors Interventions in ICU

Although there is a substantial amount of literature that indicates the importance of using HFE to improve ICUs, most of these studies are descriptive. There are only a limited number of intervention studies reported in the literature that shows how improving ICU work system design based on HFE principles could improve patient and provider outcomes. Interventions can range from organizational (e.g., changing rotation schedules to reduce fatigue) to design changes (e.g., physical layout changes, technological support). In this section, we provide an overview of these intervention studies.

In general, multifaceted HFE interventions are more effective than single-faceted interventions in ICUs. This is probably due to the complex nature of the ICU work system that includes different elements and interactions between these elements. For example, a multifaceted intervention that incorporated principles from human factors engineering has been shown to significantly reduce CLABSI. The five interventions included the following: (1) educating care providers about central line insertion and maintenance, (2) creating a central line insertion cart to make it easy for care providers to access all the supplies necessary for complying with the guidelines, (3) asking care providers daily whether a central line can be removed by adding a question to the daily rounding form, (4) implementing a central line insertion checklist that is completed by bedside nurses while the physician is completing the procedure, and (5) empowering nurses to stop procedures if there is a violation in compliance with the guideline (Berenholtz et al. 2004). A more advanced version of this multifaceted intervention that included additional interventions such as implementing a comprehensive unit-based safety program to improve the safety culture, training ICU team leaders on the basics of human factors and safety science, and providing monthly and quarterly feedback to each participating site on CLABSI rates have reduced the CLABSI rates to a median of zero in 108 Michigan ICUs (Pronovost et al. 2006).

Another human factors intervention that was successful in the ICU setting is the standardization of routine tasks using information tools, which reduces cognitive workload imposed on care providers and therefore reduces the likelihood of making an error. For example, adding a standard order for the head-of-bed angle has been shown to significantly increase the number of patients who were placed in the semirecumbent position (Helman et al. 2003). Similarly, implementing reminder mechanisms has been shown to improve ICU processes and outcomes. For example, using a computerized reminder that prompted physicians either to remove or continue the urinary catheter 72 h after catheter insertion resulted in an approximately 30% reduction in catheter days while not affecting recatheterization rates (Cornia et al. 2003).

Improving communication is an important patient safety goal in the ICU. Communication includes transfer or handoff of information and responsibility during shift change. Efforts have focused on improving handoff communication among ICU providers by adopting human factors principles. For example, Catchpole et al. (2007) developed and implemented a new handover protocol using Formula 1 pit-stop and aviation models to improve the handover of pediatric patients from the Operating Room (OR) team to the ICU after congenital heart surgery. This protocol, which focused on clarifying responsibilities during the care transition, standardizing processes, and improving situation awareness, anticipation, and communication, significantly reduced the number of technical errors (e.g., drains not located safely) and information handover omissions (Catchpole et al. 2007).

HFE interventions have also been tested in simulated ICU environments. For example, Sharp and Helmicki (1998) designed an ecological interface to support clinicians in assessing tissue oxygenation in a NICU. The ecological interface performed better in determining the correct diagnosis when compared to the traditional interface. In addition, the least experience group performed better with the ecological interface, indicating potential opportunities for providing training through the interface. However, there may be significant differences on the impact of the intervention in a hectic ICU environment considering the real-time demands of the field environment, when compared to a simulated setting.

In summary, there are only a handful number of intervention studies that have used HFE principles to improve the ICU work design. Future studies are needed to substantiate these findings, as well as provide new knowledge on the usefulness and impact of HFE interventions on ICU processes and outcomes. In many respects, this may be the one of the most formidable challenges facing HFE in the effort to improve patient safety and quality in the ICU setting.

40.10 Conclusions

ICU is a complex and safety critical environment with a range of subsystems, tasks, organizational and physical characteristics, and tools and technologies. Each of these components plays a unique and critical role toward patient care. However, the success of the ICU environment not only depends on the individual performance of each component but also on the successful synergy among them. Therefore, any effort to understand or intervene in the ICU environment needs to consider both the individual components of the ICU work system as well as the complex interactions among them. One way to systematically understand the work system characteristics of the ICU system is to use a model or a guiding framework that provides details on the individual components (i.e., individual, tasks, technologies, organization, physical environment) as well as the interactions among these components.

The systematic understanding derived from ICU-HFE research efforts needs to be integrated with successes and failures from current knowledge to provide a clear perspective of what is still needed to improve ICU outcomes. Most importantly, human factors engineers have a formidable challenge in developing, testing, and implementing interventions in the ICU work system. Although there are several descriptive, observational, or assessment type of efforts that have used HFE principles, there is a dearth of evidence on the impact of HFE redesign/design efforts on ICU processes and outcomes. Only a handful of studies have attempted to "change" an aspect of the ICU work system based on HFE principles and methods (e.g., redesigning the user interface or creating a central line cart) and understand its impact on ICU outcomes. This may be due to several reasons including inadequacy of resources and time, production pressure for care providers, etc. One of these reasons is probably the safety critical nature of the ICU work system allowing only very robust interventions to be tested or implemented. However, robustness can be successfully built into the design by iteratively testing in simulated environments before implementation in actual ICU settings, thereby preventing intervention-related patient safety issues.

Another reason for the limited number of intervention studies in ICUs is the lack of close, high-level collaboration needed among HFE experts and clinicians. Most of the hospitals, including those that belong to some of the large health care systems, do not have in-house HFE expertise. Even if they have an in-house HFE expert, these experts are usually not adequately empowered to be able to study the system sufficiently and make improvements. In summary, there is a need to increase the use of HFE expertise in studying ICUs and implementing usable design changes.

Acknowledgments

Dr. Ayse P. Gurses's effort on this book chapter was supported by grant number K01HS018762 from the Agency for Healthcare Research and Quality. The content is solely the responsibility of the authors and does not necessarily represent the official views of the Agency for Healthcare Research and Quality. Dr. Carayon's effort on this book chapter was partially supported by grant 1UL1RR025011 from the Clinical & Translational Science Award (CTSA) program of the National Center for Research Resources National Institutes of Health [PI: M. Drezner].

References

Alameddine M, Dainty KN, Deber R, and Sibbald WJ. 2009. The intensive care unit work environment: Current challenges and recommendations for the future. *J Crit Care* 24 (2): 243–248.

Anders S, Patterson E, Woods D, and Ebright P. 2007. Projecting trajectories for a new technology based on cognitive task analysis and archetypal patterns: The electronic ICU. In *Proceedings of the 8th Annual Naturalist Decision Making Conference*, June 3–6, Asilomar, CA.

Angus DC, Barnato AE, Linde-Zwirble WT, Weissfeld LA, Watson RS, Rickert T, and Rubenfeld GD. 2004. Use of intensive care at the end of life in the United States: An epidemiologic study. *Crit Care Med* 32 (3): 638–643.

Angus DC, Shorr AF, White A, Dremsizov TT, Schmitz RJ, and Kelley MA. 2006. Critical care delivery in the United States: Distribution of services and compliance with leapfrog recommendations. *Crit Care Med* 34 (4): 1016–1024.

Ash JS, Sittig DF, Dykstra R, Campbell E, and Guappone K. 2009. The unintended consequences of computerized provider order entry: Findings from a mixed methods exploration. *Int J Med Inform* 78 (Suppl 1): S69–S76.

Baggs JG, Schmitt MH, Mushlin AI, Mitchell PH, Eldredge DH, Oakes D, and Hutsin AD. 1999. Association between nurse-physician collaboration and patient outcomes in three intensive care units. *Crit Care Med* 27 (9): 1991–1998.

Barie PS, Bacchetta MD, and Eachempati SR. 2000. The contemporary surgical intensive care unit. Structure, staffing, and issues. *Surg Clin North Am* 80 (3): 791–804.

Barnato AE, McClellan MB, Kagay CR, and Garber AM. 2004. Trends in inpatient treatment intensity among medicare beneficiaries at the end of life. *Health Serv Res* 39 (2): 363–375.

Berenholtz SM, Pronovost PJ, Lipsett PA, Hobson D, Earsing K, Farley JE, Milanovich S et al. 2004. Eliminating catheter-related bloodstream infections in the intensive care unit. *Crit Care Med* 32 (10): 2014–2020.

Beuscart-Zephir MC, Pelayo S, Anceaux F, Meaux JJ, Degroisse M, and Degoulet P. 2005. Impact of CPOE on doctor-nurse cooperation for the medication ordering and administration process. *Int J Med Inform* 74 (7–8): 629–641.

Braby CD, Harris D, and Muir HC. 1993. A psychophysiological approach to the assessment of work underload. *Ergonomics* 36 (9): 1035–1042.

Breslow MJ, Rosenfeld BA, Doerfler M, Burke G, Yates G, Stone DJ, Tomaszewicz P, Hochman R, and Plocher DW. 2004. Effect of a multiple-site intensive care unit telemedicine program on clinical and economic outcomes: An alternative paradigm for intensivist staffing. *Crit Care Med* 32 (1): 31–38.

Bria WF and Shabot MM. 2005. The electronic medical record, safety, and critical care. *Crit Care Clin* 21 (1): 55–79.

Brilli RJ, Spevetz A, Branson R, Campbell GM, Cohen H, Dasta JF, Harvey MA et al. 2001. Critical care delivery in the intensive care unit: Defining clinical roles and the best practice model. *Crit Care Med* 29 (10): 2007–2019.

Cabana MD, Rand CS, Powe NR, Wu AW, Wilson MH, Abboud PA, and Rubin HR. 1999. Why don't physicians follow clinical practice guidelines? A framework for improvement. *JAMA* 282 (15): 1458–1465.

Caplan RD, Cobb S, French JRP, Harrison RV, and Pinneau SR. 1975. *Job Demands and Worker Health: Main Effects and Occupational Differences*. Washington, DC: U.S. Department of Health, Education and Welfare.

Carayon P and Gurses AP. 2005. A human factors engineering conceptual framework of nursing workload and patient safety in intensive care units. *Intensive Crit Care Nurs* 21 (5): 284–301.

Carayon P, Hundt AS, and Wetterneck TB. 2010. Nurses' acceptance of Smart IV pump technology. *Int J Med Inform* 79 (6): 401–411.

Carayon P, Schoofs HA, Karsh BT, Gurses AP, Alvarado CJ, Smith M, and Flatley BP. 2006. Work system design for patient safety: The SEIPS model. *Qual Saf Health Care* 15 (Suppl 1): i50–i58.

Carayon P and Smith M. 2000. Work organization and ergonomics. *Appl Ergon* 31 (6): 649–662.

Carayon P, Wetterneck TB, Hundt AS, Ozkaynak M, DeSilvey J, Ludwig B, Ram P, and Rough S. 2007. Evaluation of nurse interaction with bar code medication administration technology in the work environment. *J Patient Saf* 3 (1): 34–42.

Catchpole KR, de Leval MR, McEwan A, Pigott N, Elliott MJ, McQuillan A, MacDonald C, and Goldman AJ. 2007. Patient handover from surgery to intensive care: Using Formula 1 pit-stop and aviation models to improve safety and quality. *Paediatr Anaesth* 17 (5): 470–478.

Catley C, Frize M, Walker C, and Germain L. 2003. Integrating clinical alerts into an XML-based healthcare framework for the neonatal intensive care unit. In *Proceedings of the 25th Annual International Conference of the IEEE EMBS*, September 17–21, Cancun, Mexico.

Chambrin MC, Ravaux P, Calvelo-Aros D, Jaborska A, Chopin C, and Boniface B. 1999. Multicentric study of monitoring alarms in the adult intensive care unit (ICU): A descriptive analysis. *Intensive Care Med* 25 (12): 1360–1366.

Chang SY, Multz AS, and Hall JB. 2005. Critical care organization. *Crit Care Clin* 21 (1): 43–53.

Chassin MR. 1990. Practice guidelines: Best hope for quality improvement in the 1990s. *J Occup Med* 32 (12): 1199–1206.

Cheng CH, Goldstein MK, Geller E, and Levitt RE. 2003. The effects of CPOE on ICU workflow: An observational study. *AMIA Annu Symp Proc* 2003: 150–154.

Clarke EB, Luce JM, Curtis JR, Danis M, Levy M, Nelson J, and Solomon MZ. 2004. A content analysis of forms, guidelines, and other materials documenting end-of-life care in intensive care units. *J Crit Care* 19 (2): 108–117.

Clemmer TP. 2004. Computers in the ICU: Where we started and where we are now. *J Crit Care* 19 (4): 201–207.

Colpaert K, Claus B, Somers A, Vandewoude K, Robays H, and Decruyenaere J. 2006. Impact of computerized physician order entry on medication prescription errors in the intensive care unit: A controlled cross-sectional trial. *Crit Care* 10 (1): R21.

Cornia PB, Amory JK, Fraser S, Saint S, and Lipsky BA. 2003. Computer-based order entry decreases duration of indwelling urinary catheterization in hospitalized patients. *Am J Med* 114 (5): 404–407.

Crickmore R. 1987. A review of stress in the intensive care unit. *Intensive Care Nurs* 3 (1): 19–27.

Cullen DJ, Sweitzer BJ, Bates DW et al. 1997. Preventable adverse drug events in hospitalized patients: A comparative study of intensive care and general care units. *Crit Care Med* 25 (8):1289–1297.

Dara SI and Afessa B. 2005. Intensivist-to-bed ratio: Association with outcomes in the medical ICU. *Chest* 128 (2): 567–572.

Dodek PM and Raboud J. 2003. Explicit approach to rounds in an ICU improves communication and satisfaction of providers. *Intensive Care Med* 29 (9): 1584–1588.

Donchin Y, Gopher D, Olin M, Badihi Y, Biesky M, Sprung CL, Pizov R, and Cotev S. 1995. A look into the nature and causes of human errors in the intensive care unit. *Crit Care Med* 23 (2): 294–300.

Donchin Y and Seagull FJ. 2002. The hostile environment of the intensive care unit. *Curr Opin Crit Care* 8 (4): 316–320.

Dracup K and Morris PE. 2008. Passing the torch: The challenge of handoffs. *Am J Crit Care* 17 (2): 95–97.

Dutton RP, Cooper C, Jones A, Leone S, Kramer ME, and Scalea TM. 2003. Daily multidisciplinary rounds shorten length of stay for trauma patients. *J Trauma* 55 (5): 913–919.

ECRI. 2002. Critical alarms and patient safety. ECRI's guide to developing effective alarm strategies and responding to JCAHO's alarm-safety goal. *Health Devices* 31 (11): 397–417.

Entin E and Serfaty D. 1999. Adaptive team coordination. *Hum Factors* 41 (2): 312–325.

Friesdorf W, Konichezky S, Gross-Alltag F, Federolf G, Schwilk B, and Wiedeck H. 1994. System ergonomic analysis of the morning ward round in an intensive care unit. *J Clin Monit* 10 (3): 201–209.

Garland A. 2005. Improving the ICU: Part 1. *Chest* 127 (6): 2151–2164.

Giraud T, Dhainaut JF, Vaxelaire JF, Joseph T, Journois D, Bleichner G, Sollet JP, Chevret S, and Monsallier JF. 1993. Iatrogenic complications in adult intensive care units: A prospective two-center study. *crit care med* 21 (1): 40–51.

Gorges M, Markewitz BA, and Westenskow DR. 2009. Improving alarm performance in the medical intensive care unit using delays and clinical context. *Anesth Analg* 108 (5): 1546–1552.

Graf J, von den DA, Koch KC, and Janssens U. 2005. Identification and characterization of errors and incidents in a medical intensive care unit. *Acta Anaesthesiol Scand* 49 (7): 930–939.

Gurses AP and Carayon P. 2009. Exploring performance obstacles of intensive care nurses. *Appl Ergon* 40 (3): 509–518.

Gurses AP, Carayon P, and Wall M. 2009. Impact of performance obstacles on intensive care nurses' workload, perceived quality and safety of care, and quality of working life. *Health Ser Res* 44 (2 Pt 1): 422–443.

Gurses AP, Marsteller JA, Ozok AA, Xiao Y, Owens S, and Pronovost PJ. 2010. Using an interdisciplinary approach to identify factors that affect clinicians' compliance with evidence-based guidelines. *Crit Care Med* 38(8): S282–S291.

Gurses AP, Seidl K, Vaidya V, Bochicchio G, Harris A, Hebden J, and Xiao Y. 2008. Systems ambiguity and guideline compliance: A qualitative study of how intensive care units follow evidence-based guidelines to reduce healthcare-associated infections. *Qual Saf Health Care* 17 (5): 351–359.

Gurses AP and Xiao Y. 2006. A systematic review of the literature on multidisciplinary rounds to design information technology. *J Am Med Inform Assoc* 13 (3): 267–276.

Halpern NA and Pastores SM. 2010. Critical care medicine in the United States 2000–2005: An analysis of bed numbers, occupancy rates, payer mix, and costs. *Crit Care Med* 38 (1): 65–71.

Hassan E, Badawi O, Weber RJ, and Cohen H. 2010. Using technology to prevent adverse drug events in the intensive care unit. *Crit Care Med* 38 (Suppl 6): S97–S105.

Helman DL Jr., Sherner JH III, Fitzpatrick TM, Callender ME, and Shorr AF. 2003. Effect of standardized orders and provider education on head-of-bed positioning in mechanically ventilated patients. *Crit Care Med* 31 (9): 2285–2290.

Helmreich R and Merritt A. 1998. *Culture at Work: National, Organizational and Professional Influences*. Aldershot, U.K.: Ashgate.

Imhoff M and Kuhls S. 2006. Alarm algorithms in critical care monitoring. *Anesth Analg* 102 (5): 1525–1537.

Institute of Medicine. 1990. *Clinical Practice Guidelines*. Washington, DC: National Academy Press.

Johnson V, Mangram A, Mitchell C, Lorenzo M, Howard D, and Dunn E. 2009. Is there a benefit to multidisciplinary rounds in an open trauma intensive care unit regarding ventilator-associated pneumonia? *Am Surg* 75 (12): 1171–1174.

Knaus WA, Draper EA, and Wagner DP. 1986. An evaluation of outcome from intensive care in major medical centers. *Ann Intern Med* 104: 410–418.

Koppel R, Wetterneck T, Telles JL, and Karsh BT. 2008. Workarounds to barcode medication administration systems: Their occurrences, causes, and threats to patient safety. *J Am Med Inform Assoc* 15 (4): 408–423.

Korniewicz DM, Clark T, and David Y. 2008. A national online survey on the effectiveness of clinical alarms. *Am J Crit Care* 17 (1): 36–41.

Kossman SP and Scheidenhelm SL. 2008. Nurses' perceptions of the impact of electronic health records on work and patient outcomes. *Comput Inform Nurs* 26 (2): 69–77.

Krell K. 2008. Critical care workforce. *Crit Care Med* 36 (4): 1350–1353.

Landrigan CP, Rothschild JM, Cronin JW, Kaushal R, Burdick E, Katz JT, Lilly CM, Stone PH, Lockley SW, Bates DW, and Czeisler CA. 2004. Effect of reducing interns' work hours on serious medical errors in intensive care units. *N Engl J Med* 351 (18): 1838–1848.

Lapinsky SE, Wax R, Showalter R, Martinez-Motta JC, Hallett D, Mehta S, Burry L, and Stewart TE. 2004. Prospective evaluation of an internet-linked handheld computer critical care knowledge access system. *Crit Care* 8 (6): R414–R421.

Lawless ST. 1994. Crying wolf: False alarms in a pediatric intensive care unit. *Crit Care Med* 22 (6): 981–985.

Leape LL. 2005. "Smart" pumps: A cautionary tale of human factors engineering. *Crit Care Med* 33 (3): 679–680.

Linton S, Grant C, and Pellegrini J. 2008. Supporting families through discharge from PICU to the ward: The development and evaluation of a discharge information brochure for families. *Intensive Crit Care Nurs* 24 (6): 329–337.

Lustbader D and Fein A. 2000. Emerging trends in ICU management and staffing. *Crit Care Clin* 16 (4): 735–748.

Mador RL and Shaw NT. 2009. The impact of a Critical Care Information System (CCIS) on time spent charting and in direct patient care by staff in the ICU: A review of the literature. *Int J Med Inform* 78 (7): 435–445.

Malacrida R, Bomio D, Matathia R, Suter PM, and Perrez M. 1991. Computer-aided self-observation psychological stressors in an ICU. *Int J Clin Monit Comput* 8 (3): 201–205.

Malashock C, Shull S, and Gould DA. 2004. Effect of smart infusion pumps on medication errors related to infusion devices programming. *Hosp Pharm* 39 (5): 460–469.

Manias E and Street A. 2001. Nurse-doctor interactions during critical care ward rounds. *J Clin Nurs* 10 (4): 442–450.

Manser T. 2009. Teamwork and patient safety in dynamic domains of healthcare: A review of the literature. *Acta Anaesthesiol Scand* 53 (2): 143–151.

McGlynn EA, Asch SM, Adams J, Keesey J, Hicks J, DeCristofaro A, and Kerr EA. 2003. The quality of health care delivered to adults in the United States. *N Engl J Med* 348 (26): 2635–2645.

Miller A, Scheinkestel C, and Steele C. 2009. The effects of clinical information presentation on physicians' and nurses' decision-making in ICUs. *Appl Ergon* 40 (4): 753–761.

Milstein A, Galvin RS, Delbanco SF, Salber P, and Buck CR Jr. 2000. Improving the safety of health care: The leapfrog initiative. *Eff Clin Pract* 3 (6): 313–316.

Morrison JL, Cai Q, Davis N, Yan Y, Berbaum ML, Ries M, and Solomon G. 2010. Clinical and economic outcomes of the electronic intensive care unit: Results from two community hospitals. *Crit Care Med* 38 (1): 2–8.

Needham DM, Sinopoli DJ, Thompson DA, Holzmueller CG, Dorman T, Lubomski LH, Wu AW, Morlock LL, Makary MA, and Pronovost PJ. 2005. A system factors analysis of "line, tube, and drain" incidents in the intensive care unit. *Crit Care Med* 33 (8): 1701–1707.

Oates RK and Oates PR. 1995. Stress and mental health in neonatal intensive care units. *Arch Dis Child* 72 (2): F107–F110.

Oates PR and Oates RK. 1996. Stress and work relationships in the neonatal intensive care unit: Are they worse than in the wards? *J Paediatr Child Health* 32 (1): 57–59.

Osmon S, Harris CB, Dunagan WC, Prentice D, Fraser VJ, and Kollef MH. 2004. Reporting of medical errors: An intensive care unit experience. *Crit Care Med* 32 (3): 727–733.

Penoyer DA. 2010. Nurse staffing and patient outcomes in critical care: A concise review. *Crit Care Med* 38 (7): 1521–1528.

Poon EG, Keohane CA, Yoon CS, Ditmore M, Bane A, Levtzion-Korach O, Moniz T et al. 2010. Effect of bar-code technology on the safety of medication administration. *N Engl J Med* 362 (18): 1698–1707.

Pronovost PJ, Angus DC, Dorman T, Robinson KA, Dremsizov TT, and Young TL. 2002. Physician staffing patterns and clinical outcomes in critically ill patients: A systematic review. *JAMA* 288 (17): 2151–2162.

Pronovost P, Berenholtz S, Dorman T, Lipsett PA, Simmonds T, and Haraden C. 2003a. Improving communication in the ICU using daily goals. *J Crit Care* 18 (2): 71–75.

Pronovost P, Needham D, Berenholtz S, Sinopoli D, Chu H, Cosgrove S, Sexton B et al. 2006. An intervention to decrease catheter-related bloodstream infections in the ICU. *N Engl J Med* 355 (26): 2725–2732.

Pronovost PJ, Thompson DA, Holzmueller CG, Dorman T, and Morlock LL. 2007. The organization of intensive care unit physician services. *Crit Care Med* 35 (10): 2256–2261.

Pronovost P, Weast B, Schwarz M, Wyskiel RM, Prow D, Milanovich SN, Berenholtz S, Dorman T, and Lipsett P. 2003b. Medication reconciliation: A practical tool to reduce the risk of medication errors. *J Crit Care* 18 (4): 201–205.

Ramsay J, Popp H, Thull B, and Rau G. 1997. The evaluation of an information system for intensive care. *Behav Inf Technol* 16 (1): 17–24.

Rayo M, Smith P, Weinger M, Slagle J, and Dresselhaus T. 2007. Assessing medication safety technology in the intensive care unit. In *Proceedings of the Human Factors and Ergonomics Society 51st Annual Meeting*, October 1–5, Baltimore, MD.

Reader T, Flin R, and Cuthbertson B. 2007a. Teamwork in the Scottish ICU. *Scott Med J* 52 (1): 49.

Reader TW, Flin R, Mearns K, and Cuthbertson BH. 2007b. Interdisciplinary communication in the intensive care unit. *Br J Anaesth* 98 (3): 347–352.

Reddy MC, McDonald DW, Pratt W, and Shabot MM. 2005. Technology, work, and information flows: Lessons from the implementation of a wireless alert pager system. *J Biomed Inform* 38 (3): 229–238.

Reddy MC, Shabot MM, and Bradner E. 2008. Evaluating collaborative features of critical care systems: A methodological study of information technology in surgical intensive care units. *J Biomed Inform* 41 (3): 479–487.

Rothschild JM, Keohane CA, Cook EF, Orav EJ, Burdick E, Thompson S, Hayes J, and Bates DW. 2005. A controlled trial of smart infusion pumps to improve medication safety in critically ill patients. *Crit Care Med* 33 (3): 533–540.

Sapirstein A, Lone N, Latif A, Fackler J, and Pronovost PJ. 2009. Tele ICU: Paradox or panacea? *Best Pract Res Clin Anaesthesiol* 23 (1): 115–126.

Sharp T and Helmicki A. 1998. The application of the ecological interface design approach to neonatal intensive care medicine. In *Proceedings of the Human Factors and Ergonomics Society 42nd Annual Meeting*, October 5–9, Chicago, IL.

Shortell SM, Zimmerman JE, Rousseau DM, Gillies RR, Wagner DP, Draper EA, Knaus WA, and Duffy J. 1994. The performance of intensive care units: Does good management make a difference? *Med Care* 32 (5): 508–525.

Shulman R, Singer M, Goldstone J, and Bellingan G. 2005. Medication errors: A prospective cohort study of handwritten and computerised physician order entry in the intensive care unit. *Crit Care* 9 (5): R516–R521.

Sintchenko V, Iredell JR, Gilbert GL, and Coiera E. 2005. Handheld computer-based decision support reduces patient length of stay and antibiotic prescribing in critical care. *J Am Med Inform Assoc* 12 (4): 398–402.

Sluiter JK, Bos AP, Tol D, Calff M, Krijnen M, and Frings-Dresen MH. 2005. Is staff well-being and communication enhanced by multidisciplinary work shift evaluations? *Intensive Care Med* 31 (10): 1409–1414.

Smith MJ and Carayon-Sainfort P. 1989. A balance theory of job design for stress reduction. *Int J Ind Ergon* 4: 67–79.

Smith TJ, Schoenbeck K, and Clayton S. 2009. Staff perceptions of work quality of a neonatal intensive care unit before and after transition from an open bay to a private room design. *Work* 33 (2): 211–227.

Solet D, Vorvell JM, Rutan GH, and Frankel RM. 2004. Physician-to-physician communication: Methods, practice and misgivings with patient handoffs. *J Gen Intern Med* 19 (Suppl 1): 108.

Tang Z, Weavind L, Mazabob J, Thomas EJ, Chu-Weininger MY, and Johnson TR. 2007. Workflow in intensive care unit remote monitoring: A time-and-motion study. *Crit Care Med* 35 (9): 2057–2063.

Thomas EJ, Lucke JF, Wueste L, Weavind L, and Patel B. 2009. Association of telemedicine for remote monitoring of intensive care patients with mortality, complications, and length of stay. *JAMA* 302 (24): 2671–2678.

Thomas EJ, Sexton JB, and Helmreich RL. 2003. Discrepant attitudes about teamwork among critical care nurses and physicians. *Crit Care Med* 31 (3): 956–959.

Thompson D, Duling L, Holzmueller C, Dorman T, Lubomski L, Dickman F, Fahey M, Morlock L, Wu A, and Pronovost P. 2005. Computerized physician order entry, a factor in medication errors: Descriptive analysis of events in the intensive care unit safety reporting system. *J Clin Outcomes Manag* 12 (8): 407–412.

Tissot E, Cornette C, Demoly P, Jacquet M, Barale F, and Capellier G. 1999. Medication errors at the administration stage in an intensive care unit. *Intensive Care Med* 25 (4): 353–359.

Topf M and Dillon E. 1988. Noise-induced stress as a predictor of burnout in critical care nurses. *Heart Lung* 17 (5): 567–574.

Ulrich R, Quan X, Zimring C, Joseph A, and Choudhary R. 2004. *Report to The Center for Health Design for Designing the 21st Century Hospital Project*. September 2004.

Watts R, Pierson J, and Gardner H. 2006. Critical care nurses' beliefs about the discharge planning process: A questionnaire survey. *Int J Nurs Stud* 43 (3): 269–279.

Wayne JD, Tyagi R, Reinhardt G, Rooney D, Makoul G, Chopra S, and Darosa DA. 2008. Simple standardized patient handoff system that increases accuracy and completeness. *J Surg Educ* 65 (6): 476–485.

Wheelan SA, Burchill CN, and Tilin F. 2003. The link between teamwork and patients' outcomes in intensive care units. *Am J Crit Care* 12 (6): 527–534.

Whittaker J and Ball C. 2000. Discharge from intensive care: A view from the ward. *Intensive Crit Care Nurs* 16 (3): 135–143.

Wikstrom AC and Larsson US. 2004. Technology—An actor in the ICU: A study in workplace research tradition. *J Clin Nurs* 13 (5): 555–561.

Winters BD, Gurses AP, Lehmann H, Sexton JB, Rampersad CJ, and Pronovost PJ. 2009. Clinical review: Checklists—Translating evidence into practice. *Crit Care* 13 (6): 210.

Wu AW, Pronovost PJ, and Morlock L. 2002. ICU incident reporting systems. *J Crit Care* 17 (2): 86–94.

Young MP, Gooder VJ, Oltermann MH, Bohman CB, French TK, and James BC. 1998. The impact of a multidisciplinary approach on caring for ventilator-dependent patients. *Int J Qual Health Care* 10 (1): 15–26.

Zawada ET Jr., Herr P, Larson D, Fromm R, Kapaska D, and Erickson D. 2009. Impact of an intensive care unit telemedicine program on a rural health care system. *Postgrad Med* 121 (3): 160–170.

Zheng K, Haftel HM, Hirschl RB, O'Reilly M, and Hanauer DA. 2010. Quantifying the impact of health IT implementations on clinical workflow: A new methodological perspective. *J Am Med Inform Assoc* 17 (4): 454–461.

41

Human Factors and Ergonomics in the Emergency Department

Shawna J. Perry, Robert L. Wears, and Rollin J. Fairbanks

CONTENTS

41.1 Health Care Work as Systems 709
    41.1.1 Degree of Knowledge 710
    41.1.2 Degraded State 710
    41.1.3 Homeostatic Activity 710
    41.1.4 Equifinality 710
    41.1.5 Remoteness 710
    41.1.6 Unstoppability 710
    41.1.7 Mortality 710
41.2 Emergency Department Work as System 710
    41.2.1 Open System 711
    41.2.2 Unbounded System 711
    41.2.3 Complexly Interactive System 711
41.3 Emergency Department as a Sociotechnical System 711
41.4 Human Factors Issues in Emergency Departments 712
    41.4.1 Patient 712
    41.4.2 Provider 712
        41.4.2.1 Selection and Training 712
        41.4.2.2 Fatigue and Wellness 713
    41.4.3 Task and Tools 714
    41.4.4 Work Team 715
        41.4.4.1 Teamwork 715
        41.4.4.2 Communication and Coordination beyond the Emergency Department 715
        41.4.4.3 Culture 716
    41.4.5 Work Environment 716
    41.4.6 Organization and Management 717
    41.4.7 Institutional Context 717
41.5 Summary 718
    41.5.1 Macroergonomic Balance 718
References 718

41.1 Health Care Work as Systems

It has become popular among those interested in safety and quality to take a "systems view" of health care processes (Plsek, 2001). These views of a "system," based on rationalized, engineered systems, are wishful and simplistic, quite different from the actual world of health care work "in the wild" (Hutchins, 1996). One or more physiological subsystems are embedded in every health care work system—in fact, these subsystems are the focus of the larger work system. While this in itself is not unusual (e.g., chemical plants have a chemical process at the heart of their system), there are several important ways in which the health care work systems differ from simpler systems and must be identified prior to our discussion of one medical subsystem—the emergency department (ED). These areas of distinctness include (1) degree of knowledge, (2) degraded functional states, (3) homeostatic activity, (4) equifinality, (5) remoteness, (6) unstoppability, and (7) mortality.

41.1.1 Degree of Knowledge

Compared to what is known about the fundamental processes in a nuclear power plant, a jet engine, or a refinery, we know relatively little about the physiological processes of the human body in health and disease. Engineered systems have well-documented designs, and are deterministic in at least a large part of their parameter space. In particular, after an abnormality, engineered systems can be deeply investigated if necessary (e.g., completely disassembled), to provide a detailed understanding of the anomaly. Although they may occasionally manifest emergent or unexpected properties, these systems are fundamentally knowable from scientific and engineering first principles. Physiologic systems, on the other hand, are fundamentally unknown from first principles. After an abnormality or failure, detailed investigation of the physiology is not often possible; even *postmortem* examinations sometimes fail to explain the clinical course. Thus, reasoning about what is occurring in such systems, or predicting what state they will be in next, is substantively more difficult because of this fundamental lack of knowledge about a key component of the system.

41.1.2 Degraded State

In contrast to the processes on which the systems approach was originally based, the physiologic processes in the health care setting are always functioning abnormally, that is, in a degraded state. Contributors to this are biodiversity (i.e., two patients with new onset diabetes will have different manifestations of the disease 5 years later) and the natural history of medical illness, which presents in numerous ways with variable timing (i.e., men are more likely to have acute onset of severe chest pain with a heart attack while women may present with only excessive fatigue that has lasted for hours to days). Even relatively healthy individuals, classified as the "worried well," have at the very least, malfunctioning sensors that make it difficult to detect degradation until the signals are obvious and readily apparent.

41.1.3 Homeostatic Activity

Physiological systems differ from physical or chemical processes in that they actively try to counteract perceived malfunctions. Sometimes these homeostatic processes are beneficial, and sometimes they produce further problems. The point is that in health care, there are always two "actors" involved—the patient's own physiology striving to return to a "normal" state from a degrading one, and the care team's interventions, also aimed at returning to a "normal" state. These processes, often occurring simultaneously, interact in quite unpredictable ways. Physiologic systems are therefore adaptive in ways that engineered systems are not; while this is largely a benefit, it is also an additional complicating factor in understanding and anticipating system function.

41.1.4 Equifinality

A given physiologic state can be arrived at in a variety of different ways, and the trajectory of previous events and interventions can be important for assessing stability but may be difficult to obtain. While equifinality is not unique to physiologic processes, it is a fundamental characteristic, in that it means that knowledge of the current status of the process is not always sufficient for effective or efficient intervention or control.

41.1.5 Remoteness

Uncertainty about the process itself is always high because the fundamental physiologic processes are not observed directly, but rather made manifest by a relatively small number of proxy variables (symptoms, physical signs, results of testing, etc.).

41.1.6 Unstoppability

Unlike most other operations, which can often be completely stopped, disassembled, repaired, reassembled, and restarted (albeit at great expense), core physiologic processes can never be stopped. The system must be repaired while it is still running.

41.1.7 Mortality

The ability to identify a relatively small set of unambiguous events that substantially encompass the universe of possible failures has been a critical factor in advancing safety in most other fields (Schulman, 2002). This is not possible in health care. All human beings will eventually die, and most of those deaths will occur in proximity to health care (Gaba, 2000a,b). Unlike unambiguous failures such as in an airplane crash, failures within a physiologic system can be very difficult to detect and differentiate from the natural history of a human life.

41.2 Emergency Department Work as System

Applying human factors and ergonomics principles to the ED requires that we first understand the ED as a complex sociotechnical system. At the simplest and most abstract level, the ED is a basic transformation system that accepts inputs (patients), applies transformation processes to them (care), and delivers outputs (patients, hopefully in improved states). However, there are several complexities that make ED work particularly difficult to understand, manage or improve, particularly when contrasted with other types of production or service systems. In particular, the ED system is *open*, *unbounded*, and *complexly interactive*.

41.2.1 Open System

While many health care activities can be viewed as closed, or nearly closed systems, EDs are almost completely open systems; its spatial boundaries may be well defined, but its temporal, administrative, and procedural boundaries are vague to nonexistent. EDs must be open to inputs from the outside world, without limit and without prior arrangement. The inputs are generally unrestricted, in the sense that they may span the entire spectrum of possible diseases and injuries. In addition, EDs are open to inputs from within the hospital; for example, when outpatients "crash" in radiology, or other similar areas, they are typically brought to the ED for stabilization until a suitable inpatient bed can be found. The hospital influences the ED processes in many other ways besides adding to its inputs. When inpatient beds or nurses are in short supply, a standard output channel is then blocked and patients who normally would be transferred out of the ED to a ward instead remain in the ED and receive inpatient care there. Similarly, delays in laboratory, radiology, or consulting services impact the ED's internal processes directly.

The openness of the ED system means that work in EDs is almost exclusively event-driven. Long-term planning is therein probabilistic, and short-term planning is immediate and highly contextual, with very little in between.

41.2.2 Unbounded System

In addition to being open to myriad outside influences, the ED is unbounded in several dimensions and has been dubbed the "infinitely expandable space" (personal communication, James Adams, MD). It is temporally unbounded in that it must maintain 24 × 7 × 365 operations at all times; EDs do not shut down temporarily, even for major refitting or renovations. In contrast to other units, which have physical limits on the number of patients who can be accepted, the ED is unbounded in the sense that there is no natural limit to the number of patients for which it can become responsible (Derlet et al., 2001; Richardson et al., 2002; Schull et al., 2001). In fact, when other clinical units are full, the excess is held in the ED, resulting in a new class of patients,

called "boarders" and new type of care—hallway medicine—received by ED patients who are lying and sitting in hallways as a result of overcrowding (Freeman, 2003; Viccellio et al., 2008). Finally, the ED is unbounded in the types of patients and conditions it handles, with the sole exception being major surgical procedures.

41.2.3 Complexly Interactive System

An important property of the ED as a system is that its elements, particularly its patients, interact and affect each other's processes in complicated and unpredictable ways. Since patients generally do not directly interact with one another, this is not immediately apparent, but it is clearly the case that one patient's presence can affect the care given to another in ways more complicated than simple queuing effects (Boutros & Redelmeier, 2000; Magid et al., 2004; Schull et al., 2004).

41.3 Emergency Department as a Sociotechnical System

EDs as organizations are simultaneously social (i.e., consisting of people, norms, values, culture, climate) and technical (consisting of tools, facilities, equipment, technology, procedures). The social and technical components are deeply interrelated and interdependent; every change in one area affects the other, perhaps after a variable time delay. This reciprocal determinism of the social and technical aspects of the ED is influenced further by external forces (economics, regulation, litigation) and to a limited extent, also influences them. Thus, there are three aspects of the sociotechnical system of the ED: social, technical, and external. Table 41.1 provides some examples of how these aspects of the ED affect its work.

As is typical in health care, these components have tended to be managed separately in EDs, with predictable consequences due to suboptimization and the unexpected consequences of interactions with ignored areas.

In what follows, we highlight some of the more salient human factors issues in ED care, following a framework proposed by Vincent (Vincent et al., 1998, 2000) that bears a striking resemblance to the macroergonomic framework offered by Carayon (2003). Although we use a decomposition method to simplify the issues, the reader should keep in mind how the social, technical, and external aspects of the ED affect, and are affected by, these issues. To date, little work has been done to improve these aspects of emergency care; thus, we raise more problems than solutions.

TABLE 41.1

Examples of Human Factors Issues in the ED

Issue	Example/Impact
Event-driven work process: work space and resources are difficult to predict on an hour-to-hour basis	Minor bus accident results in need to treat large number of patients who arrive simultaneously, quickly outstripping the staff, supplies, and equipment already caring for patients already in the ED
Patient type is highly variable causing difficulty identifying workload or services needed	A chief complaint of abdominal pain will require differing degrees of evaluation depending on attributes such as age, past medical history, severity of illness, cultural/language barriers, etc.
Outcomes coupled in varying degrees to other microsystems within the hospital	Throughput of all patients is slowed (halted for some) with failure of a piece of laboratory equipment resulting in an extended length of stay as blood test are sent to another hospital to be performed
Work of emergency care requires many individual tasks that are highly variable in number	Each patient is seen by a clinician who will generate a series of "orders" or actions necessary to care for the patient. Each order results in series of tasks to be performed by the staff (typically 6–7 per order). Number of orders ranges from 1 to 10 per patient
Complex sociotechnical workspace involving a number of specialties and skill levels working as a "team" to provide care	Emergency care in the ED may be provided by physicians, nurses, physician assistants, paramedics, nurse practitioners, pediatricians, and require at times the involvement of the same skill types from other specialties (i.e., surgeons, dialysis nurses, etc.); standardization of work processes and cultural expectations is difficult
Lack of standardization of work tools	Tracking of patient throughput can be performed by a wide range of cognitive artifacts including hand written notes, status boards, software programs, human memory
Transitions or turnover of care are necessary	Shift work is required due to need for 24 h of operations resulting in unstructured transfer of information, authority, and responsibility for patient care
ED workspace rarely designed for the work performed there	Many ED reside in recycled clinical treatment areas designed for the work of other specialties (e.g., outpatient clinic, x-ray suites, and inpatient wards)

41.4 Human Factors Issues in Emergency Departments

41.4.1 Patient

Patient factors affect safety in all areas of health care. Language barriers, patients' physiologic status, and their general physical condition (e.g., obesity) affect the difficulty and riskiness of the work clinicians do. An additional patient domain issue that is highly specific to emergency care is that patients and providers are generally complete strangers when the encounter begins; there is no common history or shared ground, so issues of lack of confidence or trust complicate the relationship and may affect the accuracy of historical information given to the provider, or compliance with follow-up. This interaction is further complicated by the introduction of the electronic medical record (EMR), which is assumed by both patients and providers to be accurate, complete, and up to date—which it is not (Kaboli, 2004; Wagner, 1996).

41.4.2 Provider

41.4.2.1 Selection and Training

Emergency care providers are generally rigorously selected, highly trained, and strongly motivated, so further effort on selection and training seems unlikely to yield major improvements, for they have largely been maximized already. However, there are several provider areas where emergency care falls short of what has been done in other environments.

First, while the initial training and assessment of emergency care professionals is extensive and rigorous, their continuing development has been largely ignored. Once certified, there are no routine mechanisms by which the provider's physical or mental fitness for the job is reassessed. Subsequent training is haphazard at best, and is not provided for in the routine workplace; providers are expected to accomplish this on their own time and at their own expense. Given the rapid pace of change in health care, this seems to leave much to chance.

Board certification for emergency physicians has been available since 1980, but because the specialty is so new, it has only become a de facto requirement in a few metropolitan areas. There are roughly 40,000 emergency physicians in the United States, of whom only 57% are board certified (Gindle, 2009); thus, in a practical sense, board certification cannot be a requirement for employment because there are not enough board certified physicians to fill the available positions.

Emergency medicine was one of the first specialties to require recertification of its diplomates. Initially, this was required at 10 year intervals; in 2000, this was changed to a continuous certification model wherein certain activities are required yearly in order to qualify for recertification in the 10th year.

Certification is also available for emergency nurses, but seems to be less commonly pursued. Prehospital personnel (e.g. paramedic) and technicians have their own certification levels and requirements.

41.4.2.2 Fatigue and Wellness

Gaba has pointed out that in every high-hazard industry except medicine, the default assumption has been that long work hours and circadian disruption lead to health, performance, and safety problems, and that the burden of proof is placed on organizations to demonstrate that their staffing and scheduling patterns are safe (Gaba, 1998). However in health care, the burden of proof is reversed—any staffing or scheduling practice is assumed to be safe unless convincing evidence to the contrary is presented. In general, there has been a great reluctance in health care to accept data on fatigue, shift work, and circadian disruption from other settings, despite the absence of any compelling reason to believe that doctors or nurses are immune to these issues. Despite this thinking, the general association of fatigue to decreased performance seems well-established (Frank, 2002; Jha et al., 2001); it has often been asserted that the performance level of someone who has been up all night approximates that of someone who is legally intoxicated (Bonnet, 2000).

Emergency medicine was the first specialty to begin to limit physician work hours in shifts. Originally, EDs were staffed using the traditional medical "on call" model in which physicians conducted their normal work during business hours, but were occasionally "on call" for emergency cases, particularly at night or on weekends. As the demand for emergency care grew, ED work for physicians assumed a shift-work model similar to that for nurses. This model is now dominant, and the general expectation in most EDs is that physicians and nurses will be awake and at work at all times during the night (in other words, there are no "call rooms" available for rest in slack times), and that they will go home when their shift ends. However, there are no formal limits on continuous hours of service, total hours per week, or total number of hours off between shifts, with the sole exception of residents in training, who are subject to an 80 h/week rule. The effects of this limitation are gradually becoming apparent, with unexpected effects due to decreased continuity of care and of shifting workload to other unregulated workers have been noted (Laine et al., 1993; Petersen et al., 1994; Wagner et al., 2010).

The adoption of shift work in EDs limits the effects of fatigue when shifts are limited to 8–12h or less (which is almost universally current practice) (Thomas et al., 1994). However, the effects of circadian disruption due to shift work still must be dealt with (Frank, 2002; Silbergleit et al., 2006). Research in other domains (and

some studies in health care settings) has shown rather consistently that shift work is associated with several negative effects (Cao et al., 2008; Costa, 1996), principally in the areas of sleep, performance, psychological and social health, and physical health.

Night workers get roughly 25%–30% less sleep than day or evening workers, and their sleep is of poorer quality (Smith-Coggins et al., 1994; Tepas & Carvalhais, 1990). Independent of this sleep deprivation, circadian desynchrony reduces cognitive and work performance, particularly when the work does not bring its own intrinsic arousal stimulus. Thus, while ED workers on night shifts may be able to rise to the occasion for acute, exciting, and interesting cases, their performance is likely to suffer when the work is routine, or repetitive (Folkard & Tucker, 2003; Howard et al., 2003; Monk, 1990; Monk & Carrier, 1997; Paley & Tepas, 1994; Smith-Coggins et al., 1994). This performance decrement is dramatically underscored by the number of dramatic accidents occurring on night shifts, including the Challenger explosion (Vaughan, 1996), the Bhopal chemical plant disaster (Perrow, 1994), the Exxon Valdez grounding, the Three Mile Island accident (Perrow, 1984), and the Chernobyl meltdown.

Shift work is associated with mood disturbances, irritability, and if sustained for long periods of time, increased rates of substance abuse, divorce, and suicide (Cao et al., 2008; Folkard & Monk, 1985; Tepas et al., 2004). In addition, shift workers tend be more socially isolated (Monk & Folkard, 1992). A particular concern is that shift work is often cited as the primary reason for leaving emergency medicine (Hall et al., 1992).

Finally, long-term shift work is associated with poorer physical health. Shift workers have higher rates of accidents, ulcers, hypertension, and coronary disease; in addition, conditions such as diabetes and epilepsy are exacerbated by shift work (Frank, 2002). Night shift workers have a substantially greater risk of road traffic accidents on their way home from work than the general population (Monk et al., 1996), a fact that medicine seems only recently to be learning (Barger et al., 2005).

Several factors are known to decrease workers' ability to tolerate circadian disruption (Tepas & Monk, 1987). Tolerance goes down with age, especially age over 50 (Tepas et al., 1993); moonlighting (doing a second job) is another risk factor, probably related to increased fatigue; and "morning larks" do not tolerate shift work as well as "night owls." However, individual variation in tolerance is high, and the deleterious effects of shift work in emergency care may be at least partially mitigated by self-selection; individuals who do not tolerate shift work well may discover this in their training and subsequently choose other fields (Steele et al., 1997, 2000).

Professional societies have begun to address these issues albeit in a hesitant manner (American College of

Emergency Physicians, 1995); as the physician and nursing workforce in EDs age, the effects of shift work and fatigue can be expected to become more concerning. General recommendations have been made to minimize consecutive nights; to use clockwise (forward) shift rotations; to allow 24–48h off duty after a night shift; and to avoid use of sedatives or stimulants, with the possible exception of caffeine (Frank, 2002). External regulators have not addressed this area at all.

41.4.3 Task and Tools

Emergency care suffers from the same sorts of difficulties with medical devices as do other areas of care. Devices used quite commonly in emergency care (such as monitor–defibrillators) have seldom been evaluated for usability, but what few data are available indicate persistent problems in design (Fairbanks et al., 2004, 2007b; Fairbanks & Wears, 2008; Hoyer et al., 2008; Stewart, 2008). Formal assessments within the clinical environment are generally not done (by designers or purchasers), and are thus introduced and used under loose guidelines rather than precise specifications. It is then left to the user at the sharp end of care to modify or alter their work on the fly to use the device, or to find way to "workaround" it.

Emergency care has produced several unique devices to help workers cope with the demands of emergency care. The Broselow tape was developed as an aide to nonpediatric emergency caregivers who must care for critically ill or injured children (Luten et al., 1992, 1998). Although it has not been formally assessed from a human factors point of view, there is some evidence suggesting its use can be associated with improved performance (Shah et al., 2003).

ED status boards are quite interesting developments. EDs commonly use status boards as tools for managing their clinical work. These tools are not specifically mandated by any external authority, but instead were developed spontaneously and locally as the practice of emergency medicine grew more complex over the years. Thus, they can be considered cognitive artifacts produced by ED workers themselves to help them do their job better, with little or no outside influence. Status boards frequently take the form of large, manually updated "whiteboards" where patient locations are represented as rows in a grid, and columns are used to provide a variety of relevant data. They are typically located in a centrally accessible work area, such as the nursing station, where they can be used by all ED staff (i.e., not just physicians, or just nurses). Status boards began as simple tracking devices, for example, displaying patient name and location, but because they afforded further annotation, have evolved to include a great deal of additional information (Pennathur, 2008)

This additional data might include some of any of the following:

- Patient demographic information, such as chief complaint, arrival time, or length of stay
- Staff information, such as responsible ED physician, nurse, and/or tech, private physician
- Process information, such as pending procedures, laboratory or imaging studies, completed laboratory or radiology work, whether seen by ED physician yet, partial plans (e.g., "nebs × 3," or "4 h obs")
- Patient status information, such as pending consultations, admission/discharge status, and bed status
- Additional patient information, such as pregnancy, neutropenia, isolation requirements, allergies, or name conflict alerts, and so on

Status boards are important tools for providing safe care in the ED. They do this by supporting shared memory, latent processes, collaboration, shared cognition, communication, and coordination. For example, status boards are particularly useful in reducing interruptions (Chisholm et al., 1998, 2000, 2001; Coiera et al., 2002; Fairbanks et al., 2007a) and by supporting asynchronous communication among ED staff in a manner that reduces the memory burden on both sender and receiver (Parker & Coiera, 2000).

Because the manual format is flexible, the status board has often been extended to include other types of information. For example, patients not physically in the ED are often represented on the status board; transfers expected from other hospitals, referrals from physicians' offices, or ED patients who are away from the department for prolonged procedures such as arteriography or dialysis, may still be represented on the board but not associated with any specific patient location. Similarly, information affecting the unit as a whole, but not associated with any specific patient is also commonly displayed. This might include, for example, the status of critical care beds in the hospital, the current on call consultants and their contact information, ED beds that are closed for some reason, and so on.

Because status boards were developed and have evolved locally, they are exquisitely situated in their own local environments, and have not been standardized across different EDs. In addition, they are densely encoded, making use of color, cryptic abbreviations, and symbolic markings that are known to experienced staff. Newcomers assimilate this encoding during on-the-job training, but this skill is seldom remarked and never assessed, so there is a potential for hidden misunderstandings of the more idiosyncratic elements.

As designed, artifacts produced by the workers themselves and unmodified by the demands of external forces, status boards offer a view into the implicit theories of the workers (Bisantz & Ockerman, 2003; Carroll & Campbell, 1989; Woods, 1998).

A task issue that seems fairly specific to EDs is the difficulty in maintaining sufficient experience with the most difficult and dangerous procedures that might be expected. For example, the most dangerous portions of anesthesia are induction and recovery, but anesthesiologists execute these procedures several times per day, and thus are extremely facile. In contrast, the most dangerous procedures in emergency care (rapid sequence induction, cricothyrotomy, and thrombolysis) are required on the order of weekly to yearly. It is unlikely that even the best-trained practitioners will be at their peak skill level when called upon to perform these procedures. Simulation and similar venues may offer some utility in remedying this problem, but their role remains to be established.

Another important task issue is the ubiquity of "information gaps" in emergency care (Cook et al., 2000; Stiell et al., 2003). Hospital records, especially discharge summaries, details of past medical history, and other important information, are often difficult to access in an expedient fashion. The introduction of the EMR can improve access to some degree (despite accuracy issues), but not if the patient has not received care within the network utilizing the EMR. Even paper-based referral notes sent in by family doctors with the patient may not reach the emergency physician. Emergency physicians must then sometimes act on the basis of incomplete or erroneous information, and become used to doing so, resulting in that information sometimes not even sought when it might have been retrieved.

41.4.4 Work Team

41.4.4.1 Teamwork

Patient care in the ED is not a solitary activity. Caregivers from multiple professions work together in a pattern of distributed, collaborative activity. This requires additional tasks of communication and coordination among the workers, on top of the specific clinical work to be done. Physicians, nurses, and technicians are the core professionals in virtually every ED setting. A wide variety of additional professionals may supplement them (e.g., pharmacists, respiratory therapists, clerks, radiology technicians, etc.) *depending upon the clinical situation*, for instance, a trauma patient pulls together different supplemental staff than a patient with a heart attack. It is one of the strange anomalies of health care that all these professionals are trained and evaluated in their separate silos, but are expected to work smoothly together as a "team." Consequently, in most situations, they work primarily as a group, not as a team.

Some initial studies of teamwork (or lack of it) in the ED have suggested that teamwork failures could play a substantial role in the genesis of adverse events; a chart review study of malpractice claims identified an average of almost nine teamwork failures per case (Risser et al., 1999), suggesting that improved teamwork might avoid or mitigate some adverse events. The widespread acceptance of teamwork training in aviation has led to attempts to introduce similar training in emergency care (Simon et al., 1998). The MedTeams project (Simon et al., 1997) was a multi-institutional effort to adapt the principles of aviation teamwork training to the ED setting, to develop and pilot test a training program, and finally to assess the effects of teamwork training on performance. They reported that teamwork behaviors improved and observed errors decreased after training, although perceived workload did not change (Morey et al., 2002, 2003). However, teamwork training has not been generally adopted in EDs, although considerable interest in it remains (Fenendez et al., 2008; Firth-Cozens, 2001; Firth-Cozens & Moss, 1998; Jay et al., 2000; Shapiro & Morchi, 2002; Shapiro et al., 2004; Small et al., 1999).

Shift changes have long been thought to be particularly hazardous times in emergency care (Salluzo et al., 1997), but surprisingly little has been written about them. However, there is at least some evidence that shift changes may actually contribute toward the recovery from incipient failures (Wears et al., 2003). Patterson et al. (2004) noted 21 strategies used in high-hazard industries to improve the effectiveness and efficiency of shift changes; observational studies in five North American EDs noted that eight of these strategies were used frequently, four sporadically, and nine not at all (Behara et al., 2005). Thus, there is the potential to improve the effectiveness of shift change turnovers by adopting some of these unused strategies, such as limiting interruptions during turnover, or requiring the oncoming party to verbalize the "picture" of the patient they have been given to verify shared sensemaking.

Shift changes in the ED are a specific example of a larger problem within health care: facilitating the successful transfer of information, authority, and responsibility from one set of caregivers to another to assist in the continuation of safe and effective care. Table 41.2 illustrates some different types of transitions occurring in emergency care. Transitions in general and shift changes in particular appear to be extremely complex processes that have been little studied (Eisenberg, 2005; Wears & Perry, 2010).

41.4.4.2 Communication and Coordination beyond the Emergency Department

Emergency care is by definition episodic care; the continuing care of patients must be passed on either to inpatient services or to outpatient facilities. This transition

TABLE 41.2

Types and Characteristics of Transitions of ED Care

From	To	Setting	Typical No. of Patients	Time Investment
ED physician	ED physician	Shift change	Many	High
ED nurse	ED nurse	Shift change	Many	High
Paramedic	ED staff	Ambulance arrival	One	Low
ED physician	Consulting specialist	Referral or admission	One	Moderate
ED staff	Inpatient staff	Admission	One	Moderate
ED physician or Nurse	Patient and/or family	Disposition	One	Variable
ED nurse	Staff external to hospital (nursing home, another ED/hospital, funeral home)	Disposition	One	Variable

across organizational boundaries would seem fraught, since the parties to the exchange do not share a great deal of common (Coiera, 2000). Like shift change, these interactions have barely been studied (Muphy & Wears, 2009). Matthews et al. (2002) examined transitions from the ED to the internal medicine service, and found that they contained little reference to formal guidelines or standards for care, and that their perceived quality depended heavily on the confidence one participant had in the other. Other important areas, such as the ambulance to ED transition, or the ED to home care transition, have had very little study (Bost, 2010; Chan, 2009; Patterson & Roth, 2004). Communication and coordination in the reverse direction have likewise been little investigated, although ED effectiveness depends heavily on the level of responsiveness of consulting services, laboratory, and radiology.

41.4.4.3 Culture

Health professionals have been noted to have social and cultural characteristics that sometimes work against the goal of safe care. While these studies have not focused on the ED, emergency practitioners would seem to share many of these attributes, because they are to some extent reinforced by the ED environment.

One particularly problematic attribute is the presence of what has been called "patching" behavior, and the value attributed to it (Tucker & Edmondson, 2002, 2003). When health care workers encounter problems in their daily routine (which is, on average, about once per hour), they typically use "patches" to quickly deal with it and get on with their job. (e.g., a missing device might be "borrowed" from another unit, rather than reordered from the supply room. This solves the immediate problem, but moves it to another unit; and the underlying cause remains untouched.) The problem is seldom brought to anyone else's attention, so it can be dealt with in a more permanent and proactive way. The ED, with its time pressure and unpredictability, encourages this sort of "muddling through" behavior.

Thus the unit culture tolerates, in fact expects, that things will not work as designed, and workers take some pride in the clever ways in which they can devise "workarounds" to meet their goals. Although this ability is one of the reasons that the ED can function at all, it is counterproductive in the long run. In addition, it means that attempts to engineer forcing functions into devices so they cannot be used incorrectly (e.g., so that a tube intended to supply compressed air to a blood pressure cuff cannot be mistakenly connected to an intravenous line [Institute for Safe Medication Practices, 2003]) may be defeated by workers who expect things not to work together and see part of their job as finding ways for them to do so.

41.4.5 Work Environment

Most EDs are ergonomic nightmares (Wears & Perry, 2002). Few, if any, were designed using human factors principles, and in many, the original design has undergone repeated renovation and modification over the years. In addition, overcrowding in EDs has reached epidemic proportions in the past 10 years, due to reductions in hospital capacity, economic pressures, and shortages of nurses to staff inpatient beds (Derlet et al., 2001; Richardson et al., 2002; Schull et al., 2001).

Noise is a problem in many EDs (Baevsky, 2006; Barlas et al., 2001; Brown et al., 1997; Topf, 2000; Zun & Downey, 2005), although some institutions have begun to introduce noise abatement measures in an attempt to improve patient satisfaction. In the prehospital setting, noise is an occupational hazard, as EMTs demonstrate greater than expected hearing loss (Johnson et al., 1980; Pepe et al., 1985).

Patient privacy is also an important issue in many EDs, and has become increasingly difficult (at times impossible) to maintain with worsening ED overcrowding (Barlas et al., 2001; Moskop et al., 2009). There is a clear goal conflict here, between the desire to support patients' privacy, the need to be able to visually monitor patients when not physically at their bedside, and ED

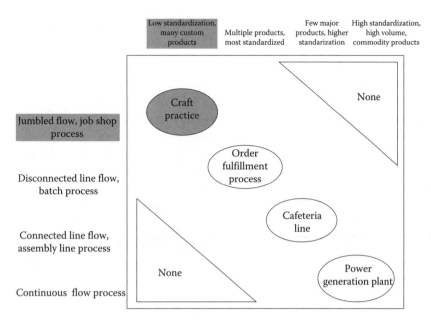

FIGURE 41.1

Product—process matrix. Product types are displayed on the horizontal axis and process types on the vertical axis. EDs seem to fit best in the upper left as a job shop model (low standardization, many custom products, jumbled flow) highlighted in gray.

volumes greater than the physical plant was designed to support.

41.4.6 Organization and Management

Effective care in EDs is much more an issue of organization and management than in other areas of care, where individual knowledge and skill play more prominent roles. Many clinical problems in emergency care are time sensitive, but demand for care is highly variable, fluctuating, and only probabilistically predictable. Thus EDs must be organized, resourced, and staffed to handle variability in not only the quantity but also the types of problems encountered. This requires long-term planning and effort at a higher level than that of the individual practitioner. It is common for ED managers to use analogies to assembly line systems (e.g., the Toyota manufacturing system [Womack et al., 1990], and Lean Six Sigma [Paul & Lin, 2007]) as models for the organization and management of EDs. However, an analysis of the "product–process" matrix (Hayes & Wheelwright, 1979) shown in the Figure 41.1 suggests that this model is not likely to fit emergency care well, which seems to have much more in common with a job shop model, with jumbled work process, a high degree of customization as a result of the biodiversity of patients and problems with challenges to standardization of work due to the openness of the system. Two implications follow from this assessment. First, EDs will necessarily require organizational slack (excess capacity) in order to function adequately, and second, assembly line models of organization and management will inevitably be mismatched to the realities of ED work.

Because of the uncertainties of their inputs, EDs typically use global outcomes to assess performance; for example, average waiting time, proportion of patients leaving before completing treatment, or the proportion of unscheduled returns to the ED. However, there are hidden difficulties in this practice. Elements of one patient's care can affect many others (Boutros & Redelmeier, 2000) in unforeseen and unpredictable ways, especially in light of growing overcrowding (Johnson, 2011). Focusing improvement efforts at the individual patient level can lead to suboptimization and be counterproductive, therein making global outcome measures more attractive. However, a focus on global optimization is naïve and raises important ethical issues. A dependence on global outcome measures has been criticized as "agricultural" (Asch & Hershey, 1995; Rothwell, 2005); an agricultural intervention is presumably concerned only with optimizing the total yield, not with the individual well-being of each individual plant or animal. Unfortunately, not only is the balance of global optimization and individual suboptimization not known for EDs, the issue is largely unrecognized and not discussed in any substantive terms.

41.4.7 Institutional Context

There are several unique aspects to the institutional milieu in which EDs operate, in addition to the general problems of economic pressures, hospital closings and mergers, litigation, etc., that affect health care generally. EDs are not allowed to make decisions to accept or not accept patients. Federal legislation (the "COBRA/EMTALA" laws) (Cross, 1992; Derlet & Nishio, 1990; Frew et al., 1988; Wanerman, 2002) mandates that any person presenting to an ED be evaluated at least to the extent necessary to determine their stability and must be stabilized before being transferred or referred elsewhere. It also mandates that a hospital/ED receiving a request to accept a transfer may not decline the transfer if they provide the sort of service the patient requires. (e.g., a head-injured patient may be transferred from one hospital to another, once stabilized, if the first hospital does not offer the services the patient requires, that is, neurosurgery. If the second hospital does offer neurosurgery, it may not refuse to accept the patient in transfer.) The effect of COBRA/EMTALA, a well-intentioned global action to protect the acutely ill and vulnerable, places substantial external pressure upon the open and unbounded work system of the ED. The constraints of COBRA/EMTALA are not widely known in other parts of the health care organization; this leads to the potential for conflict when EDs accept patients whom hospitals or specialists might prefer be sent away.

41.5 Summary

EDs are dynamic, uncertain, distributed, collaborative settings for complex technical work. A slight paraphrase from Rochlin (1997) seems an apt description of the problems they face:

> For a caregiver in an emergency, deprived of experience, unsure of context, and pressed into action only when something has already gone wrong, with varying amounts of information and few mechanisms for interpreting and assessing it, avoiding a mistake may be as much a matter of good luck as good training.

In such a complex, reciprocally interactive setting, supporting the human workers who must detect, make sense of, and repair the gaps in the system to ensure its safe and effective function would seem to be a priority. Such support, however, can be developed only with careful, meticulous research that avoids a narrow, "keyhole" view of the ED (Cook & Woods, 2003; Nemeth et al., 2004).

41.5.1 Macroergonomic Balance

Given multiple opportunities for improving work in the ED, how might managers and human factors experts

best proceed? The problem seems too large and too multifaceted to be attack on all fronts at once, even if the resources to do so were readily available.

One promising approach is to attempt to "balance" domains that are either primarily negative or relatively intractable to improvement with others in which meaningful improvements are realistically possible (Smith & Carayon, 2001; Smith & Sainfort, 1989). Of the seven domains discussed in this chapter, five are potentially amenable to improvement, balancing the positives and negatives among the domains of (1) the individual workers, (2) task and tools, (3) the team, (4) the work environment, and (5) the organization. Patient factors must be taken as a given, and although the institutional context is in principle changeable, it is beyond the scope of ED and hospital managers to directly affect. However, with a strategic focus upon improvement efforts in these five domains given their interconnectedness to internal and external forces upon the ED work system, one can influence the institutional context within which emergency care is provided.

References

American College of Emergency Physicians. (1995). Emergency physician shift work. *Ann Emerg Med*, *25*(6), 864.

Asch, D. A. & Hershey, J. C. (1995). Why some health policies don't make sense at the bedside. *Ann Intern Med*, *122*(11), 846–850.

Baevsky, R. (2006). Sound levels in the emergency department setting. *Acad Emerg Med*, *13*(2), 233.

Barlas, D., Sama, A. E., Ward, M. F., & Lesser, M. L. (2001). Comparison of the auditory and visual privacy of emergency department treatment areas with curtains versus those with solid walls. *Ann Emerg Med*, *38*(2), 135–139.

Bisantz, A. M. & Ockerman, J. J. (2003). Lessons from a focus on artifacts and implicit theories: Case studies in analysis and design. In E. Hollnagel (Ed.), *Handbook of Cognitive Task Design*. pp. 53–71, Mahwah, NJ: Lawrence Erlbaum.

Bonnet, M. H. (2000). Sleep deprivation. In M. Kryger, T. Roth, & W. C. Dement (Eds.), *Principles and Practice of Sleep Medicine*. Philadelphia, PA: W.B. Saunders.

Bost N, Crilly J, Wallis M, Patterson E, & Chaboyer W. (2010). Clinical handover of patients arriving by ambulance to the emergency department - a literature review. *Int Emerg Nurs*, *18*(4), 210–220.

Boutros, F. & Redelmeier, D. A. (2000). Effects of trauma cases on the care of patients who have chest pain in an emergency department. *J Trauma*, *48*(4), 649–653.

Brown, L. H., Gough, J. E., Bryan-Berg, D. M., & Hunt, R. C. (1997). Assessment of breath sounds during ambulance transport. *Ann Emerg Med*, *29*(2), 228–231.

Cao, C. G. L., Weinger, M. B., Slagle, J., Zhou, C., Ou, J., Gillin, S., Sheh, B., Mazzei, W. (2008). Differences in Day and Night Shift Clinical Performance in Anesthesiology. *Human Factors: The Journal of the Human Factors and Ergonomics Society, 50,* 276–290.

Carayon, P. (2003). Macroergonomics in quality of care and patient safety. In H. Luczak & K. J. Zink (Eds.), *Human Factors in Organizational Design and Management* (pp. 21–34). Santa Monica, CA: IEA Press.

Carroll, J. M. & Campbell, R. L. (1989). Artifacts as psychological theories. *Behav Inf Technol, 8*(4), 247–256.

Chan EW, Taylor SE, Marriott JL, & Barger B. (2009). Bringing patients' own medications into an emergency department by ambulance: effect on prescribing accuracy when these patients are admitted to hospital. *The Medical journal of Australia, 191*(7), 374–377.

Chisholm, C. D., Collison, E. K., Nelson, D. R., & Cordell, W. H. (2000). Emergency department work place interruptions: Are emergency physicians multi-tasking or interrupt driven? *Acad Emerg Med, 7,* 1239–1243.

Chisholm, C. D., Dornfeld, A., Nelson, D., & Cordell, W. (2001). Work interrupted: A comparison of workplace interruptions in emergency departments and primary care offices. *Ann Emerg Med, 38*(2), 146–151.

Chisholm, C. D., Pencek, A. M., Cordell, W. H., & Nelson, D. R. (1998). Interruptions and task performance in emergency departments compared with primary care offices. *Acad Emerg Med, 5*(5), 470.

Coiera, E. W. (2000). When conversation is better than computation. *J Am Med Inform Assoc, 7*(3), 277–286.

Coiera, E. W., Jayasuriya, R. A., Hardy, J., Bannan, A., & Thorpe, M. E. (2002). Communication loads on clinical staff in the emergency department. *Med J Aust, 176*(9), 415–418.

Cook, R. I., Render, M., & Woods, D. D. (2000). Gaps in the continuity of care and progress on patient safety. *BMJ, 320,* 791–794.

Cook, R. I. & Woods, D. D. (2003). The messy details: Insights from technical work studies in health care. Paper presented at the *Proceedings of the Human Factors and Ergonomics Society 47th Annual Meeting,* Oct 13–17, Denver, CO.

Costa, G. (1996). The impact of shift and night work on health. *Appl Ergon, 27*(1), 9.

Cross, L. A. (1992). Pressure on the emergency department: The expanding right to medical care. *Ann Emerg Med, 21*(10), 1266–1272.

Derlet, R. W. & Nishio, D. A. (1990). Refusing care to patients who present to an emergency department. *Ann Emerg Med, 19*(3), 262–267.

Derlet, R., Richards, J., & Kravitz, R. (2001). Frequent overcrowding in U.S. emergency departments. *Acad Emerg Med, 8*(2), 151–155.

Eisenberg, E. M., Murphy, A. G., Sutcliffe, K., Wear, R., Schenkel, S., Perry, S. J., & Vanderhoef, M. (2005). Communication in Emergency Medicine: Implications for Patient Safety. *Communication Monographs, 72*(4), 390–413.

Fairbanks, R. J., Bisantz, A. M., & Sunm, M. (2007a). Emergency department communication links and patterns. *Ann Emerg Med, 50*(4), 396–406.

Fairbanks, R. J., Caplan, S. H., Bishop, P. A., Marks, A. M., & Shah, M. N. (2007b). Usability study of two common defibrillators reveals hazards. *Ann Emerg Med, 50*(4), 424–432.

Fairbanks, R. J., Caplan, S., Shah, M. N., Marks, A., & Bishop, P. (2004). Defibrillator usability study among paramedics. Paper presented at the *Proceedings of the Human Factors and Ergonomics Society 48th Annual Meeting,* Sep. 20–24, New Orleans, LA.

Fairbanks, R. J. & Wears, R. L. (2008). Hazards with medical devices: The role of design. *Ann Emerg Med, 52*(5), 519–521.

Fernandez R, Kozlowski SWJ, Shapiro M, & Salas E. (2008). Toward a Definition of Teamwork in Emergency Medicine. *Academic Emergency Medicine, 15,* 1104–1112.

Firth-Cozens, J. (2001). Multidisciplinary teamwork: The good, bad, and everything in between. *Qual Saf Health Care, 10*(2), 65–66.

Firth-Cozens, J. & Moss, F. (1998). Hours, sleep, teamwork, and stress. Sleep and teamwork matter as much as hours in reducing doctors' stress. *BMJ, 317*(7169), 1335–1336.

Folkard, S. & Monk, T. H. (Eds.). (1985). *Hours of Work: Temporal Factors in Work Scheduling.* Chichester, U.K.: John Wiley & Sons.

Folkard, S. & Tucker, P. (2003). Shift work, safety and productivity. *Occup Med, 53*(2), 95–101.

Frank, J. R. (2002). Shiftwork and emergency medicine practice. *Can J Emerg Med, 4*(6), 421–428.

Freeman J. (2003). The emerging specialty of Hallway Medicine. *CJEM* 2003:Jul 5(4), 283–285.

Frew, S. A., Roush, W. R., & LaGreca, K. (1988). COBRA: Implications for emergency medicine. *Ann Emerg Med, 17*(8), 835–837.

Gaba, D. M. (1998). Physician work hours: The "sore thumb" of organizational safety in tertiary health care. Paper presented at the *Proceedings of the Second Annenberg Conference on Enhancing Patient Safety and Reducing Errors in Health Care,* Nov 8–10, Rancho Mirage, CA.

Gaba, D. M. (2000a). Re: Aviation fatality data. Retrieved January 21, 2000, from http://www.bmj.com/cgi/eletters/319/7203/136#EL5

Gaba, D. M. (2000b). Structural and organizational issues in patient safety: A comparison of health care to other high-hazard industries. *Calif Manag Rev, 43*(1), 83–102.

Hall, K. N., Wakeman, M. A., Levy, R. C., & Khoury, J. (1992). Factors associated with career longevity in residency-trained emergency physicians. *Ann Emerg Med, 21*(3), 291–297.

Hayes, R. H. & Wheelwright, S. C. (1979). Linking manufacturing process and product life cycle. *Harv Bus Rev, 57,* 2–9.

Hoyer, C. S., Christensen, E. F., & Eika, B. (2008). Adverse design of defibrillators: Turning off the machine when trying to shock. *Ann Emerg Med, 52*(5), 512–514.

Hutchins, E. (1996). *Cognition in the Wild.* Cambridge, MA: MIT Press.

Institute for Safe Medication Practices. (2003). Blood pressure monitor tubing may connect to IV ports. *ISMP Medicat Saf Alert, 8*(12), 1–2.

Jha, A. K., Duncan, B. W., & Bates, D. W. (2001). Fatigue, sleepiness, and medical errors. In *Making Health Care Safer: A Critical Analysis of Patient Safety Practices*. Evidence Report/Technology Assessment No. 43; AHRQ Publication 01-E058. Rockville, MD: Agency for Healthcare Research and Quality.

Johnson, D. W., Hammond, R. J., & Sherman, R. E. (1980). Hearing in an ambulance paramedic population. *Ann Emerg Med, 9*(11), 557–561.

Johnson K. D. & Winkleman C. (2011). The effect of emergency deparment crowding on patient outcomes: A literature review. *Advanced Emergency Nursing Journal, 33*(1), 39–54.

Kaboli, P. J., McClimon, B. J., Hoth, A. B., & Barnett, M. J. (2004). Assessing the accuracy of computerized medication histories. *Am J Manag Care, 10*(11 Pt 2), 872-877.

Laine, C., Goldman, L., Soukup, J. R., & Hayes, J. G. (1993). The impact of a regulation restricting medical house staff working hours on the quality of patient care. *JAMA, 269*(3), 374–378.

Luten, R. C., Broselow, J., Wears, R. L., & Blackwelder, B. (1998). Broselow-Luten system: Color-coded zones for pediatric emergencies and implications for universal application. Paper presented at the *Proceedings of the Second Annenberg Conference on Enhancing Patient Safety and Reducing Error in Health Care*, Nov 8–10, Rancho Mirage, CA.

Luten, R. C., Wears, R., Broselow, J., Zaritsky, A., Barnett, T. M., Lee, T., Bailey, A., Vally, R., Brown, R., & Rosenthal, B. (1992). Length-based endotracheal tube and emergency equipment in pediatrics [published erratum appears in Ann Emerg Med 1993 Dec; 22(12): 155]. *Ann Emerg Med, 21*(8), 900–904.

Magid, D., Asplin, B. R., & Wears, R. L. (2004). The quality gap: Searching for the consequences of emergency department crowding. *Ann Emerg Med, 44*(6), 586–588.

Matthews, A. L., Harvey, C. M., Schuster, R. J., & Durso, F. T. (2002). Emergency physician to admitting physician handovers: An exploratory study. Paper presented at the *Proceedings of the Human Factors and Ergonomics Society 46th Annual Meeting*, Sep 30–Oct 4,Baltimore, MD.

Monk, T. H. (1990). Shiftworker performance. *Occup Med, 5*(2), 183–198.

Monk, T. H. & Carrier, J. (1997). Speed of mental processing in the middle of the night. *Sleep, 20*(6), 399–401.

Monk, T. H. & Folkard, S. (1992). *Making Shiftwork Tolerable*. London, U.K.: Taylor & Francis.

Monk, T. H., Folkard, S., & Wedderburn, A. I. (1996). Maintaining safety and high performance on shiftwork. *Appl Ergon, 27*(1), 17.

Morey, J. C., Simon, R., Jay, G. D., & Rice, M. M. (2003). A transition from aviation crew resource management to hospital emergency departments: The Medteams story. Paper presented at the *Proceedings of the 12th International Symposium on Aviation Psychology*, March 5–8, 2001, Columbus, OH.

Morey, J. C., Simon, R., Jay, G. D., Wears, R. L., Salisbury, M. L., & Berns, S. D. (2002). Error reduction and performance improvement in the emergency department through formal teamwork training: Evaluation results of the MedTeams project. *Health Serv Res, 37*(6), 1553–1581.

Moskop JC, Geiderman JM, Hobgood CD, & Larkin GL. (2006). Emergency physicians and disclosure of medical errors. *Annals of Emergency Medicine, 48*(5), 523-531.

Murphy, A. G. & Wears, R. L. (2009). The medium is the message: communication and power in sign-outs. *Annals of Emergency Medicine, 54*(3), 379 - 380.

Nemeth, C. P., Cook, R. I., & Woods, D. D. (2004). The messy details: Insights from the study of technical work in health care. *IEEE Trans Syst Man Cybern A, 34*(6), 689–692.

Paley, M. J. & Tepas, D. I. (1994). Fatigue and the shiftworker: Firefighters working on a rotating shift schedule. *Hum Factors, 36*(2), 269–284.

Parker, J. & Coiera, E. (2000). Improving clinical communication: A view from psychology. *J Am Med Inform Assoc, 7*(5), 453–461.

Patterson, E. S., Roth, E. M., Woods, D. D., Chow, R., & Gomes, J. O. (2004). Handoff strategies in settings with high consequences for failure: Lessons for health care operations. *Int J Qual Health Care, 16*(2), 125–132.

Paul, J. A. & Lin, L. (2007). Simulation and parametric models for improving patient throughput in hospitals. Paper presented at the *Society for Health Systems Conference*, Feb 23–24, New Orleans, LA.

Pennathur P, Guarrera TK, Bisantz AM, Fairbanks RJ, Perry SJ, & Wears R. (2008). Cognitive artifacts in transition: an analysis of information content changes between manual and electronic patient tracking systems. Paper presented at: *Proceedings of the Human Factors and Ergonomics Society 52nd Annual Meeting*, New York, NY.

Pepe, P. E., Jerger, J., Miller, R. H., & Jerger, S. (1985). Accelerated hearing loss in urban emergency medical services firefighters. *Ann Emerg Med, 14*(5), 438–442.

Perrow, C. (1984). *Normal Accidents: Living with High-Risk Technologies*. New York: Basic Books.

Perrow, C. (1994). The limits of safety: The enhancement of a theory of accidents. *J Conting Crisis Manag, 2*(4), 212–220.

Petersen, L. A., Brennan, T. A., O'Neil, A. C., Cook, E. F., & Lee, T. H. (1994). Does housestaff discontinuity of care increase the risk for preventable adverse events? *Ann Intern Med, 121*(11), 866–872.

Plsek, P. E. & Greenhalgh, T. (2001). Complexity science: The challenge of complexity in health care. *BMJ, 323*(7313), 625–628.

Richardson, L. D., Asplin, B. R., & Lowe, R. A. (2002). Emergency department crowding as a health policy issue: Past development, future directions. *Ann Emerg Med, 40*(4), 388–393.

Risser, D. T., Rice, M. M., Salisbury, M. L., Simon, R., Jay, G. D., & Berns, S. D. (1999). The potential for improved teamwork to reduce medical errors in the emergency department. The MedTeams Research Consortium. *Ann Emerg Med, 34*(3), 373–383.

Rochlin, G. I. (1997). *Trapped in the Net: The Unanticipated Consequences of Computerization*. Princeton, NJ: Princeton University Press.

Rothwell, P. M. (2005). Treating individuals 2. Subgroup analysis in randomised controlled trials: Importance, indications, and interpretation. *Lancet, 365*(9454), 176–186.

Salluzo, R. F., Mayer, T. A., Strauss, R. W., & Kidd, P. (1997). *Emergency Department Management: Principles and Applications.* St Louis, MO: Mosby.

Schull, M. J., Szalai, J. P., Schwartz, B., & Redelmeier, D. A. (2001). Emergency department overcrowding following systematic hospital restructuring: Trends at twenty hospitals over ten years. *Acad Emerg Med, 8*(11), 1037–1043.

Schull, M. J., Vermeulen, M., Slaughter, G., Morrison, L., & Daly, P. (2004). Emergency department crowding and thrombolysis delays in acute myocardial infarction. *Ann Emerg Med, 44*(6), 577–585.

Schulman, P. R. (2002). Medical errors: How reliable is reliability theory? In M. M. Rosenthal & K. M. Sutcliffe (Eds.), *Medical Error: What Do We Know? What Do We Do?* (pp. 200–216). San Francisco, CA: Jossey-Bass.

Shah, A., Frush, K., Luo, X., & Wears, R. L. (2003). Effect of an intervention standardization system on pediatric dosing and equipment size determination. *Arch Pediatr Adolesc Med, 157*(3), 229–236.

Shapiro, M. & Morchi, R. (2002). High-fidelity medical simulation and teamwork training to enhance medical student performance in cardiac resuscitation. *Acad Emerg Med, 9*(10), 1055–1056.

Shapiro, M. J., Morey, J. C., Small, S. D., Langford, V., Kaylor, C. J., Jagminas, L., Suner, S., Salisbury, M. L., Simon, R., & Jay, G. D. (2004). Simulation based teamwork training for emergency department staff: does it improve clinical team performance when added to an existing didactic teamwork curriculum? *Qual Saf Health Care, 13*(6), 417–421.

Silbergleit, R., Kronick, S. L., Philpott, S., Lowell, M. J., & Wagner, C. (2006). Quality of emergency care on the night shift. *Acad Emerg Med, 13*(3), 325–330.

Simon, R., Morey, J., Locke, A., Risser, D. T., & Langford, V. (1997). *Full Scale Development of the Emergency Team Coordination Course and Evaluation Measures.* Andover, MA: Dynamics Research Corporations.

Simon, R., Morey, J. C., Rice, M., Rogers, L., Jay, G. D., Salisbury, M., & Wears, R. L. (1998). *Reducing errors in emergency medicine through team performance: the MedTeams project.* Paper presented at the *Proceedings of the Second Annenberg Conference on Enhancing Patient Safety and Reducing Errors in Health Care,* Nov 8–10, Rancho Mirage, CA.

Small, S. D., Wuerz, R. C., Simon, R., Shapiro, N., Conn, A., & Setnik, G. (1999). Demonstration of high-fidelity simulation team training for emergency medicine. *Acad Emerg Med, 6*(4), 312–323.

Smith, M. J. & Carayon, P. (2001). Balance theory of job design. In W. Karwowski (Ed.), *International Encyclopedia of Ergonomics and Human Factors* (pp. 1181–1184). London, U.K.: Taylor & Francis.

Smith, M. J. & Sainfort, P. C. (1989). A balance theory of job design for stress reduction. *Int J Ind Ergon, 4*, 67–79.

Smith-Coggins, R., Rosekind, M. R., Hurd, S., & Buccino, K. R. (1994). Relationship of day versus night sleep to physician performance and mood. *Ann Emerg Med, 24*(5), 928–934.

Steele, M. T., Ma, O. J., Watson, W. A., & Thomas, H. A., Jr. (2000). Emergency medicine residents' shiftwork tolerance and preference. *Acad Emerg Med, 7*(6), 670–673.

Steele, M. T., McNamara, R. M., Smith-Coggins, R., & Watson, W. A. (1997). Morningness–eveningness preferences of emergency medicine residents are skewed toward eveningness. *Acad Emerg Med, 4*(7), 699–705.

Stewart, J. A. (2008). Delayed defibrillation caused by unexpected ECG artifact. *Ann Emerg Med, 52*(5), 515–518.

Stiell, A., Forster, A. J., Stiell, I. G., & van Walraven, C. (2003). Prevalence of information gaps in the emergency department and the effect on patient outcomes. *CMAJ, 169*(10), 1023–1028.

Tepas, D. I., Barnes-Farrell, J. L., Bobko, N., Fischer, F. M., Iskra-Golec, I., & Kaliterna, L. (2004). The impact of night work on subjective reports of well-being: An exploratory study of health care workers from five nations. *Rev Saude Publica, 38*(Suppl), 26–31.

Tepas, D. I. & Carvalhais, A. B. (1990). Sleep patterns of shiftworkers. *Occup Med, 5*(2), 199–208.

Tepas, D. I., Duchon, J. C., & Gersten, A. H. (1993). Shiftwork and the older worker. *Exp Aging Res, 19*(4), 295–320.

Tepas, D. I. & Monk, T. H. (1987). Work schedules. In G. Salvendy (Ed.), *Handbook of Human Factors* (pp. 819–843). New York: Wiley & Sons.

Thomas, H., Jr., Schwartz, E., & Whitehead, D. C. (1994). Eight-versus 12-hour shifts: Implications for emergency physicians. *Ann Emerg Med, 23*(5), 1096–1100.

Topf, M. (2000). Hospital noise pollution: An environmental stress model to guide research and clinical interventions. *J Adv Nurs, 31*(3), 520–528.

Tucker, A. L. & Edmondson, A. C. (2002). When problem solving prevents organizational learning. *J Organ Change Manag, 15*(2), 122–137.

Tucker, A. L. & Edmondson, A. C. (2003). Why hospitals don't learn from failures: Organizational and psychological dynamics that inhibit system change. *Calif Manag Rev, 45*(2), 55–72.

Vaughan, D. (1996). *The Challenger Launch Decision: Risky Technology, Culture and Deviance at NASA.* Chicago, IL: University of Chicago Press.

Viccellio, P., Schneider, S. M., Adplin, B., Blum, F., Broida, R., Bukata, R. et al. (2008). *Emergency Department Crowding: High-Impact Solutions.* American College of Emergency Physicians (ACEP) publishing, Dallas, TX.

Vincent, C., Taylor-Adams, S., Chapman, E. J., Hewett, D., Prior, S., Strange, P. et al. (2000). How to investigate and analyse clinical incidents: Clinical risk unit and association of litigation and risk management protocol. *BMJ, 320*(7237), 777–781.

Vincent, C., Taylor-Adams, S., & Stanhope, N. (1998). Framework for analysing risk and safety in clinical medicine. *BMJ, 316*(7138), 1154–1157.

Wagner MM & Hogan WR. (1996). The accuracy of medication data in an outpatient electronic medical record. *Journal of the American Medical Informatics Association, 3*(3),234–244.

Wagner MJ, Wolf S, Promes S et al. (2010). Duty hours in emergency medicine: balancing patient safety, resident wellness, and the resident training experience: a consensus response to the 2008 institute of medicine resident duty hours recommendations. *Academic emergency medicine : official journal of the Society for Academic Emergency Medicine, 17*(9),1004–1011.

Wanerman, R. (2002). The EMTALA paradox. Emergency medical treatment and labor act. *Ann Emerg Med*, 40(5), 464–469.

Wears, R. L. & Perry, S. J. (2002). Human factors and ergonomics in the emergency department. *Ann Emerg Med*, 40(2), 206–212.

Wears, R. L., Perry, S. J., Shapiro, M., Beach, C., & Behara, R. (2003). Shift changes among emergency physicians: Best of times, worst of times. Paper presented at the *Proceedings of the Human Factors and Ergonomics Society 47th Annual Meeting*, Oct 13–17, Denver, CO.

Wears, R. L. & Perry, S. J. (2010). *Discourse and process analysis of shift change handoffs in emergency departments*. Paper presented at the 54th Human Factors and Ergonomics Society, San Francisco, CA.

Womack, J. P., Jones, D. T., & Roos, D. (1990). *The Machine that Changed the World*. New York: HarperCollins.

Woods, D. D. (1998). Designs are hypotheses about how artifacts shape cognition and collaboration. *Ergonomics*, 41(2), 168–173.

Zun, L. S. & Downey, L. (2005). The effect of noise in the emergency department. *Acad Emerg Med*, 12(7), 663–666.

42

Human Factors and Ergonomics in Pediatrics

Matthew C. Scanlon and Paul Bauer

CONTENTS

42.1 Introduction 723
42.2 How Children Are Different 724
42.3 Normal Pediatric Growth and Development 724
    42.3.1 Changes in Weight and Length 724
    42.3.2 Change in Head Size 724
    42.3.3 Other Physiological Changes during Development 728
    42.3.4 Pharmacokinetic Changes during Development 729
    42.3.5 Additional Motor Considerations in Children 729
    42.3.6 Cognitive and Speech Development in Children 730
    42.3.7 Obesity in Children 730
42.4 Pediatric Patient Safety 730
    42.4.1 Pediatric Patient Safety: Is It Really Different? 731
    42.4.2 Epidemiology of Pediatric Health Care Error and Injury 731
    42.4.3 Medication Errors in Pediatric Patients 732
    42.4.4 Strategies for Improving Pediatric Patient Safety—Part 1 733
42.5 Human Factors, Ergonomics, and Pediatrics 733
    42.5.1 Case for Applying Human Factors Concepts to Children 733
    42.5.2 Approach to Applying Pediatric Considerations to Human Factors and Ergonomics 734
42.6 Human Factors, Ergonomics, and Pediatric Patient Safety 734
    42.6.1 Human Factors, Technology, and Pediatric Patient Safety 735
    42.6.2 Need for Human Factors When Introducing HIT in a Pediatric Setting: The Pittsburgh Experience 736
    42.6.3 Additional Developments in Pediatrics and Human Factors 738
42.7 Strategies for Improving Pediatric Patient Safety—Part 2 739
42.8 Conclusions 739
References 740

42.1 Introduction

The U.S. Census Bureau estimates that 25% of the U.S. resident population is children less than 18 years of age (Yax 2004). Despite this fact, a corresponding quarter of U.S. resources devoted to research do not go to furthering an understanding of this segment of the population. As a result, one traditional (but flawed) approach has been to generalize concepts learned from the adult population to children and adolescents. Pediatric providers have countered with the much-used cliché that children are not simply small adults. Children have many distinctive characteristics that require specific consideration separate from adult patients. Just as an appreciation and understanding of the unique characteristics of children is critical to providing safe medical care, so is the same understanding necessary when considering application of human factors and ergonomic principles to this population.

Children differ from adults in a multitude of ways. These differences extend to their dependency on adults, their cognitive and maturational development, as well as physiologic differences. Pediatric patient safety differs from that of patient safety in adults as a result of the aforementioned differences from adults. Again, any attempt to apply concepts of human factors to the issues of pediatric patient safety requires an understanding of how children are different.

Since the first edition of this publication, disproportionately little remains published on the topic of human factors as it relates to children. One exception

723

is *Ergonomics for Children* (Lueder and Berg Rice 2008), along with a small number of articles discussed later in this chapter. Therefore, this chapter will identify the unique characteristics of both children and pediatric patient safety in an attempt to help fill the void.

42.2 How Children Are Different

The study of pediatrics is based on the central concept that from birth through early adulthood, children are dynamic. The extent of change that children experience includes size, gross and fine motor development, cognitive development, speech and psychological development, as well changes in physiology. As a result of the dynamic nature of children, the concept of "normal" is a moving target. Appropriate weight, heart rate, or speech is all age dependent.

Beyond developing an understanding of the dynamic nature of healthy children, the study of pediatrics further explores the range of illness in children. The impact of a given illness on a child may vary dramatically depending on where they are in their growth and development. Consequently, the leading causes of illness and mortality in children also vary by the age group. Unfortunately, and of potential interest to a human factors audience, accidental or unintentional injury is the leading cause of death in all children (Kochanek and Smith 2004).

In addition to the constantly changing nature of children, another commonality in pediatrics is the dependency of children on adults for care. This point is important because until adolescence (and arguably beyond), children require the provision of nutrition and shelter from others. Access to medical care is dependent on a care provider seeking care for the children in their care. This care is independent of any attempt to nurture or socialize a child. From a pediatric perspective, a patient safety perspective, and a human factors perspective, the implications of this dependency cannot be overemphasized. Children are unable to effectively advocate for themselves, make choices about their diet or lifestyle, or select products or environments, which may decrease their risk of harm. For example, motor vehicle accidents are the leading cause of death by unintentional injury in children ages 1–14 years of age (Kochanek and Smith 2004). These are not children who are electing to drive recklessly or unrestrained. Similarly, the national trend toward pediatric obesity largely reflects care decisions of adults. Obesity prevalence has increased from 15% to 22.5% in 6 to 11 year olds (Troiano et al. 1995). While there are genetic influences on the susceptibility to obesity, a significant contribution to the development

of obesity is the consumption of calories in excess of energy expenditure. Adult caretakers dictate food purchasing and portion decisions, as well as behaviors such as exercise or television viewing. In neither the case of motor vehicle accidents nor childhood obesity are children in a position to make decisions that could directly improve their health and reduce risk. The importance of a child's dependency on adults is critical when considering the challenges and solutions of patient safety, particularly from a human factors perspective.

42.3 Normal Pediatric Growth and Development

42.3.1 Changes in Weight and Length

The dramatic changes witnessed from before birth through adolescence fall into the category of growth and development. A discussion of abnormal development is beyond the scope of this work. Additionally, the biological and psychosocial determinants of normal growth and development are the subjects of other chapters and textbooks. To this end, interested readers are directed to recommended readings listed at the end of this chapter.

From the time of delivery, dramatic growth and physiologic change occur in children. Parallel to the physical changes are psychosocial changes that allow for interaction with their environment and others. The rate of growth and development is not steady state. While newborns gain an average of 30 g of weight a day and 3.5 cm of length a month, this growth begins to slow around 3 months of age (Needlman 2004). A graphic illustration of the changes in rates of growth can be seen using the invaluable tool of growth charts used to document the growth of children (Figures 42.1 and 42.2). The growth charts illustrate the ranges of growth that can be seen by weight and length or height. By following serial measurements over time, health care providers can screen for poor growth or abnormal growth patterns, which may indicate the development of acute but subtle illness, chronic illnesses, poor nutrition, or psychosocial deprivation.

42.3.2 Change in Head Size

Beyond weight and length, a child's head circumference normally demonstrates significant growth during the first 3 years of life (Figure 42.3). This growth in head circumference normally reflects normal brain growth. Not surprisingly, failure of normal brain development can lead to abnormally small head size or microcephaly. Derangements of brain development or cerebrospinal

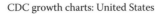

CDC growth charts: United States

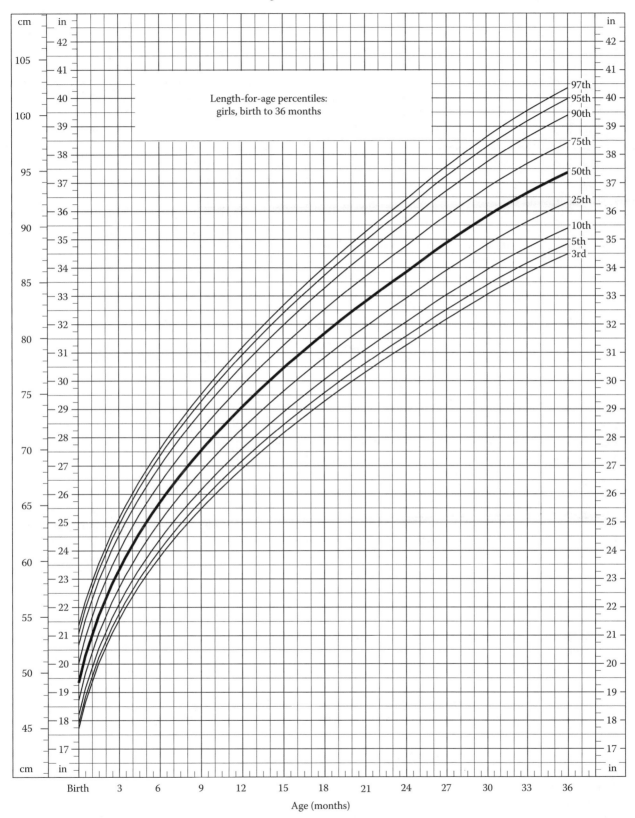

FIGURE 42.1

Growth chart illustrating normal range of changes in length with aging among girls aged birth to three years of age. These changes in length explain the need to anticipate and design for age-specific products, which account for changes in reach, size and center of gravity.

CDC growth charts: United States

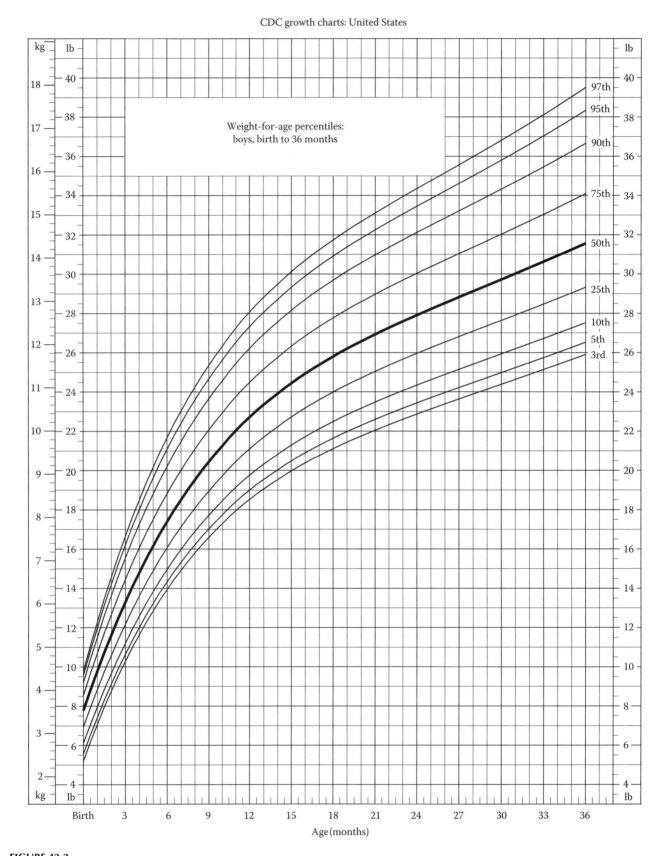

FIGURE 42.2

Growth chart illustrating normal range of changes in weight with aging among boys aged birth to three years of age. These changes in weight explain potential hazards related to medication delivery in children due to weight-based dosing.

CDS growth charts: United States

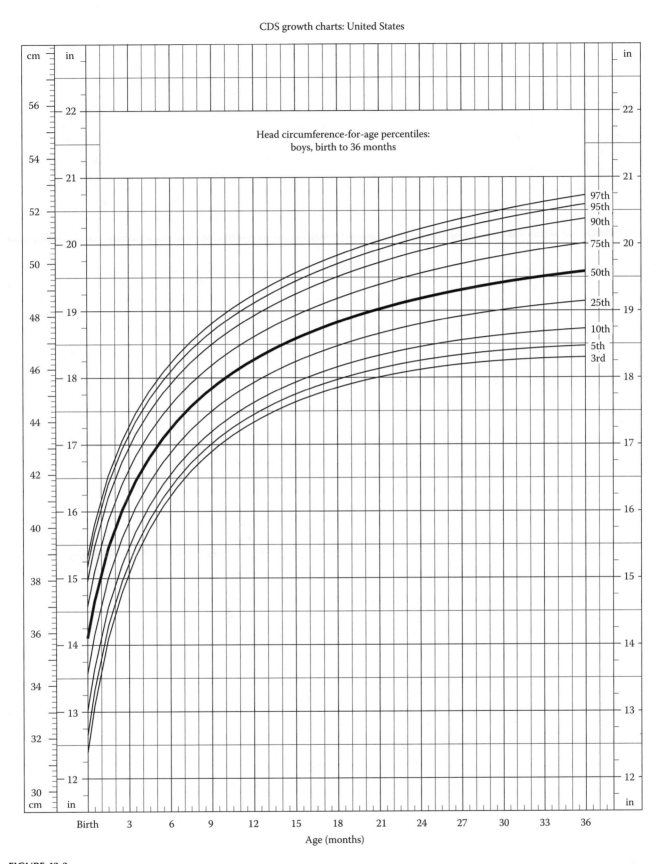

Head circumference-for-age percentiles:
boys, birth to 36 months

FIGURE 42.3
Growth chart illustrating normal range of changes in head circumference with aging among boys aged birth to three years of age. These changes, coupled with the changes in length, illustrate the potential hazards associated with higher center of gravity in young children.

fluid flow can also lead to inappropriately large head size (macrocephaly). Under normal circumstances, an infant or child's head is disproportionately large relative to the rest of its body when compared with adults. For example, an infant's head may account for 15% of its total body mass. In contrast, an adult's head typically accounts for 3% of its total body mass (Kauffman 1997). This disproportionate mass, coupled with relatively weak neck musculature, places infants and young children at greater risk for traumatic acceleration/deceleration injuries to the brain.

42.3.3 Other Physiological Changes during Development

Just as a child's growth is reflected by externally visible changes, similar changes happen at a physiologic level. Maturation of an individual can be detected in bone growth and ossification, as well as dental development. More dramatic, particularly to health care providers, are the dramatic changes in "vital signs" that normally occur in children. Vital signs are measurements of physiologic parameters that are widely used to assess the health or lack of health of a patient (adult or pediatric). Specifically, body temperature, blood pressure, and heart and respiratory rates are assessed in both health screening and ongoing patient care.

The use of vital signs is hardly unique to pediatrics; adult care providers are acutely aware of the importance of these data elements. However, unlike adult values, the pediatric values change dramatically with age changes (Figure 42.4). Additionally, within a given age-based "normal," there can be dramatic variations based on variables including sleep, level of pain or agitation, and activity. Pathologic conditions may also dramatically influence a pediatric patient's vital signs. Of note, the typical causes for an abnormality of a vital sign in a child may be quite different from that of an adult. For example, abnormally low heart rates (bradycardia) are

most commonly the result of intrinsic heart disease in adult populations. However, the leading cause in children is hypoxia or inadequate oxygen resulting from some respiratory compromise. Only through understanding the normal age-based variation in vital signs can a health care provider begin to accurately assess the status of a pediatric patient.

A child's metabolism undergoes significant changes during development. Metabolic changes, combined with changes in body size and composition that alter distribution of substances, and changes in organ development and function that may affect absorption and elimination, all influence a child's response to medications and toxins. A consequence of these changes is that medication dosing in children is either by weight or body surface area. The need for weight-based calculations to accurately dose medications carries significant implications for both patient safety and the design of technological solutions.

Another important consideration is the nature of skin in children. The skin is a person's largest organ, providing numerous functions including fluid homeostasis, temperature regulation, and protection from infection. Like the other parameters identified in this chapter, the skin of children both differs from adults and changes with age. First, the relatively high ratio of body surface area to body mass in children when compared to adult's results in the potential for much greater fluid loss. For example, a representative 5 year old child has a surface to mass ratio of 380 cm²/kg compared to an adult with a ratio of 280 cm²/kg (Malina 2004). As a result, fluid management is often more complex in pediatric patients with skin injuries. A second important difference between adults and children is the composition of skin. Children less than 2 years of age have less subcutaneous fat and thinner skin layers, which contribute to both heat and fluid loss, and vulnerability to thermal injury (O'Connor and Besner 2003). Brief immersion (5 s) in hot water that is tolerable to adults (140°F) will result in a full

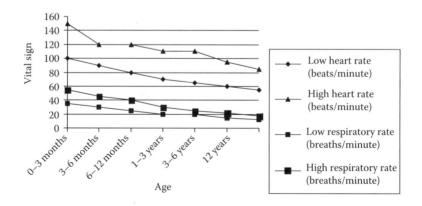

FIGURE 42.4
Changes in range of normal heart rate and respiratory rate values with age. Implications include the added challenge to recognizing abnormal values, as well as designing monitoring technology and alarm algorithms that will vary with patient age.

thickness burn to a child that would otherwise require a 2 min immersion at 126°F (Lucchesi 2004). As a result of these differences, a design with heat exposure tolerance thresholds that would be viewed as acceptable in an adult setting might lead to significant injury in children.

42.3.4 Pharmacokinetic Changes during Development

We have discussed the consideration of surface area to mass changes in the growing pediatric patient and its import on medication dosing, heat exchange, and particular vulnerabilities. Just as important is the consideration of developmental changes in organ function and composition, which lead to changing rates of drug absorption, distribution, metabolism, and excretion. Pharmacokinetics is the discipline of understanding these processes. Body surface area calculations represent an improvement over straight weight-based dosing, particularly for children over the age of 8–12 where drug metabolism approaches adult patterns. However, absolute drug dosages in young children and neonates may be lower than predicted by body surface area calculations because of differences in renal metabolism and clearance.

Bartelink et al. published a comprehensive review of pediatric dosing guidelines in *Clinical Pharmacokinetics* in 2006. They review principles of absorption, distribution, metabolism, and excretion and the changes that occur with development of the individual.

Absorption of medication in the stomach is influenced by stomach pH. The newborn will have a neutral gastric pH at birth. In fact, it is not until age 2 years that the pH of the stomach lining approaches the acidic character of the adult stomach (Rennie and Roberton 2005). Stomach emptying and movement of the bowels is slower in newborns (Strolin et al. 2005). There are other practical issues in newborn and infant children that impact on the delivery of medication through the enteral route. For example, frequent low volume feeds make medication timing difficult for medications that need to be administered without food (Bartelink et al. 2006).

Intramuscular administration of some medications in the newborn period is unreliable because of large differences in regional circulation to skeletal muscle beds. The transdermal route of medication administration can be very effective in neonates, infants, and children due to relative absence of subcutaneous fat deposits. However, assessment of individual patients informs such therapeutic decisions. Sweating, large body mass index, and wide variance in subcutaneous fat deposits may make the transdermal route difficult in some children (Strolin et al. 2005).

As the individual grows, body composition changes. The relative abundance of water (80%–90% of body weight) gives way to increases in fat and protein mass,

and may decrease in the adult to as little as 55% of total body weight (McLeod et al. 1992). These changes make a difference for drugs that are not bound in the fat compartment, but instead are highly protein bound.

As the individual matures, so do enzyme levels responsible for the metabolism of medications. Birth represents the first great change in enzymatic activity. Shortly after birth, liver blood flow increases and the baby is exposed to a remarkable increase in oxygen tension. By about 6 months of age, in the healthy liver, enzymatic clearance rates of drug approach adult levels and is linearly related to the size of the liver, which depends on body surface area, not weight (Bartelink et al. 2006).

Changes in excretion occur in the kidney as well. Renal excretion matures rapidly after birth (Kearns 2000). Creatinine levels as an indicator of glomerular filtration are often used to guide dosing for drugs that are toxic to the kidney, but some difficulty is present here as well. It is important to know that creatinine depends on the ratio of total muscle mass to glomerular filtration. Newborns have relatively less skeletal muscle mass, and values of 0.5 mg/dL can represent significant renal dysfunction where 1 mg/dL might be normal in a young male athlete. Knowing the normative values for age, sex, and size, as well as the trend in creatinine proves important in dosing for drugs that are excreted by the kidney.

The described developmental changes and variance in drug pharmacokinetics can make drug dosing in children across ages difficult. Bartelink et al. (2006) describe a flowchart for pediatric dosing based on age, with consideration of metabolic and renal pathways with eight different decision points.

42.3.5 Additional Motor Considerations in Children

Dramatic gross motor changes occur in children during the first year of life. These include the ability to roll over at approximately 4–6 months of age, the ability to crawl at 8–9 months of age, and the ability to walk at around 1 year of age. The constantly and often surprising changes in mobility have implications for both accidental injury and exposure to things previously believed to be inaccessible to children. Beyond a year of life, a child's ability to ambulate improves along a learning curve. It is not unusual for a child to run at 16 months of age, however, this running is often associated with many falls. Fine motor changes parallel the gross motor changes with children moving from crude grasping at 3 months, through reaching for objects and transferring objects between hands at 4 and 6 months to a finger-thumb "pincer" grasp between 8 and 9 months.

Children have additional challenges relating to imbalance. Imbalance is influenced by the poor gross motor coordination and high centers of gravity because of relatively large head to body mass ratios and high upper segment to lower segment ratios at young ages. The upper to lower segment ratio looks at the proportion of one's body above the pubic symphysis to that below this landmark. The ratio decreases with time from 1.7 as a newborn, 1.3 at 3 years of age until reaching a ratio of 1 at approximately 7 years of age (Needlman 2004). The higher center of gravity in younger children is another contributing factor to unintentional injuries, which may be aggravated by the design of a child's environment.

42.3.6 Cognitive and Speech Development in Children

In the setting of a stimulating environment, children flourish with rapid cognitive development. Infants begin with habituation to environmental stimulation and quickly learn to identify both the faces and voices of caretakers. By 4 months of age, children discover the world beyond their care providers. This discovery includes the phenomenon of discovering self, as well as an appreciation of cause and effect relationships. Another component of world discovery is the placement of any object encountered by an infant in their mouth. The oral exploration marks only the start of intense exploration of the surrounding environment that continues for years. By 18 months, the ability to realize an object still exists though out of sight, or object constancy, is permanent. Cause–effect relationships are better understood at this age.

Between 2 and 5 years, children experience magical thinking with egocentrism. The former leads to flawed views of causality, often as a result of the observation of coincidence. The latter is an inability to take on another's perspective. By 5 years of age, children transition through this prelogical phase to cognition that is heavily rule based and concrete. Concrete thinking continues up to adolescents when a child may begin developing formal logical thinking. However, the development of formal thinking, which includes logical evaluation, consideration of abstract processes and the application of principles, is not a forgone conclusion. Some adolescents never make this transition and remain concrete thinkers throughout life.

Of note, the majority of these cognitive milestones are often associated with emotional turmoil and behavioral changes. For example, sleep disruption is common with the development of new skills in the preschool age groups.

Normal speech development occurs in relation to cognitive development. Sound recognition begins within the first month of life, with generation of cooing noises occurring by 4 months of age. By 1 year of age, children usually have several words, typically "mama" or "dada." Vocabulary continues to grow with up to 200 words and multiword phrases between 2 and 3 years of age. At 2 years old, usually 50% of a child's words are intelligible to family, increasing to 75% at 3 years old. By 4 years, most speech is intelligible to strangers. From this point, the complexity of language skills continues to advance (Dixon 2000).

Speech development has several implications. First, the relative unintelligibility of children at young ages limits their ability to seek assistance or articulate concerns, discomforts, or fears. Second, by 4 and 5 years of age, children are able to communicate with family members. However, the ability to generate phrases and sentences that can appear relatively well thought out is misrepresentative of their ability to reason. The incongruity between "adult-like speech" and "adult thought processes" can create significant tension between children and adults.

42.3.7 Obesity in Children

The problem of obesity in children has been viewed as an epidemic. Children are now experiencing diseases like hypercholesterolemia, high blood pressure, and type 2 diabetes mellitus, which had previously been considered adult diseases (Donahue 2004). A frequent consequence of these illnesses in adults is the development of coronary artery disease. Other significant morbidities of obesity include respiratory diseases, cancer, orthopedic and joint injury, and poor self-esteem. Beyond the patient level morbidity and mortality, there are significant economic costs associated with obesity. Cost evaluation in 1995 placed the combined direct and indirect cost of obesity at $99 billion in the United States (Wolf and Colditz 1998). This figured has been increased to $117 billion based on year 2000 estimates (U.S. Department of Health and Human Services 2001).

Obesity is a multifactorial problem. There are genetic components, which are not entirely understood. Another critical determinant in the development of obesity is a mismatch between caloric intake and energy expenditure. Other factors found to be positively predictive of pediatric obesity include parental obesity and time spent viewing television (Crespo et al. 2001).

The long-term impact of this epidemic is not yet understood. However, as the problem of obesity leads to major changes in the physical characteristic and health of children, any evaluation of patient safety or human factors must consider this dilemma.

42.4 Pediatric Patient Safety

Pediatric patient safety shares many commonalities with patient safety efforts focused on adult health care. There remains a dearth of fundamental "basic science" research, particularly related to pediatrics (Perrin and Bloom 2004). Issues of culture change, communication, leadership, reporting, error identification and reduction, and human factors are important to both adult and pediatric safety work. However, many of the same unique characteristics of children, which led to a separate medical specialty, support the need for separate considerations in patient safety. This section will explore what is known about pediatric patient safety with special attention to where the unique characteristics of children might have more importance.

42.4.1 Pediatric Patient Safety: Is It Really Different?

While the reader may agree that children are not simply small adults, they may be unconvinced that children have unique patient safety needs. However, application of an established model quickly eliminates doubt. Specifically, the Systems Engineering Initiative for Patient Safety (SEIPS) model of system thinking (Carayon et al. 2006) provides a conceptual model for understanding the unique aspects of pediatric patient safety. In the SEIPS Model, there are *people* who use *tools and technology* to perform *tasks* within a specific *environment*, all in the larger context of an *organization*. In this model, safety is an emergent property that is more than the sum of the five components (italicized) and their interactions. Applying this model to the issue of pediatric patient safety, it is easy to understand the unique needs. For instance, if the task is reconciling a list of medications, then typically a physician or nurse reviews, updates and verifies medications with an adult patient at the time to presentation to the emergency department in a busy community hospital. However, simply by changing the adult patient in the scenario to a young child brought by ambulance to the same emergency department, suddenly the task of reconciling medications (and the potential added safety resulting from this task) is dramatically different. Specifically, a young child is very unlikely to know an accurate medication list. While this risk may be mitigated if an adult knowledgeable of the child's medications is present, simply changing the patient from an adult to a child illustrates the specific safety risks. Substituting children for adults or pediatric tasks for adult tasks can readily generate additional examples of the specific safety needs of children.

42.4.2 Epidemiology of Pediatric Health Care Error and Injury

There have been few systematic investigations of the scope of pediatrics errors and injuries. A study of the utilization of medical care by children referenced finding 0.8% of pediatric discharges associated with a complication (McCormick et al. 2000). One study using administrative data to assess the risk to hospitalized children found 1.81–2.96 medical errors per 100 discharges (Slonim et al. 2003). This work used International Classification of Diseases, 9th Revision (ICD-9) coding of medical errors to identify medical errors. The authors identify many of the limitations of using such administrative data including under reporting, identifying cause and effect related to variables like length of stay and a lack of clinical and physiological data. These findings are comparable to other published work, which estimated the rate of adverse events in children less than 5 years of age at 2.7 per 100 discharges (Brennan et al. 1991).

A second study applied the Agency for Healthcare Research and Quality's (AHRQ) patient safety indicators (PSIs) to administrative pediatric hospitalization data (Miller et al. 2003). This work found rates of patient safety events identified using the AHRQ tool comparable to that of adults. Aside from the previously discussed issues of using administrative datasets, the work is limited by the facts that the AHRQ tool was not designed for pediatrics and that subsequent changes in the AHRQ PSI tool may make these findings invalid (Remus 2004, personal communication).

Even less is known about pediatric medical errors in an ambulatory setting. A review of current knowledge confirms the lack of research (Miller et al. 2004). The authors identified several medial errors that are common in pediatric ambulatory settings. These errors tend to fall into one of three categories of medical errors: underuse, misuse, and overuse of medical care. One specific error identified was the lack of immunization of children. U.S. statistics demonstrate that immunization rates in children in 2002 were at only 78% (National Immunization Program 2003). The result of this is preventable illness, which can be life threatening and add to the burden of morbidity and cost. Another medical error identified in ambulatory settings was misinterpretation of pediatric echocardiograms by adult providers (Stanger et al. 1999; Ward and Purdie 2001). These could be viewed as misuse errors that occur because the appropriate individuals are not performing and interpreting studies. Another source of ambulatory errors important to the pediatric population is the over prescription of antibiotics for viral illnesses. Antibiotics are only effective in treating infections caused by bacteria. Research has extrapolated that 6.5 million prescriptions are written annually for children who were diagnosed with a viral upper respiratory infection or the common cold (Nyquist et al. 1998). This figure accounts for 12% of prescriptions written each year for children. These medications are unnecessary, can lead to adverse drug reactions and, independent of side effects or errors in the prescribing, dispensing, and administration process, likely add dramatically to the cost of health care, and encourage the emergence of resistant bacterial strains. What may aggravate this incredible overuse of medications is the fact that large numbers of medications are administered to children in school or child care settings where there are often no providers skilled in delivering medications.

42.4.3 Medication Errors in Pediatric Patients

Of all the possible medical errors that occur in children, the most is known about medication errors. A study of medication errors in children found that the rate of preventable adverse drug events (ADEs) was similar to adult studies (Kaushal 2001b). However, the rate of potential ADEs (errors that occurred yet did not reach the patient) was three times higher. These potential ADEs were most common in neonatal intensive care units, a finding that is consistent with previous work demonstrating a higher number of medication errors in neonatal and pediatric intensive care units (PICUs; Folli et al. 1987; Raju et al. 1989).

The provision of medications depends on three main steps: medication prescription, medication dispensing, and medication administration. Each of these steps is associated with errors; in pediatrics, each step has additional risks unique to delivering medications for children.

Correct prescribing is the first step to safe medication delivery. A major factor in pediatric prescription errors is the fact that medication dosing is weight based. Prescribing medication requires the correct identification of the patient's weight. An incorrect weight can lead to miscalculation of medication dosing. Similarly, confusing the units of weight, pounds, and kilograms, can lead to 2× over- or underdosing. The identification of accurate weight takes on added importance as patients often experience changes in weight over time. A single weight used as if it is a constant value can lead to dosing errors. Additional factors that have an adverse impact on pediatric medication prescriptions include decimal place errors and computational errors. Because of the weight-based calculation, which may involve small weights, decimals take on added importance. Misplacing a decimal by one place in either direction can lead to a 10-fold under- or overdose. Research reveals that pediatric resident physicians are prone to making computational errors when prescribing medications (Potts and Phelan 1996; Rowe et al. 1998). As a result, recent efforts have targeted the element of calculation in prescribing infusions of medications (see Section 42.7). The net effect is that pediatric patients are at an increased risk for prescription errors. Kaushal's previously cited work found nearly 80% of potential ADEs occurred at the ordering phase.

In addition to the inpatient concerns, ambulatory prescription is also a problem. Beyond the previously discussed dilemma of over prescription of antibiotics, a study of 1532 pediatric emergency department records found 10.1% of the charts reflected a prescribing error (Kozer et al. 2002). In the year 1998, there were 124.3 million pediatric office visits, with 52.9% resulting in prescriptions. If the pediatric emergency department numbers apply, there is likely a significant medication prescription error rate with associated potential injury rate in the pediatric outpatient setting.

Medication dispensing creates another set of challenges. Unit doses can rarely be used for medications because of the weight-based nature of dosing. Consequently, pharmacists and nurses are required to draw up correct amounts from standard vials or containers of medication. Drawing up nonstandard doses manually introduces a new step for errors to occur. Potential safeguards built into manufacturer's packaging (bar coding or otherwise) are lost when a medication is drawn up into a separate container. Another dispensing hazard occurs because the medication may be administered in both an enteral and parenteral route. Repackaging or drawing up medications may result in using the wrong syringe, and thus lead to parenteral administration of an enteral medication. Finally, repackaging in the dispensing phase creates a potential for mislabeling a medication, resulting in the wrong patient receiving the medication.

Administering medications is also a source of errors in pediatrics. The potential for mixing up patients or routes of administration can occur. Additionally, pumps designed primarily for adults must be used in pediatrics. Many of these devices are not designed for the volumes or doses that are used in pediatrics, and they create hazards through features that seem inconsequential in the relatively standardized adult patient population for which they are designed. The majority of pediatric patients are unable to recognize and intervene if they are about to receive a wrong medication or dose. While adults may be able to prevent an error by asking "Why am I getting a red pill? Mine are normally blue," most pediatric patients are unable cognitively and verbally to protect themselves.

Regardless of where a failure occurs in the medication process, children remain at greater risk. The weight-based nature of dosing increases the likelihood of factor-of-ten errors and thus larger overdoses than an adult might see. Children are considered to have less physiologic reserve than most adults; a given medication error in a child may yield worse consequences than a proportionately equivalent error in adults. This lower reserve reflects both less physiologic tolerance to stress and a limited ability to "buffer" the effect.

The challenges of delivering medications to children are many and, unfortunately, not readily aided by technological solutions available. Few, if any, computerized physician order entry (CPOE) systems provide "off-the-shelf" rules for pediatric dosing, much less safety checks. Similarly, bar coding technology has added challenges of creating bar codes that fit a wide range of patient sizes that may be encountered in pediatrics. For instance, many wristbands that are marked with bar codes may be "unreadable" to bar code scanning devices because of dressings or the curvature bands when placed on small extremities (Piehl 2003, personal communication). Also, repackaging medications for pediatric use often circumvents safeguards offered by pharmaceutical company bar coding. The need to repackage doses and thus relabel with bar codes, introduces yet another source of potential errors. These multiple challenges, and their limited technological solutions, create an environment of increased risk to pediatric patients.

42.4.4 Strategies for Improving Pediatric Patient Safety—Part 1

The unique challenges of pediatric patient safety have not been lost on all in the patient safety field. Various authors have identified specific strategies that might lead to improved safety. One strategy identified five specific topics that required concentrated attention (Perrin and Bloom 2004). These topics included the following: (1) understanding and improving communication as it relates to safety, (2) understanding the value of technology with attention to implementation, (3) advancing pediatric patient safety as a topic of importance, (4) acquiring additional knowledge beyond basic epidemiology of safety issues, and (5) further defining the true priorities in pediatric patient safety.

Another group has advocated for three main steps focusing on pediatric residency training programs as follows: (1) education related to patient safety, (2) increased reporting of errors, and (3) emphasize fixing systems rather than people (Napper et al. 2003). These authors admit that their goals may appear overly simplistic. However, an emphasis on these goals would be a major change in the current thinking of many training programs.

A third set of strategies has been described for preventing medical errors in pediatrics (Fernandez and Gillis-Ring 2003). These strategies include systems analysis, critical incident root cause analysis, the use of clinical pharmacists and computer-assisted decision making, changes in the production and distribution of pharmaceuticals, and education for families and providers. Much of this work further centers on reporting, analysis, and categorization of errors as a basis for designing solutions. It remains unclear how counting and classifying errors actually will lead to safety improvements.

Unfortunately, since the time of the first publication of this chapter, there remains little evidence for pediatric patient safety improvements. While efforts have demonstrated methods of identifying potential harm through triggers (Sharek et al. 2006; Takata et al. 2008), or error reduction through technology (Potts et al. 2004), actual improvement in safety as demonstrated by measurable reduction in harm remains the Holy Grail of pediatric safety. The one exception is the introduction of central line insertion bundles, which has resulted in measurable decreases in central venous line associated bloodstream infections (Bhutta et al. 2007; Miller et al. 2010).

In light of the absence of safety "breakthroughs," one can make a case that specific consideration of the role of human factors and ergonomics is necessary to better frame any solutions to the challenges of pediatric patient safety. What little is known related to human factors and pediatrics is considered in Section 42.5.

42.5 Human Factors, Ergonomics, and Pediatrics

To continue a theme underlying this chapter, disappointingly little has been written about human factors and ergonomics as it relates to children. With some imagination, work on design for people with functional limitations might be applied to the pediatric population (Vanderheiden 1997). While the author does not specifically identify children in his work, he addresses the need for attention to physical, cognitive, and language impairments in design. This work focused on limitations as a result of a cause (older age, disease, injury, or genetic abnormalities) rather than limitations as a preexisting though transient, state. In light of the physical, cognitive, and language developmental limitations that children may face at different ages, some of Vanderheiden's concepts bear consideration. Of note, the solution he advocates for is a concept of universal design, or the design of products that "can be effectively and efficiently used by people with no limitations as well as those operating with functional limitations" (Vanderheiden 1997).

42.5.1 Case for Applying Human Factors Concepts to Children

Before extrapolating the work of Vanderheiden and others to a pediatric population, it is worth considering whether such effort is necessary. One of the most vivid illustrations of a lack of attention to human factors when designing for children is in the toy industry.

The U.S. Consumer Product Safety Commission (CPSC) is a federal body charged with protecting the public from unreasonable risks of death and injury because of consumer products (U.S. Consumer Product Safety Commission 2004). The CPSC Web site holds a wealth of data illustrating the hazards of poor design for children.

One category tracked is deaths from toys. While the death of a child from any cause is tragic, a death because of a toy seems even more egregious. In the CPSC report for 2001, 25 deaths were reported in children less than 15 years of age (McDonald 2002b). Of these, nine deaths were linked to choking or asphyxiation from a toy. Recognizing that young children have a developmental drive to explore their world by placing new objects in their mouth, while also having narrow airways prone to obstruction, the CPSC has created a 313-page document to guide manufacturers in preventing these deaths (Smith 2002). Other work by the CPSC details over 69,500 emergency room visits in children less than 5 years of age because of injuries from nursery products (McDonald 2002a).

The medical literature also provides additional evidence for addressing the human factors needs of children. A study of playground injuries found that nearly 50% could have been prevented through changes to the equipment and playground structure (Petridou et al. 2002). The incidence of playground injuries among boys and girls was 7 per 1000 and 4 per 1000, respectively, with a total of 777 injuries identified. The number of preventable injuries would be decreased through use of protective equipment and shoes in this setting.

Aside from the known morbidity and mortality of children from play-related activity, the increasing use of computers by younger children creates another setting where design of interfaces and recognition of ergonomic implications may be more important.

42.5.2 Approach to Applying Pediatric Considerations to Human Factors and Ergonomics

Wickens et al. (1998) identify the critical areas where a child-focused consideration could drive research in human factors, ergonomics, and children. In addition to identifying the cognition, variation in human size and shape, the authors also touch on the issues of biomechanics and work (or play) physiology. Applying the differences described in Sections 42.2 and 42.3, Wickens' outline serves as a template for identifying central pediatric issues that either requires application of existing knowledge or research to acquire the needed knowledge. This, in turn, can provide a basis for extending Vanderheiden's universal design concept in the opposite direction of the aged to the young (Table 42.1).

42.6 Human Factors, Ergonomics, and Pediatric Patient Safety

Some of the existing strategies for improving patient safety were identified previously (see Section 42.4.3). While implied at times, none of the strategies expressly identified a need for applied human factors concepts to efforts to improve pediatric patient safety. The need for such consideration is based on several observations.

In 1984, a published report described a case series of what initially was viewed as neonatal sepsis (Solomon et al. 1984). On further investigation, the illnesses were the result of a repeated medication error in a neonatal hospital unit. Because of nearly identical labels, racemic epinephrine was incorrectly administered rather than vitamin E. Labeling confusion is not a problem restricted

TABLE 42.1

Human Factors and Pediatrics: Possible Design Implications

Human Factors Consideration	Pediatric-Specific Issues	Design Implications for Children[a]
Sensory changes	Development of sight and hearing in young children	Use existing knowledge of imperfect senses
Variation in size and shape	Small size can circumvent safety features and limit use of interfaces	Consider smaller extremes of size for design
Variation in biomechanics	Limited strength and leverage can limit use	Consider consequences of users with developmentally impaired strength
Work (play) physiology	Developmentally varied vital signs and energy stores limit endurance	Consider broader range of users and endurance
Cognitive considerations	Magical vs. concrete vs. abstract thinking and innate curiosity	Consider consequences of misuse due to cognitive factors
Language considerations	Varied ability to read, comprehend, and express self	Consider limited abilities in design of interface, directions, and safety labeling

[a] Design considerations predicated on desired use of device/process by children. Alternatively, consider safety features to prohibit use/misuse by children.

to pediatrics. However, the issues of how medications are administered to pediatric patients with small doses often being administered to multiple patients from a single container, rather than unit doses that are more widely used in adults. Furthermore, the case is important because pediatric differences in metabolism and the relative dose to patient size were potential factors in the poisoning.

Two papers from the Great Ormond Street Hospital for Children National Health Service (NHS) Trust in London, England address specific safety implications in the performance of surgery for neonatal congenital heart defects. The first study described findings of 243 surgeries over an 18-month time period (de Leval et al. 2000). Central to the work is the importance of both major and minor human failures independent of patient risk factors. Major failures had great potential for patient death yet could be compensated for with prevention of poor outcome. Minor human failures, however, were often unrecognized and thus no compensation occurred. Independent of how significant any given minor failure was for any specific patient's outcome, the authors found a link between these minor failures and negative outcomes. Specific factors noted as critical for mitigating the risk of major failures included the diagnostic skill and problem-solving skills of a surgeon and ability to communicate with the operative team. The second work focused on the importance of reporting and near miss identification in the performance of neonatal cardiac surgery (Carthey et al. 2001). Again, the work identifies strategies for identifying human factors in a pediatric operative setting with the intent of preventing or compensating for human errors. Neither work is limited to pediatrics in terms of its findings. To that end, the studies could have been performed in adult settings with no attention to pediatric implications. However, in the complex environment of the repair of congenital heart defects, the studies at credence for the need to apply human factors in pediatrics.

Another publication examined poor outcomes from pediatric procedural sedation with attention to human factors (Cote et al. 2000). Using critical incident analysis, the work found several important factors that could have led to improved outcomes. Identified variables associated with poor outcome (death or permanent neurologic injury) included sedation in nonhospital settings, inadequate and inconsistent monitoring of physiologic status (vital signs and oxygenation), and inadequate resuscitation. Other factors included inadequate staff to monitor the child postsedation, inadequate equipment, and medication errors. The significance of this work to a consideration of human factors in pediatric patient safety is the number of variables that could have been corrected with correct design. Monitoring procedures, equipment, and staff can be standardized, but a

presedation recognition of the age-based variation in children is essential. Similarly, as children vary in size and weight, so do resuscitation equipment and medication doses. Inattention to the need for a range of equipment and insufficient familiarity and training with the equipment creates unnecessary hazards.

A final example of the need for human factor considerations in ensuring pediatric patient safety is the U.S. Food and Drug Administration (FDA) alert describing preventable risk to children from excess radiation exposure during computed tomography (CT) scans (Fiegal 2001). Citing literature that described an increased lifetime risk of cancer from the radiation exposure during CT scans (Brenner et al. 2001), the FDA identifies how a "one-size-fits-all" approach to imaging creates risk. The unnecessary radiation exposure could be addressed through readily available interventions including reduction of tube current, adjusting scanning axes, and more aggressive screening for the appropriate use of CT imaging. Additionally, knowledge of human factors could be used to develop charts or reference tools of ideal pediatric CT settings for a range of weights, body diameters, and anatomic regions. None of this is possible without first recognizing the needs specific to children and then applying concepts of human factors to this problem.

42.6.1 Human Factors, Technology, and Pediatric Patient Safety

There is little evidence to suggest that principles of human factors and user-centered design have been effectively applied to the information technology solutions offered to address medical errors. Some of the identified challenges to implementation of CPOE include the need for weight-based dosing and age-dependent laboratory normal values (Kaushal 2001). A systematic review of information technology and its potential for pediatric patient safety identified significant potential despite little specific pediatric evidence for numerous technologies (Johnson and Davison 2004). Included in this work is a proposed research agenda for pediatric applications of safety information technology. However, nowhere in the review (including in the research agenda) is mention given to human factors consideration.

The need for consideration human factors in the design and implementation is highlighted by case studies reported from a pediatric center with a 3 year history of CPOE (Scanlon 2004). Two cases of medical errors that resulted from a lack of human factors consideration in the design and implementation are described. Specific human factor principles that were not addressed in the design of the commercially available CPOE system include an overreliance on memory for correct performance, inattention to facilitating discrimination of similar data elements, failure to recognize the implications of mental models, inadequate decision support, and insufficient feedback to users of the system.

The lack of attention to human factors is further illustrated by the flowchart describing the first time dose of a single IV medication using a CPOE system (Figure 42.5). This high-level flowchart outlines the steps required using a commercially available CPOE system in a pediatric center beginning with a provider conceiving of prescribing a medication up to a nurse administering the medication. The actual administration process is not included in the flowchart because of its extensive complexity and variation The sheer complexity of this process, coupled with the case studies of error involving the same system (Scanlon 2004), suggests that the proposed benefits of CPOE may not be readily realized without careful application of human factor principles.

The observation that a neglect of human factors may lead to imperfect technology that, in turn, may create rather than reduce risk of patient harm should not be surprising. The significance of human factors to technology has been previously described (Gosbee 2004). One important potential effect of introducing technology, patient safety or other, to a health care environment is the change to the way work is performed. Consistent with human factor concepts, technology should facilitate the performance of work. Instead, technology may introduce new cognitive requirements, disrupt workflow, cause distractions, and decrease safety.

42.6.2 Need for Human Factors When Introducing Healthcare Information Technology (HIT) in a Pediatric Setting: The Pittsburgh Experience

In 2005, there were two publications from the Children's Hospital of Pittsburgh (CHP) describing very different experiences with CPOE (Han 2005; Upperman 2005).

The first of these two manuscripts described the findings of a pre-, post-CPOE implementation study of ADEs. Using voluntary reporting as a detection method for ADEs, they described a reduction from 0.05 ADEs per 1000 medication doses to 0.03 ADEs per 1000 medication doses. Additionally, they related a complete elimination of transcription errors as a result of removing the ability to write orders. The authors concluded that CPOE led to a reduction in harmful ADEs and improved patient safety.

In contrast, Han and colleagues examined observed mortality in the CHP PICU before, during, and after the same CPOE implementation. They described an increase in the observed mortality as a percent of predicted mortality, resulting in an increased odds ratio of mortality of 3.28 attributable to CPOE.

These studies are both fascinating when viewed from a human factors perspective. First, the reduction

in ADEs (and associated harm) by Upperman and colleagues is based on a decrease in voluntary reports during the post-CPOE period. The authors believe this decrease resulted from added safety created by the CPOE system. However, when viewed from the standpoint of human performance in the face of system changes resulting in new and changed tasks created by the implementation of CPOE, it is entirely possible that the decrease in reported ADEs might instead reflect that employees were too busy (or even frustrated) to report ADEs during the post implementation time period. Because there was no alternative measure of reduced harm other than a decrease in reported events, understanding the interface of tools and technology with people and their tasks leaves readers questioning the validity of the conclusions.

The Han publication is even more interesting from a human factors standpoint. The authors describe how children being brought to the PICU from an outside facility during the pre-CPOE time period would have orders prewritten and medications prepared *even before the patient's physical arrival*. In contrast, the post-CPOE period required that patients would have to physically arrive and be registered before a single order could be written. As a result, there were documented delays in ordering, dispensing, and administering potentially life-sustaining medications, and it was this delay that was perceived to be a source of increased mortality. What is striking from a human factors standpoint is how clearly Han's work illustrates the unintended consequences of making changes to a system without understanding the potential impact. Using the SEIPS model to illustrate this, a technology change (CPOE) was made, which changed the interaction between people (PICU providers) and tasks (ordering, dispensing, and administering medications) with the context of an environment (the PICU at CHP). Not surprisingly, the changes to the task of providing medications to a critically ill child in this context created unforeseen (but potentially predictable) problems. Though this assessment risks introducing hindsight bias, it is plausible that the providers in this setting (the PICU at CHP) could have work with the information technology staff to understand how CPOE would impact clinical care. Even if this was not done, it is also plausible to believe that a work around process could be created to provide supportive care in the context of the admission of a critically ill child who needs medications at the time of admission.

The application of a human factors perspective to the CHP PICU scenario also may explain how the Han findings differ from other pediatric findings (Jacobs 2006; Longhurst 2010). In each of these reports, the interaction between people (patients and providers), tools and technology, tasks, environment, and organization

First time IV ordering/administration process

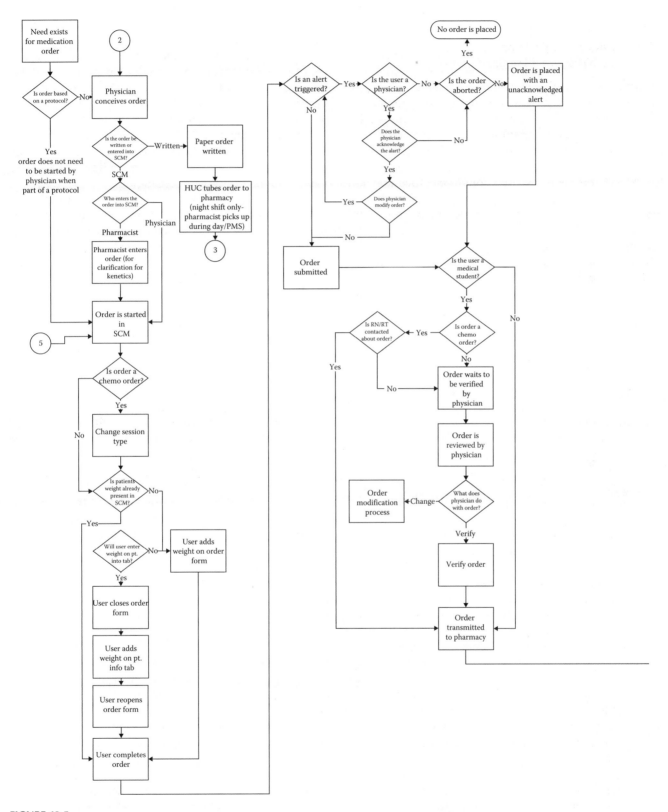

FIGURE 42.5

Flowchart illustrating high-level overview of steps in entering an order into a computer order entry system (the "SCM" in the chart). Steps in dispensing medications at pharmacy-level (not shown) are similarly complex.

is different. As a result, the same CPOE system could be implemented in multiple different pediatric settings with very different outcomes.

42.6.3 Additional Developments in Pediatrics and Human Factors

While the principles of ergonomics have not specifically changed in the field of pediatrics, the study of the relationship between human beings and their tools has led to a number of transformations in the health care industry with respect to pediatrics. These transformations can be seen in areas of tools and technology, the work environment, organizational structure and culture, and the tasks that are performed. These changes have driven alterations in care and other processes with improvements in both quality and patient safety. Other important outcomes relating to the work environment and organizations are in a period of flux.

Tools and technologies can be broadly understood to encompass medical devices, information technologies and the electronic health record, bar coding and medication administration tools, as well as the study of these tools and technologies (usability). On the regulatory end, there is now a growing human factors program at the federal governmental level with the FDA pushing for the involvement of human factors consideration in the design of medical devices (Food and Drug Administration 2010).

The work environment encompasses both the larger structure where care is delivered (hospital or clinic or home) and the layout of the room where the care is given, ambient qualities such as noise, lighting, temperature, etc. Because the pediatric patient can range in size (as described earlier) and have a variety of developmental and medical needs not routinely encountered in an adult care facility, more attention is being given to design of spaces and environments where children are cared for that facilitate care with unique technologies and tools and coordination of care among many pediatric specialties and subspecialties. As an example, the National Association of Children's Hospitals and Related Institutions (NACHRI) published a compendium of recommendations for the design of physical structures where children are cared for after an extensive architectural review (NACHRI and Center for Health Design 2008). This is but one example of how consideration of human factors is influencing change in the way the environment is constructed around the care of children.

Organizational considerations of human factors around care can be seen both at the macro- and microorganizational levels. As a collaborative of academic children's hospitals, the macroorganizational structure NACHRI is paying attention to pediatric-specific differences in quality measurement and pediatric-specific safety initiatives for its member institutions. The development of analytics and research support in NACHRI is aimed at supporting improvement in care at individual children's hospitals. On a regulatory level, the Joint Commission for the Accreditation of Healthcare Organizations has adopted a one-size-does-not-fit-all approach, recognizing that different patient populations have different needs, and that human factors are critical to improving the safety and quality of care (Gosbee 2005).

Tasks performed in the practice of pediatric vary considerably from adult care. Indeed, as indicated earlier, we have demonstrated how consideration of the pediatric patient in her developmental breadth as she grows older, has motivated unique expertise in nearly every field of medicine. This is true for the surgeon just as much as it is true for the pharmacist, developmental psychologist, or floor nurse. The number of fields dedicated to pediatric health care tasks has exploded in the last century since much of the pioneering work by Abraham Jacobi (Nichols et al. 1991). Truly, a dedication to the specific care of children has resulted in a great increase in training regarding the tasks for the care of children.

The interaction between people, tools and technology, the physical work environment, organization and culture, and the specific tasks shapes processes that result in outcomes. In medicine, since the publication of the landmark studies of the 1990s about safety (Brennan and Leape 1991; Kohn et al. 2000), there has been a dramatic shift in attention to the problem of harm in health care. Understanding the issue has driven the development of measures related to error and harm with the development of quality indicators. However, it has been shown that these measures do not carry the same meaning in pediatrics (Scanlon et al. 2008). Many adult indicators such as decubitus ulcers might be present on admission in pediatric hospitals, and prolonged ventilator dependence is not uncommon in critically ill children who have had extensive surgery. Appropriate health care quality measures in children continue to be developed.

In summary, consideration of human factors specific to the care of children are dramatically shaping the approach to interactions between tools and technology, the physical environment, and the organizational structure of care delivery. Although it has not been historically framed as such, the development of pediatric specialties and subspecialties represents the greatest human factors advance in the domain of tasks directed at the care of children. A human factors approach continues to inform and remain vital to the development of our ability to measure the quality and safety of pediatric health care.

42.7 Strategies for Improving Pediatric Patient Safety—Part 2

It is tempting to lay charges of "child neglect" on both the medical, patient safety, and human factors communities when considering the relative lack of resources dedicated to pediatrics relative to population distribution. Beyond any brief satisfaction that complaining might bring, the larger issue of how human factors relate to pediatric patient safety remains largely unanswered.

In fact, it is just as easy to suggest the pediatric patient safety community is not attending to the importance of human factors. The Institute of Medicine's landmark work *To Err Is Human* (Kohn et al. 2000) makes a point of identifying human factors as a critical aspect of improving patient safety. Unfortunately, the pediatric patient safety strategies identified to date (see Section 42.4.3) have failed to emphasize the need for more attention to pediatric human factors and its intersection with patient safety. All of the identified strategies have value and can be necessary but insufficient without further appreciation of human factors.

Instead, the pediatric patient safety community is following its adult counterparts in focusing on errors and technology without understanding the importance of human factors and design. The patient safety community has largely ignored the intersection of safe practice, economic factors, and workload (Rasmussen 1997; Cook 2003). Without an understanding of the interrelationship between how work is done and the presence or absence of patient safety, a meaningful application of human factors principles in pediatric patient safety is unlikely to be realized.

Central to this thought is that *how work is designed and performed* is critical to making health care safe. For instance, designing a CPOE system to meet the needs of the end user, as well as the environment in which it will be used, could eliminate many of the problems described in Section 42.6. The converse implication holds true. Lack of understanding of how work is performed may lead to patient safety solutions, technological or other, that actually introduce risk.

A series of publications by Karsh and colleagues further illustrates the need for better understanding of how work is done (Holden et al. in press; Karsh et al. 2009; Scanlon and Karsh 2010). Specifically, studies of implementation of bar-coded medication administration and a bidirectional interface between a computer order entry and a pharmacy system led to the findings supporting several major themes. One theme was that health care information technology changes the nature of clinical work with new and changed requirements for problem solving, as well as changed demands in clinical work.

Additionally, violations in organizational rules were found to not be behavioral deviance but instead adaptations to complex workflow (Alper et al. 2008).

Two additional themes that emerged in this work touch on mental workload and workflow. Mental workload was identified as relating not to staffing ratios or other external factors but instead to the specific demands of a task. Meanwhile, the seemingly straightforward workflow of administering medications was found to be astonishingly complex. In just one nursing unit, a flow diagram of the work in administering a medication resulted in a process map over 60 pages in length. Instead of being a clean, linear process, medication administration was nonlinear, full of interruptions, and driven by events, which led to the need for better information streams, less "noise" in systems and frequent reprioritization of work. Sadly, it appeared that the studied technologies were designed based on simple linear assumptions and thus did little to support the clinical work.

These findings, while described in pediatric settings, have not been reproduced in adult settings and thus may or may not be pediatric specific. While pediatric medication management has many steps not required for adult patients, it is not unreasonable to suspect that similar themes may be found in adult care. Thus, the need to apply human factors principles and tools to enhancing the understanding and design of work performance is arguably a global patient safety need and not solely one limited to the pediatric population.

42.8 Conclusions

The care of children is complex and challenging due, in no small part, to the variation inherent in the patients because of age-based changes in physical and intellectual development. Consequently, improving the safety of children has added complexity when compared to adult care. The numerous unique characteristics in children limit their ability to protect themselves or increase their vulnerability to medical errors and harm. Health care systems and processes, when designed, rarely consider the needs of pediatric patients but instead focus on adult care.

Existing efforts to improve pediatric patient safety have focused on improving communication, education, and reporting; developing a better understand of the epidemiology of errors; and introducing technological solutions. All are important components that may yet ignore the importance of human factors in pediatric patient safety. This oversight is largely understandable; the dearth of literature on human factors likely reflects a more global lack of appreciation for much needed knowledge. Increased recognition of the need to understand work in both pediatric and adult health care, and then design safety solutions that either is consistent with or improves the way work is performed is critical to lasting improvement.

The pediatric patient safety and human factor communities have a clear challenge ahead: develop a recognition and understanding of each other's disciplines and then apply it in a multidisciplinary manner. Only then can the work of truly improving pediatric patient safety take place.

References

Alper SJ, Holden RJ et al. (2008) Violation prevalence after introduction of a bar-coded medication administration system. *Proc 2nd Int Conf on Healthcare Systems and Ergonomics and Patient Safety*, June 2008, Strasbourg, France.

Bartelink IH, Rademaker CMA, Schobben AF, van den Anker JN. (2006) Guidelines on paediatric dosing on the basis of developmental physiology and pharmacokinetic considerations. *Clin Pharmacokinet*. 45, 1077–1097.

Bhutta A, Gilliam C, Honeycutt M, Schexnayder S, Green J, Moss M, Anand KJ. (2007) Reduction of bloodstream infections associated with catheters in paediatric intensive care units: A stepwise approach. *BMJ*. 334, 362–365.

Brennan TA, Leape LL, Kaird NM, Herbert L, Localio AR, Lawthers AG, Newhouse JP, Weiler PC, Hiatt HH. (1991) Incidence of adverse eventsand negligence in hospitalized patients. *N Engl J Med*. 324, 370–376.

Brenner DJ, Elliston CD, Hall EJ, Berdon WE. (2001) Estimated risks of radiation-induced fatal cancer from pediatric CT. *AJR*. 176,289–296.

Carayon P, Hundt AS , Karsh BT, Gurses AP, Alvarado CJ, Smith M, Flatley Brennan P. (2006) Work system design for patient safety: The SEIPS model. *Qual Saf Health Care*. 15,50–58.

Carthey J, de Leval MR, Reason JT. (2001) The human factor in cardiac surgery: Errors and near misses in high technology medical domain. *Ann Thorac Surg*. 72, 300–305.

Cook RI. (2003) Lessons from the war on cancer: The need for basic research on safety. Testimony submitted to the Agency for Healthcare Research and Quality 2nd National Summit on Patient Safety Research, pp. 1–6. Arlington, VA. www.ctlab.org. Accessed December 5, 2003.

Cote CJ, Notterman DA, Karl HW, Weinberg JA, McCloskey C. (2000) Adverse sedation in pediatrics: A critical incident analysis of contributing factors. *Pediatrics*. 105, 805–814.

CPSC: United States Consumer Product Safety Commission at http://www.cpsc.gov/. Accessed 7/31/2011.

Crespo CJ, Smit E, Troiano RP, Macera CA, Andersen RE. (2001) Television watching, energy intake, and obesity in US children: Results from the Third National Health and Nutrition Examination Survey, 1988–1994. *Arch Pediatr Adolesc Med*. 155, 360–365.

de Leval MR, Carthey J, Wright DJ, Farewell VT, Reason JT. (2000) Human factors and cardiac surgery: A multicenter study. *J Thorac Cardiovasc Surg*. 4, 661–672.

Dixon SD. (2000) Two years: Language emerges. In Dixon SD, Stein MT (eds.) *Encounters with Children: Pediatric Behavior and Development*, 3rd edn. MOSBY, St. Louis, MO.

Donahue PA. (2004) Obesity. In Behrman RE, Kliegman RM, Jenson HB (eds.) *Nelson Textbook of Pediatrics*, pp. 173–177, 17th edn. Saunders, Philadelphia, PA.

Fernandez CV, Gillis-Ring J. (2003) Strategies for the prevention of medical error in pediatrics. *J Pediatr*. 143, 155–162.

Fiegal DW. (2001) FDA Public Health Notification: Reducing radiation risk from computed tomography for pediatric and small adult patients. U.S. Food and Drug Administration–Center for Devices and Radiological Health. www.fda.gov/cdrh/safety/110201-ct.html. Accessed June 3, 2003.

Folli HL, Poole RL, Benitz WE, Russo JC. (1987) Medication error prevention by clinical pharmacists in two children's hospitals. *Pediatrics*. 79, 718–722.

Food and Drug Administration. (2010) About human factors. http://www.fda.gov/MedicalDevices/Device RegulationandGuidance/PostmarketRequirements/ HumanFactors/ucm119185.htm#2. Accessed July 30, 2010.

Gosbee LL. (2005) *Using Human Factors Engineering to Improve Patient Safety*. Joint Commission Resources, Oakbrook Terrace, IL.

Han YY, Carcillo JA, Venkataraman ST, et al. (2005). Unexpected increased mortality after implementation of a commercially sold computerized physician order entry system. *Pediatrics*. 116, 1506–1512.

Holden RJ, Scanlon MC, Patel NR, Kaushal R, Escoto KH, Brown RL, Alper SJ, Arnold JM, Shalaby TM, Murkowski K, Karsh BT. (2011) A human factors framework and study of the effect of nursing workload on patient safety and employee quality of working life. BMJ Qual Saf 2011;20:15-24 doi:10.1136/bmjqs.2008.028381

Jacobs BR, Brilli RJ, Hart KW. (2006) Perceived increase in mortality after process and policy changes implemented with computerized physician order entry. *Pediatrics*. 117(4), 1451-1452.

Johnson KB, Davison CL. (2004) Information technology: It's importance to child safety. *Ambul Pediatr*. 4, 64–72.

Karsh B, Escoto KH, Alper SJ, Holden RJ, Scanlon MC, Murkowski K, Patel N, et al. (2009) Do beliefs about hospital technologies predict nurse perceptions of their ability to provide quality care? A study in two pediatric hospitals. *Int J Human–Computer Interactions*. 25, 374–389.

Kauffman BA. (1997) Head injury and intracranial pressure. In Oldham KT, Colombani PM, Foglia RP (eds.) *Surgery of Infants and Children: Scientific Principles and Practice*, pp. 417–27. Lippincott Raven Publishers, Philadelphia, PA.

Kaushal R, Barker KN, Bates DW. (2001a) How can information technology improve patient safety and reduce medication errors in children's health care? *Arch Pediatr Adolesc Med.* 155, 1002–1007.

Kaushal R, Bates DW, Landrigan C, McKenna KJ, Clapp MD, Federico F, Goldmann DA. (2001b) Medication errors and adverse drug events in pediatric inpatients. *JAMA.* 285, 2114–2120.

Kearns GL. (2000) Impact of developmental pharmacology on pediatric study design: Overcoming the challenges. *J Allergy Clin Immunol.* 106, S128–S138.

Kochanek KA, Smith BL. (2004) *Deaths: Preliminary Data for 2002.* National Vital Statistic Report. Vol 52:13, pp. 1–48. National Center for Health Statistic, Hyattsville, MD. http://www.cdc.gov/nchs/data/nvsr/nvsr52_13.pdf. Accessed March 15, 2004.

Kohn LT, Corrigan J, Donaldson MS (eds.). (2000) *To Err is Human: Building a Safer Health System.* National Academy Press, Washington, DC.

Kozer E, Scolnik D , Macpherson A, Keays T, Shi K, Luk T, Koren. (2002) Variables associated with medication errors in pediatric emergency medicine. *Pediatrics.* 110, 737–742.

Longhurst CA, Parast L, Sandborg CI, et al. (2010). Decrease in hospital-wide mortality rate after implementation of a commercially sold computerized physician order entry system. *Pediatrics.* 126, 14–21.

Lucchesi M. (2004) Burns, thermal. *Emedicine.* http://www.emedicine.com/ped/topic301.htm. Accessed March 19, 2004.

Lueder R, Berg Rice VJ. (2008) *Ergonomics for Children: Designing Products and Places for Toddlers to teens,* 1st edn. CRC Press, Taylor & Francis, Boca Raton, FL.

Malina RM. (2004) *Growth, Maturation and Physical Activity,* p. 268, 2nd edn. Human Kinetics, Champaign, IL.

McCormick MC, Kass B, Elixhauser A, Thompson J, Simpson L. (2000) Annual report on access to and utilization of health care for children and youth in the United States-1999. *Pediatrics.* 105, 219–230.

McDonald J. (2002a) Nursery product-related injuries and deaths to children under age five. U.S. Consumer Product Safety Commission, Washington, DC. www.cpsc.gov/library/nursry02.pdf. Accessed April 1, 2004.

McDonald J. (2002b) Toy related deaths and injuries, calendar year 2001. U.S. Consumer Product Safety Commission, Washington, DC. www.cpsc.gov/library/toydth01.pdf. Accessed April 1, 2004.

McLeod HL, Relling MV, Crom WR, Silverstein K, Groom S, Rodman JH, Rivera GK, Crist WM, Evans WE. (1992) Disposition of antineoplastic agents in the very young child. *Br J Cancer Suppl.* 18, S23–S29.

Miller MR, Elixhauser A, Zhan C. (2003) Patient safety events during pediatric hospitalizations. *Pediatrics.* 111, 1358–1366.

Miller MR, Griswold M, Harris JM 2nd, Yenokyan G, Huskins WC, Moss M, Rice TB, Ridling D, Campbell D, Margolis P, Muething S, Brilli RJ. (2010) Decreasing PICU catheter associated bloodstream infections: NACHRI's quality transformation efforts. *Pediatrics.* 125, 206–213.

Miller MR, Pronovost PJ, Burstin HR. (2004) Pediatric patient safety in the ambulatory setting. *Ambul Pediatr.* 4, 47–54.

Napper C, Battles JB, Fargason C. (2003) Pediatrics and patient safety. *J Pediatr.* 142, 359–360.

National Association of Children's Hospitals and Related Institutions (NACHRI) and Center for Health Design. (2008) *Evidence for Innovation: Transforming Children's Health through the Physical Environment.* NACHRI, Alexandria, VA.

National Immunization Program. (2003) Immunization coverage in the U.S.: Results from National Immunization Survey. http://www.cdc.gov/vaccines/stats-surv/nis/default.htm. Accessed 8/6/2011.

Needlman RD. (2004) Growth and development. In Behrman RE, Kliegman RM, Jenson HB (eds.) *Nelson Textbook of Pediatrics,* pp. 23–66, 17th edn. Saunders, Philadelphia, PA.

Nichols BL, Ballabriga A, Kretchmer N, Nestlé Nutrition SA. (1991) *History of Pediatrics, 1850–1950.* Nestlé Nutrition workshop series, vol. 22. Nestlé Nutrition, Vevey, Switzerland.

Nyquist AC, Gonzales R, Steiner JF, Sande MA. (1998) Antibiotic prescribing for children with colds, upper respiratory tract infections, and bronchitis. *JAMA.* 279, 875–877.

O'Connor A, Besner GE. (2003) Burns: Surgical perspective. *Emedicine.* http://www.emedicine.com/ped/topic2929.htm. Accessed March 7, 2004.

Perrin JM, Bloom SR. (2004) Promoting safety in child and adolescent health care: Conference overview. *Ambul Pediatr.* 4, 43–46.

Petridou E, Sibert J, Dedoukou X, Skalkidis I, Trichopouos D. (2002) Injuries in public and private playgrounds: The relative contribution of structural, equipment and human factors. *Acta Paediatr.* 9, 691–697.

Potts AL, Barr FE, Gregory DF, Wright L, Patel NR. (2004) Computerized physician order entry and medication errors in a pediatric critical care unit. *Pediatrics.* 113, 59–63.

Potts MJ, Phelan KW. (1996) Deficiencies in calculation and applied mathematics skills in pediatrics among primary care interns. *Arch Pediatr Adolesc Med.* 150, 748–752.

Raju TNK, Kecskes S, Thornton JP, Perry M, Fieldman S. (1989) Medication errors in neonatal and paediatric intensive-care units. *Lancet.* 2, 374–376.

Rasmussen J. (1997) Risk management in a dynamic society: A modeling problem. *Saf Sci.* 27, 183–213.

Rennie JM, Roberton NRC. (2005) *Roberton's Textbook of Neonatology.* Elsevier/Churchill Livingstone, Edinburgh, U.K.

Rowe C, Koren T, Koren G. (1998) Errors by paediatric residents in calculating drug doses. *Arch Dis Child.* 79, 56–58.

Salvendy G (ed.). (1997) *Handbook of Human Factors and Ergonomics,* 2nd edn. John Wiley and Sons, Inc., New York.

Scanlon MC. (2004) Use of failure modes and effects analysis to evaluate the introduction of variation to a medication infusion preparation process through regulatory "Standardization." Accepted research poster. 6th Annual National Patient Safety Congress, Boston, MA.

Scanlon MC, Harris JM 2nd, Levy F, Sedman A. (2008) Evaluation of the agency for healthcare research and quality pediatric quality indicators. *Pediatrics.* 121(6), e1723–e1731.

Scanlon MC, Karsh B. (2010) Value of human factors to medication and patient safety in the intensive care unit. *Crit Care Med.* 38, s90–s96.

Sharek PJ, Horbar JD, Mason W, Bisarya H, Thurm CW, Suresh G, Gray JE, Edwards WH, Goldmann D, Classen D. (2006) Adverse events in the neonatal intensive care unit: Development, testing and findings of an NICU-focused trigger tool to identify harm in North American NICUs. *Pediatrics.* 118, 1332–1340.

Slonim AD, LaFleur BJ, Ahmed W, Joseph JG. (2003) Hospital-reported medical errors in children. *Pediatrics.* 111, 617–621.

Smith TP (ed.). (2002) Age determination guidelines: Relating children's ages to toy characteristics and play behavior. U.S. Consumer Product Safety Commission, Washington, DC. www.cpsc.gov/businfo/adg.pdf. Accessed April 1, 2004.

Solomon SL, Wallace EM, Ford-Jones EL, Baker WM, Martone WJ, Kopin IJ, Critz AD, Allen JR. (1984) Medication errors with inhalant epinephrine mimicking an epidemic of neonatal sepsis. *New Engl J Med.* 310, 166–170.

Stanger P, Silverman NH, Foster E. (1999) Diagnostic accuracy of pediatric echocardiograms performed in adult laboratories. *Am J Cardiol.* 83, 908–914.

Strolin BM, Whomsley R, Baltes EL. (2005) Differences in absorption, distribution, metabolism and excretion of xenobiotics between the paediatric and adult populations. *Expert Opin Drug Metab Toxicol.* 1, 447–471.

Takata GS, Mason W, Taketomo C, Logsdon T, Sharek PJ. (2008) Development, testing and findings of a pediatric-focused trigger tool to identify medication-related harm in US children's hospitals. *Pediatrics.* 121, e927–e935.

Troiano RP, Flegal KM, Kuczmarski RJ, Campbell SM, Johnson CL. (1995) Overweight prevalence and trends for children and adolescents. *Arch Pediatr Adolesc Med.* 149, 1085–1091.

United States Census Bureau. Population Estimates. Available at http://www.census.gov/popest/estimates.html. Accessed 8/6/2011.

Upperman JS, Staley P, Friend K, et al. (2005). The impact of hospitalwide computerized physician order entry on medical errors in a pediatric hospital. *J Pediatr Surg.* 40, 57–59.

U.S. Department of Health and Human Services. (2001) The Surgeon General's call to action to prevent and decrease overweight and obesity. U.S. Department of Health and Human Services, Public Health Service, Office of the Surgeon General, Rockville, MD. http://www.surgeon-general.gov/library. Accessed March 1, 2004.

Vanderheiden GC. (1997) Design for people with functional limitations resulting from disability, aging or circumstance. In Salvendy G (ed.) *Handbook of Human Factors and Ergonomics*, pp. 2010–2052, 2nd edn. John Wiley & Sons, Inc., New York.

Ward CJ, Purdie J. (2001) Diagnostic accuracy of paediatric echocardiograms interpreted by individuals other than paediatric cardiologist. *J Paediatr Child Health.* 37, 331–336. Wickens CD, Gordons SE, Liu Y. (1998) *An Introduction to Human Factors Engineering.* Addison Wesley Longman, Inc., New York.

Wolf AM, Colditz GA. (1998) Current estimates of the economic cost of obesity in the United States. *Obes Res.* 6, 97–106.

43

Human Factors and Ergonomics in Home Care

Teresa Zayas-Cabán and Rupa S. Valdez

CONTENTS

43.1 Background: Health and Health Care in the Home 744
    43.1.1 What Kind of Health Work Is Done at Home? 744
    43.1.2 Who Is Involved in Health and Health Care in the Home? 745
    43.1.3 Homes and Households 745
43.2 Why Human Factors? 746
    43.2.1 Goals of Human Factors and Ergonomics in Home Care 746
    43.2.2 Human Factors Domains and Health Care in the Home 746
        43.2.2.1 Cognitive Ergonomics 749
        43.2.2.2 Physical Ergonomics 750
        43.2.2.3 Macroergonomics 752
43.3 Gathering Information: How to Apply Human Factors and Ergonomics Methods in the Home 753
    43.3.1 Transition from the Workplace to the Home 753
    43.3.2 Assessment Techniques 753
        43.3.2.1 Define the System and Its Objectives 753
        43.3.2.2 Analysis Techniques 754
    43.3.3 Case Study 757
        43.3.3.1 Family Profile 757
        43.3.3.2 Distribution of Health Information 757
        43.3.3.3 Health Information Task: Checking Blood Pressure Levels 757
        43.3.3.4 Implications for Design 758
43.4 Conclusion 758
Acknowledgments and Disclaimer 759
References 759

The household is fast becoming the central site for health and health care. Although much of health care occurs in institutions, such as clinics and hospitals, since the mid-1980s, the episodes of institution-based care have become shorter and shorter, leaving much of the health care process to occur in the home (Aliotta and Andre 1997; Smith et al. 1991; Stone 2000; U.S. Bureau of Labor Statistics 2010; Youngblut et al. 1994). This shift from hospital to home-based locus of care is evidenced by the fact that employment growth in hospitals is expected to be the industry's slowest at 10%, while employment growth in smaller home health care services is expected to grow by 46% (U.S. Bureau of Labor Statistics 2010). Indeed, contemporary health care relies more often on what happens in the home environment for achieving the goals of the U.S. Government's "Healthy People 2010" (U.S. Department of Health and Human Services 2000) than on anything that occurs in hospitals, clinics, or long-term care facilities.

However, current migration of health care to the home is haphazard. Judicious application of human factors engineering principles may make this migration more successful, insuring optimal well-being for all individuals (i.e., patients, families and friends, and providers). Paying particular attention to home care is necessary since such care encompasses both social and environmental needs and is, therefore, broader than the models that dominate acute care (Stone 2000). Human factors and ergonomics, also referred to in this chapter as human factors and human factors engineering, has much to offer to further the understanding of health and health care in the home, and to ensure the appropriate design and engineering of the physical environments, information systems, and goods and services that could be used to foster health and health care goals.

Home care may be conceptualized as having two distinct components. First, professional caregivers, such as

743

nurses, physical therapists, social workers, and home health aides, make episodic visits to some people at home. Patients who are unable to leave their homes may have visiting nursing care, in which professionals come to the home to minister to the patient's needs. The work of these health professions and the human factors engineering strategies to support them, although similar to those in institutional care settings, have some special characteristics due to the physical environment of the home. Second, laypeople are also increasingly engaged in health-related activities in the home. From maintaining basic nutrition and exercise routines to managing complex disease processes, laypeople use the home and its resources to carry out the practices, activities, and tasks needed to promote health and recover from disease. Clinicians and health professionals rely on a self-motivated, well-resourced patient to do what is needed to ensure maximum benefits from modern medicine. The human factors considerations of the work of professionals in the home is well addressed in many publications (Brulin et al. 1998, 2000; Wunderlich and Kohler 2001), including other chapters in this book. Thus, this chapter will focus more directly on the needs, experiences, and human factors considerations of the work of laypeople.

43.1 Background: Health and Health Care in the Home

Health is a state of optimal well-being, encompassing physical, psychological, spiritual, and social dimensions. All people engage in behaviors and habits that are health-promoting (e.g., exercise, meditation) as well as some that are health-destroying (e.g., smoking, excess alcohol consumption). Individuals in their optimal states of well-being function to the extent possible as enabled by their innate resources coupled with the support offered by informal caregivers, pharmaceuticals, and medical assistive devices such as walkers, canes, or prosthetics.

Health care is the totality of goods and services provided by professionals in the health industry in such a way as to promote wellness, mitigate harm, or foster recovery from disease and injury. Health care is provided by trained professionals, such as nurses, physicians, social workers, pharmacists, and other clinicians who bring their talents of assessment and therapeutics to bear on the problems of individuals. Laypeople—"patients"—participate as partners in care having rights to adequate information, shared roles in decision-making, and critical responsibilities to carry out health and healing practices in collaboration with professionals.

Health care in the home represents a purposeful application of the talents of home residents and resources to accomplish health care goals. It is useful to explicitly characterize the work involved, the actors involved, and the home setting to understand where and how human factors engineering methods can ensure optimum use of the environment, the home dwellers, and the associated devices to achieve health care goals.

43.1.1 What Kind of Health Work Is Done at Home?

Home care activities can be grouped into three major areas: maintaining physical and social well-being, managing health information, and carrying out therapeutics as needed for the care and treatment of illnesses and injuries.

Contemporary health care encompasses a range of activities from self-help through self-care to disease management. *Self-help* activities include the attitudes and practices individuals initiate and follow in attempts to promote well-being. These include following a balanced diet, maintaining a balance of rest and activity, and engaging in the mental stimulation necessary to promote cognitive function. Self-help activities generally fall out of the purview of health professionals, but their importance in ensuring health and avoiding disease is well-documented.

Individuals engage in *self-care* activities as a part of a complete health care regime. Self-care activities include that which an individual does in concert with professional advice and intervention with the goal of promoting health. Self-care includes primary and secondary prevention activities recognized and endorsed by health professionals (U.S. Preventive Services Task Force 2009). Primary prevention addresses the avoidance of disease processes and includes such things as obtaining vaccines and prophylaxis on a timely cycle. Secondary prevention activities target screening for early detection of disease. While the boundary between self-help and self-care is quite indistinct, the fundamental difference arises from the critical role played by professions, who not only possess the essential knowledge base regarding self-care activities but also control access to many of the services needed for this. What brings self-help and self-care under the rubric of home care is the fact that the tasks and activities encompassed under these efforts take place in the home and these practices, although essential to health and well-being, fall largely under the volition and control of the individual.

What most individuals think of when they hear the phrase "home care" would fall under *disease management*, defined by the U.S. Preventive Services Task Force (2009) as tertiary prevention, the treatment of diseases, and mitigation of their consequences. Disease management activities are largely directed by professionals and, although they rely on the volition and participation of the layperson, these activities are directed toward the handling or recovery from a specific disease process. Changing bandages, using sensors and monitors to record physiological status, and adhering to drug management routines are all types of disease management activities. Some individuals are able to accomplish disease management on their own; others rely on the informal assistance of family members and friends. In this chapter, the household, also referred to as family, includes all individuals living within the same residence (Agency for Healthcare Research and Quality 2008). Individuals managing disease at home may have conditions such as heart disease, diabetes, or other conditions related to the circulatory, respiratory, or endocrine systems (The National Association for Home Care & Hospice 2008). As *self-help*, *self-care*, and *disease management* activities represent types of nonpaid work, the concepts and methods from human factors and ergonomics may be evoked in a manner to ensure that this work occurs in an optimal fashion.

43.1.2 Who Is Involved in Health and Health Care in the Home?

There are many people involved in health care in the home. First and foremost are individuals, sick or well, who engage in health behaviors. That is, home care participants include those traditionally thought of as ill, needing assistance for physical or mental disabilities. It is useful, however, to consider as participants in health care in the home all persons who take deliberate action in support of achieving or protecting health. So it is important to consider traditional health behaviors, such as dressing changes and exercises, as well as more common household tasks, such as grocery shopping, which take on new meaning when they are conducted in a purposeful manner to address a health need.

The physical and mental capabilities of these individuals vary widely, due in part to the developmental stage of the person (i.e., child, adult, elder) as well as to their health status and changes occurring following illness or injury. For example, when conducting a study to understand which demographic, health, and disabilities of Americans over 70 years of age were associated with living in a home with certain modifications, Tabbarah et al. (2000) identified subgroups with varying functionality within the aging population and distinct opportunities for home modifications, which could lead to reframing how this particular user group is characterized.

Family members or friends sometimes provide personal care assistance. Approximately 29% of individuals in the United States serve as informal caregivers (National Alliance for Caregiving in collaboration with AARP 2009). These informal caregivers aid others by cooking meals, assisting in personal hygiene, and aiding in mobility. Although rarely trained for the many tasks of caregiving (Donelan et al. 2002; Raphael and Cornwell 2008), these individuals develop robust strategies to manage everything from communicating with physicians and pharmacies, managing assistive equipment in the home like respirators or infusion pumps, and providing the social and emotional support needed by the ill person. Informal caregiving may be a positive experience (Donelan et al. 2002); however, informal caregivers are at risk for physical burden (e.g., fatigue, musculoskeletal injuries from improper positioning and body mechanics during personal care activities), psychosocial burden (e.g., a sense of isolation that arises from the difficulties of maintaining social contacts while managing the burdens of care), and occupational and financial burden (e.g., inability to continue working while caring for a loved one) (Canam 1993; Dokken and Sydnor-Greenberg 1998; Donelan et al. 2002; Grunfeld et al. 2004; Langa et al. 2001; Smith et al. 1991; Youngblut, Brennan, and Swegart 1994; Yun et al. 2005). Such burden may be particularly difficult when the ill person has a disability or chronic condition (Raphael and Cornwell 2008).

43.1.3 Homes and Households

Hindus (1999) emphasizes the importance of including the home in engineering research. She comments that the home is too economically important to ignore and provides a rich research field that could improve daily living for many users. She explains, however, that homes are fundamentally different from workplaces. Houses are not designed for technology and home technologies must be designed so as to not present a hazard for babies, pets, and elders. She suggests that consumers' "motivations, concerns, resources, and decisions" (p. 201) can differ from workers'. Decision-making and value setting, she comments, are different in households than those in corporate organizations.

Home care occurs in residential living spaces that differ from formal health care facilities, such as hospitals and nursing homes, in several important ways. First, contemporary culture ascribes meaning to the home as a personal space of residence where access and egress are controlled by the individuals who live there. Homes may include apartment-type dwellings as well as free-standing units. Homes generally provide places for shelter and leisure, and are increasingly becoming places of work. Homes are, for all people, sick or well, sites where health care practices occur. For some people managing chronic or complex illnesses, homes become a place where durable medical goods, supplies, medications, and assistive devices must be stored and integrated within a specific environment.

Venkatesh (1996) and Venkatesh and Mazumdar (1999) identify three environments of interest in the home: (a) the social environment, in which family behavior occurs, configured in terms of family members who adopt and use technology and made up of the social structure of the household and activities performed within it; (b) the physical environment; and (c) the technical environment, the space in which household technologies are used, which consists of the configuration of household technologies, involves family attitudes and levels of satisfaction with technologies and represents the nature of the technological environment within the household. Characterizing these three environments provides a reasonable starting place for understanding the dimensions of the home care space.

The *social* environment of the home is described in social network terms. Households may consist of individuals or families—individuals who share legal or biological relationships—or unrelated individuals with varying degrees of social ties. Additional dimensions of the social structure of households include their permeability—the number and variety of persons who interact with household members—and their network adequacy—the ability of the social contacts to help household members meet instrumental and emotional needs.

The *physical* environment of the home is characterized by a circumscribed physical space, which is devoted to the use of an individual or set of household dwellers. Like other environments more familiar to human factors engineers, such as factories or offices, the physical environment of the home can be described by space, objects held within the environment, temperature, lighting, and odor. Understanding the physical environment of the home requires awareness of the community context where the house is located. Aspects such as density, proximity to resources and health services, and accessibility of infrastructure such as sanitation, communication, and transportation also describe the physical environment of the household.

The *technical* environment of the home is important in so far as it provides the infrastructure on which technical solutions can be built. The technical environment of the home includes the mechanical, electrical, and communications framework, which governs the ability of technical solutions to "fit into" a household.

43.2 Why Human Factors?

This book addresses human factors and ergonomics in health care and patient safety, including issues of errors, safety, devices, workflow, and organization. In this chapter, we consider the full range of application of human factors to the full range of health care activities that occur in the home, exploring the progress from "Do solutions useful in other environments work in the home?" to the creation of solutions that might work best in the home. There are few human factors studies in the home, yet we will present examples that will illustrate how human factors can be used to assess and design for the home environment. This chapter should help the reader to design de novo approaches or technologies to create novel solutions for the home as well as to identify current solutions that might be extensible to the home environment.

It is important not to just take methodologies that are helpful in understanding a situation and apply them to the home; it is essential to make the bridge between these understandings and some design direction. The purpose needs to be more than simply understanding what people actually do and why they do it; human factors engineers also need to think about how it might be done better and look for an opportunity for design. This means not simply understanding how a task done in a lab or office transfers to a home environment but how the physical, social, and technical environment of the home shapes the need for the task and the ways it can be done.

43.2.1 Goals of Human Factors and Ergonomics in Home Care

The field of human factors and ergonomics identifies how to best bring together people, environments, technologies and organizations to ensure optimal outcome of work. Although historically focused on industrial systems, human factors and ergonomics has much to offer to home care. Specifically, human factors and ergonomic frameworks and methods can be applied to design optimally functioning home care systems, to reengineer the home from a place of social intercourse to one more purposefully directed toward accomplishing home care goals. Additionally, many of the issues of concern to human factors and ergonomics, including safety, quality, and efficiency, are also relevant to the home care environment.

43.2.2 Human Factors Domains and Health Care in the Home

In the field of human factors, *work* is commonly used to describe paid work conducted by professionals in industrial settings. However, as described by Hendrick and Kleiner (2002), a broader definition of work includes "any form of human effort or activity, including recreation and leisure pursuits" (p. 1), suggesting that home care can be conceptualized as a type of work.

Consequently, some human factors concepts and theories can be applied to understand home care. In particular, the home can be thought of as a work system. As defined by Hendrick and Kleiner (2001), a work system is one that "involves two or more people interacting with some form of (a) hardware and/or software, (b) internal environment, (c) external environment, and/or (d) an organizational design" (p. 1).

Work system models are a useful way to discuss the elements of the household that are relevant to human factors in home care and to draw parallels to health care organizations. For example, hardware or software in a clinic or hospital can include devices used in the provision of care to patients such as x-ray machines and stethoscopes. In the home, these can include the tools individuals use to take care of themselves or others such as medical devices (e.g., pedometers, gluco meters) or file cabinets to keep track of medical history. The internal environment in a primary care clinic typically includes a waiting area, registration desk, rooms with varying layouts and sizes, and a bathroom. The physical characteristics of homes can also vary greatly in size and layout. However, in contrast to health care settings, they are unlikely to be designed specifically for health care activities. In the provision of health care, there are many aspects of the external environment that affect an organization's ability to deliver health care services, such as the regulatory environment, availability of medical

goods, or availability of needed expertise within the organization's geographic location. Similarly, a household's ability to carry out home health care related tasks can be affected by elements of external environment such as the providers and the health care resources available in the community. Finally, the organizational design of a health care providing organization can include the number and expertise of providers, reporting structures, and organizational policies, all of which influence how health care is delivered to the patient. Household structures can range in complexity from a one-person household where the individual is responsible for all aspects of her health to an extended family residing across different locations with shared responsibility for home care activities. These varying structures will affect how home care tasks are carried out. The similarity between industrial work systems and home work systems suggests that a human factors perspective could be used to conduct in-home assessments related to home health care. Work system models can provide efficient guidance for design and evaluation.

One work system model, the balance model, gives not only awareness of what to study in the home but how to begin making modifications. Smith and Carayon-Sainfort (1989) proposed a theory that conceptualizes job design based on the balance among job elements. The balance model is composed of five elements, shown in Figure 43.1, that offer greater detail about these

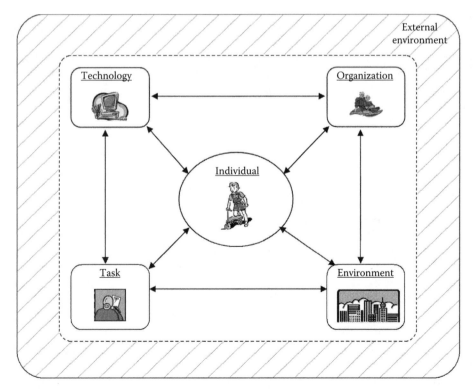

FIGURE 43.1
The household as a work system. (Adapted from Smith, M., and Carayon-Sainfort, P., *Int. J. Ind. Ergon.*, 4(1), 67, 1989.)

elements than the environments proposed by Venkatesh and Mazumdar (1999). This model provides a holistic approach to analyzing the elements that comprise jobs within the home. At the center of the model is the *individual* with his or her physical and cognitive abilities, perceptions, personality, and behavior. The individual has a number of *tools and technologies* available to perform job *tasks*. These tasks are carried out in a "work" setting, which includes the physical and social *environment*, that affect the manner in which the tasks are carried out. Also the *organizational* structure, which defines the nature and level of individual involvement, interaction, and control, affects the way in which these tasks are performed. In addition to the original elements of the balance model, the *external environment*, which can include contextual elements such as regulation, infrastructure, and economic issues, shapes the way a work system functions (Henriksen et al. 2009).

Although the balance model was initially used to analyze the job of an individual, it may also be used to analyze the work of an organization. Carayon and Smith (2000) explain this level of analysis by recognizing that an organization is composed of a group or system of individual work systems. The balance model's elements are all present in the household when carrying out any "job." For instance, human factors engineers can conceptualize the job to be studied as that of home health information management, which is comprised of a number of tasks. Thus one health information management task could be "writing down medical appointments." One or more household residents or individuals, for example the mother, may carry out these tasks. The person who carries out these tasks and the reasons for which they carry them out is a function of the household's organizational aspects (i.e., household rules that drive the way activities are carried out) as well as the individual's own characteristics (i.e., his or her ability to carry out the task). In some households, for example, a mother who has good time management skills and is adept at record keeping might be responsible for all health coordination activities in the household. Thus, all aspects of the mother's role history and acceptance by the household may influence the way in which health information management occurs.

Furthermore, the manner in which tasks are carried out is constrained by the home and community environment and will either be aided or constrained by the tools or technologies available to the household. For example, because the mother wants to easily remember when appointments are scheduled, she writes them on a calendar kept in the kitchen cupboard where she sees it every morning and where it is also far from the reach of small children. Because she depends on her sister for transportation to the appointment, she also makes sure to call her to give her the time and date and to make sure that her sister writes it in her appointment book.

The use of human factors and ergonomics methods to better understand and support home care will be especially critical as information and communication technologies (ICTs) become more common in the home. ICT solutions play an increasingly larger role in making health work in the home more feasible, efficient, and effective. Information systems link individuals in the home to health professionals and to relevant and timely health information and help individuals manage the "business" of health care (e.g., claims processing, scheduling appointments). In addition, more devices are being used in the home environment, either to monitor or provide care to an ill individual (e.g., glucose meters) or as part of self-help activities (e.g., pedometers). Sometimes, these devices are part of a larger self-care or self-management intervention that also includes the use of ICTs.

The International Ergonomics Association (IEA) (2000) defines three domains of the field of human factors and ergonomics. These domains are cognitive ergonomics, physical ergonomics, and organizational or macroergonomics. Each of these has a primary area of focus, for example, cognitive ergonomics focuses explicitly on mental processes, such as mental workload, which significantly impact individuals' ability to complete home health care tasks. Physical ergonomics focuses on very explicit physical tasks, a significant aspect of home health activities. Finally, macroergonomic approaches focus on the optimization of work systems, including their organizational structures, policies, and processes. Clearly, examining the human factors and ergonomics aspects of home health activities requires a broader human factors approach that encompasses the people who live in the home, the physical aspects of the home, the roles of the residents, the tasks they are trying to accomplish, and the resources they use to do so. All the domains are needed to examine home health care work and, as human factors engineers, it is important to be cognizant of interactions among them and how they each contribute to a more holistic view of the home environment (Zayas-Cabán and Marquard 2009; Zayas-Cabán et al. 2009). As Henriksen et al. (2009) describe, components of home health care "need to be designed in a way that takes into account their interactions with one another and with the capabilities and limitations of patients and their providers" (p. 229).

It is important to note that whereas the balance model conceptualizes the elements of the work system, the IEA describes the three domains of the field of human factors and ergonomics including the focus area and tools and methods for each domain. These domains and their tools and methods can be mapped to the elements of the work system to which they are typically applied. Thus cognitive ergonomics is primarily used to address the balance model elements of the *individual, tasks*, and *tools and technologies*. Physical ergonomics is primarily used to address the balance model elements of the *individual, environment,* and *tools and technologies.* Finally, macroergonomics is primarily used to address the balance model element of the *organization* and the *external environment.*

The following sections briefly address each of the three human factors domains, describe their relevance to human factors and ergonomics in home care through the lens of the balance model, and discuss some unique considerations for each domain when studying the home environment. These sections also contain a brief review of relevant work done applying the methods of each domain to understanding health care work done in the home.

43.2.2.1 Cognitive Ergonomics

Cognitive ergonomics is concerned with "mental processes, such as perception, memory, reasoning, and motor response, as they affect interactions among humans and other elements of a system" (International Ergonomics Association 2000). Cognitive ergonomics includes studies of mental workload, decision-making, human–computer interaction (HCI), human reliability, and work stress and training, among others. This human factors domain is most related to the *individual, tasks,* and *tools and technologies* elements of the household work system as described by the balance model. For more information, also see Chapters 26, 28, and 32.

There are several important cognitive ergonomic considerations that make the home environment unique and distinct from clinical settings. First, laypeople's health work is unpaid and motivations for conducting health-related tasks may be very different from those of a health care professional. While health care professionals receive formal training and continuing education, laypeople receive little to no training in completing home health care tasks. Instead many individuals suddenly "become" caregivers (Raphael and Cornwell 2008). In addition, there may be less standardization of skills in and across the home environment than across institutional environments and any acquisition of skill is dependent on materials provided by professionals or self-motivated.

Human factors engineers may encounter different challenges when trying to understand and support laypeople's health work at home. Consumers, in particular elderly patients and caregivers, may have diminished cognitive function, likely affecting their ability to complete or understand certain tasks and requiring additional cognitive support. Individuals have different levels of literacy; the 2003 National Assessment of Adult Literacy found that 22% of the adult population in the United States have *basic* health literacy, 14% have *below basic* health literacy, 53% have *intermediate health literacy,*

and 12% have *proficient* health literacy (Kutner 2007; Kutner et al. 2006; Paasche-Orlow et al. 2005). These characteristics need to be taken into account when designing interventions that deliver health-related content to laypeople to help them take care of themselves or a relative. Finally, health management tasks such as making decisions regarding a loved one's treatment or medications can be very complex, requiring management of different and complicated information and coordination with multiple providers (Donelan et al. 2002; Lang et al. 2009; Lippa et al. 2008; Raphael and Cornwell 2008). Cognitive ergonomics methods can be used to understand these tasks, their cognitive workload, and how to support them with different interventions.

Most of the cognitive ergonomics research that has been conducted in the home environment with a focus on health has mainly focused on (a) understanding and detecting the cognitive decline of residents and how to support individuals who may have limited or reduced cognitive abilities using technology or (b) understanding HCI issues of, primarily, information technology (IT) interventions designed to be used within the home.

One recent study examining self-care at home using complex medical equipment found that some individuals with little training and formal education were able to manage complex self-care activities (Fex et al. 2009). However, these individuals experienced some challenges in using the technology, such as remembering how to carry out procedures to use the technology in the correct order. Cognitive aids and easy to follow instructions may assist individuals in accomplishing such procedures. Furthermore, the researchers found that cognitive capacity included the ability to balance disease and technology-related tasks with other daily activities important, though not directly related, to overall health and well-being, such as interacting with friends and family.

Most of the human factors work conducted to support cognitive activity or detect cognitive decline has been implemented through smart home projects that are part of overall health monitoring. Many of these smart homes include functions that provide residents with decision support, reminder, or prompting technologies, facilitate communication between the residents and caregivers who may live outside the home, and explicitly assess cognitive status (Chan et al. 2009; Gentry 2009). However, there is recognition that additional studies are needed with regards to user needs, which is an area where human factors engineers can significantly contribute by better understanding and designing for laypeople's health tasks and cognitive needs.

HCI is concerned with the design, evaluation, and implementation of interactive computing systems for humans (Hewett et al. 1996). HCI methods such as usability testing and evaluation have commonly been used to conduct cognitive ergonomic evaluations of the person–machine interface of both medical devices and IT applications.

Kaufman et al. (2003) conducted a usability evaluation of a telemedicine application for diabetic patients. The application features use of an interactive video and computer monitoring system that helps people with diabetes monitor their blood sugar, communicate with their physicians, and review Web sites with nutritional advice. The evaluation consisted of a cognitive walk-through and field usability testing. The purpose of a cognitive walk-through is to evaluate the cognitive processes of users performing a task. For this study, the evaluation involved identifying sequences of actions and goals needed to accomplish a given task. The cognitive walk-through was conducted by the experts and was used to inform the tasks to be conducted during usability testing. Usability testing employs participants who are representative of a particular target population to evaluate the degree to which a product or system satisfies basic usability criteria. The researchers found some problems with the original interface and system. For example, the screen display of the telemedicine unit included extensive amounts of text, some of which was in a small font and was difficult to read for some participants. The telemedicine evaluation also uncovered problems arising from a lack of familiarity with the device and the need for fine eye–hand coordination to perform some of the tasks. They also found significant differences between the underserved rural and urban populations; one of the main differences was related to level of education and health literacy, including basic literacy and numeracy, which may have influenced both the acceptability of the technical intervention and the participant's familiarity with general computer skills. To address some of these barriers to use, the researchers redesigned the intervention by providing a touch screen interface that allowed the use of a finger or stylus. Screen displays were changed to make it easier to interact with the system by increasing text font size, adding spacing between links, using larger buttons, and providing better visual contrasts (Kaufman et al. 2006).

Another study of a phone-based telemedicine system used to monitor patients after ambulatory surgery showed that initial usability testing allowed the developers to identify barriers to technology use such as small keypads or difficulty in typing due to proximity of buttons (Martínez-Ramos et al. 2007). Initial testing and redesign lead to higher acceptance of the intervention and increased perceived ease of use and usefulness of the application (Martínez-Ramos et al. 2009). Importantly, the overall development process for this intervention included usability and feasibility testing of the intervention before pilot testing it in patients' homes. This allowed the developers to focus on issues

of effectiveness and context of use when pilot testing the application.

Although some efforts have been made to understand how to best design interfaces for IT solutions to be used for health management in the home, much work remains to be done. The field of human factors can help to characterize the home health care tasks, their cognitive load on individuals, and the decision-making processes laypeople undertake to take care of themselves and friends and families as well as determine how to best support those processes with appropriate solutions.

43.2.2.2 Physical Ergonomics

Physical ergonomics is concerned with understanding both the physical characteristics of the *individuals* performing the work as well as those of their work *environment*. In addition, physical ergonomics may focus on the physical requirements for using the *tools and technologies* required to perform needed tasks. This human factors domain allows intense attention to the psychomotor dimensions of specific activities and tasks common to health in the home. Relevant to this chapter are concepts arising from the design of equipment, biomechanics, and occupational health and safety.

The home environment consists primarily of private residences, which, in contrast to institutional health care settings, may not have been designed with health care work in mind. These environments can vary in shape and size and modifications to the home may or may not be under the control of its residents. The constraints of the home environment have implications for the placement of devices. For human factors engineers, it is important to evaluate the co-occurrence of tasks and type of technology or tool being used to perform it as well as what is needed to use a particular tool (e.g., power, telecommunications, lighting, or adequate flooring). The individuals performing health tasks include the patient or layperson engaging in self-help, self-care, or disease management as well as their informal caregivers, who may or may not reside in the home.

The physical conditions of the home will constrain or facilitate which interventions can be used in a home environment. While health care environments are usually under the control of management, the physical redesign (e.g., railings, ramps) of a home is dependent upon individuals' willingness and ability to make changes and adopt an intervention. While some households may have the means to purchase needed equipment, remodel their homes, and hire needed help, others may have limited financial means to do so (Lang et al. 2009). In addition, individuals may possess a variety of diminished physical characteristics such as limited vision, hearing, or motor function, which will need to be taken into consideration when designing home solutions or introducing new technologies and devices for health care (Zayas-Cabán and Dixon 2010; Zayas-Cabán and Marquard 2009).

There have been many studies that focus on physical ergonomics of the home and how these fit with its residents. Most of these address issues confronted by vulnerable populations, such as the elderly, disabled, or homebound patients (Smith 1990). Pinto et al. (2000) argue that the use of an ergonomics approach to home design may "develop an integrated strategy aimed at the well being and satisfaction of aging people" (p. 317). They describe three areas of physiological decline in the elderly: muscular strength, posture, and movements. The authors suggest that the home environment presents several risk factors, which are compounded by the elderly's "psycho-physical conditions" (p. 318). Examples of practical recommendations for entrances and kitchens in home environments include placing furniture in corners of the room or along walls to avoid collisions, and placing shelves and furniture so that they may be easily reached by this specific population.

In addition, many assistive technologies have been developed to help elderly or disabled individuals carry out activities of daily living (e.g., bathing, eating) and/ or with mobility (Chan et al. 2009). For example, Voorbij and Steenbekkers (2002) examined the twisting force and hand configuration used to open a jar. The authors found that all participants used both hands while applying force: one to twist the lid and another to hold the jar. The results showed that with increased age, difficulty with completing the task also increased. The authors found that if the torque required to open a jar was reduced to 2 Nm then only 2.4% of users 50 years of age or older would have some difficulty with this task.

Smith (1990) provided an overview of research needs and design opportunities in human factors and aging. He used the areas of work, retirement, mobility, and the home environment to illustrate applications all of which highlighted the themes of health, safety, technology, and the capabilities and limitations of age. Smith discussed several models proposed in this area that focused on a variety of issues including identifying specific tasks and task components, adaptation, changing physical and mental capabilities, and accommodating to needs.

Other in-home ergonomics studies focus on design of equipment and how they interface with both patients and caregivers to support care and day-to-day household activities. Lathan et al. (1999), for example, describe human factors issues in the use of telerehabilitation for persons with disabilities. The authors mention the need for human-centered design and point to specific design criteria with which the devices or systems must comply. The system operation must account for various impairments, including perceptual, cognitive, and motor (Jianwu et al. 1996; Lathan et al. 1999). Lathan et

al. comment that in order to build an effective device, the design process must take into account the goals, preferences, and abilities of the intended users.

Studies have found that lack of attention to the physical constraints of the home environment or the physical capabilities of its residents can lead to challenges in using interventions in the home. For example, Zayas-Cabán and Dixon (2010) found that an intervention designed to automatically transmit weight data from a scale connected to a telephone using a cable did not fit with the physical environment within some homes. This resulted in either unsafe device arrangements, by having cables crossing rooms, or in workarounds, where users would manually write down information and relay it to project staff, which could lead to additional cognitive workload and errors in the information being transmitted to providers. In addition, Huang et al. (2008) examined the use of assistive devices by children with cerebral palsy. They found that assistive devices were generally used less in the home than in other environments (e.g., school), in part, due to barriers encountered within the physical constraints of the home. For example, some of the homes did not have sufficiently wide doorways or hallways to enable use of the devices. The authors also found that in some cases, the devices themselves had barriers to use. Some devices caused physical discomfort or made the children feel unsafe, further discouraging their use. In another study, Geidl et al. (2009) conducted an evaluation of the use of ventricular assist devices in the home. To do so, they had patients take the device and use it at home and rate their appropriateness for use in daily life such as bag size, safety when being used in different environments (e.g., adequate protection against water, dust, overheating), and convenience. Other aspects of the device were tested such as battery capacity, the appropriateness of the monitor screen, and the effectiveness of alerts in waking patients up. By having study participants test the device at home, the researchers were able to uncover some issues that could be improved to make the device more useful and effective.

Empirical studies and industry publications also exist that examine occupational health and safety issues surrounding home health work. However, these publications focus on paid workers that come into the home to provide support to patients and/or caregivers (Askew and Walker 2008; Delano and Hartman 2001; Sitzman and Bloswick 2002). A more recent study, while still focusing on home care services provided by professional caregivers, examined both patients' and informal caregivers' as well as professionals' perceptions of home care safety (Lang et al. 2009). The authors found that patients and their families have different perspectives than providers of what safety means. In particular, the phrase "home care safety" did not resonate with patients and caregivers and, while they could recall unsafe experiences or situations, they did not think of those as home care safety issues. The authors also recognized that informal caregivers may agree to take care of a loved one without realizing they may not have the capacity to do so, which can lead to fatigue and stress and can impact both the caregiver and the ill person being cared for. While information about paid workers is relevant to home care, it is beyond the scope of this chapter and is better addressed in Chapter 15.

A great deal of human factors research and development has been conducted in the area of assistive devices and trying to support the physical capabilities of frail or elderly individuals. Human factors engineers can further contribute to the field of home care by conducting additional work to better characterize the physical environment of the home, how it may support or inhibit home health care tasks, and how to design appropriate solutions to support home care activities within the physical constraints of the home environment. Additional human factors studies are also needed to better understand occupational health and safety issues for informal caregivers; such caregivers may face some similar health and safety issues as those encountered by home health care workers while trying to care for ill friends or relatives, but may not have the skills or resources to be able to address or mitigate the effects of these issues.

43.2.2.3 Macroergonomics

Macroergonomics focuses on the analysis, design, and evaluation of work and work systems (Hendrick and Kleiner 2002). The term "system" refers to a sociotechnical system, which includes a technological subsystem (i.e., the tools, techniques, procedures, skills, knowledge, and devices used by the members to accomplish the tasks of the organization), personnel subsystem (i.e., the people who "work" within the organization and the relationships among them), and the relevant external environment. While all elements of the work system are relevant to macroergonomics, of particular importance are the *organizational* and the *external environment*, in which the home exists. Macroergonomics has many key concepts and dimensions that make it readily applicable to home care situations since laypeople also "engage in health work to maintain and manage their health" (Zayas-Cabán and Marquard 2009). For more information, see Chapter 5.

As human factors engineers, it is important to keep in mind that consumers' health work, workflows, and work system are distinct from that of professional health care workers. Moreover, implementing a new intervention or device within the home is akin to implementing a new health care device to support a health care process in a formal health care setting. Similar to how purchasing a new MRI machine will impact how clinicians and technicians do their work and will require training, giving a patient and his or her caregiver medical equipment to support home health management will impact the household work system. Therefore, macroergonomic work system design and evaluation must include both understanding the household to guide the design of appropriate solutions and evaluating the impact of chosen solutions on the household.

Different than industrial settings, organizations in the home setting are much less formally arranged than in industries, and the negotiation of the division of labor and authority arises more from household dynamics than from organizational hierarchies. The concepts of organizational design and structure, generally driven by the business purposes of the industry, must be modified and expanded in the home to be cognizant that the interpersonal structures exist for a variety of purposes (e.g., household dynamics, social exchange) and are rather immutable. Household decision-making strategies, degree of integration, and formalization of procedures may be much less obvious although no less fixed than what one might discover in industry. Differentiation of roles and levels is much less in the home than in industry, so the human factors strategies employed must have the capacity to absorb this high level of complexity. Thus, for the purposes of home care, the extant structure of the household must serve as the basis for assessment and redesign, but is rarely the target of reengineering activities.

In addition, areas within the home serve many purposes—for social interaction, leisure, exercise, nutritional support, storage of household goods, and safe haven for personal activities. Homes meet many physical and social needs. Two dimensions of spatial dispersion are relevant for human factors engineers: (a) health activities occur throughout the physical environment, with very few places restricted for a single purpose and (b) the organizational structure and physical environments may not be completely aligned. The nonalignment of the organizational structure and the physical environment appears in two ways: multiple families housed within a single household and single families living in distinct households but uniting to manage health concerns at home.

There are limited macroergonomics studies with regards to health at home. Most studies of work and workflow in health care focus specifically on health care organizations, while only a few human factors studies look at organizational aspects in home environment, such as household dynamics. For example, one study examining computer use in the home uncovered differences in computer use between children and parents (Kimmerly and Odell 2009). Children were more likely to use a computer in their bedrooms than parents and some families used shared computers in their kitchens. In addition, the authors found that some parents wanted to have computers in public spaces to monitor computer use more easily. These findings point to potentially implicit rules about computer use that affect where computers are placed and used and also suggest that depending on the target user of a health IT intervention, it is important to consider issues of privacy and sharing of information in its design.

In addition, there have been more recent studies that have tried to better understand home care, some with a particular focus on health information management, by applying macroergonomics methods and concepts. The study conducted by Huang et al. (2008), examining the use of assistive devices in the home by children with cerebral palsy, also found that home life and household dynamics influenced the use of the devices. The combination of children's and mothers' attitudes did not encourage the children's use of the devices in the home. The children thought of their homes as a place where they could live how they wanted and where they could rely on others for help instead of relying on the assistive devices. The mothers influenced the children's attitudes by not encouraging their use and physically assisting the children, in part, due to concern that their children would develop negative attitudes toward the devices. In addition, the mothers did not always have enough time to learn how the devices should be used and did not have the necessary knowledge to provide their children with assistance when using the devices.

Another study, the Advanced Technologies for Health@Home project is the first study that has systematically documented behaviors associated with information handling in the home. The overall goal of the project was to explore home health information management. Moen et al. (2007) presented study results that show a variety of information resources are in use to assist health management in these households. The three most common types of health information kept across participants' homes were doctor's appointment information, health insurance information, and doctor's contact information. The three most common sources of health information were doctor's visits, family and friends, and clinics or hospitals. The resources they found were mostly paper-based and had a physical presence as visible tools to assist the household members. The authors also noted that there is great variation in the different strategies used to store health information. These strategies were categorized by Moen and Brennan (2005) as just-in-time, just-because, just-in-case, and just-at-hand, which reflected the anticipated need to retrieve the information and the location, within the home, of the devices used to store the information.

There are many opportunities to apply macroergonomics concepts and methods to study home care. Additional studies are needed to better understand and characterize the work and workflow of health care in the home, in particular by understanding each element of the household work system and how these elements interact in the production of health.

43.3 Gathering Information: How to Apply Human Factors and Ergonomics Methods in the Home

43.3.1 Transition from the Workplace to the Home

Most application of human factors in organizational and job design has been carried out in industrial work settings, which, as discussed earlier, are different from homes in several key aspects. These differences will have an effect on how human factors and ergonomics methods are applied in the home. The three key areas of modification of the analysis activities are as follows: (a) involving household members in design activities; (b) design under severe space, time, and financial limitations; and (c) the need to fit design choices into the existing functional system of the household. These areas must be taken into consideration when conducting in-home human factors studies.

43.3.2 Assessment Techniques

43.3.2.1 Define the System and Its Objectives

Before any application of an assessment technique, it is essential to determine the system's objectives and define the system (Sanders and McCormick 1993). Here, the balance model is used as a framework to give an overall picture of the home environment. For each of the elements, there are several aspects that human factors engineers assess in order to better understand the job being carried out and potential for job redesign. These aspects are assessed using a number of human factors tools and methodologies. Figure 43.2 illustrates the criteria that are typically examined and the tools used to assess them. The tools or methods used by human factors engineers can be divided into three broad categories: (a) review of existing documentation; (b) direct observations, questionnaires, or interviews; and (c) use of specific techniques to assess certain aspects of work (i.e., case studies, time studies, work sampling, usability testing, and ergonomics assessment).

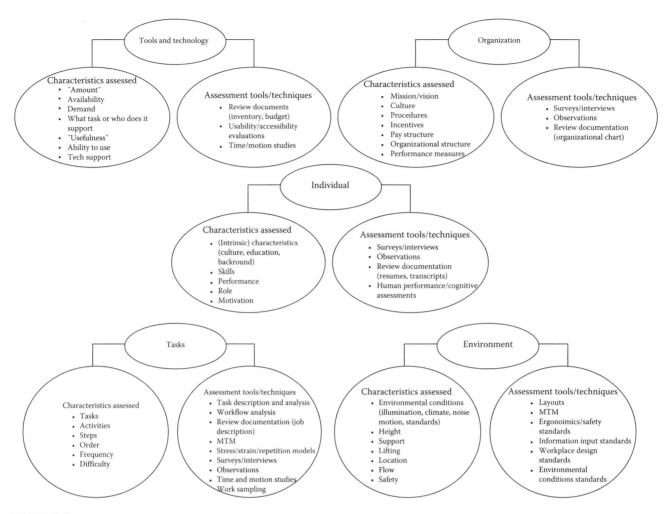

FIGURE 43.2
Human factors concepts and their assessment methodologies organized around the balance model. (Adapted from Smith, M., and Carayon-Sainfort, P., *Int. J. Ind. Ergon.*, 4(1), 67, 1989.)

43.3.2.2 Analysis Techniques

43.3.2.2.1 Review of Documentation

In traditional work settings, designers can also review supporting documentation, such as job applications, resumes, and letters of recommendation, to understand the worker's background, preparation, and skill set, and they can use performance measures to assess each individual's execution. Households generally have very few formal documentation systems. However, in terms of health concerns, there are many types of documentation that may exist. These include prescriptions, exercise routines, appointment reminders, contact information for clinicians, and health-related references. These documents tend to be stored in many locations around the house. The storage choices are generally driven by the anticipated future use of the information.

Reviewing documentation will most likely occur through interviews and observations. It is critical that the engineer remain mindful that health information takes on very personal meaning for individuals, and any exploration of documents should be done in a manner that is sensitive to the private, personal nature of the information. For example, Moen and Brennan (2005) reported on the types of health information individuals kept in their homes. Information about the different artifacts used was collected through a combination of interviews, drawing of home layouts, and taking photographs of the artifacts used. The combination of methods allowed the researchers to further characterize the artifacts by where in the home different types of health information are typically stored as well as how privately different types of health information are stored (Moen and Marquard 2004).

43.3.2.2.2 Direct Observations, Questionnaires, and Interviews

Human factors engineers have a wealth of techniques or methods they can use to assess individual characteristics. These commonly include the use of surveys, interviews, or observations. Relating specifically to health care in the home, then, human factors engineers will benefit mostly with the use of *observations*, *surveys*, and *interviews* to obtain information on the individuals that make up the home in order to assess aspects such as household composition, education, and roles and responsibilities.

The *individuals* are responsible for performing the tasks associated with their job. Traditionally, job designers are interested in understanding the nature of these tasks, the steps needed to carry them out, the frequency of the tasks, and any difficulties that workers encounter when carrying out tasks. In order to fully understand the worker's responsibilities, job designers frequently review documentation about his or her position, such as a job description, to comprehend what the job entails. In the household, exploring the dimensions of these roles and the strategies used to carry out tasks can occur through interviews with household dwellers.

In Kiesler et al. (2000), families were given a computer and Internet access. The researchers later used a variety of data sources that included questionnaires, logs of requests for support, Internet use, and home interviews to examine the dynamics of technical support for computer use at home. The results showed that family members with high technical skills became the family technical guru and were also able to influence the household's adoption of technology.

One of the most common techniques used in basic human factors design is *task* description and analysis. This process is used to list in sequence all the tasks that must be performed; each task is broken down into steps that it requires, and each step is then further analyzed. Workflow analysis is also often used to understand how processes within organizations are carried out and which individuals are involved. When analyzing the ergonomics of more repetitive, manual labor, designers also use methods-time measurement (MTM) and time and motion studies (Niebel and Freivalds 1999). Work sampling is also a common technique used to assess the nature of the work and the tasks that compose it. Job designers and human factors engineers also make use of surveys, interviews, and observations to elicit from workers their responsibilities and how their daily tasks are carried out. These last tools lend themselves more to a home environment, where, for example, documents such as job descriptions do not exist.

Denham (2002) summarizes three ethnographic studies that identify the ways family health is defined and practiced in three Appalachian communities. The study results showed that in the home, health management roles were assigned on the basis of skill level and knowledge of family members. The study aimed at understanding the kinds of family routines that led to the production of family health. These included a number of routines or health behaviors such as dietary practices, sleep and rest patterns, and avoidance behaviors. The routines, though, were not evaluated to determine whether they led to the production of health.

There have also been studies that examine the total workload of household members, which includes both paid and nonpaid work. In their study of the physical activities of female farmers in India, Nag and Chintharia (1985) examined how work changed during farming and nonfarming seasons. The study included 17 women who were interviewed and also observed. The researchers found that household activities (i.e., gathering firewood, cooking, sweeping) required 4.7 h in the farming season and 7.1 h in nonfarming season.

In another study that examined parents' total workload, Mardberg et al. (1991) aimed to develop and evaluate a questionnaire that assessed characteristics, perceived load, and positive and negative aspects from paid work and household work. The researchers mailed the questionnaire to 2050 potential participants. The questionnaire covered the following areas: (a) general background, (b) job characteristics, (c) perception of job characteristics, (d) household, (e) child care, and (f) other duties. They used responses to conduct a factor analysis of the data that identified factors within the areas of paid job descriptors, personal control, workload, and qualification. The authors then developed what the researchers found to be a reliable tool for measuring stress-related aspects of the total workload of white-collar workers.

Another study examined household work by administering a survey electronically. Marut and Hedge (1999) wanted to explore the nature of household tasks and products. Participants were asked about general demographic information and the range of household tasks, including the equipment used for the tasks, time spent on each task, how tiring each task was, and sources of help for the tasks. The results showed that scrubbing, mopping, tidying, vacuuming, and doing laundry were perceived to be the five most tiring tasks. The authors suggested that the results could serve to identify opportunities for the application of ergonomics design principles to the design of consumer products.

Workers must use *tools and technology* to complete their required tasks. Human factors engineers are interested in understanding the extent to which these tools support workers in completing their tasks. This includes the amount, availability, and demand for the resources and also the technology's usefulness, accessibility and usability, as well as technology and repair support available to workers. Again engineers could review documents on inventory and budgets to understand the availability of the technology. Usability and accessibility evaluations could help determine how good the human–machine fit is. Time and motion studies and work sampling can also be used to obtain information on technology use, frequency of use, and ease of use and its potential impact on performance. In the home, household members should be interviewed, observed, or both to gain an initial understanding of what technologies or tools are used to support health management.

Hindus et al. (2001) conducted studies to test the prototypes of the Casablanca study. The study's goal was to develop communication technologies for the home. The researchers wanted to gain a better understanding of the usability, ease of use, and overall usefulness of the prototypes. Their pilot study entailed 16 in-depth interviews on where and how the technology was used. Two of these households were later revisited for longer observations. This study was then followed up with 35 additional in-home interviews and observations to understand technology use and attitudes toward technology. Both studies yielded important themes related to use of home technology as follows: (a) households allow its residents express their identities, (b) households are private places, (c) family life is the household priority, (d) women are the household communicators, and (e) personal interaction is preferred to telephone communication.

Norris et al. (1998) conducted a study to understand the causes of accidents with stepladders in Great Britain and identify potential improvements to their design. Their methods combined the use of surveys of accidents and the stepladder market. They also conducted observations of the use of the stepladders. Participants were asked to complete two domestic tasks. The results showed that few people followed the rules of safe stepladder use. The study results were then used in a nationwide publicity campaign in the United Kingdom to raise consumer awareness about safe stepladder use.

In another study conducted to understand family utilization of the personal computer and how computer use affects the relocation of other family applications and services, Frohlich et al. (2001) conducted family in-home interviews. The authors explained that the interviews allowed them to explore all the family members' opinions on computer use and record the precise location of the computer and other technology inside the homes. This kind of study could then be complemented with more traditional human factors methods to assess either frequency of use or usability of the technology.

Jobs must be carried out in a physical *environment*, which may work to support or deter the worker's tasks. Ergonomists have mostly focused on assessing several environmental conditions such as illumination, climate, noise, and motion standards. Also when thinking of workstation design, human factors engineers focus on the task being carried out and the physical aspects of the workstation such as height, physical support, and lifting and how these may contribute to potential job-related

stress or strain injuries. Safety also falls within the human factors environmental assessments focusing on safeguarding for potentially hazardous conditions. Finally, when thinking about the physical environment from a macroergonomics view, it is important to understand where tasks are carried out and overall flow of work within the organization. Human factors engineers have developed a number of standards with respect to motion and repetitive tasks, safety, information input, workplace design, and environmental conditions that are used to evaluate the current state of any physical environment within a workplace. Also, the use of layouts may aid in understanding overall flow of work within an organization.

In their study, Pinto et al. (2000) wanted to improve the relation between the aging user and the environment. When offering practical recommendations, the illustrations used in the article showed the layout of the area of interest and illustrated the common habits or patterns of flow and use within the area. Their analysis not only recorded tasks for the tasks' sake but also aimed to understand the "artifact's role" in the task or job to be accomplished.

Similarly, the use of layouts in households can serve to obtain an understanding of location of care activities. Because of the private nature of home spaces, layouts may seem to be a very invasive tool and people may be threatened by the use of such a technique. This could, again, be complemented by the use of surveys and/or interviews to gain an overall picture of health workflow within the home. Also, depending on the overall goal of the human factors assessment, other techniques related to workplace design could then be used.

Finally, the *organization* has a strong influence on overall functioning and how jobs are organized and carried out. Organizational characteristics that have been traditionally assessed in the field of human factors are the organization's mission or vision, organizational culture, policies and procedures, incentives, pay structure, organizational structure, and performance measures. These characteristics can be assessed with the use of surveys or interviews, direct observations, and review of documentation relating to organizational structure, policies, and procedures.

As previously discussed, it is unlikely that there is documentation that describes the organizational structure of the home. Yet this is a very important aspect to be examined because it can influence all facets of home life. For example, Denham (2002) found that spiritual values and family traditions drive organizational aspects of the family concerning implementation strategies of health behaviors. Similar to the workplace, these are the ones that will drive the way that roles are assigned and tasks are carried out. In terms of organizational elements, there is a formal task and work structure within the family. The mother assumes a prominent role in health issues and there are member roles that determine who, why, where, how, and when to seek medical advice or services. These roles are tied to the family organizational structure as well as their norms of behavior. Furthermore, the roles will also be assigned on the basis of skill level and knowledge. The primary output in this case is the production of family health. This includes a number of routines or health behaviors such as dietary practices, sleep and rest patterns, avoidance behaviors, and so forth.

In a separate study, Honold (2000) defined the construct of culture within the HCI paradigm in order to develop a framework to elicit the influence of culture in product use. The researcher conducted a qualitative study of 35 households and used observations and interviews to identify eight factors that should be taken into consideration when defining product requirements in different cultures. The factors identified by the author are (a) objectives of the users, (b) characteristics of the users, (c) environment, (d) infrastructure, (e) division of labor, (f) organization of work, (g) mental models based on previous experience, and (h) tools. Human factors engineers may be better served, then, by using interviews or surveys as well as observations to assess household culture and structure to understand how they affect home health care work.

43.3.3 Case Study

The following summarizes a case study conducted to gain a deeper understanding of the job of health information management and the tasks that support it in order to improve and maintain health. Data collection encompassed three stages: (a) in-home interview of the primary health information manager or person who makes most of the transactions related to health care within the household; (b) household interviews on home health information management tasks; and (c) observation of three of the tasks mentioned during the interviews. The following description includes a family profile, a brief description of the distribution of health information throughout the home, and an example of a health information management task.

43.3.3.1 Family Profile

This extended family of seven is composed of three major household "subunits." These subunits are (a) the primary health information manager and her spouse; (b) the health information manager's sibling; and (c) the health information manager's son along with his family. For this family, blood pressure was a major health concern and although the view of the family's health is very good, the primary health information manager considers herself to be in fair health. The family uses traditional as well as home remedies, which include a parsley and lime blend used to help control high blood pressure. Although some of the family members had used a computer, the primary health information manager did not feel comfortable using e-mail, only used it "sometimes," and felt uncomfortable accessing personal information over the Internet. For this household, Spanish was their first, and for some members, only language, and this sometimes created communication problems during doctor visits. The primary health information manager mentioned that if she had an interpreter during doctor visits she would be able to understand all of the information. Also, the health information manager, who was a nurse, called her doctor and colleague and friend nurses in Mexico when she had questions about a health concern. Even though each household subunit manages their own health information, other household members were aware of what information they keep and where they keep it. Furthermore, some health information management tasks, such as managing diets and grocery shopping, were done independently by each subunit, whereas others, such as attending doctor's appointments or measuring blood pressure levels, required support from other household members as well as other family members who did not currently live at home.

43.3.3.2 Distribution of Health Information

This family reported five different locations throughout the household in which health information was stored. They also made use of 12 different storage artifacts to store this information. Table 43.1 shows the number of locations and artifacts for two different types of health information.

43.3.3.3 Health Information Task: Checking Blood Pressure Levels

Consuelo, her husband, and her brother Roberto all live in the same home. They all also happen to have high blood pressure. The three of them must check their blood pressure, but they do this in different ways. At the

TABLE 43.1

Locations and Artifacts by Health Information Type

Health Information Type	Locations	Storage Artifacts
Doctor's contact information	Kitchen	Post-it notes Refrigerator door
Prescription information/ medicines	Bedroom 1 Bedroom 2 Kitchen	Pill box Wardrobe Cookie jar Cupboard

beginning of the study, Consuelo would check her blood pressure in the morning before breakfast. She checked it monthly or every 2 months. She used her brother's blood pressure monitor and he assisted her in the task.

During the study period, Consuelo changed doctors and health insurance. This caused a change in her blood pressure medication and she had to monitor her blood pressure more closely. The doctor asked her to check her pressure daily to see how it varied. She began checking her pressure every day around breakfast time in the kitchen. Consuelo then wrote down each reading on a piece of paper that she stored in a kitchen cupboard. She then took the paper to her doctor whenever she had an appointment.

Roberto, on the other hand, kept a logbook of his readings in his bedroom, along with his monitor. He changed his logbook every 7 months. He wrote down each reading along with the average. When he had high or low blood pressure readings, Roberto wrote down circumstances that might have caused it, such as tension or exercise. He also took medication every day to control his blood pressure.

43.3.3.4 Implications for Design

The results described here have implications for human factors engineers to consider in the design of technology to support home health information management. For example, in this household, there were several subunits that manage their own health information as well as depend on other family members for support. Technology solutions, therefore, should take this into account and try to support these tasks and not create additional work. If not, the person might not be able to use it.

While the family made use of technology and resources to manage its health information, these resources were generally paper-based. This may have implications for design, because, as illustrated by the results, these resources were distributed across the household and varied in levels of complexities. They were generally suited for the different kinds of information they stored as well as took into consideration local family context like, for example, storing medications were children could not reach them. Careful consideration must be given then to the kinds of technology introduced into the home and what implications they have for ease of use.

The environment also played a crucial role in health information storage and use. Though limited by the household's layout, there may be ownership associated with different spaces within the house. Storage in one location versus another may also be associated with anticipated level of use and privacy issues. For example,

doctor appointments and calendars were kept in more visible places like kitchens or in portable devices like planners. This creates multiple levels of concern for designers in terms of distribution of use as well, access to information and privacy.

43.4 Conclusion

The home has become a relevant place for health care activities and is now an extension of the health care system. Many self-care practices such as diet, exercise, and complex disease management occur in the home. Many routine activities, such as shopping and household interaction, take on new meaning when there is a health care goal associated with them. In addition, as more and more patients manage complex conditions at home, it is becoming the primary site of care for many persons. Yet, the complexities of health work at home are not well understood. Health care represents a purposeful application of the skills of home residents and resources to accomplish health care goals. Household members involved in health care activities at home must balance these activities with professional work, additional home work, and the needs and demands of other household members. The household can be seen as a complex organization that works together to achieve household goals. This intersection of individuals, tasks, and environments pose an interesting challenge for human factors engineers to rethink the way they conceptualize and design for work, organizations, and environments.

The use of a human factors perspective provides a framework, which incorporates the analysis of the cognitive, physical, and macroergonomic aspects of home health care work and the context in which home health care activities occur. The goals of each specific human factors study in the home determine which assessment and analysis techniques will be more appropriate to achieve its objectives. For example, exploratory research conducted to better understand household roles in home health management might be better served by conducting in-depth, in-home observations. On the other hand, if the purpose of the study is to understand human factors issues in the use of a glucose meter, then designers could make use of a survey complemented with task observations to validate survey findings.

Home care is a central aspect of health care that can have implications for primary care as well as public health. Human factors has much to offer in designing for home environments and understanding how these environments fit within the larger health care system. The assessment techniques and examples presented in this chapter provide guidance for human factors engineers to begin to conceptualize the differences between home health care work and work done in industrial settings and how this may influence human factors assessment, analysis, and design.

Acknowledgments and Disclaimer

The authors thank Janice L. Genevro, PhD, Amy Helwig, MD, MS, and David Meyers, MD, from the Agency for Healthcare Research and Quality for their valuable advice on this document. The opinions expressed in this paper are those of the authors and do not reflect the official position of the Agency for Healthcare Research and Quality or the U.S. Department of Health and Human Services.

References

Agency for Healthcare Research and Quality. 2008. Medical Expenditure Panel Survey Household Component Glossary. http://www.meps.ahrq.gov/mepsweb/survey_comp/hc_ques_glossary.pdf (accessed March 9, 2010).

Aliotta, S. and J. Andre. 1997. Case management and home health care: An integrated model. *Home Health Care Manag Pract* 9 (2):1–12.

Askew, R. and J. T. Walker. 2008. Ergonomics for home care providers. *Home Health Care Nurse* 26 (7):412–417.

Brulin, C., B. Gerdle, B. Granlund, J. Hoog, A. Knutson, and G. Sundelin. 1998. Physical and psychological work-related risk factors associated with musculoskeletal symptoms among home care personnel. *Scand J Caring Sci* 12 (2):104–110.

Brulin, C., A. Winkvist, and S. Langendoen. 2000. Stress from working conditions among home care personnel with musculoskeletal symptoms. *J Adv Nurs* 31 (1):181–189.

Canam, C. 1993. Common adaptive tasks facing parents of children with chronic conditions. *J Adv Nurs* 18 (1):46–53.

Carayon, P. and M. J. Smith. 2000. Work organization and ergonomics. *Appl Ergon* 31 (6):648–662.

Chan, M., E. Campo, D. Estève, and J. Fourniols. 2009. Smart homes—Current features and future perspectives. *Maturitas* 64 (2):90–97.

Delano, K. T. and M. Hartman. 2001. Home health care: A needs assessment with design implications. *Proc Hum Factors Ergon Soc Annu Meet* 4:1289–1292.

Denham, S. A. 2002. Family routines: A structural perspective for viewing family health. *ANS Adv Nurs Sci* 24 (4):60–74.

Dokken, D. L. and N. Sydnor-Greenberg. 1998. Helping families mobilize their personal resources. *Pediatr Nurs* 24 (1):66–69.

Donelan, K., C. A. Hill, C. Hoffman, K. Scoles, P. H. Feldman, C. Levine, and D. Gould. 2002. Challenged to care: Informal caregivers in a changing health system. *Health Aff (Millwood)* 21 (4):222–231.

Fex, A., A. C. Ek, and O. Söderhamn. 2009. Self-care among persons using advanced medical technology at home. *J Clin Nurs* 18 (20):2809–2817.

Frohlich, D. M., S. Dray, and A. Silverman. 2001. Breaking up is hard to do: Family perspectives on the future of the home PC. *Int J Hum Comput Stud* 54 (5):701–724.

Geidl, L., P. Zrunek, Z. Deckert, D. Zimpfer, S. Sandner, G. Wieselthaler, and H. Schima. 2009. Usability and safety of ventricular assist devices: Human factors and design aspects. *Artif Organs* 33 (9):691–695.

Gentry, T. 2009. Smart homes for people with neurological disability: State of the art. *NeuroRehabilitation* 25 (3):209–217.

Grunfeld, E., D. Coyle, T. Whelan, J. Clinch, L. Reyno, C. C. Earle, A. Willan, R. Viola, M. Coristine, T. Janz, and R. Glossop. 2004. Family caregiver burden: Results of a longitudinal study of breast cancer patients and their principal caregivers. *CAMJ* 170 (12):1795–1801.

Hendrick, H. W. and B. M. Kleiner. 2001. *Macroergonomics: An Introduction to Work System Design*. Vol. 2. Santa Monica, CA: Human Factors and Ergonomics Society.

Hendrick, H. W. and B. M. Kleiner, eds. 2002. *Macroergonomics: Theory, Methods, and Applications*. Mahwah, NJ: Lawrence Erlbaum Associates, Inc.

Henriksen, K., A. Joseph, and T. Zayas-Cabán. 2009. The human factors of home health care: A conceptual model for examining safety and quality concerns. *J Patient Saf* 5 (4):229–236.

Hewett, T. T., R. Baecker, S. Card, T. Carey, J. Gasen, M. Mantei, G. Perlman, G. Strong, and W. Verplank. 1996. Human–computer interaction. In *ACM SIGCHI Curricula for Human–Computer Interaction*, Chapter 2. New York: ACM Press.

Hindus, D. 1999. The importance of homes in technology research. In *Proceedings of the Second International Workshop on Cooperative Buildings (CoBuild&rlenis;99)*, N. Streitz, J. Siegel, V. Hartkopf, and S. Konomi, eds. Heidelberg, Germany: Springer.

Hindus, D., S. D. Mainwaring, N. Leduc, A. E. Hagstrom, and O. Bayley. 2001. Casablanca: Designing social communication devices for the home. In *Proceedings of the ACM CHI 2001 Human Factors in Computing Systems Conference*, M. Beaudouin-Lafon and R. J. K. Jacob, eds. New York: ACM Press.

Honold, P. 2000. Culture and context: An empirical study for the development of a framework for the elicitation of cultural influence in product usage. *Int J Hum Comput Interact* 12 (3 & 4):327–345.

Huang, I.-C., D. Sugden, and S. Beveridge. 2008. Assistive devices and cerebral palsy: Factors influencing the use of assistive devices at home by children with cerebral palsy. *Child Care Health Dev* 35 (1):130–139.

International Ergonomics Association. 2000. What is ergonomics? http://www.iea.cc/01_what/What is Ergonomics.html (accessed February 8, 2010).

Jianwu, L., C. Jian, and B. Jing. 1996. Using human factors engineering as the basis for developing medical human-computer systems. *IEEE Trans Syst Man Cybern C Appl* 2:1202–1207.

Kaufman, D. R., V. L. Patel, C. Hilliman, P. C. Morin, J. Pevzner, R. S. Weinstock, R. Goland, S. Shea, and J. Starren. 2003. Usability in the real world: Assessing medical information technologies in patients' homes. *J Biomed Inform* 36 (1–2):45–60.

Kaufman, D. R., J. Pevzner, C. Hilliman, R. S. Weinstock, J. S. Teresi, and J. Starren. 2006. Redesigning a telehealth diabetes management program for a digital divide seniors population home. *Home Health Care Manag Pract* 18 (3):223–234.

Kiesler, S., B. Zdaniuk, V. Lundmark, and R. Kraut. 2000. Troubles with the Internet: The dynamics of help at home. *Hum Comput Interact* 15 (4):323–352.

Kimmerly, L. and D. Odell. 2009. Children and computer use in the home: Workstations, behaviors and parental attitudes. *Work* 32 (3):299–310.

Kutner, M. 2007. *Literacy in Everyday Life: Results from the 2003 National Assessment of Adult Literacy*, U.S. Department of Education, ed. Washington, DC: National Center for Education Statistics.

Kutner, M., E. Greenberg, Y. Jin, and C. Paulsen. 2006. *The Health Literacy of America's Adults*, U.S. Department of Education, ed. Washington, DC: National Center for Education Statistics.

Lang, A., M. Macdonald, J. Storch, K. Elliott, L. Stevenson, H. Lacroix, S. Donaldson, S. Corsini-Munt, F. Francis, and C. G. Curry. 2009. Home care safety perspectives from clients, family members, caregivers and paid providers. *Healthc Q* 12 (Special Issue):97–101.

Langa, K. M., M. W. Chernew, M. U. Kabeto, A. R. Herzog, M. B. Ofstedal, R. J. Willis, R. B. Wallace, L. M. Mucha, W. L. Straus, and A. M. Fendrick. 2001. National estimates of the quantity and cost of informal caregiving for the elderly with dementia. *J Gen Intern Med* 16 (11):770–778.

Lathan, C. E., A. Kinsella, M. J. Rosen, J. Winters, and C. Trepagnier. 1999. Aspects of human factors engineering in home telemedicine and telerehabilitation systems. *Telemed J* 5 (2):169–175.

Lippa, K. D., H. A. Klein, and V. L. Shalin. 2008. Everyday expertise: Cognitive demands in diabetes self-management. *Hum Factors* 50 (1):112–120.

Mardberg, B., U. Lundberg, and M. Frankenhaeuser. 1991. The total workload of parents employed in white-collar jobs: Construction of a questionnaire and a scoring system. *Scand J Psychol* 32 (3):233–239.

Martínez-Ramos, C., M. T. Cerdán, and R. S. López. 2009. Mobile phone-based telemedicine system for the home follow-up of patients undergoing ambulatory surgery. *Telemed J E Health* 15 (6):531–537.

Martínez-Ramos, C., M. T. Cerdán-Carbonero, R. S. López, and J. Normand Barron. 2007. Development of a mobile telemedicine system for the home follow-up of patients undergoing ambulatory surgery [in Spanish]. *Cir May Amb* 12 (4):148–156.

Marut, M. and A. Hedge. 1999. Ergonomic survey of household tasks and products. *Proc Hum Factors Ergon Soc Annu Meet* 1:506–510.

Moen, A. and P. F. Brennan. 2005. Health@Home: The work of health information management in the household (HIMH): Implications for consumer health informatics (CHI) innovations. *J Am Med Inform Assoc* 12 (6):648–656.

Moen, A., J. Gregory, and P. F. Brennan. 2007. Cross-cultural factors necessary to enable design of flexible consumer health informatics systems (CHIS). *Int J Med Inform* 76S (Suppl 1):S168–S173.

Moen, A. and J. Marquard. 2004. What do consumers do with health information at home? Paper presented at *Tromsø Telemedicine and eHealth Conference*, June 21–23, 2004, Tromsø, Norway.

Nag, A. and S. Chintharia. 1985. Physical activities of women in farming and non-farming seasons. *J Hum Ergol (Tokyo)* 14 (2):65–70.

National Alliance for Caregiving in collaboration with AARP. 2009. Caregiving in the U.S. 2009. http://www.aarp.org/research/surveys/care/ltc/hc/articles/caregiving_09.html (accessed March 9, 2010).

Niebel, B. W. and A. Freivalds. 1999. *Methods, standards, and work design*. 10th edn. Boston, MA: McGraw-Hill Companies, Inc.

Norris, B., K. Lawrence, N. Hopkinson, and J. R. Wilson. 1998. Using ergonomics investigations to improve stepladder safety. *Int J Inj Consum Saf* 5 (2):75–83.

Paasche-Orlow, M. K., R. M. Parker, J. A. Gazmararian, L. T. Nielsen-Bohlman, and R. R. Rudd. 2005. The prevalence of limited health literacy. *J Gen Intern Med* 20:175–184.

Pinto, M. R., S. De Medici, C. Van Sant, A. Bianchi, A. Zlotnicki, and C. Napoli. 2000. Ergonomics, gerontechnology, and design for the home-environment. *Appl Ergon* 31 (3):317–322.

Raphael, C. and J. L. Cornwell. 2008. Influencing support for caregivers: Lessons from home health care. *Am J Nurs* 108 (9 Suppl):78–82.

Sanders, M. S. and E. J. McCormick. 1993. *Human Factors in Engineering and Design*. 7th edn. New York: McGraw-Hill, Inc.

Sitzman, K. and D. Bloswick. 2002. Creative use of ergonomic principles in home care. *Home Health Care Nurse* 20 (2):98–103.

Smith, D. B. D. 1990. Human factors and aging: An overview of research needs and application opportunities. *Hum Factors* 32 (5):509–526.

Smith, M. and P. Carayon-Sainfort. 1989. A balance theory of job design for stress reduction. *Int J Ind Ergon* 4 (1):67–79.

Smith, C. E., L. S. Mayer, C. Parkhurst, S. B. Perkins, and S. K. Pingleton. 1991. Adaptation in families with a member requiring mechanical ventilation at home. *Heart Lung* 20 (4):349–356.

Stone, R. I. 2000. *Long-Term Care for the Elderly with Disabilities: Current Policy, Emerging Trends, and Implications for the Twenty-First Century*. New York: Milbank Memorial Fund. http://www.milbank.org/reports/008stone/LongTermCare_Mech5.pdf (accessed March 9, 2010).

Tabbarah, M., M. Silverstein, and T. Seeman. 2000. A health and demographic profile of noninstitutionalized older Americans residing in environments with home modifications. *J Aging Health* 12 (2):204–228.

The National Association for Home Care and Hospice. 2008. Basic statistics about home care. http://www.nahc.org/facts/08HC_Stats.pdf (accessed February 17, 2010).

U.S. Bureau of Labor Statistics. 2010. Career guide to industries, 2010–11 edition. http://www.bls.gov/oco/cg/cgs035.htm#outlook (accessed March 9, 2010).

U.S. Department of Health and Human Services. 2000. *Healthy People 2010: Understanding and Improving Health*. 2nd edn. Washington, DC: U.S. Government Printing Office.

U.S. Preventive Services Task Force. 2009. The guide to clinical preventive services. http://www.ahrq.gov/clinic/pocketgd09/pocketgd09.pdf (accessed March 9, 2010).

Venkatesh, A. 1996. Computers and other interactive technologies for the home. *Commun ACM* 39 (12):47–54.

Venkatesh, A. and S. Mazumdar. 1999. New information technologies in the home: A of study uses, impacts, and design strategies. Paper presented at *30th Annual Conference of the Environmental Design Research Association*, June 2–6, 1999, Orlando, FL.

Voorbij, A. I. M., and L. P. A. Steenbekkers. 2002. The twisting force of aged consumers when opening a jar. *Appl Ergon* 33 (1):105–109.

Wunderlich, G. S. and P. O. Kohler, eds. 2001. *Improving the Quality of Long-Term Care*. Washington, DC: National Academy Press.

Youngblut, J. M., P. F. Brennan, and L. A. Swegart. 1994. Families with medically fragile children: An exploratory study. *Pediatr Nurs* 20 (5):463–468.

Yun, Y. H., Y. S. Rhee, I. O. Kang, J. S. Lee, S. M. Bang, W. S. Lee, J. S. Kim, S. Y. Kim, S. W. Shin, and Y. S. Hong. 2005. Economic burdens and quality of life of family caregivers of cancer patients. *Oncology* 68 (2–3):107–114.

Zayas-Cabán, T. and B. Dixon. 2010. Considerations for the deisgn of safe and effective consumer health IT applications in the home. Qual Safe Health Care 19(3):i61–7.

Zayas-Cabán, T. and J. L. Marquard. 2009. A holistic human factors evaluation framework for the design of consumer health informatics interventions. *Proc Hum Factors Ergon Soc Annu Meet* 53:1003–1007.

Zayas-Cabán, T., J. L. Marquard, K. Radhakrishnan, N. Duffey, and D. L. Evernden. 2009. Scenario-based User Testing to Guide Consumer Health Informatics Design. *AMIA Annu Symp Proc* 719–723.

44

Human Factors and Ergonomics in Primary Care

Tosha B. Wetterneck, Jamie A. Lapin, Ben-Tzion Karsh, and John W. Beasley

CONTENTS

44.1 Introduction 763
44.2 What Is Primary Care? 763
    44.2.1 What Happens in Primary Care? 764
    44.2.2 Summary 767
44.3 Patient Safety in Primary Care 767
44.4 Macroergonomic Issues in Primary Care 768
44.5 Patient-Centered Medical Home 768
    44.5.1 Personal Physician, Physician-Directed Medical Practice—Leading a Team Who Cares for
        Patients, Whole Person Orientation, and Coordinated/Integrated Care 769
    44.5.2 Quality and Safety: As Hallmarks of Care 770
    44.5.3 Enhanced Access to Care (Open Scheduling and New Communication Options) 771
44.6 Conclusion 771
References 771

44.1 Introduction

Our original 2006 version of this chapter (Beasley et al., 2006) explored the nature of primary care, its complexity, and a dozen human factors and ergonomics (HFE) topics relevant to primary care. In the current version, we provide new information and a new direction for the chapter. We briefly review what primary care is for those unfamiliar with it, and we update what is known about the safety of primary care. Those interested in a more thorough treatment of "what is primary care and how is it different from other forms of healthcare" are referred to our original chapter. We have added two new sections: one on macroergonomic issues in primary care and one on the patient-centered medical home (PCMH). The national movement in the United States to create the PCMH in primary care has significant HFE considerations that must be addressed if the PCMH model is to succeed. These issues are discussed.

44.2 What Is Primary Care?

The Institute of Medicine (Donaldson et al., 1996) provided a formal definition of primary care:

Primary Care is the provision of integrated accessible health care services by clinicians who are accountable for addressing a large majority of personal health care needs, developing a sustained partnership with patients and practicing in the context of family and the community.

Captured in the definition are the four components of primary care that capture its essential nature: first-contact care ("accessible"), longitudinal care ("sustained partnership"), comprehensive care ("majority of personal health care needs"), and coordinated care ("context of family and the community") (Beasley et al., 1983; Donaldson et al., 1996; Starfield, 1993). *First-contact* care is not entirely unique to primary care as not all first contact events represent primary care—for example, cholesterol screening at a health fair or asking a friend with medical expertise for health advice are not examples of primary care encounters (Donaldson et al., 1996; Kovner & Jonas, 2002). However, primary care is the entry way into, as well as the patient's home in, the health care system. *Longitudinal* care indicates that continuity over time defines the care of the patient, rather than care limited to a specific disease process (e.g., cancer) or disease episode (e.g., appendicitis) (Starfield, 1993). Care continues through different stages of a patient's life and in various settings, ranging from hospital nurseries to clinics to nursing homes. The focus of care is on the individual, regardless of the type of care needed, and is provided by a single individual or team of health professionals who must also act as advocates for their patients (Donaldson et al., 1996).

Comprehensive care is integrated care that is provided for most of the common problems in the population (Starfield, 1993). Primary care clinicians must be able to utilize other health professionals and resources when this would be helpful for evaluation and treatment. The *coordination function* is extremely important and consists of integrating care that takes place through referrals, with different procedures, or various therapies (Starfield, 1993). The necessity for coordination and comprehensiveness places additional demands on the primary care clinician. For example, extra attention is required to ensure medical records are inclusive of pertinent information generated from other areas of care (Starfield, 1998). Well-coordinated care is critical if care is to be achieved in a cost-effective and safe manner (De Maeseneer et al., 2003).

Put simply, primary care is the system of care to which patients can bring all their health problems and expect to receive care or referral, be given guidance through the health care system, develop a continuous relationship with clinicians, and obtain information on disease prevention and health promotion. But these very aspects of primary care make providing services challenging. The need for clinicians and support staff to cope with a wide range of problems can lead to more chance for diagnostic and therapeutic errors. On the other hand, the flexibility of these systems allows for comprehensive management, the integration of preventive care with acute care, and "one stop shopping" for patients who can get most of their needs met by one clinician in one location.

Primary care is provided by several types of health care professionals, including physicians, nurse practitioners, midwives, and physician assistants (Kovner & Jonas, 2002). The physicians are family and general practitioners (Starfield, 1994) and include physicians in internal medicine, general pediatrics, general geriatrics, and obstetrics and gynecology (to the extent this latter group of physicians provides patients with primary care). Primary care clinicians practice in both outpatient and inpatient settings, though most patient encounters occur in outpatient settings. It is worth noting that the quality and costs of health care in the United States and elsewhere correlate most directly with the number of primary care physicians rather than the number of specialists (Baicker & Chandra, 2004; Macinko et al., 2003; Starfield, 2001). In this chapter, we consider outpatient primary care only.

44.2.1 What Happens in Primary Care?

The following section describes some of the user needs (clinician, patient, caregiver), contextual and organizational issues, and work characteristics, that must be

taken into account as efforts to improve patient safety in primary care are undertaken. The vignettes that follow are based one of the author's (J.W.B.) 35+ years of experience as a family physician and reflect the substance of actual encounters. These vignettes are intended to help those unfamiliar with primary care to develop a mental model of it. These vignettes are the same as those from the previous edition, though the analysis that follows each is abbreviated.

Patients require the integration of care for multiple problems.

Mr. J. is 84 years old and has anemia, diabetes, chronic renal failure, hypertension, atherosclerotic heart disease, arthritis, and fatigue. He takes eight different medications. He is in the clinic for a 15 minute follow-up visit.

There is a common, although incorrect, notion that most primary care patients come for care of a single, well-defined, acute medical problem. In fact, patients in primary care have, on average, slightly over three discrete problems addressed at each visit (Beasley et al., 2004).

The task of the clinician is to set management priorities and integrate care for existing conditions, to address new problems as they arise, and to integrate preventive care. The issue of the complexity engendered by patients having multiple problems is compounded by the issue of multiple agendas for the prioritization of care—agendas which may compete with each other. These agendas may arise from the caregiver (what he or she thinks is best), the patient (what's bothering him or her the most), the managed care organization (the expense of treatments), guidelines (often promulgated as the "ideal" for care for a single disease), or the threat of malpractice allegations. Adding to the complexity is the fact that between 7% and 20% of visits include discussion and medical decisions about somebody other than the identified patient—the "secondary" patient (Orzano et al., 2001; Zyzanski et al., 1998).

Diversity of clinical needs make standardization of care processes difficult.

Mrs. K. has diabetes and is in for her annual exam. She will have a pap smear and a flexible sigmoidoscopy. She has had a cough for 3 months. She has problems with insomnia. She also has a small skin lesion on her face that needs to be removed.

As seen in the aforementioned example, the issues the clinician must address are diverse and range from medical to surgical to psychosocial. This requires that more types of in-clinic equipment be available for diagnostic testing and therapeutic interventions. A range of different types of laboratory and other testing procedures is required as is a large set of consultative and support resources. This diversity of problems demands a broad range of knowledge and skills among both clinical and support personnel who need to have knowledge of these diverse medical problems and the relevant diagnostic and therapeutic strategies. This diversity also requires flexibility in scheduling and makes the establishment and use of clinical protocols and routines difficult.

Time limitations are important.

As the doctor goes to leave the room at the close of a routine 15 minute visit to monitor a 75 year old patient's hypertension, insomnia, and arthritis, his wife mentions, "By the way, when he gets up in the morning, he is very confused and sees things that aren't there."

Given the medical complexity—and the need to negotiate priorities—time management is difficult. As in the aforementioned vignette, this is not uncommonly accentuated by what is termed the "hand-on-the-doorknob" phenomenon where a major problem suddenly appears at the end of the allocated time. The need to accomplish medical tasks, listen to the patient, and complete associated administrative tasks (e.g., record keeping, billing, telephone calls) all require significant time. However in one sense, there is more, rather than less, time for patients in primary care since, in contrast to many other specialties, the work with patients and their problems occurs in continuity over a period of time, which is often measured in years.

Multiple places, systems, and people are involved in the care of patients.

Mr. J. has vascular disease. He was living at his home until he came to the clinic with an acute loss of circulation to his leg and was transferred to the hospital for an amputation. He was then admitted to a nursing home and subsequently back to his own home where he receives support consisting of family and home nursing care. His insurance has just changed and he needs to switch to a new physician and new home health caregivers as well as use different brands of medication that are on the new formulary. If he is rehospitalized, it will be at a different hospital.

Primary care clinicians must coordinate care with an average of over 200 other physicians and over 100 organizations (Pham et al., 2009). These organizations include multiple institutions such as hospitals and nursing homes, a variety of support agencies such as laboratories and x-ray units, home health agencies, and family and community resources as well as multiple payors for care. All of this takes significant time and attention on the part of clinicians and their staff.

Communication with and about patients occurs in many ways.

Mr. J. sends an e-mail to his doctor regarding his son's visit to an urgent care clinic for his asthma. The doctor gets it at home that evening. The doctor responds with a call from her cell phone while on her way to the hospital the next morning and then telephones in a prescription to the pharmacy. Later, when Mr. J. is in the clinic for his own health care, much of the time is taken up by further discussion of his son's problems.

It is typical for primary care clinicians to have multiple modes of contact with patients in addition to the direct "in the office" contact. This vignette illustrates the multiple methods of communication to which could be added communication through other members of the primary care clinical team (e.g., a medical assistant or pharmacist) or consultants. As in the aforementioned example, it is not uncommon for the medical record to be unavailable and for no record of the discussions to be created as was the case with the phone call and the clinic visit for Mr. J.

Information–it's feast and famine.

Mrs. T is a long-standing patient whose chart is in two 2″ thick volumes and contains physicians' notes, laboratory and x-ray reports, consultation reports, and hospital summaries in multiple formats scattered in various places in the chart. She was recently discharged from a regional hospital, but the clinical summary, laboratory reports, and medication changes from that hospitalization are not available when she comes to the office for follow-up. She is upset and crying and her husband who is also in the room is frustrated.

One study found that primary care clinicians reported missing clinical information in nearly 14% of visits (laboratory results [6.1% of all visits], letters/dictation [5.4%], radiology results [3.8%], history and physical examination [3.7%], and medications [3.2%]) (Smith et al., 2005). As the vignette illustrates, there are many circumstances when important medical record information is not available as well as many opportunities for information to be lost. However, it is also common for the clinician to be confronted with too much data that are often poorly organised (Schiff and Bates, 2010). The clinician often obtains data from several sources before and during the visit. The data may be in the patient's chart (electronic or paper), in memories from the last few visits, or on the scrap paper that the medical assistant wrote on when the patient checked in. The physician then collects more data while the patient is in the room by reviewing the patient's chart, by listening to the patient, picking up nonverbal cues, and perhaps listening to a relative who is also in the room as well. The dilemma of having both excess data, which may or may not be integrated into information, and at the same time potentially having important data missing is a combination that impairs medical decision making.

Context and family are important.

Mr. S. is a dairy farmer who is alcoholic. He is hospitalized for withdrawal but wants to leave to "take care of my farm." His neighbors agree to keep his herd milked and the barn cleaned out if he undergoes long-term treatment. His wife threatens to leave him unless he stays. He decides to stay for treatment.

Primary care takes place in the context of the family and the larger community and both can either help or hinder care. The involvement of the patient's family and community is especially important when the patient is a child or very elderly where issues of infirmity or cognitive impairment may play a role. This involvement may take the form of transporting the patient to the office, providing history to the clinician, helping the patient understand treatment, assisting the patient in reporting problems, and monitoring medications or the response to treatment. A lack of family support may lead to more tests and hospitalizations (Pantell et al., 2004). At the same time, family members may inhibit communication as might be the case where a parent accompanies an adolescent, thus making the discussion of issues such as sexual behavior more difficult. While the development of formal "family profiles" has been touted as a way for clinicians to recognize and act on these contextual issues, in reality much of the clinician's knowledge of contextual issues is developed over time through the care of the patient and his family (Beasley & Longenecker, 1983).

First-contact care is different.

Sammy N. is a 2 month old child whose mother brings him in because he has a fever.

Many acutely presenting problems in primary care are either minor or self-limited such as common viral infections or minor injuries, which require minimal diagnostic or therapeutic intervention. However, there is a continuing need to be alert for rare but serious diseases such as meningitis. Even when a patient has a rare but serious condition he or she tends to present to the primary care setting in the early stages when the typical signs and symptoms are not fully developed. For example, early meningitis can present without the typical physical finding of a stiff neck.

Relationships and continuity are important.

Ms. T. is a poorly controlled diabetic who is schizophrenic, lives alone, and does not like to leave her apartment. She does not trust her physician and is worried that her medications are "poisoning" her and so is not reliable in taking them.

For a patient in an intensive care unit, adherence to a prescribed medication is not an issue. He or she gets his or her medications with little or no input on their part. However, for a patient who is in a primary care setting, he or she is in control and a variety of factors may either facilitate or impede the care process. The development of a therapeutic and trusting relationship may well take precedence over the need to address other problems such as diabetes and this makes prioritization of care important. In the aforementioned vignette, the first priority has to be establishing some sort of working relationship with the patient. This prioritization according to the needs of the individual is an integral part of concurrent management of multiple problems and is often termed "treating the whole patient." It takes time for the caregiver to understand where the patient "is coming from" and what their priorities are for diagnosis and treatment. These are critical steps if there is to be a successful therapeutic alliance.

There are limitations on the use of therapeutics, procedures, and technology.

Mrs. B. is 78 year old, and in generally good health. However, she has chronic atrial fibrillation and should, according to guidelines, be on Coumadin to reduce her risk of a stroke. However, she does not want to take that "rat poison" or come in for the laboratory tests that would be required as she feels "just fine."

The omission of recommended care can also be considered to be an "error." The ability of primary care clinicians, upon whom most of the responsibility for providing care rests, to meet the guidelines for recommended care is not good (McGlynn et al., 2003). There are multiple reasons for this (Parchman et al., 2007). Some relate simply to a failure to organize a practice to deliver the needed care. Others relate to patients themselves and their multiple problems, which can result in competing demands. In some cases, there are limitations posed by the patient's physiology such as drug allergies or intolerance to certain categories of medications such as medications to treat coronary artery disease. The need to be reasonable, not only in terms of costs but also in terms of the time required, places limits on what can be accomplished both for prevention and therapy and may make following guidelines impractical (Ostbye et al., 2005; Yarnall et al., 2003).

There is a great variability of patients and their needs—one size does not fit all.

During a day's clinic session, Dr. K. sees four patients with coronary artery disease. The first has good insurance, is highly motivated, is exercising regularly, and is following his diet and taking his medications. The second patient is depressed and living alone. She cannot sleep. She continues to smoke. She has severe arthritis and cannot exercise. She is on Medicare and would like to use medications but cannot afford them. The third also has lung cancer. The lung cancer will limit her life to the point that treatment of the heart disease is no longer important. The final patient is a 63 year old immigrant from the African country of Chad who speaks no English; her husband who accompanies her has limited English skills. She is seen on an emergent basis because of an episode of chest pain and is admitted immediately to the hospital.

Even when the underlying medical problem is the same, the approach to the patient cannot be easily standardized. In the aforementioned vignette, the management of the coronary artery disease will be entirely different for each patient. In the case of the first patient, his care will meet the guidelines for the care of coronary artery disease. In the second patient, the depression is probably the most critical issue and will appropriately consume most of the caregiver's attention. In the third patient, the focus will be on the cancer and end of life issues and the coronary artery disease will not enter into the picture. For the fourth patient, there may not only be a potential language barrier, but also subtle (and perhaps unperceived) differences in concepts of disease, the nature of care, and the expectations of the caregiver. The variety of needed responses and interventions, even when a single major disease is present, makes standardization of protocols and procedures seem impractical at times.

Slack (loose coupling) in the system can allow for error recovery.

Mrs. J. is a hypertensive on metoprolol among her other medications. By accident, the physician clicks on a different dose of the medication than the one he intended. When Mrs. J. goes to the pharmacy to pick up the pills, she notices that they are different from the ones she has been getting. While she is wondering about this, the pharmacist says, "I see your doctor changed the dose." Knowing that they had not

discussed a dose change, Mrs. J calls the clinic and the doctor resends the prescription with the right dose this time.

While health care delivery is complex, and therefore arguably prone to failure, systems elements (e.g., clinicians, clinic staff, lab technicians, etc.) often do work together to prevent or recover from failures. In this case, the wrong medication dose was ordered. However, the pharmacist and patient caught the error through a process at the pharmacy, which set in motion a process for correcting the problem.

44.2.2 Summary

Dozens of HFE issues were identified in the vignettes illustrating primary care, including those related to memory, information processing, standardization, simplification, forcing functions, work pressure and work load, organizational design, uncertainty, information access, technology acceptance, usability, and error recovery. In the original first edition of this chapter (Beasley et al., 2006), some of these issues were expanded upon in more detail. We do not repeat that exercise in the current edition. Instead, having presented an overview of primary care, we now transition to a brief introduction to primary care patient safety, followed by an HFE analysis of the PCMH.

44.3 Patient Safety in Primary Care

Though initial patient safety research focused mostly on inpatient settings, it is now clear that significant patient safety problems exist in primary care, affecting children, adults, and the elderly (Gandhi et al., 2003; Goulding, 2004; Kaushal et al., 2007; Thomsen et al., 2007; Zandieh et al., 2008). The prevalence of preventable incidents or adverse events in primary care may be quite high (Thomsen et al., 2007), and evidence suggests that over half of them may be preventable (Bhasale et al., 1998; Fischer et al., 1997; Sandars & Esmail, 2003).

Ghandi et al. (2003) found that 25% of surveyed adult outpatients over the age of 18 had an adverse drug event (ADE). Of those, nearly 40% were considered preventable or at least ameliorable. The number of medications that a patient took was associated with the risk of having an event related to a medication. In a similar study of 20–75 year olds, 18% of patients reported a drug complication, though chart review of those same patients suggested that only 3% had drug complications (Gandhi et al., 2000). Patient safety problems such ADEs are likely to increase, at least among the elderly, given the ever increasing number of available and prescribed drugs (Tierney, 2003).

The causes and categories of errors in primary care are similar to those in inpatient care. Medication errors include such categories as prescribing errors, medication administration errors, documentation errors, dispensing errors, and monitoring errors (Kuo et al., 2008). Dovey and her colleagues have identified a large range of medical errors including administrative failures, investigation failures, treatment delivery lapses, miscommunication, payment system problems, errors in the execution of a clinical task, wrong treatment decision, and wrong diagnosis (Dovey et al., 2002; Phillips et al., 2006). Literature reviews have found basically the same error categories and causes (Elder & Dovey, 2002; Kuzel et al., 2004; Sandars & Esmail, 2003; Wetzels et al., 2008), though recently the problems of missing clinical information (Smith et al., 2005) and diagnostic errors have been highlighted as well (Kostopoulou et al., 2008).

Over the past 10 years, much attention has focused on the safety of transitions of care (Coleman, 2007; Joint Commission Resources, 2010), or the movement of a patient within health care between clinicians and locations, for example, a patient is hospitalized for a medical illness or sees a subspecialist for care and then returns to the primary care office for follow-up and ongoing care. Transitions of care around hospitalizations have increased greatly due to the hospitalist movement (Wachter & Goldman, 1996). Hospitalists are physicians who spend the majority of their clinical time caring for patients in the hospital setting. Most hospitalists train in internal medicine or family medicine and then choose to practice exclusively in the inpatient (hospital) setting. This movement grew out of the desire to improve the quality and safety of hospital care and the work life of primary care physicians. Because many primary care physicians have elected to handover care of their hospitalized patients to hospitalists, patient discharge from the hospital now involves a transfer of information from the hospitalist to the primary care physician. These transitions are considered significant safety vulnerabilities because of the ease with which important information can get lost in transition (Coleman & Berenson, 2004).

44.4 Macroergonomic Issues in Primary Care

The number of problems dealt with by primary care physicians per patient visit has increased over time with little increase in the average visit length (Abbo et al., 2008; Bodenheimer, 2008; Grumbach & Bodenheimer, 2002). As the population ages, primary care utilization and complexity will increase; elderly patients seek outpatient care twice as often as younger patients (National Center for Health Statistics, 2005). Electronic health records are perceived to further compress the time available (Bloom & Huntington, 2010). This time pressure may also lead to poor physician outcomes like job stress, burnout, and early retirement (Linzer et al., 2009) or decreasing work hours (The Physician's Foundation, 2008). Though the demand for primary care physicians is predicted to increase 38% by 2020, in part due to aging baby boomers living longer and requiring more health care as they develop chronic conditions (American College of Physicians, 2006), the supply of primary care physicians is not expected to meet demands given the aging of the current pool of primary care physicians (Bodenheimer, 2008) and a dwindling pipeline of primary care residents (Hauer et al., 2008).

44.5 Patient-Centered Medical Home

Clearly, transformation is needed in the way that primary care is delivered in the United States to make primary care physicians' jobs doable again; and to deliver higher quality care, maximize resource utilization, reduce waste, and improve the working life of primary care clinicians. This transformation centers around a team-based and patient-centered care approach known as the PCMH. The PCMH model, a team-based model of care, endorsed by more than 500 professional and business organizations, is proposed as the future model of primary care delivery (American College of Physicians, 2006; Patient Centered Primary Care Collaborative, 2008). The joint principles of the PCMH, agreed upon by collaborating organizations in the Patient Centered Primary Care Collaborative (PCPCC) include

1. Personal physician
2. Physician-directed medical practice—leading a team who cares for patients
3. Whole person orientation (team provides/coordinates all health care needs)
4. Care is coordinated/integrated—across the health care system and patient community. Care is facilitated using technologies such as health information technologies, registries, and health information exchanges
5. Quality and safety—as hallmarks of care
   a. Patient outcomes driven by shared decision making and a strong physician–patient relationship

b. Evidence-based medicine (EBM) and clinical decision support (CDS) tools guide decision making

c. Physician accountability for continuous quality improvement (QI)

d. Patient engagement in decision making

e. Health information technology (HIT) use to support patient care, performance measurement, education, and enhanced communication

f. Practice is recognized as PCMH (accreditation)

g. Patients and family participate in QI

6. Enhanced access to care (e.g., open scheduling, new communication options)

7. Payment restructuring

The PCMH model is currently being studied in numerous pilot and demonstration projects across the United States (Carrier et al., 2009; Nutting et al., 2009) and is being widely implemented across the Veteran Affairs Administration primary care clinics and across large primary care practices/organizations. The goals of the PCMH are to "revitalize" and "transform" primary care by improving the quality of patient care as well as the work life of primary care clinicians. Indeed, team-based care has been shown to achieve these goals (Bodenheimer, 2007; Bodenheimer & Laing, 2007). HIT and specifically electronic health records (EHRs) are crucial parts of the PCMH model and are expected to improve information flow, communication between clinicians (both clinic teams and outside providers), and record keeping. Also needed is the ability to manage panels of patients with chronic conditions using the team model of caring for patients at the practice level and the supporting of decision making with appropriate CDS tools (Bodenheimer & Grumbach, 2003).

Unfortunately, the initial lessons learned from qualitative analysis of early PCMH demonstration projects show that transforming a practice into the PCMH model is much more complicated than initially thought even though many primary care practices have previously embraced components of the PCMH model (Nutting et al., 2009). Typical changes in health care usually follow a QI model or incremental changes over time but these methods of change were not sufficient for implementing the PCMH model due to the tremendous, interdependent, ongoing change it requires. Specifically, practices found that the current information technology did not have the functionalities needed for PCMH or the functionalities were not user friendly (Nutting et al., 2009). Also, the model requires physicians to change the way they work by

moving to a team-based approach of care involving nurses and office staff and ultimately changing to a workflow that maximizes the patient experience, which may significantly impact the workflow of clinic staff. Finally, financial resources were often inadequate for health IT implementation as were operational changes like hiring new staff.

We will discuss each component of PCMH model from an HFE perspective to address some of the constructs that are helpful to consider for the new PCMH model and what is known from the HFE and other literature with regards to the design and implementation of such expansive change.

44.5.1 Personal Physician, Physician-Directed Medical Practice—Leading a Team Who Cares for Patients, Whole Person Orientation, and Coordinated/Integrated Care

From an HFE standpoint, these four principles of the PCMH pose tremendous challenges, though they have been core tenets of primary care for a long time (Donaldson et al., 1996). These principles solidify the core concepts of primary care in the PCMH model. What is new is the notion of the physician directing a team and coordinating and integrating care through the use of health information technologies. All 12 vignettes illustrated the complexity of the primary care physician being the hub of medical care, treating the whole person, and coordinating care across the health system and the community. The HFE issues of these central elements of primary care were discussed in the first edition of this chapter (Beasley et al., 2006). Here, we focus on what is new among these four principles—team-based care and the role of HIT.

The PCMH model moves primary care to a team-based model that collectively cares for individual patients and populations of patients and integrates a patient's care across the health care system and community. To do this, EHRs must support enhanced communication between the members of the primary care team, external clinicians via health information exchange, and the patient; provide tools to optimize patient care such as CDS tools based on EBM that guide decision making; and provide patient registries for monitoring population quality of care. However, existing EHRs are not well designed to support the cognitive work of teams (Nutting et al., 2009). Teams require special design consideration for communication and coordination (Burke et al., 2004; Salas et al., 2004, 2006, 2007a), which go beyond the cognitive requirements of the individuals who make up the teams. In addition, to the extent that care is provided by a team, the distribution of where information resides will change (Hutchins, 1996), and this will impact situation awareness (Salmon et al., 2009) for any given team member. For primary care to transform itself to the PCMH model, it will require

- Primary care teams to be designed. Teams do not emerge from groups of people told to work together. There is extensive HFE research about elements of effective teams and the knowledge, skills, and attitudes required by team members (Salas et al., 2004, 2007a, b). Research on effective teams will need to be used to design and implement primary care teams.

- Primary care teams will require extensive training in teamwork if the PCMH model is to succeed.

- Successful teamwork will require thoughtful redistribution of work and increased methods of integrating team members to maintain communication and coordination.

- Health information technologies will need to be designed and implemented to support the emerging needs of teams. Consideration will need to be given for how HITs are designed to support the members of the teams doing their own work, and also the needs of the team for communication, coordination, and control.

- Careful consideration will need to be given to how HITs are used by teams. Using HIT to replace synchronous communication may reduce team effectiveness (Hess et al., 2010).

- Determination of how to appropriately share information among team members at the appropriate times in order to maintain appropriate individual and team situation awareness (Prince et al., 2007).

44.5.2 Quality and Safety: As Hallmarks of Care

The quality and safety transformation principles have many subcomponents. First, true quality outcomes for a patient can only be achieved by involving the patient in decision making, a process called shared decision making (Coulter, 1997). In shared decision making, the patient and physician reach consensus on a plan of action through joint sharing of information on the options for care and the risks and benefits of the options and the patient's preferences for treatment. A strong physician–patient relationship is helpful for this process as physicians get to know their patient preferences over time (see vignettes). However, shared decision making requires attention to a variety of HFE considerations. For example, to answer the question—how do we get patients to participate in decision making?—will require attention to research on participatory ergonomics. There also needs to be attention paid to what cues physicians

rely upon to determine if a patient is actually participating versus only acquiescing. Furthermore, there is the question of how to present or describe complicated medical decisions in ways that accommodate the ranges of patient knowledge and education such that they can meaningfully participate.

Second, the evidence base for patient diagnosis and treatment is growing exponentially. Thus, physicians have access to more knowledge than ever before on how to provide quality care, but it can take decades for new evidence to become part of clinical practice (AHRQ, 2001). Clinical practice guidelines (CPGs), which are systematically developed statements to guide care based on the best evidence, have proliferated to educate and aid physicians in providing quality care but CPGs alone rarely improve care by themselves (Yarnall et al., 2003). To help physicians incorporate new knowledge into practice, CDS tools have become available to guide physician decision making. Attention has turned toward incorporating CDS into HIT, namely, EHRs and computer provider order entry (CPOE). For example, medication ordering CDS in CPOE can provide links to CPGs and medication information for medication use, default dose, frequency, and route information for medications and allergy and drug–drug interaction checking. The PCMH calls for the implementation and use of CDS tools to support patient care and patient education; however, the evidence about the effectiveness of CDS is mixed, owing in large part to usability and workflow integration problems (Karsh, 2009). If the PCMH model is to succeed, CDS systems will need to be developed that require little effort, have high effectiveness, and support both individual and team member (including patient) needs.

Third, there is a focus on continuous QI and physician performance. The PCMH calls for accountability of physicians for active involvement in clinic QI efforts and using health IT to measure physician performance on quality measures and guide improvement efforts. To make QI more patient-centered, primary care practices are encouraged to involve patients and family members in QI efforts. Professional organizations like the American Medical Association and the Ambulatory Quality Association as well as national quality and accreditation agencies like the National Committee for Quality Assurance (NCQA) have identified quality measures (e.g., The Healthcare Effectiveness Data and Information Set [HEDIS]) that primary care physicians and practices can use to monitor the quality of care provided for common diseases like diabetes, hypertension, depression, and hyperlipidemia and preventive measures like screening tests and immunizations. However, performance measures can have perverse effects on performance (Kerr, 1975). If the wrong outcomes are measured, it may drive clinicians to pay less attention to important performance outcomes such as patient outcomes in favor of those that are measured (Karsh et al., 2010). Careful attention will need to be paid to what is measured and what impact that measure will have on actual clinician performance.

44.5.3 Enhanced Access to Care (Open Scheduling and New Communication Options)

Patient-centered care also takes into consideration convenient times for patients to make appointments, like nights and weekend hours in addition to the typical daytime hours. Convenient office visit scheduling is an important means of removing patient barriers to health care access. Open scheduling allows for patients to make same-day appointments without regard to the type of appointment. Typically, same-day or same-week appointments are reserved for acute care needs but introducing open scheduling for all patient care needs gives flexibility to patients to schedule appointments. Last, new communication options should be explored and implemented to meet patients' needs. Examples include more asynchronous types of communication that are convenient for patients such as online scheduling, e-mail, and electronic communication via the patient health record as well as greater access to a primary care team member, like a nurse or medical assistant, on the telephone to immediately answer questions. Both enhanced access and new modes of communication come with HFE risks that need to be carefully managed.

Open access, while highly beneficial to the patient, may give the physician little if any time to prepare for the visit, depending on scheduling practices. If there is no time to prepare because the patient arrives on the day of his or her call, and if such visits are still scheduled for only the typical 15–20 minutes, there is a risk that the physician will not have time to fully assess the patient situation and effectively cover all the problems the patient needs to have addressed. This is often the case even with scheduled visits, but the risk is higher with open access visits. Physicians operating in open access environments may need to be allowed to schedule lengthier visits for same-day patients so that they have sufficient time for evaluation.

Likewise, the use of electronic communication may be beneficial to patients, but unless carefully managed, may be a significant burden on already burdened clinicians (Bloom & Huntington, 2010). Patient expectations for how rapidly physicians should respond to e-mails may be very high (Couchman et al., 2005), and this may translate into additional time demands for the physician. Health care organizations will need to answer HFE questions in order to incorporate electronic communication into practice, such as

- Is the system design and implementation strategy supportive of the work of primary care clinicians' work?
- Are all of our clinicians computer trained, and if not, when and how will we train them?
- How will we build in time for physicians to respond to e-mail in a timely fashion? How will we integrate this time into clinical workflow?
- Have we created proper workstation ergonomics to support this additional computer work?

44.6 Conclusion

The complexity of primary care results in many factors that can affect individual clinician and clinician team performance. These include knowledge factors related to solving problems, factors governing control of attention, and trade-off decisions that must be made when conflicting goals are present (Bogner, 1994). They also include patient factors, payor factors, clinic design, workflow, and technological factors. The different factors must integrate at both the individual and clinical team level for successful system operation, and exist in the context of situational demands, resources, and constraints that are posed by the organization (Bogner, 1994). The current national movement in the United States toward the PCMH is supposed to transform primary care in a way that better supports clinicians and patients. However, if the human factors and ergonomic details of the components and the interactions among the components are ignored, there is a risk that the very goals the PCMH is designed to achieve will be set back.

References

Abbo, E. D., Zhang, Q., Zelder, M., & Huang, E. S. (2008). The increasing number of clinical items addressed during the time of adult primary care visits. *Journal of General Internal Medicine*, 23(12), 2058–2065.

AHRQ. (2001). Translating research into practice (TRIP)-II fact sheet. Retrieved from http://www.ahrq.gov/research/trip2fac.htm (accessed July 12, 2011)

American College of Physicians. (2006). The impending collapse of primary care medicine and its implications for the state of the nation's health care: A report from the American college of physicians. Available at: http://www.acponline.org/advocacy/events/state_of_healthcare/statehco61.pdf (accessed July 12, 2011)

Baicker, K., & Chandra, A. (2004). Medicare spending, the physician workforce, and beneficiaries' quality of care. *Health Affairs*, W4-184–W4-197, doi:*10.1377/hlthaff.w4.184*.

Beasley, J. W., Hamilton-Escoto, K., & Karsh, B. (2006). Human factors in primary care. In P. Carayon (Ed.), *Handbook of Human Factors and Ergonomics in Patient Safety* (pp. 921–936). Mahwah, NJ: Lawrence Erlbaum Associates.

Beasley, J. W., Hankey, T. H., Erickson, R., Stange, K., Mundt, M., Elliott, M. et al. (2004). How many problems do family physicians manage at each encounter? A WReN study. *Annals of Family Medicine*, 2, 405–410.

Beasley, J. W., Hansen, M. F., Ganiere, D. S., Currie, B. F., Westgard, D. E., Connerly, P. W. et al. (1983). Ten central elements of family practice. *Journal of Family Practice*, 16(3), 551–555.

Beasley, J. W., & Longenecker, R. (1983). The patient/family profile. In R. B. Taylor (Ed.), *Fundamentals of Family Medicine*. New York: Springer Verlag.

Bhasale, A. L., Miller, G. C., Reid, S. E., & Britt, H. C. (1998). Analysing potential harm in Australian general practice: An incident-monitoring study [see comment]. *Medical Journal of Australia*, 169(2), 73–76.

Bloom, M. V., & Huntington, M. K. (2010). Faculty, resident, and clinic staff's evaluation of the effects of EHR implementation. *Family Medicine*, 42(8), 562–566.

Bodenheimer, T. (2007). *Building Teams in Primary Care: 15 Case Studies*. Oakland, CA: California HealthCare Foundation.

Bodenheimer, T. (2008). Coordinating care—A perilous journey through the health care system [editorial material]. *New England Journal of Medicine*, 358(10), 1064–1071.

Bodenheimer, T., & Grumbach, K. (2003). Electronic technology—A spark to revitalize primary care? *JAMA*, 290(2), 259–264.

Bodenheimer, T., & Laing, B. Y. (2007). The teamlet model of primary care. *Annals of Family Medicine*, 5(5), 457–461.

Bogner, M. (1994). *Human Error in Medicine*. Hilldale, NJ: Lawrence Erlbaum.

Burke, C. S., Salas, E., Wilson-Donnelly, K., & Priest, H. (2004). How to turn a team of experts into an expert medical team: Guidance from the aviation and military communities. *Quality & Safety in Health Care*, 13, I96–I104.

Carrier, E., Gourevitch, M. N., & Shah, N. R. (2009). Medical homes challenges in translating theory into practice. *Medical Care*, 47(7), 714–722.

Coleman, E. A. (2007). *Charting a Course for High Quality Care Transitions*. Philadelphia, PA: Haworth Press.

Coleman, E. A., & Berenson, R. A. (2004). Lost in transition: Challenges and opportunities for improving the quality of transitional care. *Annals of Internal Medicine*, 141(7), 533–535.

Couchman, G. R., Forjouh, S. N., Rascoe, T. G., Reis, M. D., Koehler, B., & van Walsum, K. L. (2005). E-mail communications in primary care: What are patients' expectations for specific test results? *International Journal of Medical Informatics*, 74(1), 21–30.

Coulter, A. (1997). Partnerships with patients: The pros and cons of shared clinical decision-making. *Journal of Health Services Research and Policy*, 2, 112–121.

De Maeseneer, J. M., De Prins, L., Gosset, C., & Heyerick, J. (2003). Provider continuity in family medicine: Does it make a difference for total health care costs? [see comment]. *Annals of Family Medicine*, 1(3), 144–148.

Donaldson, M., Yordy, K., & Lohr, K. (1996). *Primary Care America's Health in a New Era*. Washington, DC: National Academy Press.

Dovey, S. M., Meyers, D. S., Phillips, R. L., Jr., Green, L. A., Fryer, G. E., Galliher, J. M. et al. (2002). A preliminary taxonomy of medical errors in family practice. *Quality & Safety in Health Care*, 11(3), 233–238.

Elder, N. C., & Dovey, S. M. (2002). Classification of medical errors and preventable adverse events in primary care: A synthesis of the literature [erratum appears in *J Fam Pract*. 2002;51(12):1079]. *Journal of Family Practice*, 51(11), 927–932.

Fischer, G., Fetters, M. D., Munro, A. P., & Goldman, E. B. (1997). Adverse events in primary care identified from a risk-management database [see comment]. *Journal of Family Practice*, 45(1), 40–46.

Gandhi, T. K., Burstin, H. R., Cook, E. F., Puopolo, A. L., Haas, J. S., Brennan, T. A. et al. (2000). Drug complications in outpatients. *Journal of General Internal Medicine*, 15(3), 149–154.

Gandhi, T. K., Weingart, S. N., Borus, J., Seger, A. C., Peterson, J., Burdick, E. et al. (2003). Adverse drug events in ambulatory care [see comment]. *New England Journal of Medicine*, 348(16), 1556–1564.

Goulding, M. R. (2004). Inappropriate medication prescribing for elderly ambulatory care patients. *Archives of Internal Medicine*, 164, 305–312.

Grumbach, K., & Bodenheimer, T. (2002). A primary care home for Americans—Putting the house in order. *JAMA*, 288(7), 889–893.

Hauer, K. E., Durning, S. J., Kernan, W. N., Fagan, M. J., Mintz, M., O'Sullivan, P. S. et al. (2008). Factors associated with medical students' career choices regarding internal medicine. *JAMA*, 300(10), 1154–1164.

Hess, D. R., Tokarczyk, A., O'Malley, M., Gavaghan, S., Sullivan, J., & Schmidt, U. (2010). The value of adding a verbal report to written handoffs on early readmission following prolonged respiratory failure. *Chest*, 138, 1475–1479.

Hutchins, E. (1996). *Cognition in the Wild*. Cambridge, MA: MIT Press.

Joint Commission Resources. (2010). *Improving Communication during Transitions of Care*. Chicago, IL: Joint Commission Resources.

Karsh, B. (2009). *Clinical Practice Improvement and Redesign: How Change in Workflow Can Be Supported by Clinical Decision Support* (No. AHRQ Publication No. 09-0054-EF). Rockville, MD: Agency for Healthcare Research and Quality.

Karsh, B., Weinger, M., Abbott, P., & Wears, R. (2010). Health information technology: Fallacies and sober realities. *JAMIA*, 17, 617–623.

Kaushal, R., Goldmann, D. A., Keohane, C. A., Christino, M., Honour, M., Hale, A. S. et al. (2007). Adverse drug events in pediatric outpatients. *Ambulatory Pediatrics*, 7(5), 383–389.

Kerr, S. (1975). On the folly of rewarding A, while hoping for B. *Academy of Management Journal, 18*, 769–783.

Kostopoulou, O., Delaney, B. C., & Munro, C. W. (2008). Diagnostic difficulty and error in primary care—A systematic review. *Family Practice, 25*(6), 400–413.

Kovner, A., & Jonas, S. (2002). *Health Care Delivery in the United States*. New York: Springer.

Kuo, G. M., Phillips, R. L., Graham, D., & Hickner, J. M. (2008). Medication errors reported by US family physicians and their office staff. *Quality & Safety in Health Care, 17*(4), 286–290.

Kuzel, A. J., Woolf, S. H., Gilchrist, V. J., Engel, J. D., LaVeist, T. A., Vincent, C. et al. (2004). Patient reports of preventable problems and harms in primary health care. *Annals of Family Medicine, 2*(4), 333–340.

Linzer, M., Manwell, L. B., Williams, E. S., Bobula, J. A., Brown, R. L., Varkey, A. B. et al. (2009). Working conditions in primary care: Physician reactions and care quality. *Annals of Internal Medicine, 151*(1), U28–U48.

Macinko, J., Starfield, B., & Shi, L. (2003). The contribution of primary care systems to health outcomes within organization for economic cooperation and development (OECD) countries 1970–1998. *Health Services Research, 38*(3), 831–865.

McGlynn, E. A., Asch, S. M., Adams, J., Keesey, J., Hicks, J., DeCristofaro, A. et al. (2003). The quality of health care delivered to adults in the United States [see comment]. *New England Journal of Medicine, 348*(26), 2635–2645.

National Center for Health Statistics. (2005). Ambulatory health care data. Retrieved from http://www.cdc.gov/nchs/about/major/ahcd/officevisitcharts.htm (accessed July 12, 2011)

Nutting, P. A., Miller, W. L., Crabtree, B. F., Jaen, C. R., Stewart, E. E., & Stange, K. C. (2009). Initial lessons from the first national demonstration project on practice transformation to a patient-centered medical home. *Annals of Family Medicine, 7*(3), 254–260.

Orzano, A. J., Gregory, P. M., Nutting, P. A., Werner, J. J., Flocke, S. A., & Stange, K. C. (2001). Care of the secondary patient in family practice. A report from the ambulatory sentinel practice network. *Journal of Family Practice, 50*(2), 113–116.

Ostbye, T., Yarnall, K. S., Krause, K. M., Pollak, K. I., Gradison, M., & Michener, J. L. (2005). Is there time for management of patients with chronic diseases in primary care? *Annals of Family Medicine, 3*, 209–214.

Pantell, R. H., Newman, T. B., Bernzweig, J., Bergman, D. A., Takayama, J. I., Segal, M. et al. (2004). Management and outcomes of care of fever in early infancy [see comment]. *JAMA, 291*(10), 1203–1212.

Parchman, M. L., Pugh, J. A., Romero, R. L., & Bowers, K. W. (2007). Competing demands or clinical inertia: The case of elevated glycosylated hemoglobin. *Annals of Family Medicine, 5*, 196–201.

Patient Centered Primary Care Collaborative. (2008). *The Patient-Centered Medical Home: A Purchaser Guide— Understanding the Model and Taking Action*. Washington, DC: Patient Centered Primary Care Collaborative.

Pham, H. H., O'Malley, A. S., Bach, P. B., Saiontz-Martinez, C., & Schrag, D. (2009). Primary care physician's links to other physicians through medicare patients: The scope of care coordination. *Annals of Internal Medicine, 150*, 236–242.

Phillips, R. L., Dovey, S. M., Graham, D., Elder, N. C., & Hickner, J. M. (2006). Learning from different lenses: Reports of medical errors in primary care by clinicians, staff, and patients. A project of the American Academy of Family Physicians National Research Network. *Journal of Patient Safety, 2*(3), 140–146.

Prince, C., Ellis, E., Brannick, M. T., & Salas, E. (2007). Measurement of team situation awareness in low experience level aviators. *International Journal of Aviation Psychology, 17*(1), 41–57.

Salas, E., Baker, D., King, H., Battles, J., & Barach, P. (2006). On teams, organizations and safety: Of course.... *Joint Commission Journal on Quality and Patient Safety, 32*, 112–113.

Salas, E., Kosarzycki, M. P., Tannenbaum, S. I., & Carnegie, D. (2004). Principles and advice for understanding and promoting effective teamwork in organizations. In R. J. Burke & C. L. Cooper (Eds.), *Leading in Turbulent Times: Managing in the New World of Work* (pp. 95–120). Malden, MA: Blackwell Publishing.

Salas, E., Rosen, M. A., Burke, C. S., Nicholson, D., & Howse, W. R. (2007a). Markers for enhancing team cognition in complex environments: The power of team performance diagnosis. *Aviation Space and Environmental Medicine, 78*(5), B77–B85.

Salas, E., Wilson, K. A., Murphy, C. E., King, H., & Baker, D. (2007b). What crew resource management training will not do for patient safety unless.... *Journal of Patient Safety, 3*(2), 62–64.

Salmon, P. M., Stanton, N. A., Walker, G. H., Jenkins, D., Ladva, D., Rafferty, L. et al. (2009). Measuring situation awareness in complex systems: Comparison of measures study. *International Journal of Industrial Ergonomics, 39*(3), 490–500.

Sandars, J., & Esmail, A. (2003). The frequency and nature of medical error in primary care: Understanding the diversity across studies. *Family Practice, 20*(3), 231–236.

Schiff, G. D., & Bates, D. W. (2010). Can electronic clinical documentation help prevent diagnostic errors? *New England Journal of Medicine, 362*, 1066–1069.

Smith, P. C., Rodrig, A. G., Bublitz, C., Parnes, B., Dickinson, L. M., Van Vorst, R. et al. (2005). Missing clinical information during primary care visits. *JAMA, 293*, 565–571.

Starfield, B. (1993). Primary care. *Journal of Ambulatory Care Management, 16*(4), 27–37.

Starfield, B. (1994). Is primary care essential? *The Lancet, 344*(8930), 1129–1133.

Starfield, B. (1998). *Primary Care Balancing Health Needs, Services, and Technology*. New York: Oxford University Press.

Starfield, B. (2001). New paradigms for quality in primary care. *British Journal of General Practice, 51*(465), 303–309.

The Physician's Foundation. (2008). *The Physicians' Perspective: Medical Practice in 2008*. The Physician's Foundation. Available online at: http://www.physiciansfoundation.org/FoundationReportDetails.aspx?id=78 (accessed July 12, 2011)

Thomsen, L. A., Winterstein, A. G., Sondergaard, B., Haugbolle, L. S., & Melander, A. (2007). Systematic review of the incidence and characteristics of preventable adverse drug events in ambulatory care. *Annals of Pharmacotherapy, 41*(9), 1411–1426.

Tierney, W. M. (2003). Adverse outpatient drug events—A problem and an opportunity [see comment]. *New England Journal of Medicine, 348*(16), 1587–1589.

Wachter, R. M., & Goldman, L. (1996). The emerging role of "hospitalists" in the American healthcare system. *New England Journal of Medicine, 335*, 514–517.

Wetzels, R., Wolters, R., van Weel, C., & Wensing, M. (2008). Mix of methods is needed to identify adverse events in general practice: A prospective observational study. *BMC Family Practice, 9*, 35.

Yarnall, K. S., Pollak, K. I., Ostbye, T., Krause, K. M., & Michener, J. L. (2003). Primary care: Is there enough time for prevention? *American Journal of Public Health, 93*(4), 635–641.

Zandieh, S. O., Goldmann, D. A., Keohane, C. A., Yoon, C., Batfs, D. W., & Kaushal, R. (2008). Risk factors in preventable adverse drug events in pediatric outpatients. *Journal of Pediatrics, 152*(2), 225–231.

Zyzanski, S. J., Stange, K. C., Langa, D., & Flocke, S. A. (1998). Trade-offs in high-volume primary care practice [see comment]. *Journal of Family Practice, 46*(5), 397–402.

45

In Search of Surgical Excellence: A Work Systems Approach

Douglas A. Wiegmann, Sacha Duff, and Renaldo Blocker

CONTENTS

45.1 Human Error Perspectives 775
    45.1.1 Traditional View of Errors and Safety 776
    45.1.2 Systems Approach to Safety and Error Reduction 776
45.2 Work System Factors and Surgical Performance 776
45.3 Teamwork: The Bad and the Good 778
45.4 Improving Teamwork and Communication during Surgery 779
    45.4.1 Increase Awareness and Attitudes about Teamwork 779
    45.4.2 Provide Didactic and Simulation-Based Teamwork Training 780
    45.4.3 Standardize Communication Phraseology and Timing 780
45.5 Conduct Preoperative Briefings 781
45.6 Enhance Shared Mental Models 782
45.7 Conclusion 783
References 783

Surgical excellence is marked by the ability to manage errors and unexpected events during an operation (de Leval et al., 2000; Wiegmann et al., 2007). However, even experienced surgical teams can be negatively impacted by minor problems that disrupt the flow of a surgical procedure. Specifically, as the number of minor events increases, the ability of a surgical team to cope with major problems decreases significantly (Reason, 2001). The accumulation of minor events appears to diminish the compensatory resources of the surgical team, increasing their vulnerability and susceptibility to committing errors (Carthey et al., 2003). Unfortunately, little is known about the nature and frequency of surgical flow disruptions that impact surgical performance. As a result, developing interventions that improve patient safety is onerous, and errors with serious ramifications continue to occur at high rates in many surgical specialties, including cardiac surgery, vascular surgery, and neurosurgery (Kohn et al., 1999; Gawande et al., 2003).

The goal of the present chapter is to discuss the growing body of literature describing surgical errors and their apparent causes. In doing so, we will first present different viewpoints for understanding errors and why they happen. We will then focus on one particular view, namely, the work systems perspective. We will describe what is known about work system factors in the operating room that disrupt surgical flow and predispose surgical teams to making errors that could potentially cause patient harm. Following this discussion, we then focus on breakdowns in teamwork and communication in the operating room, which is one of the most common work systems factors impacting surgical performance. Finally, we will present several methods or techniques for improving teamwork and communication among surgical teams and evidence of their effectiveness. We conclude the chapter with a brief discussion of the future challenges of addressing work system factors in the operating room.

45.1 Human Error Perspectives

There are two prevailing approaches to addressing human error in complex systems: the traditional- or individual-focused approach and the work systems approach (Reason, 2000). Each approach has its own model of error causation, and each model gives rise to quite different philosophies of error prevention and management. Understanding these differences has important practical implications for developing programs for improving patient safety and surgical care.

45.1.1 Traditional View of Errors and Safety

The traditional approach to patient safety in health care generally focuses on the errors made by a specific individual, blaming them for inattentiveness, forgetfulness, or even incompetence. Historically, the medical community has tended to espouse this individual-centered approach to safety (Reason, 2000; Vincent et al., 2004). Specifically, the culture within most health care settings has been one of censuring individuals for the errors they make. This situation is not too surprising given that the culture of medicine has long focused on perfection and individual accountability for one's actions. For example, within surgical specialties, the primacy of technical skills is the underlying assumption driving rankings of surgical performance across institutions or among one's surgical colleagues. As illustrated in Figure 45.1, patient outcomes, once adjusted for patient risk factors such as disease severity or comorbidities, are presumed to be explained solely by an individual surgeon's skill (Vincent et al., 2004). Hence, when things go wrong or surgical errors are made, it is logical from this perspective to question the particular surgeon's skill or aptitude. Resultant safety interventions, therefore, often take the form of "blame you, shame you, train you" programs that are rarely effective in preventing recurrence, because they fail to focus on the underlying systemic problems that often cause errors (Reason, 2000).

45.1.2 Systems Approach to Safety and Error Reduction

In contrast, a system safety approach suggests that human error is often caused by a combination of work system factors rather than solely the inability of the individual surgeon. Specifically, the work system model (e.g., Carayon et al., 2006) indicates that in addition to surgical skill and patient conditions, performance and outcomes are also impacted by such factors as teamwork and communication, the physical working environment,

technology/tool design, task and workload factors, and organizational variables (see Figure 45.2). According to this perspective, errors are the natural consequences, not causes, of the systemic breakdown among the myriad work system factors impacting performance (ElBardissi et al., 2007). Consequently, patient safety programs are likely to be most effective when they target specific failure points within the system rather than focusing exclusively on the competency of the individual who committed the error (Carthey et al., 2001; Wiegmann et al., 2007).

Table 45.1 provides a more complete and direct comparison between the traditional and work system approaches to safety. Of particular importance is the *focus* of safety interventions that result from each approach. As would be expected, the traditional approach focuses on ways of changing the individual through training, selection, awareness campaigns, and incentives. In contrast, the focus of the work system approach is on the contextual and situational factors that impact behavior and performance. The essence of the work system approach is captured in the adage "you can't change the human condition but you can change the condition under which humans work" (Reason, 2001).

45.2 Work System Factors and Surgical Performance

The work system approach is relatively new to the general health care community; however, there is an increasing awareness of the impact that systemic factors can have on shaping human performance. For example, the role that poorly designed medical devices can play in producing errors that cause patient harm is becoming increasingly apparent (Gosbee & Gosbee, 2005). Roughly half of all medical devices recall result

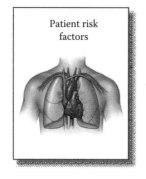  

FIGURE 45.1
The traditional safety approach assumes that outcomes are due to patient risk factors and surgical skill.

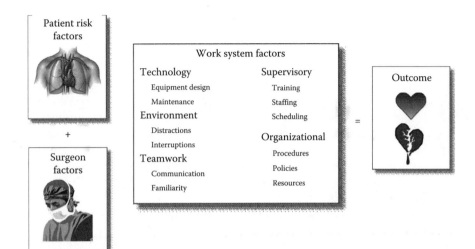

FIGURE 45.2
The system safety approach assumes that work system factors also contribute to outcomes.

TABLE 45.1

Comparison between the Traditional and System-Centered Approaches to Patient Safety

Traditional Perspective	Systems Perspective
Focuses on errors and procedural violations of individuals	Humans are fallible and errors are to be expected, even in the best organizations
Unsafe acts arise from aberrant mental processes such as inattentiveness, lack of good judgment, forgetfulness, recklessness, or even negligence	Errors are consequences of deficient processes and/or system failures rather than simply causes of bad outcomes
Interventions are directly aimed at reducing the "unwanted variability" in human behavior	Consistency in performance is important but flexibility is also invaluable during dynamic operations
Common methods include retraining, recurrent proficiency tests, disciplinary measures, or even termination and litigation	Countermeasures are based on the assumption that "though we cannot change the human condition, we can change the conditions under which humans work"
When taken to the extreme, errors are viewed as a moral issue, with the notion that bad things (i.e., errors) happen to bad people	The best people can make the worst mistakes. The important issue is not who blundered but how and why the defenses failed

Source: Based on Reason, J., *BMJ*, 320, 768, 2000.

from design flaws, with specific types of devices being associated with unusually high use-error rates, such as infusion delivery devices. Social variables, such as teamwork and communication factors, are also of growing concern. The Joint Commission on Health Care Quality and Safety recently reported "communication" as the number one root cause (70%) of reported sentinel events (Greenberg et al., 2007). Other studies also indicate that supervisory and organizational factors, such as shift work scheduling and unit staffing, can impact errors, such as medication errors within intensive care units (Carlton & Blegen, 2006). These findings clearly indicate the importance of the work system approach to understanding human error and patient safety issues across the health care system as a whole.

Within many surgical specialties, the system-centered approach to understanding surgical errors is also

becoming more widely accepted, as illustrated by the following quote:

To put it crudely, good surgical skills coupled with basic team performance and the basic equipment may enable a surgeon to achieve a 90% success rate in a high-risk operation. However, refinements in surgical skill may be a relatively small element in the drive to reduce mortality from 10% to 1%. Optimizing the surgical environment, attention to ergonomics and equipment design, understanding the subtleties of decision making in a dynamic environment, enhancing communication and team performance may be more important than skill when reaching for truly high performance. Poor team performance and inadequate equipment can cause anger and have detrimental effects on performance. Conversely, an excellent team and supportive environment will enable the surgeon to "raise his or her game" with a considerable benefit to the patient.

(Vincent et al., 2004, p. 481)

A growing number of studies have been conducted to identify the nature and frequency of work system factors in the operating room that negatively impact surgical performance or cause surgical flow disruption (deviations from the natural progression of an operation that potentially compromising safety). For example, Wiegmann et al. (2007) found several work system factors in the operating room that significantly impacted surgical flow during cardiac surgery. These "surgical flow disruptions" included miscommunication and poor coordination among the surgical team, distractions and interruptions caused by telephones and pagers or people entering or leaving the operating room, equipment malfunctions, lack of available or accessible resources, and on the job training of surgical staff. Similar findings have been reported in studies of other types of surgical procedures including laparoscopic cholecystectomy (camera-assisted gallbladder removal) and open inguinal and umbilical hernia repair procedures (Duff et al., 2010).

These studies also demonstrate that surgical flow disruptions due to work system factors are not rare. In the study by Wiegmann et al. (2007), surgical flow disruptions occurred 341 times across 31 operations (8.1/h, 11.0/case). Even higher rates of flow disruptions during cardiac surgery were observed by Blocker et al. (2010), who found a total of 898 flow disruptions in 127 h of observations for an average of 8.49 flow disruptions per hour and 42.45 flow disruptions per case. Duff et al. (2010) found that for laparoscopic cholecystectomies, which lasted an average of only 29.25 min, there was an average of 13.29 flow disruptions per minute, or one flow disruption every 2.20 min. Hernia repairs, which lasted roughly 33 min, had an average of 15 flow disruptions per procedure, for a rate of one flow disruption every 2.24 min (Duff et al., 2010).

There is clear evidence that as the number of surgical flow disruptions increases, so does the probability of surgical error. Furthermore, several studies have found that teamwork and communication appear to be the most common cause of surgical flow disruptions and predispose surgical teams to commit errors more than any other work system factors in the operating room (e.g., Wiegmann et al., 2007; Duff et al., 2010). Consequently, the remainder of this chapter will focus on teamwork and communication among surgical teams. We will briefly review what is known about breakdowns in teamwork and communication during surgery, as well as the benefits of good teamwork. We will then discuss methods or techniques for improving teamwork and performance among surgical teams.

45.3 Teamwork: The Bad and the Good

Wiegmann et al. (2007) found that teamwork factors alone accounted for roughly 45% of the variance in the errors committed by surgeons during cardiac cases. Teamwork issues generally clustered around issues of miscommunication, lack of coordination, failures in monitoring, and lack of team familiarity. Duff et al. (2010) also found that the vast majority of all flow disruptions during general surgery (62%) were due to breakdowns in communication and coordination. These findings are not specific to our studies. Another study by Lingard et al. (2004) found that 36% of communication errors in the operating room resulted in visible effects on system processes, which include inefficiency, team tension, resource waste, work-around, delay, patient inconvenience, and procedural error. Poor staff communication has been linked to poor surgical outcomes in general (de Leval et al., 2000; Carthey et al., 2001). For example, a study by Gawande et al. (2003) reported on the dangers of incomplete, nonexistent, or erroneous communication in the operating room, indicating that such miscommunication events were causal factors in 43% of errors made during surgery. Breakdown in teamwork and communication is also the most common cause of sentinel events and wrong-site surgeries (Makary et al., 2006).

Communication errors can manifest in a variety of different ways. Information might be misunderstood or misinterpreted because of wrong assumptions on the part of the receiver. On the other hand, information might simply be ignored or even withheld. In any case, communication errors can lead to delayed or improper treatment (Rosenstein & O'Daniel, 2006). A study by Lingard et al. (2004) identified the most prominent types of communication failures during surgery, which included occasion 45.7%, where timing of an exchange was requested or provided too late to be useful; content 35.7%, during which communication was missing or inaccurate; purpose 24%, where issues were not resolved; and audience 20.9%, in which key individuals were excluded. Lingard et al. (2004) suggest that these weaknesses in communication may derive from a lack of standardization in team integration, or in other words, incomplete and skewed shared mental models. A study by Greenberg et al. (2007) examined the nature of communication breakdowns during surgery. The results of this study showed that 92% of communication breakdowns during surgical care were verbal, and the majority occurred between a single transmitter and a single receiver. However, in verbal communication, spoken words account for only a small fraction of the information exchanged. Other paralinguistic cues such as volume and intonation of speech as well as body language and facial expressions are often more important in conveying the actual meaning of a message (O'Conner et al., 2010). This can present a challenge for health care workers who frequently work with unfamiliar teammates who are wearing gowns and surgical masks that conceal the majority of their facial expressions and body language.

In the most extreme situations, breakdowns in teamwork and communication can occur as a result of conflict among team members. For example, Rosenstein and O'Daniel (2006) found that the most frequent types of disruptive behavior on the part of surgeons were yelling/raising voice (79%), disrespectful interaction (72%), abusive language (62%), berating in front of peers (61%), condescension (55%), and insults (52%). Such disruptive behavior often increased stress (93%), caused frustration (92%), prevented concentration (84%), reduced collaboration (89%), hindered information transfer (86%), impeded communication (89%), and impaired relationships (87%). Disruptive behavior was observed across a variety of care settings. However, not surprisingly, it occurred most often in stressful situations. Unfortunately, it is under stressful situations that good teamwork and communication may be most important to ensuring patient safety.

Clearly, there are many aspects of poor teamwork and communication, and there is a growing body of literature describing ways in which teamwork can breakdown in the operating room. However, there is somewhat less information about what constitutes a "good" team. Still, it is clear that good teamwork is not simply the opposite of bad teamwork. In other words, effective teams do not just avoid the problems associated with bad teams, but they also exhibit specific teamwork skills and behavior that foster good team performance (Westli et al., 2010). These skills include effective task coordination (managing synchronous and/or simultaneous activities to align the pace and sequencing of others' contributions with goal accomplishment), useful information exchange (giving and receiving the knowledge and data necessary for team coordination and task completion), appropriate use of authority (observable behavior of leading the team and/or the task as required or accepting a non-leading role when appropriate), good assessment of capabilities (providing physical, cognitive, and emotional help to team members and seeking help from others when necessary), and positive supporting behaviors (providing physical, cognitive, and emotional help to teammates, and seeking help from others when necessary).

Medical teams that have been observed to exhibit these skills have been found to be more effective. Effective teams experience fewer minor problems during surgery, have better intraoperative performance, and shorter operating times (Catchpole et al., 2007). Good teamwork also results in effective error detection and

management during surgery. For example, Wiegmann et al. (2007) found that nearly half of all errors made during cardiac surgery were detected by another team member who did not make the error, resulting in successful error management processes. This finding is consistent with Catchpole et al.'s (2007) definition of effective teamwork, which is "the ability to reduce the impact of otherwise unavoidable problems." According to Makary et al. (2006), teamwork can actually be used as a meaningful surrogate measure to patient safety because it is such an integral component of the safety culture within the operating room. Clearly, improving teamwork and communication is key to optimizing outcomes of care and preventing adverse events.

45.4 Improving Teamwork and Communication during Surgery

There are several strategies for improving teamwork and performance in the operating room (Wiegmann et al., 2010). We will focus on five foundational approaches. These include increasing awareness and positive attitudes about teamwork, providing didactic and simulation-based teamwork training, standardizing communication phraseology and timing, conducting preoperative briefings, and enhancing teams' shared mental models (Wiegmann et al., 2007; ElBardissi et al., 2008).

45.4.1 Increase Awareness and Attitudes about Teamwork

Surprisingly, research indicates that surgeons generally report being satisfied with the teamwork within the operating room (Makary et al., 2006). Results of a study by Flin et al. (2006) revealed that surgical team members generally demonstrated positive attitudes toward behaviors associated with effective teamwork and safety. However, during a study by Mills et al. (2008), surgeons, nurses, and anesthesiologists were asked separately to rate their perception of teamwork on the dimensions of organizational culture, communication, and teamwork. Surgeons rated these dimensions much more highly than nurses or anesthesiologists.

Results of a study by Makary et al. (2006) revealed a large discrepancy in teamwork ratings between physician and nonphysician caregivers. Physicians were given the lowest overall teamwork ratings; nurses and surgical technologists were rated the highest. Most intragroup ratings were high: each group gave their own group the highest score they received, except Operating Room (OR) nurses who actually received a higher score from the surgeons than they rated themselves, even higher than surgeons rated other surgeons; however, nurses did not reciprocate; in fact, they gave surgeons the lowest rating out of all the caregiver types. Results were similar when anesthesiologists rated OR nurses. Nurses gave the lowest ratings across the board, rating fellow registered nurses (RNs) highest followed by surgical technologists (Makary et al., 2006). An interesting observation by Makary et al. (2006) was the difference between physicians' and nurses' ideas of what constitutes effective teamwork. For nurses, having their input respected was the key to good collaboration. For physicians, however, good collaboration was reflected by the ability of nurses to anticipate their needs and follow instructions (Makary et al., 2006).

Discrepant attitudes about teamwork have been suggested to be a considerable source of nurses' dissatisfaction with their profession, which has contributed to the critical nursing shortage (Makary et al., 2006; Mills et al., 2008). It is difficult, however, to change attitudes and teamwork behavior if surgeons and other team members are unaware of the problem! Consequently, as rudimentary as it might seem, considerable effort is still needed to spread the message that poor teamwork in the operating room remains a major problem that significantly impacts surgical care and patient safety.

45.4.2 Provide Didactic and Simulation-Based Teamwork Training

Simply increasing awareness or changing attitudes about teamwork is not enough, however, if individuals do not possess the skills needed to communicate and work together effectively. Teamwork training, often done as crew resource management (CRM) training, must also be provided. The main principles that CRM are based upon are transparency, standardization, communication, and team skills (Hammon, 2004). Typically, CRM training involves a combination of didactic and simulation-based instruction. The didactic component includes an introduction to the concepts of communication including topics such as maintaining situational awareness, team leadership and role clarification, communication, conflict resolutions, error management, and how to divide task responsibilities during high workload situations.

A series of scenario-based simulations are also utilized that require participants to apply key CRM concepts and function as a team to successfully address problems and ensure safe patient outcomes. Following each simulation exercise, an instructor typically facilitates the group's debriefing of the event, which often includes viewing a videotape of the group's behavior and a discussion of the good and bad aspects of the team's performance. One example of an advanced full team simulator is team-oriented medical simulation

(TOMS)—the training simulation involves performing a laparoscopic procedure on an instrument-controlled mannequin with sensors to assess proficiency (Helriech & Davies, 1996). This training is performed as a full OR simulation with orderlies, surgical and anesthetic consultants and registrars, and nurses. The training session includes a 1 h briefing, a 1 h simulated laparoscopic surgical procedure, and a 1 h debriefing. According to the developers, one of the most important aspects of TOMS and other successful training methods is the performance-based feedback. Proper development of full team simulator-based training methods can help to synchronize the knowledge of everyone involved in patient care (Helriech & Davies, 1996).

Team training in other domains, such as commercial aviation, shows that when didactic teamwork training is combined with dynamic, scenario-based simulations, positive reactions to teamwork concepts occur as well as an increase in knowledge of teamwork principles and improved teamwork performance (Baker et al., 2008). Furthermore, there is growing scientific evidence indicating that the gains achieved during training in critical teamwork-related competencies transfer to workplace environments, provided the application of learned skills by the trained individuals is encouraged and reinforced (Salas et al., 2006). In general, however, the impact of CRM on reducing surgical errors is difficult to establish because of the relatively low rate of occurrence of adverse events that actually cause patient harm. As a result, researchers instead have generally relied on surrogate measures, such as improvements in teamwork-related knowledge, skills, and attitudes (KSA) and demonstrations of CRM skills during surgical simulations as measures of CRM effectiveness.

45.4.3 Standardize Communication Phraseology and Timing

Standardized phraseology and communication protocols are common in high-risk industries where communication is vital to safe operations. For example, within the aviation industry, standard phraseology and protocol-drive communication (e.g., the repetition of verbal instructions or "call backs") are required to be used by pilots and air traffic controllers to decrease communication errors. Wadhera et al. (2010) sought to assess the impact of similar communication protocols and standardized phraseology on communication failures between surgeons and perfusionists during cardiac surgery. These two surgical team members were selected because they must communicate effectively to successfully operate the technologically complex cardiopulmonary bypass machine (heart–lung machine) during an operation. According to Wadhera et al., "along this communication axis, there is potential for catastrophic miscommunication as a result of divergent interpretations of colloquial language, vocal accents, differing levels of experience, and lack of familiarity."

The standardized communication protocol that was implemented focused on several critical steps in operating the heart–lung machine such as the establishment of the activated clotting time (ACT), initiation of cardioplegia, and the initiation and termination of cardiopulmonary bypass. The protocol required communication during the performance of these activities to be specific to the case, for example, the communication of numeric levels of ACT rather than vague or generic statements such as "ACT adequate for bypass." Callbacks of all communication exchanges was also instituted as a standard procedure during surgery. To assess the effectiveness of this new communication protocol, the researchers observed communication problems between cardiac surgeons and perfusionists before and after implementation. Results of the study found a clear, major reduction in communication breakdowns per case after the implementation of this standardized communication protocol (Wadhera et al., 2010).

Establishing policies that limit noncase-related discussions among surgical team members may be another strategy for improving communication. Such rules generally prohibit casual discussions during critical phases of an operation when workload is the highest and the ability to mentally focus is most critical. In aviation, this policy is referred to as the "sterile cockpit rule." It was imposed by the Federal Aviation Administration (FAA) after reviewing a series of accidents that were caused by flight crews who were distracted from their flying duties by engaging in nonessential conversations and activities during critical parts of the flight (FAA, 1981). However, there are potential "problems" with applying the sterile cockpit rule to surgery. Specifically, a study of mental workload during cardiac surgery found considerable amount of variability across surgical staff and phases of operations in terms of the timing of tasks that imposed high mental workloads (Wadhera et al., 2010). For example, during the time of intubation, workload is high for the anesthesiologist and nurse anesthetist, but relatively low for the surgeon and the perfusionist. In contrast, during the time of placing the patient onto the cardiopulmonary bypass machine, workload is high for surgeons and perfusionists but not for other surgical staff. Consequently, identifying critical phases of the operation for applying the sterile cockpit rule may be difficult. Specifically, if the sterile cockpit rule were applied consistently across tasks performed by all surgical staff, the result of the policy may be no different than imposing restrictions throughout the entire operation.

45.5 Conduct Preoperative Briefings

The use of preoperative briefings or "team huddles" as an intervention to improve teamwork and safety is a long tradition within a many high-risk industries including aviation, nuclear power, and military operations. Briefings are also becoming more common across multiple health care settings. Team meetings, such as preoperative briefings, are conducted prior to an operation and provide team members the opportunity to ask questions, clarify uncertainties, and become better acquainted with the rest of the surgical team. Briefings also help team members gain confidence in raising concerns prior to the operation, so that they are also more willing to express concerns during surgery if they identify a threat to patient safety. Briefings are more than the traditional universal protocol or presurgical pause that is designed to ensure the right patient, right site, and right procedure. Rather, briefings are meetings that are often conducted prior to the patient entering the operating room and involve a more in depth review of the case. Preoperative briefings therefore have the potential to address a greater breadth of communication and teamwork issues.

Einav et al. (2010) recently reported the results of a study on preoperative briefings and patient care during gynecological and orthopedic surgery. They found a 25% reduction in the average number of errors or "nonroutine events" that occurred within surgical cases and a significant increase in the number of surgical cases in which no nonroutine events occurred. These results are similar to those reported by DeFontes and Surbida (2004) who developed a preoperative briefing protocol for use by general surgical teams. The protocol was similar to a preflight briefing used by the airline industry and was evaluated over a 6 month period. Results revealed a significant reduction in wrong-site surgeries after implementation of preoperative briefings. Within cardiac surgery, Henrickson et al. (2009) observed significant reductions in the frequency of miscommunication events and delays waiting for appropriate instrumentation as well as a noticeable decrease in waist following the implementation of preoperative briefings. Lingard et al. (2008) found similar benefits of preoperative briefings within the context of laparoscopic surgery (Lingard et al., 2008).

The World Health Organization (WHO) recently endorsed the use of preoperative briefings as a method for improving communication and patient safety. Nonetheless, the utilization of briefings remains relatively low within many surgical specialties. This is likely due to the fact that there are no standardized protocols for conducting preoperative briefings that accommodate all types of surgery. Other barriers impeding the utilization of preoperative briefings also exist. These include individual's negative attitudes or resistance to change by surgical staff, as well as organizational barriers such as case schedules, lack of facilities, and limited resources. Consequently, the successful development of a preoperative briefing protocol takes several months to complete. In particular, considerable amount of research needs to be done to clearly understand the multiple needs and views of key stakeholders (i.e., surgical staff) as well as the nuances of the organization in which such briefings are to take place (DeFontes & Surbida, 2004).

45.6 Enhance Shared Mental Models

Carroll and Olsen (1988) define mental models as the knowledge of the components of a system, their interconnection, and the processes that change them; this forms the knowledge that provides the basis for users to be able to construct reasonable actions and explanations about why a set of actions is appropriate. An individual holds mental models for each component in a system, including devices, software systems, processes, procedures, roles, and responsibilities. The collective understanding of the components and how they function together is the individual's "mental model," or understanding of the system, why it does what it does, the contribution of each of its components, and its expected response based on the given conditions and inputs. These models develop as each individual forms knowledge, experience, and beliefs about the system. The closer a person's mental model represents the actual system the better equipped an individual is to diagnose and remedy problems within the system when they occur.

The role of mental models in team performance serves to provide a common base to which each dimension of teamwork is tethered. Cannon-Bowers and Salas (1990) describe shared mental models as a set of organized expectations for performance, from which accurate, timely predictions can be drawn. In effective teams, members have similar or compatible knowledge that they use to guide their coordinated behavior. When team members share knowledge, they can interpret cues in a similar manner, make compatible decisions, and take appropriate actions during complex tasks. The degree to which the team's mental models are congruent can be described as shared or overlapping, meaning that some information is shared between two or more team members. When teams exhibit behaviors consistent with this, such as anticipating needs and coordinating without verbal communication, they are demonstrating the affordances that shared mental models permit.

There is an empirical relationship between the quality of a team's shared mental model and effective team performance. For example, the ability to adapt quickly can be considered a marker of proficient teams (Mathieu et al., 2000). Shared mental models facilitate adaptability by allowing team members to anticipate the needs of others and to effectively coordinate their actions (Mathieu et al., 2000). During periods of high workload, shared mental models can be especially important because communication becomes more difficult, and adaptability becomes a necessity (Mathieu et al., 2000).

In the case of complex tasks, the amount of information necessary to complete the task is too great to be held by any one individual. Each team member has a highly enriched individual mental model of the tasks that they are personally responsible for, and a less complete mental model of their teammates' roles and responsibilities. In a complex system, team members will have multiple mental models activated at any given time, some are likely to be shared, others will not (Mathieu et al., 2000). These mental models overlap at the points where team coordination is necessary to achieve a successful outcome. However, it is unrealistic and impractical for all team members to have identical mental models; generally, the more understanding the team shares about the task and one another, the better they will be at coordinating their respective skill sets to achieve successful surgical outcomes.

Sharing information and verifying the correctness of held beliefs are important for effective shared mental model development. Preoperative briefings and the standardization of communication discussed previously are also techniques for building shared mental models among surgical teams. However, new health information technology (HIT) has also been recently developed to improve shared mental models and teamwork in settings that involve dynamic, collaborative interactions among multidisciplinary health care teams such as surgery. This new technology is called the "wall of knowledge" (WOK) or the OR dashboard, consisting of large wall-mounted flat-panel monitors that integrate critical case-related data, patient information, and video streams for simultaneous viewing by an entire surgical team. The WOK is designed to create a visually integrated operating room (VIOR) that facilitates the development and synchronous updating of a "shared mental model," or a common understanding of the OR team regarding the dynamic changes occurring during a procedure (Kurtz, 2008).

To date, however, there have been no empirical studies to evaluate the actual impact that the VIOR has on team cognition, coordination, or surgical performance. Albeit, anecdotally, there are reports of several patient safety improvements through the use of the WOK, including enhanced communication, fewer errors, and a reduced number of sentinel events (e.g., retained foreign bodies, wrong-site surgeries, and allergic reactions). Furthermore, related research on other new display technologies, such as the electronic white board, suggests that errors that impact patient care can be reduced when information is integrated and viewed by the entire health care team (Xiao et al., 2007). To the contrary, however, the introduction of new technology can also have unexpected consequences that can potentially induce new forms of error, which negatively affect patient safety (Ash et al., 2004; Harrison et al., 2007). New technology often requires adjustments in team communication, the development of new procedures, and altered roles of personnel that may prove problematic, at least initially. Consequently, efforts need to be made to better understand how collaborative work and surgical performance may be affected by the VIOR in order to inform decisions regarding the effective design and implementation of this new technology.

45.7 Conclusion

Within complex systems, there are several "safety nets" or "checkpoints" (Reason, 2001), which are intended to be "layered barriers to error" (Wan et al., 2009). Occasionally, these barriers are not upheld due to some failure in the system including breakdowns in teamwork and communication. Considering how the entire work system contributes and supports teamwork is important to improving patient safety more so than simply blaming the person who makes an error. We have provided only a few examples in this chapter of providing support for surgical teams and modifying the system to enhance performance. Our hope is that through these examples, we have successfully conveyed the notion that "the systems approach can lead to greater and more sustained benefit than by paying yet more attention to obvious, but often irreversible, human error" (Catchpole et al., 2007).

References

Ash, J.S., Berg, M., & Coiera, E. (2004). Some unintended consequences of information technology in health care: The nature of patient care information system-related errors. *Journal of the American Informatics Association*, 11(2), 104–112.

Baker, D.P., Gustafson, S., Beaubien, J., Salas, E., & Barach, P. (2008). Medical teamwork and patient safety: The evidence-based relation. Agency for Healthcare Research and Quality. Retrieved July 1, 2008, from www.ahrq.gov/qual/medteam

Blocker, R.C., Eggman, A., Zemple, R., Wu, C.E., & Wiegmann, D.A. (2010). Developing an observational tool for reliably identifying work system factors in the operating room that impact cardiac surgical care. *Human Factors and Ergonomics Society Annual Meeting Proceedings, Health Care*, 5, 879–883.

Cannon-Bowers, J.A. & Salas, E. (1990). Cognitive psychology and team training: Shared mental models in complex systems. Paper presented at the 5th *Annual Meeting of the Society of Industrial and Organizational Psychology*, April 19–22, Miami, FL.

Carayon, P., Hundt, A.S., Karsh, B.-T., Gurses, A.P., Alvarado, C.J., & Smith, M. (2006). Work system design for patient safety: The SEIPS model. *Quality & Safety in Health Care*, 15(Suppl. I), i50–i58.

Carlton, G. & Blegen, M.A. (2006). Medication-related errors: A literature review of incidence and antecedents. *Annual Review of Nursing Research*, 24, 19–38.

Carroll, J.M. & Olson, J.R. (1988). Mental models in human–computer interaction: Research issues about what the user of software knows. In M. Helander (Ed.), *Handbook of Human–Computer Interaction*, North Holland, Amsterdam, the Netherlands, pp. 45–65.

Carthey, J., de Leval, M.R., & Reason, J.T. (2001). The human factor in cardiac surgery: Errors and near misses in a high technology medical domain. *Annual of Thoracic Surgery*, 72, 300–305.

Carthey, J., de Leval, M.R., Wright, D.J., Farewell, V.T., & Reason, J.T. (2003). Behavioural markers of surgical excellence. *Safety Science*, 41(5), 409–425.

Catchpole, K.R., Giddings, A.E., Wilkinson, M., Hirst, G., Dale, T., & de Leval, M.R. (2007). Improving patient safety by identifying latent failures in successful operations. *Surgery*, 142, 102–110.

DeFontes, J. & Surbida, S. (2004). Preoperative safety briefing project. *Permanente Journal*, 8, 21–27.

de Leval, M.R., Carthey, J., Wright, D.J., Farewell, V.T., & Reason, J.T. (2000). Human factors and cardiac surgery: A multicenter study. *Journal of Thoracic and Cardiovascular Surgery*, 119(4), 661–672.

Duff, S.N., Windham, T.C., Wiegmann, D.A., Kring, J., Schaus, J.D., Malony, R., & Boquet, A. (2010). Identification and classification of flow disruptions in the operating room during two types of general surgery procedures. *Human Factors and Ergonomics Society Annual Meeting Proceedings, Health Care*, 5, 884–888.

Einav, Y., Gopher, D., Kara, I., Ben-Yosef, O., Lawn, M., Laufer, N., Liebergall, M., & Donchin, Y. (2010). Preoperative briefing in the operating room: Shared cognition, teamwork, and patient safety. *Chest*, 137, 443–449.

ElBardissi, A.W., Wiegmann, D.A., Dearani, J.A., & Sundt, T.M. (2007). Application of the human factors analysis and classification system methodology to the cardiovascular surgery operating room. *Annals of Thoracic Surgery*, 83, 1412–1419.

ElBardissi, A.W., Wiegmann, D.A., Henrickson, S., Wadhera, R., & Sundt, T.M. (2008). Identifying methods to improve heart surgery: An operative approach and strategy for implementation on an organizational level. *European Journal of Cardiothoracic Surgery*, 34(5), 1027–1033.

Federal Aviation Administration (FAA). (1981). *Flight Crew Duties. Federal Aviation Regulations Part 121.54*. Author, Washington, DC.

Flin, R., Yule, S., Mckenzie, L., Paterson-Brown, S., & Maran, N. (2006). Attitudes to teamwork and safety in the operating theatre. *Surgeon*, 4(3), 145–151.

Gawande, A.A., Zinner, M.J., Studdert, D.M., & Brennan, T.A. (2003). Analysis of errors reported by surgeons at three teaching hospitals. *Surgery*, 133, 614–621.

Gosbee, J.W. & Gosbee, L.L. (2005). *Using Human Factors Engineering to Improve Patient Safety*, Joint Commission Resources, Oak Brook, IL.

Greenberg, C.C., Roth, E.M., Sheridan, T.B., Gandhi, T.K., Gustafson, M.L., Zinner, M.J., & Dierks, M.M. (2007). Making the operating room of the future safer. *The American Surgeon*, 72(11), 1102–1108.

Hammon, W.R. (2004). The complexity of team training: What we have learned from aviation and its applications to medicine. *Quality & Safety in Health Care*, 13, i72–i79.

Harrison, M., Koppel, R., & Bar-Lev, S. (2007). Unintended consequences of information technologies in health care—An interactive sociotechnical analysis. *Journal of the American Medical Informatics Association*, 14(5), 542–549.

Helriech, R.L. & Davies, J.M. (1996). Human factors in the operating room: Interpersonal determinants of safety, efficiency and morale. *Bailliere's Clinical Anaesthesiology*, 10(2), 277–295.

Henrickson, S.E., Wadhera, R.K., Elbardissi, A.W., Wiegmann, D.A., & Sundt, T.M., III. (2009). Development and pilot evaluation of a preoperative briefing protocol for cardiovascular surgery. *Journal of the American College of Surgeons*, 208(6), 1115–1123.

Kohn, L., Corrigan, J., & Donaldson, M. (1999). *To Err Is Human: Building a Safer Health System* (Institute of Medicine Report), National Academy Press, Washington, DC.

Kurtz, R. (2008). *Wall of Knowledge Informs OR Team*, OR Manager, Inc., Santa Fe, NM.

Lingard, L., Espin, S., Whyte, S., Regehr, G., Baker, G.R., Reznick, R., Bohnen, J., Orser, B., Doran, D., & Grober, E. (2004). Communication failures in the operating room: An observational classification of recurrent types and effects. *Quality & Safety in Health Care*, 13, 330–334.

Lingard, L., Regehr, G., Orser, B., Reznick, R., Baker, G.R., Doran, D., Espin, S., Bohnen, J., & Whyte, S. (2008). Evaluation of a preoperative checklist and team briefing among surgeons, nurses, and anesthesiologists to reduce failures in communication. *Archives of Surgery*, 143(1), 12–17.

Makary, M.A., Sexton, B.J., Freischlag, J.A., Holzmueller, C.G., Millman, E.A., Rowen, L., & Pronovost, P.J. (2006). Operating room teamwork among physicians and nurses: Teamwork in the eye of the beholder. *Journal of the American College of Surgeons*, 202(5), 746–752.

Mathieu, J.E., Heffner, T.S., Goodwin, G.F., Salas, E., & Cannon-Bowers, J.A. (2000). The influence of shared mental models on team process and performance. *Journal of Applied Psychology*, 85(2), 273–283.

Mills, P., Neily, J., & Dunn, E. (2008). Teamwork and communication in surgical teams: Implications for patient safety. *Journal of the American College of Surgeons*, 206(1), 107–112.

O'Conner, T., Papenikolaou, V., & Keogh, I. (2010). Safe surgery, the human factors approach. *The Surgeon*, 8, 93–95.

Reason, J. (2000). Human error: Models and management. *British Medical Journal*, 320, 768–770.

Reason, J. (2001). Heroic compensations: The benign face of the human factor. *Flight Safety Australia*, 5(1), 29–31.

Rosenstein, A.H. & O'Daniel, M. (2006). Impact and implications of disruptive behavior in the perioperative area. *Journal of the American College of Surgeons*, 203(1), 96–105.

Salas, E., Wilson, K.A., Burke, C.S., & Wightman, D.C. (2006). Does crew-resource management training work? An update, an extension, and some critical needs. *Human Factors*, 48(2), 392–412.

Vincent, C., Morrthy, K., Sarker, S.K., Chang, A., & Darzi, A.W. (2004). Systems approaches to surgical quality and safety from concept to measurement. *Annals of Surgery*, 239(4), 475–482.

Wadhera, R.A., Henrickson, P.S., Burkhart, H.M., Greason, K.L., Neal, J.R., Levenick, K.M., Wiegmann, D.A., & Sundt, T.M. (2010). Is the "sterile cockpit" concept applicable to cardiovascular surgery critical intervals or critical events? The impact of protocol-driven communication during cardiopulmonary bypass. *Thoracic and Cardiovascular Surgery*, 139, 312–319.

Wan, W., Le, T., Riskin, L., & Macario, A. (2009). Improving safety in the operating room: A systematic literature review of retained surgical sponges. *Current Opinion in Anaesthesiology*, 22, 207–214.

Westli, H.K., Johnsen, B.H., Eid, J., Rasten, I., & Brattebo, G. (2010). Teamwork skills, shared mental models, and performance in simulated trauma teams: An independent group design. *Scandinavian Journal of Truman, Resuscitation and Emergency Medicine*, 18, 47.

Wiegmann, D., Eggman, A., ElBardissi, A., Henrickson-Parker, S., & Sundt, T., III. (2010). Improving cardiac surgical care: A work systems approach. *Applied Ergonomics*, 41(5), 701–712.

Wiegmann, D.A., Elbardissi, A.W., Dearani, J.A., Daly, R.C., & Sundt, T.M. (2007). Disruptions in surgical flow and their relationship to surgical errors: An exploratory investigation. *Surgery*, 142, 658–665.

World Health Organization's patient-safety checklist for surgery. (2008). *Lancet*, 372(9632), 1. PubMed Index: 18603137.

Xiao, Y., Schenkel, S., Faraj, S., Mackenzie, C., & Moss, J. (2007). What whiteboards in a trauma center operating suite can teach us about emergency department communication. *Annals of Emergency Medicine*, 50(4), 387–395.

46

Human Factors and Ergonomics in Medication Safety

Elizabeth Allan Flynn

CONTENTS

46.1 Medication System...785
46.2 Review of Significance of the Error Problem..786
46.3 Medication Safety Problems Involving Automation 787
46.4 Human Factors Affecting Medication Safety: Work Environment 788
46.5 Application...790
46.6 Conclusion 790
References...790

The application of human factors principles to medication safety is the focus of this chapter. Medication safety has been defined as "freedom from accidental injury during the course of medication use; activities to avoid, prevent, or correct adverse drug events (ADEs) which may result from the use of medications" (American Hospital Association 2002). Accurate performance of work is one critical outcome that human factors engineers strive for as they design or modify systems. Medication safety experts and human factors engineers should join forces and work together to protect patients from "a series unfortunate events" that can harm them.

46.1 Medication System

The medication distribution system is a "target-rich" environment for human factors engineers. Health care providers are expected to manage a patient's medication therapy using technologies that do not always consider their capacities and limitations in an environment that is hardly serene. Physicians diagnose patients and then may prescribe one of thousands of different medications selected by considering the patient's concurrent illnesses, calculating doses, and anticipating side effects. Mistakes at any of these steps could harm a patient. The environments in which physicians practice include noisy intensive care units, clinics, and a long-term care facility. Prescribers can consult with the pharmacist regarding the optimal drug regimen, or use an electronic drug information source. Laboratory test results may help determine the best drugs and doses—the test

results provided to the prescriber should be in a format that highlights abnormalities.

Pharmacists review and evaluate the prescriptions written by prescribers and typically enter each order into a computer system. The system helps organize the patient's medication therapy, so that all health care providers use the same list of medications in the hospital setting. Computer system software in hospital and community pharmacy settings compares the drug to the patient's other medications and predicts the risk and severity of drug interactions and evaluates the dose prescribed for appropriateness. The pharmacist can (and usually does) override the interaction due to time pressures (Murphy et al. 2004). Labels are printed for the new order, and a technician retrieves the medication for a pharmacist to inspect prior to delivering to the nursing unit. A common alternative system is for the nurse to retrieve the medication from an automated drug-dispensing device in the patient care area. In the community pharmacy setting, the prescription is filled and dispensed directly to the patient with a label that instructs the patient how to take the medication.

Nurses prepare and administer medications to inpatients and are recognized as the last defense in the system against errors. Hospitals typically store medications on the nursing unit in automated devices. Nurses retrieve medications based on orders that have been approved by pharmacists and are available to administer to the patient—the automated devices are programmed to allow access only to drugs that have been reviewed and approved by the pharmacist. Some emergency medications may be retrieved by overriding system alerts when an order has not been approved by a pharmacist. The design of effective alerts is a challenge for human factors engineers because of time pressures placed on nurses by demanding workload and the need to balance patient safety with limited time.

Ambulatory patients with a wide variety of educational backgrounds and abilities attempt to adhere to medication regimens that can be complex. The need for labeling and packaging that displays instructions in an easy-to-read manner is critical. Target pharmacy supported the reengineering of the prescription vial and label in order to help the patients comply with their medications (Institute for Safe Medication Practices 2005).

46.2 Review of Significance of the Error Problem

One of the earliest studies of medication errors in hospitals was by Miriam A. Safren and human factors pioneer Alphonse Chapanis (Safren and Chapanis 1960). Using questionnaires, they used the critical incident technique to study the causes of errors reported by nurses (primarily student nurses). A total of 178 incidents were reported on special questionnaires over a 7 month period. Common causes cited by hospital staff who discovered the error or were familiar with the circumstances were (frequency of cause in parentheses) as follows:

1. Failure to follow required checking procedures (68)
2. Misreading or misunderstanding written communication (41)
3. Transcription errors (34)
4. Medicine tickets misfiled in ticket box (15)
5. Calculational errors (11)
6. Errors due to improvisational errors (5)
7. Inaccurately labeled drugs (4)
8. Errors in assigning patients (3)
9. Misunderstood verbal communication (2)
10. Miscellaneous (4)

Those familiar with current medication error causes may recognize some or all of these issues as current problems that remain 50 years after they were initially reported. The authors examined the causes and proposed four sets of recommendations to address these problems involving written communications (including sound-alike, look-alike drug names and legibility of orders), medication procedures (including patient identification and checks by a second person), working environment (especially interruptions and distractions),

TABLE 46.1

System Failures Associated with ADEs

System Failure	Attributed Errors No. (%)
Drug knowledge dissemination	98 (29)
Dose and identity checking	40 (12)
Patient information availability	37 (11)
Order transcription	29 (9)
Allergy defense	24 (7)
Medication order tracking	18 (5)
Interservice communication	17 (5)
Device use	12 (4)

Source: Leape, L.L. et al., *JAMA*, 274, 35, 1995.

and training and education (regarding causes of errors and prevention methods) (Part II). The effect of tall-man lettering on errors with look-alike drug names has been studied by Filik et al. (2006). Standards for work environments where medications are managed have been set by the U.S. Pharmacopeia (USP—discussed at the end of this chapter).

Thirty-five years after the Safren and Chapanis study, Leape et al. (1995) conducted a systems analysis to determine which aspects of the medication system are related to ADEs. The system failures identified as being related to one or more ADEs are listed in Table 46.1. The proximal causes of the errors are listed in Figure 46.1. Human factors engineers can use this information to identify ways to prevent errors from occurring as a result of such system failures. For example, the reasons for "Faulty drug identity checking" can be identified by observing the process of checking medications and solutions developed to optimize accurate drug identity checking.

Medication errors can occur at any stage of the medication system. Prescribing errors were measured at a rate of 6% of orders in a 700-bed academic medical center (Bobb et al. 2004). Pharmacist prescription-filling error rates in observational studies range from 2% to 24% in several studies (Flynn et al. 2009). A national study of

- Lack of knowledge of the drug
- Lack of information about the patient
- Rule violations
- Slips and memory lapses
- Transcription errors
- Faulty drug identity or dose checking
- Faulty interaction with other services
- Infusion pump and parenteral delivery problems
- Inadequate monitoring
- Drug stocking and delivery problems
- Preparation errors
- Lack of standardization

FIGURE 46.1

Proximal causes of errors in a systems analysis. (Adapted from Leape, L.L. et al., *JAMA*, 274, 35, 1995.)

prescription dispensing accuracy in 50 pharmacies in six large cities found a 98% accuracy rate using disguised shoppers, which may make it seem like there is not a problem until it is multiplied by the 3 billion prescriptions that are filled each year in the United States (Flynn et al. 2003). Nurse medication administration error rates in hospitals have been measured between 10% and 18% using the observation method of error detection. This translates to approximately 1 error per patient per day in the hospital (Barker et al. 2002). Patient harm or potential harm from inpatient ADEs occurs at a rate of approximately 6% of errors (Barker et al. 2002) and 6.5% of nonobstetrical admissions (Bates et al. 1995).

Medication errors by patients with their prescribed drug regimen are described in the medication compliance and adherence literature. Patients are being treated at home as often as feasible, and complex regimens place demands on patients that require special training and can benefit from tools that make it easier for them to avoid making mistakes and being hospitalized. One example of a tool is an electronic pill box that alerts patients when it is time to take a dose. Studies have found that patients are 30%–60% compliant with their medication regimens (Meichenbaum and Turk 1987).

46.3 Medication Safety Problems Involving Automation

Technology is viewed as providing protection against errors, and large capital investments are made thinking that safety will be improved (along with charge capture and efficiency). However, design problems limit the impact of technology on medication safety. For example, a prescription verification system in a community pharmacy alerted a pharmacist that the medication she was inspecting was different from the one prescribed. A photograph of one of the medication containers was displayed on the computer screen, but this did not provide enough information to the pharmacist to determine what the difference was, so the pharmacist performed an override and approved the wrong form of the medication. Metrocream was approved instead of Metrogel. The difference should be highlighted in a way that makes it readily apparent to the inspector what the problem is—in this situation, a photograph of the correct medication package next to the incorrect one. The manufacturer has distinctive packaging for each form of medication that makes it easy to determine the form of medication (gel or cream) (Flynn and Barker 2006).

Attempts to decrease medication errors have been made, particularly through the use of technology. Has the implementation of new technology had a significant effect on medication errors? There is evidence supporting a positive effect as well as the lack of an effect. In a review of the impact of computerized prescriber order entry (CPOE) on medication safety compared to handwritten orders, 8 of 10 studies reported a significant decrease in total prescribing errors, 3 of 7 studies (43%) reported decrease in dosing errors, and 3 of 8 studies (38%) found a decrease in ADEs (Shamilyan et al. 2008).

Overrides of alerts regarding potentially harmful drug-related problems by prescribers in the ambulatory setting were studied by Isaac et al. (2009). Electronic prescriptions entered into a computer system generated alerts on 6.6% of orders; clinicians accepted (i.e., did not override and ignore)

- 9.2% of drug interaction alerts (high-severity interactions = 61.6% of alerts)
- 2%–43% of high-severity alerts accepted (depending on drug classes involved)
- 23.0% of allergy alerts

There is a need to develop and study effective signals that will ensure patient safety. It appears it may be too easy to override warnings in current systems, and clinicians report that the excessive number of alerts is annoying (Murphy et al. 2004).

The problem of nurse overrides of automated drug dispensing system warnings has also been evaluated. The Joint Commission, which accredits health care organizations, states that overrides can be performed if a delay in administering the medication would harm the patient, or if there is a sudden change in clinical status.

Oren et al. (2002) detected 10 medication errors per day that were associated with nurse overrides of automated alerts and 21% of overrides resulted in errors, but no ADEs were detected. Kowiatek et al. (2006) found that 34 overrides led to errors in a 6 month period in 2001; 2 serious errors and 13 errors required additional patient monitoring. Miller et al. (2008) found that 9 out of 59 overrides were involved in inappropriate medication administrations, and the rate of scanning bar codes to verify medication accuracy during emergency situations was lower than nonemergent (60% compared to 97%). What can be changed in the design of the devices and systems to increase their appropriate application during emergencies?

Nine percent of hospitals use bar code verification technology to double-check medication accuracy at the patient's bedside prior to nurse administration of the medication (Pedersen et al. 2006). Bar code medication verification by nurses prior to administering the medication has been evaluated using observation to detect errors (Paoletti et al. 2007). Medication accuracy rates were measured on three nursing units, one of which

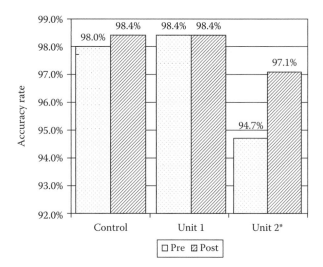

FIGURE 46.2
Effect of a bar code medication verification system on medication administration errors excluding wrong time and wrong technique errors. *$p=0.04$. (Data from Paoletti, R.D. et al., *Am. J. Health Syst. Pharm.*, 64, 536, 2007.)

served as a control group during the postimplementation phase of the study and did not install the bar code system. Figure 46.2 displays the results. The medication system on one nursing unit improved its accuracy rate while the rate on the control unit and second study unit stayed the same. Scanning compliance over an 18 month period averaged 90% and the medication administration error rate averaged 5%. Nurse managers followed up with nurses with low scanning compliance, which was a cited reason for maintaining a low error rate.

Cummings and colleagues of the University Hospitals Consortium (UHC) published recommendations for the use of bar code systems in the medication process. They recommended the use of an ongoing

noncompliance or exception report to maximize scanning, determining problem areas, and improving the overall process (Cummings et al. 2005). These noncompliance reports can reveal potential errors. Patterson and colleagues found that electronic reports cannot capture all noncompliance with bar code procedures and documented workarounds using observation that can be employed to circumvent safety-related tasks (Patterson et al. 2006).

Patterson et al. (2004) published 15 best practice recommendations for successful use of bar code medication administration systems. The recommendations are based on observations by human factors engineers of nurses using bar code systems. The recommendations are listed in Figure 46.3. The publication describes the recommendations in detail, along with the rationale and supporting data.

46.4 Human Factors Affecting Medication Safety: Work Environment

The effect of each aspect of the work environment on medication safety is recognized, but there are few studies that evaluate the relationship between environmental factors and a measurable indicator of medication safety. The key studies that are available are described here.

Illumination has been studied to determine the impact of different levels on prescription dispensing accuracy. Buchanan et al. (1991) evaluated three illumination levels in a busy outpatient pharmacy using randomized assignment of levels to each day for 21 days of observation. A total of 10,888 were analyzed during the study. The prescription dispensing error rates and

1. Put in place a standing interdisciplinary committee.
2. Train all users. Cross-train pharmacists and certain physicians.
3. Communicate known problems.
4. Display contact information for resources to resolve different types of problems.
5. Do not employ a double-documentation system.
6. Schedule planned downtimes to minimize disruptions.
7. Replace malfunctioning equipment during its servicing.
8. Develop a procedure for cleaning BCMA-related equipment.
9. Scan wristbands and medications prior to medication administration.
10. Caregivers should personally document at the time of medication administration.
11. Verify allergy information displayed in BCMA prior to administration.
12. Support staff personnel should print a report at the beginning of a shift for nurses to use as an overview worksheet.
13. Nurses should print missed medication reports once a shift.
14. Alert nurses to new stat (urgent) orders.
15. Replace wristbands as needed and periodically in long-term care.

FIGURE 46.3
Best practice recommendations for successful bar code medication administration. (Adapted from Patterson, E.S. et al., *Jt Comm. J. Qual. Saf.*, 30, 355, 2004.)

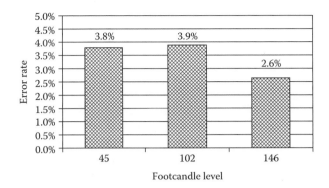

FIGURE 46.4
The effects of three levels of illumination on prescription dispensing error rates in a high-volume outpatient pharmacy (n = 10,888 prescriptions). (Data from Buchanan, T.L. et al., *Am. J. Hosp. Pharm.*, 48, 2137, 1991.)

their corresponding illumination levels are displayed in Figure 46.4. A significant decrease in dispensing error rate was not achieved until the illumination level was increased to 146 foot candles—the error rate decreased from 3.9% to 2.6%.

Illumination levels below 94 foot-candles were associated with significantly more errors than higher levels in a study comparing prescription dispensing error rates measured on 1 day (Flynn et al. 2002). This study was conducted for 1 day at each of 50 pharmacies in 6 cities. The overall dispensing error rate was 1.7% for all 50 pharmacies combined on 4481 prescriptions checked.

The Illuminating Engineering Society of North America (IESNA) has published a guide for architects to assist with the design of lighting in hospitals and health care facilities (IESNA 2006). The illumination levels mentioned in medication-related areas are lower than those identified in the research described here, but were not established based solely on error prevention.

Another interesting factor explored for an association with dispensing errors was the spacing of drug stock bottles on shelves (Flynn et al. 2002). If there was less than 1 in. of space in between containers, the pharmacy was classified as having "tightly packed shelves"; if 1 in. or greater, the pharmacy stock was "spaced." The results found a significant difference between spacing categories ($p = 0.048$)—66% of content errors were made in pharmacies with tightly packed shelves (Flynn et al. 2002). This may also be an indicator of clutter.

Workload pace can affect work environment and subsequently affect the worker performance. When workload becomes excessive in either prescription filling or nurse medication administration, the workers may skip tasks that can affect the quality of the product, such as not scanning the bar code on a medication package to verify

it is the correct item. Research indicates that prescription dispensing error rates are linearly related to prescription workload across lighting levels (Buchanan et al. 1991). Some research supports a relationship between increasing workload and dispensing errors (Guernsey et al. 1983; Flynn et al. 1999), while other studies do not support it (Kistner et al. 1994; Grasha 2000, 2002; Grasha and Schell 2001).

Interruptions and distractions were analyzed for a possible effect on prescription dispensing accuracy in an ambulatory care pharmacy over a 23 day period. Fourteen pharmacists were tested for distractibility using the group embedded figures test (GEFT) as well as for visual acuity and hearing. Interruptions and distractions were detected by reviewing videotapes of the pharmacy staff as they prepared prescriptions— these data were correlated with the prescriptions that were being filled at the time of the interruption or distraction. A pharmacist investigator compared each filled prescription with the physician's written order, noted details of deviations between the two. There was a significant association between distractibility (GEFT scores) and error rates: the less distractible the pharmacist, the fewer dispensing errors were made. A total of 5072 prescriptions were analyzed, and 164 errors were detected, for an overall error rate of 3.23%. A total of 2022 interruptions (mean ± S.D. per half hour per subject, 2.99 ± 2.70) and 2457 distractions (mean ± S.D. per half hour per subject, 3.80 ± 3.17) were detected. The error rate for sets of prescriptions with one or more interruptions was 6.65% and for sets during which there were one or more distractions, 6.55%; both of these values were a percentage point greater than the error rates for uninterrupted and undistracted prescription sets. Interruptions and distractions were both significantly associated with errors (Flynn et al. 1999). The most frequent reason for an interruption or distraction was a question from a coworker.

Ambient sound was evaluated for an association with prescription dispensing errors in the same study. Sounds detected from the videotape were categorized based on their predictability and controllability. A within-subjects case-control study design was used to determine if prescriptions with errors were filled under different sound conditions than error-free prescriptions. Loudness in terms of equivalent sound levels was also analyzed for an association with dispensing errors. Unpredictable, controllable sounds and noise were associated with prescriptions with fewer errors. The results suggest that the accuracy of pharmacist performance is not adversely affected by ambient sound as predicted. As loudness increased, so did the dispensing error rate, but only up to a point, when it began to decrease (Flynn et al. 1996).

46.5 Application

Health care professionals that process medication orders and manage medications use equipment that can benefit from intelligent design directed at improving efficiency and accuracy. Pharmacists spend a significant amount of time inspecting medications in community pharmacies, comparing what was used to fill the prescription with what the physician ordered. There is an evidence base in human factors research supporting the design of an optimal inspection station from a medication safety perspective. Inspection is a visual task involving looking at a dosage form and determining if the typically tiny imprint code matches that of the prescribed drug and strength. Incorporating magnifying glasses in the design of inspection stations helps optimize inspection accuracy. The color of the dosage form is also used to verify that the correct medication has been used to fill a prescription—this has implications for the amount and type of lighting that provides optimal color rendering. General Chapter 1066 in the USP presents guidance for lighting levels and lights for optimal color rendering. Nurses also need to read small letters on unit dose (single dose) medication packages in hospitals and could benefit from magnifying glasses, as well, especially with the increasing average age of the nurse work force. Despite the use of bar code medication verification systems in both hospitals and community pharmacies, the need for redundancy is important because technology may not be available 100% of the time it is needed.

How can health care providers working with medications be protected from interruptions and distractions? It may not always be possible to seclude them in a room away from telephones, patients, and other workers. Educating staff about the importance of interruption and distraction prevention is one step toward safer medication use. Pape proposed having nurses involved in medication administration in a hospital wear a brightly colored vest as a reminder to others not to interrupt the process (Pape et al. 2005).

The use of checklists has also been proposed as a measure to protect against errors (Pape et al. 2005) just as they have been used in airplane cockpits. Incorporating checklists into processes in a way that is efficient is critical to promoting their use with a subsequent improvement in safety. The medication administration protocol was evaluated by Pape et al. (2005) for nurse compliance—only two steps achieved 78% compliance (the highest rates): administering the medication and documenting administration.

A "medication safety zone" is a location where medications are handled in hospitals, community pharmacies, and homes, for example. USP General Chapter 1066 (2008) describes design details for medication safety zones with respect to the work environment. The illumination level standard is between 900 and 1500 lx because these levels promote accurate inspection of medications. USP also provides evidence for an interruption-free work station that limits the sound level the pharmacist is exposed to 50 dB (the level of normal conversation). Consultation with sound engineers can achieve this goal. A clutter-free workspace is proposed because of evidence that clutter can be a distraction during work.

The application of human factors standards and creativity to the design of hardware and software for health care professionals can help them efficiently and accurately retrieve drug information, order medication therapy, and manage medications. User interface and anthropometric standards are established for computers (ANSI/HFES 2007), personal digital assistants, and other technologies. Smart phones can be used for drug information but the limited screen size can be a barrier to accurate review of the data presented. Designers should consider the capacities and limitations of the users and the environment in which the tool will be used in order to optimize user accuracy and efficiency. An example of a software design problem involves the procedure for entering a medication for a prescription into a computer—a brand name drug can be selected by searching the name, or by scanning the bar code on the medication bottle. Scanning the bar code assumes that the correct medication and strength were retrieved from the storage shelf. If the original prescription is not compared to the medication that was entered in the computer, an error may be committed and the patient could receive the incorrect medication (Flynn et al. 2003).

46.6 Conclusion

Human factors principles described in this textbook need to be applied to the medication system at all levels in order to prevent medication errors. Research establishing the impact of human factors–based interventions on medication error reduction is still needed to expand the evidence base for this critical area that affects patients.

References

American Hospital Association (AHA), Health Research & Educational Trust (HRET), and the Institute for Safe Medication Practices (ISMP). 2002. Evaluation tool for pharmacists. Pathways for medication safety. http://www.medpathways.info/medpathways/tools/tools.html (accessed August 4, 2011)

American National Standards Institute—Human Factors and Ergonomics Society, ANSI/HFES 100–2007. Human factors engineering of computer workstations.

Barker, K. N., Flynn, E. A., Pepper, G. A., Bates, D. W., & Mikeal, R. L. 2002. Medication errors observed in 36 health care facilities. *Arch Intern Med* 162:1897–1903.

Bates, D. W., Cullen, D. J., Laird, N. et al. 1995. Incidence of adverse drug events and potential adverse drug events. Implications for prevention. ADE prevention study group. *JAMA* 274:29–34.

Bobb, A., Gleason, K., Husch, M. et al. 2004. The epidemiology of prescribing errors—The potential impact of computerized prescriber order entry. *Arch Intern Med* 164:785–792.

Buchanan, T. L., Barker, K. N., Gibson, J. T., Jiang, B. C., & Pearson, R. E. 1991. Illumination and errors in dispensing. *Am J Hosp Pharm* 48:2137–2145.

Cummings, J., Bush, P., Smith, D., & Matuszewski, K. 2005. Bar-coding medication administration overview and consensus recommendations. *Am J Health Syst Pharm* 62:2626–2629.

Filik, R., Purdy, K., Gale, A., & Gerrett, D. 2006. Labeling of medicines and patient safety: Evaluating methods of reducing drug name confusion. *Hum Factors* 48:39–47.

Flynn, E. A. & Barker, K. N. 2006. Effect of an automated dispensing system on errors in two pharmacies. *J Am Pharm Assoc* 46:613–615.

Flynn, E. A., Barker, K. N., Berger, B. A., Braxton, L. K., & Brackett, P. D. 2009. Dispensing errors and counseling quality in community pharmacies. *J Am Pharm Assoc* 49:48–57.

Flynn, E. A., Barker, K. N., & Carnahan, B. J. 2003. National observational study of prescription dispensing accuracy and safety in 50 pharmacies. *J Am Pharm Assoc* 43:191–200.

Flynn, E. A., Barker, K. N., Gibson, J. T., Pearson, R. E., Berger, B. A., & Smith, L. A. 1999. Impact of interruptions and distractions on dispensing errors in an ambulatory care pharmacy. *Am J Health Syst Pharm* 56:1319–1325.

Flynn, E. A., Barker, K. N., Gibson, J. T., Pearson, R. E., Smith, L. A., & Berger, B. A. 1996. Relationships between ambient sounds and the accuracy of pharmacists' prescription-filling performance. *Hum Factors* 38:614–622.

Flynn, E. A., Dorris, N. T., Holman, G. T., Carnahan, B. J., & Barker, K. N. 2002. Medication dispensing errors in community pharmacies: A nationwide study. Paper presented at the *46th Annual Meeting of the Human Factors and Ergonomics Society*, October 2, Baltimore, MD.

Grasha, A. F. 2000. Into the abyss: Seven principles for identifying the causes of and preventing human error in complex systems. *Am J Health Syst Pharm* 57:554–564.

Grasha, A. F. 2002. Psychosocial factors, workload, and risk of medication errors. *US Pharm* 27:HS32, HS35–HS36, HS39, HS43–HS44, HS47–HS48, HS52.

Grasha, A. F. & Schell, K. 2001. Psychosocial factors, workload, and human error in a simulated pharmacy dispensing task. *Percept Motor Skills* 92:53–71.

Guernsey, B. G., Ingrim, N. B., Hokanson, J. A. et al. 1983. Pharmacists' dispensing accuracy in a high-volume outpatient pharmacy service: Focus on risk management. *Drug Intell Clin Pharm* 17:742–746.

IESNA Committee for Healthcare Facilities. 2006. Lighting for hospitals and healthcare facilities. New York: Illuminating Engineering Society of North America.

Institute for Safe Medication Practices. 2005. Right on target with safer labels. http://www.ismp.org/newsletters/ambulatory/archives/200505_1.asp

Isaac, T., Weissman, J. S., Davis, R. B. et al. 2009. Overrides of medication alerts in ambulatory care. *Arch Intern Med* 169:305–311.

Kistner, U. A., Keith, M. R., Sergeant, K. A., & Hokanson, J. A. 1994. Accuracy of dispensing in a high-volume, hospital-based outpatient pharmacy. *Am J Hosp Pharm* 51:2793–2797.

Kowiatek, J. G., Weber, R. J., Skledar, S. J., Frank, S., & DeVita, M. 2006. Assessing and monitoring override medications in automated dispensing devices. *Jt Comm J Qual Patient Saf* 32:309–317.

Leape, L. L., Bates, D. W., Cullen, D. J. et al. 1995. Systems analysis of adverse drug events. ADE prevention study group. *JAMA* 274:35–43.

Meichenbaum, D. & Turk, D. C. 1987. *Facilitating Treatment Adherence: A Practitioner's Guidebook*. New York: Plenum.

Miller, K., Shah, M., Hitchcock, L. et al. 2008. Evaluation of medications removed from automated dispensing machines using the override function leading to multiple system changes. In *Advances in Patient Safety: New Directions and Alternative Approaches. Volume 4, Technology and safety*, eds. K. Henriksen, J. B. Battles, M. A. Keyes et al. pp. 1–7. http://www.ahrq.gov/downloads/pub/advances2/vol4/Advances-Miller_93.pdf (accessed February 17, 2011).

Murphy, J. E., Forrey, R. A., & Desiraju, U. 2004. Community pharmacists' responses to drug–drug interaction alerts. *Am J Health Syst Pharm* 61:1484–1487.

Oren, E., Griffiths, L. P., & Guglielmo, B. J. 2002. Characteristics of antimicrobial overrides associated with automated dispensing machines. *Am J Health Syst Pharm* 59:1445–1448.

Paoletti, R. D., Suess, T. M., Lesko, M. G. et al. 2007. Using bar-code technology and medication observation methodology for safer medication administration. *Am J Health Syst Pharm* 64:536–543.

Pape, T. M., Guerra, D. M., Muzquiz, M. et al. 2005. Innovative approaches to reducing nurses' distractions during medication administration. *J Contin Educ Nurs* 36:108–116.

Patterson, E. S., Rogers, M. L., Chapman, R. J., & Render, M. L. 2006. Compliance with intended use of bar code medication administration in acute and long-term care: An observational study. *Hum Factors* 48:15–22.

Patterson, E. S., Rogers, M. L., & Render, M. L. 2004. Fifteen best practice recommendations for bar-code medication administration in the Veterans Health Administration. *Jt Comm J Qual Saf* 30:355–365.

Pedersen, C. A., Schneider, P. J., & Scheckelhoff, D. J. 2006. ASHP national survey of pharmacy practice in hospital settings: Dispensing and administration—2005. *Am J Health Syst Pharm* 63:327–345.

Safren, M. A. & Chapanis A. 1960. A critical incident study of hospital medication errors. Part 1. *Hosp JAHA*. 34:32–34, 57–66.

Shamilyan, T. A., Duval, S., Du, J., & Kane, R. L. 2008. Just what the doctor ordered. Review of the evidence of the impact of computerized physician order entry systems on medication errors. *Health Serv Res* 43:32–53.

U. S. Pharmacopeia, Safe Medication Use Expert Committee. 2008. General Chapter 1066. Physical environments that promote safe medication use. *Pharm Forum* 34:1549–1558.

47

Human Factors and Ergonomics in Infection Prevention

Carla J. Alvarado

CONTENTS

47.1 Introduction 793
47.2 History of HAIs 793
47.3 History of Human Factors and Ergonomics 794
47.4 Infection Prevention and Human Factors/Systems Engineering 794
47.5 Current Infection Prevention and Human Factors 795
    47.5.1 Challenges of Applying Human Factors to Infection Prevention 796
    47.5.2 Work System Approach of Infection Prevention 796
    47.5.3 Example: Analysis of CVC Insertion 797
    47.5.4 Need for Broad Systems Approach 798
47.6 Human Factors and Infection Prevention Solutions 799
47.7 Conclusion 800
References 801

47.1 Introduction

Human factors and ergonomics (HFE) is a scientific discipline concerned with understanding interactions among humans and other elements of a system (IEA, 2000). The principles of human factors are a valuable resource for the challenge of improving infection prevention across the continuum of patient care. This chapter reviews the history of health care–acquired infections (HAIs), human factors and systems, and provides an example of a human factors work system task analysis of an infection prevention challenge. The discipline of human factors can very much contribute to the reduction care providers' physical and cognitive workload and improve the design of the built environment, therefore helping to decrease the complexity of infection prevention and patient safety.

47.2 History of HAIs

In 1771, orders were issued at the Manchester Infirmary that every new patient has clean sheets upon admission, additionally they have more clean sheets at least once in 3 weeks and that "two patients be not suffered to be in the same bed except that there is no spare bed in the house" (Woodward, 1974). This appears to be the first published account of an engineered control for infection prevention. At this time, British hospitals were considered far cleaner and less a risk for infection than their continental counterparts. The French hospital Hôtel-Dieu Paris had 1000 beds that were occupied by never fewer than 2000–3000 patients, and the River Seine often flooded the basement and first floor, filling the wards with contaminated water and garbage (Wagensteen & Wagensteen, 1978). But, by 1850, due to a number of wars, bad economy, poor housing and disease, all the European hospitals were stretched to the limit with diminishing resources and increased patient populations; hospital-related mortality from infection increased dramatically, especially on surgical and obstetric services (LaForce, 1993). Adding to this risk pool was that in the eighteenth and nineteenth centuries, wealthy and middle class women had their children at home and only the poor and unwed women were sent to these overcrowded and dirty hospitals to deliver. Mortality rates in these maternity hospitals were notoriously high, some reporting as much as 95% mortality during outbreaks from what appeared to be contagious puerperal fever (Nightingale, 1863).

Over 150 years ago, a Hungarian physician schooled in Vienna, Dr. Ignaz Phillip Semmelweis, wrote his observations on the transmission puerperal fever (Semmelweis, 1861). In a powerful monograph, he argued that contact with cadaveric or necrotic material could account for virtually all cases of puerperal fever and that the disease spread to patients on the hands of physicians and medical students or linens soiled with purulent material. To stop the spread of this contagion, Semmelweis insisted medical students wash their hands with chlorine compounds between examining patients and that soiled hospital linens not be reused between patients. Though Joseph Lister's work with disinfection is at least two decades in the future and the discipline of hospital infection control epidemiology will not be defined for nearly a century more, Semmelweis clearly applied epidemiological/systems analysis principles in his work and used engineered controls/technology to halt the infections.

47.3 History of Human Factors and Ergonomics

The history of human factors notes that humans have designed tools to fit their needs and abilities since the time of the cave man; but, the technology of the twentieth century and World War II ushered in more new tools and jobs than all of the previous centuries combined. As would be expected, new problems for humans' involvement with these technologies and systems also accompanied the twentieth-century inventions and innovations. These problems, never experienced in the past, required new data and new principles from the research of human sciences—physiologists, physicians, and psychologists. And, in 1946, the term "Human Factors" was first used in the title of a book describing the findings of these scientists (Chapanis, 1996). As the discipline progressed, the way in which systems were designed and how human factors contributed at each stage of the system process also developed. Human factors focuses on the interaction between the human and the work system, fitting the work to the person.

47.4 Infection Prevention and Human Factors/Systems Engineering

Debatably, Semmelweis is considered the father of modern hospital infection prevention and epidemiology programs; but, could he also be the first to use human factors engineering and a systems approach to solve an infection prevention problem? The answer to this question is found in his original works. As previously mentioned, the poor and unwed women were admitted to hospitals to deliver, whereby they were cared for at no

cost in exchange for being used as professors' "teaching material" for medical students and midwives. In 1847, Dr. Semmelweis joined the staff of the Vienna Lying-In Hospital, at that time the largest obstetrics service in the world. It was divided into two hospital wings, one for medical students and one for midwives. While rounding at the hospital, Semmelweis observed mortality was 10% or more in the medical students' hospital wing and only 3% or less in the midwives' wing. Dr. Semmelweis immediately undertook what today is described as the first published epidemiological study of a HAI outbreak.

In reviewing Semmelweis's original work, it is apparent he uses a systems or human factors approach to investigate this disease and considers the following elements currently described as the work system in the Systems Engineering Initiative for Patient Safety (SEIPS) model (Figure 47.1):

- Societal/*organization* factors such as the infected patients' socioeconomic class causes placement in the hospital vs. remaining home
- Physical *environment* factors such as water quality, patient bed linens, ventilation, and patient crowding
- *Task* factors such as difficult or extended labors requiring more physician/medical student attention than normal deliveries left to the midwives
- *Tools and technology* in use—comparing hospital and at home deliveries
- And the *people* involved in the delivery—physician, medical student, midwife, and patients

Using this approach, Semmelweis concentrated his study on the task analysis and the humans that perform it, noting that physicians and medical students had far higher patient incidence rates of puerperal fever than the midwives. What he found differed between the two work groups' tasks was physicians/medical students were required to perform autopsies on the deceased patients. Therefore, they had many opportunities to contact the purulent discharge and necrotic tissues of patients that died of puerperal sepsis. Midwives were not in medical school and did not get autopsy training. Rarely, if ever, did midwives have exposure to the deceased anatomy.

In "Tayloristic" (Taylor, 1911) fashion, Semmelweis deconstructed the physician/medical students' jobs, discovering in this task analysis that physicians and students often went directly from the autopsy theater to patient delivery without any process for decontamination of their hands. Thus, he concluded (correctly) that transmission of puerperal fever could be interrupted by scrubbing the hands with chlorine compounds after autopsy and prior to delivery. Without the knowledge of the "germ theory"

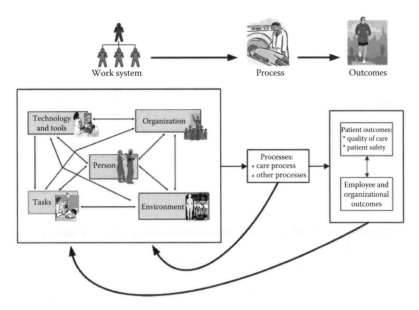

FIGURE 47.1
SEIPS model of work system and patient safety. (From Carayon, P. et al., *Qual. Saf. Health Care*, 15(suppl. 1), 50, 2006.)

of disease, Semmelweis initiated a workplace intervention (engineering control) and reduced the risk of HAI infection in the patients. He went on to describe the problem as one of advancement in medical technology, explaining that in years prior, incidence of puerperal fever had been equally low in both wings of the hospital. But, with emergence of forensic pathology and the opening of the new Institute of Anatomy, instructional autopsies performed on this indigent patient population came into academic fashion. The hospital physician administrators required all physicians and their medical students, along with the newly created staff physician/ anatomist (later known as the anatomical pathologist), to perform autopsies on their dead patients. His analysis clearly describes a latent error in the organization/ system, as the required autopsy attendance placed the medical staff at the "sharp" of the infection transmission or medical error (Reason, 2000). Unfortunately, Semmelweis, not aware of the concept of latent error in a system, did not win favor with colleagues by accusing them of causing their patients' deaths. Colleagues committed him to a mental asylum, where he died of an HAI post admission (Nuland, 2003).

47.5 Current Infection Prevention and Human Factors

Fast-forward a century and a half, infection prevention and health care epidemiology are well-established disciplines in every accredited U.S. hospital and most

European hospitals. HAI rates are trending zero in many procedures. But still these infections exist. In 2001, as a response to the need for patient safety, the Agency for Healthcare Research and Quality (AHRQ) issued the report "Making health care safer: A critical analysis of patient safety practices" (AHRQ, 2001). The report identified about 70 patient safety practices— defined as those that reduce the risk of adverse events related to medical care. The practices were then rank-ordered by strength of evidence, projected impact, effectiveness, and cost. Eleven practices were highlighted as the strongest on the basis of this rank order process. Although infection prevention was not mentioned in the first Institute of Medicine patient safety report (Institute of Medicine Committee on Quality of Health Care in America, 2001), based on the strength of evidence in the scientific literature, 4 of the 11 highlighted, practices in the 2001 AHRQ report (36%) are infection prevention practices:

- Use of maximum sterile barriers while placing central intravenous catheters to prevent infections

- Appropriate use of antibiotic prophylaxis in surgical patients to prevent perioperative infections

- Continuous aspiration of subglottic secretions (CASS) to prevent ventilator-associated pneumonia

- Use of antibiotic-impregnated central venous catheters (CVCs) to prevent catheter-related infections

This clearly acknowledges that prevention of HAI is a patient safety issue, and one that has been waiting for a human factors and systems approach since Ignaz Semmelweis.

Effective work system design is important for all aspects of a health care organization, especially in infection prevention. Ineffective working conditions can negatively impact the health care providers, patients, families, the organization, and even society (Cohen-Mansfield, 1997). The literature shows patient care time pressure and high workload are associated with increased risk of patient infection (Hugonnet et al., 2007). Because of extreme time pressure and workload, the health care providers often see infection prevention efforts as interruptions to their "real tasks" rather than integral parts of those tasks. Insufficient time for hand hygiene, excessive workload, and understaffing are all associated with HAI (Archibald et al., 1997; Harbarth et al., 1999). Hugonnet et al. (2007) found low staffing levels were followed only a few days later by the occurrence of HAIs. This suggests that under the pressure of increased workload, health care workers do not comply with infection control measures, such as hand hygiene, due to time constraints (Archibald et al., 1997; Hugonnet et al., 2002). However, many factors such as organizational support, job dissatisfaction, burnout, absenteeism, intention to leave the current position, and turnover interact and predetermine the occurrence of negative outcomes, including HAIs (Aiken et al., 2002a, 2000b).

47.5.1 Challenges of Applying Human Factors to Infection Prevention

So why has it taken since Semmelweis to adopt a systems/HFE approach to infection prevention? There are key HFE-related challenges to adopting this approach in patient care. First, if providers have an infected patient, they may ask "Is it my problem? I didn't cause this." Health care providers, through lack of infection prevention knowledge, often take no ownership as to the cause of the infection (Ward, 2010). And, more often, the providers have no sense of temporal association between their actions and patient infection. There is no clear and timely feedback associated with caregiver action and the acquired infection. If the surgeon cuts through a patient's artery in surgery, there is immediate feedback with a spurt of blood; however, if that same physician does not properly perform hand hygiene and the patient becomes infected, there is no timely feedback on the physician's (lack of) actions and this patient's outcome—patient infection occurs in days, not seconds. The cause and effect relationship in HAI is unclear (Anderson et al., 2010). Fortunately, patient care tasks are most often performed without HAI and most patients have positive experiences and

outcomes. To the individual health care provider, HAI is a rare event. The provider does not see a reward for proper performance of prevention tasks and rarely experiences a bad outcome if prevention is not done. This lack of tangible positive feedback may result in less motivation to continue time-consuming infection prevention tasks. Tasks that take more time or are less convenient to perform are often dropped, delayed, or forgotten (Anderson et al., 2010).

47.5.2 Work System Approach of Infection Prevention

Taking advantage of systems safety lessons adopted from other industries (i.e., aviation, nuclear power plants), human factors and ergonomics are currently applied in some areas of health care improvement (i.e., anesthesia and pharmacy). Unfortunately, infection prevention lags behind these specialties and a systems approach to infection prevention is often overlooked, in favor of the individual approach to HAI improvement—more training, more personal responsibility, etc. If we apply the SEIPS model (Figure 47.1) to infection prevention, this work system model provides a way of describing all of the elements of the task that affect care providers and patient infection outcomes. According to the work system model, tasks are performed by an individual who uses tools and technologies; the tasks are performed in a physical environment and under organizational conditions. The emphasis of this model is on the systemic aspects of work: the work system elements interact with each other. Whenever a change is introduced in any element of the work system, systemic changes occur in other work system elements (Carayon et al., 2006). For example, nurses have long known that manually lifting and transferring dependent patients is a high-risk activity, for both nurse and patient. Patient lifting devices are well established as engineering controls for reducing care provider back injuries (Owen et al., 1992). Research has shown this technology allows for the vertical transfers of patients without lifting manually and the ceiling lifts are currently popular installations in operating rooms (Miller et al., 2006; Silverwood & Haddock, 2006). However, no study has addressed the infection prevention requirements of these ceiling-mounted devices. The change to ceiling-mounted lifts likely reduces staff injury but does it increase infection risks to patients in the operating room? How can operating room staff physically clean and disinfect these devices between patients? What is the risk for blood and other infectious body fluids and/or multidrug resistant organisms transferring from patient to patient via the device? The installation of ceiling lifts may decrease risk of a staff or patient lifting injury; but, also may introduce the risk of patient exposures to infectious agents and at the very least increase equipment reprocessing workload for other hospital staff.

47.5.3 Example: Analysis of CVC Insertion

To illustrate the use of the work system model for infection prevention, we shall consider a human factors and ergonomics task analysis of ultrasonic guidance and infection prevention for placement of CVCs (Alvarado et al., 2006, 2008). In this initiative, rather than a full HFE task analysis (focus groups, video analysis, etc.), a series of individual observations of CVC insertions were performed to gain a better understanding of the current CVC insertion practices in the hospital. In this SEIPS work system model (Figure 47.1) task analysis, we describe a work systems approach to analyze an infection prevention issue—risk of CVC infection.

The work system can be described as the combination and interactions of the individual, the task, in use tools and technologies, the physical environment, and the current organizational conditions. In addition to using the work system model, the task analysis included an observational central line bundle checklist based on the Institute for Healthcare Improvement (IHI) best practices (Pronovost et al., 2006). Using a standardized data collection form and central line bundle checklist, the observer recorded the number and type of people (resident and staff physicians, medical students, nurses, etc.) in the room, action in the task being performed, adherence to the central line bundle practices, interruptions during these tasks, the physical environment (furniture/equipment placement, light, heat, noise, etc.) and the number of central venous cannulation attempts performed on the patient prior to central venous access.

Study observations demonstrated that required hand hygiene was not regularly performed prior to donning sterile gowns and sterile gloves. However, most nursing staff and medical staff did perform hand hygiene post procedure. Maximal sterile barrier precautions were attempted during every observation but not always in strict compliance with the prescribed standard of care. The application of the sterile, antimicrobial line dressing was performed in every observation but occasionally placed wrong side down, providing no antimicrobial protection to the catheter site. Often this error occurred when lighting was turned down for ultrasound screen confirmation of line placement and dressing the insertion site immediately followed.

At no point in the observations were clinical proficiency judgments made as to the skills of the *persons* performing the medical procedure. It was assumed the medical professional had the physical and cognitive abilities to perform the CVC placement. However, in several observations, the physician was left-hand dominant, making the entire set up and CVC insertion more difficult.

What *tasks* are being performed, and what characteristics of those tasks may contribute to unsafe patient care? What in the nature of the tasks allows the individual to perform them safely or assume risks in the process?

Several inefficiencies in the CVC line insertion process (task) were observed, especially with regard to number of interruptions and in the organization of supplies. Though the physicians were mostly experienced residents or staff, quite familiar with the ICU setting, they frequently opened sterile CVC kits and discarded the enclosed sterile disposable drape, preferring sterile operating room fabric towels and sterile cotton sheets. Physician and nurse time was spent looking for correct size or type sterile gloves, and asking for more of various items such as chlorhexidine prep swabs, sterile port covers, saline flush solutions, etc. all while performing the task under time pressure and a high cognitive workload. Required supplies often seemed inaccessible, making it difficult for the physician "to do the right thing."

The physicians used a variety of *tools and technology*, such as ultrasound machines, sterile drapes, gowns, gloves, CVC kits, etc. On occasion, the tool (i.e., correct catheter introducer) was not available on the sterile field. Rather than break sterile technique and retrieve the necessary tool, the physician inserting the CVC was shown by a more senior physician, innovative techniques to complete the task by modifying the available sterile supplies originally brought to the bedside. Additionally, the patient was being monitored by a myriad of technology (cardiac monitors, pulse oximeters, pressure monitors, respirators, IV pumps), all simultaneously transferring critical information, creating a high cognitive workload for all involved in the line insertion.

The physical *environment* was crowded with patient care equipment (e.g., ventilator, IV pumps, monitors), the patient's bed, side tables, visitor chairs, and medical students not involved in the CVC insertion but observing for teaching purposes. There is no space for a sterile-draped prep table accessible to the bedside; therefore, the sterile field was frequently set up on the sterile-draped patient's chest. The ambient lights and task light, although adequate for the task, were often dimmed to view the ultrasound image on CRT displays and, if left low for real-time ultrasonic imagery, became inadequate to perform the actual cannulation task. The environment was noisy with conversation and equipment alarms both at the patient bedside and outside the patient cubical in the general ICU ward area. Ambient room temperature was sufficiently comfortable for the covered patient; but, quite warm for the medical professionals garbed in sterile maximum barrier gowns, hats, gloves, eye protection, and masks. Poor visibility, difficulty of access, wide separation of accessories and tools used in sequence (e.g., portable ultrasound machine being positioned in an already overcrowded room) allowed breaches in infection prevention or nonoptimal practices to occur.

As is the case in the teaching hospital *organization*, the patient care area is filled with medical students, physician trainees, and medical staff. Too often the nurse was told others would assist in the procedure and exited the room to perform other nursing duties. The less-experienced medical students assumed the nurse patient advocacy role and/or fetched supplies for the physician inserting the CVC. Unfortunately, the medical students were inexperienced and unfamiliar with the needed supplies and often could not identify the requested item. The HFE analysis showed the care providers were repeatedly interrupted and many of these interruptions resulted in breaks in the primary task. While some interruptions were important (e.g., correcting insertion technique, alerting a change in vital signs), other interruptions did not concern the task or even that particular patient. Often other physicians, nurses, or therapists would come into the patient space and interrupt the task performed by the physician, asking both related and unrelated questions. In posttasks interviews, individuals were asked why they interrupted the physician during a critical task. Often, they responded that they knew where to find him/her and at that moment they could communicate all the particulars they needed to share with that person. Interruptions appear to be an *organizational* issue associated with poor or few communication opportunities to share information. There was no regard for disrupting the workflow and possibly causing the physician to overlook or skip a task, contributing to a higher chance for other errors to occur. Human beings find it difficult to stay attentive, vigilant, and productive, particularly when they are being interrupted, fatigued, or in an work overload situation (Chisholm et al., 2000, 2001). The physician whose attention is constantly shifting from one item to another may not be able to formulate a complete and coherent picture of the task at hand (Cook & Woods, 1994). It became clear that knowledge of the physician's immediate whereabouts (inserting a CVC) provided an opportunity to communicate with that physician, albeit with interruption and possible error to the task.

Using the work system as a framework for CVC insertion task analysis, we identified various areas for system interventions or improvements for patient safety (Carayon et al., 2003). The results of this task analysis suggest that the ICU work environment presently does not adequately support efficient and safe CVC insertion. The environment should be redesigned to include the space for a central line supply cart that can be wheeled to the bedside, providing area for sterile prep space, stocked with human factors/ergonomically designed central line kits, and required supplies. If the patient area has a door, it should be closed to reduce distracting ambient noise from the hallway and other patient care areas. The most surprising finding in the task analysis was the frequent interruptions of the physician inserting the catheter. To this end, signage could be designed stating "Procedure—Do Not Enter." Signage of course can be ignored, but hopefully with reinforcement and timely feedback, insertion interruptions will decline and necessary physician communications will be improved in other venues.

47.5.4 Need for Broad Systems Approach

The ability to improve infection prevention is dependent, to a large extent, on whether we can change attitudes and behaviors. Task, analysis and timely feedback provided to patient care providers are methods to support this change. In tandem with the CDC National Healthcare Safety Network (NHSN) http://www.cdc.gov/nhsn/, infection prevention programs have developed the most accurate surveillance system in the health care setting and use this data for developing specific, evidence-based infection prevention interventions. Although the lowest ever, overall nosocomial infection rates are only trending zero and still need improvement, especially in areas such as intensive care units where incidence rates of HAI still often reach 30% (Pronovost et al., 2006). In developing interventions to lower this incidence of infection, prevention programs, as in other safety programs, have focused on the individual care provider and the tasks they perform, providing more training or more technology to assist the individual in their performance of work duties. Although the health care professional, the tasks, the technology, and training are very important factors in the work system, they are not the only work organization factors infection prevention professionals should consider for prevention interventions. The physical environment—not as it relates as a fomite—but its various aspects of noise, lighting, temperature, air quality, and workplace layout and how it affects health care providers provide possible avenues of exploration to reduce infections or medical errors. Finally, infection prevention must look beyond its program boundaries and consider the organization itself, both the internal organization of the facility and the outside organization of society at large and the possible factors within the organizational context that might influence infection prevention and patient safety. Unfortunately, most infection prevention programs are grossly under-staffed and even if the infection preventionist wants to explore the use of HFE analysis, they do so in an unstructured and unassisted fashion (Anderson et al., 2010). Little to no systems/human factors expertise is available at most health care facilities. The challenges of infection prevention are many, but human factors/ergonomics and systems analysis offer infection prevention new knowledge of human abilities and design of systems already used in other industries to make products safer, jobs less stressful, safer, and more meaningful, and improve human performance. Just as Semmelweis experienced, infection prevention professionals and human factors experts may meet resistance from health care peers to change and innovative prevention processes based in human factors and ergonomics. But we have no choice but to continue to improve the patient care system, for in our lifetime we shall all enter that system and, hopefully, experience a better outcome than our mentor Ignaz Semmelweis.

47.6 Human Factors and Infection Prevention Solutions

Although infection preventionists are familiar with statistics and processes to investigate HAI outbreaks, most may be less familiar with the use of so-called process engineering in the health care center. Although process systems engineering is actually a relatively young area in chemical engineering, the title has somehow slipped into the medical jargon when referring to engineering techniques for health care improvement. The following is a brief discussion of the techniques/tools encountered in health care quality improvement engineering and infection prevention.

The Six Sigma approach to improving the quality of health care quality has been defined as "…the extent to which health services for individuals and populations increase the likelihood of desired health outcomes and are consistent with current professional knowledge" (Chassin, 1998). The term "Six Sigma" is a statistical measure of variation, the standard deviation of a normal distribution (Chassin, 1998). Implementing the Six Sigma process in the health care setting (i.e., service industry) is believed to be equally effective in improving quality of care. Eldridge et al. (2006) utilized the Six Sigma process in an effort to improve hand hygiene in intensive care units. The researchers' goal was to use Six Sigma to examine hand hygiene practices and increase compliance with the CDC guidelines (CDC, 2002) in four intensive care units at three Veterans' Affairs Hospitals. The outcome measurements were compliance with 10 required hand hygiene practices and the amount of alcohol-based hand rubs (ABHR) used per month per 100 patient days. They also assessed ICU staff attitudes and perceptions about hand hygiene by questionnaire. The results of this study found that ABHR use almost doubled in two of the ICUs ($P < .001$) and increased by 70% in the third. Increases in compliance with hand hygiene practices were statistically significant at all four

ICUs (range from 55% to 95%). The study showed 80% compliance after interventions, a rate that exceeded most other studies that tried to improve hand hygiene (CDC, 2002). Overall, the researchers concluded that the Six Sigma program's first three steps of design, measure, and analyze were helpful in permitting the identification of baseline data, an important step in improving infection prevention.

Root cause analysis (RCA)—a method of identifying "the basic and causal factor(s) that underlie variation in performance"—was introduced into the health care system in the 1990s (Wu et al., 2008). Basically, RCA answers three questions: (1) what happened, (2) why did it happen, and (3) what can be done to prevent it from happening again (Wu et al., 2008). Health care institutions are required by the Joint Commission to perform an RCA for every "sentinel" event. However, RCA often falls short of effectively preventing the problem from occurring again. Wu et al. (2008) provide an example of a preventable mistake and the subsequent RCA conducted by the institution. A nurse mistakenly infused patient-controlled analgesia (PCA) into a patient's intravenous catheter. The hospital's RCA team found there were flaws in the catheter's design that most likely contributed to the nurse's error, but believed that addressing the design flaw with the manufacturer and attempting to have the design changed was beyond the scope of the team. Instead, they developed and implemented a staff re-education program about proper use of the catheter and identified re-education programs or writing a policy as the two most common recommended actions resulting from RCAs in the health care setting (Wu et al., 2008). Unfortunately, these actions rarely reduce risk of a repeated mistake and are not disseminated to other institutions. In the example of the PCA error, the RCA team's original thought of improving the design of the catheter would have been the best solution, because that action would have positive consequences in all hospitals throughout the country that use the catheter. Health care institutions are now recognizing the limitations of the RCA process. In this example, redesign of the catheter could best be accomplished by a national organization or professional society, which would allow widespread use of the solution. Efforts by individual health care institutions to reduce mistakes of this nature have limited success and can be time consuming; developing and conducting a re-education program about PCA could only benefit the hospital(s) that implement the program.

Value stream analysis (VSA) is a planning process developed originally at Toyota (Toyota Motor Corporation, Japan). The process uses three Value Stream Maps to guide the organization to improved performance: (1) current conditions map, (2) ideal state map (what is perfection), and (3) future state map (what the value stream is to be 6–12 months from beginning)

(Rath, 2008). The value stream maps help organizations to communicate, identify waste and sources of waste, visualize the future, and create plans for action. Recently an infection prevention study presented a VSA on redesign of care and use of CVCs using Lean engineering and Six Sigma framework (Richmond, 2008). The study's findings demonstrated post VSA standardization in CVC supplies, 100% compliance with CVC insertion carts on all appropriate units, decreased nursing assistance in gathering supplies, and a projected 33% reduction in catheter-related infection predicting a $500,000 savings associated with HAI reduction (Richmond, 2008).

Failure mode and effects analysis (FMEA) is an industrial engineering–based tool that has been used by medical device and pharmaceutical companies to reduce risks in their products and processes (Rath, 2008). It is a tool that lends itself well to the health care industry and patient safety concerns. It is recommended that an individual trained as a facilitator for the FMEA process lead the team conducting the analysis. FMEA takes a proactive approach to detecting possible causes of errors and developing and implementing processes to prevent them from occurring (Duwe et al., 2005). In 2001, the Joint Commission revised the Standards in Support of Patient Safety and Medical/Health Care Error Reduction and required all acute care hospitals to perform FMEA regularly. To begin using FMEA in analyzing a process, the process must first be defined. Creating a process flow chart is an easy and effective method of defining the process to be studied. Once the possible failures are determined, the next step is to identify the possible cause(s). The next phase of this FMEA is to identify "process controls" associated with each possible failure. Process controls are methods or procedures/policies that prevent failure modes from happening, identify failure when it occurs, and minimize/avoid the severity of the failure mode. Process controls in infection prevention might include staff training, preventive equipment availability, staffing patterns, etc. The effectiveness of the process controls should then be evaluated using rating scales. What results from this scoring process is a risk priority number (RPN) for each failure mode and cause. Whenever there is a high severity score, FMEA team members should thoroughly evaluate that process step and, next, determine the process controls necessary to reduce the risk of failure and injury. The FMEA process should be re-evaluated on a regular basis to account for technology and process changes (Rath, 2008).

Task analysis is a method to describe human involvement in a system, which can help to ensure that system goals, human capabilities, and the organization operate in a way that achieves systems goals (Kerwin & Ainsworth, 1992). In the context of health care, task analysis is one method that can increase safety for staff and patients. Task analysis can identify potential hazards in the system, improve design of tasks, help analyze situations for potential human errors, and help determine what changes need to be made to prevent hazards from recurring. There are many different task analysis techniques for evaluating a variety of tasks in organizational settings; however, human factors task analysis is an appropriate type of task analysis to review for the health care setting. A previous section of this chapter describes task analysis of CVC insertion. Through observation and interviews, the reality of how tasks are actually being performed is revealed and often differs from actual task-related policies and procedures. The task analysis may identify "work arounds" or procedure deviance developed because of location of equipment, patient isolation requirements (Gurses & Carayon, 2007), or placement of critical items such as hand hygiene sinks and products (Suresh & Cahill, 2007). Task analysis within a hospital unit can highlight deficiencies in staffing, the built environment, lighting, noise reduction, and a myriad of other risk factors for HAI. A thorough task analysis is the corner stone to HFE and infection prevention efforts for reducing HAI.

47.7 Conclusion

In conclusion, human factors and systems engineering principles should and can be applied to infection prevention. Infection prevention requires determination for change. This will for change can be articulated by setting aims that specify the level of system performance with respect to errors and adverse events (Nolan, 2000). Because of the gravity of HAI, improvements should always focus on targeting zero. Error rates in any system must ultimately be determined empirically. Most processes in health care contain many steps and tasks. The error rate for a process increases as the steps in the process increase, and, therefore, its complexity increases (Nolan, 2000). Currently HAI rates have continued to improve and some approach zero in many procedures (Pronovost, 2008). Improvements to a system with error rates in the parts per thousand will necessitate more sophisticated systems analysis for improvement. We are beyond the ease of "the low hanging fruit" identified in earlier quality improvement efforts. Much is yet to be learned about how to apply human factors principles for designing infection prevention systems that are effective, safe, and sustainable. However, a solid foundation of knowledge and methods exist on which to build these systems that both patients and clinicians deserve.

References

AHRQ. (2001). Making healthcare safer: A critical analysis of patient safety practices. Evidence Report/Technology Assessment: Number 43. Rockville, MD.

Aiken, L. J., Clarke, S. P., and Sloane, D. M. (2002a). Hospital staffing, organization, and quality of care: Cross-national findings. *Int J Qual Health Care, 14,* 5–13.

Aiken, L. J., Clarke, S. P., Sloane, D. M., Sochalski, J. A., and Silber, J. H. (2002b). Hospital nurse staffing and patient mortality, nurse burnout, and job dissatisfaction. *JAMA, 288*(16), 1987–1993.

Alvarado, C., Wood, K., and Carayon, P. (2006). Human factors and ergonomics task analysis in ultrasonic guidance and infection control for CVC cannulation. In R. N. Pikaar, E. A. P. Koningsveld, & P. J. M. Settels (Eds.), *Proceedings of the IEA2006 Congress* 10–14 July 2006, Maastricht, The Netherlands. New York: Elsevier.

Alvarado, C., Wood, K., and Carayon, P. (2008). Pre-implementation technology assessment of ultrasonic placement of central venous catheter insertion. In L. Sznelwar, F. Mascia, & U. Montedo (Eds.), *Human Factors in Organizational Design and Management—IX* pp. 81-86. Santa Monica, CA: IEA Press.

Anderson, J., Gosbee, L., Bessesen, M., and Williams, L. (2010). Using human factors engineering to improve the effectiveness of infection prevention and control. *Crit Care Med, 38*(8 suppl), S269–S281.

Archibald, L. K., Manning, M. L., Bell, L. M., Banerjee, S., and Jarvis, W. R. (1997). Patient density, nurse-to-patient ratio and nosocomial infection risk in a pediatric cardiac intensive care unit. *Pediatr Infect Dis J, 16,* 1045–1048.

Carayon, P., Alvarado, C., Hsieh, Y., and Hundt, A. S. (2003). A macroergonomic approach to patient process analysis: Application in outpatient surgery. Paper presented at the *XVth Triennial Congress of the International Ergonomics Association,* August 24–29, 2003, Seoul, Korea.

Carayon, P., Hundt, A., Karsh, B. -T., Gurses, A., Alvarado, C., Smith, M. et al. (2006). Work system design for patient safety: The SEIPS model. *Qual Saf Health Care, 15*(suppl. 1), 50–58.

Centers for Disease Control and Prevention (CDC). (2002). Guideline for hand hygiene in health-care settings. Recommendations of the Healthcare Infection Control Practices Advisory Committee and the HICPAC/SHEA/APIC/IDSA Hand Hygiene Task Force. *Morb Mortal Wkly Rep, 51*(RR-16), 1–45.

Chapanis, A. (1996). *Human Factors in Systems Engineering.* New York: John Wiley & Sons.

Chassin, M. (1998). Is health care ready for six sigma quality? *Milbank Q, 76*(4), 565–591.

Chisholm, C. D., Collison, E. K., Nelson, D. R., and Cordell, W. H. (2000). Emergency department workplace interruptions: Are emergency physicians "interrupt-driven" and "multitasking"? *Acad Emerg Med, 7*(11), 1239–1243.

Chisholm, C. D., Dornfeld, A. M., Nelson, D. R., and Cordell, W. H. (2001). Work interrupted: A comparison of workplace interruptions in emergency departments and primary care offices. *Ann Emerg Med, 38*(2), 146–151.

Cohen-Mansfield, J. (1997). Turnover among nursing home staff. *Nurs Manag, 28*(5), 59–64.

Cook, R. I. and Woods, D. D. (1994). Operating at the sharp end: The complexity of human error. In M. S. Bogner (Ed.), *Human Error in Medicine* pp. 255–310. Hillsdale, NJ: Lawrence Erlbaum.

Duwe, B., Fuchs, B. D., and Hansen-Flaschen, J. (2005). Failure mode and effects analysis application to critical care medicine. *Crit Care Clin, 21*(1), 21–30.

Eldridge, N., Woods, S., Bonello, R., Clutter, K., Ellingson, L., Harris, M. et al. (2006). Using the six sigma process to implement the centers for disease control and prevention guideline for hand hygiene in 4 intensive care units. *J Gen Intern Med, 21,* S35–S42.

Gurses, A. P. and Carayon, P. (2007). Performance obstacles of intensive care nurses. *Nurs Res, 56*(3), 185–194.

Harbarth, S., Sudre, P., Dharan, S., Cadenas, M., and Pittet, D. (1999). Outbreak of *Enterobacter cloacae* related to understaffing, overcrowding and poor hygiene practices. *Infect Control Hosp Epidemiol, 20,* 598–603.

Hugonnet, S., Chevrolet, J. C., and Pittet, D. (2007). The effect of workload on infection risk in critically ill patients. *Crit Care Med, 35,* 76–81.

Hugonnet, S., Perneger, T. V., and Pittet, D. (2002). Alcohol based handrub improves compliance with hand hygiene in intensive care units. *Arch Intern Med, 162,* 1037–1043.

IEA. (2000). The discipline of ergonomics. Retrieved from http://www.iea.cc/ (accessed 15 January, 2011)

Institute of Medicine Committee on Quality of Health Care in America. (2001). *Crossing the Quality Chasm: A New Health System for the 21st.* Washington, DC: National Academy Press.

Kerwin, B. and Ainsworth, L. K. (Eds.). (1992). *A Guide to Task Analysis.* London, U.K.: Taylor & Francis, Inc.

LaForce, F. M. (Ed.). (1993). *The Control of Infections in Hospitals: 1750 to 1950.* Baltimore, MD: Williams & Wilkins.

Miller, A., Engst, C., Tate, R., and Yassi, A. (2006). Evaluation of the effectiveness of portable ceiling lifts in a new long-term care facility. *Appl Ergon, 37*(3), 377–385.

Nightingale, F. (1863). *Notes On Hospitals* 3rd edition. London: Longman, Green, Longman, Roberts, and Green.

Nolan, T. W. (2000). System changes to improve patient safety. *BMJ, 320*(7237), 771–773.

Nuland, S. B. (2003). *The Doctors' Plague: Germs, Childbed Fever and the Strange Story of Ignac Semmelweis.* New York: W.W. Norton.

Owen, B., Garg, A., and Jensen, R. C. (1992). Four methods for identification of most back-stressing tasks performed by nursing assistants in nursing homes. *Int J Ind Ergon, 9,* 213–220.

Pronovost, P. (2008). Interventions to decrease catheter-related bloodstream infections in the ICU: The keystone intensive care unit project. *Am J Infect Control, 36*(10), S171 e171–e175.

Pronovost, P., Needham, D., Berenholtz, S., Sinopoli, D., Chu, H., Cosgrove, S. et al. (2006). An intervention to decrease catheter-related bloodstream infections in the ICU. *N Engl J Med, 355*(26), 2725–2732.

Rath, F. (2008). Tools for developing a quality management program: Proactive tools (process mapping, value stream mapping, fault tree analysis, and failure mode and effects analysis). *Int J Radiat Oncol Biol Phys, 71*(1), S187–S190.

Reason, J. (2000). Human error: Models and management. *BMJ, 320*(7237), 768–770.

Richmond, A. (2008). The road to zero BSIs: How performance improvement methods can serve as your roadmap. Retrieved February 28, 2011, from http://www.apic.org/downloads/clabsi/Amy_The_Road_to_Zero_BSIs.ppt

Semmelweis, I. P. (1861). *The Etiology, the Concept and the Prophylaxis of Childbed Fever,*. K. Codell Carter translator (1983). Madison, WI: University of Wisconsin Press.

Silverwood, S., and Haddock, M. (2006). Reduction of musculoskeletal injuries in intensive care nurses using ceiling-mounted patient lifts. *Dynamics, 17*(3), 19–21.

Suresh, G., and Cahill, J. (2007). How "user friendly" is the hospital for practicing hand hygiene? An ergonomic evaluation. *Jt Comm J Qual Patient Saf, 33*, 171–179.

Taylor, F. (1911). *The Principles of Scientific Management.* New York: Norton and Company.

Wagensteen, O. H., and Wagensteen, S. D. (1978). *The Rise of Surgery.* Minneapolis, MN: University of Minnesota Press.

Ward, D. J. (2010). Infection control in clinical placements: Experiences of nursing and midwifery students. *J Adv Nurs, 66*(7), 1533–1542.

Woodward, J. (1974). *To Do the Sick No Harm: A Study of the British Voluntary Hospital System to 1875.* London, U.K.: Routledge & Kegan Paul.

Wu, A., Lipshutz, A., and Pronovost, P. (2008). Effectiveness and efficiency of root cause analysis in medicine. *JAMA, 299*, 685–687.

48

Human Factors in Anesthesiology

Matthew B. Weinger

CONTENTS

48.1 Background...803
  48.1.1 Anesthesiology and Patient Safety...803
48.2 Adverse Event Analysis and Nonroutine Events..804
  48.2.1 Event Detection Methods in Anesthesiology 804
  48.2.2 Comprehensive Open-Ended Nonroutine Event Surveys..805
  48.2.3 Prospective Video-Based Assessment of NRE 806
    48.2.3.1 NRE Analysis..806
    48.2.3.2 Relationship of NRE to Significant Physiological Disturbances....................808
48.3 Performance Shaping Factors in Anesthesiology..808
  48.3.1 Individual Factors 809
    48.3.1.1 Experience and Expertise 809
    48.3.1.2 Fatigue and Sleep Deprivation..809
    48.3.1.3 Burnout...810
  48.3.2 Tasks and Workload 811
    48.3.2.1 Measuring Clinical Workload..811
  48.3.3 Situation Awareness and Vigilance 812
    48.3.3.1 Vigilance...812
    48.3.3.2 Interruptions and Distractions 813
    48.3.3.3 Situation Awareness in Anesthesiology 813
  48.3.4 Equipment Issues...814
    48.3.4.1 Equipment Usability..814
    48.3.4.2 How Do Anesthesiologists Learn to Use New Equipment?..........................815
    48.3.4.3 Equipment Maintenance Failures..815
  48.3.5 Teamwork and Communication in Anesthesiology 816
    48.3.5.1 Handovers...816
    48.3.5.2 Organizational Issues...817
  48.3.6 Interventions to Improve Anesthesia Patient Safety 817
48.4 Conclusions...817
Acknowledgments 817
References..819

48.1 Background

48.1.1 Anesthesiology and Patient Safety

Since the dawn of modern medicine, anesthesiologists have been on the forefront of health care innovation with key involvement in the development of hemodynamic monitoring, cardiopulmonary resuscitation, intravenous (IV) fluid management, blood banking (Moore 2005), mechanical ventilation, arterial blood gases, respiratory therapy, critical care medicine, and many types of acute and chronic pain management, to name but a few. Anesthesiology has been put forward as the "poster child" for patient safety (Kohn et al. 1999) as well as leading health care in the application of human factors engineering (HFE) concepts and methods (Weinger and Englund 1990; Weinger and

Slagle 2002b). For example, over the last three decades, anesthesiologists have been instrumental in the creation and deployment of mannequin-based simulators (Gaba and DeAnda 1988; Good and Gravenstein 1989), full-scale clinical simulation (Gaba and DeAnda 1988), and experiential teamwork training (Gaba et al. 2001) across medicine, nursing, pharmacy, and the allied professions.

Reducing perioperative morbidity and mortality has been of keen interest to anesthesiologists since John Snow began deliberate evaluations of the use of ether and chloroform in the late 1800s. However, scientific study of anesthesia adverse events did not attain wide appreciation prior to Beecher and Todd's seminal (1954) study that found 1:1560 deaths related to anesthesia during 599,548 anesthetics in 10 different institutions. The soaring cost of malpractice insurance in the 1970s and 1980s provided further impetus to the specialty's efforts to decrease the incidence of preventable adverse events. By the 1990s, the incidence of preventable anesthesia deaths had improved almost 10-fold overall (Forrest et al. 1990) and 100-fold for relatively healthy surgical patients (Lienhart et al. 2006; Warner et al. 1993).

Anesthesiologists work in a complex, rapidly changing, time-constrained, and stressful work environment, most commonly in the operating room (OR) but increasingly throughout the hospital and in ambulatory care settings. The anesthesia job requires managing a highly interactive system composed of the patient, sophisticated clinical equipment, myriad medications and supplies, surgeons, nurses, and other personnel, the OR environment, and other hospital microsystems (e.g., blood bank, pharmacy, medical records, and radiology). During administration of anesthesia, practitioners are required to be vigilant, multitask, and be able to rapidly make decisions and take actions with life-or-death consequences (Gaba et al. 1994, 1995; Weinger and Englund 1990). Technology plays a large role in daily anesthesia practice.

Because of strong national leadership and an evolving culture of safety, anesthesiology has made critical contributions to patient safety (Gaba 2000). Deliberate efforts include capturing, publicizing, and analyzing adverse events (e.g., the American Society of Anesthesiologists [ASA] Closed Claims Project), creating and promulgating the use of new technologies (e.g., pulse oximetry), iteratively improving medical technology (e.g., breathing circuits and gas delivery systems), changing health care processes (e.g., intensivist care), improving provider training (particularly in crisis event management), and creating and supporting an independent foundation whose sole purpose was to advance anesthesia patient safety (the Anesthesia Patient Safety Foundation or APSF).

48.2 Adverse Event Analysis and Nonroutine Events

48.2.1 Event Detection Methods in Anesthesiology

In 1978, Cooper and colleagues applied the critical incident technique first described by Flanagan to the study of anesthesia incidents (Cooper et al. 1978). In his studies, a critical incident was defined as a human error or equipment failure that could have led (if not discovered or corrected in time) or did lead to an undesirable outcome (Cooper et al. 1978, 1982, 1984). This work demonstrated, for example, the important contributions of equipment to anesthesia critical events. The ASA Closed Claims Project, first initiated in 1985, continues to collect valuable information about anesthesia-related events from malpractice claims that have already been settled or litigated (Cheney 1999). Anesthesiology researchers have analyzed the occurrence and etiology of critical events using direct observation (Fraind et al. 2002; Lingard et al. 2004) and review of videotapes of real cases (Mackenzie et al. 1996a,b; Weinger et al. 2004a).

The most popular method currently used in hospital quality assurance (QA) programs is to examine adverse events that have already occurred and "trace back" the problem attempting to pinpoint a "root cause." While this approach has a role, especially as a hypothesis-generating activity, it is limited for several reasons. Retrospective analysis of adverse events is contaminated by cognitive (especially hindsight, outcome, and attribution) bias making an accurate assessment of causality often impossible (Reason 1990; Woods and Cook 1999). Typical QA event analyses yield specific recommendations to prevent the same event from reoccurring. However, the cause of adverse events is often multifactorial and complex. The "same event" (i.e., identical sequence of subevents and failures) is actually unlikely to reoccur. As well, reporting and analysis of adverse events is problematic because of clinicians' reluctance to report mistakes due to concerns of social, legal, and regulatory retribution (Barach and Small 2000). Finally, traditional QA systems may bias clinicians toward reporting only specific types of events (e.g., sentinel or predefined event categories) and thus fail to capture many aspects of suboptimal care that might facilitate understanding of deficiencies in health care system performance and opportunities for improving patient safety (Weinger et al. 2003).

An increasingly well-accepted model of patient safety (Reason 1990, 1997) distinguishes between active errors, which are most commonly linked with frontline providers' actions or inactions, and latent conditions, which are systemic issues that predispose to subsequent active errors. Common examples of latent conditions are dysfunctional organizational structure and policy, inadequate training, faulty communication, and poorly designed medical devices (see also Chapter 22 on Human Error in Health Care; Reason, 1997; Weinger 2003; Weinger et al. 1998). An adverse outcome is believed to result primarily from the rare coincidence of multiple events evolving from latent conditions, triggered by more readily apparent active errors. Identification and correction of latent conditions are crucial to improving safety, yet it is unclear how to best identify these potential pathways to patient harm.

Safety experts have desired an event discovery tool that is nonjudgmental, has high compliance, facilitates discovery of latent conditions, and yields sufficient data to inform intervention strategies. One such approach may be the collection of "nonroutine events" (NREs; Oken et al. 2007; Weinger and Slagle 2002b; Weinger et al. 2003, 2004). An NRE is defined as any aspect of clinical care perceived by clinicians or observers as a deviation from optimal care based on the context of the clinical situation (Figure 48.1). Unlike approaches that focus on specific "adverse events," NRE identification is open-ended. Observers or clinicians identify all events that are unusual, unexpected, or deviate from optimal care.

The NRE construct extends the definition of noteworthy safety information beyond the occurrence or near occurrence of patient injury (Weinger et al. 2003). Thus, while NREs include "near misses" (Barach and Small 2000) and "critical incidents" (Cooper et al. 1978, 1984), they also encompass events that may not have an immediately obvious link to adverse outcomes but still could provide early clues to important latent conditions in health care systems. For example, the failure of the results of routine preoperative laboratory test values to be available just prior to surgery would *not* generally be considered by most clinicians or hospital quality improvement personnel to be a "near miss." Yet detection and analysis of several of these NRE could bring to light process problems in specimen delivery, clinical laboratory operations, or results reporting that might otherwise remain unappreciated until a patient injury event occurred as a consequence.

As shown in Figure 48.1, when an event occurs, the patient's care may be shifted to a care path that deviates from "optimal." If that event (and its consequences) is not detected and mitigated in a timely manner, then there can be an increased risk of an adverse event. Moreover, the occurrence of an NRE may increase the risk that a subsequent event will occur, or that if a subsequent event does occur, then it will be more difficult to detect or mitigate.

48.2.2 Comprehensive Open-Ended Nonroutine Event Surveys

A Comprehensive Open-ended Nonroutine Event Survey (CONES) was developed to elicit NRE immediately after care episodes (Oken et al. 2007). Trained nonclinician observers used this standardized interview script to elicit possible NRE based on the clinicians' response to nine YES/NO questions. After identifying a "suboptimal" event, additional, more open-ended, questions solicit data about the event's etiology and potential contributor factors (Weinger et al. 2003). CONES interviews rarely take more than 5 min. Thus, the CONES approach appears to meet many of the criteria for an effective event detection tool—it is efficient, nonjudgmental (focusing on any NRE and not errors or

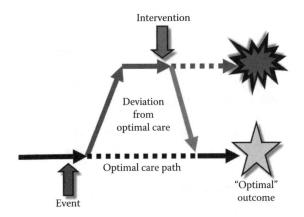

FIGURE 48.1

An optimal care path can be disrupted by an NRE. This figure shows the optimal care path predicted when a patient has a health care encounter. Often, events occur that produce a deviation from this optimal care path. If the event is not detected and its consequences mitigated through some kind of intervention by the clinician, the risk of a less than optimal outcome may be increased.

injuries), has high compliance, and yields ample data to facilitate discovery of latent condition and failure modes.

CONES data have been compared with data from the same institution's existing anesthesia QA process (Oken et al. 2007). The hospital's QA process relied on self-reporting of predefined adverse events. CONES interviews were conducted after 183 cases of varying patient, anesthesia, and surgical complexity. Fifty-five cases had at least one NRE (30.4% incidence). Over the same 30-month period, the hospital's anesthesiology QA process captured a total of 159 events (96.8% of which contained at least one NRE) out of the 8303 anesthetic procedures performed (1.9% overall incidence). The surgical patient and other clinical characteristics of the NRE-containing cases detected by the CONES methodology were more representative of the overall surgical population than were the demographics of the QA-reported cases. There were significant differences in NRE incidence ($p < 0.001$), patient impact (74.5% vs. 96.2%, $p < 0.001$), and injury (23.6% vs. 60.3%) between the CONES and QA data. Outcomes were more severe in the QA group ($p < 0.001$). Extrapolation of the CONES data suggested a significantly higher overall incidence of anesthesia-related patient injury (7.7% vs. only 1.0% from the QA method). However, the bulk of the additional injuries detected by the CONES methodology would be less severe than that detected by traditional QA self-reporting. This study showed that active surveillance using CONES can yield more events than are obtained using QA methods. The CONES approach also captured a wider variety of events than those identified by the QA process, thereby providing more information on which to base system understanding and improvement. Thus, the CONES data, being more representative of the overall surgical population, provided a more accurate picture of the true incidence of different types of events and their contributory factors. Potential limitations of this approach include the cost of observers and of data collation/analysis as well as the possibility of underreporting of NREs due to the normalization of deviance. The CONES methodology has been successfully applied to prospective studies of entire surgical teams (Slagle et al. 2009) and medication events by critical care nurses (Rayo et al. 2007).

48.2.3 Prospective Video-Based Assessment of NRE

We developed a system to facilitate the observation, collection, management, analysis, review, and archiving of multimedia data collected from the OR (Weinger et al. 2004). Audio, video, clinical information, workload, and behavioral task data are collected in real time, encoded with a common time code signal, processed, and archived to permanent media. Mackenzie, Xiao, and colleagues pioneered video data collection and analysis methods in trauma surgery and anesthesia that lead to a rich literature that has fostered our understanding of critical event etiology and the factors that affect clinical decision making of individuals and teams (Mackenzie et al. 1996b, 2004; Xiao et al. 1996).

In our video system, output from two digital cameras are directed to independent videotape recorders (VTRs). The patient's physiologic data as displayed on the vital signs monitor are recorded on a third VTR. A free-running time code generator sends frame-by-frame time code to each VTR as well as the data collection computer (Figure 48.2). Clinician participants wear wireless lavaliere microphones. Behavioral task analysis (BTA) (where clinical and nonclinical tasks are performed moment to moment) (Slagle et al. 2002), measurement of psychological and procedural workload (Weinger et al. 2004b), and vigilance assessment are performed concurrently and synchronized to the video streams. Subject mood and demographics are obtained pre- and post-case. The CONES instrument is administered after each video observation.

In our current system, videos are archived to formatted DVDs using stand-alone recorders. Video segments of interest are imported into a postproduction computer station for nonlinear video editing and deidentification (e.g., obscuration of patient and provider faces and other identifying features). Using custom software, computer-driven DVD players, each connected to its own monitor, replay the videos synchronously. Temporally related events (such as annotations of an expert reviewing the case) can also be controlled, presented, or recorded synchronously.

48.2.3.1 NRE Analysis

All identifiable patient and provider information are redacted from all datasets. Each case is assigned a unique randomly generated case number, as is each NRE identified. Data, including demographic and medical summaries, are entered into a custom password-protected database.

Two clinical experts independently review each event and determine whether the reported event meets the definition of an NRE. If there is disagreement, a consensus is reached after an open discussion of each disputed case. If agreement is not obtained, a third expert reviews the case, and the majority opinion is selected. To reduce hindsight bias, reviewers are shielded, to the extent possible, from prior knowledge of the case or its clinical course until the end of their review (Oken et al. 2007; Weinger et al. 2004).

Each NRE identified is categorized according to its patient impact (categorized as patient impact events

FIGURE 48.2

The operating room AV data collection system, and the video archiving, editing, and review stations. This composite figure shows (on the right) the components of our audio–video data collection system including video monitors, recorders, wireless microphones, data-logging computer, and two video cameras (inset) positioned in the OR, one behind the anesthesia machine and one directed down from behind the anesthesia cart. A trained assistant controls the camera views and audio levels while concurrently recording task and event data on a laptop computer. The bottom image on the left shows the video archiving station, where video data streams are transferred directly to three DVDs. The computer to the left in this bottom left image runs nonlinear video editing software to, for example, deidentify faces and voices. The video review station is depicted in the upper left image. Here, custom review and annotation software, running on a laptop computer, controls three DVD players to permit frame-synchronous playback of the three video data streams and time-stamped event annotation.

[PIE]), patient injury type, outcome severity, and putative contributing factors (Figure 48.3). Patient injury is defined as any unanticipated side effect or complication of anesthesia that affects the patient's postoperative course or quality of life. Thus, with this patient-centric definition, "injuries" included events that required the need for a higher level or prolonged postoperative care (e.g., intensive care unit [ICU] admission instead of discharge home after surgery) or emotional distress (e.g., memory of surgical events during general anesthesia). For anesthesiology events, a comprehensive structured taxonomy of 337 event descriptors (patient impact and injury events), organized hierarchically by clinical manifestation, was developed (Oken et al. 2007). Reviewers also assess whether any of 21 discrete contributory factors appear to play a role in the occurrence of each NRE. This postevent analysis is used to delineate differences between types of NRE and potential common sources or mechanisms of event etiology deserving of intervention.

We have prospectively collected video, task analysis, patient demographic, clinician variables, from hundreds of elective surgical cases representing a cross-section of anesthetic techniques, surgical procedures,

FIGURE 48.3

Still photos of videotaped anesthesia cases containing NRE. This composite figure shows two single images ("frame grabs") from videos containing NRE in the existing database of anesthesia cases. The image on the left shows a desaturation (low oxygen) event after extubation. The right image shows hypertension and bradycardia due to inadvertent intravascular injection of epinephrine by the surgical resident on the far left. A nurse anesthetist is seated and a medical student is on the left.

TABLE 48.1

Demographic Attributes of Videotaped Cases

Case Attributes	No NRE Cases (n=265)	NRE Cases (n=139)	Significance
Patient age	59.0 ± 13.2 (23–86)	59.7 ± 11.3 (30–84)	NS
Patient gender (% male)	92.1%	92.1%	NS
Patient ASA status (1/2/3/4 in %)	4.9/44.9/40.0/9.4	4.3/43.2/41.7/10.8	NS
% Performed by CRNA	41.9% (111)	35.3% (49)	NS
% By resident or fellow (CA1/CA2/CA3/fellow)	58.1% (12.8/24.2/20.7/0.4)	64.7% (18.7/30.2/15.8/0.0)	NS
Type of surgery: general/urology/OB/GYN/vascular	18.5%/15.1%/1.5%/7.2%	11.5%/20.9%/2.2%/11.5%	
Cardiothoracic/neurosurgery/ENT	10.9%/8.7%/12.8%	10.1%/10.8%/11.5%	NS
Orthopedics/plastics/other	17.0%/5.3%/2.6%	12.9%/7.2%/1.4%	Pearson χ^2

and patient complexity (Weinger et al. 2004a). In an analysis of 404 cases, 141 of these contained at least one NRE—an incidence of 35%. Thirty-three cases (23%) had more than one NRE (overall total of 190 NRE). The NRE and routine cases did not differ appreciably on high-level demographic variables (Table 48.1). There was patient impact in 145 of the prospective NRE (76%) although only 36 led to documented patient injury (usually mild and including psychological injuries). Clinician subjects rated 55 cases (39% of NRE cases) as having a 50% or greater chance of serious patient injury if the event had not been detected and/or managed properly (similar to Cooper's "critical incidents" [Cooper et al. 1978]). Table 48.2 shows the distribution of patient "impact events" associated with all NRE. Traditional adverse drug events (ADE) occurred in 28 (14.7%) of NREs.

In this dataset, there is a wide range of anesthesia events including failed, endobronchial and esophageal intubations, laryngospasm and bronchospasm, obstructed airways, profound oxygen desaturation, premature extubation, acute hypotension and hypertension, arrhythmias, myocardial ischemia, difficult or failed regional anesthetics, unanticipated movement, and surgical misadventures. A sample of typical NREs is shown in Table 48.3. Expert review suggested that the

most common factors contributing to NRE are patient factors; inadequate anesthesia provider's (AP) supervision, knowledge, experience, and/or judgment; equipment; and surgical factors (Table 48.4).

Multivariate logistic regression on an early subset of 367 cases showed an increased risk of NRE occurrence if the clinician reported, *before the case began*, difficulty sleeping the night previously (odds ratio 2.0 [95% CI of >1.0–3.9]). Additionally, decreased (less) clinical experience was an independent predictor of NRE occurrence (odds ratio 1.6 (>1.0–2.6)). Neither patient age nor physical status, as indicated by the ASA health status score, was a significant predictor of NRE occurrence.

48.2.3.2 Relationship of NRE to Significant Physiological Disturbances

Intraoperative disturbances in patient physiology predict more serious perioperative adverse events. We hypothesized that NRE-containing cases would have more episodes of significant physiological disturbance (SPD) than "routine" (i.e., non-NRE containing) cases, and that there would be an undesirable incidence of SPD in routine cases. The result of a preliminary study found that many NREs were associated with SPD. In addition, a small but significant incidence of NRE-unrelated SPDs occured in both NRE-containing and routine cases. This approach seems complementary to other patient safety efforts and could be used to validate more automated methods based on electronic health record data, the creation of more intelligent alerts, and predictive models to improve patient care.

TABLE 48.2

Distribution of Impact Events in Videotaped Anesthesia NREs

Event Category	No. of Events[a]	Percentage of NRE
Airway events	60	31.6
Cardiovascular events	53	27.9
Pulmonary events	30	15.8
Anesthetic procedures	22	11.6
Other organ systems	20	10.5
Surgical events	16	8.4
Neurological events	13	6.8
Patient disposition	9	4.7
Staff (clinician) impact	3	1.6

[a] There can be more than one impact event per NRE.

48.3 Performance Shaping Factors in Anesthesiology

A critical concept in HFE is that human performance is strongly affected by individual traits and myriad contextual factors including interpersonal/team,

TABLE 48.3

Representative Sample of NRE from Cases in the Database

No. of NRE	Brief Description of NRE
1	Desaturation to 74% after patient extubated. Required Narcan.
1	After induction, discovered that surgical attending was on vacation. Case canceled.
1	Clinician providing break gave opiate to a spontaneously breathing patient who then hypoventilated and desaturated. Required controlled ventilation until drug effect gone.
4	Hypotension with induction. Inadvertently treated with epinephrine instead of ephedrine (syringe swap). Muscle rigidity due to inadvertent bolus of opiate from poorly flowing IV. Drug concentration was double what was intended. Surgery almost done on the wrong side.
2	MI during induction of anesthesia. Pulmonary edema ensued. During treatment, RN distracted clinicians by insisting on doing a "time-out." Case canceled and patient transfered to ICU.
1	A-line tubing caught on OR table and was disconnected from the transducer during patient movement to gurney at the end of the case.
1	Accidental patient extubation by surgeon during rigid oral endoscopy. Table turned away 90°. Patient difficult to ventilate. Anesthesia attending called to help reintubate.
1	Esophageal intubation by junior resident. Distracted by research study. Reintubated.
1	Bloody nose due to failure to use nasal vasoconstrictor during awake fiberoptic nasal intubation. Resident sprayed with patient's blood upon successful intubation.

task/process, environmental, technology, and organizational factors (see Table 48.5; Weinger and Englund 1990). In this section, I discuss human factors research on the effects of these so-called performance shaping factors (or PSFs) on AP job performance. PSFs are most fruitfully considered within the context of a system-based approach to patient safety as popularized by Reason (1990), Rasmussen, Woods and Cook (2002), and others (Weinger and Englund, 1990). Due to space constraints, this section will be neither comprehensive (i.e., not all PSFs are discussed) nor complete (i.e., not all research that has been done is cited). Moreover, there is an emphasis on our own team's work, primarily since that is the work I know best.

TABLE 48.4

Putative Contributory Factors Identified in 156 Prospective Anesthesia NREs[a]

Contributory Factor Category	No. of Events	Percentage of NRE
Patient disease or unexpected response	104	54.7
AP knowledge, skill, experience, judgment, or supervision	87	45.8
Equipment failure or usability issues	48	25.3
Surgical factors, surgeon's actions	32	16.8
Other factors	30	15.8
Communication or coordination issues	28	14.7
Inadequate preoperative preparation or assessment	25	13.2
Patient positioning issues	22	11.6
Logistical or system issues	20	10.5

[a] Obtained by expert review and consensus. There is often more than one contributory factor per NRE.

48.3.1 Individual Factors

48.3.1.1 Experience and Expertise

A clear relationship between years of practical experience and performance has not been consistently demonstrated in anesthesiology research, particularly when clinicians are asked to manage unusual or difficult crisis situations. Simulation studies have consistently demonstrated significant performance gaps in *both* experienced and novice anesthesiologists (DeAnda and Gaba 1991; Murray et al. 2007; Schwid and O'Donnell 1992). For example, in a study of 35 practicing anesthesiologists and 33 anesthesia residents performing a series of 5 min simulated clinical scenarios, although performance improved with the amount of training, experienced physicians performed similarly to senior residents and both groups exhibited serious deficiencies managing some clinical events. An individual performing well in one event did not necessarily do well in others (Murray et al. 2007). Ericssen, Weinger, and others have similarly noted that years of experience do not necessarily correlate with expertise (Ericsson 2004; Weinger 2007). Studies of performance during realistic simulation have uncovered weaknesses even in highly protocolized activities such as Advanced Cardiac Life Support (ACLS; Kurrek et al. 1998; Wayne et al. 2008).

48.3.1.2 Fatigue and Sleep Deprivation

In round-the-clock industries where vigilance is critical (e.g., air traffic control or anesthesiology), the effects of fatigue and sleep deprivation on performance are of substantial concern. Fatigue impairs

TABLE 48.5

PSFs in Anesthesiology

Individual

    Experience and expertise—knowledge, skills, attitudes, and behaviors

    Demographic attributes—age, gender, ethnic, and cultural factors.

    Motivation and mood

    Fatigue/sleep deprivation/circadian effects

    State of health

    Substance use/abuse

Task

    Complexity

    Uncertainty and risk

    Task demands/workload

    Interruptions and distractions

Technology

    Design attributes (i.e., poor usability)

    Failure

Environment

    Temperature and humidity

    Lighting

    Physical ergonomics

    Crowding and disorganization

    Noise

Team

    Teamwork and interpersonal communication

    Conflict

Organizational

    Policies and procedures

    Incentives and disincentives

    Production pressure

attention, memory, and decision making; prolongs reaction times; and disrupts effective communication (Howard et al. 2002b; Weinger and Ancoli-Israel 2002a). Concerns about the dangers of sleep-deprived physicians taking care of patients led state legislatures and regulatory bodies to mandate limits to the work hours of physicians in training (Gaba and Howard 2002; Weinger and Ancoli-Israel 2002a), but as of yet no such restrictions are in place for experienced clinicians. In an early study, more than 60% of the almost 3000 APs who responded to an anonymous survey attributed at least one error in their practice to fatigue (Gravenstein et al. 1990).

Studies on sleep-deprived anesthesia residents show impairment of psychomotor performance and mood during actual cases (Cao et al. 2008) and in simulation studies (Denisco et al. 1987; Howard et al. 2003). Anesthesia residents appear to be chronically sleep deprived (Howard et al. 2002a). Using standard sleep lab techniques, Howard and colleagues demonstrated that

undisturbed anesthesia residents, even if they had *not* been on-call for two full days, had daytime sleep latencies comparable to those of patients with narcolepsy. Furthermore, often these residents denied falling asleep during the sleep tests, despite objective EEG evidence to the contrary. Sleep deprivation and fatigue appears to contribute to a higher incidence of needle stick injuries (Ayas et al. 2006) and automobile accidents (Barger 2005; Barger et al. 2006) in residents postcall compared with the overall U.S. population.

Routine performance may not be as adversely affected as is the management of critical events, but this remains a hypothesis to be tested. Fatigued APs at work use tactics to maintain an alert state including moving around, adding other duties, conversation, or reading (Cao et al. 2008; Howard et al. 2003).

In a recent study (Cao et al. 2008), we examined how fatigue affected anesthesia residents while performing clinical duties in the OR. Using a repeated-measures design, 15 pairs of day–night matched anesthesia cases were studied. Performance measures obtained in the OR included task times, workload ratings, and secondary vigilance. Participant's mood and psychomotor performance were also assessed before and after each case. Residents spent significantly higher proportions of case time on observing tasks, such as observing their physiological monitors, patient's airway, and IV fluids, during the maintenance phase of the operation ($p = 0.02$) at night than during the day. In addition, residents reported more negative mood at night than during the day, both before ($p = 0.02$) and after the case ($p < 0.01$). Workload assessments were not significantly different between day and night cases. The residents also appeared to be much more vigilant during the day than at night, especially during the induction ($p < 0.01$) and emergence phases ($p < 0.01$). When fatigued, residents altered their allocation of attention to primary task requirements, which may act to maintain their perceived workload at a manageable level.

48.3.1.3 Burnout

Burnout is a syndrome characterized by high emotional exhaustion (EE), high depersonalization (DP), and a low sense of personal accomplishment (LPA). Burnout is most often measured with the Maslach Burnout Index (MBI). Burnout can lead to health and psychological problems. The incidence of clinician burnout may be significant. We recently completed a study of burnout in perioperative clinicians using an anonymous web-based survey (Hyman et al. 2011). Our instrument consisted of 77 items including questions data, a modified form of the MBI for Human Services (MBI-HSs) and a Social Support and Personal

Coping (SSPC) survey. A summary global (burnout) score was calculated as the weighted average of the EE, DP, and LPA scores. One hundred forty-five subjects completed the survey (58% response rate). Of these, 46.2% were physicians (23.4% resident), 43.4% were nurses or nurse anesthetists, and 10.3% other health care personnel. After adjusting for gender and age, residents scored modestly higher than older MDs on the global score, EE and DP constructs, respectively. When compared to the non-MD roles, residents were significantly higher on the global score, EE, and DP items ($p < 0.05$ in all cases). Age had a strong negative relationship with DP and LPA. Finally, better health, personal support, and work satisfaction scores were related to a higher Maslach global score ($p < 0.05$). Thus, physicians (particularly residents) had the highest global burnout scores (i.e., at greater risk) compared to other health personnel. Neither age nor gender was associated with the global burnout score, but age was associated with the DP and LPA items. Three SPCC items (health, work satisfaction, and personal support) appeared protective against burnout.

48.3.2 Tasks and Workload

Gilbreth (1916) was the first to apply task analysis methods to the study of clinical performance. His time-and-motion studies showed that surgeons spent an inordinate amount of time looking for the correct instruments as they picked them off a tray. These findings led to the still current practice of the surgeon requesting instruments from a "scrub" nurse, who places the instrument in the surgeon's hand. Task analysis involves the structured decomposition of work activities and/or decisions and the classification of these activities as a series of tasks, processes, or classes. These kinds of techniques have been used to study clinical workflow, for example, in ICUs (Marasovic et al. 1997; Pierpont and Thilgen 1995; Wong et al. 2003), emergency departments (Taylor et al. 1997), and pharmacies (Guerrero et al. 1995), and of ward nurses (Paxton et al. 1996), radiologists (Gay et al. 1997), and APs (see the following text).

In one of the first task analysis studies in anesthesia, Drui et al. (1973) investigated how anesthesiologists spent their time in the OR. The practice of anesthesia was divided into a number of discrete activities, and the frequency and sequence of each activity were measured. They found that filling out the anesthesia record occupied a large proportion of the anesthesiologists' time but was rated as relatively unimportant and easy to perform. An unexpected finding was that anesthesiologists directed their attention away from the patient 42% of the time. The authors recommended automating

the task of creating the anesthesia record and redesigning the anesthesia machine to increase productivity and decrease the amount of distraction away from the patient-surgical field. Subsequent studies have corroborated and expanded these early results (Allard et al. 1995; Boquet et al. 1980; Kennedy et al. 1976; McDonald and Dzwonczyk 1988; McDonald et al. 1990; Weinger et al. 1994, 1997). More recent task analysis research has attempted to create a more comprehensive model of anesthesiology work so as to foster understanding of event etiology and the design of new devices and processes (Phipps et al. 2008).

Through a series of studies over the last 15 years, my colleagues and I have refined and applied behavioral task analysis methods to study anesthesiology care (Fraind et al. 2002; Weinger et al. 1994, 1997, 2004b). Custom software was developed to facilitate both data collection and analysis. The data collection application, typically run on a laptop or tablet PC, can be easily configured for any task list and facilitates highly efficient and reliable task recordings by observers. The team has developed robust methods for observer training and credentialing to assure high intra- and interrater reliability (Slagle et al. 2002). The task analysis application completely automates the process of collating and calculating the collected data, substantially reducing analysis time.

This methodology has been used successfully to study the impact of clinical experience (Weinger et al. 1994, 2001, 2004b; Weinger and Slagle 2002b), the effects of medical technology such as information systems (Schultz et al. 2006; Weinger et al. 1997; Wong et al. 2003), IV drug and fluid administration processes (Fraind et al. 2002), and smart infusion pumps (Rayo et al. 2007). We have also studied the effects of distracting activities such as reading (Slagle and Weinger 2009) and fatigue on AP workload and performance in both actual (Cao et al. 2008) and in realistically simulated (Howard et al. 2003) cases. For example, we investigated the task distribution and workload of experienced residents during the administration of anesthesia for cardiac surgery (Weinger et al. 1997). The use of electronic record keeping modestly decreased the time spent on record keeping during the pre-bypass maintenance phase of anesthesia care but without improving either workload or vigilance.

48.3.2.1 Measuring Clinical Workload

A major factor in the effect of additional tasks on performance appears to be what perceptual or cognitive resources are required for each new task and whether those resources are already taxed. Workload assessment is important both for evaluating the cognitive requirements of new workplace designs and equipment and for predicting workers' cognitive capacity for additional tasks. Workload can have important effects on clinical performance. For example, recovery from critical events may be impaired during high workload situations.

Different workload measurement techniques may be more sensitive and/or specific for different types of workload (Hicks and Wierwille 1979; Wierwille et al. 1985). Psychological workload measures include psychological tests and survey instruments (either retrospective or prospective). With subjective workload assessment, either an observer or the subjects themselves rate the participants' workload (or some component of it) on a predefined scale (Hill et al. 1992; Reid and Nygren 1988). For example, we assessed subjective workload asking both an observer and the subjects to rate the subjects' workload using a visual analog scale at random intervals during general anesthesia cases (Weinger et al. 1994, 1997). There is usually a very high ($R \sim 0.90$) correlation between participants' self-reported and observers' paired workload ratings. Subjective workload was significantly higher during induction of anesthesia compared with the rest of the case.

Increased workload activates the sympathetic nervous system producing physiological changes including increased heart rate and respiratory rate, decreased heart rate variability and galvanic skin response, and changes in pupil size and vocal patterns (Kantowitz and Casper 1988; Weinger and Englund 1990). The heart rate of APs increases significantly while they are administering anesthesia (Kain et al. 2002; Weinger et al. 2004b; Weinger and Slagle 2002b). More experienced individuals manifest less of an effect, although even experienced anesthesiologists' heart rates increase above baseline values during the induction and emergence phases of routine general anesthesia in healthy patients.

Procedural assessment techniques use alterations in primary and/or secondary task performance as an indirect measure of workload (Harris et al. 1982; Kantowitz and Casper 1988). For example, Gaba and Lee (1990) examined the ability of residents to answer paced arithmetic problems while also administering anesthesia. Secondary task performance was compromised in 40% of the samples (i.e., the problem was skipped or there was a more than 30s response lag). Similarly, Loeb and colleagues employed a simple vigilance probe as an indirect measure of workload (Loeb 1993, 1994, 1995). We have shown that the response latency to the illumination of an alarm light placed within the anesthesia monitors correlates with subjective workload ratings (Weinger et al. 1994, 1997, 2004b). Not only is the response time slower during induction than during maintenance, but novice residents consistently showed slower responses than did more experienced residents. This suggests that inexperienced clinicians may have less spare capacity to respond to new task demands, particularly during high workload conditions.

Measurement of the pace and difficulty of the tasks performed in a job is another way to assess workload. We described using time-and-motion data to generate a novel workload metric called "task density" (i.e., the number of tasks performed per unit time) (Weinger et al. 1994). Although task density correlated with subjective workload, its potential value was limited by the fact that the demands imposed by different tasks were weighted equally. Subsequently, "workload density" was proposed as a more accurate workload measure that incorporates the subjective workload associated with individual clinical tasks (Weinger et al. 1997, 2004b; Weinger and Slagle 2002b). Workload values for common anesthesia tasks were determined from the results of a questionnaire on which APs rated the difficulty of specific clinical tasks in terms of their mental and physical workload, and psychological stress. Workload density, calculated by multiplying the amount of time spent on each task by that task's workload rating, correlated with HR variability, alarm light response latency, and subjective workload (Vredenburgh et al. 2000; Weinger et al. 2000).

48.3.3 Situation Awareness and Vigilance

APs must be able to anticipate, detect, and aptly respond to events or problems. Situation awareness (SA) is a comprehensive and coherent representation of the current system state that is continuously updated based on repetitive assessment (Endsley 1995a,b). SA appears to be an essential prerequisite for safe operation of complex dynamic systems. SA requires the recognition of events that are caused or influenced by past precursors as well as the anticipation and prevention of potential future problems based upon an analysis of current conditions. Adequate "mental models" of the system and its critical elements are essential to effective SA. SA can be divided into three levels (Endsley 1995a,b): *Level 1—*perception of changes (this is analogous to vigilance); *Level 2*—diagnosis of the current state of the system (e.g., revised mental model); and *Level 3*—projection of future system state and choice of appropriate intervention(s) to maintain homeostasis. SA has been assessed in complex task domains using a variety of techniques including embedding extra tasks into the domain, asking participants to "think aloud," or querying specific data within the context of routine care.

48.3.3.1 Vigilance

Anesthesia vigilance can be adversely affected by many factors including clinician knowledge, skills, attitude or motivation, task complexity, workload, defective processes or equipment, and competing task demands (Weinger and Englund 1990; Weinger and Smith 1993). Environmental factors, limited cognitive resources, and psychological processes (e.g., distractions, memory lapses, fixation error, or inattention) can adversely impact vigilance, clinical decision making, and patient outcomes.

Researchers have examined the anesthesiologist's vigilance to auditory (Cooper and Cullen 1990) and visual (Loeb 1993; Weinger et al. 1994, 1997) alarm cues as well to changes in clinical variables (Blike 2000; Blike et al. 1999; Gurushanthaiah et al. 1995), in both the laboratory and during actual cases. For example, we demonstrated that novice anesthesia residents were slower to detect the illumination of an alarm light placed within their monitoring array (Weinger et al. 1994). Response rate was further impaired during periods of high workload such as the induction of anesthesia.

The maintenance of anesthesia is frequently uneventful, being characterized by lengthy periods of low cognitive or physical demands. Early task analysis studies found that, during routine maintenance periods, the clinician may perform few observable clinically relevant tasks. Such "idle" periods may occupy substantial parts (up to 40%) of routine cases (Drui et al. 1973; Loeb 1993; Weinger et al. 1994, 1997). During idle periods, many anesthetists add secondary tasks to their routine—these can be clinically relevant (rechecking the composition or organization of the anesthesia workspace, preparing meds for the next case, etc.) or unrelated to patient care (e.g., intraoperative reading, e-mail, personal tasks, phone conversations, responding to pages, perusing the internet, or casual conversations with other OR personnel).

48.3.3.2 Interruptions and Distractions

Distractions can have adverse consequences beyond purported decreases in vigilance. While performing non-patient care activities, new or ongoing clinical tasks often demand the providers' attention, representing an interruption and requiring reorientation of attention. When switching from one task to another, the cognitive readjustment between tasks, referred to as a "switch cost," results in slower responses that are more error-prone (Monsell 2003). It is more difficult and takes twice as long to be reoriented to (resume) more complex than routine tasks (Czerwinski et al. 2004). Task disruptions can have undesirable psychological effects (including more negative mood) and increase required cognitive effort (Zijlstra et al. 1999). According to the "cognitive fatigue model," cognitive disruptions increase cognitive demands and are an unexpected and uncontrollable source of mental stress, fatigue, and

information overload. Additional effort is required to ignore distractions and remain attendant to current task and goals (Cohen 1978, 1980). Task disruptions occur frequently in clinical domains (Chisholm et al. 2001; Dearden et al. 1996; Flynn et al. 1999; France et al. 2005; Potter et al. 2005) and may be important contributors to adverse events (Beso et al. 2005; Ely et al. 1995; Hicks et al. 2004; Page and Committee on the Work Environment for Nurses and Patient Safety 2004; Potter et al. 2005; Santell et al. 2003).

In one study, we sought to ascertain the incidence of intraoperative reading and measure its effects on clinicians' workload and vigilance (Slagle and Weinger 2009). In 172 selected general anesthetic cases in an academic medical center, a trained observer categorized the AP's activities into 37 possible tasks. Vigilance was assessed by the response time to a randomly illuminated alarm light. Observer- and subject-reported workload were scored at random intervals. Data from reading and nonreading periods of the same cases were compared to each other and to matched cases that contained no observed reading. The cases were matched before data analysis on the basis of case complexity and anesthesia type. Reading was observed in 35% of cases. In these 60 cases, providers read during 25% ± 3% of maintenance but not during induction or emergence. While nonreading cases (n = 112) and nonreading periods of reading cases did *not* differ in workload, vigilance, or task distribution, they both had significantly higher workload than reading periods. Vigilance was not different among the three groups. When reading, clinicians spent less time performing manual tasks, conversing with others, and record keeping. APs, even when being observed, read during a significant percentage of the maintenance period in many cases. However, reading occurred when workload was low and did not appear to affect a measure of vigilance.

48.3.3.3 Situation Awareness in Anesthesiology

Since Gaba et al. (1995) published a review of SA in anesthesiology 15 years ago, very little empirical research has been conducted on SA Levels 2 and 3. In a 2004 review, Wright et al. (2004) recommended the development and use of objective SA measures to assess performance during medical simulation—such studies have proven challenging for many of the reasons highlighted by these authors. In 2006, Drews and Westenskow reviewed the literature on the relationship between data display modalities and SA. While most studies have focused on the effects of display modality on event detection (SA Level 1, e.g., Gurushanthaiah et al. 1995), some have examined effects on diagnosis (SA Level 2, e.g., Blike et al. 1999) and even treatment decisions (e.g., Agutter et al. 2003). In fact, data display design and evaluation has been an active area of human factors research in anesthesiology (Sanderson 2006; see chapter on Human–Computer Interaction). There is also a rich literature in anesthesia on the design and limitations of auditory and visual alarms (Loeb et al. 1992; Sanderson 2006; Weinger and Smith 1993) including interesting research by Sanderson et al. (2008) on the use of vital signs sonifications that both alert and inform.

48.3.4 Equipment Issues

The interface between human and machine, or more specifically, between anesthesiologist and anesthesia equipment, plays a crucial role in anesthesia patient safety (Figure 48.4). Anesthesia equipment can be characterized by those devices that are primarily for the delivery of substances (gases, drugs, and fluids) and those used to monitor the outcome of that delivery and the physiological state of the patient. Yet these two groups increasingly share common attributes (e.g., microprocessor control) and associated problems. Although the percentage of anesthesia mishaps that are primarily due to equipment failure appears to be relatively small (Beydon et al. 2001; Cooper et al. 1984), the contribution of poor equipment design to use error may be significant. For example, the use of transesophageal echocardiography (a very complex

and cognitively demanding technology) while providing anesthesia care may, under some circumstances, impair vigilance for other clinical events (Weinger et al. 1997).

48.3.4.1 Equipment Usability

"Use error" occurs when a user does not interact with a technology in the manner intended by the designer. Accident investigations and formal studies suggest that most use error is due to deficiencies in user interface (UI) design (Weinger et al. 1998; Weinger et al. 2011). The UI includes all aspects of a device that interacts with users including the on-device controls and displays, labeling, quick reference cards, user and service manuals, etc. The users of a medical device include myriad clinicians, patients, cleaners, and clinical engineers.

At least two factors make the design of anesthesia equipment particularly difficult. First, the OR places appreciable physical and environment stresses on equipment. Second, it can be difficult to elicit from users precise or optimal design requirements. Block et al. (1985) described the installation of one of the first computerized monitoring systems specifically designed for anesthesiologists. Despite a careful predesign evaluation of the intended users to ascertain their needs and preferences, after the system was built, the users decided that they really wanted something different. This change

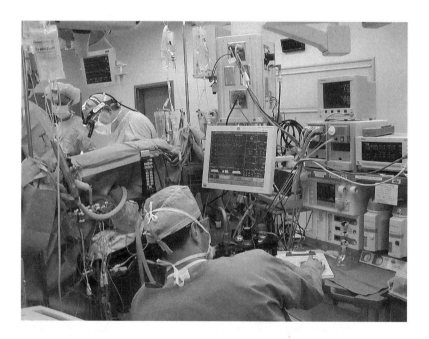

FIGURE 48.4

Anesthesiology workspace showing typical technology. The figure shows a typical anesthesiology workspace including anesthesia workstation for delivery of gases and anesthetic agents (right), integrated physiological monitor (middle), and multiple tubes, wires, and other display/control elements. The interface between the AP and the anesthesia equipment plays a crucial role in anesthesia patient safety. (With permission of Frank Painter, PhD.)

may less reflect capriciousness than it does changing experience and evolving expectations. Nonetheless, it makes the designer's task more difficult. At the very least, in the design and development of new equipment, continuous direct contact between the designer and the intended users is essential.

Poor design can manifest as too many, disorganized, or inappropriate control mechanisms. Software-related design problems include confusing displays or difficulty navigating between menus and functions. Other interface design issues include poor handling of artifact, disruptive display–control relationships, or irrational alarms.

Optimal design of complex microprocessor-based equipment requires a delicate balance between developing a device that is too complicated for the operator to understand versus one that becomes deceptively simple. If there are too many displays, or if the displays are confusing, performance may be suboptimal, and use errors can result, especially during emergent or unusual use situations. On the other hand, well-intended attempts to simplify a device can also produce poor results. Cook et al. (1991) described a humidifier that had been redesigned from a manual device to an automatic one. While the clinicians liked the newer device because it was "simpler," they did not understand the device's underlying operation. For example, they could not discern how to reactivate the device after an alarm, nor did they understand that the heating elements were turned off after an alarm. To "reset" the device after an alarm, users invariably turned the device off and then on again.

Each individual component of a system may be well thought out, but if the system design as a whole is faulty, the result will be unsatisfactory. Often the design is appropriate for one context but is transferred to another without taking into account the unique attributes of the new environment. For example, most of the original "integrated" physiological (vital signs) monitors were designed for use in the ICU and provided trend displays using a scale of six or more hours. For most anesthetic cases, the resulting display was far too compressed. The needs of anesthesiologists are substantially different than those of intensivists. New monitoring features and functions evolved as the requirements of other users became apparent.

Mosenkis (1994) suggested that excellent design should allow the clinician to use a device *correctly* the very first time they interact with it, preferably without reading the manual, whereas, to use that device *well*, practice may be required. The use of simulation may facilitate many of the design goals required for safer and more usable anesthesia equipment. By testing UI design options under almost real clinical situations, manufacturers are able to efficiently determine the strengths and weaknesses of each option.

48.3.4.2 How Do Anesthesiologists Learn to Use New Equipment?

A device manufacturer may assume that the user will depend on the user manual to become acquainted with a device during its initial use, learn its finer points, and troubleshoot problems. In fact, for the majority of anesthesia equipment, the clinicians never *see*, much less read, the user manuals. This is not due to laziness or even a busy work schedule. Rather, it is due to personality, tradition, and experience (with prior similar devices and with poorly designed manuals). Most health care providers use medical devices in the same way they use automobiles; they *expect* that a new device will work more or less the same as equivalent older devices (Mosenkis 1994). They then try to "figure out" any differences from their expectations. From a practical standpoint, most anesthesia equipment manuals are locked up in the clinical engineering offices. To address these issues, anesthesia equipment manufacturers have made greater effort to incorporate real-time help and other quick reference materials into their UIs.

How do anesthesiologists first learn to use their equipment? Typically, before their first use, they have a very brief "how to" in-service from a colleague or device company representative. This has proved problematic, especially with the introduction in the last 10 years of more complex "next-generation" anesthesia workstations and other technologies. It remains ambiguous as to whether technology training and competency is the responsibility of the manufacturers, the Food and Drug Administration, hospitals, the clinicians, or professional organizations.

48.3.4.3 Equipment Maintenance Failures

Maintenance errors often follow the same pattern as other types of errors: a chain of events, each one by itself insufficient to cause a disaster, contribute to the adverse outcome. One of the most notorious maintenance errors, first reported 40 years ago, is when the OR medical gases are switched (Mazze 1972). In this early case, a hospital engineer repaired the oxygen lines in one OR during the night so that he would not disrupt the daily routine. In a classic example of a chain of mishaps: (1) an oxygen connector was attached to one end of each of two hoses and a nitrous oxide connector to the other end; (2) both hoses were black, rather than color coded; and (3) he or she carefully and neatly twisted the hoses around each other. In these days before pulse oximetry, the first patient to be anesthetized with this machine died. It was soon realized that the second patient was undergoing the same problem, the anesthesia machine was completely disconnected, and the patient ventilated with an independent, portable, source of oxygen. The notoriety this case achieved was responsible both for an increased awareness of the possibility of OR gas switches and for the realization that continuous measurement of inspired oxygen was necessary. Sadly, gas switches still occur albeit under more insidious conditions, most typically during in-wall maintenance or construction (Holland 1991; Sato 1991). However, a more common gas supply error is the loss of central oxygen (Weller et al. 2007). Simulation has recently been used to better understand how anesthesiologists respond to these complex events (Mudumbai et al. 2010).

48.3.5 Teamwork and Communication in Anesthesiology

The anesthesiologist must function as an integral part of the OR team (Smith and Mishra 2010). OR team communication is affected by traditions, unspoken expectations, general assumptions regarding task distribution, chain-of-command hierarchies, as well as individual emotional and behavioral components. Alterations in any of these factors can impair effective team function (Kanki et al. 1989).

Failures of communication and teamwork have been identified as critical contributors to medical adverse events (Kohn et al. 1999). Even though most medical professionals work in multidisciplinary teams, training has traditionally occurred in discipline-specific silos. OR team performance can also be adversely affected by dysfunctional interpersonal interactions among team members. To assure the safest patient care, health care workers need to learn to work together in multidisciplinary teams. Thus, the anesthesiologist who is confronted with a new surgeon, OR nurse, or anesthesia resident will need to be sensitive to the "new interpersonal environment," making a special effort to communicate clearly and unambiguously, particularly under stressful conditions, such as high workload, fatigue, sleep deprivation, or unusual events.

Well-described programs to improve health care teamwork, such as TeamSTEPPS (Weaver et al. 2010), MedTeams (Morey et al. 2002), and crisis resource management (CRM) training (Howard et al. 1992; Knudson et al. 2008), have been introduced although a definitive outcome study is still lacking. Several observer-based assessment tools have been used to quantify team performance during actual patient care and in realistic simulations (Flin et al. 2003; Wright et al. 2009; see various chapters on simulation, training, and teamwork in this handbook).

48.3.5.1 Handovers

The care of hospitalized patients is marked by numerous transitions in care, including handovers of patient care responsibility at shift changes. Care transitions involving interprofessional handovers appear to be particularly vulnerable to communication failures. Poorly performed handovers can lead to subsequent errors, care lapses, inefficiencies, and patient harm (Horwitz et al. 2008; Kitch et al. 2008). Anesthesia Providers (APs) will hand over care to each other (e.g., when cases run long and duty shifts end) as well as postoperatively to Post-Anesthesia Care Unit (PACU) and ICU nurses. Successful handovers avoid unwarranted shifts in goals, decisions, priorities, or plans, including missing tasks or tasks unnecessarily duplicated by two or more clinicians (Patterson et al. 2004). Thus, handovers can be challenging, particularly when communication must cross professional boundaries (e.g., nurse and physician), where differences in culture, training, norms, attitudes, perspectives, goals, expectations, status, gender, and socioeconomic factors can predispose to misunderstandings.

We developed and deployed a simulation-based curriculum to improve actual handovers between APs and nurses (RN) in an adult (VUH) and a pediatric (VCH) PACU (Weinger et al. 2009). A simulation-based curriculum was developed that focused on communication, interpersonal skills, and strategies to manage common obstacles to effective handovers. To improve PACU handovers, APs and RNs all did web-based didactics then participated together in 2 h participatory simulation-based training sessions using standardized patients and clinicians, mannequin simulators, and facilitated video debriefing. Six months after the initial course, a 1 h simulation-based refresher course was offered only to VUH PACU clinicians. All trainees evaluated the course using a 10-point (0–9) scale. Trainees' self-evaluations (on a 5-point scale) of their actual PACU handovers were collected throughout the study period.

We trained a total of 452 clinicians (211 APs and 241 RNs). The interdisciplinary simulation-based handover training curriculum was very well received by experienced clinicians and favorably affected real-world clinician perceptions of handover interactions and effectiveness. Trained nurse observers scored the effectiveness of 981 actual PACU handovers over 16 months using a validated assessment tool. Baseline (pretraining) data were stable in both PACUs with the majority of handovers being rated as "ineffective" (global score of <3 out of 5). Posttraining, handover quality improved significantly with most handovers being rated as "effective" (≥3) in both units.

48.3.5.2 Organizational Issues

In the last decade, patients presenting for anesthesia have a greater burden of disease, and perioperative technology (both devices and drugs) is more complicated. With increasing demand for surgical services but reduced financial capital for new OR construction, surgical schedules are busier than ever. The relative shortage of experienced OR nurses in many parts of the country further constrain OR scheduling. Thus, APs are typically under appreciable pressure to maximize surgical productivity, primarily through timely case starts and minimal "turnover" times (Singer et al. 2003). Pressures to get work done (i.e., efficiency) can conflict with safety concerns (Landau and Chisholm 1995). Organizational issues affecting anesthesia patient safety in this way include production pressure (e.g., avoidance of elective case cancelation on the day of surgery), staffing decisions (e.g., reduction of "after hours" pharmacy and anesthesia technician support), misplaced priorities (i.e., OR efficiency to the detriment of safer care), inappropriate incentives or disincentives (e.g., rewarding providers with short turnover times and punishing those with case cancelations), and normalization of deviance (making routine efficiency shortcuts that degrade safety margins). As a result, while perioperative care is safer overall, having surgery after hours or the occurrence of NREs can predispose to catastrophe.

Moreover, in spite of increasing admonitions to the contrary, the "bad apple" approach to patient safety (Woods and Cook 2002) is still alive and well in many health care organizations. While most health care managers can be heard talking about how they take "a systems approach to safety," when an adverse event occurs, "blame-shame-and-train" remain a common response. More generally, there remains in health care an underemphasis on systems-based process and technology redesign and an overemphasis on policies, procedures, and training. While the specialty of anesthesiology has demonstrated 40 years of steady progress in patient safety and the resulting knowledge, tools, and attitudes have spread to other medical specialties (see Chapters 40 to 45), there is still plenty of work ahead before health care becomes an "ultrasafe" (Amalberti et al. 2005) and "high-reliability" (Carroll and Rudolph 2006) industry.

48.3.6 Interventions to Improve Anesthesia Patient Safety

As have been elucidated throughout this chapter, there are a variety of methods, tools, and techniques that have been shown or are strongly suggested to be important to assuring patient safety during anesthesia care.

The most critical intervention is to maintain conscious awareness of the risks to safety during each patient care interaction. Some of the more important actions that anesthesia professionals (and their patients) can take to improve perioperative patient safety are listed in Table 48.6.

48.4 Conclusions

In this chapter, I survey a cross-section of human factors research and relevant findings in the medical specialty of anesthesiology. Anesthesiology is a highly technological specialty where any adverse event is considered unacceptable. Thus, starting about 40 years ago, the specialty turned to engineering and psychology to understand the nature of adverse events and technology problems. A variety of HFE methods have now been employed to study anesthesia care with demonstrable improvements in systems, processes, and technologies. This chapter summarizes some of our research on NREs, a broad concept of care deviation borrowed from the process-control industries. While HFE methods and findings on cognitive factors in anesthesiology are less developed, a growing body of literature has begun to address the role in anesthesia safety of fatigue, burnout, interruptions and distraction, and situation awareness. From a patient safety standpoint, perhaps the most important area of investigation is improved understanding of teamwork failures and the design of training and related interventions. In addition to naturalistic studies, high-fidelity simulation has proven to be an effective controlled setting for studying many human factors issues in anesthesia care.

Acknowledgments

Dr. Weinger's writing efforts were supported in part through a grant from the Veterans Affairs Health Services Research and Development Service (VA HSR&D). The research described in this chapter would not have been possible without grant support over the last 10 years from the VA HSR&D, Agency for Healthcare Research and Quality (AHRQ), National Institute of Health (NIH), Anesthesia Patient Safety Foundation, National Patient Safety Foundation, United States Food and Drug Administration, and Vanderbilt University. Many dedicated individuals have contributed to our

TABLE 48.6

Important Interventions to Improve Anesthesia Patient Safety

Individual patients
Know who are your doctors and what they propose to do to care for you
Keep track of your care plan—if you are too sick, ask a family member or colleague
Ask questions especially if something is not right
Call for help early or if you have any doubt about how you are doing
Individual APs
Stay current in your knowledge and skills
Be on the hunt for potential cognitive biases (your own and others)
Participate in simulation-based training
Do not work when you are not well or are sleep deprived
Know your limitations
Ask for help early
Teamwork and communication
Perform a preoperative briefing with the responsible surgeon and conduct a preincision time-out
Communicate clearly and concisely
When in doubt, ask
Speak up if you have concerns (and listen to others)
Use read-back to "close the loop"
Participate in simulation-based team training
Take the time to do a complete and effective handover
When there are conflicts, keep the patient's safety and care foremost in everyone's mind
Tasks and workload
Maintain your and the team's situation awareness
Anticipate and balance workload
If workload becomes excessive, call for help
Avoid interruptions and distractions
Technology
Know your technology and its limitations
Use the best tools for the specific task or job at hand
Double-check all questionable sources of information
Seek disconfirming evidence as well as confirming evidence
When equipment is problematic or fails, stay focused on the most important tasks; Do not get trapped in troubleshooting; instead call for help
Participate in your facility's purchase decisions—look for devices that are easy to use and include critical safety features.
Environment
Keep your work environment organized and clean.
Avoid overcrowding and physical impediments to doing your job
Maintain lighting to assure that you can effectively monitor the patient and the workspace
Adjust your monitor alarms to the right thresholds and audible level—always leave them on
Stay in control of the sound levels in the OR—you must be able to hear the monitors, alarms, and conversations
Insist on a "sterile cockpit" during the time-out, induction, high workload periods, and critical events
Organizational
Know and follow your organizations' rules, policies, and procedures, except when they are clearly unsafe for this particular patient—then discuss and document your decisions/actions
Avoid internal and external pressures to do something that is not safe
Report all safety events even if no harm comes to the patient
Participate in your organization's quality improvement efforts and patient safety initiatives
Speak up to facility management when you discover unsafe conditions

research, but a special thanks goes to Jason Slagle, PhD, who continues in the trenches after nearly 10 years.

References

Agutter, J., F. Drews, N. Syroid, D. Westneskow, R. Albert, D. Strayer, J. Bermudez, and M. B. Weinger. 2003. Evaluation of graphic cardiovascular display in a high-fidelity simulator. *Anesth Analg* 97 (5):1403–1413.

Allard, J., D. Y. Dzwonczyk, F. E. Block Jr., and J. S. McDonald. 1995. Effect of automatic record keeping on vigilance and record keeping time. *Br J Anaesth* 74:619–626.

Amalberti, R., Y. Auroy, D. Berwick, and P. Barach. 2005. Five system barriers to achieving ultrasafe health care. *Ann Intern Med* 142 (9):756–764.

Ayas, N. T., L. K. Barger, B. E. Cade, D. M. Hashimoto, B. Rosner, J. W. Cronin, F. E. Speizer, and C. A. Czeisler. 2006. Extended work duration and the risk of self-reported percutaneous injuries in interns. *JAMA* 296 (9):1055–1062.

Barach, P. and S. D. Small. 2000. Reporting and preventing medical mishaps: Lessons from non-medical near miss reporting systems. *Br Med J* 320 (7237):759–763.

Barger, L. K. 2005. Extended work shifts and the risk of motor vehicle crashes among interns. *N Engl J Med* 352 (2):125–134.

Barger, L. K., N. T. Ayas, B. E. Cade, J. W. Cronin, B. Rosner, F. E. Speizer, and C. A. Czeisler. 2006. Impact of extended-duration shifts on medical errors, adverse events, and attentional failures. *PLOS Med* 3 (12):2440–2448.

Beecher, H. K. and D. P. Todd. 1954. A study of the deaths associated with anesthesia and surgery: Based on a study of 599,548 anesthesias in ten institutions 1948–1952, inclusive. *Ann Surg* 140 (1):2–35.

Beso, A., B. D. Franklin, and N. Barber. 2005. The frequency and potential causes of dispensing errors in a hospital pharmacy. *Pharm World Sci* 27:182–190.

Beydon, L., F. Conreux, R. Le Gall, D. Safran, J. B. Cazalaa, and "Sous-commission de Materiovigilance" for Anaesthesia and Intensive Care. 2001. Analysis of the French health ministry's national register of incidents involving medical devices in anaesthesia and intensive care. *Br J Anaesth* 86:382–387.

Blike, G. T. 2000. The challenges of human engineering research. *J Clin Monit Comput* 15:413–415.

Blike, G. T., S. D. Surgenor, and K. Whalen. 1999. A graphical object display improves anesthesiologists' performance on a simulated diagnostic task. *J Clin Monit Comput* 15 (1):37–44.

Block, F. E. Jr., L. W. Burton, M. D. Rafal, K. Burton, C. Newey, L. Dowell, F. F. Klein, D. A. Davis, and M. H. Harmel. 1985. Two computer-based anesthetic monitors, the Duke automatic monitoring equipment (DAME) system and the MICRODAME. *J Clin Monit* 1:30–51.

Boquet, G., J. A. Bushman, and H. T. Davenport. 1980. The anaesthetic machine—A study of function and design. *Br J Anaesth* 52:61–67.

Cao, C. G., M. B. Weinger, J. Slagle, C. Zhou, J. Ou, S. Gillin, B. Sheh, and W. Mazzei. 2008. Differences in day and night shift clinical performance in anesthesiology. *Hum Factors* 50 (2):276–290.

Carroll, J. S., and J. W. Rudolph. 2006. Design of high reliability organizations in health care. *Qual Saf Health Care* 15 (Suppl 1):i4–i9.

Cheney, F. W. 1999. The American Society of Anesthesiologists Closed Claims Project: What have we learned, how has it affected practice, and how will it affect practice in the future? *Anesthesiology* 91 (2):552–556.

Chisholm, C. D., A. M. Dornfield, D. R. Nelson, and W. H. Cordell. 2001. Work interrupted: A comparison of workplace interruptions in emergency departments and primary care offices. *Ann Emerg Med* 38 (2):146–151.

Cohen, S., ed. 1978. Environmental load and the allocation of attention. In *Advances in Environmental Psychology, Vol. 1. The Urban Environment*, A. Baum, J. E. Singer, and S. Valins, eds. pp. 1–29. Hillsdale, NJ: Lawrence Erlbaum & Associates.

Cohen, S. 1980. After effects of stress on human performance and social behavior: A review of research and theory. *Psychol Bull* 88:82–108.

Cook, R. I., S. S. Potter, D. D. Woods, and J. S. McDonald. 1991. Evaluating the human engineering of microprocessor controlled operating room devices. *J Clin Monit* 7:217–226.

Cooper, J. O. and B. F. Cullen. 1990. Observer reliability in detecting surreptitious random occlusions of the monaural esophageal stethoscope. *J Clin Monit* 6:271–275.

Cooper, J. B., C. D. Long, R. S. Newbower, and J. H. Philip. 1982. Critical incidents associated with intraoperative exchanges of anesthesia personnel. *Anesthesiology* 56:456–461.

Cooper, J. B., R. S. Newbower, and R. J. Kitz. 1984. An analysis of major errors and equipment failures in anesthesia management: Considerations for prevention and detection. *Anesthesiology* 60:34–42.

Cooper, J. B., R. S. Newbower, C. D. Long, and B. McPeek. 1978. Preventable anesthesia mishaps: A study of human factors. *Anesthesiology* 49:399–406.

Czerwinski, M., E. Horvitz, and S. Wilhite. 2004. A diary study of task switching and interruptions. Paper read at *Conference on Human Factors in Computing Systems*, April 24–29, at Vienna, Austria.

DeAnda, A. and D. M. Gaba. 1991. The role of experience in the response to simulated critical incidents. *Anesth Analg* 72:308–315.

Dearden, A., M. Smithers, and A. Thapar. 1996. Interruptions during general practice consultations—The patients' view. *Fam Practice* 13 (2):166–169.

Denisco, R. A., J. N. Drummond, and J. S. Gravenstein. 1987. The effect of fatigue on the performance of a simulated anesthetic monitoring task. *J Clin Monit* 3:22–24.

Drews, F.A. and D.R. Westenskow. 2006. Display design in Anesthesia. *Human Factors* 48:59–71.

Drui, A. B., R. J. Behm, and W. E. Martin. 1973. Predesign investigation of the anesthesia operational environment. *Anesth Analg* 52:584–591.

Ely, J., W. Levinson, N. Elder, A. Mainous, and D. Vinson. 1995. Perceived causes of family physicians' errors. *J Fam Prac* 40 (4):337–344.

Endsley, M. R. 1995a. Measurement of situation awareness in dynamic systems. *Hum Factors* 37:65–84.

Endsley, M. R. 1995b. Toward a theory of situation awareness in dynamic systems. *Hum Factors* 37:32–64.

Ericsson, K. A. 2004. Deliberate practice and the acquisition and maintenance of expert performance in medicine and related domains. *Acad Med* 79 (Suppl. 10):S70–S81.

Flin, R., R. Patey, R. Glavin, and N. Maran. 2003. Anaesthetists' non-technical skills. *Br J Anaesth* 105(1): 38–44.

Flynn, E. A., K. N. Barker, J. T. Gibson, R. E. Pearson, and B. A. Berger. 1999. Impact of interruptions and distractions on dispensing errors in an ambulatory care pharmacy. *Am J Health Syst Pharm* 56:1319–1325.

Forrest, J. B., M. K. Cahalan, K. Rehder, C. H. Goldsmith, W. J. Levy, L. Strunin, W. Bota et al. 1990. Multicenter study of general anesthesia. II. Results. *Anesthesiology* 72:262–268.

Fraind, D. B., J. M. Slagle, V. A. Tubbesing, S. A. Hughes, and M. B. Weinger. 2002. Reengineering intravenous drug and fluid administration processes in the operating room: Step one: Task analysis of existing processes. *Anesthesiology* 97 (1):139–147.

France, D. J., S. Levin, R. Hemphill, K. Chen, D. Rickard, R. Makowski, I. Jones, and D. Aronsky. 2005. Emergency physicians' behaviors and workload in the presence of an electronic whiteboard. *Int J Med Inform* 74 (10):827–837.

Gaba, D. M. 2000. Anaesthesiology as a model for patient safety in health care. *BMJ* 320 (7237):785–788.

Gaba, D. M. and A. DeAnda. 1988. A comprehensive anesthesia simulation environment: Recreating the operating room for research and teaching. *Anesthesiology* 69:387–394.

Gaba, D. M., K. J. Fish, and S. K. Howard. 1994. *Crisis Management in Anesthesiology*. New York: Churchill Livingstone.

Gaba, D. M. and S. K. Howard. 2002. Patient safety: Fatigue among clinicians and the safety of patients. *N Engl J Med* 347 (16):1249–1255.

Gaba, D. M., S. K. Howard, K. J. Fish, B. E. Smith, and Y. A. Sowb. 2001. Simulation-based training in anesthesia crisis resource management (ACRM): A decade of experience. *Simul Gaming* 32:175–193.

Gaba, D. M., S. K. Howard, and S. D. Small. 1995. Situation awareness in anesthesiology. *Hum Factors* 37 (1):20–31.

Gaba, D. M. and T. Lee. 1990. Measuring the workload of the anesthesiologist. *Anesth Analg* 71:354–361.

Gay, S. B., A. H. Sobel, L. Q. Young, and S. J. Dwer. 1997. Processes involved in reading imaging studies: Workflow analysis and implications for workstation development. *J Digit Imaging* 10:40–45.

Gilbreth, F. B. 1916. Motion study in surgery. *Can J Med Surg* 40:22–31.

Good, M. L. and J. S. Gravenstein. 1989. Anesthesia simulators and training devices. *Int Anesth Clin* 27:161–166.

Gravenstein, J. S., J. B. Cooper, and F. K. Orkin. 1990. Work and rest cycles in anesthesia practice. *Anesthesiology* 72:737–742.

Guerrero, R. M., N. A. Nickman, and J. N. Bair. 1995. Work activities of pharmacy teams with drug distribution and clinical responsibilities. *Am J Health Syst Pharm* 52:614–620.

Gurushanthaiah, K., M. B. Weinger, and C. E. Englund. 1995. Visual display format affects the ability of anesthesiologists to detect acute physiologic changes. A laboratory study employing a clinical display simulator. *Anesthesiology* 83:1184–1193.

Harris, R. L., J. R. Tole, A. T. Stephens, and A. R. Ephrath. 1982. Visual scanning behavior and pilot workload. *Aviat Space Environ Med* 53:1067–1072.

Hicks, R. W., S. C. Becker, D. Krenzischeck, and S. C. Beyea. 2004. Medication errors in the PACU: A secondary analysis of MEDMARX findings. *J Perianesth Nurs* 19:18–28.

Hicks, T. G. and W. W. Wierwille. 1979. Comparison of five mental workload assessment procedures in a moving-base driving simulator. *Hum Factors* 21 (2):129–143.

Hill, S. G., H. P. Iavecchia, J. C. Byers, A. C. Bittner, A. L. Zaklad, and R. E. Christ. 1992. Comparison of four subjective workload rating scales. *Hum Factors* 34 (4):429–439.

Holland, R. 1991. Another "wrong gas" incident in Hong Kong. *APSF Newsl* 6:9.

Horwitz, L. I., T. Moin, H. M. Krumholz, L. Wang, and E. H. Bradley. 2008. Consequences of inadequate sign-out for patient care. *Arch Intern Med* 168 (16):1755–1760.

Howard, S. K., D. M. Gaba, K. J. Fish, G. Yang, and F. H. Sarnquist. 1992. Anesthesia crisis resource management training: Teaching anesthesiologists to handle critical incidents. *Aviat Space Environ Med* 63 (9):763–770.

Howard, S. K., D. M. Gaba, M. R. Rosekind, and V. P. Zarcone. 2002a. Excessive daytime sleepiness in resident physicians: Risks, intervention, and implication. *Acad Med* 77 (10):1019–1025.

Howard, S. K., D. M. Gaba, B. E. Smith, M. B. Weinger, C. Herndon, S. Keshavacharya, and M. R. Rosekind. 2003. Simulation study of rested versus sleep-deprived anesthesiologists. *Anesthesiology* 98 (6):1345–1355; discussion 5A.

Howard, S. K., M. R. Rosekind, J. D. Katz, and A. J. Berry. 2002b. Fatigue in anesthesia. *Anesthesiology* 97 (5):1281–1294.

Hyman, S. A., D. Michaels, J. M. Berry, J. S. Schildcrout, N. D. Mercaldo, and M. B. Weinger. (2011). Risk of burnout in perioperative clinicians: A survey and literature review. *Anesthesiology* 114(1):194–204.

Kain, Z. N., K.-M. Chan, J. D. Katz, A. Higam, L. Fleisher, J. Dolev, and L. E. Rosenfeld. 2002. Anesthesiologists and acute perioperative stress: A cohort study. *Anesth Analg* 95:177–183.

Kanki, B. G., S. Lozito, and H. C. Foushee. 1989. Communication indices of crew coordination. *Aviat Space Environ Med* 60:56–60.

Kantowitz, B. H. and P. A. Casper. 1988. Human workload in aviation. In *Human Factors in Aviation*, E. L. Wiener and D. C. Nagel, eds. pp. 157–187. San Diego, CA: Academic Press.

Kennedy, P. J., F. Fiengold, E. L. Wiener, and R. S. Hosek. 1976. Analysis of tasks and human factors in anesthesia for coronary-artery bypass. *Anesth Analg* 55:374–377.

Kitch, B. T., J. B. Cooper, W. M. Zapol, J. E. Marder, A. Karson, M. Hutter, and E. G. Campbell. 2008. Handoffs causing patient harm: A survey of medical and surgical house staff. *Jt Comm J Qual Patient Saf* 34 (10):563–570.

Knudson, M. M., L. Khaw, M. K. Bullard, R. Dicker, M. J. Cohen, K. Staudenmayer, J. Sadjadi, S. Howard, D. Gaba, and T. Krummel. 2008. Trauma training in simulation: Translating skills from SIM time to real time. *J Trauma* 64 (2):255–263.

Kohn, L. T., J. M. Corrigan, and M. S. Donaldson, eds. 1999. *To Err Is Human: Building a Safer Health System*. Washington, DC: National Academy Press.

Kurrek, M. M., J. H. Devitt, and M. M. Cohen. 1998. Cardiac arrest in the OR: How are our ACLS skills? *Can J Anaesth* 45:130–132.

Landau, M. and D. Chisholm. 1995. The arrogance of optimism: Notes on failure-avoidance management. *J Conting Crisis Manag* 3(3):67–80.

Lienhart, A., Y. Auroy, F. Péquignot, D. Benhamou, J. Warszawski, M. Bovet, and E. Jougla. 2006. Survey of anesthesia-related mortality in France. *Anesthesiology* 105 (6):1087–1097.

Lingard, L., S. Espin, S. Whyte, G. Regehr, G. R. Baker, R. Reznick, J. Bohnen, B. Orser, D. Doran, and E. Grober. 2004. Communication failures in the operating room: An observational classification of recurrent types and effects. *Qual Saf Health Care* 13 (5):330–334.

Loeb, R. G. 1993. A measure of intraoperative attention to monitor displays. *Anesth Analg* 76:337–341.

Loeb, R. G. 1994. Monitor surveillance and vigilance of anesthesia residents. *Anesthesiology* 80:527–533.

Loeb, R. G. 1995. Manual record keeping is not necessary for anesthesia vigilance. *J Clin Monit* 11:9–13.

Loeb, R. G., B. R. Jones, R. A. Leonard, and K. Behrman. 1992. Recognition accuracy of current operating room alarms. *Anesth Analg* 75 (4):499–505.

Mackenzie, C. F., N. J. Jefferies, W. A. Hunter, W. N. Bernhard, and Y. Xiao. 1996a. Comparison of self-reporting of deficiencies in airway management with video analyses of actual performance. *Hum Factors* 38:623–635.

Mackenzie, C. F., P. Martin, and Y. Xiao. 1996b. Video analysis of prolonged uncorrected esophageal intubation. *Anesthesiology* 84:1494–1503.

Mackenzie, C. F., Y. Xiao, and R. Horst. 2004. Video task analysis in high performance teams. *Cognition Technol Work* 6:139–147.

Marasovic, C., C. Kenney, D. Elliott, and D. Sindhusake. 1997. A comparison of nursing activities associated with manual and automated documentation in an Australian intensive care unit. *Computers Nurs* 15:205–211.

Mazze, R. I. 1972. Therapeutic misadventures with O_2 delivery systems: The need for continuous in-line O_2 monitors. *Anesth Analg* 51:787–792.

McDonald, J. S., and R. R. Dzwonczyk. 1988. A time and motion study of the anaesthetist's intraoperative time. *Br J Anaesth* 61:738–742.

McDonald, J. S., R. Dzwonczyk, B. Gupta, and M. Dahl. 1990. A second time-study of the anaesthetist's intraoperative period. *Br J Anaesth* 64:582–585.

Monsell, S. 2003. Task switching. *Trends Cogn Sci* 7:134–140.

Moore, S. B. 2005. A brief history of the early years of blood transfusion at the Mayo Clinic: The first blood bank in the United States (1935). *Transfus Med Rev* 19 (3):241–245.

Morey, J. C., R. Simon, G. D. Jay, R. L. Wears, M. Salisbury, K. A. Dukes, and S. D. Berns. 2002. Error reduction and performance improvement in the emergency department through formal teamwork training: Evaluation results of the MedTeams project. *Health Serv Res* 37:1553–1581.

Mosenkis, R. 1994. Human factors in design. In *Medical Devices*, C. W. D. van Gruting, ed. pp. 41–51. Amsterdam, the Netherlands: Elsevier.

Mudumbai, S. C., R. Fanning, S. K. Howard, M. F. Davies, and D. M. Gaba. 2010. Use of medical simulation to explore equipment failures and human-machine interactions in anesthesia machine pipeline supply crossover. *Anesth Analg* 110 (5):1292–1296.

Murray, D. J., J. R. Boulet, M. Avidan, J. F. Kras, B. Henrichs, J. Woodhouse, and A. S. Evers. 2007. Performance of residents and anesthesiologists in a simulation-based skill assessment. *Anesthesiology* 107 (5):705–713.

Oken, A., M. D. Rasmussen, J. M. Slagle, S. Jain, T. Kuykendall, N. Ordonez, and M. B. Weinger. 2007. A facilitated survey instrument captures significantly more anesthesia events than does traditional voluntary event reporting. *Anesthesiology* 107 (6):909–922.

Page, A., and Committee on the Work Environment for Nurses and Patient Safety. 2004. *Keeping Patients Safe: Transforming the Work Environment for Nurses. Institute of Medicine, Committee on Quality Health Care in America*. Washington, DC: National Academy Press.

Patterson, E. S., E. M. Roth, D. D. Woods, R. Chow, and J. Gomes. 2004. Handoff strategies in settings with high consequences for failure: Lessons for health care operations. *Int J Qual Health Care* 16 (2):125–132.

Paxton, F., M. Poter, and D. Heaney. 1996. Evaluating the workload of practice nurses. *Nurs Standard* 10:33–38.

Phipps, D., G. H. Meakin, P. C. Beatty, C. Nsoedo, and D. Parker. 2008. Human factors in anaesthetic practice: Insights from a task analysis. *Br J Anaesth* 100 (3):333–343.

Pierpont, G. L. and D. Thilgen. 1995. Effect of computerized charting on nursing activity in intensive care. *Crit Care Med* 23:1067–1073.

Potter, P., L. Wolf, S. Boxerman, D. Grayson, J. Sledge, C. Dunagan, and B. Evanoff. 2005. Understanding the cognitive work of nursing in the acute care environment. *J Nurs Adm* 35 (7–8):327–335.

Rayo, M., P. Smith, M. B. Weinger, and J. S. Slagle. 2007. Assessing medication safety technology in the intensive care unit. In *Proc Hum Factors Ergon Soc* Healthcare: 692–696.

Reason, J. 1990. *Human Error*. Cambridge, MA: Cambridge University Press.

Reason, J. 1997. *Managing the Risks of Organizational Accidents*. Aldershot, U.K.: Ashgate.

Reid, G. B. and T. E. Nygren. 1988. The subjective workload assessment technique: A scaling procedure for measuring mental workload. In *Human Mental Workload*, P. A. Hancock and N. Meshkati, eds. pp. 185–218. Amsterdam, the Netherlands: North-Holland.

Sanderson, P. 2006. The multimodal world of medical monitoring displays. *Appl Ergon* 37 (4):501–512.

Sanderson, P. M., M. O. Watson, W. J. Russell, S. Jenkins, D. Liu, N. Green, K. Llewelyn, P. Cole, V. Shek, and S. S. Krupenia. 2008. Advanced auditory displays and head-mounted displays: Advantages and disadvantages for monitoring by the distracted anesthesiologist. *Anesth Analg* 106 (6):1787–1797.

Santell, J. P., R. W. Hicks, J. McMeekin, and D. D. Cousins. 2003. Medication errors: Experience of the United States Pharmacopeia (USP) MEDMARX reporting system. *J Clin Pharmacol* 43 (7):760–767.

Sato, T. 1991. Fatal pipeline accidents spur Japanese standards. *APSF Newsl* 6:14.

Schultz, K., R. Brown, S. Douglas, B. Frederick, M. Lakhani, J. Scruggs, J. Slagle, B. Slater, M. B. Weinger, K. E. Wood, and P. Carayon. 2006. Development of a job task analysis tool for assessing the work of physicians in the intensive care unit. In *Proc Hum Factors Ergon Soc* Macroergonomics:1469–1473.

Schwid, H. A. and D. O'Donnell. 1992. Anesthesiologists' management of simulated critical incidents. *Anesthesiology* 76:495–501.

Singer, S. J., D. M. Gaba, J. J. Geppert, A. D. Sinaiko, S. K. Howard, and K. C. Park. 2003. The culture of safety: Results of an organization-wide survey in 15 California hospitals. *Qual Saf Health Care* 12 (2):112–118.

Slagle, J. M. and M. B. Weinger. 2009. Effects of intraoperative reading on vigilance and workload during anesthesia care in an academic medical center. *Anesthesiology* 110 (2):275–283.

Slagle, J., M. B. Weinger, M. T. Dinh, V. V. Brumer, and K. Williams. 2002. Assessment of the intrarater and inter-rater reliability of an established clinical task analysis methodology. *Anesthesiology* 96 (5):1129–1139.

Slagle, J., M. Weinger, R. Greevy, S. Nwosu, and K. Wallston. 2009. Operating room clinicians' individual & team workload, quality of care & non-routine events (abstract). In *Academy Health Annual Research Meeting*. Held 6/28–6/30: Chicago, IL.

Smith, A. F. and A. Mishra. 2010. Interaction between anaesthetists, their patients, and the anaesthesia team. *Br J Anaesth* 105 (1):60–68.

Taylor, C. J., F. Bull, C. Burdis, and D. G. Ferguson. 1997. Workload management in A&E: Counting the uncountable and predicting the unpredictable. *J Accid Emerg Med* 14:88–91.

Vredenburgh, A. G., M. B. Weinger, K. J. Williams, M. J. Kalsher, and A. Macario. 2000. Developing a technique to measure anesthesiologists' real-time workload. *Proc IEA/HFES Congress* 44:4241–4244.

Warner, M. A., S. E. Shields, and C. G. Chute. 1993. Major morbidity and mortality within 1 month of ambulatory surgery and anesthesia. *JAMA* 270 (12):1437–1441.

Wayne, J. D., R. Tyagi, G. Reinhardt, D. Rooney, G. Makoul, S. Chopra, and D. A. Darosa. 2008. Simple standardized patient handoff system that increases accuracy and completeness. *J Surg Educ* 65 (6):476–485.

Weaver, S. J., M. A. Rosen, D. DiazGranados, E. H. Lazzara, R. Lyons, E. Salas, S. A. Knych, M. McKeever, L. Adler, M. Barker, and H. B. King. 2010. Does teamwork improve performance in the operating room? A multilevel evaluation. *Jt Comm J Qual Patient Saf* 36 (3):133–142.

Weinger, M. B. 2003. Anesthesia incidents and accidents. In *Misadventures in Health Care: The Inside Stories*, M. S. Bogner, ed. pp. 89–104. Mahwah, NJ: Lawrence Erlbaum Associates.

Weinger, M. B. 2007. Experience ≠ expertise: Can simulation be used to tell the difference? *Anesthesiology* 107 (5):691–694.

Weinger, M. B. and S. Ancoli-Israel. 2002a. Sleep deprivation and clinical performance. *JAMA* 287 (8):955.

Weinger, M. B. and C. E. Englund. 1990. Ergonomic and human factors affecting anesthetic vigilance and monitoring performance in the operating room environment. *Anesthesiology* 73:995–1021.

Weinger, M. B., D. C. Gonzales, J. Slagle, and M. Syeed. 2004a. Video capture of clinical care to enhance patient safety. *Qual Saf Health Care* 13:136–144.

Weinger, M. B., O. W. Herndon, and D. M. Gaba. 1997. The effect of electronic record keeping and transesophageal echocardiography on task distribution, workload, and vigilance during cardiac anesthesia. *Anesthesiology* 87:144–155.

Weinger, M. B., O. W. Herndon, M. H. Zornow, M. P. Paulus, D. M. Gaba, and L. T. Dallen. 1994. An objective methodology for task analysis and workload assessment in anesthesia providers. *Anesthesiology* 80 (1):77–92.

Weinger, M. B., C. Pantiskas, M. E. Wiklund, and P. Carstensen. 1998. Incorporating human factors into the design of medical devices. *JAMA* 280 (17):1484.

Weinger, M. B., S. B. Reddy, and J. M. Slagle. 2004b. Multiple measures of anesthesia workload during teaching and nonteaching cases. *Anesth Analg* 98 (5):1419–1425.

Weinger, M. B. and J. Slagle. 2002b. Human factors research in anesthesia patient safety: Techniques to elucidate factors affecting clinical task performance and decision making. *J Am Med Inform Assoc* 9 (6):S58–S63.

Weinger, M. B., J. Slagle, S. Jain, and N. Ordonez. 2003. Retrospective data collection and analytical techniques for patient safety studies. *J Biomed Inform* 36:106–119.

Weinger, M. B., J. Slagle, R. Kim, and D. Gonzales. 2001. A task analysis of the first weeks of training of novice anesthesiologists. *Proc Hum Factors Ergon Soc* 45:404–408.

Weinger, M. B., J. Slagle, A. Kuntz, D. France, J. Schildcrout, and T. Speroff. 2009. A handoff training and improvement initiative significantly improved the effectiveness of actual clinical handoffs (http://www.academyhealth.org/files/arm/ARM-2009-Call-for-Papers-Abstracts-by-Session.pdf). Paper read at *Academy Health Annual Research Meeting*, Chicago, IL.

Weinger, M. B. and N. T. Smith. 1993. Vigilance, alarms, and integrated monitoring systems. In *Anesthesia Equipment: Principles and Applications*, J. Ehrenwerth and J. B. Eisenkraft, eds. pp. 350–384. Malvern, PA: Mosby Year Book.

Weinger, M. B., A. G. Vredenburgh, C. M. Schumann, A. Macario, K. J. Williams, M. J. Kalsher, B. Smith, P. Troung, and A. Kim. 2000. Quantitative description of the workload associated with airway management procedures. *J Clin Anesthesiol* 12:273–282.

Weinger, M. B., M. Wiklund, and D. Gardner-Bonneau, eds. (2011). *Human Factors in Medical Device Design: A Handbook for Designers*. Boca Raton, FL: CRC Press/Taylor & Francis.

Weller, J., A. Merry, G. Warman, and B. Robinson. 2007. Anaesthetists' management of oxygen pipeline failure: Room for improvement. *Anaesthesia* 62 (2):122–126.

Wierwille, W. W., M. Rahimi, and J. G. Casali. 1985. Evaluation of 16 measures of mental workload using a simulated flight task emphasizing mediational activity. *Hum Factors* 27 (5):489–502.

Wong, D. H., Y. Gallegos, M. B. Weinger, S. Clack, J. Slagle, and C. T. Anderson. 2003. Changes in intensive care unit nurse task activity after installation of a third-generation intensive care unit information system. *Crit Care Med* 31:2488–2494.

Woods, D. D. and R. I. Cook. 1999. Perspectives on human error: Hindsight biases and local rationality. In *Handbook of Applied Cognition*, F. T. Durso, R. S. Nickerson, R. W. Schvaneveldt, S. T. Dumais, D. S. Lindsay, and M. T. H. Chi, eds. New York: John Wiley & Sons.

Woods, D. D. and R. I. Cook. 2002. Nine steps to move forward from error. *Cognition Technol Work* 4:137–144.

Wright, M. C., B. G. Phillips-Bute, E. R. Petrusa, K. L. Griffin, G. W. Hobbs, and J. M. Taekman. 2009. Assessing teamwork in medical education and practice: Relating behavioural teamwork ratings and clinical performance. *Med Teach* 31 (1):30–38.

Wright, M. C., J. M. Taekman, and M. R. Endsley. 2004. Objective measures of situation awareness in a simulated medical environment. *Qual Saf Health Care* 13 (Suppl 1):i65–i71.

Xiao, Y., W. A. Hunter, C. F. Mackenzie, N. J. Jefferies, and R. L. Horst. 1996. Task complexity in emergency medical care and its implications for team coordination. *Hum Factors* 38:636–645.

Zijlstra, F. R. H., R. A. Roe, A. B. Leonora, and I. Krediet. 1999. Temporal factors in mental work: Effects of interrupted activities. *J Occup Organ Psychol* 72:163–185.

Index

A

Accreditation Council for Graduate
        Medical Education (ACGME)
    duration between shifts, 204
    implemented regulations, 202
    IOM recommendations, 206
    PGY-1 residents, 199
    physician work schedules, 199
ACGME, *see* Accreditation Council for
        Graduate Medical Education
Acquisition and storage, video
    analog VHS tapes, 537
    core support group
        ATLS protocols, 537
        basic elements, 536
        ergonomic analysis, 536
        insertion instrument
            content, 536
        punitive surveillance tool, 536
        quality management, 536
        template-based analysis, 537
        VAASNET system, 536
    design requirement, 537
    DVD recorders, 537
    SMEs, 537
    SMPTE time code, 537
    software, 536
    technical performance, 538
    VAASNET®, 533–536
    video buffer, mobile video and video
        whiteboard
        data collection, 538
        extraction and analysis
            techniques, 538
        human performance, 538
        operating rooms (OR), 538
        phased scale-up strategy, 538
ACRM, *see* Anesthesia Crisis Resource
        Management Program
ACSNI, *see* Advisory Committee on the
        Safety of Nuclear Installations
Activated clotting time (ACT), 781
ADEs, *see* Adverse drug events
Advanced trauma life support (ATLS)
        protocols, 527, 562
Adverse drug events (ADEs)
    CPOE implementation, 736
    medication system, 786
    patient harm, 787
    preventable and potential, 732
    primary care, 767–768
    system failures, 786

Adverse event analysis
    CONES, 805–806
    critical incident technique, 804
    latent conditions, 804–805
    NREs, 805–809
    QA programs, 804
Adverse events
    definition, 401
    recommendations, 402
Advisory Committee on the Safety of
        Nuclear Installations (ACSNI),
        135, 136
Agency for Healthcare Research and
        Quality (AHRQ)
    and DoD, 191
    hospital survey, patient safety
        culture, 192
    initial efforts and report, 186
    responsibility, 186
    TeamSTEPPS®, 190
AHRQ, *see* Agency for Healthcare
        Research and Quality
Alert reports (ARs)
    IAs, 679
    learning ratio, 680
    NRLS, 674
American Recovery and Reinvestment
        Act (ARRA), 249
Anatomical models/systems
    Advance Trauma Life Support
        (ATLS®) training, 562
    female pelvic examination, 562
    health care workers, 561–562
    prostate models, 562
    TraumaMan® mannequin
        simulator, 562
Anesthesia Crisis Resource
        Management Program
        (ACRM)
    curriculum, 189
    death scenario, 190
    description, 189
    instructors, 189
    simulated OR, 189
    teaching institutions, 189
    teamwork performance, 189–190
Anesthesiology, human factors
    adverse event analysis, 804–809
    patient safety
        adverse events, technologies and
            provider training, 804
        anesthesiologists, 803–804

        perioperative morbidity and
            mortality, 804
        PSFs, *see* Performance shaping
            factors (PSFs)
Anesthetists' nontechnical skills
        (ANTS) behavioral marker
        system
    behavioral data, 525
    complex medical settings, 525
    data collection, 525
    individual and collaborative
        performance, 526
    real-life clinical performance, 526
    skill categories, 525
    structured and unstructured
        observations, 525
    teamwork communications, 525
    video recording, 525–526
ARRA, *see* American Recovery and
        Reinvestment Act
Artichoke systems approach
    check list, 85–86
    error, 85
    precursor events, 85
    remediation
        evaluation, effectiveness, 86–87
        identified factors and events, 86
        target determination, 86
Assessment and evaluation tools
    IET, *see* Intervention evaluation tool
    NIOSH, 237–238
    OWAS, 236–237
    PE, health care
        decision-making, 240
        hospital organizational
            structure, 239
        musculoskeletal problems, 239
        participatory approaches, 239
        physical workload measures,
            235, 236
        postural analysis and biomechanical
            tools, 236
    QEC, 239
    REBA, 238
    RULA, 238
Association for the Advancement of
        Medical Instrumentation
        (AAMI), 29, 218, 545, 547
Automatic external defibrillator (AED)
    FMEA, 485, 491, 493
    nonquantitative fault tree, 491
    possible harms, 479
    quantitative fault tree, 491–496

B

Bar coded medication administration
(BCMA)
design, 254
eMAR and hardware, 253
HFE
benefits, 254
negative side effects, 254
task distributions and self-
reports, nurses, 253–254
medication error reduction, 253
Bar-coding medication administration
(BCMA) system, 67, 72, 75,
450, 515
Behaviorally anchored ratings scale
(BARS), 190, 586
Behavioral task analysis (BTA), 806
BODY™, 560
Brainstorming
feasible use-errors, 499
investigation methods, 499
quality improvement tools, 499
Broad systems approach
HAI incidence rates, 798
HFE analysis, 798
human abilities, 798–799
infection prevention and patient
safety, 798
physical environment, 798
BTA, *see* Behavioral task analysis

C

Cardiopulmonary resuscitation (CPR),
559, 803
Care-delivery organizations (CDOs)
clinicians, 314
defining specifications, 313
high-reliability systems, 316
informatics team
adult-learning theory, 315
cost-effectiveness, 314
CPOE, 314
integrated optimization, 315
phased HIT implementation, 315
production-support team, 315
"shadow" trainers, 315
software-safety principles, 314
test process-automation
tools, 315
training program, 315
integrated process redesign, 316
nonsoftware project costs, 313
quality and dependability, 313
resource-intensive task, 316
safety, 314
software design and safety
engineering, 313–314

specific organizational
competencies, 313
technical capabilities, 316
Caregiver Support Program, 127
CASE system, *see* Comprehensive
anesthesia simulation
environment system
Cave automatic virtual environment
(CAVE), 562–563
CBI, *see* Confederation of British
Industry
CCHIT, *see* Certification Commission
for Health IT
CDM, *see* Critical decision method
CDOs, *see* Care-delivery organizations
CDSS, *see* Computerized decisions
support systems
CDS systems, *see* Clinical decision
support systems
Center for Medicare and Medicaid
Innovation (CMMI), 3
Centralization, work system
structures, 55
Central venous catheters (CVCs)
displays, 797
frequent interruptions, 798
HFE analysis, 798
infections, 681, 795
insertion task analysis, 798
kits, 797
line insertion process, 797
medical procedure, 797
physical environment, 797
placement, 795
SEIPS work system model, 795, 797
teaching hospital organization, 798
tools and technology, 797
Certification Commission for Health IT
(CCHIT)
role, 665
usability, 665
Cew resource management (CRM)
definition, 564
developed, 564
efficacy, 560
principles, 564
Challenges, patient safety
contextual factors
core intervention, 24
therapeutic agent, 23
defining methods, 23
interventions and traditional
approach, 23
principles, 22
problem
human factors investigators, 22
problem-solving mode, 22
solutions, 23
self-examination areas, 24

CHIT, *see* Consumer health information
technology
CIPP, *see* Context input process product
CIT, *see* Critical incident technique
Climate and thermal environments
clothing effects, heat exchange,
223–224
emergency medical service
(EMS), 223
factors, 219
human performance, 223
medical devices and instruments, 223
mental and physical performance, 223
OR health-care providers, 223
task stress and cultural variables,
222–223
Clinical decision support (CDS)
systems
adherence, 516
computerized alerts and clinical
reminders, 255
decision support, 516
description, 254–255, 450
functionalities, 459, 516
HFE issues, 255
macroergonomics issues, 255–256
screening rate, 509
technology, 458
tool, 516
word processing programs, 456
CME, *see* Continuing medical
education
Cognitive ergonomics
complex self-care activities, 749
HCI methods, 749–750
laypeople's health work, 749
mental processes, 749
research, 749
telemedicine system
phone-based, 750
usability evaluation, 750
Cognitive walkthrough method
(CWM), 61
Cognitive work analysis (CWA)
conceptual frameworks
contextual/social factors, 466
descriptive models, 467
drug-drug interactions, 467
naturalistic setting, 466
normative analysis, 466
written procedures, 466
description, 465
human performance
complexity, operative
elements, 466
recognition, 466
social and political climate, 466
knowledge elicitation methods
artifact analysis, 468
critical decision method, 468

description, 467
ethnographic observations, 467–468
triangulation, 468–469
usability testing, 468
performing
intrinsic work constraints, 469
phases, 469
socio-technical systems, 469
phases descriptions
control task analysis, 470–471
social organizational analysis, 472
strategies analysis, 471–472
work domain analysis, 469–470
worker competencies analysis, 472
Colorado program, 426
Common user interface (CUI) project, 664, 666
Comprehensive anesthesia simulation environment (CASE) system, 560
Comprehensive Open-ended Nonroutine Event Survey (CONES)
interviews, 805–806
vs. QA process, data, 806
Comprehensive unit-based safety program (CUSP), 574
Computed tomography (CT) scans, 20, 194, 735
Computer-integrated manufacturing, organization and people (CIMOP), 60
Computerized decisions support systems (CDSS)
and CPOE, 650
patient treatments display, 656
Computerized physician order entry (CPOE) system
drug databases, 654
and EPR system, 653
flowchart, 736, 737
human factor principles, 735–736
implementation challenges, 735
medications, 656, 659–662, 668
patient treatments display and CDSS functions, 656
pediatric dosing, 733–734
Pittsburgh experience, 736, 738
Computerized provider order entry (CPOE) system
benefits and limitations, 250–251
description, 250
drug alerts, 456
functionalities, 459
health care, 450
HFE issues
communication and coordination, 252

nonphysician workers' time, 251–252
ordering process, 252
vs. paper orders, 251
portable computers, 252
sociotechnical impacts, 251
time-motion study, 251
"illusion of communication" problem, 253
prescription errors reduction, ICUs, 699
safety-shaping mechanisms, 251
technology, 459
tools and technologies, 694
usability problems and invalid design, 252–253
workflow issues, ICUs, 699
CONES, *see* Comprehensive Open-ended Nonroutine Event Survey
Confederation of British Industry (CBI), 136
Conservation of resources (COR) theory
depersonalization component, 121
depletion, 121
description, 120
energy, 120–121
management strategies, 121
objective, personal and conditional, 120
principles, 121
Consumer health information technology (CHIT)
chronically ill patients, 256
computer skills and confidence level, patients, 256
description, 256
design and implementation challenges, 257
physical environment, patients' homes, 257
usefulness and usability, 256–257
Context input process product (CIPP), 584
Continuing medical education (CME), 563, 640, 641
Corrective and preventative action (CAPA) systems, 501
COR theory, *see* Conservation of resources theory
CPOE, *see* Computerized physician order entry; Computerized provider order entry
CPR, *see* Cardiopulmonary resuscitation
Crew resource management (CRM)
impact, surgical errors reduction, 780
teamwork training, 780

Critical decision method (CDM)
deepening, 468
incident retelling, 468
incident selection, 468
preparation, 468
timeline verification and decision point identification, 468
Critical incident technique (CIT), 140
CRM, *see* Cew resource management
CUI, *see* Common user interface
CUSP, *see* Comprehensive unit-based safety program
CVCs, *see* Central venous catheters
CWA, *see* Cognitive work analysis
CWM, *see* Cognitive walkthrough method

D

Debrief facilitation
diagnostic, 590
educate trainers and team members, 590
individual and team feedback, 590
learning climate, 589–590
outcomes, 590
processes, performance, 590
team phases, 588–589
teamwork processes, 590
video recordings, 590
Delphi techniques
description, 500
general steps, 500
risk control measures, 500
Department of Defense (DoD)
AHRQ report, 186
HCTCP, 190
TeamSTEPPS, 190
Differentiation, work system structures
departmentalization, 54
horizontal, 54
spatial, 54
vertical, 53–54
DoD, *see* Department of Defense
Dynamic approach
HE reduction strategies
first tier, 396
second tier, 396
third tier, 396–397
professional and technical system, 395
safety improvement solutions, 395

E

EBD, *see* Evidence-based design
Edinburgh taxonomy
categories, 367
error coding system, 367
intensive care unit (ICU) errors, 367
proximal and distal causes, 367–368

Efficiency–thoroughness trade-off
        (ETTO) principle, 326
EHRs, *see* Electronic health records
Eindhoven model, 371, 450
Eindhoven taxonomy
    applications, 361
    approach and underpinning
            assumptions
        human behavior, 360
        organizational factors, 358–359
        PRISMA, 358
        SMART, 358
    directive properties and limitations
        cultural problems, 362
        Eindhoven classification, 361
        remedial training, 362
    labyrinth/pinball approaches,
            360–361
Electronic health records (EHRs)
        system
    barriers
        acute care settings, 269
        cognitive support, clinicians, 270
        data types and standards, 269
    certification applicants, 665
    clinician's workflow, 268–269
    control panel display, 655
    data storage, 268
    decision support technologies, 650
    description, 507
    development and use, 268
    features, 516
    implementation phase, 508
    information, applications, 651
    nurses, 699
    nursing screens, 514
    redesign workflow, 508
    sociotechnical benefits, 699
    usability, 657
Electronic medical records (EMR)
    description, 507
    implementation, 508
    patients and providers, 712
    sociotechnical benefits, 699
    workflow, 508
Electronic medication administration
        record (eMAR), 253
Electronic patient record (EPR)
    and CPOE system, 653
    health IT, 650
Elicitation and form design
    active monitoring, 434
    anonymous schemes, 434
    classification process, 435
    colloquial terms and natural
            language, 435
    computer-based
        based tools, 435
        reporting systems, 436
    drug misadministration, 437

electronic and paper-based forms,
        435
    "free-text" fields, 435
    hardware and software resources,
            434–435
    incident reporting forms, 434
    incomplete taxonomy, 436
    NPSA, 436
    reliable self-reports, 436
    reporting forms, 435
    self-help initiative, 436
    submission techniques, 435
eMAR, *see* Electronic medication
        administration record
Emergency department (ED), *see* HFEs,
        emergency department (ED)
Emergency medical service (EMS), 223
EMR, *see* Electronic medical records
EMS, *see* Emergency medical service
Endoscopy AccuTouch® simulator, 565
Environment components
    arrangements
        clearance problems, 221
        environmental data, 220
        failure modes and effects
                analysis, 221
        furniture placement, 222
        links, definition, 221
        medical equipment, 221–222
        mobile care units, 221
        nursing station, 220
        reach requirements, 221
        task analysis data, 220
    health-care-acquired infection
        ethylene oxide (ETO)
                sterilization, 222
        MDROs, 222
        patient care procedures, 222
        patient infections, 222
    space and physical constraints
        frequency-of-use, 220
        function principle, 220
        human error, 219
        importance principle, 220
        optimum location, 220
        risk of errors, 219
        sequence-of-use, 220
        workspace design, 219–220
EPC, *see* Error-producing conditions
EPR, *see* Electronic patient record
Ergonomics program resource
        documents, 600
Ergonomics programs and effective
        implementation
    approach
        employers, 603
        goals, 603
        hazard analysis and levels, 604
        information, reporting and
                employees, 603–604

    injury data, use, 605–606
    inspiration and information,
            employees, 604
    occupational ergonomics/safety
            program, 604–605
    OSHA, 603
    periodic evaluation, 604
    risk factors, 606–607
    WMSDs risks, 604
    worker's compensation, 604
description, 598
"Holistic" approach, 600–603
importants, 597–598
matches capabilities, people, 597
musculoskeletal injuries workplace
    carpal tunnel syndrome (CTS), 598
    incremental changes, 598
    jobs, 598–599
    risk factors, 599
    scientific literature, 598
    WMSDs, 599–600
recommendations
    changes, 608
    evaluation progress, 608
    organizational design, 607–608
    processes, 608
Error
    Artichoke, lens
        reactive approach, 91
        supra-ordinate system, 90
    Artichoke systems approach, *see*
            Artichoke systems approach
    care provider, product and
            conditions
        Artichoke approach, error
                prevention, 90
        Artichoke orientation,
                observation, 90
        infusion pumps, 89–90
        monitor alarms, 89
        prevention, product design, 90
        snooze alarms, 89
    definitions, 82
    designing, real world, 93–94
    focus expansion, product design
        contexts of use, 88
        hard-wired human
                characteristics, 88
        medical devices, 87–88
        proactive error prevention, 88
    human error, 81–82
    information, health care, 82–83
    Murphy's law and Artichoke, 87
    nature
        behavior, definition, 84
        James Bond syndrome, 84–85
    problem
        cause, 83
        remedial skill training, 83
        Stop Rule, 83

product design
  device testing, 93
  excessive therapeutic radiation,
      91–93
  real-world conditions, 91
  real-world health care
      simulation, 93
Error correction
  description, 454
  pharmacist reviews, 454
  process, 455
Error detection
  definition, 452
  direct error hypothesis, 452
  error-modeling system, 452
  error suspicion, 452
  human performance stage, 453
  medication reconciliation, 453
  outcome based, 452
  self-detection mode, 452
  standard checking, 452
  taxonomy based, 452
Error disclosure
  care and support, 416
  clinicians
      disclosure conversation, 404
      harm-causing errors, 403
      health care system, 404
      missed diagnoses, 404
      ubiquity of mistakes, 404
  emotional trauma, 416
  financial assistance and practical
      help, 416
  general principles, 408
  inform patients, changes, 416
  landscape
      health care systems, 402
      impact, 401–402
      nature, 401
      standards and recommendations,
          401
  longer-term needs, patients, 415
  long-term follow-up and support
      anesthetic awareness, 418
      cardiac arrhythmia, 417–418
      health care organization, 417
      ineffective approach, 417
      medical errors, 416
      surgical procedure, 417
  medical malpractice environment
      ambulance-chasing
          attorneys, 406
      barriers, 405
      contingency-fee-based
          system, 406
      doctor-patient relationships,
          406–407
      health care providers, 406
  medical treatment, injury, 405
  overall evidence, 408

patient/family responses
    adverse events, 405
    long-term consequences, 405
    post traumatic symptoms, 405
  process
    complex cases, 410–415
    description, 410
    obvious harm, 408–410
Error explanation
  closed-loop communication, 454
  computer interfaces, 453–454
  error reporting systems, 454
  first-order problem solving, 454
  preventive solutions, 453
  spelling and grammar function, 454
  standard checking, 453
  wrong medication error, 454
Error-producing conditions (EPC)
  health care
    device criticality, 329
    error-producing factors, 329
    level of agreement, 329
    operator–designer mismatch, 329
    ratings and devices, 329
  human error, 328
  work–sleep cycles, 329
Error recovery
  conceptualizing
    "close call," 451
    diagnosis/explanation, 450
    Eindhoven model, 450
    Norman's seven stages, 450–451
    organizational learning and
        safety improvement, 451
    patient and preventive
        treatment, 451
    patient safety and quality
        efforts, 452
    three-step process, 450–451
  correction, 454–455
  description, 450
  detection, 452–453
  explanation, 453–454
  health care community
    BCMA technology, 459
    CDS functionalities, 459
    comparing rates, 459
    CPOE technology, 459
    description, 459
  health information technology, 459
  HROs, 449
  human performance issues, 455
  importance
    errors and failures, 450
    HROs characteristics, 449–450
    workers experience, 450
  negative consequences, 455
  nurse's role
    critical care environments, 457
    description, 457

emergency departments (EDs),
      456–457
    medication management
        process, 457
    proactive risk assessment
        approach, 457
    recovery, 457
  pharmacist's role
    drug-drug interactions, 458
    organizational culture, 458
    pre-post study, 458
  physician's role
    description, 458
    error reporting studies, 458
    knowledge-based errors, 458
  prevention strategies, 449
  recovery efforts, 455
  work system redesign, *see* Work
      system redesign
Evidence-based design (EBD)
  bathroom, 19
  environmental safety, 19
  quality of care, 18
  single-bed rooms, 18

F

FAA, *see* Federal Aviation
      Administration
Failure mode effects analysis (FMEA)
  detectability
    advantages and disadvantages,
        488–489
    human factors, 486
    manufacturing process
        settings, 486
  human factors formation
    brainstorm potential use-errors,
        482–483
    effectiveness ratings, 485
    failure mode/operator error,
        483–484
    high-priority failure modes, 485
    occurrence ratings, 484
    prioritization, risk, 484–485
    risk index, 484
    risk priorities, 485–486
    severity ratings, 484
    task analysis, 481–482
    team formation, 481
    worksheet, 482
  industrial engineering–based
      tool, 800
  numeric *vs.* descriptive ratings, 486
  process, 800
  range, numeric values, 486
  and RCA, 679
  risk analysis tools, 480
  team members, 800
  use-error risk analysis, 479

Fatigue Countermeasures Program, 207
Fault tree analysis (FTA)
    advantages and disadvantages
        bottom-up nature, 488
        hazards, 489–490
    aforementioned events, 487
    assign probabilities, event, 487
    combinatorial probabilistic rules, 487
    events lead to failure, 487
    fault tree events, 487
    hazard analysis and critical control
        points, 490
    healthcare failure modes, 489–490
    health hazard assessment (HHA),
        490
    human factors applications, 486–487
    probability calculation, 487
    probability estimation
        critical path analysis, 487–488
        hazard, 487
        sensitivity analysis, 488
    root cause analysis (RCA), 490
    software programs, 487
    system reliability and safety, 486
    top-level hazards, 487
Federal Aviation Administration (FAA),
    191, 576, 781
Fidelity
    designers and educators, 565
    efficacy, 565
    issues, 564–565
    physical terms, 565
    principles, 565
    semantical/phenomenal mode, 565
FLS, *see* Fundamentals of laparoscopic
    surgery
FMEA, *see* Failure mode effects analysis
Formalization, work system structures,
    54–55
Form content and delivery mechanisms
    causes and prevention
        adverse health care events,
            443–444
        causal analysis, 444
        causation and counterfactual
            arguments, 444
        CIRS reporting form, 444
        potential preventative measures,
            444
        pragmatic implications, 444
        reporting systems, 444
    computer based forms, 437
    consequences and mitigating factors
        CIRS system, 443
        computerized monitoring
            systems, 443
        immediate and long-term
            outcomes, 443
        reporting systems, 442
        voluntary reporting systems, 443

detection factors and key events
    CIRS, 442
    emergency procedure, 442
    irritation and fatigue, 441
    patient monitoring application,
        441
    proximal/immediate events, 441
    reporting, 440
    stereotypical mishaps, 442
    "tick-boxes," 442
electronic forms, 437
electronic watermarks, 437
hybrid approaches, 438
identification information
    electronic incident report, 439
    "gatekeeper" approach, 438
    NPSA's frameworks, 438
    reporting system, anonymity, 439
online reporting systems, 437
potential contributors, 437
preamble and definitions
    adverse incident, definition, 438
    reporting behavior, 438
time and place information
    analysis and retrieval
        software, 440
    hand-over procedures, 439
    health care system, 439–440
    location information, 439
    monitoring mechanisms, 440
    reporting systems, 439
Framings, 168–169
Fundamentals of laparoscopic surgery
    (FLS), 567

G

Gainesville anesthesia simulator
    (GAS), 560
GAS, *see* Gainesville anesthesia
    simulator
"Gatekeeper" approach, 438
Global domestic product (GDP), 28
Graphical user interfaces (GUIs)
    based systems, development,
        266–267
    data model, 668
    description, 266
    visibility principle, 270
Group embedded figures test (GEFT),
    789
GUIs, *see* Graphical user interfaces

H

HAIs, *see* Health–care acquired
    infections
*Handbook of Human Factors and
    Ergonomics in Healthcare and
    Patient Safety,* 4

Handoff Communication Assessment
    (HCA), 168
Handoffs and transitions, care, *see*
    Patients care handoffs and
    transitions
Haptics and multimodal perception
    collision algorithms, 565
    Endoscopy AccuTouch® simulator,
        565
    force feedback system, 565–566
    interface use, 565
    LLC, 565
    orthopedic bone pinning, 565
    participants, use, 566
    Phantom Omni®, 565
    sensation, 566
    SimQuest, 565
    surgical simulators,
        development, 565
Harvey system, 559–560
Hazards identification, risk
    management
    analysis
        AED, 478, 479
        medical device, 478
        physical injury, 478
    development process
        FMEA and FTA, 478–479
        human factors engineering
            process, 479–480
HCA, *see* Handoff Communication
    Assessment
HCTCP, *see* Healthcare Team
    Coordination Program
Head-mounted display (HMD)
    technology, 568
Head-mounted video camera (HMVC),
    534
Health care–acquired infections (HAIs)
    European hospitals, 793
    hospital-related mortality, 793
    mortality rates, 793
Health care failure mode and effects
    analysis (HFMEA), 105
Health care reporting systems
    adverse events, 424
    cost estimation, 424
    elicitation and form design, *see*
        Elicitation and form design
    feedback and analysis
        CDRH and NPSA, 445
        CIRS web-based system, 445
        electronic tracking systems, 445
        environment, increasing
            attention, 445
        national and regional
            schemes, 445
        paper-based dissemination
            techniques, 445

form content and delivery
mechanisms, *see* Form content
and delivery mechanisms
human factors problems, patient
safety
Colorado program, 426
local and national systems, 427
long-term and deep-seated
issues, 427
NYPORTS program, 426
voluntary reporting system, 427
incident reporting
causal analysis problems, 427–478
recursive problem, 427
medical device
Canadian reporting system, 426
device-related deaths, 425
device-related reporting
systems, 426
incident reporting
applications, 424
learning from errors, 426
malfunction/deterioration, 425
medication measurements, 425
monitoring potential
problems, 424
national health care system, 426
restraint-related injuries, 426
wide-ranging mandatory
system, 426
underreporting, *see* Underreporting
Healthcare systems ergonomics and
patient safety (HEPS), 672
Healthcare Team Coordination
Program (HCTCP), 190
Health-care work schedules
direct-care nursing staff, 201
groups, health-care providers, 201
laboratory and field studies, 207
patient safety, fatigue and
performance
accident rates, 203
ACGME implemented
regulations, 202
baseline data, 202
cognitive skills, 201
conflicting findings and
confounding issues, 203
cross-over observational study,
201–202
extended work hours, 203
handoffs and continuity issues,
202
intoxication levels, 202–203
neurobehavioral tests, 203
polysomnography (PSG), 202
risk of error, 201
shortened sleep durations, 202
sleep deprivation, 201

working nights and rotating
shifts, 203
physician
Institute of Medicine
Committee, 200
PGY-1 residents, 199–200
residents, 199
specialties survey, 200
U.S. Bureau of Labor
Statistics, 200
recommendations
breaks, workday, 206
clockwise and counterclockwise
rotations, 205
critical elements, 203, 204
Fatigue Countermeasures
Program, 207
IOM, 206
night shift, 203–204
nurse managers and leaders, 206
participants, staff nurse fatigue,
205–206
profound sleepiness, 204
progressive sleep, 204
risk, accidents and injuries,
204, 205
shift rotations, 204
shifts starting duration, 204
shiftwork education programs,
207
Texas Board of Nursing, 206
working permanent night shift,
205
workload, 206–207
registered nurses, *see* Registered
nurses
settings, 199
transportation industry, 207
Health information technologies (HITs)
accident, 305
adverse effect, 306
care-delivery organizations
clinicians, 314
defining specifications, 313
high-reliability systems, 316
informatics team, 314–316
integrated process redesign, 316
nonsoftware project costs, 313
quality and dependability, 313
resource-intensive task, 316
safety, 314
software design and safety
engineering, 313–314
specific organizational
competencies, 313
technical capabilities, 316
care-process compromise, 306
consumer, 306
CPOE implementation, ICU, 699
definitions, 305–307

description, 305
EHR and EMR, 699–700
electronic health record (EHR), 306
error, 306
hazard, 306
human factors engineering, 306
individual, 316
information security, 306
manufacturers
hazard-communication
plan, 313
hazard control, 312
safety-critical decision, 312
system level properties, 312
usability, 312
user-centered design, 312
non-CDO, service providers
analogous public principles, 317
medical professionalism, 316
patient's healthcare team, 317
patient, 306–307
process, 307
quality and safety, 307
safety and effectiveness
description, 307
NNT, 307
software-safety principles, *see*
Software-safety principles
societal approach
accountability, safety, 311
certification, 311
data security, 311
purchasing policies, 311
systematic reporting, 311
tools characteristics, 311–312
widespread adoption, 311
software dependability, 307
tele-ICU, 700
workflow, 307
Heterogeneous, handoffs
interaction probability, 165
patients number, 165
point, patient's care trajectory, 165
HFEs, *see* Human factors and
ergonomics
HFEs, emergency department (ED)
COBRA/EMTALA, 718
complexly interactive system, 711
health care work, systems
degraded state and homeostatic
activity, 710
degree, knowledge, 710
equifinality and remoteness, 710
vs. simpler systems, 709
unstoppability and mortality, 710
HF issues examples, 711, 712
macroergonomic balance, 718
open and unbounded system, 711
organization and management, 717
patient, 712

provider
    fatigue and wellness, 713–714
    selection and training, 712–713
task and tools
    emergency care, 714
    information gaps, 715
    status boards, data, 714
work environment, 716–717
work team
    communication and
        coordination, 715–716
    culture, 716
    teamwork, 715
HFEs, intensive care unit (ICU)
    description, 693–694
    HITs, 699–700
    interventions
        communication improvement, 703
        multifaceted, 702
        standardization, routine
            tasks, 703
    low-tech tools, 700
    medical devices and alerting
        systems
        infusion pumps and pager alerts,
            698
        patient safety, 697–698
        smart IV pump and bar coding
            medication administration
            technologies, 698
    organizational characteristics
        handoffs, 697
        MDRs, 697
        teamwork and communication,
            696–697
    patients and safety, 693–695
    physical environment
        NICU, 700
        noise levels, 701
    processes, 702
    providers
        engineers, 696
        physicians, 695–696
    safety critical nature, 704
    tasks
        measures categories, nursing
            workload, 701–702
        quality, work life, 701, 702
    work system, 694
High-reliability organizations (HROs)
    characteristics, 174, 455
    command and control systems
        effectiveness, 176
        formal rules and procedures,
            adhesion, 180–181
        migration, decision authority,
            179–180
        redundancy, 180
        situational awareness, 180
        training, 181

description, 173
effective reward systems, creation,
    175–176
error detection, 455
health care organizations
    description, 176
    patient safety, enhancement, 177
    terminology, 177
health care system features, 671
hospitals, 686
hypothetical petroleum company, 174
key characteristics, structures and
    processes, 175
manmade catastrophes, 173
micro and macro organizational
    theory, 174
mindfulness and sense making,
    174–175
perceiving risk appropriately, 176
principles, 456
process auditing
    complex mental models, 175
    error reporting systems, 177
    health care organizations, 177
    system complexities and
        interconnections, 177
quality degradation,, avoidance, 176
quality review, health care
    AHRQ, 178
    bench marking, 178
    improvement practices, 178
    substantial challenges, 178
research, 174
rewards/sanctions
    broader organizational effort, 178
    commercial aviation, 177
    error reporting systems, 177–178
risk awareness
    description, 179
    error detection and prevention
        techniques, 179
    high-reliability theory, 179
    standardized survey, 179
HIT, *see* Health information
    technologies
HIT applications
    advantages, 515
    blood pressure readings, 516
    CDS systems, 516
    digital radiology image, 516
    EHR features, 516
    electronic prescribing tool, 515
    human resources, 515
    measuring usage, 515
    non-electronic data, 515
HIT implementation, HFE
    ARRA, 249
    BCMA, 253–254
    CDS, 254–256
    CHIT, 256–257

clinician cognitive work, 250
CPOE, 250–253
incentive payments, 249–250
"meaningful use," 249–250
SHARP, 250
HMD technology, *see* Head-mounted
    display technology
"Holistic" approach
    balance model, 601
    and human factors, 600
    machinery technology, and
        materials, 601–602
    macroergonomics factors, 601
    microergonomics factors, 601
    organizational structure, 602–603
    person, 601
    problems, 600–601
    task factors, 602
    work environment, 602
Home care, HFE
    assessment and analysis techniques
        balance model, 753–754
        Casablanca study, 756
        definition and objectives, system,
            753–754
        documentation review, 754
        household culture and structure,
            757
        individual characteristics, 754–755
        job designers, 755
        organizational characteristics,
            756–757
        physical environment, 756
        task description and analysis, 755
        tools and technology, 755–756
        total workload, 755
    components, 743–744
    data collection, 757
    domains and health care
        cognitive ergonomics, 749–750
        definition, work, 746
        IEA, 748–749
        macroergonomics, 752–753
        physical ergonomics, 750–752
        work system models, 747–748
    employment growth, 743
    family profile, 757
    goals, 746
    health and health care
        description, 744
        disease management, 744–745
        homes and households, 745–746
        participants, 745
        self-help and self-care activities,
            744
    health information
        distribution, 757
        management, 758
        task, 757–758
    workplace to home, transition, 753

Hospitals and secondary care
environments
AAMI, 29
cognitive psychology variables, 30
high workload
accessing and recording
information difficulties, 34
communication and information
transfer, 34–35
insufficient work structure, 34
lack of standardization and
equipment consistency, 33
performance obstacles, 33
physical layout consistency and
lack of standardization, 33
spacey and congested work
spaces, 33–34
magnesium sulfate administration
errors, 30–31
low levels, 30
steps, 30
medical devices, 29
multiprocess and multistage medical
tasks
cognitive human performance
examination, 33
guiding principles, 32–33
training and memory research, 32
surgical procedure performance
direct medical operations, 31
formulated model, 31
functional task models, 32
integral comprehensive design,
31–32
interaction and global
performance, task, 31
operating room (OR), 31
technology, 32
Hospital Survey on Patient Safety
(HSOPS), 146
HROs, *see* High-reliability
organizations
HSOPS, *see* Hospital Survey on Patient
Safety
Human-computer interaction (HCI)
methods, 749–750
Human-computer interface design,
health care
character-based interfaces, 265–266
computer-controlled devices, 268
contextual factors
collaboration and distributed
cognition, 276
distractions and interruptions,
276
multidisciplinary collaboration,
276–277
EHR systems, 268–270
environment, 265
error-free performance, 267

GUIs, 266
hyperlinks, 266
patient monitoring
cardiovascular display, 267–268
definition, 267
elements and relationships, 267
SSSI approach, 267
task
analysis, 275
infusion pump interface, 274–275
performance, factors, 265, 266
requirements, identification,
275–276
user
affordance, 270–271
consistency, 272–273
constraints, 273–274
error tolerant systems and
recovery, 274
feedback and predictive
information, 273
gestalt principles, 271
individual user characteristics, 274
mapping and direct
manipulation, 273
memory, 271–272
mental model, 271
pattern recognition, 272
proximity compatibility principle,
271
response alternatives, 273
search processes and visual
clutter, 271
stress and sleep deprivation, 274
task-related terminology, 273
visibility, 270
web and internet-enabled
applications, 266–267
Human error, health care
accident-proneness, 324
anesthesia, studies
breathing circuit disconnections,
337
description, 337
EPC/VPC, 337–338
equipment failure, 337
human contributors, 337
interaction-related problems, 338
cognitive architecture, 324
complex sociotechnical systems, 324
context-driven behavior, 324
critical care, studies
adverse events and medical error,
335
defenses, 335
intensive care unit (ICU), 334
medicine, 334
nurse errors, 334–335
urban teaching hospital, 335
wrong medication dosages, 335

description, 323–324
factors, 324
integrative model
adaptability and flexibility, 333
cognitive architecture, 333
performance shaping factors, 333
safety-relevant dimensions, 333
theory, 333–334
variability levels, 333
integrative perspective, 324
models
adverse event, 332–333
defenses, 332
error in context, 328
hazards, 332
individual contribution, 325–328
situational and systemic
influences, 328–330
violation-producing conditions
(VPC), 331–332
violations, 330–331
performance levels, 324
surgery, studies
communication breakdowns,
336
contributing human factors,
335–336
environmental factors, 336
HFACS, 336–337
latent conditions, 337
multiple regression models, 335
OR, 335
skill-based errors, 336
Human error reduction strategies
celebrated safety models, 386
characterize behaviors, 385
coping, medicine
heterogeneity, 388
industrial field, 388
industry standards and facilitate
errors, 389
PAE, 389
professions, 388
safety management (SM), 389
decision-making, 386
description, 385
generic solution
design flaws, 389–390
drug events, 389
dynamic approach, 395–397
James Reason's concepts, 390
person approach, 390–393
system approach, 393–395
nature and extent, risk
adverse event (AE), 386–387
contextual reasons, 386
error glossary, 386–387
monitor medical practice, 386
PAEs, 387–388
patient safety, 386

PSA and SRA, 385
situation awareness, 386
Human factors (HFs), *see* Human
   factors, health IT design and
   development
Human factors and ergonomics (HFEs);
   see also HFEs, emergency
   department (ED); HFEs,
   intensive care unit (ICU)
CMMI, 3
defined, 4
description, 794
design and life cycle, system
   project, 11
   project types, 11
   updating/redesigning, 11
diversity, people
   healthcare workers, 9
   participatory ergonomics
      programs, 9
   systems and processes, 9
domains
   human error, 5
   organizational issues, 5
   physical ergonomics, 4–5
EBD research, 19
healthcare transformation,
   contribution
   "biculturals," 12
   "human factors" curriculum,
      surgical clerk students, 12
   innovations, 12
   long-term collaborations, 12
HIT implementation
   ARRA, 249
   BCMA, 253–254
   CDS, 254–256
   CHIT, 256–257
   clinician cognitive work, 250
   CPOE, 250–253
   incentive payments, 249–250
   "meaningful use," 249–250
   SHARP, 250
home care
   assessment techniques, 753–757
   case study, 757–758
   components, 743–744
   domains and health care, 746–753
   employment growth, 743
   goals, 746
   health and health care, 744–746
   workplace to home, transition, 753
human characteristics, 7
infection prevention, *see* Infection
   prevention
macroergonomic model, patient
   safety, 10–11
medical errors, 3
micro-to macroergonomics
   hardware ergonomics, 5

human–software interface
   issues, 6
"human-system interface
   technologies," 5
patient-centered medical home, 3
pediatrics, *see* Pediatrics
physical, cognitive and psychosocial
   characteristics
   BCMA, 7
   BLS, 7
   errors, 8
   information processing and
      decision making models, 8
   mental models, 8
   PDA, 7–8
   psychosocial work stress and
      burnout, 7
   user involvement, 8
primary care, 763–771
system boundaries, interactions and
   levels
   elements and characteristics, 10
   2001 IOM (Institute of Medicine)
      report, 10
   medication reconciliation, 10
work system design
   analysis, work system, 75
   continuous design, 75
   implementation, redesign, 74–75
   process, 73–74
World Alliance for Patient Safety,
   WHO, 4
Human factors engineering (HFE)
   process
concepts, 549
design deficiencies, 553
development cycle, 549
guidance, 547
health care setting, 545
learned intuition, 553
methods, 555
portable medical devices, 553
principles, 546
usability testing small group
   exercise, 553
Human factors (HFs), health IT design
   and development
authorities and administrators,
   initiatives, 665
characteristics and health care work
   applications and systems, 651
   collaborative nature, 651–652
   heterogeneous organizations,
      652–653
   implemented, multiple levels, 651
   interruptive tasks and time
      constraints, 653
   mandatory interoperability, 651
   paradigm change, information
      exchanges, 651, 652

safety critical process, 653–654
   uncertainty and criticality, 653
description, 650
design problems
   severe safety issues, 658
   usage problems, medication
      CPOE, 659–662
engineering, usability
   "design metaphors," 664
   heuristics set, 663–664
   star life cycle model, 664, 665
   testing, 664
guidance, usability standards,
   666–667
initiatives, improvement process,
   667–668
integration, usability criteria, 665–666
iterative process
   phases, 659
   SDLC, 659, 663
mandatory interoperability
   support, 654
safe and easy-to-use
   parameterization/
   customization functions
   heterogeneous organizations and
      users, 657
   usability problem, 656
support, safety critical features, 658
system making
   clinician's partner, 655–656
   team player, 654–655
and usability, 663
user-centered design methods,
   iterative safety-oriented
   usability evaluations
   designers, developers and
      vendors, 657
   safety critical health IT tools, 657
user-centered process, application
   encouragement and
   verification, 666
Human performance failures
database and root cause analysis
   event-report database, 342
   risk analysis, 342
   root cause analysis (RCA), 342
generic error modeling system,
   344–345
knowledge based, 344
natures of failures
   definition, 343
   vs. errors and mistakes, 343
Rasmussen's model, 344, 346
rule based, 344
skill based, 344
taxonomy, definition, 342
types, 344
Hybrid simulators, 561

I

ICTs, *see* Information and communication technologies
IEA, *see* International Ergonomics Association
IET, *see* Intervention evaluation tool
Incidence rate (IR), 606, 677, 794, 798
Incident reports (IRs)
    information flow, 676
    quantity, Tuscany, 679
    regional centre, 679
    worker/patient claims, 674
Industrial approach
    approach and underpinning assumptions, 350–353
    classifications, 353–354
    incredible proactive investment, 349–350
    medical settings, limitations, 354
    Rasmussen's system, 350
Infection prevention
    chlorine compounds, 794
    forensic pathology, 795
    germ theory, 794–795
    HAI, 795
    and human factors
        AHRQ report, 795
        CDC guidelines, 799
        CVCs, 795, 797–798
        challenges, 796
        FMEA, 800
        health care workers, 796
        medical care, 795
        patient infection, 796
        PCA, 799
        RCA, 799
        Six Sigma approach, 799
        systems engineering, 794–795
        VSA, 799
        work system approach, 796–797
        work system design, 796
    latent error concept, 795
    mortality, 794
    Systems Engineering Initiative for Patient Safety (SEIPS), 794–795
Information and communication technologies (ICTs), 684, 748
Infusion pump
    description, 89
    programming, 89–90
INR, *see* International Nationalized Ratio
INSAG, 134
Institute of Medicine (IOM)
    AHRQ, 186
    health system, 186
    medication errors, 185–186
    PGY-1 residents, 199
    recommendations, 206

RRR, 676–677
    statistics, 188
Integration, work system structures, 54
Intensive care units (ICUs), *see* HFEs, Intensive care unit (ICU)
International Ergonomics Association (IEA)
    ergonomics/human factors, defined, 4
    humans, 8–9
    system design, 9–10
International Nationalized Ratio (INR)
    abnormal values, 653
    CPOE systems, 656
    test, 656
International Standards Organization (ISO)
    MDs design, 668
    usability, 663
Intervention evaluation tool (IET)
    absence/staff health, 241
    accident numbers, 242
    competence compliance, 241
    ergonomics, health-care design, 243
    MSD
        exposure measures, 242
        financial impact, 242
    musculoskeletal health (MS) measure, 241
    patient
        condition, 242
        injuries, 242
        perception, 242
    psychological well-being, 242
    quality of care, 241–242
    safer lifting policies, 242–243
    safety culture, 241
IOM, *see* Institute of Medicine
IR, *see* Incidence rate
ISO, *see* International Standards Organization

J

James Bond syndrome
    Artichoke model, environmental context, 84–85
    hierarchical categories, health care terms, 84
    misadministration, 84
James Reason's concepts, 390

K

Kansai ergonomics, 61
Knowledge elicitation methods
    artifact analysis, 468
    CDM, 468
    description, 467

ethnographic observations
    behaviors, 467
    data collection, 467
    internal cognition, 467
    process tracing method, 467
    reliability and generalizability, 467
    subject sampling, 468
    traditional human factors, 467–468
triangulation
    accurate analysis, 468–469
    actual process, 469
    definition, 469
usability testing
    description, 468
    software, 468
    work environment, 468
Knowledge, skills and attitudes (KSAs)
    ACRM, 190
    team training programs, 188
    teamwork
        competencies, 629, 630
        skills, 187
KSAs, *see* Knowledge, skills and attitudes

L

Labyrinth/pinball approaches
    definition, 360
    patient related factors, 360–361
    technical problems, 360
Laparoscopy, 566–567
Latent problems
    description, 354–355
    errors, 355–358
    general approach, 355
Learning goal orientation (LGO), 633
Length of stay (LOS)
    hospital, 694
    ICU, 693, 701
    patient, 699
LGO, *see* Learning goal orientation

M

Macroergonomic analysis and design (MEAD), 60
Macroergonomic organizational questionnaire surveys (MOQS), 60
Macroergonomics
    assistive devices, 753
    cognitive ergonomics, 46
    computer use, 752–753
    description, 752
    designing, new university college, 59
    health care and patient safety, 61
    health information management, 753

human factors engineers, 752
implementation
    industrial and organizational
        (I/O) psychology, 51
    "picking the low hanging fruit," 51
    work system change, 51
intervention, petroleum company, 59
macro-to microergonomic design
    decisions, 58
    optimal ergonomic
        compatibility, 58
methodology
    anthropotechnology, 60
    CIMOP, 60
    CWM, 61
    HITOP, 60
    Kansai ergonomics, 61
    MEAD, 60
    MOQS, 60
    PE, 61
    SAT, 60
    top MODELER, 60
microergonomics, description, 46
ODAM, integration, *see*
        Organizational design and
        management
pitfalls, work system design
    criteria, design approach, 53
    "left-over" approach, function
        and task allocation, 52
    sociotechnical characteristics,
        52–53
    technology-centered design,
        51–52
Select Committee, Human Factors
        Future
    changes, industrialized societies,
        47
    ergonomics-based litigation,
        47–48
    formal development, 46
    ODAM, 48
    technology and demographic
        shifts, 47
    traditional (micro) ergonomics,
        failure, 48
    value changes, 47
    world competition, 48
sociotechnical system considerations
    degree of professionalism, 56
    environment, 57–58
    Perrow's knowledge-based
        model, 55–56
    personnel subsystem, 56
    psychosocial characteristics, 56–57
    separate assessments, integration,
        58
spatial dispersion, 752
structural analysis
    centralization, 55

differentiation, 53–54
    formalization, 54–55
    integration, 54
synergism, *see* Synergism and work
        system performance
theoretical basis
    general systems theory, 51
    joint causation and subsystem
        optimization, 50
    joint optimization *vs.* human-
        centered design, 50
    sociotechnical systems theory
        and Tavistock studies, 49
    Tavistock studies, 49–50
Madison taxonomy
    action factors, 371
    decision-making sequence, 369
    Eindhoven system, 371
    enabling categories, 372
    enabling factors
        environmental, 372
        organizational, 372–373
    failure
        analysis, 369
        classification, 374, 378–381
    general characterization, 370
    hardware failures, 371
    human performance failures,
        371–372
    medical settings, 369
    pathological/destructive actions, 372
    radiotherapy, 369
    semi-hierarchical organization, 372
    software failures, 371
Maintenance of work ability (MWA),
        617–618
Mannequins
    Aerojet General Corporation, 559
    BODY™, 560
    CASE system, 560
    function, 560
    Harvey system, 559–560
    Resusci-Anne, 559
    Sierra Engineering, 559
    SimMan system, 560
    Sim One, 559
Medical devices (MDs)
    CE, 666, 668
    design and development, 666
    health care professionals, 652
    usability engineering and risk
        management process
        relationship, design, 666, 667
Medical error
    definition, 401
    recommendations, 402
Medical failure taxonomies
    approaches to taxonomies
        characteristics, 368–369
        Edinburgh taxonomy, 367

industrial approach, 349–354
    latent problems, 354–358
    Madison taxonomy, 369–373
    NCCMERP, 363–367
    Nursing Practice Breakdown
        Research Advisory Panel,
        367–368
    psychological viewpoint, 347–349
    scope, 362–363
    selection, 347
    systemic approach, 358–362
classifications, activity and error
    fault tree, 373
    process tree, 373
error coding *vs.* taxonomies
    classification, 347
    description, 346
    New York Patient Occurrence and
        Tracking System (NYPORTS),
        346–347
human performance failures
    database and root cause analysis,
        342
    natures of failures, 343
    taxonomy, 342
integrated remedial actions
    information on failures, 376
    root causes, prevention, 376–377
nature of failures
    environments factors, 345–346
    human performance failures,
        344–345
remedial actions, classifications
    categories, 374
    failure, 374
    hierarchy, 374–376
    preventative actions, 373–374
Medical malpractice environment
    ambulance-chasing attorneys, 406
    barriers, 405
    contingency-fee-based system, 406
    doctor-patient relationships, 406–407
    health care providers, 406
    high-award cases, 406
    medical-legal dispute resolution, 406
    morbidity and mortality
        conferences, 406
    patient/family error
        communications, 406
    poor communication, 407–408
    role models/confidants, 407
    salutary effects, 407
Medical modeling and simulation
        database (MMSD), 558
Medical simulation
    anatomical models/systems
        Advance Trauma Life Support
            (ATLS®) training, 562
        female pelvic examination, 562
        health care workers, 561–562

prostate models, 562
TraumaMan® mannequin
simulator, 562
application
goals, 563–564
method, implementation, 564
training and education, 563
user characteristics, 564
body mannequins
Aerojet General Corporation,
559
BODY™, 560
CASE system, 560
function, 560
Harvey system, 559–560
Resusci-Anne, 559
Sierra Engineering, 559
SimMan system, 560
Sim One, 559
efforts, 557
head-mounted display (HMD), 568
health care
advantages, 558
disadvantages, 558
program, 558–559
human factors
anesthesiologists, affects, 566
clinicians, 566
description, 564
fidelity, 564–565
fundamentals of laparoscopic
surgery (FLS), 567
haptics and multimodal
perception, 565–566
laparoscopy, mental workload,
566–567
novices practice, 567
secondary task measures, 567
team training, 564
Wickens' multiple resource
theory, 567
hybrid simulators, 561
maternal fetal heart rate, 568
models, use, 559
research programs, 568
standardized patients, *see*
Standardized patient
technology, 557
training benefits, 557–558
virtual reality (VR) systems, *see*
virtual reality (VR) systems
Medication errors, pediatric patients
ADEs, 732
ambulatory prescription, 732
CPOE systems and bar coding
technology, 732–733
decimal place and computational
errors, 732
dispensing and administering, 732
weight based, 732

Medication safety
application
health care providers, 790
human factors, 790
inspection accuracy, 790
software design problem, 790
work environment, 790
definition, 785
distribution system
ambulatory patients, 786
nurses, 785–786
physicians and pharmacists, 785
error problem
ADEs, 786
critical incident technique, 786
electronic pill box, 787
proximal causes, 786
stages, 786–787
human factors
ambient sound, 789
environmental factors, 788
illumination levels, 789
interruptions and distractions,
789
levels assignment, 788
workload pace, 789
problems and automation
CPOE *vs.* handwritten orders, 787
double-check medication
accuracy, 787
electronic reports, 788
human factors, 788
prescription verification system,
787
scanning compliance, 788
MedTeams™
BARS, 190
purpose, 190
quasi-experimental research design,
190
train-the-trainer approach, 190
MMSD, *see* Medical modeling and
simulation database
Models, human error
adverse event, 332–333
defenses, 332
error in context, 328
hazards, 332
individual contribution
activity type, 326–327
attention focus, 327
behavior and cognitive processes
levels, 325–326
control mode, 327
definition, 325
detectability, 327
error constitute instances, 326
error likelihood, 327
ETTO principles, 326
expertise level, 327

generic error modeling system
(GEMS), 325
human cognitive architecture,
325
knowledge-based behavior, 325
natural resources, 325
psychological and organizational
contributors, 325
Reason's approach, 326
relationship to change, 327–328
rule-based performance, 325
situational influences, 327
skill-based behavior, 325
situational and systemic influences,
328–330
violation-producing conditions
(VPC), 331–332
Modern health care and proactive
human factors
"blame culture," 29
cognitive and human performance
constituents
asynchronous team, 36
collaborative involvement,
teams, 35
common knowledge and
sharing, 36
patient record file,
radiotherapy, 36
synchronous teams, 35–36
contributing factors, 27
functional consequences, 28
GDP, healthcare costs, 28
"healthcare medical paradox," 28
high investments, 28
hospitals and secondary care
environments
AAMI, 29
cognitive psychology
variables, 30
high workload
medical devices, 29
medical tasks
task analysis and system
evaluation, 30–32
technology
human factors engineering, 39–40
medical and human error, 29
reference databases
absence, comparative base, 37
compulsory reporting system, 37
data collection, 36
errors and adverse events, 36–37
information basis, 39
potential problem areas, 38
source determinants, 37
traditional data collection and
task analysis, 37
"wisdom of hindsight," 37

MOQS, *see* Macroergonomic organizational questionnaire surveys
Mortality and morbidity (M&M) reviews, 679–681, 685, 686
MTS, *see* Multiteam system
Multidisciplinary rounds (MDRs)
 communication mechanisms, 697
 ICUs, 697
Multiteam system (MTS)
 definition, 640
 performance, 640
Murphy's law and Artichoke
 human factors/ergonomics, 87
 real-world validity, 87
 recurrence, error, 87
Musculoskeletal injuries workplace
 carpal tunnel syndrome (CTS), 598
 incremental changes, 598
 jobs, 598–599
 risk factors, 599
 scientific literature, 598
 WMSDs, 599–600
MWA, *see* Maintenance of work ability

N

National Committee for Quality Assurance (NCQA), 3
National Coordinating Council for Medication Error Reporting and Prevention (NCCMERP)
 contributing factors codes, 366
 description, 363
 harm level, 364
 human factors codes, 366
 medical facility types, 363–364
National Health Service (NHS)
 design metaphor, 664
 document, 666
National Health Service Litigation Authority (NHSLA), 682
National Institute for Occupational Safety and Health (NIOSH)
 ceiling-mounted and floor lifts, 237–238
 radiography/x-ray technology, 237
National Patient Safety Agency (NPSA)
 computer-based systems, 436
 frameworks, 438
 guidance, 435
 improved and safe patient care, 683
 incident reporting, 431
 longer-term changes, 429
 and NRLS, 674
 safety culture, 428–429
National Reporting and Learning System (NRLS)
 and NPSA, 674

NCCMERP, *see* National Coordinating Council for Medication Error Reporting and Prevention
Neonatal intensive care unit (NICU)
 HFE evaluation, 700
 tissue oxygenation, 703
New York patient occurrence and tracking system (NYPORTS), 346–347
Noise
 alarms, 228–229
 disruptive effects, 227
 electronic gadgets, 226
 health-care settings, 226
 heart rate, 226
 hospital atmosphere, 226
 loudness, 227
 and performance, 227–228
 sensory organ, 227
Nonroutine events (NREs), anesthesiology
 definition, 805
 incidence, 806
 near misses and critical incidents, 805
 video-based assessment
  BTA, workload and vigilance, 806
  clinical experts, role, 806
  data collection system and stations, 806, 807
  demographic attributes, 807–808
  distribution, impact events, 808
  multivariate logistic regression, 808
  patient injury, 807
  postevent analysis, 807
  putative contributory factors, 808, 809
  relationship, SPDs, 808
  representative sample, 808, 809
  videotaped anesthesia cases, 806–807
Norman's seven stages, 450–451
NPSA, *see* National Patient Safety Agency
Nursing Practice Breakdown Research Advisory Panel
 behavior types, 367
 nursing-related events, 367
 taxonomy, 367–368
Nursing Worklife Model, 122
NYPORTS, *see* New York patient occurrence and tracking system
NYPORTS program, 426

O

Observational Teamwork Assessment for Surgery (OTAS) tool, 524–525

Occupational Safety and Health Administration (OSHA)
 central elements, 603
 employees, 604
 ergonomics standard, 605
 examine, 604
 guidance, 603
 periodic evaluation, 604
 system uses, 606
ODAM, *see* Organizational design and management
Operating room (OR)
 nurses, 779–780
 training, 780
 WOK, 782
Operating room management attitudes questionnaire (ORMAQ), 146
Opportunities, patient safety
 adaptive environments
  EBD research, 21
  homes, 20
  providers and caregivers, 20–21
  socio-technical factors, 20
 diagnostic error
  economic theory, 19
  heuristics/cognitive dispositions, 19
  inherent response dispositions, 20
  perceptual dispositions, 20
  premature closure, 19
  system-based failures, 19
 domain selection, 17
 evidence-based design (EBD)
  bathroom, 19
  environmental safety, 19
  quality of care, 18
  single-bed rooms, 18
 human-system interface
  levels, 21
  microsystem-device interface, 22
  patient-device interface, 21
  provider-device interface, 21–22
  sociotechnical-device interface, 22
 simulation
  advantages, 17–18
  high fidelity, 18
  knowledge acquisition to performance accountability, 17
Organizational design and management (ODAM)
 IEA Triennial Congress, 48
 macroergonomics
  definition, 49
  purpose, 49
 recognition, importance, 48
 Technical Committee (TC), 48
Organizational learning
 health care
  leadership, 104
  learning (or not), failure, 103

structure considerations and levels, 102–103

research
assumption sharing, 99
business decision making, 98
change, 98
contextual factors, 100
graduate students, 99
individual and organizational learning, 101
learning organizations, 100–101
single-loop and double-loop learning, 99
stress types, 99
superstitious learning, 100
synthesis and expansion, 99–100
theory of action framework, 99
two-dimensional typology, 99

risk assessments
industries, 105–106
requirements, 105

ORMAQ, *see* Operating room management attitudes questionnaire

OSHA, *see* Occupational Safety and Health Administration

Ovako working posture analysis system (OWAS)
physiotherapy, 236–237
postural risks, 237
posture types, 236
working postures, 236

P

Participatory ergonomics (PE); *see also* Assessment and evaluation tools
HFE programs, 9
patient handling
Cochrane approach, 240
intervention strategies, 240
quality score, 240

Participatory organizational intervention, phases, 614

Patient-centered medical home (PCMH) model
EHRs, 769
implementation challenges, 769
open scheduling and communication options, 771
primary care transformation, 769–770
principles, 768–769
quality and safety
evidence base, diagnosis and treatment, 770
improvement and physician performance, 770–771
shared decision making, 770

Patient-controlled analgesia (PCA), 483, 799

Patient safety center of inquiry (PSCI), 146–147

Patient safety indicator (PSI), 677

Patient safety management (PSM)
and complexity
and clinical risk manager role, 685–686
"patient safety officer," 685
components, 673
CVC, 681
and decision making centralization
development, 686
organizational culture and practices, 687
sense making, 686
function, 678
governance
centralization *vs.* distribution, 684
ICT, enabler, 684
and quality relationship, 683
systemic macroergonomic approach, 682–683
health care services, HF and ergonomics
organizations *vs.* HROs, 671, 672
role, 672
HFs, ergonomics professionals and researchers, 677
IOM report, 682
model, 673
organizational readiness, 684
organization formalization and standardization, 686
patient claim, 687
proactive and reactive methods, 677–678
risk analysis and assessment, Tuscany case
ARs and RCA, 680
vs. Danish, 679
and Denmark, 679, 680
learning and improvement ratios, 680
methods, 679
regional centre activities, axes, 678
SEA and M&M reviews, 679
sentinel events, 680
risk identification and measurement
ergonomics approach, 674
factors, Danish system, 675
health care, practices dissemination, 674
IRs, 675–676
NRLS and NPSA, 674
PSI and DVT, 677
record review limits and requirements, 677, 678

RLS, Denmark, 675
sentinel/never events, 676
safety management, 672
Tuscany, risk anticipation and control, 682
WHO and IHI, 681

Patients care handoffs and transitions
challenges, conventional wisdom
complexity, 166
heterogeneous, 165
preparatory activity, 166
purpose, 164
reconstruction, communication, 166
requisite variety *vs.* mindless standardization, 167
residency, 167
responsibility and authority, 166
sources of rescue, 165–166
transfer, understanding, 166–167
conventional wisdom and unspoken assumptions
hazardous, 164
information transfers, 164
"low hanging fruit," 164
"silver bullet" solution, 164
standardization, 164
training settings, 164
framings
accountability and social interaction, 168–169
cultural norms, 169
description, 168
types, description, 168, 169
HCA, 168
healthcare oriented databases, 167–168
improvement, mundane processes, 169–170
intervention, 168
modern healthcare requirements, 163
observational studies, 168

PE, *see* Participatory ergonomics

Pediatrics
vs. adults, 723–724
CPSC, 734
design implications, 734
dynamic nature, 724
functional and developmental limitations, 733
illness and mortality, 724
medical care, 724
motor vehicle accidents and obesity, 724
normal growth and development
cognitive and speech, 730
head size, 724, 727–728
heart rate and respiratory rate values, 728
metabolism, 728

motor considerations, 729–730
obesity, 730
pharmacokinetic changes, 729
skin, 728–729
vital signs, 728
weight and length, 724–726
patient safety
adult indicators, 738
basic science research, 730–731
congenital heart defects, 735
CPOE system, 735–737
CT scans, 735
health care error and injury,
epidemiology, 731–732
improvement strategies, 733, 739
labeling confusion, 734–735
medication errors, 732–733
organizational structures, 738
Pittsburgh experience, 736, 738
procedural sedation and critical
incident analysis, 735
SEIPS model, 731
tools and technologies, 738
pediatric providers, 723
playground injuries, 734
toy industry, 733
U.S. Census Bureau, 723
Performance goal orientation
(PGO), 633
Performance shaping factors (PSFs)
burnout
description, 810–811
global scores, 811
description, 808–810
equipment issues
design, 814–815
maintenance failures, 815–816
microprocessor-based equipment,
815
simulation and testing, 815
use error, 814
user manual, 815
workspace, 814
experience and expertise, 809
fatigue and sleep deprivation
physicians and anesthesia
residents, 809–810
repeated-measures design, 810
interruptions and distractions, 813
interventions, 817, 818
SA
levels, 812
reviews, 813–814
tasks and workload
assessment, 811–812
data collection and analysis, 811
procedural assessment
techniques, 812
psychological and subjective
workloads, 812

task density, 812
time-and-motion studies, 811
teamwork and communication
failures, 816
handovers, 816
organizational issues, 817
programs and tools, 816
vigilance
affecting factors, 812–813
task analysis studies, 813
Perrow's knowledge-based model
task
analyzability, 56
variability, 55–56
technology
classes, 56
engineering and craft, 55
knowledge-based definition, 55
routine and nonroutine, 56
Person approach
detection and recovery, 390
HE reduction strategies
simplify, 392
staff and share, 392–393
standardization, 392
supervise, 393
human error, 390
violations
first phase, 391
second phase, 391–392
third and last phase, 392
PGO, *see* Performance goal orientation
Phantom Omni®, 565
Physical environment
air quality
air-conditioning systems, 224
draught, 224
eye irritation level, 224
health-care facilities, 225
hospital environments, 225–226
humidity, 224
medical device storage
requirements, 224
weather influences moods, 224
care devices and procedures, 217
civilian health-care
institutions, 215
climate and thermal
clothing effects, heat exchange,
223–224
EMS, 22
factors, 219
human performance, 223
medical devices and
instruments, 223
mental and physical
performance, 223
OR health-care providers, 223
task stress and cultural variables,
222–223

description, 215–216
HFE involvement, 215
illumination
description, 229
inadequate lighting conditions,
230
light sources, 230
luminous intensity, 230
patient care setting, 230
presbyopia, 231
recommendations, health-care, 230
visual discomfort and eyestrain,
230
visual performance, 231
visual spectrum, 229–230
individual built environment
components
arrangements, 220–222
health-care-acquired infection,
222
space and physical constraints,
219–220
noise, *see* Noise
SEIPS model, 216
social factors and organizational
factors, 216
staff's quality, 217
standards, guidelines, rules and
reference texts
intensive care units (ICU), 219
multioccupancy patient rooms,
219
originators, 218
privacy and control, lack, 218
societal guidelines, 217
tools and technology, 217
vibration
adversely effect, 229
cumulative trauma disorders
(CTD), 229
health-care providers, 229
low frequency, 229
nonmoving/isometric controls,
229
performance degradation, 229
well-established standards and
guidelines, 217
Physical ergonomics
assistive technologies, 751
home environment, 750–751
human factors engineers, 752
occupational health and safety
issues, 751–752
psycho-physical conditions, elderly,
751
Physician professionalism
accountability
advantages, 114
health care systems, 113
root cause analysis, 114

description, 110
disease management programs
  health care system level, 113
  physician worklife, 113
  quality improvement framework, 113
  system level analyses, 113
health care systems, 112
medical approach, systems change
  health-related outcomes, 112
  HFE model, 112
  quality improvement, 111–112
  SPO model, 111–112
  structural elements, 112
multidisciplinary patient care teams, 113
organizational incentives and agency theory
  double agency, 110
  evidence-based practice, 110
  professional responsibility, 110
  quality improvement initiatives, 110
systems perspective
  balance theory, 111
  health-related outcomes, 119
  human factors engineering, 111
  knowledge and technology, medicine, 111
  quality improvement programs, 111
  well-designed systems, 111
Postanesthesia care unit (PACU), 534
Postgraduate year (PGY)-1 residents, 199
Preventable adverse events (PAEs), 387
Primary care, HFE
  clinical information, 765–766
  communication methods, 765
  comprehensive, 764
  context and family, 766
  coordination function, 764
  definition, 763
  error recovery, 767
  first-contact, 763
  longitudinal, 763–764
  macroergonomic issues, 768
  multiple problems, 764
  patient safety
    ADEs, 767–768
    medication errors, 768
    preventable incidents/adverse events, 767
    transitions, care, 768
  PCMH model, 768–771
  process standardization, challenges, 764–765
  professionals, 764
  relationships and continuity, 766

therapeutics, procedures and technology, 766–767
time limitations, 765
Probability safety assessment (PSA), 385
Process, error disclosure
  complex cases
    adverse outcome, 413
    general principles, 410
    health care systems, 410
    outcome, 412–413
    patient/family unaware, 414–415
    unexpected outcome, 410–412
  description, 410
  obvious harm
    disclosure conversations, 409–410
    general guidelines, 409
    patient/family reactions, 410
PSCI, *see* Patient safety center of inquiry
PSFs, *see* Performance shaping factors
Psychological viewpoint, failures
  complexity, 349
  performance and types, 348–349

Q

Quality assurance (QA), 428, 499, 804, 806
Quick exposure check (QEC), 239

R

Rapid entire body assessment (REBA), 238
Rapid upper limb assessment (RULA), 236, 238
Rasmussen's model, 344, 346, 350
Rasmussen taxonomy, 363
RCA, *see* Root cause analysis
Reason's approach, 326
Redundancy, 180
Registered nurses
  breaks and lunch periods, 200
  12h shifts, 200
  legislation limiting/banning mandatory, 200–201
  significance, 200
  survey, 201
  VA Health System, 201
Remedial actions, classifications
  categories, 374, 376, 378–382
  failure, 374
  hierarchy
    actions to prevent errors, 375, 382
    attention and masking noise, 375–376
    environmental problems, 375
    Institute for Safe Medical Practices (ISMP), 375

KCl syringes, 374
Madison taxonomy, 375
integrated
  information on failures, 376
  root causes, prevention, 376–37
preventative actions
  a-priori approaches, 373–374
  proactive error prevention, 374
Reporting and learning systems (RLSs)
  indicator, 680
  WHO draft guidelines, 675
Resusci-Anne, 559, 560
Retrospective record review (RRR), 676–677
Risk assessments, organizational learning
  failure, 104
  industries
    feedback mechanisms, 105
    nuclear, 105–106
    unintended consequences, 106
  requirements
    cause and effect diagrams, 105
    group's effort, 105
    interactions and outcomes, identification, 105
    team composition, 105
Risk management, medical products
  corrective and preventative action (CAPA) systems, 501
  description, 476–477
  design safety, trade-offs, 500–501
  environment use
    auditory alarms, 501
    electrical connections, 501
    user interface standards, 503
  establishing acceptable risk
    acceptable with review, 496
    ALARP, 496–497
    FMEA methods, 496
    ISO 14971 standard, 497
  expanded applications, 501
  failure modes and effects analysis, 478–479
  fault tree analysis (FTA)
    advantages and disadvantages, 488–489
    aforementioned events, 487
    assign probabilities, event, 487
    combinatorial probabilistic rules, 487
    events lead to failure, 487
    hazard analysis and critical control points, 490
    healthcare failure modes, 489–490
    health hazard assessment (HHA), 490
    human factors applications, 486–487
    probability calculation, 487

probability estimation, 487–488
RCA, 490
software programs, 487
system reliability and safety, 486
top-level hazards, 487
FMEA variations
advantages and disadvantages,
488–489
detectability, 486
numeric *vs.* descriptive ratings, 486
range, numeric values, 486
hazards identification
analysis, 478
development process, 478–479
human reliability, definition, 477
implementation considerations
achieving consensus, 499–500
brainstorming, 499
Delphi techniques, 500
team membership, 499
labeling and training, 501
medical error data, 503
patient safety, awareness, 501
regulations
FDA guidance, 498
medical device submissions, 499
software tools, 503
standards
human factors, 498
ISO, 498
usability testing, 501
use-error
definition, 477–478
medical device, 477–478
use-error FMEAs and FTAs
automatic external defibrillator
(AED), 491–496
quantitative comparison, 496
Therac-25 radiation therapy
system, 490–491
use-error risk analysis
blaming, user, 497
development tool, 497
document low-level risks, 498
overlooking critical sources,
failure, 498
probability and severity, 497–498
training/labeling, overreliance,
498
validate modes, control, 498
Root cause analysis (RCA)
heuristic evaluation, 548
HFE, 548
improvement plan, 679
interpersonal communication, 548
process, 799
regional centre, 680

S

SA, *see* Situation awareness
SADS, *see* Seasonal affective disorders
Safety attitudes questionnaire
(SAQ), 146
Safety climate survey (SCS), 146
Safety culture correlations, incident
reporting rates
comparisons
interpretation, 157
power distance, 155–156
description, 154
error reporting, 154–155
recognition, human error,
156–157
indices, 155
and performance level
estimation, 157
'guess', reporting system, 157–158
level 3+ incidents, 157
transitions, 157
staff's safety awareness/sensitivity,
155
statistical summary, 155
Safety culture, healthcare
assessment tools
description, 143–145
HSOPS, 146
ORMAQ, 146
SAQ, 146
SCS, 146
Stanford/PSCI, 146–147
association, 143
audit
comprehensive picture, 141
description, 140
questionnaire-based survey,
140–141
CIT, 140
vs. climate
distinction, 135
organizational factors, 134–135
Schein's three levels, 135
workforce's attitudes and
perceptions, 135
climate changes
activity, organizations, 153
aggregated data, 153
comparative result, 5-year
interval, 152–154
staff motivation and morale, 153
clinical specialties *vs.* wards
attitudes and perceptions, 149
dimensions, 150
doctor's specialty-based
comparisons, 149
nurse's work unit–based
comparisons, 149, 150
realistic recognition, 150

work conditions, nurses, 151
work unit group, nurses, 150, 151
conducting surveys
key stakeholders, involvement,
147
web-based tools, 147
correlations, incident reporting rates
comparisons, 155–157
description, 154
error reporting, 154–155
indices, 155
and safety performance level,
157–158
staff's safety awareness/
sensitivity, 155
statistical summary, 155
definition
ACSNI, 135
CBI, 136
holistic/shared aspect, 136
INSAG, 134
organization, 135–136
dimensions of
comparable organizations,
138–139
description, 138
eight-dimensional model, 138
instruments, 139
questionnaire responses, 138
variation, 138
focus group interviews, 140
healthcare professionals
dimensions, agreements and
disagreements, 148
ORMAQ, 147
psychological distance, 148–149
staff responses, 147–148
management
commitment, 137
objectives, 137
PSF, 137
weak points, 137
multinational comparisons
communication, 152
cross-national, 152, 153
Japanese staff's, 152
ORMAQ type, 151–152
operational safety, 134
organizational differences
US hospitals, 151
variations, dimensions, 151, 152
organizational factors, 134
outcome measurements
accident/incident rate, 142–143
self-reported rate, 143
overlapping types
domains, 136
local cultures/subcultures,
136–137
local variations, 136

national culture, 136
organizational units, 136
patient safety, improvement, 158
publication, INSAG reports, 134
questionnaire survey, 139
study and assessment, methods, 139
tools, quality requirements
description, 141
documentation, 142
job orientation and setting, 142
practicality and nonlocality, 142
relevance and
comprehensiveness, 142
reliability, 142
validity, 141–142
SAQ, *see* Safety attitudes questionnaire
SAT, *see* Systems analysis tool
SBTT, *see* Simulation-based training for
teams
Schein's three levels of culture, 135
Science of system reliability assessment
(SRA), 385
SCOPE, *see* System for the classification
of operator performance
events
SCS, *see* Safety climate survey
SEA, *see* Significant event audits
Seasonal affective disorders (SADS),
230–231
SEIPS work system model, 795, 797
SESAM, *see* Society in Europe for
Simulation Applied to
Medicine
SHARP, *see* Strategic Health IT
Advanced Research Projects
Significant event audits (SEA)
and IR, 685
M&M reviews, 679–681, 685, 686
SimMan system, 559, 560
SimQuest, 565
Simulation-based training for teams
(SBTT)
connections, 574
curriculum, scenario, 575
debrief facilitation
diagnostic, 590
educate trainers and team
members, 590
individual and team feedback,
590
learning climate, 589–590
outcomes, 590
processes, performance, 590
team phases, 588–589
teamwork processes, 590
video recordings, 590
developing event sets
and associated teamwork
responses, 581

incidents/situations objectives,
580–581
pilot test, refine, implement and
continuously evaluate, 581
teamwork competency model,
577–580
teamwork learning objectives, 580
evaluation and assessment, 574
event-based measures, 585–588
event-based methods, 575–576
general and interrelated,
applications, 574–575
goals, 574
initial rationale, 574
innovation and exploration, 574
learning activity, 575
patient safety movement, 574
performance support, 574
procedural tasks, 574
sophistication simulators, 574
systems approach, 575
team performance measurement, SBT
education training and program
evaluation frameworks, 584
location, 585
method, 585
purpose, 583
selecting training and supporting
raters, 585
systems models, 585
team science frameworks, 584
training and education, 574
Single sensor single indicator (SSSI)
approach, 267
Situation awareness (SA)
anesthesiology
levels, 812
reviews, 813–814
errors in health care organizations,
180
ICU team members, 700
Six Sigma approach, 127, 799
SMEs, *see* Subject matter experts
SMPTE time code, 537
Society in Europe for Simulation
Applied to Medicine (SESAM),
558
Software-safety principles
adverse effects, 308
automate cautiously, 309
certification, 310
characteristics, 307–308
deviance normalization, 310
HIT, 307
proactive hazard control, 308
retrospective hazard analysis, 308
software testing, 309–310
systems and software simplicity,
308–309
task-focused information, 310

SP, *see* Standardized patient
SPO model, *see* Structure-process-
outcomes model
Standardized patient (SP)
augmentation, 563
certify and license physicians, 563
medical students, physical
examination, 563
physician-patient interactions, 563
problem, 563
stethoscope, 563
Sterile cockpit rule, 781
Stop Rule, 83
Strategic Health IT Advanced Research
Projects (SHARP), 250
Stress and burnout, workplace health
care worker
COR theory, 120–121
description, 119
implications, future research
elements, behavioral
examination, 123–124
intervention mechanisms, 123
linkage, 124
organizational level, 124–125
implications, management
Caregiver Support Program, 127
employee problem-solving
committees, 126–127
interpersonal interaction, 126
intervention model, 125
medical home demonstration
project, 127
mutual aid group, 125
natural intervention, 127
organizational interventions, 126,
127
participatory decision-making,
126
personality hardiness, 126
philophonetics counseling,
nurses, 126
reflexivity encouragement, 126
six sigma approach, 127
workshop, 125
patient satisfaction
anesthesiologists and
anesthesiology, 121–122
description, 121
physicians, 122
prevalence
dentists, 120
description, 119
growth, intensive care, 120
job satisfaction, pharmacists, 120
occupational therapists, 120
rates, burnout, 120
quality of care
depersonalization component, 123
error report, 122

job performance, 122–123
medical resident, survey, 122
nurse perceptions, 122
Nursing Worklife Model, 122
physician availability, 123
zero order correlations, 123
Structure-process-outcomes (SPO)
model
components, 112
HFE model, 112
vs. human factor engineering
perspective, 112
human resources, 112
Subject matter experts (SMEs)
CDs, 537
video recording, 537
Surgical excellence
communication errors, 778
description, 775
disruptive behavior types, 779
errors and safety, traditional
view, 776
medical teams, 779
preoperative briefings, 781–782
shared mental models
enhancement
role, 782
VIOR and WOK, 782–783
systems approach, safety and error
reduction, 776
teamwork and communication
improvement
awareness and attitudes,
779–780
didactic and simulation, 780
standardize phraseology and
timing, 780–781
work system factors and
performance
medical devices, 776–777
surgical flow disruptions, 778
system-centered approach, errors,
777–778
Survey feedback process, 615
Synergism and work system
performance
incompatible organizational
designs, 59
macroergonomics, 59
System approach
HE reduction strategies
adapt governance, 394
macro-level system design, 394
safety culture, 394–395
share responsibilities and
professionalism, 394
patient safety, 393
regulations, 393
safety culture, 393
time management, 393

System for the classification of operator
performance events (SCOPE)
approach and underpinning
assumptions
error reduction, 362–363
expectation biases, 362
human performance failures, 362
interpretation failures, 362
knowledge-based action, 362
layout, 362
structure, 362
limitations
Rasmussen taxonomy, 363
SMART, 363
Systems analysis tool (SAT), 60
Systems development life cycle (SDLC)
health IT design, 659, 663
usability engineering, 650, 664
"waterfall" approach, 659
Systems engineering initiative for
patient safety (SEIPS) model
applications
BCMA technology, 72
health information technology, 73
medication administration and
outpatient surgery process, 72
HFE issues, ICUs, 694
HFs analysis, 697
work system and patient safety
care process, 71–72
healthcare leaders and policy
makers, 72

T

TACT, *see* Team adaptation and
coordination training
Tavistock studies, macroergonomics
composite work system design, 50
determinism, technological, 50
mining coal, Welsh mines, 49
sociotechnical systems theory, 49
technological change, 49–50
Team adaptation and coordination
training (TACT), 634
Team, health care
AHRQ TeamSTEPPS program,
188–189
characteristics, 186
description, 186
key features, 186
meta-analytic findings, 189
performance
confirmatory factor analysis,
187–188
inputs, processes, and outputs,
186–187
KSAs competency, 187
situation monitoring, 188

skill competencies and
definitions, 187
taskwork and teamwork, 187
T-TPQ, 188
validated model, 188
training
comprehensive analysis, 188
definition, 188
organizational factors, 188
strategies and techniques, 188
Team-oriented medical simulation
(TOMS), 780
TeamSTEPPS®, *see* Team Strategies and
Tools to Enhance Performance
and Patient Safety
Team Strategies and Tools to Enhance
Performance and Patient
Safety (TeamSTEPPS®), 574,
629, 639
curriculum, 191–192
description, 190
DoD, 190–191
evidence-based relation, 190
FAA, 191
implementation
AHRQ hospital survey, 192
organizational readiness, 192–193
Phase I, 192–193
Phase II, 193
Phase III, 193
plan, 192
sessions, 192
team performance, 192
instructional model and framework,
191
skill competencies, teamwork, 187,
191
Team task analysis (TTA)
primary steps, 632
purpose, 632
teamwork processes, 632
Team training, patient safety
analysis, 631–632
communication, 627
competencies and definitions
description, 629
teamwork, 629, 630
complex and dynamic systems, 629
daily care processes, 630
definition, 628
developing expert teams,
methods, 639
improvement science and
implementation science,
639–640
KSAs, 627
MTS, 640
optimizing teamwork, strategies and
methods

adaptation and coordination
training, 634
assertiveness training, 637
categories, 633–634
cross-training, 634, 637
error management training, 638
guided team self-correction, 637
metacognition training, 637
methods, 638
strategies relevant, 634, 635–636
team building, 634
organizational analysis, 633
PCMH, 640
person analysis, 632–633
potential behaviors, range,
630–631
purpose and structure, 628
sense, 629
SQUIRE, 640
systems approach, 631
team task analysis (TTA), 632
trainees step, 631
transfer and sustainment, 638–639
undergraduate graduate and
continuing medical education,
640, 641
Teamwork and communication,
surgery
awareness and attitudes increment
description, 779
physician and nonphysician
caregivers, 779–780
phraseology and timing
standardization
description, 780
heart-lung machine, 781
sterile cockpit rule, 781
providing didactic and simulation,
780
Teamwork and patient safety
AHRQ
CRM and DoD, 186
initial efforts and report, 186
responsibility, 186
future directions, research
implementation, team
training, 194
training and measurement,
alignment, 1945
medication errors, IOM, 185–186
structure, 186
team
AHRQ TeamSTEPPS program,
188–189
characteristics, 186
description, 186
key features, 186
meta-analytic findings, 189
performance, 186–188
training, 188

training, health care
ACRM, 189–190
cluster-randomized controlled
trial, 193–194
fundamental standard, 193
MedTeamsT, 190
pretraining/posttraining design,
194
quasi-experimental research
design, MedTeams, 193
simulations and behavior, 194
TeamSTEPPSʳ, 190–193
tools, TeamSTEPPS, 194
Telemedicine, human factors
applications
tele-consultation, 294
tele-education, 294–295
tele-home care, 295
tele-monitoring, 295
tele-surgery, 295
barriers and facilitators,
implementation
advantages, 297
communication and feedback
loops, 300
costs, 298
digital technologies, 299
human factors, 299–300
information technology systems,
297
process redesign, 299
redesigning, work processes, 301
resistance to change, 299
risks, 298
socio-technical approach, 299
special meetings, 300
technology and interconnectivity,
298
telecommunication, 297–298
usability, 298–299
work organization and workflow,
300
cost-effectiveness, 295–296
definition, 294
digital technologies and internet,
293
patient safety, 297
quality of care
Cochrane literature analysis, 296
remote populations, 296–297
telecommunication development,
293, 294
types, 293
United States, 294
Telemedicine system
phone-based, 750
usability evaluation, 750
Therac-25 radiation therapy system
AECL, 491
beam-flattening filter, 491

FMEA, 491–492
minicomputer-controlled system,
490–491
technical failures, 491
x-ray energy mode, 491
Top MODELER decision support
system, 60
Total quality management (TQM)
contradictory, 620
detrimental effects, 619–620
health care, 620
implementation, 620
nursing approach, 620
participative interventions, 615
positive impact, 619
products quality, 619
TQM, *see* Total quality management
TraumaMan® mannequin simulator,
562
Trust, health technologies
description, 281–282
face-to-face interactions, 282
interpersonal trust, 289
labor shortages and capable
technology, 282
organization, society and
technology
electronic fetal monitors, 288
implantable defibrillators, 288
mammography technologies,
288–289
vaccination technologies, 289
patients and care providers, 283
relationships
description, 281
health care systems, 286
patient and care provider, 286
provider, technology and patient,
287–288
worker and worker, 286–287
worker, technology and worker,
288
sociotechnical system
definition, 282
relationships, 282–283
system designers, 289–290
types
automation, 285
health care systems, 283–284
individual level, 284
organizational, 284
societal, 284–285
technological, 285
workers, 283
TTA, *see* Team task analysis

U

Underreporting
adverse events, 429

anesthesia quality assurance, 428
attitudes and opinions, 430
communication and coordination
    issues, 429
contemporary video analysis, 428
cultural and systemic barriers, 429
detection techniques, 428
device-related incidents, 434
exhaustive video analysis, 428
funding profile and equipment
    provision, 428
health care systems, 428
incident reporting (IR1) system, 434
in-service training sessions, 433–434
lack of certainty, 430
lessons learned applications, 431
low-criticality events, 432
mandatory reporting requirements,
    432
mandatory/voluntary schemes, 432
medical malpractice, 431
monitoring systems, 433
nonpunitive systems, 429–430
nurses error, 430
patient safety organizations, 434
"policing" reporting requirements,
    432
sentinel schemes, 434
Unexpected outcome
    definition, 401
    recommendations, 402
Usability evaluation
    architecture/facility design,
        547–548
    code cart
        boxed injection systems, 551
        dynamic settings, 551
        human factors issues, 551
        test plan, 551
        verbal orders, 551
    devices, 547
    facility design, 551
    failure modes and effects analysis
        proactive risk assessment
            technique, 548
        robust, 548
    heuristic evaluation
        description, 546
        feedback, 547
        maintain consistency, 547
        memory load, 547
        simple and natural dialog, 547
    HFE principles and methodologies,
        547
    human factors and health care
        standards documents, 547
    in-house software design and
        development
        computerized provider order
            entry (CPOE) systems, 550

design concepts, 549
functional and information
    requirements, 550
user-centered design framework,
    549
IV pumps
    ICU simulator, 552
    nurses, 552
    objective and subjective data, 552
    task correctness, 552
patient-controlled analgesia (PCA)
    pump
    design principles, 552
    human factors analysis
        techniques, 552
    post-experiment interviews, 553
    users, 552
policies and protocols
    additional problems, 550
    designer/developer, 550
    HFE, 550
    time-critical activities, 550
procurement
    biomedical engineer, 549
    factors, 549
    HFE methods, 549
RCA, 548
teaching and selling tool
    HFE, 552
    learned intuition, 553
    observed behavior, 553
testing, 546
training and education development
    developers, 550
    HFE evaluation methods, 550
    maintain competency, 551
work areas, 547
Use-error risk analysis
    blaming, user, 497
    development tool, 497
    document low-level risks, 498
    overlooking critical sources,
        failure, 498
    probability and severity, 497–498
    training/labeling, overreliance, 498
    validate modes, control, 498

V

VAASNET®, *see* Video acquisition
    system network
Value stream analysis (VSA), 799
Video acquisition system network
    (VAASNET®)
    camera systems
        acquisition display, 534–535
        anesthesiology team, 534
        extracorporeal work, 536
        head-mounted video camera
            (HMVC), 534

installation, 534–535
pan-tilt-zoom capability, 534
patient safety measures, 535
remote monitoring, 534
features, 533
mobile audio communication, 534
postanesthesia care unit (PACU), 534
Video analysis
clinical domain, communication
    crisis situations, 529
    noncommunication activities,
        528–529
    verbal, 528
critical incident
    audio–video recordings, 527
    team communication pattern,
        527–528
    template-based analysis, 527
description, 523–524
direct observation
    ANTS behavioral marker system,
        525–526
    OTAS tool, 524–525
macrocognition access
    adaptive coordination, 530
    performance omissions, 530
    scope, 530
    team process and performance
        measurement, 529–530
    visual scanning patterns, 530
qualitative data extraction
    coordination breakdowns, 527
    cultural elements, 528
    expertise coordination, 528
    knowledge sharing, 528
    patient safety, 528
    unique structuring process, 527
recording
    acquisition and storage, 533–538
    consent, 532
    privacy and confidentiality, 533
    sociolegal issues, 532–533
strengths, 531
template-based
    performance modeling, 527
    real-life performance, 527
    SMEs, 527
    task template, 527
tools
    "black box" recorder, 531
    Guerlain's software, 531
    MacSHAPA, 530
    OCS, 530
    open-source toolkit, 531
    VANNA, 530
    VINA, 530
weakness, 531–532
Video recording
    acquisition and storage, *see*
        Acquisition and storage, video

benefits, 532
consent
    ceiling-mounted camera, 532
    clinical workplace, 532
    human subjects, 532
    research hypothesis and plan
        implementation, 532–533
privacy and confidentiality
    technical approaches, 533
    trauma resuscitation unit
        (TRU), 533
sociolegal issues
    advantages, 532
    formal reports and informal
        communications, 532–533
Violation-producing conditions
description, 331
health care
    contextual factors, 332
    medication administration, 331
    patient identification
        strategies, 331
    procedures, 331–332
    technology-based defenses, 331
Violations, human error
definition, 330
health care
    negative consequences, 330–331
    surgical procedure, 330
human performance, 329–330
novel conditions, 330
types, 330
Virtual reality (VR) systems
development, 560
instruments, surgeon, 560–561
laparoscopic instruments, 561
LapSim® system, 561
MIST VR system, 561
part-task/whole-task trainers, 561
procedures, 560
surgery, 560
training systems, 562–563
Visually integrated operating room
    (VIOR)
WOK, 782

W

Wall of knowledge (WOK)
uses, 782–783
VIOR, 782
Wickens' multiple resource theory,
    567
Workflow analysis
approach
    CDS application affects, 509
    emerge depending, 510
    equipment and information, 509
    flow components, 509
    health IT application, 509

    higher-level steps, 510
    human factors perspective, 509
    levels, 509
    SEIPS model, 508–509
    temporal element, 509
BCMA, 515
clinic-based prescription writing
    process, 514
description, 510
EHR technology, 508
electronic system, 514–515
flowcharts, 514
health IT applications, *see* HIT
    applications
process stakeholders, 514
redesign, 508
tools
    analyzing and redesigning
        workflow, 511–513
    categories, 510
    description, 510
usability evaluation methods, 514
Work organization interventions
challenges, 611
characteristics, 611
conference method, 616
employer-employee discussion,
    619–620
evaluation, 621–622
goals, 612
and handling, negative emotions at
    work, 618–619
job stressors and lowered well-
    being, 612–613
managing structural changes and
    mergers, 617
multilevel organizational health
    promotion, 617–618
organizational interventions, 611
participatory phases, 613–615
planning and implementing,
    factors, 622
prevention stages, 612
quality management, 619–620
reduce stress at work, 616–617
reorganizing nursing work,
    620–621
stress management, negative
    emotions at work, 618–619
survey feedback method, 615–616
target groups, 612
TQM, 611–612
Work system approach
ceiling-mounted lifts, 796
elements, 796
SEIPS model, 796
systems safety lessons, 796
Work system design, health care
American hospital patients, medical
    error, 65

healthcare providers, 65
human factors and ergonomics
analysis, work system, 75
continuous design, 75
implementation, redesign,
    74–75
process, 73–74
model
    designing/redesigning, 66
    elements, 66–68
    ICU nurses, 68
    interactions, 69–70
    work-educational system,
        medical residents, 68–69
nursing shortage, 66
patient safety
    human error and safety
        models, 70
    pathways, 70–71
    SEIPS model, 71–73
Work system design pitfalls
criteria, effective system
    joint design and humanized task
        approach, 53
    macroergonomics, 53
    participatory ergonomics, 53
    sociotechnical characteristics,
        organization, 53
"left-over" approach, function and
    task allocation, 52
sociotechnical characteristics
    elements, 52
    empirical models, 52
technology-centered design
    hardware/software
        incorporation, 51
    human-system interfaces, 52
    reengineering efforts, 52
Work system models
balance model, 747–748
designing/redesigning, 66
elements
    health information technologies,
        67
    organizational conditions, 68
    person, 66
    physical environment, 67–68
    task, 66
    tools and technologies, 66–67
hardware/software, 747
household structures, 747
ICTs, 748
ICU nurses, 68
interactions
    central venous cannulation,
        ICU, 69
    intravenous pump users, 70
    physician workload, 69
    systems thinking, 70

work-educational system, medical
  residents, 68–69
Work system redesign
  CPOE, 456
  error recovery supporting
    strategies, 455
  HRO, 455

patient monitoring equipment,
  456
system redesign, 456
taxonomy, strategies, 456
team communication and
  supervision, 456
World Health Organization (WHO)

communication and patient safety
  improvement, 781
draft guidelines, RLSs, 675
ICPS, 673
and IHI, 681
world alliance, patient
  safety, 681